Composition and Nature of the Culicidae (Mosquitoes)

A mountain is not nearly as much a wonder as a mosquito.

Mokokoma Mokhonoana

The problem of explaining how an animal in its evolution has become adapted structurally to its environment and a special way of living is complicated in an insect such as the mosquito … it is hard to understand how evolution by means of natural selection has brought it about. On the other hand, the technique of "special creation" is entirely incomprehensible.

Robert Evans Snodgrass (Snodgrass, 1959)

Kenneth L Knight
1915–2001

Bruce A Harrison
1937–2018

E L Peyton
1929–1999

This book is dedicated with respect and admiration to the memories of the three prominent mosquito taxonomists whose insightful knowledge and infectious enthusiasm for mosquito taxonomy inspired and shaped my vocation.

Composition and Nature of the Culicidae (Mosquitoes)

Ralph E. Harbach

CABI is proud to present this book as a co-publication with the Royal Entomological Society (RES).

The RES is devoted to the understanding and development of insect science. It was established in 1833 and is dedicated to supporting international collaboration, research, and the publication of insect science discoveries and understanding. Its aim is to show every person how remarkable and valuable insects are.

Find out more about CABI's partnership with the RES here:
https://www.royensoc.co.uk/news/res-handbooks-now-available-via-cabi/

Royal Entomological Society

CABI

CABI is a trading name of CAB International

CABI
Nosworthy Way
Wallingford
Oxfordshire OX10 8DE
UK

Tel: +44 (0)1491 832111
E-mail: info@cabi.org
Website: www.cabi.org

CABI
200 Portland Street
Boston
MA 02114
USA

Tel: +1 (617)682-9015
E-mail: cabi-nao@cabi.org

A catalogue record for this book is available from the British Library, London, UK.

ISBN-13: 9781800627994 (hardback)
9781800628007 (ePDF)
9781800628014 (ePub)

DOI: 10.1079/9781800628014.0000

Commissioning Editor: Ward Cooper
Editorial Assistant: Emma McCann
Production Editor: Rosie Hayden

Typeset by Straive, Pondicherry, India
Printed and bound in the UK by Bell & Bain, Glasgow

Preface

I began my career as an insect morphologist. I always had a strong fascination for the diversity of form and structure exhibited by insects, and I was fortunate to have the opportunity to conduct comparative anatomical studies of mosquitoes prior to entering on a career as a mosquito taxonomist. The late Bruce Harrison, who was completing his PhD while I was conducting comparative morphological studies of mosquitoes as a postdoctoral student at North Carolina State University, remarked some years later, when I was undertaking mosquito taxonomic work, that I was unusual because I had learned the morphology before doing taxonomy whereas taxonomists learn the morphology while conducting taxonomic study. My fascination with morphology led me into a career as a mosquito morphologist and taxonomist that has endured for the past 49 years.

The mission of the mosquito taxonomist is to make known all species and how to identify and classify them. The purpose beyond discovery and identification is to acquire knowledge of their bionomics and evolutionary relationships. This book is my personal attempt to unify the core components of the natural history of mosquitoes.

Morphology, the discipline of biology that deals with the form, structure and specific anatomical features of organisms, has provided the basis for mosquito taxonomic study and classification since the official starting point of zoological nomenclature (Linnaeus, 1758). This book is the culmination of the morphotaxonomic studies that were completed during the past 266 years. Its purpose is to provide a full and up-to-date account of what is known about the morphology, taxonomy and classification of the global diversity of mosquitoes.

This work had its beginning as the online Mosquito Taxonomic Inventory, the main source of information that has mutated into this compendium. I created the Mosquito Taxonomic Inventory in 2007 to provide the international community with an up-to-date authoritative resource on the systematics of the Culicidae. With its impending demise, I initiated its transformation into the principal parts of this book. Some information has not been retained and much additional information has been added. The result is a much more complete coverage of mosquito classification and the taxonomic history of the currently recognized family, genus and species groups. It is my hope that this book will supersede the utility of the Mosquito Taxonomic Inventory, which in the words of William J. Sames IV (Medical Entomology Consultant, Leakey, TX, USA) 'has been a valuable go-to resource to get the latest and most accurate status of mosquito taxonomy' (personal communication, 5 November 2023).

Several themes run through this compendium, of which morphology and classification are central. I consider that classical morphology and anatomical-based classification provide the essential lens through which data can be viewed from diverse disciplines and interpreted within the landscape of molecular and phylogenetic analyses. Access to almost all other resources of mosquito information depends on scientific names and taxonomic infrastructure. In this volume, I have attempted to provide a full and up-to-date account of the current taxonomic infrastructure of the Culicidae. It is my hope that the information provided herein will stimulate students and researchers to critically examine the groupings of species based on morphological similarity to achieve a natural classification of the family. To put this in perspective, it has yet to be seen whether molecular genetics may adequately reveal what occurred during evolution to account for the amazing similar and dissimilar anatomical structures that are used by taxonomists to gauge the degree of relatedness as the basis for classifying mosquitoes into presumed natural groups. In any case, the categories and groups of classification now recognized are the stepping stones toward realizing a stable classification based on natural relationships. The substance of this book provides a platform of data that others can use, and upon which more detailed studies can be based. It is likely that much of the information presented herein will be modified or superseded by revised information, but it will have served its purpose if it induces students and researchers to undertake comparative morphological studies with the aim of discovering unknown facts about the anatomical diversity of mosquitoes.

I am grateful to all persons whose morphological and taxonomic studies have contributed to the information contained in this book. Those individuals, too numerous to list here, are the authors of the references listed at the end of the book. I am supremely grateful to the three esteemed mosquito taxonomists to whose memories this book is dedicated. My career-long accomplishments and this book would not have been possible without their invaluable tutelage, inspiration and collaboration during the early years of my work with mosquitoes. The work conducted by many students, assistants and colleagues over so many years has also contributed morphological and taxonomic information that is enshrouded throughout the taxonomic text. I am particularly indebted to Theresa Howard, who worked tirelessly as my personal Research Assistant in the Natural History Museum to produce numerous line drawings, obtain scanning electron micrographs and photographic images, and recorded morphometric data that significantly enhanced my published works.

Over the years I have been privileged to have had the opportunity to supervise the research of a number of students, the interactions with whom enriched my knowledge of mosquitoes and related subjects. I ardently thank them: **Anthony, Thomas** (MSc, PhD, Imperial College London, *Phylogeny of the Pyretophorus Series of* Anopheles *subgenus* Cellia and *Genetic structure and vector partitioning in* Plasmodium falciparum, respectively); **Azari-Hamidian, Shahyad** (PhD, University of Tehran, *Mosquitoes* *of Iran, with emphasis on the vectors of* Dirofilaria); **Cook, Shelley** (postdoctoral student, *The epidemiology of viral emergence: the molecular evolution of flaviviruses, their vectors and hosts*); **Culverwell, Lorna** (PhD, University of Helsinki, *Mosquitoes and their viruses in Finland*); **Dusfour, Isabelle** (PhD, Universite Montpellier, Anopheles sundaicus s.l., *vecteur majeur du paludisme sur le littoral sud-est asiatique. De la caractérisation du complexe d'espèces à sa surveillance dans le sud Vietnam*); **Garros, Claire** (PhD, Universite Montpellier, *The Minimus Group, a main malaria vector group in Southeast Asia. From the molecular scale to the spatial analysis*); **Lennon, Zoë** (MRes, University College London, *Molecular barcoding of medically important archival mosquito specimens (genus:* Culex*)*); **Mallampalli, Varuni** (PhD, University of Maryland, *Phylogenetics and coevolution of arboviruses and mosquitoes of the Spissipes Group of* Culex *subgenus* Melanoconion *in the Neotropics*); **Pedro, Pedro** (PhD, University of Leeds, *The influence of habitat fragmentation on* Sabethes *mosquitoes*); **Sedagaht, Mehdi** (PhD, University of Tehran, *Identification of species of the* Anopheles maculipennis *complex in Iran using morphological and molecular methods*); **Sharpe, Rosemary** (PhD, University of Leeds, *The status of cryptic species within* Anopheles minimus, *a major vector of malaria in Thailand*); **St John, Oliver** (MRes, Imperial College London, *Phylogeny of Culicini*); **Taai, Kritsana** (PhD, University of Chiang Mai, *Integrated systematics of the* Anopheles barbirostris *sibling species complex of malaria vectors*); **Thelwell, Nicola** (PhD, University of Leeds, *Gene exchange between sympatric populations of the malaria vectors* Anopheles bwambae, An. gambiae *and* An. arabiensis *in Uganda using microsatellite markers*).

I am grateful to my many contemporary collaborators and colleagues who contributed directly, indirectly or unknowingly to the information compiled in this work. I express my sincere thanks to you all for your informative discussions, hard work and efforts to enhance our knowledge of mosquito systematics.

I am a perfectionist by nature and in practice, but try as I might to make sure everything is perfect, experience has shown that this is not possible. I would like to think that this compendium is perfect in every detail, error and omission free, but surely there are errors that I have failed to notice and correct, which will come to light at a later date, much to my chagrin. My hope is that there are only a few minor errors or omissions. I hasten to add that any lapses and errors are strictly my responsibility.

Work on this volume commenced at the start of my retirement in the summer of 2018 and was completed in 2024. All taxa of the family, genus and species groups that were recognized as valid before 15 August 2024 are the substance of this book.

Ralph E. Harbach
London
August 2024

Contents

Introduction

Culicidae are a large and morphologically diverse family of familiar and mostly unfamiliar species. It is difficult to comprehend the incremental changes that must have occurred through tens of millions of years to produce the diversity of form exhibited by the extant species. It is unlikely that the morphological diversification of anatomical structures is related strictly to differences in the mitochondrial genes and genomes that are used predominantly to reconstruct evolutionary relationships. The nuclear genome, which contains nearly 2000 times more DNA base pairs than the mitochondrial genome and must have played a major role in the evolution of morphological diversity, has largely been neglected in phylogenetic studies. Surely interactions among genes, species and the environment are responsible for effecting morphological differentiation and evolution. Although molecular genetics has made important contributions to mosquito systematics, molecular studies should not overshadow studies of morphology, which provide vital information for the recognition, description and classification of taxa – the foundation for phylogenetic enquiry.

Belkin (1962) believed that hybridization has been an important mechanism in the speciation of mosquitoes and the evolution of the mosquito fauna. I share with Belkin the belief that hybridization is a plausible explanation for the many polythetic supraspecific taxa that are diagnosed by unique combinations of characters, rather than one or more characters that are exclusive of a particular group. Annectant groups, *i.e.* groups that show evidence of transitional traits, are found at all levels of classification. It is well known that hybridization can lead to introgression of genes from one species to another, providing a mechanism for recombining and shaping key traits during evolution. Hybridization between two closely related species is actually a common occurrence in nature. Indeed, examples of hybridization are well known in extant mosquitoes, *e.g. Culex pipiens/Culex quinquefasciatus* (Pipiens Complex), *Anopheles gambiae/Anopheles arabiensis* (Gambiae Complex), *Aedes mariae/Aedes zammitii* (Mariae Complex) (Rozeboom & Kitzmiller, 1958; Besansky *et al.*, 1994; Urbanelli *et al.*, 2014, respectively). Moreover, as observed and explained by Zavortink (1968, 1972, 1979), morphological similarities shared by species in areas of contact or near contact and not elsewhere can only be explained by the phenomenon of introgressive hybridization.

The current system of culicid classification, the division of mosquitoes into groups (taxa, singular taxon), is based on seemingly apparent and presumed anatomical homologies. A major problem is the question of homology. This particularly applies to structures of the male genitalia and elements of the larval and pupal chaetotaxy. Some elements of homology, which pinpoints the continuity of structures in phylogeny, must be regarded as questionable if they are based solely on comparison of extant species. Nevertheless, most homologies in mosquitoes are reasonably sound, especially those associated with innervation and musculature. Innervation has been used in determining ontogenetic homologies of many larval and pupal setae, but not for all setae and not comparatively for diverse taxa, and little is known about the actual transformation of larval structures into pupal structures and pupal structures into adult structures. It is important to realize that a structural development in one group of species is more likely to be homologous with similar structures in more closely related groups than with similar structures in more distantly related groups. The only way to increase the certainty of homologies is to collect evidence from different types of studies, including the use of physiological, biochemical, behavioural and molecular genetic methods. In comparison with comparative morphology, these other types of studies are pretty much in their infancy. It is generally understood that certain types of evidence are more reliable than others, which happens to be the case of using DNA sequence data to elucidate relationships. This is not the same as elucidating homologies. I do not know of any use of molecular applications to investigate the homology of anatomical structures. Here lies the dilemma. Is the application of molecular methods any better or more reliable in resolving relationships? To rephrase what Borkent (2018) aptly stated, analyses of phylogenetic relationships inferred from differences in DNA sequences, despite a history of inconsistent and questionable results based on limited taxon sampling, have increasingly replaced the assessment and hypothesis of relationships based on homologous structural data; until DNA sequence data can be shown to produce phylogenies in the absence of the morphology-based foundation of classification, those phylogenies should not be regarded with certainty and the application of molecular phylogenetics should not be seen as the panacea for establishing a natural classification.

Extensive changes have taken place in the classification of the Culicidae since the incipient studies of the 18th century. Greatest progress was made in the 20th century with a gradual phenomenal increase and a more abrupt decline in classical mosquito taxonomic research near the end of the century. Sadly, with the disappearance of traditional mosquito taxonomists and the rise of molecular phylogeneticists, it seems unlikely that any major formal changes will be made to the existing, largely unnatural classification of the family, the subject of this book. Unfortunately, the 20th-century era of morphotaxonomic research ended less auspiciously than it began. Fortunately, the morphologically defined categories and groups of classification now recognized are the stepping stones toward realizing a stable classification based on natural relationships.

In this volume, I have attempted to provide a full and up-to-date account of what is known about the morphology, taxonomy and classification of the currently formally recognized groups of mosquitoes. It is a thorough and complete amalgamation of the

DOI: 10.1079/9781800628014.intr

voluminous mosquito taxonomic literature, much of which is widely dispersed through numerous journals and other publications and is not readily available to individuals interested in the diversity, structure and taxonomy of the family. It must be stressed that this book only concerns formally recognized taxa. Unlike catalogues, it does not include the names of nominal groups and species that are presently regarded as synonyms. The lexicon of Harbach (2018b) and the catalogue of Wilkerson *et al.* (2021) should be consulted for detailed taxonomic information pertaining to species, higher-level taxa and their synonyms.

Background Information and Arrangement

Mosquitoes are holometabolous insects. They undergo complete metamorphosis, a process of growth, development and transformation through four life stages: egg, larva, pupa and adult. The first three stages of mosquitoes develop in aquatic habitats; the adults emerge from the pupal stage to become terrestrial insects. After emerging from an egg, the larva passes through four increments of growth (instars), each marked by notable increases in size, setal branching, number of cuticular processes and the extent of sclerotization. The fourth-instar (mature) larva transforms and moults (sheds its cuticular 'skin') into a pupa. The pupa is actually a very short period of development because most of the conventional 'pupal stage' is a developing adult enclosed in the remains of the pupal cuticle. Finally, the adult emerges leaving an empty pupal 'skin'. Larval and pupal 'skins' are the remains of the exoskeleton, properly known as 'exuviae' (the word is both singular and plural). When larvae are reared to obtain adults, the larval and pupal exuviae are retained, mounted on microscope slides and used for morphological study, knowing with certainty that the larva, pupa and adult belong to the same species. Ideally, taxonomic descriptions of species should include the morphological characterization of each of the four life stages. However, species descriptions have rarely included the egg stage and in many instances only the adult female or male have been described, often in the absence of associated larval and pupal stages. It should be borne in mind that descriptions of the larval and pupal stages are based on fourth-instar larvae (and exuviae) and pupal exuviae, respectively.

Zoological nomenclature, governed by provisions of the *International Code of Zoological Nomenclature*, readily available on the internet (https://www.iczn.org/the-code/the-code-online/), is intended to achieve stability and worldwide acceptance of scientific names. The *Code* recognizes three groups (categories, ranks) of taxa (singular taxon), the family group (families, subfamilies, tribes and subtribes), genus group (genera and subgenera) and species group (species and subspecies). No subtribes are currently recognized for mosquitoes, and based on the cogent explanation and taxonomic actions of Harbach & Wilkerson (2023), subspecies are no longer accepted as valid taxa in the classification of the Culicidae. To be validly published and to be accepted in nomenclature, the name of a newly introduced taxon must be accompanied by a verbal description of taxonomic characters that distinguish the taxon, designation of a type genus for a family-group name, designation of a type species for a genus-group name and selection of at least one specimen (holotype) to serve as the name-bearing type of a species name (see Harbach & Snow, 2003 for definitions of various kinds of type specimens). Additionally, in the case of a new species, the description must be accompanied by a statement indicating the location of the collection in which the holotype or type specimens have been or will be deposited. Because this book focuses on supraspecific taxa, type information is only provided for family- and genus-group names. The catalogue of Wilkerson *et al.* (2021) should be consulted for type information for species-group names.

The formal name of a species consists of a genus (generic) and a species (specific) name (binomen), often with a subgeneric name placed in parentheses between the two names and the species name followed by the author of the species, and sometimes also the year in which the species was named. The names of genera and subgenera, like the names of ranks above genus, must be written with the first letter capitalized, and the genus, subgenus and species names should be printed in a style of typeface, usually *italics* or **boldface**, which is different from the surrounding text. When subgenus names are enclosed in parentheses, the parentheses should not be italicized because they are elements of the surrounding text. The name of a mosquito species may be written in different formations, as illustrated by the following examples: *Culex pipiens*; *Culex* (*Culex*) *pipiens*; *Culex pipiens* Linnaeus; *Culex* (*Culex*) *pipiens* Linnaeus; *Culex pipiens* Linnaeus, 1758; *Culex* (*Culex*) *pipiens* Linnaeus, 1758. The names of authors should not be abbreviated; however, an abbreviation is allowed if the author is eminently known for his or her work and will be unmistakably recognized by the abbreviated name, *e.g.* *Culex pipiens* L. No abbreviations are used for author's names in this book. It should also be noted that family-group names can stand alone. For example, the endings of such names (-idae for family names, -inae for subfamily names and -ini for the names of tribes) denote the category of classification; hence, it is not always necessary to preface the name with 'the family', 'the subfamily' or 'the tribe'. Likewise, in many cases it is not necessary to preface the name of a genus with 'the genus'. In the case of family-group names, it is often appropriate to use the definite article 'the' before the name, *e.g.* the Culicidae, to indicate that the name is specific or particular.

The classification of the Culicidae is far from natural. Attempts to establish a natural classification based on unambiguous evolutionary relationships have largely failed. This is not surprising given that major changes to classification are problematical for non-taxonomists and individuals working in applied fields. To promote stability, the practice of recognizing few genera and many subgenera, as explained by Edwards (1932a), followed by most medical entomologists and culicidologists, and systematized in recent major taxonomic compilations (Harbach, 2018b; Wilkerson *et al.*, 2021), is embraced here.

Information in this book, following a brief History of Classification, is organized into three sections: Anatomy, Classification and Systematics, and Anatomical Terminology. Section 1 is principally an account of the external anatomy of the four life stages of mosquitoes, with emphasis on structures that are used to characterize and distinguish taxa. Little is mentioned about internal anatomy, which, perhaps unfortunately, is virtually excluded from taxonomic study and the recognition of taxa. Section 2 is by far the largest and foremost section, comprised of six parts that treat the taxonomic groupings of the extant and fossil Culicidae. Parts I–IV provide information on the typification, systematics, morphological characteristics, bionomics and principal literature sources for all currently recognized family and genus groups of mosquitoes. Parts V and VI provide information about the taxonomic history and informal (internal) classifications (where applicable) of the genera and subgenera, and a list of all currently recognized species and their placement in formal and informal classifications. Section 3 is a glossary of the morphological terminology most commonly used for describing, distinguishing and classifying mosquitoes. It includes definitions and illustrations of the chitinized anatomical structures of the adult, egg, larval and pupal stages. The sections and parts are relevant to one another and together form a coherent body of information. Most of the figures that accompany the glossary in Section 3 have been modified from illustrations in published works, principally from *Taxonomists' Glossary of Mosquito Anatomy* (Harbach & Knight, 1980).

History of Classification

Culicid nomenclature and classification had its beginning in the latter half of the 18th century (1758–1799) with the naming of 22 species of the genus *Culex* Linnaeus, 1758, 10 of which are valid today (see Knight, 1972). Seven additional genera were introduced during the 19th century (1800–1899), including *Aedes*, *Anopheles*, *Haemagogus*, *Psorophora*, *Sabethes* and *Uranotaenia*. Mosquito taxonomic activity increased rapidly during the first three decades of the 20th century (1900–1999) following the discovery at the turn of the 19th century that mosquitoes transmit the causal agents of malaria and yellow fever.

Theobald (1905e), in what was effectively the first mosquito catalogue, recognized seven subfamilies and 60 genera of Culicidae (excluding the subfamily Corethrinae). Five years later, Theobald (1910c) expanded the classification to include 10 subfamilies and 149 genera. In an updated reference catalogue, Edwards (1932a) introduced a significantly revised classification that included the subfamilies Culicinae ('true' mosquitoes), Dixinae and Chaoborinae. His Culicinae consisted of three tribes: Anophelini with three genera, Megarhinini with one genus and Culicini with 31 genera. Taxonomists waited another three decades for the substantially updated catalogue of Stone *et al.* (1959), in which the family Culicidae was restricted to the 'true' mosquitoes (Culicinae of Edwards), in which they recognized three subfamilies, Anophelinae, Culicinae and Toxorhynchitinae (Megarhinini of Edwards), with the Culicinae divided into two tribes, Culicini and Sabethini. The family as a whole included 31 genera and 95 subgenera.

The classification put forward by Stone *et al.* (1959) was not immediately accepted by all researchers. Belkin (1962) preferred to retain the subfamilies Dixinae and Chaoborinae recognized by Edwards (1932a), but restructured the classification of 'true' mosquitoes (Edwards's Culicinae) to include 12 tribes. He retained Anophelini and Toxorhynchitini and established 10 tribes in place of Edwards's Culicini. Some authors, *e.g.* Belkin *et al.* (1970), continued to treat chaoborids and dixids as subfamilies of Culicidae until they were decisively excluded from the family by Knight & Stone (1977), with the concomitant recognition of the subfamilies Anophelinae, Culicinae and Toxorhynchitinae, and the division of the subfamily Culicinae into the 10 tribes established by Belkin (1962). Surprisingly, Mattingly (1969, 1971a, 1981) did not accept the division of Culicinae into 10 tribes and recognized only two tribes, the Culicini and Sabethini of Stone *et al.* (1959). Except for Service (1993b), who continued to recognize the two tribes for convenience, the 10 tribes established by Belkin were universally accepted by mosquito taxonomists. As a consequence of phylogenetic studies based on morphological and molecular data, the subfamily Toxorhynchitinae was formally reduced to tribal rank within the subfamily Culicinae (Harbach & Kitching, 1998; Mitchell *et al.*, 2002). That action resulted in the currently accepted family-group taxa: Anophelinae, without tribal subdivision, and Culicinae, divided into 11 tribes – Aedeomyiini, Aedini, Culicini, Culisetini, Ficalbiini, Hodgesiini, Mansoniini, Orthopodomyiini, Sabethini, Toxorhynchitini and Uranotaeniini.

Knight & Stone (1977) recorded changes to the taxonomy of mosquitoes that were published prior to the end of 1973. As of that date, the Culicidae included 2960 species, 109 subspecies, 64 varieties and four infrasubgeneric forms. Three supplements to the 1977 catalogue were published before the end of the 20th century: (1) Knight (1978b), (2) Ward (1984), with the subsequent publication of an index and several additions by Gaffigan & Ward (1985), and (3) Ward (1992). At the end of January 1992, the family included 37 genera, 144 subgenera, 3273 species, 125 subspecies, 49 varieties, three infrasubspecific forms and 50 *nomina dubia* (Ward, 1992; Harbach, 2018b).

Of the 11 tribes of the subfamily Culicinae, the classification of Aedini has produced the most controversy. The traditional classification of the tribe (Knight & Stone, 1977) included nine genera, the largest of which was *Aedes*. The first revision of aedine mosquitoes was conducted by Edwards (1932a), who recognized an *Aedes* group of seven genera – *Aedes* (the largest with 29 subgenera, many of which had previously been described as genera), *Armigeres*, *Eretmapodites*, *Haemagogus*, *Heizmannia*, *Opifex* and *Psorophora*. Aedini was defined by Belkin (1962) to consist of the *Aedes* group of Edwards and genera *Ayurakitia*, *Udaya* and *Zeugnomyia*. *Ayurakitia* was later reduced to subgeneric status in *Aedes* by Mattingly (1971a); hence, the nine genera recognized by Knight & Stone (1977). Belkin recognized that some subgenera of *Aedes* were heterogeneous groups and noted that the genus would likely have to be revised into smaller more natural groups of species.

The generic classification of Aedini became much more diversified as a result of studies conducted in the first and second decades of the present century. Radical changes to the classification of the tribe were introduced in a series of studies which began with the restoration of *Verrallina* and *Ayurakitia* to generic status (Reinert, 1999c, 2000b, respectively) and were followed by the predominantly objectionable study of Reinert (2000c), who split *Aedes* into two genera, *Aedes* and *Ochlerotatus* (originally introduced as a genus), with 22 of the 43 recognized subgenera placed in *Aedes* and 21 placed in *Ochlerotatus*. This was followed by a series of four morphology-based phylogenetic studies (Reinert *et al.*, 2004, 2006, 2008, 2009), which resulted in the formal recognition of 46 genera within the tribe.

Wilkerson *et al.* (2015) reanalysed the morphological data set of Reinert *et al.* (2009) using a different method of character weighting (equal weighting as opposed to implied weighting) to 'gauge the stability' of the generic groupings of the latter authors. The result was the retention of all the genera as terminal taxa within an otherwise largely collapsed phylogenetic tree. Although deeper relationships were lost, all of the generic-level taxa of Reinert *et al.* survived, and were consequently recognized as subgenera within a very large and highly polyphyletic genus *Aedes*, which was accepted as the sinews of the traditional classification. However, close scrutiny clearly shows that the way in which Wilkerson *et al.* (2015) translated their phylogeny into a classification is problematic. Despite the various criteria discussed by the authors, their classification seems to have been based on a determination to retain the traditionally recognized genera of Aedini. If the huge polytomy is taken to be the genus *Aedes*, then some branches in their phylogenetic tree should not have been afforded generic rank while others were recognized as subgenera. For greater consistency in ranking taxa within a large genus *Aedes*, the seven traditional genera included in the polytomy, *i.e. Armigeres*, *Eretmapodites*, *Haemagogus*, *Heizmannia*, *Opifex*, *Udaya* and *Zeugnomyia*, should be ranked as subgenera. Because their generic status is retained, a reasonable alternative course of action would have been to recognize all the generic-level taxa as genera, thus preserving the genera and subgenera of Reinert *et al.* (2009). However, in as much as the results of Wilkerson *et al.* (2015) essentially confirm the results of Reinert *et al.* (2009), and the crux of disagreement seemingly only involves opposing opinions and choices about taxonomic ranking, their updated traditional classification is followed herein, with two caveats: (1) little distinction is made between the ranks of genus and subgenus – they are simply generic-level taxa, and (2) taxonomic groups (ranks) are merely hypotheses to be tested and modified as more information becomes available.

In a catalogue of the mosquitoes of Japan, Tanaka (2014, 2018) ranked the tribes Aedini, Culicini, Culisetini, Ficalbiini, Mansoniini and Orthopodomyiini as subtribes of a more encompassing tribe Culicini. In addition to subtribes Aedina, Culicina, Culisetina, Ficalbiina, Mansoniina and Orthopodomyiina, Tanaka created subtribe Heizmanniina for the genus *Heizmannia*. His subtribe Aedina included the genera *Aedes*, *Aedimorphus*, *Armigeres*, *Bruceharrisonius*, *Collessius*, *Downsiomyia*, *Edwardsaedes*, *Empihals*, *Geoskusea*, *Hopkinsius*, *Hulecoeteomyia*, *Neomelaniconion*, *Ochlerotatus*, *Phagomyia*, *Stegomyia*, *Tanakaius* and *Verrallina* for aedine species that occur in Japan. Newly recognized subgeneric taxa included *Aedimorphus* subgenera *Ecculex* and *Lepidotomyia* (currently synonyms of *Aedimorphus*), *Bruceharrisonius* subgenera *Bruceharrisonius* and *koreicoides*-group (informal groups were treated as subgenera without formal names); *Phagomyia* subgenera *Phagomyia* and *oreophila*-group; *Ochlerotatus* subgenera *communis*-group, *dorsalis*-group, *excrucians*-group, *impiger*-group, *punctor*-group, *sticticus*-group and *Woodius*; and *Stegomyia* subgenus *Quasistegomyia* (currently a synonym of *Stegomyia*). The subtribe Culicina included the genera *Culex* and *Lutzia*, and *Culex* included the subgenera *Barraudius*, *Culex*, *Culiciomyia*, *Lophoceraomyia*, *Mochthogenes* (currently a synonym of *Eumelanomyia*), *Neoculex*, *Oculeomyia*, *Protomelanoconion* (also currently a synonym of *Eumelanomyia*) and *Sirivanakarnius* for aedine species that occur in Japan. As expected, the subtribes Culisetini, Ficalbiina, Mansoniina and Orthopodomyiina included the genera that are currently included in the coordinate tribal groups Culisetini, Ficalbiini, Mansoniini and Orthopodomyiini, respectively. However, the validity (availability) of the subtribal names is doubtful. The availability of two of the names, Aedina Lahille, 1904 and Culisetina Maslov, 1964, was not mentioned, and no action was taken to reinstate and define them. The other subtribal names were introduced without explanation or justification, and are herewith treated as *nomina nuda* because family-group names published after 1930 must be accompanied by a verbal description or definition of characters that purport to distinguish them (Articles 13.1 and 13.2 of the *International Code of Zoological Nomenclature*).

Realizing that Article 45.6.4 of the *Code* prohibited the recognition of infrasubspecific taxa after 1961, Harbach & Howard (2007) critically assessed the varietal and infrasubspecific forms of mosquitoes that were recognized at the end of 2006 to determine whether they should be considered as valid subspecies or invalid infrasubspecific forms. As a result, 31 nominal forms were deemed to be subspecies, which at the time brought the number of formally recognized subspecies to 130 (incorrectly specified to be 134 in Harbach, 2018b). Ten of those subspecies were eliminated prior to the landmark publication of Harbach & Wilkerson (2023) in which the remaining 120 subspecies, deemed to be insupportable concepts, were critically assessed and recognized as either separate species or synonymous names. As a result, 96 subspecies were formally elevated to specific rank, 22 were synonymized with nominotypical forms, one was deemed to be a *nomen dubium*, one was considered a *species inquirenda* and four nominal species previously placed in synonymy were revalidated. The bottom line is that the rank of subspecies was excluded as a valid category of culicid classification.

Shortly after the publication of Harbach & Wilkerson (2023), apparently unaware that subspecies were no longer recognized as valid taxa, a new subspecies, *Aedes leptolabis talangayensis* Cornel, Kowo & Mayi, was published in Kowo *et al.* (2023). Because the authors did not explicitly indicate how the proposed subspecies differs from the presumed nominotypical form, it was not distinguished from *leptolabis sensu stricto* in their key to males of species of the 'Domesticus group' and the larva and pupa are unknown and could not be compared with those of the nominate form, *talangayensis* Cornel, Kowo & Mayi, 2023 must be regarded as a synonym of *leptolabis* Edwards, 1936.

Taking into account the additions and taxonomic changes published before 15 August 2024, the classification of the extant Culicidae includes 3726 species classified in two subfamilies and 217 genus groups (41 genera and 168 subgenera, excluding the nominotypical subgenera, 191 including the nominotypical subgenera). The taxonomic history and informal subdivision of these groups is the subject of Part V in Section 2.

Section 1: Anatomy

DOI: 10.1079/9781800628014.0001

Anatomy

Despite the advent of molecular methods of systematics research, mosquito identification and formal taxonomy relies principally on differences in features of the sclerotized anatomy. Although extremely useful, molecular techniques do not represent a substitute for morphology because they cannot be used by themselves to solve formal taxonomic problems. The reason for this is that the entire faunal classification system is based on morphology and the type specimen concept. This does not mean that species cannot be described and named on genetic or molecular data alone, but it means that molecular techniques must be integrated with classical morphological analysis to determine the correct application of available names for species concepts previously based solely on anatomical criteria. In most cases, molecular techniques merely represent a useful supplement to morphology, or novel means of analysing or utilizing anatomical data. Comparative anatomy is the foundation of mosquito systematics.

As mentioned previously, mosquitoes undergo four developmental stages – egg, larva, pupa and adult. Taxonomists have never considered whether the chicken or egg came first and have always described the adults before the other life stages in taxonomic works, and that is the case here, the order of treatment being the adult, egg, larva and pupa. Emphasis has historically been placed on the adult females due to their involvement in the transmission of pathogens of diseases in humans, and the classification of the family is rooted in anatomical studies of the adult habitus.

The account of the anatomy of mosquitoes given here is not intended to be complete. The purpose is mainly to give sufficient information about the sclerotized anatomical structures used to characterize the genus- and family-group taxa that comprise the family Culicidae. Certain structures that are not normally used in taxonomic descriptions are mentioned in the text for clarification and illumination of the principal taxonomic structures and are not, for the most part, included with the morphological terminology defined and illustrated in Section 3.

Adults

The general appearance of the adult female mosquito is illustrated in Fig. 1. As in other insects, the body has three principal regions or **tagma** (plural **tagmata**), the **head**, **thorax** and **abdomen**. The head is the anterior section (tagma) of the body bearing the compound eyes, antennae and mouthparts. It is separated from the second or intermediate division of the body, the thorax, by a membranous tube called the **cervix** (neck). The thorax is comprised of three segments, each bearing a pair of **legs**. The largest or middle segment of the thorax bears the **wings** and the reduced third or posterior segment bears the **halteres** (singular **halter**). The abdomen is the third or posterior division of the body. It consists of 10 apparent segments and the terminal genital structures involved in reproduction. Except for sclerotized elements of the alimentary canal and the reproductive system, the internal anatomy is not used in mosquito taxonomy.

Head

The generalized structure of the head of a mosquito is shown in Figs 2 and 3. The head consists of a sclerotized skull-like capsule (**cranium**) bearing the **mouthparts** and **antennae** (singular **antenna**) in front. Much of the anterolateral surface of the head is occupied by the large, convex, kidney-shaped **compound eyes**. Each compound eye consists of an aggregation of several hundred ommatidia that are collectively bounded by a line of inflection known as the **ocular suture**. Much of the anterior dorsal surface of the head behind the compound eyes constitutes the **vertex**. As in many other insects, the vertex is not differentiated from the **occiput**, **genae** and **postgenae** (singular **gena** and **postgena**). The anterior part of the vertex forms a narrow **interocular space**. In anophelines in particular, a prominent **frontal tuft** of setae and scales arises from the interocular space and the immediately adjacent portion of the vertex proper. In addition to the frontal tuft and some variously developed decumbent or semi-erect scales behind it, the main portion of the vertex is marked by a line of **ocular setae** along the dorsoposterior margins of each compound eye. By definition, the occiput is the posterior dorsal part of the cranium bearing the **erect** (forked) **scales** in front of the **postocciput**. The postocciput is a small posterior area of the cranium located at the dorsal and lateral margins of the cervix. It is separated from the anterior areas by the **postoccipital suture**, and articulates with a supporting sclerite (**cervical sclerite**) on either side of the cervix. The postgenae comprise the lateral and ventral parts of the head behind the compound eyes. The posterior margins of these areas bear a linear group of **postgenal setae**. The limits of the vertex, occiput and postgenae probably occur where the ocular setae, erect forked scales and postgenal setae are juxtaposed at the sides of the head.

Erect forked scales are interpreted as vestiture of the occiput that 'may extend [on]to the vertex' (Belkin, 1962). When the erect scales are arranged in a transverse row or narrow band at the back of the head, they are considered to be borne on the occiput. Such scales may be small or otherwise difficult to see unless the head is bent downward. The presence or absence of erect scales on the occiput appears to be of importance in delimiting genus-group taxa, *e.g.* erect scales are present in species of four of the subgenera of *Sabethes* and absent in species of the subgenus *Sabethinus*. When erect scales are more numerous and cover more of the dorsal surface of the head, they are considered to be borne on the vertex as well as the occiput. Erect scales may be numerous and extend to the orbital margins or they may be entirely absent or confined to the occiput in species of species-rich genera, such as *Uranotaenia*.

The head is supported internally by a pair of endoskeletal struts formed chiefly of the fused **anterior** and **posterior arms** of the **tentorium** (the right and left halves of the tentorium are not connected as they are in generalized insects). Each strut extends from an **anterior tentorial pit** located below the antennal socket at the front of the head to a **posterior tentorial pit** located at the lower end of the postoccipital suture at the back of the head.

Most of the front or facial area of the head circumscribed by the compound eyes is occupied by the bases of the antennae, the prominent **clypeus** and the projecting mouthparts. The clypeus is the conspicuous and projecting central lobe between the antennae and mouthparts that gives origin to the dilator muscles of the cibarium. The clypeus is bare in most mosquito species but scales or setae are present in some species and groups, *e.g.* lateral patches of scales are present in females of the Oriental Hyrcanus Group of *Anopheles* and setae are present in many species of *Trichoprosopon*. The facial area also includes the much reduced **frons**, which is largely obscured by the convergent antennal sockets, and the paired genae, which extend ventroposteriorly from the anterior tentorial pits behind the clypeus and are united ventrally between the compound eyes.

The antennae are appendicular sense organs. Each antenna consists of a hidden ring-like basal segment, the **scape**, a large globular and cup-shaped second segment, the **pedicel**, and a filamentous third segment, the **flagellum** (pleural **flagella**). The flagellum is divided into 13 false segments termed **flagellomeres**. The flagellomeres bear whorls of long curved setae which impart the typical verticillate (not plumose) appearance to the antennae of most males. The **flagellar whorls** are less conspicuous in females, and the males of certain genera, because they contain fewer and shorter setae. The pedicel often bears small scales and setae, and the first few flagellomeres often bear a few scales. Several types of sensilla are also present on the flagellomeres. The number and distribution of these sensilla, especially the sensilla coeloconica, may be of taxonomic importance.

One or both (usually) of the two apical flagellomeres (12 and 13) of male mosquitoes may be disproportionately long compared with the other flagellomeres. When they are disproportionately long, the preceding flagellomeres are shorter and more strongly verticillate than the corresponding flagellomeres in females of the same species. With the exception of *Aedeomyia* and *Bironella* (subgenera *Bironella* and *Neobironella*), when the two terminal flagellomeres are not disproportionately long, the preceding flagellomeres are relatively longer with a concomitant reduction in the development of the flagellar whorls. In the latter case, the antennae of males and females are often so similar that it is not possible to distinguish the sexes based on flagellar development. *Aedeomyia* and *Bironella* (except subgenus *Brugella*) are unique in having strongly verticillate antennae with short terminal flagellomeres. In *Brugella*, the flagella of males are weakly verticillate and closely resemble those of females. The males of some *Wyeomyia* have terminal flagellomeres that are relatively longer than usual, a somewhat intermediate condition.

The mouthparts of mosquitoes are highly modified for piercing (females) and sucking (both sexes) and form a greatly elongate **proboscis**, which projects from the lower facial region of the head below the clypeus. The proboscis includes the **labrum**, two **mandibles**, the **hypopharynx**, two maxillary **laciniae** (singular **lacinia**) and the **labium**. The labium forms the stout outer part of the proboscis. It consists of an elongate flexible **prementum**, which ensheaths the other components, an apical pair of short articulated **labella** (singular **labellum**) and a short **ligula** between the labella. The prementum is covered with scales. It may be entirely dark-scaled or have patches of pale scales that are characteristic of some species and species groups. The other components of the proboscis of females form a compact **fascicle** of needle-like piercing **stylets**. The distal ends of the maxillary laciniae have recurved teeth on their outer margins. The number of teeth on each lacinia is termed the **maxillary tooth index**, and this may exhibit differences between species. The length of the proboscis, measured from the **labial basal setae** to the tip of the labella, compared to the length of the forefemur (the **forefemur/proboscis ratio**) is sometimes a useful taxonomic character. The labrum and labium of males are developed as well as they are in females, but the other mouthparts are less developed. The hypopharynx is fused with the inner wall of the labium. Maxillae are present in all males, but they are devoid of teeth and are more weakly developed than in females. Mandibles are present in some males of some groups and absent in others; when present, they are much shorter than the labium and vary greatly in length in species of different genera and subgenera (Marshall & Staley, 1935; Snodgrass, 1959). It appears that atrophy of the maxillae, and in particular the mandibles, is an evolutionary consequence of feeding exclusively on plant juices.

As noted by Harbach & Kitching (1998), the labella have received little attention in mosquito taxonomy. In the majority of mosquitoes, each labellum consists of two segments, which are indicated by proximal and distal sclerites (Fig. 4a–c). However, the labella of toxorhynchitine and anopheline mosquitoes each consists of a single sclerite (Fig. 4a). In *Chagasia* and the nominotypical subgenus of *Anopheles*, the sclerite is divided dorsally in a manner that clearly suggests that either the one-segmented condition of anophelines has resulted from the fusion of two segments or the two-segmented condition of culicines has evolved through the cleavage of a single segment. The one-segmented labellum of *Toxorhynchites* is very long and covered with scales, a unique condition among Culicidae. Scales are absent from the labella of anophelines, cover the labella of toxorhynchitines and only occur on the proximal segment in certain species and genera of culicine mosquitoes. The elongate and densely scaled proximal segment characteristic of sabethines (except *Maorigoeldia* and *Topomyia*) may be a development (Fig. 4c) toward the condition found in *Toxorhynchites*. The condition of the proximal segment of *Haemagogus*, some *Armigeres*, a few *Aedes* (*e.g. Christophersiomyia*) and *Culex* (*Anoedioporpa*, *Carrollia*), and especially *Udaya* and *Zeugnomyia*, approaches the condition in sabethines. In specimens that have not been cleared, the elongate and densely scaled proximal sclerite of many sabethines has the appearance of being a distal extension of the prementum.

Other parts of the maxilla are present that are associated with but do not form part of the proboscis. These include the reduced **cardo** and **stipes** and the prominent **maxillary palpus** (plural **palpi**). The cardo and stipes are fused to form an internal **stipitocardinal rod**, which articulates with the lacinia and provides support for the maxillary palpus. These structures are not important in mosquito taxonomy.

The maxillary palpi arise on either side of the face immediately below the clypeus and lateral to the base of the proboscis. They may be long or short in one or both sexes and generally extend parallel to the proboscis. The palpi are composed of false segments called **palpomeres**. Prior to Tanaka *et al.* (1979) and Harbach & Kitching (1998), the palpal segmentation of female mosquitoes mistakenly included the small lobe at the base of the maxillary palpus, the **palpifer**, which is an anatomical part of the stipes. True palpomeres always bear scales or setae or both and the palpifer is always bare. When cleared specimens mounted on microscope slides are studied, it is obvious that five palpomeres is the maximum number present in both female and male mosquitoes, and shorter palpi have evolved through the reduction or loss of distal palpomeres. Thus, various groups of related species are characterized by the number of palpomeres present: one, two, three, four or five depending on the taxon.

In descriptions of *Aedes*, the maxillary palpi of females are regarded as being composed of either three, four or five palpomeres, but it is now apparent that five palpomeres do not occur in species of this genus. Palpi comprised of three palpomeres, often bearing a vestigial fourth palpomere, are characteristic of *Aedes* females observed in species representing 29 subgenera. Only two palpomeres occur in *Cancraedes*, which is regarded here as an aberration within the genus. The palpi of *Culex* females are similar to those of *Aedes* except that a vestigial fourth palpomere is not so common. Females of some subgenera, *e.g. Anoedioporpa* and probably *Acalleomyia*, have palpi composed of only two palpomeres. However, *Culex*, as presently defined, contains more than 800 species, and this condition is considered to be an aberration within this genus as well.

When fully developed, the maxillary palpus of male mosquitoes is five-segmented with clear evidence of ankylosis between the second and third palpomeres. As in the case of females, palpi with fewer than five segments are due to the loss of apical palpomeres. On the basis of the variation exhibited in the development of the palpus in males of *Aedes*, *Culex*, *Haemagogus* and *Tripteroides*, it is difficult to imagine that these genera are monophyletic. The palpus in most *Aedes* has either four or five palpomeres, but in certain subgenera it is composed of three palpomeres (*e.g. Aedes*), three palpomeres with a vestigial fourth palpomere (*e.g. Stegomyia*) or four palpomeres with a vestigial fifth palpomere (*e.g. Levua*). Males of *Belkinius* are characterized as having 'very short' palpi (Reinert, 1982) that outwardly resemble the palpus consisting of one palpomere in *Ae.* (*Cancraedes*) *indonesiae*. The latter condition is regarded as an anomaly for this subgenus as currently defined. Two disparate conditions are known to occur within *Haemagogus*. In some species the palpus is composed of five palpomeres and in others it consists of three palpomeres with an apical vestigial fourth palpomere. The majority of *Culex* males have palpi composed of five palpomeres, but four-segmented palpi, sometimes bearing a vestigial fifth palpomere, occur in some subgenera (*Aedinus*, some *Anoedioporpa* and some *Micraedes*). The male of *Cx.* (*Belkinomyia*) *eldridgei* is described as having the palpus composed of three palpomeres (Adames & Galindo, 1973), and the 'short' palpus of species of the subgenus *Acalleomyia* mentioned by Bram (1968) probably consists of three or fewer palpomeres. *Tripteroides* is clearly a heterogeneous collection of species which should be divided into two or more genera. In the majority of males, the maxillary palpus has two palpomeres as in females. Some species have fully developed palpi consisting of five palpomeres, including the type species of the subgenus *Rachisoura*, while others have reduced palpi comprised of only four palpomeres.

The palpi are covered with scales except along their inner (mesal) surface. The scales may be appressed or suberect. When they are appressed, the palpi appear thin and smooth; when they are suberect, the palpi appear thick and shaggy. The palpi may be entirely smooth or shaggy, or the proximal palpomeres may be shaggy and the distal ones smooth. The palpi are often ornamented with pale scales which frequently form pale bands (usually at the joints and apex of the terminal palpomere) and patches. The occurrence, position and extent of the pale scaling provide important characters for recognizing species.

The preoral food cavity located between the base of the hypopharynx and the under surface of the clypeus forms the primary sucking organ of mosquitoes known as the **cibarium**. This organ sometimes contains a transverse collection of teeth and other cuticular processes known as the **cibarial armature**. These structures are present in certain subgenera of *Anopheles* and, as far as known, are only developed elsewhere in the Culicini (Barraud & Covell, 1928; Belkin, 1962; Sirivanakarn, 1978). The cibarial armature was first investigated in *Anopheles* by Sinton & Covell (1927), and the divisions of the subgenus *Cellia* are based largely on characters of the **cibarial teeth**. The position and appearance of the cibarial teeth in anophelines is shown in Fig. 4d. The teeth are borne on a crest located ventrally at the posterior margin of the cibarium. There may be a single row of teeth as in the subgenus *Nyssorhynchus*, or two rows of teeth, one of **rods** and one of **cones**. When a single type is present they have more or less the character of cones. There are no cibarial teeth in species of the subgenus *Anopheles*. However, some species have a small heavily pigmented area on the **posterior hard palate** near the **ventral setae** (when present). The structures in species of Culicini are simply characterized as teeth. The finer details of the teeth are not easy to examine and appearance depends on the angle of view. The cibarial armature may be of some value in the diagnosis of certain species with extremely similar external characters. Scanning electron microscopy has been used to elucidate taxonomic differences in the cibarial armature in some species of *Anopheles* and the subgenus *Melanoconion* of *Culex*.

Thorax

The thorax (Figs 1 and 5) consists of three segments, the anterior **prothorax**, the central **mesothorax** and the posterior **metathorax**, each bearing a pair of legs ventrally. In mosquitoes, as in other Diptera, the mesothorax bears the functional **wings**, and is greatly enlarged to accommodate the flight muscles. Consequently, most of the thorax belongs to the mesothorax and the prothorax and metathorax are much reduced. The metathorax bears the small club-like **halteres** (singular **halter**), which represent the hind or second pair of wings in four-winged insects. The halteres are vibratory organs for maintaining stability during flight.

Each of the thoracic segments is divided into a dorsal **notum** (thoracic **tergum**) and a ventral **sternum**. The lateral sclerotization of the three segments comprises the thoracic **pleura** (singular **pleuron**). In many insects, a vertical **pleural suture** separates the pleuron of each segment into an anterior **episternum** and a posterior **epimeron**, and each of these sclerites may be further subdivided horizontally into upper and lower divisions, a dorsal **anepisternum** and **anepimeron** and a ventral **katepisternum** and **katepimeron**. The specific sclerites of individual segments are denoted by prefixing pro-, meso- or meta- as appropriate.

The prothorax and metathorax are so greatly reduced in mosquitoes that the individual parts are not immediately distinguishable. The notum of the prothorax, the **pronotum** (plural **pronota**), is seemingly absent medially and divided into laterally displaced halves. It is divided transversely on each side into an anterior setose lobe, the **antepronotum** (plural **antepronota**), and a larger posterior plate termed the **postpronotum** (plural **postpronota**). The setae and especially the colouration of scales on these structures sometimes provide useful taxonomic characters. The two parts of the pronotum taper downward to the **propleuron**, which consists of a small **proepimeron** and a **proepisternum**. The proepimeron is not clearly demarcated from the postpronotum but a distinct groove (**propleural suture**) above the articulation of the forecoxa clearly separates it from the proepisternum, which extends from the antepronotum to the base of the coxa and expands anteroventrally around the coxa to meet the proepisternum of the other side. Setae occurring on the proepisternum above the forecoxa are termed **upper proepisternal setae**, and these have some taxonomic importance. They may be numerous, reduced in number or entirely absent.

The development of the antepronota is generally indicative of genus groups. They are always narrowly connected but are rather small and widely separated in most mosquitoes, *e.g. Trichoprosopon*, but are large and approximated at their mesal margins in some taxa, *e.g. Haemagogus* and *Sabethes*, so as to nearly close the space between them. Polymorphism in their development appears to follow subgeneric lines in *Armigeres* and *Heizmannia*, with large and approximated antepronota occurring in the nominotypical subgenera and smaller, separated lobes in subgenera *Leicesteria* and *Mattinglyia*, respectively. The antepronota are somewhat variable in *Haemagogus*, *Malaya* and *Wyeomyia*, but are generally large and close together. They are rather large and moderately separated in *Toxorhynchites*.

The mesothorax has retained the structure more or less typical of a thoracic segment. The dorsal **mesonotum** consists of an **acrotergite**, **scutum** (including a **prescutum**), **paratergite** and **scutellum**. The scutum is the principal dorsal area of the thorax that actually represents the combined prescutum and true scutum because these areas in mosquitoes are only indistinctly separated laterally by an obsolescent **prescutal suture**. Scales are always present on the scutum, and although they are easily rubbed off, their development and colour are sometimes useful for species determination. The disc of the scutum may bear a full or partial compliment of setae or no setae at all, which is usually taxon dependent. The acrotergite is an obscure flange beneath the anterior end of the scutum, the **anterior promontory**, which projects over the cervix. The paratergite is a narrow lateral part of the mesonotum located anterior to the attachment of the wing. It is separated from the scutum by the **paranotal suture**. The scutellum is the posterior area of the mesonotum separated from the scutum by a distinct line of inflection, the **scutoscutellar suture**. The scutellum has the form of a transverse lobe, which is usually evenly rounded in species of *Anopheles*, *Chagasia* and *Toxorhynchites*, and trilobed in species of *Bironella* and Culicinae (except *Toxorhynchites*). The large convex sclerite borne below the scutellum, morphologically the acrotergite of the **metanotum**, has become the **mesopostnotum**.

Ornamentation

Characters of the thorax of use in classification and identification include the presence and absence of **scales** and **setae** (singular **seta**) (Figs 6 and 7, respectively). Various types of scales defined by their shape (see Fig. 27 and **Vestiture** below) occur on different areas of the integument (including the head and abdomen), but all setae of adult mosquitoes, with the exception of certain specialized setae that are borne on parts of the male genitalia, are simple single setae of different sizes. Setae occur in groups and occupy definite positions on the scutum and thoracic sclerites. The presence, absence or development of rows of **acrostichal** and **dorsocentral setae** on the scutum is of taxonomic importance. Setae and scales may be present on the mesopostnotum. Among the various groups of setae, the **proepisternal setae** are of some importance – they may be numerous, reduced to one or two, or absent altogether. Care must be taken not to confuse the proepisternal setae with setae of the antepronotum or the **forecoxa**, especially in pinned specimens that are liable to be somewhat shrunken or distorted in this region. Other pleural groups of setae are the **prespiracular**, **postspiracular**, **prealar**, **upper mesokatepisternal**, **lower mesokatepisternal**, **upper mesepimeral** and **lower mesepimeral setae**. A lower mesepimeral seta, infrequently two or more, is present in some group taxa and absent in others.

The following illustrates the taxonomic significance of thoracic setae and scales. Acrostichal setae are either present or absent in species of all subgenera of *Mimomyia*, absent in some *Uranotaenia* and often weakly developed and sometimes absent in *Mansonia*. Dorsocentral setae are usually present in *Aedes*, but are apparently absent in a few species of subgenera *Diceromyia*, *Finlaya*, *Leptosomatomyia*, *Pseudarmigeres*, *Scutomyia* and *Stegomyia*, and completely absent only in the species of subgenus *Christophersiomyia*. A few adventitious setae sometimes occur in *Udaya*. They are generally well developed in most *Mimomyia*, but are sometimes weak and seemingly absent in some species. Within *Tripteroides*, dorsocentral setae are absent in species of the subgenus *Rachionotomyia*, present or absent in species of the subgenera *Polylepidomyia* and *Tricholeptomyia*, and usually present in species of the subgenus *Tripteroides*.

The occurrence of mesopostnotal scales and setae in mosquitoes was examined in detail by Harrison & Bickley (1990). Scales/setae are reported to occur in only two species of *Aedes*, one species of *Armigeres* and four species of *Culex*. Only seven species of *Deinocerites* have mesopostnotal setae, but their presence does not seem to reflect evolutionary groups. Most species of the nominotypical subgenus of *Heizmannia* possess setae, and often scales, but they are absent in species of the subgenus *Mattinglyia*. The presence or absence of setae/scales seems to correspond to species groups within *Haemagogus* and *Tripteroides*, but this requires further study. Mesopostnotal setae are present in the majority of *Eretmapodites*.

Scales are present on the paratergite of *Chagasia*, and more easily distinguish species of that genus from those of *Anopheles* in the New World than the shape of the scutellum. *Culiseta* often have scales or setae below (and sometimes on) the lateral crest of the paratergite, but comparison with *Chagasia* suggests these are homologous with the paratergal scales/setae that occur on the upper (lateral) surface in other mosquitoes.

Postpronotal setae are normally absent in *Bironella*. These setae (often only one seta) are usually present in species of all subgenera of *Tripteroides*, and a single seta is present in species of *Trichoprosopon*. The setae are absent in two species of the nominotypical subgenus of *Psorophora*, but are otherwise present in species of the genus.

Belkin (1962) and Harrison & Scanlon (1975) stated that prespiracular setae (their spiracular setae) are present or absent in *Anopheles*, but they are present in the majority of species. Prespiracular setae are characteristic of species of Culisetini, Sabethini and Uranotaeniini, but are sometimes absent in *Malaya* and *Uranotaenia*.

Postspiracular setae are rarely present in species of *Culex* and *Hodgesia*. They are absent in species of the subgenera *Ayurakitia* and *Kompia* of *Aedes* and most species of *Heizmannia*. Their presence or absence coincides with subgeneric divisions in *Armigeres* and *Coquillettidia*: absent in species of the subgenus *Leicesteria* of *Armigeres* and in species of the nominotypical subgenus and *Austromansonia* of *Coquillettidia*; present in species of the nominotypical subgenus of *Armigeres* and species of the subgenus *Pseudotaeniorhynchus* of *Coquillettidia*. Species of *Johnbelkinia* sometimes have weakly developed postspiracular setae.

Prealar setae are occasionally absent in species of *Aedes* and *Anopheles*. They are present in the majority of species of *Orthopodomyia* and absent in all but one species of the subgenus *Davismyia* of *Sabethes*. Upper mesokatepisternal setae are present in most species of *Aedes* and the majority of species of *Orthopodomyia*, absent only in species of the Albipes Group. Species of the nominotypical subgenus of *Haemagogus* lack these setae, but they are present in species of the subgenus *Conopostegus*. They are rarely present in species of *Hodgesia* and *Tripteroides*.

A lower mesepimeral seta is absent in all but one species of *Armigeres* subgenus *Leicesteria* and present in all species of the nominotypical subgenus. One (usually) or rarely two or three setae are generally present in species of *Culex*, but are apparently absent in species of the subgenera *Lasioconops*, *Oculeomyia*, *Sirivanakarnius* and *Tinolestes*, and sometimes absent in species of certain other subgenera, *e.g.* *Anoedioporpa*, *Culiciomyia*, *Neoculex*, the Duttoni and Sitiens Groups of the subgenus *Culex* and the Bisulcatus Group of the subgenus *Micraedes*. Four or more lower mesepimeral setae are present in the species of *Lutzia*. The seta is present in only one species of *Orthopodomyia*, rarely absent in species of *Mimomyia* and often absent in species of *Toxorhynchites*.

Scales may be present anteriorly on the antepronotum, forming the **pronotal tuft**. The **anterior promontory** of the scutum may bear distinctive scales. The halteres usually bear scales and the colour of these is occasionally of taxonomic value. There are often small patches of scales on the pleura, usually associated with the setal groups, particularly with the mesokatepisternal and mesepimeral setae.

Metepisternal scales (and often setae) are located immediately below and behind the metathoracic spiracle. They were first noted in species of *Malaya* by Belkin (1962), in species of *Topomyia* by Marks (1971) and in two species of the subgenus *Rachisoura* of *Tripteroides* by Marks (1974). The scales are often small, translucent and difficult to see, especially in species of *Malaya*. According to Steffan & Evenhuis (1985), metepisternal scales are not considered to be homologous with the small patch of scales located near the base of the halter in species of *Toxorhynchites*. Some species of *Toxorhynchites* actually have two small patches of metapleural scales (see Fig. 13b in Harbach & Knight, 1980), one located on the metepisternum immediately below the metapleural sensilla (Harbach & Howard, 1997) and the other borne on the metepimeron posterior to the sensilla. The occurrence of metapleural scales in species of *Tripteroides* and their relation to subgeneric groupings in *Toxorhynchites* requires investigation.

The colour pattern of scales on the scutum is often a group character. The scales may be largely uniform in colour or show a contrasting pattern of light and dark stripes or patches. In some groups there may be a less definite pattern of light and dark areas, the appearance depending on the angle of observation, particularly in the case of metallic-coloured scales. The dark areas may form longitudinal stripes or distinct dark spots. In addition to any pattern of light and dark areas, the scutum may bear patches of pale scales.

Wing

The wing is one of the elongate bilaterally paired organs of flight attached to the mesothorax. The surface of the wing consists of a thin transparent or translucent sheet of cuticle (**wing membrane**) that supports a network of tubular thickenings (**veins**), which are clothed with scales on both the dorsal and ventral surfaces and often contain blood, tracheae and nerves.

The wings of mosquitoes have the venation shown in Fig. 8a. The venation is fairly uniform but some differences are useful for distinguishing genera. The terminology currently in use for the wing veins is adopted from Colless & McAlpine (1970). The principal longitudinal veins of the wing (from anterior to posterior when the wing is outstretched perpendicular to the thorax) are the **costa, subcosta, radius, media, cubitus** and **anal vein**. The branches of these veins are indicated by numbers, thus the anterior branch of the media is **media-one** (M_1) and the posterior branch is **media-two** (M_2). Short, usually transverse veins between two longitudinal veins are known as **crossveins**. In mosquitoes, the conspicuous crossveins are the **humeral crossvein** between the costa and subcosta, the **radiomedial crossvein** between the radius and media, and the **mediocubital crossvein** between the media and cubitus. An inconspicuous crossvein, the **arculus**, is located between the radius and cubitus near the base of the wing. The short thickened part of the radius basal to the arculus is known as the **remigium**. The remigium may be with or without conspicuous setae, which may be useful for distinguishing species and groups. Areas of wing membrane delimited by veins or by veins and the wing margin are termed **cells**. Each cell is named after the vein immediately anterior to it or after the posterior element if the vein is formed by the fusion of two vein branches, *e.g.* **cell R_5** is posterior to **radius-four-plus-five (vein R_{4+5})**. The surface of the wing membrane, *i.e.* the surface of the cells, is covered with minute spicules, the **microtrichia** (singular **microtrichium**). The microtrichia are similar in all mosquitoes with the exception of species of *Uranotaenia*, where they are extremely minute, numerous and not visible at low magnification.

The costa forms the thickened anterior border of the wing from the base to the apex where it continues into a thickening on the posterior margin. The posterior margin extends from the apex to a notch, the **axillary incision**, near the base of the wing. The posterior section of the wing basal to the axillary incision comprises three membranous lobes, the **alula, upper calypter** and **lower calypter**. The alula is a small lobe bounded proximally by the upper calypter and distally by the axillary incision. It covers the upper calypter when the wing is folded at rest over the abdomen. The lower calypter connects the wing to the thorax.

The alula of most mosquitoes possesses a projecting fringe of narrow (piliform) scales, but broad recumbent scales occur on the dorsal margin in species of *Eretmapodites*, *Heizmannia* and some species of *Aedes* (*e.g.* species of the subgenera *Cancraedes* and *Lorrainea*), *Mimomyia*, *Shannoniana*, *Uranotaenia* and *Wyeomyia*. Some species apparently have both types of scales, but whether or not this coincides with phyletic lines is unknown. Within *Psorophora*, the alula bears vestiture in species of *Janthinosoma* and *Psorophora*, but it is normally bare in species of *Grabhamia*.

The upper calypter bears one or more setae or hair-like scales in species of most genera of Culicinae. These setae are absent in species of *Malaya*, *Hodgesia*, *Toxorhynchites* and *Uranotaenia*, generally absent in species of *Udaya* and *Zeugnomyia*, and are sometimes reduced or absent in species of *Aedes*. The margin of the upper calypter in species of other genera may bear one or two setae, a short row of setae or the entire border may be lined with setae.

Two interesting formations characterize the wings of certain genus groups. These are the **precubital furrow** and the **cubito-marginal ridge**. The phylogenetic significance of the precubital furrow was noted by Colless (1979). The furrow is a depression with a vein-like thickening of variable distinctness running along the anterior cubitus in cell M_{3+4}. It is often at least partially hidden by scales projecting from the **cubitus anterior** (CuA), but it is usually easily seen when viewed at an angle from a distal or anterior direction. The furrow has only been observed in species of *Anopheles*, *Bironella* and *Toxorhynchites*. The cubito-marginal ridge has the appearance of a submarginal fold in the wing membrane running across the apex of the cubitus anterior. It is most easily observed when the posterior margin of the wing is viewed obliquely. The ridge of the fold is an inverted V-shaped formation characteristic of species of *Toxorhynchites*.

Scaling

The veins of the wing are clothed with scales (which are specialized flattened setae without innervation) on both the dorsal and ventral surfaces, and the apex and posterior margin of the wing bear a fringe of scales known as the **wing fringe**. The scales vary considerably in size and shape on different veins of the wing. The wings are usually entirely dark-scaled, but many species have some pale scales either intermixed or in patches. Distinct patches of pale and dark scales are known as **wing spots**. The wing fringe consists of four rows of scales (Fig. 8c), a row of long fusiform **fringe scales** (**primary fringe scales**) along the dorsal surface of the wing margin, a row of shorter **secondary fringe scales** along the ventral margin and two rows of small linear **tertiary fringe scales**, a row along the dorsal margin alongside the row of primary fringe scales and the other along the ventral margin alongside the row of secondary fringe scales. The development of the **dorsal tertiary fringe scales** appears to be diagnostic for well-defined genus-group taxa (Reinert, 2007).

Wing spots

The wings of most species of *Anopheles* and certain other mosquitoes bear a pattern of dark and pale **wing spots** (Fig. 8d,e) that correspond to areas with dark and pales scales on the veins. The spots on the costa and anterior veins are usually regular and distinct and are identified using the standardized nomenclature established by Wilkerson & Peyton (1990) based on morphological reference points, which include the humeral crossvein, the short crossvein-like **subcosta-two** (Sc_2) and the splitting of the radius into **radius-one** (R_1) and the **radial sector** (R_s). The names are those which have been used for the wing spots of anopheline and *Finlaya* species but there should be little difficulty in applying them to the wing spots of species of *Aedeomyia*, *Culex* (Mimeticus Group), *Lutzia* (subgenus *Lutzia*), *Orthopodomyia*, *Psorophora* and *Uranotaenia*.

Dark and pale spots occurring on the stem and branches of posterior veins can usually be designated by the adjectives basal, median, preapical or apical preceded by the abbreviation of the vein, *e.g.* **M_1 median pale spot**. Pale spots of the wing fringe are termed **pale fringe spots** and are named after the vein at the end of which they occur, *e.g.* **1A pale fringe spot**. Since the spots on the costa and anterior veins of species of *Anopheles* are usually constant and distinct, as are also the number and position of pale fringe spots, they provide important taxonomic characters for identification and classification. The standardized nomenclature for the costal wing spots of *Anopheles* is shown in Fig. 8e.

In fresh specimens, the darkest areas of the wing are often jet black and the palest areas are white to yellow. In some species the light and dark scales are mixed on some of the veins, giving a speckled appearance, as in species of *Mansonia*. The pale scaling is highly reduced in some species and species groups whereas in others the wings are entirely dark-scaled.

Halter

The halteres are small knobbed appendages of the metathorax. They are undoubtedly highly modified, reduced wings, which develop in non-functional metathoracic wings of the pupa (Figs 15b and 25a,b). The halter (Fig. 8b) consists of a thick base, the **scabellum**, a slender stem-like **pedicel** and a knob-like **capitellum**. The halteres are important appendages for maintaining the equilibrium and coordinated movement during flight. The capitellum usually bears scales, the colour of which is sometimes used in taxonomic descriptions to distinguish species.

Legs

The legs (Figs 1 and 9) are long, slender paired appendages of the pro-, meso- and metathoracic segments. From anterior to posterior, *i.e.* from pro- to metathorax, the legs are identified as **forelegs**, **midlegs** and **hindlegs**. The hindlegs are the longest. Each leg consists of six segments, a short basal segment, the **coxa**, which articulates with the thorax, followed by a small **trochanter**, a long **femur**, **tibia**, **tarsus** and the terminal **posttarsus** bearing a pair of **ungues** (singular **unguis**, claw). The bases of the meso- and metacoxae bear a lateral postarticular area known as a **meron**, the **mesomeron** and **metameron**, respectively. The presence or absence of scales on the metameron is sometimes a useful taxonomic character. The positional relationship of the mesomeron and the hindcoxa is also of importance. The base of the hindcoxa is usually but not always well below the dorsal margin of the mesomeron in species of *Aedes*. It is sometimes slightly below the dorsal margin of the mesomeron in species of *Heizmannia*, *Malaya* and the nominotypical subgenus of *Tripteroides*. It is more or less in line with or slightly above the dorsal margin of the mesomeron in most other mosquitoes.

The tarsus is composed of five subsegments termed **tarsomeres**. The first tarsomere is the longest and the fifth, which bears the posttarsus, is the shortest, except on the foreleg of the male where the fifth tarsomere may be longer than the fourth. The posttarsus of each leg ends in a small pair (usually) of ungues (claws) and a pair of pad-like **pulvilli** (singular **pulvillus**), one arising below the base of each unguis. The pulvilli are weakly developed and inconspicuous in most mosquitoes and distinctly developed and conspicuous in mosquitoes of Culicini (Fig. 9d). The ventral sclerite of the posttarsus, the **unguitractor plate**, on which the depressor muscle of the posttarsus is inserted, is produced apically as a pad-like or spiculose structure termed the **empodium**. The empodium may be developed differently in different taxa, but it has not been studied comparatively to determine its potential taxonomic value. The ungues may be simple or one or both of one or more legs may be toothed. In males, one unguis on the fore- and midlegs is usually distinctly larger than its mate and usually bears a tooth, and sometimes two teeth in certain taxa. Within the genus *Anopheles*, the **foreungues** of the male seem to show little variation. In some groups, in both sexes, the **forefemur** is swollen in the basal half. The coxae and trochanters usually bear some large setae, and there are shorter setae at the ends of the femora and tibiae. There is a posteromedian projection on the midcoxa which is sometimes prominent in dried specimens.

For descriptive purposes, the coxa is described as though it is in normal resting position, with the surfaces being the inner (mesal), outer (lateral), anterior and posterior. The other segments of the leg are described as though the leg is outstretched and perpendicular to the thorax, with the surfaces being anterior, dorsal, posterior and ventral. The individual segments of the legs of each thoracic segment, like the legs themselves, are identified by adding the prefixes fore-, mid- or hind- as appropriate, *e.g.* forecoxa, midtrochanter, hindfemur, *etc.*

Ornamentation

The coxae and trochanters usually bear some large setae, and there are shorter setae at the ends of the femora and tibiae. Except for the coxae, and sometimes the trochanters, and rarely the femora, the legs are entirely clothed with small appressed scales. The coxae usually bear patches of scales associated with the setae, but these are absent in certain groups.

The ornamentation of the legs is also of great importance in identification and, as with the wings, it depends on the distribution of pale and dark scales. Many species and groups, however, have entirely dark-scaled legs. Quite often though, the femora and tibiae have apical pale marks (**knee spots** on the femora) and are more or less pale beneath. The femora of some species have subapical pale bands, and the femora, tibiae and some of the tarsomeres may bear small scattered patches of pale scales referred to as speckling. The tarsi, especially the hindtarsi, are often ornamented with conspicuous bands of white or yellowish scales at the joints of the tarsomeres. These bands may be very narrow and confined to the apices of individual tarsomeres, or broader and spread across the joints onto the bases of the following tarsomeres. In some species and groups, one or more of the tarsomeres of the mid- or hindtarsi may be entirely pale-scaled.

Abdomen

The abdomen (Figs 1 and 10) is composed of eight readily visible segments, plus two more, the ninth and tenth, which are much modified in connection with the **genitalia** (plural, no singular form). The visible segments are each composed of a dorsal plate (**tergum**, plural **terga**) and a ventral plate (**sternum**, plural **sterna**). The segments and the terga and sterna are identified from anterior to posterior by the Roman numerals I–X. The tergum and sternum of each segment are jointed at the sides by the **pleural membrane**, and the segments are joined to one another anteriorly and posteriorly by the **intersegmental membranes**. When the abdomen is empty, the pleural and intersegmental membranes are folded and invisible, and each segment is retracted slightly into the one in front. When a mosquito is freshly engorged with blood (most females) or plant exudates (females and males), the abdomen is greatly distended, with the membranes fully expanded and the terga and sterna all separated from one another. The eight visible segments are similar, except for the first, which articulates with the metathorax and is shorter and has a membranous sternum. The tergum of the first segment bears a small lateral lobe or wing on either side, the **laterotergite**, separated from the median dorsal part by a longitudinal furrow. After males emerge from the pupa, the eighth segment with the genitalia rotates through 180° so that the sternum is dorsal and the tergum is ventral in position.

Ornamentation

The terms dorsum and venter are sometimes used to refer to the dorsal and ventral surfaces of the abdomen. The terga and sterna are lightly clothed with setae in anophelines whereas setae are normally only present on the lateral and posterior margins in culicines. Setae are longest and more numerous on **tergum I**.

The abdomen is completely clothed with broad decumbent scales in all culicines, except sometimes on the first and last segments. There is rarely a complete covering of scales in anopheline mosquitoes, but some scales may be present, especially on the last few segments. They may be scattered over the terga with a tendency to form **tufts** at the posterolateral corners. Scales are common on the posterior margins of the sterna of some anophelines where they may form tufts on the posterior midline. Scales are usually present on the cerci of females and the lateral surfaces of the gonocoxites of males. The terga, mainly posterior terga, of some species and groups bear tufts of scales at the posterolateral corners. Terga VI and VII of *Toxorhynchites* bear colourful caudolateral scale-tufts of long narrow outstanding scales.

Tergum I of culicines shows differences in scaling between genus groups. The laterotergite may be either bare, as in species of *Culex* and *Culiseta*, or bear scales, as in species of *Toxorhynchites* and most *Aedes*. The pattern of pale scales on the terga, and to some extent the occurrence of dark scaling on the sterna, is of taxonomic importance.

Female genitalia

The genitalia of female mosquitoes (Figs 10b and 11) are considered to include the structures posterior to abdominal segment VII. The tergum and sternum of segment VIII are included because they often exhibit differences in shape and the development of setae. In comparison with male genitalia, morphological characters of the female genitalia have been used far less in distinguishing taxa even though they often exhibit distinctive generic, subgeneric and specific differences.

The eighth abdominal segment (**segment VIII**) is narrower than the preceding segments and is normally at least partially retractile. **Tergum VIII** is variable in shape but is usually more or less trapezoidal. It bears setae posteriorly and scales may be present. In females of the subgenus *Mansonioides* of *Mansonia*, the posterior margin of the tergum bears a row of strong spiniform setae. **Sternum VIII** is also variable in shape, but it is usually wider than long and the posterior margin may have a median notch or indentation. It bears setae and sometimes scales.

Unlike males, the terminal abdominal segments of female mosquitoes do not invert (see below), *i.e.* the structures retain their original dorsal and ventral positions. Segment IX is much reduced – the sternum is essentially absent and the tergum (**tergum IX**) is usually a small narrow sclerite with or without a few setae. However, it may be bilobed, divided mid-dorsally into two plates as in species of *Eretmapodites*, *Opifex* and *Udaya*, or absent as in species of *Aedeomyia*.

Paired **cerci** (singular **cercus**) are situated posterior to tergum IX. The cerci vary in size and shape in different genera and species, but are generally cylindrical in anophelines and flap-like in culicines. They bear variously developed setae, and scales may be present or absent.

Another flattened setose lobe, the median **postgenital lobe**, is located immediately below (ventral to) the cerci. The postgenital lobe varies in length and shape, and usually has an emarginated terminal border. It varies greatly in size, shape and vestiture, providing characters of specific and higher rank. The lobe bears rather numerous setae in culicines and a pair of prominent terminal setae in species of *Anopheles*.

A membranous **proctiger**, located between the bases of the cerci and the postgenital lobe, surrounds the terminal opening (anus) of the alimentary canal. The postgenital lobe and **sternum VIII** enclose a cavity (atrium, genital pouch, **vagina**), into which the common oviduct and the spermathecal and accessory gland ducts open. The sperm storage organ (**spermatheca**, plural **spermathecae**), an ectodermal caecum of abdominal segment VIII, is attached to the anterodorsal wall of the vagina via the opening of the common oviduct (the gonopore). The spermatheca may consist of one or three sperm storage reservoirs (**spermathecal capsules**), each connected separately to the common oviduct by a **spermathecal duct**. The spermathecal capsules vary in size, shape and number, with differences in number being generally supraspecific in nature. The capsules are typically spherical, darkly pigmented and have minute 'pores' on their surface. Three capsules are normally present, with one slightly larger than the other two. A single large capsule is present in species of Anophelinae and *Aedeomyia*, *Ficalbia*, *Uranotaenia* and some groups of *Aedes*, *Mimomyia*, *Topomyia*, *Tripteroides* and *Wyeomyia* of Culicinae. One of the three capsules may be rudimentary, as in species of *Hodgesia*. **Spermathecal capsule pores** are usually located near the orifice of the spermathecal capsule and appear as small clear spots. They are larger and scattered over most of the surface of the single large capsule in anophelines and species of *Aedeomyia*.

The outer edges of the vagina are sclerotized, forming the **upper** and **lower vaginal lips**, which articulate at a small anterior thickening on either side referred to as a **hinge**. The roof of the vagina may bear a pair of sclerites (**upper vaginal sclerites**) that extend medially from either side of the upper vaginal lip adjacent to the hinges. Additionally, although usually absent, the floor of the vagina may bear a pair of **lower vaginal sclerites**.

The upper vaginal lip adjoins the postgenital lobe and is fused with it in some taxa, *e.g.* species of *Aedes* and *Orthopodomyia*. The lower vaginal lip is often weak and indefinite. In some female mosquitoes, it is associated with a small median sclerite, termed the **insula**. The insula is not discernible and seems to be absent in some mosquitoes. When distinct, it is a variously developed plate attached to the median area of the lower vaginal lip. The insula usually has the appearance of a tongue- or lobe-like enlarged portion of the lip. It usually bears a patch or double row of setae, which differ in number in different species or may be absent. The insula may also bear one or more **tuberculi** (singular **tuberculus**), which are small button-like structures having the appearance of campaniform sense organs, sometimes bearing a minute seta. The insula has received little attention in most taxa, but its form is known to correspond with two suprageneric groups within the composite genus *Aedes*: the *Aedes* group, in which the insula is tongue-like and void of setae, and the *Ochlerotatus* group, where it is lip-like and bears well-developed setae (Reinert, 2000c).

Male genitalia

The terminal abdominal structures (terminalia) of male mosquitoes involved in copulation are very important in classification and identification (Figs 10c, 12 and 13). In many groups of culicines, specific identification depends largely on characters of these structures, which have to be dissected and mounted on microscope slides for careful examination. This is normally not necessary in the case of anophelines, but since the subgenera of *Anopheles* are based mainly on certain features of the male terminalia, and since certain structures often show differences between species groups, they cannot be ignored even if they are not used for routine identification.

As for females, the genitalia of male mosquitoes are considered to include the structures posterior to abdominal segment VII, which rotate 180° a few hours after emergence of the male from the pupa, so that structures that were originally dorsal become ventral and vice versa (Fig. 10c). This torsion has caused much confusion in the morphological and taxonomic literature. In an effort to resolve this, the terms 'tergal' and 'sternal' have sometimes been used. This usage is equally confusing because many parts of the genitalia are neither tergal nor sternal in origin. For this reason, the terms 'dorsal' and 'ventral' are used here in the prerotation sense, *i.e.* with their true morphological connotation, to describe position and indicate structures which were originally dorsal and ventral, respectively, the terga being actually ventral and the sterna dorsal in mature males. The terminology adopted here is the same used by Harbach & Knight (1980), except that 'basal piece' is replaced by 'gonocoxal apodeme' (Sallum *et al.*, 2000). Likewise, the term **genitalia** is used in preference to terminalia because most of the appendages posterior to segment VIII play a role in copulation.

Although segment VIII is not usually considered part of the genitalia, it is treated in association with them. Tergum VIII (sometimes incorrectly referred to as sternum VIII) often bears distinctive morphological features, *e.g.* shape and the development and distribution of setae and scales. In many species, the posterior margin has a deep median indentation and in some aedines, *e.g.* species of *Eretmapodites*, it is divided into two sclerites. The tergum bears variously developed setae, which are generally long and project over the proximal parts of the genitalia. In contrast, sternum VIII normally lacks distinctive features.

The male genitalia comprise the following main parts: (1) tergum and sternum of the ninth segment (segment IX), (2) paired appendages (**gonocoxopodites**, forceps) of the ninth segment, (3) anal segment (**proctiger**, harpago, tenth segment) and (4) the **phallosome** surrounding the gonopore between the gonocoxites, proctiger and sternum IX, comprising the **parameres**, **prosophallus**, **aedeagus** (phallus) and **opisthophallus**.

The tergum of the ninth abdominal segment (**tergum IX**), ventral in post-rotation position, is usually a narrow bilobed sclerite bearing caudally directed setae on each of the lobes. The tergum is sometimes a much broader plate with or without weakly developed setae, *e.g.* males of *Toxorhynchites* and some *Uranotaenia*. The sternum of the ninth segment (**sternum IX**) sometimes bears a few setae, but this is usually not mentioned in species descriptions. Segment IX of anophelines is poorly developed and the tergum and sternum both lack setae.

The **gonocoxopodite** is one of a bilateral pair of appendages of the genital segment (abdominal segment IX). The gonocoxopodite is a two-segmented appendage consisting of a bulbous proximal coxopodite, the **gonocoxite** (basimere, basistyle, sidepiece), and a variously developed but usually slender distal **gonostylus** (clasper, distimere, dististyle, style). The gonostylus is articulated with the gonocoxite and is flexed by muscles that originate in the gonocoxite. The copulatory (intromittent, phallic) organ, the **aedeagus** (mesosome, phallosome), lies between the bases of the gonocoxites and is connected proximally with a basomesal apodeme of each gonocoxite (**gonocoxal apodeme**, basal piece) by a pair of small parameral apodemes (**parameres**). Each gonocoxite bears setae and sometimes scales, principally laterally, and various lobes and processes on its mesal surface that provide useful taxonomic characters, most notably the development of the **subapical lobe** and its specialized vestiture in species of *Culex*. Each gonocoxite bears a variously shaped basomesal lobe that may or may not be united with the lobe of the opposite gonocoxite. The lobe may be developed in the form of a **claspette** or as a more or less setose plate-like structure referred to as the **basal mesal lobe**. According to Christophers (1922), secondary incisions develop and separate the claspettes from the gonocoxopodites. As Edwards (1941) implied, the subapical lobe of *Culex* males is probably homologous with the claspette and basal mesal lobe. In the most primitive form, the only modification of the gonocoxite is a small setose basal mesal lobe, which is found in species of the genera *Aedeomyia*, *Culiseta*, *Ficalbia*, *Hodgesia*, *Malaya*, *Orthopodomyia*, *Toxorhynchites*, *Uranotaenia* and some genus-group taxa of Sabethini. This lobe is sometimes separated from the rest of the inner (mesal) face of the gonocoxite by a membranous area. The lobe is a prominent more distal protrusion (subapical lobe) in *Culex*, which bears specialized setae and is divided into two lobes in some subgeneric taxa. In some genera, the lobe is partly or entirely developed as an appendage known as the claspette. In many taxa of the composite genus *Aedes*, the claspettes form a pair of appendages borne at the sternobasal margin of each gonocoxite. The gonocoxites of some species of *Aedes* bear basal and apical lobes, and those of species of *Eretmapodites* may bear two or more lobe-like protuberances or appendages and collections of uniquely developed scales.

The **gonostylus** is most often a slender, simple, flexible arm-like structure borne near or at the apex of the gonocoxite, bearing a number of minute setae and a specialized terminal seta known inappropriately as the **gonostylar claw**. In species of some subgenera of *Aedes* (*e.g. Aedimorphus* and *Pseudarmigeres*), the gonostylus is strongly modified in shape and the claw is borne preapically. The gonostylus is attached distinctly before the apex of the gonocoxite in males of the subgenus *Neomelaniconion* and it usually bears scales and long setae in males of the genus *Eretmapodites*. In males of *Aedeomyia*, the claw is developed as a peculiar comb-like structure. In species of some sabethine genera, notably *Sabethes* and *Wyeomyia*, the gonostylus bears variously developed highly complex lobes which provide reliable diagnostic characters for species recognition.

The **proctiger** is a largely membranous flattened lobe with an expanded base. It consists of a pair of elongate lateral sclerites (**paraprocts**) and a more or less distinctly sclerotized dorsal area or pair of dorsal sclerites (**cercal sclerites**) between them, each usually bearing one or more small setae (**cercal setae**, absent in some taxa). **Tergum X**, usually developed as a pair of dorsolateral sclerites (basolateral sclerotizations), is located at the base of the proctiger. The dorsolateral sclerite often articulates with tergum IX and the gonocoxal apodeme, and is usually united laterally with the base of the paraproct. The paraprocts tend to exhibit generic- and tribal-level distinctions. For example, the presence of a crown of spicules at the apex of the paraproct is a unique feature of species of Culicini. The paraprocts of other mosquitoes, *e.g.* the males of *Culiseta* and *Toxorhynchites*, may have a few strong apical 'teeth', but never a collection of spicules. In most males of *Aedes*, the paraprocts are simple, without teeth. In males of *Aedeomyia*, *Anopheles* and *Uranotaenia*, the proctiger is usually, or at least apparently, completely membranous, *i.e.* without paraprocts. Cercal setae are absent in males of *Anopheles*.

The **aedeagus** (mesosome, phallosome) is the central body of the phallosome surrounding the opening of the genital duct and serving as the intromittent organ; however, in certain male mosquitoes, *e.g.* males of *Verrallina*, the aedeagus is the distal part of the phallus. In most male mosquitoes, the aedeagus is formed chiefly of a pair of lateral sclerotizations (**aedeagal sclerites**), but in males of *Verrallina* and certain other mosquitoes, the aedeagus consists of a supporting sclerite on either side of the phallus.

The **opisthophallus** is a sclerotized element of the phallosome located between the aedeagus and the proctiger. It has not been studied or recognized in most mosquitoes but it is developed in species of Culicini, *Maorigoeldia* of Sabethini and *Verrallina* of Aedini. It usually consists of a transverse dorsal bridge supported on each side by a sclerite (**opisthophallic sclerite**) that is homologous with the **lateral plate** of *Culex* males and the dorsal paramere of *Deinocerites* males. Unlike males of *Verrallina*, opisthophallic sclerites of males of Culicini are not always joined dorsally (prerotation sense) by a sclerotized transverse bridge. A sclerotized bridge is apparently absent in males of all exclusively New World genus groups of Culicini, except in males of the subgenera *Allimanta*, *Microculex* and *Neoculex* of *Culex*. The lateral plates often bear variously developed 'teeth' and processes that provide characters of specific importance.

The **prosophallus** is only known in males of *Verrallina* where it is a lobe-like structure located between the bases of the gonocoxites. It consists of a pair of sclerotized arms or plates (**prosophallic sclerites**), each of which is united basally with the base of the phallus and a paramere. The homology of the structure is unknown. Belkin (1962) stated that the prosophallus of dixids was probably homologous with the claspette of mosquitoes, but this seems unlikely since males of *Verrallina* have a well-developed, narrowly connected pair of basal mesal lobes which are very probably homologous with the claspettes of certain other aedine mosquitoes. If the prosophallus of dixids is homologous with the claspette of aedines, it would seem that the prosophallus of *Verrallina* males is an accessory structure of the phallus.

In current mosquito morphotaxonomic nomenclature, the **paramere** is one of a pair of lateral sclerites of the phallosome that articulate with the aedeagus and the gonocoxal apodemes. Parameres are not differentiated in males of *Hodgesia*.

Egg

The egg (Fig. 14) is the first life stage of an insect. The importance of the egg stage in taxonomy was realized when Hackett (1934) showed that features of the eggs provided the best means of delimiting species (as varieties) of the Maculipennis Complex of *Anopheles* in Europe.

The orientation of the egg is described relative to the orientation of the developing embryo. The **anterior end** is the end bearing the micropyle, which is adjacent to the head of the embryo; the **posterior end** is the extremity opposite the end bearing the micropyle, the end opposed to the hindmost part of the embryo; the **dorsal surface** is opposite the dorsum of the embryo, which is the lower surface of eggs that rest horizontally on the water; and the **ventral surface** is the surface of the egg opposite the venter of the embryo, the upper surface of those eggs which rest horizontally on the surface of water.

Culicid eggs exhibit a wide array of structural modification. A fairly typical egg is shown in Fig. 14f. The structures illustrated in this figure are developments of the **outer chorion**. The outer chorion is the outermost of two layers of chorion secreted by the follicular epithelium of the female mosquito. For this reason, external egg morphology is actually a phenotypic character of the adult female, but for taxonomic purposes this is unimportant.

The surfaces of the egg, *i.e.* the outer chorion, usually appear to be formed of fine **tubercles** that may differ in shape and relief and may give rise to characteristic patterns. The ventral (upper) surface of anopheline eggs (Fig. 14c) usually bears a ridge-like **frill** which commonly encloses a markedly differentiated area termed the **deck**. The frill may (1) completely or incompletely enclose an undivided deck, (2) in essence be two frills, one surrounding the anterior part and the other the posterior part of a divided deck, or (3) have the form of a collar around the anterior or posterior end of the egg. The deck is formed of tuberculate outer chorion and usually bears a variable number of large **lobed tubercles** at the anterior and posterior ends. Either side of the egg usually bears a longitudinal outgrowth of the outer chorion called a **float**. Each float contains a hollow space and bears a series of transverse ridges on the outer wall. The surface of culicine eggs (Fig. 14f,g) is normally a uniform network or pattern of impressed and raised elements of the outer chorion (**outer chorionic sculpture**), often consisting of cells (**outer chorionic cells**) bounded by a reticulum (**outer chorionic reticulum**). The **micropyle** at the anterior end of mosquito eggs is surrounded by a thin area of outer chorion called the **micropylar disc**. A thickened annular ridge of outer chorion, the **micropylar collar**, encircles the micropylar disc.

Studies of egg morphology are usually accomplished with the aid of scanning electron microscopy (SEM). Specific differences in anopheline eggs may be found in the shape and size of the eggs, the ratio of length to width, the shape and size of the floats and the number of **float ridges**, and various characteristics of the deck and frill. Culicine eggs differ in features of the outer chorionic sculpture, especially the size and shape of cells and associated **outer chorionic tubercles**, the small rounded or irregularly shaped prominences of the outer chorion.

Larva

The larva is the second stage in the life cycle of a holometabolous insect. Growth of the mosquito larva is punctuated by a series of three moults, one between each of the four stadia (*i.e.* there are four instars). Development from one instar to the next is marked by conspicuous discontinuous changes (increases) in body size, setal branching, the number of fixed cuticular processes (*e.g.* pecten spines) and the extent of sclerotization (*e.g.* the saddle). In general, the form of features which first appear in the newly emerged larva become increasingly elaborate in structural details in subsequent instars, with the most striking changes occurring between the first and second instars. The definitive form of the body parts and the relative positions of the setae are already evident in the first instar, except that certain setae of the thorax and abdomen are not yet present. There are certain characters that are specific to each instar, yet the third and fourth instars are sometimes difficult to distinguish. Since keys for the identification of species in the larval stage are based on characteristics of the mature larva, it is important to be able to recognize fourth-instar larvae. Fourth instars can be distinguished from third instars of the same species by their larger heads, and usually by the proportionately broader posterior opening of the head. Older fourth-instar larvae can also be recognized by the presence of the developing pupal setae and trumpets beneath the larval integument.

The general appearance of the fourth-instar larva is shown in Fig. 15a, a culicine larva. The body of the larva has three principal regions, the **head**, **thorax** and **abdomen**. The head is a sclerotized capsule bearing the **mouthparts**, eyes and **antennae**. It is connected to the thorax by a membranous **cervix**, which bears a lateral pair of setae and a small mid-ventral sclerite, the **ventromedian cervical sclerite**. The cervix allows the head to be rotated, which in the case of anophelines can be rotated 180° to bring the mouthparts in a position for feeding on the surface film of the water. The thorax is composed of three fused segments, the **prothorax**, **mesothorax** and **metathorax**, and the abdomen is composed of nine apparent segments.

Head

The head of the mosquito larva (Figs 15a and 16) is a more or less globular structure, usually fairly heavily sclerotized and to a greater or lesser extent flattened dorsoventrally and broadened laterally. The central area of the dorsum of the head, the **dorsal apotome** (clypeus, frons), is bounded by the **frontal ecdysial lines**. The large lateral areas of the head between the frontal ecdysial lines and the ventral **hypostomal sutures** are collectively known as the **lateralia** (plural, no singular form). The hypostomal sutures are usually either complete from the **postcoila** at the anterior margin of the head to the **posterior tentorial pits** or incomplete posteriorly, *i.e.* they end before reaching the posterior tentorial pits. Both character states occur in species of a few genera. The sutures are often complete in larvae of *Anopheles*, usually complete but indistinct in larvae of *Ficalbia* and generally absent in larvae of *Mimomyia*. Among the subgenera of *Tripteroides*, the sutures are generally complete in larvae of *Polylepidomyia* and *Tripteroides*, incomplete in larvae of *Rachisoura* and nearly absent in larvae of *Tricholeptomyia*. They reach the posterior tentorial pits in larvae of *Orthopodomyia* and *Coquillettidia*, but are often incomplete or interrupted anteriorly, respectively.

A narrow dorsal sclerite borne on the anterior margin of the head is distinct in most mosquito larvae. This sclerite, the **median labral plate**, is separated from the dorsal apotome by the **clypeolabral suture**. The suture is absent when the median labral plate is unrecognizably fused with the dorsal apotome, which is characteristic of larvae of *Lutzia*, *Maorigoeldia* and the subgenera *Carrollia*, *Lasioconops* and *Oculeomyia* of *Culex*.

The appendages of the head include the **antennae** (singular **antenna**), the **lateral palatal brushes** (mouthbrushes) and the **mouthparts** (see below). The antennae are paired anterior appendicular appendages. Each antenna is attached anterolaterally to the head capsule and consists of a narrow ring-like basal scape and a large distal shaft comprised of the unrecognizably fused pedicel and flagellum. The shaft varies in length but is usually quite long and may be finely or coarsely spiculate or lack spicules. It may be more or less cylindrical or straight or bent with the distal part narrower than the proximal part. The distal part is elongate and articulated with the proximal part in larvae of *Coquillettidia* and *Mimomyia*. It is also elongate in larvae of *Mansonia* but it is shorter and fused with the proximal part. *Aedeomyia* larvae have distinctive antennae that are greatly swollen, strongly curved and longer than the head. The lateral palatal brushes are borne anteriorly on either side of the mouth. In most larvae, the brushes consist of numerous long specialized filaments for extracting food particles from the water, but in predatory larvae these structures are reduced to about 40 much thicker and strongly pectinate filaments that aid in grasping prey.

The posterior opening of the head into the cervix is known as the **occipital foramen**. The occipital foramen of most mosquito larvae is bounded by a rim-like sclerite known as the **collar**. A collar is absent in certain taxa, *e.g.* larvae of *Deinocerites*, the subgenus *Kitzmilleria* of *Culex* and many genera of Sabethini. The posterior opening of the head is generally circular, but it bears dorsolateral slit-like extensions in sabethine taxa which lack a collar. The 'slit-like' development of the occipital foramen in larvae of *Sabethes*, *Limatus* and *Wyeomyia* was first noted by Belkin *et al.* (1970) and subsequently used by Zavortink (1979) to distinguish larvae of *Runchomyia* (including *Isostomyia*) from those of *Trichoprosopon*, *Shannoniana* and *Johnbelkinia*. Zavortink pointed out that the phylogenetic significance and distribution of the slit-like state within the Culicidae was unknown. Judd (1996) later recorded the presence of dorsolateral slits in larvae of all sabethine genera except *Maorigoeldia*, *Tripteroides* and the three noted by Zavortink (1979). Judd also noted the absence of slits in the larva of *Wyeomyia personata*, one of several species recognized by Heinemann & Belkin (1977a, and later publications) as members of an informal 'Subgenus B', which was later described as the genus *Onirion* by Harbach & Peyton (2000). All species currently included in *Wyeomyia* appear to have dorsolateral slits. In addition to various sabethine genera, a transverse occipital foramen also occurs in larvae of *Aedeomyia* and a number of species of the subgenus *Ingramia* of *Mimomyia*, which appears to be polyphyletic as currently defined. The slits do not extend as far laterad in *Aedeomyia* larvae as they do in most sabethine larvae, but they are otherwise of a similar form.

Mouthparts

Because the head of mosquito larvae is basically prognathous, the mouthparts are considered as anterior appendages. The mouthparts consist of the paired **mandibles**, paired **maxillae** (singular **maxilla**) and the composite **labiohypopharynx**. Comparative anatomical studies of the mandible (Knight, 1971a; Harbach, 1977), maxilla (Knight & Harbach, 1977) and labiohypopharynx (Harbach, 1978) provide detailed information on their structural diversity within the Culicidae. Tanaka *et al.* (1979) also provided anatomical details of the mandibles and maxillae, but their terminology differs in large part from the terminology used here.

The mouthparts of mosquito larvae have been of interest to morphologists and taxonomists since Meinert (1886) first described the mouthparts of a species of *Anopheles* and two species of *Culex*, but in general little use has been made of their considerable structural diversity in mosquito taxonomy. Structural features of the mouthparts of importance principally include modifications of the mandibles and maxillae. The modifications of the maxilla seem to be particularly numerous and significant.

Mandibles

The mandibles (Fig. 17a,b) are typically dorsoventrally flattened appendages implanted somewhat obliquely with the mesal ends, the part of each mandible nearest the mouth, bearing strongly sclerotized teeth. In resting position, the teeth of the mandible on each side rest on the labiohypopharynx, which lies between and posterior to them. The structural diversity of the mandible provides differential characters at tribal, generic and specific levels.

The mandibles bear seven principal composite structures and a number of minor features. The principal structures include the **mandibular teeth**, the spicule-bearing **mandibular lobe**, the **sellar setae** and several structures comprised of variously developed cuticular projections, the **mandibular comb**, **mandibular brush**, **mandibular rake** and **mandibular sweeper**. Differences in the development of these structures is associated with different methods used by larvae to obtain food. Most of the mandibular structures are present in larvae of all higher-level taxa, but there is a tendency for certain structures to be atrophied or hypertrophied in relation to feeding habits. In general, the mandibles of most culicine larvae are structurally more complex than the mandibles of anopheline, sabethine and toxorhynchitine larvae.

The mandibular teeth are greatly varied in structure throughout the Culicidae and potentially provide differential characters at all levels of classification. The teeth comprise a cluster of heavily sclerotized cuticular projections that includes **dorsal teeth**, **ventral teeth**, **accessory teeth** and an **auxiliary ventral tooth**. The accessory teeth and the auxiliary ventral tooth are often absent. Some species have more numerous teeth that seem to be an adaptation for crushing large food particles between the teeth and the labiohypopharynx (Schremmer, 1950). The mandibular teeth are innervated (Yin, 1970) and may be pressure sensitive and capable of informing the larva about the nature of the food.

The mandibular lobe is a lightly sclerotized, rounded protuberance borne on the mesal margin of the mandible immediately posterior to the mandibular teeth. The lobe varies considerably in form and bears as many as five groups of variously developed spicules, **mandibular lobe spicules 1–5**, which apparently function in brushing debris from the surfaces of the labiohypopharynx (Gardner *et al.*, 1973). One or more of the groups may be absent in varying combinations. The spicules are highly reduced or absent in predacious larvae.

The mandibular comb is a linearly arranged group of variously developed spicules borne on the ventroanterior margin of the mandible. The comb is not developed in *Uranotaenia* larvae and the predatory larvae of other genus groups; however, a single, branched spicule closely associated with the lateralmost spicule of the mandibular brush is present in some anopheline larvae.

The mandibular brush is a well-developed linearly arranged group of prominent spicules borne in positions from the dorsoanterior to the dorsomesal margin of the mandible. The brush is borne dorsoanteriorly in most larvae and dorsomesally in predacious larvae, in which it has a reduced number of spicules. In general, all elements of the brush are similar, but in anopheline and *Uranotaenia* larvae the spicules are noticeably different in the anterior and posterior parts of the brush, and in anopheline larvae the most lateral spicules may be flattened and plumose. In filter feeders, the mandibular brush is highly developed and the numerous flexible spicules usually bear processes to aid the removal of food particles from the lateral palatal brushes and direct them toward the mouth. In general, the spicules are reduced in number and length in larvae that feed on larger food.

The mandibular rake is a group of variously developed spicules borne ventroposterior to the ventral teeth. The anterior spicules are typically blade-like, **mandibular rake blades**, whereas the posterior spicules, **mandibular rake spicules**, are variously developed. The rake apparently serves to free food particles entangled among spicules of the maxilla. Larvae which have a single or reduced rake blade probably have no need to clean food from the maxilla – the mandibles are developed to capture and grasp large food particles.

The mandibular sweeper comprises a linear group of filamentous spicules along the dorsobasal margin of the mandible. The sweeper spicules are often divided into an adoral group, **mandibular sweeper 1**, and an oral group, **mandibular sweeper 2**. When present, the latter is located mesal to the former and arises within a loop-like band of cuticle, the **arcuate thickening**. The sweeper spicules project inward toward the mouth (Nuttall & Shipley, 1901) and apparently function to sweep food particles into the pharynx (Schremmer, 1950).

In addition to the various non-sensory spicules, a group of usually four prominent innervated setae (Yin, 1970) arise from a saddle-shaped depression, the **sella**, located on the laterodistal margin of the mandible. The sellar setae (**seta 2-Mn**) of *Anopheles* larvae are capable of capturing large food particles and directing them toward the mouth (Schremmer, 1950), and they may be able to detect the size and nature of food particles. The sellar setae are reduced in larvae that feed by grasping and breaking down large pieces of food.

Maxillae

The maxillae (Fig. 17c,d) are also typically flattened lobes, borne obliquely on the transverse margins of the lateralia ventral to the mandibles. For simplicity, the surface of the maxilla opposite a mandible is considered to be dorsal, the side bearing the

maxillary palpus is lateral, the side near the midline of the head is mesal and the surface adjacent to the lateralia is ventral (Knight & Harbach, 1977). In most mosquito larvae, the maxillary palpus is a cylindrical or fusiform lobe articulated or fused basally with the maxillary body. In many culicine larvae, the palpus bears a narrow basolateral band of cuticle, **merostipital sclerite**, which articulates with the base of the adjacent mandible. The sclerite is unrecognizably incorporated into the maxillary palpus of anopheline, toxorhynchitine and many culicine larvae, but its articulation with the mandible has been retained. The apex of the maxillary palpus bears as many as seven specialized sensory setae. The setae are all typically peg-like but three are usually foliform in anopheline larvae.

The principal part of the maxilla, the **maxillary body**, is comprised of the fused lacinia, galea and mesal part of the stipes. The mesal half of the maxillary body, the **laciniastipes**, represents the fused lacinia and stipes and bears one to three rows of variously developed spicules, the **laciniarastra** (singular **laciniarastrum**). The laciniastipes is sometimes distinctly demarcated from the lateral half of the maxillary body, the **galeastipes**, by a seam on the dorsal surface and a grove or furrow on the ventral surface. When developed, the former, termed the **dorsal maxillary suture**, extends from the apical **maxillary brush** to the base of the maxillary body; and the latter, the **ventral maxillary suture**, extends from a point immediately mesad of the maxillary brush to a small gap or slit (**basal notch**) located at the basal margin of the maxilla. In larvae of most genera, the laciniastipes and galeastipes are indistinguishably fused; they are separate and distinct only in larvae of certain aedine genera, *e.g. Opifex* and *Haemagogus*. When the laciniastipes and galeastipes are indistinct, the ventral maxillary suture is absent. The dorsal maxillary suture is usually present except in predatory species, *e.g.* larvae of *Aedes* (*Mucidus*) and *Toxorhynchites*, and species where the maxillary palpus is fused with the maxillary body (see below).

Thus, the principal components of the maxilla are the maxillary body, maxillary palpus and the **hypostomal sclerite** (Harbach & Knight, 1977, 1980). In most mosquito larvae, a cluster of variously developed setae, the **maxillary brush**, arises distally from the apex of maxillary body, or the galeastipes when distinguishable. The maxillary brush and the most mesal laciniarastrum are modified in various ways for feeding on different kinds of food. These modifications are most striking in larvae of certain genera and subgenera of Sabethini where either a modified maxillary brush plays a primary role in feeding (*e.g.* larvae of *Malaya* and *Maorigoeldia*) or the laciniarastrum has become the primary functional feature, leading toward the large toothed condition of *Sabethes* larvae.

In most mosquito larvae, the hypostomal sclerite ('cardo' of Harbach & Knight, 1980) is a small, independent plate bearing seta 6-Mx at the base of the maxilla. Knight & Harbach (1977) referred to this structure as the hypostomal sclerite because it appears to be a disassociated fragment of the cranium. The developmental state of the sclerite is easily determined by using seta 6-Mx as a landmark. In larvae of *Armigeres*, *Ficalbia* and *Toxorhynchites*, seta 6-Mx is borne on the anteroventral margin of the head without a trace of the sclerite, in most species of Aedini, Culicini, Culisetini and Mansoniini it is borne on a triangular strip of cuticle that is narrowly continuous with the edge of the cranium, and in certain other taxa, mainly sabethines, it may be fused with either the maxillary palpus or the maxillary body. Seta 6-Mx appears to be absent in *Aedeomyia* larvae. In the absence of an independent sclerite, there is no way of knowing whether it is united with the cranium or the maxilla in larvae of these species. In larvae of certain sabethine genera, the hypostomal sclerite may be fused with the base of the maxillary palpus (*Trichoprosopon*, *Tripteroides*), fused with the maxillary body (*Limatus*, some *Suaymyia* of *Topomyia* and some *Wyeomyia*) or fused with the maxillary palpus and maxillary body to form a composite maxilla (*Sabethes*, *Malaya*, *Johnbelkinia*). In larvae of other genus groups of Sabethini, the hypostomal sclerite is weakly developed and seta 6-Mx arises from membrane at the proximal margin of the maxillary body. Among non-sabethine genera, the hypostomal sclerite and maxillary palpus are fused in larvae of *Aedeomyia* and *Hodgesia*, and all three components are fused together in larvae of *Lutzia* and the nominotypical subgenus of *Psorophora*.

Most sabethine larvae have a maxillary brush similar to that in larvae of most non-sabethine species of Culicinae. The typical maxillary brush consists of a collection of independent, articulated, flexible elements that function to collect food particles from the palatal brushes. In certain sabethine larvae, all or most of the elements are partially or completely coalesced into a single articulated structure. When the elements are partially fused, bound or ensheathed, with clear evidence of individual elements internally or at least apically, the composite flexible structure is termed the **maxillary bundle**. When the elements are completely fused and unrecognizable, the structure is developed as a rigid claw-like structure termed the **maxillary claw**. The bundle and claw are distinctly hinged basally and partially opposed obliquely toward the mouth, indicating that they are adaptations for collecting food by scraping or clasping.

The presence of an **apical tooth** at the apex of the maxillary body is typical of sabethine larvae. The structure may be an elongation of the apex or the modified and enlarged distal element of the laciniarastrum. In most sabethine larvae, it is developed as a small bump or protuberance; in certain others it is a prominent tooth- or claw-like process. With the exception of larvae of *Sabethes* and a few *Wyeomyia*, a strongly developed apical tooth appears to have evolved in concert with the maxillary bundle and maxillary claw. The maxilla of *Sabethes* larvae has a reduced maxillary brush comprised of simple spicules, and the predominant feature is a strongly developed claw-like apical tooth, which probably plays a role in the acquisition of food. It seems obvious that the specialized structural modifications of the larval maxillae are indicators of phylogenetic relationships and shared feeding habits.

Labiohypopharynx

In generalized insects, the labium consists of two parts, a proximal **postlabium** (**submentum** and **mentum**) and a distal **prelabium** (**prementum**, glossae, paraglossae) and **labial palpi**. As in most nematocerous larvae, the prelabium of mosquito larvae is united with the **hypopharynx** to form the labiohypopharynx (Fig. 17e,f), which is attached to the head between the mouth and the **dorsomentum**. The hypopharynx lies dorsal to the prelabium. They are joined and supported by cibarial bars that extend from lateral areas of the cranium. The dorsomentum is the well-sclerotized transverse distal part of the mentum borne entad to the anterior border of the labiogula where it forms the protruding posterior wall of the **preoral cavity**. It is variously developed and its formation may be important in distinguishing certain group taxa, *e.g.* the subgenus *Oculeomyia* of *Culex*, as described by Harbach *et al.* (2012).

The prelabial part of the labiohypopharynx is larger than the hypopharynx and is comprised chiefly of the prementum. The central region of the prementum bears bilaterally paired assemblages of tooth-like projections, which encompass the membranous vestigial labial palpi. The palpi in anopheline and certain culicine larvae are separated by a tongue-like **ligula** that corresponds to the fused glossae and paraglossae.

In the basic condition, the tooth-like projections occur in three groups: (1) **premental teeth** consisting of a group of **lateral premental teeth** at the side of each labial palpus and a **premental dental arch** that curves around the dorsal margin of the palpi and is united with the base of the ligula; (2) a cluster of **premental cusps** located centrally between the premental dental arches; and (3) a patch of peg-like processes, **premental mala**, borne dorsolaterally beside the lateral premental teeth on each side of the labiohypopharynx.

The premental teeth are weakly developed in anopheline larvae where they are typically represented by the cusped premental dental arch. The premental teeth of culicine larvae are normally well developed and consist, except in most sabethine larvae, of three interconnected groups, which from dorsal to ventral are designated **lateral premental teeth 1–3**. The groups are quite distinct in larvae of *Hodgesia* and Mansoniini, but in larvae of certain other genera the most dorsal group is weakly developed and the lateral premental teeth of each side are reduced and coalesced with the lateral portion of the premental dental arch. In toxo-rhynchitine and most sabethine larvae, the lateral premental teeth of each side are developed as a crescentic dentate structure.

In culicine larvae, the premental dental arches tend to be separated from the lateral premental teeth and fused with the premental cusps and the ligula. In many sabethine larvae, the cusps are unrecognizably fused with the ligula to form a roughly triangular mass of **prementoligular teeth**. In anopheline larvae, a median row of **premental ridge teeth** extends dorsally from the premental teeth and separates a pair of denticulate surfaces that appear to be largely homologous with the culicine premental cusps.

The labiohypopharynx bears two groups of variously developed spicules, the **lateral premental spicules** borne on either lateral surface of the prementum next to lateral premental teeth and the **ventral premental spicules** that arise near the ventral margin. The former mainly occur in aedine larvae but are also present in larvae of *Culex*, *Culiseta* and *Orthopodomyia*. They are generally numerous and scale-like in aedine and *Culiseta* larvae and reduced in number and are spine- or lobe-like in most other culicines. They are weakly developed in *Anopheles* and *Chagasia* larvae and absent in larvae of *Bironella* and *Toxorhynchites*. A distinctly heart-shaped flap, the **premental cordate process**, is borne immediately ventral to the ligula in anopheline larvae. The process is distinctly bilobed in *Bironella* larvae and similar in appearance to the ventral premental spicules in larvae of *Uranotaenia*, *Aedeomyia*, *Hodgesia* and Mansoniini, with which it could be homologous.

Thorax

The thorax (Figs 15a, 18a and 19a) is composed of three fused segments, the **prothorax**, **mesothorax** and **metathorax**, which are distinguished by series of bilaterally paired, circumferentially arranged setae. The prothorax bears a maximum of 15 pairs of setae (numbered 0–14), the mesothorax usually has 14 pairs (1–14) and the metathorax has 13 pairs (1–13). The developmental characteristics of various setae on one or more of the segments are used in distinguishing and classifying species (see **Chaetotaxy** below).

A bilobed oval mitten-shaped structure projects dorsally from either side of the prothorax of anopheline larvae. These membranous structures, **Nuttall and Shipley's organs**, hold the thorax to the surface of the water during feeding. They are retracted into the thorax upon submersion, and although always present, they are essentially invisible in killed specimens. The organs are of no known importance in anopheline taxonomy; they are absent in culicine larvae.

Most of the setae of each segment arise directly from the integument, but certain setae or groups of setae emanate from a small sclerite termed a **setal support plate**, which may or may not be developed as a prominence or tubercle. Setae that arise from a sclerotized protuberance rather than a simple plate are said to arise from a tubercle. A setal support plate may bear a distinct spine-like process referred to as a **setal support plate spine**. Such spines are useful for distinguishing some *Anopheles* larvae and are particularly well developed on the thorax of *Stegomyia* larvae.

Abdomen

The abdomen (Figs 15a, 18a and 19a) is composed of nine apparent segments, which are denoted from anterior to posterior by Roman numerals. Since **abdominal segment VIII** actually comprises the embryonic eighth and ninth abdominal segments, the last or terminal segment is correctly recognized as **abdominal segment X** rather than IX. Rows of spicules are present on the ventral surface of the abdomen in all first-instar larvae.

Segment VIII of anopheline larvae bears a dorsal **spiracular apparatus** (Fig. 21a), which encompasses the postabdominal spiracles. The spiracular apparatus of culicine larvae is borne at the apex of an elongate sclerotized tube, inappropriately known as the **siphon**, which structurally comprises parts of embryonic abdominal segments VIII and IX. Differences in the length and shape of the siphon provide useful characters for species identification. In most larvae, the spiracular apparatus is a five-lobed valvular structure (Fig. 21a,b) consisting of a single median anterior lobe (**anterior spiracular lobe**), a pair of small lateral lobes (**anterolateral spiracular lobes**) and a pair of large posterior flap-like lobes (**posterolateral spiracular lobes**). Each lobe usually bears small simple or branched setae. The spiracular apparatus in larvae of *Coquillettidia*, *Mansonia* (Fig. 21c) and several species of *Mimomyia* is highly modified for piercing the submerged parts of aquatic plants to obtain oxygen.

Anopheline larvae bear a sclerotized **tergal plate** anteriorly on the dorsal midline of abdominal segments I–VIII, and one to three **accessory tergal plates** sometimes occur posterior to this plate in some species. The size and shape of the tergal plate can be of taxonomic significance. Fourth-instar larvae of *Orthopodomyia* develop large tergal plates on the dorsal surfaces of segments VI–VIII during growth (Zavortink, 1968).

The terminal segments of the abdomen of two anopheline larvae (Figs 18a and 20a) and several culicine larvae (Figs 19b and 20b–d) are shown in lateral view. The spiracular apparatuses of an anopheline and a culicine larva are shown in dorsal view on a larger scale in Fig. 21a and b for comparison. The spiracular apparatus sometimes exhibits characters of taxonomic value. The **anterior median process**, a variously developed dome- or knob-like membranous area situated posteriorly on the midline of the anterior spiracular lobe, is produced into a long process (filament) in some anopheline species (Fig. 20a). Anopheline larvae also bear a sclerite below the spiracular apparatus on the side of segment VIII, the **pecten plate**, which is connected posteriorly to the one on the opposite side by a **U-shaped band** extending underneath the spiracular apparatus. The posterior margin of the pecten plate bears a comb-like row of variously developed spines (**pecten spines**), which collectively comprise the **pecten**. The pecten plates and U-shaped band are at least partially homologous with the siphon of culicine larvae, which bears a posterolateral row of variously developed pecten spines. The number and structure of the pecten spines can be of taxonomic importance in both anopheline and culicine larvae. In some sabethine larvae, *e.g.* those of the genus *Sabethes*, the pecten consists of a posterior row of fine flexible filaments (Fig. 20c). A more or less intermediate condition in *Malaya* larvae and certain larvae of *Tripteroides* and *Wyeomyia* provide evidence that the pecten spines and pecten filaments are homologous. Pecten are absent in larvae of *Orthopodomyia* and *Toxorhynchites*; they are sometimes reduced or absent in larvae of *Aedes* and are nearly always present in larvae of *Culex*, *Tripteroides* and *Uranotaenia*. Pecten are usually present in larvae of *Ficalbia*, present or absent in larvae of the subgenera *Mimomyia* and *Etorleptiomyia* of *Mimomyia*, but are apparently absent in those of the subgenus *Ingramia*. The pecten is reduced to a few spines in most species of *Eretmapodites* for which the larva is known.

Culicine and first-instar anopheline larvae bear a group of specialized spicules on either side of abdominal segment VIII known collectively as the **comb**. The spicules, whether squamiform or spine-like, are termed **comb scales**. The function of the comb is not definitely known, but evidence suggests it is employed to clean the palatal brushes. The comb can comprise a single row of a few to many scales or consist of several rows or a large patch of numerous scales. The arrangement, size and shape of the scales are of taxonomic value. The scales may be flattened and bear an apical and lateral fringes of spinules of more or less equal length or the mid-apical spinule may be distinctly larger, in which case they are recognized as fringed spine-like scales. In some species, the comb consists of both evenly fringed and laterally fringed spine-like scales. Comb scales sometimes emanate from the posterior area of a sclerotized **comb plate**. A comb plate is usually present in larvae of *Aedeomyia*, *Psorophora*, *Uranotaenia*, the subgenus *Ctenogoeldia* of *Runchomyia* and the subgenus *Sabethinus* of *Sabethes*. The comb is absent in *Toxorhynchites* larvae and larvae of most species of *Trichoprosopon*.

Abdominal segment X is highly modified. It bears a prominent sclerotized **saddle**, four pairs of setae and two pairs (usually) of terminal **anal papillae**. The segment is largely formed from embryonic segment X but probably also includes parts of the true ninth and eleventh segments and the telson. Most of the dorsal surface of the segment is covered by the large sclerotized saddle. The saddle may be confined to the dorsal surface or completely encircle the segment. It is almost always complete in *Orthopodomyia* and *Culiseta* larvae, usually complete in larvae of *Culex*, *Psorophora* and *Uranotaenia*, and usually incomplete in larvae of *Aedes* and subgenus *Ingramia* of *Mimomyia*. A small attached or detached sclerite known as the **acus** is located posterolaterally at the base of the siphon in many taxa. The acus is rarely a complete basal ring as in species of *Mansonia*. Four pairs of setae (see **Chaetotaxy** below) include a lateral seta near the posterior margin of the saddle, two usually very large setae that project posteriorly, one above the other, from their insertions slightly posterior to the mid-dorsal edge of the saddle, and a linear series of more or less paired or unpaired setae, known collectively as the **ventral brush**, borne on the ventral longitudinal midline of the segment. In some taxa, the brush is divided into two groups. Setae that comprise the posterior group, which normally arise from a sclerotized **grid**, are known as **cratal setae**, and setae of the anterior group, which arise cephalad of the grid, are referred to as **precratal setae**. In certain larvae, mainly aedines, the posterior group emanates from a sclerotized elevated **boss** rather than a grid, in which case the setae of the ventral brush borne on the boss are known as **boss setae** and those borne anterior to it are the **preboss setae** (Reinert, 2002a). The **anal papillae** (singular **papilla**) are osmoregulatory organs. They are typically digiform structures, the length and shape of which vary greatly in different species depending on the salt and mineral content of the aquatic environment in which they reside. For this reason, characteristics of the anal papillae are not very useful or reliable for identifying species.

Chaetotaxy

The characters most frequently utilized in mosquito larval taxonomy involve the **chaetotaxy**, the arrangement and development of setae. Beginning with Martini (1923), a number of individuals participated in the development of a well-grounded system of nomenclature for the larval setae of mosquitoes. These efforts culminated with the detailed studies of Belkin (1950b, 1952, 1953b, 1954a, 1954b, 1960, 1962), and his system has been in general use ever since. As many as 222 pairs of setae are recognized on the fourth-instar larvae of at least some Culicidae. In the setal nomenclatural system developed by Belkin, individual setae are given a number followed with a hyphen and a capital letter or a Roman numeral to indicate the area of the body from which the seta arises. The following symbols (mouthparts omitted) are used in this system: A, antenna; C, cranium (head capsule); P, prothorax; M, mesothorax; T, metathorax; I–VIII and X, abdominal segments 1–8 and 10; and S, spiracular apparatus (anopheline larvae) and siphon (culicine larvae). The setae of the head are numbered sequentially beginning anteriorly on the dorsal surface and proceeding posteriorly and then ventrolaterally to the ventral midline. The antennal setae are numbered from beyond the base to the apex. The setae of the thorax and abdominal segments I–VIII are numbered sequentially from the dorsal midline circumferentially to the ventral midline. Setae of the pecten plate (anophelines) and siphon (culicines), the spiracular apparatus and abdominal segment X are numbered as indicted below and in the figures. It should be noted that serially homologous setae are treated as single entities, indicated by the singular 'seta', *e.g.* seta 12-P–T and seta 5-II–VII.

The consistent pattern of chaetotaxy in larvae throughout the Culicidae is an indication of its monophyletic origin, and its ontogenetic and phylogenetic homology, as well as the serial homology of setae borne on the thoracic and abdominal segments. Nerve connections between the external setae of one instar and the internal developing setae of the next instar or pupa provide absolute evidence of ontogenetic homology (Belkin, 1962). As shown by Harbach (1991b), phylogenetically homologous setae are more uniform in appearance and position on first-instar larvae, and since first instars have fewer setae, phylogenetic homologies are more readily apparent than in later instars. Tracing the development and positional relationships of setae through the four instars of *Sabethes chloropterus* revealed ontogenetic evidence for the phylogenetic homology of setae that first appear in later instars.

The arrangement and development of setae are subject to variation, but in general structural differences in setae provide useful specific characters while positional differences in setae represent supraspecific group characters (Belkin, 1952; Harbach, 1991b). In other words, the appearance and branching of setae are useful in distinguishing species while absolute differences in setal positions are useful in defining supraspecific groups.

Specific structural characteristics of setae of importance in delimiting both species and groups are too numerous to be mentioned here, but in general they include strength, length, number of branches, position (insertion) and presence or absence of a supporting sclerite. Setae may be single (unbranched) or branched. Single setae may be simple or bear short lateral or apical processes (not true branches). Branched setae may have branches emanating at or some distance above the base, or borne along the length of the main stem. The descriptors used to name single setae include aciculate, barbed, brush-tipped, foliform, lanceolate, pectunculate, peg-like, simple, spiniform and spinulate; descriptors used to name branched setae include dendritic, fan-like, forked, palmate, pectinate, plumose and stellate. The names of single setae are used to describe the branches of branched setae. The principal types of setae are illustrated in Fig. 26.

The complete chaetotaxy of the fourth-instar larva of *Anopheles subpictus* and *Culex pipiens* is shown in Figs 22 and 23, respectively. It must be mentioned that the positional relationships of the setae in these two species, although similar in related species, will be different in other groups of species, especially those that are most distantly related to the specific groups (defined by specific shared characteristics) in which they are classified. Variation in the locational relationships of the thoracic setae and the setae of abdominal segments I–VI in species of the Anopheleinae, Aedini, Culicini, Sabethini and Uranotaeniini is masterfully shown in the composite drawings of Harbach & Knight (1980: figs 69–71).

Head

The antenna of fourth-instar larvae bears six setae, **setae 1–6-A**, and the cranium bears a maximum of 20 pairs of setae, designated **setae 0–19-C**. The principal antennal seta, **seta 1-A**, varies in position on the shaft and is usually large with multiple branches but is single or two-branched in some species. **Setae 2,3-A** are inserted apically in association with **setae 4–6-A** on the antennae of most mosquito larvae, but are noticeably subapical in some species. They are always located a considerable distance from the apex and divide the antenna into proximal and distal parts in species of Mansoniini.

All of the cranial setae are potentially of taxonomic importance. **Setae 2–4-C** are taxonomically the most important setae on the head of anopheline larvae. Seta 2-C is usually well developed in anopheline larvae and absent in most other mosquito larvae. It is present in some species of *Culex* and *Orthopodomyia*, rarely developed in species of *Coquillettidia* and is weakly to well-developed in species of *Deinocerites*. Seta 3-C is inserted immediately posterior to the median labral plate when it is distinctly delineated from the dorsal apotome. In *Toxorhynchites* larvae, seta 3-C is associated with **seta 0-C** on the median labral plate. When the plate is indistinguishably fused with the dorsal apotome, seta 3-C either arises from the adoral (dorsal) surface or oral (ventral) surface near the anterior margin of the head. In sabethine larvae, seta 3-C is associated with seta 0-C in a ventral position, indicating an anterior migration of the seta from the dorsal apotome to the median labral plate, concomitant with loss of the clypeolabral suture. Seta 3-C is absent in species of *Lutzia* and the subgenera *Aedinus*, *Anoedioporpa* and *Tinolestes* of *Culex*, and it is usually represented by a spicule when present in species of *Deinocerites*. The development of **setae 4–7-C** and their positions (insertions) relative to one another are important in delimiting both species and groups. The other dorsal and ventral setae of the head are used infrequently in anopheline taxonomy, but often exhibit useful characteristics and positional relationships for distinguishing culicine larvae, *e.g.* the development and relative positions of **setae 14-** and **15-C** in certain *Sabethes* larvae. **Setae 16-** and **17-C** are minute setae borne at the intersection of the lateralia and the collar in certain *Culex* larvae. **Setae 18-** and **19-C** are small ventrolateral setae of the cervix that are inserted at the level of the ventromedian cervical sclerite.

Thorax

The thoracic chaetotaxy provides important characters for the classification and identification of species and groups. In fourth-instar larvae, the prothorax bears a maximum of 15 pairs of setae (**setae 0–14-P**), the mesothorax usually has 14 pairs (**setae 1–14-M**) and the metathorax has 13 pairs (**setae 1–13-T**). With the exception of **seta 0-P**, presumptive serial homologies have been established for all of the thoracic setae. Most of the largest setae occur dorsoanteriorly and laterally.

Setae 1–3-P, the three usually close-set setae nearest the dorsal midline immediately behind the head, may arise independently from the integument or two or all three may be inserted on a common support plate. When two are borne on a plate (usually setae 2,3-P), they may or may not arise from a sclerotized tubercle. The development and insertional relationships of the three setae are often used to delimit species of *Aedes*, *Anopheles* and *Culex*. The length of seta 3-P relative to the length of seta 1-P, for example, has been used as a character to distinguish the subgenera of *Culex*.

The pleural setal groups (**setae 9–12-P,M,T**) are of great importance in the identification of anopheline larvae. Each group is a close-set cluster of four setae inserted on a common setal support plate, often in close apposition to a distinct spine (setal support plate spine). The absence and development of the spine when present are useful in distinguishing species, especially anophelines. Three setae of the propleural group (setae 9,10,12-P) and two of the meso- and metapleural groups (setae 9,10-M,T) are usually much longer than the other seta(e) of the group. Characteristics of the longer setae, *e.g.* simple, aciculate, branched or plumose, have been used for classifying and identifying species of *Anopheles*.

Seta 13-P in anopheline larvae arises anterior to the tubercle bearing the pleural setal group; when present in culicine larvae it is usually borne mesal to these setae. This seta is present in larvae of *Maorigoeldia* and *Tripteroides*, and occurs sporadically in species of *Aedes*, *Psorophora*, *Uranotaenia* and *Wyeomyia*.

In general, fewer setae of the meso- and metathorax have been used for distinguishing and diagnosing species and groups than those of the prothorax. The absence of **seta 8-M** in all species of *Kimia* and *Trichoprosopon*, in concert with similar development of the occipital foramen and insertion of seta 13-T (see below), is phylogenetically significant (Zavortink, 1979; Harbach *et al.*, 2007a). Loss of seta 8-M is only known elsewhere in one species of *Sabethes* (*Sa. gymnothorax*) and six species of the subgenus *Tricholeptomyia* of *Tripteroides*.

Seta 13-T is variously developed. It is normally branched with few to numerous branches, and ranges in size from small and inconspicuous, as in most aedine larvae, to moderately large in species of many groups, to extremely long in certain sabethine larvae, *e.g.* species of the subgenus *Sabethinus* of *Sabethes*. When it is large, it is often borne on a setal support plate. Larvae of *Kimia* and *Trichoprosopon* are unique in having seta 13-T inserted on a plate with the pleural setal group (setae 9–12-T).

Abdomen

In fourth-instar larvae, the chaetotaxy of abdominal segment I is reduced and somewhat unusual, that of segments II–VII is similar in overall pattern and segment VIII exhibits a highly modified pattern. Abdominal segment I bears a maximum of 13 pairs of setae (**setae 1–13-I**), abdominal segments II–VII have 15 pairs (**setae 0–14-II–VII**), segment VIII has seven pairs (**setae 0–5,14-VIII**), there are 13 pairs on the siphon (pecten plate of anophelines) and the spiracular apparatus (**setae 1–13-S**), and segment X has four pairs (**setae 1–4-X**), with seta 4-X being any seta of the ventral brush. Setae 0, 8 and 14 are absent from segment I.

The arrangement and form of the abdominal setae are extremely important in classification and identification. The setae of segments I–VII are divided into three groups based on general location: setae 0–5 are the dorsal setae, setae 6–9 are the lateral setae and setae 10–14 are the ventral setae. The following setae are of particular importance in distinguishing and identifying species and species groups.

Seta 0-II–VII is the most anterior of the dorsal setae. It is usually minute and inconspicuous but it is prominent and stellate in some sabethine larvae. **Seta 1-I–VII** is the usually well-developed dorsal seta nearest to the longitudinal midline and posterior margin of the respective segment. Seta 3 and sometimes seta 2 on segment VI may be nearer to the midline in some taxa but they are normally more weakly developed than seta 1. In anopheline larvae, **seta 1** may be a small branched seta or it may be developed into a **palmate seta** on some or all of the abdominal segments except VIII and X. The **leaflets** of a palmate seta may be simple and lanceolate, but they are more often differentiated into a **blade** with a terminal **filament**. The number of palmate setae present and the shape of their individual leaflets are very important in the classification and identification of *Anopheles*.

Setae 4,5,6-III–V are usually associated with an empty alveolus, inappropriately referred to as an **abdominal puncture**. A similar alveolus sometimes occurs anomalously on segment II. The occurrence of the punctures has largely been ignored in studies of mosquito larvae, but the presence or absence of their homologues in pupae suggests that they may provide characters for distinguishing species and groups.

Seta 6-I–VII is the lateral seta often borne on a prominent sclerite located approximately midway between the dorsal and ventral longitudinal midlines of the abdomen. It is strongly developed on segments I and II, normally slightly less well developed on segments III–VI and smaller on segment VII. **Seta 7-I–VII** is the lateral seta borne immediately ventral or slightly posteroventral to seta 6. It is generally strongly developed and similar to seta 6 on segments II and III and usually much smaller on segments III–VII. It is uniquely stellate in some taxa. **Seta 9-I–VII** arises immediately anterior, posterior or ventral to seta 7. It varies in development from small and single to moderately large and branched, and is stellate in some taxa.

Setae 10–13-I–VII comprise a generally transverse group of ventral setae. The positional relationships and development of these setae vary considerably on different segments and in different taxa, and thus provide numerous important characters for delimiting species and groups. Except for seta 12, which is absent on segment I in sabethine larvae and some aedine larvae, the four setae are routinely present in a variety of relationships. The presence or absence of seta 12-I is a distinctive characteristic of species of the subgenera of *Aedes*, present in species of approximately half of the subgenera and absent in the others. Seta 13 is inserted nearest to the ventral longitudinal midline on segments I and II (and also on at least some of the other segments in certain species and groups) and is usually the most strongly developed of the group on all of the segments. It is a well-developed stellate seta in some taxa.

Seta 14-II–VIII is the ventral seta inserted close to the longitudinal midline at the anterior margin of the segment. It is incorrectly placed on the posterior margins of the segments in may published illustrations. Seta 14 is absent from segment I, usually absent from segment II and absent from all of segments I–VII in some sabethine groups. It is minute when present except in certain sabethines.

Segment VIII bears five usually well-developed generally vertically aligned posterior setae (**setae 1–5-VIII**) and two minute anterior setae (**setae 0,14-VIII**). Setae 1–5 are inserted ventral to the pecten in anopheline larvae and posterior to the comb in culicine larvae. Setae 3 and 5 are generally the most strongly developed of the posterior setae. The distance between setae 4 and 5 in relation to the dorsal margin of segment X is useful in distinguishing some sabethine taxa.

Seta 1-S is normally a well-developed seta located below the posterolateral spiracular lobe on either side of segment VIII in anopheline larvae and is borne posterolaterally on either side of the siphon in culicine larvae. Larvae of Culicini are unique in having three or more setae along the posterolateral margin, which are designated as **seta 1a-S**, **seta 1b-S**, *etc.*, from the most proximal to the most distal seta. With the exception of Culicini and species of the subgenus *Rusticoidus* and some species of *Aedes sensu stricto*, the siphon of all other culicines bears a single bilateral pair of seta 1-S. The single seta 1-S may be inserted at or beyond the base of the siphon. It is inserted at the base of the siphon in larvae of *Hodgesia* and *Culiseta*, with the exception of larvae of the monotypic subgenus *Austrotheobaldia*. The seta is inserted near the base of the siphon in some species of *Toxorhynchites*. When inserted beyond the base of the siphon, its attachment before or after the most distal pecten spine provides a useful taxonomic character.

Segment X bears four pairs of setae (**1–4-X**). **Seta 1-X** is normally borne on the lower posterior area of the saddle, but arises from the integument at the margin of the saddle in some taxa. **Setae 2,3-X**, commonly referred to as the **dorsal brush**, are attached immediately posterior to the dorsal midline of the saddle. **Seta 2-X**, also known as the upper caudal seta, may be long or short, unbranched (single), branched with many branches mainly on one side of the main stem or fan-like (plumose) with branches borne more or less equally on both sides of the main stem. **Seta 3-X**, or lower caudal seta, is normally much longer and has fewer branches than seta 2-X. It is often single. **Seta 4-X (ventral brush)** provides taxonomic characters that are important in delimiting species and groups, including the number of setae, their attachment to a grid or boss, and the presence or absence of precratal setae. The absence of these setae in two African species of *Culex*, *Cx.* (*Kitzmilleria*) *moucheti* and *Cx.* (*Culiciomyia*) *harleyi*, is a unique feature within the Culicidae. A single pair of seta 4-X is present in all species of the Australasian subgenus *Acallyntrum* of *Culex* and species of Sabethini, except those of the subgenus *Sabethinus* of *Sabethes*, which possess a second pair of small accessory setae of uncertain homology. Two pairs of seta 4-X characterize species of the subgenus *Ingramia* of *Mimomyia* whereas four pairs are present in all other members of the genus, subgenera *Etorleptiomyia* and *Mimomyia*. The ventral brush of most mosquito larvae consists of more than three pairs of setae, and some larvae have an unpaired seta, *i.e.* one more or one less than the usual number of paired setae that characterize a particular group.

Pupa

The pupa (Fig. 15b) is the third or penultimate stage in the life cycle of holometabolous insects. In mosquitoes, the pupal–adult apolysis occurs shortly after the larval–pupal ecdysis. Hence, there is a very short exposed pupal period, perhaps 1–5 hours, followed by a long pharate adult phase, perhaps about 44–47 hours of the 2 days that so many species are conventionally said to be in the 'pupal stage'. Thus, most of the conventional 'pupal stage' is an adult, not a pupa, enclosed by the remains of the pupal integument. For this reason, most authors prior to Harbach & Knight (1980) adopted terms for parts of the pupal exuviae ('skin'), which include the words sheath, case or plate because the parts cover or enclose the corresponding structure of the developing adult, *e.g.* scutal plate, which overlies the adult scutum. To avoid ambiguity, cuticular areas of the pupa are named without including the words sheath, shield, case, plate, pouch or sac.

The free-living pupa is comma-shaped (shaped like a filled-in number 9) (Fig. 15b), the enlarged upper end of the comma being the large unsegmented **cephalothorax** and the prolonged extension being the segmented **abdomen**. Dorsally the cephalothorax bears a pair of usually movable, dorsolateral appendages, the **trumpets**, which contain the mesothoracic spiracles, and the developing eyes, mouthparts, legs and wings of the adult are evident ventrally. The abdomen has eight obvious segments, an insignificant ninth segment and terminates in a **genital lobe** and a large pair of **paddles**.

Although pupae are seldom used for routine identification of species, the study of pupal exuviae has shown that they are of considerable value in anopheline and sabethine taxonomy. Many taxonomic studies have shown that the pupal stage frequently offers the best characters for differentiating some species and groups. The most useful taxonomic characters and their significance are indicated below.

Cephalothorax

The cephalothorax (Figs 15b and 24a,b) is a remarkable section of the mosquito pupa. It consists of the combined head and thorax of the developing adult. The head is attached to the lower anterior area of the thorax with the dorsal surface of the developing adult directed anteriorly. The segmented **antenna** of either side curves upward and posteriorly along the edge of the **mesothoracic wing**. The **compound eye** is conspicuous below the proximal portion of the antenna on either side of the elongate **proboscis**, which projects posteriorly between the **maxillary palpi** (singular **palpus**) and the **legs** beneath the thorax. Lying between the bases of the antennae and the **postpronota** (singular **postpronotum**) is a clearly demarcated sclerite, the **dorsal apotome**, which abuts the cephalic margin of the **mesonotum**. The dorsal apotome is easily examined in pupal exuviae that have been split open along the dorsal ecdysial line, spread out and mounted on a microscope slide. When flattened, the dorsal apotome projects from the median anterior margin of the cephalothorax, revealing its size and form. Reinert (1999d) conducted a comparative study of the dorsal apotome and concluded that major differences in its formation were of phylogenetic significance. It is a variously developed single oblong sclerite in anopheline pupae and consists of two elongate sclerites of various shapes connected by a medial membrane in culicine pupae.

The mesonotum of the pupa consists of a large **scutum** and a small **postscutal area**. The extent (length) of the scutum is delimited by the anterior and posterior ends of a distinct median longitudinal ridge, the **median keel**, which bears the dorsal ecdysial line where the pupal cuticle splits open to allow the adult to emerge from the pupal exuviae. The postscutal area lies behind the scutum. It is composed of the usually ill-defined scutellum and mesopostnotum. The metathorax is represented by a small sclerite, the **metanotum**, which is closely associated with and appears to be a part of the abdomen. The metanotum bears the **metathoracic wings** laterally. The wing is an elongate triangular lobe lying behind the mesothoracic wing with the apex directed posteriorly along the lateral margin of abdominal segment I. The wing is not clearly differentiated from the metanotum and one or two of the metanotal setae usually appear to be borne on the wing.

The cephalothorax of some species bears pigmented areas or patterns of pigmentation. Patches of darker pigmentation may be present on the antenna, maxillary palpus, scutum and mesothoracic wing. Some aedine pupae have clear unpigmented spots on the scutum and metanotum that distinguish species and groups (Reinert *et al.*, 2009), and the wings of members of the *Anopheles hyrcanus* group have a lattice pattern of spots that is useful in identifying species (Harrison & Scanlon, 1975).

Trumpet

The trumpet (Figs 15b and 24) is a double-walled usually movable appendage borne dorsolaterally on either side of the cephalothorax, articulated directly with the scutal integument or supported by a distinct basal tubercle. The outer wall of the trumpet is hydrophilic; the inner wall is hydrophobic, being lined with a network of specialized spicules that trap air and resist wetting. In the majority of mosquito species, the inner and outer walls are closely adherent so that the double-walled structure of the trumpet is not obvious. This common type of trumpet, **angusticorn trumpet**, is shown in Fig. 24d,g. It is characterized by having the longest axis vertical more or less in line with the stem, it is narrowly funnel-shaped when closed and has a split (**meatal cleft**) down one side that allows it to open widely at the surface of the water. The stem of the trumpet below the meatal cleft is termed the **meatus** and the upper part is known as the **pinna**. The narrow basal part of the meatus in most species bears a more or less distinct series of annular ridges and furrows known as the **tracheoid area**. The tracheoid area may be weakly developed, indistinct or absent. In some species, especially those of the subgenus *Anopheles*, there is a different and more complex type of trumpet, the **laticorn trumpet** (Fig. 24c,e). This type of trumpet has the longest axis transverse to the stem, it often has a secondary cleft in the pinna opposite the meatal cleft and it is sometimes less wide-mouthed but with a more or less elaborate lobe, the **tragus**, on the rim of the pinna (Fig. 24e).

It is interesting that pupae of *Coquillettidia*, *Mansonia* and several species of *Mimomyia*, the larva of which have a highly modified siphon for piercing the roots of aquatic plants to obtain air, have long horn-like trumpets (Fig. 24f) that serve the same purpose. The trumpets in these species are borne on a more or less distinct tubercle nearer the mid-dorsal line and farther forward on the scutum than in other species. The apex of the trumpet is narrowed and sharply pointed to foster insertion into the roots of plants.

Abdomen

The abdomen of the pupa (Figs 15b and 25a,b) is more or less compressed and expanded laterally. The eight obvious segments are similarly sclerotized, except segment I, which has a membranous ventral surface. Segment I has a mainly sclerotized dorsal surface whereas segments II–VIII each have a well-developed dorsal **tergum** and ventral **sternum**. The formation of segments II–VIII is largely intermediate between their structures in the larva and adult. The terga and sterna are fused to one another laterally – the dorsal and ventral integument is contiguous in larvae whereas the terga and sterna of adults are separated by a flexible **pleural membrane** – and the segments are connected, as in adults, by flexible **intersegmental membranes**, which enable the pupa to flex the abdomen and tumble ('swim') through the water when disturbed.

Abdominal segment IX (Fig. 15c) is represented chiefly by a transverse lobe, the **median caudal lobe**, associated with the caudal margin of tergum VIII. Although the lobe appears to be a part of tergum VIII, and probably includes a small part of tergum VIII, it is largely formed of **tergum IX**. The median caudal lobe is located dorsal to the **proctiger**, which is formed of abdominal segments X and XI, with its **cerci** (singular **cercus**), and the telson. In female pupae, each cercus is a more or less distinct lateral lobe of the proctiger. Structures associated with the proctiger include the flanking **paddles** and the ventral **genital lobe**. The genital lobe is usually small, rounded and spiculate in females. It is much larger and partially bilobed in males because it must provide sufficient space for the developing gonocoxopodites of the adult. This difference provides an easy means of distinguishing male and female pupae. The genital lobe of males of some species ends in a knob or constricted terminus, and the size and shape of the lobe sometimes differs between species. **Sternum IX** is sometimes present as a reduced and indistinct transverse sclerite at the base of the genital lobe.

Paddle

The paddle (Figs 15b and 25a,b) is one of a pair of movable or immovable appendages of abdominal segment IX, but they appear to be borne on the caudolateral areas of abdominal segment VIII. The paddle is a flattened, largely transparent natatorial organ, strengthened by a thickened basolateral border, the **buttress**, and a central vein-like **midrib**. Each paddle has inner and outer parts and inner and outer margins. The **outer part** is the area lateral to the midrib and the **inner part** is the area mesad of the midrib. The **outer margin** is the lateral edge of the paddle extending from the basolateral area of the outer part to the apex of the midrib, or its projection, and the **inner margin** is the mesal edge of the paddle extending from the basomesal area of the inner part to the apex of the midrib. In anopheline larvae, the proximal part of the outer margin usually bears serrations, the **paddle marginal serrations**, and the distal part and the inner margin bear a fringe of spicules, the **paddle marginal spicules**. The margins of the paddle of most culicine larvae are smooth or bear only minute serrations, but the paddle of some taxa, *e.g.* various groups of aedines, have a fringe of hair-like spicules.

The shape of the paddle sometimes serves as a useful character for distinguishing species. The paddle may be relatively slender or broad. These variations are expressed by the **paddle index**, the ratio of maximum **paddle length** to maximum **paddle width**. Sometimes there is a pigmented or differentiated membranous area at the base of the paddle sufficiently constant to be of diagnostic use. The extent of the serrations on the outer margin of the paddle varies in anopheline species, and since the part bearing the serrations is refractile to light it is called the **refractile border**. The length of the refractile border is expressed as a fraction of the paddle length. This fraction is called the **refractile index**. The size and shape of the serrations on the refractile border may show specific differences, and the paddle marginal spicules beyond the serrations may also vary in character.

Although pupae are seldom used for routine identification of species, the study of pupal exuviae has shown that they are of considerable value in anopheline taxonomy (Harrison & Peyton, 1984). During the latter part of the 20th century, it become apparent that the pupal stage frequently offers the best characters for differentiating some of the series, groups and species of *Anopheles*.

Chaetotaxy

The body of the pupa bears a definite arrangement of bilaterally paired setae. The arrangement and nomenclature of the setae (**chaetotaxy**) developed slowly until the pupae of sufficient numbers of species were collected and described. The discovery that the developing pupal setae are connected by sensory neurons to the setae of the fourth-instar larva allowed Belkin (1960, 1962) to establish the ontogenetic homologies of the setae. Attempts to create a system of nomenclature for pupal setae since names were first applied to most of them by Macfie (1920) culminated in the soundly based system established by Belkin (1962). This system recognizes a total of 119 pairs of pupal setae. Although most pupal setae are easily identified, the positions of two or three setae sometimes make it necessary to arbitrarily decide which is which.

The cephalothoracic and abdominal chaetotaxy of the anopheline and culicine pupae (composites) are shown in Figs 24a, 25a and 24b, 25b, respectively. The more important elements of the pupal chaetotaxy are treated below. Setae of the cephalothorax are designated with a number connected by a hyphen to the suffix CT, the use of which was initiated by Knight (1971b) as the replacement for the letter 'C' that was previously in use for both pupal cephalothoracic and larval cranial setae. The apparent absence of ontogenetic homology between the setae of the two life stages (one exception is now known) provided the rationale for using a different suffix for the cephalothoracic setae. The terminology for setae of the abdominal segments is the same for larvae and pupae (a number followed with a hyphen and a Roman numeral) because it is based principally on established ontogenetic homologies.

Cephalothorax

The cephalothorax bears 13 pairs of setae, three laterally on the head (**setae 1–3-CT**), four on the prothorax (**setae 4–7-CT**), two on the mesothorax (**setae 8,9-CT**) and usually three, sometimes four, on the metathorax (**setae 10–13-CT**). The setae of the cephalothorax are shown in Fig. 24a,b. Most of the cephalothoracic setae are of limited taxonomic use. **Setae 1-CT** and **11-CT** have received the most attention. Seta 1-CT is the most dorsal of the three head setae; it is ontogenetically homologous with larval seta 10-C. Species of some genus groups have a large seta 1-CT. It is normally considerably stronger than the other two head setae in those taxa, and exceptionally stronger in pupae of Sabethini and Toxorhynchitini. In many species of Sabethini, seta 1-C is double (two-branched), sigmoidally curved and apically hooked, a specialization that apparently allows pupae to anchor to objects below the surface of the water (Iyengar & Menon, 1948). The development of seta 11-CT, whether single or branched, is useful for distinguishing some species. When present, **seta 13-CT** occurs consistently in very few groups and sporadically in species of various groups. The occurrence of seta 13-CT in the family and the tribe Aedini was reviewed by Reinert (1981a, 2000c, respectively).

Abdomen

The abdominal segments (Fig. 25a,b), except the first and last, have a maximum of 15 pairs of setae (**setae 0–14-II–VII**), the serial homologies of which are usually determined without difficulty. Seven or eight setae are normally borne on each side of the terga and six or seven are usually borne on either side of the sterna. Segment I has eight pairs of dorsal setae (**setae 1–7,9-I**) and one or two minute setae may occur on the membranous ventral surface (**setae 10,11-I**). Segment VIII bears a maximum of five setae (**setae 1,4,9,13,14-VIII**). A seta borne near the lateral margin of the median caudal lobe is arbitrarily named **seta 1-IX**. An anomalous seta borne laterally on the cercus in females and the proctiger in males is designated **seta 1-XI**. For the most part, the numbers given to the abdominal setae of segments I–VIII indicate ontogenetic homologies with the abdominal setae of fourth-instar larvae.

Abdominal segment I bears a conspicuous pair of usually dendritic setae, **seta 1-I**, which function to suspend the pupa from the surface of the water. They may also detect disturbances to the surface of the water and elicit a swimming response to possible danger. It is noteworthy that this seta is usually very small and single in pupae that remain below the surface the water, *e.g.* pupae of species of *Coquillettidia*, *Ficalbia* and *Mansonia*, and tends to have fewer simpler branches in pupae which are submerged much of the time while feeding, *e.g. Eretmapodites*.

Seta 0-II–VIII is the most anterior of the dorsal setae on segments II–VII, and also on segment VIII except in species of *Anopheles* and *Bironella* and most species of *Coquillettidia* where it is inserted at or usually slightly posterior to the mid-length of the segment. Seta 0 is absent from segment I.

Setae 1,2,3-II–VII are the usually more or less closely associated setae borne on the mesodorsal areas of the terga. Seta 1 is usually inserted nearest to the posterior margin of the terga. However, if seta 2 is equally near or nearer to the posterior margin, it is also usually nearer to the longitudinal midline and typically shorter than seta 1; if seta 3 is equally near or nearer to the posterior margin, it is also nearer to the lateral margin, but mesad of seta 1 in species of *Anopheles*. Seta 2 may be inserted at positions anywhere from lateral to mesal of seta 1, but it is nearly always shorter than setae 1 and 3 and otherwise identifiable based on serial resemblance. Seta 3 may stand alone or be more closely associated with setae 4 and 5 in some taxa.

Setae 4,5,6-III–V, as in fourth-instar larvae, are usually associated with an empty alveolus (**abdominal puncture**) borne on the dorsolateral area of the segments, and anomalously on segment II. The presence or absence of one or more of the punctures is of taxonomic importance in sabethine taxa (see, *e.g.* Zavortink, 1979, as dorsal sensillum, and Harbach, 1994b).

Seta 5-II–VII is a dorsoposterior seta that along with **seta 4-II–VII** arises lateral to setal group 1-2-3 and mesad of setae 6 and 9. However, the positions of setae 5 and 4 vary in relationship to one another and in their associations with setae 1, 2 and 3 in various taxa, and are important in distinguishing species and groups. The development of seta 5 on segments IV–VI is also important in recognizing particular taxa. It may be weakly developed, barely reaching the middle of the following tergum; slightly shorter to slightly longer than the following tergum or much longer than the following tergum. Seta 5 is absent from abdominal segment VIII, except in species of the subgenus *Kenknightia* of *Aedes*.

The insertion of **seta 6-VII** relative to a longitudinal line through the insertion of seta 9-VII reflects supraspecific relationships. It may be inserted more or less in line directly posterior to seta 9, on the dorsal surface of the segment well mesad of that line, more or less directly in line anterior to seta 9 or on the ventral surface mesad of the line.

Seta 9-II–VII is usually inserted posterolaterally at the very edge of the ventral surface but it is clearly inserted dorsally at or removed from the caudolateral angle of the tergum in some culicine pupae. The seta is normally minute in culicine pupae and increases progressively in length from short and peg-like to long and spine-like from segment II to segment VII in anopheline pupae. **Seta 9-VIII** is normally large and branched in mosquito pupae, but its insertion (and development) relative to the posterolateral corner of the segment differs between groups and species, from being on the corner, slightly anterior or mesad of the corner to noticeably anterior to the corner. Seta 9-VIII is variable and somewhat difficult to characterize in many *Toxorhynchites* and *Uranotaenia* pupae where it is borne at the caudolateral angle of the segment in positions ranging from clearly dorsal to slightly ventral.

Seta 14-III–VII is a minute seta borne close to its mate at the longitudinal midline on the anterior margin of the abdominal sterna. These setae are present in New World species of *Mansonia* (subgenus *Mansonia*) and absent in Old Word species (subgenus *Mansonioides*). They are apparently always absent from segments III and IV of *Toxorhynchites* pupae, but are sometimes present on segment VI and/or VII. The setae are absent in all New World sabethine pupae, as well as pupae of *Malaya* and many *Topomyia* in the Old World. The removal of species from *Topomyia* to *Kimia* suggests that the presence or absence of these setae in species of *Topomyia* occurs along phyletic lines. The setae are apparently present in *Maorigoeldia* pupae and the pupae of all species of *Tripteroides*. **Seta 14-VIII** in anopheline pupae occurs in the same position as seta 14 on the other segments, but it is widely separated from its mate in culicine pupae.

Seta 1-IX is an anomalous seta that occurs in some species of *Aedes*. The seta is rarely absent in species of *Culex* but often absent in species of *Uranotaenia*. In *Mimomyia*, seta 1-IX is absent in the species of subgenera *Etorleptiomyia* and *Ingramia*. In pupae of the subgenus *Mimomyia*, it is a small peg-like seta distinctly removed from the median caudal lobe and located on membrane between the lobe and the base of the paddle.

Seta 1-XI is a lateral seta of the cercus in females and a corresponding area of the proctiger in males. This seta is represented by an alveolus in some *Culiseta* pupae. The seta occurs anomalously in a few species of *Aedes* and *Anopheles*.

Seta 1-Pa, the terminal or outer seta of the paddle, may be very short or long and single or multi-branched. **Seta 2-Pa**, the subterminal or inner seta, is ventral in position. One or the other or both paddle setae may be present or absent, usually in all (but not always) species of various supraspecific groups. Seta 2-Pa is absent more often than seta 1-Pa. Paddle setae are present in the majority of *Culex* species, most *Uranotaenia* species and all species of *Culiseta* except those of the subgenus *Theomyia*. Paddle setae are present in pupae of the subgenus *Mimomyia*, present or absent in those of the subgenus *Etorleptiomyia* and usually absent or replaced by specialized spicules in pupae of the subgenus *Ingramia*.

Vestiture

Vestiture is an archaic word for clothing. The vestiture of mosquitoes (Figs 26–28) consists of two types of external cuticular projections, articulated processes that arise from a basal **alveolus** (plural **alveoli**) and non-articulated processes (**spicules**) that are continuous outgrowths of the integument. There are two basic classes of articulated processes: **setae** (singular **seta**) and **scales**. Typical setae are innervated sensory organs (mechanical receptors) whereas scales are basically flattened non-innervated setae with longitudinal ridges. Setae generally arise from rather large alveoli, the presence of which should be checked to determine whether setae of taxonomic importance have been lost. In comparison, scales usually arise from extremely small alveoli that are difficult to see. Scales of different shapes (Fig. 27) and colours form the principal body covering (ornamentation) of adult mosquitoes. Scales are absent in larvae and pupae, in which the vestiture consists of multiple forms of setae and spicules (Figs 26 and 28).

In addition to setae and patches of scales, areas of the thoracic integument of adult mosquitoes often bear a fine covering of **tomentum** that imparts a pollinose or pruinose appearance to the surface. Tomentum is technically a covering of minute cuticular processes known as **aculeae** (singular **aculea**).

The distribution and development of setae and spicules are important taxonomic features of the larval and pupal stages. Adjectives are used to describe and name these elements of vestiture, *i.e.* an adjective is used to characterize the structural forms of setae and spicules. The adjectival names of the common types of setae and spicules are listed, described and illustrated in Section 3.

Section 2: Classification and Systematics

DOI: 10.1079/9781800628014.0002

Classification and Systematics

Insects (Insecta), a Class of the Phylum Arthropoda, are divided into Orders, the names of most of which are compound words ending with -ptera (plural of the Greek *pteron*, meaning 'wing'). Flies, including mosquitoes, comprise the order Diptera ('two-wings'), so named because they only have a single pair of wings. Orders are divided into Families. Mosquitoes constitute the family Culicidae, which is composed of two Subfamilies, the Anophelinae and Culicinae. The subfamily Culicinae is further divided into 11 Tribes: Aedeomyiini, Aedini, Culicini, Culisetini, Ficalbiini, Hodgesiini, Mansoniini, Orthopodomyiini, Sabethini, Toxorhynchitini and Uranotaeniini. The subfamily Anophelinae and six of the tribes of Culicinae are further divided into two or more Genera (Culisetini, Hodgesiini, Orthopodomyiini, Toxorhynchitini and Uranotaeniini are monogeneric, each includes only one genus), and many genera are divided into Subgenera. Subgenera, and genera without subgenera, are the taxonomic groups in which one or more species are classified.

The present classification of mosquitoes is the result of more than 266 years of taxonomic work conducted by numerous individuals. It must be borne in mind that the classification is a hypothesis of evolutionary lineages – the subordinate groups at each level of classification are considered to represent natural groups of species, but only until new information becomes available. Thus, it must be emphasized that many of the categories of classification treated below are probably polyphyletic groups composed of species that exhibit morphological similarities but are actually descendants of two or more evolutionary lineages.

The supraspecific units of the Culicidae treated here include the subfamilies, tribes, genera and subgenera recognized by Harbach (2018b), Wilkerson *et al.* (2021) and more recent authors. The taxonomic actions of Reinert *et al.* (2004, 2006, 2008, 2009), division of the 'traditional' concept of *Aedes* into many genus-group taxa, and Foster *et al.* (2017), generic status for the Neotropical subgenera of *Anopheles*, are eschewed herein.

The current hierarchical classification of the Culicidae is treated here in six parts: (I) family groups (family, subfamilies and tribes), (II) genus groups of Anophelinae (genera and subgenera), (III) genus group of Burmaculicinae (single genus), (IV) genus groups of Culicinae (genera and subgenera), (V) genus groups and informal group taxa, and (VI) species and their classification. Parts I–IV provide information on the typification, systematics, morphological characteristics, bionomics and principal literature sources for the family, subfamilies, tribes and all currently recognized genera and subgenera. The taxa in those parts are arranged alphabetically in hierarchical order, each superior taxon is followed by its subordinate taxon or taxa. Part V consists of detailed accounts of the taxonomic history and informal subdivision (where applicable) of the genera and subgenera, and Part VI presents a comprehensive hierarchical tabulation of the informal (internal) classifications of the genera and subgenera, and a list of all currently recognized species and their taxonomic placement in the formal and informal groups of the Culicidae. From a pragmatic point of view, there is no need for lineages regarded as subgenera, including the nominotypical subgenus, to be classified as components of a higher category (genus) rather than separate genera. Since taxonomic ranks are arbitrary constructs, little distinction is made between the ranks of genus and subgenus in the disquisitions on genus groups in part IV. The Latin term *sensu stricto* (meaning 'in the strict sense') is used herein for the nominotypical subgenus of a genus to distinguish it from the genus, *e.g. Anopheles* (the genus) and *Anopheles sensu stricto* (the type subgenus).

The ages provided for fossil taxa follow the geological timescale determined by the International Commission on Stratigraphy (ICS). These ages generally differ slightly from those provided with the published descriptions of fossil mosquitoes.

Part I: Family Groups

Family-group names are plural nouns in the nominative case. They are based on the name of a genus, the type genus of the group. A name proposed for a family-group name at any rank simultaneously establishes it for all higher and lower ranks in the family group, each having the same type genus. The family Culicidae, for example, and the coordinate subfamily name Culicinae and the coordinate tribal name Culicini are based on the genus *Culex* Linnaeus, which is the type genus of all three.

The family Culicidae includes 14 subordinate family groups – three subordinate subfamily names, Anophelinae, Burmaculicinae and Culicinae, and the last subfamily includes 11 subordinate tribal groups, Aedeomyiini, Aedini, Culicini, Culisetini, Ficalbiini, Hodgesiini, Mansoniini, Orthopodomyiini, Sabethini, Toxorhynchitini and Uranotaeniini. It is apparent from these names that the endings of family-group taxa are added to the stem of the type genus. The suffix -idae is added to the stem for family names, -inae for subfamily names and -ini for the names of tribes. Should the tribes be subdivided into subtribal groups, the names would be formed by adding -ina to the stem name. A few subtribes were established for mosquitoes in the past, but none are recognized today.

Culicidae Meigen, 1818

The family name Culicidae, typified by mosquitoes of the genus *Culex* Linnaeus, 1758, was established by Meigen, 1818 – as Culiciformes, the coordinate family-group name formed from the genitive singular of the type genus *Culex* (*Culicis*, combining stem *Culici-*); thus, Culiciformes, Culicidae (family), Culicinae (subfamily) and Culicini (tribe).

****

Systematics: Culicidae, true mosquitoes, comprise a large and abundant group that occurs throughout temperate and tropical regions of the world, and well beyond the Arctic Circle. Mosquitoes are most diverse in tropical forest environments. The family includes 3726 formally recognized extant species and 28 extinct species. The extant and some extinct species are classified in two subfamilies: Anophelinae, with three genera, and Culicinae, which has 38 genera segregated into 11 tribes. One genus with no living species belongs to the extinct subfamily Burmaculicinae. The family is monophyletic but deeper relationships are largely unresolved. Burmaculicinae is the sister of all other extinct and extant Culicidae. Anophelinae is a monophyletic lineage basal to all other Culicidae, and *Chagasia* is a monophyletic lineage basal to other Anophelinae (Harbach & Kitching, 1998; Harbach, 2007). Molecular data appear to support the monophyly of Culicinae and at least nine of the tribes (Aedeomyiini, Aedini, Culicini, Culisetini, Mansoniini, Orthopodomyiini, Sabethini, Toxorhynchitini and Uranotaeniini) (da Silva *et al.*, 2020a; Soghigian *et al.*, 2023; Chen *et al.*, 2024), but the monophyly and evolutionary relationships of these family groups will not be definitely known until their monophyly has been specifically tested with data obtained from the type species and sufficient numbers of representative species of all genus groups of the subfamily.

****

Characteristics: Adult mosquitoes are slender, long-legged insects that are easily recognized by their long proboscis and the presence of scales on most parts of the body. Larvae are distinguished from other aquatic insects by the absence of legs, the presence of a distinct head bearing mouthbrushes and antennae, a bulbous thorax that is wider than the head and abdomen, posterior anal papillae and either a pair of respiratory openings (subfamily Anophelinae) or an elongate siphon (subfamily Culicinae) borne near the end of the abdomen. Mosquitoes are usually, and most reliably, identified as mature (fourth-instar) larvae and adults.

****

Bionomics: Since mosquitoes are delicate insects, they are always found where the air is relatively cool and the humidity is high. Many species live within a few metres of the ground, whereas many forest species occur primarily in the forest canopy. Vertical distribution is largely dependent on feeding preferences. All males and the females of many species feed exclusively on plant liquids, including nectar, honeydew, fruit juices and exudates. Females of a great number of species feed on the blood of living animals, but some that are normally blood feeders may produce eggs without a blood meal. Warm-blooded vertebrates are a common source of blood for most species, but many species also attack cold-blooded animals such as snakes, turtles, toads, frogs and other insects, including nymphal cicadas, lepidopterous larvae and mantids. A species of *Uranotaenia* is known to feed on earthworms and leeches. The time of flight and feeding activity is usually quite specific for most species. Some species are active at night (nocturnal) or twilight (crepuscular) while others are active during the daylight hours (diurnal).

Mosquito larvae occupy a spectrum of aquatic environments. The majority of larvae feed on suspended particulate matter and microorganisms which they extract from the water with filamentous mouthbrushes. Other species are obligatory or facultative predators which capture and feed largely on the immature stages of other mosquitoes by means of modified mouthbrushes or grasping mandibles or maxillae. Some larvae resort to scavenging or cannibalism when food is scarce. The larvae of most mosquitoes obtain oxygen from the atmosphere by coming to the surface of the water. All species of *Mansonia* and *Coquillettidia* and some species of *Mimomyia* obtain oxygen from the air vessels of aquatic plants, which they pierce with a specialized siphon. *Aedeomyia* species apparently use their enlarged antennae for respiration. Some species have greatly enlarged anal papillae that are well supplied with tracheae, and these species seldom come to the surface and probably obtain dissolved oxygen from the water.

Mosquitoes are important because the females of many species are bloodsucking, they annoy humans and other animals, and they may transmit the pathogens that cause human and animal diseases. The pathogens transmitted by mosquitoes include viruses (arboviruses), filarial worms (helminths) and protozoa. Mosquitoes are the indirect cause of more morbidity and mortality among humans that any other group of organisms.

****

Principal references: Barraud, 1934 (southern Asia, Culicinae, as Culicini and Megarhinini, keys, taxonomy, descriptions, distributions, bionomics); Bates, 1949 (biology); Edwards, 1941 (Afrotropical Region, keys, descriptions of adults and pupae, distributions); Hopkins, 1952 (Afrotropical Region, keys, descriptions of larvae, distributions, bionomics); Lane, 1953 (Neotropical Region, taxonomy, keys, descriptions, distributions); Carpenter & LaCasse, 1955 (North America, taxonomy, keys, descriptions, bionomics); Mattingly & Knight, 1956 (Arabia); Stone *et al.*, 1959 (catalogue); Belkin, 1962 (South Pacific, as subfamily Culicinae, taxonomy, keys, descriptions, distributions, bionomics); DuBose & Curtin, 1965 (Mediterranean area, keys,); Cova-Garcia *et al.*, 1966 (Venezuela, keys, species descriptions, distributions); Belkin, 1968a (New Zealand, taxonomy, keys, descriptions, bionomics); Belkin *et al.*, 1970 (Jamaica, taxonomy, keys, descriptions, bionomics); Gillett, 1971 (biology); Mattingly, 1971a (world, keys for genera); Horsfall, 1972 (bionomics); Gutsevich *et al.*, 1974 (former USSR, taxonomy, keys, descriptions, distributions, bionomics); Knight & Stone, 1977 (catalogue); Knight, 1978b (catalogue supplement); Tanaka *et al.*, 1979 (Japan, taxonomy, keys, descriptions, distributions, bionomics); Wood *et al.*, 1979 (Canada, taxonomy, keys, descriptions, distributions, bionomics); Lee *et al.*, 1980, 1982, 1984, 1987a, 1987b, 1988a, 1988b, 1988c, 1989a, 1989b, 1989c (Australasian Region, taxonomy, keys, literature, distributions, bionomics); Darsie & Ward, 1981, 2005 (North America, keys); Lu & Li, 1982 (China, keys); Clark-Gil & Darsie, 1983 (Guatemala, keys, distributions, bionomics); Ward, 1984, 1992 (catalogue supplements); Danilov, 1985a, 1985b (Afghanistan, keys); Darsie, 1985 (Argentina, keys); Gaffigan & Ward, 1985 (catalogue supplement); Lounibos *et al.*, 1985 (bionomics); Mitchell & Darsie, 1985 (Argentina, distributions, literature); Cranston *et al.*, 1987 (Britain, keys, distributions, bionomics); Laird, 1988 (larval bionomics); Debenham *et al.*, 1989 (Australasian Region, taxonomy, species list, geographies, disease relationships); Darsie & Pradhan, 1990 (Nepal, keys, distributions, bionomics); Clements, 1992, 1999, 2012 (biology); Service, 1990 (Afrotropical Region, keys, descriptions, bionomics); Service, 1993a, 1993b (bionomics, taxonomy); Rattanarithikul & Panthusiri, 1994 (Thailand, keys, medically important species); Harbach & Sandlant, 1997 (genera, keys); Harbach & Kitching, 1998 (morphology, phylogenetic relationships); Becker *et al.*, 2003, 2010, 2020 (Europe, keys, descriptions, distributions, bionomics); Rattanarithikul *et al.*, 2005a, 2005b, 2006a, 2007b, 2007, 2010 (Thailand, keys, distributions, bionomics); Harbach, 2007 (taxonomy, classification, phylogenetic relationships); Thielman & Hunter, 2007 (Canada, key to females); González *et al.*, 2008 (Chile, key to genera with species of medical importance); Azari-Hamidian & Harbach, 2009 (Iran, keys); Reidenbach *et al.*, 2009 (phylogenetic relationships); Borkent, 2012 (pupal morphology); Jeffery *et al.*, 2012 (Peninsula Malaysia, keys to adults, references, checklist, taxonomic notes); Tuten *et al.*, 2012 (pyloric armature, phylogenetic relationships); Bhattacharyya *et al.*, 2014 (India, checklist, classification, vector species); Rozo-Lopez & Mengual, 2015 (Colombia, checklist); Rossi, 2015 (Argentina, checklist, distributions); Rueda *et al.*, 2015 (Laos, checklist, distributions); Talaga *et al.*, 2015 (French Guiana, checklist, taxonomic notes); Tyagi *et al.*, 2015a (India, species list, keys for larvae and adults of vector species); Borkent & Grimaldi, 2016 (Burmaculicinae, morphology, phylogenetic position); Tantely *et al.*, 2016 (Madagascar, checklist); Webb *et al.*, 2016 (Australia, species, descriptions, bionomics, distributions); Lemine *et al.*, 2017 (Mauritania, species list, distributions, medical importance); Coetzee, 2017 (Afrotropical Region, morphology, bionomics, keys to genera); Namgay *et al.*, 2018 (Bhutan, checklist); Ruzzante *et al.*, 2019 (evolutionary genomics); Tsuda, 2019 (Japan, illustrated keys, distributions, bionomics); Yeo *et al.*, 2019 (Singapore, adult females, key to genera); da Silva *et al.*, 2020a (Neotropics, family groups, phylogenetic relationships); Mahgoub *et al.*, 2020 (Sudan, checklist, identification, distributions, bionomics); Talaga *et al.*, 2020a (French Guiana, species described from the country, taxonomic notes); Talaga *et al.*, 2020b (French Guiana, taxonomic review of indigenous Culicidae); van der Beek *et al.*, 2020 (Dutch Leeward Islands, key for adults and larvae, ecology, distributions); Bromley-Schnur, 2021 (Israel, neighbouring areas, keys for adults and larvae, taxonomy, bionomics, distributions); Ebejer & Gatt, 2021 (Maltese Islands, species list); Maquart *et al.*, 2021 (Cambodia, checklist, distributions, bionomics); Merabti *et al.*, 2021 (Algeria, checklist, taxonomy); Ndiaye *et al.*, 2021 (Burkina Faso, Chad, Gambia, Mauritania, Niger, Senegal, species distributions); Nugroho & Mujiyono, 2021 (Indonesia, keys to genera); Wilkerson *et al.*, 2021 (World, biology, taxonomy, systematics, morphology, medical importance, catalogue); Yee *et al.*, 2021 (Puerto Rica, Vieques, species, collection records, bionomics); Robert *et al.*, 2022 (Burkina Faso, Cape Verde, Gambia, Mali, Mauritania, Niger, Senegal, Chad, keys for adults and larvae); Dawah *et al.*, 2023 (Saudi Arabia, distributions, notes, medical importance, keys for adult females); Harbach & Wilkerson, 2023 (World, elimination of subspecies); Russell & Burkot,

2023 (Pacific Island countries and territories, species lists, distributions, bionomics); Soghigian *et al.*, 2023 (World, family groups, molecular markers, phylogenetic relationships); Chen *et al.*, 2024 (Eastern Asia, family groups, mitochondrial protein sequences, phylogenetic relationships); Ortega-Morales *et al.*, 2024 (Mexico, Morelos State, checklist, bionomics, distributions, *COI* sequences, phylogenetic relationships).

Anophelinae Grassi, 1900

The subfamily Anophelinae is typified by mosquitoes of the genus *Anopheles* Meigen, 1818. The International Commission on Zoological Nomenclature (1959a: Opinion 547) placed 'Anophelinae (correction of Anophelina) Theobald, 1901' on the *Official List of Family-Group Names in Zoology*. However, the use of Anophelinae by Grassi (1900) has historical precedence for priority.

****

Systematics: Anophelinae includes 524 formally recognized extant species and three extinct species. Many genetic species of contemporary sibling species complexes await formal names. The extant species are classified in three genera: *Anopheles*, *Bironella* and *Chagasia*. Two fossil species are placed in *Anopheles* and one is the nominotypical member of the extinct genus *Priscoculex*. Mosquitoes belonging to these genera are referred to as 'anophelines'. Most species of the subfamily belong to *Anopheles*, the species of which occur in temperate, subtropical and tropical areas of the world, except for island groups in the Pacific and isolated islands in the Atlantic. *Bironella* and *Chagasia* are small genera, restricted to the Australasian and Neotropical Regions, respectively. The anophelines comprise a monophyletic lineage basal to all other Culicidae, and *Chagasia* is a monophyletic lineage basal to other genera. *Anopheles* is not demonstrably monophyletic with regard to *Bironella* (Harbach & Kitching, 1998; Sallum *et al.*, 2000, 2002; Krzywinski *et al.*, 2001a, 2001b; Harbach & Kitching, 2005, 2016; Harbach, 2007; Foster *et al.*, 2017). More recent molecular phylogenetic studies indicate that the ancestors of the Anophelinae and Culicinae diverged during the Jurassic Period (201.3–145.0 Mya) of the Mesozoic Era (da Silva *et al.*, 2020a; Soghigian *et al.*, 2023; Chen *et al.*, 2024).

****

Characteristics: Adult anophelines are easily recognized by their appearance. Most species stand with the body inclined at an angle of 30–45° to the surface and have dark and pale spots of scales on the veins of the wings. Some species have the wing veins entirely covered with dark scales. The maxillary palpi of both sexes are about as long as the proboscis (except in *Bironella*). The palpi of females sometimes have semi-erect scales that give them a rather shaggy appearance. The scutellum is evenly rounded in *Anopheles* (with a few exceptions) and *Bironella* and trilobed in *Chagasia*. The abdominal sterna, and usually the terga, are completely or nearly devoid of scales. Anopheline larvae lack a respiratory siphon, the head is longer than wide and pairs of palmate setae are normally present on some or all of abdominal segments I–VII.

****

Bionomics: Anopheline larvae are found principally in groundwater, but some species are restricted to plant cavities. All species rest at the water surface and feed by rotating the head trough 180° to draw food particles at the surface of the water into the mouth with their mouthbrushes. Most species utilize stagnant or relatively still water, and even those that are associated with fast-flowing streams actually occur in niches where currents are substantially reduced by vegetation, debris or other objects. Most anopheline adults are active at night and during twilight periods. They rest in protected places during the day. Females feed on warm-blooded vertebrates, primarily mammals, but some species feed largely on birds. Many species readily attack humans. The hum from the wings of anophelines is low-pitched and almost inaudible to humans unless the mosquito is hovering close to the ears.

Species of *Bironella* and *Chagasia* are of no medical or economic importance. *Anopheles* contains the mosquitoes that transmit malarial parasites to humans and other animals. Some species of *Anopheles* also transmit microfilariae and arboviruses. Because of the importance of *Anopheles* to human health, it has been the subject of more scientific study than any other genus of mosquitoes.

****

Principal references: Lane, 1953 (Neotropical Region, as Anophelini); Belkin, 1962 (South Pacific, as Anophelini); Cova-Garcia, 1961 (Venezuela, as Anophelini); Forattini, 1962 (Neotropical Region, as Anophelini, descriptions, distributions, bionomics); Reid, 1968 (Malaysia, Borneo, classification, keys, descriptions, distributions, bionomics); Tanaka *et al.*, 1979 (Japan, keys, descriptions, distributions, bionomics); Lee *et al.*, 1987b, 1988a (Australasian Region, taxonomy, keys, literature, distributions, bionomics); Sallum *et al.*, 2000, 2002 (phylogenetic relationships); Becker *et al.*, 2003, 2010, 2020 (Europe, keys, descriptions, distributions, bionomics); Harbach & Kitching, 2005, 2016 (phylogenetic relationships); Foster *et al.*, 2017 (mitochondrial protein sequences, phylogenetic relationships); Hao *et al.*, 2017 (phylogenetic relationships); Poinar *et al.*, 2019 (*Priscoculex*, diagnosis, type species description, evolutionary notes and comments); da Silva *et al.*, 2020a (phylogenetic relationships); Guo *et al.*, 2021 (phylogenetic relationships); Soghigian *et al.*, 2023 (molecular markers, phylogenetic relationships, host use); Chen *et al.*, 2024 (mitochondrial protein sequences, phylogenetic relationships).

Burmaculicinae Borkent & Grimaldi, 2016

The subfamily Burmaculicinae includes the extinct genus *Burmaculex* Borkent & Grimaldi, 2004.

****

Systematics: Burmaculicinae has no living species and only one extinct species of the type genus is currently known. Phylogenetic analysis based on morphological data has shown that Burmaculicinae is the sister of all other fossil and extant species of Culicidae (Borkent & Grimaldi, 2004).

****

Characteristics: ADULTS: Clypeus of the head with a few setae; proboscis moderately elongate, about 0.3–0.4 the length of the antenna; maxillary palpus longer than the proboscis, without scales; wing largely without scales, present only on the costa and posterior margin, vein R_1 with a distinct cephalad curvature anterior to vein R_{2+3}, anal vein (vein 1A) sharply curved distally and ending at the wing margin proximal to the furcation of the mediocubital crossvein and vein CuA; abdomen apparently without scales. IMMATURE STAGES: Unknown.

****

Bionomics: The only known species of Burmaculicinae existed during the early part of the Upper Cretaceous of the Mesozoic Era approximately 90–95 Mya. The climate was relatively warm, there were numerous shallow inland seas, and dinosaurs were the dominant land animals. Flowering plants and many insects evolved during the Cretaceous Period. Potential hosts for species of the subfamily would have included reptiles and birds, and perhaps other vertebrates that existed in the mid-Cretaceous.

****

Principal references: Borkent & Grimaldi, 2004 (type genus and species description, phylogenetic position, bionomics); Borkent & Grimaldi, 2016 (Burmaculicinae, diagnosis, type genus description, phylogenetic position, comparative morphology).

Culicinae Meigen, 1818

The subfamily Culicinae is typified by mosquitoes of the genus *Culex* Linnaeus, 1758. Meigen (1818) introduced the first family-group name (Culiciformes) that established the date of priority for Culicidae and its coordinate subfamily and tribal names, Culicinae and Culicini, respectively.

****

Systematics: Culicinae is the largest subfamily of mosquitoes, containing 3202 extant species classified in 38 genera. Culicinae also includes 24 fossil species, 20 in six extant genera, eight in six extinct genera. Fossils are also known for four extant species and a *nomen nudum*. Members of Culicinae are found in all zoogeographic regions of the world, but the majority of species occur in tropical regions. The genera are grouped into 11 tribes: Aedeomyiini, Aedini, Culicini, Culisetini, Ficalbiini, Hodgesiini, Mansoniini, Orthopodomyiini, Sabethini, Toxorhynchitini and Uranotaeniini. Species belonging to this subfamily are referred to as 'culicines'; however, species of Aedini are referred to as 'aedines', species of Culicini are sometimes referred to as 'culicinines' and species of Sabethini are known as 'sabethines'. Morphological and molecular data appear to support the monophyly of Culicinae and the 11 tribes (Harbach & Kitching, 1998; Harbach, 2007; Reidenbach *et al.*, 2009; da Silva *et al.*, 2020a; Soghigian *et al.*, 2023; Chen *et al.*, 2024), but the monophyly and evolutionary relationships of these groups must be considered uncertain until their monophyly has been tested with data obtained from the type species and sufficient numbers of all subordinate groups of each family group. The molecular phylogenetic studies conducted by da Silva *et al.* (2020a), Soghigian *et al.* (2023) and Chen *et al.* (2024) indicate that the ancestors of the Culicinae and Anophelinae diverged during the Jurassic Period (201.3–145.0 Mya) of the Mesozoic Era.

****

Characteristics: Culicine females have maxillary palpi that are usually much shorter than the proboscis. Males have long palpi with numerous long setae, but they are not swollen apically like those of anophelines. The scutellum normally has three lobes with setae confined to each lobe. The wing veins are usually entirely dark-scaled, but speckles or patches of white or yellow scales are present in some species. Some species of some genus groups, *e.g. Aedeomyia*, *Culex sensu stricto* (Mimeticus Group), *Finlaya*, *Lutzia sensu stricto*, *Orthopodomyia* and *Uranotaenia*, have spots of pale scales similar to those of many *Anopheles*. An emargination on the posterior margin of the wing opposite vein CuA (cubitus anterior) is found only in species of Toxorhynchitini. The abdominal terga and sterna are densely covered with scales, which distinguishes them immediately from almost all anophelines.

Adult culicines stand with the body parallel to the surface on which they are resting. All culicine larvae have a respiratory siphon, the dorsal and ventral abdominal setae of most species arise separately and usually without basal sclerites, and the mouthbrushes are composed of numerous, usually slender filaments (except *Lutzia*, *Mucidus* and *Toxorhynchites*).

****

Bionomics: Culicine larvae occupy a wide variety of aquatic habitats ranging from small, temporary to large, permanent bodies of water. Most species occur in collections of groundwater, but many species inhabit a variety of plant cavities (phytotelmata) and artificial containers of different sizes. Most larvae use their siphons to suspend from the surface of the water, but larvae of *Mansonia*, *Coquillettidia* and some *Mimomyia* insert the modified tips of their siphons into plants to obtain air. The larvae of most species feed by extracting suspended particulate matter from the water with filamentous mouthbrushes but grazing on submerged surfaces commonly occurs and the larvae of some species are facultative or obligate predators of other mosquito larvae and other insects. Culicine adults are active during the day or night, but most species are probably only active during the daylight hours, including crepuscular periods.

Species of the genus groups *Culex*, *Mansonia* and *Stegomyia* are the most important from a medical point of view because they contain the majority of culicine mosquitoes that transmit helminths and arboviruses. Females of species of many genera are annoying pests of humans and domestic animals.

****

Principal references: Barraud, 1934 (southern Asia, as Culicini and Megarhinini, keys, taxonomy, descriptions, distributions, bionomics); Edwards, 1941 (Afrotropical Region, keys, descriptions of adults and pupae, distributions); Hopkins, 1952 (Afrotropical Region, keys, descriptions of larvae, distributions, bionomics); Lane, 1953 (Neotropical Region, taxonomy, keys, descriptions, distributions); Thurman, 1959 (Thailand, as family Culicidae, taxonomy, descriptions, distributions); Belkin, 1962 (South Pacific, as subfamily Culicinae less Anophelini, taxonomy, keys, descriptions, distributions, bionomics); Cova Garcia *et al.*, 1966 (Venezuela, Culicinae + Toxorhynchitinae, keys, descriptions, distributions); Delfinado, 1966a (Philippines, in part as Culicini, keys, taxonomy, descriptions, distributions, bionomics); Tanaka *et al.*, 1979 (Japan, keys, descriptions, distributions, bionomics); Ibrahim *et al.*, 1983 (Iraq, larvae, keys); Lee *et al.*, 1984, 1987a, 1988a, 1988b, 1988c, 1989a, 1989b, 1989c (Australasian Region, genus-group descriptions, taxonomy, literature, distributions and bionomics for species); Hudson & Abul-Hab, 1987 (Iraq, adults, keys); Service, 1990 (Afrotropical Region, Culicinae in part + Toxorhynchitinae, keys, descriptions, distributions, bionomics); Lu *et al.*, 1997a (China, Culicinae + Toxorhynchitinae, keys, taxonomy, descriptions, distributions, bionomics); Harbach & Kitching, 1998 (phylogenetic relationships); Becker *et al.*, 2003, 2010, 2020 (Europe, keys, descriptions, distributions, bionomics); Azari-Hamidian & Harbach, 2009 (Iran, keys); Reidenbach *et al.*, 2009 (phylogenetic relationships); da Silva *et al.*, 2020a (phylogenetic relationships); Soghigian *et al.*, 2023 (molecular markers, phylogenetic relationships, host use); Chen *et al.*, 2024 (mitochondrial protein sequences, phylogenetic relationships).

Aedeomyiini Theobald, 1901

The tribe Aedeomyiini Theobald, 1901a is typified by mosquitoes of the genus *Aedeomyia* Theobald, 1901a. The subfamily name Aedeomyiinae (as Aedeomyina) is attributed to Theobald (1901a), and he is considered to have simultaneously established the coordinate tribal name Aedeomyiini.

****

Systematics: Aedeomyiini is a subordinate taxon of the subfamily Culicinae. It includes seven extant species in a single genus, *Aedeomyia*. Members of the tribe are found in the Afrotropical, Australasian, Neotropical and Oriental Regions. The phylogenetic relationships of the tribe are uncertain. Based on morphology and distribution, Belkin (1962) suggested that '*Aedeomyia* is probably a very ancient primitive but highly specialized segregate, which originated in the Old World and reached the New World prior to or very early in the Tertiary' (see Harbach & Kitching, 1998; Reidenbach *et al.*, 2009). In support of this, in a phylogenetic analysis of amino acid sequences of 709 genes Aedeomyiini was recovered in a basal relationship to all other culicine taxa included in the study (Soghigian *et al.*, 2023). It is interesting to note that in the analysis of mitogenomes of Neotropical mosquitoes conducted by da Silva *et al.* (2020a), Aedeomyiini and Uranotaeniini were recovered as sisters in a basal relationship to the other culicine taxa included in the study; however, the result was based on sequence data obtained from a single species of each of those tribes. Based on the results of the phylogenetic analysis of amino acid sequences of 709 genes (Soghigian *et al.*, 2023), it appears that the ancestor of the Aedeomyiini diverged from the ancestor of the other tribes of Culicinae during the Lower Cretaceous Period (145.0–100.5 Mya) of the Mesozoic Era.

****

Characteristics: Adults of species of the tribe are small, highly ornamented mosquitoes with very broad scales, superficially resembling some aedine mosquitoes, but they are readily distinguished by the structure of the antenna – all antennal flagellomeres of females and the apical two flagellomeres of males are unusually short and thick. Larvae have distinctive antennae that are greatly swollen, strongly curved and longer than the head.

****

Bionomics: Larvae of species of the tribe are found primarily in dense swamps and ground pools with abundant aquatic vegetation, but they have also been collected from river margins. They are known to stay submerged for long periods of time, and are suspected of being capable of cuticular respiration through their greatly enlarged antennae. Females apparently feed predominantly on birds. Females of one species, *Aedeomyia africana*, have been collected from human bait stationed on platforms in forest canopy, but members of the tribe normally are not attracted to humans. A few arboviruses and avian malarial protozoa have been isolated from species of the tribe, but none of the species are considered to be of importance to human health.

****

Principal references: Belkin, 1962 (South Pacific, taxonomy, description, systematics, bionomics, distribution); Tyson, 1970a (World, taxonomy, keys, descriptions, distributions, bionomics); da Silva *et al.*, 2020a (phylogenetic relationships); Soghigian *et al.*, 2023 (molecular markers, phylogenetic relationships, host use).

Aedini Neveu-Lemaire, 1902

The tribe Aedini Neveu-Lemaire, 1902c is typified by mosquitoes of the genus *Aedes* Meigen, 1818. The subfamily name Aedinae (as Aedeinae) was established by Neveu-Lemaire (1902c), and he is considered to have simultaneously established the coordinate tribal name Aedini.

****

Systematics: Aedini is a subordinate taxon of the subfamily Culicinae. Species of Aedini are referred to as 'aedines'. Aedini is the largest tribe of mosquitoes with 1298 extant species classified in 10 genera: *Aedes*, *Armigeres*, *Eretmapodites*, *Haemagogus*, *Heizmannia*, *Opifex*, *Psorophora*, *Udaya*, *Verrallina* and *Zeugnomyia*. Four fossil aedine species are known, two are recognized as species of *Aedes* (*Ochlerotatus*), two are members of extinct genera, *Aetheapnomyia* and *Eoaedes*. Aedines are worldwide in distribution, but are better represented in the Old World and the Nearctic Region. Morphological and molecular evidence indicates that Aedini is a monophyletic taxon (see Harbach, 2007; Reidenbach *et al.*, 2009; da Silva *et al.*, 2020a; Lorenz *et al.*, 2021; Soghigian *et al.*, 2023; Chen *et al.*, 2024). Based on the assumption that annectant forms may have been derived by hybrid origin, Belkin (1962) believed that Aedini differentiated in the Indomalayan area of the Old World where the majority of annectant forms presently exist. Based on morphology, the genera of Aedini comprise two genus groups, the *Ochlerotatus* Genus Group and the more highly derived *Aedes* Genus Group. This division is supported by the molecular phylogenetic study of Soghigian *et al.* (2017). Aedini was recovered as a paraphyletic lineage relative to a monophyletic Mansoniini in the phylogenetic study of Harbach & Kitching (1998), as the sister of *Mansonia* in the studies of Reinert *et al.* (2004, 2006, 2009) and as sister to *Orthopodomyia* + *Mansonia* in the study of Reinert *et al.* (2008) based on morphological data. Aedini was recovered as the sister to *Orthopodomyia* + *Coquillettidia* in the study of Reidenbach *et al.* (2009) based on combined molecular (six nuclear genes) and morphological data. Aedini was not recovered as a monophyletic group in the analyses of Wilkerson *et al.* (2015) based on equal weighting of the data set of Reinert *et al.* (2009) – *Mansonia* + *Psorophora* was recovered in a sister relationship to the other generic-level taxa of the tribe. Aedini was recovered in a sister relationship with Culicini in a phylogenetic analysis of mitogenomes derived from Neotropical representatives of eight tribes of Culicinae (da Silva *et al.*, 2020a), as the sister of a clade comprised of 17 species of *Culex* and a single species of *Uranotaenia* in an analysis of 13 mitochondrial protein-coding genes (Lorenz *et al.*, 2021) and as the sister to a clade comprised of five other tribes (Culicini, Mansoniini, Sabethini, Toxorhynchitini and Uranotaeniini) in a tree generated from an analysis of the protein-coding gene 123 (Chen *et al.*, 2024) and as the sister to a lineage comprised of six tribes (Culicini, Culisetini, Mansoniini, Orthopodomyiini, Sabethini and Toxorhynchitini) in a study based on analysis of amino acid sequences of 709 genes (Soghigian *et al.*, 2023). Unlike the study of Chen *et al.*, Uranotaeniini was recovered in a basal relationship to Aedini and the other tribes, with Aedeomyiini basal to them all. The results of the phylogenetic studies conducted by da Silva *et al.* (2020a), Soghigian *et al.* (2023) and Chen *et al.* (2024) indicate that the ancestral lineage of the Aedini arose during the latter part (120–100 Mya) of the Lower Cretaceous Period of the Mesozoic Era.

****

Characteristics: Species of Aedini are extremely varied, and many are difficult to identify to genus because of overlapping suites of shared anatomical features. Hence, combinations of characters are required to define most of the genera, subgenera and species. General features of the tribe include the presence of toothed ungues (tarsal claws) and a pointed abdomen in most females. Although toothed ungues are not universally present, they are not found in any other tribe of Culicinae. Larvae have relatively short, stout siphons with a single pair of seta 1-S (except for species of *Aedes sensu stricto* and the Rusticoidus Group of *Ochlerotatus*) inserted well above the base, usually beyond the middle of the siphon. A comb is always present on segment VIII and the ventral brush is usually represented by five or more pairs of setae.

The heterogeneity of the tribe is illustrated by the following description of principal taxonomic features (excluding those above) found in the adult and larval stages. ADULTS: Compound eyes contiguous or separated above antennae; erect scales of head numerous, restricted to a single posterior row, or absent; clypeus with or without scales, never setae; proboscis variable in length and development; antennae distinctly shorter to slightly longer than proboscis; scutal scales and setae varied; mesopostnotum with or without setae; paratergite with or without scales; antepronota large or small, sometimes close together; pleural setae and scales varied; base of hindcoxa in line with or below base of mesomeron; pulvilli very small, not evident; tarsi normal; wing membrane with distinct microtrichia; cell R_2 longer than vein R_{2+3}; vein R_s without basal spur; vein 1A ending distal to base of mediocubital crossvein; base of subcosta without setae ventrally except in *Opifex*; scales of alula varied; upper calypter usually with complete line of setae or hair-like scales, sometimes reduced in number or absent; laterotergite varied, completely or partially covered with scales or bare. LARVAE: Occipital foramen normal, circular; collar varied, usually distinct; hypostomal suture complete; seta 2-C absent; seta 3-C dorsal in position; seta 13-P rarely present; no plumose or palmate setae on thorax or abdomen; seta 12-I present or absent; pecten usually present, sometimes absent; saddle usually incomplete.

****

Bionomics: Aedine larvae inhabit a diversity of aquatic habitats. Some species utilize temporary groundwaters such as pools, puddles, hoof prints, wallows and the margins of swampy or marshy areas. Other species utilize rock holes, crab holes, tree holes, leaf axils and flower bracts, pitcher plants and artificial containers. The females of many species readily attack humans. Both daytime and night-time biters are known. Certain members of the tribe are of great importance in the transmission of viruses and helminths to humans and other animals.

****

Principal references: Belkin, 1962 (South Pacific, taxonomy, description, systematics, bionomics, distribution); Belkin *et al.*, 1970 (Jamaica, taxonomy, keys, descriptions, bionomics); Lee *et al.*, 1980, 1982, 1984, 1987a, 1988b, 1988c (Australasian Region, taxonomy, keys, literature, distributions, bionomics); Lu & Su, 1987 (China, keys); Huang, 2001 (Afrotropical Region, keys to subgenera of *Aedes*); Reinert *et al.*, 2004, 2006, 2008, 2009 (phylogenetic relationships, classification); Reidenbach *et al.*, 2009 (phylogenetic relationships); Rattanarithikul *et al.*, 2010 (Thailand, keys); Wilkerson *et al.*, 2015 (phylogenetic relationships, classification); Mohamed *et al.*, 2017 (Republic of the Sudan, bionomics, keys); da Silva *et al.*, 2020a (phylogenetic relationships); Lorenz *et al.*, 2021 (mitochondrial protein-coding genes, phylogenetic relationships); Soghigian *et al.*, 2023 (molecular markers, phylogenetic relationships, host use); Chen *et al.*, 2024 (mitochondrial protein sequences, phylogenetic relationships).

Culicini Meigen, 1818

The tribe Culicini is typified by mosquitoes of the genus *Culex* Linnaeus, 1758. Meigen (1818) introduced the first family-group name (Culiciformes) that established the date of priority for Culicidae, Culicinae and Culicini.

****

Systematics: Culicini is a subordinate taxon of the subfamily Culicinae. It includes 847 extant species classified in four genera: *Culex*, *Deinocerites*, *Galindomyia* and *Lutzia*. Fossils of seven extinct species and four extant species of *Culex* are known. *Culex* has a cosmopolitan distribution, *Deinocerites* and *Galindomyia* occur in the Neotropical Region and *Lutzia* is represented by species in all zoogeographical regions except the western Palaearctic and Nearctic Regions. The monophyly of Culicini has never been disputed. The tribe was recovered as the sister of Sabethini in the cladistic analysis of Harbach & Kitching (1998) based on morphological data. Mattingly (1981) noted that certain *Culex* share characters with sabethines, but there is no substantial evidence to support a relationship between Culicini and Sabethini. Belkin (1962) considered Culicini to be a natural and ancient group that shares some similarities in the immature stages with certain groups of *Culiseta*. Adames (1971) considered *Deinocerites* and *Galindomyia* as a monophyletic group based on shared characteristics of the antennae and male genitalia, and regarded this clade as the sister group of *Culex*. These relationships were supported in the generic analysis of Harbach & Kitching (1998). A number of published and unpublished phylogenetic studies based on morphological and molecular data support the monophyly of all of the generic-level groups of Culicini except subgenera *Culex*, *Eumelanomyia* and *Neoculex* (Mallampalli, 1995; Miller *et al.*, 1996; Isoe, 2000; Navarro & Liria, 2000; Juthayothin, 2004; St John, 2007; Deus, 2009; Vesgueiro *et al.*, 2011; Demari-Silva *et al.*, 2011; Harbach *et al.*, 2012; Laurito & Almirón, 2013; Harbach *et al.*, 2017a). The results of these studies, however, provide conflicting support for the generic status of *Lutzia* and the subgeneric statuses of *Oculeomyia*, *Phenacomyia*, *Phytotelmatomyia* and *Sirivanakarnius*. With the exception of Mallampalli (1995), analyses based solely on morphological data support the generic status of *Lutzia* (Navarro & Liria, 2000; St John, 2007; Harbach *et al.*, 2012; Laurito & Almirón, 2013) by placing it outside the clade comprised of the subgenera of *Culex*, whereas the analyses based on nucleotide sequences of ITS1 and ITS2 of rDNA and *COI* of mtDNA do not, because *Lutzia* is placed among species or subgenera of *Culex* (Miller *et al.*, 1996; Vesgueiro *et al.*, 2011; Demari-Silva *et al.*, 2011). In stark contrast, however, the phylogenetic analysis of Sun *et al.* (2019) based on the nucleotide sequences of 13 protein-coding genes obtained from the complete mitochondrial genomes of 10 species of *Culex* and two species of *Lutzia* provided strong support for the generic status of *Lutzia* as the sister of *Culex*. The morphological analysis of Navarro & Liria (2000) recovered *Phenacomyia* in a basal relationship relative to all other subgenera of *Culex* (a clade that also included *Deinocerites*). In contrast, *Phenacomyia* and *Phytotelmatomyia*, and also subgenera *Oculeomyia* and *Sirivanakarnius*, have been placed among species of *Culex sensu stricto* in other studies, both morphological and molecular, that included species of these subgenera (Deus, 2009; Vesgueiro *et al.*, 2011; Demari-Silva *et al.*, 2011; Harbach *et al.*, 2012; Laurito & Almirón, 2013; Harbach *et al.*, 2017a). More recent, more in-depth molecular studies support the monophyly of Culicini (da Silva *et al.*, 2020a; Soghigian *et al.*, 2023; Chen *et al.*, 2024). The tribe was recovered as the sister to a clade comprised of Mansoniini, Sabethini, Toxorhynchitini and Uranotaeniini in a phylogenetic analysis of the protein-coding gene 123 (Chen *et al.*, 2024) and as the sister to a lineage comprised of Culisetini, Mansoniini, Orthopodomyiini, Sabethini and Toxorhynchitini in a study based on the analysis of amino acid sequences of 709 genes (Soghigian *et al.*, 2023). The results of the phylogenetic studies conducted by da Silva *et al.* (2020a), Soghigian *et al.* (2023) and Chen *et al.* (2024) indicate that the ancestral lineage of the Culicini existed as early as the start (about 100 Mya) of the Upper Cretaceous Period of the Mesozoic Era.

****

Characteristics: The immature stages of Culicini are very similar to the immature stages of some groups of *Culiseta*. As far as known, the cibarial armature of the females is developed elsewhere only in the Anophelinae. The absence of prespiracular setae and postspiracular setae is rather distinctive of Culicini, but the latter are present in at least two species of *Culex*. The tribe is unique in the development of more or less distinct pulvilli at the base of the ungues, but this character is rather difficult to observe. The crown of spicules on the paraprocts of the male genitalia is also a unique feature of Culicini, but the spicules are reduced in some subgenera of *Culex*.

****

Bionomics: Species of *Culex* are found primarily in more or less permanent bodies of groundwater, but many species inhabit the leaf axils of plants, tree holes, rock holes and crab holes. A few species that normally occur in groundwater or tree holes also utilize artificial containers. The eggs of most *Culex* species are laid in rafts on the water surface, but species that inhabit the axils of plants are likely to lay their eggs individually, possibly in individual gelatinous coverings like species of *Microculex*. The females of the majority of species attack humans, other mammals and birds. Some species appear to feed primarily on birds, and some are known to feed on frogs and lizards. Several species, primarily of *Culex sensu stricto*, are more or less closely associated with humans. Several species of the subgenera *Culex* and *Melanoconion* are involved in the transmission of agents that cause human diseases, particularly filariasis and arboviral encephalitis.

****

Principal references: Belkin, 1962 (South Pacific, taxonomy, description, systematics, bionomics, distribution); Lee *et al.*, 1980, 1988b, 1989a, 1989b (Australasian Region, taxonomy, keys, literature, distributions, bionomics); Harbach *et al.*, 2012 (morphology, phylogenetic relationships); da Silva *et al.*, 2020a (phylogenetic relationships); Soghigian *et al.*, 2023 (molecular markers, phylogenetic relationships, host use); Chen *et al.*, 2024 (mitochondrial protein sequences, phylogenetic relationships). See the references listed for the genera *Culex*, *Deinocerites*, *Galindomyia* and *Lutzia*.

Culisetini Belkin, 1962

The tribe Culisetini is typified by mosquitoes of the genus *Culiseta* Felt, 1904. Maslov (1964) introduced the family-group name Culisetina, but the Culisetini of Belkin (1962) has priority. The older 'Culisetini' of Stone (1957) is a *nomen nudum*.

****

Systematics: Culisetini is a subordinate taxon of the subfamily Culicinae. It only includes the genus *Culiseta*, which is comprised of 41 extant and five extinct species. Species of *Culiseta* are found in all zoogeographic regions, but none occur in South America. The tribe is largely confined to the temperate regions of North America, Europe and Asia. The affinities of the tribe are uncertain. The only included genus is extremely generalized and may consist of several phyletic lines from which some of the other genera of Culicidae were derived. Similarities of some species have been noted with species of Aedini, Culicini, Ficalbiini and Orthopodomyiini. Culisetini was paired with Toxorhynchitini in the implied weighting analysis of Harbach & Kitching (1998), but the authors admitted that this relationship was problematic. Culisetini was recovered in a weakly supported sister relationship with Orthopodomyiini in the phylogenetic analysis of amino acid sequences of 709 genes conducted by Soghigian *et al.* (2023), with the pair as the sister to Mansoniini. The results of the study indicate that the lineage that gave rise to the Culisetini existed during the Upper Cretaceous Period (100.5–66.0 Mya) of the Mesozoic Era.

****

Characteristics: Species of the tribe are characterized by the presence of setae at the base of the subcosta on the lower surface of the wing, but these setae are not unique to members of the tribe. They also occur in *Coquillettidia* (*Austromansonia*) *tenuipalpis* and the two species of *Opifex*, all indigenous to New Zealand. The larvae are remarkably diverse in the chaetotaxy of the head, the length of the antennae and details of the terminal abdominal segments. ADULTS: Maxillary palpus short in females, variable in length in males but usually as long as the proboscis; prespiracular setae present; postspiracular setae absent; pulvilli not evident; abdomen of females bluntly rounded, eighth segment not retractile; gonocoxite of male genitalia rather long, basal mesal lobe present, apical lobe small in some species, absent in others; gonostylus simple, with terminal claw; claspettes absent. LARVAE: Head wider than long; comb and pecten present; single pair of seta 1-S present at base of siphon.

****

Bionomics: The immature stages of most species of the tribe develop in bogs, marshes, ponds, streams, and small ground and rock pools. Several species are sometimes found in artificial containers and occasionally in tree holes. An African species, *Culiseta fraseri*, is restricted to tree holes. Several Australian species are found underground. Species generally live in colder regions or at higher elevations. Little is known of the biology of the adults. Several species feed on domestic animals and some, especially those of *Culiseta sensu stricto*, readily attack humans, but the preferred hosts of most species are unknown. Eggs are usually laid on the surface of the water in boat-shaped masses, but there are exceptions. The eggs of species of the subgenera *Climacura* and *Culicella* lay their eggs singly in captivity. Several North American species vector encephalitis viruses.

****

Principal references: Belkin, 1962 (South Pacific, taxonomy, description, systematics, bionomics, distribution); Belkin, 1968a (New Zealand, taxonomy); Maslov, 1967, 1989 (World, systematics, keys, descriptions, distributions, bionomics); Soghigian *et al.*, 2023 (molecular markers, phylogenetic relationships, host use).

Ficalbiini Belkin, 1962

The tribe Ficalbiini, established by Belkin (1962), is typified by mosquitoes of the genus *Ficalbia* Theobald, 1903a.

****

Systematics: Ficalbiini is a subordinate taxon of the subfamily Culicinae. It includes 54 extant species divided between two genera, *Ficalbia* and *Mimomyia*. Ficalbiini is an Old World taxon. The species mainly occur in the tropical areas of western Africa and the Oriental and Australasian Regions. Three species extend into the eastern Palaearctic. A number of similarities between the larvae and pupae of some species of Ficalbiini and Mansoniini, *e.g.* specializations for obtaining oxygen from submerged plant parts, are suggestive of a phylogenetic relationship, but this is not supported by the results of the morphology-based cladistic analysis of mosquito genera conducted by Harbach & Kitching (1998). Edwards (1932a) and Mattingly & Grjebine (1958) suggested a relationship with Hodgesiini, Orthopodomyiini and Uranotaeniini but the cladistic analyses of Harbach & Kitching (1998) indicated a possible close affinity with only Hodgesiini and Uranotaeniini. The morphological data included in the latter study do not support the monophyly of the tribe, *i.e.* the two genera of the tribe were not recovered as sister taxa. This supports the opinion of Mattingly (1981) that *Ficalbia* and *Mimomyia* are more distantly related than their inclusion in the same tribe would suggest. No species of the tribe have been included in any of the molecular phylogenetic studies conducted to date.

****

Characteristics: Adults of species of the tribe exhibit some overlapping features with other groups, but they are quite characteristic and fairly easy to recognize. The larvae and pupae vary considerably but also share a number of characters with other tribes. Larvae exhibit similarities to *Mansonia* and *Culiseta*, but they are readily separated from the larvae of both in the complete or almost complete absence of a hypostomal suture and by the median position of seta 1-S on the siphon (except for *Ficalbia*).

****

Bionomics: Species of Ficalbiini are found in bodies of groundwater with dense vegetation or in various types of plant containers, such as tree holes, leaf axils and pitchers plants. Several species of *Mimomyia* have a piercing siphon in the larva and piercing trumpets in the pupa that are used to obtain oxygen from the submerged parts of aquatic plants. Very little is known about the bionomics of the adults. Several species occasionally bite humans both indoors and outdoors, but none of the species are serious pests. Most species appear to be active at night. Members of the tribe are not known to be of medical or economic importance.

****

Principal references: Belkin, 1962 (South Pacific, as *Ficalbia*, taxonomy, description, systematics, bionomics, distribution). See the references listed for *Ficalbia* and *Mimomyia*.

Hodgesiini Belkin, 1962

The tribe Hodgesiini, established by Belkin (1962), is typified by mosquitoes of the genus *Hodgesia* Theobald, 1904a.

****

Systematics: Hodgesiini is a subordinate taxon of the subfamily Culicinae. It includes a single genus, *Hodgesia*, with 11 extant species. Species of Hodgesiini occur in the Afrotropical, Australasian and Oriental Regions. The phylogenetic relationships of the tribe are uncertain. The morphological data used in the cladistic analysis of mosquito genera by Harbach & Kitching (1998) indicate a close relationship with Ficalbiini. No species of the tribe have been included in any of the molecular phylogenetic studies conducted to date.

****

Characteristics: Species of Hodgesiini exhibit a unique combination of characters in the adult and larval stages that distinguish them from members of other tribes. The adults are small mosquitoes with silver markings and distinctive outstanding scales on the wings, but they otherwise resemble members of the tribe Uranotaeniini in the development of the anal vein and the absence of scales or setae on the alula of the wing. Larvae bear a superficial resemblance to larvae of Culisetini and Ficalbiini.

****

Bionomics: The immature stages of species of *Hodgesia* are usually found in dense vegetation in swamps, marshes and pools in marshy areas, often in association with species of Ficalbiini. The feeding habits of most species are not known. *Hodgesia sanguinae* in Africa and *Ho. solomonis* of the Solomon Islands are known to attack humans. Adult females are so small that they may be mistaken for ceratopogonid midges. Species of the tribe do not appear to be of economic or medical importance.

****

Principal references: Belkin, 1962 (South Pacific, taxonomy, description, systematics, bionomics, distribution). See the references listed for *Hodgesia*.

Mansoniini Belkin, 1962

The tribe Mansoniini, established by Belkin (1962), is typified by mosquitoes of the genus *Mansonia* Blanchard, 1901.

****

Systematics: Mansoniini is a subordinate taxon of the subfamily Culicinae. It includes 83 extant and five fossil species. Species of Mansoniini occur in the Old and New Worlds. The majority of species of the two included genera, *Coquillettidia* and *Mansonia*, are found in the Old World, mainly in tropical areas. Mansoniini appears to share an affinity with Aedini. It was found to be a strongly supported monophyletic group in a derived relationship to Aedini in the cladistic analyses of morphological data conducted by Harbach & Kitching (1998), and *Mansonia titillans* (as an outgroup taxon) was recovered as the sister of Aedini in the morphology-based cladistic analyses of Reinert *et al.* (2004, 2006, 2009). Harbach & Kitching (1998) listed six synapomorphies that support the monophyly of *Coquillettidia* + *Mansonia*: larval seta 5-VIII inserted close to seta 4, usually near or above the level of the dorsal margin of segment X; larval siphon without pecten; pupal trumpet supported by a tubercle; tracheoid area of trumpet present; pupal seta 9-VIII inserted dorsally; paddle without setae. Mansoniini was not associated with Aedini in the molecular phylogenetic studies of do Nascimento *et al.* (2021), Lorenz *et al.* (2021), Soghigian *et al.* (2023) and Chen *et al.* (2024). In the analysis of 13 mitochondrial protein-coding genes conducted by Lorenz *et al.* (2021), Mansoniini, represented by *Coquillettidia chrysonotum* and *Mansonia amazonensis*, was recovered in a basal relationship to a clade comprised of species of Aedeomyiini, Sabethini and Toxorhynchitini; however, with the addition of *Cq. nigricans* to the same data set used in the study of do Nascimento *et al.* (2021), Mansoniini was recovered as the sister of Sabethini + Toxorhynchitini. In the study of Chen *et al.* (2024), based on sequences of mitochondrial genomes, Mansoniini was also recovered in a sister relationship with Sabethini + Toxorhynchitini. In the study of Soghigian *et al.* (2023), Mansoniini was placed in a basal relationship to a weakly supported clade comprised of Culisetini + Orthopodomyiini. The results of the phylogenetic studies conducted by da Silva *et al.* (2020a), Lorenz *et al.* (2021), Soghigian *et al.* (2023) and Chen *et al.* (2024) indicate that the ancestral lineage of the Mansoniini existed during the early part of the Upper Cretaceous Period (100.5–66.0 Mya) of the Mesozoic Era.

****

Characteristics: Species of the tribe are most strikingly characterized in the larval stage and, to a lesser extent, in the pupal stage. The adults appear to be quite heterogeneous and may be easily mistaken for species of *Culex* or aedine mosquitoes.

****

Bionomics: Females of several species of the tribe are aggressive nocturnal biters. Biting activity and seasonal prevalence varies considerably with different species. The larvae and pupae of Mansoniini attach to the submerged parts of aquatic plants to obtain oxygen from pockets of air. Some species use a wide variety of plants whereas other species appear to be host specific. *Pistia* (water lettuce) is a common host plant for species of *Mansonia*, particularly those of *Mansonioides*. The immature stages generally occur in more or less permanent bodies of water. Larvae readily detach and re-attach to host plants. Some pupae are capable of changing host plants whereas others are not. Pupae detach and float to the surface of the water prior to emergence of the adults. Several species of the tribe have been found naturally infected with arboviruses and microfilariae, and have been shown to transmit them under experimental conditions. Some species are annoying pests and vicious biters.

****

Principal references: Belkin, 1962 (South Pacific, taxonomy, description, systematics, bionomics, distribution). See the references listed for *Coquillettidia* and *Mansonia*.

Orthopodomyiini Belkin, Heinemann & Page, 1970

The tribe Orthopodomyiini is typified by mosquitoes of the genus *Orthopodomyia* Theobald, 1904c. According to Sabrosky (1999), 'Orthopodomyiini' of Belkin (1962) is a *nomen nudum* because family-group names published after 1930 must be accompanied by a verbal description or definition of characters that purport to distinguish them. The requirement of availability for the tribal name Orthopodomyiini was first met by Belkin *et al.* (1970).

****

Systematics: Orthopodomyiini is a subordinate taxon of the subfamily Culicinae. It includes 36 extant species in a single genus, *Orthopodomyia*. Orthopodomyiini has a worldwide distribution, but the species are largely restricted to tropical and warm temperate areas. The phylogenetic relationships of the tribe are uncertain (see Harbach & Kitching, 1998). Orthopodomyiini was recovered in a weakly supported sister relationship with Culisetini in the phylogenetic analysis of amino acid sequences of 709 genes conducted by Soghigian *et al.* (2023). The results of the study indicate that the ancestral lineages of Orthopodomyiini and Culisetini diverged from a common ancestor during the Upper Cretaceous Period (100.5–66.0 Mya) of the Mesozoic Era.

****

Characteristics: The adults of Orthopodomyiini are very similar in habitus. The species and species groups differ principally in the ornamentation of the thorax and wings. The ornamentation of the proboscis, maxillary palpus, legs and abdomen is extremely variable. The male genitalia are also very similar. The larvae and pupae are extremely similar and distinguishing characters are restricted, for the most part, to the length and number of setal branches. The siphon of larvae lacks a pecten, the comb is uniquely developed and abdominal segments VII and VIII have large dorsal plates. The pupae resemble those of species of Aedini that lack a tracheoid area on the trumpet.

****

Bionomics: Very little is known about the biology of the species of Orthopodomyiini. Larvae are filter feeders that principally inhabit tree holes. Adults inhabit forests and appear to be active only after dark. Adult females appear to feed principally on birds. Only two species in the Oriental Region, *Orthopodomyia albipes* and *Or. andamanensis*, are known to feed on humans, but neither is known to vector pathogens of human disease.

****

Principal references: Zavortink, 1968 (World, systematics); Soghigian *et al.*, 2023 (molecular markers, phylogenetic relationships, host use). See the references listed for *Orthopodomyia*.

Sabethini Blanchard, 1905

The tribe Sabethini is typified by mosquitoes of the genus *Sabethes* Robineau-Desvoidy, 1827. The subfamily name Sabethinae (as Sabettinae) established by Blanchard (1905) has priority over Sabethinae Dyar, 1906; thus, Blanchard (1905) is considered to have simultaneously established the coordinate tribal name Sabethini.

****

Systematics: Sabethini is a subordinate taxon of the subfamily Culicinae. It includes 448 currently recognized extant species that comprise 14 genera: *Isostomyia*, *Johnbelkinia*, *Kimia*, *Limatus*, *Malaya*, *Maorigoeldia*, *Onirion*, *Runchomyia*, *Sabethes*, *Shannoniana*, *Topomyia*, *Trichoprosopon*, *Tripteroides* and *Wyeomyia*. Species of Sabethini are referred to as 'sabethines'. Sabethines are found mainly in the Neotropical, Oriental and Australasian Regions. Only a few species occur in temperate regions. Nine genera are almost entirely restricted to the Neotropical Region. These include *Isostomyia*, *Johnbelkinia*, *Limatus*, *Onirion*, *Sabethes*, *Shannoniana*, *Runchomyia*, *Trichoprosopon* and *Wyeomyia*. A few species of *Wyeomyia* occur in eastern North America. Only five genera are recognized in the Old World. *Malaya* occurs in Africa as well as the Oriental and Australasian Regions, *Topomyia* is known from the Oriental Region and New Guinea, *Maorigoeldia* is restricted to New Zealand, *Kimia* occurs in eastern areas of the Oriental Region, and *Tripteroides* occurs in the Oriental, Australasian and the southern part of the Palaearctic. The morphology-based cladistic analyses of Judd (1996), Harbach & Kitching (1998) and Harbach & Peyton (2000) and the molecular analyses of mitogenomes (da Silva *et al.*, 2020a; Chen *et al.*, 2024) and amino acid sequences (Soghigian *et al.*, 2023) bolstered the monophyly of the tribe. In addition, the analyses of morphological data indicated that the Old World genera comprise a paraphyletic assemblage relative to the monophyletic New World taxa. Contrary to these findings, the New World genera of Sabethini were not recovered as a monophyletic clade in a derived relationship to the Old World genera when Harbach *et al.* (2007a) included the newly recognized *Kimia* in the cladistic analysis of Harbach & Peyton (2000). The New World *Trichoprosopon* was paired with the Old World *Tripteroides* in a sister relationship with *Kimia*, and the Old World *Malaya*, which was sister to *Topomyia* in the analyses of Harbach & Peyton, was placed as sister to the New World *Limatus*. The results of the molecular phylogenetic studies conducted by da Silva *et al.* (2020a), Soghigian *et al.* (2023) and Chen *et al.* (2024) indicate that the Sabethini arose and began to diversify near the end of Upper Cretaceous Period (100.5–66.0 Mya) of the Mesozoic Era. Muñoz-Gamba *et al.* (2021) included species of *Johnbelkinia* (two species), *Limatus* (two species), *Sabethes* (10 species), *Trichoprosopon* (four species) and *Wyeomyia* (21 species) in a phylogenetic analysis of mtDNA *COI* gene sequences in which only *Johnbelkinia* and *Limatus* were recovered as monophyletic groups.

****

Characteristics: Sabethine adults are difficult to characterize. Most species have the base of the hindcoxa in line with or slightly above the base of the mesomeron, the mesomeron is generally distinctly smaller than in non-sabethine mosquitoes, prespiracular setae or scales are virtually always present (absent in one species of *Malaya*; scales replace setae in *Limatus*) and the mesopostnotum usually has a cluster of setae or scales. The position of the hindcoxa relative to the mesomeron is not universal among sabethines and is not unique to the tribe. It is below the base of the mesomeron in *Maorigoeldia*, some *Malaya* and some *Tripteroides*, and many species of Aedini, including *Armigeres*, *Udaya*, *Zeugnomyia* and certain *Aedes* and *Heizmannia* resemble the majority of sabethines in having the base of the hindcoxa in line with the base of the mesomeron. Of the culicine genera, prespiracular setae are found elsewhere only in species of *Culiseta* and *Psorophora*, but these species are easily distinguished from sabethines based on other characteristics. Mesopostnotal setae are absent in *Kimia*, *Malaya*, *Maorigoeldia*, *Topomyia* and some *Tripteroides*, and are present in many species of Aedini and Culicini. The reduction of the larval ventral brush (seta 4-X) to one or rarely two pairs of setae is more or less diagnostic for the tribe (the homology of these setae with the ventral brush of non-sabethines is questionable). The position of seta 3-C on the ventral side of the head and the lateral displacement of the anterior labropalatal setae appear to be diagnostic of sabethine larvae, but these characters are not always easy to see.

The following descriptions of principal taxonomic features (excluding those above) are provided to illustrate the varied development of the tribe. ADULTS: Eyes contiguous or narrowly separated above the antennae; erect scales of the head restricted to a single posterior row or absent; clypeus sometimes with setae or scales; proboscis variable in length and development; antennae shorter than proboscis; scutal scales varied, setae reduced or absent; mesopostnotum with setae or scales or both, sometimes bare; paratergite bare; antepronota often large and close together; postpronotum with or without posterior setae; postspiracular setae absent; prealar setae few or absent; lower mesepimeral setae absent; hindtibia markedly shorter than hindtarsomere 1; tarsi normal, pulvilli not evident; wing membrane with distinct microtrichia; cell R_2 longer than vein R_{2+3}; vein R_s often with basal line of scales; vein 1A usually ends distal to the base of the mediocubital crossvein (not so in *Malaya* and some *Topomyia*); subcosta without setae at the base ventrally; upper calypter varied, with or without setae or hair-like scales, laterotergite varied, with complete or partial covering of scales or bare; tip of abdomen blunt in females. LARVAE: Occipital foramen circular with distinct collar or transverse and slit-like without collar; hypostomal suture complete, incomplete or absent; maxillae or mandibles often developed for grasping, maxillary brush unmodified or represented by a maxillary bundle or maxillary claw for grasping; seta 2-C absent; seta 13-P usually absent (present in *Maorigoeldia*, *Tripteroides* and a few *Wyeomyia*); no plumose or palmate setae on the thorax or abdomen; seta 12-I absent; comb rarely absent; seta 5-VIII usually inserted near or above dorsal margin of segment X; siphon with at least one pair of setae, usually more, in addition to setae 1,2-S; pecten developed as spines, filaments or absent; saddle incomplete.

****

Bionomics: Sabethine larvae are almost entirely restricted to water contained in dead and living plant material, including leaf axils, bromeliads, bamboo, tree holes, flower bracts and spathes, and pitcher plants. A few species have been collected from snail shells and rock holes, and several are capable of developing in artificial containers. Some species require a specific host plant. Predacious larvae are known in several genera. The mandibles or maxillae of those species are modified for capturing prey. Most if not all sabethine larvae lie on their backs at the bottom of the cavity and seldom come to the surface. Sabethine adults are forest mosquitoes. The females of most species suck blood and sometimes bite humans. Many species are active during the day. The majority of species fly with the hindlegs bent upward and over the back. Sabethines as a whole are not important as vectors of disease agents. Various species of *Limatus*, *Johnbelkinia*, *Sabethes*, *Trichoprosopon* and *Wyeomyia* have been implicated in the transmission of arboviruses in the Neotropical Region, but none of the Old World species are even suspected of transmitting pathogens to humans or other animals.

****

Principal references: Belkin, 1962 (South Pacific, taxonomy, description, systematics, bionomics, distribution); Lane, 1953 (Neotropical Region, taxonomy, keys, descriptions, distributions); Forattini, 1965b (Neotropical Region, descriptions, distributions, bionomics); Belkin *et al.*, 1970 (Jamaica, taxonomy, keys, descriptions, bionomics); Tanaka *et al.*, 1979 (Japan, keys, descriptions, distributions, bionomics); Harbach & Peyton, 1993 (larval mouthparts); Judd, 1996 (phylogenetic relationships); Harbach & Peyton, 2000 (phylogenetic relationships); Harbach *et al.*, 2007a (phylogenetic relationships); da Silva *et al.*, 2020a (phylogenetic relationships); Muñoz-Gamba *et al.*, 2021 (mtDNA *COI* sequences, phylogenetic relationships); Soghigian *et al.*, 2023 (molecular markers, phylogenetic relationships, host use); Chen *et al.*, 2024 (mitochondrial protein sequences, phylogenetic relationships).

Toxorhynchitini Lahille, 1904

The tribe Toxorhynchitini is typified by the genus *Toxorhynchites* Theobald, 1901a. The subfamily name Toxorhynchitinae (as Toxorhynchina) is attributed to Lahille (1904), not Theobald (1905e) (as Toxorhynchitinae); thus, Lahille (1904) is considered to have simultaneously established the coordinate tribal name Toxorhynchitini.

****

Systematics: Toxorhynchitini is a subordinate taxon of the subfamily Culicinae. The tribe includes only one genus, *Toxorhynchites*, with 101 extant and one fossil species. Species of Toxorhynchitini are referred to as 'toxorhynchitines'. Toxorhynchitine species are largely tropical, with a few species occurring in the far eastern areas of Asia and North America and three introduced species occurring on Fiji and Samoa in the South Pacific. The tribal status of the mosquitoes belonging to *Toxorhynchites* is based on the morphological and molecular phylogenetic analyses of Harbach & Kitching (1998) and Mitchell *et al.* (2002), respectively, which did not support the traditional recognition of a separate subfamily for this taxon. The monophyly of Toxorhynchitini and its evolution from a common ancestor with Sabethini is supported by molecular analyses of mitogenomes (da Silva *et al.*, 2020a; Chen *et al.*, 2024) and amino acid sequences (Soghigian *et al.*, 2023). The results of the studies conducted by Chen *et al.* and Soghigian *et al.* indicate that the ancestral linages of the Toxorhynchitini and Sabethini diverged from a common ancestor during the Upper Cretaceous Period (100.5–66.0 Mya) of the Mesozoic Era. In the more taxon impoverished study of da Silva *et al.*, Toxorhynchitini was recovered in a sister relationship with Mansoniini + Sabethini.

****

Characteristics: Toxorhynchitine adults are large and colourful mosquitoes with the body covered with iridescent green, purple or red scales. The distal half of the proboscis is slender and bent sharply downwards and backwards. The scutellum is evenly rounded (as in *Anopheles* and *Bironella*) and the posterior margin of the wing is distinctly emarginate opposite the termination of vein CuA. Larvae have mouthbrushes composed of about 10 broad, flat filaments that are used to capture prey, and the dorsal and ventral abdominal setae occur in groups on large sclerotized plates. A comb and pecten are both absent.

****

Bionomics: Toxorhynchitine larvae are found in plant cavities, mainly tree holes and bamboo stumps. The larvae of all species are predators. They feed mainly on other mosquito larvae, including their own kind when other species are few or absent. The adults are basically forest mosquitoes. Males and females both feed exclusively on nectar and other sugary substances. The adults are active during the day. Since toxorhynchitines do not suck blood, they are not of medical importance. The larvae of a few species have been used with some success to control economically important mosquitoes whose larvae inhabit plant cavities and artificial containers.

****

Principal references: Belkin, 1962 (South Pacific, taxonomy, description, systematics, bionomics, distribution). See the references listed for *Toxorhynchites*.

Uranotaeniini Lahille, 1904

The tribe Uranotaeniini is typified by mosquitoes of the genus *Uranotaenia* Lynch Arribálzaga, 1891b. The subfamily name Uranotaeniinae (as Uranotaenina) is attributed to Lahille (1904), not Coquillett (1906d) (as Uranotaeniinae); thus, Lahille (1904) is considered to have simultaneously established the coordinate tribal name Uranotaeniini.

****

Systematics: Uranotaeniini is a subordinate taxon of the subfamily Culicinae. It includes 276 extant species assigned to a single genus, *Uranotaenia*. Species of *Uranotaenia* occur in all zoogeographic regions but are absent from New Zealand, New Caledonia and numerous oceanic islands. The majority of species are found in the Old World tropics. The affinities of Uranotaeniini with the other tribes of the subfamily Culicinae are uncertain. The tribe is usually regarded as one of the most primitive groups of Culicinae. It shares morphological features with Anophelinae, Aedeomyiini, Toxorhynchitini and certain other groups. Uranotaeniini was paired with Aedeomyiini in a basal relationship to other tribes of Culicinae in the analysis of mitogenomes (da Silva *et al.*, 2020a), placed in a paraphyletic position between Aedeomyiini (basal) and other tribes of the subfamily (Soghigian *et al.*, 2023) and recovered as the sister to a clade composed of Mansoniini, Sabethini and Toxorhynchitini in an analysis of the protein-coding gene 123 (Chen *et al.*, 2024). None of those studies included sequence data for species of Ficalbiini and Hodgesiini. That aside, the results of the three studies suggest that the ancestral lineage of the Uranotaeniini originated during the Upper Cretaceous Period (100.5–66.0 Mya) of the Mesozoic Era.

****

Characteristics: Species of Uranotaeniini are small mosquitoes that bear the following characteristics. ADULTS: Maxillary palpus very short, consisting of a single palpomere; wing membrane with extremely minute and numerous microtrichia that are not visible at lower magnification, scales of the veins usually all broad and small, truncate or rounded apically, cell R_2 shorter than vein R_{2+3} and never longer than cell M_2, vein M usually without scales on basal 0.75 of dorsal surface (rarely with a few scattered scales), vein 1A sharply curved distally and ending before junction of the mediocubital crossvein and vein CuA; female with a single large spermathecal capsule. LARVAE: Seta 1-C arises from a distinct apical process of the median labral plate; seta 14-C inserted near anterior margin at base of maxilla; hypostomal suture absent; segment VIII usually with a sclerotized plate bearing a single row of comb scales on the posterior margin; siphon with a single pair of posterolateral setae (seta 1-S) arising beyond proximal 0.25; siphon usually with few to many pecten spines. PUPAE: Paddle with inner part deeply excavated toward the base, toothed or with filamentous fringe on inner and outer margins.

****

Bionomics: The immature stages of most species of the tribe are found in swamps, marshes, stream margins and temporary ground pools. A large number of species also inhabit crab holes, rock holes, tree holes, bamboo, fallen plant parts, leaf axils, inflorescences and pitcher plants. A few species are sometimes found in artificial containers. Species that utilize ground pools lay their eggs in rafts. The larvae rest parallel to the surface and are often mistaken for anopheline larvae. Species that utilize small collections of water in plant containers apparently lay their eggs individually. Little or nothing is known about the bionomics of most species. The feeding preferences of females are largely unknown. At least some species apparently feed on amphibians and birds, and the females of a few species will bite humans. Several species are known to be active in shaded places during daylight hours. Species of the tribe are not of medical or economic importance to humans.

****

Principal references: Belkin, 1962 (South Pacific, taxonomy, description, systematics, bionomics, distribution). See the references listed for *Uranotaenia*.

Part II: Genus Groups of Anophelinae

The traditional classification of Anophelinae includes three contemporary genera, *Anopheles* (cosmopolitan), *Bironella* (Australasian) and *Chagasia* (Neotropical), and one extinct genus, *Priscoculex*, known from a single specimen in Burmese amber. The majority of extant anopheline species belong to *Anopheles*.

Anopheles Meigen, 1818

Anopheles Meigen, 1818, originally proposed and currently recognized as a genus of Anophelinae, is typified by *Anopheles maculipennis* Meigen, 1818. Genus abbreviation – *An.*

****

Systematics: *Anopheles* is a subordinate taxon of the subfamily Anophelinae. It includes 511 formally recognized extant species and more than 50 members of contemporary species complexes without formal names. Two fossil species are recognized as species of *Anopheles*, but one is a questionable member of the genus. *Anopheles* has an almost worldwide distribution. Species occur in temperate, subtropical and tropical areas, but are absent from the majority of the Pacific Islands, including the large ones of New Zealand, Fiji and New Caledonia, and isolated islands in the Atlantic. *Anopheles* species are found at elevations from coastal areas to mountainous terrain. The formally named species are placed in eight subgenera: *Anopheles sensu stricto*, *Baimaia*, *Cellia*, *Christya*, *Kerteszia*, *Lophopodomyia*, *Nyssorhynchus* and *Stethomyia*. Foster *et al.* (2017) proposed the elevation of *Kerteszia*, *Lophopodomyia*, *Nyssorhynchus* and *Stethomyia* to generic status based on phylogenetic analyses of mitochondrial protein-coding genes, but their justification for that action is weak and problematic. Because the four groups are fully diagnosable and a change of status does not add stability to the classification of Anophelinae, their elevation to generic status is unnecessary and has not been widely accepted. For these reasons, traditional subgeneric status of the four groups is retained here.

Cladistic analyses of morphological data indicate that *Anopheles* is not demonstrably monophyletic with regard to *Bironella* and the subgenera *Lophopodomyia* and *Stethomyia*. Subgenera *Kerteszia*, *Nyssorhynchus* and *Cellia* are monophyletic, and *Kerteszia* and *Nyssorhynchus* are sister taxa. Some molecular studies add further support for the sister relationship between *Kerteszia* and *Nyssorhynchus* (Freitas *et al.*, 2015; da Silva *et al.*, 2020a) while others do not (Foster *et al.*, 2017; Chen *et al.*, 2024). The monophyly of the Sections and most Series of the subgenera *Anopheles*, *Cellia* and *Nyssorhynchus* are doubtful. Based on analysis of a data set consisting of *COI* and *COII* mtDNA and 2.8S rDNA sequences (Freitas *et al.*, 2015), *Anopheles sensu stricto* (10 species), as well as *Cellia* (25 species), *Kerteszia* (three species), *Nyssorhynchus* (six species) and *Stethomyia* (two species), were recovered as monophyletic clades. These groups were also recovered as monophyletic clades in the studies of Foster *et al.* (2017) and Guo *et al.* (2021) based on analyses of mitochondrial protein-coding genes. See the review of Harbach (2007) and the evolutionary hypotheses of Harbach (2013b) and Harbach & Kitching (2016). Karimian *et al.* (2014) used sequences of *COI* mtDNA and ITS2 rDNA to investigate the phylogenetic relationships of the *Anopheles* species (15 and 28, respectively) that occur in the region of the Iranian plateau. The study included 16 species of *Anopheles sensu stricto*, 12 from the Anopheles Series and four from the Myzorhynchus Series; and 18 species of *Cellia*, eight from the Neocellia Series, six from the Myzomyia Series, three from the Paramyzomyia Series and one from the Pyretophorus Series. The monophyly of the subgenera was supported by the ITS2 data but not by the *COI* data when these were subjected to maximum parsimony analyses. The ITS2 data supported the integrity of the Series for the most part, but was not able to separate the one species of the Pyretophorus Series from the members of the Paramyzomyia Series. Neafsey *et al.* (2015) constructed a maximum likelihood phylogeny for 18 species of the subgenera *Anopheles* (two species), *Cellia* (14 species) and *Nyssorhynchus* (two species) from aligned protein sequences of 1085 single-copy orthologs. The results, although taxon limited, support the monophyly of the three subgenera and indicate that the ancestral stem-group gave rise to *Nyssorhynchus* about 100 Mya and a lineage which later gave rise to *Anopheles* and *Cellia*. In contrast, Freitas *et al.* (2015) (see above) inferred that *Anopheles* diverged from a common ancestor approximately 110 Mya, which is supported by the results of the molecular phylogenetic studies of da Silva *et al.* (2020a), Soghigian *et al.* (2023); Chen *et al.* (2024). The results of Neafsey *et al.* convincingly support the monophyly of the Pyretophorus Series of *Cellia* and its emergence about 30 Mya. In agreement with the supposition of Reid & Knight (1961) that the monobasic *Christya* (as the Christya Series) is the more primitive group of *Anopheles sensu stricto*, the phylogenetic analyses of Harbach & Kitching (2016) indicate that this taxon is sister to all Anophelinae except *Chagasia* (*Christya* has not been included in any molecular study conducted to date that could provide support or refute its ancestral relationship to all other extant *Anopheles*). As noted above, Foster *et al.* (2017) proposed (suggested?) the elevation of the subgenera *Kerteszia*, *Lophopodomyia*, *Nyssorhynchus* and *Stethomyia* to generic status based on phylogenetic analyses of mitochondrial protein-coding genes. They also suggested the elevation of the Myzorhynchella Section of *Nyssorhynchus* to subgeneric status in their proposed genus *Nyssorhynchus*. The suggested actions are questionable in view of relationships portrayed in the featured phylogeny of the authors that was rooted on *Chagasia*, where the other groups of anophelines formed a polytomy with three branches: (1) *Stethomyia*, (2) *Lophopodomyia* + *Kerteszia* and (3) *Bironella* + (*Nyssorhynchus* + (*Cellia* + *Anopheles*)). In view of the results of previous studies, both morphological and molecular (see above), and the paucity of Old World taxa included in the analyses, the relationships should be regarded with scepticism until significantly more molecular data are available to test, corroborate and further resolve relationships of the global diversity of Anophelinae. Also see the relationships of *Anopheles* generated in the phylogenetic analyses of Wang *et al.* (2015) based on 2007 single-copy genes, and Hao *et al.* (2017), Chu *et al.* (2018), Guo *et al.* (2021) and Lorenz *et al.* (2021) based on mitochondrial genomes. The results of the studies of Hao *et al.* and Guo *et al.* support the monophyly of the Myzomyia, Neocellia, Neomyzomyia and Pyretophorus Groups of the subgenus *Cellia* and the Albitarsis, Argyritarsis and Oswaldoi Groups of the subgenus *Nyssorhynchus*. The results of Guo *et al.* also provide support the monophyly of the Albimanus, Argyritarsis and Myzorhynchella Sections of *Nyssorhynchus*.

****

Characteristics: Most species of *Anopheles* are immediately recognized by their overall appearance. The majority of the adults are slender insects with the head and abdomen oriented in a straight line at an angle to the surface when at rest. The maxillary palpi are about as long as the proboscis, and in males they are clubbed apically; the posterior margin of the scutellum is evenly rounded (distinction from *Chagasia*); veins M_{1+2} and M_{3+4} are straight or evenly curved, not sinuate (distinction from *Bironella*); and the abdominal segments have only small patches of scales or none at all (distinction from culicines). Larvae are immediately recognized by the absence of a siphon (distinction from culicines), the absence of palmate setae on the mesothorax (distinction from *Bironella*) and the absence of both uniquely developed palmate setae on abdominal segments III–V and a fringe of setae on the posterolateral lobes of the spiracular apparatus (distinction from *Chagasia*).

****

Bionomics: *Anopheles* larvae are adapted to a variety of aquatic habitats, but they occur predominantly in groundwaters. Some species require aerated water, others brackish water and some inhabit cavities such as tree holes (Plumbeus Group, *Anopheles sensu stricto*) and the axils of epiphytic plants (*Kerteszia*, except for *An. bambusicola* which inhabits bamboo). Specific habitats contain stagnant water or water that is slowed down by vegetation or objects in specific niches occupied by the larvae. The larvae of all species feed at the water surface, where they attach to the surface film by the spiracular apparatus, palmate setae and special notched organs of the prothorax. They rotate the head 180° so that particles of food at the surface can be swept into the mouth by currents produced by the mouthbrushes. The larvae generally rest with the end of the abdomen against objects and are therefore found in greatest numbers in areas with emergent vegetation at the margins of the habitats. The adults of most anophelines are active at night (nocturnal) or during twilight periods (crepuscular), and rest in cool, damp places during the day. Bloodfeeding is largely restricted to warm-blooded animals. Information on host specificity primarily pertains to those species that feed on humans and domestic animals. Females bite humans inside or outside houses. They normally fly no farther than 1–3 km from the larval habitats. Mosquitoes of *Anopheles* are the sole vectors of human malarial parasites. Some species are effective vectors of microfilariae and some may be involved in the transmission of encephalitis viruses. *Anopheles* are vectors of numerous animal pathogens, including species of malaria protozoa that do not affect humans. A review of arboviruses harboured by anophelines (Minkeu & Vernick, 2018) revealed that at least 51 viruses have been identified in various species of *Anopheles sensu stricto*, *Cellia* and *Nyssorhynchus*, including viruses which have the potential to cause febrile disease.

****

Principal references: Lane, 1953 (Neotropical Region, taxonomy, keys, descriptions, distributions); Mattingly & Knight, 1956 (Arabia, taxonomic notes, keys for adults and larvae, distributions); Cova-Garcia, 1961 (Venezuela, descriptions, keys); Belkin, 1962 (South Pacific, taxonomy, keys, descriptions, distributions, bionomics); Forattini, 1962 (Neotropical Region, classification, descriptions, bionomics, distributions); DuBose & Curtin, 1965 (Mediterranean area, keys); Grjebine, 1966 (Madagascar, classification, taxonomy, keys, descriptions, bionomics, distributions); Gillies & de Meillon, 1968 (Afrotropical Region, keys, taxonomy, descriptions, bionomics, distributions); Reid, 1968 (Malaysia, Borneo, classification, keys, descriptions, distributions, bionomics); Belkin *et al.*, 1970 (Jamaica, taxonomy, keys, descriptions, bionomics); Cagampang-Ramos & Darsie, 1970 (Philippine Islands, keys); Zavortink, 1970b (World, tree-hole species, keys, descriptions, systematics, bionomics, distributions); Zavortink, 1973 (*Kerteszia*, keys, descriptions, systematics, bionomics, distributions); Gutsevich *et al.*, 1974 (former USSR, keys, descriptions, distributions, bionomics); Harrison & Scanlon, 1975 (*Anopheles sensu stricto*, Thailand, classification, keys, taxonomy, descriptions, distributions, bionomics); Klein, 1977a (Cambodia, species, bionomics); Faran, 1980 (Albimanus Section of *Nyssorhynchus*, keys, taxonomy, descriptions, bionomics, distributions); Tanaka *et al.*, 1979 (Japan, taxonomy, keys, descriptions, distributions, bionomics); Wood *et al.*, 1979 (Canada, taxonomy, keys, descriptions, distributions, bionomics); Harrison, 1980 (Thailand, Myzomyia Series of *Cellia*, keys, taxonomy, descriptions, distributions, bionomics); Darsie & Ward, 1981, 2005 (North America, keys); Faran & Linthicum, 1981 (Amazonia, *Nyssorhynchus*, keys, diagnoses, bionomics, distributions); Lu & Li, 1982 (China, keys); Clark-Gil & Darsie, 1983 (Guatemala, keys, distributions, bionomics); Rao, 1984 (India, keys, keys, descriptions, bionomics, distributions); Lee *et al.*, 1987b (Australasian Region, taxonomy, keys, literature, distributions, bionomics); Gillies & Coetzee, 1987 (Afrotropical Region, keys, supplemental information to Gillies & de Meillon, 1968); Linthicum, 1988 (Argyritarsis Section of *Nyssorhynchus*, keys, classification, taxonomy, descriptions, bionomics, distributions); Das *et al.*, 1990 (India, keys); Darsie & Pradhan, 1990 (Nepal, distributions, bionomics, taxonomic notes); Wilkerson & Strickman, 1990 (Central America and Mexico, keys); Glick, 1992 (southwestern Asia and Egypt, keys); Peyton *et al.*, 1992 (*Kerteszia* and *Nyssorhynchus*, morphological distinctions); Harbach, 1994, 2004a (informal classification); Rattanarithikul & Panthusiri, 1994 (Thailand, keys, medically important species); Nagpal & Sharma, 1995 (India, identification, species information); Lu *et al.*, 1997b (China, keys, taxonomy, descriptions, distributions, bionomics); Oo *et al.*, 2004 (Myanmar, malaria vectors, bionomics, collection sites); Harbach *et al.*, 2005 (*Baimaia*, description, systematics, bionomics, distribution); Oo *et al.*, 2006 (Myanmar, morphology, keys for larvae and adults); Rattanarithikul *et al.*, 2006b (Thailand, classification, bionomics, distributions, keys); Harbach, 2007 (phylogenetic relationships); Azari-Hamidian & Harbach, 2009 (Iran, keys); Harbach, 2013b (classification, phylogenetic relationships); Harbach & Kitching, 2016 (subgeneric status of *Christya*, phylogenetic relationships); Freitas *et al.*, 2015 (phylogenetic relationships); Gunathilaka *et al.*, 2014 (Sri Lanka, key, larvae); Gunathilaka, 2017 (Sri Lanka, key, adults); Foster *et al.*, 2017 (mitochondrial protein sequences, phylogenetic relationships); Hao *et al.*, 2017 (mitochondrial protein sequences, phylogenetic relationships); Chu *et al.*, 2018 (mitochondrial genomes, mtDNA *COI* sequences, phylogenetic relationships); Minkeu & Vernick, 2018 (arboviruses in anophelines, global distribution); Poinar *et al.*, 2019 (*Priscoculex*, description, origin), Coetzee, 2020 (Afrotropical Region, keys); Irish *et al.*, 2020 (Afrotropical Region, country distributions); Sallum *et al.*, 2020a (South America, classification, bionomics, distributions); Sallum *et al.*, 2020b (South America, fourth-instar larvae, morphology, key); Sallum *et al.*, 2020c (South America, male genitalia, morphology, key); Sallum *et al.*, 2020d (South America, adult females, morphology, key); Adugna *et al.*, 2021 (Ethiopia, list of species, abundance, distributions); Guo *et al.*, 2021 (mitochondrial protein sequences, phylogenetic relationships); Lorenz *et al.*, 2021 (mitochondrial protein-coding genes, phylogenetic relationships); Santos *et al.*, 2022 (phylogenetic relationships of genus and species groups); da Silva *et al.*, 2020a (phylogenetic relationships of genus groups); Russell & Burkot, 2023 (Pacific Island countries and territories, species lists, distributions, bionomics, vector status); Chen *et al.*, 2024 (mitochondrial protein sequences, phylogenetic relationships of genus groups).

Anopheles Meigen, 1818 *sensu stricto*

Anopheles sensu stricto, the nominotypical subgenus of *Anopheles*, is typified by *Anopheles maculipennis* Meigen, 1818. Subgenus abbreviation – *Ano.*

****

Systematics: *Anopheles sensu stricto* includes 207 formally recognized extant species divided between two informal Sections that are further subdivided into a number of Series comprised of Groups and Subgroups that are believed to represent phylogenetically related assemblages of species based principally on morphological similarity (Harbach, 1994a, 2004). *Anopheles sensu stricto* has an almost worldwide distribution. Species are found at elevations from coastal areas to mountainous terrain in temperate, subtropical and tropical areas, but are absent from the majority of the Pacific Islands, including the large ones of New Zealand, Fiji and New Caledonia. The morphological and molecular phylogenetic studies of Sallum *et al.* (2000, 2002) and Harbach & Kitching (2005, 2016) indicate that the taxon is not demonstrably monophyletic with regard to *Bironella* and the subgenera *Lophopodomyia* and *Stethomyia*, and the monophyly of the Sections and most Series are doubtful. In contrast, the results of a cladistic analysis of morphological data by Collucci & Sallum (2007) and relationships recovered in the phylogenetic analyses of *COI* and *COII* mtDNA and 2.5S rDNA sequences by Freitas *et al.* (2015) suggest that *Anopheles sensu stricto* is monophyletic. The results of Collucci & Sallum (2007), however, were biased by the selection of outgroup taxa whose interrelationships with the ingroup taxa were unresolved in the earlier studies of Sallum *et al.* (2000, 2002). A molecular phylogeny of 18 species constructed from aligned protein sequences of 1085 single-copy orthologs indicates that *Anopheles sensu stricto* is sister to *Cellia* (Neafsey *et al.*, 2015). *Anopheles* was also recovered as sister to *Cellia* in the phylogenies of Foster *et al.* (2017), Hao *et al.* (2017), da Silva *et al.* (2020a), Ghassemi-Khademi *et al.* (2021), Guo *et al.* (2021), Lorenz *et al.* (2021) and Chen *et al.* (2024) based on analyses of mitogenomic data, but representatives of major species groups, *e.g.* the Australasian Stigmaticus Group, were not included the studies.

****

Characteristics: *Anopheles sensu stricto* includes species with and without spotted wings. Those with pale markings have no more than two pale areas on the costal margin. Pale scales of other veins, when present, usually do not form distinct pale areas. Cibarial armature is absent in females and the genitalia of males have one to three (usually two) specialized setae (generally one internal and two parabasal setae) on the gonocoxites, with at least the inner parabasal borne on a distinct prominence. Proepisternal setae are usually numerous and prespiracular setae are usually present. Larvae, with few exceptions, have seta 1-A branched and usually inserted on the mesal (inner) side of the antenna. Seta 2-C are closely approximated (closer, rarely equal, than the distance between seta 2-C and seta 3-C on either side) and setae of each pleural setal group (setae 9–12-P,M,T) are generally all simple.

Anopheles sensu stricto is divided into two informal Sections based on the shape of the pupal trumpet. The Laticorn Section includes species with a wide funnel-shaped trumpet having the longest axis transverse to the stem (laticorn), and the Angusticorn Section includes species with a semi-tubular trumpet having the longest axis vertical more or less in line with the stem (angusticorn). Pupal seta 9 on abdominal segments V–VII is usually short and blunt, and seta 1-Pa of the paddle is usually short and straight, seldom long and bent.

****

Bionomics: Larvae of species of *Anopheles sensu stricto* are adapted to a variety of aquatic habitats, but occur predominantly in groundwaters. Some species require aerated water, others brackish water and some inhabit cavities such as rock holes and tree holes (*e.g.* the Plumbeus Group). Adults are active at night and during twilight periods, but the females of some species will accost potential hosts in dimly lit places during daylight hours. Many species are known to feed on humans and domestic animals, but the host preferences of most species are unknown.

Anopheles sensu stricto is divided into six Series, but only females of the Myzorhynchus and Anopheles Series include vectors of human malarial parasites. The Hyrcanus and Maculipennis Groups include principal vectors of historical and contemporary importance in the Holarctic Region: *An. sinensis* in southeastern areas of the Palaearctic Region and *An. atroparvus* in Europe and eastern Asia (Hyrcanus Group); *An. freeborni* in western North America and *An. pseudopunctipennis* at higher elevations in Central and South America (Maculipennis Group). Some members of *Anopheles sensu stricto* are effective vectors of microfilariae and some may be involved in the transmission of encephalitis viruses.

****

Principal references: Lane, 1953 (Neotropical Region, taxonomy, keys, descriptions, distributions); Mattingly & Knight, 1956 (Arabia, taxonomic notes, keys for adults and larvae, distributions); Cova-Garcia, 1961 (Venezuela, descriptions, keys); Forattini, 1962 (Neotropical Region, classification, descriptions, bionomics, distributions); Gillies & de Meillon, 1968 (Afrotropical Region, keys, taxonomy, descriptions, bionomics, distributions); Reid, 1968 (Malaysia, Borneo, classification, keys, descriptions, distributions, bionomics); Belkin *et al.*, 1970 (Jamaica, taxonomy, keys, descriptions, bionomics); Gutsevich *et al.*, 1974 (former USSR, keys, descriptions, distributions, bionomics); Harrison & Scanlon, 1975 (Thailand, classification, keys, taxonomy, descriptions, distributions, bionomics); Klein, 1977a (Cambodia, species, bionomics); Tanaka *et al.*, 1979 (Japan, taxonomy, keys, descriptions, distributions, bionomics); Wood *et al.*, 1979 (Canada, taxonomy, keys, descriptions, distributions, bionomics); Darsie & Ward, 1981, 2005 (North America, keys); Lu & Li, 1982 (China, keys); Clark-Gil & Darsie, 1983 (Guatemala, keys); Rao, 1984 (India, keys, keys, descriptions, bionomics, distributions); Lee *et al.*, 1987b (Australasian Region, taxonomy, keys, literature, distributions, bionomics); Gillies & Coetzee, 1987 (Afrotropical Region, keys, supplemental information to Gillies & de Meillon, 1968); Das *et al.*, 1990 (India, keys); Darsie & Pradhan, 1990 (Nepal, distributions, bionomics, taxonomic notes); Wilkerson & Strickman, 1990 (Central America and Mexico, keys); Glick, 1992 (southwestern Asia and Egypt, keys); Rattanarithikul & Panthusiri, 1994 (Thailand, keys, medically important species); Nagpal & Sharma, 1995 (India, identification, species information); Lu *et al.*, 1997b (China, keys, taxonomy, descriptions, distributions, bionomics); Becker *et al.*, 2003, 2010, 2020 (Europe, keys, descriptions, distributions, bionomics); Rattanarithikul *et al.*, 2006b (Thailand, classification, bionomics, distributions, keys); Collucci & Sallum, 2007 (phylogenetic relationships); Gómez *et al.*, 2015 (Neotropics, Arribalzagia Series, molecular taxonomy); Foster *et al.*, 2017 (mitochondrial protein sequences, phylogenetic relationships); Hao *et al.*, 2017 (mitochondrial protein sequences, phylogenetic relationships); Namgay *et al.*, 2020 (Bhutan, Lindesayi Complex, morphology, rDNA ITS2 sequences, phylogenetic relationships); Ghassemi-Khademi *et al.*, 2021 (mitochondrial genomes, phylogenetic relationships); Guo *et al.*, 2021 (mitochondrial protein sequences, phylogenetic relationships); Lorenz *et al.*, 2021 (mitochondrial protein-coding genes, phylogenetic relationships); Coetzee, 2022 (Afrotropical Region, adult females, descriptions, biology, distributions); Russell & Burkot, 2023 (South Pacific, country lists of species, distributions, bionomics, vector status).

Baimaia Harbach, Rattanarithikul & Harrison, 2005

Baimaia Harbach *et al.*, 2005, originally proposed and currently recognized as a subgenus of *Anopheles*, is typified by *Anopheles kyondawensis* Abraham, 1947. Subgenus abbreviation – *Bmi.*

****

Systematics: *Baimaia* is monobasic. It only includes the extant type species. The species has been found only in forested hilly and mountainous areas on either side of the Thai-Myanmar border located roughly between 14° and 17° N (Kanchanaburi and Tak Provinces, Thailand; Mon State, Myanmar), and at one other place approximately 400 km northeast of these localities near the Thai–Laos border in Nan Province, Thailand. The phylogenetic study of Anophelinae conducted by Harbach & Kitching (2005) based on morphological characters revealed that *Baimaia* is the earliest taxon derived from an ancestor that gave rise to all other *Anopheles*. This was not supported in the more rigorous later analysis of the data set by Harbach & Kitching (2016) in which *Baimaia* was recovered in a heterogeneous clade that also included subgenera *Anopheles*, *Lophopodomyia*, *Stethomyia* and

Bironella. Molecular data have not been available for *An. kyondawensis*; thus, how *Baimaia* may affect the relationships of the genus groups of Anophelinae are unknown.

****

Characteristics: *Baimaia* appears to be related to *Anopheles sensu stricto*, especially the Aitkenii, Alongensis and Culiciformis Groups, but differs principally in features of the male genitalia, which are unique within *Anopheles*. The gonocoxite does not have differentiated parabasal setae or an inner seta, and the gonostylus is flattened and mitten-shaped distally and lacks a gonostylar claw. Males also have uniquely developed maxillary palpi, which are straight and very nearly cylindrical with palpomeres 4 and 5 barely swollen and only slightly flattened. Adults lack thoracic scaling, and females have a dense covering of long sensilla between the antennal whorls that depart a fuzzy appearance to the antennae. Larvae have strongly inwardly curved antennae, setae 5–7-C are reduced, long single compressed or flattened somewhat lanceolate setae are present on the thorax and abdomen, and seta 6 on abdominal segments IV–VI is as long as seta 6-I–III. Pupae have a trumpet that appears undifferentiated and intermediate between the angusticorn and laticorn types, the spiracular scars of abdominal segments II–VII and seta 9-II–VIII are borne ventrally, seta 1-III is plumose and the inner and outer margins of the paddle bear a long fringe of spicules.

****

Bionomics: The immature stages of *An. kyondawensis* apparently inhabit crab holes associated with streams or streamlets and are occasionally washed out of these habitats. Larvae have been found only in crab holes and small pools along the sides of streams shaded by forest in hilly and mountainous areas. Adults of this species have never been collected in the wild, and nothing is known about their biology or behaviour. The species is not of medical importance to humans.

****

Principal references: Harrison & Scanlon, 1975 (*An. kyondawensis* as species of subgenus *Anopheles*, larva, taxonomy, distribution); Harbach *et al.*, 2005 (*Baimaia*, taxonomy, adults, larva, pupa, bionomics, distribution); Rattanarithikul *et al.*, 2006b (Thailand, classification, bionomics, distributions, keys).

Cellia Theobald, 1902

Cellia, originally proposed as a genus by Theobald (1902b), currently classified as a subgenus of *Anopheles*, is typified by *Anopheles pharoensis* Theobald, 1901b [see Theobald, 1902b]. Subgenus abbreviation – *Cel.*

****

Systematics: *Cellia* includes 233 formally recognized extant species and a growing number of unnamed members of contemporary sibling species complexes that are divided between six informal Series comprised of Groups that are believed to represent phylogenetically related assemblages based principally on morphological similarity (Harbach, 1994a, 2004). Most species of *Cellia* have distributions in the Afrotropical, Australian and Oriental Regions, but some species occur in southern areas of the Palaearctic Region. Species of *Cellia* are conspicuously absent from the majority of the Pacific islands, including the large islands of New Zealand, Fiji and New Caledonia. Phylogenetic analyses of morphological and molecular data (Krzywinski *et al.*, 2001a, 2001b; Sallum *et al.*, 2000, 2002; Harbach & Kitching, 2005, 2016; Mohanty *et al.*, 2009; Wang *et al.*, 2015; Norris & Norris, 2015; Freitas *et al.*, 2015; Hao *et al.*, 2017; da Silva *et al.*, 2020a; Lorenz *et al.*, 2021; Chen *et al.*, 2024) have shown that *Cellia* is a monophyletic group. Two unique, shared characters, cibarial armature of females with cones only and the genital lobe of pupae with a finger-like projection, support a sister-group relationship between *Cellia* (Old World) and *Kerteszia* + *Nyssorhynchus* (New World) (Harbach & Kitching, 2005). It is noteworthy, however, that a molecular phylogeny of 18 species constructed from aligned protein sequences of 1085 single-copy orthologs indicates that the *Nyssorhynchus* lineage diverged prior to the lineages that gave rise to the subgenera *Anopheles* and *Cellia* (Neafsey *et al.*, 2015). Dixit *et al.* (2010) inferred the phylogeny and divergence times of the Myzomyia, Neocellia and Pyretophorus Series based on *COII* sequences of the mitochondrial genome. The molecular phylogeny of Neafsey *et al.* (2015) convincingly supports the monophyly of the Pyretophorus Series and its emergence about 30 Mya. The molecular studies of Sallum *et al.* (2002), Sum *et al.* (2014) and Wang *et al.* (2015) provide evidence that the Funestus Group (Myzomyia Series), Maculatus Group (Neocellia Series) and the Leucosphyrus Group (Neomyzomyia Series) are monophyletic assemblages; however, the results of morphological studies (Sallum *et al.*, 2000; Harbach & Kitching, 2005, 2016) and other molecular studies (Sallum *et al.*, 2002; Mohanty *et al.*, 2009; Norris & Norris, 2015; Freitas *et al.*, 2015; Munawar *et al.*, 2020) indicate that a significant part of the current internal classification of *Cellia* does not reflect evolutionary relationships. The molecular phylogenies of Neafsey *et al.* (2015), Foster *et al.* (2017), Hao *et al.* (2017), da Silva *et al.* (2020a), Ghassemi-Khademi *et al.*, 2021; Guo *et al.* (2021), Lorenz *et al.* (2021) and Chen *et al.* (2024) based on analyses of mitochondrial data indicated that *Cellia* is sister to *Anopheles sensu stricto*. The mitogenomic studies of Hao *et al.* and Guo *et al.* support the monophyly of the Myzomyia, Neocellia, Neomyzomyia and Pyretophorus Groups of the subgenus.

****

Characteristics: *Cellia* includes species with four or more pale spots on the wings involving the costa and veins R–R_1, and the other veins usually also have distinct pale markings. With few exceptions, the cibarial armature is well developed in females and the genitalia of males have a cluster of 4–6 (usually 5) parabasal setae (no internal seta) on the gonocoxites that are usually not borne on raised tubercles. Larvae usually have a small, single seta 1-A inserted on the lateral (outer) side of the antenna. Seta 2-C are normally well separated, at least as far apart as the distance between seta 2-C and seta 3-C on either side. Each pleural setal group (setae 9–12-P,M,T) is variable, but one or both long mesopleural and metapleural setae are often plumose. Pupae have a semi-tubular trumpet (angusticorn) with the longest axis vertical more or less in line with the stem, seta 9 on abdominal segments V–VII is usually relatively long and sharply pointed, the paddle seta 1-Pa is commonly long and hooked or curved, and the genital lobe of male pupae has an apical pair of rounded points or knobs.

****

Bionomics: The immature stages of species of *Cellia* are adapted to a wide variety of groundwater habitats, ranging from small collections of water such as animal footprints and puddles that lack vegetation to large bodies of water such as rice fields, swamps and river margins that have shelter provided by vegetation. A number of species thrive in both fresh and brackish water habitats and some normally inhabit only brackish water pools. Whereas the bionomics of species involved in malaria transmission is generally well known, comparatively little is known about the biology of most species. Each of the six informal Series contains vectors of malarial protozoa and microfilariae. The principal vectors of malaria include members of sibling species complexes: the Gambiae and Funestus Complexes in the Afrotropical Region; the Culicifacies, Dirus; Fluviatilis, Leucosphyrus, Maculatus, Minimus and Sundaicus Complexes in the Oriental Region; the Farauti and Punctulatus Complexes in the Australasian Region. *Anopheles sergentii* and *An. stephensi* are also major vectors in the Middle East and the Indian Subcontinent. Over the past decade or so, the latter species has extended its range to the Horn of Africa (Djibouti, Ethiopia, Somalia, Sudan). It has also been found in Ghana, Kenya and Nigeria, and is likely to have spread to other African countries. O'nyong'nyong virus in Africa is primarily transmitted by *An. funestus* and *An. gambiae*.

****

Principal references: Mattingly & Knight, 1956 (Arabia, taxonomic notes, keys for adults and larvae, distributions); Belkin, 1962 (South Pacific, taxonomy, keys, descriptions, distributions, bionomics); DuBose & Curtin, 1965 (Mediterranean area, keys); Grjebine, 1966 (Madagascar, classification, taxonomy, keys, descriptions, bionomics, distributions); Gillies & de Meillon, 1968 (Afrotropical Region, keys, taxonomy, descriptions, bionomics, distributions); Reid, 1968 (Malaysia, Borneo, classification, keys, descriptions, distributions, bionomics); Cagampang-Ramos & Darsie, 1970 (Philippine Islands, keys); Gutsevich *et al.*, 1974 (former USSR, keys, descriptions, distributions, bionomics); Klein, 1977a (Cambodia, species, bionomics); Tanaka *et al.*, 1979 (Japan, taxonomy, keys, descriptions, distributions, bionomics); Harrison, 1980 (Thailand, Myzomyia Series, keys, taxonomy, descriptions, distributions, bionomics); Lu & Li, 1982 (China, keys); Rao, 1984 (India, keys, keys, descriptions, bionomics, distributions); Lee *et al.*, 1987b (Australasian Region, taxonomy, keys, literature, distributions, bionomics); Gillies & Coetzee, 1987 (Afrotropical Region, keys, supplemental information to Gillies & de Meillon, 1968); Das *et al.*, 1990 (India, keys); Darsie & Pradhan, 1990 (Nepal, distributions, bionomics, taxonomic notes); Glick, 1992 (southwestern Asia and Egypt, keys); Rattanarithikul & Panthusiri, 1994 (Thailand, keys, medically important species); Nagpal & Sharma, 1995 (India, identification, species information); Lu *et al.*, 1997b (China, keys, taxonomy, descriptions, distributions, bionomics); Rattanarithikul *et al.*, 2006b (Thailand, classification, bionomics, distributions, keys); Beebe *et al.*, 2015 (southwestern Pacific, Punctulatus Group, molecular species identification, evolution, population genetics, distributions, medical importance); Hao *et al.*, 2017 (mitochondrial protein sequences, phylogenetic relationships); Coetzee, 2020 (Afrotropical Region, keys); Irish *et al.*, 2020 (Afrotropical Region, country distributions); Ghassemi-Khademi *et al.*, 2021 (mitochondrial genomes, phylogenetic relationships); Guo *et al.*, 2021 (mitochondrial protein sequences, phylogenetic relationships); Lorenz *et al.*, 2021 (mitochondrial protein-coding genes, phylogenetic relationships); Russell & Burkot, 2023 (South Pacific, country lists of species, distributions, bionomics, vector status).

Christya Theobald, 1903

Christya, originally proposed as a genus by Theobald (1903b), currently classified as a subgenus of *Anopheles*, is typified by *Anopheles implexus* (Theobald, 1903b), original combination: *Christya implexa*. Subgenus abbreviation – *Chr.*

****

Systematics: *Christya* includes two extant species in sub-Saharan Africa. *Anopheles implexus* is most abundant in highland forested areas of eastern Africa from Ethiopia through Sudan to Kenya, Uganda, Burundi and Zambia. It is uncommon and occurs in localized areas in western Africa from Burkina Faso and Ivory Coast through Ghana, Nigeria, Central African Republic and Cameroun to Angola and the Congo basin. Isolated populations are reported from Tanzania and South Africa. *Anopheles okuensis* is currently only known from highlands in Cameroun. In the first phylogenetic study of Anophelinae by Sallum *et al.* (2000),

An. implexus was recovered as sister to *Kerteszia* + *Nyssorhynchus*, and this clade was sister to *Cellia* + all other anophelines except *Chagasia*. Based on revision and the addition of two taxa to the data set of Sallum *et al.*, *An. implexus* was recovered as the sister to *Cellia* + (*Kerteszia* + *Nyssorhynchus*) in the study of Harbach & Kitching (2005). With further revision of the data set, addition of another species and application of more stringent methods of assessing clade support (Harbach & Kitching, 2016), *An. implexus* was recovered as the sister to all Anophelinae except those of *Chagasia*.

****

Characteristics: Based on the morphology of *An. implexus*, *Christya* shares a number of features with *Chagasia*, including speckled legs, tarsi with unusually broad pale bands and a simple pupal trumpet, and is easily recognizable by having long tufts of narrow scales projecting from the abdominal segments. *Anopheles implexus* is a large mosquito with the following distinctive features: maxillary palpus of females with a peculiar knotted appearance due to groups of semi-erect scales; scutum with sparse covering of yellowish setae and yellowish piliform scales; meso- and metathoracic pleura with conspicuous dark scale-patches sharply delimited by silvery lines and pale scaling; about 6 lower mesepimeral setae present; wing with variable pale scaling; femora and tibiae speckled, spots sometimes united to form bands; hindtarsomeres 3 and 4 usually entirely pale, hindtarsomere 5 entirely dark or more often with the apical 0.5 or less pale; abdominal segments 1–7 with prominent posterolateral tufts of long narrow mainly dark scales projecting perpendicular to the segments; sterna 1–7 each with a pair of conspicuous round clearly defined pale spots. The gonocoxite of the male genitalia has 1 short parabasal seta borne on a prominent tubercle; the claspette is bilobed. Larvae have seta 1-A large, brush-like; the pair of seta 2-C inserted close together, simple or split into 2 or 3 apical branches; seta 3-C multi-branched, nearly as long as seta 2-C; seta 3-T and seta 1-I,II,VII multi-branched, not palmate, branches arranged on both sides of a rachis (pinnate), seta 1-III–VI palmate with simple transparent lanceolate leaflets (not differentiated as blades with an apical filament); seta 6-I–VI all plumose and borne on tubercles. Pupae have the cephalothorax with a median keel forming a conspicuous crest posteriorly; the trumpet is unusually broad and widely open.

****

Bionomics: Species of *Christya* are forest mosquitoes. The immature stages of *An. implexus* are found in streams, seepage, swamps, marshes, pools, ponds and elephant tracks, principally in small heavily shaded residual collections of water. Females of the species are active during the day in some areas and at night in other areas. They commonly bite humans who enter their haunts and are found resting on tree trunks. Larvae of *An. okuensis* have been found in a mountain stream. The species of *Christya* are not of medical importance to humans.

****

Principal references: Evans, 1938 (as Group Christya, *An. implexus*, taxonomy, description, bionomics, distribution); Gillies & de Meillon, 1968 (*An. implexus*, taxonomy, description, bionomics, distribution); Gillies & Coetzee, 1987 (as Series Christya, *An. implexus*, bionomics, distribution); Harbach & Kitching, 2016 (subgeneric status, phylogenetic relationships).

Kerteszia Theobald, 1905

Kerteszia, originally proposed as a genus by Theobald (1905f), currently classified as a subgenus of *Anopheles*, is typified by *Anopheles boliviensis* (Theobald, 1905f), original combination: *Kerteszia boliviensis*. Subgenus abbreviation – *Ker.*

****

Systematics: *Kerteszia* includes 12 extant species. It has a Neotropical distribution: south from Veracruz State in Mexico through Central America and Atlantic South America, along the Andes and along the coast, to the States of Misiones in Argentina and Rio Grande do Sul in Brazil, and also extends south along the Pacific Coast of South America to the State of El Oro, Ecuador. Species of *Kerteszia* are absent from all islands of the West Indies except Trinidad, and from most of the vast expanse of the Amazon basin in South America (Zavortink, 1973). Phylogenetic studies of anopheline mosquitoes based on both morphological and molecular data support the monophyly of *Kerteszia* (Sallum *et al.*, 2000, 2002; Collucci & Sallum, 2003; Harbach & Kitching, 2005, 2016; Foster *et al.*, 2017; do Nascimento *et al.*, 2021; Ghassemi-Khademi *et al.*, 2021; Guo *et al.*, 2021; Lorenz *et al.*, 2021). Except for the studies of Freitas *et al.*, Foster *et al.*, do Nascimento *et al.*, Ghassemi-Khademi *et al.* and Guo *et al.*, all of the studies that support the monophyly of *Kerteszia* also support its sister relationship with *Nyssorhynchus*. *Kerteszia* was recovered as the sister to *Stethomyia* in the analyses of *COI* and *COII* mtDNA and 2.5S rDNA sequences conducted by Freitas *et al.* and in the analysis of mitochondrial protein sequences in the study of Guo *et al.* (2021); it was recovered in a sister relationship with *Lophopodomyia* in the study of Foster *et al.* based on analyses of mitochondrial protein-coding genes; and as sister to a clade comprised of *Nyssorhynchus* + (*Anopheles sensu stricto* + *Cellia*) in the analyses of mitochondrial data undertaken in the studies of do Nascimento *et al.* (2021) and Ghassemi-Khademi *et al.* (2021). Collucci & Sallum (2003) explored the phylogenetic relationships of 12 species of *Kerteszia*, and Kirchgatter *et al.* (2020) examined the relationships of four species, two collected in Brazil and two collected in Peru.

****

Characteristics: *Kerteszia* is most closely related to *Nyssorhynchus*. The two groups are very similar in overall morphology, but the following features easily distinguish the adults, larvae and pupae of *Kerteszia*. The scutum of adults has four distinct longitudinal lines of dark pollinose integument and a few scales in patches on the lateral and/or anterior areas. The scutellum is normally void of scales (a few rarely occur medially) and the mid- and hindfemora have a stripe of white scales on the distal 0.33 of their anterior surfaces. The wings are distinct in having crossveins sc-r and r_1-r_s relatively far apart (distance about 0.5 the length of the sector dark spot on the costa and radius-one), a spot of dark scales where the humeral crossvein intersects the radius, the subcosta ending before the subcostal pale spot, vein R_{4+5} dark-scaled apically, vein CuA mostly dark-scaled with a basal dark spot and the anal vein normally entirely dark-scaled (rarely with one or two small pale spots). The gonocoxite, claspette and aedeagus of the male genitalia exhibit a number of differential and diagnostic features. The gonocoxite lacks a subapical seta, two subequal accessory setae are inserted beyond mid-length, the internal seta is very stout and inserted far basad of the accessory setae, and the parabasal seta is considerably longer than the internal seta. The ventral lobes of the claspette are narrowly joined separate structures, the dorsal lobes each bear two distinct groups of specialized setae, and the aedeagus is slender and uniformly tapered to a narrowly rounded apex. Larvae exhibit many distinctive features, which include, but are not limited to, seta 1-A single and inserted dorsolaterally, seta 2-C inserted on same level or slightly posterior to seta 3-C, seta 4-C inserted posterolateral to seta 3-C, hypostomal suture undeveloped or very short, absence of a sclerotized tubercle at the base of seta 4-C, long seta 4-M (about as long as seta 6-M), weakly plumose setae 8-M,T and 6,7-I,II, single seta 2 on segments I–IV, absence of seta 14-III, seta 2-IV inserted lateral to seta 4-IV, weakly developed seta 1-S (single, with 2 or 3 branches or weak terminal branches or aciculae) and 9 pairs of seta 4-X. Salient features of *Kerteszia* pupae include a trumpet with a short pinna (less than 0.5 as long as the trumpet) and no meatal cleft, seta 9-I shorter than 0.33 the length of seta 6-I, weakly developed setae 1-I–VII and 5-V–VII (less than 0.5 as long as the segment), seta 14-III absent, seta 2-IV,V inserted anterior and nearly in line with seta 3, seta 3-VI inserted lateral to seta 1-VI, seta 10-VI present and seta 1-Pa stout and shorter than seta 2-Pa (absent in *An. neivai*).

****

Bionomics: The immature stages of species of *Kerteszia*, except those of *An. bambusicola*, are found in water that accumulates in the leaf axils of terrestrial and epiphytic bromeliads. The immature stages of *An. bambusicola* are normally found in unbroken bamboo internodes. The adults of *Kerteszia* are strongly anthropophilic. Peak biting activity occurs during the evening hours, but females are known to bite throughout the night, in the morning and in the shade during the daylight hours. Six species of *Kerteszia* are known to transmit malarial protozoa, but only *An. bellator* in Trinidad and *An. cruzii* in Brazil are important vectors. *Anopheles bellator* also transmits the helminths that cause Bancroftian filariasis. *Anopheles neivai* and *An. boliviensis* have been found naturally infected with arboviruses, including yellow fever virus in the former species, and eggs of *Dermatobia hominis* have been found in the latter species.

****

Principal references: Komp, 1937b (species descriptions, taxonomy, keys); Cova-Garcia, 1961 (Venezuela, descriptions, keys); Zavortink, 1973 (revision, keys, descriptions, systematics, bionomics, distributions); Peyton *et al.*, 1992 (*Kerteszia* and *Nyssorhynchus*, morphological distinctions); Collucci & Sallum, 2003 (morphology, phylogenetic relationships); Harrison *et al.*, 2012 (morphology, key to females); Freitas *et al.*, 2015 (*COI* and *COII* mtDNA and 2.5S rDNA sequences, phylogenetic relationships); Lorenz *et al.*, 2015 (*An. cruzii* and *An. homunculus*, characterization, evolutionary relationships); Oliveira *et al.*, 2016 (mitochondrial genomes); Foster *et al.*, 2017 (mitochondrial protein sequences, phylogenetic relationships); Hao *et al.*, 2017 (mitochondrial protein sequences, phylogenetic relationships); do Nascimento *et al.*, 2021 (mitochondrial protein-coding genes, phylogenetic relationships); Guo *et al.*, 2021 (mitochondrial protein sequences, phylogenetic relationships); Lorenz *et al.*, 2021 (mitochondrial protein-coding genes, phylogenetic relationships).

Lophopodomyia Antunes, 1937

Lophopodomyia Antunes, 1937, originally proposed and currently recognized as a subgenus of *Anopheles*, is typified by *Anopheles squamifemur* Antunes, 1937. Subgenus abbreviation – *Lph.*

****

Systematics: *Lophopodomyia* includes six formally recognized extant species, which are known to occur in areas of Panama and northern South America (Brazil, Colombia, Ecuador, French Guiana, Venezuela). *Lophopodomyia* appears to be a monophyletic taxon, but its phylogenetic relationship with *Anopheles sensu stricto*, as currently defined, and its subgeneric status are uncertain (Sallum *et al.*, 2000, 2002; Harbach & Kitching, 2005, 2016; Freitas *et al.*, 2015). *Lophopodomyia* was recovered in a sister relationship with *Kerteszia* in the study of Foster *et al.* (2017) based on analyses of mitochondrial protein-coding genes. The results of their study stand in stark contrast to the results of morphological and molecular studies that support the sister relationship of *Kerteszia* and *Nyssorhynchus* (Sallum *et al.*, 2000, 2002; Collucci & Sallum, 2003; Harbach & Kitching, 2005, 2016; Lorenz *et al.*,

2021). Surprisingly, *Lophopodomyia* was recovered as the sister to all other subgenera of *Anopheles* in the analysis of mitochondrial protein sequences conducted by Guo *et al.* (2021). Not so surprisingly, *Lophopodomyia* was recovered in a sister relationship with *Stethomyia*, and the pair was recovered in a basal position relative to all other *Anopheles* included in the phylogenetic analysis of 13 mitochondrial protein-coding genes conducted by Lorenz *et al.* (2021).

****

Characteristics: The species of *Lophopodomyia* have not been studied in detail. The adults resemble species of *Anopheles sensu stricto*, and differ principally in features of the male genitalia. The gonocoxite bears 4 or 5 specialized setae: one basal and two parabasal setae with an external accessory seta not inserted on tubercles, and usually one internal seta. The dorsal claspette is rudimentary. In larvae, the long setae of the pro- and mesothoracic pleural groups are predominantly simple and seta 1 of the abdominal segments is distinctly palmate. Pupae have fringed paddles bearing a straight seta 1-Pa.

****

Bionomics: Species of *Lophopodomyia* are forest mosquitoes. Little is known about their bionomics, but females are apparently active during twilight periods and will feed on humans and other animals that enter their realm. The immature stages are found in shaded stream pools that are rich in organic matter. Species of *Lophopodomyia* are not of medical importance to humans.

****

Principal references: Lane, 1953 (taxonomy, keys, descriptions, distributions); Cova-Garcia, 1961 (Venezuela, descriptions, keys); Forattini, 1962 (classification, descriptions, bionomics, distributions); Wilkerson & Strickman, 1990 (Central America and Mexico, keys); Foster *et al.*, 2017 (mitochondrial protein sequences, phylogenetic relationships); Guo *et al.*, 2021 (mitochondrial protein sequences, phylogenetic relationships); Lorenz *et al.*, 2021 (mitochondrial protein-coding genes, phylogenetic relationships).

Nyssorhynchus Blanchard, 1902

Nyssorhynchus, originally proposed as a genus by Blanchard (1902), currently classified as a subgenus of *Anopheles*, is typified by *Anopheles argyritarsis* Robineau-Desvoidy, 1827. Subgenus abbreviation – *Nys.*

****

Systematics: *Nyssorhynchus* includes 45 formally recognized extant species, a growing number of unnamed members of contemporary sibling species complexes and a single fossil species. The extant species are divided between three informal Sections, the two largest of which are further subdivided into Series and Groups that are believed to represent phylogenetically related assemblages based principally on morphological similarity (Harbach, 2004). Species of *Nyssorhynchus* are restricted to the Neotropical Region, except for *An. albimanus*, which extends into the Nearctic Region (northern Mexico and along the Rio Grande River in Texas, USA). Phylogenetic studies of anopheline mosquitoes based on both morphological and molecular data support the monophyly of *Nyssorhynchus* (Sallum *et al.*, 2000, 2002; Collucci & Sallum, 2003; Harbach & Kitching, 2005, 2016; Foster *et al.*, 2017; Guo *et al.*, 2021; Lorenz *et al.*, 2021). Except for the studies of Freitas *et al.*, Foster *et al.*, Guo *et al.* and Chen *et al.*, all of the studies that support the monophyly of *Nyssorhynchus* also support its sister relationship with *Kerteszia*. *Nyssorhynchus* was recovered as (1) the sister to *Stethomyia* in the analyses of *COI* and *COII* mtDNA and 2.5S rDNA sequences conducted by Freitas *et al.* (2015), (2) as the sister to *Cellia* + *Anopheles* in the studies of Foster *et al.* (2017) and Guo *et al.* (2021) based on analyses of mitochondrial protein-coding genes, (3) as the sister to a clade comprised of *Bironella* + (*Kerteszia* + *Stethomyia*) in the study of Chen *et al.*, also based on mitochondrial protein sequences, and (4) as the sister to a clade comprised of *Nyssorhynchus* + (*Anopheles sensu stricto* + *Cellia*) generated in the analyses of mitochondrial data conducted by do Nascimento *et al.* (2021) and Ghassemi-Khademi *et al.* (2021). The analysis of morphological data by Harbach & Kitching (2005) indicated that *Cellia* (Old World) is sister to *Nyssorhynchus* + *Kerteszia*, but this relationship was not recovered in their later study (Harbach & Kitching, 2016). It is noteworthy that a molecular phylogeny of 18 species constructed from aligned protein sequences of 1085 single-copy orthologs indicated that the *Nyssorhynchus* lineage diverged prior to the lineages that gave rise to the subgenera *Anopheles* and *Cellia* (Neafsey *et al.*, 2015). Phylogenetic relationships among species of *Nyssorhynchus* based on nuclear and mitochondrial gene sequences were elucidated by Bourke *et al.* (2010). Similarly, relationships among five of the six species belonging to the Myzorhynchella Section of *Nyssorhynchus* were investigated by Bourke *et al.* (2011) based on the ITS2 region of rDNA and the nuclear *white* gene. Foster *et al.* (2013) analysed combined data for the nuclear *white* and *CAD* genes and the *COI* barcode region of mtDNA obtained from representatives of the Albimanus, Argyritarsis and Myzorhynchella Sections and only found support for the monophyly of the last section. Similarly, their analyses did not support the monophyly of the Strodei Subgroup of the Oswaldoi Series or the monophyly of the Albitarsis Series of the Argyritarsis Section. The analyses of Foster *et al.* (2017) did not support the monophyly of the Myzorhynchella Section, which excluded *An. parvus*, and the Argyritarsis Series, which excluded *An. atacamensis*. Motoki *et al.* (2021) used mtDNA *COI* sequences to investigate the composition of the Albitarsis Group and establish that divergence within the group occurred 0.58–2.25 Mya during the Pleistocene Epoch.

****

Characteristics: *Nyssorhynchus* is most closely related to *Kerteszia*. The two groups are very similar in overall morphology, but the following features easily distinguish the adults, larvae and pupae of *Nyssorhynchus*. The scutum of adults lacks dark longitudinal lines and has scales on the acrostichal and dorsocentral areas in addition to a few scales in patches on the lateral and/or anterior areas. The scutellum has numerous scattered white, yellow or silver scales along the posterior border. The wings differ in having crossveins sc-r and r_1-r_s closer together (distance about 0.30 or less the length of the sector dark spot on the costa and radius-one), no spot of dark scales where the humeral crossvein intersects the radius, the subcosta terminates at the proximal end or the middle of the subcostal pale spot, vein R_{4+5} is narrowly pale-scaled apically with a corresponding isolated apical pale fringe spot, vein CuA is mostly pale-scaled or equally pale- and dark-scaled without a basal dark spot and the anal vein has three pale-scaled areas setting off two dark areas between the basal and an apical pale spot (a greater amount of dark scaling occurs in species of the Myzorhynchella Section). The gonocoxite, claspette and aedeagus of the male genitalia exhibit a number of distinctions. The gonocoxite has a subapical seta, two subequal accessory setae are inserted before mid-length (about 0.4 from the base), the internal seta is as long but much weaker than the accessory setae, and the parabasal seta is shorter than the internal seta (about 0.6 its length). The ventral lobes of the claspette are fused to form a single structure, the dorsal lobes each bear a single group of two or three specialized setae, and the aedeagus is broad and not noticeably tapered (at least in the basal 0.8). Larvae exhibit differential and diagnostic features, which include, but are not limited to, seta 1-A branched and inserted dorsomesally, seta 2-C inserted anterior to seta 3-C, seta 4-C inserted posteromesal to seta 3-C, hypostomal suture well developed (but not reaching the posterior tentorial pit), presence of a sclerotized tubercle at the bases of setae 1,2-P, seta 4-M short (not more than 0.5 as long as seta 6-M), setae 8-M,T and 6,7-I,II strongly plumose, seta 3-T palmate or fan-like, seta 2-I–IV multi-branched, seta 14-III present, seta 2-IV inserted mesad to seta 4-IV, seta 1-S well developed (with multiple branches) and ventral brush with eight pairs of seta 4-X. Salient features of *Nyssorhynchus* pupae include a trumpet with a long pinna (more than 0.5 as long as the trumpet) and a meatal cleft, seta 9-I more than 0.5 the length of seta 6-I, setae 1-I–VII and 5-V–VII well developed (nearly as long or longer than the segment), seta 14-III present, seta 2 on segments IV and V inserted mesad of seta 3, seta 3-VI inserted mesad to or in line with seta 1-VI, seta 10-VI absent, and paddle seta 1-Pa slender, similar to seta 2-Pa or sinuous and apically hooked.

****

Bionomics: The immature stages of *Nyssorhynchus* species occur in a variety of sunlit or partially shaded habitats, including ponds, lakes, swamps, stream and river margins, canals, ditches, flood land, ground pools, borrow pits, spring water, wheel ruts and animal tracks. The immature stages of *An. aquasalis* mainly occur in brackish water habitats such as mangrove swamps and coastal ground pools, and those of *An. albimanus*, in addition to ground pools and stream margins, have been found in crab holes, tree holes and artificial containers. Females of most species feed at dusk or during the night. They feed on a variety of large mammals and infrequently on birds. Some species readily attach humans. *Anopheles albimanus*, *An. aquasalis*, *An. argyritarsis*, *An. darlingi*, *An. nuneztovari* and *An. oswaldoi s.l.* are vectors of malarial protozoa. *Anopheles albimanus*, *An. albitarsis* and *An. aquasalis* also transmit arboviruses, and some species also transmit *Wuchereria bancrofti*.

****

Principal references: Lane, 1953 (taxonomy, keys, descriptions, distributions); Cova-Garcia, 1961 (Venezuela, descriptions, keys); Forattini, 1962 (classification, descriptions, bionomics, distributions); Faran, 1980 (Albimanus Section, classification, keys, taxonomy, descriptions, bionomics, distributions); Faran & Linthicum, 1981 (Amazonia, keys, diagnoses, bionomics, distributions); Clark-Gil & Darsie, 1983 (Guatemala, keys); Linthicum, 1988 (Argyritarsis Section, keys, classification, taxonomy, descriptions, bionomics, distributions); Wilkerson & Strickman, 1990 (Central America and Mexico, keys); Peyton *et al.*, 1992 (*Nyssorhynchus* and *Kerteszia*, morphological distinctions); Bourke *et al.*, 2010 (phylogenetic relationships); Foster *et al.*, 2017 (mitochondrial protein sequences, phylogenetic relationships); Hao *et al.*, 2017 (mitochondrial protein sequences, phylogenetic relationships); do Nascimento *et al.*, 2021 (mitochondrial protein-coding genes, phylogenetic relationships); Guo *et al.*, 2021 (mitochondrial protein sequences, phylogenetic relationships); Lorenz *et al.*, 2021 (mitochondrial protein-coding genes, phylogenetic relationships); Motoki *et al.*, 2021 (Albitarsis Group, mtDNA *COI* sequences, phylogenetic network, geographic data); Chen *et al.*, 2024 (mitochondrial protein sequences, phylogenetic relationships).

Stethomyia Theobald, 1902

Stethomyia, originally proposed as a genus by Theobald (1902b), currently classified as a subgenus of *Anopheles*, is typified by *Anopheles nimbus* (Theobald, 1902b), original combination: *Stethomyia nimbus*. Subgenus abbreviation – *Ste.*

****

Systematics: *Stethomyia* includes only five extant species. Species of *Stethomyia* principally occur in southern Central America (Costa Rica, Panama) and northern South America (Brazil, Colombia, French Guiana, Guyana, Suriname, Venezuela), but one or two species are known to occur on the Islands of Trinidad and Tobago and as far south as Peru and Bolivia. *Stethomyia* appears to be a monophyletic taxon, but its phylogenetic relationship with *Anopheles sensu stricto*, as currently defined, and its subgeneric status are uncertain (Sallum *et al.*, 2000, 2002; Harbach & Kitching, 2005, 2016). *Stethomyia* was recovered as the sister to *Kerteszia* in (1) the analyses of *COI* and *COII* mtDNA and 2.5S rDNA sequences conducted by Freitas *et al.* (2015) and (2) the analyses of mitochondrial protein sequences conducted by Guo *et al.* (2021) and Chen *et al.* (2024). It is noteworthy that *Stethomyia* + *Kerteszia* was recovered as the sister of *Bironella* in the study of Chen *et al.* (2024) and was recovered as one of three branches of a polytomy comprised of 30 species of *Anopheles* and a species of *Bironella* in the study of Foster *et al.* (2017), both studies based on analyses of mitochondrial protein-coding genes. In contrast to the results of those studies, *Stethomyia* was recovered as the sister of *Lophopodomyia* in a basal relationship to all other *Anopheles* in the study of Lorenz *et al.* (2021) based on the analysis of 13 mitochondrial protein-coding genes.

****

Characteristics: *Stethomyia* are dark mosquitoes. There is no pale scaling on the maxillary palpi, wings or legs but the scutum bears a median longitudinal silvery stripe. Prealar setae are absent. The gonocoxite of the male genitalia has an internal seta and a single long parabasal seta that is not borne on a tubercle. Larvae have widely separated seta 2-C, as in species of *Cellia*, the long pleural setae have thorn-like branches on one side only, abdominal setae 1 (float setae) are absent or vestigial, each anterolateral spiracular lobe has a narrow finger-like process and the spiracles are prominent and widely spaced. Pupae do not seem to bear subgeneric characters.

****

Bionomics: *Stethomyia* are forest mosquitoes. Very little is known about their bionomics. Females are known to feed on humans and other animals. Immature stages have been collected in shaded marshy and swampy areas along streams in forest. Species of *Stethomyia* are not of medical importance to humans.

****

Principal references: Lane, 1953 (taxonomy, keys, descriptions, distributions); Cova-Garcia, 1961 (Venezuela, descriptions, keys); Forattini, 1962 (classification, descriptions, bionomics, distributions); Wilkerson & Strickman, 1990 (Central America and Mexico, keys); Foster *et al.*, 2017 (mitochondrial protein sequences, phylogenetic relationships); Guo *et al.*, 2021 (mitochondrial protein sequences, phylogenetic relationships); Lorenz *et al.*, 2021 (mitochondrial protein-coding genes, phylogenetic relationships); Chen *et al.*, 2024 (mitochondrial protein sequences, phylogenetic relationships).

Bironella Theobald, 1905

Bironella Theobald, 1905f, originally proposed and currently recognized as a genus of Anophelinae, is typified by *Bironella gracilis* Theobald, 1905f. Genus abbreviation – *Bi.*

****

Systematics: *Bironella* is a subordinate taxon of the subfamily Anophelinae. It includes eight extant species divided between three subgenera: *Bironella sensu stricto*, *Brugella* and *Neobironella*. Species of *Bironella* occur in the Australasian Region, principally in the Papuan Subregion. The phylogenetic relationships of *Bironella* are uncertain. It may be the sister group of *Anopheles* (Harbach & Kitching, 1998; Besansky & Fahey, 1997; Foley *et al.*, 1998; Krzywinski *et al.*, 2001a, 2001b; Foster *et al.*, 2017; Lorenz *et al.*, 2021) or a species group of *Anopheles sensu stricto* (Sallum *et al.*, 2000, 2002; Harbach & Kitching, 2005, 2016). *Bironella* has been placed in different relationships in various molecular studies based on different methods of aligning data sets (Lorenz *et al.*, 2021). The single species of *Bironella* (*Bi. hollandi*) included in the study of Foster *et al.* based on analyses of mitochondrial protein-coding genes was recovered as the sister of a clade parenthetically expressed as *Nyssorhynchus* + (*Cellia* + *Anopheles*). The evolutionary relationships of the subgenera and species of *Bironella* have not been investigated.

****

Characteristics: The adults and immature stages of species of *Bironella* are similar to those of *Anopheles*. Species of *Bironella* are delicate insects that vary from light to dark brown in overall colouration. They are easily distinguished from *Anopheles* by wing venation and larval thoracic chaetotaxy. Wing veins M_{1+2} and M_{3+4} are conspicuously wavy; cell R_2 may be complete and nearly as long as vein R_{2+3}, complete but distinctly shorter than vein R_{2+3} (about 0.20–25 its length) or incomplete with vein R_3 absent basally. Larval thoracic seta 1-M and seta 3-T are both palmate and seta 4-P is inserted closer to setae 1–3-P than to setae 5–7-P. In *Anopheles*, veins M_{1+2} and M_{3+4} are both straight, cell R_2 is complete and approximately as long as vein R_{2+3}; only larval thoracic seta 3-T is palmate and seta 4-P is inserted closer to setae 5–7-P than to setae 1–3-P.

****

Bionomics: The immature stages of species of *Bironella* are usually found in undisturbed areas, most frequently in dense shade in swamp margins and the margins of slow-moving streams. They also inhabit rock pools in and along margins of streams, backwaters of streams, seepages, river banks and temporary pools. The adults are seldom collected or seen in nature, but some have been captured in Malaise traps set up in virgin forest. The females of *Bi. gracilis* have been reported to bite humans, and have been induced to bite in the laboratory, but species of *Bironella* do not habitually attack people. Species of *Bironella* are not known to transmit any pathogens of human diseases.

****

Principal references: Tenorio, 1977 (revision, keys, taxonomy, descriptions, bionomics, distributions); Lee *et al.*, 1988a (taxonomy, keys, literature, distributions, bionomics); Foster *et al.*, 2017 (mitochondrial protein sequences, phylogenetic relationships); Lorenz *et al.*, 2021 (mitochondrial protein-coding genes, phylogenetic relationships).

Bironella Theobald, 1905 *sensu stricto*

Bironella sensu stricto, treated as the nominotypical subgenus of the genus *Bironella*, is typified by *Bironella gracilis* Theobald, 1905f. Subgenus abbreviation – *Bir.*

****

Systematics: *Bironella sensu stricto* includes only two extant species. The type species is the most widely distributed species of *Bironella*, ranging from New Britain (Bismarck Archipelago) through New Guinea to Ceram (Moluccas). *Bironella simmondsi* is only known to occur in Queensland, Australia. The evolutionary relationships of *Bironella sensu stricto* and its species have not been investigated.

****

Characteristics: The following combinations of characters distinguish species of *Bironella sensu stricto* from those of *Brugella* and *Neobironella*. ADULTS: Vertex with pale scales and setae, some scales forked apically; occiput with dense dark erect forked scales; interocular space with long pale scales and setae; antenna strongly verticillate in males, weakly verticillate in females, flagellar whorls attached near base of the flagellomeres; maxillary palpus nearly as long as the proboscis in males, 0.50–0.75 the length of the proboscis in females, with dense dark brown scales; anterior promontory with a patch of pale piliform scales; upper proepisternum setose; cell R_2 of wing short, 0.20–0.25 the length of vein R_{2+3}; radiomedial crossvein distal to the origin of vein M_{3+4}; hindfemur pale-scaled on ventral and most of posterior surfaces to near apex, hindtibia pale on anterior surface from basal 0.10–0.50. MALE GENITALIA: Gonocoxite without scales, basotergal area with well-developed projection; aedeagus long, slender, rod-like, with leaflets at apex; claspette absent. LARVAE: Seta 2-C long, single; seta 3-C as long or nearly as long as seta 2-C, single to 5-branched; setae 5–7-C multi-branched, 0.67–0.75 the length of head capsule. PUPAE: Trumpet notched at apical margin; seta 13-CT absent, alveoli present; seta 9-II–VII short, stub-like; paddle with fringe of spicules on outer and inner margins.

****

Bionomics: The immature stages of species of *Bironella sensu stricto* are normally found among floating debris and vegetation in densely shaded clear-water pools, swamps and rivers in forest. Tenorio (1977) listed seepage and flower bracts as larval habitats for *Bi. gracilis*. Females of *Bi. gracilis* are known on rare occasions to bite humans and enter human dwellings. Species of *Bironella sensu stricto* are not considered to be involved in the transmission of pathogens to humans or domestic animals.

****

Principal references: Tenorio, 1977 (revision, keys, taxonomy, descriptions, bionomics, distributions); Lee *et al.*, 1988a (taxonomy, keys, literature, distributions, bionomics).

Brugella Edwards, 1930

Brugella Edwards, 1930a, originally proposed and currently recognized as a subgenus of *Bironella*, is typified by *Bironella travestita* (Brug, 1928b), original combination: *Anopheles* (*Bironella*) *travestitus*. Subgenus abbreviation – *Bru.*

****

Systematics: *Brugella* includes only three extant species. The distribution of *Brugella* includes the Bismarck Archipelago, Ceram (Moluccas), New Guinea and the Solomon Islands. *Bironella hollandi* is known from the Bismarck Archipelago and Solomon Islands; *Bi. obscura* is only known from localities in northeastern and northwestern New Guinea; *Bi. travestita* is recorded from Ceram and New Guinea. *Bironella hollandi* was included in the study of Foster *et al.* (2017) based on analyses of mitochondrial protein-coding genes, and was recovered as the sister of a clade parenthetically expressed as *Nyssorhynchus* + (*Cellia* + *Anopheles*). The evolutionary relationships of *Brugella* and its species have not been investigated.

****

Characteristics: The following combinations of characters distinguish species of *Brugella* from those of *Bironella* and *Neobironella*. ADULTS: Vertex, occiput and interocular space essentially as in species of *Bironella sensu stricto*; antenna weakly verticillate in males as well as females, flagellar whorls attached near base of the flagellomeres; maxillary palpus short, 0.17–0.20 the length of the proboscis, dark brown- to black-scaled; anterior promontory with a patch of pale piliform scales as in species of *Bironella sensu stricto*; upper proepisternum bare; cell R of wing about as long as vein R_{2+3}; origin of vein M_{3+4} proximal to the radiomedial crossvein; hindfemur pale-scaled on anterior surface except apical 0.17, anterior surface of hindtibia pale between proximal 0.20 and apical 0.25. MALE GENITALIA: Gonocoxite with scales on basolateral area, basotergal area with no more than short rounded lobe with apical setae; claspette present; aedeagus similar to that of species of *Bironella sensu stricto*. LARVAE: Seta 3-C single, 0.50–0.67 the length of seta 2-C; setae 5–7-C multi-branched, about 0.50 the length of head capsule. PUPAE: Trumpet without notch at apical margin; seta 13-CT rarely present; seta 9-II,III short, stub-like but slightly longer than in species of *Bironella sensu stricto*; seta 9-VII moderately long, spine-like, seta 9-IV–VI shorter than seta 9-VII; paddle with fringe of spicules on outer margin, absent from inner margin.

****

Bionomics: The immature stages of species of *Brugella* are found mainly in cool, clean water among floating debris and vegetation in shaded bodies of water such as streams, swamps, ground pools and lagoons at the mouths of rivers, and smaller collections of water such as road ruts, seepage pools, springs and pig wallows. They are sometimes found in natural and artificial containers. Adults of *Bi. obscura* and *Bi. travestita* have been collected in Malaise traps; the latter species has also been collected in carbon dioxide-baited traps and biting humans. Adults of *Bi. hollandi* have never been collected in the wild. Species of *Brugella* are not known to transmit pathogens of human diseases.

****

Principal references: Tenorio, 1977 (revision, keys, taxonomy, descriptions, bionomics, distributions); Lee *et al.*, 1988a (taxonomy, keys, literature, distributions, bionomics); Foster *et al.*, 2017 (mitochondrial protein sequences, phylogenetic relationships).

Neobironella Tenorio, 1977

Neobironella Tenorio, 1977, originally proposed and currently recognized as a subgenus of *Bironella*, is typified by *Bironella confusa* Bonne-Wepster, 1951. Subgenus abbreviation – *Nbi.*

****

Systematics: *Neobironella* includes only three extant species. Species of *Neobironella* are recorded from islands of the Bismarck Archipelago (*Bi. confusa*) and the Moluccas (*Bi. papuae*), New Guinea (*Bi. confusa*, *Bi. papuae*, *Bi. derooki*) and northern Queensland, Australia (*Bi. papuae*). The evolutionary relationships of *Neobironella* and its species have not been investigated.

****

Characteristics: The following combinations of characters distinguish species of *Neobironella* from those of *Bironella* and *Brugella*. ADULTS: Vertex without scales, setae brown; occiput with sparse dark brown to black weakly forked piliform scales; interocular space without scales, with light to dark brown setae; antenna strongly verticillate in males, weakly verticillate in females, flagellar whorls attached near apex of the flagellomeres; maxillary palpus as long as the proboscis and palpomere 5 club-like in males, 0.20–0.67 the length of the proboscis in females, clothed in light brown to beige scales; anterior promontory without a patch of pale scales, with brown setae; upper proepisternum bare; vein R_3 incomplete, proximal part absent; radiomedial crossvein distal to the origin of vein M_{3+4}; proximal 0.67 of hindfemur and apical 0.20 of hindtibia pale-scaled. MALE GENITALIA: Gonocoxite without scales, basotergal area with well-developed projection or small rounded lobe with apical setae; aedeagus not differentiated, phallosome simple, bottle-like or with three apical lobes; claspette absent. LARVAE: Seta 3-C minute, shorter than 0.10 the length of seta 2-C, single or forked; setae 5–7-C multi-branched, shorter than 0.50 the length of head capsule. PUPAE: Trumpet without notch at apical margin; seta 13-CT sometimes present, its alveolus present if seta absent; seta 9-II,III short, seta 9-IV–VII long, progressively longer from segment IV to segment VII; paddle with fringe of spicules on outer and inner margins.

****

Bionomics: The immature stages of species of *Neobironella* are mainly found among debris or vegetation along the shaded margins of streams, rivers and ponds. They have also been found in pools, rock pools and pot holes associated with streams. Larvae of *Bi. confusa* were once found in a tin can near a steam and once in a tree hole (Tenorio, 1977). Females of *Bi. confusa* have been collected in Malaise traps, but nothing is known about the behaviour and bionomics of the adults. Species of *Neobironella* are not known to transmit pathogens of human diseases.

****

Principal references: Tenorio, 1977 (revision, keys, taxonomy, descriptions, bionomics, distributions); Lee *et al.*, 1988a (taxonomy, keys, literature, distributions, bionomics).

Chagasia Cruz, 1906

Chagasia Cruz, 1906, originally proposed and currently recognized as a genus of Anophelinae, is typified by *Chagasia neivae* Cruz, 1906 [subjective synonym of *Chagasia fajardi* (Lutz, 1904, in Bourroul, 1904)]. Genus abbreviation – *Ch.*

****

Systematics: *Chagasia* is a subordinate taxon of the subfamily Anophelinae. It includes only five extant species. Four species of *Chagasia* are restricted to South America and one (*Ch. bathana*) extends from Ecuador, Colombia and Venezuela through Central America into southern Mexico. *Chagasia* has been considered an ancient group showing affinities with non-anophelines. Cladistic analyses of morphological data and DNA sequences of various ribosomal, mitochondrial and nuclear genes strongly support the monophyly of *Chagasia* and its placement in an ancestral relationship to all other anophelines (Harbach & Kitching, 1998; Sallum *et al.*, 2000, 2002; Krzywinski *et al.*, 2001a, 2001b; Harbach & Kitching, 2005, 2016; Foster *et al.*, 2017; Lorenz *et al.*, 2021).

****

Characteristics: The adults of *Chagasia* are similar to those of *Anopheles*, but the resting posture is like culicine mosquitoes with the head and abdomen at angles to the thorax, and the scutellum is trilobed with setae in three distinct groups. The wings have dark scales or a mixture of dark and pale scales, but there are no distinct spots as there are in most *Anopheles*. Eggs have a circumferential covering of longitudinal floats and the micropylar apparatus is borne dorsally at the anterior end. Larvae have three pairs of exceptionally long, forward-projecting broom-like setae (setae 2,4,6-C) at the dorsoanterior margin of the head, uniquely shaped palmate setae (seta 1) on abdominal segments III–V, and the spiracular apparatus bears a long median process and a fringe-like row of setae on either side. In addition to the strong lateral spine-like seta 9-V–VIII that is usually present in anopheline pupae, seta 2-III–VII of *Chagasia* pupae is developed as a strong dorsal spine, and the paddle seta 1-Pa is also stout and spine-like.

****

Bionomics: *Chagasia* larvae are usually found in shaded streams among the roots of trees and in grassy margins or dead leaves and other debris. They sometimes occur in clear rock pools along shaded streams. Adults remain in vegetation near the larval habitats or enter nearby forest canopy. Females bite during the day and night, but seldom feed on humans. Species of *Chagasia* are not known to transmit any pathogens of human diseases.

****

Principal references: Lane, 1953 (taxonomy, keys, descriptions, distributions); Harbach & Howard, 2009 (review, descriptions, distributions, literature summaries, keys); Foster *et al.*, 2017 (mitochondrial protein sequences, phylogenetic relationships); Lorenz *et al.*, 2021 (mitochondrial protein-coding genes, phylogenetic relationships).

Priscoculex Poinar, Zavortink & Brown, 2019

Priscoculex Poinar *et al.*, 2019, originally proposed and currently recognized as a genus of Anophelinae, is typified by *Priscoculex burmanicus* Poinar, Zavortink & Brown, 2019. Genus abbreviation – *Pr.*

****

Systematics: *Priscoculex* is a subordinate taxon of the subfamily Anophelinae. It has no living species. It only includes the extinct type species, which existed 98.79 ± 0.62 Mya during the Cenomanian Age of the Early Upper Cretaceous of the Mesozoic Era. The species is only known from a female preserved in Burmese amber collected in the Hukawng Valley in Kachin State of Myanmar. The evolutionary relationships of *Priscoculex* have not been investigated. Poinar *et al.* (2019) proposed that *Priscoculex* is a primitive lineage of Anophelinae which evolved as the result of an earlier diversification that occurred in the Lower Cretaceous some 130–140 Mya after Gondwana had separated from Laurasia but had not yet split into continents.

****

Characteristics: *Priscoculex* mainly exhibit anopheline features, but some traits are not typical of extant anophelines and are regarded as ancestral characters. ADULT FEMALES: Vertex of the head with scales; antenna longer than the proboscis, flagellum with 13 flagellomeres (as in extant mosquitoes); maxillary palpus also longer than the proboscis, with 5 palpomeres and long setae, palpomere 3 with a thick-walled scalloped and apically pointed sensillum chaeticum projecting from a socket; proboscis straight, subequal in length to the maxillary palpus, setaceous, with long setae; prespiracular setae absent; scutellum with a rounded margin; wing with 10 veins reaching the margin, veins with scales and setae, costa continues around the apex of the wing but is weaker along the posterior margin, subcosta intersects the costa beyond mid-length of the wing; tarsi scaled; pulvilli absent; abdomen truncate, without scales, cercus oval.

****

Bionomics: The climate during the Cretaceous Period was warm and humid. Dinosaurs were the dominant land animals, but many modern groups of animals, including placental mammals, evolved during the Period. The Crocodylia, the true crocodiles, first appeared during the Cenomanian Age when the type female of *Pr. burmanicus* was trapped and fossilized in tree resin. The presence of stylets in the proboscis of *Pr. burmanicus*, as in modern anophelines, indicates that females probably fed on vertebrates.

****

Principal reference: Poinar *et al.*, 2019 (genus diagnosis, type species description, evolutionary notes and comments).

Part III: Genus Group of Burmaculicinae

Forty seven fossil species, excluding fossils of three extant species and a *nomen nudum*, have been described as members of Culicidae, but only 28 can be placed in the family with confidence. Twenty-two of the fossil mosquitoes are from the Cretaceous Era, including species of *Anopheles* (see above), *Coquillettidia*, *Culex*, *Toxorhynchites* and six extinct genera. The discovery of the two cretaceous species confirms the view of Edwards (1923c) that the evolution of the Culicidae must have begun in the Mesozoic Era. Burmaculicinae includes a single genus and a single species from the mid-Cretaceous (89.3–99.6 Mya) of the Mesozoic.

Burmaculex Borkent & Grimaldi, 2004

Burmaculex Borkent & Grimaldi, 2004, originally proposed and currently recognized as a genus of Burmaculicinae, is typified by *Burmaculex antiquus* Borkent & Grimaldi, 2004. Genus abbreviation – *Bu.*

****

Systematics: *Burmaculex* is monobasic. It currently only includes the extinct type species. The species is only known from a female preserved in Burmese amber collected near Tanai, Kachin Province, Myanmar. The age of the fossil is estimated to be between the Turonian and the Late Albian (89.8–113.0 Mya) of the Mesozoic Era. Phylogenetic analysis showed that *Bu. antiquus* is the sister to all other fossil and extant species of the Culicidae.

****

Characteristics: The legs of *Bu. antiquus* are covered with well-developed scales, a characteristic (synapomorphy) shared by all fossil and extant species of the Culicidae. ADULT FEMALES: Head with the eyes narrowly if at all separated above the antennal pedicels; antenna with 13 flagellomeres (as in extant mosquitoes), flagellomeres 2–11 (possibly also 12 and 13) each with a basal whorl of setae arising from a ring of pale cuticle, flagellomeres 2–13 with 4 types of setae: a subbasal whorl of very long setae, a somewhat scattered subapical circle of shorter setae, numerous pointed sensilla scattered on the surface and a few blunt-tipped sensilla, apex of flagellomere 13 bifid; clypeus with a few setae; maxillary palpus longer than the proboscis, with 5 palpomeres, palpomere 1 very short, palpomere 3 elongate, bearing numerous sensilla basiconica; proboscis short, about 0.33 the length of the antenna, expanded laterally, bent downward, labella well developed; body without scales; thorax with numerous setae on the anterior area of the scutum, at least 6 mesokatepisternal setae; wing length 1.9 mm, with slender setae on many veins, microtrichia on the membrane, scales restricted to the alula and the fringe on the posterior margin; mid- and hindlegs with numerous setae, midleg with the apices of tarsomeres 1–4 each with a pair of short spines, hindleg with the apices of tarsomeres 2–4 each with a pair of short spines; mid- and hindungues present; midleg with an empodium.

****

Bionomics: As noted by Borkent & Grimaldi (2004), the presence of a stylet-like labrum and laciniae in *Burmaculex antiquus*, which is shared with extant Culicidae, indicates the capability of biting and feeding on vertebrates.

****

Principal references: Borkent & Grimaldi, 2004 (genus and species description, comparative morphology, possible bionomics, phylogenetic position); Borkent & Grimaldi, 2016 (species description, phylogenetic position).

Part IV: Genus Groups of Culicinae

The practice established by Edwards (1932a) of recognizing few genera and many subgenera is evident in the present subgenus-rich classification of the subfamily Culicinae. The subfamily comprises 218 extant genus-group taxa (including the nominotypical subgenera of 21 genera), 38 currently recognized at the rank of genus and 180 of which are classified as subgenera. Many of the 38 genera are undoubtedly polyphyletic taxa, which in a truly phylogenetic classification would see numerous subordinate groups raised to generic status. A natural classification of the subfamily cannot be realized until all groups have been thoroughly studied and compared using both morphological and molecular data. Unfortunately, many groups are at present very poorly known and are rarely collected and investigated, and consequently the currently accepted classification is far from being based on evolutionary history and natural relationships. Ranks within the genus group represent arbitrary constructs and, in reality, do not contribute to an easier understanding of phylogenetic relationships. In this treatise, little distinction is made between the ranks of genus and subgenus. The current status of each genus-group taxon is indicated in the treatments below, but it should be borne in mind that removal of the 21 genera and retention of their nominotypical subgenera would leave 181 genus-level groups of equal rank. Culicinae also includes four extinct genus groups.

Aedeomyia Theobald, 1901

Aedeomyia Theobald, 1901a, originally proposed and currently recognized as a genus, is typified by *Aedeomyia squamipennis* (Lynch Arribálzaga, 1878), original combination: *Aedes squamipennis*. Genus abbreviation – *Ad.*

****

Systematics: *Aedeomyia* is a subordinate taxon of the subfamily Culicinae, tribe Aedeomyiini. *Aedeomyia* is the only genus of Aedeomyiini. It includes seven extant species classified in two subgenera: *Aedeomyia sensu stricto* and *Lepiothauma*. *Aedeomyia* is pantropic. Four species of *Aedeomyia* are Afrotropical, one is Australasian, one is Neotropical and one occurs in the Oriental and Australasian Regions. The phylogenetic origins of *Aedeomyia* are uncertain (Reidenbach *et al.*, 2009). Belkin (1962) surmised that *Aedeomyia* was probably an ancient and primitive group with a number of highly specialized features. The cladistic analyses of morphological data conducted by Harbach & Kitching (1998) indicated that either *Aedeomyia* alone or *Aedeomyia* + *Uranotaenia* is the basal group of the Culicinae. Interestingly, the latter relationship was recovered in the molecular analysis of mitogenomes conducted by da Silva *et al.* (2020a), which included data for only one species of each of the two genera, whereas the former (*Aedeomyia* alone) was supported by the results of the phylogenetic analysis of amino acid sequences conducted by Soghigian *et al.* (2023), which included data for two species of *Aedeomyia* and three species of *Uranotaenia*. Oddly, *Ad. squamipennis* was paired with *Toxorhynchites portoricensis* in a sister relationship to species of Sabethini in an analysis of 13 mitochondrial protein-coding genes conducted by Lorenz *et al.* (2021) whereas in a different analysis of the same molecular data (do Nascimento *et al.*, 2021) *Ad. squamipennis* was recovered in a basal relationship to species of the tribes Aedini, Culicini, Mansoniini, Sabethini and Toxorhynchitini. Evolutionary relationships within *Aedeomyia* have not been investigated.

****

Characteristics: The adults of *Aedeomyia* are small, scaly mosquitoes of very similar appearance. They are generally easily recognized by a combination of characters. All antennal flagellomeres of females and the apical two flagellomeres of males are unusually short and thick, the clypeus bears a patch of scales, prespiracular setae and postspiracular setae are absent, lower mesepimeral setae are present, the wings are densely covered with broad yellow, white and brown scales, and the mid- and hindfemora have large apical scale-tufts. Larvae are distinctive. They share a transverse occipital foramen with various sabethine genera, but the slits do not extend as far laterad in *Aedeomyia*. The antennae are greatly swollen, strongly curved and longer than the head. Seta 6-Mx of the maxilla (hypostomal sclerite) appears to be absent. They possess strongly developed setae which give them a 'hairy' appearance, the lateral thoracic setae are remarkably long, the siphon is uniformly pubescent and lacks a pecten, and the spiracular apparatus bears a pair of large hook-like setae (seta 9-S). The comb of segment VIII consists of a single row of scales on a small sclerotized plate. Pupae have a large distinctive paddle that is narrow at the base and broad distally with a deep emargination at the insertion of seta 1-Pa, which is unusually long and simple or weakly aciculate.

****

Bionomics: *Aedeomyia* larvae are found primarily in dense swamps and ground pools with abundant aquatic vegetation, but they have also been collected from river margins. They are known to stay submerged for long periods of time, and are suspected of being capable of cuticular respiration through their greatly enlarged antennae. Females of *Aedeomyia* apparently feed predominantly on birds. Females of *Ad. africana* have been collected from human bait stationed on platforms in forest canopy, but members of *Aedeomyia* normally are not attracted to humans. A few arboviruses and avian malarial protozoa have been isolated from species of *Aedeomyia*, but none of the species are considered to be of medical importance to humans.

****

Principal references: Lane, 1953 (Neotropical Region, taxonomy, description, distribution); Belkin, 1962 (South Pacific, taxonomy, description, distribution); Tyson, 1970a (revision, taxonomy, description, distribution, bionomics, keys); Lee *et al.*, 1982 (Australasian Region, taxonomy, keys, literature, distributions, bionomics); Service, 1990 (Afrotropical Region, descriptions, bionomics, keys); Rattanarithikul *et al.*, 2006a (Thailand, keys); Brunhes *et al.*, 2011 (Afrotropical and Malagasy Regions, taxonomy, descriptions, keys, bionomics).

Aedeomyia Theobald, 1901 *sensu stricto*

Aedeomyia sensu stricto, the nominotypical subgenus of *Aedeomyia*, is typified by *Aedeomyia squamipennis* (Lynch Arribálzaga, 1878), original combination: *Aedes squamipennis*. Subgenus abbreviation – *Ady.*

****

Systematics: *Aedeomyia sensu stricto* includes six extant species. One species is found in Australia, three are found in the Afrotropical Region, one is Neotropical and one occurs in the Oriental and Australasian Regions. The evolutionary relationships of the species of *Aedeomyia sensu stricto* have not been investigated.

****

Characteristics: Species of *Aedeomyia sensu stricto* bear the following distinctive features. ADULTS: Hindtarsus with decumbent scales, without semi-erect clumps of scales. Males: last flagellomere without scales; foreungues simple. LARVAE: Mesal surface of the antenna without filamentous spicules (except *Ad. catasticta*); branches of setae 1,2,5,9,13-I–IV aciculate (weakly so or simple in *Ad. venustipes*), apex of most branches simple (*Ad. africana*, *Ad. pauliani*, *Ad. squamipennis*, *Ad. venustipes*) or fimbriated (*Ad. catasticta*). PUPAE: Seta 9-II–VI (and sometimes seta 9-VII,VIII) more or less spine-like, single or forked.

****

Bionomics: Larvae of species of *Aedeomyia sensu stricto* are nearly always found in association with aquatic vegetation, especially *Pistia*, *Salvinia*, *Eichhornia* and *Potamogeton*. Little is known about the feeding habits of females, but they seem to feed at night, mainly on birds above ground level. Females of *Ad. africana* have been collected from human bait on the ground and stationed on platforms in forest canopy. Females of *Ad. catasticta* commonly bite humans in Thailand, but species of *Aedeomyia* are not considered to be of medical importance to humans.

****

Principal references: Lane, 1953 (Neotropical Region, *Ad. squamipennis*, taxonomy, description, distribution); Belkin, 1962 (South Pacific, *Ad. catasticta*, taxonomy, description, bionomics, distribution); Tyson, 1970a (Southeast Asia, *Ad. catasticta*, *Ad. venustipes*, taxonomy, descriptions, distributions); Lee *et al.*, 1982 (Australasian Region, *Ad. catasticta*, *Ad. squamipennis*, *Ad. venustipes*, literature, distributions, bionomics); Brunhes *et al.*, 2011 (Afrotropical and Malagasy Regions, *Ad. africana*, *Ad. madagascarica*, *Ad. pauliani*, taxonomy, descriptions, keys).

Lepiothauma Enderlein, 1923

Lepiothauma, originally proposed as a genus by Enderlein (1923), currently classified as a subgenus of *Aedeomyia*, is typified by *Aedeomyia furfurea* (Enderlein, 1923), original combination: *Lepiothauma furfurea*. Subgenus abbreviation – *Lpi.*

****

Systematics: *Lepiothauma* is monobasic. It only includes the extant type species. The species is widely distributed in the Afrotropical Region, including Madagascar. The evolutionary relationships of *Lepiothauma* have not been investigated.

****

Characteristics: The only species of *Lepiothauma* is distinguished as follows. ADULTS: Hindtarsomeres 1–4 with clumps of semi-erect black scales; males with few white scales on proximal (first) flagellomere and one foreunguis with an apically rounded basal tooth. LARVAE: Mesal surface of the antenna with numerous filamentous spicules; branches of setae 1,2,5,9,13-I–IV aciculate, apex of branches strongly brush-tipped. PUPAE: Seta 9-II–VIII strongly spiniform, broad and heavily sclerotized.

****

Bionomics: Larvae mainly inhabit swamps, ponds and borrow pits with aquatic plants, usually *Pistia* or *Potamogeton*. They are sometimes found in the margins and backwaters of rivers with floating vegetation, and have been found in rice fields in Madagascar. Females probably feed predominantly on birds, but they are attracted to humans at ground level and especially in forest canopy. They reportedly feed mainly during the early hours of the night. Females of *Ad. furfurea* may approach and land on humans, but the species is not considered to be of medical importance.

****

Principal references: Edwards, 1941 (*Ad. furfurea*, taxonomy, adults, distribution); Hopkins, 1952 (*Ad. furfurea*, larva, bionomics); Tyson, 1970a (*Ad. furfurea*, key characters of adults, larva, pupa); da Cunha Ramos & Ribeiro, 1975 (*Ad. furfurea*, Angola, taxonomic notes, bionomics); Service, 1990 (*Ad. furfurea*, adults, larva, bionomics, distribution); Brunhes *et al.*, 2011 (*Ad. furfurea*, taxonomy, description).

Aedes Meigen, 1818

Aedes Meigen, 1818, originally proposed and currently recognized as a genus in the traditional classification of Culicidae, is typified by *Aedes cinereus* Wiedemann, 1818 (in Meigen, 1818). [*Aedes cinereus* is traditionally credited to Meigen (1818) but the heading '*Aedes cinereus* Hoffmgg,' and the description are both in quotation marks, followed by a statement which clearly shows that Meigen did not know or describe the species. Wiedemann apparently described the species and used a Hoffmansegg manuscript name (Sabrosky, 1999).] Genus abbreviation – *Ae.*

****

Systematics: *Aedes* is a subordinate taxon of the subfamily Culicinae, tribe Aedini. The traditional *Aedes*, *sensu* Wilkerson *et al.* (2015), is a very large genus with 968 extant species, 28 which are not placed in subgenera and 930 which are divided among 80 subgenera (all but seven were recognized as genera in the classification of Aedini established by Reinert *et al.*, 2009). *Aedes* also includes two fossil species currently placed in *Ochlerotatus*. *Aedes* has a worldwide distribution, but it is not so well represented in the Neotropical Region. Most of the subgenera occur in the Oriental and Australasian Regions. Little can be said about the phylogenetic relationships of *Aedes* because it is a highly polyphyletic group. In the phylogeny of mosquito genera generated in the study of Harbach & Kitching (1998), *Aedes* was recovered in an unresolved assemblage involving *Eretmapodites* and a clade comprised of *Opifex* in a basal relationship to *Psorophora* + Mansoniini. Soghigian *et al.* (2017) included two or more species for each of 26 subgenera of *Aedes* in their molecular phylogenetic study of Aedini: *Acartomyia*, *Aedes sensu stricto*, *Aedimorphus*, *Bothaella*, *Bruceharrisonius*, *Catageiomyia*, *Collessius*, *Diceromyia*, *Dobrotworskyius*, *Downsiomyia*, *Georgecraigius*, *Halaedes*, *Howardina*, *Hulecoeteomyia*, *Jarnellius*, *Lorrainea*, *Macleaya*, *Mucidus*, *Neomelaniconion*, *Ochlerotatus*, *Paraedes*, *Phagomyia*, *Rhinoskusea*, *Rusticoidus*, *Stegomyia* and *Tanakaius*. All but six of these genus-group taxa were recovered as monophyletic. *Aedimorphus*, *Catageiomyia* and *Neomelaniconion* in Clade A (corresponding to genus *Aedes* of Reinert, 2000d) and *Collessius*, *Ochlerotatus* and *Phagomyia* in Clade B (corresponding to genus *Ochlerotatus* of Reinert, 2000d) were not monophyletic. However, the monophyly of *Collessius*, *Neomelaniconion* and *Phagomyia* was violated by species with only a single genetic marker. The molecular phylogenies of Muñoz-Gamba *et al.* (2021), based on mitochondrial *COI* gene sequences, Soghigian *et al.* (2023), based on amino acid sequences, and Chen *et al.* (2024), based of mitochondrial genome sequences, also indicated that *Aedes* is a polyphyletic group.

****

Characteristics: *Aedes* are often difficult to distinguish from other aedine genera because of the great diversity of subgenera and species which are defined by various combinations of characters that overlap with other members of the tribe. The spectrum of polymorphism that occurs in the tribe is also found in the traditional concept of *Aedes*. Adults are usually differentiated from those of other aedine genera by the following combination of characters: antepronota usually rather small and widely separated, prespiracular setae absent, postspiracular setae nearly always present, mesopostnotal setae usually absent, base of hindcoxa usually below base of mesomeron, base of subcosta without ventral patch of setae, and apex (tip) of the abdomen of females usually, but not always, more or less pointed. Larvae are difficult to distinguish from those of *Haemagogus* in the New World and *Heizmannia*, *Udaya* and *Zeugnomyia* in the Oriental Region. The treatments of these genera should be consulted for features that distinguish them from *Aedes*. The following combination of characters is usually used to recognize *Aedes* larvae: metathoracic pleural group of setae (setae 9–12-M) well developed, a single pair of seta 1-S inserted near or beyond the middle of the siphon (species of *Aedes sensu stricto* and *Rusticoidus* have more than one pair of seta 1-S) and pecten apparently always present (represented by one or more spines on each side of the siphon).

****

Bionomics: *Aedes* is the dominant group of mosquitoes from the tropics to the arctic. Some species also occur at high altitudes. The immature stages of *Aedes* develop exclusively in temporary collections of water, including ground pools, snow pools, rock holes, tree holes, bamboo, leaf axils, artificial containers, and brackish water swamps and pools. Females of all species lay eggs on surfaces subject to being flooded or inundated by water. Those that oviposit in cavities such as tree holes, leaf axils or domestic containers attach their eggs to the walls of the cavity just above the water. Eggs have a capacity for resisting desiccation for months or a year or longer. Mating of all species probably takes place in the immediate vicinity of the larval habitats. Females of many species readily attack humans, some during the daylight hours and others at night. *Aedes* includes species that transmit yellow fever virus, dengue viruses and other arboviruses. A few species also transmit the helminths that cause Brugian and Bancroftian filariasis. *Stegomyia* is the most important subgenus of *Aedes* from a medical point of view. It includes *Ae. aegypti*, the classic vector of urban yellow fever and dengue fever viruses. *Aedes aegypti* is widely distributed through the tropical and subtropical areas of the world. Other important vectors include *Ae. africanus* and *Ae. bromeliae*, which transmit yellow fever virus in Africa, and *Ae. albopictus*, an important vector of dengue fever virus in the Oriental Region. The latter species has been introduced into the USA and countries in Central and South America, Europe and Africa. Other important vectors of dengue fever virus include *Ae. scutellaris*, *Ae. pseudoscutellaris* and *Ae. polynesiensis* in the Pacific Islands. The latter two species also transmit the helminths which cause Bancroftian filariasis. *Aedes vexans* of *Aedimorphus*, which has an extensive distribution in tropical Africa, Central America, Southeast Asia and temperate regions of the Nearctic and Palaearctic Regions, is a vector of *Dirofilaria immitis* (dog heartworm) and Tahyna virus, which affects humans in Europe. The females of many species of *Aedes* are annoying and persistent biters which can be involved in the transmission of arboviruses. A few species of *Finlaya*, including *Ae. poicilius*, *Ae. togoi*, *Ae. kochi*, *Ae. niveus* and *Ae. harinasutai*, are important vectors of arboviruses and microfilariae in the Oriental Region. Two species of *Diceromyia*, *Ae. taylori* and *Ae. furcifer*, are implicated in the transmission of yellow fever virus and the spread of Chikungunya virus in Africa. A few medically important species belong to *Ochlerotatus*. *Aedes vigilax* transmits the agents of Bancroftian filariasis in New Caledonia and several viruses (*e.g.* Ross River virus) in Australia. *Aedes sierrensis* and *Ae. atlanticus* are important vectors of dog heartworm in North America. A few other species transmit various arboviruses, mainly those that cause encephalitis. *Aedes triseriatus* of *Protomacleaya* is an important vector of the California group of encephalitis viruses. See table 1 of Wilkerson *et al.* (2015) for a detailed list of medically important species.

****

Principal references: Barraud, 1934 (southern Asia, keys, taxonomy, descriptions, distributions, bionomics); Lane, 1953 (Neotropical Region, taxonomy, keys, descriptions, distributions); Muspratt, 1956a (South Africa, *Stegomyia*, keys, descriptions, distributions, bionomics); Belkin, 1962 (South Pacific, taxonomy, keys, descriptions, distributions, bionomics); Dobrotworsky, 1965 (Australia, classification, keys, descriptions, bionomics, distributions); DuBose & Curtin, 1965 (Mediterranean area, keys); Forattini, 1965a (Neotropical Region, diagnosis, classification, species, bionomics, distributions); Cova-Garcia *et al.*, 1966 (Venezuela, keys, species descriptions, distributions); Belkin, 1968a (New Zealand, taxonomy, keys, descriptions, bionomics); Huang, 1968b (Papuan Subregion, *Huaedes*, key to subgenera); Berlin, 1969a (*Howardina*, classification, keys, taxonomy, descriptions, bionomics, distributions); Belkin *et al.*, 1970 (Jamaica, taxonomy, keys, descriptions, bionomics); Reinert, 1970a, 1972a, 1973b, 1974, 1976b, 1976c, 1981b, 1982, 1984, 1985, 1987, 1990b, 1993, 1999a, 1999b, 2000a, 2000c, 2003 (various subgenera, various geographic regions, taxonomy, keys, descriptions, distributions, bionomics); Schick, 1970a (former *Ae. terrens* group in the Neotropical Region, see Zavortink, 1972); Tyson, 1970b (*Mucidus*, classification, taxonomy, keys, distributions, bionomics); McIntosh, 1971 (*Neomelaniconion*, southern Africa, keys, descriptions, distributions, bionomics); Huang, 1972 (*Ae. scutellaris* group, Southeast Asia, keys, descriptions, distributions, bionomics); Zavortink, 1972 (various subgenera, New World, classification, keys, taxonomy, descriptions, bionomics, distributions); Gutsevich *et al.*, 1974 (former USSR, taxonomy, keys, descriptions, distributions, bionomics); McIntosh, 1975 (*Aedimorphus*, southern Africa, classification, keys, taxonomy, descriptions, distributions, bionomics); Arnell, 1976 (*Ae. scapularis* group, *Ochlerotatus*, New World, classification, keys, taxonomy, descriptions, bionomics, distributions); Abercrombie, 1977 (*Christophersiomyia*, taxonomy, keys, descriptions, distributions, bionomics); Huang, 1977a, 1979a (*Stegomyia*, Oriental Region, classification, keys, taxonomy, descriptions, distributions, bionomics); Tanaka *et al.*, 1979 (Japan, taxonomy, keys, descriptions, distributions, bionomics); Wood *et al.*, 1979 (Canada, taxonomy, keys, descriptions, distributions, bionomics); Darsie & Ward, 1981, 2005 (North America, keys); Lee *et al.*, 1982, 1984, 1987a (Australasian Region, taxonomy, keys, literature, distributions, bionomics); Lu & Li, 1982 (China, keys); Clark-Gil & Darsie, 1983 (Guatemala, keys, distributions, bionomics); Darsie, 1985 (Argentina, keys); Darsie & Pradhan, 1990 (Nepal, keys, distributions, bionomics); Huang, 1990 (*Stegomyia*, *Ae. africanus* group, Afrotropical Region, keys, taxonomy, descriptions, distributions, bionomics); Rattanarithikul & Panthusiri, 1994 (Thailand, keys, medically important species); Jupp, 1996 (southern Africa, subgenus descriptions, keys to species); Lu *et al.*, 1997a (China, keys, taxonomy, descriptions, distributions, bionomics); Reinert, 2000e, 2002e (female genitalia); Huang, 2004 (*Stegomyia*, Afrotropical Region, classification, species-group diagnoses, keys, new species descriptions, distributions, bionomics); Becker *et al.*, 2003, 2010, 2020 (Europe, keys, descriptions, distributions, bionomics); Huang *et al.*, 2010 (*Coetzeemyia*, delineation, phylogenetic relationships); Huang *et al.*, 2012 (*Levua*, taxonomy); Wilkerson *et al.*,

2015 (phylogenetic relationships, classification); Huang & Rueda, 2015b (*Pseudalbuginosus*, taxonomy, morphology, affinities); Mohamed *et al.*, 2017 (Republic of the Sudan, bionomics, keys); Brunhes *et al.*, 2017 (*Paulianius*, as a genus, taxonomy, bionomics, keys); Soghigian *et al.*, 2017 (phylogenetic relationships); Muñoz-Gamba *et al.*, 2021 (mtDNA *COI* sequences, phylogenetic relationships); Russell & Burkot, 2023 (South Pacific, country lists of species, bionomics, vector status).

Abraedes Zavortink, 1970

Abraedes Zavortink, 1970c, originally proposed and currently recognized as a subgenus of *Aedes*, is typified by *Aedes papago* Zavortink, 1970c. Subgenus abbreviation – *Abr.*

****

Systematics: *Abraedes* is monobasic. It only includes the extant type species, which occurs in the southwestern area of the USA. *Abraedes* was recovered as the sister of *Lewnielsenius*, and this pair was sister to *Aztecaedes*, in the phylogenetic analysis of Aedini conducted by Reinert *et al.* (2009). *Abraedes* was sister to *Gymnometopa* in the phylogeny of Wilkerson *et al.* (2015).

****

Characteristics: The adults of *Ae. papago* are highly ornamented and bear the following distinctive features: integument of the head and thorax dark brown to black, strongly contrasting with yellowish to amber integument of the abdomen; compound eyes broadly separated above the antennae, interocular space with broad flat silvery scales; vertex with few erect scales; decumbent scales broad and flat; proboscis conspicuously longer than the forefemur; maxillary palpus short in females, less than 0.3 the length of the proboscis, apparently comprised of 4 palpomeres, long in males, with 5 palpomeres, palpomeres 4 and 5 bent downward; scutum of the thorax with long rows of acrostichal and dorsocentral setae, narrow dark scales and conspicuous lines of white and silvery scales; scutellum with patches of large broad flat silvery scales; paratergite with broad flat silvery scales on lower anterior area; pleura with broad flat silvery scales in small dense patches or lines; few setae present on mesokatepisternum, lower mesepimeral setae absent; legs disproportionately short; tarsi marked with white bands or rings; ungues simple in females, anterior unguis of the fore- and midlegs larger and toothed in males; wing with plume scales on dorsum of veins R_s, R_{2+3}, R_2, R_3, middle of M and sometimes apical portion of 1A; caudal terga and sterna with dense lateral patches of outstanding black and iridescent silvery scales. The genitalia of males have tergum IX poorly developed, without lobes, with 1 or 2 fine setae on each side; gonocoxite without lobes and specialized scales; claspette with long simple filament; proctiger unusually long, paraproct without basolateral lobe. Larvae have normal palatal brushes; setae 5,7-C inserted about level of antennal base; setae 5,6,7-P on separate tubercles; seta 2-III–V usually inserted anteromesal to seta 1; seta 3-VII relatively short, not reaching base of the siphon; seta 12-I absent; seta 13-VI relatively long, usually double or triple, inserted laterad and usually caudad of seta 10-VI; comb with 4 or 5 large spine-like scales; siphon short, without acus; saddle without spines on caudal margin; ventral brush (seta 4-X) weakly developed, with 5 pairs of setae inserted on boss, seta 4a-X short, multi-branched. Pupae are distinctive in having the trumpet short, tracheoid virtually absent; cephalothoracic setae short; seta 1-II farther from midline than seta 1-I; seta 2-I–VII relatively long and strong, usually inserted anteromesal to seta 1 on segments III–V; seta 5-IV–VI shorter than corresponding tergum; seta 9-II–VI relatively long and strong, becoming longer on the posterior segments, usually inserted anterior to the insertion of seta 6-II–IV, anterior or posterior to the insertion of seta 6-V, posterior to the insertion of seta 6-VI; seta 9-VIII relatively far mesad of caudolateral angle of the segment; paddle deeply emarginate, outer margin distad of external buttress, inner margin with long filamentous spicules.

****

Bionomics: Larvae of the type species have been reared from eggs collected in a rot hole in an oak tree, but this may not be the normal habitat of the immature stages. Females have been captured while attacking and attempting to bite humans. The species is unlikely to be of medical or economic importance to humans.

****

Principal references: Zavortink, 1970c (new subgenus, new species, descriptions, bionomics, distribution); Zavortink, 1972 (subgenus and species, taxonomy); Reinert *et al.*, 2004, 2006, 2009 (as a genus, morphology, phylogenetic relationships); Wilkerson *et al.*, 2015 (phylogenetic relationships, classification).

Acartomyia Theobald, 1903

Acartomyia, originally proposed as a genus by Theobald (1903a), currently recognized as a subgenus of *Aedes*, is typified by *Aedes zammitii* (Theobald, 1903a), original combination: *Acartomyia zammitii*. Subgenus abbreviation – *Acy.*

****

Systematics: *Acartomyia* includes only three extant species that comprise the Mariae Complex (Coluzzi *et al.*, 1976). The species inhabit coastal areas of the Mediterranean Region. The phylogenetic affinities of *Acartomyia* are uncertain. In the phylogenetic study of Reinert *et al.* (2009) based on morphological data, *Acartomyia* was recovered as the basal member of a clade comprised of *Acartomyia* + (*Jarnellius* + (*Halaedes* + *Opifex*)), and this clade was sister to *Ochlerotatus*. It was recovered as sister to *Halaedes* + *Opifex* in the phylogenetic analysis of Wilkerson *et al.* (2015) and as a monophyletic group in a sister relationship to species of *Ochlerotatus* in the molecular phylogeny of Soghigian *et al.* (2017).

****

Characteristics: The following combinations of characters distinguish species of *Acartomyia* from other genus groups of Aedini in the Palaearctic Region. ADULTS: Vertex of head with both broad and narrow decumbent scales; erect scales on occiput and vertex; antennal pedicel bare; maxillary palpus of females dark-scaled, with 4 palpomeres, 4th vestigial, maxillary palpus of males about as long as the proboscis, apical palpomeres swollen; proboscis with pale scales or band near mid-length; scutum with diffuse pattern of narrow golden and brown scales; scutellum with narrow curved scales on all lobes; anterior and posterior acrostichal setae and anterior and posterior dorsocentral setae present; paratergal and lower proepisternal scales present; scale-patches present on hypostigmal area, subspiracular area, postspiracular area, prealar area (upper and lower) and postprocoxal membrane; wing with a mixture of dark and pale scales, remigium without dorsal setae; tarsi with conspicuous basal and apical pale bands, hindtarsomere 4 pale-scaled; ungues of females equal, both foreungues toothed, fore- and midungues of males unequal, larger foreunguis with 2 teeth, smaller foreunguis and both midungues each with 1 tooth, hindungues equal and simple in both sexes. FEMALE GENITALIA: Intersegmental membrane long between segments VII and VIII; cercus without scales; upper vaginal sclerite absent; insula tongue-like, setae in lateral patches. MALE GENITALIA: Sternum IX with setae; gonocoxite without scales, dorsomesal apical lobe absent, dorsomesal basal lobe weakly developed, basomesal area of dorsal surface with several stout setae, mesal surface entirely membranous; gonostylus borne at apex of the gonocoxite, simple; gonostylar claw spiniform, inserted at apex of the gonostylus; claspette comprised of single stem with apical claspette filament, filament nearly as long as the stem, narrow, slightly curved and flattened; aedeagus tube-like. LARVAE: Antenna shorter than the head, with numerous spicules; setae 5,6-C single, seta 4-C inserted posterior to seta 6-C; seta 6-C inserted anterior to seta 7-C; seta 19-C present; seta 12-I present; comb with 15–45 scales in 2 or 3 irregular rows; siphon short, siphon index ≤ 2.0; pecten spines closely and evenly spaced, long, simple or with proximal denticles; seta 1-S long, ≥ 0.67 width of the siphon at the point of insertion, inserted distal to the pecten; saddle weakly developed (small dorsal sclerite); seta 2-X short, fan-like, multi-branched (> 10 branches); ventral brush (seta 4-X) usually with 4–6 precratal setae, sometimes more, 11–13 cratal setae on grid with lateral and transverse grid bars; anal papillae very short, spherical. PUPAE: Trumpet with weakly developed tracheoid area; seta 2-II inserted lateral to seta 1-II; seta 3-I as long or longer than seta 6-I, seta 3-II inserted lateral to seta 2-II; seta 5-II longer than seta 3-II; seta 9-VIII inserted on posterolateral corner of the segment; paddle without fringe of hair-like spicules; seta 1-Pa single (rarely 2-branched); seta 2-Pa absent.

****

Bionomics: The immature stages of species of *Acartomyia* are found in coastal rock pools containing sea water. Adult females of *Ae. mariae* are known to feed on birds and annoy humans in coastal areas. Little else in known about the bionomics of the species, which are not known to be of medical or economic importance to humans. *Aedes mariae* is a vector of *Plasmodium relictum*, the causal agent of bird malaria.

****

Principal references: Coluzzi & Sabatini, 1968 (as *Aedes mariae* complex, *Ae. mariae*, morphology, cytogenetics, bionomics, distribution); Reinert *et al.*, 2008 (as a subgenus of *Ochlerotatus*, morphology, phylogenetic relationships); Reinert *et al.*, 2009 (as a genus, morphology, phylogenetic relationships); Becker *et al.*, 2003, 2010 (as Ochlerotatus Mariae Complex, in keys, descriptions, bionomics, distributions); Becker *et al.*, 2020 (in keys, descriptions, bionomics, distributions); Wilkerson *et al.*, 2015 (phylogenetic relationships, classification).

Aedes Meigen, 1818 *sensu stricto*

Aedes sensu stricto, the nominotypical subgenus of the genus *Aedes*, is typified by *Aedes cinereus* Wiedemann, 1818 (in Meigen, 1818). Subgenus abbreviation – *Aed.*

****

Systematics: *Aedes sensu stricto* includes 12 extant species. *Aedes cinereus* occurs in the Nearctic and Palaearctic Regions; the other 11 species are confined to areas of the Palaearctic Region. *Aedes* was recovered as the sister of *Paraedes* + *Verrallina* in a paraphyletic relationship with *Neomelaniconion* and *Edwardsaedes* in the morphology-based phylogenies of Reinert *et al.* (2009) and Wilkerson *et al.* (2015). Species of *Aedes sensu stricto* comprised a monophyletic group in a sister relationship to species of *Paraedes* in the molecular phylogeny of Soghigian *et al.* (2017).

****

Characteristics: Species of *Aedes sensu stricto* exhibit the following distinctive features. ADULTS: Vertex of head with narrow curved scales, occiput with broad flat scales; maxillary palpus very short in both sexes, males with 3 palpomeres; scutum concolourous, paratergite with scales, lateral lobes of scutellum with narrow scales; scale-patches of thoracic pleura weakly developed, postspiracular scales present, subspiracular area bare, mesepimeral setae absent; both hindungues toothed in females, simple in males; laterotergite with scales, abdominal terga of males with numerous long apically curved setae; females with distal 0.7 or more of tergum VIII covered with numerous setae. MALE GENITALIA: Phallosome divided into a pair of plates or rods; gonocoxite short and stout, gonostylus bifurcate and attached subapically, gonostylar claw absent; claspette stem and filament absent. LARVAE: Siphon (as in species of *Rusticoidus*) with accessory anterolateral seta 2a-S in addition to seta 1-S.

****

Bionomics: Species of *Aedes sensu stricto* appear to be univoltine, over wintering in the egg stage and hatching in spring in pools formed by melted snow. One species, *Ae. yamadai*, is reported to hibernate as adults. The larvae are found in woodland ground pools, ditches, ponds, stream margins, sunlit temporary pools and occasionally in swamps and marshes. Females sometimes bite humans but species of *Aedes sensu stricto* are not medically important.

****

Principal references: Lu *et al.*, 1997a (China, keys, taxonomy, descriptions, distributions, bionomics); Reinert, 2000e (female genitalia); Becker *et al.*, 2003, 2010, 2020 (Europe, in keys, descriptions, bionomics, distributions); Gornostaeva, 2003 (principally Russia, keys, male genitalia, distributions); Gornostaeva, 2005 (Palaearctic Region, male genitalia, distributions); Reinert *et al.*, 2004, 2006, 2008, 2009 (as a genus, morphology, phylogenetic relationships); Wilkerson *et al.*, 2015 (phylogenetic relationships, classification); Soghigian *et al.*, 2017 (phylogenetic relationships).

Aedimorphus Theobald, 1903

Aedimorphus, originally proposed as a genus by Theobald (1903a), currently recognized as a subgenus of *Aedes*, is typified by *Aedes domesticus* (Theobald, 1901d), original combination: *Uranotaenia domestica*. Subgenus abbreviation – *Adm.*

****

Systematics: *Aedimorphus* includes 76 extant species. Most species occur in the Afrotropical Region, some occur in the Oriental Region, a few occur in the Australasian Region and one, *Ae. vexans*, occurs in the Holarctic and Oriental Regions, Central America, South Africa and the Papuan area. The affinities of *Aedimorphus*, as currently defined, are indefinite. It appears to be a polyphyletic assemblage of species – it was not recovered as a monophyletic group in the maximum likelihood phylogeny of Soghigian *et al.* (2017) based on seven molecular markers. A clade containing nine species of *Aedimorphus* also included two species of *Catageiomyia* and a species of *Polyleptiomyia*. A tenth species of *Aedimorphus* formed a disparate lineage with a species of *Neomelaniconion*. *Aedimorphus* was recovered as the sister of a number of subgeneric taxa in the phylogenetic analysis of Aedini by Reinert *et al.* (2009) based on morphological data recorded for 20 species of *Aedimorphus*. The sister group included subgenera divided between two clades, one comprised of *Polyleptiomyia* + (*Bifidistylus* + (*Albuginosus* + (*Tewarius* + (*Christophersiomyia* + (*Huaedes* + *Leptosomatomyia*))))) and the other comprised of *Neomelaniconion* + (*Edwardsaedes* + (*Aedes* + (*Paraedes* + *Verrallina*))). Three species of *Aedimorphus* were recovered in a paraphyletic series in a sister relationship to *Edwardsaedes* + (*Neomelaniconion* + (*Aedes* + (*Paraedes* + *Verrallina*))) in the phylogeny of Wilkerson *et al.* (2015). Thirteen nominal taxa now in synonymy for species groups within *Aedimorphus* may require subgeneric recognition.

****

Characteristics: Species of *Aedimorphus* are distinguished from species of other generic-level taxa of Aedini by the following combinations of characters. ADULTS: Vertex of head with narrow decumbent scales medially, erect forked scales usually numerous, on vertex and occiput; eyes usually narrowly separated above the antennal pedicels; antenna of females long, normally as long or longer than the proboscis, shorter than the proboscis in males, with numerous flagellar setae directed mainly dorsally and ventrally; maxillary palpus of females short, about 0.15–0.30 the length of the proboscis, with 5 palpomeres, palpomere 5 vestigial if present; maxillary palpus of males usually longer than the proboscis, with 5 palpomeres, palpomeres 4 and 5 usually thickened, setose and turned downward, rarely slender, bare and turned upward; proboscis usually distinctly longer than the forefemur; scutum with narrow scales; acrostichal setae and dorsocentral setae well developed, numerous; scutellum with narrow, broad or both narrow and broad scales; paratergite with or without scales; antepronotum and postpronotum usually with narrow scales; upper proepisternum with numerous setae; pleural scaling varied, postspiracular area, subspiracular area and prealar area with or without scales; lower mesepimeral setae absent; hindtarsomeres entirely dark-scaled or with basal pale bands; both ungues of fore- and midlegs of females toothed, fore- and midungues of males enlarged, toothed, hindungues usually simple in both sexes but sometimes toothed; abdominal segment I usually extensively scaled, laterotergite with large scale-patch, terga of males with numerous lateral setae. FEMALE GENITALIA: Segment VIII largely or completely retracted into preceding segment; sternum VIII with deep posteromedian emargination separating broadly rounded lobes; cercus long, narrow, usually without scales; postgenital lobe with median apical indentation; insula tongue-like, with tuberculi on distal 0.25, each usually with minute seta; 3 spermathecal capsules. MALE GENITALIA: Tergum VIII sometimes widened posteriorly; tergum IX usually narrowed medially, ninth tergal lobes more or less distinct, with or without setae; sternum IX usually with postmedian row of setae; gonocoxite without distinct apical and basal tergomesal lobes, mesal surface membranous from base to apex; gonostylus usually expanded distally, with 1 or more gonostylar claws; claspette represented by short process or setose plaque at base of the gonocoxite; aedeagus comprised of 2 lateral plates, each with distolateral teeth; paraproct with blunt or rounded apex, sternal arm present; cercal setae usually absent, present in some African species. LARVAE: Extremely varied; antenna usually long, spiculate, seta 1-A strongly developed, usually multi-branched; seta 4-C small, multi-branched, inserted closer to seta 5-C than to seta 6-C; setae 5,6-C strongly developed, usually inserted close together on level or posterior to seta 7-C; thoracic and abdominal chaetotaxy not studied for most species; seta 12-I absent; seta 6-III branched; setae 1,2-VIII inserted on common setal support plate; comb of segment VIII varied, comb scales few in single row or numerous in large patch, scales squamiform or spine-like; siphon usually long, siphon index ≥ 3.0, pecten varied, distal pecten spines usually simple, always more widely spaced than proximal spines; seta 1-S multi-branched, shorter than 0.5 the diameter of siphon, inserted distal to pecten; saddle incomplete, acus present or absent; seta 1-X often single, single to 4-branched; seta 2-X usually multi-branched; ventral brush (seta 4-X) with 3 or 4 precratal setae and 8–14 setae on grid with lateral and transverse grid bars. PUPAE: Trumpet short, only slightly widened distally, tracheoid area present or absent, weakly developed if present; seta 8-CT inserted cephalad to or at level of trumpet base; setae 10,11-CT approximated, 10-CT multi-branched, 11-CT single; genital lobe of females usually prominent; paddle ovoid, marginal spicules usually small or inconspicuous; seta 1-Pa short, single to triple.

****

Bionomics: The immature stages of species of *Aedimorphus* principally inhabit temporary and semi-permanent freshwater ground pools, but have been found in a variety of habitats, including swamps, marshes, bogs, pits, wells, ground pools, puddles, flood pools, stream pools, seepage, rice fields, animal footprints, wheel tracks, rock pools, rock holes and artificial containers. Females of many species readily attack and feed on humans, and are sometimes serious pests. Females of other species prefer non-human hosts and only occasionally feed on humans. Some species feed in the daytime, chiefly outdoors toward dusk, and some feed mainly during the night. A number of species have been implicated in the transmission of pathogens that cause diseases in humans and domestic animals. *Aedes vexans* has been found naturally infected with a number of arboviruses, including Eastern and Western equine encephalitis viruses, viruses of the California encephalitis group (see details and references in Reinert, 1973c) and Japanese encephalitis virus (Burke & Leake, 1988). This species is involved in the transmission of Tahyna virus in Europe (see references in Becker *et al.*, 2003, 2010, 2020), it is a vector of *Dirofilaria immitis* (dog heartworm) (Reinert, 1973c) and is considered to be a potentially important bridge vector of West Nile virus (Turell *et al.*, 2005). Other pathogens associated with various species of *Aedimorphus* include Batai, Bunyamwera, Middelburg, Rift Valley fever, Sindbis, Shokwe, Spondweni and Wesselsbron viruses, and microfilariae of *Brugia malayi*.

****

Principal references: Reinert, 1973c (Southeast Asia, taxonomy, keys, descriptions, distributions, bionomics); McIntosh, 1975 (southern Africa, classification, keys, taxonomy, descriptions, distributions, bionomics); Lee *et al.*, 1982 (Australasian Region, taxonomy, keys, literature, distributions, bionomics); Lu *et al.*, 1997a (China, keys, taxonomy, descriptions, distributions, bionomics); Reinert, 2000e (female genitalia); Reinert *et al.*, 2004 (morphology, phylogenetic relationships); Reinert *et al.*, 2009 (as a genus, morphology, phylogenetic relationships); Rattanarithikul *et al.*, 2010 (as a genus, Thailand, keys, bionomics); Wilkerson *et al.*, 2015 (phylogenetic relationships, classification).

Alanstonea Mattingly, 1960

Alanstonea Mattingly, 1960, originally proposed and currently recognized as a subgenus of *Aedes*, is typified by *Aedes treubi* (de Meijere, 1910), original combination: *Scutomyia treubi*. Subgenus abbreviation – *Ala.*

****

Systematics: *Alanstonea* includes two extant species in Southeast Asia, recorded from Brunei, Indonesia (Java, Sumatra), Malaysia (Peninsular, Sarawak) and Thailand. The past inclusion of the species in *Armigeres* and *Stegomyia* suggested affinities with those two taxa. However, in the comprehensive cladistic study of Aedini conducted by Reinert *et al.* (2009), *Alanstonea* and the Afrotropical *Pseudarmigeres* were recovered as sister taxa in a clade that comprised *Petermattinglyius* + (*Heizmannia* + (*Alanstonea* + *Pseudarmigeres*)), and this clade in a sister relationship to a clade comprised of six genus groups: *Armigeres, Belkinius, Eretmapodites, Lorrainea, Udaya* and *Zeugnomyia*. A clade comprising *Armigeres* +

(*Alanstonea* + *Eretmapodites*) was recovered in the phylogenetic analyses of Wilkerson *et al.* (2015). Additional study is needed to resolve the affinities of *Alanstonea*.

****

Characteristics: The adults of species of *Alanstonea* generally resemble the adults of *Armigeres* in size and colouration, in having the proboscis laterally compressed and curved downward, and in the absence of dorsocentral setae. They differ from those of *Armigeres* in the absence of a lower mesepimeral seta and features of the male genitalia. Differences are more striking in the immature stages. The paddles of pupae lack a midrib and a marginal fringe, which are present in *Armigeres*. The siphon of *Alanstonea* larvae is narrow and bears a pecten and the anal papillae are narrow and pointed whereas the pecten is absent, the siphon is very broad and stumpy and the anal papillae are broad with rounded apices in *Armigeres*. The similarity of *Alanstonea* species with species of *Stegomyia* lies mainly with the male genitalia and with several larval characters. However, *Alanstonea* adults differ from *Stegomyia* adults in having a curved proboscis and legs without white bands. Larvae are distinguished from other Aedini by the combination of single setae 2- and 3-X and the absence of a grid or boss at the base of the ventral brush. Other characteristics include the following. ADULTS: Pedicel of the antenna with scales on the lateral surface; maxillary palpus of females shorter than 0.25 the length of the proboscis; maxillary palpus of males longer than the proboscis, slender, almost bare, palpomere 5 no thicker than palpomere 4; paratergite with numerous scales; mesopleuron with hypostigmal and upper and lower mesokatepisternal scales, patches more or less contiguous; upper mesokatepisternal setae absent; postspiracular area with few or no scales; mesepimeral scale-patch extended into the upper mesepimeral setae; anteprocoxal scales present; alula of the wing with broad scales, dorsal tertiary fringe scales absent; hindcoxa more or less in line with the dorsal margin of the mesomeron; fore- and midungues of males simple. FEMALE GENITALIA: Insula without setae. MALE GENITALIA: Gonocoxite with scales; gonostylus with a single apical claw; basal mesal lobe with numerous relatively short stout setae; aedeagus with the apices of the lateral plates fused. LARVAE: Setae 4–6-C inserted well anterior to seta 7-C; setae 1,2-P on a common tubercle; seta 1-I reduced, with 2 or 3 main branches and some secondary branching or feathering; seta 7-I–V small and inconspicuous; seta 1-S inserted near apex of the siphon; seta 8-S very long; ventral brush (seta 4-X) comprised of 8 or 9 setae that are all or mostly single. PUPAE: Trumpet without tracheoid area; seta 6-I shorter than seta 7-I; seta 9-VII,VIII very long; paddle very small, shorter than seta 9-VIII, subtriangular; seta 1-Pa well developed.

****

Bionomics: The immature stages of species of *Alanstonea* are found in *Nepenthes* pitcher plants. The larvae are carnivorous and feed on other mosquito larvae (Mogi & Yong, 1992; Mogi & Chan, 1996). Nothing is known about the biology of the adults. The species are of no medical or economic importance to humans.

****

Principal references: Mattingly, 1960 (new subgenus, taxonomy, description, bionomics, distribution); Mattingly, 1971b (eggs); Reinert, 2000e (female genitalia); Ramalingam & Ramakrishnan, 1971 (*Ae. brevitibia*, taxonomy, description, bionomics, distribution); Reinert *et al.*, 2004 (as a genus, morphology, phylogenetic relationships); Reinert *et al.*, 2009 (as a genus, morphology, phylogenetic relationships); Rattanarithikul *et al.*, 2010 (as a genus, Thailand, keys, bionomics); Wilkerson *et al.*, 2015 (phylogenetic relationships, classification).

Albuginosus Reinert, 1987

Albuginosus Reinert, 1987, originally proposed and currently recognized as a subgenus of *Aedes*, is typified by *Aedes marshallii* (Theobald, 1901b), original combination: *Stegomyia marshallii*. Subgenus abbreviation – *Alb.*

****

Systematics: *Albuginosus* includes nine extant species in the Afrotropical Region. *Albuginosus* fell within a pectinate series of taxa in the cladistic analysis of Reinert *et al.* (2009) based on extensive morphological data: *Polyleptiomyia* + (*Bifidistylus* + (*Albuginosus* + (*Tewarius* + (*Christophersiomyia* + (*Huaedes* + *Leptosomatomyia*))))). The relationships of these seven subgenera were weakly supported by suites of homoplastic characters, and were lost in the reanalysis of Wilkerson *et al.* (2015).

****

Characteristics: The following distinctive features distinguish species of *Albuginosus* from species of other aedine taxa. ADULTS: Vertex and occiput of the head with numerous long erect forked scales; maxillary palpus and proboscis of females dark-scaled; maxillary palpus of males long, apical 2 palpomeres short and downturned, ventral surface at apex of palpomere 3 and ventral surface of palpomere 4 with several long setae, palpomere 5 with a basal pale band; anterolateral margins of scutum usually with broad silvery-white scales; acrostichal setae (anterior and posterior) present and well developed; scutellum usually with flat silvery-white scales at least on lateral lobes; subspiracular area with a patch of silvery-white scales; wing with or without a spot of silvery-white scales at the base of the costa; hindtarsomeres 1–4 each with an apical white band, hindtarsomere 5 all white. FEMALE GENITALIA: Tergum VIII with basal 0.55–0.90 retracted into segment VII, proximal margin nearly straight or slightly concave, caudal margin broad and straight or very slightly concave; proximal margin of sternum VIII moderately broad and more or less straight, distal margin broad and convex with a moderately deep median emargination, upper vaginal lip narrow, upper lateral area sharply convex, posterior margin flat with squared lateral angles; cercus short, without scales; postgenital lobe with a moderately deep to deep median apical emargination; upper vaginal sclerite large, moderately wide throughout its length and fenestrated; insula tongue-like, setae absent, with 3–11 tuberculi. MALE GENITALIA: Tergum IX bilobed, each lobe broadly rounded with 3–8 stout apical setae; gonostylus approximately 0.44–0.66 the length of the gonocoxite, tapering from about mid-length to a narrow bluntly pointed apex; gonostylar claw spiniform, long; basal mesal lobe with several short thin setae and 4–9 short stout lanceolate setae; aedeagus divided into 2 lateral plates, each plate with 5–8 posteriorly directed teeth on the apical 0.5, dorsal flap present. LARVAE: Head about as broad as long; seta 1-S inserted immediately beyond the pecten; saddle incomplete; seta 2-X long, with 1–5 branches; seta 3-X single, significantly longer than seta 2-X; ventral brush (seta 4-X) with 3 or 4 pairs of setae on a grid and 2 or 3 precratal setae. PUPAE: Known for only 4 of the 9 species of *Albuginosus*. Setae 11-CT, 3-I,II and 6-III single; seta 6-I longer than seta 7-I; seta 3-II as thick or thicker than seta 1-II; midrib of paddle strong, extending to or near the apex of the paddle; seta 1-Pa present, single, about 0.33 the length of the paddle.

****

Bionomics: The usual habitats of the immature stages of species of *Albuginosus* are plant containers, including tree holes, rot holes, bored and cut bamboo, axils of bananas and cavities in palm and fig trees. Species of *Albuginosus* are not known to be of medical or economic importance to humans. However, Bunyamwera virus has been isolated from a mixed pool of aedine species in Cameroun that included *Ae. capensis*, and Middleburg virus has been isolated from specimens of *Ae. marshallii* captured in Natal, South Africa (references listed in Huang & Rueda, 2015a).

****

Principal references: Reinert, 1987 (new subgenus, taxonomy, description, distribution, bionomics); Reinert, 2000e (female genitalia); Reinert *et al.*, 2004 (as a genus, morphology, phylogenetic relationships); Reinert *et al.*, 2009 (as a genus, morphology, phylogenetic relationships); Huang & Rueda, 2015a (key to adults); Wilkerson *et al.*, 2015 (phylogenetic relationships, classification).

Ayurakitia Thurman, 1954

Ayurakitia, originally proposed as a genus by Thurman (1954a), currently classified as a subgenus of *Aedes*, is typified by *Aedes griffithi* (Thurman, 1954a), original combination: *Ayurakitia griffithi*. Subgenus abbreviation – *Ayu*.

****

Systematics: *Ayurakitia* includes only two extant species in the Oriental Region (Malaysia, Myanmar and Thailand). *Ayurakitia* was recovered as the sister to *Diceromyia* in the morphology-based phylogeny of Reinert *et al.* (2009). It fell within an eight-clade polytomy, including *Diceromyia*, in the phylogeny of Wilkerson *et al.* (2015).

****

Characteristics: Adults of *Ayurakitia* lack postspiracular setae and scales. The vestiture of the thoracic pleura is restricted to small patches of broad silvery scales on the upper proepisternum, upper and lower mesokatepisternum and upper mesepimeron. The midungues of males are unequal and simple. The genitalia of females are unique among Aedini in having a large tergum VIII that is nearly covered with broad scales. Tergum IX and the gonostylus of the male genitalia are uniquely developed. The posterior margin of tergum IX has a wide, deep median emargination that separates broadly rounded lobes, each bearing a few thin, posterolaterally directed setae. The gonostylus is large and distinctively mitten-shaped. Larvae differ from other Aedini in the development of setae 4–7-C, which have multiple aciculate branches, and the placement of setae 4–6-C, which are inserted in an equilateral triangular pattern. Additionally, the location of seta 8-C is unusual in being inserted well mesad of the frontal ecdysial line. Pupae of *Ayurakitia* are unique in having the outer part of the paddle ending before the apex of the inner part, with seta 1-Pa inserted subapically adjacent to the apex of the outer part. Pupae are also unusual in having seta 7-CT very long and double.

****

Bionomics: Very little is known about the bionomics of species of *Ayurakitia*. Larvae have been collected from bamboo stumps and internodes, and the axils of bananas and *Pandanus*. Adults have been found resting on tree trunks and vegetation in jungle in mountain valleys. The species of *Ayurakitia* are of no medical or economic importance to humans.

****

Principal references: Thurman, 1954a (as new genus, new species, diagnosis, description); Lu *et al.*, 1997a (China, *Ae. peytoni*, taxonomy, descriptions, distribution, bionomics); Reinert, 2000b (as a genus, description, distribution, bionomics); Reinert *et al.*, 2004, 2009 (as a genus, morphology, phylogenetic relationships); Rattanarithikul *et al.*, 2010 (as a genus, Thailand, keys, bionomics); Wilkerson *et al.*, 2015 (phylogenetic relationships, classification).

Aztecaedes Zavortink, 1972

Aztecaedes Zavortink, 1972, originally proposed and currently recognized as a subgenus of *Aedes*, is typified by *Aedes ramirezi* Vargas & Downs, 1950. Subgenus abbreviation – *Azt.*

****

Systematics: *Aztecaedes* is monobasic. It only includes the extant type species, which is only known from elevations between 1600 and 1900 m in Jalisco and Morelos States of Mexico. Zavortink (1972) stated that '*Aztecaedes* shows no definite relationship to any other New World group' of aedine mosquitoes. The morphology-based phylogenies of Reinert *et al.* (2008, 2009) support this observation. *Aztecaedes* was recovered in a basal relationship to *Gymnometopa* and *Kompia* in a clade comprised of *Lewnielsenius* + (*Abraedes* + (*Aztecaedes* + (*Gymnometopa* + *Kompia*))) in the study of Reinert *et al.* (2008) and a clade of similar topology and relationships was recovered in the more comprehensive study of Reinert *et al.* (2009): *Howardina* + (*Gymnometopa* + (*Kompia* + (*Aztecaedes* + (*Abraedes* + *Lewnielsenius*)))). *Aztecaedes* is clearly distinct and not definitely related to other aedine generic-level taxa (see also Wilkerson *et al.*, 2015).

****

Characteristics: *Aztecaedes* is distinguished from all other New World generic-level taxa of Aedini by the following distinctive features. ADULTS: Vertex of the head with numerous erect scales and few to numerous moderately broad decumbent scales; antennal pedicel with a large patch of scales on the mesal surface; maxillary palpus and proboscis entirely dark-scaled; maxillary palpus of males with a long palpomere 3, ratio of its length to the length of the proboscis > 0.21; scutum with broad decumbent pale scales in a narrow acrostichal line, a narrow posterior outer dorsocentral line and a narrow marginal line; scutal fossa without setae; paratergite bare; upper area of the pleura with 2 nearly horizontal lines of pale scales; postpronotum with dark falcate scales along the upper margin and a lower line of pale spatulate scales; hypostigmal scales absent; tibiae and tarsi of all legs entirely dark-scaled. FEMALE GENITALIA: Upper vaginal sclerite absent. MALE GENITALIA: Sternomesal surface of the gonocoxite with numerous long strong setae, base of the tergomesal surface without a single differentiated seta or setal tuft; claspette with a pair of strongly developed setae on the dorsoapical area of the claspette stem; aedeagus with a median constriction. LARVAE: Head with very large ocular lobes; antenna without spicules; setae 5-C, 7-C and 5-P single; seta 7-C inserted posterior to seta 5-C; setae 14-M, 13-T and 13-I with few strongly developed branches; seta 12-I absent; siphon very short; seta 1-S shorter than the diameter of the siphon; pecten short, conspicuously curved dorsad distally or oblique, pecten spines with numerous basal denticles or serration; saddle incomplete ventrally; seta 1-X strongly developed, single; ventral brush (seta 4-X) borne on a boss, usually comprised of 8 pairs of setae, seta 4a-X (the posteriormost seta) very short relative to seta 4c-X; anal papillae very large, inflated and tapered. PUPAE: Terga II–VI or VII with a strong irregular transverse ridge slightly caudad of seta 0; seta 3-I single, seta 3-II inserted posteromesal to seta 2-II and longer than seta 6-II; seta 6-I shorter than seta 7-I; paddle broad, midrib restricted to approximately proximal 0.6.

****

Bionomics: The immature stages of *Ae. ramirezi* have been found in holes in volcanic rock and in a rot hole of a tree. A female was captured landing-biting a human during mid-afternoon. The species is of no medical or economic importance to humans.

****

Principal references: Zavortink, 1972 (new subgenus, diagnosis, *Ae. ramirezi*, description, bionomics, distribution); Reinert, 2002e (as a subgenus of *Ochlerotatus*, female genitalia); Reinert *et al.*, 2004, 2006, 2009 (as a genus, morphology, phylogenetic relationships); Wilkerson *et al.*, 2015 (phylogenetic relationships, classification).

Belkinius Reinert, 1982

Belkinius Reinert, 1982, originally proposed and currently recognized as s subgenus of *Aedes*, is typified by *Aedes aurotaeniatus* (Edwards, 1922a), original combination: *Stegomyia aureostriata*. Subgenus abbreviation – *Blk.*

****

Systematics: *Belkinius* is monobasic. It only includes the extant type species, which is only known from Negros Occidental and Mindoro Oriental Provinces of the Philippine Islands. It appears to have closest affinity with *Zeugnomyia*, as indicated in the clade recovered in the morphological phylogeny of Reinert *et al.* (2009) that comprises *Lorrainea* + ((*Eretmapodites* + *Armigeres*) + (*Udaya* + (*Belkinius* + *Zeugnomyia*))). This sister-group association of *Belkinius* and *Zeugnomyia* was also recovered in the study of Reinert *et al.* (2004) based on an analysis of fewer morphological characters. *Belkinius* was recovered within a polytomy with *Bothaella*, *Catatassomyia*, *Udaya* and *Zeugnomyia* in the phylogeny reconstructed by Wilkerson *et al.* (2015).

****

Characteristics: *Belkinius*, like most generic-level taxa of Aedini, is diagnosed and distinguished by a combination of characters. ADULTS: Somewhat similar to the adults of *Stegomyia*. Vertex of the head with broad decumbent scales; few short erect forked scales confined to the occiput; antenna distinctly shorter than the proboscis; flagellomeres 1–12 of males with scales on the proximal side of the flagellar whorls; maxillary palpus very short in both sexes; scutal scales narrow, entirely pale; acrostichal setae absent; few anterior dorsocentral setae present, posterior dorsocentral setae absent; entire length of the dorsocentral area with a stripe of pale scales; scutal fossa with contrasting areas of pale scales; prescutellar space covered with scales; upper proepisternum with broad scales and 2–4 setae, lower area bare; mesepimeron with 1–4 setae above the scale-patch, lower mesepimeral setae absent; anal vein (vein 1A) of the wing ends at or slightly distad of the intersection of the cubitus anterior (vein CuA) and the mediocubital crossvein; hindtibia entirely dark-scaled; both ungues of fore- and midlegs toothed in females, equal and without tooth in males. FEMALE GENITALIA: Tergum IX short, band-like, without setae; postgenital lobe moderately long and wide with concave lateral margins; insula tongue-like with few tuberculi on the apical 0.35. MALE GENITALIA: Tergum IX uniquely trilobed, setae absent; gonocoxite short and broad; basal mesal lobe short, borne near the mesal surface of the gonocoxite; gonostylus large with blade-like spines; gonostylar claw absent; aedeagus with 2 lateral plates, each plate with apical teeth. LARVAE: Unknown. PUPAE: Trumpet with a long pinna; metanotum with clear unpigmented spots, seta 13-CT present; seta 2-I short; seta 3-I very long; seta 5-II inserted posterior or posterolateral to seta 4-II; paddle with a fringe of long hair-like spicules; seta 1-Pa single with recurved tip.

****

Bionomics: *Aedes aurotaeniatus* occurs in hilly, mountainous areas. Females are known to bite humans in areas of high altitude. The immature stages have only been found in the axil of a wild banana tree. The species is of no medical or economic importance to humans.

****

Principal references: Reinert, 1982 (new subgenus, taxonomy, descriptions, distribution, bionomics); Reinert, 2000e (female genitalia); Reinert *et al.*, 2004, 2009 (as a genus, morphology, phylogenetic relationships); Wilkerson *et al.*, 2015 (phylogenetic relationships, classification).

Bifidistylus Reinert, Harbach & Kitching, 2009

Bifidistylus, originally proposed as a genus by Reinert *et al.* (2009), currently classified as a subgenus of *Aedes*, is typified by *Aedes lamborni* Edwards, 1923a. Subgenus abbreviation – *Bif.*

****

Systematics: *Bifidistylus* includes only two extant species. Species of *Bifidistylus* are known to occur in Cameroon, Democratic Republic of the Congo, Equatorial Guinea, Kenya, Malawi, South Africa and Zambia. Although weakly supported, the cladistic analysis of Reinert *et al.* (2009) placed *Bifidistylus* in a position basal to a clade comprised of *Albuginosus* + (*Tewarius* + (*Christophersiomyia* + (*Huaedes* + *Leptosomatomyia*))), with *Polyleptiomyia* basal to *Bifidistylus*. Although the internal branches of the clade were weakly supported and only nine homoplastic characters diagnosed the *Bifidistylus* branch (the clade was not recovered when the data set was analysed in the study of Wilkerson *et al.*, 2015), *Bifidistylus* exhibits clear distinctions from the other generic-level taxa within the clade. Characters of the females and the male and female genitalia that were not included in the analysis support the monophyly of *Bifidistylus*.

****

Characteristics: *Bifidistylus* is named for the distinctive bifurcated gonostylus of the male genitalia. Like most generic-level taxa of Aedini, *Bifidistylus* is distinguished by a unique combination of characters. ADULTS: Occiput and vertex of the head with numerous erect forked scales; eyes above the antennal pedicels contiguous or separated by less than the diameter of 1 corneal facet; antennal pedicel with several broad scales on the mesal surface; clypeus bare; maxillary palpus of females dark-scaled with pale scales at the apex, maxillary palpus of males shorter than the proboscis, comprised of 5 palpomeres; scutum with dark and pale falcate scales, prescutellar area with some broad pale scales mesal to the setae on each side; acrostichal, dorsocentral and prescutellar setae well developed; all lobes of the scutellum with broad silvery scales; paratergite, upper proepisternum, subspiracular area, postspiracular area, upper and lower prealar areas, upper and lower areas of mesokatepisternum and upper to middle area of mesepimeron with patches of broad pale scales; lower mesepimeral setae absent; wing dark-scaled with pale patch at the base of the costa; upper calypter with numerous marginal setae; ante- and postprocoxal membranes bare; hindfemur, hindtibia and hindtarsomeres 1–4 with pale apical bands, hindtarsomere 5 pale-scaled; fore-, mid- and hindungues of females equal, each with 1 tooth, fore- and midungues of males unequal, each with 1 tooth; laterotergite of tergum I with broad pale scales; terga II–VI with basal pale areas. FEMALE GENITALIA: Tergum and sternum VIII wider than long, tergum with numerous broad scales distally and few scattered scales proximally; tergum IX comprised of 2 lateral sclerites separated by a mesal membrane, each with 5–13 setae; postgenital lobe with median emargination; lower vaginal sclerite absent; insula tongue-like, without setae; cercus with or without scales; 1 large and 2 slightly smaller spermathecal capsules. MALE GENITALIA: Tergum IX lobes moderately large, each with several setae; sternum IX with 1–4 posteromedian setae; gonocoxite with numerous broad scales on the lateral and ventral surfaces; gonostylus attached at the apex of the gonocoxite, bifurcate near mid-length, outer lobe longer with a short subapical seta, inner lobe with a short flattened gonostylar claw at the apex; claspette comprised of a small basal plaque bearing a few slender setae proximally and several stout setae distally; aedeagus comprised of 2 lateral sclerites, each with a few elongate teeth on approximately the distal 0.5; proctiger without cercal setae. LARVAE: Seta 1-C slender, single; seta 4-C short, with 4 or 5 branches, inserted more or less directly mesal to seta 6-C; seta 5-C with 8 or 9 branches, inserted posteromesal to setae 6,7-C; seta 6-C inserted close to seta 5-C and mesal and slightly posterior to seta 7-C; seta 12-C shorter and inserted mesal to seta 13-C; seta 14-C single; seta 19-C absent; setae 1–3-P not inserted on a common setal support plate, seta 1-P longer than seta 2-P, seta 2-P longer than seta 3-P; setae 5,6-P long, single, seta 5-P longer than setae 6,7-P; seta 7-P long, with 2 or 3 branches; seta 4-M short, with 2 or 3 branches; setae 5,7-M long, single, seta 5-M longer than seta 7-M; seta 2-T with 4 or 5 branches; seta 6-T single; seta 6-I–VI long, stout, double; seta 7-I long, stout, with 2 or 3 branches; seta 12-I absent; seta 1-VII very long, stout, single, noticeably longer than the saddle; setae 2,4-VIII single; comb comprised of numerous scales in a triangular patch; siphon moderately long, acus absent; pecten comprised of numerous spines; seta 1-S inserted distal to or within the pecten; saddle of segment X incomplete ventrally, acus absent; seta 1-X short, single or double, inserted on the saddle; seta 2-X moderately long, with 7–10 branches; seta 3-X long, single; ventral brush (seta 4-X) inserted on a grid with both transverse and lateral grid bars, with 1 (rarely 2) precratal setae. PUPAE: Tracheoid area of the trumpet weakly developed; setae 1,3-CT similarly developed; setae 4,5-CT branched and similar in length; setae 10,12-CT branched; seta 11-CT normally single, longer than setae 10,12-CT; seta 3-I long, stout, single, longer than setae 6,7-I; seta 1-II with numerous slender branches; seta 2-II inserted lateral to setae 1,3-II; seta 3-II,III long, stout, single; seta 3-III longer than seta 5-III; seta 5-V longer than tergum VI; seta 6-VII inserted posterior and slightly mesal to seta 9-VII, seta 9-VII branched, longer than seta 6-VII; seta 9-VIII with 3–5 branches; paddle rounded apically with the midrib extending to near the apex, without hair-like marginal spicules; seta 1-Pa short, single or double.

****

Bionomics: Larvae of species of *Bifidistylus* principally inhabit rock pools under heavy shade. Females of *Ae. lamborni* have been collected biting humans in forest. Species of *Bifidistylus* are not medically or economically important to humans.

****

Principal references: Reinert *et al.*, 2009 (as a new genus, phylogenetic relationships, description, distribution, bionomics); Reinert, 2010g (as a genus, female genitalia); Wilkerson *et al.*, 2015 (phylogenetic relationships, classification).

Borichinda Harbach & Rattanarithikul, 2007

Borichinda, originally proposed as a genus by Rattanarithikul & Harbach, 2007 (in Harbach *et al.*, 2007b), currently classified as a subgenus of *Aedes*, is typified by *Aedes cavernicola* (Rattanarithikul & Harbach, 2007) (in Harbach *et al.*, 2007b). Subgenus abbreviation – *Bor.*

****

Systematics: *Borichinda* is monobasic. It only includes the extant type species, which is only known from Borichinda Cave in northern Thailand. Based on the inclusion of morphological data for *Borichinda* in the parsimony analysis of Reinert *et al.* (2004), *Borichinda* appears to be more closely related to *Isoaedes* and *Ayurakitia* than to other generic-level taxa of Aedini, with a closer affinity with *Isoaedes*, which is reflected in the relationships of *Fredwardsius*, *Isoaedes*, *Borichinda* and a clade of other subgenera, each on a separate branch, recovered in the cladistic analysis of more extensive morphological data by Reinert *et al.* (2009): *Fredwardsius* + (*Isoaedes* + (*Borichinda* + other subgenera)). *Borichinda* was recovered as the sister of *Isoaedes* in the phylogeny of Wilkerson *et al.* (2015), and as the sister of *Nyctomyia* in the maximum likelihood phylogeny of Soghigian *et al.* (2017) based on molecular markers.

****

Characteristics: The following salient anatomical differences distinguish *Borichinda* from *Ayurakitia*, *Isoaedes* and *Nyctomyia*. ADULTS: Compound eyes narrowly separated; scutum with a pattern of dark and pale scaling; scutellum with broad scales on the midlobe; postspiracular scales and subspiracular scales present; remigial setae absent; only one foreungues of males toothed (distinction from *Isoaedes* and *Nyctomyia*). MALE GENITALIA: Apex of the claspette expanded with numerous setae. LARVAE: Seta 4-C short, with 7–11 branches; seta 14-C branched (distinction from *Isoaedes*); seta 5-P double; seta 5-T large, stellate, multi-branched; seta 1-I–VII large, stellate, inserted far anterolateral to seta 1; seta 9-II–VI inserted far anterior to seta 7. PUPAE: Seta 6-III single; seta 6-VII inserted anterior to seta 9; seta 9-IV–VI inserted anterior to seta 8; apex of the paddle slightly concave.

****

Bionomics: *Aedes cavernicola* is a cave-dwelling species that has only been found and reared from larvae that inhabit a single rimstone pool in Borichinda Cave of Chiangmai Province in northern Thailand. The cave harbours a population of bats that is likely to be the primary source of food for adult females. Nothing else is known about the biology of the species. It is not a medically or economically important species.

****

Principal references: Harbach *et al.*, 2007b (as a new genus, descriptions, phylogenetic relationships, DNA sequences, bionomics, distribution); Reinert *et al.*, 2009 (as a genus, morphology, phylogenetic relationships); Rattanarithikul *et al.*, 2010 (as a genus, Thailand, keys, bionomics); Wilkerson *et al.*, 2015 (phylogenetic relationships, classification); Soghigian *et al.*, 2017 (phylogenetic relationships).

Bothaella Reinert, 1973

Bothaella Reinert, 1973f, originally proposed and currently recognized as a subgenus of *Aedes*, is typified by *Aedes helenae* Reinert, 1973f. Subgenus abbreviation – *Bot.*

****

Systematics: *Bothaella* includes six extant species. The species occur in the Oriental Region, principally in Southeast Asia. The monophyly of *Bothaella* was confirmed in the phylogenetic study of Aedini by Reinert *et al.* (2009), and was closely associated with *Ae. meronephada* of *Catatassomyia* based on a unique combination of 12 homoplastic characters that are shared by the two taxa: two characters of adults, two characters of the female genitalia, one character of the male genitalia, four setal characters of larvae and three setal characters of pupae (Reinert *et al.*, 2009: fig. 2H). *Bothaella* fell within a polytomy with *Belkinius* and *Catatassomyia*, and the genera *Udaya* and *Zeugnomyia*, in the phylogeny reconstructed by Wilkerson *et al.* (2015). In the absence of species of *Catatassomyia*, *Bothaella* was recovered in a sister relationship with *Nyctomyia* + *Borichinda* in the maximum likelihood phylogeny of Soghigian *et al.* (2017) based on analysis of seven molecular markers.

****

Characteristics: Adults of *Bothaella* are dark brown to black with distinctive patches of silvery-white scales anteriorly on the vertex, the mesal side of the antennal pedicel, midlobe of the scutellum, antepronotum, upper proepisternal, postspiracular, upper mesokatepisternal, lower mesokatepisternal and upper mesepimeral areas of the thoracic pleura, the laterotergite of abdominal segment I and lateral areas of abdominal terga II–VII. The erect scales of the head are arranged in a transverse posterior row, the postpronotum lacks scales and a lower anterior mesepimeral seta is present. The hindfemur is largely or completely pale-scaled proximally, hindtarsomeres 1–4 have basal and apical pale bands (except *Ae. kleini*, in which the hindtarsus is entirely dark-scaled) and both ungues of the fore- and midlegs bear a tooth. The gonostylus of the male genitalia is distinctly enlarged distally, the basal mesal lobe is expanded and heavily setose distally, and the proctiger lacks cercal setae. *Bothaella* larvae have a well-developed seta 13-P, setae 4- and 6-C inserted far anterior to seta 7-C, setae 5- and 6-C long and single, comb scales arranged in two or more irregular rows, evenly spaced pecten spines, a branched seta 1-S borne distal to the pecten, an incomplete saddle and a ventral brush (seta 4-X) comprised of 8 or 9 setae. Pupae have setae 4,6-CT, 3-II,III and 5-IV,V very long and single, seta 6-VII inserted anterior to anteromesad of seta 9-VII and no seta 2-Pa.

****

Bionomics: *Bothaella* larvae have been collected from rock pools and rock holes, occasionally from tree holes, bamboo stumps and split bamboo. Females have been collected biting humans. Nothing else is known about the bionomics of these mosquitoes. Species of *Bothaella* are not known to be of medical or economic importance to humans.

****

Principal references: Reinert, 1973f (new subgenus, descriptions, keys, distributions, bionomics); Lu *et al.*, 1997a (China, *Ae. eldridgei*, taxonomy, description, distributions, bionomics); Reinert *et al.*, 2004, 2009 (as a genus, morphology, phylogenetic relationships); Rattanarithikul *et al.*, 2010 (as a genus, Thailand, keys, bionomics); Cook *et al.*, 2010 (as a genus, phylogenetic relationships, DNA sequence); Wilkerson *et al.*, 2015 (phylogenetic relationships, classification); Soghigian *et al.*, 2017 (phylogenetic relationships).

Bruceharrisonius Reinert, 2003

Bruceharrisonius Reinert, 2003, originally proposed and currently recognized as a subgenus of *Aedes*, is typified by *Aedes greenii* (Theobald, 1903a), original combination: *Howardina greenii*. Subgenus abbreviation – *Brh.*

****

Systematics: *Bruceharrisonius* includes nine extant species. Species of *Bruceharrisonius* occur principally in the Oriental Region, with extensions into the Australasian Region (New Guinea) and southeastern areas of the Palaearctic Region (China, Maritime and Ussurii Provinces of Russia, Korea and Japan). Records for countries of the Oriental Region include India, Indonesia, Malaysia, Nepal, Philippine Islands, Sri Lanka, Taiwan and Thailand. Species of *Bruceharrisonius* appear to be most closely related to species of *Himalaius*, *Kenknightia*, *Vansomerenis* and *Zavortinkius*, which comprised a clade parenthetically expressed as (*Himalaius* + *Bruceharrisonius*) + (*Vansomerenis* + (*Zavortinkius* + *Kenknightia*)) in the morphology-based phylogenetic study of Reinert *et al.* (2009). The sister relationship of *Bruceharrisonius* and *Himalaius* was first observed in the phylogenetic analysis of morphological data by Reinert *et al.* (2006) and confirmed in the successively more comprehensive analyses of morphological data by Reinert *et al.* (2008) and Reinert *et al.* (2009), and reanalysis of the data set by Wilkerson *et al.* (2015), and in the maximum likelihood phylogeny of Soghigian *et al.* (2017) based on analysis of seven molecular markers.

****

Characteristics: Species of *Bruceharrisonius* are distinguished from species of other aedine generic-level taxa that occur in their realm by one or more or a combination of the following principal features. ADULTS: Vertex of the head with a median patch of pale falcate scales and several erect forked scales; eyes contiguous above the antennal pedicels; maxillary palpus dark-scaled in both sexes, palpus of males slender, nearly straight and shorter than the proboscis, palpomeres 4 and 5 fused, few short to moderately long setae on palpomeres 3–5; scutum with acrostichal and dorsocentral lines of pale scales contrasting with general covering of dark falcate scales, pale scaling also on the scutal fossa, supraalar area and either side of the prescutellar area; acrostichal and dorsocentral setae present; scutellum with narrow scales only; subspiracular and lower prealar areas with broad pale scales; lower mesepimeral seta absent (except in *Ae. okinawanus*). FEMALE GENITALIA: Sternum VIII nearly entirely covered with broad spatulate scales; tergum IX a single sclerite with more lightly pigmented median area and 3–7 setae distally on either side of the midline; insula lip-like with 2 close-set setae on either side of the midline; cercus moderately long and relatively narrow, sometimes with 1 or 2, rarely 4–6, scales. MALE GENITALIA: Tergum IX lobes short with 3–6 stout somewhat flattened setae; ventral surface of the gonocoxite with a row of setae extending from the base to the apex; gonostylus long and narrow, attached to the apex of the gonocoxite; gonostylar claw attached apically; claspette comprised of a columnar stem with a terminal large flattened filament and a narrow lateral arm arising from the base of the stem that bears 2 to several distal setae, one or more of which are flattened. LARVAE: Antenna long and slender; seta 1-A usually single; seta 4-C short and branched; seta 5-C long and branched; seta 6-C exceptionally long, single; insertions of setae 4,5,6-C form a small triangle with seta 4-C mesad of setae 5,6-C, anterior to seta 5-C and at more or less the same level as seta 6-C, and seta 5-C inserted posteromesad to seta 6-C; seta 7-C inserted posterior to the base of the antenna at the level of setae 4–6-C; seta 6-I,II relatively short, seta 6-II shorter than seta 6-I,III; seta 7-I longer than seta 6-I; seta 1-VII very long, single; setae 2,4-VIII normally single; comb with many scales in a patch. PUPAE: Seta 3-CT very long, much longer than seta 1-CT; seta 3-I long, normally double; paddle relatively narrow with midrib extending from base to apex; seta 1-Pa relatively long, single.

****

Bionomics: The immature stages of species of *Bruceharrisonius* are usually found in water contained in tree holes and the stumps and internodes of bamboo. Some species have been found in rock pools, bromeliads, ferns and artificial containers. Adults may be active during day or night. Some species have been collected in light traps and some are known to bite humans during the day. The species of *Bruceharrisonius* are of no medical or economic importance to humans.

****

Principal references: Reinert, 2003 (as a subgenus of *Ochlerotatus*, *Ae. greenii*, taxonomy, descriptions, distribution, bionomics); Reinert *et al.*, 2004, 2008, 2009 (as a genus, morphology, phylogenetic relationships); Rattanarithikul *et al.*, 2010 (as a genus, Thailand, keys, bionomics); Wilkerson *et al.*, 2015 (phylogenetic relationships, classification); Soghigian *et al.*, 2017 (phylogenetic relationships).

from the apex of the gonostylus; claspette well developed, with numerous setae; aedeagus strongly toothed; paraproct with a sternal arm; cercal setae absent. LARVAE: Seta 4-C well developed, single, simple, inserted closer to seta 6-C than to seta 5-C, inserted anteromesad of seta 6-C; seta 12-I absent; seta 6-II as long as seta 6-III; seta 1-VIII single; seta 2-VIII inserted far from seta 1; comb scales in a single row; seta 1-S single, long, inserted distal and anterior to the pecten; seta 3-X branched; ventral brush (seta 4-X) with 5 pairs of setae on a grid, without precratal setae. PUPAE: Seta 6-CT single, long, stout; seta 7-CT much shorter than seta 6-CT; seta 3-II thinner than seta 1-II; seta 3-III shorter than seta 5-III; seta 5-V shorter than the following tergum; seta 2-VI inserted lateral to seta 1-VI; seta 9-VI,VII single, long, stout, much longer and stouter than seta 9-I–V; paddle margins with fringe of fine hair-like spicules, apex with moderately deep emargination; seta 1-Pa present, single and long.

****

Bionomics: Females of *Ae. cozi* have been collected biting humans in gallery forest in eastern Senegal. The specimens that comprise the type series of this species were reared from eggs obtained from a biting female; hence, nothing is known about the bionomics of the immature stages. The species is not known to be of medical or economic importance to humans.

****

Principal references: Huang, 2005 (new subgenus, *Ae. cozi*, taxonomy, descriptions, distribution, bionomics); Reinert *et al.*, 2009 (as a genus, morphology, phylogenetic relationships); Wilkerson *et al.*, 2015 (phylogenetic relationships, classification).

Dahliana Reinert, Harbach & Kitching, 2006

Dahliana, originally proposed as a genus by Reinert *et al.* (2006), currently classified as a subgenus of *Aedes*, is typified by *Aedes geniculatus* (Olivier, 1791), original combination: *Culex geniculatus*. Subgenus abbreviation – *Dah.*

****

Systematics: *Dahliana* includes three extant species in the Palaearctic Region. *Dahliana* was placed in a sister-group relationship to a pair of terminal clades in the cladistic analyses of Reinert *et al.* (2009) based on anatomical characters of all life stages. The two clades of the sister group comprised 10 generic-level taxa: *Howardina* + (*Gymnometopa* + (*Kompia* + (*Aztecaedes* + (*Abraedes* + *Lewnielsenius*)))) and (*Finlaya* + *Danielsia*) + (*Downsiomyia* + *Haemagogus*). Relationships of *Dahliana* with other generic-level taxa of Aedini were not resolved in the phylogeny of Wilkerson *et al.* (2015).

****

Characteristics: Species of *Dahliana* are easily distinguished from those of other generic-level taxa of Aedini that occur in the Palaearctic Region by one or more of the distinctive features listed below. ADULTS: Vertex of the head with narrow pale falcate scales; numerous erect scales on the vertex and occiput; ocular line with narrow pale scales; eyes narrowly separated above the antennal pedicels; maxillary palpus and proboscis dark-scaled in both sexes, proboscis longer than the forefemur; scutum with narrow falcate scales, narrow longitudinal pale stripes on the entire length of the acrostichal area, the posterior part of the dorsocentral area extending on the sides of the prescutellar setae and on the anterior and lateral margins of the prescutellar bare area; antealar area with pale scales extending as a narrow stripe on the supraalar area; scutal fossa covered with pale scales; acrostichal and dorsocentral setae present; scutellum with narrow falcate scales on the lateral lobes; broad pale scales on the paratergite, antepronotum, postpronotum (also some dark scales), postspiracular and subspiracular areas, upper proepisternum, upper and lower mesokatepisternum, prealar area and mesepimeron; lower mesepimeral setae absent; wing dark-scaled, upper calypter with several marginal setae; ante- and postprocoxal membranes bare; hindfemur with a well-developed apical pale band; fore- and midungues of females equal, each with 1 tooth, fore- and midungues of males unequal, larger unguis with 2 teeth, smaller one with 1 tooth, hindungues equal and simple in both sexes; laterotergite of abdominal segment I with patch of pale scales. FEMALE GENITALIA: Cercus moderately long, broadly rounded apically, dorsal surface without scales; setae on distal 0.6–0.7 of dorsal surface; postgenital lobe moderately wide, apex rounded; proctiger membranous; upper vaginal lip curved outward, caudal part moderately wide and somewhat flattened; upper vaginal sclerite moderately large; lower vaginal lip narrow; lower vaginal sclerite absent; insula lip-like, flat transversally, with 3–5 (usually 4 or 5) moderately long setae on each side; 3 spermathecal capsules, 1 large, 2 slightly smaller. MALE GENITALIA: Tergum IX comprised of 2 broad lateral plates each with a posteromesal lobe bearing 2–6 flattened setae; sternum IX with setae; gonocoxite with scales on the dorsal, lateral and ventral surfaces; gonostylus attached at the apex of the gonocoxite, with a single gonostylar claw at the apex; claspette with a columnar stem, claspette filament long, somewhat flattened, slightly curved and pointed; aedeagus simple, tube-like; proctiger with few to several cercal setae. LARVAE: Antenna without spicules; seta 1-A single or double; seta 1-C single; seta 4-C multi-branched, inserted mesal to seta 6-C; seta 5-C single, inserted posterior to setae 4,6,7-C; seta 6-C with 1–3 branches, inserted more or less directly anterior to seta 5-C and mesal to seta 7-C; seta 7-C double or triple; seta 12-C branched, inserted mesal to seta 13-C; seta 13-C single, longer than seta 12-C; seta 14-C short, branched; seta 19-C absent; setae 1–3-P inserted on a common setal support plate; seta 5-P double, longer than seta 6-P; seta 8-P multi-branched; seta 4-M single; seta 5-M longer than seta 7-M; seta 1-T borne on a setal support plate; setae 2,6-T, 7-I,II, 8-II, 12-VII and 2,4-VIII single; seta 1-I,II similarly developed; seta 12-I present; seta 1-VII stellate; seta 3-VII inserted at same level or posterior to seta 1-VII; comb comprised of several scales in a single row; siphon with attached acus; seta 1-S multi-branched, inserted distal to the pecten; saddle incomplete, acus absent; seta 1-X branched, inserted on the saddle; setae of the ventral brush (seta 4-X) usually with 2–4 branches arising from a long stem, inserted on a grid with both transverse and lateral grid bars, 2 precratal setae. PUPAE (based on *Ae. geniculatus*): Setae 1,3-CT similar, normally double; seta 11-CT single; seta 3-I single, shorter than seta 6-I; seta 6-I longer than seta 7-I; seta 2-II inserted lateral to setae 1,3-II; seta 3-II normally single, shorter than seta 6-II; seta 5-II longer than seta 3-II, slightly shorter than seta 6-II, inserted lateral to seta 4-II; setae 3,6-III long, single; seta 5-V longer than tergum VI; seta 2-VI inserted anterolateral to seta 1-VI and mesal to seta 3-VI; seta 6-VII short, inserted posteromesal to seta 9-VII; seta 9-VII,VIII moderately long, aciculate, seta 9-VII usually double, seta 9-VIII with 2–5 branches, usually 2 or 3; paddle much longer than wide, margins without hair-like spicules, midrib complete to apex; seta 1-Pa single.

****

Bionomics: The immature stages of species of *Dahliana* normally inhabit tree holes, but they are also found in open tree stumps, root holes and pools in rocks. The species are of no medical or economic importance to humans.

****

Principal references: Reinert *et al.*, 2006 (as a new genus, morphology, phylogenetic relationships, description, bionomics); Reinert, 2008c (as a genus, female genitalia); Reinert *et al.*, 2009 (as a genus, morphology, phylogenetic relationships); Becker *et al.*, 2003, 2010 (as species of *Ochlerotatus* subgenus *Finlaya*, in keys, descriptions, bionomics, distributions); Wilkerson *et al.*, 2015 (phylogenetic relationships, classification); Becker *et al.*, 2020 (in keys, descriptions, bionomics, distributions).

Danielsia Theobald, 1904

Danielsia, originally proposed as a genus by Theobald (1904c), currently classified as a subgenus of *Aedes*, is typified by *Aedes albotaeniatus* (Leicester, 1904) (in Theobald, 1904c), original combination: *Danielsia albotaeniata*. Subgenus abbreviation – *Dan.*

****

Systematics: *Danielsia* includes three extant species in the Oriental Region, recorded from localities in China, India, Indonesia, Malaysia, Nepal, Philippines, Sri Lank and Thailand. *Danielsia* appears to be most closely related to *Haemagogus* and the subgenera *Downsiomyia* and *Finlaya* based on the cladistic analysis of extensive morphological data by Reinert *et al.* (2009) that generated relationships of these genera expressed parenthetically as (*Finlaya* + *Danielsia*) + (*Downsiomyia* + *Haemagogus*). The genealogical affinities of *Danielsia* require further study as the character support for its sister relationship with *Finlaya* was not very strong. However, some degree of affinity with *Finlaya* is not unlikely as the species of the two subgenera were previously included in *Finlaya*. *Danielsia* was also recovered as the sister of *Finlaya* in the study of Wilkerson *et al.* (2015). The type species of *Danielsia* included in the maximum likelihood phylogeny of Soghigian *et al.* (2017) based on seven molecular markers was recovered as the sister of the monobasic *Luius*.

****

Characteristics: *Danielsia*, like most generic-level taxa of Aedini, is a polythetic taxon that is diagnosed by a unique combination of characters. ADULTS: Occiput of the head with numerous erect forked scales that may extend slightly onto the vertex; eyes contiguous, separated by less than 1 corneal facet; maxillary palpus and proboscis of both sexes dark-scaled; palpus of males about as long as the proboscis, slender, palpomeres 4 and 5 short; scutum with narrow falcate scales, a patch of pale scales on anterior part of acrostichal area or covering both acrostichal and dorsocentral areas; antealar and supraalar areas dark-scaled; acrostichal setae, paratergal scales and lower mesepimeral seta absent; wing with patch of pale scales at the base of the costa; mid- and hindtarsi with pale bands on at least tarsomeres 1 and 2; fore- and midungues toothed, anterior and posterior unguis of the fore- and midlegs of males unequal in size, larger unguis with 2 teeth; abdominal segments VII and VIII of females laterally compressed. FEMALE GENITALIA: Sternum VIII mostly covered with scales, posterior margin gently sloping anteromesad from posterolateral corners; insula lip-like with 3–5 setae on each side; lower vaginal lip lightly tanned in contrast with the more darkly tanned insula; upper vaginal lip small; posterior margin of the postgenital lobe straight. MALE GENITALIA: Tergum IX narrow, broadened laterally, with 1 or 2 short setae inserted on a small lobe on each side; gonocoxite with a compact basal patch of slender setae on the dorsomesal margin, ventromesal surface with patch of large lanceolate scales near mid-length; claspette comprised of a columnar stem bearing an apical flattened and postbasally expanded claspette filament; aedeagus simple, somewhat scoop-like with distal 0.25 noticeably narrower; cercal setae present. LARVAE: Antenna relatively short, without spicules; seta 1-A short, not longer than 3 times the diameter of the antenna at the point of insertion; seta 4-C branched, inserted anteromesal to seta 6-C; seta 6-C moderately long, inserted anteromesal to seta 7-C; seta 13-C single; setae 1–3-P borne on a common setal support plate; seta 5-P longer than seta 6-P; setae 2,6-T single; many thoracic and abdominal setae stellate; seta 6-II,III branched; seta 7-II long, stout, single or double; seta 8-II branched; seta 12-VII and setae 2,4-VIII single; siphon with attached acus; seta 1-S inserted distal to the pecten, with 2–6 branches; distal 1–3 pecten spines more widely spaced; saddle incomplete, with well-developed spicules on posterior margin; seta 1-X long, stout, single (rarely double); seta 2-X double. PUPAE: Seta 1-CT similar to seta 3-CT (but shorter); seta 4-CT longer than seta 5-CT; seta 6-CT noticeably shorter than seta 7-CT; seta 1-II double, branches slender, inserted mesal to seta 3-II; seta 3-I very long, stout, longer than seta 6-I; seta 3-VI inserted lateral to seta 1-VI; seta 9-VII shorter than seta 6-VI; seta 9-VIII with 6–12 stout aciculate branches; paddle rounded apically, without fringe of hair-like spicules, midrib complete to apex.

****

Bionomics: The immature stages of species of *Danielsia* are found principally in bamboo habitats, *e.g.* bamboo pots, bamboo stumps, fallen and split bamboo, but the type species has been found in tree holes in Malaysia (Macdonald & Traub, 1960) and a freshwater reservoir at an elevation of 2200 m in Nepal (Darsie *et al.*, 1993). Females of the type species are known to bite humans in forested areas during daylight hours (Macdonald, 1957). Nothing else is known about the ecology and behaviour of members of *Danielsia*. Species of *Danielsia* are not known to be of medical or economic importance to humans.

****

Principal references: Reinert *et al.*, 2008, 2009 (as a genus, morphology, phylogenetic relationships); Reinert, 2008e (as a genus, female genitalia); Reinert, 2009a (as a genus and type species, taxonomy, descriptions, distributions, bionomics); Rattanarithikul *et al.*, 2010 (as a genus, Thailand, keys, bionomics); Wilkerson *et al.*, 2015 (phylogenetic relationships, classification); Soghigian *et al.*, 2017 (phylogenetic relationships).

Dendroskusea Edwards, 1929

Dendroskusea Edwards, 1929b, originally proposed and currently recognized as a subgenus of *Aedes*, is typified by *Aedes micropterus* (Giles, 1901a), original combination: *Culex micropterus*. Subgenus abbreviation – *Dsk.*

****

Systematics: *Dendroskusea* includes five extant species in the Oriental Region, recorded form Bangladesh, India, Nepal, Pakistan and Sri Lanka. The relationships of *Dendroskusea* are uncertain. It was placed in a large clade in the phylogeny of Aedini reconstructed by Reinert *et al.* (2009). The first group of the clade to branch off comprised *Skusea* + (*Indusius* + *Cancraedes*), followed by *Fredwardsius*, *Isoaedes* and *Borichinda* on separate branches, and then *Diceromyia* + *Ayurakitia*, *Dendroskusea* and *Scutomyia* + (*Catatassomyia* + *Bothaella*). In the phylogeny generated by Wilkerson *et al.* (2015), *Dendroskusea* was placed within a polytomy that included *Ayurakitia*, *Belkinius*, *Borichinda*, *Bothaella*, *Catatassomyia*, *Cornetius*, *Diceromyia*, *Isoaedes*, *Petermattinglyius*, *Pseudarmigeres*, *Scutomyia* and *Stegomyia*, as well as the genera *Heizmannia*, *Udaya* and *Zeugnomyia*.

****

Characteristics: *Dendroskusea* is a polythetic taxon that is diagnosed by a unique combination of characters. Species of *Dendroskusea* are similar to species of *Diceromyia*, but should be distinguishable from species of *Diceromyia* by the following combination of characters. ADULTS: Vertex of the head with a pair of whitish scale-patches; proboscis not longer than the forefemur; antenna of males with the setae of the flagellar whorls directed mainly dorsally and ventrally; maxillary palpus of males as long as the proboscis, palpomeres 4 and 5 short and turned downwards; scutal setae very strongly developed, posterior acrostichal setae present or absent, usually present; 1 or 2 lower mesepimeral setae present in both sexes (except *Ae. periskelatus*); paratergite with pale scales; ungues of females all simple, ungues of males toothed; wing with broad scales on anterior veins. FEMALE GENITALIA: Sternum VIII with seta 2-S inserted lateral and on or near the same level as seta 1-S; cerci short and broad; 3 spermathecal capsules. MALE GENITALIA: Tergum IX small; gonocoxite without definite lobes; gonostylus without setae on the distal 0.33, with a single short apical or preapical gonostylar claw; aedeagus divided into 2 lateral plates, each bearing conspicuous teeth; proctiger with well-developed paraprocts; cercal setae absent. LARVAE: Seta 4-C usually inserted at the same level but sometimes anterior to seta 6-C; seta 4-P branched, as long as or shorter than seta 3-P; seta 3-VII usually single. PUPAE: Seta 2-V inserted posterior or directly mesal to seta 3-V; seta 9-VIII inserted slightly anterior or mesal to the posterolateral corner of the segment; seta 9-VIII single or double.

****

Bionomics: The immature stages of *Ae. kanarensis* and *Ae. ramachandrai* are unknown; larvae of the other three members of *Dendroskusea* have only been found in tree holes. Females of *Ae. ramachandrai* were collected in forest in all seasons of the year, either in vegetation or coming to feed on humans (Reuben, 1967). Nothing else is known about the bionomics of the species of *Dendroskusea*. The species are not known to be of medical or economic importance to humans.

****

Principal references: Reinert *et al.*, 2009 (as a genus, morphology, phylogenetic relationships); Wilkerson *et al.*, 2015 (phylogenetic relationships, classification).

Diceromyia Theobald, 1911

Diceromyia, originally proposed as a genus by Theobald (1911c), currently classified as a subgenus of *Aedes*, is typified by *Diceromyia africana* Theobald, 1911c [subjective synonym of *Aedes furcifer* (Edwards, 1913b)]. Subgenus abbreviation – *Dic.*

****

Systematics: *Diceromyia* includes eight extant species in sub-Saharan Africa divided between four species groups. *Diceromyia* was recovered as sister to *Ayurakitia* in the phylogenetic study of Aedini by Reinert *et al.* (2009) based on extensive morphological data of adults, larvae and pupae. In the phylogeny generated by Wilkerson *et al.* (2015), *Diceromyia* was placed within a polytomy that included *Ayurakitia*, *Belkinius*, *Borichinda*, *Bothaella*, *Catatassomyia*, *Cornetius*, *Dendroskusea*, *Isoaedes*, *Petermattinglyius*, *Pseudarmigeres*, *Scutomyia* and *Stegomyia*, as well as the genera *Heizmannia*, *Udaya* and *Zeugnomyia*. As the Malagasy species were transferred to *Paulianius*, *Diceromyia* may share affinities with *Paulianius*. In the absence of species of *Paulianius*, *Diceromyia* was recovered as sister to a clade comprised of *Petermattinglyius* + (*Cornetius* + *Lorrainea*) in the maximum likelihood phylogeny of Soghigian *et al.* (2017) based on seven molecular markers.

****

Characteristics: *Diceromyia* is a polythetic taxon that is diagnosed by a unique combination of characters. ADULTS: Vertex of the head with broad pale decumbent scales; maxillary palpus about as long as the proboscis, usually with a median pale band and a small basal pale spot in females, maxillary palpus of males straight with pale bands at the bases of palpomeres 3–5; scutum with areas of narrow (sometimes rather board) pale scales; prescutellar area with median and/or posterior scales; acrostichal and dorsocentral setae strong and numerous; paratergite and all lobes of the scutellum (except in *Ae. flavicollis*) with broad pale scales; postpronotum, postspiracular area, subspiracular area and upper prealar area with broad pale scales; lower mesepimeral setae present; costa of the wing entirely dark-scaled or speckled with pale scales, scales very broad in some species, alula with broad scales on the margin and dorsal surface; postprocoxal scales present; femora, tibiae and hindtarsomere 1 speckled (dark in *Ae. fascipalpis*); tarsomeres 1–4 of all legs and hindtarsomere 5 with basal pale bands; foreungues toothed in both sexes; abdominal terga usually with basal pale bands or pale median spot and basolateral pale patches, bands becoming pre-basal laterally in all species except *Ae. fascipalpis*; sterna dark-scaled or speckled with basal pale band. FEMALE GENITALIA: Segment VIII not completely retracted into segment VII; sternum VIII with a straight posterior margin; sternum IX a single transversely crescent-shaped sclerite, anterior margin convex; postgenital lobe not emarginate at the apex. MALE GENITALIA: Gonocoxite with a conspicuous apical scale-tuft (except in *Ae. fascipalpis*), with a ventromesal row or patch of long broad scales, basal dorsomesal lobe poorly formed except in *Ae. fascipalpis*, mesal surface membranous; gonostylus attached subapically, broad proximally, with elongate lobe on the lateral surface, gonostylar claw inserted at or before the middle, flattened and leaf-like; claspette long, basally enlarged, number and nature of setae variable; aedeagus comprised of 2 lateral plates, apparently fused apically, with numerous lateral and apical teeth. LARVAE: Antenna spiculate, seta 1-A with a few branches; seta 4-C inserted posterior to seta 6-C; seta 6-C inserted at the same level or posterior to seta 7-C; seta 4-P branched, as long as or shorter than seta 3-P; seta 5-P longer than seta 6-P; the setal support plate of the meso- and metapleural setal groups with a spine; seta 1-VII inserted on posterior 0.45 of the tergum near seta 1-VII; seta 3-VIII single; comb scales in a single row; ventral brush (seta 4-X) poorly developed, precratal setae absent; siphon short; pecten spines evenly spaced. PUPAE: Seta 1-CT weakly developed, considerably shorter than seta 3-CT; seta 3-III as long as or shorter than seta 5-III; seta 2-VI inserted lateral to seta 1; seta 1-Pa ≥ 0.8 the length of the paddle.

****

Bionomics: Females of some species of *Diceromyia* bite humans, at ground level and in or just below forest canopy, mainly during the night hours. The immature stages are found mainly in tree holes, occasionally in artificial containers and rarely in axils of plants. *Aedes furcifer* and/or *Ae. taylori* are involved in the monkey-to-human transmission of yellow fever virus in the Gambia (Germain *et al.*, 1980). They are vectors of Chikungunya and Dengue type 2 viruses in Senegal (Diallo *et al.*, 1999; Diallo *et al.*, 2003); they have been incriminated as vectors of yellow fever virus in Burkina Faso, Mali and the Sudan and Chikungunya virus in Zimbabwe and South Africa; and are known to harbour Zika, Bouboui and Bunyamwera viruses in Senegal and Dengue type 2 virus in the Ivory Coast (Roche *et al.*, 1983; Robert *et al.*, 1993; and references cited in Jupp, 1996). However, the species previously identified as *taylori* in eastern Africa (Uganda, Kenya and Tanzania) and South Africa (Natal) is *Ae. cordellieri*; consequently, the identity of the species associated with arboviruses in the Sudan, Zimbabwe and South Africa requires confirmation (Huang, 1986a).

****

Principal references: Mattingly, 1959 (Indomalayan area, taxonomy, keys, descriptions, bionomics, distributions); Reinert, 1970b (in part, Afrotropical Region, distributions); Huang, 1986a (Furcifer Group, diagnosis, bionomics, *Ae. cordellieri* and *Ae. taylori*, descriptions, distributions, bionomics); Jupp, 1996 (southern Africa, keys); Reinert, 2000e (in part, female genitalia); Reinert *et al.*, 2004, 2009 (as a genus, morphology, phylogenetic relationships); Wilkerson *et al.*, 2015 (phylogenetic relationships, classification); Huang & Rueda, 2016 (medical importance, key to adults); Soghigian *et al.*, 2017 (phylogenetic relationships).

Dobrotworskyius Reinert, Harbach & Kitching, 2006

Dobrotworskyius, originally proposed as a genus by Reinert *et al.* (2006), currently classified as a subgenus of *Aedes*, is typified by *Aedes tubbutiensis* Dobrotworsky, 1959. Subgenus abbreviation – *Dob.*

****

Systematics: *Dobrotworskyius* includes seven extant species in Australia. It was recovered in a sister relationship to *Georgecraigius* + *Patmarksia* in the phylogenetic studies of Reinert *et al.* (2006, 2009), and these three taxa were sister to a clade comprised of *Tanakaius* + (*Hulecoeteomyia* + (*Gilesius* + *Collessius*)) in the latter study. *Dobrotworskyius* was recovered in a polytomy that included the sister pair *Georgecraigius* + *Patmarksia* and four subgenera (*Collessius*, *Gilesius*, *Hulecoeteomyia* and *Tanakaius*) in an unresolved branch in the study of Wilkerson *et al.* (2015). *Dobrotworskyius* was recovered as sister to a clade comprised of *Bruceharrisonius* + *Himalaius* in the maximum likelihood phylogeny of Soghigian *et al.* (2017) based on seven molecular markers. Although their study included species of *Georgecraigius* (no species of *Patmarksia* were included), there was no evidence of a relationship with that group.

****

Characteristics: *Dobrotworskyius* is characterized by the following combinations of characters, which distinguish it from all other generic-level taxa of Aedini. ADULTS: Vertex of the head with pale and dark narrow decumbent scales, vertex and occiput with numerous erect forked scales; eyes above the antennal pedicels contiguous (but separated by approximately the width of 3.5 corneal facets in *Ae. rubrithorax*); antennal pedicels with broad scales on the mesal surface, flagellar whorls of males directed dorsally and ventrally; maxillary palpus of females with pale scales at the apices of palpomeres 3 and 4, maxillary palpus of males comprised of 5 palpomeres, palpomeres 4 and 5 slightly downturned; proboscis longer than the forefemur, dark-scaled (few scattered pale scales in *Ae. alboannulatus* and *Ae. tubbutiensis*); scutum with narrow pale and dark scales, pale scales on the lateral margins of the prescutellar space (*Ae. alboannulatus* and *Ae. milsoni* also with a few broad pale scales at this location) and paratergite; anterior and posterior acrostichal setae and anterior and posterior dorsocentral setae present; scutellum with narrow scales on all lobes; antepronotal scales all narrow; postpronotum with both narrow and broad scales; postspiracular area with scales; broad pale scales on upper proepisternum, subspiracular area, upper and lower mesokatepisternum, lower and upper prealar areas, and upper and middle mesepimeron; lower mesepimeral setae absent; wing dark-scaled, sometimes with 1–3 pale scales at the base of the costa; remigium with 1–3 dorsal setae; upper calypter with setae; alula with row of narrow dark scales on margin; hindtarsomeres 1–4 with broad basal pale bands; both fore-, mid- and hindungues of females with 1 tooth, fore- and midungues of males unequal, larger unguis with 2 teeth, smaller one with 1 tooth, hindungues each with 1 tooth; laterotergite of abdominal segment I with pale scales; terga dark-scaled with basomesal and basolateral pale areas not forming complete bands. FEMALE GENITALIA: Tergum VIII broadly rounded posteriorly, broad scales on distal 0.5; sternum VIII as long as to slightly longer than wide, caudal margin with an emargination separating 2 lobes, broad scales on lateral areas, seta 2-S inserted posterior to seta 1-S; tergum IX comprised of 2 sclerites, setae absent (*Ae. milsoni* occasionally with 1 seta distally on 1 or both sclerites); cercus broadly rounded apically, without scales; postgenital lobe narrower distally and truncate or with minute emargination; lower vaginal sclerite absent; insula lip-like, depressed medially, setae in lateral patches; 3 spermathecal capsules. MALE GENITALIA: Tergum IX lobes short, each with few setae; sternum IX with posteromedian setae; gonocoxite with small dorsobasal lobe bearing a few setae, without dorsoapical lobe, mesal surface membranous, lateral and ventral surfaces with broad scales, gonostylus attached apically, single gonostylar claw at the apex; claspette with a single columnar stem and a long apical claspette filament; aedeagus simple, tube-like. LARVAE: Seta 4-C short, with few branches, inserted mesad to seta 5-C and noticeably posterior to seta 6-C; seta 5-C inserted posteromesal to seta 6-C; seta 6-C inserted at the same level or slightly anterior to seta 7-C; seta 7-C inserted anterolateral to seta 5-C; seta 12-C branched, inserted mesal to seta 13-C; seta 13-C single, longer than seta 12-C; seta 14-C single; seta 19-C absent; setae 4,8-P branched; seta 2-T branched; seta 6-T normally single; seta 3-II branched; seta 6-I,II long, double, 6-II longer than seta 6-III; seta 7-I long, single; seta 12-I present; setae 7-II short, branched; seta 1-VII 0.48–0.85 the mid-dorsal length of segment X; seta 12-VII normally single; setae 2,4-VIII single or double; comb comprised of numerous scales in a patch; siphon with attached acus, pecten with numerous close-set spines, seta 1-S inserted slightly distal to or on level with the distal 1 or 2 spines; saddle incomplete ventrally, acus absent, seta 1-X inserted on the saddle; ventral brush (seta 4-X) comprised of multi-branched setae inserted on a grid with well-developed transverse and lateral grid bars, anterior 2 setae shorter and 1 of these occasionally inserted anterior to the grid. PUPAE: Setae 1,3-CT similarly developed; seta 11-CT single or double; seta 3-I,II branched, longer than seta 6-I,II, respectively; seta 6-I longer than seta 7-I; seta 1-II multi-branched; seta 5-II inserted lateral to seta 4-II; seta 3-III single, longer than seta 5-III; seta 5-V single, longer than the following tergum; seta 2-VI inserted mesal to setae 1,3-VI; seta 6-VII inserted posteromesal to seta 9-VII; seta 9-VIII inserted on the posterolateral corner of the segment; paddle rounded apically, midrib extending to or near the apex, without hair-like spicules on the margins; seta 1-Pa single.

****

Bionomics: Larvae of species of *Dobrotworskyius* are found in rock pools in river and stream beds and occasionally in ground pools. Some species have been found in old wells, collections of water around houses, artificial containers, a rot hole in a tree and cavities in logs. Females bite humans during the day, evening and night. Species of *Dobrotworskyius* are of no known medical or economic importance to humans.

****

Principal references: Reinert *et al.*, 2006, 2009 (as a genus, morphology, phylogenetic relationships); Reinert, 2008k (as a genus, female genitalia); Wilkerson *et al.*, 2015 (phylogenetic relationships, classification); Soghigian *et al.*, 2017 (phylogenetic relationships).

Downsiomyia Vargas, 1950

Downsiomyia Vargas, 1950, originally proposed and currently recognized as a subgenus of *Aedes*, is typified by *Aedes niveus* (Ludlow, 1903) (in Theobald, 1903a), original combination: *Stegomyia nivea*. Subgenus abbreviation – *Dow.*

****

Systematics: *Downsiomyia* includes 31 extant species in the Oriental Region and adjoining areas of the Australasian and Palaearctic Regions. It appears to be most closely related to *Haemagogus* and the subgenera *Danielsia* and *Finlaya* based on the cladistic analysis of extensive morphological data by Reinert *et al.* (2009) that generated relationships of these taxa expressed parenthetically as (*Finlaya* + *Danielsia*) + (*Downsiomyia* + *Haemagogus*). The genealogical affinities of *Downsiomyia* require further study as the character support for its sister relationship with *Haemagogus* was not very strong. It is interesting, however, that *Downsiomyia* appears to be more closely related to *Haemagogus* than to either *Danielsia* or *Finlaya*, which includes species that were previously included with the species of *Downsiomyia* in the traditional polyphyletic concept of *Finlaya*. *Downsiomyia* formed a single branch in the large unresolved polytomy recovered in the study of Wilkerson *et al.* (2015). In contrast with the finding of Reinert *et al.* (2009), *Downsiomyia* was recovered as sister to a clade comprised of *Danielsia* + *Luius* in the maximum likelihood phylogeny of Soghigian *et al.* (2017) based on seven molecular markers.

****

Characteristics: The adults of species of *Downsiomyia* are easily recognized by the presence of broad decumbent scales on the vertex and erect scales on the occiput of the head, the presence of patches of pale scales covering the scutal fossae and by the absence of acrostichal and dorsocentral setae and postspiracular scales. Patches of pale scales also occur on the scutal fossae of members of the Gubernatoris Group of *Finlaya*, and a similar condition occurs in some species of *Stegomyia*, but those species differ in having pale scaling on the supraalar area, which is dark-scaled in *Downsiomyia*. The simple aedeagus with small apical teeth of *Downsiomyia* males is somewhat similar to the aedeagus of some species of *Ochlerotatus*, but otherwise differ in numerous other features. Larvae of *Downsiomyia* resemble those of *Ayurakitia* in having setae 4,5,6-C multi-branched and inserted slightly anterior or posterior to seta 7-C, but differ in having seta 12-I and fewer comb scales arranged in a single row.

Like other generic-group taxa of Aedini, the life stages of *Downsiomyia* are diagnosed and distinguished by combinations of characters, as follow. ADULTS: Vertex of the head with only broad decumbent scales, erect scales confined to the occiput; eyes contiguous; antennal pedicel with scales and setae on the mesal surface; maxillary palpus and proboscis dark-scaled; acrostichal and dorsocentral setae absent; scutal fossa with pale falcate scales; supraalar area dark-scaled; all lobes of the scutellum with broad scales; paratergite, postspiracular area, subspiracular area and metameron without scales; wing entirely dark-scaled; hindleg, except proximal part of the femur, dark-scaled; fore- and midungues of males unequal, each with 1 tooth. FEMALE GENITALIA: Tergum VIII and sternum VIII with numerous broad scales; tergum IX comprised of 2 narrow lobes each with few apical setae; cercus short and broad, with 5–7 stout distolateral setae; upper vaginal sclerite small; insula lip-like with few long setae in lateral patches; 1 large (Niveus Group) or 1 large and 2 smaller (Alboniveus Group) spermathecal capsules. MALE GENITALIA: Gonocoxite with a ventromesal cluster of several to numerous broad fusiform scales (except *Ae. mikrokopion*) and a narrow thumb-like setose lobe borne dorsomesally at the base; gonostylus short with a long gonostylar claw at the apex; claspette columnar with long longitudinally striated claspette filament at the apex; aedeagus simple, scoop-like with 1 or usually 2 or more small apical teeth on each side. LARVAE: Antenna spiculate; seta 1-A multi-branched; seta 1-C single, attenuate distally; setae 4–7-C well developed, multi-branched; seta 12-I present; seta 7-I large and single, seta 7-II short and multi-branched; seta 6-II shorter than seta 6-III; setae 2,4-VIII both single; spines of pecten evenly spaced; seta 1-S branched, inserted distal to the pecten; saddle incomplete; seta 1-X inserted on the saddle. PUPAE: Seta 1-CT long, stout and normally single; setae 2,3-CT short and slender; seta 5-CT distinctly longer than seta 4-CT, often long and single; seta 11-CT long, stout and single; seta 3-I long and stout, longer than seta 6-I; seta 6-VII short, with a few branches, inserted posterior to seta 9-VII; seta 9-VII long, stout and normally double or triple; seta 9-VIII long, stout, multi-branched, inserted on the margin of the segment; paddle without hair-like marginal spicules, midrib well developed.

****

Bionomics: *Downsiomyia* are forest mosquitoes. The immature stages normally inhabit water in tree holes but they are found occasionally in stump holes, bamboo stumps, bamboo internodes, bamboo cups and split bamboo on the ground (Macdonald & Traub, 1960; Rattanarithikul *et al.*, 2010). Two species, *Ae. niveoides* and *Ae. novoniveus*, have been found in an artificial container and a rock hole, respectively (Rattanarithikul *et al.*, 2010). Adult females of some species, *e.g. Ae. albolateralis*, *Ae. harinasutai*, *Ae. niveoides* and *Ae. vanus*, are known to bite humans at ground level and in forest canopy during the daylight, early morning and evening hours (Macdonald, 1957; Wharton, 1962; Macdonald *et al.*, 1965). *Aedes niveus* and *Ae. harinasutai* are vectors of *Wuchereria bancrofti* in the Philippines (Rozeboom & Cabrera, 1964) and Thailand (Gould *et al.*, 1982). Zagaria & Savioli (2002) also listed these two species as vectors of filarial parasites. Shriram *et al.* (2005) found females of *Ae. niveus* (as *Ochlerotatus niveus*) in the Andaman and Nicobar Islands of India to be naturally infected with microfilariae of *W. bancrofti*, but the identification of the species is questionable (Huang & Rueda, 1998). Rudnick *et al.* (1986) reported that species of *Downsiomyia* (as members of the Niveus Subgroup of *Finlaya*) were sylvatic vectors of dengue fever virus in Malaysia.

****

Principal references: Reinert *et al.*, 2004, 2006, 2009 (as a genus, morphology, phylogenetic relationships); Reinert & Harbach, 2006 (as a genus and type species, taxonomy, descriptions, bionomics, distributions); Reinert, 2008h (as a genus, female genitalia); Rattanarithikul *et al.*, 2010 (as a genus, Thailand, keys, bionomics); Wilkerson *et al.*, 2015 (phylogenetic relationships, classification); Soghigian *et al.*, 2017 (phylogenetic relationships).

Edwardsaedes Belkin, 1962

Edwardsaedes Belkin, 1962, originally proposed and currently recognized as a subgenus of *Aedes*, is typified by *Aedes imprimens* (Walker, 1860), original combination: *Culex imprimens*. Subgenus abbreviation – *Edw.*

****

Systematics: *Edwardsaedes* includes three extant species with distributions in the Australasian Region (Australia, Papua New Guinea, Solomon Islands), eastern Palaearctic (China, Japan, Korea, Russia) and the Oriental Region (Bangladesh, Cambodia, India, Indonesia, Malaysia, Philippines, Thailand, Vietnam). Belkin (1962) noted that *Edwardsaedes* bears similarities to *Aedimorphus*, *Neomelaniconion* and *Verrallina*. Based on the cladistic analysis of extensive morphological data by Reinert *et al.* (2009), *Edwardsaedes* appears to be related to those generic-level taxa as well as to *Aedes sensu stricto* – the five groups comprised a clade expressed parenthetically as *Neomelaniconion* + (*Edwardsaedes* + (*Aedes* + (*Paraedes* + *Verrallina*))). *Edwardsaedes* was recovered as the sister to a clade comprised of *Neomelaniconion* + (*Aedes* + (*Parades* + *Verrallina*)) in the phylogeny generated by Wilkerson *et al.* (2015). The only species of *Edwardsaedes*, *Ae. bekkui*, included in the maximum likelihood phylogeny of Soghigian *et al.* (2017) based on seven molecular markers was recovered as the sister of a clade comprised of five species of *Verrallina*.

****

Characteristics: Species of *Edwardsaedes* exhibit a number of unique features, as noted in the following diagnosis. ADULTS: Eyes narrowly separated; decumbent scales of the vertex largely narrow, erect scales numerous, extending to the ocular setae; antennal pedicel with few broad scales and fine setae, flagellar whorls of males with dorsoventrally directed setae; maxillary palpus of females slender, about 0.20–0.25 the length of the proboscis, with 5 palpomeres, palpomere 5 small, palpomeres 1 and 2 ankylosed; maxillary palpus of males longer than the proboscis, palpomeres 1 and 2 short, palpomeres 3–5 long, palpomere 5 upturned; scutum and scutellum with narrow falcate scales only, scales on median and/or posterior area of the scutellum; acrostichal and dorsocentral setae short, numerous, in complete rows; paratergite bare; scales on the anterior part of the antealar area all pale; antepronotum and postpronotum with narrow falcate scales; postspiracular scales present, reduced in male; subspiracular area without or with short setae intermixed with broad scales; prealar area without scales; mesepimeron with a median scale-patch; lower mesepimeral setae absent; femora with knee spots; tibiae and tarsi with basal pale markings; both ungues of all legs toothed in both sexes, fore- and midungues of males unequal; tergum I and laterotergite extensively scaled; terga II–VII with lateral pale markings, reduced in males, terga II–V or VI with basal pale bands, terga I–VII of males with numerous long lateral setae. FEMALE GENITALIA: Segment VIII completely retracted into segment VII, tergum VIII with setae on the distal 0.7 or more, anterior and posterior margins of sternum VIII with a deep emargination; cercus long and slender, without setae; upper vaginal sclerite present, lower vaginal sclerite absent; insula long, tongue-like, without setae or tuberculi; 3 spermathecal capsules, unequal. MALE GENITALIA: Short, inconspicuous, deeply retracted; tergum IX comprised of 2 plates separated by a median membrane, each plate with a conspicuous projecting lobe, lobes close together; gonocoxite very short and broad, with a few broad scales, setae absent from dorsomesal area of the dorsal surface, basal mesal lobe a raised plate with numerous setae, basal dorsomesal lobe not developed, gonostylus attached at the apex, with large mesal lobe; gonostylar claw absent; aedeagus relatively simple, slender, apex sagittate, denticles present on the distal 0.7; paraproct with a strongly developed apical claw-like projection, sternal arm absent; cercal setae absent. LARVAE: Antenna long, slender with numerous spicules; seta 1-A multi-branched; seta 4-C inserted mesal and cephalad of setae 5,6-C; setae 5,6-C and 12,13-C approximated, 12,13-C removed from the cephalic margin of the head; ventromedian cervical sclerite well developed; setae 9–12-M,T borne on a large plate with a long stout spine; seta 12-I absent; seta 1-VII short, multi-branched; setae 1,2-VIII borne on a common setal support plate; seta 3-VII inserted anterior to seta 1-VII; seta 4-VIII branched; comb scales in an irregular patch or double row, with basolateral fringes and long apical spine; siphon relatively long, acus present; pecten of numerous spines, distal spines more widely spaced; seta 1-S inserted distal to the pecten, small, multi-branched; saddle complete or incomplete, without acus, caudal margin with short heavy spicules dorsally; seta 1-X single; seta 2-X multi-branched; seta 3-X single or branched, longer than 2-X; ventral brush (seta 4-X) with 5 pairs of setae borne on a poorly developed grid and a number of precratal setae borne on the saddle. PUPAE: Trumpet short, tracheoid area absent; seta 8-CT inserted cephalad of the trumpet; seta 10-CT inserted caudad and mesal or laterad of seta 11-CT; setae 2,3-I moderately to widely separated; seta 9-VII multi-branched, inserted directly cephalad to and longer than seta 6-VII; genital lobe of females prominent, long; paddle without a marginal fringe of hair-like spicules; seta 1-Pa short.

****

Bionomics: Species of *Edwardsaedes* are known from locations at sea level to 1200 m. The immature stages usually inhabit shaded temporary pools and flowing water in flooded areas in forest. Eggs are laid individually and require drying for 4–6 weeks prior to hatching. Larvae mature within 3 days. Adults have been found resting in houses, stables and on vegetation, and have been captured in light traps. Females viciously attack and bite humans during daytime and are capable of biting through heavy clothing. Nothing is known of the disease relations of species of *Edwardsaedes*, but available evidence suggests they are unlikely to vector pathogens to humans.

****

Principal references: Belkin, 1962 (new subgenus and type species, taxonomy, descriptions, bionomics, distribution); Reinert, 1976c (subgenus and type species, taxonomy, descriptions, distribution, bionomics); Mogi, 1977 (*Ae. bekkui*, comparative morphology); Lu *et al.*, 1997a (China, keys, taxonomy, descriptions, distributions, bionomics); Reinert, 2000e (female genitalia); Reinert *et al.*, 2004, 2009 (as a genus, morphology, phylogenetic relationships); Rattanarithikul *et al.*, 2010 (as a genus, Thailand, keys, bionomics); Wilkerson *et al.*, 2015 (phylogenetic relationships, classification); Soghigian *et al.*, 2017 (phylogenetic relationships).

Elpeytonius Reinert, Harbach & Kitching, 2009

Elpeytonius, originally proposed as a genus by Reinert *et al.* (2009), currently classified as a subgenus of *Aedes*, is typified by *Aedes apicoannulatus* (Edwards, 1912a) [replacement name for *Aedimorphus alboannulatus* Theobald, 1905b], original combination: *Ochlerotatus apicoannulatus*. Subgenus abbreviation – *Elp.*

****

Systematics: *Elpeytonius* includes only two extant species in sub-Saharan Africa. The type species is recorded only from Nigeria and Sierra Leone; *Ae. simulans* is recorded from Cameroon, Central African Republic, Democratic Republic of the Congo, Ghana, Liberia, Nigeria, Sierra Leone, Sudan and Uganda. *Elpeytonius* is sister to *Catageiomyia* in the morphological phylogeny of Reinert *et al.* (2009). These two taxa appear to be rather distantly related to *Aedimorphus* in which they were previously assigned. *Elpeytonius* formed a single branch in the large unresolved polytomy recovered in the study of Wilkerson *et al.* (2015).

****

Characteristics: *Elpeytonius* is characterized by the following combinations of characters, which distinguish it from all other generic-level taxa of Aedini. ADULTS: Decumbent scales of the vertex all narrow, numerous erect forked scales on the occiput and vertex; eyes above the antennal pedicels contiguous or separated by the diameter of 2 corneal facets; antennal pedicel with a few setae and sometimes a few scales on the mesal surface; maxillary palpus of females dark-scaled, maxillary palpus of males with 5 palpomeres, palpomeres 4 and 5 somewhat downturned, palpomeres 4 and 5 and distal part of 3 with ventrolateral setae; proboscis pale-scaled ventrally near mid-length; scutum with narrow falcate scales; acrostichal (anterior and posterior) and dorsocentral (anterior and posterior) setae present; scutellum with broad silvery scales on all lobes; paratergite broad, with pale scales; antepronotum with broad silvery scales; postspiracular area without scales; upper proepisternum with 1−4 setae; lower mesepimeral setae absent; wing dark-scaled with pale scales at the base of the costa; hindfemur and hindtibia with apical pale scales; hindtarsomeres 1–4 with a wide apical white band, hindtarsomere 5 entirely or nearly entirely white-scaled; fore- and midungues of females equal, each with a tooth, fore- and midungues of males unequal, both foreungues with a tooth, larger midunguis with a tooth; laterotergite of tergum I with board pale scales, terga I−VI dark-scaled dorsally; segment VII of females laterally compressed. FEMALE GENITALIA: Tergum VIII with setae on the distal 0.6 or less; sternum VIII wider than long, apex with a median emargination, scales absent or few in number, seta 2-S inserted posterior to seta 1-S; cercus without scales; insula tongue-like with 4–6 distal tuberculi; 1 large spermathecal capsule (shared with *Catageiomyia*). MALE GENITALIA: Gonocoxite with broad scales, gonostylus attached at the apex; distal part of the gonostylus with an elongate lobe on the lateral surface (shared with *Catageiomyia*), apex with broad leaf-like gonostylar claw; claspette a short narrow plaque at the base of the gonocoxite, with a few basal setae; aedeagus comprised of 2 sclerites. LARVAE: Seta 4-C with 3–9 branches, inserted mesal to seta 6-C; seta 6-C branched (shared with *Catageiomyia*), inserted close to seta 5-C; setae 1–3-P not inserted on a common setal support plate; seta 7-I long, double; seta 12-I absent; seta 5-II longer than seta 3-II (shared with *Catageiomyia*); comb scales in a patch; siphon acus present; spines of the pecten evenly spaced; seta 1-S with 2–5 branches, inserted distal to the pecten; saddle incomplete, acus absent; seta 1-X single to triple, inserted on the saddle; ventral brush (seta 4-X) with branched fan-like setae borne on a grid with both transverse and lateral grid bars, several precratal setae. PUPAE: Seta 1-CT triple, shorter than seta 3-CT; seta 11-CT single; seta 3-I very long, stout, single; seta 6-I longer than seta 7-I; seta 3-II inserted mesal or anterior to seta 2-II; seta 5-V longer than tergum VI; seta 9-VII branched, inserted anterolateral to seta 6-VII; paddle without hair-like marginal spicules, apical margin rounded, midrib complete; seta 1-Pa short, double or triple.

****

Bionomics: The immature stages of species of *Elpeytonius* inhabit cavities in trees, stumps and axils of plants of the *Dracaena* family. Females of *Ae. simulans* have been captured occasionally biting humans in forest and plantations, mainly during daylight hours. Species of *Elpeytonius* are not known to be of medical importance to humans or domestic animals.

****

Principal references: Reinert *et al.*, 2009 (as a genus, morphology, phylogenetic relationships, description, distribution, bionomics); Reinert, 2010e (as a genus, female genitalia); Wilkerson *et al.*, 2015 (phylogenetic relationships, classification).

Finlaya Theobald, 1903

Finlaya, originally proposed as a genus by Theobald (1903a), currently classified as a subgenus of *Aedes*, is typified by *Aedes kochi* (Dönitz, 1901), original combination: *Culex kochi*. Subgenus abbreviation – *Fin.*

****

Systematics: *Finlaya* includes 36 extant species in southern areas of the Oriental Region and northern and eastern areas of the Australasian Region. *Finlaya* appears to be most closely related to *Danielsia* and *Downsiomyia*, which comprise species previously included in the traditional polyphyletic concept of *Aedes* (*Finlaya*), as well as the Neotropical *Haemagogus*, based on the cladistic analysis of morphological data by Reinert *et al.* (2009) that generated relationships of these groups expressed parenthetically as (*Finlaya* + *Danielsia*) + (*Downsiomyia* + *Haemagogus*). *Finlaya* was also recovered as the sister of *Danielsia* in the study of Wilkerson *et al.* (2015). Pao & Knight (1970) recognized the Marshallii Group of *Aedes* (*Aedimorphus*) to include *Ae.* (*Elp.*) *apicoannulatus*, *Ae.* (*Elp.*) *simulans*, *Ae.* (*Pseudalbuginosus*) *grjebinei* and eight species of *Finlaya*, but there is no evidence of a close phylogenetic relationship between the first three species and *Finlaya*. The internal relationships of *Finlaya* and its genealogical association with other aedine genera require further study.

****

Characteristics: The following combinations of characters easily distinguish species of *Finlaya* from those of other aedine taxa. ADULTS: Maxillary palpus and proboscis with pale-scaled areas; maxillary palpus of males long with the distal part of palpomere 3 slightly swollen, upturned and bearing several long setae on the ventrolateral area, palpomeres 4 and 5 short and downturned; a few setae on upper proepisternum; postpronotum with a large patch of broad scales; postspiracular area with a few setae and normally no scales; veins of the wing covered with broad scales arranged in pale and dark areas (spots), costa with 3 or more pale areas, wing fringe with pale areas at the ends of the veins; femora and tibiae of all legs with pale bands and/or spots (pale bands or spots, or both bands and spots, also occur on the femora and tibiae of some species of *Diceromyia* and *Mucidus*); tarsi of all legs with extensive pale areas, hindtarsomere 1 with basal, median and apical pale bands (median band rarely reduced to a pale spot), hindtarsomeres 2 and 3 with apical pale bands (nearly completely pale in some species), hindtarsomere 5 always white-scaled; larger fore- and midunguis of males with 2 teeth, smaller fore- and midunguis with 1 tooth; abdominal terga with white-scaled areas. FEMALE GENITALIA: Distinctive; tergum and sternum VIII with large patches of broad scales; tergum IX composed of 2 long narrow lateral sclerites connected basally by a lightly pigmented strip, bearing 1–5 short setae apically (absent in *Ae. knighti*); upper vaginal sclerite well developed and heavily pigmented; cercus short with a broadly rounded or flattened apex bearing a few moderately long and long stout setae, several short lanceolate setae and a number of broad scales on the dorsal surface. MALE GENITALIA: Distinctive; aedeagus scoop-shaped with distal 0.33 expanded; dorsal surface of the gonocoxite with a large patch of numerous tightly packed setae on the basomesal area, ventral surface with a patch of long to very long fusiform scales on the dorsomesal margin (males of *Downsiomyia*, *Haemagogus* and *Ae.* (*Och.*) *albotaeniatus* also have long fusiform scales on the ventral surface but they extend more distad in *Downsiomyia* and *Haemagogus*); claspette with a columnar stem usually bearing a single short seta on both the proximal and distal 0.33, claspette filament long, foliform and normally annulated transversely (an annulated claspette filament is also found in some species of the Gubernatoris and Papuensis Groups of *Ochlerotatus*); tergum IX consists of broad lateral plates, each with a very small subapical lobe on the posterior margin bearing 1–3 moderately long setae, lateral plates connected mesally by a narrow membranous strip. LARVAE: Fourth-instar larvae display a number of unusual features, including seta 1-C stout proximally and branched distally (a branched seta 1-C also occurs in *Ae.* (*Dow.*) *axitiosus*, *Ae.* (*How.*) *walkeri* and *Ae.* (*Stg.*) *pandani*); seta 3C absent; the development and positions of setae 4–7-C are distinctive; seta 13-C inserted mesad of seta 12-C; seta 14-C displaced caudally, longer than and posterior or lateral to seta 15-C; seta 2-A noticeably longer than setae 3–6-A; seta 0-P displaced laterally; seta 2-P very long and single; setae 1,3-P short, stellate and multi-branched; setae 5,10,12-M and 10-T very long; seta 5-T inserted mesad and slightly caudad of seta 6-T; seta 8-T displaced anteriorly; seta 11-M,T relatively moderately long and usually with 5 or more branches; seta 5-I–VI large, stellate and inserted noticeably anterior to seta 6-I–VI; seta 5-III–VI inserted anterior to seta 8-III–VI; seta 9-II–VI long, displaced anteriorly and multi-branched; setae 6,8-S very long, single; spiracular apodeme of the siphon with a distinctive shape; ventral brush (seta 4-X) comprised of long pectinate setae borne on a boss; comb scales and pecten spines distinctively developed (larvae of *Ae.* (*Cts.*) *meronephada* have setae 14,15-C, 5-I–VI and 9-II–IV similarly developed to larvae of *Finlaya*, and larvae of *Lorrainea* have seta 5-I–VI inserted slightly anterior to seta 6-I–VI but they are small and simple). PUPAE: Cephalothorax and/or abdomen with clear unpigmented spots (present also in *Ae.* (*Ala.*) *brevitibia* and *Ae.* (*Blk.*) *aurotaeniatus* but otherwise unique within Aedini); other unusual or unique features of *Finlaya* pupae include seta 2-II inserted anterior and usually slightly mesad of seta 1-II, seta 5-II inserted mesad of seta 4-II, seta 8-II present, seta 9-VI inserted anterior to seta 8-VI and setae 8,9-VII inserted dorsal and far anterior on the tergum.

****

Bionomics: Larvae of *Finlaya* are found almost exclusively in small collections of water in leaf axils of plants, especially species of Pandanaceae and Araceae. Belkin (1962) noted that at least four species on Guadalcanal may be found in a single host plant (*Pandanus*) at the same time but each species occurred predominantly in certain leaf axils and generally at a specific level of the plant. He also noted that the immature stages are sometimes found in extremely shallow water, and the larvae can crawl out of the water and move into a more suitable niche. Females of *Ae. fijiensis*, *Ae. kochi*, *Ae. oceanicus* and *Ae. samoanus* are known to bite humans. They appear to be predominantly nocturnal feeders but there are some records of diurnal attacks (Belkin, 1962). Species of *Finlaya* do not appear to be of medical or economic importance to humans.

****

Principal references: Stone & Bohart, 1944 (Philippines, as *Aedes* (*Finlaya*) *kochi* group, keys, taxonomy, species descriptions); Lu *et al.*, 1997a (China, *Finlaya sensu* Knight & Stone, 1977, keys, species groups, descriptions, distributions, bionomics); Reinert *et al.*, 2004, 2006, 2008, 2009 (as a genus, morphology, phylogenetic relationships); Reinert & Harbach, 2005 (as a genus and type species, taxonomy, descriptions, distribution, bionomics); Rattanarithikul *et al.*, 2010 (as a genus, Thailand, keys, bionomics); Wilkerson *et al.*, 2015 (phylogenetic relationships, classification).

Fredwardsius Reinert, 2000

Fredwardsius Reinert, 2000a, originally proposed and currently recognized as a subgenus of *Aedes*, is typified by *Aedes vittatus* (Bigot, 1861), original combination: *Culex vittatus*. Subgenus abbreviation – *Fre.*

****

Systematics: *Fredwardsius* is monobasic. It only includes the extant type species, which is recorded from many countries in the Afrotropical, Oriental and Palaearctic Regions. The closest relatives of *Fredwardsius* are uncertain. In the morphological phylogeny of Reinert *et al.* (2009), *Fredwardsius* was the basal branch of a large clade that was sister to *Skusea* + (*Indusius* + *Cancraedes*), and was followed by a pectinate series of five lineages: *Isoaedes*, *Borichinda*, *Diceromyia* + *Ayurakitia*, *Dendroskusea*, and *Scutomyia* + (*Catatassomyia* + *Bothaella*) in a sister relationship to a group of other generic-level taxa. *Fredwardsius* was recovered as the sister of *Indusius* + *Skusea* in the phylogenetic analyses of Wilkerson *et al.* (2015). *Aedes vittatus* was variously classified as a species of *Stegomyia*, *Scutomyia* and *Aedimorphus*, but it bears a number of unique and unusual features that are not shared with members of those groups. *Aedes vittatus* was recovered as sister to a clade comprised of *Diceromyia* + (*Petermattinglyius* + (*Cornetius* + *Lorrainea*)) in the maximum likelihood phylogeny of Soghigian *et al.* (2017) based on seven molecular markers.

****

Characteristics: The following unique and unusual features distinguish *Fredwardsius* from all other generic-level taxa of Aedini. ADULTS: Eyes narrowly to moderately separated above the antennal pedicels; decumbent scales of the vertex both broad and narrow; antennal pedicel with separate mesal and lateral patches of broad silvery scales; clypeus of females with a patch of pale scales on either side; maxillary palpus of females with some pale scales, palpomeres 1–3 short, palpomere 4 long and palpomere

5 small and bud-like; maxillary palpus of males longer than the proboscis by about the length of palpomere 5, palpomeres 2–5 each with a broad basal pale band, palpomere 3 with a few short apical setae, palpomeres 4 and 5 narrow and upturned, palpomere 4 with a nearly complete ventrolateral row of setae, palpomere 5 with a few ventral and apical setae; scutum and antepronotum with both broad and narrow scales; upper and lower areas of the prealar area with scales; 2–6 lower mesepimeral setae; metameron usually with scales; remigium of the wing without dorsal setae; postprocoxal scales present; femora with complete subapical pale bands; hindtarsomere 2 with basal pale scales. FEMALE GENITALIA: Sternum VIII without scales, heavily pigmented, posterior margin with a deep, narrow, median emargination; postgenital lobe narrow, ventral width of the distal $0.2 \leq 0.65$ the width of the cercus at mid-length; ventral surface of the postgenital lobe with setae inserted mesally (unique); upper vaginal sclerite well developed, with jagged posterior margin. MALE GENITALIA: Posterior margin of tergum VIII strongly convex, with numerous long stout setae; tergum IX lobes small and relatively narrow; sternum IX without setae and scales; gonocoxite relatively narrow, ventral surface with a small apicomesal lobe bearing several moderately long setae; distal 0.5 of the gonostylus greatly expanded; gonostylar claw long, narrow, strongly curved and borne on the ventral surface of an expanded area; basal mesal lobe oblong, mostly detached from the mesal surface of the gonocoxite, with only very short setae. LARVAE: Seta 4-C minute, inserted anteromesad of seta 5-C and near and slightly posteromesad of seta 6-C; setae 5,6-C single; seta 5-C well removed posterior to setae 4,6-C; seta 7-C moderately long, branched, inserted anterolateral to seta 5-C and posterolateral to seta 6-C; seta 8-P relatively long, stout, aciculate, with 2 or 3 branches; seta 3-VII inserted at the same level or slightly posterior to seta 1-VII; seta 4-VII long, single, noticeably longer than setae 1,3-VII; seta 1-VIII short, branched; seta 2-VIII long, single; seta 5-I–VI short, seta 5-II–V normally single; seta 1-X short, single, inserted on the posteroventral margin of the saddle; ventral brush (seta 4-X) usually composed of 8 cratal and 3–5 precratal setae. PUPAE: Trumpet without tracheoid area; seta 6-I as long as or shorter than seta 7-I; seta 1-II short, with multiple fine branches; seta 3-II branched, longer than seta 6-II; seta 4-III,IV short, single; seta 4-VIII noticeably longer than seta 9-VIII; seta 5-II longer than seta 3-II; seta 6-II as long as or shorter than seta 7-II; seta 5-V shorter than tergum VI; seta 6-IV–VI inserted at about the same level or slightly dorsoposterior to seta 9; paddle broad, apical margin slightly rounded, outer margin with very few minute spicules, without fringe of long hair-like spicules, midrib weak, terminating well before the apex of the paddle.

****

Bionomics: The immature stages of *Ae. vittatus* usually inhabit rock pools and rock holes, but they have been found in tree holes, bamboo pots, hoof prints, wells and artificial containers. Females are known to feed on humans. For additional bionomical information, see Mattingly (1965), Service (1970b), Huang (1977b) and Becker *et al.* (2003, 2010, 2020). *Aedes vittatus* is a potential vector species. It has been shown to transmit yellow fever virus from monkey to monkey in the laboratory (Philip, 1929) and has been suspected as a vector in the Nuba Mountain epidemic of yellow fever in Sudan (Lewis, 1943) (also see Clements, 2012).

****

Principal references: Reinert, 2000a (new subgenus, taxonomy, description, distribution, bionomics); Reinert *et al.*, 2004, 2009 (as a genus, morphology, phylogenetic relationships); Becker *et al.*, 2003, 2010, 2020 (in keys, description, bionomics, distribution); Rattanarithikul *et al.*, 2010 (as a genus, Thailand, keys, bionomics); Wilkerson *et al.*, 2015 (phylogenetic relationships, classification); Soghigian *et al.*, 2017 (phylogenetic relationships).

Georgecraigius Reinert, Harbach & Kitching, 2006

Georgecraigius, originally proposed as a genus by Reinert *et al.* (2006), currently classified as a subgenus of *Aedes*, is typified by *Aedes atropalpus* (Coquillett, 1902d), original combination: *Culex atropalpus*. Subgenus abbreviation – *Grg.*

****

Systematics: *Georgecraigius* includes three extant species with distributions in South, Central and North America. In the morphology-based phylogeny of Reinert *et al.* (2009), *Georgecraigius* was included in a clade in which *Rampamyia* + (*Molpemyia* + (*Luius* + *Macleaya*)) were sister to a clade that was divided into two clades comprised of *Dobrotworskyius* + (*Patmarksia* + *Georgecraigius*) and *Tanakaius* + (*Hulecoeteomyia* + (*Gilesius* + *Collessius*)). In the phylogeny recovered in the study of Wilkerson *et al.* (2015), *Georgecraigius* was the sister of *Patmarksia* in a polytomy that also included *Collessius*, *Gilesius*, *Hulecoeteomyia* and *Tanakaius* in an unresolved clade, and *Dobrotworskyius*. *Georgecraigius* was recovered as sister to *Gymnometopa* in the maximum likelihood phylogeny of Soghigian *et al.* (2017) based on seven molecular markers. There was no evidence of a relationship between *Georgecraigius* and the genera associated with it in the phylogenies of Reinert *et al.* and Wilkerson *et al.*

****

Characteristics: The following combinations of characters easily distinguish the species of *Georgecraigius* from those of other aedine taxa. ADULTS: Vertex of the head with both narrow central and broad lateral scales, with numerous erect forked scales mainly or entirely on the occiput; ocular line with narrow pale scales; eyes separated above the antennal pedicels; antennal pedicel with a few scales and setae on the mesal surface, flagellar whorls of males with setae directed dorsally and ventrally; clypeus bare;

maxillary palpus and proboscis dark-scaled, palpus of males with 5 palpomeres, shorter than the proboscis, palpomeres 4 and 5 nearly straight relative to palpomere 3; anterior and posterior acrostichal setae present; anterior dorsocentral setae absent, posterior dorsocentral setae present; supraalar area with a longitudinal patch of pale scales; prescutellar area with pale scales on the lateral margins of the bare space; scutellum with narrow scales on all lobes; scales present on the antepronotum and postpronotum, broad scales on the subspiracular, upper proepisternal, upper and lower mesokatepisternal, lower prealar and upper mesepimeral areas; lower mesepimeral setae absent; metameron bare; wing dark-scaled with a pale patch at the base of the costa; upper calypter with several setae; alula with narrow scales on the margin; remigial setae absent; hindfemur with pale scales at the apex; hindtarsomeres 1 and 2 with a narrow pale basal band, both fore- and midungues of females with 1 tooth, hindungues simple in both sexes, fore- and midungues of males unequal, larger unguis with 2 teeth, smaller one with 1 tooth. FEMALE GENITALIA: Tergum and sternum VIII broadly rounded posteriorly, with numerous scales on much of the surface, sternum VIII slightly wider than long; seta 2-S inserted posterior to seta 1-S; tergum IX comprised of 2 oval sclerites, without setae; postgenital lobe truncate or with a shallow median emargination; upper vaginal sclerite small; lower vaginal sclerite absent; insula lip-like, with a few setae in lateral patches; cercus with broadly rounded apex, without scales; 3 spermathecal capsules. MALE GENITALIA: Tergum IX lobes narrowly connected, without setae; sternum IX with setae; gonocoxite with an apicodorsal lobe, with a setose basodorsal lobe; gonostylus attached apically, long, with single short gonostylar claw at the apex; claspette comprised of a narrow columnar stem with a long somewhat flattened and slightly curved apical claspette filament; aedeagus simple, tube-like, distal 0.5 slightly wider, apex convex; cercal setae present. LARVAE: Antenna with a few to several spicules; seta 1-C single, thin, distally attenuate; seta 4-C short, branched, inserted mesal and slightly posterior to seta 6-C; seta 5-C stout, inserted posterior to seta 6-C; seta 6-C stout; seta 7-C stout, branched, inserted about the same level or slightly posterior to seta 5-C; seta 12-C branched, shorter than and inserted mesal to seta 13-C; seta 13-C single; seta 14-C short, single; seta 19-C absent; seta 4-P single; seta 5-M longer than seta 7-M; setae 2,6-T single; seta 12-I present; seta 6-I–III long, aciculate, branched; seta 7-I long, stout, aciculate; setae 7,10,12-VII single; seta 1-VIII single; seta 2-VIII longer than seta 1-VIII, setae 2,4-VIII single; comb comprised of numerous scales in a patch; siphon with attached acus; pecten with numerous spines, evenly spaced or the distal 1 or 2 spines more widely spaced; seta 1-S moderately long, aciculate, multi-branched; saddle relatively small, incomplete ventrally, acus absent; seta 1-X single, inserted ventral to the saddle (occasionally on the margin of the saddle in *Ae. fluviatilis*); ventral brush (seta 4-X) borne on a grid with transverse and lateral grid bars, precratal setae absent. PUPAE: Seta 11-CT single, longer than setae 10,12-CT; seta 3-I stout, single (rarely 2-branched on 1 side of *Ae. atropalpus*), noticeably longer than seta 6-I; setae 6-I,II shorter than seta 7-I,II, respectively; seta 1-II multi-branched; seta 2-II short, inserted lateral to setae 1,3-II; seta 3-III single, long, stout, inserted on level with seta 1-III; seta 5-III inserted mesal to seta 4-III; seta 6-VII short, slender, single or double, inserted mesal and posterior to seta 9-VII; seta 9-VII stout, branched, longer than seta 6-VII; seta 9-VIII stout, multi-branched; paddle wide; without hair-like spicules on the lateral margin; midrib distinct to the apex; seta 1-Pa single.

****

Bionomics: The immature stages of species of *Georgecraigius* are usually found in rock holes and rock pools, but they are also found in artificial containers and occasionally in tree holes, ground pools, stream pools and axils of *Agave* leaves. Other than the type species, little is known about the bionomics of the adults. Laboratory studies have shown that *Ae. atropalpus* is capable of transmitting La Crosse, West Nile and other encephalitic viruses (Freier & Beier, 1984; Turell *et al.*, 2001), but its importance as a vector of pathogens of human diseases is unknown.

****

Principal references: Reinert *et al.*, 2006 (new genus, morphology, phylogenetic relationships, description, distribution, bionomics); Reinert *et al.*, 2009 (as a genus, morphology, phylogenetic relationships); Reinert, 2008a (as a genus, female genitalia); Wilkerson *et al.*, 2015 (phylogenetic relationships, classification); Soghigian *et al.*, 2017 (phylogenetic relationships).

Geoskusea Edwards, 1929

Geoskusea Edwards, 1929b, originally proposed and currently recognized as a subgenus of *Aedes*, is typified by *Aedes fimbripes* Edwards, 1924b. Subgenus abbreviation – *Geo.*

****

Systematics: *Geoskusea* includes 10 extant species. Species of *Geoskusea* are recorded from Indonesia (Kabaena, Timor and Saparoea), Philippine Islands, New Guinea and the Bismarck Islands, Solomon Islands and New Hebrides in the South Pacific. Belkin (1962) noted that adults of *Geoskusea* share a number of characters with the adults of *Levua* and *Rhinoskusea*, and the pupae are very similar to those of *Levua*. Based on the cladistic analysis of morphological data by Reinert *et al.* (2009), *Geoskusea* appears to be related to three other subgenera of *Aedes* – the four subgenera comprised a clade expressed parenthetically as *Sallumia* + (*Geoskusea* + (*Levua* + *Rhinoskusea*)). In the phylogeny of Wilkerson *et al.* (2015), *Geoskusea* formed a polytomy along with *Levua*, *Rhinoskusea* and *Sallumia*, and this group was sister to *Cancraedes*. A single species of *Geoskusea* (*Ae. baisasi*) was recovered in a sister relationship with a single species of *Kenknightia* (*Ae. dissimilis*) in the maximum likelihood phylogeny of Soghigian *et al.* (2017) based on a single molecular marker.

****

Characteristics: The presence of fine setae on the metameron of *Geoskusea* adults is a particularly distinctive feature of this group of aedine mosquitoes. ADULTS: Vertex of the head with broad decumbent scales; erect scales numerous, scattered over the vertex in females, usually concentrated on the occiput in males; eyes contiguous above the antennal pedicels; antenna of females about as long as the proboscis, antenna of males distinctly shorter than the proboscis, flagellomeres 12 and 13 together about as long as the rest of the flagellum; proboscis dark-scaled, about 1.3 the length of the forefemur; maxillary palpus of females about 0.15 the length of the proboscis, palpomere 5 minute; maxillary palpus of males 0.2–0.7 the length of the proboscis; scutum with narrow scales; acrostichal setae and dorsocentral setae strongly developed; scutellum with narrow or broad scales or both; paratergite bare; antepronotum usually without scales; postpronotum with broad or narrow scales; pleural scaling restricted to upper mesokatepisternal and upper mesepimeral patches, the latter may extend posteriorly onto the lower 0.5 of the sclerite; lower mesepimeral setae absent but numerous fine setae borne posteriorly near middle of the sclerite; metameron with fine setae (distinctive feature); wing dark-scaled, alula with fringe of hair-like or moderately broad scales; tarsi dark-scaled, ungues of females simple, fore- and midungues of males toothed, midungues both toothed, larger one with 1 tooth; tergum I dark-scaled, laterotergite without scales, other terga entirely dark-scaled, with basolateral pale spots or basal pale bands; lateral abdominal setae of males sparse, short. FEMALE GENITALIA: Segment VIII largely retracted into segment VII, without scales, tergum with long setae on the lateral margins of the proximal 0.4; cercus long, without scales; insula with 2 setae on either side of the midline; 1 large and 2 smaller spermathecal capsules. MALE GENITALIA: Segment VIII reduced, tergum with a long median sclerotized lobe extending into the caudal membrane; tergum IX poorly sclerotized, some or all setae stout, with a pair of close-set posterior lobes; gonocoxite without dorsomesal basal lobe, without a distinct mesal membrane; claspette developed as a single basal setose plaque, columnar lobe absent; aedeagus small, simple, poorly sclerotized; proctiger small, paraproct with a simple short apical spine; cercal setae usually in 2 or 3 pairs. LARVAE: Antenna long, spiculate; seta 1-A branched; palatal brushes filamentous; dorsomentum with numerous narrow denticles; seta 4-C very small, inserted cephalad of seta 6-C; setae 5,6-C very long, single to triple; setae 4–7-C inserted caudad of the antennal base; seta 8-C long, single; setae 12,13-C widely separated; seta 19-C present; seta 13-P absent; seta 7-M as long as or longer than seta 5-M; seta 12-I present; comb comprised of small fringed scales in a large patch; siphon short, acus present; pecten long, spines evenly spaced; seta 1-S short, length 0.5 the diameter of the siphon, inserted beyond the pecten; saddle small, incomplete; seta 1-X not inserted on the saddle; seta 2-X multi-branched; ventral brush (seta 4-X) with 6 pairs of setae, all on a grid. PUPAE: Seta 8-C inserted at the level of the bases of the trumpets, with numerous delicate terminal branches; setae 2,3-I approximated; seta 9-VI small, similar to seta 9-II–V; seta 5-V very long, single; seta 6-VII very small, single, inserted posterior to seta 7; genital lobe of males very long; paddle short, broadly ovate; seta 1-Pa very long, single.

****

Bionomics: The immature stages of all species of *Geoskusea* appear to be associated with crab holes, where resting adults can also be found. Larvae and pupae may be found in ground pools in the immediate vicinity of crab holes. Females of some species attack humans readily, mainly at night. Species of *Geoskusea* are not known to be of medical importance.

****

Principal references: Knight & Hull, 1952 (Philippine Islands, description, bionomics, distribution); Mattingly, 1959 (Indomalayan area, taxonomy, keys, descriptions, bionomics, distributions); Belkin, 1962 (South Pacific, taxonomy, keys, descriptions, bionomics, distributions); Lee *et al.*, 1984 (Australasian Region, taxonomy, keys, literature, distributions, bionomics); Reinert, 2002e (female genitalia); Reinert *et al.*, 2004, 2006, 2008, 2009 (as a genus, morphology, phylogenetic relationships); Wilkerson *et al.*, 2015 (phylogenetic relationships, classification); Soghigian *et al.*, 2017 (phylogenetic relationships).

Gilesius Reinert, Harbach & Kitching, 2006

Gilesius, originally proposed as a genus by Reinert *et al.* (2006), currently classified as a subgenus of *Aedes*, is typified by *Aedes pulchriventer* (Giles, 1901a), original combination: *Culex pulchriventer*. Subgenus abbreviation – *Gil.*

****

Systematics: *Gilesius* includes two extant species with distributions in China, India, Nepal, Pakistan and Taiwan. *Gilesius* was recovered as sister to *Collessius* in the cladistic analysis of Reinert *et al.* (2009) based on morphological data of the Aedini. These groups were the terminal taxa in a clade of subgenera parenthetically expressed as *Tanakaius* + (*Hulecoeteomyia* + (*Gilesius* + *Collessius*)). *Gilesius* and *Collessius*, *Hulecoeteomyia* and *Tanakaius* comprised a polyphyletic clade in the phylogeny of Wilkerson *et al.* (2015).

****

Characteristics: The following combinations of characters distinguish species of *Gilesius* from those of other aedine taxa. ADULTS: Vertex of the head with narrow decumbent scales, broad scales present or absent anterolaterally; erect forked scales numerous on the occiput and vertex; eyes narrowly separated; flagellar whorls of males directed dorsally and ventrally; maxillary palpus of females with or without postbasal pale scales; maxillary palpus of males 0.48–0.80 the length of the proboscis, palpomeres 4 and 5 slightly downturned; proboscis longer than the forefemur, dark-scaled; scutum with narrow falcate scales, scales pale on the anterior acrostichal and anterior dorsocentral areas (confluent), the posterior acrostichal and posterior dorsocentral areas, the lateral margins of the prescutellar area and forming a stripe on the supraalar area; scales present medially on the prescutellar area; acrostichal (anterior and posterior) and dorsocentral (anterior and posterior) setae present; scutellum with narrow curved scales on all lobes, median lobe with or without a few broad scales; paratergite with pale scales; antepronotal scales both narrow and broad; thoracic pleura with broad pale scales on the hypostigmal area, postspiracular area, subspiracular area, upper proepisternum, upper and lower mesokatepisternal areas, upper and lower prealar areas and the upper and middle areas of the mesepimeron; lower mesepimeral setae absent; wing dark-scaled, with or without a small pale spot at the base of the costa; alula with narrow dark scales on the margin; upper calypter with several marginal setae; remigium with 2 or 3 dorsal setae; legs mainly dark-scaled, hindfemur with a knee spot, both fore- and midungues of females with 1 tooth, larger fore- and midunguis of males with 2 teeth, the smaller unguis with 1 tooth, hindungues equal and simple in both sexes; laterotergite of abdominal segment I with pale scales, terga with basal pale areas, segment VII dorsoventrally flattened in females. FEMALE GENITALIA: Tergum VIII nearly covered with broad scales, setae on posterior 0.7 or more; sternum VIII slightly wider than long, posterior margin with a small median emargination, several broad scales on the lateral areas, seta 2-S inserted posterior to seta 1-S; tergum IX comprised of 2 sclerites, each with several distal setae; lower vaginal sclerite absent; insula lip-like, with a few setae on either side of the midline; 3 spermathecal capsules. MALE GENITALIA: Tergum IX with a median bilobed pincushion-like structure bearing numerous setae, lateral area developed as a large oblong lobe projecting cephalad; sternum IX with setae and scales; gonocoxite without apical and basal lobes on the mesal surface, with a row of stout close-set comb-like setae on the mesal margin of the dorsal surface, ventral surface with a patch of numerous fusiform scales on the distal 0.5 of the mesal margin, mesal surface membranous, lateral and ventral surfaces with broad scales; gonostylus attached apically, broad proximally, median part noticeably narrower, with single spiniform gonostylar claw; claspette comprised of a narrow columnar stem with a small subapical lobe and a long curved apical claspette filament; aedeagus simple, tube-like; paraproct with apical teeth. LARVAE: Antennal seta 1-A with 3 or 4 (rarely 2) branches; seta 4-C branched, inserted posterior to seta 6-C; seta 6-C inserted more or less directly lateral to seta 5-C; setae 4–6-C inserted noticeably anterior to seta 7-C; seta 12-C inserted mesal to seta 13-C; seta 13-C single, longer than 12-C; seta 19-C absent; setae 1–3-P borne on a common setal support plate, seta 3-P multi-branched; seta 8-P stellate, shorter than seta 4-P; seta 1-II with 3 or more branches, stellate; seta 6-I,II long, double, seta 6-II longer than seta 6-III; seta 7-I,II long, single; seta 12-I present; seta 1-VII long, single; seta 12-VII single; comb scales in a patch; siphon with attached acus; pecten spines numerous, close-set; seta 1-S inserted slightly distal to the pecten; saddle incomplete, with numerous stout spicules on the posterior margin dorsal to seta 1-X, acus absent; seta 1-X single, inserted on the saddle; ventral brush (seta 4-X) comprised of 6 pairs of setae attached to a grid with both transverse and lateral grid bars. PUPAE: Setae 1,3-CT similarly developed; seta 11-CT single, longer than seta 10-CT; seta 3-I,II single, longer than seta 6; seta 6-I shorter than seta 7-I; seta 1-II long, with 3 branches; seta 5-II inserted lateral to seta 4; seta 3-III single, about as long as seta 5-III; seta 5-V single, shorter than the following tergum; seta 2-VI inserted mesal to setae 1 and 3; seta 6-VII single, inserted slightly posterior and mesal to seta 9-VII; seta 9-VIII inserted on the posterolateral corner of the segment; paddle without hair-like spicules on the margin; seta 1-Pa short, single.

****

Bionomics: The immature stages of species of *Gilesius* commonly inhabit rock pools and tree holes but are also found in freshwater ground pools, seepage pools, mud pools, barrels, stump holes, holes in fallen trees and artificial containers. The larvae of *Ae. alius* can survive under ice. Females of *Ae. pulchriventer* have been collected landing on humans at dusk. Species of *Gilesius* are not known to be of medical importance to humans.

****

Principal references: Reinert *et al.*, 2006 (new genus, morphology, phylogenetic relationships, description, distribution, bionomics); Reinert *et al.*, 2008, 2009 (as a genus, morphology, phylogenetic relationships); Reinert, 2009e (as a genus, female genitalia); Rattanarithikul *et al.*, 2010 (as a genus, Thailand, keys, bionomics); Wilkerson *et al.*, 2015 (phylogenetic relationships, classification).

Gymnometopa Coquillett, 1906

Gymnometopa, originally proposed as a genus by Coquillett (1906a), currently classified as a subgenus of *Aedes*, is typified by *Aedes mediovittatus* (Coquillett, 1906d), original combination: *Stegomyia mediovittata*. Subgenus abbreviation – *Gym.*

****

Systematics: *Gymnometopa* is monobasic. It only includes the extant type species, which is found in the West Indies, from Cuba and the Cayman Islands eastward and southward through Jamaica, Hispaniola and Puerto Rico to the Virgin Islands. The species is also recorded from north-central Venezuela. Dyar (1928) and Zavortink (1972) stated that *Gymnometopa* is definitely related to *Howardina*. Zavortink also suggested that *Gymnometopa* is possibly related to *Abraedes*, *Aztecaedes* and *Kompia* that occur north of the tropics in Mexico and the southwestern USA. *Gymnometopa* was recovered as sister to a clade comprised of those three groups and *Lewnielsenius* in the phylogenetic analysis of Reinert *et al.* (2009) based on morphological data of all life stages: *Gymnometopa* + (*Kompia* + (*Aztecaedes* + (*Abraedes* + *Lewnielsenius*))). *Gymnometopa* was recovered as the sister of *Abraedes* in the phylogenetic analysis of Wilkerson *et al.* (2015). In contrast, *Gymnometopa* was recovered as sister to *Georgecraigius* in the maximum likelihood phylogeny of Soghigian *et al.* (2017) based on seven molecular markers.

****

Characteristics: *Gymnometopa* is easily separated from the other New World generic-level taxa of Aedini by the following combinations of characters. ADULTS: Vertex of the head, the paratergite and the scutellum with broad decumbent scales; compound eye bordered by 3 patches of broad decumbent silvery scales, scales of interocular space all broad, setae absent; antennal pedicel with a patch of broad overlapping silvery scales on the mesal surface, flagellomeres 2–10 of females with 8 setae in basal whorls; maxillary palpus of males with palpomeres 4 and 5 upturned and sparsely setose; proboscis with pale scales; scutum with lines and patches of golden and silvery-white scales, scutal fossal scales sparse, dark scales anteriorly on antealar area; scales on the lateral lobes of the scutellum all broad; parascutellar seta absent; postpronotum with narrow dark falcate scales along the upper margin and broad decumbent silver scales in front of the mesothoracic spiracle; 1–4 upper proepisternal setae; lower prealar scales absent; mesepimeral scales in 2 patches; alula of the wing with broad scales on the margin and the dorsal surface; postprocoxal scales present; midfemur and hindfemur with a long pale stripe on the anterior surface; hindtibia dark with pale-scaled areas. FEMALE GENITALIA: Tergum IX comprised of 2 plates separated by a lightly pigmented area; cercal scales present, distal part of the cercus narrowly rounded; insula lip-like. MALE GENITALIA: Gonocoxite with a large membranous basal sternomesal area, specialized setae (sinuous and coarse to near apices) along the distal sternomesal margin and a strongly differentiated seta borne on a basal tergomesal tubercle; gonostylar claw spiniform; claspette with a long stem, claspette filament with a retrorse angle on the convex side; aedeagus expanded distally. LARVAE: Antenna without spicules; labiogula about as wide as long; setae 1–3-P borne on a common support plate; seta 5-P branched; seta 12-I absent; seta 2-VII branched; seta 3-VII long, strong, single; seta 9-III–V stellate; seta 10-VII single; seta 12-VII inserted at approximately the same level as seta 13-VIII; comb scales with 1 or 2 long and usually 1–3 shorter simple spines borne on a large basal plate; siphon acus absent; pecten spines simple or apically laciniate, borne in a dorsally curved row; seta 1a-S single; siphon acus absent; saddle incomplete ventrally, with large spines on the caudal margin; seta 3-X with 3 or 4 branches; ventral brush (seta 4-X) borne on strongly sclerotized boss, seta 4d-X with 5 or more branches, not plumose. PUPAE: Trumpet without tracheoid area; seta 6-I shorter than seta 7-I; seta 3-II longer than seta 6-II; seta 2-III–V inserted far laterad of seta 1; seta 5-IV,V short, 0.5–0.8 the length of the tergum; seta 6-VII strongly developed, long; seta 9-III–VI usually elongate, strong, seta 9-VII shorter than seta 6-VII, seta 9-VIII with 6–16 long strong primary branches; seta 1-Pa 0.4–0.6 the length of the paddle.

****

Bionomics: The immature stages of *Ae. mediovittatus* are found in a wide variety of habitats, including tree holes, broken and cut bamboo internodes, bromeliad axils, rock holes, ground pools and artificial containers. Adults have been found resting in tree holes and crab holes, and females have been captured while feeding on humans and donkeys. Poole-Smith *et al.* (2015) demonstrated that *Ae. mediovittatus* maintained in colony derived from eggs collected in Puerto Rico is a competent vector of dengue fever virus.

****

Principal references: Zavortink, 1972 (in keys, taxonomy, taxonomy, bionomics, distribution); Reinert, 2002e (as a subgenus of *Ochlerotatus*, female genitalia); Reinert *et al.*, 2004, 2006, 2009 (as a genus, morphology, phylogenetic relationships); Wilkerson *et al.*, 2015 (phylogenetic relationships, classification); Soghigian *et al.*, 2017 (phylogenetic relationships).

Halaedes Belkin, 1962

Halaedes Belkin, 1962, originally proposed and currently recognized as a subgenus of *Aedes*, is typified by *Aedes australis* (Erichson, 1842), original combination: *Culex australis*. Subgenus abbreviation – *Hal.*

****

Systematics: *Halaedes* includes three extant species with distributions in the southern part of the Australasian Region, including Lord Howe and Norfolk Islands in the South Pacific. The male genitalia of species of *Halaedes* have a generalized but distinct small claspette that is only found elsewhere in some species of *Aedimorphus*. The immature stages share several features with species of *Opifex*. *Halaedes* was recovered in a clade comprising *Acartomyia* + (*Jarnellius* + (*Halaedes* + *Opifex*)) in a sister-group relationship with *Ochlerotatus* in the phylogenetic study of Reinert *et al.* (2009). Cook *et al.* (2005) investigated the relationships among 20 species of Aedini – *Halaedes* (three species), *Aedes* (one species), *Aedimorphus* (one species), *Diceromyia* (two species),

Ochlerotatus (10 species) and *Stegomyia* (three species) – based on *COI* and *COII* sequences of mtDNA. The three species of *Halaedes* formed a strongly supported clade in the *COII* phylogeny (*COI* sequence was not available for these species) comprised of *Ae. ashworthi* + (*Ae. wardangensis* + *Ae. australis*). Brust & Mahon (1997) surmised that the anautogenous *Ae. ashworthi* in Western Australia may have diverged from *Ae. australis* in eastern Australia. Belkin (1962) believed that *Halaedes* is a relict taxon with its species preserved in a highly specialized habitat. *Halaedes* was recovered as the sister of *Opifex*, and this pair was sister to *Acartomyia*, in the phylogeny of Wilkerson *et al.* (2015). In the maximum likelihood phylogeny of Soghigian *et al.* (2017) based on seven molecular markers, species of *Halaedes* comprised a distinct lineage within a clade that included species of *Geoskusea*, *Kenknightia*, *Ochlerotatus* and *Phagomyia*.

****

Characteristics: Species of *Halaedes* are unornamented mosquitoes with dark tarsi, long setae on the antennal pedicel, numerous postspiracular and prealar setae, and prominent patches of broad scales on the postspiracular area, prealar area, mesokatepisternum and mesepimeron. ADULTS: Eyes distinctly separated above the antennae; decumbent scales of the vertex mainly narrow; erect scales numerous, seta-like in front; maxillary palpus of females about 0.2 the length of the proboscis, with 5 palpomeres, palpomere 5 small; maxillary palpus of males slightly shorter than the proboscis, palpomeres 4 and 5 long, widened, setae numerous from the middle of palpomere 3; antenna shorter than the proboscis; scutal scales narrow; acrostichal and dorsocentral setae present; scutellum with narrow scales; paratergite with scales; pleural scaling extensive, scales broad, mesokatepisternal scales in a single large patch; pleural setae numerous, 20 or more on the upper proepisternum; mesepimeron with a lower anterior line of setae; tarsi dark-scaled; all ungues of both sexes toothed, larger midunguis of males with 2 teeth; remigium of the wing with conspicuous setae; alula with an extensive marginal fringe; tergum I largely bare, laterotergite with a small scale-patch. FEMALE GENITALIA: Segment VIII partially retracted into segment VII; tergum VIII largely covered with setae; cercus short and broad. MALE GENITALIA: Segment VIII narrowed at base; tergum IX long, poorly sclerotized, lobes small, with setae; sternum IX well developed, with a median patch of setae; gonocoxite elongate, simple, mesal membrane from the base to the apex; claspette short, rounded, bearing numerous specialized setae; gonostylus long, slender, widened in the basal 0.6; gonostylar claw apical, slender; aedeagus simple; ventral paramere long; proctiger strongly developed, basolateral sclerotization long; tergal area with 2 membranous lobes; paraproct with a strong apical tooth; cercal setae present. LARVAE: Head broad; seta 4-C inserted between and on level with setae 5,6-C; setae 12,13-C approximated; antenna short, simple, without spicules, seta 1-A inserted near mid-length; seta 13-P absent; seta 5-M poorly developed; seta 7-M as long as seta 5-M; setae 9–12-T removed cephalad and dorsad close to the mesonotal setae; seta 6-I–V strongly developed, branched, seta 6-VI short, multi-branched; seta 7-I strongly developed; seta 12-I present; seta 2-III–VI inserted near seta 1-III, seta 2-VIII larger than seta l-VIII; seta 4-VIII multi-branched, inserted near seta 3-VIII; comb comprised of small fringed scales in a large patch; siphon rather swollen, acus absent; pecten spines closely spaced except those at the base of the siphon; seta 1-S large, multi-branched; lobes of spiracular apparatus small; trachea strongly swollen; saddle incomplete, very small, acus absent; seta 1-X small, not inserted on the saddle; seta 2-X multi-branched; seta 3-X single; ventral brush (seta 4-X) with 7 pairs of setae on a grid; anal papillae represented by slight lobes. PUPAE: Trumpet short, broad, tracheoid area developed at base; seta 8-CT inserted caudad to the attachment of the trumpet; setae 2,3-I approximated; seta l-II short, removed from the caudal margin, seta l-III–VII short; seta 2-II as long as seta 1-II, seta 2-III–V inserted only slightly mesad of seta 1, seta 2-VI,VII inserted laterad of seta 1; seta 3-I longer than seta 6-I; seta 5-II longer than seta 3-II; seta 9-VI small, similar to 9-II–V, slightly dorsal; segment VIII very short, seta 4-VIII removed mesad; genital lobe of females only slightly projecting; paddle very large, wider than segment VIII; margins without distinct spicules.

****

Bionomics: Females of *Ae. ashworthi* and *Ae. wardangensis* are anautogenous whereas those of *Ae. australis* are generally autogenous in the first ovarian cycle (Brust, 1997). The immature stages are found in saline rock pools located at or slightly above the high tide mark along rocky coastlines. Females of *Ae. australis* are known to feed on humans and are reported to be pests in eastern Tasmania. Females of this species from Tasmania and New Zealand have been infected experimentally with Ross River virus, and the New Zealand population is capable of transmitting the virus (see references in Brust, 1997).

****

Principal references: Belkin, 1962 (new subgenus and type species, South Pacific, taxonomy, descriptions, distributions, bionomics); Lee *et al.*, 1984 (Australasian Region, taxonomy, keys, literature, distributions, bionomics); Reinert *et al.*, 2004, 2008, 2009 (as a genus, morphology, phylogenetic relationships); Wilkerson *et al.*, 2015 (phylogenetic relationships, classification); Soghigian *et al.*, 2017 (phylogenetic relationships).

Himalaius Reinert, Harbach & Kitching, 2006

Himalaius, originally proposed as a genus by Reinert *et al.* (2006), currently classified as a subgenus of *Aedes*, is typified by *Aedes gilli* (Barraud, 1924d), original combination: *Finlaya gilli*. Subgenus abbreviation – *Him.*

****

Systematics: *Himalaius* includes two extant species. The distributions of the species are poorly known as they have only been collected in India, Nepal and Thailand. *Himalaius* was recovered as the sister to *Bruceharrisonius* in the phylogenies of Aedini recovered in the studies of Reinert *et al.* (2006, 2008, 2009) and Wilkerson *et al.* (2015). These groups were recovered in a sister relationship to *Vansomerenis* + (*Zavortinkius* + *Kenknightia*) in the study of Reinert *et al.* (2009). In agreement with the findings of Reinert *et al.* and Wilkerson *et al.*, one of the two known species of *Himalaius*, *Ae. gilli*, was recovered in a fairly strongly supported sister relationship with two species of *Bruceharrisonius* in the maximum likelihood phylogeny of Soghigian *et al.* (2017) based on several molecular markers.

****

Characteristics: The following combinations of characters distinguish species of *Himalaius* from those of other aedine taxa. ADULTS: Vertex of the head with narrow (medially) and broad (laterally) decumbent scales; occiput with numerous erect forked scales; maxillary palpus and proboscis of both sexes dark-scaled, proboscis approximately equal to the length of the forefemur; maxillary palpus of males with 5 palpomeres, palpomeres 4 and 5 short, slightly downturned; scutum with pale scales covering approximately the anterior 0.6, posterior acrostichal area with a stripe of similar scales extending and forking to the margins of the prescutellar bare area, similar pale stripe on the posterior dorsocentral area extending to the lateral margin of the prescutellar setae, scutal fossa with a patch of pale scales; supraalar area with a longitudinal pale stripe; anterior dorsocentral setae absent; scutellum with narrow pale scales on all lobes; paratergite without scales; antepronota widely separated, with both narrow and broad scales; postpronotum mostly with broad pale scales, narrow scales dorsally; upper proepisternum, upper and lower mesokatepisternum, lower prealar area and mesepimeron with broad pale scales; lower mesepimeral setae absent; wing dark-scaled; upper calypter with setae; alula with narrow dark scales on the margin; remigium with a few dorsal setae; ante- and postprocoxal membranes without scales; hindfemur without a knee spot; hindtibia with a basal pale band; hindtarsomeres 1–3 with broad basal pale bands; fore- and midungues of females equal, each with 1 tooth, fore- and midungues of males unequal, larger unguis of each with 2 teeth, smaller unguis with 1 tooth, hindungues of females and males equal and simple; laterotergite of abdominal segment I with a pale scale-patch; segment VII laterally compressed in females. FEMALE GENITALIA: Tergum VIII with broad scales on all but a narrow proximal strip; sternum VIII slightly wider than long, some broad scales proximally, posterior margin nearly straight; seta 2-S inserted posterior to seta 1-S; tergum IX with a median posterior emargination; postgenital lobe with apex rounded; lower vaginal sclerite absent; cercus without scales, apex narrowly rounded; insula lip-like, with a few setae in lateral patches; 3 spermathecal capsules. MALE GENITALIA: Tergum IX comprised of 2 sclerites; gonocoxite without dorsal apicomesal and dorsal basomesal lobes, scales present, ventral surface without specialized setae or scales, mesal surface membranous; gonostylus attached apically, without setae on the distal 0.33, with a single gonostylar claw at the apex; claspette with a single columnar stem and 1 stout spiniform; aedeagus simple, tube-like; proctiger with 2 or 3 apical teeth; cercal setae present. LARVAE: Head with a scoop-like pouch located anteroventral to the antenna; seta 1-A single, extending beyond the apex of the antenna; seta 1-C spiniform; seta 4-C inserted near, mesal and slightly posterior to seta 5-C; seta 5-C single, inserted near, mesal and slightly posterior to seta 6-C; seta 6-C single; seta 7-C single, rarely double, inserted anterolateral to setae 4–6-C and posterior to the antenna; seta 12-C inserted mesal to seta 13-C; seta 14-C branched; seta 19-C absent; seta 2-P longer than setae 1,3-P; setae 6,7-P longer than seta 5-P; seta 6-P longer than seta 7-P; seta 7-P single; seta 1-M approximately 2.9 times as long as seta 2-M; seta 7-M longer than seta 5-M; seta 6-T branched; seta 12-I present; seta 7-I only slightly shorter than seta 6-I; seta 1-VII very long, with 2 or 3 branches; seta 3-VII inserted lateral and slightly anterior to seta 1-VII; setae 7,10,12-VII and 2,4-VIII branched; comb comprised of 80–94 scales in a patch; siphon with attached acus; pecten comprised of 24–36 evenly spaced spines; seta 1-S short, with 6–8 branches, inserted distal to the pecten; saddle incomplete, acus absent; seta 1-X branched, inserted on the saddle; seta 3-X single; ventral brush (seta 4-X) inserted on a grid with both transverse and lateral grid bars, 2 precratal setae. PUPAE: Seta 1-CT very long, double; seta 3-CT also very long, with 3–5 branches; seta 4-CT shorter than seta 5-CT; seta 7-CT long, with 3–5 branches; seta 11-CT normally with 2 branches; seta 3-I long, with 3 or 4 branches; seta 6-I single or double, longer than seta 7-I; seta 2-II inserted lateral to seta 1-II and mesal to seta 3-II; seta 3-II with 2 or 3 branches, longer than seta 6-II; seta 4-II inserted mesal to seta 5-II; seta 6-II long, single to 3-branched, longer than seta 7-II; seta 3-III long, with 2 branches; seta 5-IV,V exceptionally long; seta 2-IV inserted mesal to setae 1,3-IV; seta 6-VII inserted mesal and slightly anterior to seta 9-VII; seta 9-VIII inserted noticeably anterior to the posterior margin of the segment; paddle ovoid, without hair-like spicules on the margins; seta 1-Pa single, 0.65–1.02 the length of the paddle.

****

Bionomics: The immature stages of *Ae. gilli* have been found in tree holes and bamboo. Nothing is known about the bionomics of the adults. Species of *Himalaius* are unlikely to be of medical and economic importance to humans.

****

Principal references: Reinert *et al.*, 2006 (new genus, morphology, phylogenetic relationships, description, distribution, bionomics); Reinert *et al.*, 2008, 2009 (as a genus, morphology, phylogenetic relationships); Rattanarithikul *et al.*, 2010 (as a genus, Thailand, keys, bionomics); Wilkerson *et al.*, 2015 (phylogenetic relationships, classification); Soghigian *et al.*, 2017 (phylogenetic relationships).

Hopkinsius Reinert, Harbach & Kitching, 2008

Hopkinsius, originally proposed as a genus by Reinert *et al.* (2008), currently classified as a subgenus of *Aedes*, is typified by *Aedes ingrami* Edwards, 1930a. Subgenus abbreviation – *Hop.*

****

Systematics: *Hopkinsius* includes seven extant species, five placed in the Hopkinsius Group and two in the Yamada Group. Species of *Hopkinsius* have distributions in sub-Saharan Africa, China, India, Japan, South Korea and Taiwan. *Hopkinsius* was placed as sister to a large clade comprised of *Dahliana* in a basal relationship to 10 other generic-level taxa – (*Howardina* + (*Gymnometopa* + (*Kompia* + (*Aztecaedes* + (*Abraedes* + *Lewnielsenius*)))) + ((*Finlaya* + *Danielsia*) + (*Downsiomyia* + *Haemagogus*))) – in the phylogeny of Aedini recovered in the study of Reinert *et al.* (2009). *Hopkinsius* was recovered as the sister of *Ae. crossi* (subgenus uncertain) in the phylogenetic analyses of Wilkerson *et al.* (2015).

****

Characteristics: The following combinations of characters distinguish species of *Hopkinsius* from those of other aedine taxa. ADULTS: Decumbent scales of the vertex all broad or with narrow scales along the coronal suture; erect forked scales on the occiput; eyes more or less contiguous; flagellar whorls of males directed dorsally and ventrally; maxillary palpus dark-scaled, palpus of males with palpomeres 4 and 5 slightly downturned; proboscis longer than the forefemur, dark-scaled; approximately anterior 0.5 of the scutum pale-scaled or pale scales present on the anterior promontory, antedorsocentral area, scutal fossa extending over the anterior dorsocentral area, a stripe on posterior dorsocentral area, entire acrostichal area, antealar and supraalar areas and margins of the prescutellar area; anterior dorsocentral setae present or absent, posterior dorsocentral setae present; scutellum with falcate scales on all lobes (*Ae. ingrami* with broad scales); broad pale scales on the paratergite (scales absent in *Ae. albocinctus*), subspiracular area, upper proepisternum, lower prealar area, upper and lower mesokatepisternal areas and upper mesepimeral area; lower mesepimeral setae absent; costa of the wing dark-scaled with a basal pale spot (occasionally absent in *Ae. ingrami*), remigium with a few dorsal setae, upper calypter with numerous marginal setae, alula with narrow scales on the posterior margin; ante- and postprocoxal membranes bare; base of the hindcoxa below the dorsal margin of the mesomeron; legs mainly dark-scaled, hindfemur without apical pale scales, hindtarsomeres 1 and 2 and sometimes hindtarsomere 3 with basal pale bands, fore- and midungues of males unequal, larger unguis with 2 teeth, smaller unguis with 1 tooth; hindtarsus with 2 equal, simple ungues; laterotergite of abdominal segment I with broad scales, terga II–VI with basolateral pale patches, *Ae. albocinctus* and *Ae. seoulensis* also with narrow basal pale bands (incomplete in *Ae. albocinctus*). FEMALE GENITALIA: Tergum and sternum VIII wider than long, largely covered with broad scales, posterior margin of the tergum broadly rounded, sternum with a posterior emargination separating broadly rounded lobes; tergum IX with a deep posterior emargination, sternum IX with a few posteromesal setae; postgenital lobe with a small apical emargination; lower vaginal sclerite absent; insula lip-like, with 2 or 3 setae in lateral patches; 3 spermathecal capsules. MALE GENITALIA: Tergum IX with 2 blunt lobes on the posterior margin, each lobe bearing a few flattened setae; gonocoxite with a small basomesal lobe, with a few setae (Hopkinsius Group) or a few short and 1 longer stouter seta (Yamada Group), lateral and ventral surfaces with broad scales, mesal margin membranous; gonostylus attached apically, distal part curved mesad, with 1 gonostylar claw at the apex; claspette with a narrow columnar stem and a long claspette filament with recurved apex; aedeagus tube-like. LARVAE: Antenna slightly curved mesad, spiculate; seta 1-A single or with a few branches; seta 4-C multi-branched, inserted posterior to seta 6-C; seta 5-C branched, inserted posteromesal to seta 6-C; seta 7-C inserted directly lateral to setae 4–6-C; seta 13-C branched; 2 or 3 of setae 1–3-P normally inserted on a setal support plate; comb comprised of a few to several scales in a single row (Hopkinsius Group) or numerous scales in a patch (Yamada Group); siphon with attached acus, pecten with numerous evenly spaced spines; seta 1-S multi-branched, inserted distal to the pecten; saddle of segment X incomplete ventrally; seta 1-X inserted on the saddle; ventral brush (seta 4-X) with several normally 2-branched setae on a grid with transverse grid bars and 2 precratal setae. PUPAE: Seta 7-CT about 3 times as long as seta 6-CT (Hopkinsius Group) or greater than 6 times the length of 6-CT (Yamada Group); seta 6-I longer than seta 7-I; seta 2-II inserted lateral to setae 1 and 3; seta 2-VI inserted mesal or directly anterior to seta 1; seta 5-II inserted lateral to seta 4; seta 5-IV and normally 5-V,VI longer than the following tergum; seta 6-VII inserted posteromesal to seta 9-VII; seta 9-VIII inserted on the posterolateral corner of the segment; paddle without fringe of hair-like spicules.

****

Bionomics: The immature stages of species of *Hopkinsius* have been collected from tree holes and bamboo. Nothing is known about the bionomics of the adults. Species of *Hopkinsius* are unlikely to be of medical and economic importance to humans.

****

Principal references: Reinert *et al.*, 2008 (new genus, morphology, phylogenetic relationships, description, distribution, bionomics); Reinert *et al.*, 2009 (as a genus, morphology, phylogenetic relationships); Reinert, 2009b (as a genus, female genitalia); Rattanarithikul *et al.*, 2010 (as a genus, Thailand, keys, bionomics); Wilkerson *et al.*, 2015 (phylogenetic relationships, classification); Huang & Rueda, 2017 (as Wellmanii and Barnardi Groups of *Aedes* subgenus *Finlaya*, Afrotropical Region, key to adults).

Howardina Theobald, 1903

Howardina, originally proposed as a genus by Theobald (1903a), currently classified as a subgenus of *Aedes*, is typified by *Aedes walkeri* (Theobald, 1901b), original combination: *Culex walkeri*. Subgenus abbreviation – *How.*

****

Systematics: *Howardina* includes 34 extant species in the Neotropical Region. Prior to the series of phylogenetic studies of Aedini by Reinert *et al.* (2004, 2006, 2008, 2009) it was generally assumed that *Howardina* shared affinities with *Finlaya* (in the broad traditional sense) based on similarities in all life stages (Edwards, 1932a; Berlin, 1969a; Belkin *et al.*, 1970). Berlin (1969a) also noted that the immature stages of *Howardina* are very similar to those of some species of *Haemagogus*. However, with the recognition that species traditionally grouped in *Finlaya* do not comprise a monophyletic taxon, it became apparent that the affinities of *Howardina* are uncertain. In the comprehensive study of Reinert *et al.* (2009), *Howardina* was placed as sister to a clade that included five other New World groups: *Howardina* + (*Gymnometopa* + (*Kompia* + (*Aztecaedes* + (*Abraedes* + *Lewnielsenius*)))). Interestingly, this group of generic-level taxa was sister to a clade that included *Haemagogus* and three subgenera that were previously included in the broad traditional concept of *Finlaya*: (*Finlaya* + *Danielsia*) + (*Downsiomyia* + *Haemagogus*). *Howardina* was not associated with other generic-level taxa in the phylogeny of Wilkerson *et al.* (2015). Contrary to the findings of Reinert *et al.* (2009), *Howardina* was strongly supported as sister to *Haemagogus* in the maximum likelihood phylogeny of Soghigian *et al.* (2017) based on a number of molecular markers.

****

Characteristics: The following morphological features distinguish *Howardina* from other generic-level taxa of Aedini in the Neotropical Region. ADULTS: Vertex of the head with a median stripe of narrow scales extending forward to the eyes, erect scales restricted to the occiput; mesonotum with a linear pattern of silvery or golden scales; scutellar scales narrow; thoracic pleura with silvery scales; maxillary palpus of males with palpomeres 4 and 5 nearly straight relative to palpomere 3; ungues of females simple (without teeth); abdominal terga of males with a few relatively short lateral setae. FEMALE GENITALIA: Tergum VIII with straight posterior margin; sternum VIII rather inconspicuous; insular setae inserted in lateral patches. MALE GENITALIA: Claspette small, formed as a single basal plaque with a stout spiniform seta; aedeagus simple; proctiger without a sternal arm. LARVAE: Antenna spiculate; seta 5-C single; setae 1–3-P borne on a common setal support plate; seta 4-P longer than seta 3-P; seta 13-P present; seta 12-I absent; seta 2-VII branched; seta 12-VII single, inserted anterior to seta 13-VII in most species; seta 1-VIII ≥ 1.1 the length of seta 2-VIII; comb scales squamous, without median apical spine; length of seta 1-S 1.1–2.0 times the width of the siphon; saddle incomplete ventrally with the lateral margins near either side of the ventral brush. PUPAE: Paddle with the midrib strongly development, extending to or near the apex of the paddle.

****

Bionomics: The immature stages of *Howardina* species are found in containers, including rock holes, tree holes, broken bamboo, fallen plant parts, axils of bromeliads and aroids, and flower bracts of *Heliconia*. A few species have been found in various types of artificial containers. Adults of *Howardina* are active during the daylight hours. Females of a number of species are known to bite humans in forested areas, but their natural hosts are vertebrates that inhabit various strata from ground level to canopy in forests. Some species of *Howardina* readily attack humans but little is known about the potential of these mosquitoes to harbour and transmit arboviruses. Experimental studies have shown that *Ae. fulvithorax* and other species of *Howardina* can be infected with and maintain yellow fever virus, but they are not able to transmit the virus (Davis & Shannon, 1931; Whitman & Antunes, 1937; Boshell-Manrique & Osorno-Mesa, 1944; Bugher *et al.*, 1944; Aitken, 1960). Bates (1949) suggested that species of *Howardina* may be able to transmit the virus if incubation is prolonged at favourable temperatures.

****

Principal references: Berlin, 1969a (classification, keys, taxonomy, descriptions, bionomics, distributions); Reinert, 2002e (as a subgenus of *Ochlerotatus*, female genitalia); Reinert *et al.*, 2004, 2006, 2008, 2009 (as a genus, morphology, phylogenetic relationships); Wilkerson *et al.*, 2015 (phylogenetic relationships, classification); Soghigian *et al.*, 2017 (phylogenetic relationships).

Huaedes Huang, 1968

Huaedes Huang, 1968b, originally proposed and currently recognized as a subgenus of *Aedes*, is typified by *Aedes wauensis* Huang, 1968b. Subgenus abbreviation – *Hua.*

****

Systematics: *Huaedes* includes three extant species that are only known to occur on the island of New Guinea. Huang (1968b) observed that adults of *Huaedes* are superficially similar in colouration to the adults of some species of *Finlaya*, and males seem to share some basic genitalic characters with species of *Christophersiomyia*, suggesting affinities with that group. In the phylogeny of Aedini recovered in the cladistic analyses of morphological data conducted by Reinert *et al.* (2009), *Huaedes* (based on characters coded for *Ae. wauensis*) was placed in a clade comprised of *Polyleptiomyia* + (*Bifidistylus* + (*Albuginosus* + (*Tewarius* + (*Christophersiomyia* + (*Huaedes* + *Leptosomatomyia*))))), indicating an affinity with *Leptosomatomyia* as well as *Christophersiomyia*. *Huaedes* was recovered as the sister to *Leptosomatomyia* in the phylogeny of Wilkerson *et al.* (2015), and the sister pair was sister to a clade comprised of *Armigeres* + (*Alanstonea* + *Eretmapodites*).

****

Characteristics: Species of *Huaedes* are ornate mosquitoes that have a dark-scaled proboscis which is longer than the forefemur, narrow scales covering the scutum, patches of broad pale scales on the thoracic pleura, including a large V-shaped patch on the mesepimeron, femora sprinkled with yellow scales, tarsi with narrow basal pale bands and abdominal segment VIII of females is partially retracted into segment VII. These and the following features distinguish *Huaedes* from other generic-level taxa of Aedini. ADULTS: Decumbent scales of the vertex largely pale and narrow; erect scales numerous, not restricted to the occiput; eyes separated in both sexes; antenna less than 0.8 the length of the proboscis, verticillate in males, pedicel and flagellomere 1 with pale scales on the mesal side; proboscis dark-scaled, longer than the forefemur; maxillary palpus less than 0.25 the length of the proboscis in both sexes; clypeus bare; scales of the scutum all narrow; acrostichal and dorsocentral setae present; scutellar scales all narrow or narrow and broad; paratergite and postspiracular area with scales; proepisternum, subspiracular area and upper and lower areas of the mesokatepisternum with patches of broad pale scales, large V-shaped patch on the mesepimeron; lower prealar area without scales; upper mesokatepisternal setae absent; lower mesepimeral seta present; femora and tibiae speckled with yellowish scales; tarsomeres with narrow basal pale bands; ungues of females simple, anterior fore- and midungues toothed in males. FEMALE GENITALIA: Abdominal segment VIII partially retracted into segment VII; posterior margin of tergum VIII rounded; posterior margin of sternum VIII with a median emargination; tergum IX comprised of 2 sclerites connected by membrane; postgenital lobe with a median emargination; cercus short and rounded apically; 1 large and 2 smaller spermathecal capsules. MALE GENITALIA: Gonocoxite long, more than twice as long as wide, scales on the dorsolateral, lateral and ventral surfaces; gonostylus simple, elongate, with a conspicuous apical spiniform gonostylar claw; claspette a single setose plaque at the base of the gonocoxite; aedeagus with several teeth on the distal ≥ 0.68 of each side; paraprocts sclerotized at the apex; cercal setae absent. LARVAE: Setae 4,5-C inserted caudad of the antennal base; seta 4-C small, branched, inserted anterolateral and close to seta 5-C; seta 6-C inserted well anterior and slightly mesad of seta 5-C; seta 12-I absent; seta l-VIII small, branched; comb scales in a triangular patch, scales with lateral spicules, sometimes with a small apical spine; siphon short, siphon index less than 4.0, acus present, small; saddle complete; seta 1-X single, removed from the caudal margin of the saddle; seta 2-X multi-branched; seta 3-X single; ventral brush (seta 4-X) with more than 6 pairs of plumose setae on a grid, no precratal setae; anal papillae lanceolate. PUPAE: Setae 2,3-I not widely separated, distance between them the same as the distance between setae 4,5-I; seta 1-II branched; seta 2-II inserted laterad of seta 3-II; seta 2-V inserted caudad of seta 3-V; seta 9-VII relatively large, fan-like with aciculate branches, markedly different than seta 9-II–IV, seta 9-VIII inserted mesal to the posterolateral corner of the segment; paddle margins distinctly spiculate, midrib weak, terminating well before the apex of the paddle; seta 1-Pa branched.

****

Bionomics: The immature stages of species of *Huaedes* inhabit plant cavities, including bamboo, tree holes and pitcher plants. A female of *Ae. variepictus* was captured while attempting to bite, but nothing else is known about the bionomics of the species of *Huaedes*. The species are unlikely to be of medical and economic importance to humans.

****

Principal references: Huang, 1968b (new subgenus, type species, descriptions, bionomics, distribution); Lee *et al.*, 1984 (Australasian Region, taxonomy, keys, literature, distributions, bionomics); Reinert, 2000e (female genitalia); Reinert *et al.*, 2004, 2009 (as a genus, morphology, phylogenetic relationships); Wilkerson *et al.*, 2015 (phylogenetic relationships, classification).

Hulecoeteomyia Theobald, 1904

Hulecoeteomyia, originally proposed as a genus by Theobald (1904c), currently classified as a subgenus of *Aedes*, is typified by *Hulecoeteomyia trilineata* Leicester, 1904 [subjective synonym of *Aedes chrysolineatus* (Theobald, 1907), original combination: *Howardina chrysolineata*]. Subgenus abbreviation – *Hul.*

****

Systematics: *Hulecoeteomyia* includes 17 extant species. The species are native to the Oriental and eastern Palaearctic Regions. *Aedes japonicus* is established in the USA and central Europe. *Hulecoeteomyia* appears to share affinities with several subgenera of *Aedes* based on phylogenetic analyses of morphological data. In the study of Reinert *et al.* (2006), the taxon was recovered as the sister to *Tanakaius* and *Collessius* in a clade comprised of *Georgecraigius* + (*Hulecoeteomyia* + (*Tanakaius* + *Collessius*)). In the more comprehensive study of Reinert *et al.* (2009), *Hulecoeteomyia* was recovered as the sister to *Gilesius* and *Collessius* in a clade comprised of *Tanakaius* + (*Hulecoeteomyia* + (*Gilesius* + *Collessius*)). *Hulecoeteomyia*, *Collessius*, *Gilesius* and *Tanakaius* comprised a polyphyletic clade in the phylogeny of Wilkerson *et al.* (2015). Contrary to what was stated by Soghigian *et al.* (2017), *Hulecoeteomyia* was not recovered as a monophyletic lineage in their maximum likelihood phylogeny based on seven molecular markers. Two species of *Collessius* formed a branch within a clade that included five species of *Hulecoeteomyia*, and that clade was sister to a clade comprised of species of *Rhinoskusea* + *Collessius*. In a phylogenetic analysis of mtDNA *COI* sequences (Somboon *et al.*, 2023b), *Hulecoeteomyia* was recovered, albeit weakly, as the sister to a clade comprised of species of 12 other subgenera of *Aedes*, including *Collessius* and *Tanakaius*.

****

Characteristics: The adults and larvae of *Hulecoeteomyia* resemble the adults and larvae of the *Collessius*, *Orohylomyia* and *Tanakaius*. The adults differ conspicuously in having basal pale bands on hindtarsomeres 1–3; the larvae differ in having spine-like comb scales with lateral fringes or strong denticles (occasionally rounded with an even fringe in *Ae. harveyi* and *Ae. pallirostris*) and the siphon usually has one or more simple pecten spines inserted adjacent or distal to the insertion of seta 1-S. Species of *Hulecoeteomyia* exhibit the following characteristics. ADULTS: Vertex of the head with both broad and narrow scales, narrow scales medially, broad scales laterally; erect forked scales dorsally from the compound eyes to the occiput; maxillary palpus of females dark-scaled, usually with some pale scales apically, maxillary palpus of males shorter than the proboscis, more or less straight, setae absent or few on palpomeres 3 and 4; proboscis dark-scaled, some species with areas of pale scaling; scutum with narrow pale scales arranged in narrow longitudinal lines, a long median line, long submedian line interrupted at the scutal angle, line or diffused area over the wing base, small patch just before the wing base and a patch along the anterior margin of the scutum from the end of the submedian line to the level of the postpronotal scales; scutellum with narrow scales, broader and appressed laterally on the middle lobe; antepronotum with a dense cover of broad pale scales, sometimes a few narrow scales also present; postpronotum variably scaled; proepisternum with a patch of broad pale scales continuing onto the anterior surface; thoracic pleura with patches of broad white scales on the postspiracular area, lower prealar area, upper and lower mesokatepisternal areas and upper mesepimeral area, some species with paratergal and subspiracular scales; wing dark-scaled, a line of pale scales nearly always present at the base of the costa; tarsi with basal pale bands on some tarsomeres; fore- and midungues of females equal, each with 1 tooth, fore- and midungues of males unequal, larger unguis with a median tooth and a smaller basolateral tooth, smaller unguis with a postbasal tooth, hindungues equal and simple in both sexes; abdominal terga usually with basal pale markings, with prominent basolateral pale spots, laterotergite of tergum I with scales. FEMALE GENITALIA: Tergum VIII wider than long, with dense covering of broad scales on the posterior 0.6–0.8, caudal margin broadly rounded; sternum VIII with a caudal emargination separating broadly rounded setose lobes; tergum IX comprised of 2 lateral sclerites; postgenital lobe truncate with a shallow apical emargination; lower vaginal sclerite absent; insula lip-like with setae in lateral patches; 1 large and 2 smaller spermathecal capsules. MALE GENITALIA: Ninth tergal lobes small but prominent, each bearing strong setae; sternum IX with setae and sometimes scales; gonocoxite densely scaled, mesal surface membranous, dorsal surface without an apical lobe, usually without a basal lobe, with 30–60 short setae along the mesal margin, variable number of these setae more or less clustered in the basal area; gonostylus simple with a simple terminal claw; claspette with a basal stem and an apical sickle-shaped claspette filament; aedeagus short, simple; paraproct with a single apical tooth; cercal setae present. LARVAE: Antenna cylindrical with sparse spicules, seta 1-A with 1–5 branches (usually 2 or 3); labiogula short, length shorter than width; seta 5-C branched; seta 4-C very small, with 2–7 branches; setae 4–6-C usually in a more or less even transverse line; setae 5,6-C fan-like with stiff branches; seta 14-C branched; abdominal seta 1 is the most prominent dorsal seta on segments III–VII; seta 2-II branched, seta 2-VIII single; seta 3-VII long to very long; seta 4-VIII single; seta 5-II with 2 or more branches, not stellate; seta 13 is the most prominent ventral seta on segments III–VII, 13-III–V with 1–6 branches, 13-VI single, 13-VII with 1–4 branches; comb scales in a patch; seta 1-S inserted at or just beyond mid-length of the siphon, with 3–7 plumose branches; segment X with pilose integument, saddle incomplete, posterior margin with variously developed spicules; ventral brush (seta 4-X) comprised of 10–13 setae, normally all on a grid. PUPAE: Seta 1-CT very strongly developed, considerably longer than seta 3-CT; seta 10-CT inserted mesad of seta 11-CT; seta 11-C greatly elongate, single; seta 1-II usually well developed, with 2 or more branches; setae 2,3-I approximated; seta 2-VI inserted well mesad of seta 1-VI; setae 3-I–III and 5-IV–VII long, single; seta 3-VII moderately long; seta 6-VII with 2–5 simple or barbed branches; seta 9-VII similar to seta 6-VII but usually larger with more branches; seta 1-Pa elongate, similar to seta 5-VII.

****

Bionomics: The immature stages of species of *Hulecoeteomyia* commonly occur in ground pools, rock holes, crab holes, artificial containers, rock pools, hollow logs, tree stumps, tree holes, bamboo stumps, bamboo internodes, split bamboo, axils of *Pandanus* and *Nipa* and other plants, ginger inflorescence, pitcher plants, banana stumps, fallen leaves and coconut shells. The immature stages of some species are found occasionally in artificial containers. Little is known about the bionomics of the adults. Species of *Hulecoeteomyia* are not known to be of medical importance, but have been collected biting humans and are thus regarded as potential vectors of arboviruses.

****

Principal references: Knight, 1968 (as Chrysolineatus Subgroup of subgenus *Finlaya*, Southeast Asia, classification, keys, taxonomy, descriptions, distributions, bionomics); Reinert, 2002e (as Chrysolineatus Assemblage of *Ochlerotatus* subgenus *Finlaya*, female genitalia); Reinert *et al.*, 2006, 2008, 2009 (as a genus, morphology, phylogenetic relationships); Reinert, 2008d (as a genus, female genitalia); Rattanarithikul *et al.*, 2010 (as a genus, Thailand, keys, bionomics); Wilkerson *et al.*, 2015 (phylogenetic relationships, classification); Soghigian *et al.*, 2017 (phylogenetic relationships).

Indusius Edwards, 1934

Indusius Edwards, 1934 (in Barraud, 1934), originally proposed and currently recognized as a subgenus of *Aedes*, is typified by *Aedes pulverulentus* Edwards, 1922a. Subgenus abbreviation – *Ind.*

****

Systematics: *Indusius* is monobasic. It only includes the extant type species, which is only known from Pakistan. Based on the cladistic analysis of morphological data by Reinert *et al.* (2009), *Indusius* appears to be related to *Skusea* and *Cancraedes*. *Skusea* and *Indusius* were recovered in a paraphyletic relationship with respect to *Cancraedes*: *Skusea* + (*Indusius* + *Cancraedes*). *Indusius* was sister to *Skusea* and this pair was sister to *Fredwardsius* in the phylogenetic analysis of Wilkerson *et al.* (2015).

****

Characteristics: *Indusius* is principally distinguished by the distinctive form of the maxillary palpi, foretarsi and genitalia of males. ADULTS: Vertex of the head with broad decumbent scales; few erect forked scales confined to the occiput; compound eyes widely separated; interocular space with broad scales and 6 or more well-developed setae; antennal pedicel with a few scales and setae mesally; maxillary palpus with pale scales, maxillary palpus of males with 4 palpomeres, palpomere 3 about 0.5 the length of palpomere 4, palpomere 4 long, swollen and setose; proboscis of males shorter than the forefemur, swollen and somewhat compressed distally, with numerous long setae on the lateral and ventral areas; scutum with a large patch of pale scales on the anterior 0.7, scutal fossa with contrasting pale scales in a large patch; few anterior acrostichal setae, no posterior acrostichal setae; few anterior dorsocentral and numerous posterior dorsocentral setae; paratergite bare; scutellum with broad and narrow scales on the midlobe; scales present on the antepronotum, postpronotum, upper proepisternum, subspiracular area, postspiracular area, upper prealar area and the mesepimeron; lower anterior mesepimeral setae absent; alula of the wing with long narrow scales on the margin, similar to setae on the upper calypter; remigium without setae; femora short, somewhat thickened; tibiae with numerous stout erect setae, very noticeable on the hindtibia; hindtarsomere 1 with apical pale scales, hindtarsomere 2 with basal and apical pale scales; fore- and midungues of females equal, each with 1 tooth, hindungues equal, simple, fore- and midungues of males large, foreungues equal and simple, larger midunguis with 1 tooth, smaller midunguis simple; abdominal terga of males with short setae on the lateral and posterior margins. FEMALE GENITALIA: Segment VIII completely retracted into segment VII; tergum VIII without scales; sternum VIII with a deep median posterior indentation and a median longitudinal non-sclerotized area, scales absent or only a few present, setae 1–3-S all inserted proximad; tergum IX comprised of 2 long sclerites, each expanded caudally, with setae on the distal and much of the median area; insula tongue-like with a few small tuberculi; lower vaginal sclerite absent; upper vaginal sclerite well developed; postgenital lobe with a median caudal indentation; cercus without scales; 3 spermathecal capsules, 1 large, 2 smaller. MALE GENITALIA: Tergum IX with a convex posterior margin, several setae on each side of the midline; sternum IX narrow, without setae; gonocoxite somewhat conical, scales absent, lateral surface with mostly short setae, dorsomesal surface with numerous long setae between the base and apex; gonostylus attached subapically, exceptionally short, narrow proximally, bulbous distally with file-like ridges and stout spicules; gonostylar claw absent; phallosome with moderately long and broad parameral apodeme with a long curved projection on its caudomesal margin; proctiger short, sternal arm present, paraproct divided distally into 2 narrow arms; cercus membranous, without setae. LARVAE: Unknown. PUPAE: Unknown.

****

Bionomics: Females of *Ae. pulverulentus* are known to feed on cattle, but nothing else is known about the bionomics of the species. It is unlikely to be of medical and economic importance to humans.

****

Principal references: Edwards, 1934 (in Barraud, 1934) (new subgenus, type species, description, distribution); Reinert, 1976c (subgenus and type species, taxonomy, descriptions, distribution, bionomics); Reinert, 2000e (female genitalia); Reinert *et al.*, 2004, 2009 (as a genus, morphology, phylogenetic relationships); Wilkerson *et al.*, 2015 (phylogenetic relationships, classification).

Isoaedes Reinert, 1979

Isoaedes Reinert, 1979, originally proposed and currently recognized as a subgenus of *Aedes*, is typified by *Aedes cavaticus* Reinert, 1979. Subgenus abbreviation – *Isa*.

****

Systematics: *Isoaedes* is monobasic. It only includes the extant type species, which is only known from Thailand. Based on the parsimony analysis of Harbach *et al.* (2007b), *Isoaedes* appears to be more closely related to *Borichinda* than to other generic-level taxa of Aedini. This is also reflected in the relationships of generic-level taxa recovered in the cladistic analysis of more extensive morphological data by Reinert *et al.* (2009): *Fredwardsius* + (*Isoaedes* + (*Borichinda* + other generic-level taxa)).

****

Characteristics: The following salient anatomical differences distinguish *Isoaedes* from other generic-level taxa of Aedini in the Oriental Region, particularly the morphologically similar *Borichinda* and *Ayurakitia*. ADULTS: Decumbent scales of the vertex and ocular scales narrow; interocular space narrow, interocular setae present; eyes contiguous; maxillary palpus of males straight, with palpomere 5 $\leq$ 0.55 the length of palpomere 4; scutal scales all pale; scutal fossal scales sparse, contrasting with surrounding scales; paratergite bare; scales on mid and lateral lobes of the scutellum all narrow; antepronotal scales and postpronotal scales present; subspiracular scales and postspiracular scales absent; postspiracular setae present; lower anterior mesepimeral setae present; remigial setae present; hindtarsomere 1 without pale-scaled area; both foreungues and both midungues of males toothed; abdominal terga of males with relatively few relatively short lateral setae. FEMALE GENITALIA: Upper vaginal sclerite present; insula tongue-like, insular setae absent; 3 spermathecal capsules. MALE GENITALIA: Tergum IX with setae; sternum IX relatively short, without setae; gonostylus long, cylindrical; claspette with a narrow apex and few setae; aedeagus comprised of 2 lateral plates, plates apparently fused distally but each with numerous long stout curved teeth laterally and apically; proctiger with sternal arm; paraproct with apical teeth. LARVAE: Antenna moderately long to long, $\geq$ 0.4 the median length of the dorsal apotome; length of seta 1-A $\leq$ 3.0 times the width of the antenna at the point of attachment; seta 4-C short, with 3–6 branches; setae 5,6-8,13,14-C single; seta 7-C single (rarely double); cervical sclerite present; seta 1-P single, longer than seta 2-P; setae 3,5-P single; seta 5-T small, single; seta 2-I–VII small, single, inserted near seta 1; seta 11-I small, single or double; seta 12-I absent; seta 5-II–VI small, single; seta 7-II long, single or double; seta 9-II–VI near seta 7; seta 10-II–V inserted lateral to setae 11 and 12; seta 6-III single; seta 13-III–V single; seta 5-VII small, single or double; seta 1-VIII also single; comb scales scale-like, in a patch; siphon acus present, pecten with the distal spine more widely spaced; seta 1-X small, single; seta 2-X single or double (rarely triple); ventral brush (seta 4-X) with 5 pairs of setae on a grid, precratal setae absent. PUPAE: Seta 1-CT weakly developed, considerably shorter than seta 3-CT; seta 5-CT > 1.3 the length of seta 4-CT; seta 7-CT longer than seta 6-CT; setae 3-I,III and 6-III branched; seta 3-II as long as or shorter than seta 6-II; seta 6-VII inserted posterior to seta 9; seta 9-IV–VI inserted posterior to seta 8; paddle produced apically.

****

Bionomics: *Isoaedes* is a cavernicolous taxon. The immature stages of *Ae. cavaticus* are found in shallow temporary cold-water pools in complete darkness. Nothing in known about the bionomics of the adults. The species is of no medical or economic importance to humans.

****

Principal references: Reinert, 1979 (new subgenus and type species, descriptions, bionomics, distribution); Reinert, 2000e (female genitalia); Reinert *et al.*, 2004, 2006, 2009 (as a genus, morphology, phylogenetic relationships); Harbach *et al.*, 2007b (as a genus, comparative morphology); Rattanarithikul *et al.*, 2010 (as a genus, Thailand, keys, bionomics); Wilkerson *et al.*, 2015 (phylogenetic relationships, classification).

Jarnellius Reinert, Harbach & Kitching, 2006

Jarnellius, originally proposed as a genus by Reinert *et al.* (2006), currently classified as a subgenus of *Aedes*, is typified by *Aedes varipalpus* (Coquillett, 1902d), original combination: *Culex varipalpus*. Subgenus abbreviation – *Jar.*

****

Systematics: *Jarnellius* includes five extant species with distributions in western areas of the Nearctic Region – Canada (British Columbia), Mexico (Baja California Sur, Sonora) and the USA (Arizona, California, Idaho, Montana, Nevada, New Mexico, Oregon, Utah, Washington). Arnell & Nielsen (1972) noted that the relationships of *Jarnellius* (their Varipalpus Group) and other aedine taxa were obscure. Based on distributions, rather simple male genitalia and probable tropical origin, Arnell & Nielsen stated that the group seemed to be an ancient lineage, with *Ae. laguna* 'probably closest to the ancestral stock'. In the phylogeny of Aedini recovered in the study of Reinert *et al.* (2009) based on morphological data, *Jarnellius* was recovered in a clade comprised of *Acartomyia* + (*Jarnellius* + (*Halaedes* + *Opifex*)), which was sister to *Ochlerotatus*. *Jarnellius* was not associated with other generic-level taxa in the phylogeny of Wilkerson *et al.* (2015). In the maximum likelihood phylogeny of Soghigian *et al.* (2017) based on seven molecular markers, species of *Jarnellius* comprised a monophyletic lineage in a sister relationship with two species of *Ochlerotatus*.

****

Characteristics: Adults of species of *Jarnellius* are distinguished from the adults of all other New World aedine taxa by the presence of white patches on the maxillary palpus and apical white rings on the tarsi; larvae are characterized by a poorly developed ventral brush comprised of six pairs of usually double setae that are gradually stronger distally. *Jarnellius* is characterized by the following combinations of characters. ADULTS: Maxillary palpus of females dark-scaled with pale-scaled areas, maxillary palpus of males with palpomeres 4 and 5 nearly straight; scutal fossa with white lines of scales on the lateral and posterior margins (the lines are not expanded mesally); scutellum with broad scales on the lateral lobes; postpronotum with narrow scales dorsally and broad scales ventrally; hypostigmal scales absent; mesepimeron with 2 patches of white scales; metameral scales present; hindtibia dark-scaled with white scales basally; hindtarsomere 1 with broad basal and apical white bands, hindtarsomeres 2 and 3 with broad apical white bands; fore- and midungues (both sexes) each with 1 tooth. FEMALE GENITALIA: Cercus moderately long and wide, apex sharply rounded, without scales; upper vaginal sclerite absent. MALE GENITALIA: Ninth tergal lobes with 4–8 setae; gonocoxite with strong setae and scales on the lateral 0.33 of the dorsal surface, short setae becoming stronger toward the apex on the mesal 0.67 of the dorsal surface, basal tergomesal lobe with a strongly differentiated apically curved seta and 1–3 smaller slender setae, gonostylus strongly curved apically, gonostylar claw spiniform, about 0.3 the length of the gonostylus; claspette filament slightly longer than the stem, flattened beyond the middle, apex sharply recurved. LARVAE: Antenna without spicules; seta 1-C spiniform; seta 6-C double; seta 4-P as long as or shorter than seta 3-P; seta 8-P multi-branched; setae 2-T and 3-I,II,VII single; seta 7-I,II similar, single (occasionally double in *Ae. sierrensis*), moderately long to long; seta 1-S 1.2–2.0 the width of the siphon; ventral brush (seta 4-X) attached to a grid with only transverse grid bars, seta 4a-X long, seta 4d-X single or double (rarely triple), not plumose. PUPAE: Seta 2-CT very strongly developed, considerably longer than seta 3-CT; seta 1-II with 3 or fewer branches (rarely 4-branched); seta 3-II as thick or thicker than seta 1-II; seta 5-II inserted lateral to seta 4-II; seta 6-III single; seta 2-V inserted anterior to seta 3; seta 5-V shorter than the following tergum; seta 6-VII inserted posterior and mesal to seta 9-VII.

****

Bionomics: The immature stages of the species of *Jarnellius* are usually found in tree holes. Females are known, with the exception of *Ae. deserticola*, to avidly bite humans who enter their haunts during daylight hours and at dusk. Species of *Jarnellius* are of no known medical or economic importance to humans.

****

Principal references: Arnell & Nielsen, 1972 (as Varipalpus Group of *Aedes* subgenus *Ochlerotatus*, systematics, keys, descriptions, bionomics, distributions); Reinert *et al.*, 2006 (new genus, morphology, phylogenetic relationships, description, distribution, bionomics); Reinert *et al.*, 2008, 2009 (as a genus, morphology, phylogenetic relationships); Reinert, 2010b (as a genus, female genitalia); Wilkerson *et al.*, 2015 (phylogenetic relationships, classification); Soghigian *et al.*, 2017 (phylogenetic relationships).

Jihlienius Reinert, Harbach & Kitching, 2006

Jihlienius, originally proposed as a genus by Reinert *et al.* (2006), currently classified as a subgenus of *Aedes*, is typified by *Aedes chungi* Lien, 1968b. Subgenus abbreviation – *Jih.*

****

Systematics: *Jihlienius* includes three extant species recorded from areas of China, India, Nepal and Taiwan. *Jihlienius*, based on *Ae. chungi*, was recovered as the sister of *Molpemyia* in the phylogenetic studies of Reinert *et al.* (2006, 2008), but morphological differences not included in the analyses indicate that the species is not closely related to *Molpemyia*. *Jihlienius* + *Molpemyia* was sister to *Phagomyia* in the 2008 study, and *Jihlienius* was recovered as the sister of *Phagomyia* in the studies of Reinert *et al.* (2009) and Wilkerson *et al.* (2015).

****

Characteristics: *Jihlienius* is characterized by the following combinations of characters. ADULTS: Vertex, ocular area and interocular space of the head with broad decumbent scales, erect forked scales numerous on the occiput, few semi-erect scales on the vertex; compound eyes more or less contiguous above the antennal pedicels; maxillary palpus and proboscis dark-scaled, proboscis longer than the forefemur, palpus of males with 5 palpomeres in a nearly straight line, shorter than the proboscis; scutum with narrow scales except on the median prescutellar area and a strip between the supraalar and prescutellar setae, mainly pale-scaled with variable areas of dark scaling; paratergite with broad scales, postpronotal scales narrow and broad; upper and lower areas of the mesokatepisternum, prealar area and the upper to middle areas of the mesepimeron with broad pale scales; lower mesepimeral setae absent; wing dark-scaled with a pale spot at the base of the costa; hindfemur mainly pale-scaled, with basal and preapical dark bands; hindtarsomere 1 with a broad basal pale band; fore- and midungues equal in females, each with 1 tooth, unequal in males, larger unguis with 2 teeth, smaller unguis with 1 tooth, hindungues equal and simple in both sexes; abdominal terga dark-scaled with small basolateral pale patches, laterotergite of segment I with pale scales, segment VII laterally compressed in females. FEMALE GENITALIA: Tergum VIII largely covered with broad scales, posterior margin broadly rounded; sternum VIII wider than long, with a median caudal emargination, without scales, seta 2-S inserted posterior to seta 1-S; tergum IX comprised of 2 sclerites; cercus without scales; lower vaginal sclerite absent; insula lip-like with a few setae in lateral patches; 3 spermathecal capsules. MALE GENITALIA: Setae of the ninth tergal lobes all slender; gonocoxite without apical and basal lobes on the mesal margin of the dorsal surface, with a small basomesal patch of stout setae, ventral surface with a patch of lanceolate setae on the distal area of the mesal margin, mesal surface membranous, lateral surface with broad scales; gonostylus attached at the apex of the gonocoxite, with 1 apical gonostylar claw; claspette with a narrow columnar stem and a long striated somewhat flattened apical claspette filament; aedeagus simple, tube-like, with small distal spicules; proctiger with a few cercal setae. LARVAE: Antenna without spicules, short, $\leq$ 0.4 the length of the dorsal apotome; seta 4-C multi-branched, inserted slightly mesal and approximately in line with seta 6-C; seta 5-C single, inserted posterior to setae 4,6,7-C; seta 6-C single, shorter than seta 5-C, inserted anterior and at approximately the same level as 5-C, inserted mesal and anterior to seta 7-C; seta 7-C with 3–5 branches, inserted lateral and slightly anterior to seta 5-C; seta 12-C inserted mesal to seta 13-C; seta 13-C single, longer than seta 12-C; seta 14-C single; seta 19-C absent; setae 1–3-P borne on a common setal support plate; seta 4-P single or double, longer than seta 3-P; seta 5-P normally double; seta 8-P shorter than seta 4-P; setae 1,4-M single; seta 12-I present; seta 1-S multi-branched, inserted slightly distal to the pecten; saddle of segment X incomplete ventrally, with stout spicules on the posterior margin dorsal to the insertion of seta 1-X, seta 1-X single or double, inserted on the saddle; seta 2-X branched (normally double); ventral brush (seta 4-X) inserted on a grid with both transverse and lateral grid bars, with 2 precratal setae. PUPAE: Setae 1,3-CT similarly developed; seta 10-CT single; seta 11-CT double, shorter than seta 10-CT; seta 3-I,II single, shorter than seta 6-I,II; seta 6-I longer than seta 7-I, seta 6-III branched, seta 6-VII inserted slightly posterior and mesal to seta 9-VII; seta 5-II inserted lateral to seta 4-II, seta 5-V single, longer than tergum VI; seta 2-VI inserted mesal to setae 1,3-VI; seta 9-VIII inserted on the posterolateral corner of the segment, with 3–5 branches; paddle with rounded apical margin, midrib extends to the apex, without hair-like marginal spicules; seta 1-Pa single or forked.

****

Bionomics: The immature stages of species of *Jihlienius* have been found in tree holes and bamboo stumps. The species are unlikely to be of medical and economic importance to humans.

****

Principal references: Reinert *et al.*, 2006 (new genus, morphology, phylogenetic relationships, description, distribution, bionomics); Reinert *et al.*, 2009 (as a genus, morphology, phylogenetic relationships); Reinert, 2009d (as a genus, female genitalia); Rattanarithikul *et al.*, 2010 (as a genus, Thailand, keys, bionomics); Wilkerson *et al.*, 2015 (phylogenetic relationships, classification).

Kenknightia Reinert, 1990

Kenknightia Reinert, 1990b, originally proposed and currently recognized as a subgenus of *Aedes*, is typified by *Aedes dissimilis* (Leicester, 1908), original combination: *Stegomyia dissimilis*. Subgenus abbreviation – *Ken.*

****

Systematics: *Kenknightia* includes 12 extant species recorded form localities in Bangladesh, India, Malaysia, the Philippines, Thailand and Vietnam. In the morphology-based phylogeny of Aedini recovered in the study of Reinert *et al.* (2009), *Kenknightia* was placed in a clade comprised of *Himalaius* + *Bruceharrisonius* and *Vansomerenis* + (*Zavortinkius* + *Kenknightia*), followed by two pairs of species of uncertain genus-group assignment, *Ae. biocellatus* + *Ae. crossi* and *Ae. candidoscutellum* + *Ae. keefei*. In the results of Reinert *et al.* (2006, 2008), these four groups were arranged in a paraphyletic series (with *Hopkinsius* included with *Ae. biocellatus* and *Ae. crossi* in the study of Reinert *et al.*, 2006). *Kenknightia* was not associated with other generic-level taxa in the phylogeny of Wilkerson *et al.* (2015). A single species of *Kenknightia* (*Ae. dissimilis*) was recovered in a sister relationship with a single species of *Geoskusea* (*Ae. baisasi*) in the maximum likelihood phylogeny of Soghigian *et al.* (2017) based on a single molecular marker.

****

Characteristics: The following combinations of distinctive features distinguishes species of *Kenknightia* from species of other aedine generic-level taxa. ADULTS: Interocular scales of the head all broad; maxillary palpus, proboscis, tibiae and tarsi dark-scaled; maxillary palpus of males noticeably shorter than the proboscis, slender throughout, palpomeres 4 and 5 nearly straight, short with few setae; vertex with broad recumbent scales; ocular line with broad silvery scales; acrostichal and dorsocentral setae absent; scutellum with only broad scales on the mid and lateral lobes; postpronotum, paratergite, prealar knob and upper proepisternum with broad silvery scales; 1 or 2 upper proepisternal setae usually present; subspiracular area, postspiracular area and lower prealar area without scales; mesokatepisternum with 1 (rarely 2) upper, 1 (rarely 2) median posterior and 1–5 lower posterior setae; mesokatepisternal scales in a single large patch; mesepimeron with 1–11 (usually 1–6) upper posterior setae; hindtarsomere 2 without basal pale scales; abdominal terga of males with few relatively short lateral setae. FEMALE GENITALIA: Caudal margin of sternum VIII with a minute median depression and short curved lanceolate setae intermixed with stout setae on the margin; tergum IX comprised of 2 elongate lateral plates connected by membrane; insula lip-like with 4–9 setae; upper vaginal sclerite absent or very small; postgenital lobe narrow, apex rounded; cercus moderately long, wide, distal 0.33 tapered, flattened and gently oblique, this area with stout setae intermixed with curved lanceolate setae. MALE GENITALIA: Tergum IX lobes moderately broad to broad, widely spaced, setae all slender; gonocoxite long and narrow, with more or less evenly spaced setae along the mesal margin, basal mesal lobe absent; gonostylus moderately long, with a simple long and narrow gonostylar claw, ≥ 0.4 the length of the gonostylus; aedeagus trough-like, apex without teeth and flattened. LARVAE: Antenna long, narrow, with spicules proximal to the insertion of seta 1-A, seta 1-A single, stout, inserted 0.4–0.6 from the base; setae 4–6-C inserted at the angles of a small triangle posterior to the level of seta 7-C; seta 4-C short, inserted more or less on level with or posterior to seta 6-C; setae 5,6-C long and stout, seta 5-C single, seta 6-C shorter than seta 5-C; seta 13-C branched; setae 1–3-P usually not inserted on a common setal support plate; seta 6-I,II stout and relatively long; seta 12-I present; seta 2-VII single; seta 3-VII branched, inserted at the same level or posterior to seta 1-VII; comb with squamiform or weakly spine-like scales in a triangular patch; seta 6-S noticeably longer than the distal width of the siphon; ventral brush (seta 4-X) with 5 or 6 pairs of setae, caudal 8 setae with 5–12 long branches. PUPAE: Seta 5-CT noticeably longer than seta 4-CT, both with 2–9 branches; seta 6-CT short; seta 7-CT very long, with 2–7 branches; seta 1-II with 3 or fewer branches, rarely 4-branched; seta 3-II longer than seta 6-II, 3-III branched, rarely single; seta 5-IV–VI very long and stout; branches of setae 1-I and 9-VIII with long aciculae; paddle with well-developed buttress on the proximal 0.8 or less, apicolateral margin more or less flattened; seta 1-Pa single.

****

Bionomics: The immature stages of species of *Kenknightia* usually inhabit tree holes and stumps, but occasionally occur in rock holes, bamboo stumps, coconut shells, axils of *Colocasia*, palm stubs and artificial containers such as cans, pots and buckets. Adult females have been collected biting humans and water buffalo. Species of *Kenknightia* are not of medical and economic importance to humans.

****

Principal references: Reinert, 1990b (new subgenus, classification, keys, descriptions, distributions, bionomics); Reinert *et al.*, 2004, 2006, 2009 (as a genus, morphology, phylogenetic relationships); Rattanarithikul *et al.*, 2010 (as a genus, Thailand, keys, bionomics); Wilkerson *et al.*, 2015 (phylogenetic relationships, classification); Soghigian *et al.*, 2017 (phylogenetic relationships).

Kompia Aitkin, 1941

Kompia Aitkin, 1941, originally proposed and currently recognized as a subgenus of *Aedes*, is typified by *Aedes purpureipes* Aitken, 1941. Subgenus abbreviation – *Kom.*

****

Systematics: *Kompia* is monobasic. It only includes the extant type species, which is found in Mexico and the southwestern USA (Arizona). Zavortink (1972) stated that *Kompia* 'shows no definite relationship to any other New World group' of Aedini, but noted that some larval features, such as the absence of seta 12-I, the development and position of seta 13-VI, the single seta 3-VII and the boss supporting the ventral brush suggest that the group 'may have been derived from the same stock as *Abraedes*, *Aztecaedes*, *Gymnometopa* and *Howardina*'. *Kompia* was recovered as sister to a clade comprised of *Aztecaedes* + (*Abraedes* + *Lewnielsenius*) in the morphology-based phylogenetic analysis of Aedini conducted by Reinert *et al.* (2009). *Kompia* was not associated with other generic-level taxa in the phylogeny of Wilkerson *et al.* (2015). However, *Ae. purpureipes* was recovered in a sister relationship with the monobasic *Lewnielsenius* in the maximum likelihood phylogeny of Soghigian *et al.* (2017) based on seven molecular markers, and the pair was sister to a large clade comprised of 37 species of *Ochlerotatus*.

****

Characteristics: The following combinations of features distinguish *Kompia* from other generic-level taxa of Aedini. ADULTS: Dark scaling of the head, proboscis, maxillary palpus, legs and abdomen with metallic blue, violet or purple reflections; vertex and interocular space with broad decumbent scales; maxillary palpus and proboscis entirely dark-scaled; scutum with golden longitudinal lines on black background; midlobe of the scutellum with broad decumbent scales; paratergite bare; postspiracular setae absent; yellowish scales at the bases of the femora, tibiae and tarsi entirely dark-scaled; abdominal sterna with yellowish scales. FEMALE GENITALIA: Tergum VIII with setae on the distal 0.7 or more; posterior margin of sternum VIII more or less uniformly sloping cephalad from the apicolateral corners to the midline. MALE GENITALIA: Tergum IX lobes prominent, each with a series of strongly developed setae; basal tergomesal area of the gonocoxite with a patch of numerous fine setae and 1 strong stout seta, apical area of the sternomesal surface with numerous strong elongate setae; gonostylar claw a moderately broad spiniform with the apex bluntly pointed; aedeagus expanded apically. LARVAE: Seta 7-C $\leq$ 0.19 median length of the dorsal apotome; ventromedian cervical sclerite absent; integument of the thorax and abdomen spiculose; seta 7-M as long as or longer than seta 5-M; seta 12-I absent; comb scales (3–7) large, spine-like, in a single row; ventral brush (seta 4-X) comprised of 7 pairs of setae borne on a boss, posteriormost seta very short and multi-branched; anal papillae sausage-shaped, dorsal pair much longer than the ventral pair and curved ventrad. PUPAE: Seta 2-III–V inserted laterad of seta 1; seta 5-IV,V short, 0.4–0.7 the length of the segment; paddle with a moderately deep apical emargination.

****

Bionomics: The immature stages of *Ae. purpureipes* are found in tree holes. Adults are known to rest in tree holes and are attracted to artificial lights. Females have been collected landing on and biting humans during the day and at dusk. The species is not known to be of medical and economic importance.

****

Principal references: Zavortink, 1972 (*Ae. purpureipes*, in keys, description, bionomics, distribution); Reinert *et al.*, 2004, 2006, 2008, 2009 (as a genus, taxonomy, classification, phylogenetic relationships); Wilkerson *et al.*, 2015 (phylogenetic relationships, classification); Soghigian *et al.*, 2017 (phylogenetic relationships).

Leptosomatomyia Theobald, 1905

Leptosomatomyia, originally proposed as a genus by Theobald (1905f), currently classified as a subgenus of *Aedes*, is typified by *Aedes aurimargo* Edwards, 1922d. Subgenus abbreviation – *Lep.*

****

Systematics: *Leptosomatomyia* is monobasic. It only includes the extant type species, which is known from localities in Australia (Cape York Peninsula and Torres Strait Islands) and New Guinea. *Leptosomatomyia* was recovered as the sister of *Huaedes* in the phylogenetic study of Aedini, based on morphological data, conducted by Reinert *et al.* (2009). The two subgenera were the terminal taxa of a clade comprised of *Polyleptiomyia* + (*Bifidistylus* + (*Albuginosus* + (*Tewarius* + (*Christophersiomyia* + (*Huaedes* + *Leptosomatomyia*))))). *Leptosomatomyia* was also recovered as the sister of *Huaedes* in the phylogenetic analysis of Wilkerson *et al.* (2015), where this pair was sister to a clade comprised of *Armigeres* + (*Alanstonea* + *Eretmapodites*).

****

Characteristics: *Leptosomatomyia* is characterized by the following combinations of characters. ADULTS: Dark mosquitoes; vertex of the head with broad and narrow scales; eyes narrowly separated above the antennal pedicels; antenna not strongly verticillate in males, flagellar whorls with several setae directed more or less dorsally and ventrally and several directed laterally; maxillary palpus very short in both sexes, shorter than 0.25 the length of the proboscis in males; scutum with pale-scaled margins, median pale stripe and curved posterolateral pale lines ending at the base of the wing, pale scaling sometimes reduced in males; acrostichal setae absent; dorsocentral setae present; scutellum with broad and narrow scales; paratergite and postspiracular area with narrow pale scales; upper proepisternum, mesokatepisternum and mesepimeron with large patches of broad pale scales, 1 large patch on the mesokatepisternum, 2 patches on the mesepimeron; several lower mesepimeral setae present; base of the hindcoxa more or less in line with the dorsal margin of the mesomeron, hindfemur without dorsal and apical pale scales, hindtibia entirely dark-scaled; hindtarsomere 1 as long as or slightly longer than the hindtibia; ungues of all legs simple in both sexes; abdomen of males very elongate. FEMALE GENITALIA: Abdominal segment VIII small, retracted into segment VII, seta 2-S inserted lateral and approximately on level with seta 1-S; tergum IX with a deep median posterior emargination and wrinkled lateral areas; cercus short, wider distally than at mid-length; postgenital lobe covered with small spicules, apex emarginate; insula tongue-like with tuberculi near the apex. MALE GENITALIA: Sternum IX without setae and scales; gonocoxite without an apical lobe, with a large setose sternomesal lobe, setae on the basomesal area of the dorsal surface all slender; gonostylus simple, noticeably broader at mid-length than proximally; gonostylar claw short, broad, inserted at the apex of the gonostylus; claspette absent; aedeagus comprised of 2 lateral plates, each with several teeth; paraproct long, with apical teeth. LARVAE: Antenna without spicules, seta 1-A single; labiogula wider than long; setae 4–7-C multi-branched, seta 4-C inserted on level with seta 5-C, seta 7-C inserted approximately on level with seta 5-C; seta 1-M single; seta 12-VII inserted approximately on level with seta 13-VII; comb with 4–6 scales in a single row; siphon short, relatively broad, pecten on the proximal 0.5, seta 1-S inserted at the distal end of the pecten; saddle almost complete; anal papillae very long, sausage-shaped. PUPAE: Trumpet without a tracheoid area; seta 5-CT $\leq$ 1.2 the length of seta 4-CT.

****

Bionomics: Larvae of *Ae. aurimargo* have been found in coconut shells, bamboo, tree holes and tree cavities. Adult females are known to attack humans who enter their dwellings. The species is not known to be of medical or economic importance.

****

Principal references: Penn, 1947 (larva); Penn, 1949 (pupa); Lee *et al.*, 1984 (taxonomy, keys, literature, distribution, bionomics); Reinert, 2000e (female genitalia); Reinert *et al.*, 2004, 2009 (as a genus, morphology, phylogenetic relationships); Wilkerson *et al.*, 2015 (phylogenetic relationships, classification).

Levua Stone & Bohart, 1944

Levua Stone & Bohart, 1944, originally proposed and currently recognized as a subgenus of *Aedes*, is typified by *Aedes suvae* Stone & Bohart, 1944 [synonym of *Aedes geoskusea* Amos, 1944]. Subgenus abbreviation – *Lev.*

****

Systematics: *Levua* is monobasic. It only includes the extant type species, which is only known from the Fiji Islands. Belkin (1962) stated that '*Levua* is undoubtedly an ancient derivative of *Ochlerotatus*'. Stone & Bohart (1944) noted that *Levua* closely resembles *Geoskusea*. In the morphology-based phylogenetic study of Aedini conducted by Reinert *et al.* (2009), *Levua* and *Rhinoskusea* were recovered as sister taxa in a clade that comprised *Sallumia* + (*Geoskusea* + (*Levua* + *Rhinoskusea*)). *Levua* was recovered as sister to *Rhinoskusea* + *Geoskusea* in the phylogenetic analyses of Huang *et al.* (1910) based solely on adult morphology. In the phylogeny of Wilkerson *et al.* (2015), *Levua* formed a polytomy along with *Geoskusea*, *Rhinoskusea* and *Sallumia*, and this group was sister to *Cancraedes*.

****

Characteristics: The following combinations of features distinguish *Levua* from other generic-level taxa of Aedini. ADULTS: Dark unornamented mosquitoes; decumbent scales of the vertex all narrow; antenna of males not strongly verticillate; proboscis longer than the forefemur; maxillary palpus of females scarcely longer than the clypeus, maxillary palpus of males about 0.2 the length of the proboscis; scutellum with narrow scales on the mid and lateral lobes; mesokatepisternal scales in 2 patches; lower mesepimeral setae absent; hindtarsomere 1 shorter than the hindtibia; ungues of female all simple. FEMALE GENITALIA: Cercus long. MALE GENITALIA: Tergum IX lobes prominent; gonocoxite with a basal dorsomesal lobe, no apicodorsal lobe, basomesal area of the dorsal surface with 1 or more stout or flattened setae, lateral setae mostly short; gonostylus rather short, proximal part broad; 2 gonostylar claws, apex of the proximal claw bluntly pointed; claspette present, with slender terminal filament; aedeagus simple. LARVAE: Seta 1-C spiniform; seta 6-C inserted anterior to seta 7-C; seta 6-T branched; seta 6-II single; seta 3-VII inserted at the same level or posterior to seta 1-VII; seta 12-VII branched; seta 1-X not inserted on the saddle. PUPAE: Seta 5-II longer than seta 3-II; seta 6-III single; seta 2-V inserted anterior to seta 3-V; seta 2-VI inserted lateral to seta 1-VI; seta 1-Pa 0.4–0.6 the length of the paddle.

****

Bionomics: The immature stages of *Ae. geoskusea* inhabit 'crab holes' made by mud lobsters and mangrove crabs and are sometimes found in discarded household containers. Nothing is known about the bionomics of the adults. Females apparently do not attack humans; hence, *Ae. geoskusea* is not suspected to be of medical or economic importance.

****

Principal references: Stone & Bohart, 1944 (new subgenus, type species, descriptions); Lee *et al.*, 1984 (taxonomy, keys, literature, distribution, bionomics); Reinert *et al.*, 2004, 2006, 2008 (morphology, phylogenetic relationships); Reinert *et al.*, 2009 (as a genus, morphology, phylogenetic relationships); Huang *et al.*, 2010 (morphology, phylogenetic relationships); Huang *et al.*, 2012 (taxonomy, morphology); Wilkerson *et al.*, 2015 (phylogenetic relationships, classification).

Lewnielsenius Reinert, Harbach & Kitching, 2006

Lewnielsenius Reinert *et al.*, 2006, originally proposed as a subgenus of *Jarnellius* and currently recognized as a subgenus of *Aedes*, is typified by *Aedes muelleri* Dyar, 1920g. Subgenus abbreviation – *Lew.*

****

Systematics: *Lewnielsenius* is monobasic. It only includes the extant type species, which is only known from localities in Mexico and the USA (Arizona, New Mexico, Texas). Zavortink (1972) suggested a close relationship between *Lewnielsenius* (as the Muelleri Group) and *Jarnellius* (as the Varipalpus Group) based on complementary distributions and some morphological resemblance, particularly with *Ae. laguna*. In the morphology-based phylogenetic study of Reinert *et al.* (2009), *Lewnielsenius* was recovered as the sister of *Abraedes* in a clade comprised of *Howardina* + (*Gymnometopa* + (*Kompia* + (*Aztecaedes* + (*Abraedes* + *Lewnielsenius*)))), indicating that it is not closely related to *Jarnellius*, in which it was originally classified. Although *Lewnielsenius* was paired with *Jarnellius* in the study of Wilkerson *et al.* (2015) (omitted from their fig. 3, shown only in supplementary trees), the relationship seems unlikely based on the pairing of *Lewnielsenius* with *Kompia* in the maximum likelihood phylogeny of Soghigian *et al.* (2017) based on seven molecular markers.

****

Characteristics: *Lewnielsenius* is distinguished by the following combinations of characters. ADULTS: Vertex of the head with only narrow falcate scales; interocular space with both narrow and broad scales; maxillary palpus and proboscis entirely dark-scaled, males with numerous long setae on palpomeres 3 and 4; scutum with narrow scales only, white to dirty-white scales in distinct acrostichal and lateral prescutellar lines and either with additional pale scaling in distinct anterior dorsocentral, posterior outer dorsocentral, lateral marginal and posterior fossal lines or covering 0.33 of the scutum; scutellum with only narrow scales; paratergite (sometimes) and postpronotum with narrow scales; antepronotal scales all narrow; thoracic pleura with extensive covering of scales and setae, ≥ 20 upper proepisternal setae; metameral scales present; tibiae and tarsi dark-scaled; ungues of females simple; abdominal terga of males with long lateral setae. FEMALE GENITALIA: Intersegmental membrane between segments VII and VIII long; tergum VIII with setae on the distal 0.7 or more. MALE GENITALIA: Gonocoxite with a dorsomesal basal lobe bearing a very stout seta, tergal surface with numerous scales; claspette filament without retrorse apical expansion; aedeagus barrel-shaped, widest in the proximal 0.7. LARVAE: Antenna ≥ 0.4 the length of the dorsal apotome, with spicules; seta 4-C 0.2–0.4 and seta 6-C 0.4–0.8 the length of the dorsal apotome; seta 1-P no longer than seta 2-P; seta 12-I absent; seta 3-V ≥ 1.8 the length of seta 5-V; seta 3-VII short, usually single; comb comprised of 8–12 fringed scales with an apical spine; seta 1a-S 1.2–2.0 width of the siphon; seta 2-X single or double (rarely triple); ventral brush (seta 4-X) usually with 7 pairs of setae on a grid with lateral and transverse grid bars, seta 4a-X short with 3–5 branches. PUPAE: Seta 7-CT ≥ 6.0 the length of seta 6-CT; seta 1-II–IV and sometimes 1-V well developed, 1-VI,VII and sometimes 1-V usually conspicuously more weakly developed; seta 3-II no longer than seta 6; seta 5-II inserted mesal to seta 4; setae 6,9-III–VI more or less strongly developed, long; seta 2-V inserted on a line mesal to seta 3; seta 5-V longer than the following tergum; seta 9-VIII single or double; paddle rather long and ovate.

****

Bionomics: The immature stages of *Ae. muelleri* are usually found in tree holes, but they have also been found in artificial containers and leaf axils of maguey. Females are known to attack and bite humans during the day and brief periods before darkness, but the species is not known or suspected to be of medical or economic importance.

****

Principal references: Zavortink, 1972 (as Muelleri Group of *Aedes* subgenus *Ochlerotatus*, *Ae. muelleri*, taxonomy, description, bionomics, distribution); Reinert *et al.*, 2004 (morphology, phylogenetic relationships); Reinert *et al.*, 2006 (as a subgenus of *Jarnellius*, morphology, phylogenetic relationships); Reinert *et al.*, 2008 (as a subgenus of *Ochlerotatus*, morphology, phylogenetic relationships); Reinert *et al.*, 2009 (as a genus, morphology, phylogenetic relationships); Reinert, 2010c (as a genus, female genitalia); Wilkerson *et al.*, 2015 (phylogenetic relationships, classification); Soghigian *et al.*, 2017 (phylogenetic relationships).

Lorrainea Belkin, 1962

Lorrainea Belkin, 1962, originally proposed and currently recognized as a subgenus of *Aedes*, is typified by *Aedes dasyorrhus* King & Hoogstraal, 1946c. Subgenus abbreviation – *Lor.*

****

Systematics: *Lorrainea* includes five extant species, which are found in countries from Thailand and the Philippines southward to northern Australia and eastward to the Solomon Islands. Belkin (1962) noted that *Lorrainea* shows affinities with *Stegomyia* but differs markedly in all life stages, with a very differently developed aedeagus in males. In the morphology-based phylogeny of Aedini recovered by Reinert *et al.* (2009), *Lorrainea* was placed in a clade comprised of *Lorrainea* + ((*Eretmapodites* + *Armigeres*) + (*Udaya* + (*Belkinius* + *Zeugnomyia*))), and this clade was sister to another clade comprised of *Petermattinglyius* + (*Heizmannia* + (*Alanstonea* + *Pseudarmigeres*)). In the phylogeny of Aedini recovered in the study of Wilkerson *et al.* (2015), *Lorrainea* was sister to a clade comprised of *Cancraedes* + (*Geoskusea* + *Levua* + *Rhinoskusea* + *Sallumia*). In stark contrast, *Lorrainea*, with species in the Australasian and Oriental Regions, was recovered as sister to the Afrotropical *Cornetius* in the maximum likelihood phylogeny of Soghigian *et al.* (2017) based on seven molecular markers.

****

Characteristics: The following combinations of features distinguish *Lorrainea* from other generic-level taxa of Aedini. ADULTS: Vertex of the head with decumbent scales very broad; erect scales restricted to the occiput; eyes apparently contiguous above the antennal pedicels; interocular scales absent; antenna of females longer than the proboscis, shorter than the proboscis in males; maxillary palpus of females about 0.2 the length of the proboscis, with 5 palpomeres, palpomere 5 very small and hidden by scales, maxillary palpus of males 0.75 the length of the proboscis to slightly longer than the proboscis, palpomeres 4 and 5 with no more than a few setae; scutal scales all narrow and dark; acrostichal and dorsocentral setae present; scutellum with broad scales; paratergal and subspiracular scales absent; remigium of the wing without distinct setae, alula with broad scales; hindfemur without pale scales at the apex; tarsi dark-scaled; ungues of females all simple, larger fore- and midunguis of males simple, smaller of these apparently with a small basal tooth; abdominal tergum I and the laterotergite almost completely covered with scales. FEMALE GENITALIA: Segment VIII apparently largely retractile, seta 2-S inserted lateral and more or less on level with seta 1-S; tergum IX without setae; cercus short, broad; 3 spermathecal capsules. MALE GENITALIA: Tergum IX poorly developed, without setae; sternum IX with a patch of setae; gonocoxite complex, mesal area of the ventral surface with a row or patch of lanceolate setae, a conspicuous basal sternal lobe with apical spiniform setae present; gonostylus attached subapically, with scales; gonostylar claw a relatively narrow spiniform with a truncate apex, long, ratio of its length to the length of the gonostylus ≥ 0.4; claspette comprised of 2 basal lobes, columnar lobe absent; aedeagus simple, without teeth; proctiger strongly developed, paraproct with apical teeth; cercal setae absent. LARVAE: Head with a poorly developed collar; antenna short, without spicules, seta 1-A single, inserted beyond mid-length; setae 4,6,7-C inserted cephalad of the antennal base, seta 4-C strongly developed, multi-branched, inserted cephalad of seta 6-C; seta 14-C single; seta 6-I–VI strongly developed, long and usually branched; seta 12-I absent; seta 2-III–VII inserted far mesad of seta 1; comb scales small, fringed, spatulate, in a large patch; seta 4-VIII inserted close to seta 3; siphon relatively short, incomplete at the base ventrally, acus absent; pecten spines fringed; seta 1-S long, single; seta 8-S longer than the distal width of the siphon; saddle incomplete, very small, acus and marginal spicules absent; seta 1-X long; seta 2-X multi-branched; seta 3-X single; ventral brush (seta 4-X) with 5 pairs of setae on a poorly developed grid, seta 4d-X with 5 or more branches (rarely 4-branched), not plumose. PUPAE: Seta 1-CT long; setae 2,3-CT widely separated, 2-CT short, 3-CT long; seta 5-CT > 1.3 the length of seta 4-CT; setae 2,3-I approximated; seta 1-II inserted near the midline, with 5 or more branches, seta 1-III–VII strongly displaced laterad; seta 2-III–VII inserted distinctly mesad of seta 1 near the caudal border of the tergum; seta 3-II inserted lateral to seta 2; seta 3-VI inserted mesal or directly anterior to seta 1; seta 9-VI small, inserted laterally, similar to seta 9-II–V; external and internal margins of the paddle with a fringe of long hair-like spicules; seta 1-Pa 0.4–0.6 the length of the paddle.

****

Bionomics: The immature stages of species of *Lorrainea* have been found in phytotelm habitats, such as coconut shells, palm axils and tree holes, and artificial containers. Bohart & Ingram (1946b) found that larvae of *Ae. lamelliferus* were abundant in a mangrove swamp but few females were biting, indicating that either humans are not preferred hosts or that feeding may normally occur at other times of the day or night. Little else is known about the bionomics of the species of *Lorrainea*. They are of no medical or economic importance.

****

Principal references: Belkin, 1962 (new subgenus, type species, taxonomy, descriptions, bionomics, distribution); Lee *et al.*, 1984 (taxonomy, keys, literature, distributions, bionomics); Reinert *et al.*, 2004, 2009 (as a genus, morphology, phylogenetic relationships); Rattanarithikul *et al.*, 2010 (as a genus, Thailand, keys, bionomics); Wilkerson *et al.*, 2015 (phylogenetic relationships, classification); Soghigian *et al.*, 2017 (phylogenetic relationships).

Luius Reinert, Harbach & Kitching, 2008

Luius, originally proposed as a genus by Reinert *et al.* (2008), currently classified as a subgenus of *Aedes*, is typified by *Aedes fengi* Edwards, 1935a. Subgenus abbreviation – *Lui.*

****

Systematics: *Luius* is monobasic. It only includes the extant type species, which is only known from localities in China (Chekiang and Hunan Provinces) and Taiwan. *Luius* was recovered as the sister of *Finlaya* in the morphology-based phylogenetic study of Reinert *et al.* (2008), and the *Luius* + *Finlaya* clade was sister to *Macleaya*. In the more comprehensive study of Reinert *et al.* (2009), *Luius* was recovered as the sister of *Macleaya* in a clade comprised of *Rampamyia* + (*Molpemyia* + (*Luius* + *Macleaya*)). *Luius* was also recovered as the sister to *Macleaya* in the study of Wilkerson *et al.* (2015). In contrast, *Luius* was recovered in a sister relationship with the only species of *Danielsia* (*Ae. albotaeniatus*) included in the maximum likelihood phylogeny of Soghigian *et al.* (2017) based on seven molecular markers.

****

Characteristics: *Luius* is characterized and distinguished from other generic-level taxa of Aedini by the following combinations of characters. ADULTS: Vertex of the head with narrow decumbent scales; erect scales on the occiput and vertex; eyes contiguous; antennal pedicel with a patch of broad pale scales on the mesal surface; maxillary palpus of females dark-scaled, with 4 palpomeres, maxillary palpus of males with 5 palpomeres, about 0.8 the length of the proboscis, palpomeres 4 and 5 nearly straight, with a few ventrolateral setae; proboscis dark-scaled, about as long as the forefemur; scutum with a pattern of narrow pale scales interposed with dark scales, including a pale stripe on the anterior dorsocentral area; acrostichal and dorsocentral setae present; scutellum with narrow scales on all lobes; antepronota widely separated, with broad pale scales; broad pale scales on the postpronotum, postspiracular area, subspiracular area, upper proepisternum, mesokatepisternum, lower prealar area and the mesepimeron; lower mesepimeral setae absent; wing dark-scaled, upper calypter with numerous marginal setae, alula with narrow scales on the margin; remigium with 1–3 dorsal setae; ante- and postprocoxal scales absent; base of hindcoxa below the dorsal margin of the mesomeron; hindfemur with a dorsoapical pale patch; tibiae dark-scaled; hindtarsomeres 1–3 each with a narrow basal pale band or dorsobasal pale spot; anterior and posterior fore- and midungues of females equal, each with 1 tooth, fore- and midungues of males unequal, larger unguis with 2 teeth, smaller unguis with 1 tooth, hindungues equal and simple in both sexes; abdominal terga II–VIII dark-scaled with basolateral pale patches, terga of male with a few short lateral setae, segment VII of females laterally compressed. FEMALE GENITALIA: Tergum VIII wider than long, posterior margin straight, broad scales on the posterior 0.6; sternum VIII wider than long, posterior margin with a median emargination, scales absent; tergum IX with a median posterior emargination separating narrow pointed lobes; postgenital lobe long, narrow, apical margin straight or slightly emarginate; lower vaginal sclerite absent; insula lip-like with a few setae; 3 spermathecal capsules. MALE GENITALIA: Tergum IX with a few stout (somewhat flattened) setae; gonocoxite with a small dorsomesal lobe near the base, lateral and ventral surfaces with numerous broad scales; gonostylus narrow, with a narrow terminal gonostylar claw; claspette with a relatively stout columnar stem bearing a broad flattened claspette filament apically; aedeagus tube-like, broader proximally. LARVAE: Antenna about 0.45 the length of the dorsal apotome, narrow, spicules absent, seta 1-A double or triple; seta 1-C single, thin, distally attenuate; seta 4-C inserted mesal and slightly posterior to seta 6-C; seta 5-C single, stout, inserted posterolateral to seta 6-C; seta 6-C with 2 stout branches; seta 7-C with 2 or 3 stout branches, inserted posterolateral to setae 4,6-C and anterolateral to seta 5-C; seta 13-C single, inserted anterior and slightly mesal to seta 12-C; seta 19-C absent; setae 1–3-P borne on a common setal support plate, 1-P very long, stout; seta 6-I–V double; seta 7-I single, shorter than seta 6-I; seta 12-I present; comb scales in a patch; seta 1-VII 0.5–0.9 the length of segment X; siphon with attached acus, pecten comprised of numerous evenly spaced spines, seta 1-S inserted distal to the pecten; saddle incomplete, with numerous long stout spicules on the dorsoposterior margin; seta 1-X single, inserted on the saddle; ventral brush (seta 4-X) borne on a grid with transverse grid bars, with 2 precratal setae. PUPAE: Cephalothorax with clear unpigmented spots; seta 1-CT single, long, stout, noticeably longer than setae 2,3-CT; setae 4,5-CT more or less equally developed; seta 3-I shorter than seta 6-I; seta 2-II inserted lateral to seta 1-II and slightly mesal to seta 3-II, seta 2-VI inserted mesal to seta 3-VI; seta 5-IV–VII single, long, stout; seta 6-VII inserted mesal and slightly anterior to seta 9-VII; seta 9-VII single to triple, seta 9-VIII with a few branches, inserted on the posterolateral corner of the segment; paddle rounded apically, without hair-like marginal spicules, midrib extends to the apex; seta 1-Pa single (rarely forked at tip).

****

Bionomics: The immature stages of *Ae. fengi* have only been found in bamboo stumps. Adult females fiercely bite humans in bamboo forest. The species is not known to be of medical and economic importance.

****

Principal references: Lu & Ji, 1997 (*Ae. fengi*, as a species of the *Ae. albotaeniatus* group of *Finlaya*, in keys, taxonomy, description); Reinert *et al.*, 2008 (new genus, morphology, phylogenetic relationships, description, distribution, bionomics); Reinert *et al.*, 2009 (as a genus, morphology, phylogenetic relationships); Reinert, 2009c (as a genus, female genitalia); Wilkerson *et al.*, 2015 (phylogenetic relationships, classification); Soghigian *et al.*, 2017 (phylogenetic relationships).

Macleaya Theobald, 1903

Macleaya, originally proposed as a genus by Theobald (1903e), currently classified as a subgenus of *Aedes*, is typified by *Aedes tremulus* (Theobald, 1903e), original combination: *Macleaya tremula*. Subgenus abbreviation – *Mac.*

****

Systematics: *Macleaya* includes 11 extant species divided between two species groups, the Chaetocruiomyia and Macleaya Groups (subgenera *Chaetocruiomyia* and *Macleaya*, respectively, of Reinert *et al.*, 2006). Species of *Macleaya* occur in the Australasian Region, principally Australia. The phylogenetic relationships of *Macleaya* are uncertain. *Macleaya* was recovered as the sister of *Luius* in a clade comprised of *Rampamyia* + (*Molpemyia* + (*Luius* + *Macleaya*)) in the phylogeny of Reinert *et al.* (2009) based on morphological data. *Macleaya* was also recovered as the sister to *Luius* in the study of Wilkerson *et al.* (2015). In the absence of the sole species of *Luius*, it was recovered in a sister relationship with 17 species of *Ochlerotatus* in the maximum likelihood phylogeny of Soghigian *et al.* (2017) based on seven molecular markers.

****

Characteristics: Species of *Macleaya* are characterized and distinguished from the species of other generic-level taxa of Aedini by the following combinations of characters. ADULTS: Small thick-set mosquitoes with strong scutal setae; vertex with broad decumbent scales; proboscis short; maxillary palpus of males about as long as the proboscis; antepronotal scales all narrow; subspiracular scales present; lower mesepimeral setae absent; legs mainly dark-scaled; hindtibia with strong setae; tarsi with pale bands, hindtarsomere V usually all white; all ungues of females equal and simple, fore- and midungues of males unequal, larger anterior fore- and midunguis with a strong tooth, hindungues equal and simple; abdominal segments VII and VIII of females large, laterally compressed. FEMALE GENITALIA: Sternum VIII large, posterior margin with a median emargination separating broadly rounded lateral lobes; insula lip-like; cerci very short, sometimes hidden. MALE GENITALIA: Gonostylus broader in the middle than proximally, without distal setae, with a terminal gonostylar claw, ≥ 0.4 the length of the gonostylus; phallosome simple. LARVAE: Antenna without spicules; ventromedian cervical sclerite absent; setae 4–7-C inserted anterior to the base of the antenna; seta 5-P branched; seta 1-M ≥ 3.5 the length of seta 2-M; seta 4-M single; seta 7-M as long as or longer than seta 5-M; seta 3-II branched; seta 5-II single; seta 3-V ≤ 1.6 the length of seta 5-V; seta 1-VIII ≥ 1.1 the length of seta 2-VIII; seta 1-S inserted beyond the pecten, single; saddle incomplete. PUPAE: Seta 5-II inserted mesal to seta 4-II; seta 5-II longer than seta 3-II; seta 6-II no longer than seta 7-I; seta 2-V inserted anterior to seta 3-V; seta 9-VIII inserted slightly anterior to the posterolateral corner of the segment; seta 1-Pa single.

****

Bionomics: The immature stages of species of *Macleaya* develop in tree holes and hollow stumps; those of *Ae. tremulus* have been found in artificial containers. Adult females of most species are known to attack and bite humans and domestic animals during daylight hours, but some are known to bite in the evening and early morning hours, or during the night. Two species, *Ae. moloiensis* and *Ae. wattensis*, sometimes bite indoors. Species of *Macleaya* are not of medical and economic importance.

****

Principal references: Marks, 1963a (as *Aedes* subgenus *Chaetocruiomyia*, keys, classification, descriptions, bionomics, distributions); Dobrotworsky, 1965 (as *Aedes* subgenera *Chaetocruiomyia* and *Macleaya*, in keys, descriptions, bionomics, distributions); Lee *et al.*, 1982 (as *Aedes* subgenus *Chaetocruiomyia*, key, classification, literature, distributions, bionomics); Lee *et al.*, 1984 (key, literature, distributions, bionomics); Reinert, 2002e (as *Aedes* subgenera *Chaetocruiomyia* and *Macleaya*, female genitalia); Reinert *et al.*, 2006, 2008, 2009 (as a genus, morphology, phylogenetic relationships); Wilkerson *et al.*, 2015 (phylogenetic relationships, classification); Soghigian *et al.*, 2017 (phylogenetic relationships).

Molpemyia Theobald, 1910

Molpemyia, originally proposed as a genus by Theobald (1010c), currently classified as a subgenus of *Aedes*, is typified by *Aedes purpureus* (Theobald, 1910c), original combination: *Molpemyia purpurea*. Subgenus abbreviation – *Mol.*

****

Systematics: *Molpemyia* includes three extant species which are found in Australia, the states of Queensland, Northern Territory and Western Australia. In the morphology-based phylogeny of Aedini recovered by Reinert *et al.* (2009), *Molpemyia* was placed in a clade comprised of *Rampamyia* + (*Molpemyia* + (*Luius* + *Macleaya*)). *Molpemyia* was not associated with other generic-level taxa in the phylogeny of Wilkerson *et al.* (2015).

****

Characteristics: The following combinations of features, especially characteristics of the female and male genitalia and the uniquely developed ventral brush and boss of the larvae, distinguish *Molpemyia* from other generic-level taxa of Aedini. ADULTS: Vertex and occiput of the head with narrow curved decumbent scales and numerous long erect forked scales; maxillary palpus and proboscis dark-scaled; scutum with complete rows (anterior and posterior groups) of acrostichal and dorsocentral setae; scutal fossa with contrasting pale scales in a large patch; paratergite with broad pale scales; upper proepisternum with 5–12 setae; upper prealar scales present; costa with a pale spot near the base of the wing, anal vein terminates approximately in line with the intersection of the mediocubital crossvein and the cubitus; tibiae of all legs dark-scaled, tarsomeres 1 and 2 of all legs and tarsomere 3 of the hindleg with a broad basal white band; abdominal segment VII of females dorsoventrally flattened; abdominal terga II–VII with a large basomedian pale patch or band; larger unguis of the fore- and midlegs of males with 2 teeth, smaller unguis with 1 tooth, hindungues equal and simple. FEMALE GENITALIA: Sternum VIII large, with a subapical line of 4 long setae extending toward the midline from the apicolateral corners; caudal margin of tergum IX with a U-shaped median indentation and 5–8 long setae on either side of the midline; postgenital lobe broad basally, narrower distally; caudal margin of the upper vaginal lip broadly rounded, upper vaginal sclerite large and heavily pigmented; insula large, lip-like, with a cluster of 6–9 long setae on either side of the midline. MALE GENITALIA: Tergum IX lobes small, each with 4–7 setae; gonocoxite long and narrow, mesal membrane extends to the base of the gonostylus, lateral and ventral surfaces with numerous setae and broad scales; gonostylar claw bluntly pointed; basal mesal lobe comprised of a long narrow laterally curved stem bearing short setae distally and a long distally tapered apical foliform; aedeagus simple, trough-like, distal 0.5 divided and produced laterad, widest in the distal 0.33, teeth absent; paraproct with a short curved beak-like apex bearing 2–4 small teeth; cercal sclerite with 3–7 setae. LARVAE: Antenna short, ratio of its length to the median length of the dorsal apotome ≤ 0.4; seta 1-A single; setae 5–7-C single; seta 4-C inserted mesad and slightly caudad of seta 6-C; seta 5-C inserted directly caudad of seta 6-C, ratio of its length to the median length of the dorsal apotome 0.2−0.4; seta 7-C inserted on about the same level as seta 6-C, ratio of its length to the median length of the dorsal apotome ≤ 0.2; seta 13-C single; seta 4-P longer than seta 3-P; seta 7-P single; seta 2-T branched; setae 9,10-T single, stout, strongly aciculate; seta 1-II single; ratio of the length of seta 1-VII to the mid-dorsal length of segment X ≤ 0.4; seta 6-I–VI single, long, stout; seta 8-II single; comb with 5–12 short broad scales in a single row; siphon short, broad at the base, index 1.1–1.6, saddle/siphon index 1.1–1.3; pecten comprised of 8–22 slender evenly spaced spines, ratio of the length of seta 1-S to the width of the siphon ≤ 0.4; saddle large, incomplete ventrally; seta 1-X small, with 2–7 branches; setae 2,3-X single, long, stout, ratio of the length of seta 2-X to the length of seta 3-X ≥ 0.9; ventral brush (seta 4-X) comprised of 17–22 stout usually apically split plumose setae in an irregular row along the ventral and caudal margins of an exceptionally large caudally projecting rudder-like boss. PUPAE: Setae 1–3-CT moderately long, seta 2-CT slightly shorter than setae 1,3-CT; seta 4-CT single, about as long as seta 5-CT; seta 2-II–V inserted mesad and slightly cephalad of seta l, seta 2-VI,VII inserted laterad of seta 1; seta 3-II inserted lateral to seta 2-II; seta 5-V shorter than the median length of the following tergum; seta 6-VII inserted mesad and caudad of seta 9; seta 9-III–V apparently inserted ventrally; seta 9-VII,VII single, long, stout, spinulate, inserted dorsally near the caudolateral margin of the segment; paddle with a broad weakly developed midrib terminating well before the apex.

****

Bionomics: The immature stages of species of *Molpemyia* have been found in cavities in trees, which sometimes contain foul water. Larvae of the three known species of *Molpemyia* are reportedly predacious on other mosquito larvae. Females are known to bite humans, but the species are not known to be of medical or economic importance. Attempts to recover virus from *Ae. purpureus* produced negative results (cited in Lee *et al.*, 1982).

****

Principal references: Reinert, 1993 (taxonomy, descriptions, distributions, bionomics); Reinert *et al.*, 2006, 2008, 2009 (as a genus, morphology, phylogenetic relationships); Wilkerson *et al.*, 2015 (phylogenetic relationships, classification); Soghigian *et al.*, 2017 (phylogenetic relationships).

Mucidus Theobald, 1901

Mucidus, originally proposed as a genus by Theobald (1901a), currently classified as a subgenus of *Aedes*, is typified by *Aedes alternans* (Westwood, 1835), original combination: *Culex alternans*. Subgenus abbreviation – *Muc.*

****

Systematics: *Mucidus* includes 15 extant species, divided between two species groups, the Mucidus and Pardomyia Groups (subgenera *Mucidus* and *Pardomyia*, respectively, of Reinert, 2006b). Species of *Mucidus* are found in the Afrotropical, Australasian and Oriental Regions. The phylogenetic studies of Reinert *et al.* (2008, 2009) based on cladistic analysis of morphological data corroborate the monophyly of *Mucidus*, its distinction from *Aedes sensu stricto* and *Ochlerotatus*, and the recognition of the Mucidus and Pardomyia Groups. The results of their studies indicate that *Psorophora* is the most ancestral clade of Aedini and the sister to *Mucidus* + all other aedine taxa. *Mucidus* was not associated with other generic-level taxa in the phylogeny of Wilkerson *et al.* (2015). In contrast to the results of Reinert *et al.*, two species of the Mucidus Group were recovered as the sister to a clade comprised of species of *Haemagogus*, *Howardina*, *Macleaya* and *Ochlerotatus* in the maximum likelihood phylogeny of Soghigian *et al.* (2017) based on seven molecular markers.

****

Characteristics: The following combinations of features distinguish *Mucidus* from other generic-level taxa of Aedini. ADULTS: Large mosquitoes; eyes narrowly separated above the antennae; frons with scales; decumbent scales on the vertex broad and narrow or all narrow, erect scales numerous; antenna slightly shorter than the proboscis; maxillary palpus of females 0.3–0.6 the length of the proboscis, palpomere 5 large, maxillary palpus of males longer than the proboscis, palpomeres 4 and 5 subequal; scutal scales narrow, acrostichal and dorsocentral setae present; scutellum with narrow scales; paratergite with or without scales; postpronotal setae numerous; pleural scaling varied, pleural setae numerous, mesepimeron with anterior median setae; legs with conspicuous ornamentation, tarsi banded, ungues of females toothed on at least the fore- and midlegs; males with ungues of the fore- and midlegs enlarged, teeth varied; wing membrane darkly pigmented at the base of vein R_s, the basal segment of vein M_{3+4} and around the radiomedial crossvein; alula with a complete marginal fringe; scaling restricted on tergum I and the laterotergite. FEMALE GENITALIA: Segment VIII completely retracted or protruding for 0.5 or more of its length from segment VII; cercus long, slender; 1 large and 2 smaller spermathecal capsules. MALE GENITALIA: Segment VIII strongly developed, narrowed at base; tergum IX poorly sclerotized, lobes poorly differentiated but with setae; sternum IX large, with a group of setae; gonocoxite with a distinct basomesal lobe bearing differentiated setae; gonostylus long, simple, with subapical setae; gonostylar claw slender, attached at the apex of the gonostylus; aedeagus simple, trough-like; claspette strongly differentiated, with a slender stem and an apical seta in the form of a long appendage; proctiger strongly developed, with strong basolateral sclerotization; paraproct with a strong apical spine; cercal setae present. LARVAE: Median labral plate very wide and large; mouthparts modified for predation, filaments of each lateral palatal brush in a compact group on a prominent lateral lobe, filaments (25–30) rather short, curved and with strong apical pectination; mandibular teeth very large and strong; seta 1-C strongly displaced laterad; setae 0,3-C inserted close together; seta 4-C cephalad of the antennal base, long; setae 5–7-C inserted far caudad of the antennal base, 6-C short, 5,7-C long; setae 12–15-C strongly displaced cephalad; seta 13-P absent; seta 12-I present; seta 6-I–VI long, usually single or double; seta 7-I,II large, multi-branched; seta 1-III–V strongly developed; setae 1,2-VIII inserted on a small common basal plate; comb scales fringed, in a patch; siphon long, acus strongly developed, attached; segment X elongate; saddle incomplete, marginal spicules present or absent, acus absent; seta 1-X single, inserted on the saddle; setae 2,3-X single; ventral brush (seta 4-X) strongly developed, with 13 or 14 pairs of setae extending nearly the entire length of the segment, more caudal setae borne on a grid or boss. PUPAE: Trumpet elongate, tracheoid area strongly developed; setae 2,3-CT closely approximated, removed from caudal margin of the sclerite; setae 2,3-I approximated; seta l-II single, inserted very close to the midline; seta 9-VI small, similar to seta 9-II–V; seta 2-VI,VII inserted laterad of seta 1.

****

Bionomics: Larvae of species of *Mucidus* are found in temporary ground pools of various types, and are predacious on other mosquito larvae. Females of several species of the Mucidus Group are reported to attack humans at least occasionally, chiefly at night. Females carry the abdomen curved downward beneath the thorax. Because of a lack of succession of generations in larval habitats, Hopkins (1952) suggested that females deposit eggs while flying, and the eggs are resistant to desiccation. *Mucidus* species primarily inhabit freshwater, but larvae of *Ae. alternans* are said to develop successfully in saline marshes (Knight, 1947) and *Ae. chrysogaster* has been found in highly polluted water (Bick, 1951, as a subspecies of *Ae. aurantius*). Species of *Mucidus* are not of medical and economic importance.

****

Principal references: Mattingly, 1961 (Indomalayan area, taxonomy, keys, descriptions, bionomics, distributions); Tyson, 1970b (classification, taxonomy, keys, distributions, bionomics); Lu *et al.*, 1997a (China, *Ae. scatophagoides*, taxonomy, descriptions, distribution, bionomics); Reinert *et al.*, 2004, 2006, 2008, 2009 (as a genus, morphology, phylogenetic relationships); Reinert, 2006b (as a genus, *Pardomyia*, description, distribution, comparative morphology); Rattanarithikul *et al.*, 2010 (as a genus, Thailand, keys, bionomics); Wilkerson *et al.*, 2015 (phylogenetic relationships, classification); Soghigian *et al.*, 2017 (phylogenetic relationships).

Neomelaniconion Newstead, 1907

Neomelaniconion, originally proposed as a genus by Newstead, 1907 (in Newstead *et al.*, 1907), currently classified as a subgenus of *Aedes*, is typified by *Aedes palpalis* (Newstead, 1907), original combination: *Neomelaniconion palpale*. Subgenus abbreviation – *Neo.*

****

Systematics: *Neomelaniconion* includes 29 extant species. Species of *Neomelaniconion* predominantly occur in the Afrotropical Region. One species, *Ae. lineatopennis*, extends through the Oriental Region into the Australasian Region. In the Australasian Region, it is known from Timor, Amboina and northern Australia. Mattingly (1961) observed that *Neomelaniconion* 'shows marked affinities' with *Aedes sensu stricto* and *Paraedes* in all life stages and surmised a close relationship with the former based on similarities of the male genitalia, particularly the subapical attachment of the gonostylus to the gonocoxite. In the morphology-based phylogeny of Aedini recovered by Reinert *et al.* (2009), *Neomelaniconion* was placed in a clade comprised of *Neomelaniconion* + (*Edwardsaedes* + (*Aedes sensu stricto* + (*Paraedes* + *Verrallina*))). *Neomelaniconion* was recovered as the sister of a clade comprised of *Aedes* + (*Paraedes* + *Verrallina*) in the study of Wilkerson *et al.* (2015). It was not recovered as a monophyletic lineage in the maximum likelihood phylogeny of Soghigian *et al.* (2017) based on seven molecular markers because the clade with nine species of *Neomelaniconion* included a branch with *Ae.* (*Adm.*) *dentatus*. It seems likely, however, that the specimen of *Ae. dentatus* was a misidentified specimen of a species of *Neomelaniconion*; thus, it comprises a monophyletic group.

****

Characteristics: Species of *Neomelaniconion* are comparatively darkly ornamented species, usually with bright yellow or whitish scales on the vertex and sides of the scutellum. The following features, for the most part, distinguish *Neomelaniconion* from other generic-level taxa of Aedini. ADULTS: Decumbent scales of the vertex all narrow; erect scales fairly numerous but not reaching the front of the vertex; ocular scales both narrow and broad; 6 or more interocular setae present; eyes contiguous or nearly so; antenna of females normal, antenna of males strongly verticillate with the flagellar whorls directed mainly dorsally and ventrally; maxillary palpus of females about 0.2 the length of proboscis, maxillary palpus of males slightly longer than the proboscis, distinctive, palpomere 4 large, upturned and densely setose, palpomere 5 minute or absent; proboscis usually entirely dark-scaled, sometimes pale below (with a complete pale ring in *Ae. taeniarostris*); scutum and scutellum with narrow scales; scutal fossa with contrasting pale scales in a large patch; dorsocentral and acrostichal setae present; scutellar scales all narrow; paratergite bare; antepronotal scales absent; prealar scales absent; mesokatepisternal scale-patches well separated; lower anterior mesepimeral setae present; wing with narrow dark scales, remigium without dorsal setae, alula with a narrow fringe scales; tarsi dark-scaled; foreungues toothed in females and males; abdominal terga with basal pale bands, sometimes also with apical pale scaling. FEMALE GENITALIA: Abdominal segment VIII small, completely retractile, tergum VIII longer than wide; cercus long, narrow; postgenital lobe deeply emarginate apically. MALE GENITALIA: Tergum IX lobes shallowly separated, setose; gonocoxite without mesal membranous area, basal dorsomesal lobe a short process bearing 2 stout setae, ventromesal surface with a longitudinal row of a few to several short blunt-tipped spiniforms (unique character); gonostylus inserted subapically, narrow basally, enlarged medially; claspette absent; phallosome with toothed lateral plates; proctiger triangular, with a sternal arm; apex of the paraprocts sclerotized and pointed. LARVAE: Antenna spiculate; seta 1-A small with a few branches; seta 4-C small, inserted slightly anterior to seta 6-C; setae 4–6-C inserted caudad of seta 7-C; dorsomentum with numerous small teeth (except in *Ae. bolensis*); comb scales stout, in 1 or 2 irregular rows; pecten extends beyond mid-length of the siphon, with 1 or more distal spines more widely separated; seta 1-S small, inserted distal to the pecten; saddle incomplete; ventral brush (seta 4-X) with precratal setae. PUPAE: Seta 5-CT > 1.3 the length of seta 4-CT; seta 10-CT inserted laterad and markedly posterior to seta 11-CT; setae 2,3-I widely separated, seta 3-I with 3 or more branches; seta 1-II with at least slightly expanded base and 4 or more primary branches; seta 7-VI minute, seta 7-VII small, at least somewhat darkened and thickened; paddle with several rows of minute spicules on the distal edge; midrib weakly developed; seta 1-Pa small, single or branched.

****

Bionomics: The immature stages of species of *Neomelaniconion* are found in various groundwater habitats, such as stream pools, ground pools, flood pools, marshy ground and ponds. Adult females of several species are avid blood-suckers and readily attack humans. In general, very little is known about the bionomics of the species of *Neomelaniconion*. The group includes a number of species that have been implicated in the transmission of numerous arboviruses, including viruses of the families Flaviviridae, Togaviridae and Bunyaviridae. Specific isolations have been documented from *Ae. mcintoshi*, *Ae. circumluteolus*, *Ae. luridus* and other species (see, *e.g.* Worth *et al.*, 1961; McIntosh, 1971; McIntosh *et al.*, 1972).

****

Principal references: Edwards, 1941 (as subgenus *Banksinella*, Afrotropical Region, taxonomy, key, descriptions, distributions); Mattingly, 1961 (Indomalayan area, taxonomy, keys, descriptions, bionomics, distributions); McIntosh, 1971 (southern Africa, species descriptions, keys); Lee *et al.*, 1984 (characteristics, literature, bionomics); Jupp, 1996 (southern Africa, keys); Lu *et al.*, 1997a (China, *Ae. lineatopennis*, taxonomy, descriptions, distributions, bionomics); Reinert *et al.*, 2004, 2009 (as a genus, morphology, phylogenetic relationships); Rattanarithikul *et al.*, 2010 (as a genus, Thailand, keys, bionomics); Wilkerson *et al.*, 2015 (phylogenetic relationships, classification); Soghigian *et al.*, 2017 (phylogenetic relationships).

Nyctomyia Harbach, 2013

Nyctomyia, originally named as a genus by Harbach (2013a), currently classified as a subgenus of *Aedes*, is typified by *Aedes pholeocola* (Linton & Harbach, 2013) (in Harbach *et al.*, 2013). *Nyctomyia* is a replacement name for *Nyx* Harbach & Linton, 2013 (in Harbach *et al.*, 2013) (*non* Heppner, 1982). Subgenus abbreviation – *Nyc.*

****

Systematics: *Nyctomyia* includes two extant species only known from two caves in southern Thailand: *Ae. pholeocola* from Wat Tham Phanturat in Surat Thani Province and *Ae. biunguiculatus* from Wat Tham Sumano in Phatthalung Province. Based on the inclusion of morphological data for *Nyctomyia* in the parsimony analysis of Reinert *et al.* (2009) (as *Nyx*), this taxon appears to be more closely related to *Borichinda* and *Isoaedes* than to other generic-level taxa of Aedini (Harbach *et al.*, 2013). The addition of *Nyctomyia* resulted in its interposition on a separate branch between *Isoaedes* and *Borichinda* with an almost equal division of the characters that supported this branch on either side of the node leading to *Nyctomyia*, yielding the following branching pattern: (*Skusea* (*Indusius*, *Cancraedes*)) (*Fredwardsius* (*Isoaedes* (*Nyctomyia* (*Borichinda* ((*Diceromyia*, *Ayurakitia*) (*Dendroskusea*, remainder of the *Stegomyia*-group)))))). In the absence of *Isoaedes*, *Nyctomyia* was recovered as the sister of *Borichinda* in the maximum likelihood phylogeny of Soghigian *et al.* (2017) based on molecular markers.

****

Characteristics: The following combinations of features distinguish *Nyctomyia* from other generic-level taxa of Aedini. ADULTS: Compound eyes narrowly separated; maxillary palpus of males very short, 0.3 the length of the proboscis, comprised of 4 palpomeres, palpomeres 4 vestigial; scutum with some pale scaling, scales on anterior part of the antealar area all pale; mid-lobe of the scutellum with narrow scales; postpronotal scales all narrow; postspiracular scales present; subspiracular scales absent; costal scales of the wing all dark; remigial setae absent; hindtibia dark-scaled; foreungues of males both toothed. FEMALE GENITALIA: Intersegmental membrane between segments VII and VIII short to intermediate in length, segment VIII fully retracted into segment VII; tergum VIII with setae on the distal 0.6 or less, more than 14 scales present; sternum VIII with more than 10 scales, covering much of the surface; insula without setae. MALE GENITALIA: Sternum IX long, setae present; gonostylus long, enlarged distally; apex of claspette narrow with a few setae; aedeagus widest in the distal 0.33. LARVAE: Seta 4-C short, with 3 or 4 branches; seta 14-C single or double, longer than seta 15-C; seta 1-P longer than seta 2-P; setae 1,3-P usually double, infrequently triple; seta 5-P double, as long as or shorter than seta 6-P; seta 5-T small, single; seta 2-I,II large, with 2 or 3 branches, seta 2-III–VI small, single or double, inserted anterior to seta 1; seta 11-I large, with 2 or 3 branches; seta 3-II single; seta 5-II–VI small, usually single; seta 7-II single, short, about 0.5 the length of seta 6-II; seta 9-III–VI inserted anterior and close to seta 7; seta 10-II–V inserted lateral to setae 11 and 12; seta 13-III–V single; seta 5-VII small, single; comb scales spine-like, in a single row; distal 1 or 2 pecten spines more widely spaced than the others; seta 1-X small, single; ventral brush (seta 4-X) with 1 precratal seta, cratal setae borne on a grid with transverse grid bars, seta 4d-X single or double (rarely triple), not plumose. PUPAE: Seta 7-CT longer than seta 6-CT; seta 3-III single; seta 6-III single or double; seta 6-VII inserted posteromesal to seta 9; seta 9-IV–VI inserted posterior to seta 8, seta 9-VII single; apex of the paddle slightly produced.

****

Bionomics: Species of *Nyctomyia* are cave-dwelling mosquitoes that have only been found and reared from larvae and pupae collected from rimstone pools in caves in southern Thailand. The water in the pools was clear, cold and devoid of vegetation and plant matter. The caves harbour populations of bats that are likely to be the primary source of food for adult females. Females of *Ae. pholeocola*, however, are known to bite individuals who enter their cavernicolous environment; thus, species of *Nyctomyia* could potentially transmit pathogens from bats to humans. Nothing else is known about the biology of the species.

****

Principal references: Harbach *et al.*, 2013 (as genus *Nyx*, type species, descriptions, DNA sequences, phylogenetic relationships, bionomics, distribution); Harbach, 2013 (*Nyctomyia*, replacement name); Harbach & Taai, 2014 (as a genus, *Ae. biunguiculatus*, description, troglobitic mosquito species); Wilkerson *et al.*, 2015 (subgeneric status, phylogenetic relationships); Soghigian *et al.*, 2017 (phylogenetic relationships).

Ochlerotatus Lynch Arribálzaga, 1891

Ochlerotatus, originally proposed as a genus by Lynch Arribálzaga (1891a, b), currently classified as a subgenus of *Aedes*, is typified by *Ochlerotatus confirmatus* Lynch Arribálzaga, 1891b [subjective synonym of *Aedes scapularis* (Rondani, 1848)]. Subgenus abbreviation – *Och.*

****

Systematics: *Ochlerotatus* includes 194 extant species and two fossil species. Twenty-two species groups are recognized for the extant species. *Ochlerotatus* is a large and abundant group with species in all temperate, subtropical and tropical areas of the world. The greatest diversity of species occurs in the Australasian, Nearctic and Neotropical Regions. Comparatively few species are found in the Afrotropical and Oriental Regions. The monophyly of *Ochlerotatus* is doubtful. Many species are questionably placed in the currently recognized species groups. Consequently, the affinities of *Ochlerotatus* are very uncertain. *Ochlerotatus* was recovered as the sister group of a clade comprised of *Acartomyia* + (*Jarnellius* + (*Halaedes* + *Opifex*)) in the phylogeny generated in the study of Reinert *et al.* (2009) based on morphological data of all life stages. *Ochlerotatus* was not associated with other generic-level taxa in the phylogeny of Wilkerson *et al.* (2015). It was recovered as a polyphyletic group in the maximum likelihood phylogeny of Soghigian *et al.* (2017) based on molecular markers for 85 species.

****

Characteristics: *Ochlerotatus* is a large highly varied group of species that generally exhibit the following characteristics. ADULTS: Vertex of the head largely with narrow decumbent scales, erect forked scales numerous; interocular space with or without scales; compound eyes more or less separated above the antennal pedicels; antenna of females shorter to longer than the proboscis, distinctly shorter than the proboscis in males, flagellar whorls of males with numerous long setae normally directed dorsally and ventrally; maxillary palpus of females 0.2–0.4 the length of the proboscis, with 5 palpomeres, palpomere 5 minute, maxillary palpus of males as long as or longer than the proboscis, palpomeres 4 and 5 subequal, downturned and setose; scutal scales usually narrow; anterior and posterior acrostichal and dorsocentral setae normally present; scutellum with narrow scales only; paratergite usually (not always) with scales; pleural scaling and setation varied; lower mesepimeral setae present or absent; remigium of the wing usually with conspicuous setae; alula with an extensive marginal fringe of long piliform scales; scaling of the tarsi varied; ungues varied, at least fore- and midungues usually toothed, fore- and midungues of males enlarged; abdominal scaling varied, especially tergum I and the laterotergite; terga with numerous lateral setae and reduced scaling in males. FEMALE GENITALIA: Segment VIII usually completely retracted within segment VII; tergum VIII longer than wide, with long setae on the proximal 0.4 of the lateral margins; upper vaginal sclerite absent; insula lip-like, bearing well-developed setae; cercus usually long and narrow, distal part narrowly rounded; 3 spermathecal capsules. MALE GENITALIA: Segment VIII usually narrowed proximally; ninth tergal lobes usually more or less distinct, with setae; gonocoxite usually elongate, with a more or less strongly developed basal dorsomesal lobe, sometimes with an apical dorsomesal lobe, mesal surface membranous from the base to the apex; gonostylus simple, with an apical usually long spiniform gonostylar claw; claspette comprised of an elongate stem with a single apical blade-like, sickle-like or otherwise modified claspette filament; aedeagus simple; paraproct with a strong usually simple apical spine; cercal setae present. LARVAE: Head broad; antenna short to quite long, usually distinctly spiculate, seta 1-A varied, usually inserted near mid-length; seta 1-C spiniform; seta 4-C usually small, varied in position; setae 5,6-C inserted caudad of the antennal bases; seta 13-P absent; seta 12-I present; setae 6-I,II and 7-I strongly developed, seta 6-III–VI at least moderately developed; comb scales varied, in a large patch or a single row; siphon varied in length, pecten spines evenly spaced or distal spines more widely separated; seta 1-S inserted within or beyond the pecten; saddle complete or incomplete, acus present or absent; seta 1-X usually inserted on the saddle; seta 2-X multi-branched; ventral brush (seta 4-X) usually with at least 7 pairs of setae on a grid and 2 or more precratal setae. PUPAE: Trumpet short, broad, tracheoid area indistinct; seta 8-CT widely separated from seta 9-CT; setae 2,3-I usually approximated, seta 2-III–VII varied in position; seta 9-VI small, similar to seta 9-II–V; paddle with or without marginal spicules; seta 1-Pa single or branched.

****

Bionomics: The immature stages of most species of *Ochlerotatus* are found in various types of temporary freshwater ground pools, but several species inhabit rock holes, natural and artificial container habitats, and brackish water pools. Females of many species viciously attack humans and are often very troublesome pests, particularly in northern temperate regions. The majority of species bite in the daytime, particularly during crepuscular periods, but some species also feed during the night. A number of species of *Ochlerotatus* are known to harbour natural infections of arboviruses and microfilariae (see Wilkerson *et al.*, 2015: table 1).

****

Principal references: Belkin, 1962 (South Pacific, taxonomy, keys, descriptions, distributions, bionomics); Lu *et al.*, 1997a (China, keys, taxonomy, descriptions, distributions, bionomics); Reinert, 2000d (as a subgenus of *Ochlerotatus*, classification, morphology); Reinert, 2002e (as a subgenus of *Ochlerotatus*, female genitalia); Reinert *et al.*, 2008, 2009 (as a genus, morphology, phylogenetic relationships); Rattanarithikul *et al.*, 2010 (as a genus, Thailand, keys, bionomics); Huang & Rueda, 2014 (Afrotropical Region, key to adults); Wilkerson *et al.*, 2015 (phylogenetic relationships, classification); Soghigian *et al.*, 2017 (phylogenetic relationships); Ferreira de Freitas & Bartholomay, 2021 (taxonomic history, type species).

Orohylomyia Somboon & Harbach, 2023

Orohylomyia Somboon & Harbach (in Somboon *et al.*, 2023b), originally proposed and currently recognized as a subgenus of *Aedes*, is typified by *Aedes yunnanensis* (Gaschen, 1934). Subgenus abbreviation – *Oro.*

****

Systematics: *Orohylomyia* is monobasic. It only includes the extant type species, which has only been collected in mountainous localities in Guizhou, Sichuan and Yunnan Provinces of China. The single species of *Orohylomyia* is most similar to species of *Collessius*, *Gilesius*, *Hulecoeteomyia* and *Tanakaius*. In a phylogenetic analysis of mtDNA *COI* sequences (Somboon *et al.*, 2023b), *Ae. yunnanensis* formed a strongly supported lineage in an unresolved relationship with species of *Collessius* (Alloeomyia Group) and *Gilesius*.

****

Characteristics: The single species of *Orohylomyia* is most similar to species of *Collessius*, *Gilesius*, *Hulecoeteomyia* and *Tanakaius*. Significant similarities with species of *Collessius*, *Gilesius*, *Hulecoeteomyia* and *Tanakaius* include the scutal ornamentation of adults and the linear arrangement of setae 4–6-C of the larva. Adults of *Orohylomyia*, *Gilesius* and *Ae.* (*Tanakaius*) *savoryi* differ from adults of *Collessius* and *Hulecoeteomyia* in having hindtarsomeres 1–3 entirely dark-scaled (these tarsomeres have basal pale bands in species of *Hulecoeteomyia* and *Ae.* (*Tanakaius*) *togoi* and basal and apical pale bands in species of *Collessius*). The adults (both sexes) of *Ae. yunnanensis* and *Tanakaius* differ from those of *Collessius*, *Gilesius* and *Hulecoeteomyia* in having the hindungues with a median tooth (the hindungues are simple, without a tooth, in species of *Collessius*, *Gilesius* and *Hulecoeteomyia*). Species of *Orohylomyia* are characterized and distinguished from species of other generic-level taxa of Aedini by the following combinations of characters. ADULTS: Antennal pedicel with scales on the mesal and dorsal surfaces; flagellomere 1 with a mesal patch of scales; maxillary palpus of females usually with some pale scales apically, maxillary palpus of males with pale scales on the dorsal surface of palpomeres 1 and 2; proboscis dark-scaled, longer than the forefemur in females, shorter than the forefemur in males; scutum with acrostichal setae, acrostichal and anterior dorsocentral areas with pale scales, acrostichal line forked posteriorly at the sides of the prescutellar bare space, pale scales laterally from the anterior promontory to the margins of the scutal fossa and along the inner side of the supraalar area to the posterior margin of the scutum, supraalar area with narrow pale falcate scales; paratergite with narrow pale scales; patches of broad pale scales on the postspiracular area, subspiracular area, prealar area and upper mesokatepisternum; lower mesepimeral seta absent; wing dark-scaled, with a few pale scales at the extreme base of the costa; midfemur with anterior stripe of pale scales on the proximal 0.8, tarsi of all legs entirely dark-scaled; females with the ungues of all legs equal, each with a median tooth, males with unequal fore- and midungues, larger unguis with a median tooth, smaller unguis with a basolateral tooth, hindungues as in females; abdominal terga dark-scaled with basal pale bands on terga II–VIII, females with segment VII dorsoventrally flattened and segment VIII not retracted into segment VII. FEMALE GENITALIA: Tergum IX 1.3 times wider than long, with convex posterior margin and about 18 more or less stiff simple setae on the posterolateral areas; cerci broad, moderately long; postgenital lobe with slight caudal emargination; insula pad-like, with 2 irregular rows of 4 or 5 setae; 1 large and 2 smaller spermathecal capsules. MALE GENITALIA: Ninth tergal lobes small, widely separated by a narrow faintly developed bridge, lobes with short simple setae; gonocoxite with a row of long narrow lanceolate setae along the distal 0.5 of the ventromesal margin; gonostylus elongate, curved, tapered toward the apex, proximal portion distinctly stouter than the distal portion, with long slender gonostylar claw at the apex; claspette filament long, slender, cylindrical, sickle-shaped, slightly acuminate apically, borne at the apex of an elongate tubular stem, about twice the length of the stem; paraproct long, slender, strongly sclerotized, with more strongly sclerotized beak-like apex. LARVAE: Antenna about 0.33 the length of the dorsal apotome, moderately spiculate; seta 1-A with 2–4 (usually 4) aciculate branches, inserted more or less at mid-length of the shaft; setae 4–6-C arranged in a transverse row, 4-C small, weak, inserted about the same level as 5-C, 5,6-C fan-like with strong aciculate branches, 6-C longer than 5-C; seta 7-C inserted on level posterior to 5-C; setae 5,10,12-M and 10-T long, double, aciculate; comb scales in a patch, scales evenly fringed at the sides and apex; siphon short, stout, pecten slightly longer than 0.5 siphon length, seta 1-S inserted at or slightly beyond distal pecten spine; saddle incomplete, posterior margin with short needle-like spicules; seta 1-X single; seta 2-X with 3 or 4 branches; seta 3-X usually double; ventral brush (seta 4-X) comprised of 12 setae inserted on a grid with lateral and transverse grid bars. PUPAE: Seta 1-CT usually single, longer than seta 3-CT; seta 11-CT well developed, single, long, stiff, inserted immediately lateral to seta 10-CT; seta 1-II well developed, multi-branched; seta 2-VI inserted mesad of seta 1-VI; setae 3-II,III and 5-IV–VII long, single; seta 6-VII usually single, inserted immediately posteromesad to seta 9-VII; seta 9-VII dissimilar to seta 6-VII, much stronger and longer than 6-VII, single or with 2 or 3 apical branches, seta 9-VIII more strongly developed than 9-VII, branched at base, with 4–8 strong aciculate branches; paddle broad, asymmetrical, inner part narrower proximally and broader distally than outer part, emarginate at the termination of the midrib, distal inner and distal outer margins spiculate; seta 1-Pa single, inserted on the edge of the outer part immediately adjacent to the apex of the midrib; seta 2-Pa absent.

****

Bionomics: The immature stages of the type species of *Orohylomyia* have been found in small groundwater habitats and artificial containers, including buckets, cement tanks, crevices in rocks, iron pans, oil drums, ponds, puddles, rock pools, rocky caves, stagnant pools, stone pits, stone troughs and tyres in mountainous terrain at elevations above 1000 m. Nothing is known about the biting behaviour of females and whether or not they may vector pathogenic agents.

****

Principal references: Knight & Marks, 1952 (as a species of Group H (Geniculatus-group: Protomacleaya) of *Finlaya*); Lei, 1989b (as species of *Finlaya*, in key, description, bionomics, distribution); Lu *et al.*, 1997a (as species of *dissimilis* group of *Finlaya*, in key, description, bionomics, distribution); Dong *et al.*, 2009 (as species of *Ochlerotatus yunnanensis* group of *Finlaya*, in keys, description, bionomics, distribution); Somboon *et al.*, 2023b (new subgenus, China, taxonomy, description of subgenus and type species, bionomics, distribution, phylogenetic relationships).

Paraedes Edwards, 1934

Paraedes, originally proposed as a genus by Edwards, 1934 (in Barraud, 1934), currently classified as a subgenus of *Aedes*, is typified by *Aedes barraudi* (Edwards, 1934), original combination: *Paraedes barraudi*. Subgenus abbreviation – *Par.*

****

Systematics: *Paraedes* includes nine extant species with distributions in the Oriental Region from India and Sri Lanka eastward to the Philippines. *Paraedes* appears to be most closely related to *Aedes sensu stricto* and *Verrallina*. In the phylogenetic study of Reinert *et al.* (2004) based on morphological data, *Paraedes* was recovered as the sister of a clade comprised of *Aedes* + *Verrallina*; it was recovered as the sister of *Verrallina* in a clade comprised of *Aedes sensu stricto* + (*Paraedes* + *Verrallina*) in the studies of Reinert *et al.* (2009) and Wilkerson *et al.* (2015). In agreement with their results, two species of *Paraedes* were recovered in a sister relationship to five species of *Aedes* in the maximum likelihood phylogeny of Soghigian *et al.* (2017) based on seven molecular markers. The *Aedes* + *Paraedes* clade, however, was sister to a clade consisting of *Verrallina* + *Edwardsaedes*.

****

Characteristics: Species of *Paraedes* are characterized and distinguished from species of other generic-level taxa of Aedini by the following combinations of characters. ADULTS: Maxillary palpus of males less than 0.1 the length of the proboscis; scutum with numerous anterior and posterior dorsocentral setae; anterior acrostichal area with a stripe of pale scales, acrostichal setae usually absent; prescutellar area without scales; paratergite usually bare (with scales in males of *Ae. thailandensis* and females of *Ae. pagei*); lower mesepimeral seta absent. FEMALE GENITALIA: Tergum and sternum VIII usually without scales; cercus usually without scales; upper vaginal sclerite well developed, narrow; insula tongue-like, distal 0.5 with 2–7 tuberculi, each with a minute spicule; only 1 large spermathecal capsule (3 in *Ae. menoni*). MALE GENITALIA: Claspette (basal mesal lobe of Reinert, 1981b) comprised of 2 elongate lobes, one long, narrow, curved, with 3 stout apical setae, the other broader with 3–5 flattened apical setae or with long narrow spines; gonostylus with 2 or 3 arms, 1 or 2 arms with spicules forming file-like ridges, 1 arm with fine setae; gonostylar claw absent; aedeagus comprised of 2 approximated lateral plates, each with 1–3 lateral teeth. LARVAE: Scape of antenna spiculate; seta 1-A stout, multi-branched; setae 4,6-C inserted at the same level, both more or less equally distant from seta 5-C, seta 6-C inserted laterad and slightly cephalad of seta 5-C; seta 7-C inserted caudomesad of the base of the antenna; seta 6-I–V long, stout, barbed; seta 4-VIII single (rarely with 2 or 3 branches in *Ae. chrysoscuta*), noticeably longer than seta 3-VIII; comb with 7–18 spine-like scales in a single curved row; pecten with the distal 1–3 spines more widely spaced than the proximal spines; seta 1-S $\leq$ 0.4 the width of the siphon; saddle incomplete, acus absent; seta 2-X moderately long, multi-branched; seta 3-X very long, single; ventral brush (seta 4-X) borne on a grid with transverse grid bars. PUPAE: Seta 5-CT > 1.3 the length of seta 4-CT; seta 9-CT longer than seta 8-CT; seta 1-II with 5–22 branches; seta 2-II long, stout, only slightly shorter than seta 3-II, as long as or longer than seta 1-II; seta 3-I nearly as long as or longer than seta 6-I, seta 3-II longer than seta 6-II; seta 9-VIII long, single (sometimes double in *Ae. bonneae*); paddle without a fringe of hair-like spicules; seta 1-Pa long, single.

****

Bionomics: The immature stages of species of *Paraedes* are usually found in freshwater in crab holes, but they have also been found in holes near streams, elephant footprints, fallen *Nipa* palm fronds, ground pools and wheel tracks. Habitats are located in partially to heavily shaded forest, plantations, bamboo grooves and scrub areas in valleys, hills and mountainous terrain. Adults of most species are known to feed on humans, sometimes viciously biting during daytime, but species of *Paraedes* are not of medical and economic importance.

****

Principal references: Mattingly, 1958a (Indomalayan area, keys, classification, taxonomy, descriptions, bionomics, distributions); Reinert, 1981b (taxonomy, keys, descriptions, distributions, bionomics); Reinert, 2000e (female genitalia); Reinert *et al.*, 2004, 2009 (as a genus, morphology, phylogenetic relationships); Rattanarithikul *et al.*, 2010 (as a genus, Thailand, keys, bionomics); Wilkerson *et al.*, 2015 (phylogenetic relationships, classification); Soghigian *et al.*, 2017 (phylogenetic relationships).

Patmarksia Reinert, Harbach & Kitching, 2006

Patmarksia, originally proposed as a genus by Reinert *et al.* (2006), currently classified as a subgenus of *Aedes*, is typified by *Aedes papuensis* (Taylor, 1914d), original combination: *Leucomyia australiensis* variety *papuensis*. Subgenus abbreviation – *Pat.*

****

Systematics: *Patmarksia* includes 13 extant species in the Australasian Region: Australia, Bismarck Archipelago, Moluccas, New Guinea and Solomon Islands. Most species are only known from New Guinea. The phylogenetic relationships of *Patmarksia* are uncertain. The known species resemble adults of the Scutellaris Group of *Stegomyia*, and were previously recognized, along with two unrelated species, as a species group (Papuensis Group/Subgroup) of *Finlaya* (King & Hoogstraal, 1946d; Knight & Marks, 1952; Lee *et al.*, 1982, respectively). In the phylogenetic study of Reinert *et al.* (2009) based on extensive morphological data, *Patmarksia* was recovered as the sister of *Georgecraigius* in a clade comprised of *Dobrotworskyius* + (*Patmarksia* + *Georgecraigius*), and this clade was sister to a clade comprised of *Tanakaius* + (*Hulecoeteomyia* + (*Gilesius* + *Collessius*)). In the phylogeny recovered in the study of Wilkerson *et al.* (2015), *Patmarksia* was the sister of *Georgecraigius* in a polytomy that also included the *Collessius*, *Gilesius*, *Hulecoeteomyia* and *Tanakaius* in an unresolved clade, and *Dobrotworskyius*.

****

Characteristics: Species of *Patmarksia* are characterized and distinguished from species of other generic-level taxa of Aedini by the following combinations of characters. ADULTS: Vertex of the head with narrow scales; forked scales on the occiput and vertex; eyes contiguous or nearly so above the antennal pedicels; maxillary palpus and proboscis dark-scaled, maxillary palpus of males with 5 palpomeres, slightly shorter to slightly longer than the proboscis; scutum with median longitudinal patch of pale scales on the anterior 0.5–0.7 of the acrostichal and extending laterally over mesal part of dorsocentral areas, posterior dorsocentral area without a pale stripe, supraalar area without pale scales; acrostichal setae absent, anterior dorsocentral area normally without setae, posterior dorsocentral area with several setae; scutellum with narrow scales on all lobes; paratergite with broad scales; postpronotum normally with broad scales ventrally and narrow scales dorsally; mesokatepisternum with upper and lower patches of broad scales; costa of the wing dark-scaled; hindfemur with a dorsoapical patch of pale scales; tibiae dark-scaled; hindtarsomeres 1–4 with broad basal pale bands; both fore- and midungues of males toothed. FEMALE GENITALIA: Tergum VIII with numerous broad scales distally, setae inserted on the posterior ≥ 0.7; sternum VIII longer than wide, with broad scales in lateral patches, posterior margin with a median emargination, seta 2-S inserted posterior to seta 1-S; tergum IX comprised of 2 lateral plates, posteromesal margins fused to the anterolateral margins of the cerci, without setae; lower vaginal sclerite absent; insula lip-like, with setae in lateral patches; cercus usually without scales; 3 spermathecal capsules. MALE GENITALIA: Tergum IX comprised of 2 narrow lateral plates narrowly joined mesally, without setae; gonocoxite without a dorsoapical lobe, with a small basal lobe bearing setae, mesal surface membranous; gonostylus long and narrow, attached at the apex of the gonocoxite, gonostylar claw inserted at the apex; claspette comprised of a rather long slender columnar stem with a striated claspette filament borne at the apex; aedeagus simple, tube-like. LARVAE: Antenna with a few to numerous spicules; seta 1-A normally branched; seta 4-C branched, inserted slightly posterior and mesal to seta 5-C; setae 5,6,7-C multi-branched, displaced far anteriorly, seta 6-C inserted lateral and slightly anterior to seta 5-C; seta 7-C inserted lateral and posterior to setae 4–6-C; seta 12-C inserted mesal to seta 13-C; seta 19-C absent; setae 1–3-P inserted on a common setal support plate; seta 12-I present; seta 1-II single; seta 7-II short and branched; comb of segment VIII normally with numerous scales in a patch (some species with several scales in 1 or 2 irregular rows); siphon with the pecten spines evenly spaced, seta 1-S inserted distal to the pecten, seta 6-S noticeably longer than the distal diameter of the siphon; saddle of segment X incomplete ventrally, with moderately well-developed spicules on the posterior margin; seta 1-X inserted on the saddle; ventral brush (seta 4-X) borne on a grid with both transverse and lateral grid bars. PUPAE: Seta 11-CT normally single; seta 3-I long, stout, single, longer than seta 6-I; seta 6-I single, longer than seta 7-I; seta 2-II inserted lateral to seta 1; seta 3-II,III long, stout, single, 3-II longer than seta 6-II, 3-III longer than seta 5-III; seta 5-IV,V long, stout, longer than the following tergum; paddle without hair-like marginal spicules; seta 1-Pa single.

****

Bionomics: Species of *Patmarksia* are apparently confined to forests. The immature stages develop in phytotelm habitats such as fallen leaves, bracts, palm fronds and coconut husks, tree holes, log holes, cut bamboo and cup fungi. They are also found in small puddles, ruts, borrow pits, rock pools in stream beds, crevasses in rocks and artificial containers, *e.g.* metal cans, drum heads and flower bowls. Adults are rarely encountered and females, with the exception of *Ae. palmarum* and *Ae. papuensis*, which have been captured biting during the day in forest, are not known to attack humans. Species of *Patmarksia* are not known or suspected of being of medical and economic importance.

****

Principal references: Lee *et al.*, 1982 (*papuensis* subgroup [excluding *alticola* Bonne-Wepster, 1948 and *toxopeusi* Bonne-Wepster, 1948], keys, taxonomy, bionomics, distribution, literature); Reinert *et al.*, 2006 (new genus, description, distribution, bionomics); Reinert *et al.*, 2008, 2009 (as a genus, morphology, phylogenetic relationships); Reinert, 2008g (as a genus, female genitalia); Wilkerson *et al.*, 2015 (phylogenetic relationships, classification).

Paulianius Brunhes & Boussès, 2017

Paulianius, originally proposed as a genus by Brunhes & Boussès (in Brunhes *et al.*, 2017), currently classified as a subgenus of *Aedes*, is typified by *Aedes tiptoni* Grjebine, 1953a. Subgenus abbreviation – *Pln.*

****

Systematics: *Paulianius* includes eight extant species, which are divided between three species groups, the Tiptoni Section (*Ae. hirsutus*, *Ae. tiptoni*), the Niveus Section (*Ae. zethus*) and the Coulangesi Section (*Ae. ambremontis*, *Ae. coulangesi*, *Ae. grassei*, *Ae. madagascarensis*, *Ae. rodhaini*). *Paulianius* includes species in Madagascar (seven species) and southern Africa (*Ae. niveus*, known from Zambia and Zimbabwe). It appears to be most closely related to *Tewarius*, which are sylvatic species that occur in the mountainous western areas of India. *Paulianius* also shares affinities, to a lesser degree, with the African *Diceromyia* (Brunhes *et al.*, 2017).

****

Characteristics: The following features are most distinctive of species of *Paulianius*. ADULTS: Black mosquitoes adorned with numerous spots, rings or bands of large white scales; vertex of the head covered with separate areas of black and white scales, white scales generally forming longitudinal patches; pedicel of the antenna with white scales; proboscis black-scaled; scutum black-scaled with longitudinal stripes of white scales or white scaling on the anterior 0.5; dorsocentral setae strongly developed, acrostichal setae absent; large white scales present at least on the midlobe of the scutellum; thoracic pleura with numerous spots of silvery scales; lower mesepimeral seta present (absent in *Ae. niveus*); fore- and midungues of males unequal; wing entirely black-scaled or at most with a few pale scales at the base of the costa; legs black-scaled with stripes and rings of white scales; forefemur, and sometimes midfemur, with a ventral white patch or strip; tarsi entirely black-scaled or with white rings; abdominal terga mainly black-scaled, terga II–VI with a pale median basal band and lateral patches of white scales. MALE GENITALIA: Gonostylus formed of 2 equal arms (distinction from the African *Diceromyia*); dorsal arm often flattened and spatulate at the apex, bearing very small sensory setae; ventral arm slightly shorter than the dorsal arm and sometimes cylindrical, bearing a short broad more or less spatulate claw; claspette always present, often long (poorly developed in *Tewarius*). LARVAE: Antenna very slightly spiculate at the base or glabrous; seta 4-C with about 15 short flexible branches; setae 5,6-C single; seta 15-C with 3 or 4 very long flexible branches; seta 1-P very long; support plates of the pleural setal groups without a strong spine (distinction from *Diceromyia*); seta 4-X with 5 pairs of setae. PUPAE: Inner and outer margins of the paddle lined with denticles or long filaments, midrib clearly visible to near the apex; seta 1-Pa single, relatively short.

****

Bionomics: Species of *Paulianius* are forest inhabitants. The larval habitats of species of the Coulangesi Section are unknown, but the immature stages of the other species of *Paulianius*, like those of *Tewarius*, develop in plant cavities, frequently in tree holes. Little is known about the bionomics of the adults. Females of *Ae. tiptoni* feed mainly during twilight hours, and are known to bite humans outdoors and more frequently indoors. Females of *Ae. madagascarensis* bite during the day and those of *Ae. coulangesi* have been captured on human bait but are reluctant to bite. Females of *Ae. madagascarensis* have been found naturally infected with West Nile virus, but it is unlikely that species of *Paulianius* are of medical and economic importance (see references in Brunhes *et al.*, 2017).

****

Principal reference: Brunhes *et al.*, 2017 (new genus, classification, taxonomy, descriptions, bionomics, distributions, keys).

Petermattinglyius Reinert, Harbach & Kitching, 2009

Petermattinglyius, originally proposed as a genus by Reinert *et al.* (2009), currently classified as a subgenus of *Aedes*, is typified by *Aedes iyengari* Edwards, 1923b. Subgenus abbreviation – *Pet.*

****

Systematics: *Petermattinglyius* includes five extant species classified in two species groups, the Aglaonotus and Petermattinglyius Groups (subgenera *Aglaonotus* and *Petermattinglyius*, respectively, of Reinert *et al.*, 2009). Species of *Petermattinglyius* are recorded from localities in Bangladesh, India, Indonesia, Malaysia, Myanmar, Singapore and Thailand. *Petermattinglyius* was recovered as the sister to (*Alanstonea* + *Pseudarmigeres*) + *Heizmannia* in the morphology-based phylogeny of Reinert *et al.* (2009), and that clade was sister to another comprised of *Lorrainea* + ((*Eretmapodites* + *Armigeres*) + (*Udaya* + (*Belkinius* + *Zeugnomyia*))). *Petermattinglyius* was recovered as the sister of *Heizmannia* + *Pseudarmigeres* in the study of Wilkerson *et al.* (2015).

****

Characteristics: Species of *Petermattinglyius* are characterized and distinguished from species of other generic-level taxa of Aedini by the following combinations of characters. ADULTS: Vertex of the head with broad decumbent scales, erect scales mainly on the occiput; eyes contiguous above the antennal pedicels; maxillary palpus of females with pale scales distally (pale scales absent in *Ae. whartoni* and *Ae. punctipes*), maxillary palpus of males with palpomeres 4 and 5 slightly downturned, palpomere 5 with pale scales at least basally (dark-scaled in *Ae. whartoni*); proboscis longer than the forefemur, entirely dark-scaled in females, with a narrow pale band distal to mid-length in males (entirely dark-scaled in *Ae. whartoni*); scutum with at least the posterior 0.25 covered with broad scales that extend anteriorly over the supraalar areas; acrostichal and dorsocentral setae absent; scutellum covered with broad dark scales; paratergite with pale scales; ante- and postpronotum with broad scales, sometimes also with narrow scales; postspiracular area without scales, subspiracular area, upper proepisternum, upper and lower areas of the mesokatepisternum and mesepimeron with broad pale scales; prealar scales present or absent; lower mesepimeral setae present (absent in *Ae. whartoni*); wing dark-scaled with a pale patch at the base of the costa, remigium with dorsal setae, upper calypter with marginal setae, alula with dark scales on the posterior margin and a few broad dark scales above the marginal scales; postprocoxal membrane with broad pale scales (absent in *Ae. whartoni*); all femora with a preapical pale band and pale scales at the apex; tibiae with several median pale bands or spots (absent in *Ae. whartoni*); hindtarsomere 1 with pale basal and pale apical spots and 2 median bands (latter absent in *Ae. whartoni*); ungues all equal and simple in females, fore- and midungues unequal in males, larger unguis with 1 tooth (midunguis of *Ae. franciscoi* without a tooth); abdominal tergum I with a basomedian pale patch (absent in *Ae. whartoni*) and broad pale scales on the laterotergite; segment VII laterally compressed in females. FEMALE GENITALIA: Tergum and sternum VIII with numerous broad scales, sternum with seta 2-S inserted lateral to seta 1-S; tergum IX wider than long, with a median caudal emargination separating rounded lobes, each with 1–6 setae (rarely 1 lobe without setae); postgenital lobe with a distal emargination; lower vaginal sclerite absent; insula tongue-like, with 2–4 distal tuberculi, without setae; 3 spermathecal capsules. MALE GENITALIA: Tergum IX lobes setose; sternum IX with 1 or more posterior setae; gonocoxite with numerous broad scales on the lateral and ventral surfaces, dorsal surface with several short flattened setae distally on the mesal margin (absent in *Ae. whartoni*), ventral surface with a small patch of long setae proximally on the mesal margin (less well developed in *Ae. franciscoi*), gonostylus attached apically, a single preapical gonostylar claw inserted some distance from the apex; claspette a small basal plaque with several short setae; aedeagus comprised of 2 lateral sclerites, each with elongate lateral teeth extending from the base to the apex; cercal setae absent. LARVAE: Antenna short, without spicules (long with few spicules in *Ae. whartoni*); seta 1-C single; seta 4-C multi-branched, inserted mesal and slightly posterior to seta 6-C; seta 5-C single, inserted posterior to setae 4,6,7-C; seta 12-C branched, inserted mesal to seta 13-C; seta 13-C single, longer than seta 12-C; seta 14-C single; seta 19-C absent; setae 1–3-P not borne on a common setal support plate; seta 6-P single, longer than branched setae 5,7-P; seta 5-M single, noticeably longer than setae 6,7-M; seta 6-I–V double, aciculate; seta 7-I single or double; seta 12-I absent; seta 8-II with 2 or 3 branches; seta 1-VII noticeably longer than the saddle; siphon relatively short (long in *Ae. whartoni*), seta 1-S single, inserted distal to the pecten; saddle incomplete ventrally; seta 1-X single, inserted on the saddle; ventral brush (seta 4-X) with several setae with 2–4 branches inserted on a grid with transverse grid bars and 2 branched precratal setae. PUPAE: Trumpet with a weakly developed tracheoid area; seta 6-CT single, much shorter than seta 7-CT; seta 11-CT longer than seta 12-CT, both single; seta 1-II short, double, rarely triple; seta 3-II,III long, single; seta 6-VII inserted mesal and posterior to seta 9-VII; seta 9-VIII with 2–5 branches; paddle with a very shallow apical emargination and a fringe of hair-like spicules on the outer and inner margins (absent in *Ae. whartoni*); seta 1-Pa short, single (branched in *Ae. whartoni*).

****

Bionomics: The immature stages of species of *Petermattinglyius* are normally found in bamboo and occasionally in holes in trees or logs. Females have been collected biting humans, in human-baited traps and light traps, and resting in houses. Species of *Petermattinglyius* are not known to be of medical importance.

****

Principal references: Reinert, 1970a (Southeast Asia, as *Diceromyia* in part, *Ae. franciscoi*, *Ae. iyengari*, *Ae. scanloni*, *Ae. whartoni*, in keys, descriptions, distributions, bionomics); Reinert *et al.*, 2009 (new genus, morphology, phylogenetic relationships, description, distribution, bionomics); Rattanarithikul *et al.*, 2010 (as a genus, Thailand, keys, bionomics); Reinert, 2010h (as a genus, female genitalia); Wilkerson *et al.*, 2015 (phylogenetic relationships, classification).

Phagomyia Theobald, 1905

Phagomyia, originally proposed as a genus by Theobald (1905e), currently classified as a subgenus of *Aedes*, is typified by *Aedes gubernatoris* (Giles, 1901a), original combination: *Culex gubernatoris*. Subgenus abbreviation – *Phg.*

****

Systematics: *Phagomyia* includes 16 extant species. Species of *Phagomyia* primarily occur in the Oriental Region. One or two species extend into far eastern areas of the Palaearctic, one occurs in Southeast Asia and Sulawesi and at least two species are known from the Australasian Region (northern Australia and New Guinea). Knight & Marks (1952) noted that *Phagomyia* (as the Gubernatoris Subgroup of *Finlaya*) seemed to be most closely related to *Downsiomyia* (as the Niveus Subgroup of *Finlaya*). Rattanarithikul *et al.* (2010) indicated, without specifically suggesting phylogenetic relationships, that adults of *Phagomyia* generally resemble those of *Christophersiomyia*, *Downsiomyia* and *Jihlienius* in having the anterior area of the scutum largely pale-scaled and the hindtarsomeres with basal and apical pale bands; they also stated that larvae resemble those of *Jihlienius*, but did not indicate how they resemble one another. *Phagomyia* was recovered as the sister of *Jihlienius* in the phylogenetic studies of Reinert *et al.* (2009) and Wilkerson *et al.* (2015) based on cladistic analyses of morphological data from all life stages. Species of *Phagomyia* were recovered in a paraphyletic relationship relative to *Geoskusea* + *Kenknightia* in the maximum likelihood phylogeny of Soghigian *et al.* (2017) based on seven molecular markers.

****

Characteristics: Species of *Phagomyia* are characterized and distinguished from species of other generic-level taxa of Aedini by the following combinations of characters. ADULTS: Vertex of the head with broad decumbent scales medially; maxillary palpus of males with numerous long setae arising apically on palpomere 3, all along palpomere 4, some on palpomere 5; scutum with a large anterior area covered with pale scales (covering the anterior 0.3–0.7 of the acrostichal, dorsocentral and scutal fossal areas), pale scaling sometimes divided partially or completely in the middle, with a patch of pale scales on the anterior supraalar-posterior antealar area, scales on the anterior part of the antealar area all dark; anterior dorsocentral setae absent; postpronotum with a patch of broad scales, sometimes without scales; subspiracular and postspiracular scales absent; prealar scales present or absent; remigium of the wing without dorsal setae; foretibia with a dorsal white area near or at the apex; hindtarsomere 1 with basal and apical pale bands or spots, tarsomere 2 with basal pale band or spot, sometimes with pale scaling over the joint between tarsomeres 2 and 3 or a few pale scales at the base of tarsomere 3; some abdominal sterna with outstanding or roughened scales posteriorly. FEMALE GENITALIA: Tergum and sternum VIII wider than long, tergum densely covered with scales, sternum with or without a few scales; postgenital lobe long and relatively narrow, with setae on 0.4–0.7 of the ventral surface; cercus sharply oblique distally, without scales; both upper and lower vaginal sclerites present; insula lip-like with 2–5 setae on either side of the midline. MALE GENITALIA: Gonocoxite without a tuft of specialized scales, with a row or patch of long lanceolate setae on the mesal area of the ventral surface; claspette filament curved, blade-like. LARVAE: Median labral plate indistinguishably fused with the dorsal apotome; seta 1-C slender or stout and spiniform; seta 4-C inserted posterior to seta 6-C; seta 7-C inserted far posterior to the base of the antenna, single, rarely double; seta 2-T branched; seta 12-I present; comb scales rounded and evenly fringed. PUPAE: Trumpet with a weakly developed tracheoid area; seta 2-II inserted lateral to seta 1-II; seta 6-III single; seta 9-VIII inserted on the posterolateral corner of the segment; paddle without a fringe of hair-like spicules; seta 1-Pa long, 0.4–0.6 the length of the paddle; seta 2-Pa absent.

****

Bionomics: The immature stages of species of *Phagomyia* are commonly found in rock pools, rock holes, tree holes, stump holes, bamboo stumps, bamboo pots, bamboo internodes, split bamboo and occasionally in artificial containers. Adult females are known to bite humans, but species of *Phagomyia* are not of medical and economic importance.

****

Principal references: Knight & Marks, 1952 (as Subgroup III, Gubernatoris, of subgenus *Finlaya*, in key, diagnosis); Reinert *et al.*, 2006, 2008, 2009 (as a genus, taxonomy, morphology, phylogenetic relationships); Rattanarithikul *et al.*, 2010 (as a genus, Thailand, keys, bionomics); Reinert, 2008b (as a genus, female genitalia); Wilkerson *et al.*, 2015 (phylogenetic relationships, classification); Soghigian *et al.*, 2017 (phylogenetic relationships). The adults and larvae of all but five of the species of *Phagomyia*, *Ae. iwi*, *Ae. kiangsiensis*, *Ae. melanoptera*, *Ae. plumiferus* and *Ae. watasei*, are treated by Barraud (1934), as species of *Finlaya*.

Polyleptiomyia Theobald, 1905

Polyleptiomyia, originally proposed as a genus by Theobald (1905e), currently classified as a subgenus of *Aedes*, is typified by *Aedes albocephalus* (Theobald, 1903c), original combination: *Stegomyia albocephalus*. Subgenus abbreviation – *Pmt.*

****

Systematics: *Polyleptiomyia* includes only two extant species in sub-Saharan Africa: *Ae. albocephalus* is widely distributed whereas *Ae. gandarai* is only known from Príncipe and São Tomé. The evolutionary relationships of *Polyleptiomyia* are uncertain. *Aedes albocephalus* was recovered in a basal relationship to *Bifidistylus* + (*Albuginosus* + (*Tewarius* + (*Christophersiomyia* + (*Huaedes* + *Leptosomatomyia*)))) in the morphology-based phylogenetic study of Aedini conducted by Reinert *et al.* (2009). *Polyleptiomyia* was not associated with other generic-level taxa in the phylogeny of Wilkerson *et al.* (2015).

****

Characteristics: The known species of *Polyleptiomyia* exhibit the following distinctive features. ADULTS: Vertex of the head mainly with narrow decumbent scales (almost entirely broad in males of *Ae. albocephalus*); erect scales on the vertex and occiput; maxillary palpus and proboscis entirely dark-scaled, maxillary palpus of males slightly longer than the proboscis; scutum with a variable mixture of dark and pale scales; paratergite with narrow scales; scutellum with narrow scales on the midlobe, sometimes also with some broad scales, lateral lobes with broad or a mixture of broad and narrow scales; lower proepisternal scales present; postspiracular scales present or absent; subspiracular scales present; prealar scales absent; wing dark-scaled with a line of pale scales at the base of the costa; femora each with a very small inconspicuous pale knee spot (absent on the fore- and midfemora of males?); hindtibia with a small apical pale spot; tarsi dark-scaled (tarsomere 1 sometimes pale-scaled posteriorly in *Ae. albocephalus*); fore- and midungues of females equal, both pair with 1 tooth, fore- and midungues of males unequal, larger unguis with 1 tooth, smaller unguis simple, hindungues equal and simple in both sexes; abdominal terga with basolateral pale patches, with or without narrow pale basomesal bands or spots (always absent in males?), sterna pale-scaled with narrow apical dark bands. FEMALE GENITALIA: Tergum and sternum VIII without scales, basolateral areas of the tergum expanded outward, caudal margin straight, sternum VIII wider than long, caudal margin with a median emargination; cercus widest in the proximal 0.5, scales absent; postgenital lobe with a median caudal emargination; insula tongue-like, setae or tuberculi present. MALE GENITALIA: Gonocoxite elongate, with a setose basal mesal lobe; gonostylus forked near mid-length, dorsal arm with a fine apical seta, ventral arm bearing a strong terminal gonostylar claw, no other setae present; aedeagus elongate, comprised of 2 lateral plates, each with a few apical teeth; paraproct long, tapered to a point; cercal setae absent. LARVAE: Antenna slender, curved, with a few spicules; seta 1-A multi-branched; seta 1-C spiniform; comb comprised of relatively few spine-like scales in a patch; siphon relatively short, pecten extending to about mid-length of the siphon, seta 1-S inserted well beyond the distal pecten spine; saddle incomplete; seta 1-X single, about as long as the saddle; ventral brush (seta 4-X) with 3–5 precratal setae and about 5 pairs of setae on a grid with lateral and transverse grid bars. PUPAE: Poorly known; trumpet with weakly developed tracheoid area; seta 9-VIII inserted on the corner of the segment, with 4 or more branches; paddle oval, rounded posteriorly, without a fringe of hair-like spicules, midrib well developed; seta 1-Pa single or forked; seta 2-Pa absent.

****

Bionomics: Hopkins (1952) reported that the immature stages of *Ae. albocephalus* typically inhabit grassy swamps, but are also found in ground pools of various kinds, marshy pools in small streams and drains, and crab holes; exceptionally found in tins and tree holes (van Someren *et al.*, 1955). Larvae of this species have been found in saltwater marshes and tidal pools in South Africa (Muspratt, 1951; McIntosh, 1975), confirming Hopkins supposition that 'it can tolerate some degree of salinity'. Females readily feed on domestic mammals and sometimes also on humans, but they rarely enter human dwellings (Muspratt, 1951; van Someren *et al.*, 1955; McIntosh, 1975). Nothing has been reported about the biology of *Ae. gandarai*. Worth *et al.* (1961) reported the isolation of Middelburg virus from *Ae. albocephalus* in South Africa, but species of *Polyleptiomyia* are not considered to be of medical importance.

****

Principal references: Reinert *et al.*, 2009 (as a genus, morphology, phylogenetic relationships); Reinert, 2010f (as a genus, female genitalia); Wilkerson *et al.*, 2015 (phylogenetic relationships, classification).

Protomacleaya Theobald, 1907

Protomacleaya, originally proposed as a genus by Theobald (1907), currently classified as a subgenus of *Aedes*, is typified by *Aedes triseriatus* (Say, 1823), original combination: *Culex triseriatus*. Subgenus abbreviation – *Pro.*

****

Systematics: *Protomacleaya* is a taxon of convenience for 40 extant species that do not appear to share a common ancestry. The species occur in areas from southern Canada to northern Argentina and southeastern Brazil, but they are absent from the West Indies. The Triseriatus, Kompi and Terrens Groups replace each other geographically; the Zoosophus Group occurs with the Triseriatus Group in southcentral USA and the Knabi Group is found with the Terrens Group in southern Mexico (Zavortink, 1972). *Protomacleaya* is a polyphyletic assemblage of species (Reinert *et al.*, 2006, 2008). Zavortink (1972) noted that the species of *Protomacleaya* in general resemble each other phenotypically, by virtue of what they lack; hence, it is an unnatural group whose relationships with other generic-level taxa cannot be determined.

****

Characteristics: Species of *Protomacleaya* are characterized and distinguished from other species of *Aedes* by the following combinations of characters. ADULTS: Vertex of the head with numerous erect and narrow falcate scales; eyes wholly or partly separated above the antennal pedicels; maxillary palpus and proboscis entirely dark-scaled, proboscis relatively stout; scutum usually dark-scaled with anterolateral or lateral white or silver markings; paratergite with a large patch of scales; postpronotum completely covered with decumbent spatulate scales; mesokatepisternum with 3 distinct patches of scales; mesepimeron usually with a large scale-patch; laterotergite of abdominal segment I with a large patch of scales. MALE GENITALIA: Ninth tergal lobes each with 1–9 setae in a single row; gonocoxite with a basal tergomesal patch of a few to numerous setae, apical area of the sternomesal surface without numerous long setae, median tergomesal sclerite absent to well developed, with a tuft of setae when present; gonostylus short, outer surface near the apex without setae; gonostylar claw long, spiniform; aedeagus simple, scoop-like, usually rounded and slightly expanded basally, more or less parallel-sided distally, truncate apically; paraproct with an apical spine. LARVAE: Seta 3-VII weakly to moderately developed, short, with 2–6 branches; seta 5-II–IV moderately to strongly developed; seta 12-I usually present; comb of most species with numerous fringed scales in a large patch; ventral brush (seta 4-X) moderately to strongly developed, with 6–8 pairs of setae on a grid, setae 4b,c with 3–12 branches in most species. PUPAE: Trumpet usually strongly pigmented; terga II–VII without an anterior transverse ridge; seta 1-II usually relatively strongly developed with long branches; seta 5-IV,V moderately long to long; seta 9-III–VI usually fine, sometimes stout.

****

Bionomics: The immature stages of species of *Protomacleaya* are found primarily in tree holes, but they are occasionally found in broken or cut bamboo internodes, coconut shells and pitcher plants, rock holes and artificial containers. *Aedes triseriatus* is a vector of La Crosse virus and has been shown to be capable of transmitting yellow fever, Eastern equine, Venezuelan and Western equine encephalomyelitis viruses in the laboratory. This species is also considered to be a potential vector of dog heartworm (*Dirofilaria immitis*).

****

Principal references: Schick, 1970a, 1970b (Terrens Group, as a group of *Finlaya*, taxonomy, classification, descriptions, distribution, bionomics, keys); Zavortink, 1972 (classification, keys, taxonomy, descriptions, bionomics, distributions); Reinert *et al.*, 2004, 2006, 2008, 2009 (as '*Ochlerotatus* (*Protomacleaya*)' *sensu auctorum*, morphology, phylogenetic relationships); Wilkerson & Linton, 2015 (status as a subgenus of *Aedes*).

Pseudalbuginosus Huang & Rueda, 2015

Pseudalbuginosus Huang & Rueda, 2015b, originally proposed and currently recognized as a subgenus of *Aedes*, is typified by *Aedes grjebinei* Hamon, Taufflieb & Maillot, 1957. Subgenus abbreviation – *Pal.*

****

Systematics: *Pseudalbuginosus* is monobasic. It only includes the extant type species, which is only known from localities in Cameroun and Gabon. The affinities of *Pseudalbuginosus* are unknown. Huang & Rueda (2015b) stated that *Pseudalbuginosus* 'possesses some rather important basic characters in common' with *Aedimorphus*, *Albuginosus*, *Diceromyia* and *Stegomyia*, thus clearly indicating that *Pseudalbuginosus*, like most generic-level taxa of Aedini, is a polythetic taxon that is diagnosed by a unique combination of characters.

****

Characteristics: *Pseudalbuginosus* is diagnosed and distinguished from all other generic-level taxa of Aedini based on features of the adults (immature stages unknown). ADULTS: Vertex of the head with numerous erect scales (not restricted to the occiput) and a patch of broad decumbent pale scales on either side above the margins of the compound eyes; maxillary palpus of males about as long as the proboscis, with 5 palpomeres, palpomeres 4 and 5 short, combined length shorter than the length of palpomere 3, slightly downturned and setose, palpomere 5 with a broad pale band at the base; proboscis of males with a complete pale band at mid-length; antennal pedicel with a patch of broad pale scales on the mesal surface; scutum with a large silvery patch of broad scales on the anterior scutal fossal area; acrostichal setae present; scutellum with broad pale scales on all lobes; paratergite with broad pale scales; subspiracular area without scales; postspiracular setae present; lower prealar scale-patch absent; lower mesepimeral setae absent; postprocoxal membrane without scales; hindfemur and -tibia with an apical pale spot, hindtarsomeres 1–4 with apical pale bands, hindtarsomere 5 entirely pale. MALE GENITALIA: Tergum IX short, constricted medially, with 4 or 5 setae posteriorly on either side of a shallow median emargination, lobes unapparent; gonostylus attached at the apex of the gonocoxite, long, proximal 0.6 narrow, broader distally with a prominent dorsal process bent toward the apex; gonostylar claw broad, leaf-like, inserted slightly before the apex; claspette elongate, somewhat carrot-shaped with 2 short setae at the apex; aedeagus strongly toothed; paraproct rather long, rounded apically, 3 cercal setae present.

****

Bionomics: The immature stages of *Ae. grjebinei* are unknown and nothing is known about the bionomics of the adults. The species is unlikely to be of medical or economic importance.

****

Principal references: Huang & Rueda, 2015a (as Grjebinei Group of subgenus *Aedimorphus*, in key to adults of species of *Albuginosus* and *Aedimorphus* Grjebinei and Apicoannulatus Groups); Huang & Rueda, 2015b (new subgenus, type species, descriptions, affinities).

Pseudarmigeres Stone & Knight, 1956

Pseudarmigeres Stone & Knight, 1956, originally proposed and currently recognized as a subgenus of *Aedes*, is typified by *Aedes argenteoventralis dunni* Evans, 1928, a synonym of *Ae. argenteoventralis* (Theobald, 1909), original combination: *Dendromyia argenteoventralis*. Subgenus abbreviation – *Psa.*

****

Systematics: *Pseudarmigeres* includes six extant species. *Pseudarmigeres* are Afrotropical species with distributions in Africa south of the Sahara. Edwards (1941) suggested a close relationship between *Pseudarmigeres* (as a subgenus of *Dunnius*) and the Oriental *Armigeres* based on striking morphological similarities of the adults, notably the reduction of dorsocentral setae, densely scaled proepisternum and postprocoxal membrane, general ornamentation and the presence of many setae on the male gonostylus. Based on the morphology-based phylogenetic studies of Reinert *et al.* (2004, 2009), *Pseudarmigeres* is more closely related to the Oriental *Alanstonea* and *Heizmannia*. *Pseudarmigeres* and *Heizmannia* were recovered as sister taxa in their 2004 study and in a clade comprising *Petermattinglyius* + (*Heizmannia* + (*Alanstonea* + *Pseudarmigeres*)) in their more comprehensive 2009 study. *Pseudarmigeres* was recovered as the sister of *Heizmannia*, and this pair was sister to *Petermattinglyius*, in the study of Wilkerson *et al.* (2015).

****

Characteristics: *Pseudarmigeres* is characterized and distinguished from other generic-level taxa of Aedini by the following combinations of characters. ADULTS: Ornate brown and white species; vertex with broad decumbent scales, erect scales absent; maxillary palpus and proboscis dark-scaled, proboscis longer than the forefemur, maxillary palpus of males slender, upturned, practically devoid of setae, 0.5–0.8 the length of the proboscis; acrostichal setae absent, dorsocentral setae and prescutellar setae few or absent; scutum with pale scales on the anterior promontory, posterior dorsocentral area, supraalar area and the prescutellar area; scutellum, antepronotum, postpronotum and paratergite with broad flat scales; proepisternum and usually the thoracic pleura densely scaly, hypostigmal, upper prealar and lower prealar scales present; lower mesepimeral setae present or (usually?) absent; wing dark-scaled, alula with broad scales on the dorsal surface; postprocoxal membrane largely or entirely covered with scales, hindtibia with an apical pale spot, tarsi dark-scaled, anterior fore- and midungues toothed; laterotergite of abdominal segment I with scales, terga II–V with large lateral pale patches, terga V–VII with complete basal pale bands; sterna II–V pale-scaled, sterna VI and VII dark-scaled with basal pale markings. FEMALE GENITALIA: Segment VIII mostly or completely retracted into segment VII, sternum VIII scaled, posterior margin more or less straight; tergum IX heart-shaped; cercus rather broad; postgenital lobe emarginate. MALE GENITALIA: Gonocoxite densely setose, a patch of short flat setae on the distal 0.5 of the ventromesal margin, basal dorsomesal lobe indistinct, with dense setae, mesal membrane nearly complete to the apex of the gonocoxite, gonostylus attached at the apex, noticeably narrower proximally, gonostylar claw inserted near mid-length, with a close-set row of stout setae (similar to the claw) between the claw and the apex; claspette closely appressed to the mesal membrane of the gonocoxite, with an apical tuft of setae, one seta much longer. LARVAE: Antenna smooth; seta 1-C spiniform; seta 4-C large, as long as seta 7-C; comb scales relatively few in number; elements of the ventral brush (seta 4-X) unpaired, precratal setae present. PUPAE: Setae 10–12-CT single, 10,11-CT about the same length, longer than 12-CT; most abdominal setae single; seta 3-I longer than setae 6,7-I; seta 2-VI inserted lateral to seta 1-VI; seta 5-V shorter than the median length of the tergum; seta 9-VII,VIII not strongly developed; paddles without a fringe or obvious denticles on the margin, midrib rather stout, without an external buttress.

****

Bionomics: Species of *Pseudarmigeres* are essentially forest mosquitoes. The immature stages are found in tree holes and bamboo. The species are not of medical or economic importance.

****

Principal references: Edwards, 1941 (as subgenus *Dunnius*, key, descriptions of adults and pupae, distributions); Hopkins, 1952 (as subgenus *Dunnius*, in key, descriptions of larvae, distributions, bionomics); Jupp, 1996 (southern Africa, diagnosis); Reinert, 2000e (female genitalia); Reinert *et al.*, 2004, 2009 (as a genus, morphology, phylogenetic relationships); Wilkerson *et al.*, 2015 (phylogenetic relationships, classification).

Pseudoskusea Theobald, 1907

Pseudoskusea, originally proposed as a genus by Theobald (1907), currently classified as a subgenus of *Aedes*, is typified by *Aedes multiplex* (Theobald, 1903a), original combination: *Skusea multiplex*. Subgenus abbreviation – *Psk*.

****

Systematics: *Pseudoskusea* currently includes only four extant species recorded from localities in Australia and New Guinea. The affinities of *Pseudoskusea* are uncertain. The group was recovered as the sister to a clade comprised of three species groups, *Culicelsa* + (*Buvirilia* + *Empihals*), each as a subgenus of *Ochlerotatus* in the phylogeny of Reinert *et al.* (2009) based on morphological data.

****

Characteristics: Species of *Pseudoskusea* are characterized and distinguished from other species of *Aedes* by the following combinations of characters. ADULTS: Dark mosquitoes without obvious ornamentation; eyes wholly or partly separated above the antennal pedicels, interocular scales present; vertex with broad decumbent scales, sometimes a few narrow decumbent scales on the occiput, erect scales numerous on the occiput, sparse on the vertex; antenna of males shorter than the proboscis, flagellar whorls with setae mainly directed dorsally and ventrally; maxillary palpus with 5 palpomeres, palpus of females about 0.2 the length of the proboscis, palpus of males more or less same length as the proboscis, palpomere 4 and the apex of palpomere 3 slightly swollen, bearing long setae; proboscis dark-scaled, distinctly longer than the forefemur; scutum with narrow dark or golden scales; acrostichal and dorsocentral setae moderately developed; scutellum with narrow scales on all lobes; antepronotum without scales; postpronotum with narrow scales, usually sparse, sometimes with some broad scales; upper proepisternum, subspiracular area and postspiracular area with or without scales, lower prealar area without scales; scale-patches present on the mesokatepisternum and the upper mesepimeron; lower mesepimeral setae and scales absent; metameron with or without short fine setae; wing dark-scaled; tarsi dark-scaled, fore- and midungues of females equal, with or without a tooth, fore- and midungues of males unequal, toothed; tergum I of the abdomen with a median patch of dark or pale scales, terga II–VI with either complete basal pale bands or basolateral patches with or without basomedian pale scaling; segment VIII of females usually completely retracted into segment VII. FEMALE GENITALIA: Tergum and sternum VIII normally without scales, tergum with 1 long basolateral seta, sternum with seta 2-S inserted more or less directly lateral to seta 1-S; cercus long, slender, sharply rounded apically, without scales; upper vaginal sclerite present. MALE GENITALIA: Ninth tergal lobes distinct, with setae; gonocoxite long and narrow, with a basal dorsomesal patch of setae, dorsomesal apical lobe indistinct, dorsomesal basal lobe absent; gonostylus long, simple, with a long apical gonostylar claw and several long preapical setae; claspette a single basal setose plaque, with a few inconspicuous setae, columnar lobe absent; aedeagus simple, scoop-like; paraproct with an apical spine. LARVAE: Strongly differentiated; antenna spiculate; seta 1-A branched; labiogula wider than long; seta 1-C single, thin, attenuate distally; seta 4-C inserted anterior to seta 6-C; seta 5-C ≥ 0.9 the length of the dorsal apotome; seta 6-C very long, single or double, ≥ 0.9 the length of the dorsal apotome; seta 7-C 0.4–0.8 the length of the dorsal apotome; seta 13-C branched; 2 or 3 of seta 1–3-P inserted on a common setal support plate; setae 2-T, 3-I,II,VII and 8-II all single; seta 7-II developed like seta 7-I; seta 3-VIII ≥ 0.9 the length of segment X; comb of segment VIII comprised of fringed scales in a large patch; siphon short or long; pecten spines evenly spaced; seta 1-S inserted beyond the pecten, ≥ 2.2 the width of the siphon at the point of attachment; seta 9-S stout, hook-like; saddle of segment X incomplete; seta 1-X single; ventral brush (seta 4-X) with 6 pairs of setae on a grid. PUPAE: Not studied in detail, apparently without characters of subgeneric significance.

****

Bionomics: The immature stages of three species of *Pseudoskusea*, *Ae. bancroftianus*, *Ae. multiplex* and *Ae. postspiraculosus*, are found in rain-filled ground pools; those of *Ae. culiciformis* are found in freshwater crab holes. Species of *Pseudoskusea* are not known to be of medical or economic importance.

****

Principal references: Dobrotworsky, 1961a (excluding *Ae. australis* (Erichson), descriptions, bionomics, distributions); Dobrotworsky, 1965 (descriptions, bionomics, distributions); Lee *et al.*, 1984 (taxonomy, key, literature, distributions, bionomics); Reinert, 2002e (female genitalia); Reinert *et al.*, 2004, 2008, 2009 (as a subgenus of *Ochlerotatus*, morphology, phylogenetic relationships); Reinert *et al.*, 2006 (as a genus, morphology, phylogenetic relationships); Wilkerson *et al.*, 2015 (status as a species group); Wilkerson & Linton, 2015 (status as a subgenus of *Aedes*).

Rampamyia Reinert, Harbach & Kitching, 2006

Rampamyia, originally proposed as a genus by Reinert *et al.* (2006), currently classified as a subgenus of *Aedes*, is typified by *Aedes notoscriptus* (Skuse, 1889), original combination: *Culex notoscriptus*. Subgenus abbreviation – *Ram.*

****

Systematics: *Rampamyia* includes four extant species. Species of *Rampamyia* occur in the Australasia Region (Australia, Bismarck Archipelago, New Caledonia, New Zealand, Papua New Guinea and Solomon Islands), with an extension into Indonesia (Java, Moluccas, Western New Guinea). The affinities of *Rampamyia* are uncertain. *Rampamyia* was recovered as the sister of *Molpemyia* + (*Luius* + *Macleaya*) in the phylogeny of Reinert *et al.* (2009) based on morphological data. The clade was sister to a group of two clades comprising *Dobrotworskyius* + (*Patmarksia* + *Georgecraigius*) and *Tanakaius* + (*Hulecoeteomyia* + (*Gilesius* + *Collessius*)). In the phylogeny recovered in the study of Wilkerson *et al.* (2015), *Rampamyia* was the sister of *Georgecraigius* in a polytomy that also included the *Collessius*, *Gilesius*, *Hulecoeteomyia* and *Tanakaius* in an unresolved clade, and *Dobrotworskyius*. The only species of *Rampamyia*, *Ae. notoscriptus*, included in the maximum likelihood phylogeny of Soghigian *et al.* (2017) based on seven molecular markers was recovered in a poorly supported clade with species of *Bruceharrisonius*, *Dobrotworskyius*, *Himalaius* and *Ochlerotatus*.

****

Characteristics: Species of *Rampamyia* are characterized and distinguished from species of other generic-level taxa of Aedini by the following combinations of characters. ADULTS: Vertex of the head with both broad and narrow decumbent scales; erect forked scales on the occiput and vertex; eyes contiguous; antennal pedicel with broad silvery scales on the mesal surface; maxillary palpus of females pale at the tip, palpus of males longer than the proboscis, with 5 palpomeres; proboscis with pale scales, a pale band near mid-length; acrostichal area, posterior dorsocentral area and the lateral margin of the scutal fossa with a narrow pale stripe, and also usually on the posterior margin of the scutal fossa; scales on the anterior part of the antealar area all dark; supraalar area with a narrow longitudinal pale patch; anterior and posterior acrostichal and dorsocentral setae present; scutellum with broad scales on all lobes; broad scales on the paratergite, antepronotum, postpronotum, subspiracular area, upper proepisternum, lower prealar area, upper and lower posterior mesokatepisternum, and the upper and lower mesepimeron; lower mesepimeral setae absent; wing dark-scaled, remigium with dorsal setae inserted distally; mid- and hindfemora with an anterior pale stripe extending from the base to near the apex; hindtarsomeres 1–4 with broad basal pale bands; fore- and midungues of females equal, toothed, fore- and midungues of males unequal, toothed, larger one with 1 tooth, hindungues equal and simple; laterotergite of abdominal segment I with silvery scale-patch; segment VII of females laterally compressed. FEMALE GENITALIA: Intersegmental membrane between segments VII and VII long; tergum and sternum VIII wider than long, normally without scales (tergum occasionally with 1–3 adventitious scales), sternum with seta 2-S inserted posterior to seta 1-S; tergum IX a single sclerite with a median posterior emargination; cercus without scales; postgenital lobe with a few setae distally; lower vaginal sclerite absent; insula lip-like, with 5–7 setae in lateral patches; 3 spermathecal capsules. MALE GENITALIA: Tergum IX lobes short, narrow, acute, each with a few setae; gonocoxite with basomesal and apicomesal lobes, ventral surface with a patch of long lanceolate setae distally at the mesal margin, mesal surface membranous; gonostylus attached at the apex of the gonocoxite, relatively short, gonostylar claw borne apically, ≥ 0.4 the length of the gonostylus; claspette comprised of a long columnar stem with an apical flattened and curved claspette filament; aedeagus simple, tube-like, distinctly long and narrow. LARVAE: Antenna without spicules; seta 1-A normally single, inserted on the distal 0.33 of the antenna; seta 4-C inserted slightly posteromesal to seta 6-C; setae 4,6-C displaced far anterior; seta 5-C inserted posterior and slightly lateral to setae 4,6-C and mesal and slightly anterior to seta 7-C; setae 5,6,13-C single; seta 19-C present; setae 1–3-P not on a common setal support plate; seta 8-P short, multi-branched; seta 7-I single; seta 12-I present; seta 1-VIII ≥ 1.1 the length of seta 2-VIII; comb comprised of numerous scales in a patch; pecten spines of the siphon evenly spaced; seta 1-S branched, inserted distal to the pecten; saddle incomplete, acus present, with well-developed spicules on the dorsoposterior margin; ventral brush (seta 4-X) attached to a grid with transverse and lateral grid bars; seta 1-X multi-branched, inserted on the saddle. PUPAE: Seta 1-CT weakly developed, considerably shorter than seta 3-CT; seta 5-CT > 1.3 the length of seta 4-CT; seta 6-I longer than seta 7-I; seta 3-II long, single, as long as or shorter than seta 6-II, inserted mesal to seta 2-II; seta 5-II single, longer than seta 3-II; seta 5-IV,V longer than the following tergum; seta 2-VI inserted mesal to seta 1; seta 6-VII inserted posteromesal to seta 9-VII; paddle broadly rounded apically, without a fringe of hair-like spicules; seta 1-Pa single.

****

Bionomics: The immature stages of species of *Rampamyia* are found in plant containers (phytotelmata), including tree holes, bamboo, leaf axils of various plants, palm fronds, coconut shells and husks. They also inhabit rock holes and artificial containers made of metal, cement and wood. Species of *Rampamyia* are not of medical and economic importance.

****

Principal references: Lee *et al.*, 1982 (as *notoscriptus* subgroup, excluding *Ae. mallochi* Taylor and *Ae. plagosus* Marks, of Group E of subgenus *Finlaya*, taxonomy, literature, distributions, bionomics); Reinert *et al.*, 2006, 2008, 2009 (as a genus, morphology, phylogenetic relationships); Reinert, 2008f (as a genus, female genitalia); Wilkerson *et al.*, 2015 (phylogenetic relationships, classification); Soghigian *et al.*, 2017 (phylogenetic relationships).

Reinertia Somboon, Namgay & Harbach, 2021

Reinertia Somboon *et al.*, 2021a, originally proposed and currently recognized as a subgenus of *Aedes*, is typified by *Aedes suffusus* Edwards, 1922a. Subgenus abbreviation – *Rei.*

****

Systematics: *Reinertia* is monobasic. It only includes the extant type species, which is only recorded from localities at high altitude in the Himalaya Mountains of Bhutan, India and Nepal. The larva of *Ae. suffusus* closely resembles larvae of species of *Downsiomyia*, and the adults and larva share anatomical features with those of species of *Dahliana*. In maximum likelihood and Bayesian phylogenetic analyses of the mitochondrial *COI* gene (Somboon *et al.*, 2021a), *Reinertia* was not associated with *Dahliana* or *Downsiomyia* but was recovered as the weakly supported sister of *Protomacleaya*.

****

Characteristics: The following salient anatomical differences distinguish *Reinertia* from other generic-level taxa of Aedini. ADULTS: Small blackish brown mosquitoes; vertex of the head with dark upright forked scales covering most of the dorsal surface, recumbent scales on the dorsal midline largely narrow and white; proboscis and maxillary palpus dark-scaled; scutum mostly covered with narrow pale falcate scales; postspiracular area with pale scales and setae; anterior surfaces of the fore- and midfemora mostly dark-scaled with pale scaling posteroventrally, approximately proximal 0.5 of hindfemur pale-scaled with narrow anteroventral line of pale scales connected to an apical pale patch on the anterior surface, tibiae and tarsi of all legs dark-scaled; abdominal terga III–V and all sterna with conspicuous basal pale bands. FEMALE GENITALIA: Segment VIII broad, not completely retracted into segment VII; tergum IX short, bilobed, lobes widely separated by a narrow basal strip; cercus moderately long, somewhat cylindrical in dorsal view. MALE GENITALIA: Gonocoxite without dorsomesal apical and dorsomesal basal lobes, ventromesal margin without a group of prominent scales; mesal membrane with a setose basomesal sclerite (homologous with dorsomesal basal lobe). LARVAE: Antenna very long, narrow and bowed; seta 1-A very long and branched; seta 4-C inserted mesad and slightly posterior to seta 6-C; seta 5-C inserted slightly posterior to seta 7-C; seta 6-C inserted slightly anterior to seta 7-C; seta 12-C present; seta 13-C branched; seta 14-C single; comb scales spine-like, in a single irregular row; pecten spines evenly spaced; seta 1-S inserted distal to the pecten; ventral brush (seta 4-X) with precratal setae. PUPAE: Seta 11-CT well developed, single; seta 6-VII inserted posterior to the level of seta 9-VII; paddle long, essentially symmetrical.

****

Bionomics: The immature stages of *Ae. suffusus* have been found in tree holes at high altitudes in mountain forest (Edwards, 1922a; Barraud, 1934; Darsie *et al.*, 1996; Somboon *et al.*, 2021a). Darsie *et al.* (1996) found females resting inside a guest house in northwestern Nepal, and those authors, as well as Bhat (1975) working in West Bengal State of India, observed that females are attracted to humans. Whether the single species of *Reinertia* may play a role in the transmission of pathogens is unknown.

****

Principal references: Barraud, 1934 (as species of *Finlaya*, key for adults, description, bionomics, distribution); Knight & Marks, 1952 (as species of *Finlaya*, key, classification, morphology, taxonomy, bionomics); Darsie *et al.*, 1996 (as species of *Finlaya*, Nepal, collection records, bionomics); Somboon *et al.*, 2021a (new subgenus, type species, Bhutan, taxonomy, descriptions, bionomics, distribution, phylogenetic relationships).

Rhinoskusea Edwards, 1929

Rhinoskusea Edwards, 1929b, originally proposed and currently recognized as a subgenus of *Aedes*, is typified by *Aedes longirostris* (Leicester, 1908), original combination: *Ficalbia longirostris*. Subgenus abbreviation – *Rhi.*

****

Systematics: *Rhinoskusea* includes only four extant species recorded from areas of the Oriental Region, Papua New Guinea and the northern tip of Australia. Mattingly (1958a) stated that *Rhinoskusea* is 'clearly related' to *Cancraedes* and *Paraedes* and 'shows some interesting resemblance to *Geoskusea*' in basal structures of the male genitalia. Based on the phylogenetic study of Reinert *et al.* (2009), *Rhinoskusea* appears to be more closely related to *Geoskusea* and *Levua*. *Rhinoskusea* was recovered as the sister of *Levua* in a clade parenthetically expressed as *Sallumia* + (*Geoskusea* + (*Levua* + *Rhinoskusea*)). In the phylogeny of Wilkerson *et al.* (2015), *Rhinoskusea* formed a polytomy along with *Geoskusea*, *Levua* and *Sallumia*, and this group was sister to *Cancraedes*. In contrast to the findings of Reinert *et al.* and Wilkerson *et al.* based on analyses of morphological data, two species of *Rhinoskusea* were recovered in a sister relationship with two species of *Collessius* in the maximum likelihood phylogeny of Soghigian *et al.* (2017) based on seven molecular markers.

****

Characteristics: *Rhinoskusea* is characterized and distinguished from the other generic-level taxa of Aedini by the following combinations of characters. ADULTS: Small unornamented mosquitoes; vertex of the head with broad decumbent scales, few erect scales restricted to the occiput; eyes contiguous above the antennal pedicels; maxillary palpus of males short as in females; scutum with both narrow and broad scales, the latter on the anterior promontory, scales sparse on the scutal fossa, scales present on the prescutellar area; acrostichal setae (anterior and posterior) and dorsocentral setae (anterior and posterior) numerous; scutellum with broad scales on all lobes; paratergal, antepronotal and postpronotal scales absent; thoracic pleura without scales except for a small patch on the lower area of the mesepisternum; lower mesepimeral setae absent; upper calypter of the wing of males with 0–3 setae; legs dark-scaled, hindtarsomere 1 slightly longer than the hindtibia, all ungues of females small, simple, anterior fore- and midungues of males larger, with or without a tooth. FEMALE GENITALIA: Postgenital lobe short and broad; upper vaginal sclerite absent; insula lip-like with 3–5 long setae; only 1 spermathecal capsule. MALE GENITALIA: Tergum IX lobes with or without setae; mesal surface of the gonocoxite partly sclerotized, dorsomesal apical lobe present, median and distal parts of the gonostylus much broader than the proximal part, gonostylar claw short, striated, scoop-like; claspette comprised of

2 basal lobes, columnar lobe absent, long distal portion free of the gonocoxite, without an apical filament; aedeagus simple, trough-like, wide, widest in the distal 0.33. LARVAE: Lateral palatal brushes without pectinate elements; antenna with setae 2,3-A inserted noticeably subapical; seta 6-C branched; 2 or 3 of setae 1–3-P borne on a common setal support plate; seta 4-P single; seta 6-II shorter than seta 6-III; seta 8-II single; seta 1-VII 0.5–0.9 the length of segment X; comb comprised of numerous scales in a triangular patch; seta 1-S long, ≥ 2.2 the width of the siphon, multi-branched, inserted far distad on the siphon; seta 1-X long, single, inserted on the saddle; ventral brush (seta 4-X) with 10 multi-branched setae borne on a grid with transverse grid bars; anal papillae small, knob-like. PUPAE: Trumpet with well-develop tracheoid area distal to the base; seta 11-CT long, stout, double; intersegmental membranes and membrane around the base of seta 1-I with small dark granules; seta 6-I shorter than seta 7-I; seta 2-II as long as or longer than seta 1-II; seta 3-II longer than seta 6-II, thinner than seta 1-II, seta 3-III branched; seta 5-II inserted mesal to seta 4-II, seta 5-III–VI long, with 2–6 branches, seta 5-V shorter than tergum VI; seta 4-VII much longer than seta 5-VII; seta 9-VIII multi-branched with secondary branches on the distal 0.5; seta 1-Pa with 3–5 branches.

****

Bionomics: Species of *Rhinoskusea* are confined to brackish water coastal and river areas. The immature stages usually inhabit crab holes and small ground pools associated with mangrove, but they are sometimes found in other small collections of water, natural and constructed containers, and rock pools. Adults have been collected in light traps, human- and animal-baited traps, and resting in crab holes. Females are known to feed on humans, but species of *Rhinoskusea* are not known to be of medical importance.

****

Principal references: Knight & Hull, 1952 (Philippine Islands, description, bionomics, distribution); Mattingly, 1958a (in part, Indomalayan area, key, descriptions, bionomics, distributions); Reinert, 1976b (taxonomy, keys, descriptions, distributions, bionomics); Reinert, 2002e (as a subgenus of *Ochlerotatus*, female genitalia); Reinert *et al.*, 2004, 2006, 2008 (as a subgenus of *Ochlerotatus*, morphology, phylogenetic relationships); Reinert *et al.*, 2009 (as a genus, morphology, phylogenetic relationships); Rattanarithikul *et al.*, 2010 (as a genus, Thailand, keys, bionomics); Wilkerson *et al.*, 2015 (phylogenetic relationships, classification); Soghigian *et al.*, 2017 (phylogenetic relationships).

Rusticoidus Shevchenko & Prudkina, 1973

Rusticoidus Shevchenko & Prudkina, 1973, originally proposed and currently recognized as a subgenus of *Aedes*, is typified by *Aedes refiki* Medschid, 1928. Subgenus abbreviation – *Rus*.

****

Systematics: *Rusticoidus* includes nine extant species. Species of *Rusticoidus* have distributions in the Holarctic Region. Seven species occur in the Palaearctic, principally in Europe, and two occur in the Nearctic. Within the phylogeny of Aedini generated in the study of Reinert *et al.* (2009) based on morphological data, *Rusticoidus* was recovered as the sister of *Ae. flavescens* in a clade, comprised of a large number of species, that was diagnosed by the unique presence of parascutellar scales. In the maximum likelihood phylogeny of Soghigian *et al.* (2017) based on seven molecular markers, three species of *Rusticoidus* were recovered in a sister relationship to a clade comprised of *Ae.* (*Och.*) *bancroftianus* + (*Ae.* (*Och.*) *sagax* + *Ae.* (*Och.*) *vittiger*). Interestingly, *Ae.* (*Och.*) *bancroftianus*, a member of *Pseudoskusea*, and *Ae.* (*Och.*) *sagax* and *Ae.* (*Och.*) *vittiger*, which are without subgeneric placement, are Australasian species whereas species of *Rusticoidus* have distributions in the Holarctic Region.

****

Characteristics: The species of *Rusticoidus* are distinguished from other species of *Aedes* and other Aedini in the Holarctic Region by the following combinations of characters. ADULTS: Large mosquitoes; ocular line of the head broad; antennal pedicel of females with large mesal and lateral scale-patches (patches sometimes contiguous dorsally); parascutellar area with 1–6 scales and 1–3 setae; scutum and scutellum with narrow scales only; acrostichal area with a pale-scaled stripe; membrane posteromesad of the metapostnotum with a patch of broad scales; antepronotum with both narrow and broad scales; postpronotum with only broad decumbent scales; antepronotum, upper proepisternum and prealar area with a large number of long setae; lower proepisternal scales present; hypostigmal area with broad scales; subspiracular area with an elongate patch of broad scales; mesepimeron with 1–8 long setae on the anteroventral area; metameron with broad scales; remigium of the wing with dorsal setae inserted proximally; ante- and postprocoxal scales present; hindtarsomere 2 without basal pale scales; ungues of males each toothed, larger fore- and midungues with 2 teeth. FEMALE GENITALIA: Tergum and sternum VIII largely covered with broad spatulate scales (except in *Ae. bicristatus*), posterior margin of the sternum with a median emargination separating broadly rounded lateral lobes; dorsal surface of the cercus with broad spatulate scales on the distal 0.6, ventral surface with several setae on the distal 0.5–0.6 and a few broad spatulate scales; postgenital lobe with a median apical indentation, ventral surface narrowly connected with the upper vaginal lip; upper and lower vaginal sclerites absent; insula lip-like, with 2–4 setae on either side of the midline; 3 spermathecal capsules. MALE GENITALIA: Ninth tergal lobes with stout flattened setae; sternum IX with a heavily pigmented round median caudal area with setae and similar lateral triangular strips; gonocoxite with a basal dorsomesal lobe bearing specialized setae; claspette moderately thick with short annulated filament; aedeagus simple, long and narrow, narrowest in the middle, base with a circular opening (bearing lateral triangular flaps in some species), apex with a small median lobe separating lateral flattened areas bearing minute spicules; paraproct heavily pigmented, with a curved pointed apex. LARVAE: Antenna spiculate; seta 1-A short, branched; seta 1-C single, thin, attenuate distally; seta 4-C branched, inserted slightly posterior to seta 7-C and anteromesal to seta 6-C; seta 5-C inserted caudomesad of seta 6-C and slightly lateral to seta 4-C; seta 4-P about as long as seta 3-P; seta 8-P ≥ 1.8 the length of seta 4-P; seta 12-P very long, single; seta 1-M with 3 or more branches; seta 12-I present, short; seta 3-VII 0.5–0.9 the mid-dorsal length of segment X; seta 1-VIII noticeably shorter than setae 3,5-VIII, 0.5–1.0 the length of seta 2-VIII; comb scales spine-like, borne in 1 or 2 irregular rows; siphon with an accessory seta 1-S inserted on the anterior side of the pecten proximal to the level of the usual seta 1-S and 1–4 accessory anterolateral seta 2a-S; seta 9-S stout, hook-like; pecten with 1 or the more distal spines more widely spaced; saddle incomplete, without acus; ventral brush (seta 4-X) with 2–6 precratal setae and posterior setae borne on a grid with lateral and transverse grid bars. PUPAE (only known for *Ae. quasirusticus* and *Ae. rusticus*): Seta 11-CT inserted closer to seta 10-CT than to seta 12-CT; seta 2-II,VII inserted lateral to seta 1, seta 2-III–VI inserted mesad of seta 1; seta 5-IV–VI very long; paddle noticeably longer than wide; seta 1-Pa usually single, occasionally forked or branched; seta 2-Pa absent.

****

Bionomics: The species of *Rusticoidus* are monocyclic snow-melt mosquitoes that overwinter in the egg stage. Some larvae hatch in late autumn and pass winter in the larval stage; others hatch early the following year during or shortly after snow-melt, or after heavy rainfall. Larvae are predominantly found in semi-permanent and permanent collections of water in woodlands, primarily swampy and marshy habitats, but they are also found in ditches and flooded meadows. Adults are usually encountered close to larval habitats, emerging from late April to June. The females are mainly zoophilic but several species, including *Ae. refiki*, *Ae. rusticus*, *Ae. subdiversus* and *Ae. bicristatus*, are known bite humans at dusk and during the day in shaded situations, but species of *Rusticoidus* are not known to be of medical or economic importance.

****

Principal references: Carpenter & LaCasse, 1955; Wood *et al.*, 1979 (*Ae. bicristatus* and *Ae. provocans*, respectively, description, biology, distribution); Reinert, 1999a (taxonomy, morphology); Reinert, 2002e (as a subgenus of *Ochlerotatus*, female genitalia); Melero-Alcíbar, 2004, 2005 (pupa of *Ae. quasirusticus* and *Ae. rusticus*, respectively); Reinert *et al.*, 2004, 2006, 2008, 2009 (as a subgenus of *Ochlerotatus*, morphology, phylogenetic relationships); Becker *et al.*, 2003, 2010 (as a subgenus of *Ochlerotatus*, Europe, descriptions, biology, distributions); Becker *et al.*, 2020 (Europe, descriptions, biology, distributions); Wilkerson *et al.*, 2015 (status as species group); Wilkerson & Linton, 2015 (status as a subgenus of *Aedes*); Soghigian *et al.*, 2017 (phylogenetic relationships).

Sallumia Reinert, Harbach & Kitching, 2008

Sallumia Reinert *et al.* (2008), originally proposed as a subgenus of *Ochlerotatus* and currently recognized as a subgenus of *Aedes*, is typified by *Aedes hortator* Dyar & Knab, 1907a. Subgenus abbreviation – *Sal.*

****

Systematics: *Sallumia* includes only two extant species. *Sallumia* is a Neotropical taxon, with the species occurring in countries from Bolivia northward to Trinidad. *Aedes hortator* is recorded from localities in Bolivia, Brazil, Colombia, French Guiana, Peru, Surinam, and Trinidad and Tobago. *Aedes perventor* is known from localities in Brazil and French Guiana. Reinert *et al.* (2008), in their morphology-based phylogenetic study, recovered *Sallumia* as the sister to *Geoskusea* + (*Levua* + *Rhinoskusea*) deeply nested within *Ochlerotatus*. In the more comprehensive study of Reinert *et al.* (2009), the same sister-group clade was placed in a sister relationship to a large terminal clade comprised of generic-level taxa other than *Ochlerotatus*. In the phylogeny of Wilkerson *et al.* (2015), *Sallumia* formed a polytomy along with the subgenera *Geoskusea*, *Levua* and *Rhinoskusea*, and this group was sister to *Cancraedes*.

****

Characteristics: *Sallumia* is characterized and distinguished from other generic-level taxa of Aedini by the following combinations of characters. ADULTS: Vertex of the head with broad decumbent scales, erect scales restricted to the occiput; eyes contiguous or separated by less than 1 corneal facet; antenna of males with setae directed dorsally and ventrally; maxillary palpus of females dark-scaled, with 4 palpomeres, palpomere 4 minute, bulbous, maxillary palpus of males longer than the proboscis, palpomeres 4 and 5 upturned, palpomere 5 ≤ 0.6 the length of palpomere 4; proboscis dark-scaled, longer than forefemur; scutum with narrow dark scales; anterior and posterior acrostichal setae and posterior dorsocentral setae absent, anterior dorsocentral area with 1 or 2 setae; scutellum with broad scales on all lobes; paratergite without scales; antepronota widely separated, without scales; postpronotum, hypostigmal area, postspiracular area, subspiracular area, upper and lower proepisternum, mesepimeron and metameron without scales; prealar area and lower mesokatepisternum with a few scales; lower mesepimeral setae absent; wing dark-scaled, remigium without setae, anal vein (vein 1A) terminates approximately in line with the intersection of the mediocubital crossvein and the cubitus; ante- and postprocoxal membranes bare; base of the hindcoxa well below the dorsal margin of the mesomeron; femora without knee spots; tibiae and tarsi dark-scaled; fore- and midungues of females equal, anterior and posterior unguis of each with 2 teeth, fore- and midungues of males unequal, larger unguis with 1 tooth, smaller unguis simple, hindungues of females and males both toothed; laterotergite of abdominal segment I with broad scales, terga II–VII dark-scaled with small basolateral pale patches, terga of males with a few lateral setae; segment VII of females cylindrical in cross-section. FEMALE GENITALIA: Tergum and sternum VIII longer than wide, posterior margin of the tergum straight, sternum with a median posterior emargination surrounded by darkly pigmented areas, both without scales; postgenital lobe with a posterior emargination; upper and lower vaginal sclerites absent; insula lip-like with a few setae; cercus without scales; 3 spermathecal capsules. MALE GENITALIA: Tergum IX lobes large, each lobe with a few setae; gonocoxite with scales, with a dorsomesal apical lobe, dorsal surface with a large elongate lobe narrowly connected to the basomesal area, lobe with many setae, 1 flattened; gonostylus attached at the apex of the gonocoxite, with a single long terminal gonostylar claw, ≥ 0.4 the length of the gonostylus; claspette with a long columnar stem bearing a broad claspette filament; aedeagus tube-like, broader proximally than distally, apex with a few denticles; cercal setae present. LARVAE: Antenna with spicules; seta 1-A double, extending beyond the apex of the antenna; seta 1-C stout, spiniform; seta 4-C branched, inserted mesal and slightly anterior to seta 6-C; seta 5-C inserted posteromesal to seta 6-C; seta 6-C branched, with 1 branch stouter and longer than the others; seta 7-C inserted anterolateral to setae 4–6-C; seta 13-C single, inserted anterolateral to seta 12-C; seta 14-C single; seta 19-C absent; setae 1–3-P not inserted on a common setal support plate; seta 4-P double, no longer than seta 3-P; seta 8-P ≥ 1.8 the length of seta 4-P; seta 6-I double; seta 7-I single, slightly shorter than seta 6-I; seta 12-I present; seta 6-II,III single, 6-III longer than 6-II; seta 1-VII 0.5–0.9 the length of segment X; seta 12-VII and seta 2-VIII branched; comb scales in a patch; siphon with the pecten spines evenly spaced, seta 1-S inserted distal to the pecten, multi-branched; saddle incomplete; seta 1-X single, inserted on the saddle; seta 2-X short, multi-branched; ventral brush (seta 4-X) attached to a grid with both transverse and lateral grid bars, and with 2 precratal setae. PUPAE: Setae 1–3-CT normally double, seta 1-CT weakly developed, considerably shorter than seta 3-CT; seta 3-I with 2 long stout branches, shorter than setae 6,7-I; seta 2-II inserted lateral to setae 1,3-II; seta 3-II single, longer than seta 6-II, seta 3-III single or double, both longer than seta 5; seta 2-VI inserted slightly mesal to seta 1; seta 6-VII inserted posteromesal to seta 9-VII, 9-VII no longer than seta 6-VII; seta 9-VIII inserted on the posterolateral corner of the segment; paddle rounded apically, without hair-like marginal spicules; midrib complete to the apex of the paddle; seta 1-Pa single.

****

Bionomics: The immature stages of species of *Sallumia* are found in temporary pools (*Ae. hortator*) and crab holes (*Ae. perventor*). The females of *Ae. hortator* are known to attack humans in wooded areas, but species of *Sallumia* are not known to be of medical or economic importance.

****

Principal references: Reinert *et al.*, 2008 (as a subgenus of *Ochlerotatus*, morphology, phylogenetic relationships, description, distribution, bionomics); Reinert *et al.*, 2009 (as a genus, morphology, phylogenetic relationships); Reinert, 2010a (as a genus, female genitalia); Wilkerson *et al.*, 2015 (phylogenetic relationships, classification).

Scutomyia Theobald, 1904

Scutomyia, originally proposed as a genus by Theobald (1904c), currently classified as a subgenus of *Aedes*, is typified by *Aedes albolineatus* (Theobald, 1904c), original combination: *Scutomyia albolineata*. Subgenus abbreviation – *Sct.*

****

Systematics: *Scutomyia* includes nine extant species. The type species is widely distributed in the Oriental and Australasian Regions, six species are known only from the Philippines and two are known only from Sulawesi. The phylogenetic relationships of *Scutomyia* are uncertain. *Scutomyia* appears to share affinities with some species of *Stegomyia*, and was previously recognized as a species group (Albolineatus Group) within that taxon (Knight & Rozeboom, 1946; Huang, 1979a). *Scutomyia* was recovered as the sister of *Catatassomyia* + *Bothaella* in the phylogeny of Reinert *et al.* (2009) based on morphological data. Its relationship with other generic-level taxa was unresolved in the phylogeny of Wilkerson *et al.* (2015). The one species of *Scutomyia*, *Ae. albolineatus*, included in the maximum likelihood phylogeny of Soghigian *et al.* (2017) based on seven molecular markers was recovered in a poorly supported relationship with subgenera *Bothaella* + (*Borichinda* + *Nyctomyia*).

****

Characteristics: Species of *Scutomyia* are characterized and distinguished from species of other generic-level taxa of Aedini by the following combinations of characters. ADULTS: Vertex of the head with both broad and narrow decumbent scales; erect scales confined to the occiput; ocular scales all narrow; antennal pedicel with a patch of broad silvery scales on the mesal surface; maxillary palpus dark-scaled, with 4 palpomeres in females, 5 in males, maxillary palpus of males 0.5–0.8 the length of the proboscis (longer, about the length of the proboscis in *Ae. bambusicola*), with palpomeres 4 and 5 nearly straight; scutum with or without a median longitudinal pale stripe (absent in *Ae. platylepidus*) extending from the anterior promontory to at least the posterior end of the acrostichal area, present on the posterior acrostichal area; paratergite bare; acrostichal setae absent; dorsocentral setae present or absent; scutellum with broad scales on all lobes; postpronotum without scales; upper and lower proepisternal scales present; subspiracular and postspiracular scales absent; wing dark-scaled, sometimes with a small pale spot at the base of the costa, upper calypter with 0–3 hair-like scales; hindcoxa more or less at the same level as the dorsal margin of the mesomeron, hindtarsus with a basal pale band at least on tarsomere 1; foreungues of females both simple; abdominal terga of males with a few relatively short lateral setae. FEMALE GENITALIA: Tergum VIII with setae on the caudal ≤ 0.6; sternum VIII with the caudal margin shallowly concave or with a deep U-shaped notch, setae 1–3-S in a more or less curved line from the anteromedian area to the posterolateral area, seta 1-S inserted mesally near the anterior margin, seta 2-S inserted close to seta 1-S, seta 3-S widely separated posteriorly from setae 1,2-S; insula tongue-like, without setae, with 3–9 (usually 6–9) small tuberculi on the distal 0.4, each with or without a short spicule. MALE GENITALIA: Tergum IX lobes unapparent or slightly produced; gonostylus attached at the apex of the gonocoxite, elongate, no setae on the distal 0.33, gonostylar claw spiniform, inserted near mid-length of the gonostylus; aedeagus widest in the distal 0.33, with several distolateral teeth; paraproct with apical teeth, without a sternal arm, with or without an apical lobed process. LARVAE: Antenna with or without spicules; seta 5-C branched; seta 6-C 0.4–0.8 the length of the dorsal apotome; seta 12-I absent; seta 1-VII 0.5–0.9 the length of segment X; seta 2-VII branched; seta 12-VII inserted anterior to seta 13-VII; comb with scales in a single row, comb plate present or absent; siphon short, siphon index about 2.0–3.1, acus present, pecten on the proximal 0.5 of the siphon, pecten spines evenly spaced, seta 1-S inserted distad of the pecten; segment X with inconspicuous spicules or large spines on the posterior margin; ventral brush (seta 4-X) with 5 pairs of setae. PUPAE: Seta 2-II as long as or longer than seta 1-II; seta 1-VI inserted lateral to seta 1-VI; seta 9-VIII with 3 or more branches, inserted slightly anterior or mesal to the posterolateral corner of the segment; paddle more or less oval, margins without a fringe of hair-like spicules, at most with minute spicules; seta 1-Pa branched, long, 0.4−0.6 the length of the paddle.

****

Bionomics: Species of *Scutomyia* are usually found in plant containers such as bamboo stumps and tree holes, but they also occur in numerous other phytotelmata, artificial containers, rock holes, lagoons and jungle pools. Females of at least three species are known to bite humans, but species of *Scutomyia* are not of medical and economic importance.

****

Principal references: Huang, 1979a (as Albolineatus Group of *Stegomyia*, diagnosis, keys); Reinert, 1985 (taxonomy, description, species groups, characteristics); Reinert, 2000e (female genitalia); Reinert *et al.*, 2004, 2006, 2008, 2009 (as a genus, morphology, phylogenetic relationships); Rattanarithikul *et al.*, 2010 (as a genus, Thailand, keys, bionomics); Wilkerson *et al.*, 2015 (phylogenetic relationships, classification); Soghigian *et al.*, 2017 (phylogenetic relationships).

Skusea Theobald, 1903

Skusea, originally proposed as a genus by Theobald (1903a), currently classified as a subgenus of *Aedes*, is typified by *Aedes pembaensis* Theobald, 1901c. Subgenus abbreviation – *Sku.*

****

Systematics: *Skusea* includes four extant species, which are known to occur in Madagascar (three species), the Seychelles (two species, including the type species) and Comoro Islands (one species, also found in Madagascar). The type species, in addition to its occurrence in Madagascar and the Seychelles, is recorded from Mozambique and Tanzania. Prior or Belkin (1962), *Skusea* included a collection of unrelated species whose adults exhibited similar scaling of the head and scutellum. He established *Lorrainea* for the Oriental and Australasian species previously included in the group and recognized *Skusea* as a monobasic group represented by *Ae. pembaensis* in eastern Africa and Madagascar. Edwards (1928a) noted that the position of this species relative to other aedine taxa was 'a matter of uncertainty owing to the fact that no very close allies are known'. The affinities of *Skusea*, now with four closely related species, remain uncertain. *Skusea* was recovered as the sister of *Indusius* + *Cancraedes* in the phylogeny of Reinert *et al.* (2009) based on morphological characters of all life stages of the type species. It was recovered as the sister of *Indusius*, and this pair was sister to *Fredwardsius*, in the phylogeny of Wilkerson *et al.* (2015).

****

Characteristics: Species of *Skusea* are characterized and distinguished from species of other generic-level taxa of Aedini by the following combinations of characters. ADULTS: Vertex of the head with broad decumbent scales, erect forked scales confined to the occiput; maxillary palpus of males about as long as the proboscis, palpomeres 4 and 5 short, slightly thickened and setose; scutum with narrow scales, acrostichal and dorsocentral setae present; scutellum with broad decumbent scales; paratergite bare; thoracic pleura with broad scales on the postspiracular area, upper and lower areas of the mesokatepisternum and the upper area of the mesepimeron; lower mesepimeral setae present (apparently absent in males of *Ae. pembaensis*); wing dark-scaled with a short patch of pale scales at the base of the costa; femora without knee spots, ungues of females all simple; abdominal terga largely dark-scaled with basolateral pale patches, some terga usually with narrow pale basal bands. FEMALE GENITALIA: Intersegmental membrane between segments VII and VIII short to intermediate in length; tergum VIII with setae inserted on ≤ distal 0.6; posterior margin of sternum VIII with a median emargination separating sublateral lobes; tergum IX a single sclerite, wrinkled (at least in *Ae. pembaensis*); postgenital lobe without scales; upper vaginal sclerite absent. MALE GENITALIA Segment VIII unusually large; tergum IX unusually small; sternum IX without setae; gonocoxite short and broad, mesal surface not membranous, with a double subapical lobe bearing a row of long apically hooked setae on the lateral side that extends to or beyond the apex of the gonostylus and a dense patch of short setae on the mesal side (the lobe is apparently the terminal part of a uniquely modified claspette that extends as a narrow strip along the mesal margin of the gonocoxite); gonostylus short and stout, with a distolateral fringe-like crest and a dorsal spur-like process immediately proximal to a short spine-like gonostylar claw; aedeagus widest in the distal 0.33, comprised of a pair of simple elongate lateral plates, teeth absent; paraproct very poorly developed. LARVAE: Antenna without spicules, seta 1-A inserted about 0.25 from the base; seta 5-P branched, no longer than seta 6-P; seta 7-P single; seta 7-M longer than seta 5-M; seta 7-I with 4 or more branches, shorter than 0.5 the length of seta 6-I; seta 1-VIII less than 0.4 the length of seta 2-VIII; seta 2-VIII inserted approximately midway between setae 1 and 3; seta 3-VIII simple; comb comprised of numerous scales in a patch, scales bifid and trifid (branched from the base), branches frayed at the tip; siphon nearly cylindrical, relatively short, siphon index about 3.5, acus absent; pecten more or less on the proximal 0.5 of the siphon, pecten spines evenly spaced; seta 1-S very short, multi-branched, inserted beyond the pecten; saddle very small, weakly sclerotized, with a dorsoposterior patch of coarse spicules; seta 1-X not inserted on the saddle, single; seta 2-X multi-branched, with about 8 branches; ventral brush (seta 4-X) with 6–7.5 pairs of multi-branched setae inserted on a grid. PUPAE: Trumpet without tracheoid area; seta 7-CT ≥ 6.0 times the length of seta 6-CT; setae 11-CT and 3-I,II branched; seta 5-II inserted mesal to seta 4-II; seta 5-II longer than seta 3-II; seta 2-VI inserted lateral to seta 1-VI; seta 9-VIII inserted slightly anterior or mesal to the posterolateral corner of the segment; paddle broad, round, margins smooth; seta 1-Pa single, long, ≥ 0.8 the length of the paddle.

****

Bionomics: Larvae of species of *Skusea* are common in crab holes in tidal swamps and estuaries. Females of *Ae. pembaensis* are known to bite humans from early morning until dusk, with peak activity in late afternoon. Females of *Ae. lambrechti* are known to feed on mammals and are attracted to humans. Both sexes have been collected inside houses in early morning and outdoors in the vicinity of larval habitats. Eggs of *Ae. pembaensis* have been found attached to crabs. Microfilariae have been found in wild-caught females of *Ae. lambrechti*, but it is not known if species of *Skusea* are of medical importance.

****

Principal references: Knight & Hull, 1952 (Philippine Islands, descriptions, bionomics, distributions); Mattingly, 1959 (Indomalayan area, taxonomy, keys, classification, descriptions, bionomics, distributions); Brunhes, 1977b (male genitalia, taxonomy, description of *Ae. cartroni*); Reinert, 2000e (female genitalia); Reinert *et al.*, 2004, 2009 (as a genus, morphology, phylogenetic relationships); Wilkerson *et al.*, 2015 (phylogenetic relationships, classification).

Stegomyia Theobald, 1901

Stegomyia, originally proposed as a genus by Theobald (in Howard, 1901), currently classified as a subgenus of *Aedes*, is typified by *Aedes aegypti* (Linnaeus, 1762), original combination: *Culex aegypti*. Subgenus abbreviation – *Stg.*

****

Systematics: *Stegomyia* is a large group with 133 extant species classified in 16 species groups, two groups of which are further divided into subgroups. Species of *Stegomyia* have distributions in the Afrotropical, Australasian and Oriental Regions. Two species, *Ae. aegypti* and *Ae. albopictus*, have been transported to areas of the Nearctic, Neotropical and Palaearctic Regions by human agency. *Stegomyia*, as defined by Reinert *et al.* (2009), appears to be a highly evolved monophyletic group with uncertain affinities. *Stegomyia* was recovered as sister to a clade comprised of 10 aedine genera in the phylogeny of Reinert *et al.* based on extensive morphological data, including *Alanstonea*, *Belkinius*, *Cornetius*, *Diceromyia*, *Eretmapodites*, *Heizmannia*, *Lorrainea*, *Patmarksia*, *Udaya* and *Zeugnomyia*. Huang (1979a, 2004) suggested that *Stegomyia* exhibits 'strongest affinities' with *Diceromyia* based on 'important' shared characters of the adult and immature stages. In the phylogeny of Wilkerson *et al.* (2015), *Stegomyia* was recovered as the sister of *Cornetius* within a large polytomy that included 11 other subgenera and the genera *Heizmannia*, *Udaya* and *Zeugnomyia*. *Stegomyia*, represented by 33 species, was recovered as a monophyletic group in the maximum likelihood phylogeny of Soghigian *et al.* (2017) based on molecular markers. Their findings support the monophyly of the Aegypti and Simpsoni Groups. Hide *et al.* (2024) conducted a maximum likelihood analyses of a data set consisting of 435-bp sequences of the mtDNA

COI gene to assess the relationships of 32 species of *Stegomyia*. The results, depicted in the tree generated in the study, did not support the monophyly of the Scutellaris, Unilineatus and W-albus Groups, nor the Albopictus Subgroup of the Scutellaris Group. The results did, however, support the monophyly of the Aegypti and Simpsoni Groups, the latter weakly supported, and the W-albus Subgroup of the W-albus Group.

****

Characteristics: *Stegomyia* is characterized and distinguished from other generic-level taxa of Culicinae by the following combinations of characters. ADULTS: Ornate mosquitoes, usually dark with white or silver markings; vertex of the head with broad decumbent scales, erect scales restricted to the occiput, usually few in number; clypeus bare or with scales; lateral surface of the antennal pedicel with scales; proboscis usually about as long as the forefemur; maxillary palpus with white scales, palpus of females about 0.1–0.3 the length of the proboscis, comprised of 3 or 4 palpomeres, palpomere 4 minute when present, palpus of males longer than 0.5 the length of the proboscis, comprised of 5 palpomeres, palpomeres 4 and 5 subequal, slender, upturned, with a few setae, their combined length at least 0.4 the combined length of the proximal palpomeres; scutum entirely or mainly covered with narrow scales; acrostichal setae absent; scutellum usually with broad decumbent scales; paratergite usually with scales; mesopostnotum bare; antepronotum usually with broad decumbent scales; lower proepisternal scales present; prespiracular setae absent; postspiracular setae present; lower mesepimeral setae usually absent; postprocoxal membrane without scales; wing with narrow plume scales, some pale scales near the base of the costa; hindtarsus with a basal white band on one or more tarsomeres; ungues of fore- and midlegs toothed or simple, those of the hindleg usually simple, anterior fore- and midungues of males larger than the posterior ungues; laterotergite of abdominal segment I with scales. FEMALE GENITALIA: Segment VIII retractile, somewhat compressed, sternum VIII with prominent rounded lateral lobes; segment IX with a wide and deep posterior emargination, shorter than cercus; cercus broad and rather short, with scales; postgenital lobe with a shallow caudal emargination; insula tongue-like, longer than broad, with minute setae and 2–10 larger setae on the apical 0.3–0.5; 3 spermathecal capsules, one larger than the others. MALE GENITALIA: Sternum IX bare; gonocoxite sometimes with a membranous mesal area, gonostylus usually simple, gonostylar claw present, inserted apically or subapically, developed as a relatively narrow spiniform, truncate, rarely slightly rounded; claspette well developed, with numerous setae; aedeagus strongly toothed, comprised of distinct lateral plates, each with numerous teeth; paraproct without teeth; cercal setae absent. LARVAE: Antenna without spicules; setae 4–6-C inserted far forward, cephalad of the antennal base; seta 4-C well developed, branched; setae 5,6-C seldom strongly branched; seta 4-C inserted closer to seta 6-C than to seta 5-C, inserted cephalad and mesad of seta 6-C; seta 6-C inserted cephalad of setae 5,7-C; seta 6-C branched; seta 7-C inserted at approximately the same level as seta 5-C; seta 4-P branched; seta 5-P longer than seta 6-P; seta 12-I absent; seta 2-VIII distant from seta 1-VIII; comb scales usually in a single row; saddle complete or incomplete; seta 2-X single or double, rarely triple; ventral brush (seta 4-X) with 4 or 5 pairs of setae on a grid or boss, without precratal setae. PUPAE: Subgeneric characters not evident; seta 8-C inserted well posterior to the trumpet, at least halfway between the trumpet and the mesonotum.

****

Bionomics: The immature stages of species of *Stegomyia* are found in natural and artificial containers. Typical habitats are tree holes, but many species inhabit small amounts of water contained in dead and fallen plant parts. A few species utilize rock holes, crab holes and the leaf axils of various plants. Eggs are normally resistant to desiccation and hatch when the habitat is filled with water. Females are typically diurnal and many species (approximately 50) are known to bite humans. They also feed on a variety of domestic and wild animals, including mammals, birds, reptiles and amphibians. *Stegomyia* is a medically important group. *Aedes aegypti* is the classical vector of yellow fever and dengue fever viruses and a proven vector of other viruses. Zika virus is suspected to be transmitted mainly by *Ae. aegypti* in urban environments of South America. Evidence also suggests that *Ae. albopictus*, which is also an important vector of dengue fever virus, is capable of natural vertical transmission of Zika virus in Brazil (Alencar *et al.*, 2021). Other recognized vectors of yellow fever virus include *Ae. africanus* and *Ae. luteocephalus* in areas of central and western Africa and *Ae. bromeliae* in eastern Africa. Some species of the Scutellaris Group are efficient vectors of *Wuchereria bancrofti* in the South Pacific. Various arboviruses have been isolated from other species of *Stegomyia*.

****

Principal references: Knight & Hull, 1952 (Philippine Islands, classification, descriptions, bionomics, distributions); Belkin, 1962 (South Pacific, taxonomy, keys, descriptions, distributions, bionomics); Huang, 1977a, 1979a (Oriental Region, classification, keys, taxonomy, descriptions, distributions, bionomics); Lee *et al.*, 1987a (Australasian Region, taxonomy, key, literature, distributions, bionomics); Huang, 1990 (Afrotropical Region, Africanus Group, keys, taxonomy, descriptions, distributions, bionomics); Huang, 1997 (Afrotropical Region, Dendrophilus Group, keys, taxonomy, descriptions, distributions, bionomics); Lu *et al.*, 1997a (China, keys, taxonomy, descriptions, distributions, bionomics); Reinert, 2000e (female genitalia); Huang, 2004 (Afrotropical Region, classification, species-group diagnoses, keys, new species descriptions, distributions, bionomics); Reinert *et al.*, 2004, 2006, 2008, 2009 (as a genus, morphology, phylogenetic relationships); Rattanarithikul *et al.*, 2010 (as a genus, Thailand, keys, bionomics); Wilkerson *et al.*, 2015 (phylogenetic relationships, classification); Soghigian *et al.*, 2017 (phylogenetic relationships).

Tanakaius Reinert, Harbach & Kitching, 2004

Tanakaius, originally proposed as a genus by Reinert *et al.* (2004), currently classified as a subgenus of *Aedes*, is typified by *Aedes togoi* (Theobald, 1907), original combination: *Culicelsa togoi*. Subgenus abbreviation – *Tan.*

****

Systematics: *Tanakaius* includes two extant species. The type species is known to occur naturally in eastern areas of China and Russia, and in Japan, South Korea, Malaysia (Peninsular), South Prymorye, Southern Kurile Islands, Taiwan, Thailand and Vietnam. It was introduced to British Columbia, Canada, and subsequently extended its range into Washington State, USA. *Aedes savoryi* is indigenous to Ogasawa-guntô (the Bonin Islands) of Japan. *Tanakaius* was recovered as sister to a clade comprised of *Hulecoeteomyia* + (*Gilesius* + *Collessius*) in the phylogenetic study of Reinert *et al.* (2009) based on morphological data of the adult, larval and pupal stages of Aedini. *Tanakaius* and *Collessius*, *Gilesius* and *Hulecoeteomyia* comprised a polyphyletic clade in the phylogeny of Wilkerson *et al.* (2015). *Tanakaius* was recovered as the sister to a clade comprised of species of *Collessius*, *Hulecoeteomyia* and *Rhinoskusea* in the maximum likelihood phylogeny of Soghigian *et al.* (2017) based on molecular markers. In a phylogenetic analysis of mtDNA *COI* sequences (Somboon *et al.*, 2023b), *Tanakaius* was recovered as a weakly supported clade in a polytomy with two clades comprised of 12 other subgenera of *Aedes*.

****

Characteristics: Species of *Tanakaius* are characterized and distinguished from species of other generic-level taxa of Aedini by the following combinations of characters. ADULTS: Vertex of the head with narrow decumbent scales, occiput and most of the vertex with numerous erect forked scales; eyes narrowly separated above the antennal pedicels; maxillary palpus about 0.25 the length of the proboscis in females, about 0.75 in males (comprised of 5 palpomeres); scutum with narrow scales, pale scales in lines and stripes on the acrostichal and dorsocentral areas, anterior margin of the scutal fossa, posterior margin of the scutal fossa to the middle of the supraalar area and laterally on the antealar and supraalar areas; scutal fossal scales sparse; acrostichal and dorsocentral setae present; paratergite with broad pale scales; scutellum with narrow falcate scales on all lobes; lower mesepimeral setae present or absent; wing dark-scaled, costa with a small pale spot at the base; remigial setae absent; alula with a row of narrow scales on the margin; upper calypter with numerous marginal setae; ante- and postprocoxal membranes bare; femora with knee spots and incomplete pale stripes on the anterior and posterior surfaces; tarsi dark-scaled or with narrow basal and apical pale bands on some tarsomeres; both hindungues toothed in females and males; laterotergite of abdominal tergum I pale-scaled, terga II–VII with basolateral pale patches; sterna with basolateral pale patches often connected mesally; segments VII and VIII of females laterally compressed. FEMALE GENITALIA: Intersegmental membrane long between segments VII and VIII; tergum VIII with setae on the distal 0.7 or more; sternum VIII with a shallow posterior emargination, seta 2-S inserted lateral to and more or less at the level with seta 1-S; tergum IX with a V-shaped caudal margin; upper and lower vaginal sclerites absent; insula lip-like with setae on either side of the midline; 3 spermathecal capsules, one slightly larger than the others. MALE GENITALIA: Tergum IX narrowly connected laterally to the sternum, posterior margin bilobed, lobes with setae; gonocoxite with numerous spatulate scales on the dorsolateral, lateral and ventral surfaces, ventral surface with an irregular or double row of numerous fusiform scales, gonostylus attached apically, gonostylar claw borne at the apex; claspette comprised of a columnar stem and a long strongly curved claspette filament, stem $\leq$ 0.8 the length of the aedeagus; aedeagus simple, scoop-like with a narrowed apex; paraproct narrow, apex strongly curved into a short beak-like point, 2–5 cercal setae. LARVAE: Antenna short, $\leq$ 0.4 the length of the dorsal apotome; seta 1-A with 3–6 branches (rarely double), not reaching the apex of the antenna; setae 2,3-A inserted apically; seta 4-C branched; setae 4–6-C arranged in a transverse row, inserted anterior to the level of seta 7-C; seta 13-C single, inserted laterad of seta 12-C; seta 19-C absent; seta 4-P single; seta 5-P as long as or shorter than seta 6-P; seta 7-M as long as or longer than setae 5-M; seta 3-VII inserted at the same level or posterior to seta 1; seta 6-I–III with 3 or more branches; seta 7-I,II long, with 2 or 3 branches; seta 12-I single or double; setae 2,10-VII single; comb scales in a patch; siphon short, acus attached, pecten spines evenly spaced; seta 1-S inserted distal to the pecten; saddle small, incomplete ventrally, without acus; seta 1-X single to triple, not inserted on the saddle; seta 2-X with 5–11 branches; seta 3-X single; ventral brush (seta 4-X) comprised of 12–14 multi-branched setae attached to a grid with both lateral and transverse grid bars, occasionally 1 precratal seta. PUPAE: Trumpet with a weakly developed tracheoid area; seta 1-CT longer than seta 3-CT; seta 7-CT > 3.5 the length of seta 6-CT; seta 3-I branched; seta 6-I with 1–5 branches, distinctly shorter than seta 7-I; seta 2-II single or double, inserted lateral to seta 1; seta 5-II with 1–4 branches, inserted mesad of seta 4; seta 5-V single to triple, longer than the following tergum; seta 3-VI inserted laterad of seta 1; seta 6-VII with 1–7 branches, inserted posteromesad of seta 9; paddle broad, without a fringe of hair-like spicules, midrib distinct to near the apex; seta 1-Pa single.

****

Bionomics: The immature stages of species of *Tanakaius* usually inhabit brackish water ground and rock pools in coastal areas, but *Ae. togoi* has been found on occasion in freshwater containers. Females of *Tanakaius* are known to bite humans during diurnal hours. *Aedes togoi* is a vector of *Brugia malayi*; females have been found naturally infected with Japanese encephalitis virus and transmission of the virus has been demonstrated in laboratory studies (Gutsevich *et al.*, 1974; Tanaka *et al.*, 1979; Rattanarithikul *et al.*, 2010). It is also an experimental vector of Semliki Forest virus (Nye & Lien, 1960).

****

Principal references: Reinert *et al.*, 2004 (new genus, type species, morphology, phylogenetic relationships, descriptions, bionomics, distribution); Reinert *et al.*, 2009 (as a genus, morphology, phylogenetic relationships); Reinert, 2008i (as a genus, female genitalia); Rattanarithikul *et al.*, 2010 (as a genus, Thailand, keys, bionomics); Wilkerson *et al.*, 2015 (phylogenetic relationships, classification); Soghigian *et al.*, 2017 (phylogenetic relationships); Somboon *et al.*, 2023b (comparative morphology, phylogenetic relationships).

Tewarius Reinert, 2006

Tewarius, originally proposed as a genus by Reinert (2006a), currently classified as a subgenus of *Aedes*, is typified by *Aedes agastyai* Tewari & Hiriyan, 1992b. Subgenus abbreviation – *Tew.*

****

Systematics: *Tewarius* includes four extant species, which are divided between two species groups, the Agastyai Group (*Ae. agastyai*, *Ae. reubenae*) and the Nummatus Group (*Ae. nummatus*, *Ae. pseudonummatus*). The species occur in the Oriental Region. Three species are known from India (*Ae. agastyai*, *Ae. nummatus*, *Ae. reubenae*) and one from Thailand (*Ae. pseudonummatus*). The cladistic analysis of morphological data by Reinert *et al.* (2009) placed *Tewarius* in a clade with six other Old World groups: *Polyleptiomyia* + (*Bifidistylus* + (*Albuginosus* + (*Tewarius* + (*Christophersiomyia* + (*Huaedes* + *Leptosomatomyia*))))). In the phylogeny of Wilkerson *et al.* (2015), *Tewarius* fell within a three-way polytomy that included *Christophersiomyia* and a clade comprised of (*Armigeres* + (*Alanstonea* + *Eretmapodites*)) + (*Huaedes* + *Leptosomatomyia*).

****

Characteristics: The adults of species of *Tewarius* are very similar to those of *Downsiomyia* but differ in having dorsocentral setae and pale scales on the postspiracular area. The large patch of pale scales on the anterior half of the scutum is somewhat similar in species of *Phagomyia* and *Himalaius*, but species of *Tewarius* differ in numerous other features, especially features of the male and female genitalia. The moderately long and thin maxillary palpus of *Tewarius* males bears some resemblance to certain species of *Scutomyia*, but species of *Tewarius* otherwise exhibit major differences in features of the adult, larval and pupal stages. The forked gonostylus of *Tewarius* males is somewhat similar to the condition in some Afrotropical species of *Diceromyia*, but the latter differ in having the inner arm of the gonostylus much shorter than the outer arm. The larvae of *Tewarius* are similar to those of *Petermattinglyius* but differ in the development of seta 1-S. Pupae of *Tewarius* and *Ae. gilli* of *Himalaius* have a very long single seta 1-Pa but differ noticeably in the positions of setae 6,9-VII and 9-VIII.

The following features are most distinctive of species of *Tewarius*. ADULTS: Vertex of the head covered with broad decumbent scales, erect forked scales restricted to the occiput; eyes contiguous above the antennal pedicels; maxillary palpus short, about 0.2 the length of the proboscis in females, 0.4–0.5 the length of the proboscis in males, males with 5 palpomeres; approximately the anterior 0.5 of the scutum covered with white scales; supraalar area dark-scaled; acrostichal setae absent; dorsocentral setae present; midlobe of the scutellum with broad scales; mesokatepisternum with a large patch of broad white scales extending from the upper area to the lower posterior area; lower mesepimeral seta absent; fore- and midungues equal in females and unequal in males, each with 1 tooth, hindungues equal in both sexes. FEMALE GENITALIA: Tergum and sternum VIII with numerous broad scales; insula tongue-like with 1 or 2 distal tuberculi; upper vaginal sclerite large; cercus short with a few scales on the dorsal surface; 3 spermathecal capsules. MALE GENITALIA: Tergum IX lobes broad, with several setae; gonocoxite short, with apical and basal lobes on the dorsomesal margin; gonostylus split near mid-length into 2 narrow arms, lateral arm with minute apical setae, mesal arm with a stout gonostylar claw at the apex; claspette comprised of a basal setose plaque; aedeagus comprised of 2 lateral sclerites bearing numerous elongate teeth; cercal setae absent. LARVAE: Antenna with a few spicules, setae 2,4-A long; seta 4-C short, branched, inserted mesal and slightly posterior to seta 6-C; seta 5-C long, branched, inserted posterolateral to seta 4-C and posteromesal to setae 6,7-C; seta 6-C very long, normally single, inserted lateral to setae 4,5-C and slightly posteromesal to seta 7-C; seta 13-C branched; setae 1,3-P branched, shorter than seta 2-P; setae 4-M, 6-T and 13-T branched, 13-T long; seta 6-II shorter than seta 6-III; seta 1-VII single, long; setae 2,4-VIII branched; comb with 20–32 scales in a patch; seta 1-S multi-branched, inserted distal to the pecten; pecten comprised of numerous long slender evenly spaced spines; saddle incomplete; ventral brush (seta 4-X) with several long multi-branched cratal setae and 2 or 3 short precratal setae; grid with well-developed transverse grid bars; setae 2,3-X single (2-X occasionally double). PUPAE: Setae 3-CT, 3-I–III, 6-II and 5-IV–VI single and very long, seta 3-CT noticeably longer than seta 1-CT; seta 3-I noticeably longer than seta 6-I, seta 3-II,III noticeably longer than the following tergum; seta 6-II shorter than seta 3-II; seta 5-IV–VI longer than the following 2 terga; paddle ovoid, margin without hair-like spicules; seta 1-Pa single (rarely split apically), very long, about as long as the paddle.

****

Bionomics: *Tewarius* are forest species. Immature stages have been found in tree holes, bamboo stumps, cut bamboo and once in a broken bottle in jungle. Nothing is known about the bionomics of the adults. Species of *Tewarius* are not of medical and economic importance.

****

Principal references: Reinert, 1973e (*Ae. nummatus* and *Ae. pseudonummatus* as species of *Diceromyia*, taxonomy, descriptions, distributions, bionomics); Reinert, 2006a (new genus, taxonomy, description, distribution, bionomics, diagnosis); Reinert *et al.*, 2008, 2009 (as a genus, morphology, phylogenetic relationships); Rattanarithikul *et al.*, 2010 (as a genus, Thailand, keys, bionomics); Wilkerson *et al.*, 2015 (phylogenetic relationships, classification).

Vansomerenis Reinert, Harbach & Kitching, 2006

Vansomerenis, originally proposed as a genus by Reinert *et al.* (2006), currently classified as a subgenus of *Aedes*, is typified by *Aedes pulchrithorax* Edwards, 1939a. Subgenus abbreviation – *Van.*

****

Systematics: *Vansomerenis* includes three extant species, which are known to occur in six central African countries: Angola, Democratic Republic of the Congo, Kenya, Tanzania, Uganda and Zambia. The phylogenetic relationships of *Vansomerenis* are uncertain. It was recovered as the sister of *Zavortinkius* + *Kenknightia* in a clade comprised of (*Himalaius* + *Bruceharrisonius*) + (*Vansomerenis* + (*Zavortinkius* + *Kenknightia*)) in the phylogeny of Reinert *et al.* (2009) based on morphological data. Relationships of *Vansomerenis* with other generic-level taxa of Aedini were not resolved in the phylogeny of Wilkerson *et al.* (2015).

****

Characteristics: *Vansomerenis* includes distinctive species that are characterized and distinguished from species of other generic-level taxa of Aedini by the following combinations of characters. ADULTS: Vertex of the head with a median pale patch, erect forked scales on the vertex and occiput; ocular line with broad silvery scales; eyes contiguous; antennal pedicel without vestiture on the mesal surface (unique character); maxillary palpus and proboscis dark-scaled; scutum with narrow pale stripe on the anterior and posterior acrostichal areas extending to the prescutellar area, similar stripe on the posterior dorsocentral area extending to the posterior margin of the scutum, a longitudinal pale patch on the supraalar area; acrostichal setae absent, anterior and posterior dorsocentral setae present; scutellum with narrow scales on all lobes; broad silvery scales on the paratergite, antepronotum, postpronotum, upper proepisternum, upper and lower mesokatepisternal areas, lower prealar area and the mesepimeron (a large patch covering upper and lower areas); lower mesepimeral setae absent; wing dark-scaled; hindfemur without a knee spot, hindtibia with a broad basal pale patch, hindtarsomeres 1 and 2 with pale basal bands; both sexes with ungues of all legs each with 1 tooth; laterotergite of abdominal segment I with a silvery scale-patch, segment VII of females laterally compressed. FEMALE GENITALIA: Tergum VIII with broad scales on much of the surface; sternum VIII wider than long, caudal margin nearly straight or with a shallow median emargination, with a few to several broad scales, seta 2-S inserted posterior to seta 1-S; tergum IX comprised of 2 long narrow sclerites connected on the mesal 0.7 by a pigmented band; cercus without scales, distal part narrowly rounded; upper and lower vaginal sclerites absent; insula lip-like, with a few setae in lateral patches; 3 spermathecal capsules. MALE GENITALIA: Tergum VIII with a small posteromedian lobe; tergum IX comprised of 2 small sclerites connected mesally by a tongue-like median anteriorly directed lobe of tergum X, each sclerite with a small posterior lobe bearing a few setae, lateral part of the lobe connected by a narrow band with sternum IX; gonocoxite with scales on the dorsal surface, without an apicomesal lobe, with a basal lobe, distal part of the lobe with 2 or 3 setae, proximal part expanded with a series of basal wrinkles and 1 or more stout or flattened setae, mesal surface of the gonocoxite membranous; gonostylus attached at the apex of the gonocoxite, with a single apical gonostylar claw; claspette comprised of a long slender columnar stem with a long distally expanded and flattened claspette filament; aedeagus simple, tube-like. LARVAE: Antenna with a few scattered spicules; seta 1-A single or double, occasionally triple; seta 4-C inserted mesal and slightly posterior to seta 6-C; seta 5-C with 1–3 branches, most often single, inserted posteromesal to seta 6-C and posterolateral to seta 4-C; seta 6-C longer than seta 5-C, inserted mesal and very slightly anterior to seta 7-C; seta 12-C inserted mesal to and shorter than seta 13-C; seta 19-C absent; seta 4-P not longer than seta 3-P; seta 5-P branched, shorter than seta 6-P; seta 6-I,II long, branched (6-II rarely single in *Ae. luteostriatus*), seta 6-III branched, shorter than 6-II; seta 7-I long, double or triple, 7-II short, multi-branched; comb comprised of numerous scales in a patch; seta 1-VIII ≥ 1.1 the length of seta 2-VIII; seta 2-VIII branched; pecten spines of the siphon evenly spaced; seta 1-S branched, inserted distal to the pecten; saddle of segment X incomplete ventrally, acus absent; setae 1-X branched, inserted on the saddle; ventral brush (seta 4-X) with 2 precratal setae, other setae inserted on a grid with well-developed transverse grid bars, lateral grid bars absent, represented by a faint depression. PUPAE: Seta 7-CT very long, longer than 6 times the length of seta 6-CT; seta 6-I single, very long, noticeably longer than seta 7-I; seta 7-I usually double, rarely single or triple; seta 2-II inserted lateral to seta 1 and mesal to seta 3, seta 2-VI inserted mesal to setae 1 and 3; seta 3-II,III long, stout, single; seta 5-II shorter than seta 3-II; seta 6-II longer than seta 7-II; seta 7-II longer than the following tergum; seta 2-V inserted anterior to seta 3-V; seta 6-VII inserted posteromesal to seta 9-VII; seta 9-VIII inserted near the posterolateral corner of the segment; paddle longer than wide, paddle index 1.2–1.4, without hair-like spicules on the margins; seta 1-Pa single.

****

Bionomics: The immature stages of species of *Vansomerenis* have been found in tree holes, a rot hole in the fork of a tree, a tree buttress and bored bamboo. Nothing is known about the bionomics of the adults, which are only known from larval rearings. The species of *Vansomerenis* are not of medical and economic importance.

****

Principal references: Reinert *et al.*, 2006 (new genus, morphology, phylogenetic relationships, description, distribution, bionomics); Reinert *et al.*, 2008, 2009 (as a genus, morphology, phylogenetic relationships); Reinert, 2008j (as a genus, female genitalia); Wilkerson *et al.*, 2015 (phylogenetic relationships, classification); Huang & Rueda, 2017 (as Pulchrithorax Group of *Finlaya*, Afrotropical Region, key to adults).

Zavortinkius Reinert, 1999

Zavortinkius Reinert, 1999b, originally proposed and currently recognized as a subgenus of *Aedes*, is typified by *Aedes longipalpis* (Grünberg, 1905), original combination: *Stegomyia longipalpis*. Subgenus abbreviation – *Zav.*

****

Systematics: *Zavortinkius* includes 11 extant species in the Afrotropical Region. *Zavortinkius* was recovered as sister to *Kenknightia* in a clade comprised of *Himalaius* + *Bruceharrisonius* and *Vansomerenis* + (*Zavortinkius* + *Kenknightia*) in the morphology-based phylogenetic analysis of Aedini conducted by Reinert *et al.* (2009). It is not surprising that *Zavortinkius* appears to be related to *Vansomerenis* as the two groups are endemic to the Afrotropical Region. However, relationships of *Zavortinkius* with other generic-level taxa of the tribe were not resolved in the phylogeny of Wilkerson *et al.* (2015).

****

Characteristics: The following combinations of features distinguish *Zavortinkius* from other generic-level taxa of Aedini. ADULTS: Decumbent scales of the vertex and interocular space all broad; ocular line with broad silvery spatulate scales; antennal pedicel without scales; proboscis shorter than the forefemur; acrostichal and dorsocentral setae absent; scutal scales both narrow and broad; supraalar area, postpronotum and pleural areas of the thorax with broad silvery spatulate scales; subspiracular scales absent; lower proepisternal and upper prealar scales present; scutellar scales all broad; hindungues both toothed in females and males; lateral abdominal setae of males shorter than the dorsal width of the terga and few in number; abdominal segment VII of females dorsoventrally flattened. FEMALE GENITALIA: Sternum VIII with scales, often covering much of the surface, posterior margin with a median emargination separating broadly rounded lateral lobes; tergum IX comprised of 2 narrow lateral plates connected by a basomesal strip, with apical setae on each side of the midline; postgenital lobe long, relatively narrow, base bilobed, basomesal apodeme well developed; cercus long, without scales; upper vaginal lip well developed; insula lip-like, with setae. MALE GENITALIA: Tergum IX comprised of 2 lateral sclerites, each with a single posterior row of curved setae; tergum X narrow with a tongue-like median anteriorly directed lobe separating the 2 halves of tergum IX; gonocoxite long with a basal dorsomesal lobe bearing at least 1 stout seta, ventromesal margin with a double row of stout setae; gonostylar claw long, narrow, curved distally, apex blunt; claspette narrow, stem with a long flattened filament; aedeagus simple, trough-like. LARVAE: Antenna long, narrow, with a few spicules; seta 1-A usually single; setae 4–6-C in a triangular formation (but in a diagonal line in *Ae. pollinctor*); seta 4-C inserted between and mesad of setae 5,6-C (except in *Ae. pollinctor*); seta 4-C short, branched; setae 5-C single; seta 7-C single (rarely double), inserted slightly anterolateral to slightly posterolateral to seta 6-C; setae 8–10,13-C branched (10-C rarely single in *Ae. geoffroyi*); seta 18-C present; seta 19-C and the ventromedian cervical sclerite absent; seta 1-P longer than seta 2-P, 2-P single, longer than seta 3-P; setae 4-P and 6-T branched; seta 12-I present, seta 12-VII branched; seta 1-VII long; seta 2-VIII branched, longer than seta l-VIII; comb comprised of numerous scales in a patch; siphon without acus; saddle incomplete; ventral brush (seta 4-X) with 5 or 6 pairs of fan-like setae on a grid. PUPAE: Seta 7-CT long, with 2–6 stout branches; seta 8-CT long, branched; seta 6-I,II long, seta 6-II longer than seta 7-II; seta 2-II inserted laterad of seta 1-II and mesad of seta 3-II; setae 3-II,III and 5-IV–VI stout, longer than the following tergum; seta 6-VII branched, shorter than seta 9-VII; seta 9-VII,VIII stout, branched (9-VII rarely single in *Ae. fulgens*), seta 9-VIII inserted noticeably anterior to the corner of the segment; seta 1-Pa single (very rarely split apically).

****

Bionomics: Species of *Zavortinkius* prefer forested areas, but females of some species have been captured in plantations, bush and buildings. Adults are principally active during the daylight hours, especially in forest canopy, but some species are active to a lesser degree at night. Females are known to feed on humans and monkeys. The immature stages are usually found in tree holes, but they are sometimes found in bamboo, leaf axils, banana fibre and seed pods, and occasionally in rock holes, snail shells and artificial containers such as tyres, tins and gutters. *Aedes fulgens* is regarded as a potential vector of Chikungunya virus based on laboratory experiments, and Uganda S virus appears to have been isolated from either *Ae. longipalpis* or *Ae. ingrami*.

****

Principal references: Reinert, 1999b (new subgenus, taxonomy, classification, descriptions, distributions, bionomics, keys); Reinert, 2002e (female genitalia); Reinert *et al.*, 2004, 2006, 2008, 2009 (as a genus, morphology, phylogenetic relationships); Wilkerson *et al.*, 2015 (phylogenetic relationships, classification); Huang & Rueda, 2015c (Afrotropical Longipalpis Group, pictorial key to adults of six species).

Aetheapnomyia Harbach & Greenwalt, 2012

Aetheapnomyia Harbach & Greenwalt, 2012, originally proposed and currently recognized as an extinct genus, is typified by the fossil species *Aetheapnomyia hoffeinsorum* (Szadziewski, 1998), original combination: *Aedes* (*Finlaya*) *hoffeinsorum*. Genus abbreviation – *At.*

****

Systematics: *Aetheapnomyia* is a subordinate taxon of the subfamily Culicinae, tribe Aedini. It has no living species. It only includes the extinct type species, which existed during the Eocene Epoch (33.9–56.0 Mya) of the Cenozoic Era. The species is only known from a male preserved in Baltic amber from Poland. Based on the anatomical information and illustrations provided by Szadziewski (1998) and Podėnas (1999), the affinities of *Aetheapnomyia* with other genus groups of Aedini are uncertain.

****

Characteristics: ADULT MALES: Antenna shorter than the proboscis, pedicel enlarged, terminal flagellomere with an apical stylet-like prolongation; maxillary palpus shorter than the proboscis, palpomere 5 greatly reduced, about 0.3 the length of palpomere 4; proboscis noticeably swollen distally; wing with slender scales with convex apices, cell R_2 of the wing much shorter than vein R_{2+3}; fore- and midungues long, unequal, simple, borne well before the apex of tarsomere 5, hindungues short, equal, simple; the genitalia (see Szadziewski, 1998: fig. 3a) bear some resemblance to the genitalia of certain members of the *Ochlerotatus* group of genera, *e.g. Aedes* (*Acartomyia*) *mariae*.

****

Bionomics: The climate during the Eocene Epoch was warm and humid. Forests were abundant in temperate and subtropical areas. The ancestors of most groups of the modern mammals were present. It is possible that females of *At. hoffeinsorum* fed on mammals as well as other vertebrates. Based on the phylogenetic findings of Soghigian *et al.* (2017), it is equally likely that the immature stages of *At. hoffeinsorum* may have developed in either groundwater or phytotelm habitats.

****

Principal references: Szadziewski, 1998 (species description, illustrations, keys for fossil species in Baltic amber and Europe); Podėnas, 1999 (illustrations, key to aedine males in Baltic amber); Harbach & Greenwalt, 2012 (new genus, diagnosis).

Armigeres Theobald, 1901

Armigeres Theobald, 1901a, originally proposed and currently recognized as a genus, is typified by *Armigeres obturbans* (Walker, 1859), original combination: *Culex obturbans*. Genus abbreviation – *Ar.*

****

Systematics: *Armigeres* is a subordinate taxon of the subfamily Culicinae, tribe Aedini. It includes 58 extant species divided between two subgenera: *Armigeres sensu stricto* and *Leicesteria*. Species of *Armigeres* occur in the Oriental, Palaearctic and Australasian Regions. Available evidence indicates that *Armigeres* is a monophyletic taxon that shares uncertain affinities with *Belkinius*, *Eretmapodites*, *Udaya* and *Zeugnomyia*, which exhibited the following relationships in the cladistic analysis of morphological data in the study of Reinert *et al.* (2009): (*Udaya* + (*Belkinius* + *Zeugnomyia*)) + (*Eretmapodites* + *Armigeres*). As is generally true among higher-level taxa of the Culicidae, these relationships are weakly supported, but this is not surprising since to date neither morphological nor molecular data have satisfactorily resolved deeper relationships among culicine genera (Harbach, 2007; Reidenbach *et al.*, 2009). *Armigeres* was recovered as the sister to *Alanstonea* + *Eretmapodites* in the phylogenetic analyses of Wilkerson *et al.* (2015). Six species of *Armigeres* comprised a monophyletic group in the maximum likelihood phylogeny of Soghigian *et al.* (2017) based on seven molecular markers. *Armigeres flavus* of *Leicesteria* was basal to five species of *Armigeres sensu stricto* – *Ar. durhami*, *Ar. kesseli*, *Ar. malayi*, *Ar. obturbans* and *Ar. subalbatus* – included in the analysis.

****

Characteristics: *Armigeres* adults are morphologically similar to species of other aedine generic-level taxa in the Oriental and Australasian Regions but they are generally larger and usually have the proboscis slightly curved downwards and flattened laterally. They are more reliably distinguished from the adults of other aedine genera by the following combination of characters: dorsal scales of the head mostly broad and flat; compound eyes separated ventrally by 2 long rows of scales; acrostichal setae, dorsocentral setae and prespiracular setae absent; postspiracular scales present; postspiracular setae present in species of *Armigeres sensu stricto*, absent in species of *Leicesteria*; mesomeron small with the base more or less in line with the base of the hindcoxa; alula and upper calypter of the wing with hair-like scales on the margin; pulvilli not evident. Larvae are distinguished from all other aedine genera within their range by the absence of a pecten on the siphon.

****

Bionomics: Larvae of species of *Armigeres* are found in small collections of water, particularly those containing foul water or water with a high organic content. They occur in hollow logs, rock holes, tree holes, stump holes, bamboo, *Pandanus* axils, sago palm and banana stumps, fruit shells and husks, fallen leaves and spathes, flower bracts, pitcher plants, artificial containers that contain organic matter and small collections of groundwater. Larvae appear to be at least partially carnivorous. Adults occur primarily in forested and plantation areas and are mainly active during the daytime and crepuscular periods. The females of a number of species readily attack and viciously bite humans. Several species of *Armigeres* are suspected of transmitting *Wuchereria bancrofti*.

****

Principal references: Barraud, 1934 (southern Asia, keys, taxonomy, descriptions, distributions, bionomics); Thurman, 1959 (Thailand, taxonomy, keys, descriptions, distributions); Macdonald, 1960 (*Leicesteria*, taxonomy, keys, descriptions, distributions, bionomics); Belkin, 1962 (South Pacific, taxonomy, keys, descriptions, distributions, bionomics); Delfinado, 1966a (Philippines, taxonomy, keys, descriptions, distributions, bionomics); Steffan, 1968 (Papuan Subregion of the Australasian Region, taxonomy, keys, descriptions, distributions, bionomics); Tanaka *et al.*, 1979 (Japan, genus, subgenus, *Ar. subalbatus*, keys, descriptions, bionomics, distributions); Lu & Li, 1982 (China, keys); Lee *et al.*, 1988a (Australasian Region, taxonomy, keys, literature, distributions, bionomics); Darsie & Pradhan, 1990 (Nepal, keys, distributions, bionomics); Lu *et al.*, 1997a (China, keys, taxonomy, descriptions, distributions, bionomics); Darsie, 2000 (pupae of five species of *Armigeres sensu stricto*, descriptions, key); Reinert, 2002d (female genitalia); Reinert *et al.*, 2004, 2009 (morphology, phylogenetic relationships); Rattanarithikul *et al.*, 2010 (Thailand, keys, bionomics); Wilkerson *et al.*, 2015 (phylogenetic relationships); Soghigian *et al.*, 2017 (phylogenetic relationships); Russell & Burkot, 2023 (Pacific Island countries and territories, species lists, distributions, bionomics, vector status).

Armigeres Theobald, 1901 *sensu stricto*

Armigeres sensu stricto, the nominotypical subgenus of *Armigeres*, is typified by *Armigeres obturbans* (Walker, 1859), original combination: *Culex obturbans*. Subgenus abbreviation – *Arm.*

****

Systematics: *Armigeres sensu stricto* includes 40 extant species with distributions in the Oriental, Palaearctic and Australasian Regions. Only one species, *Ar. subalbatus*, also occurs in the Palaearctic Region (China, Korea and Japan). The phylogenetic studies of Reinert *et al.* (2004, 2009) based on cladistic analysis of morphological data strongly support the sister relationship of *Armigeres sensu stricto* with *Leicesteria*. Five species of *Armigeres sensu stricto*, *Ar. durhami*, *Ar. kesseli*, *Ar. malayi*, *Ar. obturbans* and *Ar. subalbatus*, were recovered in a derived position relative to *Ar. flavus* of *Leicesteria* in the maximum likelihood phylogeny of Soghigian *et al.* (2017) based on seven molecular markers.

****

Characteristics: Species of *Armigeres sensu stricto* exhibit the following distinctions. ADULTS: The adults of species of *Armigeres sensu stricto* are easily distinguished from those of *Leicesteria* by the presence of postspiracular setae and a lower mesepimeral seta. Other distinctions include the following: clypeus bare or with lateral white scales; maxillary palpus of females < 0.33 the length of the proboscis; scutum normal, not compressed nor produced over the head; postpronotum with pale scales posteriorly (no black scales); thoracic pleura with patches of white scales; hindtibia longer or equal in length to the fore- and midtibiae. MALE GENITALIA: Gonostylus scythe-shaped with close-set claws along the inner margin of the apical 0.5 (except in *Ar. treubi*, which has a single claw on a bulbous gonostylus, and in *Ar. theobaldi*, which has 12 claws at the apex as in species of *Leicesteria*). LARVAE: Although larvae of *Armigeres sensu stricto* are easily distinguished from those of other genera, no single character reliably distinguishes the larvae from those of *Leicesteria*. Rattanarithikul *et al.* (2010) used the form of the comb scales to distinguish larvae of the two subgenera in Thailand: comb scales rather elongate, rounded or pointed and evenly fringed at the sides and the apex in species of *Armigeres sensu stricto*. PUPAE: Usually with seta 6-CT as long as or longer than seta 7-CT; seta 3-CT and usually seta 1-CT long and very stout; seta 1-Pa usually present and at least 0.3 the length of the paddle.

****

Bionomics: Females of *Armigeres sensu stricto* lay their eggs singly (Macdonald & Traub, 1960; Steffan, 1968). The eggs resist drying to some extent. The larvae generally inhabit phytotelm habitats, including tree holes, log holes, bamboo stumps, sago palm and banana stumps, flower bracts, fallen coconut, cacao and other fruit shells, *Nepenthes* pitchers, but are often found in artificial containers and small collections of surface water, *e.g.* ground pools, rock pools, drains and septic tanks. The water is generally either very foul or has a high organic content. Larvae are in some cases carnivorous. Adults are usually active in the daytime in heavily shaded areas. Females of a number of species readily attack and viciously bite humans. Females of *Ar. subalbatus* are suspected of transmitting *Wuchereria bancrofti*. Mackerras (1946) investigated the potential of *Ar. breinli* and *Ar. milnensis* for transmitting dengue fever virus. Results were negative for these two species; however, sufficient data were not obtained for *Ar. milnensis* nor possibly for *Ar. breinli* to exonerate them conclusively as vectors of this virus. They should be considered potential vectors of this and other arboviruses. *Armigeres subalbatus* is considered to be a potential vector of Zika virus in China where it has been shown to develop infections of the virus and transmit it to suckling mice (Yang *et al.*, 2022).

****

Principal references: Barraud, 1934 (southern Asia, keys, taxonomy, descriptions, distributions, bionomics); Thurman, 1959 (Thailand, taxonomy, keys, descriptions, distributions); Delfinado, 1966a (Philippines, taxonomy, keys, descriptions, distributions, bionomics); Steffan, 1968 (Papuan Subregion of the Australasian Region, taxonomy, keys, descriptions, distributions, bionomics); Tanaka *et al.*, 1979 (Japan, *Ar. subalbatus*, keys, descriptions, bionomics, distributions); Lee *et al.*, 1988a (Australasian Region, taxonomy, keys, literature, distributions, bionomics); Darsie & Pradhan, 1990 (Nepal, keys, distributions, bionomics); Lu *et al.*, 1997a (China, keys, taxonomy, descriptions, distributions, bionomics); Darsie, 2000 (pupae of five species, descriptions, key); Reinert, 2002d (female genitalia); Reinert *et al.*, 2004, 2009 (morphology, phylogenetic relationships); Rattanarithikul *et al.*, 2010 (Thailand, keys, bionomics); Soghigian *et al.*, 2017 (phylogenetic relationships).

Leicesteria Theobald, 1904

Leicesteria, originally proposed as a genus by Theobald (1904c), currently classified as a subgenus of *Armigeres*, is typified by *Armigeres longipalpis* (Leicester, 1904) (in Theobald, 1904c), original combination: *Leicesteria longipalpis*. Subgenus abbreviation – *Lei.*

****

Systematics: *Leicesteria* includes 18 extant species, which occur almost exclusively in the Oriental Region. A single Oriental species, *Ar. annulipalpis*, is recorded from Ceram Island of eastern Indonesia in the Moluccas west of New Guinea in the Australasian Region. The phylogenetic studies of Reinert *et al.* (2004, 2009) based on cladistic analyses of morphological data strongly supports the sister relationship of *Leicesteria* with *Armigeres sensu stricto*. The evolutionary relationships of the species of *Leicesteria* have not been investigated. However, *Ar. flavus* was recovered in a basal relationship to five species of *Armigeres sensu stricto* in the maximum likelihood phylogeny of Soghigian *et al.* (2017) based on seven molecular markers.

****

Characteristics: Species of *Leicesteria* exhibit the following distinctions. ADULTS: The adults are easily distinguished from those of *Armigeres sensu stricto* by the absence of postspiracular and lower mesepimeral setae (except *Ar. flavus*). Other distinctions include: maxillary palpus of females 0.5–0.8 the length of the proboscis; scutum more or less produced over the head, narrow and laterally compressed in some species; hindtibia longer than the foretibia and equal in length to the midtibia. *Armigeres flavus* is peculiar in having a few scales and minute seta on the mesopostnotum, 1 or 2 lower mesepimeral setae and an unusually short hindtibia. MALE GENITALIA: Gonocoxite with a small basal lobe on the inner side bearing spines or processes; gonostylus with apical claws (except *Ar. cingulatus*). LARVAE: Although *Armigeres* larvae are easily distinguished from those of other genera, no single character reliably distinguishes the larvae of *Leicesteria* from those of *Armigeres sensu stricto*. Rattanarithikul *et al.* (2010) used the form of the comb scales to distinguish larvae of the two subgenera in Thailand: comb scales relatively short, with strong denticles and frequently with a stout median spine. PUPAE: Seta 7-CT at least 1.6 times as long as seta 6-CT; setae 1–3-CT usually slender; seta 1-Pa usually absent, if present it is no longer than the paddle.

****

Bionomics: The oviposition strategy of species of *Leicesteria* was thoroughly discussed by Macdonald (1960). Females attach the eggs in mass to their hindlegs before depositing them on the water. The eggs apparently do not resist drying. Larvae inhabit the internodes of dead and living bamboo in lowland and dipterocarp forests, but are absent from primary forest (Macdonald, 1960; Macdonald & Traub, 1960). Adults occur primarily in forested or plantation areas and are mainly active during the daytime and crepuscular periods. Females of a number of species readily attack and viciously bite humans. None of the species of *Leicesteria* are known to transmit pathogens.

****

Principal references: Barraud, 1934 (southern Asia, keys, taxonomy, descriptions, distributions, bionomics); Thurman, 1959 (Thailand, taxonomy, keys, descriptions, distributions); Macdonald, 1960 (taxonomy, keys, descriptions, distributions, bionomics); Delfinado, 1966a (Philippines, taxonomy, keys, descriptions, distributions, bionomics); Lee *et al.*, 1988a (Australasian Region, taxonomy, keys, literature, distributions, bionomics); Darsie & Pradhan, 1990 (Nepal, keys, distributions, bionomics); Lu *et al.*, 1997a (China, keys, taxonomy, descriptions, distributions, bionomics); Reinert, 2002d (female genitalia); Reinert *et al.*, 2004, 2009 (morphology, phylogenetic relationships); Rattanarithikul *et al.*, 2010 (Thailand, keys, bionomics); Soghigian *et al.*, 2017 (phylogenetic relationships).

Coquillettidia Dyar, 1905

Coquillettidia Dyar 1905a, originally proposed and currently recognized as a genus, is typified by *Coquillettidia perturbans* (Walker, 1856); original combination: *Culex perturbans*. Genus abbreviation – *Cq.*

****

Systematics: *Coquillettidia* is a subordinate taxon of the subfamily Culicinae, tribe Mansoniini. It includes 58 extant species and five extinct species, two preserved in Baltic amber and three preserved as compression/impression fossils. The extant species are classified in three subgenera: *Austromansonia*, *Coquillettidia sensu stricto* and *Pseudotaeniorhynchus*. The single species of *Austromansonia* occurs in New Zealand, species of *Pseudotaeniorhynchus* are confined to the Neotropical Region and species of *Coquillettidia sensu stricto* are found mainly in the Afrotropical Region but some occur in the Oriental and Australasian Regions, one occurs in North America and two occur in the Palaearctic Region. Morphological and molecular data strongly support the sister relationship of *Coquillettidia* + *Mansonia* (Harbach & Kitching, 1998; da Silva *et al.*, 2020a; do Nascimento *et al.*, 2021; Lorenz *et al.*, 2021; Soghigian *et al.*, 2023). The phylogeny of the subgenera and species has not been investigated.

****

Characteristics: *Coquillettidia* are generally rather large, yellowish mosquitoes that are similar to species of *Mansonia* and resemble some species of *Culex* and Aedini. The ungues are simple and the abdomen is truncate in females (distinction from Aedini), pulvilli are not evident (distinction from *Culex*) and the dorsal wing scales are usually narrow and unicolourous, rarely broad and mixed dark and pale but never markedly asymmetrical (distinction from *Mansonia*). Furthermore, species of *Coquillettidia sensu stricto* and *Austromansonia* are distinguished from aedine species and species of *Mansonia* by the absence of postspiracular setae, and Neotropical species (*Pseudotaeniorhynchus*) are distinct in having a conspicuous preapical white band or spot on the anterior surface of the femora. The larvae of *Coquillettidia* and *Mansonia* differ from almost all other mosquito larvae in having the spiracular apparatus and siphon uniquely developed for piercing and securing air from plant tissues. Larvae of *Coquillettidia* have the distal part of the antenna elongate (usually as long as the basal part) and articulated with the basal part.

****

Bionomics: Larvae of species of *Coquillettidia* attach to aquatic plants to obtain oxygen from air cells for respiration. A variety of plants are used, particularly grasses. Larvae detach and re-attach to host plants quite readily. The females of several species readily attack humans. Both nocturnal and diurnal biters are known. Some species are notorious pests of humans and domestic animals in Africa, Europe and North America. Several species of *Coquillettidia sensu stricto* and *Pseudotaeniorhynchus* are natural vectors of various arboviruses. Most importantly, *Cq. perturbans* is a vector of Eastern equine encephalitis virus in North America.

****

Principal references: Lane, 1953 (as *Rhynchotaenia*, as a subgenus of *Mansonia*, keys, taxonomy, distributions); Belkin, 1962 (South Pacific, as a subgenus of *Mansonia*, taxonomy, keys, descriptions, distributions, bionomics); Cova Garcia *et al.*, 1966 (Venezuela, keys, species descriptions, distributions); Delfinado, 1966a (Philippines, *Coquillettidia sensu stricto*, keys, taxonomy, bionomics, distributions); Belkin, 1968a (New Zealand, *Austromansonia* and *Coquillettidia*, taxonomy, keys, descriptions, bionomics); Belkin *et al.*, 1970 (Jamaica, *Cq. nigricans*, taxonomy, description, bionomics, distribution); Tanaka *et al.*, 1979 (Japan, as a subgenus of *Mansonia*, keys, taxonomy, descriptions, bionomics, distributions); Lee *et al.*, 1988a (Australasian Region, keys, taxonomy, literature, bionomics, distributions); Service, 1990 (Afrotropical Region, keys, taxonomy, literature, bionomics, distributions); Lu *et al.*, 1997a (China, keys, taxonomy, descriptions, distributions, bionomics); Rattanarithikul *et al.*, 2006a (Thailand, keys, bionomics); Nugroho *et al.*, 2020 (Indonesia, list of species, taxonomic notes, key to adult females); da Silva *et al.*, 2020a (phylogenetic relationships); do Nascimento *et al.*, 2021 (mitochondrial protein-coding genes, phylogenetic relationships); Lorenz *et al.*, 2021 (mitochondrial protein-coding genes, phylogenetic relationships); Soghigian *et al.*, 2023 (molecular markers, phylogenetic relationships).

Austromansonia Belkin, 1968

Austromansonia Belkin, 1968a, originally proposed and currently recognized as a subgenus of *Coquillettidia*, is typified by *Coquillettidia tenuipalpis* (Edwards, 1924b); original combination: *Taeniorhynchus* (*Coquillettidia*) *tenuipalpis*. Subgenus abbreviation – *Aus.*

****

Systematics: *Austromansonia* is monobasic. It only includes the extant type species, which is native to New Zealand. Belkin (1968a) noted that '*Austromansonia* seems to be the most primitive and generalized phylad of *Coquillettidia* and of the entire tribe Mansoniini'. The evolutionary relationships of *Austromansonia* with the other subgenera of *Coquillettidia* have not been explicitly investigated.

****

Characteristics: The following features distinguish *Austromansonia* from the other groups of *Coquillettidia*. ADULTS: Antennal flagellum of males with flagellomeres 12 and 13 together distinctly longer than the remainder of the flagellum; maxillary palpus of males with palpomeres 4 and 5 slightly upturned, with sparse setae; mesokatepisternum and the base of the prealar area with a continuous patch of scales; subcosta of the wing with a large patch of setae on the lower surface; ungues of all legs with long fine spicules at the base of the outer surface. MALE GENITALIA: Gonocoxite without a mesal membrane and lobes; aedeagus long. LARVAE: Unknown. PUPAE: Paddle with only the inner margin produced at the apex; seta 2-VII present; seta 3-VII absent, frequently represented by an alveolus; seta 9-VIII inserted at the apex of the caudolateral angle of the segment; seta 1-IX absent.

****

Bionomics: The immature stages of *Cq. tenuipalpis* are known to occur in the shallow margins of ponds and small lakes choked with various types of vegetation, the roots and stems of which provide a source of attachment and respiration for the larvae and pupae. Adults rest in vegetation found in and around the larval habitats and adjacent banks. Females have been collected attempting to bite humans in the vicinity of the aquatic habitats, but few in number (Belkin, 1968a). The species is not known to be of medical or economic importance.

****

Principal references: Belkin, 1968a (New Zealand, taxonomy, keys, descriptions, bionomics); Lee *et al.*, 1988a (New Zealand, in key, taxonomy, literature, description).

Coquillettidia Dyar, 1905 *sensu stricto*

Coquillettidia sensu stricto, the nominotypical subgenus of the genus *Coquillettidia*, is typified by *Coquillettidia perturbans* (Walker, 1856); original combination: *Culex perturbans*. Subgenus abbreviation – *Coq.*

****

Systematics: *Coquillettidia sensu stricto* includes 44 extant species. Except for *Cq. perturbans*, which occurs in North America, species of *Coquillettidia sensu stricto* are confined to the Old World. Most species are found in the Afrotropical Region, but some occur in the Oriental and Australasian Regions. Two species occur in the Palaearctic Region. Belkin (1962) noted that *Coquillettidia* (as a subgenus of *Mansonia*) appears to be closely related to *Pseudotaeniorhynchus*, but its evolutionary relationships with the other subgenera of *Coquillettidia* have not been explicitly investigated.

****

Characteristics: Species of *Coquillettidia sensu stricto* exhibit the following anatomical features. ADULTS: Compound eyes narrowly separated above the antennal pedicels, interocular space with or without scales; maxillary palpus of females with a minute apical palpomere, without scales, maxillary palpus of males as long as or longer than the proboscis, apex of palpomere 3 and all of palpomere 4 somewhat broadened, palpomere 5 well developed, setose, not upturned; proboscis with or without pale scales, longer than the forefemur; scutal scales of 1 or 2 colours; acrostichal setae in a more or less complete line; scutellar scales often sparse; antepronotum without scales; postspiracular setae absent; mesokatepisternum with scales largely restricted to the lower area or, if numerous, on the upper part not continuous with the prealar sales; dorsal scales of the wing usually all narrow and unicolourous, rarely with a mixture of broader dark and pale scales; legs with or without conspicuous pale markings, tarsi with or without pale bands; scales of abdominal tergum I confined to a small central area. FEMALE GENITALIA: Abdominal segment VIII largely exposed, sternum without a lateral flap-like lobe; cercus rounded, without an upturned apical process; postgenital lobe usually more or less rounded distally, at most slightly emarginate apically; 3 spermathecal capsules. MALE

GENITALIA: Ninth tergal lobes more or less distinct, with few to many setae; gonocoxite more or less cylindrical or conical, mesal membrane present or absent; gonostylus with an apical or subapical spiniform gonostylar claw; claspette short, poorly developed, with 1 or more elongate apical spiniform setae; aedeagus variable, with or without spicules. LARVAE: Hypostomal suture usually interrupted in the anterior 0.5; antenna with the distal part usually as long as the proximal part, flexible, annulated, jointed and movable along a peculiar diagonal ventral connection to the basal part; setae 2,3-A much shorter than the distal part; seta 4-III,IV inserted laterad of seta 3; comb usually with at least 5 spine-like scales, scales sometimes with basolateral denticles and spicules; setae 3,4,5-VIII always in a compact group; seta 1-S usually on a distal lobe of the siphon; ventral brush (seta 4-X) with 4 or 5 pairs of setae on a grid, some precratal setae sometimes also present on the ventral midline of the saddle; accessory setae (x,y,z-X) sometimes present on the saddle. PUPAE: Trumpet with the apex produced into a narrow poorly sclerotized process surrounded by a 'cellular' lobe on 2 sides (usually broken off in pupae detached from plant parts); all cephalothoracic setae unbranched; seta 9-VIII inserted on the dorsal surface of the segment, removed from the caudolateral angle; all setae very small and single; seta 3-III–VI absent; seta 11-II–VII usually absent, an alveolus occasionally present; paddle long and narrow.

****

Bionomics: The bionomics of most species of *Coquillettidia sensu stricto* are poorly known. The immature stages are chiefly found among the roots of various plants, particularly grasses, in the shallow water of marshes, swamps, ponds, lakes and seepage. Adults may be active during the day or night. Some species are notorious pests of humans and domestic animals in Africa, Europe and North America. *Coquillettidia perturbans* is a vector of Eastern equine encephalitis and Tensaw viruses in North America (Turell *et al.*, 2005; White & Faust, 2014, respectively); *Cq. crassipes* is a secondary vector of *Brugia malayi* (Wharton, 1962). *Coquillettidia aurites*, *Cq. metallica* and *Cq. pseudoconopas* appear to be important vectors of bird malaria in Africa (Njabo *et al.*, 2009). White & Faust (2014) list *Cq. fuscopennata* as a natural vector of Chikungunya and Sindbis viruses, *Cq. metallica* as a natural vector of West Nile virus and *Cq. fuscopennata*, *Cq. microannulata* and *Cq. versicolor* as natural vectors of Rift Valley fever virus in Africa. They also list *Cq. richiardii* as a natural vector of Calovo virus in Europe.

****

Principal references: Belkin, 1962 (South Pacific, as a subgenus of *Mansonia*, taxonomy, keys, descriptions, distributions, bionomics); Delfinado, 1966a (Philippines, taxonomy, keys, descriptions, distributions, bionomics); Belkin, 1968a (New Zealand, *Cq. iracunda*, description, bionomics, distribution); Tanaka *et al.*, 1979 (Japan, as a subgenus of *Mansonia*, *Cq. crassipes* and *Cq. ochracea*, keys, taxonomy, descriptions, bionomics, distributions); Lee *et al.*, 1988a (Australasian Region, keys, taxonomy, literature, bionomics, distributions); Service, 1990 (Afrotropical Region, as a subgenus of *Mansonia*, keys, taxonomy, bionomics, distributions); Rattanarithikul *et al.*, 2006a (Thailand, keys, bionomics); Nugroho *et al.*, 2020 (Indonesia, list of species, taxonomic notes, key to adult females); Russell & Burkot, 2023 (South Pacific, country lists of species, vector status).

Pseudotaeniorhynchus Theobald, 1911

Pseudotaeniorhynchus, originally proposed as a genus by Theobald (1911a), currently classified as a subgenus of *Coquillettidia*, is typified by *Coquillettidia fasciolata* (Lynch Arribálzaga, 1891b); original combination: *Taeniorhynchus fasciolatus*. Subgenus abbreviation – *Pst.*

****

Systematics: *Pseudotaeniorhynchus* includes 13 extant species. Most species of *Pseudotaeniorhynchus* are recorded from countries in South America, from the countries bordering the Caribbean southward into northern Argentina and Uruguay (there are no records for Chile). Four of the South American species, *Cq. arribalzagae*, *Cq. fasciolata*, *Cq. nigricans* and *Cq. venezuelensis*, also occur in Central America. The distributions of *Cq. nigricans* and *Cq. venezuelensis* extend into Mexico, and the former is recorded from Jamaica. Belkin (1962) indicated that the adults of species of *Pseudotaeniorhynchus* (as *Rhynchotaenia*) appear to be more primitive than the adults of *Coquillettidia sensu stricto* (both as subgenera of *Mansonia*), and seem to be intermediate between *Coquillettidia* and *Mansonia*. The evolutionary relationships of *Pseudotaeniorhynchus* have not been explicitly investigated.

****

Characteristics: Species of *Pseudotaeniorhynchus* exhibit the following anatomical features. ADULTS: Antennal flagellum approximately as long as the proboscis; maxillary palpus of males longer than the proboscis, palpomeres 4 and 5 strongly setose; proboscis as long as or longer than the forefemur, with a pale ring beyond mid-length; scutum with well-developed acrostichal, dorsocentral and supraalar setae; postpronotum with scales; thoracic pleura with scales on the mesokatepisternum and upper mesepimeron; wing mostly dark-scaled; femora with a conspicuous preapical pale band. MALE GENITALIA: Ninth tergal lobes with well-developed setae; gonocoxite with a basal dorsomesal lobe bearing 1 or 2 (usually 1) long strong setae; paraproct with several strong apical teeth. LARVAE: Known for only a few species; chaetotaxy of the known species has not been studied in detail; the part of the antenna borne distal to setae 2,3-A much longer than the proximal part; setae 2,3-A short; accessory seta x-X present on the saddle in addition to seta 1-X. PUPAE: Known for few species; not studied in detail. Seta 6-I–VI absent.

****

Bionomics: Little is known about the bionomics of species of *Pseudotaeniorhynchus*. Larvae have been found in floating and emergent vegetation, such as grasses, sedges and *Pistia*, in swamps and lakes. Females of *Cq. arribalzagae*, *Cq. fasciolata* and *Cq. nigricans* have been collected landing on and biting humans who enter their haunts during daylight and evening hours (*e.g.* Heinemann & Belkin, 1977b; Heinemann *et al.*, 1980). Clark-Gil & Darsie (1983) reported that females of *Cq. venezuelensis* were attracted to a light trap in forest and were captured when attracted to humans in dwelling and forest areas during dusk. *Coquillettidia venezuelensis* is a natural vector of Bussuquara, Catu, Cotia, Guama, Mayaro, Murutucu and Oropouche viruses, and *Cq. arribalzagae* is a natural vector of Wyeomyia virus (White & Faust, 2014).

****

Principal references: Lane, 1953 (as *Rhynchotaenia*, Neotropical Region, taxonomy, keys, descriptions, distributions); Cova-Garcia *et al.*, 1966 (as *Rhynchotaenia*, Venezuela, species list, literature); Belkin *et al.*, 1970 (Jamaica, *Cq. nigricans*, in keys, taxonomy, bionomics, distribution).

Culex Linnaeus, 1758

Culex Linnaeus, 1758, originally established and currently recognized as a genus, is typified by *Culex pipiens* Linnaeus, 1758. Genus abbreviation – *Cx.*

****

Systematics: *Culex* is a subordinate taxon of the subfamily Culicinae, tribe Culicini. It is a large and important genus of mosquitoes. It includes 819 extant species. Seven species are without subgeneric placement; the remainder are divided between 28 subgenera: *Acalleomyia*, *Acallyntrum*, *Aedinus*, *Afroculex*, *Allimanta*, *Anoedioporpa*, *Barraudius*, *Belkinomyia*, *Carrollia*, *Culex sensu stricto*, *Culiciomyia*, *Eumelanomyia*, *Kitzmilleria*, *Lasioconops*, *Lasiosiphon*, *Lophoceraomyia*, *Maillotia*, *Melanoconion*, *Micraedes*, *Microculex*, *Neoculex*, *Nicaromyia*, *Oculeomyia*, *Phalangomyia*, *Phenacomyia*, *Phytotelmatomyia*, *Sirivanakarnius* and *Tinolestes*. The larger subgenera are arranged in informal classifications that variously include Sections, Series, Groups, Subgroups and Complexes. *Culex* also includes seven extinct species, four extant species, including the type species, that are known from fossils and one fossil is recognized as a *nomen nudum*. Species of *Culex* occur in all zoogeographical regions. They range from the tropics to cool temperate regions, but do not extend to the extreme northern latitudes where only *Aedes* occur.

Belkin (1962) indicated that the affinities of the subgenera of *Culex* were obscure, but surmised that *Lutzia* (as a subgenus of *Culex*) is an ancient derivative that appears to have a strong affinity with *Culex sensu stricto*. It is generally assumed that the Melanoconion Group of subgenera, *i.e. Aedinus*, *Anoedioporpa*, *Belkinomyia*, *Carrollia*, *Melanoconion*, *Micraedes*, *Microculex* and *Tinolestes*, includes the most derived mosquitoes of *Culex* (Belkin, 1968a; Berlin, 1969b; Adames & Galindo, 1973; Sirivanakarn, 1983). Sirivanakarn (1983) noted that *Melanoconion* shares several anatomical features with the 'more primitive' and widespread *Neoculex*, suggesting that an offshoot of *Neoculex* may have given rise to the Melanoconion Group. Valencia (1973) suggested that *Carrollia* and the closely related *Deinocerites* and *Galindomyia* share a similar evolutionary history as derivatives from a primitive stock of *Melanoconion*. Miller *et al.* (1996) used sequence divergence in the ITS1 and ITS2 regions of rDNA to infer relationships between 14 species representing four subgenera. Neighbour-joining analyses produced a tree (rooted on *Melanoconion*) that supported the notion that *Melanoconion* bears a number of resemblances with *Neoculex*, and *Lutzia* shares an ancestry with *Culex sensu stricto*. Navarro & Liria (2000) conducted a parsimony analysis of 30 characters of the larval mandibles and maxillae to infer the phylogenetic relationships of *Deinocerites* and seven subgenera of *Culex*. The phylogeny indicated that *Lutzia* and *Culex* formed distinct monophyletic clades, with *Lutzia* being the more primitive of the two. Although only New World species were included in the analysis, members of the Melanoconion Group formed a distinct monophyletic clade in a sister-group relationship with *Culex sensu stricto*. It is interesting to note that this clade included *Deinocerites* in an unresolved polytomy with *Anoedioporpa*, *Melanoconion* and *Microculex* in a derived position relative to *Carrollia*. Vesgueiro *et al.* (2011) used rDNA ITS2 sequences and neighbour-joining to infer the relationships of 17 Neotropical species representing *Lutzia* (one species) and *Culex sensu stricto* (nine species), *Microculex* (one species), *Melanoconion* (five species) and *Phenacomyia* (one species). In a companion study, Demari-Silva *et al.* (2011) used a 478-bp fragment of the *COI* gene and Bayesian methodology to assess the relationships of the same 17 species (and a species of *Carrollia*). The species of *Lutzia* was placed among species of *Culex* in both studies, and a similar result was obtained in the phylogenetic assessment of mitochondrial genomes conducted by Chen *et al.* (2024). In the neighbour-joining tree of Vesgueiro *et al.*, *Lutzia* and *Phenacomyia* were placed in a clade among species of *Culex sensu stricto*. In the Bayesian analysis of Demari-Silva *et al.*, *Lutzia* was recovered as the sister of a clade comprised of the subgenera *Culex* and *Phenacomyia*. Harbach *et al.* (2012) explored relationships among taxa of Culicini using 169 morphological characters from 86 exemplar species representing the four genera and 26 subgenera of Culicini, most species groups and subgroups of *Culex sensu stricto* and an outgroup of five species from five other tribes. As in previous phylogenetic studies based on morphology (Navarro & Liria,

2000; St John, 2007), *Lutzia* was recovered as a clade separate from *Culex*. However, *Culex* was not recovered as monophyletic because it included *Deinocerites* and *Galindomyia* as derived members of the New World subgenera. With the exception of *Culex sensu stricto*, *Eumelanomyia* and *Neoculex*, there was strong support for the monophyly of all subgenera. They noted that *Culex sensu stricto* would be monophyletic if it did not include the Afrotropical *Cx. duttoni*, the Neotropical *Cx. apicinus* and the Australasian Atriceps Group. Based on a series of comparative and data exclusion analyses of the data set of Harbach *et al.* (2012) and a critical examination of the results of other phylogenetic studies, Kitching *et al.* (2015) concluded that no morphological or molecular data set analysed in any previously published study had adequate signal to place *Lutzia* unequivocally with regard to other taxa in Culicini. Harbach *et al.* (2017a) critically analysed a morphological data set comprised of 286 characters of the larval, pupal and adult stages of 150 species of *Culex sensu stricto* and an outgroup of 17 species representing *Lutzia*, 11 other subgenera of *Culex* and the tribes Aedini, Culisetini and Sabethini. Unlike the results of previous analyses of morphological data that placed *Lutzia* outside *Culex* (Harbach *et al.*, 2012; Navarro & Liria, 2000; St John, 2007; Laurito & Almirón, 2013), the more stringent assessment of clade support applied in their study indicated, in agreement with Kitching *et al.* (2015), that the placement of *Lutzia* relative to *Culex* based on morphology is highly equivocal. In stark contrast to the molecular studies mentioned above, the phylogenetic analysis of Sun *et al.* (2019) based on the nucleotide sequences of 13 protein-coding genes obtained from the complete mitochondrial genomes of 10 species of *Culex* and two species of *Lutzia* provided strong support for recognizing *Lutzia* as the sister of *Culex*.

****

Characteristics: *Culex* adults are usually drab, unicolourous mosquitoes, but some species of *Culex sensu stricto* have markings on the legs and pale spots on the wings similar to *Anopheles*. *Culex* are characterized by the presence of distinct pulvilli and the absence of prespiracular and postspiracular setae (the latter are present in the Australian *Cx. postspiraculosus*). These characters, however, will not distinguish *Culex* from *Deinocerites*, *Galindomyia* and *Lutzia*. Species of *Culex* differ from those of *Deinocerites* and *Galindomyia* in having the apical flagellomere of the antenna much shorter than the first flagellomere, and differ from the species of *Lutzia* in having few (usually one) lower mesepimeral setae. *Culex* larvae are distinguished from the larvae of other genera by the following characters: seta 2-C usually absent; seta 3-C located on the dorsal side of the head, sometimes absent; palatal brushes normal, not developed for grasping prey; mandible normal, without a lateral lobe at the base; maxillary brush present, well developed; seta 12-I and comb always present; siphon with 3 or more pairs of prominent setae (seta 1-S); pecten normally present; saddle usually complete, sometimes incomplete and greatly reduced but never divided into dorsal and ventral sclerites or longer than the siphon; ventral brush (seta 4-X) usually with 3 or more pairs of setae.

****

Bionomics: *Culex* larvae occur primarily in semi-permanent or permanent bodies of groundwater, but a large number of species live exclusively in leaf axils, tree holes, rock holes and crab holes. Some utilize artificial containers as well as the normal groundwater habitats. A few species, including the filarial vector *Cx. quinquefasciatus*, are found in organically polluted waters. *Culex* females mainly bite at night. Many species attack humans and other mammals. Females of some species are known to feed on birds and some are known to feed on amphibians and reptiles. Several species of *Culex sensu stricto* and *Melanoconion* are of medical importance.

****

Principal references: Barraud, 1934 (southern Asia, keys, taxonomy, descriptions, distributions, bionomics); Edwards, 1941 (Afrotropical Region, keys, descriptions of adults and pupae, distributions); Hopkins, 1952 (Afrotropical Region, keys, descriptions of larvae, distributions, bionomics); Lane, 1953 (Neotropical Region, taxonomy, keys, descriptions, distributions); Mattingly & Knight, 1956 (Arabia, taxonomic notes, keys for adults and larvae, distributions); Belkin, 1962 (South Pacific, taxonomy, keys, descriptions, distributions, bionomics); Colless, 1965 (*Lophoceraomyia*, Peninsular Malaysia and Singapore, taxonomy, keys, description, bionomics, distributions); DuBose & Curtin, 1965 (Mediterranean area, keys); Forattini, 1965a (Neotropical Region, diagnosis, classification, species, bionomics, distributions); Cova-Garcia *et al.*, 1966 (Venezuela, keys, species descriptions, distributions); Delfinado, 1966a (Philippines, taxonomy, keys, descriptions, distributions, bionomics); Bram, 1967a (Thailand, keys, descriptions, distributions, bionomics); Bram, 1967b (*Culex sensu stricto*, New World, taxonomy, keys, descriptions, distributions); Belkin, 1968a (New Zealand, taxonomy, keys, descriptions, bionomics); Sirivanakarn, 1968 (*Lophoceraomyia*, taxonomy, classification, keys, descriptions, bionomics, distributions); Belkin *et al.*, 1970 (Jamaica, taxonomy, keys, descriptions, bionomics); Sirivanakarn, 1971, 1972, 1976, 1977a (various subgenera, Oriental Region, taxonomy, classification, keys, descriptions, bionomics, distributions); Valencia, 1973 (*Carrollia*, taxonomy, classification, keys, descriptions, bionomics, distributions); Gutsevich *et al.*, 1974 (former USSR, taxonomy, keys, descriptions, distributions, bionomics); Tanaka *et al.*, 1979 (Japan, taxonomy, keys, descriptions, distributions, bionomics); Wood *et al.*, 1979 (Canada, taxonomy, keys, descriptions, distributions, bionomics); Berlin & Belkin, 1980 (Neotropical Region, *Aedinus*, *Anoedioporpa*, *Tinolestes*, keys, descriptions, bionomics, distributions); Ribeiro & da Cunha Ramos, 1980 (Angola, keys, taxonomic notes, distributions, bionomics); Darsie & Ward, 1981, 2005 (North America, keys); Lu & Li, 1982 (China, keys); Clark-Gil & Darsie, 1983 (Guatemala, keys, distributions, bionomics); Sirivanakarn, 1983 (*Melanoconion*, classification, keys, descriptions, bionomics, distributions); Darsie, 1985 (Argentina, keys); Lee *et al.*, 1988b, 1989a, 1989b (Australasian Region, taxonomy, keys, literature, distributions, bionomics); Harbach, 1988 (southwestern Asia and Egypt, *Culex sensu stricto*, classification, taxonomy, keys, descriptions, bionomics, distributions); Darsie & Pradhan, 1990 (Nepal, keys, distributions, bionomics); Harbach & Peyton, 1992b (*Phenacomyia*, diagnosis, comparative morphology, keys); Pecor *et al.*,

1992 (*Melanoconion*, catalogue); Rattanarithikul & Panthusiri, 1994 (Thailand, keys, medically important species); Reuben *et al.*, 1994 (Southeast Asia, keys, Japanese encephalitis vectors); Sallum & Forattini, 1996 (*Melanoconion*, Spissipes Section, classification, keys, descriptions, distributions, bionomics); Lu *et al.*, 1997a (China, keys, taxonomy, descriptions, distributions, bionomics); González Broche & Rodríguez R., 2001 (*Nicaromyia*, diagnosis, description, distribution, bionomics); Tanaka, 2004 (*Oculeomyia* and *Sirivanakarnius*, pupae, taxonomy, descriptions, distributions, keys); Rattanarithikul *et al.*, 2005a, 2005b (Thailand, keys, distributions, bionomics); Rossi & Harbach, 2008 (*Phytotelmatomyia*, diagnosis, comparative morphology, distribution, bionomics, phylogenetic relationships); Harbach, 2011 (informal classification); Vesgueiro *et al.*, 2011 (rDNA ITS2 sequences, phylogenetic relationships); Demari-Silva *et al.*, 2011 (mtDNA *COI* sequences, phylogenetic relationships); Harbach *et al.*, 2012 (morphology, phylogenetic relationships); Harbach *et al.*, 2017a (*Culex sensu stricto*, morphology, phylogenetic relationships); Simsaa *et al.*, 2021 (Sudan, key to adults); González *et al.*, 2023 (Neotropical subgenera, comparative morphology, molecular associations); Russell & Burkot, 2023 (Pacific Island countries and territories, species lists, distributions, bionomics, vector status); Chen *et al.*, 2024 (mitochondrial protein sequences, phylogenetic relationships).

Acalleomyia Leicester, 1908

Acalleomyia, originally proposed as a genus by Leicester (1908), currently classified as a subgenus *Culex*, is typified by *Culex obscurus* (Leicester, 1908), original combination: *Acalleomyia obscurus*. Subgenus abbreviation – *Aca*.

****

Systematics: *Acalleomyia* is monobasic. It only includes the extant type species, which is known from localities in the Indomalayan Subregion of the Oriental Region. Belkin (1962) and Bram (1968) suggested a relationship between *Acalleomyia*, *Acallyntrum* and *Culiciomyia* based on similarities observed in larvae. While the study of St John (2007) suggested a sister relationship with *Acallyntrum*, the morphology-based phylogenetic analyses of Harbach *et al.* (2012) showed that *Acalleomyia* has a closer affinity with *Culiciomyia* than it does with *Acallyntrum*.

****

Characteristics: Species of *Acalleomyia* exhibit the following anatomical features. ADULTS: Small species without distinctive ornamentation; antenna of males sparsely verticillate; maxillary palpus of males short, 0.15 the length of the proboscis (distinction from other genus groups of Culicini in the Oriental Region); acrostichal setae absent; scutellum with broad flat scales on the midlobe; thoracic pleura without noticeable scale-patches, with prominent setae on the prealar area, a single prominent seta posteriorly on the mesokatepisternum and a single seta on the lower area of the mesepimeron; hindfemur mainly pale with an apical dark band that extends proximally on the dorsal surface; tarsi and abdominal terga entirely dark-scaled. MALE GENITALIA: Gonocoxite with scales and a unique basolateral tubercle bearing 2 large rod-like sensory structures; subapical lobe at the extreme apex of the gonocoxite, bearing 2 leaf-like setae in addition to 3 rod-like setae and several slender setae; gonostylus with an expanded distal part bearing a dorsal recurved spine; phallosome similar to the phallosome of species of *Culiciomyia*, lateral plate with a basal tooth and lateral denticles; paraproct with a short basal sternal process and a crown of fine dorsomesal spicules and ventrolateral blades; cercal setae present. LARVAE: Palatal brushes normal; seta 1-A large, inserted near mid-length of the antenna; seta 1-C exceptionally short, fine; seta 4-C single; setae 5–7-C with multiple branches, inserted in a more or less straight diagonal line; setae 1–3-P relatively short, single, seta 3-P considerably shorter than setae 1,2-P; seta 8-P very short, single; comb comprised of numerous fan-shaped scales in a patch; pecten on the proximal 0.33 of the siphon, spines uniquely developed with 3–5 large basal denticles and a series of fine distal spicules on one side; 3 pairs of posterolateral seta 1-S inserted in a row distad of the pecten; saddle incomplete; ventral brush (seta 4-X) comprised of 4 pairs of short setae inserted on a grid. PUPAE: Similar to pupae of the species of *Culiciomyia*.

****

Bionomics: Species of *Acalleomyia* occur in coastal brackish water environments. Larvae have been found in the axils and stumps of *Nipa* palms and a hole in a fallen mangrove tree. Females will apparently bite humans in the vicinity of larval habitats, but *Cx. obscurus* is not a medically important species.

****

Principal references: Bram, 1968 (*Cx. obscurus*, description, distribution, bionomics); Lien *et al.*, 1975 (*Cx. obscurus*, distribution); Harbach *et al.*, 2012 (morphology, phylogenetic relationships).

Acallyntrum Stone & Penn, 1948

Acallyntrum Stone & Penn, 1948, originally proposed and currently recognized as a subgenus of *Culex*, is typified by *Culex perkinsi* Stone & Penn, 1948. Subgenus abbreviation – *Acl.*

****

Systematics: *Acallyntrum* includes eight extant species in tropical areas of the Australasian Region. *Acallyntrum* is divided into two groups, the Bicki and Perkinsi Groups. Belkin (1962) noted that *Acallyntrum* appears to be most closely related to *Culiciomyia* based on overall morphological similarity, and hypothesized that these taxa may have evolved from a common ancestor. The study of St John (2007) suggested a sister relationship with *Acalleomyia* whereas the more detailed study of Harbach *et al.* (2012) showed that *Acallyntrum* has a closer affinity with *Culiciomyia*.

****

Characteristics: Species of *Acallyntrum* exhibit the following anatomical features. ADULTS: Rather small mosquitoes; vertex of the head with a large central area of narrow curved scales, broad flat scales laterally but not bordering the margins of the eyes above the lower edge of the central patch of narrow scales; antenna of males unmodified; maxillary palpus of females not more than 0.25 the length of the proboscis; palpomere 3 of males with a ventral row of specialized scales, palpomeres 4 and 5 setose; proboscis of males with some long setae on the ventral surface; acrostichal setae strongly developed; scutellar scales narrow; lower mesepimeral seta present; thoracic pleura without scale-patches; scales of the wing narrow, dark; hindfemur almost completely white-scaled; tarsi dark; pale markings of the terga basal if present. MALE GENITALIA: Gonocoxite without scales; subapical lobe with setae *a–c* differentiated, some setae in the group *d–f* specialized; gonostylus simple, tapered apically, with 1 or 2 small setae on the inner surface, 1 or 2 terminal setae short, broad; phallosome simple, lateral plates joined by a ventral bridge; paraproct with a crown of variable small subequal spicules in a single row, basal sternal process absent; cercal setae absent. LARVAE: Palatal brushes normal; seta 1-A large, inserted near mid-length of the antenna; seta 4-C branched; setae 5,6-C inserted on the anterior 0.5 of the head capsule, seta 5-C multi-branched; setae 6,7-C with 3 or more branches; siphon moderately long, with 3 pairs of long posterolateral seta 1-S usually inserted distal to the pecten; saddle incomplete, with distinct caudal marginal spicules; ventral brush (seta 4-X) comprised of 1 pair of short setae. PUPAE: Seta 1-CT much longer than setae 2,3-CT; seta 1-Pa present or absent, seta 2-Pa absent.

****

Bionomics: The immature stages of species of *Acallyntrum* have been found in leaf axils of *Pandanus* (principally), taro and sago palm. Adult females are not known to bite humans. The species are of no known medical importance.

****

Principal references: Belkin, 1962 (South Pacific, taxonomy, keys, descriptions, distributions, bionomics); Lee *et al.*, 1989a (Australasian Region, taxonomy, key, literature, distributions, bionomics); Harbach *et al.*, 2012 (morphology, phylogenetic relationships).

Aedinus Lutz, 1904

Aedinus, originally proposed as a genus by Lutz, 1904 (in Bourroul, 1904), currently classified as a subgenus of *Culex*, is typified by *Culex amazonensis* (Lutz, 1905), original combination: *Aedinus amazonensis*. Subgenus abbreviation – *Ads.*

****

Systematics: *Aedinus* includes four extant species in the Neotropical Region, primarily northern South America into Panama. The adults of *Aedinus* resemble those of *Tinolestes* and *Belkinomyia*, but the phyletic relationships of these three taxa are uncertain. *Aedinus* was recovered in a sequentially nested relationship with *Anoedioporpa*, *Tinolestes* and *Belkinomyia* in the morphology-based phylogenetic study of Culicini conducted by Harbach *et al.* (2012). In the molecular phylogenetic study of the Neotropical subgenera of *Culex* conducted by González *et al.* (2023), which employed sequences for the nuclear genes *CAD* and *HB* and mitochondrial *COI* gene, *Aedinus* (represented by *Cx. amazonensis*) was recovered as the sister of *Cx. apicinus* of *Phalangomyia*. In the absence of *Phalangomyia*, *Aedinus* (also represented by *Cx. amazonensis*) was recovered in a basal relationship to a clade comprised of *Maillotia* (represented by *Cx. hortensis*) and a sister group of 24 species (23 species of *Culex sensu stricto* and a species of *Phenacomyia*) in the molecular study (mitogenomes) of da Silva *et al.* (2020a).

****

Characteristics: Species of *Aedinus* exhibit the following anatomical features. ADULTS: Small, dark, unadorned mosquitoes; decumbent scales of the vertex all narrow; antenna slightly longer than the proboscis in females, subequal to the proboscis in males; maxillary palpus short in both sexes, apparently with 4 palpomeres, palpomeres 1 and 2 ankylosed; proboscis longer than the forefemur; scutum with narrow dark falcate scales; acrostichal setae absent; lower mesepimeral seta present; postpronotum with a few dark scales; thoracic pleural scaling restricted to rows of broad translucent scales along the mesokatepisternal setae; wing and legs dark-scaled; ungues of females simple, fore- and midungues of males enlarged, unequal, larger unguis of the foreleg (but not the midleg) with a tooth, hindungues as in females; abdominal terga predominantly dark-scaled. MALE GENITALIA: Lobes of tergum IX distinctly separated, digitiform, bearing a few weak apical setae; gonocoxite with scales; subapical lobe submedian in position, undivided, bearing 1 (usually) or 2 appendages; gonostylus at least 0.5 the length of the gonocoxite, relatively simple or enlarged subapically; gonostylar claw subapical in position; phallosome not columnar, lateral plate with a broadly sclerotized basal hook (usually without denticles), a caudally directed blunt apical process and 2 short sternal processes; paraproct with a crown of 5–9 teeth; cercal setae indistinct. LARVAE: Head capsule faintly imbricate; antenna with a long multi-branched seta 1-A inserted beyond mid-length; setae 2,3-C absent; seta 3-P moderately long, approaching 0.5 the length of setae 1,2-P, single or double; abdominal seta 2 inserted relatively close to seta 1 within the posterior 0.5 of the terga; setae 1,2-VIII on distinct sclerotized plates; comb scales in a patch; siphon long and slender; 5 or 6 posterolateral pairs of seta 1-S inserted beyond the pecten and 2 pairs of anterolateral seta 2a-S, proximal one inserted opposite the pecten; saddle complete, with distinct attached acus, posterior margin without spines; seta 1-X short, submarginal; seta 2-X long, with 2 short postbasal branches; seta 3-X long, single; ventral brush (seta 4-X) with 6 pairs of setae on a grid. PUPAE: Cephalothorax and abdomen with a distinct pattern of pigmentation; trumpet moderately long, tracheoid area distinct; seta 2-III inserted mesad of seta 1, seta 2-VI,VII inserted laterad of seta 1; seta 9-VII,VIII single, short, distinctly shorter than the tergum; paddle with a distal pigmented spot along the midrib, margin without spicules; setae 1,2-Pa both absent.

****

Bionomics: The immature stages of species of *Aedinus* inhabit the margins and root caves of swamps in primary forest, and are occasionally found in ground pools in partially forested areas. Females are not known to bite humans, but they harbour and apparently transmit a variety of arboviruses to forest animals.

****

Principal references: Lane, 1953 (key for males, descriptions, distributions); Berlin & Belkin, 1980 (keys, descriptions, bionomics, distributions); Harbach *et al.*, 2012 (morphology, phylogenetic relationships); da Silva *et al.*, 2020a (*Cx. amazonensis*, molecular associations); González *et al.*, 2023 (Neotropical subgenera of *Culex*, comparative morphology, molecular associations).

Afroculex Danilov, 1989

Afroculex Danilov, 1989, originally proposed and currently recognized as a subgenus of *Culex*, is typified by *Culex lineatus* (Theobald, 1912a), original combination: *Pseudohowardina lineata*. Subgenus abbreviation – *Afc.*

****

Systematics: *Afroculex* is monobasic. It only includes the extant type species, which is known only from South Africa. Danilov (1989) suggested that *Afroculex* is intermediate between *Barraudius* and *Lutzia* based on similarities of the male genitalia. The morphology-based phylogenetic studies of St John (2007) and Harbach *et al.* (2012) indicated a closer affinity with *Maillotia*, and those groups were recovered in a sister relationship with *Lasiosiphon* in the latter study, parenthetically expressed as *Lasiosiphon* + (*Afroculex* + *Maillotia*).

****

Characteristics: The single species of *Afroculex* exhibits the following principal anatomical features. ADULTS: Head black with black erect scales, decumbent pale scales present only on the dorsal midline and along the margins of the eyes; maxillary palpus of females white-scaled at the apex, palpus of males as long as the proboscis, slender, without setae; scutum uniquely marked with 5 conspicuous longitudinal lines of pale scales; thoracic pleura with 2 longitudinal lines of white scales extending from the antepronotum to the prealar area and from the upper proepisternum to the upper mesepimeral setae; lower mesepimeral seta present; subcosta of the wing white-scaled except distally, other veins dark-scaled; legs predominantly dark-scaled, fore- and midfemora with a complete anterior stripe of white scales, hindfemur with a stripe of white scales on the distal 0.5, tarsi entirely dark-scaled, hindtarsomere 1 about the length of the hindtibia; abdominal terga with narrow incomplete apical pale bands. MALE GENITALIA: Ninth tergal lobes very narrow, without setae; gonocoxite rather short, stout, with scales laterally, bearing a rather stout specialized seta between the base of the subapical lobe and the attachment of the gonostylus; subapical lobe prominent, subapical in position, undivided, bearing 2 stout spine-like setae apically and a less stout one subapically; gonostylus short, rather broad and curved, with a terminal claw; paraproct of a peculiar form, with a short spiculate thumb-like projection dorsally from the apex; cercal sclerite strongly sclerotized, with a thumb-like projection at the base, setae absent; phallosome much as in species of *Lutzia*, lateral plate a long finger-like process with a group of small spicules on the mesal side near mid-length. LARVAE and PUPAE: Unknown.

****

Bionomics: Females of *Cx. lineatus* have been collected feeding on lizards (*Pseudocordylus melanotus*) resting in cracks in rocks (Van As *et al.*, 2015), but nothing else is known about the bionomics of the species.

****

Principal references: Edwards, 1941 (as *Cx. pulchrithorax*, in key, taxonomy, description, distribution); Danilov, 1989 (new subgenus, description, distribution); Harbach *et al.*, 2012 (morphology, phylogenetic relationships).

Allimanta Casal & García, 1968

Allimanta Casal & García, 1968a, originally proposed and currently recognized as a subgenus of *Culex*, is typified by *Culex tramazayguesi* Duret, 1954. Subgenus abbreviation – *Alm.*

****

Systematics: *Allimanta* is monobasic. It only includes the extant type species, which is known only from Argentina. *Allimanta* was placed among principally Afrotropical subgenera of *Culex* in the analyses of St John (2007) and was recovered in a basal relationship to all Old World groupings of Culicini in the analysis of Mallampalli (1995). *Culex tramazayguesi* was originally described as a species of *Culex sensu stricto* with certain features shared with species of *Barraudius* and *Lasiosiphon* (Duret, 1954; Casal & García, 1968a), which occur principally in the Mediterranean Subregion of the Palaearctic. Consequently, it is interesting that *Allimanta* was recovered in a basal relationship to a large clade comprised of New World generic-level taxa in the phylogenetic study of Harbach *et al.* (2012).

****

Characteristics: The following combinations of characters distinguish *Allimanta* from other genus groups of Culicini in the New World. ADULTS: Proboscis with a ring of pale scales; acrostichal setae present; scutellum with white scales; thoracic pleura with distinct pale scale-patches; lower mesepimeral seta present; costa and subcosta of the wing with scattered pale scales; femora and tibiae with apical pale bands; tarsi with pale rings at the joints, hindtarsomere 5 entirely pale-scaled; abdominal terga with narrow basal pale bands. MALE GENITALIA: Tergum IX lobes poorly developed, widely separated, with a few small setae; gonocoxite rather short and stout, without scales; subapical lobe postmedian in position, with distinct proximal and distal divisions, stem of the proximal division with 2 apical rod-like setae, distal division with a single rod-like seta and a slender seta on the lateral side of the stem; gonostylus slightly flattened and flared distally, with a short broad claw at the apex; phallosome not columnar, lateral plate gradually tapered to a laterally bent point, with a small basal hook; paraproct with a dense crown of short needle-like spicules, basal lateral arm rather stout and straight; cercal setae present. LARVAE: Setae 2,3-C present; seta 3-P as long as setae 1,2-P; seta 8-P well developed, double; seta 7-I subequal in length to seta 6-I, seta 7-II small, similar to seta 7-III–V, seta 7-VI single; siphon with 5 or 6 seta 1-S inserted in a more or less single posterior row from the pecten to near the apex, with a single subapical pair of dorsolateral setae; saddle complete, seta 2-X with 3–5 branches, seta 3-X single; ventral brush (seta 4-X) with 5.5–6 pairs of branched setae inserted on a grid. PUPAE: Trumpet long, distally flared, pinna as long as the meatus; seta 6-I longer than seta 7-I; seta 5-IV–VI longer than the following tergum; seta 6-VII with 5–9 branches, inserted posteromesad of seta 9-VII; paddle without marginal spicules; setae 1,2-Pa both present.

****

Bionomics: Larvae of *Cx. tramazayguesi* were found in hypersaline groundwater near a thermal spring. Females were captured indoors, but the species is not of medical importance.

****

Principal references: Duret, 1954 (*Cx. tramazayguesi*, description of female and male); Casal & García, 1968a (new subgenus, *Cx. tramazayguesi*, descriptions, adults, pupa, larva); Harbach *et al.*, 2012 (morphology, phylogenetic relationships); González *et al.*, 2023 (Neotropical subgenera of *Culex*, comparative morphology).

Anoedioporpa Dyar, 1923

Anoedioporpa, originally proposed as a genus by Dyar (1923e), currently classified as a subgenus of *Culex*, is typified by *Culex conservator* Dyar & Knab, 1906d. Subgenus abbreviation – *And.*

****

Systematics: *Anoedioporpa* includes 12 extant species in the Neotropical Region. The species are divided between two groups, the Conservator and Restrictor Groups. Similarities in the male genitalia of *Anoedioporpa* and *Melanoconion* suggest that they may have shared a common ancestor. *Anoedioporpa* definitely shares affinities with other subgenera of the Melanoconion Group (Harbach *et al.*, 2012).

****

Characteristics: Species of *Anoedioporpa* exhibit the following anatomical features. ADULTS: Small, dark, unadorned mosquitoes; vertex of the head with only narrow decumbent scales; antenna slightly shorter than the proboscis in both sexes; maxillary palpus short in both sexes (except in *Cx. restrictor*), palpus of females with 4 palpomeres, palpomeres 1 and 2 ankylosed, palpus of males with 4 or 5 palpomeres; proboscis longer than the forefemur; scutum predominantly with narrow dark falcate scales; acrostichal setae absent (except in *Cx. corrigani* and *Cx. restrictor*); pleural integument of the thorax generally yellowish; lower mesepimeral seta present or absent; postpronotum with a few scales; pleural scaling restricted to a few broad translucent scales along the mesokatepisternal setae; wing and legs dark-scaled; ungues of females simple, fore- and midungues of males enlarged, unequal, each unguis with a tooth, hindungues as in females; abdominal terga dark-scaled with basolateral pale patches. MALE GENITALIA: Tergum IX lobes small, mound-like, widely separated, with a variable number of setae; gonocoxite with a few scales; subapical lobe postmedian in position, with distinct proximal and distal divisions, stem of the proximal division with 2 apical rods, distal division with specialized apical setae and with or without a foliform setae on the stem; gonostylus relatively simple, distal 0.33 with many close-set transverse ridges on the ventral surface; gonostylar claw subapical in position; phallosome not columnar, lateral plate with a broadly sclerotized basal hook, a caudally directed apical process and a shorter sternal process; paraproct with a crown of 6–13 teeth; cercal setae varied. LARVAE: Head with a conspicuous lateral expansion posterior to the antenna; antenna with a long multi-branched seta 1-A inserted about 0.75 from the base; setae 2,3-C absent; median tooth of the dorsomentum constricted (nipple-like) distally (except in *Cx. restrictor*); seta 3-P moderately long, about 0.5 the length of setae 1,2-P; abdominal seta 2 inserted relatively close to seta 1 within the posterior 0.5 of the terga; setae 1,2-VIII not on a common tubercle; comb scales in a patch; siphon long and slender, pecten on the proximal 0.25, with 2 pairs of dorsolateral (2a-S) and 4–6 pairs of ventrolateral setae (1a-S) borne distal to the pecten; seta 2-S distinct, slightly curved, with or without a submedian branch; saddle complete, without acus, slightly spiculate posteriorly; seta 1-X varied, branched; setae 2,3-X long, single; ventral brush (seta 4-X) with 5 pairs of setae (6 pairs in *Cx. restrictor*) borne on a grid. PUPAE: Cephalothorax and abdomen without a distinct pattern of pigmentation; trumpet moderately long, tracheoid area distinct; seta 2-III–VII submarginal, inserted either laterad or mesad of seta 1, 2-VI,VII inserted laterad of seta 1; seta 9-VIII long, subequal in length to the tergum; caudal lobe of tergum VIII overlying the lateral part of tergum IX; seta 1-IX present (absent in *Cx. restrictor*); paddle without marginal spicules; setae 1,2-Pa both absent.

****

Bionomics: The immature stages of species of *Anoedioporpa* inhabit phytotelm habitats, mainly tree holes and bamboo, and occasionally artificial containers. The adults are principally forest dwellers, but they are sometimes collected in peri-urban areas. The females are not known to bite humans; hence, the species are not of medical importance.

****

Principal references: Berlin & Belkin, 1980 (keys, descriptions, bionomics, distributions); Harbach *et al.*, 2012 (morphology, phylogenetic relationships); González *et al.*, 2023 (Neotropical subgenera of *Culex*, comparative morphology).

Barraudius Edwards, 1921

Barraudius Edwards, 1921d, originally proposed and currently recognized as a subgenus of *Culex*, is typified by *Culex pusillus* Macquart, 1850. Subgenus abbreviation – *Bar.*

****

Systematics: *Barraudius* includes four extant species with distributions in the Palaearctic (principally) and Afrotropical Regions. The affinities of *Barraudius* are not clear, but the group appears to be closely related to the Afrotropical *Afroculex* and *Maillotia* and the Neotropical *Allimanta*. When Edwards (1921d) established *Barraudius*, he pointed out that species of *Barraudius* 'appear to be more distinct from typical *Culex* than any other groups occurring in the Old World'. The basal placement of *Barraudius* relative to taxa traditionally included in *Culex sensu stricto* in the morphology-based phylogenetic analyses of Harbach *et al.* (2012) support Edwards's observation.

****

Characteristics: Species of *Barraudius* exhibit the following characteristics. ADULTS: Small yellowish-brown mosquitoes; vertex of the head with narrow scales only; proboscis shorter than the forefemur, entirely dark-scaled; maxillary palpus with 3 palpomeres in females (palpomere 4 vestigial or absent), 5 palpomeres in males; acrostichal and dorsocentral setae present; thoracic pleura with distinct pale scale-patches; lower mesepimeral seta present; tarsi entirely dark-scaled, foretarsomere 5 equally long or longer than tarsomere 4, hindtarsomere 1 distinctly shorter than the tibia; abdominal terga dark-scaled but basolateral pale patches may be present. MALE GENITALIA: Tergum IX lobes poorly developed, widely separated, with setae; gonocoxite with scales; subapical lobe median in position, divided, each division bearing 1–3 specialized setae of uncertain homology; gonostylus simple, tapered distally, with a rather long claw at the apex; phallosome elongate, lateral plate narrow, tapered, apically pointed and bent laterad; paraproct with a crown of numerous short thorn-like spicules, basal lateral arm not developed; cercal setae present.

LARVAE: Palatal brushes normal; seta 1-A large, inserted distad to mid-length of the antenna; setae 4,5,6-C inserted at mid-length of the head capsule, seta 4-C single; seta 3-P nearly as long as setae 1,2-P; seta 8-P well developed, double; seta 7-I subequal in length to seta 6-I, seta 7-II small, similar to seta 7-III–V, seta 7-VI single; siphon moderately long, pecten spines long and slender, seta 1-S with 8–11 long multi-branched elements in a posterior zigzag row extending from within the length of the pecten to the apex of the siphon; saddle complete; seta 2-X with 2 or 3 branches; seta 3-X single, longer than seta 2-X; ventral brush (seta 4-X) with 5.5–6.5 pairs of long setae inserted on a grid. PUPAE: Setae of the cephalothorax generally short, subequal in length; seta 6-I longer than seta 7-I; seta 5-IV–VI shorter or slightly longer than the following tergum; setae 1,2-Pa both present.

****

Bionomics: The immature stages of species of *Barraudius* inhabit a variety of groundwater habitats, especially brackish water marshes, swamps and rice fields. Females are known to bite humans and domesticated animals during the day and night. *Culex modestus* vectors arboviruses and is regarded as a vector of West Nile virus in Europe. Females have been found naturally infected with tularaemia (Gutsevich *et al.*, 1974).

****

Principal references: Gutsevich *et al.*, 1974 (eastern Palearctic, *Cx. modestus*, *Cx. pusillus*, descriptions, bionomics, distributions); Tanaka *et al.*, 1979 (Japan, subgenus, *Cx. inatomii*, descriptions, bionomics, distributions); Lu *et al.*, 1997a (China, *Cx. modestus*, taxonomy, descriptions, distribution, bionomics); Becker *et al.*, 2003, 2010, 2020 (Europe, subgenus, *Cx. modestus*, *Cx. pusillus*, descriptions, bionomics, distributions); Harbach *et al.*, 2012 (morphology, phylogenetic relationships).

Belkinomyia Adames & Galindo, 1973

Belkinomyia Adames & Galindo, 1973, originally proposed and currently recognized as a subgenus of *Culex*, is typified by *Culex eldridgei* Adames & Galindo, 1973. Subgenus abbreviation – *Bel.*

****

Systematics: *Belkinomyia* is monobasic. It only includes the extant type species, which is known only from the Pacific coast of Colombia. *Belkinomyia* shares affinities with *Tinolestes* and the genera *Galindomyia* and *Deinocerites*. The results of the phylogenetic studies of St John (2007) and Harbach *et al.* (2012) based on morphological data indicate *Belkinomyia* is the sister of *Galindomyia* + *Deinocerites*.

****

Characteristics: The single species of *Belkinomyia* exhibits the following principal anatomical features. ADULTS: Decumbent scales of the vertex and orbital lines narrow; antenna longer than the proboscis, flagellar whorls similar in both sexes, setae of the whorls slightly longer in males; proboscis slightly longer than the forefemur; maxillary palpus very short, with 3 (2?) palpomeres in both sexes; scutum and scutellum with narrow falcate scales except for bare narrow dorsocentral lines; acrostichal setae absent; postpronotum with scales; thoracic pleura with a few scales along the mesokatepisternal setae; lower mesepimeral seta present; wing, legs and abdominal terga dark-scaled; veins R_2 and R_3 of the wing with broad scales; ungues of the tarsi simple, fore- and midungues unequal in males. MALE GENITALIA: Tergum IX lobes small, globular, setose; gonocoxite without scales; subapical lobe somewhat as in *Deinocerites*, undivided, with 1 foliform and 3 spiniform setae; gonostylus uniquely developed, exceptionally short and broad with an apical forked claw as in species of *Deinocerites*; phallosome not columnar, lateral plate mostly membranous, membranous areas bearing distinct denticles; paraproct with a comb-like crown of short and heavy spicules. LARVAE: Head without a lateral expansion caudad of the antenna; setae 2,3-C absent; setae 4–6-C closely grouped together, inserted caudad of the level of seta 7-C; seta 3-P significantly shorter than setae 1,2-P, with 4–6 weak branches; seta 7-II small like seta 7-III–VI; comb scales in a patch; siphonal pecten comprised of simple spines; seta 1-S inserted beyond the pecten, with 4 posterolateral setae and a distal lateral seta; seta 2-X with 1 short postbasal branch; seta 3-X single; ventral brush (seta 4-X) with 6 pairs of setae on a grid. PUPAE: Meatus of the trumpet without a slit from the proximal part of the pinna; seta 8-CT longer than the other setae of the cephalothorax, with multiple branches; seta 1-II,III strongly developed, with numerous branches, slightly resembling seta 1-I; seta 2-III–VI inserted nearly on the posterior margin of the terga, seta 2-III,VI inserted lateral to seta 1, seta 2-IV,V inserted mesad of seta 1; seta 5-IV–VI weakly developed, barely reaching the middle of the following tergum; seta 9-VII short, distinctly shorter than seta 3-VII; caudal lobe of tergum VIII not overlying the lateral part of tergum IX; setae 1,2-Pa both present.

****

Bionomics: The immature stages of *Cx. eldridgei* are found in crab holes containing slightly saline water. The adults rest in crab holes. The species is not medically important.

****

Principal references: Adames & Galindo, 1973 (new subgenus, type species, descriptions, bionomics, distribution); Harbach *et al.*, 2012 (morphology, phylogenetic relationships); González *et al.*, 2023 (Neotropical subgenera of *Culex*, comparative morphology).

Carrollia Lutz, 1905

Carrollia, originally proposed as a genus by Lutz (1905), currently classified as a subgenus of *Culex*, is typified by *Culex iridescens* (Lutz, 1905), original combination: *Carrollia iridescens*. Subgenus abbreviation – *Car.*

****

Systematics: *Carrollia* includes 18 extant species. The species are restricted to Neotropical forests from southern Mexico to the States of São Paulo and Paraná in Brazil and Misiones Province in northeastern Argentina, but they are not known to occur south of the Guayas River Basin in Ecuador. *Carrollia* shares similarities with other subgenera of the Melanoconion Group, but it appears to be most closely related to *Tinolestes* based on the absence of dorsocentral setae. Valencia (1973), however, suggested that *Carrollia* and the closely related *Deinocerites* and *Galindomyia* share a similar evolutionary history as derivatives from a primitive stock of *Melanoconion*. *Carrollia* shared a sister relationship with *Melanoconion* in a Bayesian analysis of mtDNA *COI* sequences obtained from species of those subgenera and *Microculex* of the Melanoconion Group, as well as species of *Culex sensu stricto*, *Phenacomyia* and *Lutzia* (Demari-Silva *et al.*, 2011).

****

Characteristics: Species of *Carrollia* exhibit the following characteristics. ADULTS: Dark scales of the head, proboscis, maxillary palpus, legs and abdomen almost always with metallic iridescence; vertex of the head with a transverse band of broad decumbent scales anteriorly and narrow decumbent scales posteriorly, present or absent on the occiput, orbital line without scales; proboscis about as long as or longer than the forefemur; maxillary palpus of females scarcely 0.25 the length of the proboscis, with 3 palpomeres, palpomere 4 vestigial or absent; palpus of males about as long as the proboscis, slender, bare, with 5 palpomeres; acrostichal setae usually absent; dorsocentral setae present or absent; lower mesepimeral seta present; wing entirely dark-scaled; femora with patches of metallic scales; tarsi all dark-scaled or hindtarsomere 4 with a broad basal pale ring; ungues of females simple, subequal on fore- and midlegs; fore- and midungues of males enlarged and uneven, larger unguis of the foreleg always and that of the midleg sometimes with a tooth, both with very small pulvilli, smaller unguis of fore- and midlegs simple and seemingly without pulvilli; hindungues small, simple and equal as in females but with very short pulvilli; abdomen laterally compressed, especially in males, terga with basolateral spots of pale iridescent scales; tergum VIII of males very deeply emarginate in the middle, almost divided. MALE GENITALIA: Ninth tergal lobes approximated or widely separated, small, simple, conical or flattened and rounded, with setae largely directly laterad; gonocoxite with or without scales; subapical lobe subapical in position, proximal division always present, bearing 2 flattened apical setae and sometimes 1 subapical or apical relatively simple seta, additional setae sometimes present on the stem and the base, distal division poorly developed or absent, usually with 1 lanceolate seta when present, an accessory division frequently developed near the base of the proximal division as a simple long digitiform process, a complex lobe with a broad dorsal expansion or 2 separate columnar processes; gonostylus usually enlarged, variously shaped; phallosome not columnar, much as in species of *Melanoconion*, lateral plate with a basal hook; paraproct with a crown comprised of a single row of blunt flattened spicules; cercal sclerite usually membranous, cercal setae present or absent. LARVAE: Antenna with a small seta 1-A inserted in middle 0.33; median labral plate poorly differentiated; palatal brushes normal; seta 2-C strongly developed, inserted mesal to seta 1-C; seta 3-C nearly always absent, rarely represented by a spicule; seta 3-P shorter than setae 1,2-P, usually branched; seta 1-VII very strongly developed, extending beyond the base of the siphon; seta 7-II normally similar to seta 7-III–VI; comb scales apically fringed and in a patch or spine-like and in a single row; siphon moderately long, pecten restricted to the proximal 0.5, spines with or without denticles or marginal fringe; seta 1-S in an irregular double posterior row of 8–16 elements and 2 accessory elements, 1 lateral and 1 posterolateral; occasionally a third accessory (posterolateral) element present; saddle short, complete; seta 1-X single or branched; seta 2-X always strong, branched; seta 3-X always strong, single; ventral brush (seta 4-X) with 4–6 pairs of setae on a poorly developed grid without lateral grid bars. PUPAE: Meatus of the trumpet without a slit from the proximal part of the pinna; seta 2-II inserted mesad or laterad of seta 3; seta 2-III,VI,VII inserted mesad of seta 1; seta 9-VII distinctly larger than seta 3-VII; dorsal lobe of tergum VIII not overlying the lateral part of tergum IX; seta 1-IX not developed; seta 1-Pa present, rarely absent, seta 2-Pa absent.

****

Bionomics: *Carrollia* are sylvan mosquitoes. The immature stages are found in a variety of container habitats, including artificial containers, but occasionally occur in collections of groundwater. They are most frequently found in fallen leaves and spathes and fronds of palms, and tree holes and cut or broken bamboo. The adults are diurnal but nothing is known about the feeding habits of females. They are seldom attracted to humans. Species of *Carrollia* are not of medical importance.

****

Principal references: Valencia, 1973 (taxonomy, classification, keys, descriptions, bionomics, distributions); Harbach, 2011 (informal classification); Harbach *et al.*, 2012 (morphology, phylogenetic relationships); González *et al.*, 2023 (Neotropical subgenera of *Culex*, comparative morphology).

Culex Linnaeus, 1758 *sensu stricto*

Culex sensu stricto, the nominotypical subgenus of *Culex*, is typified by *Culex pipiens* Linnaeus, 1758. Subgenus abbreviation – *Cux*.

****

Systematics: *Culex sensu stricto*, as presently defined, includes 216 extant species. A fossil specimen of the type species is preserved in Baltic amber. The group has a cosmopolitan distribution and is almost certainly polyphyletic and will remain a 'taxon of convenience' until the monophyly, phylogeny and classification of the included taxa and their relationships with other genus groups are resolved. *Phenacomyia*, *Oculeomyia*, *Phytotelmatomyia* and *Sirivanakarnius* have been placed among species of *Culex sensu stricto* in both morphological and molecular studies that included species of these subgenera (Deus, 2009; Vesgueiro *et al.*, 2011; Demari-Silva *et al.*, 2011; Harbach *et al.*, 2012; Laurito & Almirón, 2013; da Silva *et al.*, 2020a). Analyses of the mitochondrial genomes of four species of the Coronator Group suggest that the *ATP6*, *ATP8* and *NADH5* genes may be useful for resolving phylogenetic relationships within *Culex sensu stricto* (Demari-Silva *et al.*, 2015). Harbach *et al.* (2017a) conducted a phylogenetic analysis of *Culex sensu stricto* using implied and equally weighted parsimony methods to analyse a data set comprised of 286 anatomical characters of the larval, pupal and adult stages of 150 species of *Culex sensu stricto* and an outgroup of 17 species. The analyses largely failed to resolve relationships among the species and the informal groups in which they are currently placed; thus, the internal classification of *Culex sensu stricto* based on the intuitive assessment of morphological similarities and differences does not seem, in large part, to reflect evolutionary relationships. It is notable that the phylogenetic analysis of Sun *et al.* (2019) based on the nucleotide sequences of 13 protein-coding genes obtained from the complete mitochondrial genomes of 10 species of *Culex* strongly supported the suppositions of Edwards (1932a), Sirivanakarn (1976) and Harbach *et al.* (2017a) that the Sitiens Group possibly should be recognized as a genus-group taxon. It is also interesting to note that their study did not support the monophyly of the Pipiens Group, which was recovered as a polyphyletic group in the studies of Laurito & Almirón (2013) and Harbach *et al.* (2017a), but did support the monophyly of the Coronator Group.

****

Characteristics: Adults of species currently placed in *Culex sensu stricto* are mainly drab, unicolourous mosquitoes, but some species have pale markings on the legs and species of the Mimeticus Subgroup of the Sitiens Group have pale spots on the wings as in species of *Anopheles*. The species exhibit the following characteristics. ADULTS: Generally medium-sized mosquitoes; decumbent scales of the vertex predominantly narrow; eyes narrowly separated or contiguous above the antennal pedicels; antenna slightly longer than the proboscis, antenna of males without tufts of specialized setae or scales; maxillary palpus of females with 3 palpomeres, palpus of males with 5 palpomeres, usually distinctly longer than the proboscis, palpomere 3 with a row of ventral setae and variable number of longer ventrolateral setae, palpomeres 4 and 5 upturned and usually with numerous setae; proboscis usually as long as or slightly longer than the forefemur, with or without a median pale ring or pale scaling on the ventral surface, proboscis of males with or without a median ventral cluster of setae but usually with a false joint before mid-length; scutal scaling varied; acrostichal setae present in more or less a double row, sometimes short; dorsocentral setae, scutal fossal setae, antealar setae and supraalar setae well developed; ante- and postpronota and thoracic pleura usually with extensive scaling, usually present at least on the upper proepisternum, mesokatepisternum and mesepimeron, present or absent on the lower proepisternum, anteprocoxal membrane, postspiracular area and prealar area; pleural setae varied, postspiracular setae rarely present, lower mesepimeral setae present or absent; wing most often entirely dark-scaled, sometimes with scattered pale scales or distinct pale spots; legs varied, tarsi with or without pale bands; ungues small, equal and simple in females, outer unguis of fore- and midlegs toothed and larger than the inner unguis in males; pulvilli well developed; abdominal terga usually with basal pale bands and basolateral pale spots, sometimes with apical pale bands, apicolateral pale patches or entirely dark. MALE GENITALIA: Ninth tergal lobes usually short and broad, with 1 or 2 rows of setae; gonocoxite with or without scales; subapical lobe not distinctly divided, setae *a–c* rod-like, setae *d–f* and seta *h* varied, 1 or all of setae *d–f* sometimes absent, seta *g* foliform or simple; gonostylus generally simple, curved and tapered; gonostylar claw weak, subapical, usually shoehorn-shaped; phallosome complex, varied, lateral plate usually with distinct inner and outer divisions bearing variously developed teeth and processes; paraproct with a crown of numerous spicules of varying size and character, basal sternal process varied, sometimes absent; cercal setae present. LARVAE: Antenna shorter than or as long as the head; seta 1-A usually strongly developed; setae 2,3-A single, usually subapical, sometimes apical; median labral plate usually distinctly delimited from the dorsal apotome; seta 1-C slender or stout; setae 1–3-P subequal and long, usually single, seta 3-P rarely distinctly shorter than setae 1,2-P; abdominal chaetotaxy varied; setae 6-I,II and 7-I strongly developed; seta 6-III–VI moderately to strongly developed; seta 7-II large like seta 7-I or small like seta 7-III–V; seta 1-III–VII varied; comb varied from numerous fringed scales to a few strong spines; siphon varied, short and stout to long and slender; pecten varied from a few to many spines, denticulation varied; seta 1-S in 3–7 posterolateral pairs, 1 or more distal elements sometimes out of line; seta 2-S single, sometimes hook-like and large; saddle usually complete, rarely incomplete; seta 1-X varied, not very long; setae 2,3-X strongly developed, seta 2-X branched, seta 3-X single; ventral brush (seta 4-X) comprised of 6 or 7 pairs of setae, all inserted on a grid. PUPAE: Trumpet varied in length, generally cylindrical or distally flared with a large pinna; meatus usually without a slit; tracheoid area variable; cephalothoracic setae varied in development, all shorter than the trumpet; abdominal chaetotaxy varied, setae 1-III–VII and 5-IV–VI varied in development; seta 2-III–VI single, very small, always inserted mesad of seta 1, seta 2-VII usually inserted laterad of seta 1, sometimes variable in position; seta 6-I,II usually distinctly longer than seta 7-I,II; paddle generally broad and oval, apex rounded, slightly produced or slightly emarginate, outer margin minutely spiculate proximally; setae 1,2-Pa small and usually single, seta 2-Pa sometimes absent.

****

Bionomics: The immature stages of species of *Culex sensu stricto* typically occur in groundwater habitats, but several species are frequently found in large artificial containers. A number of species inhabit rock holes, crab holes, tree holes, bamboo and the leaf axils of plants. Many species are notorious pests of humans and domesticated animals. Females are primarily active during crepuscular or nocturnal periods. The hosts of most species are unknown but probably principally include birds and mammals. *Culex sensu stricto* includes most of the medically important and pest species of the Culicini. Many species vector pathogens that cause human diseases, including Japanese encephalitis in the Oriental Region, several encephalitic diseases in North America, Murray Valley encephalitis and Ross River fever in Australia, West Nile fever in Africa, Europe, India and North America, and Rift Valley fever in Africa. A few species are important vectors of *Wuchereria bancrofti*, and some species may be involved in the transmission of *Brugia malayi*.

****

Principal references: Barraud, 1934 (southern Asia, keys, taxonomy, descriptions, distributions, bionomics); Edwards, 1941 (Afrotropical Region, keys, descriptions of adults and pupae, distributions); Hopkins, 1952 (Afrotropical Region, keys, descriptions of larvae, distributions, bionomics); Lane, 1953 (Neotropical Region, taxonomy, keys, descriptions, distributions); Carpenter & LaCasse, 1955 (North America, taxonomy, keys, descriptions, bionomics); Belkin, 1962 (South Pacific, taxonomy, keys, descriptions, distributions, bionomics); Forattini, 1965a (Neotropical Region, diagnosis, classification, species, bionomics, distributions); Cova-Garcia *et al.*, 1966 (Venezuela, keys, species descriptions, distributions); Delfinado, 1966a (Philippines, taxonomy, keys, descriptions, distributions, bionomics); Bram, 1967a (Thailand, keys, descriptions, distributions, bionomics); Bram, 1967b (Neotropical Region, taxonomy, keys, descriptions, distributions); Belkin *et al.*, 1970 (Jamaica, taxonomy, keys, descriptions, bionomics); Sirivanakarn, 1976 (Oriental Region, keys, descriptions, bionomics, distributions); Gutsevich *et al.*, 1974 (former USSR, taxonomy, keys, descriptions, distributions, bionomics); Tanaka *et al.*, 1979 (Japan, taxonomy, keys, descriptions, distributions, bionomics); Wood *et al.*, 1979 (Canada, taxonomy, keys, descriptions, distributions, bionomics); Darsie & Ward, 1981, 2005 (North America, keys); Clark-Gil & Darsie, 1983 (Guatemala, keys, distributions, bionomics); Darsie, 1985 (Argentina, keys); Lee *et al.*, 1989a (Australasian Region, taxonomy, keys, literature, distributions, bionomics); Harbach, 1988 (southwestern Asia and Egypt, classification, taxonomy, keys, descriptions, bionomics, distributions); Lee *et al.*, 1989a (Australasian Region, taxonomy, keys, literature, distributions, bionomics); Darsie & Pradhan, 1990 (Nepal, keys, distributions, bionomics); Rattanarithikul & Panthusiri, 1994 (Thailand, keys, medically important species,); Reuben *et al.*, 1994 (Southeast Asia, keys, Japanese encephalitis vectors); Lu *et al.*, 1997a (China, keys, taxonomy, descriptions, distributions, bionomics); Rattanarithikul *et al.*, 2005b (Thailand, keys, distributions, bionomics); Deus, 2009 (North and Central America, selected species, relationships based on mtDNA, rDNA and a nuclear gene); Harbach, 2011 (informal classification); Harbach *et al.*, 2012 (morphology, phylogenetic relationships); Laurito & Almirón, 2013 (morphology, phylogenetic relationships of species of the subgenus in Argentina); Laurito *et al.*, 2013 (Argentina, Brazil, 22 species, mtDNA *COI* sequences, identification, relationships); Harbach *et al.*, 2017a (World, morphology, phylogenetic relationships); Sun *et al.*, 2019 (phylogenetic relationships based on mitochondrial genomes); da Silva *et al.*, 2020a (Neotropical species, mitochondrial phylogenomics); Simsaa *et al.*, 2021 (Sudan, key to adults); González *et al.*, 2023 (Neotropical subgenera of *Culex*, comparative morphology, molecular associations); Russell & Burkot, 2023 (Pacific Island countries and territories, species lists, distributions, bionomics, vector status).

Culiciomyia Theobald, 1907

Culiciomyia, originally proposed as a genus by Theobald (1907), currently classified as a subgenus of *Culex*, is typified by *Culiciomyia inornata* Theobald, 1907 [subjective synonym of *Culex fragilis* Ludlow, 1903]. Subgenus abbreviation – *Cui.*

****

Systematics: *Culiciomyia* includes 58 extant species recorded from tropical areas of the Afrotropical, Oriental and Australasian Regions. A few species extend into the eastern Palaearctic Region (China, Japan and Korea). The affinities of *Culiciomyia* are not clear. They share similarities with species of *Acallyntrum* and *Lophoceraomyia*. Females of *Culiciomyia* and *Lophoceraomyia* are very difficult to distinguish. Belkin (1962) and Bram (1968) suggested a relationship between *Culiciomyia*, *Acalleomyia* and

Acallyntrum based on similarities observed in larvae, and surmised that *Culiciomyia* is the more primitive group. However, in the phylogenetic analysis of morphological data conducted by Harbach *et al.* (2012), which included species of *Acalleomyia*, *Culiciomyia*, *Eumelanomyia* and *Lophoceraomyia*, *Culiciomyia* was recovered as the sister of *Acalleomyia* in a clade parenthetically expressed as *Eumelanomyia* + (*Lophoceraomyia* + (*Acalleomyia* + *Culiciomyia*)).

****

Characteristics: Species of *Culiciomyia* exhibit the following characteristics. ADULTS: Small to rather large dark-coloured mosquitoes; head with 1 or 2 rows of broad decumbent scales on the ocular line, decumbent scales narrow in the centre of the vertex; eyes contiguous above the antennae; frontal tuft not developed; antenna longer than the proboscis in females, usually slightly shorter than the proboscis in males, pedicel without conspicuous scales; proboscis, maxillary palpus and tarsi dark-scaled; several long ventral setae near the middle of the proboscis in males; maxillary palpus with 3 palpomeres in females; palpus longer than the proboscis in males, with 5 palpomeres, distal 0.5 of palpomere 3 usually with a ventral row of outstanding lanceolate scales (absent in *Cx. dispectus* and *Cx. hainanensis*), palpomeres 4 and 5 upturned and setose ventrally, palpomere 4 usually noticeably shorter than palpomere 5; acrostichal setae usually absent, present in some species, sometimes rarely weakly developed near the prescutellar space; scutal scaling very dense with a smooth appearance; antepronotum with or without scales; postpronotum with narrow scales; thoracic pleura without scale-patches; 1 or 2 lower mesepimeral setae present; wing usually entirely dark-scaled, sometimes with pale scales on the remigium; abdominal terga with or without basal or apical pale bands. MALE GENITALIA: Ninth tergal lobes weakly developed; gonocoxite without scales; subapical lobe strongly developed; setae *a–c* strongly developed, spiniform, seta *a* usually removed sternad from setae *b* and *c*, setae *d–f* varied, seta *g* usually more or less foliform, seta *h* usually developed, sometimes numerous additional specialized tergal setae present, sometimes with a setose apicosternal accessory lobe developed from the subapical lobe; gonostylus varied in shape, distal portion recurved, usually with a distinct outer crest of strong spicules; phallosome simple, lateral plate with a strong apical spine and usually a series of short denticles on the lateral margin; dorsal and ventral aedeagal bridges distinct; paraproct with a large cluster of numerous fine spicules dorsomesally and heavy curved spine-like spicules in a single ventrolateral row, basal sternal process usually long, slender and curved; cercal setae varied. LARVAE: Antenna at least 0.5 the length of the head; seta 1-A inserted near mid-length or beyond; setae 2,3-A slightly removed from the apex; seta 1-C slender, filamentous or robust and gradually tapered; seta 2-C usually present; setae 4–7-C inserted on the anterior 0.5 of the cranium; seta 12-C inserted mesad of seta 13-C; seta 15-C usually with at least 4 branches, inserted on the anterior 0.5 of the labiogula; seta 1-P single or bifid; seta 2-P single; seta 3-P single or branched, about 0.5 the length of setae 1,2-P; seta 7-II only slightly larger than seta 7-III; seta 1-IV,V strongly developed; comb comprised of numerous fringed scales in a patch; siphon variable in length and shape, with 3–6 pairs of posterolateral setae inserted in a line distad of the pecten (pecten absent in *Cx. dispectus* and *Cx. hainanensis*); saddle complete; seta 1-X large, single or double; setae 2,3-X usually single; ventral brush (seta 4-X) comprised of 3.5–6 pairs of setae borne on a grid. PUPAE: Very similar to pupae of *Culex sensu stricto*; trumpet moderately long and slender, pinna short, tracheoid area varied; cephalothoracic setae well developed; abdominal setae moderately to strongly developed; seta 5-II inserted anterior or posterior to seta 4-II; seta 6-VII well developed, conspicuous; seta 1-IX present or absent; paddle normal, marginal spicules inconspicuous or absent; setae 1,2-Pa both present.

****

Bionomics: The immature stages of species of *Culiciomyia* occupy a wide range of habitats, including ground pools, swamps, running streams, tree holes, crab holes, leaf axils of plants and artificial containers, but they appear to prefer temporary accumulations of water with a high organic content. Virtually nothing is known about the bionomics of the adults. Females are not known to attack humans. Although two species, *Cx. pallidothorax* and *Cx. spathifurca*, have been experimentally infected with *Wuchereria bancrofti*, species of *Culiciomyia* are unlikely to vector pathogens of human diseases in nature.

****

Principal references: King & Hoogstraal, 1946a (New Guinea, descriptions, key); Belkin, 1962 (South Pacific, taxonomy, keys, descriptions, distributions, bionomics); Vattier & Hamon, 1962 (sub-Saharan Africa, *Cx. gilliesi*, description, key to larvae); Bram, 1967a (Thailand, keys, descriptions, distributions, bionomics); Harrison, 1987 (historical review, morphology); Lee *et al.*, 1989b (Australasian Region, taxonomy, keys, literature, distributions, bionomics); Lu *et al.*, 1997a (China, keys, taxonomy, descriptions, distributions, bionomics); Rattanarithikul *et al.*, 2005b (Thailand, keys, distributions, bionomics); Harbach, 2011 (informal classification); Harbach *et al.*, 2012 (morphology, phylogenetic relationships); Cornel *et al.*, 2020 (Afrotropical Region, descriptions, keys to males).

Eumelanomyia Theobald, 1909

Eumelanomyia, originally proposed as a genus by Theobald (1909), currently classified as a subgenus of *Culex*, is typified by *Culex inconspicuosus* (Theobald, 1908a), original combination: *Eumelanomyia inconspicuosa*. Subgenus abbreviation – *Eum.*

****

Systematics: *Eumelanomyia* includes 79 extant species in the Afrotropical and Oriental Regions, with extensions into the Papuan Subregion and South Pacific of the Australasian Region and the Manchurian Subregion of the Palaearctic Region. Sirivanakarn (1971, 1972) hypothesized a close relationship between *Eumelanomyia*, *Maillotia* and *Neoculex* based on morphological similarities of the adults and male genitalia, and suggested that *Eumelanomyia* is an ancient derivative of the *Maillotia* lineage. He also suggested that the *Eumelanomyia* lineage probably gave rise to *Lophoceraomyia*. Sirivanakarn (1972) also noted similarities between the adults of *Eumelanomyia* and *Culiciomyia*. The results of the study by St John (2007) indicate a closer relationship with *Culiciomyia* and *Lophoceraomyia* than with *Maillotia* and *Neoculex*. A limited molecular study by Juthayothin (2004) using the *COI* gene of mtDNA showed poor resolution among *Culiciomyia*, *Eumelanomyia* and *Culex*. Based on this sole molecular marker, *Eumelanomyia* seemed to be the most primitive of the three; however, substantial homoplasy was detected in the data set making this conclusion unreliable. The results of the phylogenetic study of Harbach *et al.* (2012) based on morphological data indicate that *Eumelanomyia* is not a monophyletic group.

****

Characteristics: Species of *Eumelanomyia* exhibit the following characteristics. ADULTS: Small to medium-sized dark unornamented mosquitoes; vertex with narrow to broad scales; antenna as long as or longer than the proboscis in females, shorter than or as long as the proboscis in males; maxillary palpus and proboscis dark-scaled; palpus of females with 3 or 4 palpomeres, palpomere 4 vestigial or absent; palpus of males with 3–5 palpomeres, distinctly shorter than the proboscis; proboscis as long as or slightly longer than the forefemur; proboscis of males with or without a false joint; scutal scales narrow and predominantly dark; acrostichal setae usually present anteriorly, sometimes absent or only a few present; pleural scales absent or very few present on the mesokatepisternum; lower mesepimeral seta present or absent; wing dark-scaled, cell R_2 distinctly longer than vein R_{2+3}; tibiae and tarsi dark-scaled; ungues of all legs small, equal and simple in females, fore- and midungues enlarged in males, one larger than the other, each with a small denticle; pulvilli poorly or well developed; abdominal terga usually entirely dark-scaled, sometimes with pale basal bands or basolateral spots. MALE GENITALIA: Tergum IX lobes usually small, sometimes large and prominent, setose; gonocoxite without scales; subapical lobe usually small, not clearly divided, proximal division with 3 rod-like setae; distal division with or without 1 strong basal seta and 1 foliform seta, always with a variable number of hair-like or blade-like setae mesally; gonostylus simple or strongly modified; gonostylar claw well developed, length and thickness varied, usually subapical, sometimes apical; phallosome oval to nearly spherical in shape, tergal bridge median or submedian in position, lateral plates usually with denticles on the upper tergal surface; paraproct with a crown of fine spine-like spicules (absent in *Cx. manusensis*), some coarse spicules sometimes present, basal sternal process absent; cercal setae present. LARVAE: Head broader than long; antenna as long as the head, spiculate, seta 1-A well developed, inserted distal to mid-length of the antenna, setae 2,3-A inserted at or removed from the apex; seta 1-C stout, spine-like; setae 5,6-C varied in length and position; seta 3-P much shorter than setae 1,2-P; seta 4-P varied in length from shorter than seta 3-P to as long as setae 1,2-P; seta 7-I strong, similar to seta 6-I, seta 7-II small, similar to seta 7-III–VI; comb comprised of numerous fringed scales in a patch; siphon slender, moderately long to very long, pecten restricted to the basal area, spines varied in shape; 4–6 posterolateral pairs of seta 1-S present; saddle complete, without conspicuous spicules; setae 2,3-X long, seta 2-X single or with short accessory branches, seta 3-X single; ventral brush (seta 4-X) usually with 4.5–8 pairs of setae borne on a grid. PUPAE: Trumpet usually cylindrical, pinna with or without a slit extending into the meatus; seta 6-I,II well developed, longer than seta 7; seta 2-II inserted laterad of setae 1,3-I, seta 2-III–VI inserted mesad of seta 1, seta 2-VII inserted mesad or laterad of seta 1; seta 5-IV–VI well developed, larger than seta 5-III,VII; seta 9-VIII shorter than the tergum, inserted anteromesal to the caudolateral angle of the sternum; paddle broad and oval, margins smooth; seta 1-Pa present or absent; seta 2-Pa present.

****

Bionomics: Species of *Eumelanomyia* are common in tropical forests in inland hilly and mountainous terrain. The immature stages are found in various ground pools, rock pools, rock holes, tree holes, bamboo and artificial containers, generally containing little organic matter. The adults of most species have been collected resting on vegetation in the vicinity of streams and swamps. Nothing is known about the feeding habits of females. No species of *Eumelanomyia* have been incriminated as vectors of human disease agents, but three different viruses have been isolated from a single species in South Africa.

****

Principal references: Sirivanakarn, 1971 (taxonomy, keys and descriptions of species groups and subgroups); Sirivanakarn, 1972 (Oriental Region, taxonomy, classification, keys, descriptions, bionomics, distributions); Tanaka *et al.*, 1979 (Japan, taxonomy, keys, descriptions, distributions, bionomics); Lee *et al.*, 1989b (Australasian Region, taxonomy, keys, literature, distributions, bionomics); Lu *et al.*, 1997a (China, keys, taxonomy, descriptions, distributions, bionomics); Rattanarithikul *et al.*, 2005b (Thailand, keys, distributions, bionomics); Harbach, 2011 (informal classification); Harbach *et al.*, 2012 (morphology, phylogenetic relationships).

Kitzmilleria Danilov, 1989

Kitzmilleria Danilov, 1989, originally proposed and currently recognized as a subgenus of *Culex*, is typified by *Culex moucheti* Evans, 1923a. Subgenus abbreviation – *Kit.*

****

Systematics: *Kitzmilleria* is monobasic. It only includes the extant type species, which is known only from the Equatorial zone of Africa from Kenya westward to Sierra Leone. The affinities of *Kitzmilleria* are not clear. Danilov (1989) regarded *Kitzmilleria* to be closely related to *Afroculex* and *Barraudius* based on shared anatomical features of the larvae and male genitalia. Danilov also noted that *Kitzmilleria* and *Barraudius* share similarities with *Lutzia*. However, the analysis of St John (2007) indicated that *Kitzmilleria* is probably more closely related to species of *Maillotia* than other Afrotropical groups. Based on the morphological character data analysed in the study of Harbach *et al.* (2012), *Kitzmilleria* appears to be most closely related to *Acallyntrum* and the Atriceps Group of *Culex sensu stricto* in the Australasian Region.

****

Characteristics: The single species of *Kitzmilleria* exhibits the following principal anatomical features. ADULTS: Erect scales of the vertex nearly all orange, decumbent scales white; clypeus usually with a few small pale scales on the distal margins; antennal pedicel usually orange; maxillary palpus of males longer than the proboscis, moderately setose, dark-scaled; acrostichal setae present; 2–5 lower mesepimeral setae present; wing entirely dark-scaled; hindfemur with a black dorsal stripe from base gradually widening to the apex but allowing pale scaling of the anterior surface to extend almost to the apex ventrally, tibiae and tarsi dark-scaled; abdominal terga with basolateral pale patches that are larger and visible dorsally on terga VI and VII, tergum VIII with a basal pale band. MALE GENITALIA: Gonocoxite without scales; subapical lobe undivided, with 8 more or less specialized setae generally developed as in species of *Culex sensu stricto*; phallosome somewhat as in *Barraudius* and *Lutzia*, each half divided into 2 prongs, outer prong longer than the inner prong and minutely ridged apically; paraproct as in most *Culex* except the basal arm extends obliquely instead of at a right angle. LARVAE: Antenna short, about 0.25 the length of the head; setae 5,6-C widely spaced, inserted far forward near seta 7-C; seta 4-C minute, inserted far forward; seta 1-P as long as setae 2,3-P; some dorsal setae of the meso- and metathorax and some dorsal and ventral setae of abdominal segments I–VII apparently highly reduced or absent; comb with 2–5 peculiar short broad scales; siphon exceptionally short, pecten absent, with 2 pairs of very small posterolateral seta 1a-S and 1 pair of very small subapical seta 2a-S; saddle relatively small, without spicules; seta 1-X large with multiple branches; seta 2-X multi-branched; seta 3-X single; ventral brush (seta 4-X) and grid absent. PUPAE: Setae 2,3,5,6-CT large with multiple branches; trumpets short, laticorn, with the longest axis transverse to the stem, tracheoid area absent; seta 1-I rather stout; seta 9-VII,VII large with numerous branches, 0.5 the length of the paddle; paddle large, broad, apex emarginate; setae 1,2-Pa both absent.

****

Bionomics: Females of *Cx. moucheti* are thought to feed principally on birds but are known to attack humans during day and night in forests and plantations in western Uganda. Larvae have been found in a variety of habitats that include holes in fallen papaya trees, cavities in banana and oil palm stumps, banana axils, fallen leaves, snail shells, small puddles, ditches and pit latrines. The species is not of medical importance.

****

Principal references: Edwards, 1941 (*Cx. moucheti* as a species of *Culex sensu stricto*, adults and pupa, distribution); Hopkins, 1952 (*Cx. moucheti* as species of *Culex sensu stricto*, larva, bionomics); Danilov, 1989 (new subgenus, description, distribution); Harbach *et al.*, 2012 (morphology, phylogenetic relationships).

Lasioconops Theobald, 1903

Lasioconops, originally proposed as a genus by Theobald (1903c), currently classified as a subgenus of *Culex*, is typified by *Culex poicilipes* (Theobald, 1903a), original combination: *Lasioconops poicilipes*. Subgenus abbreviation – *Lsi.*

****

Systematics: *Lasioconops* is monobasic. It only includes the extant type species, which is widely distributed in the Afrotropical Region, with populations occurring along the Nile River into Egypt, the Sinai Peninsula and the Negev Desert of southern Israel. Harbach (1988) provisionally placed *Cx. poicilipes* in the Sitiens Subgroup of the Sitiens Group of subgenus *Culex* based on similarities of the larval stage with the larva of *Cx. whitmorei*, but noted it was likely that *poicilipes* and *whitmorei* would have to be placed in separate groups after species of the Sitiens Group had been studied more thoroughly. In the morphology-based phylogenetic study of Harbach *et al.* (2012), *Cx. poicilipes* was recovered as the sister to *Cx. bitaeniorhynchus* + *Cx. infula* of the subgenus *Oculeomyia*, and that clade was sister to species of the Mimeticus Subgroup.

****

Characteristics: *Lasioconops* is distinguished by the following principal anatomical features. ADULTS: Moderately large mosquitoes easily distinguished from species of the other subgenera of *Culex* in the Afrotropical and Palaearctic Regions by the presence of rows of pale spots on the femora and tibiae; maxillary palpus of females comprised of 3 palpomeres, palpus of males with 5 palpomeres; proboscis with median pale band; scutum with variable amount of pale scaling but anterior two-thirds not mainly covered with pale scales as in species of *Oculeomyia*; proepisternum with lower patch of pale scales in addition to the usual upper patch; wing with variable pale scaling, sometimes absent or indistinct but usually scattered or in lines or spots on the anterior veins; femora and tibiae with row of pale spots on the anterior surface; tarsi with pale rings on the bases and apices of tarsomeres 1–3 and the base of tarsomere 4; abdominal terga II–VII with variable basal pale bands, generally produced medially, more so on more posterior terga, no apical pale bands. MALE GENITALIA: Tergum IX lobes scarcely produced, with a row of irregularly spaced setae; gonocoxite rather elongate, subapical lobe undivided, foliform seta (*g*) present, an unusual auxiliary seta present basally on the proximal side of the lobe; inner division of the phallosome closely fused to the ventromesal margin of the outer division, with a number of large dorsolaterally directed teeth borne between the dorsocaudal angle and the mesal surface of the outer division, outer division with a large concavo-convex flap arising from the dorsocaudal margin; proctiger large, paraproct long and slender, without a basal lateral arm, crown composed of numerous short spine-like spicules; cercal setae present. LARVAE: Antenna shorter than the head, scape developed, seta 1-A large, with numerous branches, inserted at mid-length of the shaft; median labral plate distinct, separated from the dorsal apotome by a well-developed clypeolabral suture; dorsomentum broadly triangular with relatively few distinct teeth; maxillary body elongate, somewhat shaped like an arch-shaped door in ventral view, maxillary palpus not fused with and shorter than 0.25 the length of the maxillary body, hypostomal sclerite not fused with the cranium; setae 4–6-C inserted posterior to the mid-length of the head capsule, relatively distant from seta 7-C; seta 3-P similar to setae 1,2-P, equally long and single; seta 7-I similar to seta 6-I, seta 7-II small like seta 7-III–VI; seta 1-III–VIII strongly developed, multi-branched, distinctly longer than 0.5 the length of the segment; seta 2-VI inserted lateral to seta 3-VI; comb scales large, spine-like, fringed on either side at the base; siphon short, stout, with pronounced forward curvature, spiracular lobes unusually large; pecten borne on basal 0.33 of the siphon; seta 1-S consists of 5 setae, 4 inserted at the posterior midline, considerably longer than the diameter of the siphon, fifth seta inserted far lateral to 2 middle posterior setae; saddle complete, rather long; seta 1-X short, branched; seta 2-X branched; ventral brush (seta 4-X) composed of 6 or 7 pairs of highly branched setae on a grid; anal papillae of variable length, ventral pair usually, but not always, longer than the dorsal pair. PUPAE: Trumpet more or less cylindrical, pinna short; seta 1-II with 8–15 branches; seta 2-II inserted anterolateral to seta 3-II, seta 2-III–V, and usually 2-VI,VII, inserted mesad of seta 1; seta 1-III–V strongly developed, multi-branched, about the length of the following tergum; seta 5-V,VI considerably longer than the following tergum; seta 6-III–VI branched, branches flexible; seta 9-VIII inserted on the sternum some distance from the lateral and posterior margins; caudolateral angle of segment VIII acutely produced; paddle with darkened area on the distal 0.5 or less and mainly on the inner part, outer and inner parts about the same width; setae 1,2-Pa present.

****

Bionomics: The immature stages of the type species are found in sunlit bodies of clean standing water with thick vegetation. Females have been found to enter houses and readily attack humans near larval habitats. They are not known to vector any pathogens of human disease; however, Rift Valley fever virus has been isolated from specimens collected in the Senegal River Basin, suggesting that *Cx. poicilipes* may play a role in the transmission of the virus in the Afrotropical Region.

****

Principal references: Edwards, 1941 (*Cx. poicilipes* as a species of subgenus *Culex*, adults and pupa, distribution); Harbach, 1988 (*Cx. poicilipes* as a species of subgenus *Culex*, description, bionomics, distribution); Harbach *et al.*, 2012 (morphology, phylogenetic relationships); Harbach, 2023 (subgeneric status, taxonomy, morphological distinctions).

Lasiosiphon Kirkpatrick, 1925

Lasiosiphon Kirkpatrick, 1925a, originally proposed and currently recognized as a subgenus of *Culex*, is typified by *Culex adairi* Kirkpatrick, 1926 [replacement name for *Culex pluvialis* Kirkpatrick, 1925a]. Subgenus abbreviation – *Las.*

****

Systematics: *Lasiosiphon* is monobasic. It only includes the extant type species, which is known only from the desert areas of northern Africa and southwestern Asia. The affinities of *Lasiosiphon* are uncertain, but it appears to be more closely related to *Afroculex* and *Maillotia* than to other taxa in the Palaearctic and Afrotropical Regions. Members of these groups were placed in the same clade in the equally weighted and Bayesian analyses of St John (2007) and the parsimony analyses of Harbach *et al.* (2012) based on morphological data.

****

Characteristics: The single species of *Lasiosiphon* exhibits the following principal anatomical features. ADULTS: Small yellow-brown mosquito without special ornamentation; scales of the vertex, antepronotum and scutellum all narrow; maxillary palpus of females short, less than 0.25 the length of the proboscis, maxillary palpus of males shorter than the proboscis, slender, without setae; proboscis slightly longer than the forefemur; lower mesepimeral seta present; wing with linear scales, vein R_{2+3} shorter than cell R_2 in females, about the same length in males; hindtarsomere 1 somewhat shorter than the tibia; ungues of females simple, fore- and midungues of males each with a tooth, hindungues simple; pulvilli rudimentary; abdominal terga with narrow inconspicuous apical pale bands or apical lateral pale spots only. MALE GENITALIA: Gonocoxite without scales; subapical lobe prominent, setae not highly modified, without foliform seta; gonostylus simple; paraprocts with an exceptionally long basal arm, spicules of the crown numerous but short and weak; phallosome comprised of 2 pairs of slender rods (distinction from all other Culicini). LARVAE: Antenna noticeably longer than the head capsule; setae 5,6-C minute (distinction from all other Culicini); integument of the thorax and abdomen densely covered with minute spicules; siphon moderately long with about 16 long seta 1-S in a single slightly zigzag posterior row extending from near the base to near the apex of the siphon and 2 very small setae inserted on the distal 0.5 on either side of the posterior row; saddle narrowly incomplete ventrally; ventral brush (seta 4-X) strongly developed, with 1 precratal seta. PUPAE: Seta 7-I,II very long, seta 7-I longer than seta 6-I; seta 2-II inserted far mesad of seta 1-II; seta 5-IV–VI longer than the following tergum; seta 1-IX absent; paddle very large, ovoid; seta 2-Pa absent.

****

Bionomics: Larvae of *Cx. adairi* are found in clear freshwater pools with rocky bottom. Nothing is known about the bionomics of the adults, but females are unlikely to feed on humans. The species is not of medical importance.

****

Principal references: Kirkpatrick, 1925a (new subgenus, type species [as *Cx. pluvialis*], description, bionomics); Kirkpatrick, 1925b (as *Cx. pluvialis*, description); Rioux & Quézel, 1957 (*Cx. adairi*, description, bionomics); St John, 2007 (morphology, phylogenetic relationships); Harbach *et al.*, 2012 (morphology, phylogenetic relationships).

Lophoceraomyia Theobald, 1905

Lophoceraomyia, originally proposed as a genus by Theobald (1905h), currently classified as a subgenus of *Culex*, is typified by *Culex uniformis* (Theobald, 1905h), original combination: *Lophoceraomyia uniformis*. Subgenus abbreviation – *Lop.*

****

Systematics: *Lophoceraomyia* includes 113 extant species. The group is largely confined to the tropical and subtropical areas and islands of the Oriental and Australasian Regions, with a small extension into southeastern areas (China, Japan and Korea) of the Palaearctic Region. Belkin (1962) noted that *Lophoceraomyia* bears similarities to *Culiciomyia*, *Eumelanomyia* (as *Mochthogenes*) and *Neoculex*. The results of the phylogenetic study by St John (2007) based on morphological data indicated that *Lophoceraomyia* is closer to *Culiciomyia* and *Eumelanomyia* than to *Neoculex*. Sirivanakarn (1971, 1972) stated that *Lophoceraomyia* exhibits the strongest affinity with *Eumelanomyia*. In the phylogenetic analyses of Harbach *et al.* (2012), also based on morphological data, *Lophoceraomyia* was placed in a clade with species of *Acalleomyia*, *Culiciomyia* and *Eumelanomyia*, related as *Eumelanomyia* + (*Lophoceraomyia* + (*Acalleomyia* + *Culiciomyia*)).

****

Characteristics: Species of *Lophoceraomyia* exhibit the following morphological characteristics. ADULTS: Small to medium-sized mosquitoes; vertex usually with a more or less distinct orbital line of broad decumbent scales; eyes contiguous above the antennae; frontal tuft not developed; antenna distinctly longer than (females) or about as long (males) as the proboscis, pedicel without distinct scales, pedicel of males with or without a blunt prominence on the mesal side, flagellum of males with specialized tufts of scales and setae on flagellomeres 5–9 or 10, sometimes restricted to flagellomeres 7 and 8, rarely entirely absent; proboscis, maxillary palpus, wing and tarsi dark-scaled; proboscis very slender, particularly at the base; proboscis of males with or without a false joint, dorsal surface with or without a double row of setae, ventral surface usually with a few rows of short setae, sometimes with a group of long setae at mid-length; maxillary palpus of females with 3 palpomeres; palpus of males with 5 palpomeres, variable in length, usually slightly longer than the proboscis, sometimes reduced to 0.5 the length of the proboscis, palpomere 1 with or without 2 pairs of finger-like slender processes on the lateral and ventral surfaces, palpomere 3 with 1 or 2 ventral rows of short setae, palpomeres 4 and 5 with a few to numerous setae; acrostichal setae usually absent, rarely present; scutal scales sparse, narrow and dark; postpronotum sometimes with a few scales; pleural scaling practically absent, sometimes a few scales along the mesokatepisternal setae, lower mesepimeral seta usually present; wing cell R_2 longer than vein R_{2+3}; ungues of females small, equal and simple; fore- and midungues enlarged in males, with basal denticles; pulvilli well developed; abdominal terga dark-scaled or with basolateral pale spots or basal pale bands. MALE GENITALIA: Small and inconspicuous, withdrawn into segment VIII; ninth tergal lobes poorly defined, widely separated, with 3–5 setae; gonocoxite small, without scales; subapical lobe prominent, not clearly divided, setae densely packed but proximal and distal groups discernible, proximal group usually with 3 rod-like setae (*a–c*), distal group with a mesal subgroup usually comprised of a foliform seta and 4–6 variously developed setae (*d–f*) and a lateral subgroup with 1 foliform seta (*g*?) and an accessory seta (*h*); gonostylus relatively simple; gonostylar claw small, subapical; phallosome simple, broad and bulbous at the base in dorsal aspect and with a rather broad basal tergal bridge, lateral plate varied in development, with a strong sternal process and with or without a smaller tergal process;

paraproct with a crown comprised of a variable number of small to large spicules, basal sternal process absent; cercal setae present. LARVAE: Head usually broader than long; ocular bulge usually prominent; antenna shorter than the head, seta 1-A inserted distinctly beyond mid-length, setae 2,3-A usually removed from the apex; seta 1-C moderately thickened, with or without lateral spicules; seta 2-C absent; seta 3-P much shorter than setae 1,2-P; seta 7-I large, developed similar to seta 6-I, seta 7-II usually small and developed like seta 7-III–VI, sometimes developed like seta 7-I; comb comprised of numerous fringed scales in a patch; siphon usually long and slender; pecten well developed, spines evenly fringed or spiculate on the ventral margin; seta 1-S with 3–8 pairs of weak elements inserted distad of the pecten; saddle complete, without an acus, caudolateral margin lightly to strongly spiculate; seta 1-X usually short and weak; seta 2-X with a strong primary stem and 1–4 weaker basal branches; ventral brush (seta 4-X) with 5 or 6 pairs of setae borne on a grid. PUPAE: Trumpet usually long and narrow, pinna small, with or without a slit extending to the meatus; seta 5-IV–VI strong, 1–2 times as long as the following tergum; seta 9-VIII well removed cephalad from the caudolateral angle of the segment; paddle large, generally ovoid, outer margin smooth or minutely spiculate basally; setae 1,2-Pa both usually present, minute.

****

Bionomics: The immature stages of species of *Lophoceraomyia* have been found in ground pools (swamps, streams and temporary pools), tree holes, bamboo stumps, crab holes, rock holes, pitcher plants, palm bracts, sago stumps and leaf axils of aroids, sago palms and *Pandanus*, and artificial containers. Very little is known about the bionomics of the adults. Females probably feed principally on birds, but may also feed on amphibians, reptiles and mammals. They apparently will attack humans in forest, but the species are unlikely to play a role in the transmission of pathogens to humans or domestic animals. However, based on the isolation of a number of arboviruses from species in Malaysia and northern Australia, species of *Lophoceraomyia* probably play a role in the maintenance of viruses in sylvatic cycles.

****

Principal references: Belkin, 1962 (South Pacific, taxonomy, keys, descriptions, distributions, bionomics); Colless, 1965 (Peninsular Malaysia and Singapore, taxonomy, keys, description, bionomics, distributions); Bram, 1967a (Thailand, keys, descriptions, distributions, bionomics); Sirivanakarn, 1968 (New Guinea and Bismarck Archipelago, taxonomy, classification, keys, descriptions, bionomics, distributions); Sirivanakarn, 1977a (Oriental Region, taxonomy, classification, keys, descriptions, bionomics, distributions); Lee *et al.*, 1989b (Australasian Region, taxonomy, keys, literature, distributions, bionomics); Lu *et al.*, 1997a (China, keys, taxonomy, descriptions, distributions, bionomics); Harbach, 2011 (informal classification); St John, 2007 (morphology, phylogenetic relationships); Harbach *et al.*, 2012 (morphology, phylogenetic relationships).

Maillotia Theobald, 1907

Maillotia, originally proposed as a genus by Theobald (1907), currently classified as a subgenus of *Culex*, is typified by *Maillotia pilifera* Theobald, 1907 [subjective synonym of *Culex hortensis* Ficalbi, 1889a]. Subgenus abbreviation – *Mai.*

****

Systematics: *Maillotia* includes 11 extant species, which principally occur in areas of the Afrotropical Region, northern Africa and southwestern Asia. One species (*Cx. hortensis*) extends into Spain, Portugal and Eurasia. The phylogenetic relationships of *Maillotia* are uncertain. It appears to be most closely related to *Afroculex*, *Barraudius* and *Lasiosiphon*. Sirivanakarn (1972) hypothesized a close relationship between *Maillotia* and *Eumelanomyia*, *Lophoceraomyia* and *Neoculex* based on similarities of the adults, and suggested that *Maillotia* is the more primitive of these groups. The parsimony analysis of Mallampalli (1995), based on morphological characters of larvae, pupae and adults, indicated that *Maillotia*, *Allimanta* and *Culex sensu stricto* are the most primitive taxa of *Culex*. The morphological data analysed in the studies of St John (2007) and Harbach *et al.* (2012) indicated that *Maillotia* is most closely related to *Afroculex*. In a molecular phylogenetic analysis of mitochondrial genomes representing one species each of *Maillotia*, *Aedinus* and *Phenacomyia* and 23 species of *Culex sensu stricto* (da Silva *et al.*, 2020a), *Maillotia* was recovered in a position between *Aedinus* and a clade comprised of the other species, which does not shed light on the true likely affinities of the group.

****

Characteristics: Species of *Maillotia* exhibit the following morphological characteristics. ADULTS: Medium-sized mosquitoes; decumbent scales of the vertex narrow and pale; ocular line with narrow scales; maxillary palpus of females short, palpus of males shorter to longer than the proboscis, slender, almost devoid of setae; proboscis longer than the forefemur; scutum with rather coarse predominantly pale scales; thoracic pleura with extensive scale-patches, prealar and postspiracular scales usually present; lower mesepimeral seta present; wing dark-scaled; legs mainly dark-scaled, femora with distinct or inconspicuous pale knee spots; hindtarsomere 1 as long as or only slightly shorter than the hindtibia; abdominal terga with apical pale bands. MALE GENITALIA: Ninth tergal lobes widely separated, small, bearing setae; gonocoxite with or without scales; subapical lobe median or postmedian in position, divided or undivided (usually), 2 or 3 rod-like setae always present, setae *d–f* and a foliform seta usually present; phallosome short, broad, oval, subspherical or cup-shaped in tergal view, tergal bridge at or just above mid-length of the lateral plates, denticles not developed or sometimes only a few present on the apex; proctiger heavily sclerotized with a crown of flat blunt spicules arranged in a row or sometimes mixed with coarse pointed spicules in a tuft. LARVAE: Antenna long, seta 1-A inserted near mid-length, setae 2,3-A inserted some distance before the apex; setae 5,6-C long and conspicuous, much thicker and longer than seta 4-C; seta 3-P nearly as long as setae 1,2-P; comb scales fringed, in a patch; siphon with an anterolateral row of setae; seta 1-S in close-set rows or a single median posterior row; seta 2-S well developed; saddle complete; seta 2-X branched; seta 3-X single; ventral brush (seta 4-X) with 5–7 pairs of setae borne on a grid and 1 or more (usually) precratal setae. PUPAE: Seta 5-V,VII nearly as long or longer than 2 terga; seta 6-III–VI branched; setae 1,2-Pa both present.

****

Bionomics: The immature stages of species of *Maillotia* are found in shaded ground pools, stream pools, rock pools and other collections of groundwater. Very little is known about the bionomics of the adults. The species are not of medical importance.

****

Principal references: Sirivanakarn, 1971 (taxonomy, keys and descriptions of species groups); Lu *et al.*, 1997a (China, *Cx. hortensis*, taxonomy, descriptions, distribution, bionomics); Becker *et al.*, 2003, 2010, 2020 (Europe, subgenus and *Cx. hortensis*, description, bionomics, distribution); Harbach, 2011 (informal classification); Harbach *et al.*, 2012 (morphology, phylogenetic relationships).

Melanoconion Theobald, 1903

Melanoconion, originally proposed as a genus by Theobald (1903a), currently classified as a subgenus of *Culex*, is typified by *Culex atratus* Theobald, 1901c. Subgenus abbreviation – *Mel.*

****

Systematics: *Melanoconion* includes 179 extant species. The group ranges from the southern part of the USA southward though most of Brazil, Paraguay and northern Argentina, and to Peru on the western side of the Andes Mountains. Sirivanakarn (1983) suggested that the *Melanoconion* lineage may have given rise to *Aedinus*, *Anoediopora*, *Belkinomyia*, *Carrollia*, *Micraedes*, *Microculex* and *Tinolestes* based on morphological similarities shared with each of these groups. The results of a cladistic analysis based on implied weighting of morphological data by St John (2007) supported this hypothesis. The cladistic analyses of Harbach *et al.* (2012) supported the monophyly of the Melanoconion Group comprised of *Aedinus*, *Anoedioporpa*, *Belkinomyia*, *Carrollia*, *Melanoconion*, *Micraedes*, *Microculex* and *Tinolestes*, and indicated, as was suggested by Berlin (1969b), Adames & Galindo (1973) and Sirivanakarn (1983), that *Melanoconion*, *Deinocerites* and *Galindomyia* comprise the most derived mosquitoes of the Culicini. Torres-Gutierrez *et al.* (2018) investigated the phylogenetic relationships of 43 species of *Melanoconion* based on fragments of two single-copy nuclear genes (*CAD*, *HB*) and a mitochondrial gene (*COI*). The results supported the monophyly of the Melanoconion and Spissipes Sections and a few groups and subgroups, but the monophyly of the Vomerifer and Crybda Groups of the latter section was doubtful. The authors emphasized the need for further investigation based on additional genes and species represented by accurately identified specimens, *i.e.* males and females with associated larval and pupal exuviae.

****

Characteristics: Species of *Melanoconion* exhibit the following anatomical features. ADULTS: Small to medium-sized mosquitoes; antenna as long as the proboscis; decumbent scales of the vertex varied, narrow, linear or falcate in the centre to entirely broad; maxillary palpus and proboscis entirely dark-scaled; palpus with 3 or 4 palpomeres in females, palpomere 4 vestigial or absent; palpus of males with 5 palpomeres, usually longer than the proboscis, sometimes slightly shorter, palpomeres 4 and 5 usually strongly setose; scales on the anterior 0.5 of the scutum usually entirely dark, infrequently partially pale, producing a distinct pattern; acrostichal setae absent (except *Cx. spissipes*); antepronotum with a row of setae only, scales absent; scales present on the lower posterior surface of the mesokatepisternum, present or absent on the upper corner and upper median area of the mesanepimeron; legs usually dark-scaled, femora sometimes with pale knee spots and/or pale rings at the joints of the hindtarsomeres; wing scales varied in shape; abdominal terga usually with distinct basolateral pale spots, sometimes with complete basal pale bands, rarely entirely dark-scaled. MALE GENITALIA: Ninth tergal lobes varied in size and form, narrowly to rather widely separated; gonocoxite varied in shape, with scales; subapical lobe subapical in position, divided, proximal and distal divisions strongly developed, widely separated, each division with a basal columnar stem of varied length, stem of the proximal division distally divided or undivided, bearing 2 apical rod-like setae, stem of the distal division shorter or as long as the stem of the proximal division, bearing 1 hooked rod, 2 blade-like setae, 3 or 4 appressed foliform setae and/or 1 or more variously modified setae; gonostylus at least 0.5 the length of the gonocoxite, varied in form; phallosome not columnar, aedeagus usually without an upper tergal bridge connecting the lateral plates; lateral plate varied in development, with a prominent basal hook; paraproct with a comb-like crown of 5–12 blunt or pointed blades; cercal setae present or absent. LARVAE: Head broader than long, usually with prominent ocular bulges; antenna as long as or slightly shorter than the head; seta 1-A usually large, rarely reduced, inserted beyond mid-length of the antenna; seta 2-C usually absent, sometimes present; seta 5-C weakly to strongly developed, single or branched; seta 6-C long, always single; seta 3-P much shorter then setae 1,2-P; seta 7-I large, similar to seta 6-I, seta 7-II small, similar to seta 7-III–VI; comb usually comprised of numerous fringed scales in a patch, sometimes with relatively few rather large spine-like scales in a patch or single row; setae 1,2-VIII usually with a basal sclerotized plate; siphon tubular, distally tapering, length varied; pecten spines in a close-set row, rarely widely spaced, usually with fine barbs or numerous denticles; seta 1-S in 3–8 posterolateral pairs, length varied, usually inserted beyond the pecten, sometimes placed within the pecten, 2 (sometimes 3 or 4) dorsolateral setae; seta 2-S usually well developed, hooked and usually with a recurved tooth; saddle complete, without an acus, usually without spines, inconspicuous spicules usually present; ventral brush (seta 4-X) with 5 or 6 pairs of setae borne on a grid. PUPAE: Trumpet with a distinct meatal slit extending from the pinna, rarely absent; setae 1,5-III–VI usually with several branches; seta 9-VIII placed at or removed cephalad of the caudolateral angle of the segment; caudal lobe of tergum VIII not overlying the lateral part of tergum IX; paddle broad, ovate, outer margin without spicules; setae 1,2-Pa both present.

****

Bionomics: The immature stages of species of *Melanoconion* are found in a variety of groundwater habitats in heavy or partial shade or full sunlight, including forest pools, jungle swamps, ponds, marshes, lakes, margins of streams and canals containing plants or decaying plant material. Eggs are laid individually or in a batch. Adults are active at night, and females apparently feed on a variety of vertebrates, including humans. A number of species are important actual or potential vectors of Venezuelan equine encephalitis virus and other arboviruses.

****

Principal references: Rozeboom & Komp, 1950a (review, key to males, annotated species list); Sirivanakarn, 1983 (taxonomy, classification, morphology, bionomics, medical importance, distribution); Pecor *et al.*, 1992 (illustrated catalogue, species listings, type data, synonyms, literature, distributions); Sallum & Forattini, 1996 (Spissipes Section, classification, keys, morphology, descriptions, distributions, bionomics); Navarro & Weaver, 2004 (Vomerifer and Pedroi Groups, molecular phylogeny); Harbach, 2011 (informal classification); Torres-Gutierrez & Sallum, 2015 (catalogue of South American species, classification, species listings, type data, synonyms, distributions, literature); Harbach *et al.*, 2012 (morphology, phylogenetic relationships); González *et al.*, 2023 (Neotropical subgenera of *Culex*, comparative morphology, molecular associations).

Micraedes Coquillett, 1906

Micraedes, originally proposed as a genus by Coquillett (1906a), currently classified as a subgenus of *Culex*, is typified by *Culex bisulcatus* (Coquillett, 1906a), original combination: *Micraedes bisulcatus*. Subgenus abbreviation – *Mca.*

****

Systematics: *Micraedes* includes eight extant species, which are confined to the American Mediterranean Region *sensu* Belkin (1962). Species are recorded from the Antilles, Costa Rica, Cuba, Jamaica, Mexico, Panama, Tobago, Trinidad and the USA (Florida). *Micraedes* appears to be closely related to *Microculex* and *Carrollia*. It was recovered as the sister of *Carrollia* in a clade comprised of members of these three groups in the parsimony analyses of Harbach *et al.* (2012) based on morphological data.

****

Characteristics: Species of *Micraedes* exhibit the following morphological characteristics. ADULTS: Small mosquitoes; proboscis, maxillary palpus, wing and legs dark-scaled; decumbent scales of the vertex narrow; antenna subequal to the proboscis; proboscis distinctly longer or subequal to the forefemur; maxillary palpus of females with 3 palpomeres; palpus of males with 5 palpomeres, palpomere 5 sometimes minute; scutum with narrow curved scales; acrostichal setae present; postpronotum with a few scales on the upper margin; pleural integument of the thorax generally yellowish; pleural scaling usually absent, a few scales may be present along the mesokatepisternal setae and among the upper mesepimeral setae; lower mesepimeral seta present or absent; ungues of all legs simple in females, fore- and midungues enlarged and unequal in males, larger unguis with a submedian tooth, hindungues as in females; abdominal terga with basolateral pale patches. MALE GENITALIA: Ninth tergal lobes small, approximated, usually distinct and mound-like with a varied number of setae; gonocoxite roughly triangular in outline, moderately inflated, usually without scales; subapical lobe more or less distinctly divided in most species, proximal division usually with 2 rod-like setae, distal division with or without a prominent foliform seta; gonostylus relatively simple, attenuate distally or distal 0.33 thickened, usually with a row of fine ridges distally on the sternal margin and 2 distinct short subapical setae on the tergal surface; gonostylar claw broadened distally, inserted subapically; phallosome not columnar, lateral plate with a broadly sclerotized basal hook, a long posteriorly directed apical process and a sternal spine; paraproct strongly developed, produced into a conspicuous dorsal conical lobe below the ninth tergal lobe, with a crown of 6–13 teeth; cercal setae present. LARVAE: Antenna relatively short and slender, uniformly tapered, seta 1-A inserted near mid-length of the shaft, branched; setae 2,3-C not developed; setae 5–7-C always multi-branched, with at least 4 branches, long, subequal in length, setae 5,6-C distinctly longer than the antenna, usually with 4 branches; setae 0-P, 1,3,14-M and 1,2,13-T stellate; setae 1,3-P at least moderately long, multi-branched; setae 1,2,4,11,13-I, 1,2,5,9,13-II, 1,2,5,7,9,13-III–VI and 1,2,5,8,10,13-VII usually stellate; seta 2 always inserted on the anterior area of the tergum; seta 7-I long, double, seta 7-II developed like seta 7-I; comb scales in a patch; siphon long, pecten on proximal 0.4; seta 1-S strongly developed, inserted beyond the pecten, comprised of 4 pairs of posterolateral setae and 2 pairs of anterolateral setae; saddle complete, with long marginal spines on the caudolateral border; seta 1-X short to moderately long, single, double or forked; seta 2-X long, at least double, without postbasal branches; seta 3-X long, single; ventral brush (seta 4-X) with 5 pairs of setae borne on a grid with distinct lateral grid bars. PUPAE: Trumpet moderately long; seta 5-CT strongly developed, at least twice as long as seta 4-CT; setae 10,11-CT inserted moderately close together, removed from seta 12; seta 2-III–VII inserted cephalad or anteromesal to seta 1, seta 2-VII always shorter than 0.8 the length of the tergum; seta 9-II–VI inserted anterolateral to seta 6; seta 9-VIII long, triple, either subequal to or distinctly longer than tergum VIII; caudal lobe of tergum VIII overlies the lateral part of tergum IX; seta 1-IX absent; paddle margins with distinct hair-like spicules in some species; seta 1-Pa present, seta 2-Pa absent.

****

Bionomics: The immature stages of species of *Micraedes* are found in leaf axils, primarily in epiphytic and terrestrial bromeliads and rarely in aroids. Very little is known about the bionomics of the adults. Collections of resting adults suggest that species of *Micraedes* are active during crepuscular or nocturnal hours. Species are not known to be of medical or economic importance.

****

Principal references: Berlin, 1969b (taxonomy, descriptions, keys, bionomics, distributions); Harbach *et al.*, 2012 (morphology, phylogenetic relationships); González *et al.*, 2023 (Neotropical subgenera of *Culex*, comparative morphology).

Microculex Theobald, 1907

Microculex, originally proposed as a genus by Theobald (1907), currently classified as a subgenus of *Culex*, is typified by *Microculex argenteoumbrosus* Theobald, 1907 [subjective synonym of *Culex imitator* Theobald, 1903a]. Subgenus abbreviation – *Mcx.*

****

Systematics: *Microculex* includes 34 extant species with distributions in the Neotropical Region, from Mexico to northern Argentina. *Microculex* appears to be related to *Micraedes* and *Carrollia*. It was recovered in a paraphyletic basal relationship to the reciprocally monophyletic *Micraedes* and *Carrollia* in the parsimony analyses of Harbach *et al.* (2012) based on morphological data.

****

Characteristics: Species of *Microculex* exhibit the following morphological characteristics. ADULTS: Small mosquitoes; head with narrow scales on the vertex and orbital line; maxillary palpus of females with 3(?) palpomeres, palpus of males as long as or longer than the proboscis, with 5 palpomeres; scutum with a pattern of dark, golden or silvery scales; acrostichal setae present or absent; antepronotum with narrow scales; mesokatepisternum with scales; lower mesepimeral seta present; veins R_2 and R_3 of the wing with narrow scales; legs entirely dark-scaled (some species) or tarsi with basal pale bands (most species); abdominal terga dark-scaled. MALE GENITALIA: Tergum IX lobes distinct, widely separated; gonocoxite without scales; subapical lobe subapical in position, little if at all divided, proximal division with 2 rod-like setae, distal division weakly developed but usually bearing a number of setae; gonostylus simple, tapered distally; phallosome not columnar, lateral plate with a long basal hook; paraproct with a single row of blunt spicules. LARVAE: Antenna shorter than the head, constricted beyond the insertion of seta 1-A, seta 1-A inserted about 0.8 from the base; seta 2-C weakly developed or absent; seta 3-P shorter than setae 1,2-P; comb with many scales in a patch; siphon usually very long and slender; seta 1-S inserted distal to the pecten; saddle long, complete; seta 1-X small, branched; setae 2,3-X long, single; ventral brush (seta 4-X) with 4 pairs of setae borne on a grid. PUPAE: Trumpet long and slender, tracheoid area beyond the base; meatus without a slit from the proximal part of the pinna; seta 2-III inserted mesad of seta 1; seta 2-VI inserted laterad of seta 1; seta 9-VIII relatively long, subequal to the length of tergum VIII; caudal lobe of tergum VIII not overlying the lateral part of tergum IX; seta 1-Pa present, seta 2-Pa absent.

****

Bionomics: The immature stages of species of *Microculex* inhabit epiphytic bromeliads, and less frequently bamboo internodes. Eggs are enclosed separately in gelatinous envelopes. Lane (1953) indicated that adult females prefer to feed on birds. Species of *Microculex* are not of medical importance.

****

Principal references: Edwards, 1932a (characteristics, distribution, bionomics); Rozeboom & Komp, 1950b (taxonomy, *Cx. elongatus*, description, key to males); Lane & Whitman, 1951 (Brazil, keys, classification, species descriptions, bionomics); Lane, 1953 (descriptions, keys for adults and larvae, distributions); Coutinho & Forattini, 1962 (three species, descriptions, comparative morphology); Cotrim & Galati, 1977 (Pleuristriatus Series, taxonomy, descriptions, keys); Harbach *et al.*, 2012 (morphology, phylogenetic relationships); González *et al.*, 2023 (Neotropical subgenera of *Culex*, comparative morphology).

Neoculex Dyar, 1905

Neoculex, originally proposed as a genus by Dyar (1905a), currently classified as a subgenus of *Culex*, is typified by *Culex territans* Walker, 1856. Subgenus abbreviation – *Ncx*.

****

Systematics: *Neoculex* includes 26 extant species, predominantly distributed in the Old World but several species occur in the Nearctic Region. *Neoculex* does not appear to be a monophyletic group (St John, 2007; Harbach *et al.*, 2012). It includes several morphologically disparate lines that share various similarities with *Culiciomyia*, *Eumelanomyia*, *Lophoceraomyia* and *Melanoconion*. Sirivanakarn (1972) hypothesized a close relationship between *Eumelanomyia*, *Lophoceraomyia*, *Maillotia* and *Neoculex* based on similarities of the adults, and Sirivanakarn (1983) noted that *Melanoconion* shares several anatomical features with the 'more primitive' *Neoculex*.

****

Characteristics: Species of *Neoculex* exhibit the following morphological characteristics. ADULTS: Small to medium-sized mosquitoes; maxillary palpus of females very short, with 3 palpomeres, sometimes with a very small palpomere 4; maxillary palpus of males as long as or longer than the proboscis, with 5 palpomeres, palpomere 3 at most with short setae in a ventromesal line, palpomeres 4 and 5 with or without long setae; scutal scales usually predominantly pale, sometimes arrayed in a striking pattern of colouration or entirely dark; acrostichal setae present; thoracic pleura usually with extensive patches of scales on the proepisternum, the upper area and lower posterior border of the mesokatepisternum, the upper area of the mesepimeron and sometimes on the postspiracular and prealar areas, scale-patches rarely entirely absent; lower mesepimeral seta present or absent, middle and lower areas of the mesepimeron sometimes with numerous small setae; wing and legs dark-scaled; abdominal terga with or without basal or apical pale bands, sometimes with apicolateral pale spots. MALE GENITALIA: Tergum IX lobes varied, distinct or weakly developed, with or without setae; gonocoxite without scales; subapical lobe always with 2 long rod-like setae on the proximal division, distal division with only narrow flattened setae, foliform setae absent; phallosome columnar, H-shaped in tergal view with the tergal bridge above the midpoint or near the apex of the lateral plates, rarely at mid-length, a few denticles present or absent, without a basal hook on each side; paraproct with a crown of flat blunt spicules in a single row, sometimes also with a few coarse pointed spicules. LARVAE: Antenna long, seta 1-A inserted near mid-length or more distal; setae 2,3-A inserted a short distance from the apex; seta 3-P shorter than setae 1,2-P, less than 0.5 the length of seta 1-P; comb scales small, apically fringed, in a patch; siphon long, slender, with numerous posterolateral seta 1-S, at least 1 seta out of line; segment X long, saddle complete, posterior margin with distinct caudal spicules; seta 2-X with 2 or more branches; seta 3-X single; ventral brush (seta 4-X) comprised of 6 or 7 pairs of setae borne on a grid and 1 or more precratal setae. PUPAE: Meatus of trumpet without a slit from the proximal part of the pinna; seta 2-III,VI inserted mesad of seta 1; seta 9-VII longer than seta 3-VII; caudal lobe of tergum VIII not overlying the lateral part of tergum IX; seta 9-VIII inserted at or immediately adjacent to the caudolateral angle of the segment; setae 1,2-Pa both present.

****

Bionomics: The immature stages of species of *Neoculex* inhabit permanent and temporary bodies of groundwater, as well as rock holes, crab holes, tree holes, bamboo internodes and artificial containers. Adult females apparently do not feed on humans. Females of some species are known to feed on frogs, small mammals and birds. Egg rafts are apparently laid above the water line and the newly hatched larvae wriggle their way to the water. Species of *Neoculex* are not known to be of medical or economic importance.

****

Principal references: Belkin, 1962 (South Pacific, taxonomy, keys, descriptions, distributions, bionomics); Bram, 1967a (Thailand, keys, descriptions, distributions, bionomics); Sirivanakarn, 1971 (taxonomy, keys and descriptions of species groups); Lee *et al.*, 1988b (Australasian Region, taxonomy, keys, literature, distributions, bionomics); Lu *et al.*, 1997a (China, keys, taxonomy, descriptions, distributions, bionomics); Becker *et al.*, 2003, 2010, 2020 (Europe, subgenus, species descriptions, bionomics, distributions); Harbach *et al.*, 2012 (morphology, phylogenetic relationships).

Nicaromyia González Broche & Rodríguez R., 2001

Nicaromyia González Broche & Rodríguez R., 2001, originally proposed and currently recognized as a subgenus of *Culex*, is typified by *Culex nicaroensis* Duret, 1967a. Subgenus abbreviation – *Nic.*

****

Systematics: *Nicaromyia* is monobasic. It only includes the extant type species, which is known only from Cuba. The affinities of *Nicaromyia* are unknown. The male genitalia share similarities with members of the Melanoconion Group. The immature stages of *Nicaromyia* and *Tinolestes* are similar, and those of both inhabit crab holes. Seta 1-C and the pecten spines of larvae are developed like their homologues in members of the Sitiens Group of *Culex sensu stricto*. *Nicaromyia* was recovered in a basal relationship to species of *Anoedioporpa*, *Aedinus*, *Belkinomyia*, *Melanoconion* and *Tinolestes* in the parsimony analyses of Harbach *et al.* (2012) based on morphological data.

****

Characteristics: The single species of *Nicaromyia* exhibits the following principal anatomical features. ADULTS: Medium-sized dark mosquitoes; maxillary palpus of females short, about 3 times the length of the clypeus, palpus of males as long as the proboscis; proboscis slightly longer than the forefemur; acrostichal, supraalar and antealar setae absent; antepronotum without scales; a row of 5–7 scales beside the lower mesokatepisternal setae; wing and legs (except lower surfaces of the femora) dark-scaled; abdominal terga dark-scaled with basolateral pale patches. MALE GENITALIA: Tergum IX lobes produced, closely approximated, setose caudally; gonocoxite short, stout, with scales; subapical lobe subapical in position, divided, proximal division with 2 rod-like setae inserted apically on a long curved cylindrical stem, distal division short and stout, bearing 6 setae in 2 groups and another distally compressed and slightly bent seta between them, proximal group comprised of 3 more or less rod-like setae of different lengths, also with a pair of identical long distally curved rod-like setae and a fine slender seta laterally at their base; gonostylus uniquely developed, about 0.67 the length of the gonocoxite, enlarged in distal 0.5, with a dorsal undulated crest proximal to a prominent horn-like process, lateral surface with a membranous flap-like expansion, apex with an abruptly upturned narrow snout with the gonostylar claw projecting at its base; phallosome not columnar; lateral plate of the aedeagus with a prominent basal hook, a caudally directed apical process and a sternal spine; apex of the paraproct with a crown of 10–15 curved teeth generally in 2 rows. LARVAE: Seta 1-C short, stout, with a blunt tip; seta 2-C absent; seta 12-C inserted well posterior of the level of seta 15-C; seta 13-C inserted nearly at the level of seta 11-C, well separated from seta 12-C; seta 14-C removed from the margin of the head capsule, inserted nearly at the level of seta 15-C; setae 1,2-P single; seta 3-P branched, much shorter than setae 1,2-P; seta 4-T well developed, about as large as setae 2,3-T; seta 7-II dissimilar to seta 7-I, developed like seta 7-III–VII; comb scales numerous, in a patch; siphon long, evenly tapered; pecten comprised of long slender spines bearing a ventral fringe of numerous minute spicules; seta 1-S comprised of 4 posterolateral elements inserted beyond the pecten and 1 anterolateral element inserted near the level of the proximal end of the spiracular apodeme; saddle complete, seta 1-X shorter than the saddle; seta 2-X branched, shorter than seta 3-X; seta 3-X single; ventral brush (seta 4-X) comprised of 6 pairs of setae borne on a grid. PUPAE: Trumpet long, relatively slender, without a meatal cleft; seta 1-I,III–VII not strongly developed, seta 1-VI,VII inserted mesad of seta 2; seta 7-I as long as seta 6-I; seta 5-IV–VI equally long, longer than the tergum (seta 3-IV–VI incorrectly labelled as 5-IV–VI, and reverse, in González Broche & Rodríguez R., 2001); seta 9-VIII with 2 or 3 branches, shorter than the tergum, inserted at the caudolateral corner of the segment; paddle pointed, not rounded apically, without marginal spicules; setae 1,2-Pa both present.

****

Bionomics: The immature stages and resting adults of *Cx. nicaroensis* are found in crab holes in coastal areas with mangrove. The species is not medically important.

****

Principal references: Duret, 1967a (*Cx. nicaroensis*, description); González Broche & Rodríguez R., 2001 (new subgenus, *Cx. nicaroensis*, descriptions, bionomics, distribution); González Broche, 2006 (*Cx. nicaroensis*, description); Harbach *et al.*, 2012 (morphology, phylogenetic relationships); González *et al.*, 2023 (Neotropical subgenera of *Culex*, comparative morphology).

Oculeomyia Theobald, 1907

Oculeomyia, originally proposed as a genus by Theobald (1907), currently classified as a subgenus of *Culex*, is typified by *Oculeomyia sarawaki* Theobald, 1907 [subjective synonym of *Culex infula* Theobald, 1901b]. Subgenus abbreviation – *Ocu.*

****

Systematics: *Oculeomyia* includes 22 extant species with distributions in the Afrotropical, Australasian and Oriental Regions, and far eastern areas of the Palaearctic. The evolutionary relationships of *Oculeomyia* are not clear. Sirivanakarn (1976) suggested that *Oculeomyia* (as the Bitaeniorhynchus Subgroup of the Sitiens Group of *Culex sensu stricto*) is a very old group that may have given rise to the Sitiens Group. Its pairing with *Kitzmilleria* in the morphology-based phylogenetic study of St John (2007), although not inconceivable, is only supported by the unique development of the larval comb. Species of *Oculeomyia* were placed among species of *Culex sensu stricto* in the phylogenetic study of (Harbach *et al.*, 2012), also based on morphological data.

****

Characteristics: Species of *Oculeomyia* exhibit the following principal anatomical features. ADULTS: Rather large mosquitoes; maxillary palpus of females comprised of 3 palpomeres, palpus of males with 5 palpomeres, longer than the proboscis, palpomere 3 with numerous setae, apex of palpomere 5 with pale scales and setae; proboscis with a median pale band, proboscis of males with a prominent median ventral tuft of setae; anterior 0.67 of the scutum usually covered mainly with pale scales; ante- and postpronota with some narrow scales; postspiracular and prealar areas without scales; lower mesepimeral seta absent; wing completely dark-scaled or lightly to heavily speckled with pale scales; anterior surfaces of the femora and tibiae vary from lightly to heavily speckled with pale scales or completely dark-scaled; tarsomeres with basal and apical pale bands; abdominal terga with basal or apical pale bands or both, sometimes completely dark-scaled. MALE GENITALIA: Tergum IX lobes not produced, with relatively few setae in a row; subapical lobe undivided, usually without a foliform seta; gonostylus simple, sickle-shaped; inner division of the phallosome simple or complex, with or without a cluster of finger- or teeth-like processes, sometimes with a subapical lobe bearing 2–4 foliform processes; outer division of the phallosome simple and broad, leaf-like or with distinct median, lateral and sternal spines, sometimes largely reduced or not developed; basal sternal process of the paraproct varied from short, slender and pale to very long, thick and dark, crown composed of numerous spine-like spicules and a few blades; cercal setae present. LARVAE: Head relatively smaller than in other larvae of Culicini; antenna shorter than the head, seta 1-A inserted at mid-length; median labral plate not distinguished, completely fused with the dorsal apotome (distinction from all other genus groups of Culicini); maxilla, short and broad, maxillary palpus more than twice as long and partially fused with the maxillary body, hypostomal sclerite fused with the cranium; maxillary palpus partially fused with the maxillary body and the hypostomal sclerite fused with the cranium; dorsomentum acutely triangular with denticulate or minutely serrated edges; seta 3-P developed similar to setae 1,2-P; seta 7-I similar to seta 6-I, seta 7-II small like seta 7-III–VI; setae 2,4-VIII with 2–7 branches; comb varied, scales usually large, spine-like, few in number; siphon very long and slender, pecten reduced to a few small spines; seta 1-S comprised of 2.5–6 pairs of posterolateral elements; saddle complete; seta 1-X short, branched; seta 2-X branched; seta 3-X single; ventral brush (seta 4-X) comprised of 5 or 6 pairs of setae borne on a grid. PUPAE: Trumpet asymmetrically funnel-shaped, pinna with a wide opening; setae of the cephalothorax (except setae 10,11-CT) very short; seta 1-II usually single to triple or at most 5-branched; seta 2-II inserted anterolateral to seta 3-II, seta 2-III–VI usually inserted mesad of seta 1 and close to the posterior margin of the tergum, seta 2-VII variable in position, often inserted mesad of seta 1; seta 6-III–VI normally single, more or less straight and rigid; seta 9-VIII inserted on the sternum, removed from the lateral and posterior margins of the segment, distinctly shorter than the sternum; caudolateral angle of segment VIII acutely produced; paddle varied in pigmentation, outer part wider than the inner part, margin practically smooth; seta 1-Pa present, rarely absent, seta 2-Pa present.

****

Bionomics: The immature stages of species of *Oculeomyia* inhabit all sorts of freshwater ground pools and stagnant streams containing masses of green algae. Little or nothing is known about the bionomics of the adults of most species. Females probably feed on a variety of avian and mammalian hosts. Females of *Cx. bitaeniorhynchus* are known to bite humans. Most species of *Oculeomyia* are of no medical or economic importance. *Culex bitaeniorhynchus* is a suspected vector of encephalitic viruses and has been found naturally infected with microfilariae.

****

Principal references: Sirivanakarn, 1976 (as Bitaeniorhynchus Subgroup of Sitiens Group of *Culex sensu stricto*, descriptions, bionomics, distributions); Tanaka, 2004 (subgeneric status, larval stage, pupal stage, pupae and distributions of two species in Japan); Harbach *et al.*, 2012 (morphology, phylogenetic relationships); Harbach, 2023 (removal of *Cx. poicilipes*, diagnostic differences).

Phalangomyia Dyar & Knab, 1914

Phalangomyia, originally proposed as a genus by Dyar & Knab (1914), currently classified as a subgenus of *Culex*, is typified by *Culex apicinus* Philippi, 1865. Subgenus abbreviation – *Pha.*

****

Systematics: *Phalangomyia* is monobasic. It only includes the extant type species, which is only recorded from localities in Argentina, Bolivia, Chile and Peru. The evolutionary relationships of *Phalangomyia* are uncertain. *Culex apicinus* was recovered in a sister relationship with *Cx. tramazayguesi*, the type species of *Allimanta*, in the morphology-based phylogenetic study of Harbach *et al.* (2012), both of which share morphological features with Old World taxa, *e.g.* the presence of a sclerotized opisthophallic bridge. However, they were recovered in a basal relationship to a large clade that consisted, with one exception, of New World generic-level taxa. *Culex apicinus* was recovered as the sister of *Cx. scheuberi* in the morphology-based phylogenetic study of species of the subgenus *Culex* conducted by Laurito & Almirón (2013); however, the relationship is unlikely because data were only available for features of the male genitalia of *Cx. scheuberi* and the study was limited to species of *Culex sensu stricto*. In a subsequent morphology-based study of *Culex sensu stricto* conducted by Harbach *et al.* (2017a), phylogenetic analyses largely failed to resolve relationships among the species. In a later study of the Neotropical subgenera of *Culex* conducted by González *et al.* (2023), *Cx. apicinus* and the Atacama form of the species were recovered in a strongly supported sister relationship with *Cx. amazonensis* of *Aedinus* based on a Bayesian and Maximum Likelihood analysis of a data set consisting of sequences for two nuclear protein-coding genes, *CAD* and *HB*, and the mitochondrial *COI* gene.

****

Characteristics: *Phalangomyia* is diagnosed by the following principal anatomical features. ADULTS: Dark brown mosquitoes; first flagellomere of antenna with scales on the inner side; maxillary palpus of females dark brown with pale scales, comprised of 3 palpomeres, palpus of males with 5 palpomeres; proboscis longer than forefemur, with dorsal and ventral pale scaling; scutum with narrow pale scales, except on bare lines between the acrostichal and dorsocentral setae; wing with a patch of pale scales near the middle of veins M_2 and M_3; legs dark-scaled except femora pale-scaled ventrally, forefemur longer than foretarsomere 1, foretarsomere 1 as long as the combined length of foretarsomeres 2–5; terga II–VII dark-scaled with broad basal pale bands. FEMALE GENITALIA: Cerci distinctly longer than the postgenital lobe, 1.6–1.7 times longer; upper vaginal lip and upper vaginal sclerite strongly sclerotized; insula with a patch of prominent rigid setae (not in lateral rows). MALE GENITALIA: Posterior margin of tergum VIII deeply emarginate; tergum IX lobes small, widely separated, each bearing a row of slender flexible setae; gonocoxite modified, with apical cluster of 8–20 long flexible setae, nearly as long as the gonostylus, basoventral margin with a cluster of shorter flexible setae, subapical lobe not developed, setae highly modified and spread over the mesal area from near the base to near the apex, middle of the dorsomesal margin with a protrusion bearing setae *a*, *b* and *c*, foliform seta *g* (seta *h* of Laurito & Almirón, 2013) borne near the base of the gonostylus, homologies of the other setae not apparent; gonostylus slightly tapered and strongly curved, apex abruptly upturned with conical subapical gonostylar claw projecting in the opposite direction, giving the apex a bifurcate appearance; phallosome broad, with a sclerotized opisthophallic bridge (dorsal aedeagal bridge of Harbach & Knight, 1980), lateral plate (opisthophallic sclerite) stout, about as long as aedeagal sclerite, dorsal arm short, slightly shorter than the lateral arm, lateral arm with a series of laterally directed denticles on the caudal margin, ventral arm not distinctly differentiated from denticles of the lateral arm; proctiger modified, paraproct strongly produced basolaterally into a blunt dorsally directed basolateral arm, crown with a large patch of short spine-like spicules. LARVAE: Head, siphon and saddle strongly sclerotized; antenna shorter than the head, seta 1-A inserted about 0.75 from the base; seta 16-C absent; seta 8-P inserted from directly anterior to slightly anteromesal to pleural setal group 9–12-P; seta 7-I,II about the same length, both longer than seta 7-III but much shorter than seta 6-I,II; elements of siphonal seta 1-S (a–e) in approximated posterior rows, distal element (1e-S) slightly out of line (more lateral); siphon relatively short, stout, tapered, index $\leq$ 3.5. PUPAE: Seta 5-IV–VI equally long, about twice the length of the following tergum; seta 8-II sometimes present; seta 13-III–VI (at least alveolus) usually present; seta 1-IX absent; surface of paddle minutely spiculate; seta 2-Pa present.

****

Bionomics: The immature stages of species of *Phalangomyia* can be found throughout the year in temporary to permanent collections of water, including ditches, rock holes, rock pools, river pools and stream margins. Larvae of *Cx. apicinus* have also been found in water tanks and swimming pools (Rossi *et al.*, 2008). Females are not known to bite humans and probably feed on domestic animals and livestock. The species is apparently of no medical importance to humans.

****

Principal references: Rossi *et al.*, 2008 (*Cx. apicinus*, description, bionomics, distribution); Harbach *et al.*, 2012 (*Cx. apicinus*, morphology, phylogenetic relationships); Laurito & Almirón, 2013 (*Cx. apicinus*, morphology, phylogenetic relationships of species of *Culex sensu stricto* in Argentina); Harbach *et al.*, 2017a (*Cx. apicinus*, morphology, taxonomy, phylogenetic relationships); González *et al.*, 2023 (subgeneric status, molecular phylogeny of the Neotropical subgenera of *Culex*).

Phenacomyia Harbach & Peyton, 1992

Phenacomyia Harbach & Peyton, 1992b, originally proposed and currently recognized as a subgenus of *Culex*, is typified by *Culex corniger* Theobald, 1903a. Subgenus abbreviation – *Phc.*

****

Systematics: *Phenacomyia* includes three extant species in areas of the Neotropical Region from Mexico to Uruguay. The phylogenetic relationships of *Phenacomyia* are not clear. It appears to be an ancient group that shares affinities with *Culex sensu stricto* and *Lutzia*. Species of *Phenacomyia* were placed among species of *Culex sensu stricto* in phylogenetic studies of both morphological and molecular data (Deus, 2009; Vesgueiro *et al.*, 2011; Demari-Silva *et al.*, 2011; Harbach *et al.*, 2012; da Silva *et al.*, 2020a).

****

Characteristics: Species of *Phenacomyia* exhibit the following anatomical features. ADULTS: Medium-sized species with a moderate degree of ornamentation; decumbent scales of the vertex narrow; proboscis with a ventral patch of pale scaling beyond mid-length; maxillary palpus of females with 3 palpomeres, palpus of males with 5 palpomeres, longer than the proboscis; scutum with conspicuous pale scaling extending as a lateral band from the anterior dorsocentral line to the anterior part of the supraalar area, with a less conspicuous narrow extension dorsally at the scutal angle often appearing as a small spot at the termination, and on either side of the prescutellar space; upper mesepimeral scales absent; 1–5 lower mesepimeral setae present; wing dark-scaled; tarsi with narrow pale bands across the joints; abdominal terga with basal pale markings. MALE GENITALIA: Resembling those of the Pipiens Group of *Culex sensu stricto*; tergum IX lobes slightly produced, each bearing 12–29 setae in 3–5 rows; subapical lobe of the gonocoxite with setal group *d–f* represented by 1 small hair-like setae (*f*); lateral plate of the phallosome without an inner division, outer division complex, dorsal arm small or absent; paraproct with a dense crown of spicules, with a prominent basal sternal process; cercal setae present. LARVAE: Antenna short, about 0.4 the length of the head, cylindrical, slightly if at all narrowed distally, seta 1-A weakly developed, with a few short branches, inserted at mid-length of the antennal shaft; maxilla rectangular, maxillary brush reduced, brush and pilose area on the dorsal surface removed laterad of the maxillary suture, setae 4,5-Mx relatively far apart, seta 1-Mx inserted about 0.67 from the base of the maxilla; setae 4–7-C inserted in a more or less straight transverse line; seta 8-C inserted anterior to seta 9-C, usually single, longer than seta 10-C; seta 14-C inserted slightly posterior to seta 13-C; seta 3-P with 3 or 4 branches, shorter than setae 1,2-P; seta 7-II as large as seta 7-I; siphon short, with a prominent dorsal seta at mid-length, elements of seta 1-S largely inserted within the length of the pecten; saddle complete, short, usually with short spine-like marginal spicules; setae 2,3-X single; ventral brush (seta 4-X) with 7 pairs of setae borne on a grid. PUPAE: Trumpet short and strongly flared, length of the pinna less than the diameter at the apex; seta 6-I,II hardly if at all longer than seta 7-I,II; seta 2-II inserted lateral to seta 3-II, seta 2-III–VII inserted mesal to seta 1; seta 9-VII,VIII not strongly developed, distinctly shorter than the tergum; seta 1-Pa present, seta 2-Pa present or absent.

****

Bionomics: Larvae of species of *Phenacomyia* are found in a wide variety of permanent or semi-permanent groundwaters, such as puddles, ditches, stream pools and rock holes, and container habitats, including broken or cut bamboo, tree holes, leaf axils of aroids and ground bromeliads, flower bracts of *Heliconia*, fallen leaves, coconut shells and occasionally artificial containers. Females feed on mammals and birds and occasionally bite humans. Whether or not species of *Phenacomyia* are of medical importance is unknown.

****

Principal references: Strickman & Pratt, 1989 (*Cx. corniger*, *Cx. lactator*, as species of *Culex sensu stricto*, descriptions, distributions); Harbach & Peyton, 1992b (new subgenus, diagnosis, characterization); Deus, 2009 (*Cx. lactator*, relationships based on mtDNA, rDNA and a nuclear gene); Demari-Silva *et al.*, 2011 (mtDNA *COI* sequences, phylogenetic relationships); Vesgueiro *et al.*, 2011 (rDNA ITS2 sequences, phylogenetic relationships); Harbach *et al.*, 2012 (morphology, phylogenetic relationships); da Silva *et al.*, 2020a (*Cx. corniger*, phylogenetic relationships); González *et al.*, 2023 (Neotropical subgenera of *Culex*, comparative morphology).

Phytotelmatomyia Rossi & Harbach, 2008

Phytotelmatomyia Rossi & Harbach, 2008, originally proposed and currently recognized as a subgenus of *Culex*, is typified by *Culex renatoi* Lane & Ramalho, 1960. Subgenus abbreviation – *Phy.*

****

Systematics: *Phytotelmatomyia* includes four extant species in the Neotropical Region, along the Río Paraná Basin from Brazil to Argentina, including Paraguay and Uruguay. *Phytotelmatomyia* is a distinct monophyletic group that appears to have its closest affinities with *Culex sensu stricto* and *Phenacomyia*. In the cladistic analyses of morphological data conducted by Harbach *et al.* (2012), species of *Phytotelmatomyia* were placed in a clade comprised of three species of the Pipiens Group of *Culex sensu stricto* and two species of *Phenacomyia* in the relationship *Cx. chidesteri* + (*Cx. dolosus* + (*Phytotelmatomyia* + (*Cx. janitor* + *Phenacomyia*))).

****

Characteristics: Species of *Phytotelmatomyia* exhibit the following principal anatomical features. ADULTS: Scutum with a pattern of pale scales, principally on the margins (inconspicuous in *Cx. castroi* and *Cx. machadoi*); females with a patch of pale scales on the maxillary palpus (except *Cx. castroi* and *Cx. machadoi* which lack pale scales) and a more or less evident pale patch or band on the proboscis; tarsi with broad bands across the joints (except in *Cx. castroi* and *Cx. machadoi*). FEMALE GENITALIA: Postgenital lobe trapezoidal, at least in *Cx. castroi*, *Cx. hepperi* and two undescribed species. MALE GENITALIA: Differ from those of *Phenacomyia* and similar to various species of *Culex sensu stricto* in having setae *e* and *f* on the subapical lobe of the gonocoxite; otherwise without other distinctive characteristics except for the number of teeth on the lateral plate of the phallosome and the number of setae in the ninth tergal lobes. LARVAE: Antenna very short, less than 0.33 the length of the head, nearly cylindrical, seta 1-A relatively small with multiple branches, setae 2-4-A forked or bifid; seta 1-C short, stout, branched or forked; seta 4-C long, reaching the anterior margin of the head, fan-like, developed similar to setae 5,7-C; maxilla short, stout, seta 1-Mx inserted 0.6 from the base; maxillary brush relatively short, slightly shorter than the maxillary body, spiculose area lateral to seta 5-Mx poorly developed; seta 0-P large, inserted posterior to seta 5-P; seta 3-P multi-branched, short, about 0.33 the length of seta 1-P; seta 4-P similar to seta 3-P; setae 14-P, 1-M,T and 1-II usually with multiple branches; seta 1-I,III multi-branched; seta 7-II long, similar to seta 7-I; saddle largely covered with spicules that grade into long spicules posteriorly, especially dorsally; ventral brush (seta 4-X) comprised of 4 pairs of setae. PUPAE: Seta 1-CT significantly longer than setae 2,3-CT; seta 7-II as long as seta 6-II; seta 5-IV as long as seta 5-V,VI, about 1.5 the length of the following tergum; paddle spiculate on the anterior 0.4 of the surface, with distinct serration posteriorly on the margins of the inner and outer parts.

****

Bionomics: Adults of species of *Phytotelmatomyia* are attracted to humans, but have never been collected inside or outside houses or in light traps. The immature stages of *Cx. hepperi*, *Cx. castroi* and *Cx. renatoi* were found in two species of *Eryngium* in Punta Lara, Buenos Aires Province, Argentina. Eggs are deposited in small rafts of 4–8 eggs or single eggs are found in different axils. The eggs are small but generally resemble the eggs of species of *Culex sensu stricto*. Whether or not species of *Phytotelmatomyia* are of medical importance is unknown.

****

Principal references: Rossi & Harbach, 2008 (diagnosis, comparative morphology, distribution, bionomics, phylogenetic relationships); Harbach *et al.*, 2012 (morphology, phylogenetic relationships); González *et al.*, 2023 (Neotropical subgenera of *Culex*, comparative morphology).

Sirivanakarnius Tanaka, 2004

Sirivanakarnius Tanaka, 2004, originally proposed and currently recognized as a subgenus of *Culex*, is typified by *Culex boninensis* Bohart, 1957. Subgenus abbreviation – *Sir.*

****

Systematics: *Sirivanakarnius* is monobasic. It only includes the extant type species, which is known only from the Bonin Islands. The phylogenetic relationships of *Sirivanakarnius* are obscure. The type species was originally placed in *Culex sensu stricto*, but it appears to be more closely related to species of *Acalleomyia*, *Acallyntrum* and *Culiciomyia* based on the phylogenetic analysis of morphological data in the study of St John (2007). The sole species of *Sirivanakarnius* was placed among species of *Culex sensu stricto* in the phylogenetic study of Harbach *et al.* (2012), also based on morphological data.

****

Characteristics: The single species of *Sirivanakarnius* exhibits the following principal anatomical features. ADULTS: Proboscis and maxillary palpus lightly speckled with pale scales, proboscis with an ill-defined median pale ring; wing and tarsi dark-scaled; proboscis subequal to the forefemur in females, slightly longer in males; maxillary palpus with 3 palpomeres in females, 5 in males, palpomeres 3–5 not setose; acrostichal setae present; thoracic pleura with distinct scale-patches; lower mesepimeral seta absent; hindtarsomere 1 as long as or slightly longer than the hindtibia; abdominal terga with basolateral pale patches. MALE GENITALIA: Tergum IX lobes poorly developed, widely separated, each bearing 6–9 setae in a single row; gonocoxite without scales; subapical lobe undivided, borne in postmedian position, with setae *a–c* rod-like as in most *Culex sensu stricto*, setae *d*,*e* represented by 1 small hair-like seta, seta *f* distally enlarged and borne lateral to setae *d*,*e*, seta *g* a narrow cycle-shaped leaf; gonostylus short, stout, tapered distally, with annulations dorsally at the apex; lateral plate of the phallosome relatively short, bearing a caudally projecting clump of teeth; paraproct with a crown comprised of small spine-like spicules and a row of conspicuous lateral blades, basal lateral arm moderately developed; cercal setae present. LARVAE: Antenna long, about 0.8 the length of the head, slender, seta 1-A strongly developed, with numerous branches, inserted about 0.7 from the base of the antenna; insertions of setae 4,5-C form points of a regular triangle at mid-length of the head; seta 8-C inserted on level with seta 9-C, usually single; seta 3-P as long as setae 1,2-P; seta 8-P strongly developed, as long as setae 1–3-P, with 2 or 3 branches; seta 7-II much shorter than seta 7-I but distinctly longer than seta 7-III–V, seta 7-VI double; siphon moderately long, straight; pecten spines with distinct ventral denticles; seta 1-S with 5–8 (usually 6) relatively large multi-branched elements in an irregular posterior row borne distal to the pecten, seta 2a-S comprised of 1 or 2 pairs (usually 2) anterolateral setae; saddle complete; seta 2-X usually double, with the dorsal branch shorter; seta 3-X single, longer than seta 2-X; ventral brush (seta 4-X) with 5.5–6.5 pairs of setae borne on a grid. PUPAE: Trumpet cylindrical, pinna about as long as the meatus; setae of the cephalothorax relatively short, subequal; surfaces of the abdominal terga and sterna with pronounced imbrications; seta 6-I longer than seta 7-I; setae 1,2-III–VI inserted on the posterior edge of the respective tergum; seta 2-VII usually inserted anterolaterad of seta 1; seta 9-II–V inserted caudad of seta 7; setae 1,2-Pa both present.

****

Bionomics: The immature stages of *Cx. boninensis* are found in ground pools, rock holes and artificial containers. According to Bohart (1957), the larvae are predacious, feeding on mosquito larvae and small aquatic organisms. Adult females will feed on humans, but are not persistent biters. *Culex boninensis* is not of medical importance.

****

Principal references: Bohart, 1957 (*Cx. boninensis* as a species of *Culex sensu stricto*, description, bionomics, distribution); Tanaka *et al.*, 1979 (*Cx. boninensis* as a species of *Culex sensu stricto*, description, bionomics, distribution); Tanaka, 2004 (new subgenus, adult and larval stages, pupa and distribution of *Cx. boninensis*); St John, 2007 (morphology, phylogenetic relationships); Harbach *et al.*, 2012 (morphology, phylogenetic relationships).

Tinolestes Coquillett, 1906

Tinolestes, originally proposed as a genus by Coquillett (1906a), currently classified as a subgenus of *Culex*, is typified by *Culex latisquama* (Coquillett, 1906a), original combination: *Tinolestes latisquama*, original type designation by the author (see Evenhuis, 2018). Subgenus abbreviation – *Tin.*

****

Systematics: *Tinolestes* includes three extant species with distributions in Central America and Florida, USA. *Tinolestes* shares similarities with other genus-group taxa of the Melanoconion Group, but it appears to be most closely related to *Carrollia* based on the absence of dorsocentral setae. A possible sister-group relationship with *Belkinomyia*, *Galindomyia* and *Deinocerites* is supported by the shared ecological association of these taxa with crab holes. In fact, *Tinolestes* was recovered in a clade that also comprised *Anoedioporpa*, *Aedinus* and *Belkinomyia*, in a sequentially nested relationship, with *Galindomyia* and *Deinocerites* as terminal groups, in the phylogenetic analyses of Harbach *et al.* (2012) based on morphological data.

****

Characteristics: Species of *Tinolestes* exhibit the following anatomical features. ADULTS: Small dark unadorned mosquitoes; decumbent scales of the vertex narrow; antenna slightly longer than the proboscis in females, subequal to the proboscis in males; maxillary palpus of females short, apparently with 4 palpomeres, palpomeres 1 and 2 ankylosed; palpus of males approximately 0.5 the length of the proboscis, with 5 palpomeres, palpomeres 1 and 2 ankylosed; proboscis longer than the forefemur; scutum with narrow dark falcate scales; acrostichal and dorsocentral setae absent; upper proepisternal setae strongly developed, 20–25 present; mesokatepisternum almost entirely covered by a broad patch of translucent scales; mesepimeron with a distinct central patch of scales and short setae, lower mesepimeral seta absent; metameron with setae; wing and legs dark-scaled; ungues of females simple, fore- and midungues of males enlarged, unequal, larger unguis of foreleg (but not midleg) with a tooth, hindungues as in females; abdominal terga dark-scaled with basolateral pale patches. MALE GENITALIA: Tergum IX lobes distinct, flattened, with long erect setae; gonocoxite with a few scales; subapical lobe subapical in position, undivided, bearing 3 rods with cup-shaped expanded tips and 6 or 7 setae; gonostylus simple, stout, bearing an external crest; gonostylar claw simple, apical in position; phallosome not columnar, lateral plate with a broadly sclerotized basal hook, a caudally directed sinuous apical process and a sternal spine; paraproct with a crown of 22–24 teeth; cercal setae distinct. LARVAE: Head capsule faintly imbricate; antenna with a large multi-branched seta 1-A inserted about 0.75 from the base; setae 2,3-C absent; seta 3-P much shorter than setae 1,2-P, with multiple weak branches; abdominal seta 2 inserted relatively close to seta 1 within the posterior 0.5 of the terga; setae 1,2-VIII not on a common tubercle; comb scales in a patch; siphon relatively long and slender; 4 posterolateral pairs of seta 1-S and 2 pairs of anterolateral seta 2a-S inserted beyond the pecten; saddle complete, acus absent, posterior margin without spines; seta 1-X with multiple branches; seta 2-X long, with 3 postbasal branches; seta 3-X long, single; ventral brush (seta 4-X) with 6 pairs of setae borne on a grid. PUPAE: Cephalothorax and abdomen evenly pigmented; trumpet moderately long, tracheoid area distinct; seta 11-CT always double; seta 2-III inserted mesad of seta 1, seta 2-VI,VII inserted laterad of seta 1; seta 9-VII,VIII double, seta 9-VIII shorter than the tergum; tergum VIII not overlapping the base of tergum IX; paddle without marginal spicules; setae 1,2-Pa both present.

****

Bionomics: The immature stages of species of *Tinolestes* are found in crab holes. Adults apparently rest in crab holes. Females are not known to bite humans. The species are not of medical importance.

****

Principal references: Berlin & Belkin, 1980 (keys, descriptions, bionomics, distributions); Harbach *et al.*, 2012 (morphology, phylogenetic relationships); González *et al.*, 2023 (Neotropical subgenera of *Culex*, comparative morphology).

Culiseta Felt, 1904

Culiseta Felt, 1904, originally proposed and currently recognized as a genus, is typified by *Culex absobrinus* Felt, 1904 [subjective synonym of *Culiseta impatiens* (Walker, 1848)]. Genus abbreviation – *Cs.*

****

Systematics: *Culiseta* is a subordinate taxon of the subfamily Culicinae, tribe Culisetini. It includes 42 extant and three fossil species. Two of the extant species are without subgeneric placement; the other extant species are divided between seven subgenera: *Allotheobaldia*, *Austrotheobaldia*, *Climacura*, *Culicella*, *Culiseta sensu stricto*, *Neotheobaldia* and *Theomyia*. Most species occur in the Palaearctic, Nearctic and Australasian Regions, but some also occur in the Afrotropical and Oriental Regions. Two species are found in Central America, but none occur in South America. The affinities of *Culiseta* are unknown. *Culiseta* was paired with *Toxorhynchites* in the implied weighting analysis of Harbach & Kitching (1998) based on morphological data, but this relationship seems unlikely. *Culiseta* appears to be an ancient group that shares similarities with a number of other genera. From divergence times based on sequence data for nuclear protein-coding genes and fossil calibration points, it appears that *Culiseta* may have appeared as early as 191.0 Mya (credibility interval 154.7–218.8 Mya) during the Jurassic (Reidenbach *et al.*, 2009). Soghigian *et al.* (2023) included nine species representing four subgenera of *Culiseta* in their phylogenetic analysis of amino acid sequences of 709 genes, which formed a monophyletic clade, but the relationships of the species were not explicitly disclosed. However, contrary to the emergence time indicated in the study of Reidenbach *et al.*, the chronogram generated in the study of Soghigian *et al.* revealed that *Culiseta* may have appeared much more recently during the Upper Cretaceous, perhaps approximately 80 Mya.

****

Characteristics: *Culiseta* are rather large mosquitoes which bear a superficial resemblance to species of *Culex*. They lack perceptible pulvilli, possess prespiracular setae and bear a patch of setae at the base of the subcosta (absent in *Theomyia*) on the ventral surface of the wing. The subcostal setae are virtually diagnostic for *Culiseta*. They are known to occur in only three other mosquitoes, *Coquillettidia* (*Austromansonia*) *tenuipalpis* (New Zealand), *Opifex* (*Nothoskusea*) *chathamicus* (Chatham Island) and *Opifex* (*Opifex*) *fuscus* (New Zealand). *Culiseta* are distinguished from these species by the presence of prespiracular setae. With one exception (*Cs. littleri* of *Austrotheobaldia*, Australia), larvae have seta 1-S inserted at the base of the siphon. The only other species that have seta 1-S in this position belong to *Ficalbia* and *Hodgesia*. Larvae of *Culiseta* differ from the larvae of *Ficalbia* in having a complete hypostomal suture; they differ from the larvae of *Hodgesia* in having seta 5-VIII located far below the dorsal margin of abdominal segment X. Larvae of *Austrotheobaldia* are unique in having the antennae and head setae 5,6-C unusually long, with seta 6-C single and about twice as long as the head.

****

Bionomics: Most *Culiseta* are cold-adapted species which only occur in warmer climates during the colder parts of the year or at higher elevations where temperatures are low. The larvae of most species are found in groundwaters such as bogs, marshes, ponds, streams, ditches and rock pools, but an African species occurs in tree holes, a common eastern Palaearctic species occurs in wells and rock pools, and several Australian species occur in underground collections of water. Little is known about the blood-feeding habits of females. Females of most species feed on birds and mammals, but a few feed on reptiles. Females of several species attack domestic animals and occasionally humans. Three species, *Cs. inornata*, *Cs. melanura* and *Cs. dyari*, are vectors of both Eastern and Western equine encephalitis virus in North America.

****

Principal references: Belkin, 1962 (South Pacific, taxonomy, keys, descriptions, distributions, bionomics); DuBose & Curtin, 1965 (Mediterranean area, keys); Forattini, 1965b (Neotropical Region); Maslov, 1967, 1989 (World, systematics, keys, descriptions, distributions, bionomics); Dobrotworsky, 1971 (Southeast Asia, keys for all subgenera, descriptions, distributions, bionomics, medical importance); Gutsevich *et al.*, 1974 (former USSR, taxonomy, keys, descriptions, distributions, bionomics); Tanaka *et al.*, 1979 (Japan, taxonomy, keys, descriptions, distributions, bionomics); Wood *et al.*, 1979 (Canada, taxonomy, keys, descriptions, distributions, bionomics); Darsie & Ward, 1981, 2005 (North America, keys); Lu & Li, 1982 (China, keys); Lee *et al.*, 1988b (Australasian Region, taxonomy, keys, literature, distributions, bionomics); Service, 1990 (Afrotropical Region, keys, descriptions, bionomics, distributions); Lu *et al.*, 1997a (China, keys, taxonomy, descriptions, distributions, bionomics); Harbach & Kitching, 1998 (morphology, phylogenetic relationships); Becker *et al.*, 2003, 2010, 2020 (Europe, keys, descriptions, distributions, bionomics); Reidenbach *et al.*, 2009 (nuclear genes, morphology, phylogenetic relationships); Soghigian *et al.*, 2023 (molecular markers, phylogenetic relationships); Ortega-Morales *et al.*, 2024 (Mexico, keys for larvae and adults).

Allotheobaldia Brolemann, 1919

Allotheobaldia Brolemann, 1919, originally proposed as a subgenus of *Theobaldia* Neveu-Lemaire, 1902a and currently recognized as a subgenus of *Culiseta*, is typified by *Culiseta longiareolata* (Macquart, 1838a), original combination: *Culex longiareolatus*. Subgenus abbreviation – *All.*

****

Systematics: *Allotheobaldia* is monobasic. It only includes the extant type species, which is known only from localities in the southwestern Palaearctic and Afrotropical Regions. *Culiseta longiareolata* was included in the phylogenetic analysis of amino acid sequences of 709 genes conducted by Soghigian *et al.* (2023), but its relationship with species of other subgenera included in the analysis was not explicitly indicated. Maslov (1967, 1989) suggested that it may have evolved from 'ancient Mediterranean representatives' of *Culiseta sensu stricto*.

****

Characteristics: The following combinations of characters distinguish *Allotheobaldia* from all other *Culiseta*. ADULTS: Head, thorax, abdominal terga and legs with patterns of broad dark and pale scales; maxillary palpus of males shorter than the proboscis, about 0.67 as long, swollen distally, junction between palpomeres 2 and 3 usually distinctly visible; scutum with longitudinal lyre-shaped pale stripes; paratergite long, very narrow; thoracic pleura scaly, with a stripe of broad scales extending along the dorsal margin of the postspiracular area below the paratergite; costa of the wing mainly with pale scales, subcostal setae present; femora and tibiae with well-defined pale spots and stripes; abdominal tergum I mainly without scales, laterotergite with scales. FEMALE GENITALIA: Sternum VIII very large, with a large posteromedian excavation with rows of setae on either side; tergum IX rather broad, constricted medially, somewhat butterfly-shaped; upper vaginal lip with a median bilobed process; insula conspicuous, separated from the lower vaginal lip. MALE GENITALIA: Tergum IX very large, H-shaped, tergum IX lobes long, widely separated, clavate; gonocoxite broad, without an apicodorsal lobe; basal mesal lobe with rows of progressively longer setae toward the apex; gonostylus truncate, with 2 apical gonostylar claws; phallosome with a pair of twisted lateral plates. LARVAE: Antenna short, seta 1-A weakly developed; siphon short, not sclerotized at the base, siphon index < 2.5, pecten comprised of simple spines, setae 1a-S and 2a-S absent; saddle incomplete ventrally, covering approximately the dorsal 0.5 of segment X; seta 2-X multi-branched, seta 3-X single or double; ventral brush (seta 4-X) extended anteriorly on the ventral midline of the segment, with about 9 pairs of setae, setae more or less unpaired anteriorly. PUPAE: Trumpet short, broad, tracheoid area indefinite; seta 5-IV–VI longer than the following tergum; paddle large, broad, distal margin finely spiculate, buttress and midrib distinct; seta 1-Pa short, single; seta 2-Pa absent.

****

Bionomics: The immature stages of *Cs. longiareolata* are found in rock holes and various types of artificial containers, including barrels, tanks and wells, and rarely occur in pools, ditches and drainage canals. They sometimes inhabit slightly saline and highly polluted bodies of water. The species overwinters in the larval stage. Adults are found from February to November in temperate areas. Females do not enter dwellings and rarely bite humans. Birds are the main hosts, but females may produce eggs without a blood meal. *Culiseta longiareolata* is not known to be of medical importance.

****

Principal references: Edwards, 1941 (as a subgenus of *Theobaldia*, adults, pupa, identification, distribution); Hopkins, 1952 (as a subgenus of *Theobaldia*, larva, identification, bionomics); DuBose & Curtin, 1965 (Mediterranean area, keys); Dobrotworsky, 1971 (in keys for subgenera); Gutsevich *et al.*, 1974 (former USSR, identification, taxonomy, adults, larvae, distribution, bionomics); Maslov, 1967, 1989 (as a genus, taxonomy, identification, adults, larvae, distribution, bionomics); Service, 1990 (Afrotropical Region, keys, descriptions, bionomics, distributions); Becker *et al.*, 2003, 2010, 2020 (Europe, keys, descriptions, distributions, bionomics).

Austrotheobaldia Dobrotworsky, 1954

Austrotheobaldia Dobrotworsky, 1954, originally proposed as a subgenus of *Theobaldia* Neveu-Lemaire, 1902a and currently recognized as a subgenus of *Culiseta*, is typified by *Culiseta littleri* (Taylor, 1914c), original combination: *Chrysoconops littleri*. Subgenus abbreviation – *Aut.*

****

Systematics: *Austrotheobaldia* is monobasic. It only includes the extant type species, which is known only from Australia, localities in New South Wales, Victoria and Tasmania. The phylogenetic relationships of *Austrotheobaldia* have not been investigated. Maslov (1967, 1989) suggested that 'genus' *Austrotheobaldia* may have evolved from an ancestral lineage which also gave rise to *Climacura*, *Culicella* and *Neotheobaldia*.

****

Characteristics: The following combinations of characters distinguish *Austrotheobaldia* from the other genus groups of *Culiseta*. ADULTS: Vertex with upright pale scales; maxillary palpus of males as long as or shorter than the proboscis; scutum with pale golden scales; 2 or 3 prespiracular setae; postspiracular area with a few small scales; mesepimeron with a scale-patch near the middle, usually with 3 lower mesepimeral setae; wing with subcostal setae; tarsi entirely dark-scaled. MALE GENITALIA: Ninth tergal lobes each with 4–7 setae; gonocoxite long and slender, basal mesal lobe very long, about 0.67 the length of the gonocoxite, separated from the gonocoxite except basally, apex with a cluster of strong spine-like setae; gonostylus very long and slender, nearly as long as the gonocoxite; aedeagus ovoid, weakly sclerotized; paraproct with 3 apical teeth. LARVAE: Head large, broad; all cranial setae with smooth branches; antenna long, setae 2,3-A branched, inserted before the apex; comb of segment VIII with 16–30 small scales in an anterior row and 8–15 longer scales in 2 or 3 posterior rows; siphon long, slightly tapered, siphon index about 7, pecten comprised of 13–15 short triangular spines, seta 1-S inserted near mid-length of the siphon, single or double, setae 1a-S and 2a-S absent; segment X with a complete saddle, seta 1-X single, seta 2-X with 5 or 6 branches, seta 3-X with 1 very long and 2 very short branches, ventral brush (seta 4-X) with 10 or 11 long setae borne on a grid and 2 short precratal setae. PUPAE: Trumpet short, with a large oblique opening with a fringed margin; seta 9-VIII single; paddle oval, outer margin with minute denticles; setae 1,2-Pa both present, single.

****

Bionomics: The immature stages of *Cs. littleri* inhabit shaded permanent pools of clean, cold water in forest. Females occasionally attack humans, but the species is not known to be of medical importance.

****

Principal references: Dobrotworsky, 1965 (Australia, taxonomy, adults, larvae, bionomics, distribution); Dobrotworsky, 1971 (in keys for subgenera); Lee *et al.*, 1988b (Australia, taxonomy, literature, distribution, bionomics); Maslov, 1967, 1989 (as a genus, taxonomy, identification, adults, larvae, distribution, bionomics).

Climacura Howard, Dyar & Knab, 1915

Climacura Howard *et al.*, 1915, originally proposed and currently recognized as a subgenus of *Culiseta*, is typified by *Culiseta melanura* (Coquillett, 1902b), original combination: *Culex melanurus*. Subgenus abbreviation – *Cli.*

****

Systematics: *Climacura* includes six extant species with distributions in the Australasian, Oriental and Nearctic Regions, specifically Australia (*Cs. antipodea*, in Queensland and Victoria), New Zealand (*Cs. novaezealandiae*, *Cs. tonnoiri*), Peninsular Malaysia (*Cs. marchettei*) and North America (*Cs. melanura*, in Canada, Mexico and the USA). Soghigian *et al.* (2023) included *Cs. melanura* in their phylogenetic analysis of amino acid sequences of 709 genes, but its relationship with species of other subgenera included in the analysis was not explicitly indicated. Maslov (1967, 1989) hypothesized that *Climacura* evolved from an offshoot of the lineage that also gave rise to *Culicella*.

****

Characteristics: The following combinations of characters distinguish *Climacura* from the other genus groups of *Culiseta*. ADULTS: Maxillary palpus of females with palpomere 5 small, round or elongate, palpus of males as long as or longer than the proboscis, palpomeres 4 and 5 slender or more or less thickened; antepronotum without scales, with setae only; postpronotum with posterior setae and narrow scales; 1–7 prespiracular setae; scutellum with narrow scales; wing with or without darker clusters of scales, base of the subcosta with a reduced number of ventral setae, mediocubital crossvein located well before the radiomedial crossvein; tarsi entirely dark-scaled or hindtarsi with distinct basal pale bands; abdominal tergum I with setae only; terga II–VIII usually entirely dark-scaled (*Cs. marchettei* with basal pale bands on terga V–VIII). MALE GENITALIA: Ninth tergal lobes with long setae; gonocoxite without apicodorsal lobe, basal mesal lobe rather small, conical, with variously developed setae; gonostylus with a single terminal gonostylar claw; aedeagus usually simple and weakly sclerotized (strongly sclerotized laterally with 3 or 4 prominent apical denticles in *Cs. marchettei*). LARVAE: Head very large; antenna long, slender, longer than the head; seta 1-A large, inserted well beyond mid-length of the antenna, setae 2,3-A very long, usually slightly removed from the apex of the antenna; comb scales in a single row; siphon long, index 4.0–7.5, seta 1-S small, inconspicuous, single or with 2–5 branches, inserted at the extreme base of the siphon, setae 1a-S and 2a-S present, seta 1a-S in a more or less single mid-posterior row on the distal 0.5–0.75 of the siphon; saddle complete; seta 2-X multi-branched; ventral brush (seta 4-X) with or without precratal setae, small if present. PUPAE: Generally not well studied; setae 1,2-Pa usually both present (2-Pa apparently absent in *Cs. novaezealandiae*).

****

Bionomics: The immature stages of species of *Climacura* are usually found in permanent bodies of water, mainly swamps, but they may also inhabit semi-permanent pools (larvae of *Cs. marchettei* have not been found in nature). Females primarily feed on birds, but those of *Cs. melanura* and *Cs. tonnoiri* have been captured biting humans, and the former are also known to feed on small mammals and snakes. West Nile virus has been isolated from *Cs. melanura*, but it is regarded as an inefficient vector (Turell *et al.*, 2005). This species circulates Eastern and Western Equine Encephalitis viruses among birds (White & Faust, 2014). Whataroa virus has been isolated from *Cs. tonnoiri* in New Zealand (Chamberlain, 1981), and this species is capable of acquiring and transmitting Coxsackie A6 virus (Maguire, 1970).

****

Principal references: Dobrotworsky, 1965 (Australia, taxonomy, adults, larvae, bionomics, distribution); Maslov, 1967, 1989 (World, systematics, keys, descriptions, distributions, bionomics); Dobrotworsky, 1971 (in keys for subgenera, taxonomy, adults, larvae, bionomics, distribution); Gutsevich *et al.*, 1974 (former USSR, taxonomy, keys, descriptions, distributions, bionomics); Wood *et al.*, 1979 (Canada, taxonomy, keys, descriptions, distributions, bionomics); Lee *et al.*, 1988b (Australasian Region, taxonomy, keys, literature, distributions, bionomics).

Culicella Felt, 1904

Culicella, originally proposed as a genus by Felt (1904), currently classified as a subgenus of *Culiseta*, is typified by *Culex dyari* (Coquillett, 1902b) [junior synonym of *Culiseta morsitans* (Theobald, 1901c)]. Subgenus abbreviation – *Cuc.*

****

Systematics: *Culicella* includes 14 extant species. Seven species occur in Australia, six in the Palaearctic Region and two in the Nearctic Region. *Culiseta morsitans* has a Holarctic distribution. Soghigian *et al.* (2023) included three species of *Culicella* in their phylogenetic analysis of amino acid sequences of 709 genes, but the relationship of the group with species of the other subgenera included in the analysis was not explicitly indicated. Maslov (1967, 1989) hypothesized that the subgenus ‘probably originated in the lower latitudes of Europe’, but this seems highly unlikely in view of the antiquity of *Culiseta* (Harbach & Greenwalt, 2012). Morphological evidence suggests that *Culicella* and *Climacura* may be sister taxa (Maslov, 1967, 1989; Harbach & Greenwalt, 2012).

****

Characteristics: The following combinations of characters distinguish *Culicella* from the other genus groups of *Culiseta*. ADULTS: Maxillary palpus of males as long as or longer than the proboscis, dense setae on palpomeres 4 and 5 and the distal area of palpomere 3; antepronotum without scales or with very long piliform scales; postpronotum with setae and fine piliform scales; prespiracular setae numerous (few in Australian species); postspiracular setae few (not more than 5) or absent; wing without darker clusters of scales (*Cs. ochroptera* with indistinct darker cluster at the base of vein R_{4+5}); subcostal setae present (few in Australian species), mediocubital crossvein distinctly located proximal to the radiomedial crossvein, at a distance equal to or greater (usually) than the length of the radiomedial crossvein; tarsi with narrow or indistinct apical pale bands that tend to extend onto the apex of the following tarsomere. MALE GENITALIA: Ninth tergal lobes not produced; gonocoxite without an apico-dorsal lobe; aedeagus weakly sclerotized, egg-shaped. LARVAE: Head broad; palatal brushes very large; antenna nearly as long to longer than the head, seta 1-A well developed, large, inserted well beyond the middle of the shaft, usually near the apex, setae 2,3-A long, removed from the apex; comb of segment VIII with numerous scales (60 to more than 100) in a patch; siphon long, tapering, index 4.5–8.0; pecten comprised of small spines near the base of the siphon (except *Cs. fumipennis* with several stout more distal spines), seta 1-S inserted at the base, row of setae 1a-S and seta 2a-S absent; seta 3-X branched; ventral brush (seta 4-X) with precratal setae inserted on the saddle. PUPAE: Unknown or inadequately studied, subgeneric characters not apparent.

****

Bionomics: Species of *Culicella* generally overwinter in the egg or larval stage. The immature stages of the majority of species are found in bogs, marshes, ponds, streams, ditches and ground pools. The larvae of several Australian species are found in underground collections of water: crayfish tunnels (*Cs. victoriensis*) and pits (*Cs. drummondi*, *Cs. sylvanensis* and *Cs. weindorferi*). Little is known about the blood-feeding habits of adult females. They seldom enter dwellings and mainly feed on birds, but some will feed on amphibians, reptiles and mammals, including humans. *Culiseta victoriensis* is a troublesome biter in Australian forests. Females apparently deposit their eggs individually on damp substrates above water level, similar to species of Aedini. *Culiseta morsitans* is a carrier of Ockelbo virus in Sweden (Francy *et al.*, 1989).

****

Principal references: Carpenter & LaCasse, 1955 (North America, taxonomy, keys, descriptions, bionomics); Dobrotworsky, 1965 (Australia, taxonomy, adults, larvae, bionomics, distribution); Gutsevich *et al.*, 1974 (former USSR, taxonomy, keys, descriptions, distributions, bionomics); Wood *et al.*, 1979 (Canada, taxonomy, keys, descriptions, distributions, bionomics); Lee *et al.*, 1988b (Australasian Region, taxonomy, keys, literature, distributions, bionomics); Maslov, 1967, 1989 (World, systematics, keys, descriptions, distributions, bionomics); Lu *et al.*, 1997a (China, *Cs. nipponica*, in keys, taxonomy, descriptions, distributions, bionomics).

Culiseta Felt, 1904 *sensu stricto*

Culiseta sensu stricto, the nominotypical subgenus of *Culiseta*, is typified by *Culex absobrinus* Felt, 1904 [subjective synonym of *Culiseta impatiens* (Walker, 1848)]. Subgenus abbreviation – *Cus.*

****

Systematics: *Culiseta sensu stricto* includes 14 extant species in the Holarctic Region. Eight species occur in the Palaearctic Region and five in the Nearctic Region. *Culiseta alaskaensis* has a Holarctic distribution and the Nearctic *Cs. particeps* extends into the Neotropical Region as far south as Panama. Soghigian *et al.* (2023) included four species of *Culiseta sensu stricto* in their phylogenetic analysis of amino acid sequences of 709 genes, but the relationship of the group with species of the other subgenera included in the analysis was not explicitly indicated. Maslov (1967, 1989) hypothesized that the group is the 'most primitive' of the *Culiseta*, an early offshoot of an ancestral lineage in the Holarctic Region. From divergence times based on sequence data for nuclear protein-coding genes and fossil calibration points, the study of Reidenbach *et al.* (2009) indicated that *Culiseta sensu stricto* may have evolved during the Jurassic. However, the chronogram generated in the molecular phylogenetic study of Soghigian *et al.* (2023) revealed that *Culiseta* may have appeared more recently, perhaps about 80 Mya during the Upper Cretaceous.

****

Characteristics: The following combinations of characters distinguish *Culiseta sensu stricto* from the other genus groups of *Culiseta*. ADULTS: Maxillary palpus of females with palpomere 5 sometimes distinctly white-scaled, palpus of males about as long as or longer than the proboscis, usually with very long setae, palpomeres 4 and 5 more or less swollen, not upturned; scutellum usually with narrow scales (broad in *Cs. niveitaeniata*); antepronotum with setae and a few scales; postpronotum with posterior setae and broad or narrow scales; prespiracular area usually with more than 7 setae, sometimes with a few scales; wing often with darker clusters of scales, subcostal setae numerous, mediocubital and radiomedial crossveins usually approximated, if not then the distance between them is not more than the length of the mediocubital crossvein; tarsi uniformly dark-scaled or with basal pale bands; abdominal terga with more or less distinct basal pale bands. MALE GENITALIA: Ninth tergal lobes usually with long setae; gonocoxite with a distinct apicodorsal lobe or a patch of setae, basal mesal lobe relatively small, conical, with numerous setae; gonostylus with an apical usually bifid gonostylar claw; aedeagus with strongly sclerotized lateral plates, plates usually separated, egg-shaped. LARVAE: Head not very broad; antenna short to moderately long, about 0.5 the length of the head, slightly flexed, seta 1-A inserted near mid-length of the shaft, setae 2–6-A usually short, inserted apically; comb of segment VIII with about 50 scales in a triangular patch; siphon relatively short, index 2.5–4.5, pecten comprised of strong spines, seta 1-S inserted at the base of the siphon, a row of setae 1a-S present, seta 2a-S absent; saddle of segment X complete, ventral brush (seta 4-X) usually with 2–5 precratal setae inserted on the saddle. PUPAE: Not well studied; seta 1-Pa present, seta 2-Pa apparently absent.

****

Bionomics: The immature stages of species of *Culiseta sensu stricto* are found predominantly in sunlit or shaded ground pools, ditches, pond margins, swamps and marshes among vegetation or plant debris. Larvae of *Cs. annulata* and *Cs. inornata* are sometimes found in artificial containers, and larvae of the former species and *Cs. glaphyroptera* have been found in tree holes. Larvae are mainly bottom feeders and remain submerged for long periods of time. Females overwinter and feed in the spring. They deposit their eggs in rafts. Little is known about the host preferences and feeding habits of adult females. The females of most species probably feed principally on birds, but some at least occasionally feed on reptiles and mammals. *Culiseta alaskaensis*, *Cs. annulata* and *Cs. subochrea* frequently feed on humans. Most species are not known or suspected to be vectors of pathogens. However, *Cs. annulata* is considered to be a potential vector of Tahyna virus (Ribeiro *et al.*, 1988) and is known to transmit avian malarial plasmodia (Gutsevich *et al.*, 1974). *Culiseta inornata* is considered a potential vector of Western equine encephalitis and West Nile viruses (Turell *et al.*, 2005), and is probably the chief vector of Cache Valley virus (Corner *et al.*, 1980).

****

Principal references: Carpenter & LaCasse, 1955 (North America, taxonomy, keys, descriptions, bionomics); Maslov, 1967, 1989 (World, taxonomy, keys, descriptions, distributions, bionomics); Gutsevich *et al.*, 1974 (former USSR, taxonomy, keys, descriptions, distributions, bionomics); Wood *et al.*, 1979 (Canada, taxonomy, keys, descriptions, distributions, bionomics); Lu *et al.*, 1997a (China, keys, taxonomy, descriptions, distributions, bionomics); Becker *et al.*, 2003, 2010, 2020 (Europe, keys, descriptions, distributions, bionomics).

Neotheobaldia Dobrotworsky, 1958

Neotheobaldia Dobrotworsky, 1958b, originally proposed as a subgenus of *Theobaldia* Neveu-Lemaire, 1902a and currently recognized as a subgenus of *Culiseta*, is typified by *Culiseta hilli* (Edwards, 1926a), original combination: *Theobaldia hilli*. Subgenus abbreviation – *Net.*

****

Systematics: *Neotheobaldia* includes three extant species known only from the highlands of Victoria, Australia. The phylogenetic relationships of *Neotheobaldia* have not been investigated. Maslov (1967, 1989) hypothesized that *Neotheobaldia* evolved from the ancestral lineage that also gave rise to *Culicella* and *Climacura*.

****

Characteristics: The following combinations of characters distinguish *Neotheobaldia* from the other genus groups of *Culiseta*. ADULTS: Vertex with upright pale and dark scales, sometimes mostly dark; maxillary palpus and proboscis entirely dark-scaled, palpus of males as long or longer than the proboscis; scutum with pale golden scales; ante- and postpronota with pale falcate scales and at least some pale setae; prespiracular area with a few setae (5 or fewer); postspiracular area with a few setae; mesepimeron with a scale-patch and a few setae near the middle; subcostal setae present; femora each with a conspicuous knee spot, tarsi of females with tarsomeres 3–5 or 4 and 5 pale-scaled, tarsomeres of males entirely dark- or pale-scaled apically. MALE GENITALIA: Ninth tergal lobes prominent, each with 9–15 setae; basal mesal lobe very long, 0.7–0.9 the length of the gonocoxite, largely or partially separated from the gonocoxite, apex setose; gonostylus very long and slender, nearly as long as the gonocoxite; aedeagus relatively large and strongly sclerotized; paraproct with 7 or 8 apical teeth, cercal setae numerous. LARVAE: Head large, broad; antenna moderately long, usually not shorter than the head, seta 1-A inserted well beyond mid-length of the shaft, setae 2,3-A single, inserted slightly before the apex; comb of segment VIII with approximately 90 scales in a triangular patch; siphon long, slightly tapered, siphon index 5.0–5.4, pecten comprised of 7–11 hair-like spines borne on the proximal 0.5 or less of the siphon, seta 1-S inserted at the base of the siphon, single or double, setae 1a-S and 2a-S absent; segment X with a complete saddle, seta 1-X branched, seta 2-X with 9–11 branches, seta 3-X very long, single, ventral brush (seta 4-X) with 12 or 14 long setae borne on a grid, no precratal setae. PUPAE: Trumpet short with a large oblique opening with a fringed margin; seta 9-VIII with 4–6 branches; paddle oval with smooth margins; setae 1,2-Pa both present, 1-Pa with 3 or 4 branches, 2-Pa single.

****

Bionomics: Species of *Neotheobaldia* are forest mosquitoes. The immature stages are found in water in the subterranean tunnels of terrestrial crayfish. Females are active during the daytime and readily attack humans. Eggs apparently are laid individually on soil above water level. The species pass the winter months as larvae. The species are not known to be of medical importance.

****

Principal references: Dobrotworsky, 1965 (classification, keys, descriptions, bionomics, distributions); Maslov, 1967, 1989 (taxonomy, identification, adults, larvae, distribution, bionomics); Lee *et al.*, 1988b (taxonomy, in keys, literature, distribution, bionomics).

Theomyia Edwards, 1930

Theomyia Edwards, 1930a, originally proposed as a subgenus of *Theobaldia* Neveu-Lemaire, 1902a and currently recognized as a subgenus of *Culiseta*, is typified by *Culiseta fraseri* (Edwards, 1914b), original combination: *Leptosomatomyia fraseri*. Subgenus abbreviation – *Thm.*

****

Systematics: *Theomyia* is monobasic. It only includes the extant type species, which is known from countries in sub-Saharan Africa. The phylogenetic relationships of *Theomyia* have not been investigated. Maslov (1967, 1989) hypothesized that *Theomyia* is an early offshoot of the lineage that later successively gave rise to *Austrotheobaldia*, *Neotheobaldia*, *Climacura* and *Culicella*.

****

Characteristics: The following combinations of characters distinguish *Theomyia* from the other genus groups of *Culiseta*. ADULTS: Maxillary palpus and proboscis dark-scaled, palpus of males very short, about 0.5 the length of the proboscis, slender, without setae; scutum with narrow pale scales forming an indefinite pattern of broad lines; antepronotum with narrow pale scales anteriorly, broad silvery scales posteriorly; prespiracular setae rather numerous; postspiracular area without scales; subcostal setae absent; mediocubital and radiomedial crossveins approximated; hindtarsus with a pale ring at the base of the first tarsomere; laterotergite of abdominal segment I without scales. MALE GENITALIA: Ninth tergal lobes prominent, well separated, each with a cluster of apical setae; gonocoxite elongate, tapering to the apex; gonostylus long, slender; basal mesal lobe appressed to the gonocoxite, crescentiform (ridge-like), with a row of several setae on the crest; aedeagus relatively large, strongly sclerotized; paraproct with 2 or 3 apical teeth. LARVAE: Head very large, broad; antenna short, about 0.25 the length of the head, stout, seta 1-S inserted slightly beyond mid-length of the shaft, setae 2,3-A single, inserted on the apex; thorax with a pair of large thin-walled air sacs; comb of segment VIII with approximately 50 scales in a triangular patch; siphon long, slightly tapered, very strongly tanned, siphon index greater than 7; pecten comprised of about 25 bifid spines borne on the proximal 0.33 of the siphon; seta 1-S single, inserted at the base of the siphon, setae 1a-S and 2a-S absent; saddle complete, very strongly tanned, seta 1-X single, about 0.5 the length of the saddle, setae 2,3-X both very long, single; ventral brush (seta 4-X) with 8 pairs of long setae borne on a grid, no precratal setae. PUPAE: Trumpet short, with a large oblique opening; seta 9-VIII with 6–8 branches; paddle broadest distally, slightly emarginate distal to the midrib, outer part broader than the inner part, outer margin serrate; setae 1,2-Pa both absent.

****

Bionomics: *Culiseta fraseri* is a forest dweller. Females feed on reptiles, birds and mammals, including humans who enter their haunts. The immature stages normally develop in tree holes, but larvae have been found on occasion in other habitats (Service, 1990). The species is not known to be of medical importance.

****

Principal references: Edwards, 1941 (as a subgenus of *Theobaldia*, adults, distribution); Hopkins, 1952 (as a subgenus of *Theobaldia*, larvae, bionomics); van Someren, 1956 (pupa); Maslov, 1967, 1989 (as a genus, taxonomy, in keys, adults, larvae, distribution, bionomics); Service, 1990 (Afrotropical Region, in keys, adults, larvae, bionomics, distribution).

Deinocerites Theobald, 1901

Deinocerites Theobald, 1901a, originally proposed and currently recognized as a genus, is typified by *Deinocerites cancer* Theobald, 1901c. Genus abbreviation – *De.*

****

Systematics: *Deinocerites* is a subordinate taxon of the subfamily Culicinae, tribe Culicini. It includes 18 extant species that are not divided into subgenera. Species of *Deinocerites* are primarily restricted to the Neotropical Region, but a few extend northward into the southern USA. Valencia (1973) suggested that *Deinocerites*, as well as *Carrollia* and the closely related *Galindomyia*, evolved from a primitive stock of *Melanoconion*. In the cladistic analysis of morphological features of larval mandibles and maxillae conducted by Navarro & Liria (2000), *Deinocerites* fell within an unresolved polytomy with *Anoedioporpa*, *Melanoconion* and *Microculex*. Based on this relationship, Navarro & Liria proposed subgeneric status for *Deinocerites* in *Culex*, but this action has not been accepted because this group of species is diagnosed by many unique (autapomorphic) features in the immature and adult stages. Autapomorphies are a measure of (often rapid) divergence and adaptive radiation, which traditional taxonomists have intuitively taken into consideration when defining genera. *Deinocerites*, and *Galindomyia* as its sister taxon, were recovered as derived members of the New World subgenera of *Culex* in the phylogenetic analyses of Harbach *et al.* (2012) based on morphological data. Soghigian *et al.* (2023) included two species of *Deinocerites* in their phylogenetic analysis of amino acid sequences of 709 genes, but the relationship of the group with other genus groups of Culicini included in the analysis was not explicitly indicated. The phylogenetic relationships of the species of *Deinocerites* have not been investigated.

****

Characteristics: The adults of *Deinocerites* superficially resemble *Culex* mosquitoes. They differ in having the antennae much longer than the proboscis, the first flagellomere is remarkably long in most species and the maxillary palpi are very short. *Deinocerites* larvae are distinguished from the larvae of other genera as follows: the head capsule is widest at the level of the antennae, the mandibles have a ‘hairy’ lateral lobe that is unique for *Deinocerites*, the pecten spines are bifid or trifid and the saddle of segment X is represented by widely separated dorsal and ventral sclerites.

****

Bionomics: The larvae of all *Deinocerites* usually inhabit crab holes, but some species have been collected from other habitats, including post holes, rock holes, tree holes and various artificial containers. Adults appear to be active after sundown and rest in crab holes during the daylight hours. The feeding preferences of females are not well known, but some species feed on a variety of hosts, including humans and other mammals, birds, lizards, frogs and toads. Venezuelan and St Louis equine encephalitis viruses have been isolated from *De. pseudes*, and laboratory studies have shown that this species is capable of transmitting these viruses. Nevertheless, species of *Deinocerites* are not important pests of humans and probably play little if any role in the transmission of pathogens.

****

Principal references: Belkin & Hogue, 1959 (Central America and adjoining areas, classification, descriptions, keys, bionomics, phylogenetic relationships); Belkin *et al.*, 1970 (Jamaica, *De. cancer*, in keys, description, bionomics, distribution); Adames, 1971 (classification, taxonomy, phylogenetic relationships, keys, descriptions, distributions, bionomics).

Eoaedes Harbach & Greenwalt, 2012

Eoaedes Harbach & Greenwalt, 2012, originally proposed and currently recognized as an extinct genus, is typified by *Eoaedes damzeni* (Szadziewski, 1998), original combination: *Aedes* (*Finlaya*) *damzeni*. Genus abbreviation – *Eo.*

****

Systematics: *Eoaedes* is a subordinate taxon of the subfamily Culicinae, tribe Aedini. It has no living species. It only includes the extinct type species, which existed during the Eocene Epoch (33.9–56.0 Mya) of the Cenozoic Era. The species is only known from a male preserved in Baltic amber from the Bay of Gdańsk?, Poland. Based on the anatomical information and illustrations provided by Szadziewski (1998) and Podėnas (1999), the affinities of *Eoaedes* with other genus groups of Aedini are uncertain.

****

Characteristics: ADULT MALES: Darkly ornamented species; antenna longer than the proboscis; palpomere 5 of the maxillary palpus short, about 0.4 the length of palpomere 4; wing with narrow scales, vein R_{2+3} less than 0.5 the length of vein R_2; foretarsomere 5 with a prominent setose lobe ventrally at the base; fore- and midungues long, unequal, longer one with a tooth, hindungues short, simple, equal; gonocoxite slender, with a long apical spine; claspette elongate with an apical sickle-like appendage (as in extant *Ochlerotatus*); proctiger barely visible; aedeagus? (Szadziewski, 1998: fig. 3b,c) similar to the aedeagus of extant species of the *Aedes* group of genera, *e.g. Aedimorphus*.

****

Bionomics: The Eocene Epoch was warm and humid, forests were widespread in temperate and subtropical areas and the vertebrate fauna included the ancestors of most groups of modern mammals. Females of *Eo. damzeni* may have fed on mammals as well as other vertebrates. In light of the results of the molecular phylogenetic study of Aedini by Soghigian *et al.* (2017), the immature stages of *Eo. damzeni* may have developed in either groundwater or phytotelm habitats.

****

Principal references: Szadziewski, 1998 (species description, illustrations, keys for fossil species in Baltic amber and Europe); Podėnas, 1999 (illustrations, key to aedine males in Baltic amber); Harbach & Greenwalt, 2012 (new genus, diagnosis).

Eretmapodites Theobald, 1901

Eretmapodites Theobald, 1901d, originally proposed and currently recognized as a genus, is typified by *Eretmapodites quinquevittatus* Theobald, 1901d. Genus abbreviation – *Er.*

****

Systematics: *Eretmapodites* is a subordinate taxon of the subfamily Culicinae, tribe Aedini. It includes 51 extant species confined to the Afrotropical Region. No subgenera are recognized. *Eretmapodites* was recovered as the sister of *Armigeres* in the morphology-based phylogeny of Aedini generated by Reinert *et al.* (2009). It was recovered as the sister of *Alanstonea*, and this pair was sister to *Armigeres*, in the phylogeny of Wilkerson *et al.* (2015). *Eretmapodites quinquevittatus*, the only species of the genus included in the maximum likelihood phylogeny of Soghigian *et al.* (2017) based on seven molecular markers, was recovered as the sister of a clade comprised of five species of *Armigeres*. Soghigian *et al.* (2023) included a single species of *Eretmapodites*, *Er. intermedius*, in their phylogenetic analysis of amino acid sequences of 709 genes, but the relationship of the species with other Aedini included in the analysis was not explicitly indicated. The phylogenetic relationships of the species of *Eretmapodites* have not been investigated.

****

Characteristics: *Eretmapodites* are yellowish-brown mosquitoes with a very broad patch of large, flat, silver scales separating the eyes above the antennae. They possess postspiracular setae like other Aedini in the Afrotropical Region, but the paratergite is both very broad and bare. Larvae have four or fewer pecten spines and a ventral brush (seta 4-X) composed of four pairs of setae. Additionally, the antennae are short and cylindrical and the lateral setae of abdominal segments I–VI arise from large conical projections in many species.

****

Bionomics: *Eretmapodites* are forest mosquitoes, but some species have become adapted to life in banana plantations. Larvae are found in water contained in fallen leaves, fruit husks, leaf axils, snail shells, artificial containers, bamboo and rarely tree holes. The larvae of two species occur in water held in the concave upper surface of polyporaceous fungi. Larvae of most species are facultative predators. Females apparently feed during the daytime. They will attack humans but probably prefer other hosts. Some arboviruses, including Rift Valley fever virus, have been isolated from a few species. *Eretmapodites chrysogaster* has been shown to be capable of transmitting yellow fever virus.

****

Principal references: Haddow, 1946 (classification, taxonomy, bionomics, descriptions, keys); Edwards, 1941 (keys, descriptions of adults and pupae, distributions); Hopkins, 1952 (keys, descriptions of larvae, distributions, bionomics); Hamon, 1961 (Oedipodeios Group, former subspecies of *Er. oedipodeios*, descriptions); Hamon & van Someren, 1961b (Oedipodeios Group,

Er. grenieri, description, male genitalia key to species); Service, 1990 (classification, keys, descriptions of adults and larvae, bionomics, distributions); Reinert *et al.*, 2004, 2009 (morphology, phylogenetic relationships); Wilkerson *et al.*, 2015 (phylogenetic relationships); Soghigian *et al.*, 2017 (amino acid sequences, phylogenetic relationships).

Ficalbia Theobald, 1903

Ficalbia Theobald, 1903a, originally proposed and currently recognized as a genus, is typified by *Ficalbia minima* (Theobald, 1901c), original combination: *Uranotaenia minima*. Genus abbreviation – *Fi.*

****

Systematics: *Ficalbia* is a subordinate taxon of the subfamily Culicinae, tribe Ficalbiini. *Ficalbia* includes eight extant species, four in the Afrotropical Region, three in the Oriental Region (one of which extends into the Australasian Region) and one in the eastern Palaearctic (Ryukyu Archipelago). The phylogenetic relationships of *Ficalbia* are uncertain. Morphological data suggest affinities with *Hodgesia* and *Mimomyia*. Although *Ficalbia* and *Mimomyia* are the only genera of Ficalbiini, the cladistic analyses of generic relationships based on morphological data conducted by Harbach & Kitching (1998) suggest that they are not sister taxa. This is concordant with the opinion of Mattingly (1981) that *Ficalbia* and *Mimomyia* are more distantly related than their inclusion in the same tribe would suggest. The phylogeny of *Ficalbia* has not been investigated.

****

Characteristics: *Ficalbia* are small mosquitoes. The adults are rather easily distinguished from the adults of other genus groups, except *Mimomyia*, by the following combinations of characters: first flagellomere of the antenna elongate, 2–6 times as long as the second flagellomere in females; head with erect scales restricted to the occiput; acrostichal and dorsocentral setae present; prespiracular and postspiracular setae absent; upper calypter and alula of the wing with a fringe of setae or hair-like scales, anal vein (vein 1A) ends beyond the base of the mediocubital crossvein; tarsi normal, pulvilli not evident. A fringe of narrow scales on the alula distinguishes *Ficalbia* from *Mimomyia*. In *Mimomyia*, the alula is either bare or has a patch of broad, flat scales on the margin. Larvae of *Ficalbia* are characterized as follows: distal part of the antenna fused with the basal part (not articulated as in *Mimomyia*); hypostomal suture incomplete or absent; seta 12-I present; comb scales in a single row; siphon unmodified, with seta 1-S inserted near the base (distinction from *Mimomyia*); pecten with at most 2 spines; ventral brush (seta 4-X) with 4–6 pairs of setae; saddle complete, with 1 or more minute setae on the ventral midline (distinction from *Mimomyia*). The larvae share many characteristics with species of *Hodgesia*, which differ in having a complete hypostomal suture and exceptionally long thoracic setae.

****

Bionomics: Little is known about the bionomics of species of *Ficalbia*. Larvae are found in swamps, marshes, ponds, pools and river margins with abundant vegetation. Adults have been captured on lower vegetation and near the canopy in forest at night. Nothing is known about the biting habits of females. Species of *Ficalbia* are apparently of no medical or economic importance.

****

Principal references: Edwards, 1941 (Afrotropical Region, as subgenus *Ficalbia*, in keys, descriptions of adults and pupae, distributions); Hopkins, 1952 (Afrotropical Region, as subgenus *Ficalbia*, in keys, descriptions of larvae, distributions, bionomics); Mattingly, 1957a (Indomalayan area, as subgenus *Ficalbia*, keys, descriptions, bionomics, distributions); da Cunha Ramos & Ribeiro, 1975 (Angola, as subgenus *Ficalbia*, keys, description, distributions, bionomics); Grjebine, 1986 (Madagascar, classification, descriptions, bionomics, distributions); Lee *et al.*, 1988b (Australasian Region, taxonomy, literature, distributions, bionomics); Service, 1990 (Afrotropical Region, keys, descriptions, bionomics, distributions); Lu *et al.*, 1997a (China, keys, taxonomy, descriptions, distributions, bionomics); Harbach & Kitching, 1998 (morphology, phylogenetic relationships); Toma & Higa, 2004 (comparative morphology, distributions); Rattanarithikul *et al.*, 2006a (Thailand, *Fi. minima*, distribution, bionomics, key characters).

Galindomyia Stone & Barreto, 1969

Galindomyia Stone & Barreto, 1969, originally proposed and currently recognized as a genus, is typified by *Galindomyia leei* Stone & Barreto, 1969. Genus abbreviation – *Ga.*

****

Systematics: *Galindomyia* is a subordinate taxon of the subfamily Culicinae, tribe Culicini. The genus is monobasic. It only includes the extant type species, which is known only from Colombia and Ecuador. Adames (1971) considered *Galindomyia* and *Deinocerites* to be sister taxa based on shared characteristics of the antennae and male genitalia, and regarded this monophyletic group as the sister of *Culex*. These relationships were supported in the cladistic analyses of Harbach & Kitching (1998) and Harbach *et al.* (2012) based on morphology. *Galindomyia* + *Deinocerites* was recovered in a derived relationship to *Culex* in the former study and as the derived taxon of the New World subgenera of *Culex* in the latter study.

****

Characteristics: *Galindomyia leei* is very similar to species of *Culex*, but the antennae are much longer than the proboscis and the maxillary palpi are very short as they are in species of *Deinocerites*. The species differs from species of *Deinocerites* in having the proboscis swollen apically in both sexes. *Galindomyia* is characterized as follows. ADULTS: Compound eyes contiguous; head with a few erect scales on the occiput; antenna longer than the proboscis, first and last flagellomeres very long, about the same length; acrostichal, prespiracular and postspiracular setae absent; paratergite bare; 1 lower mesepimeral seta present; anal vein (vein 1A) of the wing ends beyond the base of the mediocubital crossvein; wing membrane with distinct microtrichia; tarsi normal; pulvilli distinct. LARVAE: Maxilla with a well-developed maxillary brush; hypostomal suture complete; occipital foramen circular with an incomplete collar; comb and pecten present; siphon with 3 pairs of seta 1-S; saddle poorly developed, apparently composed of weakly sclerotized dorsal and ventral sclerites; ventral brush (seta 4-X) with 5 pairs of setae. The larva of *Galindomyia* is very similar to the larvae of *Culex* and *Deinocerites*. It can be separated from *Culex* larvae by the combination of seta 3-C absent (usually present in *Culex*), collar incomplete (usually well developed in *Culex*) and saddle composed of poorly defined dorsal and ventral sclerites (distinct and usually complete in *Culex*). The larva is distinguished from those of *Deinocerites* by the absence of seta 2-C, as well as the absence of a lateral lobe at the base of the mandible.

****

Bionomics: Little is known about the bionomics of *Galindomyia*. Larvae and adults have been collected from crab holes with brackish water in Colombia and males have been collected in light traps in Ecuador. Females apparently lay their eggs singly above water level. *Galindomyia leei* is not a medically important species.

****

Principal references: Stone & Barreto, 1969 (Colombia, description, adults); Adames & Galindo, 1975 (Colombia, pupa, larva, descriptions, bionomics); Harbach & Kitching, 1998 (morphology, phylogenetic relationships); Harbach *et al.*, 2012 (morphology, phylogenetic relationships).

Haemagogus Williston, 1896

Haemagogus Williston, 1896, originally proposed and currently recognized as a genus, is typified by *Haemagogus splendens* Williston, 1896. Genus abbreviation – *Hg.*

****

Systematics: *Haemagogus* is a subordinate taxon of the subfamily Culicinae, tribe Aedini. *Haemagogus* includes 28 extant species divided between two subgenera: *Conopostegus* and *Haemagogus sensu stricto*. Species of *Haemagogus* are distributed from Argentina through Central America and into North America as far as the southern tip of Texas, but they are most abundant in the forests in the central region of South America. Morphological and molecular data (Harbach & Kitching, 1998; Reinert *et al.*, 2004, 2006, 2008, 2009; Wilkerson *et al.*, 2015; da Silva *et al.*, 2020b; do Nascimento *et al.*, 2021) support the monophyly of *Haemagogus*. Arnell (1973) postulated that '*Haemagogus* is one of the more derived genera of Aedini', 'undoubtedly derived from the genus *Aedes*'. He also noted that the male genitalia of species of *Haemagogus* and *Finlaya* (as the Kochi Group of *Aedes* subgenus *Finlaya*) exhibit 'remarkable similarities', but then stated that 'there is little in the adults or immature stages to indicate relationship'. However, the results of the phylogenetic study of Reinert *et al.* (2009) based on morphological data appear to contradict the latter statement, wherein *Haemagogus* was recovered as the sister of *Downsiomyia* in a clade comprised of (*Finlaya* + *Danielsia*) + (*Downsiomyia* + *Haemagogus*). *Haemagogus* was not associated with other generic-level taxa in the phylogeny of Wilkerson *et al.* (2015). A close affinity between *Haemagogus* and the Oriental *Heizmannia* proposed by Edwards (1922a) and Mattingly (1957b) and other authors is not supported by the phylogenetic analyses. Four species of *Haemagogus* formed a strongly supported sister relationship with two species of *Howardina* in the maximum likelihood phylogeny of Soghigian *et al.* (2017) based on seven molecular markers. The phylogenetic analysis of amino acid sequences of 709 genes conducted by Soghigian *et al.* (2023) included two species of *Haemagogus*, but their relationships with species of other groups of Aedini included in the analysis were not explicitly indicated. Contrary to the pairing of *Hg.* (*Conopostegus*) *leucotaeniatus* with *Aedes* (*Ochlerotatus*) *euiris* in the phylogeny of mitochondrial *COI* genes generated in the study of Muñoz-Gamba *et al.* (2021), the analysis of mitochondrial protein sequences (da Silva *et al.*, 2020b) supports the monophyly of *Conopostegus* and *Haemagogus sensu stricto*. Oddly, *Hg. leucotaeniatus* was recovered in a clade that included five species of *Aedes* (representatives of three subgenera) in the analysis of 13 mitochondrial protein-coding genes conducted by Lorenz *et al.* (2021).

****

Characteristics: *Haemagogus* are brightly coloured mosquitoes that resemble sabethines. The scutum is covered with smooth metallic scales in almost all species, and the acrostichal, dorsocentral and prescutellar setae are absent centrally. The antepronota are large and close together. The adults are distinguished from sabethines by the position of the hindcoxa below the base of the mesomeron and the absence of prespiracular setae. They are usually distinguished from other New World aedines without much difficulty by the character of the scutal scaling and the presence of silver scales in a single broad vertical band (*Haemagogus sensu stricto*) or three vertical arcs (*Conopostegus*) extending from the scutum to the coxae. Larvae are separated from those of other genus groups in the New World, except some other Aedini, based on one or more of the following characters: maxillary brush present; hypostomal suture complete; occipital foramen circular, with a distinct collar; comb and pecten both present; seta 5-VIII inserted well below the dorsal margin of segment X; siphon unmodified, with a single pair of seta 1-S inserted distal to the pecten; ventral brush (seta 4-X) with 5 or 6 pairs of setae. The larvae are distinguishable with some difficulty from similar species of other genus groups of Aedini by the combination of (1) setae 5–7-P borne on small separate tubercles, (2) seta 12-I present and/or seta 7-III–V much larger than seta 9 on the same segment, (3) seta 3-VII very long and usually single (similar to sabethines) and (4) saddle incomplete.

****

Bionomics: The larvae of all species of *Haemagogus* inhabit cavities. They are found mainly in tree holes and bamboo, but are often found in the axils of bromeliads, fallen fruit husks, and occasionally in ground pools, rock holes and artificial containers in urban areas. *Haemagogus* adults are active during the daylight hours. They inhabit primary and secondary tropical forests, open deciduous forests and mangroves. Species that inhabit primary forest are largely arboreal and feed primarily in the forest canopy. Many species readily attack humans in forest clearings, open secondary growth and littoral situations associated with mangroves. Several species, including *Hg. janthinomys*, *Hg. equinus*, *Hg. lucifer*, *Hg. leucocelaenus*, *Hg. mesodentatus*, *Hg. spegazzinii* and *Hg. capricornii* are vectors of sylvatic yellow fever virus. Ilheus virus has been isolated from *Hg. janthinomys* and *Hg. spegazzinii* in Panama and from *Hg. leucocelaenus* in Brazil. Evidence also suggests that *Hg. leucocelaenus* is capable of natural vertical transmission of Zika virus in Brazil (Alencar *et al.*, 2021).

****

Principal references: Lane, 1953 (Neotropical Region, taxonomy, keys, descriptions, distributions); Martínez *et al.*, 1961 (Argentina, taxonomy, literature, species treatments); Forattini, 1965b (Neotropical Region, diagnosis, classification, species, bionomics, distributions); Cova-Garcia *et al.*, 1966 (Venezuela, keys, species descriptions, distributions); Belkin *et al.*, 1970 (Jamaica, taxonomy, keys, descriptions, bionomics); Zavortink, 1972 (*Conopostegus*, taxonomy, keys, descriptions, bionomics, distributions); Arnell, 1973 (taxonomy, classification, keys, descriptions, bionomics, distributions); Clark-Gil & Darsie, 1983 (Guatemala, keys, distributions, bionomics); Darsie, 1985 (Argentina, keys); Reinert, 2002c (female genitalia); Reinert *et al.*, 2004, 2006, 2008, 2009 (morphology, phylogenetic relationships); Marcondes & Alencar, 2010 (Brazil, key to adults, distributions, bionomics, medical importance); Wilkerson *et al.*, 2015 (phylogenetic relationships); Soghigian *et al.*, 2017 (amino acid sequences, phylogenetic relationships); da Silva *et al.*, 2020b (mitochondrial protein sequences, phylogenetic relationships); Lorenz *et al.*, 2021 (mitochondrial protein-coding genes, phylogenetic relationships).

Conopostegus Dyar, 1925

Conopostegus Dyar, 1925d, originally proposed as a subgenus of *Aedes* and currently recognized as a subgenus of *Haemagogus*, is typified by *Haemagogus leucocelaenus* (Dyar & Shannon, 1924b), original combination: *Aedes leucocelaenus*. Subgenus abbreviation – *Con*.

****

Systematics: *Conopostegus* includes four formally named extant species and at least four unnamed species (Zavortink, 1972). Species of *Conopostegus* are recorded from localities that extend from Honduras to Argentina and Brazil. *Conopostegus* and *Haemagogus sensu stricto* were recovered as sister taxa in the phylogenetic studies of Reinert *et al.* (2006, 2008, 2009) based on morphological data and da Silva *et al.* (2020b) based on mitochondrial protein sequences, with *Conopostegus* in a basal relationship to *Haemagogus sensu stricto*, but this does not appear to support the suggestion of Arnell (1973) that species of *Conopostegus* are closer to the ancestral lineage that gave rise to *Haemagogus* because they appear to have retained several characters that are not present in species of *Haemagogus sensu stricto*. Zavortink (1972) noted that relationships between the formally named species of *Conopostegus* are not obvious. The only species of *Conopostegus* (*Hg. leucocelaenus*) included in the maximum likelihood phylogeny of Soghigian *et al.* (2017) based on several molecular markers was paired with *Hg. mesodentatus* of *Haemagogus sensu stricto* in a clade that included two other species of that group. The relationship, however, is doubtful because *Hg. leucocelaenus* was represented by a single molecular marker as opposed to several markers for the three species of *Haemagogus sensu stricto*. Soghigian *et al.* (2023) included *Hg. leucocelaenus*, along with a species of *Haemagogus sensu stricto*, in their phylogenetic analysis of amino acid sequences of 709 genes, but the relationships of the species with other Aedini included in the analysis were not explicitly indicated.

****

Characteristics: The following combinations of features distinguish species of *Conopostegus* from species of *Haemagogus sensu stricto*. ADULTS: Scutum with deep brown scaling, scutum and scutellum without metallic-coloured scales; acrostichal, antealar and prescutellar areas with conspicuous lines or patches of silver scales; prescutellar setae present, well developed; postspiracular setae well developed; thoracic pleura with 3 conspicuous more or less vertical bands of silver scales; midfemur with a patch of silvery-white scales on the proximal 0.5–0.7 of the anterior surface. FEMALE GENITALIA: Sternum VIII divided by a narrow median longitudinal membranous area; tergum IX V-shaped, setae absent. MALE GENITALIA: Tergum VIII without specialized scales or setae on the posterior margin; tergum IX with a pair of widely separated posterior lobes; aedeagus with a pair of sternal plates, without a tergal sclerotized process. LARVAE: Integument of the thorax and abdomen without spicules; seta 12-I present, seta 12-II slightly to noticeably weaker than seta 10-II; comb comprised of 5–9 scales in a single row; ventral brush (seta 4-X) borne on a strongly sclerotized boss. PUPAE: Seta 5-CT strongly developed, long; seta 7-CT 6 or more times as long as seta 6-CT; seta 1-Pa strongly developed, long, 0.4–0.6 the length of the paddle.

****

Bionomics: The immature stages of species of *Conopostegus* are found in tree holes, bamboo internodes and artificial containers. Females have been collected in horse-baited traps and feeding on humans. *Haemagogus leucocelaenus* is a vector of sylvatic yellow fever virus. Ilheus virus has been isolated from this species in Brazil.

****

Principal references: Zavortink, 1972 (taxonomy, keys, descriptions, bionomics, distributions); Arnell, 1973 (in keys, morphological distinctions); Reinert, 2002c (female genitalia); Reinert *et al.*, 2004, 2009 (morphology, phylogenetic relationships); Marcondes & Alencar, 2010 (Brazil, subgenus and *Hg. leucocelaenus*, in key to adults); Soghigian *et al.*, 2017 (amino acid sequences, phylogenetic relationships); da Silva *et al.*, 2020b (mitochondrial protein sequences, phylogenetic relationships).

Haemagogus Williston, 1896 *sensu stricto*

Haemagogus sensu stricto, the nominotypical subgenus of *Haemagogus*, is typified by *Haemagogus splendens* Williston, 1896. Subgenus abbreviation – *Hag.*

****

Systematics: *Haemagogus sensu stricto* includes 24 extant species in the Neotropical Region, from Paraguay, northern Argentina and Bolivia northward to Sonora, Mexico on the Pacific and southern Texas on the Atlantic of North America. *Haemagogus sensu stricto* and *Conopostegus* were recovered as sister taxa in the phylogenetic studies of Reinert *et al.* (2006, 2008, 2009) based on morphological data and da Silva *et al.* (2020b) based on mitochondrial protein sequences, with *Conopostegus* in a basal relationship to *Haemagogus sensu stricto*, but this does not appear to support the contention of Arnell (1973) that species of *Conopostegus* are more highly derived than species of *Conopostegus*. As noted for *Conopostegus*, the inclusion of *Hg. leucocelaenus* in a clade with three species of *Haemagogus sensu stricto* in the maximum likelihood phylogeny of Soghigian *et al.* (2017) based on molecular markers is doubtful. Morphological differences strongly support the recognition of *Conopostegus* and *Haemagogus sensu stricto* as separate generic-level taxa. Soghigian *et al.* (2023) included a single species of *Haemagogus sensu stricto*, *Hg. janthinomys*, along with *Hg.* (*Conopostegus*) *leucocelaenus*, in their phylogenetic analysis of amino acid sequences of 709 genes, but the relationships of the species with other Aedini included in the analysis were not explicitly indicated.

****

Characteristics: The following combinations of features distinguish species of *Haemagogus sensu stricto* from species of *Conopostegus*. ADULTS: Antepronota approximated; scutum with a dense covering of metallic green, blue, copper or bronze scales; prescutellar area with scales medially and/or posteriorly; thoracic pleura with a single broad vertical band of silver scales; postspiracular and upper mesokatepisternal setae absent; postprocoxal scales present; hindcoxa more or less at the same level as the dorsal margin of the mesomeron. FEMALE GENITALIA: Tergum IX Y-shaped, with or without setae. MALE GENITALIA: Tergum VIII with long lanceolate scales or large spine-like setae on the posterior margin, sternum VIII with seta 2-S inserted lateral and at about the same level as seta 1-S; tergum IX with or without setae; gonocoxite with or without an apical lobe, with 1 or more stout setae on the basomesal area of the dorsal surface; gonostylus simple to highly modified, gonostylar claw apical or subapical, 0.2–0.7 the length of the gonostylus; claspette filament simple or highly modified; aedeagus with a variously developed sclerotized apicotergal process. LARVAE: Seta 1-A no longer than 3 times the width of the antenna at the point of attachment; seta 4-C not longer than 0.2 the length of the dorsal apotome; comb comprised of numerous scales in a patch; thoracic and abdominal integument spiculate; seta 12-I absent or seta 12-II stronger than seta 10-II; ventral brush (seta 4-X) borne on a weakly sclerotized boss. PUPAE: Seta 5-CT relatively weakly developed, no longer than 1.2 the length of seta 4-CT; seta 5-V shorter than the length of tergum VI; seta 1-Pa relatively weakly developed.

****

Bionomics: Species of *Haemagogus sensu stricto* are inherently sylvatic. Females of most species readily attack humans at ground level, but some species tend to be principally arboreal, *e.g. Hg. janthinomys*, and are seldom encountered at ground level. The immature stages mainly inhabit tree holes and cut or broken bamboo internodes but are often found in other phytotelm habitats, *e.g.* bromeliads and fruit husks, and sometimes in ground pools and rock holes. Several species, including *Hg. janthinomys*, *Hg. equinus*, *Hg. lucifer*, *Hg. mesodentatus*, *Hg. spegazzinii* and *Hg. capricornii*, are vectors of sylvatic yellow fever virus. Ilheus virus has been isolated from *Hg. janthinomys* and *Hg. spegazzinii* in Panama.

****

Principal references: Lane, 1953 (Neotropical Region, taxonomy, keys, descriptions, distributions); Martínez *et al.*, 1961 (Argentina, taxonomy, literature, species treatments); Forattini, 1965b (as subgenera *Haemagogus*, *Longipalpifer* and *Stegoconops*, taxonomy, species, bionomics, distributions); Cova-Garcia *et al.*, 1966 (Venezuela, keys, species descriptions, distributions); Belkin *et al.*, 1970 (Jamaica, *Hg. equinus* as a species of subgenus *Longipalpifer*, taxonomy, description, bionomics, distribution); Arnell, 1973 (taxonomy, classification, keys, descriptions, bionomics, distributions); Clark-Gil & Darsie, 1983 (Guatemala, keys, distributions, bionomics); Darsie, 1985 (Argentina, keys); Reinert, 2002c (female genitalia); Reinert *et al.*, 2004, 2009 (morphology, phylogenetic relationships); Marcondes & Alencar, 2010 (Brazil, key to adults, distributions, bionomics, medical importance); Soghigian *et al.*, 2017 (amino acid sequences, phylogenetic relationships); da Silva *et al.*, 2020b (mitochondrial protein sequences, phylogenetic relationships).

Heizmannia Ludlow, 1905

Heizmannia Ludlow, 1905a, originally proposed and currently recognized as a genus, is typified by *Heizmannia scintillans* Ludlow, 1905a. Genus abbreviation – *Hz.*

****

Systematics: *Heizmannia* is a subordinate taxon of the subfamily Culicinae, tribe Aedini. It includes 40 extant species classified in two subgenera: *Heizmannia sensu stricto* and *Mattinglyia*. The species occur mainly in the Oriental Region. A few species occur in southeastern areas of the Palaearctic Region (China, Japan and South Korea). One species of *Heizmannia sensu stricto* is known only from its type locality in the Australasian Region (Halmahera, Molucca Islands and Indonesia). *Heizmannia* was recovered in a clade comprised of *Petermattinglyius* + (*Alanstonea* + *Pseudarmigeres*) + *Heizmannia*)) in the phylogenetic study of Reinert *et al.* (2009) based on morphological data, and this clade was sister to a clade comprised of *Lorrainea* + ((*Eretmapodites* + *Armigeres*) + (*Udaya* + (*Belkinius* + *Zeugnomyia*))). *Mattinglyia* was paraphyletic relative to *Heizmannia sensu stricto*. *Heizmannia* and *Pseudarmigeres* seem to be very closely related, although *Alanstonea* is included as the sister of *Pseudarmigeres*. *Heizmannia* and *Pseudarmigeres* were recovered as sister taxa in the phylogeny of Wilkerson *et al.* (2015). *Heizmannia* was not recovered as a monophyletic lineage in the maximum likelihood phylogeny of Soghigian *et al.* (2017) based on seven molecular markers. The *Heizmannia* clade included *Udaya subsimilis* as the sister of *Hz. discrepans*. It is possible, however, that the former species was misidentified as a species of *Heizmannia*.

****

Characteristics: *Heizmannia* are similar to *Haemagogus* in the New World. Adults are brightly coloured mosquitoes reminiscent of some sabethines. Many species have broad iridescent scales on the scutum, and species of *Heizmannia sensu stricto* have a small patch of setae on the mesopostnotum. *Haemagogus* shares many of the characteristics found in species of other aedine genera in the Oriental Region, but it is separable from them by the presence of one or more or various combinations of the following characters: erect scales restricted to the back of the head, occasionally absent; acrostichal and dorsocentral setae absent; prespiracular setae absent; postspiracular setae usually absent; upper mesokatepisternal setae absent; lower mesepimeral setae present; subspiracular area, postspiracular area, paratergite and laterotergite all with scales; upper calypter of the wing with a fringe of setae, alula with broad flat scales on the dorsal margin, anal vein (vein 1A) ending beyond the base of the mediocubital crossvein; tarsi normal, pulvilli not evident. The essential features distinguishing *Heizmannia* larvae from larvae of other genus groups include the following: hypostomal suture complete to the posterior tentorial pit, maxilla with a typical maxillary brush (*Heizmannia sensu stricto*) or a maxillary bundle of coalesced filaments (*Mattinglyia*); seta 12-I absent; comb and pecten present; a single pair of seta 1-S inserted well above the base of the siphon; saddle incomplete ventrally; ventral brush (seta 4-X) usually with 4 pairs of setae. Some larvae of other aedine genus groups closely resemble *Heizmannia* larvae but usually differ in having setae 4,7-C small with few branches. If these setae are large with many branches as in *Heizmannia* larvae, then seta 6-C is inserted posterior to seta 4-C and/or it does not have two unequal branches.

****

Bionomics: *Heizmannia* larvae are found mainly in tree holes and bamboo, but some species are occasionally found in small ground pools and one species has been collected from split coconuts and artificial containers. Little is known about the bionomics of the adults. They are apparently active during the daytime in forests where females readily attack humans, but none of the species are known to vector disease-causing agents.

****

Principal references: Barraud, 1934 (southern Asia, keys, taxonomy, descriptions, distributions, bionomics); Mattingly, 1957b (Indomalayan area, taxonomy, classification, keys, descriptions, bionomics, distributions); Thurman, 1959 (Thailand, taxonomy, keys, descriptions, distributions); Delfinado, 1966a (Philippines, taxonomy, keys, descriptions, distributions, bionomics); Mattingly, 1970 (Southeast Asia, keys, descriptions, distributions, bionomics); Reinert, 1973a (taxonomy); Tanaka *et al.*, 1979 (Japan, taxonomy, keys, descriptions, distributions, bionomics); Lee *et al.*, 1988b (Australasian Region, *Hz. aurea*, taxonomy, literature, distribution); Lu *et al.*, 1997a (China, keys, taxonomy, descriptions, distributions, bionomics); Reinert, 2002b (female genitalia); Reinert *et al.*, 2004, 2009 (morphology, phylogenetic relationships); Rattanarithikul *et al.*, 2010 (Thailand, keys, bionomics); Wilkerson *et al.*, 2015 (phylogenetic relationships); Soghigian *et al.*, 2017 (phylogenetic relationships).

Heizmannia Ludlow, 1905 *sensu stricto*

Heizmannia sensu stricto, the nominotypical subgenus of *Heizmannia*, is typified by *Heizmannia scintillans* Ludlow, 1905a. Subgenus abbreviation – *Hez.*

****

Systematics: *Heizmannia sensu stricto* includes 34 extant species distributed mainly in the Oriental Region, with a few species recorded from southeastern areas of the Palaearctic Region (China, Japan, South Korea). *Heizmannia aurea* is the only species known from the Australasian Region (Halmahera, Molucca Islands, Indonesia). The subgenus was found to be monophyletic in the morphology-based phylogenetic studies of Reinert *et al.* (2004, 2009), in a derived relationship relative to the paraphyletic *Mattinglyia*.

****

Characteristics: Species of *Heizmannia sensu stricto* are distinguished from species of *Mattinglyia* as follows. ADULTS: Antenna of males not strongly verticillate; antepronota nearly contiguous; mesopostnotal setae and sometimes mesopostnotal scales present; lower mesepimeral seta absent; fore- and midungues of females of most species simple, without teeth. FEMALE GENITALIA: Sternum VIII with seta 1-S inserted some distance from the anterior margin; tergum IX represented by 2 lateral plates separated by membrane; insula with setae. MALE GENITALIA: Gonocoxite with a well-developed subapical lobe bearing 1 or 2 stout setae. LARVAE: Maxillary brush normal, comprised of separate filaments. PUPAE: Generally not separable from the pupae of *Mattinglyia*.

****

Bionomics: The immature stages of species of *Heizmannia sensu stricto* are found primarily in tree holes and split, cut and bored bamboo, but also occur in other types of container habitats and sometimes in groundwater and crab holes. Adults of most species are known to readily attack and feed on humans in shaded forest habitats. As far as known, species of *Heizmannia sensu stricto* are not of medical importance.

****

Principal references: Barraud, 1934 (southern Asia, keys, taxonomy, descriptions, distributions, bionomics); Mattingly, 1957b (Indomalayan area, taxonomy, classification, keys, descriptions, bionomics, distributions); Thurman, 1959 (Thailand, taxonomy, keys, descriptions, distributions); Delfinado, 1966a (Philippines, taxonomy, keys, descriptions, distributions, bionomics); Mattingly, 1970 (Southeast Asia, keys, descriptions, distributions, bionomics); Tanaka *et al.*, 1979 (Japan); Lee *et al.*, 1988b (Australasian Region, *Hz. aurea*, taxonomy, literature, distribution); Lu *et al.*, 1997a (China, keys, taxonomy, descriptions, distributions, bionomics); Reinert, 2002b (female genitalia); Reinert *et al.*, 2004, 2009 (morphology, phylogenetic relationships); Rattanarithikul *et al.*, 2010 (Thailand, keys, bionomics).

Mattinglyia Lien, 1968

Mattinglyia, originally proposed as a genus by Lien (1968c), currently classified as a subgenus of *Heizmannia*, is typified by *Heizmannia catesi* (Lien, 1968c); original combination: *Mattinglyia catesi*. Subgenus abbreviation – *Mat.*

****

Systematics: *Mattinglyia* includes six extant species in the Oriental Region. They are recorded from localities in southern China, India, Indonesia, Malaysia, Taiwan and Thailand. Reinert *et al.* (2004, 2009) found that *Mattinglyia* was paraphyletic with respect to *Heizmannia sensu stricto* based on phylogenetic analyses of morphological data.

****

Characteristics: Species of *Mattinglyia* are distinguished from species of *Heizmannia sensu stricto* as follows. ADULTS: Antenna of males strongly verticillate; antepronota moderately separated; lower mesepimeral seta present (except in *Hz. catesi*); midlobe of the scutellum with broad silvery scales; mesopostnotal setae and scales absent; fore- and midungues of females toothed. FEMALE GENITALIA: Sternum VIII with seta 1-X inserted near the anterior margin; tergum IX a single sclerite; insula without setae. MALE GENITALIA: Gonocoxite without a subapical lobe, a spine or spines normally associated with it are represented by 2 flattened leaf-like structures with expanded apices; gonostylus relatively simple; one lobe of the proximal claspette with a fan-like array of flattened blades, the other lobe with recurved setae. LARVAE: Maxillary brush developed as a bundle of coalesced filaments. PUPAE: Generally not separable from pupae of *Heizmannia sensu stricto*.

****

Bionomics: The immature stages of species of *Mattinglyia* primarily inhabit tree holes and bamboo. Females will viciously attack humans who enter their haunts. As far as known, species of *Mattinglyia* are not of medical importance.

****

Principal references: Mattingly, 1970a (Southeast Asia, keys, descriptions, distributions, bionomics); Reinert, 1973a (taxonomy); Lu *et al.*, 1997a (China, keys, taxonomy, descriptions, distributions, bionomics); Reinert, 2002b (female genitalia); Reinert *et al.*, 2004, 2009 (morphology, phylogenetic relationships); Rattanarithikul *et al.*, 2010 (Thailand, keys, bionomics).

Hodgesia Theobald, 1904

Hodgesia Theobald, 1904a, originally proposed and currently recognized as a genus, is typified by *Hodgesia sanguinae* Theobald, 1904a. Genus abbreviation – *Ho.*

****

Systematics: *Hodgesia* is a subordinate taxon of the subfamily Culicinae, tribe Hodgesiini. It includes 11 extant species in the Old World tropics. Four species are found in Africa, two in the Oriental Region, three in the Australasian Region and two occur in both the Oriental and Australasian Regions. *Hodgesia* appears to be most closely related to *Culiseta*, *Ficalbia* and *Mimomyia*. *Hodgesia* and *Ficalbia* were recovered as sister taxa in the phylogenetic analysis of Harbach & Kitching (1998) based on a single synapomorphy, *i.e.* seta 1-S inserted at the base of the siphon, a characteristic that also occurs in species of *Culiseta*. Belkin (1962) noted that the adults also bear some similarity with *Uranotaenia* and hypothesized that *Hodgesia* may have originated through hybridization between ancient members of Ficalbiini and Uranotaeniini. Phylogenetic relationships within *Hodgesia* have not been investigated.

****

Characteristics: *Hodgesia* are very small mosquitoes that are unique in having the dorsal scales on the distal half of the wing elongate and forked at the tip. Larvae are similar to *Culiseta* and *Ficalbia* in having seta 1-S attached near the base of the siphon. Larvae are distinguished from the larvae of *Culiseta* in having seta 5-VIII inserted near the dorsal margin of segment X, and differ from the larvae of *Ficalbia* in having the hypostomal sclerite fused with the maxillary palpus and the hypostomal suture complete to the posterior tentorial pit.

****

Bionomics: Very little is known about the bionomics of *Hodgesia*. Larvae are usually found in swamps and marshes in water with very dense vegetation, and often in association with species of *Mimomyia*. The feeding habits of most species are unknown. However, females of *Ho. sanguinae* in Africa have been reported to feed on humans (Edwards, 1941); females of *Ho. cyptopus* in Uganda freely feed on humans in forest, bush and plantations, and sometimes enters houses (Haddow & Ssenkubuge, 1963); and the females of *Ho. solomonis* viciously attack humans in the vicinity of its larval habitats in the Solomon Islands (Belkin, 1962). Females are so small that they may be mistaken for biting midges (Ceratopogonidae); hence, it is likely that other species of *Hodgesia* that may feed on humans have not been recognized. None of the species of *Hodgesia* are of medical importance.

****

Principal references: Barraud, 1934 (southern Asia, keys, descriptions, distributions, bionomics); Edwards, 1941 (Afrotropical Region, keys, descriptions of adults and pupae, distributions); Hopkins, 1952 (Afrotropical Region, keys, descriptions of larvae, distributions, bionomics); Thurman, 1959 (Thailand, taxonomy, descriptions, distributions); Belkin, 1962 (South Pacific, taxonomy, description, bionomics, distribution); Lee *et al.*, 1988b (Australasian Region, taxonomy, key to females, literature, distributions, bionomics); Service, 1990 (Afrotropical Region, keys, descriptions, bionomics, distribution); Lu *et al.*, 1997a (China, as a genus of Sabethini, *Ho. bailyi*, descriptions, distribution, bionomics); Harbach & Kitching, 1998 (morphology, phylogenetic relationships); Rattanarithikul *et al.*, 2006a (Thailand, keys, distributions, bionomics).

Isostomyia Coquillett, 1906

Isostomyia Coquillett, 1906c, originally proposed and currently recognized as a genus, is typified by *Isostomyia perturbans* (Williston, 1896), original combination: *Aedes perturbans*, subsequent designation by Coquillett (1910) (see Evenhuis, 2018). Genus abbreviation – *Is.*

****

Systematics: *Isostomyia* is a subordinate taxon of the subfamily Culicinae, tribe Sabethini. It includes four extant species, which are only known to occur in the Neotropical Region. *Isostomyia* shares close affinities with other New World genera of Sabethini. It was recovered in a sister-group relationship with *Ctenogoeldia* in the morphology-based phylogenetic analysis of Judd (1996) and as the sister of *Shannoniana* in the morphology-based phylogenetic analyses of Harbach & Kitching (1998), Harbach & Peyton (2000) and Harbach *et al.* (2007a). The phylogenetic relationships of the species of *Isostomyia* are unknown.

****

Characteristics: The adults of *Isostomyia* are distinguished from the adults of other members of Sabethini in the New World by the following combinations of characters: dorsal head scales with weak to moderate green or blue reflections; proboscis distinctly longer than the forefemur; scutal scales moderately broad to broad and flat; postpronotum without posterior setae; lower mesokatepisternal setae usually not extended above the lower edge of the mesepimeron; all tarsi completely dark-scaled. The larvae of *Isostomyia* differ from those of *Johnbelkinia*, *Onirion*, *Shannoniana*, *Trichoprosopon* and some *Wyeomyia* in having a transverse slit-like occipital foramen and a filamentous pecten on the posterior surface of the siphon. They are distinguished from larvae of other New World sabethines by the presence of a maxillary claw. They also differ from otherwise very similar larvae of *Runchomyia* in having seta 3-X branched and seta 6-S strongly developed, rigid and hooked at the tip. Campos & Zavortink (2010) showed, without mention, that seta 13-T of the larva of *Is. paranensis* is inserted on a common plate with setae 9–12-T, a feature that was only previously known in species of *Kimia* and *Trichoprosopon* (Harbach *et al.*, 2007a).

****

Bionomics: What little is known about the bionomics of *Isostomyia* species is based on collections of *Is. espini*. The larvae of this species are found in the leaf axils of terrestrial plants (aroids). The females are known to enter houses and to feed on humans outdoors in domestic areas. It is unlikely that species of *Isostomyia* are involved in the transmission of pathogens to humans.

****

Principal references: Zavortink, 1979 (as a subgenus of *Runchomyia*, in key to adults, male genitalia, pupae and larvae); Harbach & Peyton, 1993 (generic status, comparative morphology of larval maxillae); Judd, 1996 (morphology, phylogenetic relationships); Harbach & Kitching, 1998 (morphology, phylogenetic relationships), Harbach & Peyton, 2000 (morphology, phylogenetic relationships); Harbach *et al.*, 2007a (morphology, phylogenetic relationships).

Johnbelkinia Zavortink, 1979

Johnbelkinia Zavortink, 1979, originally proposed and currently recognized as a genus, is typified by *Johnbelkinia longipes* (Fabricius, 1805), original combination: *Culex longipes*. Genus abbreviation – *Jb.*

****

Systematics: *Johnbelkinia* is a subordinate taxon of the subfamily Culicinae, tribe Sabethini. *Johnbelkinia* includes only three extant species recorded from localities extending from southern Mexico to coastal Ecuador, eastern Bolivia and the central coastal area of Brazil. Species of *Johnbelkinia* are absent from the West Indies except Trinidad and Tobago. The affinities of *Johnbelkinia* with other sabethine genera are uncertain. Species of *Johnbelkinia* share characteristics with species of other New World genus groups, especially *Shannoniana*, *Runchomyia* and *Trichoprosopon*. The relationship of *Johnbelkinia* to other genus groups of Sabethini was not resolved in the morphology-based phylogeny of Judd (1996), but it was recovered as the sister to *Runchomyia* in the phylogeny generated from the analysis of morphological data in the studies of Harbach & Kitching (1998) and Harbach & Peyton (2000). With the addition of *Kimia* to the data set (Harbach *et al.*, 2007a), *Johnbelkinia* was recovered in a sister relationship to a clade comprised of *Sabethes* + (*Limatus* + *Malaya*). Muñoz-Gamba *et al.* (2021) included two species of *Johnbelkinia*, *Jb. longipes* and *Jb. ulopus*, along with species of *Limatus*, *Sabethes*, *Trichoprosopon* and *Wyeomyia*, in a phylogenetic analysis of mtDNA *COI* gene sequences in which *Johnbelkinia* was recovered as the sister of an unidentified species of *Sabethes*.

****

Characteristics: The adults of *Johnbelkinia* are distinguished from the adults of other genus groups of Sabethini in the New World by the following combination of characters: dorsal head scales with brilliant silver and blue reflections; proboscis distinctly longer than the forefemur; scutal scales moderately broad and flat and dull to moderately iridescent; postpronotum with 1 or 2 posterior setae; lower mesokatepisternal setae usually not extending above the lower edge of the mesepimeron; mid- and hindtarsi with pale markings. Larvae are recognized by the presence of a normal circular occipital foramen (distinguishes *Johnbelkinia* from *Isostomyia*, *Limatus*, *Phoniomyia*, *Sabethes*, *Runchomyia* and most *Wyeomyia*) in combination with a long row of branched setae on the posterior midline of the siphon (distinguishes *Johnbelkinia* from *Onirion*, *Shannoniana*, *Trichoprosopon* and certain *Wyeomyia*). The presence of a maxillary bundle distinguishes larvae of *Johnbelkinia* from larvae of all New World sabethines except *Runchomyia*.

****

Bionomics: The larvae of *Johnbelkinia* usually occur in the leaf axils and flower bracts of plants. They have been reported to be predacious on the larvae of *Wyeomyia*. The adults are forest mosquitoes. They are mainly active during the day. The females bite humans and other mammals. Arboviruses have been isolated from *Jn. ulopus* in Trinidad and Colombia; hence, species of *Johnbelkinia* are considered to be potential vectors of pathogens of human diseases. Species of *Johnbelkinia* are known to carry the eggs of the human bot fly, *Dermatobia hominis*.

****

Principal references: Zavortink, 1979 (new genus, systematics, genus and species descriptions, keys, bionomics, distributions); Harbach & Peyton, 1993 (comparative morphology of larval maxillae); Judd, 1996 (morphology, phylogenetic relationships); Harbach & Kitching, 1998 (morphology, phylogenetic relationships), Harbach & Peyton, 2000 (morphology, phylogenetic relationships); Harbach *et al.*, 2007a (morphology, phylogenetic relationships); Muñoz-Gamba *et al.*, 2021 (mtDNA *COI* sequences, phylogenetic relationships).

Kimia Vu & Harbach, 2007

Kimia Vu & Harbach 2007 (in Harbach *et al.*, 2007a), originally proposed and currently recognized as a genus, is typified by *Kimia decorabilis* (Leicester, 1908), original combination: *Topomyia decorabilis*. Genus abbreviation – *Km.*

****

Systematics: *Kimia* is a subordinate taxon of the subfamily Culicinae, tribe Sabethini. It currently includes five extant species with distributions in eastern areas of the Oriental Region. *Kimia* formed a sister-group relationship with *Trichoprosopon* + *Tripteroides* in the cladistic analysis of Harbach *et al.* (2007a) based on the absence of larval seta 8-M, which was not contradicted. A number of morphological characters not included in the analysis provide additional evidence for a relationship with the New World *Trichoprosopon*. This relationship sheds doubt on the monophyly of the New World Sabethini. The phylogenetic relationships of the species of *Kimia* are unknown, but based on similarities of the male genitalia, *Km. decorabilis* and *Km. nemorosa* appear to be most closely related to *Km. imitata* and *Km. suchariti*, respectively. Harbach *et al.* (2007a) stated that the insertion of seta 13-T on a common plate with setae 9–12-T is a unique feature of *Kimia* and *Trichoprosopon*; however, Campos & Zavortink (2010), without mention, illustrated this condition in the larva of *Isostomyia paranensis*.

****

Characteristics: *Kimia* are relatively small, attractive mosquitoes with a median longitudinal stripe of broad silver scales on the scutum. This character, along with the absence of setae or hair-like scales on the upper calypter, distinguishes the adults from the adults of all other Old World genus groups except *Malaya* and *Topomyia*. Despite the overall similarity, the unmodified proboscis of *Kimia* could never be confused with the setose, bent and apically swollen proboscis of *Malaya*. Females of *Kimia* are distinguished from those of *Topomyia* by the absence of setae and scales on sternum IX; males are distinguished by having the inner foreunguis distinctly larger than the outer one. Males also differ in several features of their genitalia. The larvae of *Kimia* are distinguished from the larvae of all other genus groups in the Oriental Region based on the following characters: presence of a single pair of seta 4-X (distinction from non-sabethine genera); presence of an oval occipital foramen and absence of seta 8-M (distinctions from *Malaya* and *Topomyia*); insertion of seta 13-T on a plate with setae 9–12-T and absence of a pecten (distinctions from *Malaya* and *Tripteroides*). Whereas punctures are present on two or three of abdominal segments III–V in pupae of *Topomyia* and *Tripteroides*, they are entirely absent in *Kimia*. Seta 6-VII is strongly developed and inserted anterior to seta 9-VII in pupae of *Kimia* and *Malaya*. This seta, although not strongly developed, is also inserted anterior to seta 9-VII in *Polylepidomyia*, *Rachionotomyia*, *Rachisoura* and *Tripteroides sensu stricto*. The long seta 5-III and short seta 5-VI distinguish the pupae of *Kimia* from those of *Topomyia* and *Tripteroides* in which seta 5-III is short and seta 5-VI is long. Whereas seta 14-III–VII is present in *Tripteroides*, it is absent in *Kimia*, *Malaya* and *Topomyia*. The absence of pupal seta 14-VIII in *Kimia* is a unique feature within Sabethini; in fact this seta is present in all other Culicidae except for some species of *Toxorhynchites*.

****

Bionomics: Little bionomical information is available for species of *Kimia*. Adult and pupal stages have never been collected in the wild and are known only from laboratory rearings of larvae found in bamboo stumps and erect bamboo internodes bearing small holes presumably made by beetles. It is uncertain whether larvae are facultative or obligatory predators. The maxillae appear to be modified for grasping. Before reaching the pupal and adult stages, the holotype of *Km. suchariti* was seen to prey on larvae of *Armigeres* and *Tripteroides*. Larvae of *Aedes aegypti* were used as a source of food while rearing larvae of *Km. decorabilis* that were collected in northern Vietnam (Harbach *et al.*, 2007a). Species of *Kimia* are of no known medical or economic importance.

****

Principal reference: Harbach *et al.*, 2007a (new genus and type species, taxonomy, phylogenetic relationships, comparative morphology with *Topomyia*, descriptions, bionomics, distribution).

Limatus Theobald, 1901

Limatus Theobald, 1901c, originally proposed and currently recognized as a genus, is typified by *Limatus durhamii* Theobald, 1901c. Genus abbreviation – *Li.*

****

Systematics: *Limatus* is a subordinate taxon of the subfamily Culicinae, tribe Sabethini. It includes nine extant species with distributions in Central America, eastern South America and the West Indies. Whereas *Limatus* was embedded within *Wyeomyia* in the morphological phylogeny of Judd (1996), it was recovered as a monophyletic lineage outside of *Wyeomyia* in the separate morphological and allozyme phylogenies of Motta *et al.* (2007). *Limatus* was recovered in a sister relationship with *Sabethes* in the phylogenetic analyses of morphological data conducted by Harbach & Kitching (1998) and Harbach & Peyton (2000), but was paired with *Malaya* when Harbach *et al.* (2007a) included *Kimia* in the data set of the latter authors. The phylogenetic relationships of the species of *Limatus* have not been investigated; however, two species of the group, *Li. flavisetosus* and *Li. durhamii*, formed an independent lineage in the phylogeny of mtDNA *COI* gene sequences generated in the study of Muñoz-Gamba *et al.* (2021), which also included species of *Johnbelkinia*, *Sabethes*, *Trichoprosopon* and *Wyeomyia*. It is interesting to note that two studies based on analyses of 13 mitochondrial protein-coding genes generated different relationships between *Li. flavisetosus*, *Wyeomyia* (*Prosopolepis*) *confusa* and four species of *Sabethes*: the relationship *Li. flavisetosus* + (*Wyeomyia confusa* + *Sabethes*) was recovered in the study of do Nascimento *et al.* (2021) whereas the relationship *Sabethes* + (*Li. flavisetosus* + *Wyeomyia confusa*) was recovered in the study of Lorenz *et al.* (2021).

****

Characteristics: The adults of *Limatus* are unique in having a single unguis (claw) on the hindleg, and differ from all other mosquitoes of Sabethini in having scales on the prespiracular area instead of setae. They generally resemble *Sabethes* in overall ornamentation, but the scutum is distinctive in bearing a striking pattern of gold, blue and violet scales. The short and broad caudolateral slits of the occipital foramen and the absence of an apical tooth on the maxilla distinguishes the larvae of *Limatus* from those of *Sabethes* and the majority of species currently classified in *Wyeomyia*. Seta 1-C are inserted close together, but the larvae are otherwise very similar to those of *Wyeomyia*.

****

Bionomics: *Limatus* are forest mosquitoes. Larvae have been found in cavities such as bamboo, tree holes, coconut husks, cacao pods, fallen leaves and spathes, snail shells, rock holes along streams and a variety of small artificial containers. Adults have been captured during landing-biting collections made at ground level and on towers in forest. Wyeomyia virus has been isolated from a pool of *Limatus* in Trinidad, and Venezuelan equine encephalitis virus has been isolated from *Li. flavisetosus*, but species of *Limatus* are unlikely to be of medical or economic importance.

****

Principal references: Lane, 1953 (taxonomy, keys, descriptions, distributions); Forattini, 1965b (description, bionomics, distributions); Cova-Garcia *et al.*, 1966 (Venezuela, description, keys); Belkin *et al.*, 1970 (Jamaica, *Li. hoffmani*, description, bionomics, distribution); Harbach & Peyton, 1993 (comparative morphology of larval maxillae); Judd, 1996 (morphology, phylogenetic relationships); Harbach & Kitching, 1998 (morphology, phylogenetic relationships), Harbach & Peyton, 2000 (morphology, phylogenetic relationships); Harbach *et al.*, 2007a (morphology, phylogenetic relationships); Motta *et al.*, 2007 (morphology, allozymes, phylogenetic relationships); do Nascimento *et al.*, 2021 (mitochondrial protein-coding genes, phylogenetic relationships); Lorenz *et al.*, 2021 (mitochondrial protein-coding genes, phylogenetic relationships); Muñoz-Gamba *et al.*, 2021 (mtDNA *COI* sequences, phylogenetic relationships).

Lutzia Theobald, 1903

Lutzia Theobald, 1903a, originally proposed and currently recognized as a genus, is typified by *Lutzia bigoti* (Bellardi, 1862), original combination: *Culex bigoti*. Genus abbreviation – *Lt.*

****

Systematics: *Lutzia* is a subordinate taxon of the subfamily Culicinae, tribe Culicini. *Lutzia* includes nine extant species divided between three subgenera: *Insulalutzia*, *Lutzia sensu stricto* and *Metalutzia*. The genus includes two species in the Neotropical Region, five in Asia and the Australasian Region, one in the Afrotropical Region and one in the Ogasawara Islands of Japan. Belkin (1962) suggested that *Lutzia* (as a subgenus of *Culex*) is an ancient group allied with *Culex sensu stricto*. Phylogenetic analyses based of morphological data support the generic status of *Lutzia* (Navarro & Liria, 2000; St John, 2007; Harbach *et al.*, 2012) by placing it outside the clade comprised of the subgenera of *Culex*. In stark contrast, analyses based on nucleotide sequences of ITS1 and ITS2 of rDNA and *COI* of mtDNA do place *Lutzia* among species of the subgenera of *Culex* (Miller *et al.*, 1996; Deus, 2009; Vesgueiro *et al.*, 2011; Demari-Silva *et al.*, 2011). Based on a series of comparative and data exclusion analyses of the data set of Harbach *et al.* (2012) and a critical examination of the results of other phylogenetic studies, Kitching *et al.* (2015) concluded that no morphological or molecular data set analysed in any previously published study had adequate signal to place *Lutzia* unequivocally with regard to other taxa of Culicini. However, in stark contrast to the molecular studies mentioned above, the phylogenetic analyses of Sun *et al.* (2019) and do Nascimento *et al.* (2021) based on the nucleotide sequences of 13 protein-coding genes, obtained from the complete mitochondrial genomes of 10 and eight species of *Culex*, respectively, and two species of *Lutzia*, provided strong support for the generic status of *Lutzia* as the sister of *Culex*. Phylogenetic relationships within *Lutzia* have not been investigated; however, *Lutzia sensu stricto* (represented by the type species) was recovered in a basal relationship to two species of *Metalutzia* (*Lt. fuscana* and *Lt. halifaxii*) in the phylogenetic study of Harbach *et al.* (2012). Soghigian *et al.* (2023) included a single species of *Lutzia*, *Lt. halifaxii*, in their phylogenetic analysis of amino acid sequences of 709 genes, but its relationship with other taxa of Culicini included in the analysis was not explicitly indicated.

****

Characteristics: *Lutzia* are large mosquitoes. The adults have four or more lower mesepimeral setae (distinction from the adults of other Culicini). The anterior surfaces of the fore- and midfemora have numerous patches of light-coloured scales, the pulvilli are large and the radiomedial crossvein is close to the base of vein M_{3+4}. The gonocoxite of males lacks a foliform seta (seta *g*). *Lutzia* larvae are distinguished from the larvae of other culicine genera by the following characteristics: mouthparts modified for predation, lateral palatal brushes with about 40 stout and strongly pectinate filaments; siphon short, with pecten extending to or near the apex (except *Insulalutzia*); saddle complete, dorsal surface strongly extended caudad making segment X appear long and pointed. Pupae have all or most of setae 1–9-CT single and seta 9-VIII inserted distinctly cephalad of the caudolateral angle of the tergum, about 0.7 from the anterior margin.

****

Bionomics: The immature stages of *Lutzia* occur in a wide variety of groundwater habitats, but are sometimes found in tree holes and other phytotelm habitats. Several species are commonly found in artificial containers, and appear to tolerate or prefer water with a high organic content. The larvae of all species are predacious. They mainly feed on other mosquito larvae, but also on other insects. Females attack domestic animals and sometimes humans, but otherwise very little is known of the bionomics of the adults. None of the species of *Lutzia* are of medical or economic importance.

****

Principal references: Edwards, 1941 (as a subgenus of *Culex*, Afrotropical Region, keys, descriptions of adults and pupae, distributions); Hopkins, 1952 (as a subgenus of *Culex*, Afrotropical Region, keys, descriptions of larvae, distributions, bionomics); Belkin, 1962 (as a subgenus of *Culex*, South Pacific, *Lt. halifaxii*, in keys, taxonomy, descriptions, bionomics, distribution); Bram, 1967a (as a subgenus of *Culex*, Thailand, keys, descriptions, distributions, bionomics); Tanaka *et al.*, 1979 (as a subgenus of *Culex*, Japan, taxonomy, keys, descriptions, distributions, bionomics); Lee *et al.*, 1988b (as a subgenus of *Culex*, Australasian Region, taxonomy, keys, literature, distributions, bionomics); Miller *et al.*, 1996 (as a subgenus of *Culex*, molecular data, phylogenetic relationships); Lu *et al.*, 1997a (China, as a subgenus of *Culex*, keys, taxonomy, descriptions, distributions, bionomics); Navarro & Liria, 2000 (as a subgenus of *Culex*, *Lt. bigoti*, morphology of maxillae and mandibles, phylogenetic relationships); Tanaka, 2003 (genus status, Japan, taxonomy, pupae); Rattanarithikul *et al.*, 2005b (Thailand, bionomics, keys); Deus, 2009 (as a subgenus of *Culex*, molecular data, phylogenetic relationships); Demari-Silva *et al.*, 2011 (*Lt. bigoti*, mtDNA *COI* sequences, phylogenetic relationships); Vesgueiro *et al.*, 2011 (*Lt. bigoti*, rDNA ITS2 sequences, phylogenetic relationships); Harbach *et al.*, 2012 (morphology, phylogenetic relationships); Kitching *et al.*, 2015 (phylogenetic relationships); Sun *et al.*, 2019 (phylogenetic relationship with *Culex*); do Nascimento *et al.*, 2021 (mitochondrial protein-coding genes, phylogenetic relationships).

Insulalutzia Tanaka, 2003

Insulalutzia Tanaka, 2003, originally proposed and currently recognized as a subgenus of *Lutzia*, is typified by *Lutzia shinonagai* (Tanaka, Mizusawa & Saugstad, 1979), original combination: *Culex* (*Lutzia*) *shinonagai*. Subgenus abbreviation – *Ilt.*

****

Systematics: *Insulalutzia* is monobasic. It only includes the extant type species, which is known only from Ogasawara-guntô (the Bonin Islands). The affinities of *Insulalutzia* have not been investigated.

****

Characteristics: *Lutzia shinonagai* is distinguished from the species of *Lutzia sensu stricto* and *Metalutzia* as follows. ADULTS: Compound eyes contiguous above the antennal pedicels, separated dorsally (distinction from *Lutzia sensu stricto*); maxillary palpus of males without conspicuous setae; proboscis entirely dark-scaled; prealar scales absent; wing entirely dark-scaled (distinction from *Lutzia sensu stricto*); tarsi without pale bands (distinction from *Lutzia sensu stricto*). MALE GENITALIA (distinctions from *Lutzia sensu stricto*): Seta *h* of the subapical lobe of the gonocoxite not developed or differentiated; gonostylar claw borne at the apex of the gonostylus; paraproct with a prominent basal lateral arm. LARVAE: Antenna more or less cylindrical, not narrowed beyond seta 1-A (distinction from *Lutzia sensu stricto*); seta 6-C inserted posterior to seta 5-C; seta 9-C inserted posterior to seta 8-C (distinction from *Lutzia sensu stricto*); seta 10-C inserted anterior to seta 9-C (distinction from *Lutzia sensu stricto*); seta 4-P normally single (distinction from *Lutzia sensu stricto*); seta 1-III,IV large, 0.5–1.0 the length of the segment (distinction from *Metalutzia*); seta 1-V large, 0.5–1.0 the length of the segment (distinction from *Lutzia sensu stricto*); comb comprised of 18–20 scales; siphon shorter than or at most as long as the saddle, elements of seta 1-S much longer than the diameter of the siphon (distinction from *Lutzia sensu stricto*), pecten restricted to the middle 0.33 of the siphon, pecten spines closely spaced; ventral brush (seta 4-X) without precratal setae (distinction from *Lutzia sensu stricto*). PUPAE: Seta 1-III,IV single; seta 5-II inserted lateral of seta 4-II; seta 9-II–VI sharply pointed (distinction from *Metalutzia*); segment VIII with the posterolateral corner not produced.

****

Bionomics: A larva of *Lt. shinonagai* was found in a tree hole. Nothing else has been published about the bionomics of this species. *Lutzia shinonagai* is apparently rare and is unlikely to be of medical or economic importance.

****

Principal references: Tanaka *et al.*, 1979 (as a species of *Culex* subgenus *Lutzia*, Japan, in keys, description, distribution, bionomics); Tanaka, 2003 (new subgenus, Japan, pupa, description, identification).

Lutzia Theobald, 1903 *sensu stricto*

Lutzia sensu stricto, the nominotypical subgenus of *Lutzia*, is typified by *Lutzia bigoti* (Bellardi, 1862), original combination: *Culex bigoti*. Subgenus abbreviation – *Lut.*

****

Systematics: *Lutzia sensu stricto* includes two extant species with distributions in Central and South America, from Mexico southward into Argentina. In a cladistic analysis of morphological features of larval mandibles and maxillae conducted by Navarro & Liria (2000), *Lt. bigoti* was recovered in a basal relationship to a clade that consisted of representatives of six subgenera of *Culex* and a species *Deinocerites*. In contrast, phylogenetic analyses based on nucleotide sequences of the *COI* gene of mtDNA and the ITS2 locus of rDNA (Demari-Silva *et al.*, 2011; Vesgueiro *et al.*, 2011, respectively) placed *Lt. bigoti* among species of the subgenera of *Culex*. In the morphology-based phylogenetic study of Harbach *et al.* (2012), *Lt. bigoti* was recovered in a basal relationship to two species of *Metalutzia* (*Lt. fuscana* and *Lt. halifaxii*). Investigations that include species of all subgenera of *Lutzia* are needed to confirm this relationship, as well as the relationship of the genus with other genus groups of the Culicini.

****

Characteristics: The following characters distinguish species of *Lutzia sensu stricto* from the species of *Insulalutzia* and *Metalutzia*. ADULTS: Compound eyes completely separated above the antennal pedicels; maxillary palpus of males with numerous conspicuous setae (distinction from *Insulalutzia*); proboscis with pale scales in a ring (distinction from *Insulalutzia*); prealar scales present (distinction from *Insulalutzia*); wing with pale markings (distinction from *Insulalutzia* and *Metalutzia*); tarsi with pale bands. MALE GENITALIA: Subapical lobe of the gonocoxite with a distinct seta *h*; gonostylar claw inserted more or less distinctly before the apex of the gonostylus; lateral plate (opisthophallic sclerite) of the phallosome with a small (reduced) dorsal arm; basal lateral arm of the paraproct absent or inconspicuous. LARVAE: Antenna narrowed beyond seta 1-A, seta 1-A inserted near mid-length of the antenna (about halfway from the base); seta 4-C inserted at the level of seta 7-C; seta 6-C inserted far anterior to seta 5-C, at the level of setae 4,7-C; seta 9-C inserted anterior to seta 8-C; seta 10-C inserted at the level of seta 9-C; seta 4-P multi-branched; seta 1-III,IV large, 0.5–1.0 the length of the segment (distinction from *Metalutzia*); seta 1-V very large, twice or more the length of the segment; comb with more than 30 scales (distinction from *Insulalutzia*); siphon longer than the saddle, elements of seta 1-S much shorter than the diameter of the siphon; ventral brush (seta 4-X) with precratal setae. PUPAE (distinctions from *Insulalutzia*): Seta 1-III,IV branched; seta 5-II inserted mesad of seta 4-II; seta 9-II–VI sharply pointed (distinction from *Metalutzia*); segment VIII with the posterolateral corner rather sharply produced.

****

Bionomics: The immature stages of species of *Lutzia sensu stricto* typically occur in small ground pools, and have been found in wheel ruts and barrels. The larvae are voracious predators. The females apparently do not attack humans and are not hematophagous; hence, species of the group are not of medical or economic importance.

****

Principal references: Lane, 1953 (as species of *Culex* subgenus *Lutzia*, taxonomy, keys, descriptions, distributions); Navarro & Liria, 2000 (as a subgenus of *Culex*, *Lt. bigoti*, morphology of maxillae and mandibles, phylogenetic relationships); Demari-Silva *et al.*, 2011 (*Lt. bigoti*, mtDNA *COI* sequences, phylogenetic relationships); Vesgueiro *et al.*, 2011 (*Lt. bigoti*, rDNA ITS2 sequences, phylogenetic relationships); Harbach *et al.*, 2012 (*Lt. bigoti*, morphology, phylogenetic relationships); Kitching *et al.*, 2015 (phylogenetic relationships).

Metalutzia Tanaka, 2003

Metalutzia Tanaka, 2003, originally proposed and currently recognized as a subgenus of *Lutzia*, is typified by *Lutzia vorax* Edwards, 1921d. Subgenus abbreviation – *Mlt.*

****

Systematics: *Metalutzia* includes six extant species, which occur mainly in the Oriental Region. The distribution of *Lt. fuscana* extends to Timor in the Australasian Region and *Lt. halifaxii* is widely distributed in the Australasian as well as the Oriental Region. *Lutzia tigripes* is widely distributed in the Afrotropical Region. In a phylogenetic analysis conducted by Sun *et al.* (2019), based on the nucleotide sequences of 13 protein-coding genes obtained from the complete mitochondrial genomes of two species of *Metalutzia* (*Lt. fuscana* and *Lt. halifaxii*), nine species of *Culex sensu stricto* and a species of *Culiciomyia*, *Metalutzia* was recovered in a sister relationship to a clade comprised of the other species. In the morphology-based phylogenetic study of Harbach *et al.* (2012), *Metalutzia* (also represented by *Lt. fuscana* and *Lt. halifaxii*) was recovered as the sister of *Lutzia sensu stricto*. Investigations that include species of all subgenera and species of *Lutzia* are needed to confirm this relationship, as well as their affinities with other genus groups of the Culicini.

****

Characteristics: Species of *Metalutzia* are distinguished from the species of *Insulalutzia* and *Lutzia sensu stricto* as follows. ADULTS: Compound eyes partially separated, contiguous above the antennal pedicels (distinction from *Lutzia sensu stricto*); maxillary palpus of males with numerous conspicuous setae (distinction from *Insulalutzia*); proboscis usually with variable pale scaling mid-ventrally or in a ring (distinction from *Insulalutzia*); prealar scales present (distinction from *Insulalutzia*); wing entirely dark-scaled (distinction from *Lutzia sensu stricto*); tarsi without pale bands (distinction from *Lutzia sensu stricto*). MALE GENITALIA (distinctions from *Lutzia sensu stricto*): Seta *h* of the subapical lobe of the gonocoxite not developed or differentiated; gonostylar claw borne at the apex of the gonostylus; lateral plate (opisthophallic sclerite) of the phallosome without a dorsal arm; paraproct with a prominent basal lateral arm. LARVAE (distinctions from *Lutzia sensu stricto* except where indicated otherwise): Antenna more or less cylindrical, not narrowed beyond seta 1-A, seta 1-A inserted on the proximal 0.5 of the antenna (about 0.25 from the base); seta 4-C inserted well posterior to seta 7-C; seta 6-C inserted posterior to seta 5-C; seta 9-C inserted posterior to seta 8-C; seta 10-C inserted anterior to seta 9-C; seta 4-P normally single; seta 1-III,IV small, shorter than 0.5 the length of the segment (distinction from *Insulalutzia* and *Lutzia sensu stricto*); seta 1-V large, 0.5–1.0 the length of the segment; comb with more than 30 scales (distinction from *Insulalutzia*); siphon shorter than or at most as long as the saddle, elements of seta 1-S much longer than the diameter of the siphon; ventral brush (seta 4-X) without precratal setae (always?). PUPAE (distinctions from *Insulalutzia*): Seta 1-III,IV branched; seta 5-II inserted mesad of seta 4-II; seta 9-II–VI blunt-tipped (distinction from *Insulalutzia* and *Lutzia sensu stricto*); segment VIII with the posterolateral corner rather sharply produced.

****

Bionomics: The immature stages of species of *Metalutzia* occur in a wide range of habitats, which are limited only by the presence or absence of other mosquito larvae and aquatic organisms that serve as prey. They commonly inhabit various temporary and permanent bodies of stagnant groundwater, but frequently occur in all types of natural and artificial containers. Females are rarely attracted to humans. Little else is known about the bionomics of the adults. Species of the *Metalutzia* are not known or suspected of being of medical importance.

****

Principal references: Edwards, 1941 (Afrotropical *Lt. tigripes*, as a species of *Culex* subgenus *Lutzia*, in key, descriptions of adults and pupae, distribution); Hopkins, 1952 (Afrotropical *Lt. tigripes*, as a species of *Culex* subgenus *Lutzia*, in key, description of larvae, distribution, bionomics); Belkin, 1962 (South Pacific, *Lt. halifaxii*, as a species of *Culex* subgenus *Lutzia*, in key, descriptions, distribution, bionomics); Bram, 1967a (Thailand, as species of *Culex* subgenus *Lutzia*, keys, descriptions, distributions, bionomics); Tanaka *et al.*, 1979 (Japan, as species of *Culex* subgenus *Lutzia*, keys, descriptions, distributions, bionomics); Lee *et al.*, 1988b (Australasian Region, as species of *Culex* subgenus *Lutzia*, taxonomy, keys, literature, distributions, bionomics); Tanaka, 2003 (new subgenus, taxonomy, pupae, keys); Rattanarithikul *et al.*, 2005b (Thailand, bionomics, keys); Harbach *et al.*, 2012 (*Lt. fuscana*, *Lt. halifaxii*, morphology, phylogenetic relationships); Kitching *et al.*, 2015 (*Lt. fuscana*, *Lt. halifaxii*, phylogenetic relationships); Sun *et al.*, 2019 (*Lt. fuscana*, *Lt. halifaxii*, phylogenetic relationships based on mitochondrial genomes).

Malaya Leicester, 1908

Malaya Leicester, 1908, originally proposed and currently recognized as a genus, is typified by *Malaya genurostris* Leicester, 1908. Genus abbreviation – *Ml.*

****

Systematics: *Malaya* is a subordinate taxon of the subfamily Culicinae, tribe Sabethini. The group includes 12 extant species with distributions in central Africa, through most of the Oriental Region and into Australia, New Guinea, the Solomon Islands, China and the Ryukyu Archipelago. Belkin (1962) stated that '*Malaya* appears to be an ancient derivative from a generalized sabethine stock which also gave rise to *Topomyia*.' In support of this, *Malaya* was recovered as the sister of *Topomyia* in the cladistic analyses of morphological data conducted by Judd (1996), Harbach & Kitching (1998) and Harbach & Peyton (2000). However, it was paired with the New World *Limatus* when Harbach *et al.* (2007a) included *Kimia*, which was established for species previously placed in *Topomyia*, in the data set of Harbach & Peyton (2000). Phylogenetic relationships among the species of *Malaya* have not been studied.

****

Characteristics: Adults of species of *Malaya* are unique in having a setose, bent, apically swollen proboscis that is folded beneath the body when at rest. They are otherwise similar to the adults of *Kimia* and *Topomyia*, which also have a conspicuous median longitudinal white or silver stripe on the scutum and lack setae on the upper calypter, which distinguish them from other Old World sabethines (*Maorigoeldia* and *Tripteroides*). In the areas where species of *Malaya* occur, the larvae are distinguished from non-sabethine genera by the presence of a single pair of seta 4-X. *Malaya* larvae differ from the larvae of *Kimia* and *Tripteroides* in having a slit-like occipital foramen and comb scales in two or more rows. The larvae of *Malaya* closely resemble the larvae of *Kimia* and *Topomyia* but are distinct in having the maxillary palpus fused with the maxillary body and seta 4-X single rather than branched.

****

Bionomics: Larvae of species of *Malaya* occupy cavities, including the leaf axils of plants, tree holes and the water-filled nests of arboreal ants. Adults are active during the daytime. Their feeding habits are unique among mosquitoes. They feed on the regurgitation of ants. To accomplish this, the mosquito accosts an ant and brings the tip of its proboscis into contract with the mouth of the ant until a drop of liquid is produced. The regurgitated liquid is rapidly sucked up and the ant goes away unharmed. Females are incapable of taking a blood meal; hence, they are not of medical or economic importance to humans.

****

Principal references: Barraud, 1934 (southern Asia, as genus *Harpagomyia*, keys, taxonomy, descriptions, bionomics, distributions); Wharton, 1947 (Australasian and Oriental species, as genus *Harpagomyia*, descriptions, bionomics, distributions, keys); Thurman, 1959 (Thailand, taxonomy, descriptions, bionomics, distributions); Belkin, 1962 (South Pacific, *Ml. solomonis*, taxonomy, descriptions, bionomics, distribution); Tanaka *et al.*, 1979 (Japan, *Ml. genurostris*, in keys, descriptions, distribution, bionomics); Lee *et al.*, 1988b (Australasian Region, key to females, literature, descriptions, bionomics, distributions); Service, 1990 (Afrotropical Region, keys, descriptions, bionomics, distributions); Harbach & Peyton, 1993 (comparative morphology of larval maxillae); Judd, 1996 (morphology, phylogenetic relationships); Lu *et al.*, 1997a (China, keys, taxonomy, descriptions, distributions, bionomics); Harbach & Kitching, 1998 (morphology, phylogenetic relationships); Harbach & Peyton, 2000 (phylogenetic relationships); Harbach *et al.*, 2007a (phylogenetic relationships); Rattanarithikul *et al.*, 2007 (characteristics, distributions, bionomics, keys).

Mansonia Blanchard, 1901

Mansonia Blanchard, 1901, originally proposed and currently recognized as a genus, is typified by *Mansonia titillans* (Walker, 1848), original combination: *Culex titillans*. Genus abbreviation – *Ma*.

****

Systematics: *Mansonia* is a subordinate taxon of the subfamily Culicinae, tribe Mansoniini. It includes 25 extant species classified in two subgenera: *Mansonia sensu stricto* and *Mansonioides*. It also includes three fossil species. *Mansonia* has a nearly worldwide distribution. Species of *Mansonia sensu stricto* occur in the New World and those of *Mansonioides* occur in the Old World. The majority of species are tropical, but several range into the colder parts of the world. Morphological data strongly supported the sister relationship of *Coquillettidia* + *Mansonia* (Harbach & Kitching, 1998). The sister relationship was also recovered in the molecular phylogenetic analysis of mitochondrial genomes conducted by da Silva *et al.* (2020a), which included three species of *Mansonia sensu stricto* and five species of New World *Coquillettidia*. Soghigian *et al.* (2023) included three species of *Mansonia sensu stricto* and a species of *Mansonioides* in their phylogenetic analysis of amino acid sequences of 709 genes, but their relationships with other taxa of Mansoniini included in the analysis were not explicitly indicated.

****

Characteristics: Adults of *Mansonia* are generally large mosquitoes characterized by the presence of broad, asymmetrical scales on the wing veins. There is often a mixture of dark and pale scales that imparts a speckled appearance to the wings. *Mansonia* resemble some species of *Culex*, Aedini and *Coquillettidia*, but the tarsal claws are simple, the abdomen is truncate in females (distinctions from Aedini), pulvilli are not evident (distinction from *Culex*) and postspiracular setae are present (distinction from Old World species of *Coquillettidia*). New World species of *Coquillettidia* possess postspiracular setae, but differ from *Mansonia* in having a conspicuous preapical white band on the anterior surface of the femora. The larvae of *Mansonia* resemble those of *Coquillettidia* in having the spiracular apparatus and siphon distinctively modified for piercing plant tissues. They differ from *Coquillettidia* in having the distal part of the antenna fused with and much shorter than the basal part.

****

Bionomics: The larvae of species of *Mansonia* occur in permanent waters in association with aquatic plants that have roots used for attachment by the siphon to obtain oxygen from air cells for respiration. Larvae of some species burrow into debris on the bottom whereas others cling to the roots of plants in floating masses. Water lettuce (*Pistia*) is commonly used as a host plant, particularly by species of *Mansonioides*. Larvae detach and re-attach to host plants quite readily. The females of several species are vicious nocturnal biters. *Mansonia titillans* of *Mansonia sensu stricto* is an important pest in South and Central America and in the southern USA. It is known to transmit various arboviruses, including Venezuelan equine encephalitis. Some species of *Mansonioides* transmit several arboviruses, but they are mainly important as vectors of the helminths that cause Brugian filariasis in India and Southeast Asia. *Mansonia uniformis*, which is widely distributed from western Africa through southern Asia to Japan and the Australasian Region, is a vector of *Wuchereria bancrofti* in Western New Guinea.

****

Principal references: Barraud, 1934 (southern Asia, *Mansonioides*, key to females, descriptions, bionomics, distributions); Edwards, 1941 (Afrotropical Region, *Mansonioides* as subgenus of *Taeniorhynchus*, in keys, descriptions of adults and pupae, distributions); Hopkins, 1952 (Afrotropical Region, *Mansonioides* as subgenus of *Taeniorhynchus*, in key, descriptions of larvae, bionomics); Lane, 1953 (Neotropical Region, including *Pseudotaeniorhynchus* (as *Rhynchotaenia*) of *Coquillettidia*, taxonomy, keys, descriptions, distributions); Belkin, 1962 (South Pacific, including *Coquillettidia*, taxonomy, keys, descriptions, distributions, bionomics); Ronderos & Bachmann, 1963b (Neotropical Region, descriptions, keys, distribution); Forattini, 1965b (Neotropical Region, as a subgenus, species, bionomics, distributions); Cova-Garcia *et al.*, 1966 (Venezuela, keys, descriptions, distributions); Delfinado, 1966a (Philippines, taxonomy, keys, descriptions, distributions, bionomics); Belkin *et al.*, 1970 (Jamaica, taxonomy, keys, descriptions, bionomics); Darsie, 1985 (Argentina, keys); Lee *et al.*, 1988b (Australasian Region, *Mansonioides*, taxonomy, keys, literature, distributions, bionomics); Service, 1990 (Afrotropical Region, *Mansonioides*, descriptions, keys, distribution); Rattanarithikul & Panthusiri, 1994 (Thailand, keys, medically important species); Lu *et al.*, 1997a (China, keys, taxonomy, descriptions, distributions, bionomics); Harbach & Kitching, 1998 (morphology, phylogenetic relationships); Rattanarithikul *et al.*, 2006a (Thailand, bionomics, keys); da Silva *et al.*, 2020a (Neotropical species, mitochondrial phylogenomics).

Mansonia Blanchard, 1901 *sensu stricto*

Mansonia sensu stricto, the nominotypical subgenus of *Mansonia*, is typified by *Mansonia titillans* (Walker, 1848), original combination: *Culex titillans*. Subgenus abbreviation – *Man*.

****

Systematics: *Mansonia sensu stricto* includes 15 extant species, which occur predominantly in the Neotropical Region. The distributions of two species extend northward into the Nearctic Region: *Ma. dyari* is distributed throughout Central America and occurs in Florida and Georgia; *Ma. titillans* is widely distributed from Argentina northward into southeastern Texas and Florida. Species of *Mansonia sensu stricto* are very similar to those of the Old World *Mansonioides*. The two groups are probably sister taxa, but this requires phylogenetic confirmation. The evolutionary relationships of *Mansonia sensu stricto* and its species have not been investigated.

****

Characteristics: Species of *Mansonia sensu stricto* exhibit the following principal anatomical features. ADULTS: Proboscis usually as long as or slightly longer than the forefemur; maxillary palpus of males longer than the proboscis; scutal scaling sparse, not forming a pattern; scutellum with pale scales, denser on the midlobe; femora and tibiae conspicuously speckled with dark and pale scales, femora without a distinct preapical pale band on the anterior surface; tarsomeres with basal pale rings; wing veins distinctly speckled with broad pale scales. FEMALE GENITALIA: Cerci strongly developed and angled, with wedge-shaped bases projecting toward the midline; postgenital lobe somewhat cleft medially; 3 spermathecal capsules, 2 large and 1 small. MALE GENITALIA: Abdominal tergum IX membranous, without setae; gonocoxite with a membranous mesal surface, tergomesal margin not noticeably produced, with many relatively short slender setae; gonostylus irregular in shape, gonostylar claw present; claspette shorter than the gonocoxite, with 1 (usually) or 2 short apical spiniform setae; aedeagus broad, more or less oval and rounded distally, lateral plates joined at the proximal and distal ends. LARVAE: Head distinctly wider than long; antenna with the distal part beyond setae 2,3-A subequal to or shorter than the proximal part, distinctly pigmented; setae 2,3-A more or less the same length as the distal part of the antenna; comb with 3–7 spine-like scales with or without basolateral spicules, sometimes with fringed squamiform scales; ventral brush (seta 4-X) with 3 or 4 pairs of setae borne on a grid and a line of precratal setae inserted mid-ventrally on the saddle. PUPAE: Apex of the trumpet not produced into a heavily sclerotized spine projecting freely from the lobe on one side; all or some of setae 1,3,4,6-II–VII very long; seta 3-II–VI present, 3-VII absent; seta 11-II–VII absent; paddle relatively narrow, inner part usually distinctly shorter than the outer part.

****

Bionomics: Larvae and pupae of species of *Mansonia sensu stricto*, like those of *Mansonioides*, attach themselves to the roots of aquatic plants to obtain oxygen. Water lettuce (*Pistia*) and water hyacinth (*Eichornia*) that flourish in marshes, ponds and lakes are commonly used for respiration. Females avidly feed on humans outdoors, often miles from the habitats occupied by the immature stages. Females of *Ma. indubitans* and *Ma. titillans* are vectors of Venezuelan equine encephalitis virus in the New World tropics (White & Faust, 2014). The latter species is also known to vector the nematodes that cause filariasis.

****

Principal references: Lane, 1953 (Neotropical Region, including *Pseudotaeniorhynchus* (as *Rhynchotaenia*) of *Coquillettidia*, taxonomy, keys, descriptions, distributions); Ronderos & Bachmann, 1963a (subgeneric status, diagnosis, included species); Ronderos & Bachmann, 1963b (descriptions, keys, distribution); Forattini, 1965b (species, bionomics, distributions); Cova-Garcia *et al.*, 1966 (Venezuela, keys, descriptions, distributions); Belkin *et al.*, 1970 (Jamaica, taxonomy, keys, descriptions, bionomics); Clark-Gil & Darsie, 1983 (Guatemala, keys, distributions, bionomics); Darsie, 1985 (Argentina, keys).

Mansonioides Theobald, 1907

Mansonioides, originally proposed as a genus by Theobald (1907), currently classified as a subgenus of *Mansonia*, is typified by *Mansonioides septemguttata* Theobald, 1907 [junior synonym of *Mansonia annulifera* (Theobald, 1901c)]. Subgenus abbreviation – *Mnd.*

****

Systematics: *Mansonioides* includes 10 extant species with distributions in the Afrotropical Region (sub-Saharan Africa and Madagascar), Oriental Region, Manchurian Subregion of the Palaearctic and the Australasian Region (Solomon Islands and northward from Queensland, Australia). Species of *Mansonioides* are very similar to those of the New World *Mansonia sensu stricto*. The two groups are probably sister taxa, but this requires phylogenetic confirmation. The evolutionary relationships of *Mansonioides* and its species have not been investigated.

****

Characteristics: Species of *Mansonioides* exhibit the following principal anatomical features. ADULTS: Proboscis shorter than the forefemur; maxillary palpus of males distinctly longer than the proboscis, palpomere 5 minute; scutal scales numerous and of at least 2 colours, pale scaling forming spots or irregular longitudinal lines; acrostichal setae weakly developed or absent; legs with numerous pale markings, tarsi with pale rings; wing veins speckled with broad asymmetrical dark and pale scales. FEMALE GENITALIA: Abdominal segment VIII small, usually retracted into segment VII, tergum short with thick spiniform setae near the posterior border, sternum with conspicuous flap-like extensions laterally; cerci small, with pointed apices, without internal wedge-shaped extensions; postgenital lobe small, almost completely divided medially into lateral plates; 3 spermathecal capsules, 2 large, 1 rudimentary. MALE GENITALIA: Gonocoxite excavated and extensively membranous on the mesal surface, tergomesal margin produced and bearing a row of thick strongly developed setae; gonostylus irregular in shape, gonostylar claw absent; claspette very long, nearly reaching the apex of the gonocoxite, with 1 or more apical spiniform setae; aedeagus more or less elongate, with continuous broad dorsal sclerotization, usually with a median projection. LARVAE: Antenna with the distal part beyond setae 2,3-A less than 0.5 the length of the basal part; setae 2,3-A longer than the distal part of the antenna; seta 4-III,IV inserted mesad of seta 3; setae 4,5-VIII removed from seta 3-VIII; comb with only 1–3 blunt rounded scales without a fringe; siphon with seta 1-S inserted adjacent to the base of a modified spiracular lobe; ventral brush (seta 4-X) with 5 pairs of setae borne on a grid and a line of precratal setae inserted mid-ventrally on the saddle. PUPAE: Apex of the trumpet produced into a heavily sclerotized spine projecting freely from the lobe on one side; seta 1- or 5-II–VII or both very long; seta 3-II–VI present, 3-VII absent; seta 11-II–VII absent; paddle relatively broad.

****

Bionomics: The immature stages of species of *Mansonioides* are most often associated with aquatic plants of the genus *Pistia*, but the larvae and pupae of some species use the roots of other aquatic plants to obtain air. They are commonly found in ponds, lakes, swamps, marshes and ditches. Larvae usually occur in fairly deep water. Females insert the end of their abdomen under the surface of the water to deposit their eggs. Females of most species readily attack humans both indoors and outdoors and are often series pests. They are normally active during crepuscular and nocturnal periods, but some species feed during the daytime. Some species are vectors of *Wuchereria malayi* and *W. bancrofti*, and may play a role in the transmission of yellow fever virus. Viruses and *Mansonioides* vectors listed by White & Faust (2014) include Banzi (*Ma. africana*), Bunyamwera (*Ma. africana*, *Ma. uniformis*), Chikungunya (*Ma. africana*, *Ma. uniformis*), Rift Valley fever (*Ma. africana*, *Ma. uniformis*?), Sindbis (*Ma. africana*), Spondweni (*Ma. africana*, *Ma. uniformis*) and Wesselsbron (*Ma. uniformis*) viruses in Africa, and Septik virus (*Ma. septempunctata*) in Australia. Azari-Hamidian *et al.* (2020) reviewed the medical importance of *Ma. uniformis*.

****

Principal references: Barraud, 1934 (southern Asia, key to females, descriptions, bionomics, distributions); Edwards, 1941 (Afrotropical Region, as a subgenus of *Taeniorhynchus*, in keys, descriptions of adults and pupae, distributions); Carter, 1950 (Sri Lanka, as a subgenus of *Taeniorhynchus*, descriptions, keys, bionomics, distributions); Hopkins, 1952 (Afrotropical Region, as a subgenus of *Taeniorhynchus*, in key, descriptions of larvae, bionomics); Belkin, 1962 (South Pacific, subgenus and *Ma. melanesiensis*, taxonomy, description, bionomics, distribution); Ronderos & Bachmann, 1963a (subgeneric status, diagnosis, included species); Ronderos & Bachmann, 1963b (descriptions, keys, distribution); Delfinado, 1966a (Philippines, taxonomy, keys, descriptions, distributions, bionomics); Tanaka *et al.*, 1979 (Japan, *Ma. uniformis*, taxonomy, description, distribution, keys); Lee *et al.*, 1988b (Australasian Region, taxonomy, literature, keys, descriptions, bionomics, distributions); Service, 1990 (Afrotropical Region, keys, descriptions, distributions, bionomics); Rattanarithikul & Panthusiri, 1994 (Thailand, medical importance, keys); Lu *et al.*, 1997a (China, keys, taxonomy, descriptions, distributions, bionomics); Rattanarithikul *et al.*, 2006a (Thailand, description, bionomics, keys); Azari-Hamidian *et al.*, 2020 (Iran, *Ma. uniformis*, medical importance).

Maorigoeldia Edwards, 1930

Maorigoeldia Edwards, 1930a, originally proposed as a subgenus of *Rachionotomyia* and currently classified as a genus, is typified by *Maorigoeldia argyropus* (Walker, 1848), original combination: *Culex argyropus*. Genus abbreviation – *Mg.*

****

Systematics: *Maorigoeldia* is a subordinate taxon of the subfamily Culicinae, tribe Sabethini. *Maorigoeldia* is monobasic. It only includes the extant type species, which is endemic in New Zealand. *Maorigoeldia* exhibits many peculiar characteristics that led Belkin (1962) to regard it as the most primitive sabethine. Whereas this view was contradicted in the phylogenetic analysis of sabethines conducted by Judd (1996), which placed *Maorigoeldia* as the sister to the New World genera of Sabethini, it was corroborated in the phylogenetic analyses of mosquito genera (Harbach & Kitching, 1998) and sabethine genera (Harbach & Peyton, 2000; Harbach *et al.*, 2007a), in which *Maorigoeldia* was recovered as the most basal clade of the Sabethini.

****

Characteristics: The adults of *Mg. argyropus* are large, beautiful mosquitoes with light bluish and silver scales. They are unique among sabethines in having well-developed acrostichal setae. Larvae are distinguished from other Old World genus groups by the presence of a single pair of seta 4-X (distinction from non-sabethines), a circular occipital foramen (distinction from *Malaya* and *Topomyia*), seta 13-P (distinction from *Kimia*, *Malaya* and *Topomyia*) and comb scales in more than a single row (distinction from *Kimia* and *Tripteroides*).

****

Bionomics: Larvae of *Mg. argyropus* are found in tree holes and various artificial containers near forests. They were collected once in a pool with leaves in the bed of a drying stream. The larvae rest on the bottom of the cavity with the dorsal surface down and seldom come to the surface to obtain air. Females have been observed resting on the trunks of trees and approaching humans without landing or attempting to feed. *Maorigoeldia argyropus* is not a medically or economically important species.

****

Principal references: Belkin, 1962 (New Zealand, genus and species descriptions, bionomics, distribution); Belkin, 1968a (New Zealand, taxonomy, description, bionomics, distribution); Lee *et al.*, 1988c (New Zealand, taxonomy, genus and species descriptions, literature, bionomics); Harbach & Peyton, 1993 (comparative morphology of larval maxillae); Judd, 1996 (morphology, phylogenetic relationships); Harbach & Kitching, 1998 (morphology, phylogenetic relationships); Judd, 1996 (morphology, phylogenetic relationships); Harbach & Kitching, 1998 (morphology, phylogenetic relationships); Harbach & Peyton, 2000 (sabethine phylogeny); Harbach *et al.*, 2007a (sabethine phylogeny).

Mimomyia Theobald, 1903

Mimomyia Theobald, 1903a, originally proposed and currently recognized as a genus, is typified by *Mimomyia splendens* Theobald, 1903a. Genus abbreviation – *Mi.*

****

Systematics: *Mimomyia* is a subordinate taxon of the subfamily Culicinae, tribe Ficalbiini. It includes 46 extant species divided between three subgenera: *Etorleptiomyia*, *Ingramia* and *Mimomyia sensu stricto*. Species of *Mimomyia* occur in the Australasian, Oriental and Afrotropical Regions. Two species extend into the eastern Palaearctic (Ryukyu Archipelago). The phylogenetic relationships of *Mimomyia* are uncertain. Mattingly (1981) suggested that *Ficalbia* and *Mimomyia* are more distantly related than their inclusion in the same tribe would suggest, and this notion is supported, although weakly, by the pairing of *Ficalbia* with *Hodgesia* rather than *Mimomyia* in the phylogenetic analyses of Harbach & Kitching (1998) based on morphological data. The phylogeny of the subgenera and species has not been investigated.

****

Characteristics: *Mimomyia* are generally small mosquitoes. The adults are rather easily distinguished from the adults of other genus groups, except *Ficalbia*, by the following combination of characters: first flagellomere of the antenna sometimes elongate (1.5–3.0 times as long as the second flagellomere in females of *Mimomyia sensu stricto*); head with erect scales on the vertex and occiput or restricted to the occiput; acrostichal and dorsocentral setae present; prespiracular and postspiracular setae present or absent; upper calypter with a fringe of setae or hair-like scales; alula bare or with broad recumbent scales on the dorsal margin (distinction from *Ficalbia*); anal vein (vein 1A) ends beyond the base of the mediocubital crossvein; tarsi normal, pulvilli not evident. Larvae of *Mimomyia* are characterized as follows: distal part of the antenna articulated with the basal part (not fused as in *Ficalbia*); hypostomal suture incomplete or absent, usually absent; seta 12-I present; siphon usually sharply narrowed distally, with the spiracular apparatus sometimes modified and fused with the siphon for piercing plant tissues; seta 1-S inserted beyond the base of the siphon (distinction from *Ficalbia*); pecten absent or reduced, at most with 4 spines; ventral brush (seta 4-X) with 4–6 pairs of setae; saddle complete or incomplete, without minute setae on the mid-ventral area of segment X (distinction from *Ficalbia*). The larvae share many characteristics with larvae of *Hodgesia*, which differ in having a complete hypostomal suture and exceptionally long thoracic setae.

****

Bionomics: Very little is known about the bionomics of species of *Mimomyia*. The larvae of most species occur in swamps and marshes with dense vegetation. The larval siphon of several species is modified for piercing aquatic plants to obtain oxygen. Larvae of *Ingramia* are found in the leaf axils of plants. The adults of several species have been reported to feed on humans, but none are serious pests. Blood tests have revealed that the females of a number of species had fed on amphibians (Snyman, 2021). Most species appear to be active at night. Species of *Mimomyia* are not of medical importance.

****

Principal references: Barraud, 1934 (southern Asia, as a subgenus of *Ficalbia*, descriptions, bionomics, distributions); Edwards, 1941 (Afrotropical Region, as a subgenus of *Ficalbia*, species groups, adults and pupae, descriptions, distributions); Hopkins, 1952 (Afrotropical Region, as a subgenus of *Ficalbia*, larvae, descriptions, bionomics); Mattingly, 1957a (Indomalayan area, as a subgenus *Ficalbia*, keys, descriptions, bionomics, distributions); Mattingly & Grjebine, 1958 (as a subgenus of *Ficalbia*, taxonomy, keys, bionomics, zoogeography, affinities); Belkin, 1962 (South Pacific, as a subgenus of *Ficalbia*, *Mi. gurneyi*, taxonomy, description, bionomics, distribution); Delfinado, 1966a (Philippines, as a subgenus of *Ficalbia*, taxonomy, descriptions, distributions, bionomics); Tanaka *et al.*, 1979 (Japan, *Etorleptiomyia*, descriptions, keys, distributions, bionomics); Grjebine, 1986 (Madagascar, taxonomy, descriptions, distributions, bionomics, keys); Lee *et al.*, 1988c (Australasian Region, taxonomy, descriptions, literature, bionomics, keys); Service, 1990 (Afrotropical Region, descriptions, keys, bionomics, distributions); Lu *et al.*, 1997a (China, keys, taxonomy, descriptions, distributions, bionomics); Harbach & Kitching, 1998 (morphology, phylogenetic relationships); Rattanarithikul *et al.*, 2006a (Thailand, bionomics, keys).

Etorleptiomyia Theobald, 1904

Etorleptiomyia, originally proposed as a genus by Theobald (1904b), currently classified as a subgenus of *Mimomyia*, is typified by *Mimomyia mediolineata* (Theobald, 1904b), original combination: *Etorleptiomyia mediolineata*. Subgenus abbreviation – *Eto.*

****

Systematics: *Etorleptiomyia* includes six extant species in the Australasian, Oriental and Afrotropical Regions. Two species in the Oriental Region, *Mi. elegans* and *Mi. luzonensis*, extend into the eastern Palaearctic Region (Ryukyu Archipelago), two other species, *Mi. bougainvillensis* and *Mi. solomonis*, are known only from the Solomon Islands and three species, *M. martinei*, *Mi. mediolineata* and *Mi. xanthozona*, have distributions in the Afrotropical Region. *Mimomyia martinei* is the only species of the subgenus known to occur (solely) in Madagascar. The evolutionary relationships of *Etorleptiomyia* have not been investigated.

****

Characteristics: Species of *Etorleptiomyia* exhibit the following principal anatomical features. ADULTS: Scaly mosquitoes; vertex of the head with narrow falcate scales dorsally and broad scales laterally; erect forked scales numerous on the occiput and vertex; eyes well separated above the antennal pedicels; flagellomere 1 of females with broad scales, 1.0–1.5 the length of flagellomere 2; maxillary palpus of females with a pale apex, palpus of males 0.6–0.9 the length of the proboscis, with 4 palpomeres (palpomere 4 very small), only short setae present; proboscis of females distinctly swollen in the distal 0.2, proboscis of males strongly swollen in the distal 0.5; acrostichal setae present or absent; dorsocentral setae present; scutellum largely with broad scales, a few narrow scales apically; ante- and postpronota with broad scales, antepronotum sometimes with narrow falcate scales; thoracic pleura largely with broad scales, present on the upper proepisternum, postspiracular area, mesokatepisternum and sometimes the mesepimeron; lower mesepimeral setae absent; wing densely covered with large broad dark and pale scales (some heart-shaped), cell R_2 3.0–4.5 the length of vein R_{2+3}, alula with a marginal patch of broad scales, upper calypter with a fringe of narrow piliform scales; tarsi with pale bands; abdominal terga with some pale markings. FEMALE GENITALIA: 1 spermathecal capsule. MALE GENITALIA: Aedeagus simple, slightly sclerotized. LARVAE: Head much broader than long; antenna articulated distal to the insertion of setae 2,3-A, distal part shorter than 0.5 the length of the proximal part, setae 2–4-A long, single; setae 5,6,7,11-C very long, seta 11-C usually double; siphon long, pecten absent, siphon index 5.0 to > 8.0, seta 2-S trifid, setae of the spiracular apparatus short, spiracular lobes short, not modified for piercing plant tissues; ventral brush (seta 4-X) with 3 pairs of long setae and 1 or 2 pairs of short setae. PUPAE: Trumpet very long, index 20–30, not modified for piercing plant tissues; seta 1-I single, small; setae 1,5-VII very large, dark, usually with at least 3 branches; seta 9-VIII large, usually with 8–11 aciculate or plumose branches; paddle much longer than wide, index 4 to >10, margins spiculate or serrate, buttress not developed; seta 1-Pa present or absent.

****

Bionomics: The immature stages of species of *Etorleptiomyia* are found in permanent groundwater habitats, most commonly ponds, marshes and swamps, where the edge of the water or its entire surface is covered by vegetation. Practically nothing is known about the bionomics of the adults. The females of two species, *Mi. luzonensis* and *Mi. mediolineata*, are known to feed on amphibians (Braima *et al.*, 2017; Boreham *et al.*, 1975, respectively). They are apparently active at night. Species of *Etorleptiomyia* are not of medical importance.

****

Principal references: Mattingly, 1957a (as a subgenus of *Ficalbia*, Indomalayan area, keys, descriptions, bionomics, distributions); Mattingly & Grjebine, 1958 (as a subgenus of *Ficalbia*, taxonomy, keys, bionomics, zoogeography, affinities); Belkin, 1962 (as a subgenus of *Ficalbia*, South Pacific, taxonomy, bionomics, distribution, keys); Delfinado, 1966a (as a subgenus of *Ficalbia*, Philippines, taxonomy, descriptions, distributions, bionomics); Tanaka *et al.*, 1979 (Japan, descriptions, keys, distributions, bionomics); Grjebine, 1986 (Madagascar, taxonomy, descriptions, distributions, bionomics, keys); Lee *et al.*, 1988c (Australasian Region, taxonomy, literature, bionomics, distribution, keys); Service, 1990 (Afrotropical Region, descriptions, keys, bionomics, distributions).

Ingramia Edwards, 1912

Ingramia, originally proposed as a genus by Edwards (1912b), currently classified as a subgenus of *Mimomyia*, is typified by *Mimomyia fusca* (Leicester, 1908), original combination: *Dasymyia fusca*. Subgenus abbreviation – *Ing.*

****

Systematics: *Ingramia* includes 22 extant species. The distributions of the species are not well known. Five species are recorded from countries in the Oriental Region. The other 17 species are only known from Madagascar and neighbouring islands. The evolutionary relationships of *Ingramia* have not been investigated.

****

Characteristics: Species of *Ingramia* exhibit the following principal anatomical features. ADULTS: Not as scaly as species of *Etorleptiomyia*; erect scales of the head variable in extent, generally broad, restricted to the occiput or extended forward onto the vertex; eyes well separated above the antennal pedicels; flagellomere 1 of females with broad scales, 1.0–1.4 the length of flagellomere 2, males with flagellomeres 12 and 13 very long; maxillary palpus of females shorter than 0.33 the length of the proboscis, palpus of males 0.4–0.8 the length of the proboscis, with 4 palpomeres, with short setae apically; proboscis of males strongly swollen in the distal 0.5; acrostichal setae present or absent; dorsocentral setae present; scutellum with broad scales; scale-patches present or absent on the postpronotum, mesokatepisternum and mesepimeron, postspiracular scales absent; lower area of the mesepimeron sometimes with a scattering of fine setae; wing mainly with broad scales, some more or less heart-shaped, scales entirely dark, entirely pale or both dark and pale, cell R_2 normally 1.3–2.1 times as long as vein R_{2+3}, alula with broad scales on the margin, upper calypter with a fringe of broad scales; tarsi usually with pale bands (all dark in *Mi. fusca* except hindtarsomeres 4 and 5 and the apex of 3 paler); abdominal terga with some pale markings. FEMALE GENITALIA: 1 spermathecal capsule. MALE GENITALIA: Aedeagus simple, slightly sclerotized. LARVAE: Antenna articulated distal to setae 2,3-A; setae 5,6,7,11-C variable in length, often short; setae 5,6-M, 9–12-M, 7-T, 9–12-T and sometimes 9–12-P borne on lateral digit-like protuberances; siphon variable, index 2.3–10.0, pecten absent, seta 2-S single or bifid, spiracular apparatus as in species of *Etorleptiomyia*, not modified for piercing plant tissues; ventral brush (seta 4-X) with 2 pairs of setae. PUPAE: Trumpet very long, index 25 to > 50, not modified for piercing plant tissues; seta 1-I small, single or forked; setae 1,5-VII normally single or forked; seta 9-VIII large, usually with 14–22 branches; paddle much longer than wide, index 10–20.

****

Bionomics: The immature stages of species of *Ingramia* develop in phytotelm habitats, including tree holes, cut bamboo, leaf axils of *Ravenala*, *Pandanus*, *Typhonodorum* and *Colocasia*, and *Nepenthes* pitchers. Species of *Ingramia* are not of medical importance.

****

Principal references: Mattingly, 1957a (as subgenus *Ravenalites* of *Ficalbia*, Indomalayan area, keys, descriptions, bionomics, distributions); Mattingly & Grjebine, 1958 (as subgenus *Ravenalites* of *Ficalbia*, taxonomy, keys, bionomics, zoogeography, affinities); Delfinado, 1966a (as subgenus *Ravenalites* of *Ficalbia*, Philippines, taxonomy, descriptions, distributions, bionomics); Grjebine, 1986 (Madagascar, taxonomy, descriptions, distributions, bionomics, keys); Service, 1990 (Afrotropical Region, descriptions, keys, bionomics, distributions).

Mimomyia Theobald, 1903 *sensu stricto*

Mimomyia sensu stricto, the nominotypical subgenus of *Mimomyia*, is typified by *Mimomyia splendens* Theobald, 1903a. Subgenus abbreviation – *Mim.*

****

Systematics: *Mimomyia sensu stricto* includes 18 extant species with distributions in the Afrotropical, Australasian and Oriental Regions. Ten species occur in the Afrotropical Region and seven occur in the Oriental Region, one of which (*Mi. chamberlaini*) is also found in Australia and New Guinea Island. *Mimomyia gurneyi* is only known from the Solomon Islands. The evolutionary relationships of *Mimomyia sensu stricto* have not been investigated. The group is unlikely to be monophyletic. The species are morphologically and biologically diverse and are currently classified into five species groups (see Mattingly & Grjebine, 1958 and Grjebine, 1986 for taxonomic treatments).

****

Characteristics: Species of *Mimomyia sensu stricto* exhibit the following principal anatomical features. ADULTS: Not as scaly as species of *Etorleptiomyia*; vertex with variable scaling; flagellomere 1 of females 1.5–3.0 the length of palpomere 2; maxillary palpus of males 1.1–1.3 the length of the proboscis, clubbed (similar to *Anopheles*) with spiny setae; proboscis of females faintly swollen distally, proboscis of males slightly to moderately swollen in the distal 0.33; acrostichal setae present or absent; dorsocentral setae usually present; scutellum with narrow scales only; mesepimeron without scales; lower mesepimeral setae absent or represented by very fine setae; wing with reduced scaling, scales dark, mostly confined to the apices of the veins (except *Mi. parenti* which has more heavily scaled wings), sometimes with pale scales at the base of the costa, cell R_2 0.6–1.2 the length of vein R_{2+3}, alula bare, upper calypter with a fringe of narrow scales; legs with various pale markings. FEMALE GENITALIA: 1 or 3 spermathecal capsules. MALE GENITALIA: Variously developed, especially the basal mesal lobe of the gonocoxite, the aedeagus and the proctiger. LARVAE: Antenna articulated distad of setae 2,3-A, setae 2,3,4-A single; siphon variable, pecten present or absent, index 2.0–8.0, sometimes modified for piercing plant tissues; ventral brush (sea 4-X) with 4 pairs of setae. PUPAE: Trumpet index about 9.0–17.0, sometimes modified for piercing plant tissues; seta 1-I usually large, multi-branched (except *Mi. perplexens*); seta 9-VIII small, single or with 2–5 branches; paddle index 1.5–2.8; seta 1-Pa present or absent.

****

Bionomics: The immature stages of species of *Mimomyia sensu stricto* have been found in ground pools, swamp margins and small depressions in marshy areas, usually in dense vegetation. Very little is known about the bionomics of the adults. Blood tests have shown that females of *Mi. mimomyiaformis*, *Mi. plumosa* and *Mi. splendens* feed on amphibians (Boreham *et al.*, 1975; Crabtree *et al.*, 2013; Omondi *et al.*, 2015). Species of the group are not of medical importance.

****

Principal references: Mattingly, 1957a (as a subgenus of *Ficalbia*, Indomalayan area, classification, keys, descriptions, bionomics, distributions); Mattingly & Grjebine, 1958 (as a subgenus of *Ficalbia*, taxonomy, classification, keys, bionomics, zoogeography, affinities); Belkin, 1962 (as a subgenus of *Ficalbia*, South Pacific, taxonomy, bionomics, distribution, keys); Delfinado, 1966a (as a subgenus of *Ficalbia*, Philippines, taxonomy, descriptions, distributions, bionomics); Grjebine, 1986 (Madagascar, taxonomy, classification, descriptions, distributions, bionomics, keys); Lee *et al.*, 1988c (Australasian Region, taxonomy, literature, bionomics, distribution, keys); Service, 1990 (Afrotropical Region, descriptions, keys, bionomics, distributions).

Neoculicites Evenhuis, 1994

Neoculicites Evenhuis, 1994, *nomen novum* for *Culicites* Meunier, 1915, originally proposed and currently recognized as an extinct genus, is typified by *Neoculicites depereti* (Meunier, 1915), original combination: *Culicites depereti*. Genus abbreviation – *Nc.*

****

Systematics: *Neoculicites* is a subordinate taxon of the subfamily Culicinae, tribe (?). The three species assigned to *Neoculicites* are based on compression/impression fossils of females from the Oligocene Epoch (23.03–33.90 Mya) of the Cenozoic Era found in France (two fossil species) and Germany (one fossil species).

****

Characteristics: The species of *Neoculicites* are described as having the appearance of *Culex* mosquitoes. The authors provided only limited general information that cannot be used to diagnose the genus. Edwards (1923c) noted that the image of the type specimen of *Nc. depereti* provided by Meunier (1915) is that of a typical culicine mosquito. He stated that *Nc. ceyx* could possibly be a species of *Aedes*, but the fossil only consists of 'portions of the thorax and abdomen; there is nothing to prove that it is a mosquito at all'. The photographs and brief descriptions of the female and unassociated pupa of *Nc. arvernensis* do not provide diagnostic features. These fossils are so poorly known that the status of the genus must be regarded as uncertain.

****

Bionomics: The Oligocene Epoch was more temperate than today, but average global temperatures and sea levels were decreasing. Numerous species of reptiles, birds and mammals were available for blood-feeding. Groundwater and phytotelm habitats were available for the immature stages of mosquitoes.

****

Principal references: Evenhuis, 1994 (*Neoculicites nomen novum*, summary of mosquito fossils); Poinar *et al.*, 2000 (review of descriptions of fossil mosquitoes).

Onirion Peyton & Harbach, 2000

Onirion Peyton & Harbach, 2000 (in Harbach & Peyton, 2000), originally proposed and currently recognized as a genus, is typified by *Onirion personatum* (Lutz, 1904) (in Bourroul, 1904), original combination: *Dendromyia personata*. Genus abbreviation – *On.*

****

Systematics: *Onirion* is a subordinate taxon of the subfamily Culicinae, tribe Sabethini. It includes seven extant species in the Neotropical Region. The affinities of the group are uncertain. *Onirion* was recovered as a distinct entity among the New World genera of Sabethini in the phylogenetic analyses of Harbach & Peyton (2000) and Harbach *et al.* (2007a) based on morphological data. Whereas *Onirion* was recovered as the sister to a terminal clade comprised of *Johnbelkinia* + (*Sabethes* + (*Limatus* +

Malaya)), it was embedded within *Wyeomyia* in the phylogenetic analysis of morphological and allozyme data conducted by Motta *et al.* (2007). As pointed out by Harbach (2007), the placement of *Onirion* within *Wyeomyia* in the latter study may have been due to the omission of *Hystatomyia* and certain discriminating characters from the data set. Phylogenetic relationships among the species of *Onirion* have not been investigated.

****

Characteristics: Adults of *Onirion* are similar to the adults of *Wyeomyia* in overall ornamentation, but the pale markings on the mid- and hindtarsi are quite characteristic of members of the group. The midtarsus has bright white scaling on the anterior surface from the middle of tarsomere 2 through tarsomere 5 and the hindtarsus has similar scaling ventrally from the middle of tarsomere 3 through tarsomere 5. The cercal setae of the male genitalia are extremely long and distinctive of *Onirion*, and the collection of long tergal setae on the gonocoxites (except in one species) is an unusual feature that occurs in few other sabethine taxa. Larvae of *Onirion* have a circular occipital foramen (distinction from *Isostomyia*, *Limatus*, *Sabethes*, *Runchomyia* and most *Wyeomyia*) but are otherwise easily distinguished from the larvae of all New World sabethine genus groups by the predominance of aciculae on one side of the branches of setae 4-P and 7-T, the stout apically spiked seta 11-T, the placement of strongly developed seta 13-IV,V laterad of setae 11 and 12, and the presence of a dense filamentous pecten extending the length of the siphon. Pupae have punctures only on terga III and IV, a noticeable pair of scar-like depressions borne centrally on terga II–VII, and terga I–VII are largely darkly pigmented centrally with little or no pigment around the bases of many setae.

****

Bionomics: Little reliable bionomical information is available for species of *Onirion*. All species are sylvan. Adults are active during the daytime and females are attracted to humans that enter their realm. The immature stages typically inhabit bamboo internodes. One species was collected from a terrestrial bromeliad. Species of *Onirion* are not known to be of medical importance.

****

Principal references: Harbach & Peyton, 2000 (new genus, taxonomy, genus and species descriptions, bionomics, distributions, phylogenetic relationships); Harbach *et al.*, 2007a (morphology, phylogenetic relationships); Motta *et al.*, 2007 (morphology and allozymes, phylogenetic relationships).

Opifex Hutton, 1902

Opifex Hutton, 1902, originally proposed as a genus of Tipulidae and currently classified as a genus of Culicidae, is typified by *Opifex fuscus* Hutton, 1902. Genus abbreviation – *Op.*

****

Systematics: *Opifex* is a subordinate taxon of the subfamily Culicinae, tribe Aedini. It includes two extant species placed in separate subgenera: *Nothoskusea* and *Opifex sensu stricto*. The type species is found in New Zealand and the Kermadec Islands; *Op.* (*Nothoskusea*) *chathamicus* is known only from the Chatham Islands. Belkin (1962) considered many features of *Op. fuscus* to represent the most primitive conditions found within Aedini and believed that they were annectant with other tribes, but this was not corroborated in the phylogenetic analyses of Harbach & Kitching (1998), Reinert *et al.* (2004, 2006, 2008 2009) and Wilkerson *et al.* (2015) based on morphological data. In the comprehensive study of Reinert *et al.* (2009) based on more extensive morphological data, *Opifex* was recovered as the sister of *Halaedes* in a clade comprised of *Acartomyia* + (*Jarnellius* + (*Halaedes* + *Opifex*)), and this clade was sister to *Ochlerotatus*. A similar relationship, but without *Jarnellius*, *i.e. Acartomyia* + (*Halaedes* + *Opifex*), was recovered in the analysis of Wilkerson *et al.* (2015). *Opifex fuscus* was recovered in a basal relationship to the Ochlerotatus Group of generic-level taxa (clade B) in the maximum likelihood phylogeny of Soghigian *et al.* (2017) based on seven molecular markers. Soghigian *et al.* (2023) included *Op. fuscus* in their phylogenetic analysis of amino acid sequences of 709 genes, but its relationship with other taxa of Aedini included in the analysis was not explicitly indicated. It must be noted that analyses of mtDNA *COI* and rDNA ITS1 sequences obtained for the mosquito fauna of New Zealand (Cane *et al.*, 2020) indicated that *Op. chathamicus*, the type species of *Nothoskusea*, appears to be more closely related to species currently classified in *Aedes* than to *Op. fuscus*.

****

Characteristics: The adults of *Opifex* are large mosquitoes. They have a wide interocular space and frons, numerous scutal and abdominal setae, and setae are present at the base of the subcosta of the wing in females (and males of *Op. fuscus*) as in species of *Culiseta*. Males exhibit two unique features: the antennal flagellomeres have only a few short setae and the two apical flagellomeres are approximately the same length as the others. Larvae have seta 6-I–V usually with at least three branches, seta 7-I usually with at least two branches, a relatively small dorsal saddle, a strongly developed ventral brush (seta 4-X) and small rounded anal papillae. *Opifex* shares many features with *Halaedes*. The species of *Opifex* exhibit the following generally diagnostic characters. ADULTS: Eyes broadly separated above the antennal pedicels; interocular space with 6 or more setae; terminal 2 flagellomeres of the antenna about the same length as the proximal flagellomeres, flagellar whorls of males with a few short setae dispersed more or less around the flagellomeres; palpomere 4 of the maxillary palpus absent or vestigial in males; scutal fossal scales sparse; paratergal scales and subspiracular scales absent; base of the subcosta of the wing with ventral setae in females; laterotergite of abdominal tergum I without scales. MALE GENITALIA: Gonocoxite with setae on the basomesal area of the dorsal surface all slender; gonostylar claw a short claw-like spiniform, bluntly pointed. LARVAE: Seta 7-I shorter than 0.5 the length of seta 6-I; seta 1-II single. PUPAE: Seta 1-CT weakly developed, shorter than seta 3-CT.

****

Bionomics: *Opifex* larvae occur in coastal rock pools containing brackish, saline or hypersaline water. The larvae of *Op. fuscus* have also been collected from a hole in a tree trunk projecting over the seashore, a freshwater stream and a horse trough. The adults of *Op. chathamicus* have not been encountered in nature, but females of *Op. fuscus* are known to attack at night and produce a painful bite. Males of the latter species exhibit a unique mating behaviour that includes precocious coupling with females before they have fully emerged from their pupal exuviae. The species of *Opifex* are not of medical or economic importance.

****

Principal references: Belkin, 1962 (*Opifex sensu stricto*, as a genus, New Zealand area, in keys, description, bionomics, distribution); Belkin, 1968a (*Opifex sensu stricto* and *Nothoskusea*, as a genus and subgenus of *Aedes*, respectively, New Zealand area, in keys, descriptions, bionomics, distribution); Lee *et al.*, 1984 (*Nothoskusea*, as a subgenus of *Aedes*, taxonomy); Lee *et al.*, 1988c (*Opifex sensu stricto*, as a genus, taxonomy, literature, distribution, bionomics); Harbach & Kitching, 1998 (morphology, phylogenetic relationships); Reinert *et al.*, 2004, 2006, 2008, 2009 (morphology, phylogenetic relationships); Wilkerson *et al.*, 2015 (phylogenetic relationships); Soghigian *et al.*, 2017 (phylogenetic relationships); Cane *et al.*, 2020 (mtDNA *COI* and rDNA ITS1 sequences, phylogenetic relationships).

Nothoskusea Dumbleton, 1962

Nothoskusea Dumbleton, 1962, originally proposed as a subgenus of *Aedes* and currently classified as a subgenus of *Opifex*, is typified by *Opifex chathamicus* (Dumbleton, 1962), original combination: *Aedes* (*Nothoskusea*) *chathamicus*. Subgenus abbreviation – *Not.*

****

Systematics: *Nothoskusea* is monobasic. It only includes the extant type species, which is known only from the Chatham Islands. *Nothoskusea* was consistently recovered as the sister of *Opifex sensu stricto* in the series of morphology-based phylogenetic studies conducted by Reinert *et al.* (2004, 2006, 2008, 2009). Although *Nothoskusea* is currently retained as a subgenus of *Opifex*, phylogenetic analyses of New Zealand mosquito species based on mtDNA *COI* and rDNA ITS1 sequences (Cane *et al.*, 2020) showed that *Op. chathamicus* may be more closely related to species of *Aedes* than it is to *Op. fuscus*. The researchers noted, however, that 'additional research involving more extensive sampling and exploring more molecular markers' is needed to establish the placement of the species.

****

Characteristics: The only species of *Nothoskusea*, the largest mosquito in the New Zealand area, bears the following unique and distinctive features. ADULTS: Vertex of the head with numerous erect scales; eyes widely separated, essentially as in *Opifex sensu stricto*; antenna with a relatively small scape, flagellar whorls poorly developed, flagellomere 13 of females with an apical process and a few subapical setae; maxillary palpus very short, about 0.25 the length of the proboscis in both sexes; wing of females with setae at the base of the subcosta, absent or inconspicuous in males; all tarsi with rows of minute thorn-like setae among scales on the ventral surface; ungues of all legs equal, toothed, with short spicules dorsally at the base, essentially similar in males and females; abdomen of males distinctly longer than the abdomen of females, terga and sterna II–VII with numerous setae and very sparse scales in both sexes. FEMALE GENITALIA: Largely retracted into segment VIII; tergum VIII about 0.7 the length of sternum VIII, with a few scales on about the posterior 0.4 and numerous setae on most of the surface; sternum VIII with a similar distribution of scales and setae, setae 1–3-S not distinguished, basolateral seta absent; tergum IX comprised of 2 lateral plates separated by membrane; cercus without scales, setae on about the distal 0.8 of the dorsal surface; postgenital lobe poorly sclerotized; upper vaginal sclerite striated; lower vaginal sclerite absent; insula lip-like, with 2–4 setae on each side of the midline. MALE GENITALIA: Tergum IX lobes poorly developed; gonocoxite with long setae on the lateral and ventral surfaces, mesal membrane poorly developed, narrow; gonostylus long, somewhat sinuous, gonostylar claw a short stout spiniform inserted at the apex of the gonostylus; claspette a prominent elongate setose lobe; aedeagus very small, simple; cercal setae short, 4–8 present. LARVAE: Prothoracic pleural setal group (setae 9–12-P) very small, meso- and metathoracic pleural setal groups (setae 9–12-M,T) large, mesothoracic group borne far mesad of seta 8-M; seta 6-VI small, significantly shorter than seta 6-V; pecten not reduced, spines in a continuous row or with a gap near the middle of the pecten; seta 6-S inconspicuous; segment X with a very small detached acus; seta 1-X not inserted on the saddle; ventral brush (seta 4-X) usually with 9 pairs of setae. PUPAE: Cephalothorax with normally developed median keel; trumpet short, broad, trumpet index about 1.7; seta 1-I usually with fewer than 20 simple branches; seta 3-III inserted caudad of the level of seta 5; seta 5-IV,V with 2–4 branches, usually 3; seta 9-VII large, dissimilar to the smaller seta 9 on the preceding segments, inserted dorsally at the margin; seta 1-Pa usually branched.

****

Bionomics: The immature stages of *Op. chathamicus* have only been found in saline or brackish water rock pools at or just above high tide mark. Nothing is known about the adults as they have not been encountered in nature and are only known from reared larvae. The species is not known to be of medical or economic importance.

****

Principal references: Belkin, 1968a (as a subgenus of *Aedes*, *Op. chathamicus*, in keys, description, bionomics, distribution); Lee *et al.*, 1984 (as a subgenus of *Aedes*, taxonomy); Reinert, 2002e (as a subgenus of *Ochlerotatus*, female genitalia); Reinert *et al.*, 2004 (status as a subgenus of *Opifex*, morphology, phylogenetic relationships); Reinert *et al.*, 2006, 2008, 2009 (morphology, phylogenetic relationships); Cane *et al.*, 2020 (mtDNA *COI* and rDNA ITS1 sequences, phylogenetic relationships).

Opifex Hutton, 1902 *sensu stricto*

Opifex sensu stricto, the nominotypical subgenus of *Opifex*, is typified by *Opifex fuscus* Hutton, 1902. Subgenus abbreviation – *Opi.*

****

Systematics: *Opifex sensu stricto* is monobasic. It only includes the extant type species, which is known only from New Zealand and the Kermadec Islands. *Opifex fuscus* was consistently recovered as the sister of *Nothoskusea chathamicus* in the series of morphology-based phylogenetic studies conducted by Reinert *et al.* (2004, 2006, 2008, 2009). However, phylogenetic analyses of New Zealand mosquito species based on mtDNA *COI* and rDNA ITS1 sequences (Cane *et al.*, 2020) showed that *Op. chathamicus* may be more closely related to species of *Aedes*. *Opifex fuscus*, in the absence of *Op. chathamicus*, was recovered in a basal relationship to the Ochlerotatus Group of generic-level taxa (clade B) in the maximum likelihood phylogeny of Soghigian *et al.* (2017) based on seven molecular markers. *Opifex fuscus* was included in the phylogenetic analysis of amino acid sequences of 709 genes conducted by Soghigian *et al.* (2023), but its relationship with other taxa of Aedini included in the analysis was not explicitly indicated.

****

Characteristics: The only species of *Opifex sensu stricto*, a large mosquito, bears the following unique and distinctive features. ADULTS: Vertex of the head with setae instead of erect scales; antenna with a fully developed scape, flagellomeres 2–4 of males each with a dorsal basal swelling bearing a greatly enlarged spine-like seta; maxillary palpus short, about 0.5 and 0.7 the length of the proboscis in females and males, respectively; palpomere 4 of females club-shaped with an invaginated sense organ at the apex; wing of males as well as females with setae at the base of the subcosta; foreleg markedly shorter than the mid- and hindlegs; tarsi of males with a conspicuous hydrofuge fringe of setae on the ventral surface; ungues of females equal, toothed, ungues of males simple, foreungues extremely large; abdominal terga and sterna II–VII with numerous setae and very sparse scales as in *Nothoskusea*. FEMALE GENITALIA: Nearly completely retracted into segment VIII; tergum VIII about 0.7 the length of sternum VIII, with a few scales on the posterior 0.4–0.6 and numerous setae on the posterior 0.6–0.7; sternum VIII without scales, with dense hydrofuge setae distally and a densely spiculose invaginated caudal membrane, setae 1–3-S long, inserted in more or less a diagonal line from the anteromesal area to the caudolateral corner, basolateral seta absent; tergum IX comprised of 2 lateral plates separated by a narrow membrane; cercus without scales, setae on the distal 0.5–0.6 of the dorsal surface; postgenital lobe poorly sclerotized; upper vaginal sclerite a small short broad lobe extending inward and situated some distance caudally from the hinge; lower vaginal sclerite absent; insula lip-like, very short, with 2–4 setae on each side of the midline. MALE GENITALIA: Segment VIII markedly shortened; tergum IX without lobes, with lateral patches of numerous short setae; gonocoxite with dense hydrofuge setae on the dorsal, lateral and ventral surfaces, mesal surface membranous from the base to the apex; gonostylus very short, stout, with a rounded basomesal swelling, gonostylar claw stout, spiniform, deeply inserted at the apex of the gonostylus; claspette a long setose plate-like sclerite on the mesal membrane of the gonocoxite; aedeagus long, simple; cercal setae strongly developed, numerous. LARVAE: Pro- and metathoracic pleural setal groups (setae 9–12-P,T) very small, all setae normally single; mesothoracic pleural setal group (setae 9–12-M) borne immediately adjacent (close) to seta 8-M; seta 6-VI large, developed like seta 6-V; pecten reduced, represented by a few simple widely spaced spines; seta 6-S very long, aciculate; segment X without an acus; seta 1-X inserted on the margin of the saddle; ventral brush (seta 4-X) with 8 or 8.5 pairs of setae. PUPAE: Median keel of the cephalothorax very strongly developed; trumpet short, extremely broad, trumpet index about 1.1; seta 1-I reduced, with a few simple branches; seta 3-III inserted caudad of the level of seta 5; seta 5-IV,V usually single; seta 9-VII small, similar to seta 9-II–VI, inserted ventrally near the caudolateral margin; seta 1-Pa usually single.

****

Bionomics: The immature stages of *Op. fuscus* are found primarily in brackish, saline and hypersaline rock pools above high tide level, but they have also been found in a hole in a dead pohutukawa trunk, a freshwater stream and a horse trough (Belkin, 1968a). Adult females are autogenous, but are known to seek blood meals at night. They attack and feed on humans who enter their haunts. Males spend most of their time on the water surface searching for pupae, and precociously couple with females before they have fully emerged from their pupal exuviae. The species is not known or suspected of being of medical importance.

****

Principal references: Belkin, 1962 (as a genus, New Zealand area, in keys, description, bionomics, distribution); Belkin, 1968a (as a genus, New Zealand area, in keys, descriptions, bionomics, distribution); Lee *et al.*, 1988c (as a genus, taxonomy, literature, distribution, bionomics); Reinert, 2001a (as a genus, female genitalia); Reinert *et al.*, 2004 (subgeneric status, morphology, phylogenetic relationships); Reinert *et al.*, 2006, 2008, 2009 (morphology, phylogenetic relationships); Wilkerson *et al.*, 2015 (phylogenetic relationships); Cane *et al.*, 2020 (mtDNA *COI* and rDNA ITS1 sequences, phylogenetic relationships).

Orthopodomyia Theobald, 1904

Orthopodomyia Theobald, 1904c, originally proposed and currently recognized as a genus, is typified by *Orthopodomyia albipes* Leicester, 1904 (in Theobald, 1904c). Genus abbreviation – *Or.*

****

Systematics: *Orthopodomyia* is a subordinate taxon of the subfamily Culicinae, tribe Orthopodomyiini. It includes 36 extant species. Most species occur in the Neotropical and Oriental Regions. Two Neotropical species extend their ranges into the USA and Canada, a number of species occur in Madagascar (only three species are known to occur elsewhere in the Afrotropical Region, *i.e.* Cameroon), and a single species, *Or. pulcripalpis*, occurs in the Palaearctic Region. No species of the genus occur in Australia. The affinities of *Orthopodomyia* are uncertain. The genus was recovered as the sister of *Culiseta* + *Toxorhynchites* in the phylogenetic analysis of Harbach & Kitching (1998) based on morphological data. Edwards (1932a) believed that *Orthopodomyia* is most closely related to *Culiseta* (as *Theobaldia*). He noted the occurrence of a pair of thoracic air sacs in larvae that also occur in *Culiseta fraseri*, and elsewhere only in members of Mansoniini (as the genus *Mansonia*) and Toxorhynchitini (as the genus *Megarhinus*). It is particularly noteworthy that the phylogenetic analysis of amino acid sequences of 709 genes conducted by Soghigian *et al.* (2023) fully supports Edwards's hypothesis of a close relationship with *Culiseta* – two species of *Orthopodomyia* were placed as the sister of a clade comprised of nine species representing four subgenera of *Culiseta*. Phylogenetic relationships of species of the North American Signifera Group based on the ITS2 region of rDNA were investigated by Byrd *et al.* (2012).

****

Characteristics: *Orthopodomyia* are beautifully marked mosquitoes with bands and stripes of white, silver or sometimes gold scales. Some species are mottled with pale and dark scales. *Orthopodomyia* are readily distinguished from the adults of all other mosquito genus groups by having tarsomere 1 of the fore- and midlegs longer than the four distal tarsomeres combined, and tarsomere 4 shorter than tarsomere 5. Larvae lack a pecten, the comb is usually uniquely developed with a posterior row of long scales and an anterior row of shorter scales (rarely with one or three rows of scales) and abdominal segment VIII has a large dorsal sclerite (except in *Or. alba*).

****

Bionomics: Little is known about the biology of *Orthopodomyia* species. Larvae occur principally in tree holes, but some species are found in bamboo internodes, the axils of bromeliads and the spathes of *Heliconia* plants. Most species are also occasionally found in artificial containers. The larvae apparently feed by filtering microorganisms and particulate matter from the water. Adults inhabit forests and appear to be active only after dark. The feeding habits of females are largely unknown, but birds appear to be the primary hosts. Two species in the Oriental Region are known to approach and feed on humans. None of the species of *Orthopodomyia* are of medical or economic importance.

****

Principal references: Barraud, 1934 (southern Asia, keys, descriptions, bionomics, distributions); Knight & Mattingly, 1950 (classification, keys, outdated Anopheloides Subgroup, descriptions); Lane, 1953 (Neotropical Region, descriptions, keys, distributions); Thurman, 1959 (Thailand, keys, descriptions, distributions); Delfinado, 1966a (Philippines, taxonomy, keys for adults, descriptions, distributions, bionomics); Zavortink, 1968 (World, taxonomy, classification, descriptions, keys, bionomics, distributions); Zavortink, 1971 (Southeast Asia, classification, descriptions, keys, distributions, bionomics); Belkin *et al.*, 1970 (Jamaica,

Or. waverleyi, description, bionomics, distribution); Tanaka *et al.*, 1979 (Japan, *Or. anopheloides*, descriptions, distribution, bionomics); Darsie & Ward, 1981, 2005 (North America, keys); Lee *et al.*, 1988c (Australasian Region, taxonomy, literature, bionomics, distributions); Service, 1990 (Afrotropical Region, keys, taxonomy, literature, bionomics, distributions); Brunhes & Hervy, 1995 (Afrotropical Region, taxonomy, descriptions, keys, bionomics); Lu *et al.*, 1997a (China, keys, taxonomy, descriptions, distributions, bionomics); Harbach & Kitching, 1998 (morphology, phylogenetic relationships); Rattanarithikul *et al.*, 2007 (Thailand, bionomics, keys); Byrd *et al.*, 2012 (Signifera Group, morphology, rDNA ITS2, phylogenetic relationships); Azari-Hamidian *et al.*, 2017 (Iran, *Or. pulcripalpis*, notes on systematics of the genus).

Paleoculicis Poinar, Zavortink, Pike & Johnston, 2000

Paleoculicis Poinar *et al.*, 2000, originally proposed and currently recognized as an extinct genus, is typified by *Paleoculicis minutus* Poinar, Zavortink, Pike & Johnston, 2000. Genus abbreviation – *Pc.*

****

Systematics: *Paleoculicis* is a subordinate taxon of the subfamily Culicinae, tribe (?). It has no living species. It only includes the extinct type species, which existed during the Campanian Age (72.1–83.6 Mya) during the Upper Cretaceous Epoch of the Mesozoic Era. The species is only known from a male preserved in Canadian amber from near Grassy Lake village, Alberta, Canada. Unique characteristics of the specimen suggest that *Paleoculicis* may represent a major lineage (tribe?) that died out during the evolution of the extant groups of Culicinae. A number of other features of *Pc. minutus* suggest that it might possibility be an aedine mosquito.

****

Characteristics: ADULT MALES: Minute mosquitoes; vertex of the head with erect piliform scales; eyes with fewer and larger facets in relation to the size of the head than extant culicids; antenna with the pedicel greatly enlarged, without scales or setae, flagellum strongly verticillate, flagellomere 1 with scales; clypeus smaller than the antennal pedicel, without scales or setae; maxillary palpus long, dark-scaled, comprised of 5 palpomeres, palpomere 3 gradually enlarged toward the apex, with long setae ventrally at the apex, palpomere 4 with numerous long strong setae, palpomere 5 swollen in the distal 0.5, with fewer shorter and finer setae; scutellum rounded, with a uniform row of setae on the posterior edge; mesopostnotum without setae or scales; antepronota small, widely separated; thoracic pleura without scales; prespiracular and postspiracular setae absent; upper proepisternal, prealar, mesokatepisternal and upper anterior and posterior mesepimeral setae present; mesokatepisternum with a broad band of setae in 2 or 3 vertical rows extending the entire length of the sclerite, mesepimeron with 2 strong lower anterior setae and 2 or 3 posterior setae; prealar area and mesokatepisternum not separated by a suture; mesomeron large, upper edge distinctly above the base of the hindcoxa; wing with distinct microtrichia; subcostal vein with setae at the base ventrally; vein R_s without a basal spur; cell R_2 2.3 the length of vein R_{2+3}; anal vein (vein 1A) terminates beyond the base of the mediocubital crossvein; alula with marginal piliform scales; upper calypter with a row of marginal setae; ungues paired, enlarged, unequal, larger anterior unguis with a sharp basal denticle and a long blunt submedian tooth, smaller posterior unguis with a short basal spur projecting laterally; empodium present; pulvilli apparently not developed; abdominal terga and sterna with long setae, without scales; gonocoxite with few setae, with strong spicules distally; claspette (?) a large rounded lobe with setae.

****

Bionomics: The Campanian Age was generally warm and species of all vertebrate groups, including the first placental mammals, were present. Since females of *Pc. minutus* were probably blood feeders, they could have fed on reptiles, including dinosaurs, birds and mammals. It is equally likely that the immature stages of *Pc. minutus* may have developed in groundwater or phytotelm habitats.

****

Principal reference: Poinar *et al.*, 2000 (new genus, diagnosis, descriptions, illustrations, phylogenetic considerations, list of fossil mosquitoes).

Psorophora Robineau-Desvoidy, 1827

Psorophora Robineau-Desvoidy, 1827, originally proposed and currently recognized as a genus, is typified by *Psorophora ciliata* (Fabricius, 1794); original combination: *Culex ciliata* (*sic*). Genus abbreviation – *Ps.*

****

Systematics: *Psorophora* is a subordinate taxon of the subfamily Culicinae, tribe Aedini. It includes 49 extant species divided between three subgenera: *Grabhamia*, *Janthinosoma* and *Psorophora sensu stricto*. Species of *Psorophora* are associated with flood-water pools in the tropics and warmer temperate regions of North and South America. Most species occur in the southern temperate and tropical countries. *Psorophora* was recovered as the sister of all other Aedini in the phylogenetic studies of Reinert *et al.* (2004, 2006, 2008, 2009) based on morphological data of all life stages, and also in the maximum likelihood phylogenies of Soghigian *et al.* (2017), based on seven molecular markers, as well as do Nascimento *et al.*, 2021 and Lorenz *et al.* (2021) based on 13 mitochondrial protein-coding genes. Liria & Navarro (2014) conducted a phylogenetic analysis of relationships for 29 of the 48 recognized species (only 45 were indicated in the published article) of the genus based on 66 morphological characters from the adult, larval and pupal stages under equal weighting. The analysis weakly supported the monophyly of *Psorophora* and strongly supported the monophyly of *Grabhamia* and *Psorophora sensu stricto*. However, in the strict consensus tree of 11 most parsimonious trees, *Janthinosoma* was not recovered as monophyletic due to the ambiguous position of *Ps.* (*Jan.*) *cyanescens*, and within the three subgenera, the relationships among the respective included species were poorly resolved. The relationships and conclusions discussed in the paper should be regarded as provisional due to a dubious choice of characters, problematic character coding and questionable homologies. Muñoz-Gamba *et al.* (2021) included 13 species of *Psorophora* in an analysis of mitochondrial *COI* gene sequences in which five species of *Janthinosoma* comprised a monophyletic clade in a sister relationship with an intermixed group that included five species of *Grabhamia* and three species of *Psorophora sensu stricto*. *Psorophora* fell outside of Aedini in an improbable sister relationship with *Mansonia* in the phylogeny of Wilkerson *et al.* (2015), probably due to equal weighting of the data set of Reinert *et al.* (2009). *Psorophora* was included in the phylogenetic analysis of amino acid sequences of 709 genes conducted by Soghigian *et al.* (2023), but its relationship with other genus groups of Aedini included in the analysis was not explicitly indicated.

****

Characteristics: The adults of *Psorophora* are small to very large mosquitoes that are distinguished from the adults of other genus groups in the New World by the presence of both prespiracular and postspiracular setae. Larvae have a well-developed ventral brush (seta 4-X) with a row of precratal setae attached to a complete saddle, or extending to the middle of segment X if the saddle is incomplete. The latter condition occurs in many aedine species, however the combination of an incomplete saddle and a strongly swollen siphon will distinguish *Psorophora*. Additional morphological distinctions are as follow. ADULTS: Maxillary palpus of males with palpomeres 4 and 5 upturned; prescutellar area with median and posterior scales. FEMALE GENITALIA: Tergum and sternum VIII mostly membranous with heavily sclerotized lateral rod-shaped structures; tergum IX with setae distally and on much of the median area; upper vaginal sclerite absent; insula tongue-like. MALE GENITALIA: Claspette with 3 or more moderately long, stout, bluntly pointed, spiculose setae. LARVAE: Comb plate present; seta 1-S short, length not more than 0.4 the width of the siphon; seta 6-S noticeably longer than the distal width of the siphon; ventral brush (seta 4-X) borne on a grid with transverse grid bars. PUPAE: Seta 3-II longer than seta 6-II; seta 5-V shorter than the following tergum; seta 6-VII inserted anterior to seta 9-VII.

****

Bionomics: *Psorophora* are called flood-water mosquitoes because females lay their eggs on damp or dry mud and debris in fields and wooded plains where they may withstand long periods (months or years) of desiccation and hatch when the habitat is inundated by rain or flood waters. Larvae of *Psorophora sensu stricto* are predacious. The adults are diurnal and the females of many species avidly feed on humans. Several species are vectors of arboviruses. Most notably, Ilheus virus is transmitted by *Ps. ferox* and Venezuelan equine encephalitis virus is transmitted by *Ps. ferox* and *Ps. confinnis*. A number of viruses have been isolated from other species of *Psorophora*.

****

Principal references: Horsfall *et al.*, 1952 (egg, descriptions, keys); Lane, 1953 (Neotropical Region, taxonomy, keys, descriptions, distributions); Forattini, 1965a (Neotropical Region, diagnosis, descriptions, bionomics, distributions, key to adults); Cova-Garcia *et al.*, 1966 (Venezuela, keys, descriptions, distributions); Barr & Barr, 1969 (North America, Puerto Rico, pupae, keys, descriptions); Belkin *et al.*, 1970 (Jamaica, keys, descriptions, bionomics, distributions); Wood *et al.*, 1979 (Canada, keys, descriptions, bionomics, distributions); Darsie & Ward, 1981, 2005 (North America, keys); Clark-Gil & Darsie, 1983 (Guatemala, keys); Darsie, 1985 (Argentina, keys); Harbach & Kitching, 1998 (morphology, phylogenetic relationships); Reinert, 2000d (female genitalia); Reinert *et al.*, 2004, 2006, 2008, 2009 (morphology, phylogenetic relationships); Liria & Navarro, 2014 (morphology, phylogenetic relationships); Wilkerson *et al.*, 2015 (phylogenetic relationships); Muñoz-Gamba *et al.*, 2021 (mtDNA *COI* sequences, phylogenetic relationships); Soghigian *et al.*, 2017 (phylogenetic relationships); do Nascimento *et al.*, 2021 (mitochondrial protein-coding genes, phylogenetic relationships); Lorenz *et al.*, 2021 (mitochondrial protein-coding genes, phylogenetic relationships).

Grabhamia Theobald, 1903

Grabhamia, originally proposed as a genus by Theobald (1903a), currently classified as a subgenus of *Psorophora*, is typified by *Psorophora jamaicensis* (Theobald, 1901b); original combination: *Culex jamaicensis*. Subgenus abbreviation – *Gra*.

****

Systematics: *Grabhamia* includes 16 extant species with distributions in the New World, from approximately 51° N in southern Canada to northern Argentina and Uruguay in South America (there are currently no records from Chile). *Grabhamia* was recovered as the sister of *Janthinosoma* + *Psorophora sensu stricto* in the preferred phylogeny of Reinert *et al.* (2009) based on morphological data for the genus groups of Aedini. The monophyly of *Grabhamia* was strongly supported in the phylogenetic study of Liria & Navarro (2014) based on 66 morphological characters from the adult, larval and pupal stages analysed under equal weighting. However, relationships within *Grabhamia* were poorly resolved in their strict consensus tree of 11 most parsimonious trees. *Grabhamia* was recovered with rather weak support as the sister of *Psorophora sensu stricto* in the maximum likelihood phylogeny of Soghigian *et al.* (2017) based on seven molecular markers, and the pair was strongly supported as the sister to *Janthinosoma*. *Grabhamia* was included in the phylogenetic analysis of amino acid sequences of 709 genes conducted by Soghigian *et al.* (2023), but its relationship with other genus groups of Aedini included in the analysis was not explicitly indicated.

****

Characteristics: Species of *Grabhamia* are characterized and distinguished from species of *Janthinosoma* and *Psorophora sensu stricto* by the following combinations of characters. ADULTS: Small to medium-sized mosquitoes; interocular space with 6 or more setae; proboscis of females with a ring or ventral patch of pale scales; areas between the acrostichal and dorsocentral setae and between the supraalar and prescutellar setae with scales; wing with a mixture of dark and pale scales, alula without marginal scales; femora with a narrow subapical pale band, tibiae with a speckling or spots of pale scales, at least tarsomeres 1 and 2 of all legs with basal pale markings. MALE GENITALIA: Claspette without a foliform setae; aedeagus conical or cylindrical, without a lateral spine. LARVAE: Head and the lateral palatal brushes normal, antenna short, distinctly shorter than the head capsule, no more than 0.8 as long; siphon strongly inflated, pecten usually with fewer than 20 spines. PUPAE: Trumpet relatively short; setae 10-CT and 5-II usually with more than 5 branches, 5-II inserted anterior to seta 3-II; caudolateral angle of abdominal tergum IV without spines; paddles without contrasting pigmentation near the external buttress and apex.

****

Bionomics: The immature stages of species of *Grabhamia* are found primarily in open, sunlit groundwater habitats. Females of several species readily feed on humans. Venezuelan equine encephalitis virus is transmitted by *Ps. confinnis*. Cache Valley, Tensaw and Venezuelan equine encephalitis and West Nile viruses have also been isolated from this species. Females of *Ps. cingulata* have tested positive for Cache Valley virus.

****

Principal references: Lane, 1953 (Neotropical Region, taxonomy, keys, descriptions, distributions); Forattini, 1965a (Neotropical Region, diagnosis, descriptions, bionomics, distributions, key to adults); Cova-Garcia *et al.*, 1966 (Venezuela, keys, descriptions, distributions); Barr & Barr, 1969 (North America, Puerto Rico, pupae, keys, descriptions); Belkin *et al.*, 1970 (Jamaica, keys, descriptions, bionomics, distributions); Wood *et al.*, 1979 (Canada, keys, descriptions, bionomics, distributions); Darsie & Ward, 1981, 2005 (North America, keys); Clark-Gil & Darsie, 1983 (Guatemala, keys); Darsie, 1985 (Argentina, keys); Reinert, 2000d (female genitalia); Reinert *et al.*, 2004, 2006, 2008, 2009 (morphology, phylogenetic relationships); Liria & Navarro, 2014 (morphology, phylogenetic relationships); Soghigian *et al.*, 2017 (phylogenetic relationships).

Janthinosoma Lynch Arribálzaga, 1891

Janthinosoma, originally proposed as a genus by Lynch Arribálzaga (1891a, b), currently classified as a subgenus of *Psorophora*, is typified by *Psorophora discrucians* (Walker, 1856); original combination: *Culex discrucians*. Subgenus abbreviation – *Jan*.

****

Systematics: *Janthinosoma* includes 23 extant species with distributions primarily in the Neotropical Region, in countries from Mexico in Central America to northern Argentina and Uruguay in South American. Two species, *Ps. longipalpus* and *Ps. mexicana*, are only recorded from Mexico and the USA, *Ps. johnstonii* is widely distributed in the Caribbean and occurs in the extreme south of Florida, and *Ps. cyanescens* and *Ps. ferox* have distributions that extend from Argentina and Uruguay northward into the USA, with the latter species reaching southern Canada. *Janthinosoma* was the sister of *Psorophora sensu stricto*, and this pair was the sister of *Grabhamia*, in the phylogeny of Aedini recovered in the phylogenetic study of Reinert *et al.* (2009) based on morphological data of all life stages. Liria & Navarro (2014) explored relationships within *Psorophora* using 66 characters from the adult, larval and pupal stages coded for 12 species of *Janthinosoma*, five species of *Psorophora sensu stricto* and seven species of *Grabhamia*. *Janthinosoma* was not recovered as monophyletic in their strict consensus tree of 11 most parsimonious trees due to the ambiguous position of *Ps.* (*Jan.*) *cyanescens*, and the relationships of the species of *Janthinosoma* included in the study were poorly resolved. Contrary to the findings of Reinert *et al.*, *Janthinosoma* was recovered as the sister of *Grabhamia* + *Psorophora sensu stricto* in the maximum likelihood phylogeny of Soghigian *et al.* (2017) based on seven molecular markers. *Janthinosoma* was included in the phylogenetic analysis of amino acid sequences of 709 genes conducted by Soghigian *et al.* (2023), but its relationship with other genus groups of Aedini included in the analysis was not explicitly indicated.

****

Characteristics: Species of *Janthinosoma* are characterized and distinguished from species of *Grabhamia* and *Psorophora sensu stricto* by the following combinations of characters. ADULTS: Small to medium-sized mosquitoes; proboscis, legs and abdominal terga with dark violet metallic scaling, proboscis entirely dark-scaled; scutum with scales between the acrostichal and dorsocentral setae and between the prescutellar and supraalar setae; lower proepisternal scales present; wing entirely dark-scaled or with a few inconspicuous pale scales on the costa and subcosta; anteprocoxal scales present; apices of the femora without erect scales, hindfemur without a subapical ring of pale scales, tibiae uniformly dark-scaled with violet to purple reflections, tarsomeres 1–3 of all legs without basal pale markings, hindtarsomeres 4 or 4 and 5 and sometimes 3 with conspicuous white markings in most species; ungues of all legs with 1 tooth; tip of the abdomen with a tapering sharply pointed appearance. MALE GENITALIA: Claspette with 1 short simple and 2 large twisted foliform setae on the apex; aedeagus conical or cylindrical, without a lateral spine. LARVAE: Head and lateral palatal brushes normal, antenna extending well beyond the anterior margin of the head capsule; pecten reduced, usually with fewer than 10 pecten spines. PUPAE: Trumpet long; seta 7-CT ≥ 6.0 the length of seta 6-CT; seta 10-CT usually with more than 5 branches; seta 11-CT branched; seta 3-II branched; seta 5-II inserted anterior to seta 3-II, usually with more than 5 branches; posterolateral angle of abdominal segment IV with conspicuous spines; paddle without contrasting pigmentation near the external buttress and apex.

****

Bionomics: The immature stages of species of *Janthinosoma* are seldom collected because they are found in primarily shaded habitats in flooded woodland areas that are often inaccessible. Females of some species readily attack and feed on humans who enter their haunts. Ilheus and Venezuelan equine encephalitis viruses are transmitted by *Ps. ferox*. West Nile virus, as well as Cache, Kairi, Mayaro, Oriboca, St Louis and Una viruses, have also been isolated from this species. Other species of *Janthinosoma* that have tested positive for viruses include *Ps. albipes* (Ilheus and Una viruses) and *Ps. lutzii* (Guama and Ilheus viruses).

****

Principal references: Lane, 1953 (Neotropical Region, taxonomy, keys, descriptions, distributions); Forattini, 1965a (Neotropical Region, diagnosis, descriptions, bionomics, distributions, key to adults); Cova-Garcia *et al.*, 1966 (Venezuela, keys, descriptions, distributions); Barr & Barr, 1969 (North America, Puerto Rico, pupae, keys, descriptions); Belkin *et al.*, 1970 (Jamaica, keys, descriptions, bionomics, distributions); Wood *et al.*, 1979 (Canada, keys, descriptions, bionomics, distributions); Darsie & Ward, 1981, 2005 (North America, keys); Clark-Gil & Darsie, 1983 (Guatemala, keys); Darsie, 1985 (Argentina, keys); Reinert, 2000d (female genitalia); Reinert *et al.*, 2004, 2006, 2008, 2009 (morphology, phylogenetic relationships); Harrison *et al.*, 2008 (USA, descriptions, key to females); Liria & Navarro, 2014 (morphology, phylogenetic relationships); Soghigian *et al.*, 2017 (phylogenetic relationships).

Psorophora Robineau-Desvoidy, 1827 *sensu stricto*

Psorophora sensu stricto, the nominotypical subgenus of *Psorophora*, is typified by *Psorophora ciliata* (Fabricius, 1794); original combination: *Culex ciliata* (*sic*). Subgenus abbreviation – *Pso.*

****

Systematics: *Psorophora sensu stricto* includes 10 extant species, which occur predominantly in the Neotropical Region from approximately 35° S latitude northward to Mexico. The distributions of *Ps. ciliata* and *Ps. howardii* extend north of Mexico, with the former reaching southern Canada. *Psorophora sensu stricto* + *Janthinosoma* was the sister of *Grabhamia* in the phylogeny of Aedini recovered in the phylogenetic study of Reinert *et al.* (2009) based on morphological data of all life stages. The monophyly of *Psorophora sensu stricto* was strongly supported in the phylogenetic study of Liria & Navarro (2014) based on 66 morphological characters from the adult, larval and pupal stages analysed under equal weighting. However, relationships within the subgenus were poorly resolved in their strict consensus tree of 11 most parsimonious trees. Contrary to the findings of Reinert *et al.*, *Psorophora sensu stricto* was recovered as the sister of *Grabhamia* in the maximum likelihood phylogeny of Soghigian *et al.* (2017) based on seven molecular markers. *Psorophora sensu stricto*, represented by *Ps. ciliata* and *Ps. howardii*, was included in the phylogenetic analysis of amino acid sequences of 709 genes conducted by Soghigian *et al.* (2023), but its relationship with other genus groups of Aedini included in the analysis was not explicitly indicated.

****

Characteristics: Species of *Psorophora sensu stricto* are characterized and distinguished from species of *Grabhamia* and *Janthinosoma* by the following combinations of characters. ADULTS: Huge mosquitoes; scutum without scales between the acrostichal and dorsocentral setae and between the prescutellar and supraalar setae; wing entirely dark-scaled or with a few inconspicuous pale scales on the costa and subcosta; hindleg with long erect scales, shaggy, hindfemur without a subapical ring of pale scales, hindtarsomere 5 never entirely white-scaled; ungues of all legs with 1 tooth; laterotergite of abdominal segment I without scales. MALE GENITALIA: Gonocoxite without scales; aedeagus with a broad sharp lateral subapical spine, with small distal spicules. LARVAE: Head capsule more or less quadrate in dorsal view, truncate anteriorly; lateral palatal brushes reduced, filaments thickened; antenna short, not reaching the anterior margin of the head capsule, seta 1-A single or double, short, length $\leq$ 3 times the width of the antenna at the point of insertion; seta 1-C spiniform; seta 4-C inserted posterior to seta 6-C; seta 6-C inserted anterior to seta 7-C; seta 7-C inserted more or less on level with seta 5-C; seta 12-C inserted lateral to seta 13-C; seta 7-P single; pecten comprised of numerous spines, distal 1 or more spines more widely spaced; seta 1-S single, length 1.2–2.0 the width of the siphon. PUPAE: Seta 10-C usually with fewer than 5 branches; seta 5-II longer than seta 3-II, inserted posterior to seta 3-II, usually with fewer than 5 branches; paddle heavily pigmented near the external buttress and apex.

****

Bionomics: The immature stages of the species of *Psorophora sensu stricto* are found in open sunlit temporary pools and the flooded margins of swamps. The larvae are predacious. The intricacies of the prey-capture behaviour of *Ps. ciliata* were elucidated by Hancock *et al.* (2022). Females of at least some species avidly attack and feed on humans during daylight hours in the vicinity of larval habitats. *Psorophora ciliata* is known to carry Eastern, Venezuelan and Western equine encephalitis viruses, Tensaw virus and West Nile virus, but it is not known whether it is a competent vector of these viruses. Other species of *Psorophora sensu stricto* that have tested positive for viruses include *Ps. cilipes* (Venezuelan equine encephalitis virus) and *Ps. howardii* (California group viruses).

****

Principal references: Lane, 1953 (Neotropical Region, taxonomy, keys, descriptions, distributions); Forattini, 1965a (Neotropical Region, diagnosis, descriptions, bionomics, distributions, key to adults); Cova-Garcia *et al.*, 1966 (Venezuela, keys, descriptions, distributions); Barr & Barr, 1969 (North America, Puerto Rico, pupae, keys, descriptions); Belkin *et al.*, 1970 (Jamaica, *Ps. ciliata*, in keys, descriptions, bionomics, distributions); Wood *et al.*, 1979 (Canada, keys, descriptions, bionomics, distributions); Darsie & Ward, 1981, 2005 (North America, keys); Clark-Gil & Darsie, 1983 (Guatemala, keys); Darsie, 1985 (Argentina, keys); Reinert, 2000d (female genitalia); Reinert *et al.*, 2004, 2006, 2008, 2009 (morphology, phylogenetic relationships); Liria & Navarro, 2014 (morphology, phylogenetic relationships); Soghigian *et al.*, 2017 (phylogenetic relationships).

Runchomyia Theobald, 1903

Runchomyia Theobald, 1903a, originally proposed and currently recognized as a genus, is typified by *Runchomyia frontosa* Theobald, 1903a. Genus abbreviation – *Ru.*

****

Systematics: *Runchomyia* is a subordinate taxon of the subfamily Culicinae, tribe Sabethini. It includes eight extant species grouped in two subgenera, *Ctenogoeldia* and *Runchomyia sensu stricto*. The species have distributions in the Neotropical Region. *Runchomyia* was recovered as the sister of *Johnbelkinia* in the phylogenetic analysis of Harbach & Kitching (1998). This relationship was supported by a single synapomorphy: maxillary brush of the larva represented by a flexible bundle of coalesced spicules. This character only occurs elsewhere in certain *Topomyia*, hence it lends rather strong support to the relationship. However, the phylogenetic analyses of Judd (1996) provided evidence that *Runchomyia* is not a monophyletic taxon. Judd recorded four differences in larval mouthparts and one involving the pupal trumpet that distinguish the two subgenera, and apparently support a closer relationships with other New World genus groups of Sabethini. In analyses of 13 mitochondrial protein-coding genes that included species of Neotropical sabethines (do Nascimento *et al.*, 2021; Lorenz *et al.*, 2021), *Ru. reversa* was paired with *Trichoprosopon pallidiventer* in a sister relationship to a clade comprised of one species each of *Limatus* and *Wyeomyia* and four species of *Sabethes*.

****

Characteristics: The adults of *Runchomyia* are distinguished from the other members of Sabethini in the New World by the following combinations of characters: dorsal head scales with brilliant silver and blue reflections, proboscis distinctly longer than the forefemur, scutal scales narrow and curved, postpronotum with 1 or more posterior setae, lower mesokatepisternal setae not extending above the lower edge of the mesepimeron and tarsi entirely dark-scaled. *Runchomyia* larvae possess a slit-like occipital foramen and a filamentous pecten on the siphon, which distinguish them from larvae of *Johnbelkinia*, *Onirion*, *Shannoniana* and *Trichoprosopon*. The presence of a maxillary bundle distinguishes them from all other New World sabethines. *Runchomyia* are otherwise very similar to larvae of *Isostomyia*, but differ in having seta 3-X single and seta 6-S weakly developed, flexible and not hooked at the tip.

****

Bionomics: Little is known about the bionomics of species of *Runchomyia*. Larvae have been collected from flower bracts (*Calathea* and *Heliconia*), leaf axils (aroids) and both arboreal and terrestrial bromeliads. The adults of at least two species have been captured during landing-biting collections on humans in forest. Species of *Runchomyia* are unlikely to be of medical or economic importance.

****

Principal references: Lane, 1953 (as subgenera *Ctenogoeldia* and *Runchomyia* of *Trichoprosopon*, in keys, descriptions, distributions); Zavortink, 1979 (genus as currently defined less *Isostomyia*, keys to genus groups, taxonomy); Clark-Gil & Darsie, 1983 (Guatemala, keys); Darsie, 1985 (Argentina, keys); Harbach & Peyton, 1993 (comparative morphology of larval maxillae); Judd, 1996 (morphology, phylogenetic relationships); Harbach & Kitching, 1998 (morphology, phylogenetic relationships); do Nascimento *et al.*, 2021 (mitochondrial protein-coding genes, phylogenetic relationships); Lorenz *et al.*, 2021 (mitochondrial protein-coding genes, phylogenetic relationships).

Ctenogoeldia Edwards, 1930

Ctenogoeldia Edwards, 1930a, originally proposed as a subgenus of *Goeldia* Theobald, 1903a and currently classified as a subgenus of *Runchomyia*, is typified by *Runchomyia magna* (Theobald, 1905f), original combination: *Phoniomyia magna*. Subgenus abbreviation – *Cte.*

****

Systematics: *Ctenogoeldia* includes two extant species. The type species is known to occur in countries from Belize and Guatemala in Central America southward into Bolivia, with records of its occurrence in Brazil, Colombia, Costa Rica, Ecuador, Honduras, Nicaragua, Panama, Suriname and Venezuela. *Runchomyia walcotti* is only recorded from Brazil. The phylogenetic analyses of morphological data conducted by Judd (1996) indicated that *Ctenogoeldia* is more closely related to *Isostomyia* than to *Runchomyia*. The relationship was supported by two unambiguous synapomorphies, both features of the larval labiohypopharynx: (1) anterior edge of the prementoligular teeth level with the bases of seta 1-Lh and (2) the teeth not extending posteriorly beyond the labial palpus.

****

Characteristics: The immature stages and male are known only for the type species; distinguished from other members of New World Sabethini by the following combinations of characters. Characters that distinguish *Ctenogoeldia* from *Runchomyia* are indicated by an asterisk (*). ADULTS: Scales of the vertex with brilliant silver and azure blue reflections; proboscis 1.2–1.4 the length of the forefemur; scutal scales narrow and curved; *scutellum with scales on the base of the midlobe with brilliant silver reflections; postpronotum with 1 or more posterior setae; a row of lower mesokatepisternal setae usually not extending dorsad to the level of the lower edge of the mesepimeron; mid- and hindtarsi dark-scaled. MALE GENITALIA: *Gonocoxite without distinctly separated basal mesal lobe. LARVAE: Head capsule with a short to long transverse slit-like occipital foramen, not bounded by a distinct collar; *anterior margin of the median labral plate between the bases of seta 1-C concave; *mandibular spiculose area present; *maxilla with a rigid apical spine shorter than the articulated maxillary bundle; *anterior edge of the prementoligular teeth level with the base of seta 1-Lh, *teeth not extending posteriorly beyond the labial palpus; *seta 5-T strongly developed, moderately long; *most or all comb scales arise from a single large plate; siphon with a filamentous pecten; seta 3-X single; seta 6-S weakly developed, flexible, not hooked. PUPAE: *Trumpet elongate, more or less cylindrical, not noticeably wider apically, apical width at least twice the length of the internal section; *seta 5-C much longer and stronger than seta 7-C, with 3–10 branches; punctures present on terga III–V; seta 5-V,VI well developed, long; *seta 6-I more or less strongly developed but weaker than seta 3-I and/or seta 6-II, seta 6-VII single or double, inserted caudomesad of seta 9-VII; caudolateral angle of tergum VIII not produced or only slightly produced, not extending more than 0.25 the distance to the apex of segment IX; paddle oval to triangular, inner part glabrous.

****

Bionomics: Females of *Ru. magna* have been collected while landing-feeding on humans between 17:00 and 18:00 in forest (Heinemann & Belkin, 1977b, as *Trichoprosopon magnum*). Larvae have been found in the leaf axils and flower bracts of plants of the genera *Calathea* (Marantaceae) and *Heliconia* (Heliconiaceae), and the leaf axils of species of *Xanthosoma* (aroids, Araceae) (Heinemann & Belkin, 1978a, as *Tr. magnum*). Linton *et al.* (2013) reported the rearing of a female from a pupa collected in turbid ground seepage water in open sunlight. Circumstantial evidence suggests that the larvae may be predatory. Species of *Ctenogoeldia* are not known to be of medical or economic importance.

****

Principal references: Lane, 1953 (as a subgenus of *Trichoprosopon*, in keys, descriptions, distributions); Zavortink, 1979 (subgenus as currently defined, keys to genus groups, taxonomy); Harbach & Peyton, 1993 (comparative morphology of larval maxillae); Judd, 1996 (morphology, phylogenetic relationships).

Runchomyia Theobald, 1903 *sensu stricto*

Runchomyia sensu stricto, the nominotypical subgenus of *Runchomyia*, is typified by *Runchomyia frontosa* Theobald, 1903a. Subgenus abbreviation – *Run.*

****

Systematics: *Runchomyia sensu stricto* includes six extant species. With the exception of Panama, species of *Runchomyia sensu stricto* are recorded almost exclusively from countries of South America, including Argentina (*Ru. frontosa*; *Ru. reversa*), Brazil (all species of the group except *Ru. rapax*), Guyana (*Ru. frontosa*), Paraguay (*Ru. reversa*) and Trinidad and Tobago (*Ru. frontosa*, *Ru. rapax*, *Ru. theobaldi*). *Runchomyia cerqueirai* and *Ru. frontosa* are recorded from Panama. The phylogenetic analyses of morphological data conducted by Judd (1996) indicated that *Runchomyia sensu stricto* is not closely related to *Ctenogoeldia*, which was recovered as the sister of *Isostomyia*, the relationship supported by two unambiguous characters of the larval labiohypopharynx: (1) anterior edge of the prementoligular teeth level with the bases of seta 1-Lh and (2) the teeth not extending posteriorly beyond the labial palpus.

****

Characteristics: Species of *Runchomyia sensu stricto* are distinguished from other members of New World Sabethini by the following combinations of characters. Characters that distinguish *Runchomyia sensu stricto* from *Ctenogoeldia* are indicated by an asterisk (*). ADULTS: Scales of the vertex with brilliant silver and azure blue reflections; proboscis 1.2–1.4 the length of the forefemur; scutal scales narrow and curved; *scales on the midlobe of the scutellum without silver reflections; postpronotum with 1 or more posterior setae; a row of lower mesokatepisternal setae usually not extending dorsad to the level the of lower edge of the mesepimeron; mid- and hindtarsi dark-scaled. MALE GENITALIA: Gonocoxite long, *with a distinctly separated basal mesal lobe; aedeagus with a narrow sharp apical beak, apex of the submedian tergal arm narrow; paraproct with 2–6 strong apical teeth. LARVAE: Head capsule with a short to long transverse slit-like occipital foramen, not bounded by a distinct collar; *anterior margin of the median labral plate between the bases of seta 1-C straight; *mandibular spiculose area absent; *maxilla with a rigid apical spine longer than the articulated maxillary bundle; *anterior edge of the prementoligular teeth posterior to the base of seta 1-Lh, *teeth extend laterally and mesally toward the ventral premental spicules; *seta 5-T weakly developed, short; *comb plate absent, comb scales arise from hyaline integument; siphon with filamentous pecten; seta 3-X single; seta 6-S weakly developed, flexible, not hooked. PUPAE: *Trumpet widest apically, equal in length to the internal section; *seta 5-C subequal in length and not as strong as seta 7-C or at least with 40 branches; punctures present on terga III–V; seta 5-V,VI well developed, long; *seta 6-I strongly developed, as strong or stronger than setae 3-I and 6-II; seta 6-VII single or double, inserted caudomesad of seta 9-VII; caudolateral angle of tergum VIII not produced or only slightly produced, not extending more than 0.25 the distance to the apex of segment IX; paddle oval to triangular, inner part glabrous.

****

Bionomics: Very little is known about the bionomics of the species of *Runchomyia sensu stricto*. The species are known to be diurnal. Females of *Ru. ?frontosa* are attracted to and feed on humans, and their larvae inhabit epiphytic bromeliads and the leaf axils of *Heliconia wagneriana* (Heinemann *et al.*, 1980). Zavortink (1986b) found that the immature stages of *Ru. frontosa* live in fluids held by the carnivorous bromeliad *Brocchinia reducta* and noted that the larvae are facultative predators. Campos (2013) considered the larvae of *Ru. reversa*, which were collected from the internodes of two species of bamboo (*Guadua chacoensis* and *G. trinii*), to be potential facultative predators. Larvae listed as uncertain larvae of *Ru. cerqueirai* and *Ru. reversa* by Heinemann & Belkin (1979) were collected in terrestrial bromeliads (both species) and epiphytic bromeliads and cut or broken bamboo (the latter species). Species of *Runchomyia sensu stricto* are not known to be of medical or economic importance.

****

Principal references: Lane, 1953 (as a subgenus of *Trichoprosopon*, in keys, descriptions, distributions); Zavortink, 1979 (subgenus as currently defined, generic-level keys, taxonomy); Darsie, 1985 (Argentina, keys include *Ru. frontosa* and *Ru. reversa*); Harbach & Peyton, 1993 (comparative morphology of larval maxillae); Judd, 1996 (morphology, phylogenetic relationships).

Sabethes Robineau-Desvoidy, 1827

Sabethes Robineau-Desvoidy, 1827, originally proposed and currently recognized as a genus, is typified by *Sabethes locuples* Robineau-Desvoidy, 1827 [subjective synonym of *Sabethes cyaneus* (Fabricius, 1805)]. Genus abbreviation – *Sa.*

****

Systematics: *Sabethes* is a subordinate taxon of the subfamily Culicinae, tribe Sabethini. It currently includes 42 extant species classified in five subgenera: *Davismyia*, *Peytonulus*, *Sabethes sensu stricto*, *Sabethinus* and *Sabethoides*. The only known species of

Davismyia occurs in South America. The other genus groups have species in Central as well as South America. The phyletic affinities of *Sabethes* are uncertain. Based on morphological data, *Sabethes* was recovered as the sister of *Wyeomyia* in the phylogenetic analysis of Judd (1996), as the sister of *Limatus* in the analyses of Harbach & Kitching (1998) and Harbach & Peyton (2000), and was placed as the sister of *Limatus* + *Malaya* when Harbach *et al.* (2007a) included *Kimia* in the data set of Harbach & Peyton (2000). The phylogeny of the subgenera and species has not been investigated. However, an analysis of 13 mitochondrial protein-coding genes of four species conducted by Lorenz *et al.* (2021) generated a clade that translates to *Sabethes sensu stricto* + (*Sabethoides* + *Peytonulus*). Two species of *Sabethes*, *Sa.* (*Sabethes*) *cyaneus* and *Sa.* (*Sabethoides*) *chloropterus*, were included in the phylogenetic analysis of amino acid sequences of 709 genes conducted by Soghigian *et al.* (2023), but their relationships with other taxa of Sabethini included in the analysis were not explicitly indicated. *Sabethes*, represented by 10 species, was not recovered as a monophyletic group in the phylogenetic analysis of mtDNA *COI* gene sequences conducted by Muñoz-Gamba *et al.* (2021). One species (unidentified) was recovered in a sister relationship with two species of *Johnbelkinia* and the other nine species were dispersed among 21 species of *Wyeomyia*. Whether this casts doubt on the monophyly of *Sabethes* requires further investigation.

****

Characteristics: *Sabethes* are brilliantly coloured mosquitoes with a smooth covering of metallic green, blue, violet, gold and silver scales. The scutum is uniformly clothed in broad, flat scales and setae are present only on the extreme anterior margin and on the supraalar areas. The antepronota are large and close together. Prealar setae are absent in all species except *Sa. petrocchiae* of *Davismyia*. Most species of *Sabethes sensu stricto* have 'paddles' of outstanding scales on some or all of the legs. The maxillae of *Sabethes* larvae have a prominent apical tooth and the maxillary palpus is fused with the maxillary body. A few species of *Wyeomyia* have a similar apical tooth on the maxilla, but the maxillary palpus is not fused with the maxillary body.

****

Bionomics: *Sabethes* are forest mosquitoes. The larvae inhabit plant cavities, primarily bamboo and tree holes, but bromeliads reportedly harbour a few species. The larval habitats of most species are unknown. Adults are active during the daylight hours. They are attracted to humans on the ground but are most abundant in the forest canopy. Blood-seeking species of *Sabethes sensu stricto* approach humans slowly and have a predilection for landing on the nose. *Sabethes cyaneus* (*Sabethes sensu stricto*) and *Sa. chloropterus* (*Sabethoides*) are known to exhibit complex courtship and copulatory behaviours (Hancock *et al.*, 1990; Zsemlye *et al.*, 2005). *Sabethes chloropterus* appears to play a role in the transmission of yellow fever virus in Central America. Ilheus and St Louis encephalitis viruses have also been isolated from this species in Panama, and the latter virus has been isolated from *Sa. belisarioi* in Brazil.

****

Principal references: Lane & Cerqueira, 1942 (keys, descriptions, bionomics, distributions); Lane, 1953 (keys, descriptions, distributions); Forattini, 1965b (description, bionomics, distribution); Cova-Garcia *et al.*, 1966 (Venezuela, description, keys to species, literature); Clark-Gil & Darsie, 1983 (Guatemala, keys); Darsie, 1985 (Argentina, keys); Harbach, 1991a (genus groups, comparative morphology); Harbach & Peyton, 1993 (comparative morphology of larval maxillae); Harbach, 1994b (*Sabethinus*, descriptions, keys, bionomics, distributions); Judd, 1996 (morphology, phylogenetic relationships); Harbach & Kitching, 1998 (morphology, phylogenetic relationships); Lorenz *et al.*, 2021 (mitochondrial protein-coding genes, phylogenetic relationships); Muñoz-Gamba *et al.*, 2021 (mtDNA *COI* sequences, phylogenetic relationships); Neves *et al.*, 2024 (Brazil, keys to adult females, bionomics, distributions).

Davismyia Lane & Cerqueira, 1942

Davismyia Lane & Cerqueira, 1942, originally proposed as a subgenus of *Wyeomyia* and currently classified as a subgenus of *Sabethes*, is typified by *Sabethes petrocchiae* (Shannon & del Ponte, 1928), original combination: *Miamyia petrocchiae*. Huang (2002) returned *Davismyia* to its previous subgeneric status in *Wyeomyia* based on a single adult character, antepronotum with a prominent apicomesal angle, but this action has not been accepted because Huang failed to consider the adult and larval characters that are synapomorphic for *Sabethes*. Subgenus abbreviation – *Dav.*

****

Systematics: *Davismyia* is monobasic. It only includes the extant type species, which is recorded from Paraguay and neighbouring areas of Bolivia, northern Argentina and southern Brazil. The affinities of *Davismyia* are unknown.

****

Characteristics: Adults of *Davismyia* are distinguished from the adults of other *Sabethes* by the presence of prealar setae; they possess upper proepisternal setae (distinction from most species of *Sabethes sensu stricto*) and lower mesokatepisternal setae (distinction from *Sabethinus* and some species of *Sabethes sensu stricto*), and lack leg paddles (distinction from most members of *Sabethes sensu stricto*). Females have some white scaling on the ventral side of the midtarsus. Larvae have seta 4-C inserted lateral to seta 1-C, a highly reduced maxillary brush and seta 8-T inserted between seta 7-T and the pleural setal group (setae 9–12-T); seta 1-I is inserted lateral to seta 2-I (distinction from *Peytonulus*, *Sabethes sensu stricto* and *Sabethoides*), seta 10-I is inserted far lateral to seta 13-I, seta 1-VII arises from the integument and is inserted lateral to seta 2-VII (distinction from *Peytonulus*) and a comb plate and an auxiliary seta 4b-X are absent (distinctions from *Sabethinus*). Pupae have seta 5-III as long as seta 5-IV–VI (distinction from all other *Sabethes* except certain species of *Sabethes sensu stricto*), seta 1-VII is present (distinction from *Peytonulus*) and seta 8-VII is dorsal in position (distinction from *Sabethinus*).

****

Bionomics: Little information has been published on the bionomics of *Sa. petrocchiae*. Shannon & del Ponte (1928) collected females while being bitten and speculated that larvae would be found in tree holes. Lane & Causey (1955) described and illustrated the larva and pupa, but did not mention the habitat from which they were collected. Roberts *et al.* (1984) collected females feeding on humans during the daytime at canopy level in a double canopy gallery forest in southeastern Bolivia. *Sabethes petrocchiae* is not of medical and economic importance.

****

Principal references: Harbach & Peyton, 1991 (as a subgenus of *Sabethes*, taxonomy, description, bionomics); Harbach, 1991a (comparative morphology); Neves *et al.*, 2024 (Brazil, in keys to adult females, bionomics, distribution).

Peytonulus Harbach, 1991

Peytonulus Harbach, 1991a, originally proposed and currently recognized as a subgenus of *Sabethes*, is typified by *Sabethes aurescens* (Lutz, 1905), original combination: *Sabethinus aurescens*. Subgenus abbreviation – *Pey.*

****

Systematics: *Peytonulus* currently includes 13 extant species, which occur in tropical forests of Central and South America. *Peytonulus* exhibits the most divergent morphology of any group of *Sabethes*, but its phyletic affinities are unknown. Species included in *Peytonulus* were formerly placed in *Sabethinus*, but these subgenera have very little in common other than overt similarities in the ornamentation of the adults. Based on character agreement, *Peytonulus* appears to be more closely related to *Sabethoides* and *Davismyia*. These three groups, however, differ markedly in all life stages. The presence of prealar setae in the adults of *Davismyia* is particularly noteworthy. Prealar setae are plesiomorphic for *Sabethes* and symplesiomorphic with *Wyeomyia*. *Peytonulus* was placed in a sister relationship to *Sabethoides* in the phylogenetic analysis of Judd (1996) based on morphological data, which also included data for species of *Sabethes sensu stricto*.

****

Characteristics: The absence of prealar setae distinguishes the adults of *Peytonulus* from the adults of *Davismyia*; the presence of upper proepisternal setae and/or the absence of paddles of scales on the legs distinguish them from species of *Sabethes sensu stricto*; the presence of lower mesokatepisternal setae distinguishes them from the adults of *Sabethinus*; and the absence of white scaling on the midtarsus distinguishes them from the adults of *Sabethoides*. *Peytonulus* larvae are easily distinguished from all other *Sabethes* larvae by having seta 1-VII highly modified and borne on a strong tubercle. Larvae also have seta 8-T inserted anterior or anteroventral to seta 9-T (distinction from *Davismyia*, *Sabethes sensu stricto* and *Sabethoides*), seta 10-I inserted in line with or mesal (usually) to seta 13-I (distinction from *Davismyia*, *Sabethes sensu stricto* and *Sabethinus*), and lack a comb plate and an auxiliary seta 4b-X (distinctions from *Sabethinus*). *Peytonulus* pupae are easily recognized by the absence of seta 1-VII (this seta is present in all other *Sabethes*).

****

Bionomics: The immature stages of *Peytonulus* most often develop in water contained in uncut bamboo internodes, but they are sometimes found in cut and broken bamboo, at various heights above ground level. The holotype of *Sa. paradoxus* (synonym of *Sa. shannoni*) was collected as a larva found in a terrestrial bromeliad. Nothing is known about the bionomics of the adults. The species of *Peytonulus* are not known to be of medical or economic importance.

****

Principal references: Lane & Cerqueira, 1942 (as species of *Sabethinus*, keys, descriptions, bionomics, distributions); Lane, 1953 (as species of *Sabethinus*, keys, descriptions, distributions); Harbach, 1991a (new subgenus, taxonomy, comparative morphology); Judd, 1996 (morphology, phylogenetic relationships); Neves *et al.*, 2024 (Brazil, keys to adult females, bionomics, distributions).

Sabethes Robineau-Desvoidy, 1827 *sensu stricto*

Sabethes sensu stricto, the nominotypical subgenus of *Sabethes*, is typified by *Sabethes locuples* Robineau-Desvoidy, 1827 [subjective synonym of *Sabethes cyaneus* (Fabricius, 1805)]. Subgenus abbreviation – *Sab.*

****

Systematics: *Sabethes sensu stricto* currently includes 18 extant species, which occur in tropical forests in Central and South America. The phyletic affinities of *Sabethes sensu stricto* are uncertain. In the absence of *Davismyia* and *Sabethinus*, it was placed in a sister relationship to *Peytonulus* + *Sabethoides* in the phylogenetic analyses of morphological data conducted by Judd (1996).

****

Characteristics: Species of *Sabethes sensu stricto* are the *Morpho* butterflies of the mosquito fauna. Most species of the group have 'paddles' of outstanding scales on some or all of the legs, and the absence of upper proepisternal setae (except in *Sa. batesi*, which has leg paddles) distinguishes the adults from species of the other groups of *Sabethes*. The absence of prealar setae distinguishes them from the adults of *Davismyia*, and the absence of ventral white scaling on the midtarsus distinguishes them from species of *Sabethoides*. Larvae of *Sabethes sensu stricto* are distinguished from other *Sabethes* larvae by having seta 8-T inserted dorsal or anterodorsal to seta 9-T (distinction from *Peytonulus*); seta 1-I inserted mesal to seta 1-I (distinction from *Davismyia* and *Sabethinus*); seta 10-I usually inserted lateral (sometimes in line with) seta 13-I (distinction from *Peytonulus* and *Sabethoides*); seta 1-VII unmodified and not borne on a tubercle (distinction from *Peytonulus*); absence of a comb plate and an auxiliary seta 4b-X (distinctions from *Sabethinus*). Pupae of *Sabethes sensu stricto* are not readily distinguished from pupae of *Davismyia* and *Sabethoides*. The presence of seta 1-VII distinguishes them from pupae of *Peytonulus* and the ventral placement of seta 8-VII distinguishes them from pupae of *Sabethinus*.

****

Bionomics: Species of *Sabethes sensu stricto* are diurnal, principally canopy-dwelling mosquitoes. The principal hosts of females are unknown, but they are attracted to humans on the ground and have a predilection for landing on the nose or around the head. Adults of *Sa. amazonicus*, *Sa. cyaneus*, *Sa. quasicyaneus* and *Sa. schnusei* were captured during human-bait collections made at ground level and on platforms in forest canopy in Peru (Harbach, 1992; Pecor *et al.*, 2000). *Sabethes cyaneus* exhibits a complex courtship and copulatory behaviour (Hancock *et al.*, 1990), and it is likely that other species of the group exhibit similar behaviour. Larvae of *Sa. gymnothorax* are apparently predacious as they attack and consume larvae of other species in the laboratory. Larvae of this species took 4–5 weeks to develop at 25°C under laboratory conditions (Harbach & Petersen, 1992). *Sabethes cyaneus* is a facultative predator that captures prey by bending the abdomen and using the siphon to move the prey toward the head to be grasped between the toothed maxillae (Hancock *et al.*, 2022). The larval habitats of most species are unknown. Komp (1936a) found larvae of *Sa. cyaneus* on several occasions deep in the slender stumps of a species of palm. Pecor *et al.* (2000) collected larvae of *Sa. cyaneus* and *Sa. quasicyaneus* in Peru, but did not mention the habitat (probably bamboo internodes) where they were found. Dyar (1928) mentioned a specimen of *Sa. cyaneus* bred from a tree hole in Panama, and suggested that larvae of this species are probably predacious upon other tree-hole species. Larvae of *Sa. bipartipes* have been found in tree holes, where they are said to be predacious (Dyar, 1928), and fallen banana leaves (Shannon, 1931). The type series of *Sa. shannoni* was bred from larvae collected from a bromeliad, 'bananeira brava' (Bird-of-paradise) and a tree stump (Cerqueira, 1961). St Louis encephalitis virus has been isolated from *Sa. belisarioi* in Brazil, but species of *Sabethes sensu stricto* are not known to transmit this or other viruses to humans.

****

Principal references: Lane & Cerqueira, 1942 (keys, descriptions, bionomics, distributions); Lane, 1953 (keys, descriptions, distributions); Harbach, 1991a (taxonomy, comparative morphology); Judd, 1996 (morphology, phylogenetic relationships); Neves *et al.*, 2024 (Brazil, keys to adult females, bionomics, distributions).

Sabethinus Lutz, 1904

Sabethinus, originally proposed as a genus by Lutz, 1904 (in Bourroul, 1904), currently classified as a subgenus of *Sabethes*, is typified by *Sabethes intermedius* (Lutz, 1904) (in Bourroul, 1904), original combination: *Sabethinus intermedius*. Subgenus abbreviation – *Sbn.*

****

Systematics: *Sabethinus* currently includes six extant species, which occur in tropical forests in Central and South America. The phyletic affinities of *Sabethinus* and the evolutionary relationships of the included species are unknown.

****

Characteristics: Adults of *Sabethinus* lack leg paddles (distinction from most species of *Sabethes sensu stricto*) and lower mesokatepisternal setae (distinction from the other groups of *Sabethes* except some species of *Sabethes sensu stricto*). The absence of prealar setae distinguishes them from the adults of *Davismyia*, and the absence of ventral white scaling on the midtarsus distinguishes them from species of *Sabethoides*. *Sabethinus* larvae are readily distinguished from all other *Sabethes* larvae by the presence of a comb plate and an auxiliary seta 4b-X. The insertion of seta 1-I lateral to seta 2-I in species of *Sabethinus* is only shared with *Davismyia*. The strong development and dorsal position of seta 8-VII readily distinguishes pupae of *Sabethinus* from those belonging to the other groups of *Sabethes*.

****

Bionomics: Little bionomical information is available for species of *Sabethinus*. The species are all sylvan. Adults are diurnal and females are attracted to humans on the ground and in forest canopy. They are also known to approach horses and birds. Larvae inhabit uncut and cut bamboo internodes. Species of *Sabethinus* are not known to be of medical importance.

****

Principal references: Lane & Cerqueira, 1942 (including species of *Peytonulus*, keys, descriptions, bionomics, distributions); Lane, 1953 (including species of *Peytonulus*, keys, descriptions, distributions); Harbach, 1991a (taxonomy, comparative morphology); Harbach, 1994b (taxonomy, descriptions, bionomics, distributions); Neves *et al.*, 2024 (Brazil, keys to adult females, bionomics, distributions).

Sabethoides Theobald, 1903

Sabethoides, originally proposed as a genus by Theobald (1903a), currently classified as a subgenus of *Sabethes*, is typified by *Sabethoides confusus* Theobald, 1903a [subjective synonym of *Sabethes chloropterus* (von Humboldt, 1819)], original combination: *Sabethoides confusus*. Subgenus abbreviation – *Sbo.*

****

Systematics: *Sabethoides* currently includes four extant species, which occur in tropical forests in Central and South America. The phyletic affinities of *Sabethoides* are uncertain. Based on character agreement, *Sabethoides* appears to be more closely related to *Sabethinus* and *Davismyia* than to *Peytonulus* and *Sabethes sensu stricto*. The three groups, however, differ markedly in all life stages. The presence of prealar setae in the adults of *Davismyia* is particularly noteworthy. Prealar setae are plesiomorphic for *Sabethes* and symplesiomorphic with *Wyeomyia*. *Sabethoides* and *Peytonulus* were revolved as sister taxa in the phylogenetic analysis of Judd (1996) based on morphological data, which also included data for species *Sabethes sensu stricto*.

****

Characteristics: The adults of *Sabethoides* differ from those of the other groups of *Sabethes* by having the ventral surface of the midtarsi largely white-scaled. The presence of upper proepisternal setae and the absence of leg paddles distinguish them from most species of *Sabethes sensu stricto*. They lack prealar setae (distinction from *Davismyia*) and possess lower mesokatepisternal setae (distinction from *Sabethinus*). Larvae of *Sabethoides* have seta 1-I inserted mesal to seta 2-I (distinction from *Davismyia* and *Sabethinus*); seta 10-I usually inserted mesal (sometimes in line with) seta 13-I (distinction from *Davismyia* and most species of *Sabethes sensu stricto* and *Sabethinus*); seta 1-VII unmodified and not borne on a tubercle (distinction from *Peytonulus*); and they lack a comb plate and an auxiliary seta 4b-X (distinctions from *Sabethinus*). Pupae of *Sabethoides* are not readily distinguished from pupae of *Davismyia* and *Sabethes sensu stricto*. The presence of seta 1-VII distinguishes them from pupae of *Peytonulus* and the ventral placement of seta 8-VII distinguishes them from pupae of *Sabethinus*.

****

Bionomics: Mosquitoes of *Sabethoides* are primarily arboreal. Except for *Sa. chloropterus*, little bionomical information is available for species of the group. Females seek blood meals in greatest numbers during the early afternoon, which corresponds to the rest period of most arboreal mammalian hosts. *Sabethes chloropterus* is stenogamous, with mating occurring on a surface following a brief period of prenuptial activity by the male (Zsemlye *et al.*, 2005). Females of this species oviposit by hovering in front of a small opening and forcibly ejecting one or two eggs into the phytotelm habitat (Galindo, 1958). Larvae develop in plant cavities and rot holes with small openings, but virtually nothing else is known about their bionomics in the wild. *Sabethes chloropterus* appears to play a role in the maintenance and transmission of yellow fever virus in Central America. Ilheus and St Louis encephalitis viruses have also been isolated from this species in Panama.

****

Principal references: Lane & Cerqueira, 1942 (keys, descriptions, bionomics, distributions); Lane, 1953 (keys, descriptions, distributions); Harbach, 1991a (taxonomy, comparative morphology); Harbach, 1991b (*Sa. chloropterus*, larval ontogeny); Judd, 1996 (morphology, phylogenetic relationships); Neves *et al.*, 2024 (Brazil, keys to adult females, bionomics, distributions).

Shannoniana Lane & Cerqueira, 1942

Shannoniana Lane & Cerqueira, 1942, originally proposed as a subgenus of *Trichoprosopon* and currently recognized as a genus, is typified by *Shannoniana fluviatilis* (Theobald, 1903a), original combination: *Goeldia fluviatilis*. Genus abbreviation – *Sh.*

****

Systematics: *Shannoniana* is a subordinate taxon of the subfamily Culicinae, tribe Sabethini. It includes only four extant species with distributions in South and Central America. *Shannoniana* shares close affinities with other New World genera of Sabethini. It was recovered as an independent lineage between *Johnbelkinia* and *Ctenogoeldia* + *Isostomyia* in the study of Judd (1996) and as the sister of *Isostomyia* in the studies of Harbach & Kitching (1998), Harbach & Peyton (2000) and Harbach *et al.* (2007a), all based on phylogenetic analyses of morphological data. The evolutionary relationships of the species of *Shannoniana* are unknown.

****

Characteristics: The adults of *Shannoniana* are distinguished from other members of Sabethini in the New World by the following combination of characters: proboscis about as long as the forefemur, a broad complete or incomplete postmedian pale band on the hindtibia (unique to *Shannoniana*), lower mesokatepisternal setae extend above the lower edge of the mesepimeron (distinction from *Johnbelkinia*, *Runchomyia* and *Isostomyia*), postprocoxal scales present and the laterotergite of abdominal segment I is densely and completely covered with scales (distinctions from *Trichoprosopon*). *Shannoniana* larvae are recognized by the following combination of characters: normal circular occipital foramen with a distinct collar (distinguishes *Shannoniana* from *Isostomyia*, *Limatus*, *Phoniomyia*, *Sabethes*, *Runchomyia* and most *Wyeomyia*), maxillary claw present and maxillary suture incomplete or absent (distinctions from *Onirion*, *Trichoprosopon* and *Wyeomyia*), and siphon without a line of setae on the posterior midline (distinction from *Johnbelkinia*). The presence of a maxillary claw distinguishes *Shannoniana* from all New World genera of Sabethini except *Isostomyia*, and the unusually strong articulation of the maxilla with the head capsule far ventrad of the maxillary palpus is known elsewhere only in *Rachisoura*, currently classified as a subgenus of *Tripteroides*.

****

Bionomics: Little is known about the bionomics of species of *Shannoniana*. Larvae have been collected from bamboo and the leaf axils of aroids. Females have been captured while landing-feeding on a horse and humans in forest during the daytime. Species of *Shannoniana* are unlikely to be of medical or economic importance.

****

Principal references: Lane & Cerqueira, 1942 (new subgenus of *Trichoprosopon*, in keys, descriptions, bionomics, distributions); Lane, 1953 (as a subgenus of *Trichoprosopon*, in keys, descriptions, distributions); Zavortink, 1979 (genus as currently defined, keys to genus groups); Clark-Gil & Darsie, 1983 (Guatemala, keys); Harbach & Peyton, 1993 (comparative morphology of larval maxillae); Judd, 1996 (morphology, phylogenetic relationships); Harbach & Kitching, 1998 (morphology, phylogenetic relationships); Harbach & Peyton, 2000 (morphology, phylogenetic relationships); Harbach *et al.*, 2007a (morphology, phylogenetic relationships).

Topomyia Leicester, 1908

Topomyia Leicester, 1908, originally proposed and currently recognized as a genus, is typified by *Topomyia minor* Leicester, 1908. Genus abbreviation – *To.*

****

Systematics: *Topomyia* is a subordinate taxon of the subfamily Culicinae, tribe Sabethini. It includes 69 extant species. Two species are without subgeneric placement (Rattanarithikul *et al.*, 2007; Harbach & Culverwell, 2014) and the others are classified in three subgenera: *Miyagiella*, *Suaymyia* and *Topomyia sensu stricto*. Species of *Topomyia* occur in the Oriental Region, with extensions into the southern Palaearctic (Ryukyu Archipelago) and the Australasian Region (Sulawesi, Maluku Islands, Biak Island and New Guinea). *Topomyia* formed a sister-group relationship with *Malaya* in the phylogenetic analyses of Judd (1996), Harbach & Kitching (1998) and Harbach & Peyton (2000) based on morphological data. It was placed as the sister of *Limatus* when *Kimia* was distinguished from *Topomyia* by Harbach *et al.* (2007a) and included in the data set of Harbach & Peyton (2000). The internal phylogeny of the genus has not been investigated.

****

Characteristics: *Topomyia* are relatively small, attractive mosquitoes with a median longitudinal stripe of broad white or silver (occasionally brownish) scales on the scutum. This character, along with the absence of setae or hair-like scales on the upper calypter and the termination of the anal vein (vein 1A) before or slightly beyond the base of the mediocubital crossvein, distinguishes the adults from the adults of all other Old World genera except *Kimia* and *Malaya*. Despite the overall similarity, the unmodified proboscis of *Topomyia* could never be confused with the setose, bent and apically swollen proboscis of *Malaya*. In the areas where *Topomyia* occur, the larvae are distinguished from the larvae of all other genus groups based on the following characters: presence of a single pair of seta 4-X (distinction from non-sabethine genera); presence of a slit-like occipital foramen and absence of seta 13-P (distinctions from *Tripteroides*); presence of seta 8-M and seta 13-T removed from the plate bearing setae 9–12-T (distinctions from *Kimia*); maxillary palpus freely attached to the maxillary body and seta 4-X branched rather than single (distinctions from *Malaya*).

****

Bionomics: Little is known about the bionomics of species of *Topomyia*. The species are associated with forests. The immature stages have been found in leaf axils (*e.g. Allocasia*, *Ananas*, *Colocasia*, *Cordyline*, *Crinum*, *Curcuma*, *Donax*, *Heliconia*, *Homalomena*, *Hymenocollis*, *Nipa*, *Pandanus*, *Phrynium*, taro and wild banana), inflorescences (such as *Cystandra*, *Rafflesia* and Zingiberaceae), *Nepenthes* pitcher plants, bamboo internodes and stumps, and tree holes (rarely). The larvae of at least some species are predacious, with well-developed maxillae. Miyagi & Toma (2010a) noted that most species of *Suaymyia* are browsers and predators that normally occur individually in bamboo internodes, bamboo stumps or leaf axils of various plants; and that most species of *Topomyia sensu stricto* are not predacious and occur in leaf axils with many larvae of the same or different species. The feeding habits of the adults are unknown. Females do not feed on humans. Adults have been collected resting on tree trunks in deeply shaded forest. Species of *Topomyia* do not appear to be of medical or economic importance.

****

Principal references: Thurman, 1959 (Thailand, descriptions, keys, bionomics); Marks, 1971 (metapleural setae and scales); Tanaka *et al.*, 1979 (Japan, *To. yanbarensis*, descriptions, distributions, bionomics); Lee *et al.*, 1988c (Australasian Region, taxonomy, literature, bionomics, distributions); Harbach & Peyton, 1993 (comparative morphology of larval maxillae); Lu *et al.*, 1997a (China, keys, taxonomy, descriptions, distributions, bionomics); Harbach *et al.*, 2007a (comparative morphology, phylogenetic relationships); Rattanarithikul *et al.*, 2007 (Thailand, description, classification, bionomics, keys); Harbach & Culverwell, 2014 (*Miyagiella*, descriptions, comparative morphology, bionomics, distribution).

Miyagiella Harbach, 2014

Miyagiella Harbach, 2014 (in Harbach & Culverwell, 2014), originally proposed and currently recognized as a subgenus of *Topomyia*, is typified by *Topomyia discors* Harbach, 2014 (in Harbach & Culverwell, 2014). Subgenus abbreviation – *Myg.*

****

Systematics: *Miyagiella* is monobasic. It only includes the extant type species, which is known only from the holotype male collected in Sabah, Malaysian Borneo. The phylogenetic affinities of *Miyagiella* have not been studied. The male shares a number of features with the males of *Suaymyia* and *Topomyia sensu stricto*, but its affinities with these groups cannot be determined until all species and their life stages are known and have been thoroughly studied.

****

Characteristics: The subgenera of *Topomyia* are currently defined solely by characters of the forelegs and genitalia of males (Thurman, 1959); however, as shown by Somboon *et al.* (2023c), the subgeneric placement of species based on the length of foretarsomeres 2 and 3 is problematic. That aside, *Miyagiella* is clearly a distinct lineage based on the unique development and combination of the following characters: larger foreunguis of males toothed (unique); tergum IX broad, without setae (unique); sternum IX without scales (unique), setae only; gonocoxite with a few inconspicuous scales (unique), gonostylus attached to the gonocoxite subapically (distinction from *Suaymyia* and most species of *Topomyia sensu stricto*); gonostylar claw absent (unique); claspette without a dorsal lobe (distinction from *Topomyia sensu stricto*); aedeagus bulbous (unique); tergum X a narrow sclerite joined to tergum IX (unique); paraproct a stout process of tergum IX, without apical teeth (unique); cercal setae absent (distinction from *Topomyia sensu stricto* and some species of *Suaymyia*).

****

Bionomics: Nothing is known about the bionomics of this taxon. The holotype male of *To. discors* was captured in a light trap at the edge of a stand of bamboo at the side of an agricultural field. The immature stages of this species probably develop in bamboo internodes. The species is not of medical or economic importance.

****

Principal reference: Harbach & Culverwell, 2014 (new subgenus, descriptions, comparative morphology, bionomics, distribution).

Suaymyia Thurman, 1959

Suaymyia Thurman, 1959, originally proposed and currently recognized as a subgenus of *Topomyia*, is typified by *Topomyia cristata* Thurman, 1959. Subgenus abbreviation – *Sua.*

****

Systematics: *Suaymyia* includes 21 extant species recorded from eastern areas of the Oriental Region (Cambodia, China, Indonesia, Malaysia, Philippines, Thailand), one of which (*To. cristata*) is also reported from India, another (*To. yanbarensis*) that is also known to occur in the Ryukyu Islands of Japan and a third species (*To. papuensis*) is also recorded from northeastern islands of the Australasian Region, including New Guinea and the Maluku Islands (Ambon, Seram, Halmahera) of Indonesia. Two species originally included in *Suaymyia*, *To. spathulirostris* and *To. yanbarensis*, are not currently placed in a subgenus (Rattanarithikul *et al.*, 2007; Harbach & Culverwell, 2014). The phylogeny and phylogenetic relationships of the species and subgenera of *Topomyia* have not been investigated.

****

Characteristics: Thurman (1959) defined the subgenera of *Topomyia* based on characters of the forelegs and genitalia of males, but as noted by Lee *et al.* (1988c), the Australasian species cannot be placed with certainty in either *Suaymyia* or *Topomyia sensu stricto* based on those characters, and as shown by Somboon *et al.* (2023c), the subgeneric placement of species based on characteristics of the forelegs is unreliable. Consequently, the following diagnosis may not apply to all species currently included in *Suaymyia*, which is unlikely to be a monophyletic group, as indicated by the recognition of *Kimia* for five species previously included in the subgenus (Harbach *et al.*, 2007a). Diagnosis: ungues equal and simple in both sexes (distinction from males of *Miyagiella*); tergum IX narrow, with setae (unique); sternum IX with scales (distinction from *Miyagiella*) and usually setae; gonocoxite with conspicuous scales (distinction from *Miyagiella*), gonostylus attached at the apex of the gonocoxite (distinction from *Miyagiella* and some species of *Topomyia sensu stricto*), gonostylar claw present (distinction from *Miyagiella*); claspette without a dorsal lobe (distinction from *Topomyia sensu stricto*); aedeagus elongate (distinction from *Miyagiella*); tergum X not evident (distinction from *Miyagiella*); paraproct developed as an independent slender sclerite with apical teeth (unique); cercal setae present or absent.

****

Bionomics: Little is known about the biology of species of *Suaymyia*. The larvae of most species are browsers and predators that normally develop individually in bamboo internodes, bamboo stumps or leaf axils of various plants (Miyagi & Toma, 2010a). The feeding habits of the adults are unknown, but females do not feed on humans. Adults have been found resting on tree trunks in totally shaded places. Species of *Suaymyia* are of no medical or economic importance to humans.

****

Principal references: Thurman, 1959b (Thailand, new subgenus, descriptions, keys, bionomics); Tanaka *et al.*, 1979 (Japan, *To. yanbarensis*, distribution, bionomics); Lee *et al.*, 1988c (Australasian Region, taxonomy, literature, bionomics, distributions); Harbach & Peyton, 1993 (comparative morphology of larval maxillae); Lu *et al.*, 1997a (China, keys, taxonomy, descriptions, distributions, bionomics); Rattanarithikul *et al.*, 2007 (Thailand, bionomics, keys); Harbach & Culverwell, 2014 (comparative morphology).

Topomyia Leicester, 1908 *sensu stricto*

Topomyia sensu stricto, the nominotypical subgenus of *Topomyia*, is typified by *Topomyia minor* Leicester, 1908. Subgenus abbreviation – *Top.*

****

Systematics: *Topomyia sensu stricto* includes 45 extant species. Most species are recorded from southeastern areas of the Oriental Region (Cambodia, China, Laos, Indonesia, Malaysia, Philippines, Thailand, Vietnam). *Topomyia aureoventer*, *To. bifurcata* and *To. hirtusa* are also recorded from India (*To. aureoventer* also from Nepal). *Topomyia irianensis* is the only formally described species of *Topomyia sensu stricto* with a distribution in the Australasian Region, including Papua New Guinea and Biak Island, and the Maluku Islands (Ambon, Seram, Halmahera) of Indonesia. The phylogeny and phylogenetic relationships of the species and subgenera of *Topomyia* have not been investigated.

****

Characteristics: Characters of the forelegs and genitalia of males used by Thurman (1959) to place species in *Topomyia sensu stricto* are unreliable (Lee *et al.*, 1988c; Somboon *et al.*, 2023c); consequently, the following diagnosis may not apply to all species currently included in the group, which may not be monophyletic. Diagnosis: ungues equal and simple in both sexes (distinction from males of *Miyagiella*); tergum IX broad, with setae (unique); sternum IX with scales (distinction from *Miyagiella*) and setae; gonocoxite with conspicuous scales (distinction from *Miyagiella*), gonostylus usually attached apically (distinction from *Miyagiella*), attached subapically in *To. gracilis*, *To. hardini*, *To. minor*, *To. pilosa* and *To. winter*; gonostylar claw present (distinction from *Miyagiella*); claspette with a dorsal lobe (unique); aedeagus elongate (distinction from *Miyagiella*); tergum X not evident (distinction from *Miyagiella*); paraproct developed as an independent elongate sclerite with a pointed apex (unique); cercal setae present (distinction from *Miyagiella* and some species of *Suaymyia*).

****

Bionomics: Little bionomical information is available for species of *Topomyia sensu stricto*. The species are associated with forests. The larvae of several species are apparently predatory but the larvae of most species are browsers that develop in leaf axils with other larvae of the same or different species (Miyagi & Toma, 2010a). The feeding habits of the adults are unknown. Females, if they are blood feeders, do not feed on humans. Adults are known to rest in deeply shaded places in forest. Species of *Topomyia sensu stricto* are of no medical or economic importance to humans.

****

Principal references: Thurman, 1959 (Thailand, descriptions, keys, bionomics); Lee *et al.*, 1988c (Australasian Region, taxonomy, literature, bionomics, distributions); Harbach & Peyton, 1993 (comparative morphology of larval maxillae); Lu *et al.*, 1997a (China, keys, taxonomy, descriptions, distributions, bionomics); Rattanarithikul *et al.*, 2007 (Thailand, bionomics, keys); Harbach & Culverwell, 2014 (comparative morphology).

Toxorhynchites Theobald, 1901

Toxorhynchites Theobald, 1901a, originally proposed and currently recognized as a genus, is typified by *Toxorhynchites brevipalpis* Theobald, 1901b (International Commission on Zoological Nomenclature, 1959b: Opinion 548). Genus abbreviation – *Tx.*

****

Systematics: *Toxorhynchites* is a subordinate taxon of the subfamily Culicinae, tribe Toxorhynchitini. It includes 101 currently recognized extant species classified in four subgenera: *Afrorhynchus*, *Ankylorhynchus*, *Lynchiella* and *Toxorhynchites sensu stricto*. A single fossil species, *Tx. mexicanus* Zavortink & Poinar, 2008, placed in the nominate subgenus, is known from the Cenozoic (Mexican amber, Early Miocene to Late Oligocene, 16.4–28.5 Mya). Extant species of *Toxorhynchites* are largely confined to tropical regions, but a few species occur in the eastern fringes of Asia and North America. Three apparently introduced species are found in Fiji and Samoa in the South Pacific, but species of *Toxorhynchites* are otherwise absent from all oceanic islands, including New Zealand. Species of *Toxorhynchites sensu stricto*, with the exception of the fossil species, occur in the Old World, those of *Afrorhynchus* occur in Africa, and species of *Ankylorhynchus* and *Lynchiella* are confined to the New World. *Toxorhynchites* was traditionally placed in a separate subfamily. Belkin (1962) believed that this over-emphasized the peculiar developments found in species of the group and regarded it as constituting one of 12 tribes of 'true mosquitoes'. He was convinced that most of the distinctive features of *Toxorhynchites* are secondary adaptive characters related to the large size and feeding habits of the adults and larvae. Species of *Toxorhynchites* share the un-lobed condition of the scutellum and the presence of a precubital furrow on the wing with *Anopheles* and *Bironella*, characters considered to be symplesiomorphic by Colless (1979). Harbach (1977) discussed possible steps in the evolution of feeding behaviour in mosquito larvae and postulated that *Toxorhynchites* and the Anophelinae were derived from an early common ancestor of the Culicidae. In many respects, the adults resemble sabethine mosquitoes in overall ornamentation, reduction of the thoracic chaetotaxy, lack of scales on the alula and upper calypter of the wing, and structures of the male genitalia. Based on characteristics of the larval labiohypopharynx, Harbach (1978) hypothesized that *Toxorhynchites* was derived from a sabethine ancestor. Features of the larvae, including the absence of a comb and the placement of setae on common tubercles, are paralleled in certain sabethines, *e.g. Trichoprosopon*. The lateral abdominal setae of some *Eretmapodites* larvae, as well as those of *Armigeres* (*Leicesteria*) *dolichocephalus*, are also borne on conspicuous common tubercles. *Toxorhynchites* also share with Sabethini the absence of a tracheoid area at the base of the pupal trumpet, the position of setae 2,3-CT and the absence of setae at the apex of the paddle. Belkin believed that the similarities between *Toxorhynchites* and sabethines were indicative of a close relationship rather than the result of convergence. Whereas *Toxorhynchites* was paired with *Culiseta* in the implied weighting analysis of Harbach & Kitching (1998) based on morphological data, it was recovered as the sister of Sabethini when all characters in their data set were equally weighted or analysed using successive approximations character weighting. Steffan & Evenhuis (1985) and Ribeiro (2004) discussed similarities and postulated affinities with other Culicidae, but most importantly, the molecular phylogenies of Soghigian *et al.* (2023), based on amino acid sequences of 709 genes, and Chen *et al.* (2024), based on mitogenome sequences, strongly support the sister relationship between *Toxorhynchites* and Sabethini, which overshadows the results of the analysis of mitochondrial genomes conducted by da Silva *et al.* (2020a), wherein a single species of *Toxorhynchites* was recovered in a sister relationship with a clade composed of species of different genera of Mansoniini and Sabethini. However, in view of the results of the analysis of a data set consisting of six nuclear genes and 80 morphological characters revealed in the study of Reidenbach *et al.* (2009), in which *Mimomyia luzonensis* was recovered as the sister to *Tx. amboinensis* + *Aedeomyia squamipennis*, it is possible that different relationships may have been recovered in the studies of Chen *et al.* and Soghigian *et al.* if they had included data for a species of *Mimomyia*.

****

Characteristics: *Toxorhynchites* includes the largest mosquitoes. Adults are easily recognized by their large size and long strongly bent proboscis. The body is covered with brightly coloured iridescent scales and the posterior abdominal segments have lateral scale-tufts. The scutellum is evenly rounded (as in *Anopheles* and *Bironella*) and the posterior margin of the wing is distinctly emarginated opposite the termination of vein CuA. Larvae are very large and easily recognized. They vary from pink to red and purple in colour, and possess modified mouthbrushes comprised of 10 or fewer thick, flattened filaments for grasping prey. The abdominal setae occur in groups of three to five on common sclerites and a comb and pecten are both absent. Species of *Toxorhynchites* are remarkably similar in all life stages throughout the world and are difficult to identify.

****

Bionomics: *Toxorhynchites* are basically forest mosquitoes. The larval habitats are mainly tree holes and bamboo, but a few species are found in leaf axils, pitcher plants, rock pools and artificial containers. The larvae of all species are predacious. They feed mainly on the larvae of other mosquito species, but exhibit cannibalism or feed on detritus in the absence of suitable prey. Males and females both feed exclusively on nectar and other sugary substances. The adults are active during the day. Steffan & Evenhuis (1981) and Collins & Blackwell (2000) provide details on the biology of *Toxorhynchites*. Species of *Toxorhynchites* are not involved in the transmission of human or animal pathogens. The larvae of a few species have been used with some success to control economically important mosquitoes whose larvae inhabit plant cavities and artificial containers.

****

Principal references: Barraud, 1934 (southern Asia, as *Megarhinus*, keys, taxonomy, descriptions, distributions, bionomics); Edwards, 1941 (Afrotropical Region, as *Megarhinus*, keys, adults, pupae, descriptions, distributions); van Someren, 1946b (Afrotropical Region, as *Megarhinus*, classification, descriptions, key to adults); Hopkins, 1952 (Afrotropical Region, keys, descriptions of larvae, distributions, bionomics); Lane, 1953 (Neotropical Region, taxonomy, keys, descriptions, distributions); Vargas, 1953 (North America, as *Megarhinus*, keys); Thurman, 1959 (Thailand, descriptions, keys, bionomics); Belkin, 1962 (South Pacific, taxonomy, keys, descriptions, distributions, bionomics); da Costa Lima *et al.*, 1962 (Neotropical Region, as Toxorhynchitini, descriptions, key to adults); Forattini, 1965b (Neotropical Region, description, bionomics, distribution); Cova-Garcia *et al.*, 1966 (Venezuela, keys, species descriptions, distributions); Belkin *et al.*, 1970 (Jamaica, taxonomy, *Tx. portoricensis*, keys, description, bionomics); Tanaka *et al.*, 1979 (Japan, taxonomy, keys, descriptions, distributions, bionomics); Steffan *et al.*, 1980 (annotated bibliography); Steffan & Evenhuis, 1981 (biology); Clark-Gil & Darsie, 1983 (Guatemala, *Lynchiella*, keys, distributions, bionomics); Darsie, 1985 (Argentina, keys); Steffan & Evenhuis, 1985 (Australasian, eastern Palaearctic and Oriental Regions, *Toxorhynchites sensu stricto*, species groups, descriptions, keys); Evenhuis & Steffan, 1986 (Oriental Region, *Toxorhynchites sensu stricto*, Acaudatus Group, keys, descriptions, distributions, bionomics); Lee *et al.*, 1988c (Australasian Region, *Toxorhynchites sensu stricto*, taxonomy, keys, literature, distributions, bionomics); Service, 1990 (Afrotropical Region, keys, adults, larvae, bionomics, distribution); Ribeiro, 1991 (Afrotropical Region, Brevipalpis Group, classification, keys, descriptions, distributions); Ribeiro, 1992a (*Afrorhynchus*, diagnosis, adults, key to subgenera); Lu *et al.*, 1997a (China, keys, taxonomy, descriptions, distributions, bionomics); Harbach & Kitching, 1998 (morphology, phylogenetic relationships); Collins & Blackwell, 2000 (literature, biology, distributions); Ribeiro, 2004 (Madagascar, *Afrorhynchus*, Pauliani Group, keys, descriptions, diagnoses); Ribeiro, 2005b (Afrotropical Regions, key to adults); Reidenbach *et al.*, 2009 (nuclear genes, morphology, phylogenetic relationships); Tyagi *et al.*, 2015b (southeastern and south-central Asia, keys, comparative morphology, bionomics, distributions, phylogenetic relationships); da Silva *et al.*, 2020a (mitogenomes, phylogenetic relationships); Donald *et al.*, 2020 (bionomics); Soghigian *et al.*, 2023 (molecular markers, phylogenetic relationships); Chen *et al.*, 2024 (mitochondrial protein sequences, phylogenetic relationships).

Afrorhynchus Ribeiro, 1992

Afrorhynchus Ribeiro, 1992a, originally proposed and currently recognized as a subgenus of *Toxorhynchites*, is typified by *Toxorhynchites lutescens* (Theobald, 1901b), original combination: *Megarhinus lutescens*. Subgenus abbreviation – *Afr.*

****

Systematics: *Afrorhynchus* includes 19 extant species in the Afrotropical Region. The distributions of the species are poorly known. Most species are known from only one or two countries. The phylogenetic relationships of *Afrorhynchus* have not been investigated, but see Ribeiro (2004) for comments on affinities.

****

Characteristics: Species of *Toxorhynchites* are extremely similar in all life stages; adult characters are used to characterize and distinguish the subgenera. Species of *Afrorhynchus* are distinguished by the following principal features. ADULTS: Maxillary palpus of females short, about 0.25 the length of the proboscis, apex rounded; antenna of females normal; integument of the scutum mainly blackish, scales greenish; postpronotum and the thoracic pleura mainly or entirely yellow; postpronotum and forecoxa with golden scales, contrasting with the dark greenish scales of the scutum; pleural scaling mainly white, lower mesokatepisternal and prealar scales golden; mesokatepisternum with or without 1 or 2 strong dark setae and some scattered weaker setae; laterotergite of abdominal segment I with or without a few scales; terga II–VIII with golden or purplish scaling on the lateral margins, lateral tufts of terga VI–VIII golden, red or orange; halter with a yellow capitellum; sterna golden-scaled; femora golden-scaled ventrally; hindtibia with a patch of whitish scales on the distal 0.33 of the dorsal surface; hindtarsomere 1 with an indication of a narrow white ring some distance from the base, hindtarsomere 2 entirely dark-scaled, hindtarsomere 3 with a basal white ring; midungues of males small, equal, simple. MALE GENITALIA: Gonocoxite noticeably widened at mid-length, with very long and stout setae on the inner surface; gonostylus dark, gonostylar claw inserted at a right angle to the gonostylus, very long, nearly 0.33 the length of the gonostylus; aedeagus with a wide dorsal aedeagal bridge; paraproct with an unsclerotized transverse band. LARVAE and PUPAE: Poorly studied or unknown, the known larvae of some species are indistinguishable based on currently incomplete descriptions.

****

Bionomics: Larvae of species of *Afrorhynchus* have been found in cut and bored bamboo, leaf axils and tree holes. See Steffan & Evenhuis (1981) and Collins & Blackwell (2000) for details of *Toxorhynchites* biology.

****

Principal references: Edwards, 1941 (as species of *Megarhinus*, Group B *lutescens* group, descriptions of adults of *Tx. aeneus*, *Tx. erythrurus*, *Tx. lutescens* and *Tx. viridibasis*, and pupae of *Tx. aeneus* and *Tx. erythrurus*, keys, bionomics, distributions); Hopkins, 1952 (descriptions of larvae of *Tx. aeneus*, *Tx. kaimosi*, *Tx. nairobiensis*, *Tx. ruwenzori* and *Tx. viridibasis*); Steffan & Evenhuis, 1981 (as a subgenus *Toxorhynchites* in part, *Tx. aeneus*, *Tx. erythrurus*, *Tx. kaimosi*, *Tx. lutescens*, *Tx. nairobiensis*, *Tx. pauliani*, *Tx. ruwenzori* and *Tx. viridibasis*, biology); Service, 1990 (in part, keys and descriptions of adults and larvae of species listed for Steffan & Evenhuis, 1981); Ribeiro, 1992a (new subgenus, description, adult key to the subgenera of *Toxorhynchites*); Ribeiro, 2004 (Madagascar, keys, descriptions, phylogenetic relationships).

Ankylorhynchus Lutz, 1904

Ankylorhynchus, originally proposed as a genus by Lutz, 1904 (in Bourroul, 1904), currently classified as a subgenus of *Toxorhynchites*, is typified by *Toxorhynchites purpureus* (Theobald, 1901b), original combination: *Megarhinus purpureus*. Subgenus abbreviation – *Ank*.

****

Systematics: *Ankylorhynchus* includes four extant species in South America. The distribution of *Ankylorhynchus* is poorly known. The species have only been recorded from localities in Brazil (*Tx. catharinensis*, *Tx. purpureus*, *Tx. trichopygus*), Bolivia (*Tx. hexacis*), French Guiana (*Tx. trichopygus*) and Uruguay (*Tx. purpureus*). The phylogenetic relationships of *Ankylorhynchus* have not been investigated.

****

Characteristics: Species of *Toxorhynchites* are extremely similar in all life stages; adult characters are used to characterize and distinguish the subgenera. Species of *Ankylorhynchus* are distinguished by the following principal features. ADULTS: Maxillary palpus of females long, about as long as the proboscis, apical palpomere acute (pointed) and upturned; antenna of females moderately verticillate; forecoxa with white scales; mesokatepisternum without golden scales; laterotergite of abdominal segment I with a dense covering of scales; midungues of males unequal, larger unguis with a tooth. MALE GENITALIA: Gonostylar claw short, 0.1–0.2 the length of the gonostylus; aedeagus with a narrow dorsal bridge; paraproct without an unsclerotized transverse band. LARVAE and PUPAE: Not well studied, poorly described; unknown for *Tx. hexacis*.

****

Bionomics: Larvae of species of *Ankylorhynchus* have been found in cut and bored bamboo, tree holes and artificial containers. See Steffan & Evenhuis (1981) and Collins & Blackwell (2000) for details of *Toxorhynchites* biology.

****

Principal references: Lane, 1953 (taxonomy, keys, descriptions, distributions); da Costa Lima *et al.*, 1962 (as a genus, descriptions, key to adults); Forattini, 1965b (adult morphology, bionomics); Steffan & Evenhuis, 1981 (biology); Darsie, 1985 (Argentina, adult key includes *Tx. purpureus*).

Lynchiella Lahille, 1904

Lynchiella, originally proposed as a genus by Lahille (1904), currently classified as a subgenus of *Toxorhynchites*, is typified by *Toxorhynchites haemorrhoidalis* (Fabricius, 1787), original combination: *Culex haemorrhoidalis*. Subgenus abbreviation – *Lyn*.

****

Systematics: *Lynchiella* includes 20 extant species, which occur predominantly in the Neotropical Region. The distribution of *Tx. rutilus* extends northward into eastern USA and southeastern Canada. The phylogenetic relationships of *Lynchiella* have not been investigated, but see Steffan & Evenhuis (1985) and Ribeiro (2004) for discussions of similarities and postulated affinities of *Toxorhynchites* with other Culicidae. Three species of *Lynchiella* included in the analysis of amino acid sequences of 709 genes conducted by Soghigian *et al.* (2023) were recovered as a clade in a sister relationship with *Tx.* (*Tox.*) *amboinensis*, the only other species of *Toxorhynchites* included in the study.

****

Characteristics: Species of *Toxorhynchites* are extremely similar in all life stages; adult characters are used to characterize and distinguish the subgenera. Species of *Lynchiella* are distinguished by the following principal features. ADULTS: Maxillary palpus of females 0.7–0.8 the length of the proboscis, apex blunt; antenna of females normal; mesokatepisternum with golden scales; forecoxa white-scaled; laterotergite of abdominal segment I with a dense covering of scales; midungues of males unequal, larger unguis with a tooth. MALE GENITALIA: Gonostylar claw short, 0.1–0.2 the length of the gonostylus; aedeagus with a narrow dorsal aedeagal bridge; paraproct without an unsclerotized transverse band. LARVAE and PUPAE: Poorly studied or unknown, larvae of species for which the larva is known are indistinguishable based on available (largely incomplete) descriptions.

****

Bionomics: The immature stages of species of *Lynchiella* have been found in tree holes, cut and bored bamboo, bromeliads, plant axils, rock holes and artificial containers. All instars of larvae are predacious and mainly feed on larvae of other mosquito species. Females and males feed exclusively on plant secretions. Females are harmless to humans and other vertebrates. See Steffan & Evenhuis (1981) and Collins & Blackwell (2000) for details of *Toxorhynchites* biology.

****

Principal references: Lane, 1953 (taxonomy, keys, descriptions, distributions); da Costa Lima *et al.*, 1962 (as a genus, descriptions, key to adults); Forattini, 1965b (adult morphology, bionomics); Cova-Garcia *et al.*, 1966 (Venezuela, five species, key, distributions, literature); Belkin *et al.*, 1970 (Jamaica and adjacent islands, keys for three species; taxonomy, description, bionomics, distribution of *Tx. portoricensis*); Steffan & Evenhuis, 1981 (biology); Clark-Gil & Darsie, 1983 (Guatemala, *Lynchiella*, key for adults of *Tx. grandiosus*, *Tx. superbus*, as a subspecies of *haemorrhoidalis*, and *Tx. theobaldi*, key for larvae of *Tx. haemorrhoidalis superbus* and *Tx. theobaldi*, distributions, bionomics); Darsie, 1985 (Argentina, keys for adults and larvae).

Toxorhynchites Theobald, 1901 *sensu stricto*

Toxorhynchites sensu stricto, the nominotypical subgenus of *Toxorhynchites*, is typified by *Toxorhynchites brevipalpis* Theobald, 1901b (International Commission on Zoological Nomenclature, 1959b: Opinion 548). Subgenus abbreviation – *Tox*.

****

Systematics: *Toxorhynchites sensu stricto* includes 58 extant species with distributions in the Afrotropical, Australian and Oriental Regions, with extensions into the Manchurian Subregion of the Palaearctic. Interspecific relationships between Oriental and Australasian species of *Toxorhynchites sensu stricto* were explored by Moss *et al.* (1979) and Simon *et al.* (1983) using phenetic and cladistic methods, respectively. Steffan & Evenhuis (1985) proposed an internal classification of *Toxorhynchites sensu stricto* comprised of seven species groups based in large part on morphological similarities recognized by Moss *et al.* and phylogenetic trees recovered in the analyses of Simon *et al.* *Toxorhynchites* (*Tox.*) *amboinensis* was recovered as the sister of a clade comprised of three species of *Lynchiella* included in the analysis of amino acid sequences of 709 genes conducted by Soghigian *et al.* (2023).

****

Characteristics: Species of *Toxorhynchites* are extremely similar in all life stages; adult characters are used to characterize and distinguish the subgenera. Species of *Toxorhynchites sensu stricto* are distinguished by the following principal features. ADULTS: Maxillary palpus of females short, about 0.25 the length of the proboscis, apex blunt; antenna of females normal; mesokatepisternum without golden scales; forecoxa white-scaled; laterotergite of abdominal segment I with a dense covering of scales; midungues of males unequal, larger unguis with a tooth. MALE GENITALIA: Gonostylar claw short, 0.1–0.2 the length of the gonostylus; aedeagus with a narrow dorsal aedeagal bridge; paraproct without an unsclerotized transverse band. LARVAE and PUPAE: Known for many species and well described or illustrated for some, but subgeneric characters are not delineated; pupae seem to be easily distinguished based on distinctive features of the chaetotaxy, trumpet and paddle but larvae exhibit few diagnostic differences.

****

Bionomics: The immature stages of species of *Toxorhynchites sensu stricto* have been found in tree holes, stump holes, bamboo internodes, bamboo stumps, split bamboo, axils of *Allocasia* and *Colocasia*, *Nepenthes* pitchers, rock holes and artificial containers. All instars of larvae are predacious and mainly feed on larvae of other mosquito species. Details of the prey-capture behaviour of *Tx. amboinensis* were examined and described by Hancock *et al.* (2022). Females, like the males, feed exclusively on plant exudates, and are therefore not vectors of human or animal pathogens. See Steffan & Evenhuis (1981) and Collins & Blackwell (2000) for details of *Toxorhynchites* biology.

****

Principal references: Barraud, 1934 (southern Asia, as *Megarhinus*, keys, descriptions, bionomics, distributions); Edwards, 1941 (Afrotropical Region, as *Megarhinus*, also species of *Afrorhynchus*, adults, keys, descriptions, distributions; pupa of *Tx. brevipalpis*); Hopkins, 1952 (Afrotropical Region, also species of *Afrorhynchus*, larvae, descriptions, bionomics); Thurman, 1959 (Thailand, keys, taxonomy, descriptions, bionomics, distributions); Belkin, 1962 (South Pacific, three species, keys, taxonomy, descriptions, bionomics, distributions); Moss *et al.*, 1979 (phenetic similarities of Oriental species); Tanaka *et al.*, 1979 (Japan, three species, keys, taxonomy, descriptions, distributions, bionomics); Steffan & Evenhuis, 1981 (biology); Simon *et al.*, 1983 (phylogenetic relationships of Oriental and Australasian species); Steffan & Evenhuis, 1985 (Australasian, Oriental and eastern Palaearctic Regions, phylogenetic relationships, biogeography, bionomics, taxonomy, morphology, classification, keys); Evenhuis & Steffan, 1986 (Oriental Region, Acaudatus Group, keys, taxonomy, distributions, bionomics); Lee *et al.*, 1988c (Australasian Region, keys, taxonomy, literature, distributions, bionomics); Service, 1990 (Afrotropical Region, also species of *Afrorhynchus*, keys, descriptions, bionomics, distributions); Ribeiro, 1991 (Afrotropical Region, Brevipalpis Group, keys, taxonomy, descriptions, distributions); Lu *et al.*, 1997a (China, keys, taxonomy, descriptions, distributions, bionomics); Rattanarithikul *et al.*, 2007 (Thailand, keys, bionomics, keys); Tyagi *et al.*, 2015b (southeastern and south-central Asia, keys, comparative morphology, bionomics, distributions, phylogenetic relationships).

Trichoprosopon Theobald, 1901

Trichoprosopon Theobald, 1901a, originally proposed and currently recognized as a genus, is typified by *Trichoprosopon nivipes* Theobald, 1901c [subjective synonym of *Trichoprosopon digitatum* (Rondani, 1848)]. Genus abbreviation – *Tr.*

****

Systematics: *Trichoprosopon* is a subordinate taxon of the subfamily Culicinae, tribe Sabethini. It includes 17 extant species with distributions in Central and South America. *Trichoprosopon* is not divided into subgenera. The phyletic affinities of *Trichoprosopon* are not definitely known. It was recovered as the basal clade within a monophyletic assemblage of New World sabethine genus groups in the phylogenetic analysis of Judd (1996) based on morphological data. Its relationships with other New World genera were unresolved in the analysis of Harbach & Kitching (1998), also based on morphological data, and was recovered as the sister of *Johnbelkinia* + *Runchomyia* when *Onirion* was included in their data set (Harbach & Peyton, 2000). When Harbach *et al.* (2007a) included *Kimia* in the latter data set, *Trichoprosopon* was paired with the Old World *Tripteroides* in a sister relationship with *Kimia*. Speculation about a close relationship between *Trichoprosopon* and *Tripteroides* has existed since Lee (1946) could find no clear distinctions to separate them. *Trichoprosopon digitatum* was included in the phylogenetic analysis of amino acid sequences of 709 genes conducted by Soghigian *et al.* (2023), but its relationship with other taxa of Sabethini included in the analysis was not explicitly indicated. *Trichoprosopon*, represented by four species, was not recovered as a monophyletic group in the phylogenetic analysis of mtDNA *COI* gene sequences conducted by Muñoz-Gamba *et al.* (2021). *Trichoprosopon digitatum* was paired with *Sabethes* (*Sabethes*) *cyaneus*, *Tr. compressum* was paired with *Wyeomyia* (*Cruzmyia*) *forattinii* and the other two species, *Tr. evansae* and *Tr. pallidiventer*, were basal and paraphyletic to all other sabethines included in the study. In analyses of 13 mitochondrial protein-coding genes conducted by do Nascimento *et al.* (2021) and Lorenz *et al.* (2021), *Tr. pallidiventer* was paired with *Runchomyia reversa* in a sister relationship to a clade comprised of one species each of *Limatus* and *Wyeomyia* and four species of *Sabethes*.

****

Characteristics: The adults of species of *Trichoprosopon* are distinguished from the adults of other sabethines in the New World by the following combinations of characters: proboscis relatively short, about as long as the forefemur; clypeus of some species with setae; antepronota relatively small and separated; prespiracular and mesopostnotal setae present; postprocoxal scales absent;

lower mesokatepisternal setae extend above the lower edge of the mesepimeron; laterotergite of abdominal segment I with some sparse scales distally. Larvae are recognized by the presence of a circular occipital foramen with a distinct collar, maxilla with a maxillary brush and the cardo fused with the base of the maxillary palpus, a complete hypostomal suture, absence of seta 8-M, and the absence of pecten spines and mid-ventral filaments on the siphon. Larvae have seta 13-T inserted on a common plate with setae 9–12-T, a feature which is only found in species of *Kimia*, *Isostomyia* and *Trichoprosopon* (Harbach *et al.*, 2007a; Campos & Zavortink, 2010).

****

Bionomics: *Trichoprosopon* are basically forest mosquitoes. Larvae are found in small containers of water. They have been collected from bamboo, fallen leaves and spathes, cacao pods, coconut shells, nuts (monkey pods), flower bracts of *Heliconia*, leaf axils, tree holes and artificial containers. Females of a few species are known to feed on humans in shaded areas during the daytime. *Trichoprosopon digitatum* is regarded as a potential vector of arboviruses to humans: Pixuna and Wyeomyia viruses have been isolated from females, and Bussuquara, Ilheus and St Louis encephalitis viruses have been isolated from mixed pools that included the species.

****

Principal references: Lane, 1953 (in part, as subgenus *Trichoprosopon*, keys, species descriptions, distributions); Forattini, 1965b (description, bionomics, distributions); Zavortink, 1979 (genus as currently defined, keys to genus groups); Zavortink, 1981 (internal classification, notes on characteristics, bionomics, distribution); Darsie, 1985 (Argentina, keys); Harbach & Peyton, 1993 (comparative morphology of larval maxillae); Judd, 1996 (morphology, phylogenetic relationships); Harbach & Kitching, 1998 (morphology, phylogenetic relationships); Harbach & Peyton, 2000 (phylogenetic relationships); Harbach *et al.*, 2007a (phylogenetic relationships); do Nascimento *et al.*, 2021 (mitochondrial protein-coding genes, phylogenetic relationships); Lorenz *et al.*, 2021 (mitochondrial protein-coding genes, phylogenetic relationships); Muñoz-Gamba *et al.*, 2021 (mtDNA *COI* sequences, phylogenetic relationships); Rivera-García *et al.*, 2023 (keys to known species, note on internal classification).

Tripteroides Giles, 1904

Tripteroides Giles, 1904, originally proposed and currently recognized as a genus, is typified by *Runchomyia philippinensis* Giles, 1904 [subjective synonym of *Tripteroides nitidoventer* (Giles, 1904)]. Genus abbreviation – *Tp.*

****

Systematics: *Tripteroides* is a subordinate taxon of the subfamily Culicinae, tribe Sabethini. It is a large group of 126 extant species divided between five subgenera: *Polylepidomyia*, *Rachionotomyia*, *Rachisoura*, *Tricholeptomyia* and *Tripteroides sensu stricto*. The species are largely distributed in the Oriental and Australasian Regions. The range of the group extends from India and Sri Lanka to Japan, Taiwan, the Philippines, Fiji in the South Pacific and south to Tasmania. When Harbach *et al.* (2007a) included *Kimia* in the data set of Harbach & Peyton (2000), *Tripteroides* was paired with the New World *Trichoprosopon* in a sister relationship with *Kimia*. Speculation about a close relationship between *Tripteroides* and *Trichoprosopon* has existed since Lee (1946) could find no clear distinctions to separate them. However, *Tripteroides* is a complex and heterogeneous genus group that has always been difficult to characterize. Despite the study of Mattingly (1981), *Tripteroides* remains a confused and apparently polyphyletic assemblage of species (Judd, 1996). One species each of *Polylepidomyia*, *Rachionotomyia* and *Rachisoura* were included in the phylogenetic analysis of amino acid sequences of 709 genes conducted by Soghigian *et al.* (2023), but their relationships with each other and with other genus groups of Sabethini included in the analysis were not explicitly indicated. Until the included species are sorted into monophyletic groups, it will not be possible to know the affinities of *Tripteroides* or its subordinate groups with certainty.

****

Characteristics: Species of *Tripteroides* are distinguished from species of all other genus groups that occur within its distribution based on one or more of the following characters in the adult and larval stages. ADULTS: Erect scales of the head restricted to the occiput, acrostichal setae absent from the median longitudinal area of the scutum, scutellum trilobed, prespiracular setae present, upper calypter of the wing with marginal setae and the anal vein (vein 1A) ends beyond the base of the mediocubital crossvein. LARVAE: Seta 3-C ventral, occipital foramen circular with a distinct collar, seta 13-P present, comb scales in a single row and a single pair of seta 4-X.

****

Bionomics: *Tripteroides* includes domestic and forest species. The larvae inhabit small collections of water in tree holes, bamboo, coconut shells and husks, fallen leaves, leaf axils, flower bracts, pitcher plants and snail shells. A few species occur in artificial containers, and *Tr. tasmaniensis* has been collected in rock holes. Larvae usually lie upside down on the bottom of the cavity. They feed on arthropods or their remains. Larvae of *Rachisoura* are predacious. The adults appear to be active during the daytime.

A few species attack and bite humans, but nothing is known about the feeding habits of most species. Among the Oriental species only *Tr. aranoides* is known to feed on humans. Some species have become pests in northwestern Australia where they invade houses. No species of *Tripteroides* are known to be involved in the transmission of pathogenic agents.

****

Principal references: Barraud, 1934 (southern Asia, keys, descriptions, bionomics, distributions); Lee, 1946 (Australasian Region, excluding *Maorigoeldia*, classification, keys, descriptions, bionomics, distributions); Baisas & Ubaldo-Pagayon, 1953 (Philippines, distributions, bionomics, classification, keys, descriptions); Peters, 1959 (New Guinea, descriptive notes, keys); Thurman, 1959 (Thailand, keys, taxonomy, descriptions, bionomics, distributions); Belkin, 1962 (South Pacific, keys, taxonomy, genus, subgenus and descriptions, bionomics, distributions); Peters, 1963c (New Guinea, taxonomy, new species descriptions, key to pupae of known species); Tanaka *et al.*, 1979 (Japan, genus, *Tripteroides sensu stricto* and *Tr. bambusa*, descriptions, distributions, bionomics); Mattingly, 1980 (classification); Mattingly, 1981 (Oriental Region, *Rachionotomyia*, *Tricholeptomyia*, keys, descriptions, distributions, bionomics); Lee *et al.*, 1989c (Australasian Region, diagnoses, keys, literature, bionomics); Harbach & Peyton, 1993 (comparative morphology of larval maxillae); Judd, 1996 (morphology, phylogenetic relationships); Lu *et al.*, 1997a (China, keys, taxonomy, descriptions, distributions, bionomics); Harbach & Kitching, 1998 (morphology, phylogenetic relationships); Harbach & Peyton, 2000 (phylogenetic relationships); Harbach *et al.*, 2007a (phylogenetic relationships); Rattanarithikul *et al.*, 2007 (Thailand, *Rachionotomyia* and *Tripteroides sensu stricto*, keys, bionomics).

Polylepidomyia Theobald, 1905

Polylepidomyia, originally proposed as a genus by Theobald (1905f), currently classified as a subgenus of *Tripteroides*, is typified by *Tripteroides argenteiventris* (Theobald, 1905f), original combination: *Polylepidomyia argenteiventris*. Subgenus abbreviation – *Pol.*

****

Systematics: *Polylepidomyia* includes 19 extant species, which occur exclusively in the Australasian Region, including Timor, New Guinea, Bismarck Archipelago, South Pacific from the Solomon Islands to Rotuma Island (but not known in Fiji), Australia (all states) and Tasmania. A species of *Polylepidomyia*, *Tp. tasmaniensis*, and one species each of *Rachionotomyia* and *Rachisoura*, were included in the phylogenetic analysis of amino acid sequences of 709 genes conducted by Soghigian *et al.* (2023), but their relationships with each other and with other genus groups of Sabethini included in the analysis were not explicitly indicated. The phylogenetic relationships of the species and species groups of *Polylepidomyia* have not been investigated.

****

Characteristics: *Polylepidomyia* are unornamented mosquitoes which exhibit the following principal anatomical features. ADULTS: Vertex of the head with entirely or largely dark scales, ocular scales white or whitish, 3–6 ocular setae on either side; clypeus sometimes with scales; maxillary palpus of females at most 0.2 the length of the proboscis, usually shorter, palpus of males 0.7–0.9 the length of the proboscis; scutum with narrow scales; dorsocentral setae present or absent; prescutellar setae present; postpronotal setae sometimes absent; scales of the thoracic pleura white or whitish; mesopostnotal setae sometimes present; wing veins with narrow scales, sometimes boarder scales apically; femora with the anterior surfaces dark-scaled, tibiae and tarsi dark-scaled, hindungues of both sexes usually paired (1 hindunguis in a few species); pale scaling of the abdominal terga white or whitish. FEMALE GENITALIA: Postgenital lobe with an apical emargination; insula with a single row of marginal setae, usually with 2–5 basolateral setae on each side (absent in *Tp. marksae*); 3 spermathecal capsules. MALE GENITALIA: Large, conspicuous, usually fully exposed; ninth tergal lobes with slender or sometimes flattened setae; gonostylus slender, tapering toward the apex; lateral plate of the aedeagus without a sternal lobe. LARVAE: Maxilla with a short apical tooth and a typical maxillary brush; hypostomal suture strongly developed, complete to the posterior tentorial pit; seta 15-C inserted far posterior near the collar (except in *Tp. coheni*); seta 6-M varied, long and single, short and branched or very thick and rigid; seta 7-T varied, often very thick and rigid; seta 8-M,T varied, rather small with a few branches or larger with many stellate branches; seta 14-I–VI usually well developed; siphon with a lateral seta similar to seta 2a-S. PUPAE: Seta 1-CT very long, usually double, sometimes triple; seta 1-I strongly developed, multi-branched; punctures present on terga III–V or IV and V; seta 3-VII short, shorter than tergum VIII; seta 9-VII,VIII strongly developed; paddle sometimes with the apex slightly produced.

****

Bionomics: The immature stages of species of *Polylepidomyia* have been found in a variety of plant and artificial containers. Larvae and pupae of several species are known to inhabit *Nepenthes* pitchers, and some have been found in tree holes, stump holes, coconut shells, cut bamboo, cacao pods, ground holes and rock pools. Females of several species are known to feed on humans during daytime, but there is no evidence that they harbour or transmit pathogens of human or animal diseases.

****

Principal references: Lee, 1946 (as subgenus *Mimeteomyia*, excluding the *obscura*-group, classification, in keys, descriptions, bionomics, distributions); Belkin, 1955 (Melanesia, Caledonicus Group, as the *caledonica* complex, distribution, bionomics, characteristics, keys); Belkin, 1962 (South Pacific, seven species as members of *Rachionotomyia*, taxonomy, descriptions, distributions, bionomics); Peters, 1963c (New Guinea, taxonomy, new species descriptions, in key to pupae of known species of *Tripteroides*); Mattingly, 1980 (classification); Mattingly, 1981 (key characters); Lee *et al.*, 1989c (keys, taxonomy, characteristics, classification, literature, distributions, bionomics).

Rachionotomyia Theobald, 1905

Rachionotomyia, originally proposed as a genus by Theobald (1905h), currently classified as a subgenus of *Tripteroides*, is typified by *Tripteroides ceylonensis* (Theobald, 1905h), original combination: *Rachionotomyia ceylonensis*. Subgenus abbreviation – *Rah.*

****

Systematics: *Rachionotomyia* includes 14 extant species in the Oriental Region, eastward from India and Sri Lanka. A species of *Rachionotomyia*, *Tp. aranoides*, and one species each of *Polylepidomyia* and *Rachisoura*, were included in the phylogenetic analysis of amino acid sequences of 709 genes conducted by Soghigian *et al.* (2023), but their relationships with each other and with other genus groups of Sabethini included in the analysis were not explicitly indicated. The phylogenetic relationships of the species and species groups of *Rachionotomyia* have not been investigated.

****

Characteristics: *Rachionotomyia* are unornamented mosquitoes which exhibit the following principal anatomical features. ADULTS: Vertex of the head with broad dark decumbent scales, sometimes with subdued bronzy reflections; ocular scales white or whitish, sometimes with subdued blue reflections (species of the Affinis Subgroup have brilliant blue reflections); 2–4 ocular setae (usually 3) on either side (often 4 in species of the Nepenthis Subgroup); maxillary palpus of both sexes at most 0.2 the length of the proboscis (0.33 as long in *Tp. edwardsi*); scutum with broad scales except on the midline; dorsocentral setae absent; prescutellar setae present or absent; postpronotal setae absent (present in *Tp. edwardsi*); scales of the thoracic pleura white or whitish (with silvery reflections in species of the Affinis Subgroup); subspiracular scales present; wing veins mostly with narrow scales, broader distally on veins R_2 and R_3; femora pale ventrally, anterior surface entirely dark-scaled or narrowly pale ventrally at the base, tibiae and some tarsomeres often paler ventrally, hindungues paired; pale scaling of the abdominal terga white or whitish (with silvery reflections in species of the Affinis Subgroup). FEMALE GENITALIA: Postgenital lobe with or without a shallow apical emargination; insula with a single row of 8 or fewer marginal setae on either side (in a double row and more numerous in species of the Affinis Subgroup), with 2–7 basolateral setae on each side; 3 spermathecal capsules. MALE GENITALIA: Small, inconspicuous, retracted, not normally fully exposed; ninth tergal lobes with stout relatively short setae (long and slender in *Tp. rozeboomi*); gonostylus rather stout, slightly if at all tapered toward the apex (more slender in *Tp. dofleini*; highly modified in *Tp. rozeboomi*); lateral plate of the aedeagus with a spiculate sternal lobe. LARVAE: Maxilla with a short apical tooth and a typical maxillary brush; hypostomal suture conspicuous, complete to the posterior tentorial pit; seta 15-C inserted far posterior, near the collar; setae 6-M and 7-T single, short, stout and rigid; seta 8-M,T strongly developed (except in *Tp. dofleini*); seta 14-I–VI usually strongly developed (minute or absent in *Tp. dofleini* and species of the Nepenthis Subgroup); siphon with or without a lateral seta, similar to seta 2a-S when present. PUPAE: Seta 1-CT very long, double (single in *Tp. nepenthis*); seta 1-I strongly developed, multi-branched, seta 1-III–VI usually slender and delicate, sometimes multi-branched and conspicuous; seta 3-II,III strongly developed, much longer than seta 5-III, seta 3-VII much shorter than segment VIII (except in species of the Affinis Subgroup); seta 5-IV long, stout, conspicuous, much longer than seta 1-IV; seta 6-III–V short and delicate; punctures present on terga III–V or IV and V; seta 9-VII,VIII long, multi-branched; paddle with the apex at most bluntly pointed.

****

Bionomics: The immature stages of species of *Rachionotomyia* have been found in tree holes, hollow logs, bamboo and tree stumps, root holes, split bamboo, plant axils (including banana, *Allocasia*, *Colocasia*, *Nipa*, *Pandanus*, pineapple and others), *Nepenthes* pitchers, ginger florescence, fallen leaves, coconut shells and husks, rock holes and artificial containers. The feeding preferences of females are unknown. They rarely, if ever, feed on humans.

****

Principal references: Barraud, 1934 (southern Asia, *Tp. affinis*, *Tp. aranoides*, *Tp. edwardsi* and *Tp. serratus* as subspecies of *Tp. aranoides*, taxonomy, descriptions, bionomics, distributions); Thurman, 1959 (Thailand, *Tp. affinis* and *Tp. aranoides* as species of *Tripteroides sensu stricto*, taxonomy, descriptions, distributions); Mattingly, 1980 (taxonomy); Mattingly, 1981 (taxonomy, keys, descriptions, distributions, bionomics); Lu *et al.*, 1997a (China, *Tp. aranoides*, taxonomy, descriptions, distribution, bionomics); Rattanarithikul *et al.*, 2007 (Thailand, six species, bionomics, keys).

Rachisoura Theobald, 1910

Rachisoura, originally proposed as a genus by Theobald (1910c), currently classified as a subgenus of *Tripteroides*, is typified by *Tripteroides sylvestris* (Theobald, 1910c), original combination: *Rachisoura sylvestris*. Subgenus abbreviation – *Rac.*

****

Systematics: *Rachisoura* includes 28 extant species, which are found only in parts of the Australasian Region (New Guinea, northern Australia and the Solomon Islands). The phylogenetic relationships of *Rachisoura* and its species have not been investigated. Belkin (1962) indicated the relationships of *Rachisoura* were not apparent, but noted that similarities with some species of *Trichoprosopon* were indicative of 'genetic affinities'. A species of *Rachisoura*, *Tp. sylvestris*, and one species each of *Polylepidomyia* and *Rachionotomyia*, were included in the phylogenetic analysis of amino acid sequences of 709 genes conducted by Soghigian *et al.* (2023), but their relationships with each other and with other genus groups of Sabethini included in the analysis were not explicitly indicated. The phylogenetic relationships of the species and species groups of *Rachisoura* have not been investigated.

****

Characteristics: *Rachisoura* are unornamented mosquitoes which exhibit the following principal anatomical features. ADULTS: Vertex of the head entirely or largely with dark decumbent scales, ocular scales white or whitish, 3–6 ocular setae on either side; clypeus sometimes with scales; maxillary palpus of females 0.2–0.3 the length of the proboscis, palpus of males 0.2–1.0 the length of the proboscis; scutum with mainly narrow scales; acrostichal and dorsocentral setae absent; prescutellar setae reduced or absent; mesopostnotum with a cluster of a few minute setae (always?); postpronotal setae reduced or absent; scales of the thoracic pleura white or whitish; subspiracular scales present; wing scales all broad (Filipes Group) or broad only toward the apex of the wing (Vanleeuweni Group); all femora with the anterior surface usually entirely dark-scaled, tibiae and tarsi dark-scaled, hindungues always paired; pale scaling of the abdominal terga white or whitish. FEMALE GENITALIA: Insula with 10–35 setae in a longitudinal band on either side of the midline; 3 spermathecal capsules. MALE GENITALIA: Large, not retracted (fully exposed); ninth tergal lobes with slender or flattened setae; lateral plate of the aedeagus with a sternal lobe. LARVAE: Maxilla with a short apical tooth and a large articulated maxillary claw (adaptation for grasping; homologous with the maxillary brush); hypostomal suture incomplete, not reaching the posterior tentorial pit; seta 15-C inserted approximately midway between the collar and the anterior margin of the labiogula; seta 6-M single, long, slender; seta 7-T strongly developed, long, multi-branched; seta 8-M,T very small, multi-branched; seta 14-I–VI minute or absent; siphon without a lateral seta. PUPAE: Seta 1-CT normally long, single, sometimes double; seta 1-I strongly developed; punctures, when present, only on terga III and IV; seta 9-VII,VIII usually well developed; paddle usually with more or less conspicuous spicules at the apex.

****

Bionomics: The larvae of approximately half of all species of *Rachisoura* are known only to inhabit *Nepenthes* pitchers. Other species have been found in cut bamboo, coconut shells, husks and spathes, leaf axils, tree holes and artificial containers. Larvae, as far as known, use their modified maxillae to capture and feed on other mosquito larvae that share their habitat, and perhaps feed on dead or dying insects and other arthropods as well. Females of several species are known to feed on humans in the daytime, but no species of *Rachisoura* are known to harbour or transmit pathogens that cause human or animal diseases.

****

Principal references: van den Assem, 1959 (New Guinea, taxonomy, descriptions, bionomics); Belkin, 1962 (South Pacific, *Tp. mathesoni*, *Tp. stonei* and *Tp. torokinae*, taxonomy, descriptions, bionomics, distributions); Mattingly, 1981 (key characters); Lee *et al.*, 1989c (Australasian Region, keys, taxonomy, characteristics, classification, literature, distributions, bionomics); Harbach & Peyton, 1993 (comparative morphology of larval maxillae).

Tricholeptomyia Dyar & Shannon, 1925

Tricholeptomyia, originally proposed as a genus by Dyar & Shannon (1925a), currently classified as a subgenus of *Tripteroides*, is typified by *Tripteroides nepenthicola* (Banks, 1909), original combination: *Wyeomyia nepenthicola*. Subgenus abbreviation – *Tri.*

****

Systematics: *Tricholeptomyia* includes nine extant species known only from the Philippines. The phylogenetic relationships of *Tricholeptomyia* and its species and species groups have not been investigated.

****

Characteristics: *Tricholeptomyia* are unornamented mosquitoes which exhibit the following principal anatomical features. ADULTS: Vertex of the head with broad dark decumbent scales, sometimes with pale scales on the midline or posteriorly, ocular scales white or whitish with subdued blue reflections, 1 or 2 ocular setae (usually 2) on either side; maxillary palpus 0.1–0.2 the length of the proboscis; scutum with narrow scales (moderately broad in *Tp. apoensis*); dorsocentral setae present or absent; prescutellar setae present; postpronotal setae present (absent in *Tp. barraudi*); scales of the thoracic pleura white or whitish; subspiracular scales present; wing veins with narrow scales, sometimes somewhat broader distally on veins R_2 and R_3; all femora with the anterior surface entirely dark-scaled, tibiae and some tarsomeres often paler ventrally, hindlegs of both sexes with single unguis (not paired), pale scaling of the abdominal terga white or whitish. FEMALE GENITALIA: Postgenital lobe with an apical emargination; insula with 3–10 marginal setae, with 0–3 basolateral setae on each side; 3 spermathecal capsules. MALE GENITALIA: Small, retracted or exposed; ninth tergal lobes with relatively long slender attenuated setae (spine-like in *Tp. christophersi*); gonostylus rather stout, clubbed distally; lateral plate of the aedeagus without a sternal lobe. LARVAE: Maxilla with a short apical tooth and a typical maxillary brush; hypostomal suture unapparent; seta 15-C inserted approximately midway between the collar and the anterior margin of the labiogula; setae 6-M and 7-T single, very stout and rigid (6-M relatively long and sender in *Tp. delpilari* and *Tp. nepenthicola*); seta 8-M,T very small, minute or absent; seta 14-I–VI minute; siphon with a minute lateral seta, usually branched. PUPAE: Seta 1-CT short; seta 1-I weakly developed, often single (except in *Tp. delpilari*); other abdominal setae short and slender or longer and relatively stout; punctures present on terga III–V; seta 9-VII,VIII usually well developed (small in *Tp. apoensis* and *Tp. barraudi*); paddle with spiculate margins.

****

Bionomics: Little is known about the bionomics of species of *Tricholeptomyia*. The immature stages develop in pitcher plants. Nothing is known about the biology of the adults. Females are not known to feed on humans and are unlikely to be vectors of pathogenic agents.

****

Principal references: Baisas & Ubaldo-Pagayon, 1953 (as Group A of subgenus *Tripteroides* except *Tp. rozeboomi*, distributions, bionomics, classification, keys, descriptions); Mattingly, 1981 (keys, taxonomy, descriptions, distributions, bionomics).

Tripteroides Giles, 1904 *sensu stricto*

Tripteroides sensu stricto, the nominotypical subgenus of *Tripteroides*, is typified by *Runchomyia philippinensis* Giles, 1904 [subjective synonym of *Tripteroides nitidoventer* (Giles, 1904)]. Subgenus abbreviation – *Trp.*

****

Systematics: *Tripteroides sensu stricto* includes 56 extant species in the Australasian and Oriental Regions and the Manchurian Subregion of the Palaearctic. Its distribution extends from northeastern India to the Philippines, northward to Japan and Korea, and southward to the Moluccas, northern Australia, New Guinea, the Bismarck Archipelago and the South Pacific (Solomon Islands, Santa Cruz Islands and Fiji). The phylogenetic relationships of *Tripteroides sensu stricto* and its species and species groups have not been investigated.

****

Characteristics: Unlike species of the other genus groups of *Tripteroides*, species of *Tripteroides sensu stricto* are ornate mosquitoes with a bright blue or silver band of scales anteriorly on the head, silvery scales on the thoracic pleura and one or all femora have silvery spots on the anterior surface. The species also exhibit the following anatomical features. ADULTS: Head with 1–3 ocular setae (usually 2) on either side; maxillary palpus very short in both sexes, less than 0.2 the length of the proboscis; scutum with narrow piliform scales; dorsocentral and prescutellar setae usually present; mesopostnotum bare; thoracic pleural scaling sometimes reduced, subspiracular scales often absent; wing veins with narrow to moderately broad scales, broader on veins R_2 and R_3; tibiae and tarsi entirely dark-scaled, hindungues of males always paired; abdominal terga with lateral silvery spots or bands (except in some *Tp. distigma*). FEMALE GENITALIA: Postgenital lobe with a shallow apical emargination; insula with 2–8 marginal setae (usually 3–5) and 1–10 basolateral setae on each side; Oriental species usually with 1 spermathecal capsule, Australasian species with 3 (except *Tp. littlechildi*). MALE GENITALIA: Usually small and often retracted; gonostylus usually short and stout, expanded distally in some species; lateral plate of the aedeagus with an apical tergal process, projecting sternolaterad and usually ribbed or spiculate. LARVAE: Subgeneric characters not evident at present, Mabinii and Nitidoventer Groups distinguished as follows. Mabinii Group (1 species): maxilla with an exceptionally long apical tooth; hypostomal suture incomplete, not reaching the posterior tentorial pit; seta 15-C inserted far posterior, near the collar; setae 6-M and 7-T single, very stout and rigid; seta 14-I–VI minute or absent. Nitidoventer Group (*Tripteroides sensu stricto*): maxilla with a short apical tooth and a typical maxillary brush; hypostomal suture complete to the posterior tentorial pit; seta 15-C inserted far posterior, near the collar; seta 14-I–VI strongly developed. PUPAE: Seta 1-CT very long, stout, conspicuous; seta 1-I strongly developed.

****

Bionomics: The immature stages of species of *Tripteroides sensu stricto* are found primarily in tree holes, but they also occur in coconut shells, bamboo internodes, bamboo stumps, split bamboo, plant axils, *Nepenthes* pitchers, log and stump holes, and artificial containers. Females of several species are known to feed on humans during daylight hours, but no species of *Tripteroides sensu stricto* are known to harbour or transmit pathogens of human or animal diseases.

****

Principal references: Baisas & Ubaldo-Pagayon, 1953 (as Group B of subgenus *Tripteroides*, distributions, bionomics, classification, keys, descriptions); Thurman, 1959 (Thailand, *Tp. aeneus*, *Tp. hybridus* and *Tp. similis*, synonymy, adult morphology, distributions); Belkin, 1962 (South Pacific, *Tp. binotatus*, *Tp. bonneti*, *Tp. distigma*, *Tp. lipovskyi* and *Tp. purpuratus*, taxonomy, descriptions, bionomics, distributions); Tanaka *et al.*, 1979 (Japan, *Tp. bambusa*, taxonomy, descriptions, distributions, bionomics); Mattingly, 1980 (classification); Mattingly, 1981 (Oriental Region, Mabinii and Nitidoventer Groups, characterization); Lee *et al.*, 1989c (Australasian Region, keys, taxonomy, characteristics, classification, literature, distributions, bionomics); Lu *et al.*, 1997a (China, keys, taxonomy, descriptions, distributions, bionomics); Rattanarithikul *et al.*, 2007 (Thailand, nine species, bionomics, keys).

Udaya Thurman, 1954

Udaya Thurman, 1954b, originally proposed as a subgenus of *Paraedes* and currently recognized as a genus, is typified by *Udaya argyrurus* (Edwards, 1934) (in Barraud, 1934), original combination: *Paraedes argyrurus*. Genus abbreviation – *Ud.*

****

Systematics: *Udaya* is a subordinate taxon of the subfamily Culicinae, tribe Aedini. It includes only three extant species in the Oriental Region, the distributions of which are poorly known. Macdonald & Mattingly (1960) hypothesized that *Udaya* was an annectant group between *Aedes* and *Eretmapodites*. *Udaya* was recovered as the sister of *Belkinius* + *Zeugnomyia* in the phylogeny of Aedini generated in the phylogenetic analyses of Reinert *et al.* (2009) based on morphological data, and this relationship formed part of a larger clade parenthetically expressed as *Lorrainea* + ((*Eretmapodites* + *Armigeres*) + (*Udaya* + (*Belkinius* + *Zeugnomyia*))). *Udaya* fell within a polytomy with *Belkinius*, *Bothaella*, *Catatassomyia* and *Zeugnomyia* in the phylogeny reconstructed by Wilkerson *et al.* (2015). The only species of *Udaya*, *Ud. subsimilis*, included in the maximum likelihood phylogeny of Soghigian *et al.* (2017) based on molecular markers was paired with *Heizmannia discrepans* in a clade that included five other species of *Heizmannia*, suggesting that *Ud. subsimilis* was misidentified or that *Udaya* is not distinct enough from *Heizmannia* to justify generic status.

****

Characteristics: *Udaya* are small mosquitoes with conspicuous patches of silver scales on the head, thoracic pleura and abdominal terga. The adults are characterized by a spoon-shaped patch of broad overlapping silver scales that separates the eyes and extends to the antennal pedicels, the absence of acrostichal setae, a patch of broad silver scales on the antepronotum that separates the setae into two groups, the absence of scales on the postpronotum and postspiracular area, broad overlapping silver scales on the paratergite, and the alula of the wing has broad flat scales on the dorsal margin. Larvae are similar to the larvae of species of the *Aedes* genus group, *Heizmannia* and *Zeugnomyia*, but are distinguished by the following characters: setae 4,5,6-C placed far forward on the head and seta 7-C inserted far from the lateral margin of the dorsal apotome (distinctions from the *Aedes* genus group); comb with 10 or fewer scales in a single row (distinction from *Heizmannia*); saddle of abdominal segment X without strongly developed spines along the posterior margin (distinction from *Zeugnomyia*).

****

Bionomics: Very little is known about the bionomics of species of *Udaya*. Larvae of *Ud. argyrurus* and *Ud. subsimilis* have been collected from bamboo. Females of *Ud. argyrurus* have been found on occasion in catches of biting flies. Species of *Udaya* are not of medical or economic importance.

****

Principal references: Thurman, 1959 (Thailand, *Ud. argyrurus*, in keys, adults, descriptions, bionomics, distribution); Macdonald & Mattingly, 1960 (descriptions and comparative morphology of *Ud. argyrurus* and *Ud. lucaris*); Reinert, 1990a (*Ud. subsimilis*, taxonomy, description of male); Reinert, 2000d (female genitalia); Reinert *et al.*, 2004, 2009 (morphology, phylogenetic relationships); Rattanarithikul *et al.*, 2010 (Thailand, bionomics, keys); Wilkerson *et al.*, 2015 (phylogenetic relationships); Soghigian *et al.*, 2017 (phylogenetic relationships).

Uranotaenia Lynch Arribálzaga, 1891

Uranotaenia Lynch Arribálzaga, 1891a, originally proposed and currently recognized as a genus, is typified by *Uranotaenia pulcherrima* Lynch Arribálzaga, 1891b. Genus abbreviation – *Ur.*

****

Systematics: *Uranotaenia* is a subordinate taxon of the subfamily Culicinae, tribe Uranotaeniini. It includes 276 extant species divided between two subgenera: *Pseudoficalbia* and *Uranotaenia sensu stricto*. Most species occur in the Afrotropical and Oriental Regions, but several occur in Australia, one in Europe and the Middle East, two extend from the Neotropical Region into the southern USA, and one occurs in the West Indies and from Mexico to Canada. No species are known from New Zealand, New Caledonia or small oceanic islands. What is hypothesized about the phylogenetic affinities of *Uranotaenia* is interesting. The largely membranous proctiger of the male genitalia is found elsewhere only in Anophelinae, Aedeomyiini and some Aedini, and the greatly elongate ventral portion of the larval head capsule is found in Anophelinae, Toxorhynchitini and some predacious larvae of other groups (Belkin, 1962). The larval labiohypopharynx bears a strong resemblance to that of anophelines and dixids (Harbach, 1978). *Uranotaenia* was recovered in a sister-group relationship with *Aedeomyia* in the phylogenetic study of Harbach & Kitching (1998) based on an implied weighting analysis of morphological data; however, *Aedeomyia* was placed basal to *Uranotaenia* when character weights were assigned using successive approximation character weighting. The former relationship was recovered in the molecular analysis of mitogenomes conducted by da Silva *et al.* (2020a), which included data for only one species of each of the two genera, *Ur. pulcherrima* and *Ad. squamipennis*, respectively. In the study of Chen *et al.* (2024), also based of mitogenome sequences but in the absence of a species of *Aedeomyia*, *Uranotaenia* was recovered as the sister to a clade comprised of species of Mansoniini, Sabethini and Toxorhynchitini. The sister relationship of *Aedeomyia* and *Uranotaenia*, and their basal relationship to all other Culicinae, are supported by the results of the phylogenetic analysis of amino acid sequences of 709 genes conducted by Soghigian *et al.* (2023), which included data for two species of *Aedeomyia* and three species of *Uranotaenia sensu stricto*. The phylogenetic relationships within *Uranotaenia* have not been investigated.

****

Characteristics: *Uranotaenia* are small, delicate mosquitoes that are characterized by features of the wings. The microtrichia are extremely minute and not visible under ordinary magnification, cell R_2 is always shorter than vein R_{2+3} and cell M_2, and the anal vein (vein 1A) is sharply curved distally and ends before the base of the mediocubital crossvein. Some species are adorned with silver scales with bluish reflections. Larvae are recognized by the following characters: seta 1-C arises from a distinct process on the anterior margin of the head, seta 14-C is always near the anterior margin at the base of the maxilla, the hypostomal suture is absent, abdominal segment VIII usually bears a sclerite (comb plate) with a single row of comb scales on its posterior margin, the siphon has a single pair of seta 1-S inserted well above the base and a pecten is present that consists of a few to many spines.

****

Bionomics: The immature stages of species of *Uranotaenia* are found in a range of habitats. The larvae of most species inhabit groundwaters, including swamps, marshes, stream margins and temporary pools with vegetation, but many also occur in rock holes, crab holes, tree holes, bamboo, plant parts on the ground, leaf axils, flower bracts, pitcher plants and artificial containers. Females rarely feed on humans. The feeding preferences of most species are unknown, but available data indicate that amphibians, reptiles, birds and mammals serve as hosts. A number of species are attracted to the calls of frogs and feed on amphibians (Toma *et al.*, 2014; Camp *et al.*, 2018; Snyman, 2021). Females of *Ur. sapphirina* of *Uranotaenia sensu stricto* specialize in feeding on earthworms and leeches of the phylum Annelida (Reeves *et al.*, 2018). Adults rest on moist, protected surfaces near the larval habitats and may be found in large numbers in suitable shelters. Many species are attracted to light and are occasionally found resting inside houses. A few species feed on humans but none are involved in the transmission of pathogens.

****

Principal references: Barraud, 1934 (southern Asia, keys, descriptions, bionomics, distributions); Edwards, 1941 (Afrotropical Region, keys, descriptions of adults and pupae, distributions); Hopkins, 1952 (Afrotropical Region, keys, descriptions of larvae, distributions, bionomics); Lane, 1953 (Neotropical Region, taxonomy, keys, descriptions, distributions); Galindo *et al.*, 1954 (Panama, taxonomy, descriptions, keys, bionomics); Belkin, 1962 (South Pacific, keys, taxonomy, descriptions, bionomics, distributions); Delfinado, 1966a (Philippines, taxonomy, keys, descriptions, distributions, bionomics); Belkin *et al.*, 1970 (Jamaica, taxonomy, keys, descriptions, bionomics); Peyton, 1972 (classification, comparative morphology, species lists); Peyton, 1977 (*Pseudoficalbia*, Southeast Asia, classification, descriptions, keys, distributions, bionomics); Tanaka *et al.*, 1979 (Japan, taxonomy, keys, descriptions, distributions, bionomics); Clark-Gil & Darsie, 1983 (Guatemala, keys); Darsie, 1985 (Argentina, keys); Mitchell & Darsie, 1985 (Argentina, distributions, literature); Lee *et al.*, 1989c (Australasian Region, taxonomy, keys, literature, distributions, bionomics); Darsie & Pradhan, 1990 (Nepal, keys, distributions, bionomics); Service, 1990 (Afrotropical Region, keys, adults, larvae, bionomics, distribution); da Cunha Ramos, 1993 (Afrotropical Region, classification, descriptions, keys, distributions, bionomics); Harbach & Kitching, 1998 (morphology, phylogenetic relationships); da Cunha Ramos & Brunhes, 2004

(Madagascar, classification, descriptions, keys, distributions, bionomics); Lu *et al.*, 1997a (China, keys, taxonomy, descriptions, distributions, bionomics); Brunhes, 2004 (Madagascar, classification, descriptions, keys, distributions, bionomics); Rattanarithikul *et al.*, 2006a (Thailand, bionomics, keys); Rivera-García *et al.*, 2019 (Mexico, keys); da Silva *et al.*, 2020a (mitogenomes, phylogenetic relationships); Soghigian *et al.*, 2023 (molecular markers, phylogenetic relationships); Chen *et al.*, 2024 (mitochondrial protein sequences, phylogenetic relationships).

Pseudoficalbia Theobald, 1912

Pseudoficalbia, originally proposed as a genus by Theobald (1912b), currently classified as a subgenus of *Uranotaenia*, is typified by *Uranotaenia pandani* (Theobald, 1912b), original combination: *Pseudoficalbia pandani*. Subgenus abbreviation – *Pfc.*

****

Systematics: *Pseudoficalbia* includes 153 extant species, which occur principally in the tropical and subtropical regions of the Old World, with the exception of some islands in the Pacific and Atlantic Oceans, *Ur. anhydor* in the southwestern USA and Mexico, several species with a northward extension from the Oriental Region into the Palaearctic Region, and *Ur. unguiculata* in the Mediterranean Subregion and adjacent areas of the European and Siberia Subregions of the Palaearctic. Phylogenetic relationships within *Pseudoficalbia* have not been investigated. However, da Cunha Ramos (1993) employed UPGMA (Unweighted Pair Group Method with Arithmetic Mean) methodology to construct phenetic trees for Afrotropical species of *Pseudoficalbia* and *Uranotaenia sensu stricto*.

****

Characteristics: Species of *Pseudoficalbia* exhibit the following principal anatomical features. ADULTS: Without lines or patches of metallic scales; head with numerous erect scales covering most of the vertex, frontal tuft absent; posterior mesanepisternum (prealar area) not separated from the mesokatepisternum by a suture (mesanapleural suture); wing without pale-scaled areas, alula with or without a few broad dorso-marginal scales; anterior unguis of the legs of females not noticeably broader than the posterior unguis. MALE GENITALIA: Gonocoxite with a well-developed basal mesal lobe, usually produced dorsoapically into a very distinct finger-like lobe; gonostylus moderately long, usually slender and tapered to a narrow apex, sometimes rather stout and distended on the dorsal preapical margin; lateral sclerites of the aedeagus normally distinctly connected dorsally and ventrally. LARVAE: Setae 5-C (normally) and 6-C slender, usually single and simple or weakly branched, usually inserted far forward on the anterior 0.33 of the head, rarely inserted near mid-length; grid of the ventral brush (seta 4-X) normally not joined mid-ventrally to the saddle. PUPAE: Trumpets normally set far apart, nearer to the mesothoracic wing than to the median keel, usually with an indistinct tracheoid area at the base anteriorly.

****

Bionomics: The immature stages of species of *Pseudoficalbia* are generally found in phytotelm habitats, but some species have been found in ground pools, swamps and artificial containers. Very little is known about the feeding habits of females, but the females of *Ur. mashonaensis* and *Ur. montana* have been observed feeding on a species of toad in South Africa (Netherlands *et al.*, 2020). Species of *Pseudoficalbia* are not of medical or economic importance.

****

Principal references: Peyton, 1972 (characterization, classification, keys, species list); Peyton, 1977 (Southeast Asia, classification, descriptions, keys, distributions, bionomics); Tanaka *et al.*, 1979 (Japan, taxonomy, keys, descriptions, distributions, bionomics); Lee *et al.*, 1989c (Australasian Region, taxonomy, keys, literature, distributions, bionomics); Service, 1990 (Afrotropical Region, keys, adults, larvae, bionomics, distribution); da Cunha Ramos, 1993 (Afrotropical Region, classification, descriptions, keys, distributions, bionomics); Lu *et al.*, 1997a (China, keys, taxonomy, descriptions, distributions, bionomics); da Cunha Ramos & Brunhes, 2004 (Madagascar, classification, descriptions, keys, distributions, bionomics); Rattanarithikul *et al.*, 2006a (Thailand, bionomics, keys).

Uranotaenia Lynch Arribálzaga, 1891 *sensu stricto*

Uranotaenia sensu stricto, the nominotypical subgenus of *Uranotaenia*, is typified by *Uranotaenia pulcherrima* Lynch Arribálzaga, 1891b. Subgenus abbreviation – *Ura.*

****

Systematics: *Uranotaenia sensu stricto* includes 123 extant species, which occur principally in tropical areas and to some extent subtropical areas of the Afrotropical, Australasian, Neotropical and Oriental Regions. *Uranotaenia sapphirina* is primarily a

Nearctic species, extending northward from the Dominican Republic, Mexico and Puerto Rico through the USA into Canada. Phylogenetic relationships within *Uranotaenia sensu stricto* have not been investigated. However, da Cunha Ramos (1993) employed UPGMA (Unweighted Pair Group Method with Arithmetic Mean) methodology to construct phenetic trees for Afrotropical species of *Uranotaenia sensu stricto* and *Pseudoficalbia*. Three species of *Uranotaenia sensu stricto*, in the absence of species of *Pseudoficalbia*, included in the phylogenetic analysis of amino acid sequences of 709 genes conducted by Soghigian *et al.* (2023), were recovered in a sister relationship with two species of *Aedeomyia*, and the pair was positioned as the basal lineage of the Culicinae.

****

Characteristics: Species of *Uranotaenia sensu stricto* exhibit the following principal anatomical features. ADULTS: Some species have lines or patches of silver, gold or blue metallic scales; head usually without erect scales, when present usually only a few restricted to the occiput, rarely a few on the vertex, frontal tuft present; posterior mesanepisternum (prealar area) separated from the mesokatepisternum by a suture (mesanapleural suture); wing of some species with pale-scaled areas, alula without scales; females of some species with the anterior unguis of all legs much broader than the posterior unguis. MALE GENITALIA: Gonocoxite with a weakly developed basal mesal lobe, occasionally absent; gonostylus very short and stout, usually with the dorsal preapical margin distended; lateral sclerites of the aedeagus distinctly connected only dorsally. LARVAE: Setae 5,6-C very stout and normally spinulate, always inserted far back on the head, seta 6 inserted at the level or posterior to the base of the antenna and seta 5 inserted near mid-length of the head; grid of segment X usually distinctly joined mid-ventrally to the saddle. PUPAE: Trumpets normally set close together, nearer to the median keel than to the mesothoracic wing, usually with a distinct tracheoid area on the basal 0.25 or more.

****

Bionomics: Species of *Uranotaenia sensu stricto* are found mainly in groundwater habitats, including springs, stream margins, swamps, temporary pools with vegetation, rock holes and crab holes. Some species have been found in tree holes, plant axils and artificial containers. Females of some species are known to feed on frogs, birds and mammals, but are normally not attracted to humans. Species known to feed on amphibians include *Ur. lowii* in North America, which feeds on frogs and salamanders (Remington, 1945) and *Ur. sapphirina*, also in North America, which specializes in feeding on the blood of earthworms and leeches (Reeves *et al.*, 2018). Species of *Uranotaenia sensu stricto* are not of medical or economic importance.

****

Principal references: Lee *et al.*, 1989c (Australasian Region, taxonomy, keys, literature, distributions, bionomics); Service, 1990 (Afrotropical Region, keys, adults, larvae, bionomics, distribution); da Cunha Ramos, 1993 (Afrotropical Region, classification, descriptions, keys, distributions, bionomics); Lu *et al.*, 1997a (China, keys, taxonomy, descriptions, distributions, bionomics); da Cunha Ramos & Brunhes, 2004 (Madagascar, classification, descriptions, keys, distributions, bionomics); Rattanarithikul *et al.*, 2006a (Thailand, bionomics, keys).

Verrallina Theobald, 1903

Verrallina Theobald, 1903a, originally proposed and currently recognized as a genus, is typified by *Verrallina butleri* (Theobald, 1901c), original combination: *Aedes butleri*. Genus abbreviation – *Ve.*

****

Systematics: *Verrallina* is a subordinate taxon of the subfamily Culicinae, tribe Aedini. It includes 95 extant species divided between three subgenera: *Harbachius*, *Neomacleaya* and *Verrallina sensu stricto*. Species of *Verrallina* primarily occur in the Oriental Region with extensions northward to Japan (Kyushu) and south-eastward into the Caroline Islands, New Guinea, the southern Pacific islands and northeastern Australia. *Verrallina* was recovered as a strongly supported monophyletic group in a sister relationship with *Aedes* in a phylogenetic analysis of Aedini based on morphological data (Reinert *et al.*, 2004). *Harbachius* was recovered as the sister of *Neomacleaya* + *Verrallina sensu stricto*. Different relationships were recovered in the much more comprehensive analysis of Aedini conducted by Reinert *et al.* (2009), where *Verrallina* was recovered in a clade comprised of *Neomelaniconion* + (*Edwardsaedes* + (*Aedes* + (*Paraedes* + *Verrallina*))) (also recovered in the phylogeny of Wilkerson *et al.*, 2015); within *Verrallina*, *Harbachius* and *Neomacleaya* were recovered as monophyletic but *Verrallina sensu stricto* was paraphyletic with respect to *Neomacleaya*. *Verrallina* was recovered in a sister relationship with *Edwardsaedes* in the maximum likelihood phylogeny of Soghigian *et al.* (2017) based on molecular markers.

****

Characteristics: The remarkable development of the male and female genitalia and the following combinations of morphological characters distinguish *Verrallina* from all other genus groups of Aedini. ADULTS: Decumbent scales of the vertex all broad (infrequently with a few narrow curved scales on the margins of the coronal suture); antennal pedicel with several very short fine setae and usually a few small broad scales mesally (scales not overlapping or forming a dense patch); maxillary palpus of males very short, 0.1–0.2 the length of the proboscis; scutum and scutellum with only narrow curved scales, paratergite bare (except *Ve. indica*); anterior and posterior acrostichal areas with several to numerous setae (species of *Harbachius* usually with only a few setae on the posterior area); anterior and posterior dorsocentral areas with setae; mesopostnotum normally bare (two species with 1 or 2 setae); postpronotum with or without narrow curved scales dorsally; prespiracular area bare; postspiracular setae present; wings with narrow dark scales (rarely with a few pale scales at the base of the costa and/or the remigium); alula with narrow to moderately broad scales on the margin; tarsi dark-scaled. FEMALE GENITALIA: Lower vaginal sclerite present; insula ill-defined and small in *Neomacleaya* and *Verrallina sensu stricto*, absent in *Harbachius*; 3 spermathecal capsules; cerci more or less triangular, usually with numerous broad scales; tergum and sternum VIII with numerous broad scales. MALE GENITALIA: Tergum IX without setae; gonocoxite with 1 or more well-developed lobes, with fleshy projections or spiniforms dorsally and/or ventrally at the apex (except in two species); gonostylar claw absent; paraprocts usually highly modified into short or long free projections or broad contiguous plates; cercal setae absent; phallosome complex, with opisthophallus and prosophallus; phallus with a distal tergal filament. LARVAE: Seta 4-C short, multi-branched, inserted mesad and slightly cephalad of seta 6-C; seta 5-C inserted caudad and slightly mesad of seta 6-C, inserted laterad and caudad of seta 4-C; insertions of setae 5,6-C widely separated; seta 7-C usually inserted laterad and slightly cephalad of seta 6-C, inserted laterad and at the same level as seta 6-C in a few species; setae 5–7-C stout, branched, approximately equal in length or seta 7-C slightly shorter than setae 5,6-C (6-C single in *Ve. unca*); setae 5,6-P single (double in *Ve. nigrotarsis*); seta 8-P long, thickened, usually single (occasionally double, rarely triple); siphon with an acus, pecten with distal 1–5 spines more widely spaced than the others; saddle incomplete ventrally, without spines on the posterior margin, acus absent; ventral brush (seta 4-X) usually with 10 setae borne on a grid and 2 or 3 precratal setae. PUPAE: Setae 1–3-CT approximately equally developed; setae 1,3-IV branched (usually multi-branched, rarely single); insertions of setae 2,3-I widely separated; seta 3-I branched, rarely single, short to moderately long, seta 3-II,III moderately long to long, single, seta 3-II rarely double; setae 6,9-VII both relatively short, weakly developed; seta 9-VII single, rarely double; seta 9-VIII single to 6-branched; paddle without a fringe of hair-like spicules; seta 1-Pa single, occasionally double.

****

Bionomics: Very little information is available for species of *Verrallina*. As far as known, the immature stages are found in temporary ground pools. Some species seem to be adapted to breeding in jungle pools that are frequently flushed whereas others inhabit more open coastal pools that are not subjected to flushing. *Verrallina cuccioi* has only been collected in pools and tree holes at ground level in stream beds. Eggs are apparently laid in ribbon-like gelatinous strings attached to objects in locations subject to inundation by rain water. The females of several species readily attack humans, chiefly during the day and largely in shaded places or in the vicinity of forest. *Verrallina carmenti* and *Ve. lineata* are serious pests in the South Pacific. The latter seems to be a semi-domestic species that may have been dispersed by human agency. No species of *Verrallina* appear to play a role in the transmission of pathogens that cause diseases in humans.

****

Principal references: Laffoon, 1946 (Philippines, as *Aedes* subgenus *Aedes*, keys, descriptions, bionomics, distributions); King & Hoogstraal, 1947b (New Guinea, as *Aedes* subgenus *Aedes*, descriptions, keys); Belkin, 1962 (South Pacific, as a subgenus of *Aedes*, keys, descriptions, bionomics, distributions); Reinert, 1974 (as a subgenus of *Aedes*, descriptions, keys, distributions, bionomics); Delfinado, 1967 (Southeast Asia, *Neomacleaya*, as a subgenus *Aedes*, keys, descriptions, distributions, bionomics); Delfinado, 1968 (Southeast Asia, *Neomacleaya*, as a subgenus *Aedes*, keys, descriptions, distributions, bionomics); Huang, 1968a (Papuan Subregion, as a subgenus of *Aedes*, keys, descriptions, bionomics, distributions); Reinert, 1984 (Sri Lanka, as a subgenus of *Aedes*, classification, keys, descriptions, distributions, bionomics); Lee *et al.*, 1987a (Australasian Region, as a subgenus of *Aedes*, taxonomy, keys, literature, distributions, bionomics); Lu *et al.*, 1997a (China, as a subgenus of *Aedes*, keys, taxonomy, descriptions, distributions, bionomics); Reinert, 1999c (keys, genus, subgenus and type species descriptions, distribution); Reinert, 2001b (female genitalia); Reinert *et al.*, 2004, 2009 (morphology, phylogenetic relationships); Rattanarithikul *et al.*, 2010 (Thailand, keys, bionomics); Wilkerson *et al.*, 2015 (phylogenetic relationships); Soghigian *et al.*, 2017 (phylogenetic relationships).

Harbachius Reinert, 1999

Harbachius Reinert, 1999c, originally proposed and currently recognized as a subgenus of *Verrallina*, is typified by *Verrallina yusafi* (Barraud, 1931b), original combination: *Aedes yusafi*. Subgenus abbreviation – *Har.*

****

Systematics: *Harbachius* includes 13 extant species recorded mainly from countries in the Oriental Region, with occurrences in Japan and Russia. *Harbachius* was recovered as the sister of *Neomacleaya* + *Verrallina sensu stricto* in the morphology-based phylogenetic analysis of Aedini conducted by Reinert *et al.* (2004), and as sister to a clade in which *Verrallina* was paraphyletic with respect to *Neomacleaya* in the much more comprehensive analysis of Aedini conducted by Reinert *et al.* (2009). In agreement with Reinert *et al.* (2004), *Harbachius* was recovered as the sister of *Neomacleaya* + *Verrallina sensu stricto* in the maximum likelihood phylogeny of Soghigian *et al.* (2017) based on molecular markers.

****

Characteristics: The following combinations of characters distinguish species of *Harbachius* from those of *Neomacleaya* and *Verrallina sensu stricto*. ADULTS: Eyes separated dorsally; mesepimeron without fine hair-like setae posterior or ventral to the scale-patch; hindungues of both females and males with a tooth. FEMALE GENITALIA: Tergum VIII with scales distally; sternum VIII with numerous stout setae on the apical lobes, scales on a relatively small area; postgenital lobe relatively narrow, apex with a median indentation or broadly rounded; upper vaginal lip with the median posterior area produced into a large horizontal shield extending over or around the spermathecal eminence; lower vaginal sclerite comprised of a pair of ribbon-like apically fused plates, attached to the lower vaginal lip, without spicules; insula absent, apparently replaced by an apical U-shaped projection of the lower vaginal sclerite, without tuberculi; spermathecal eminence relatively simple, without spicules, with a spine-lined cephalad pouch. MALE GENITALIA: Gonocoxite with a moderately broad basotergal apodeme; prosophallus comprised of a pair of narrow apically approximated sclerites, longer than twice the length of the phallus; phallus comprised of a pair of aedeagal sclerites joined by a sternal bridge, each sclerite with a narrow projection basolaterally extending at approximately a right angle; opisthophallus usually long, with the caudal margin sharply rounded or pointed; paramere longer than 1.5 times the length of the phallus. LARVAE: Seta 1-A single or double; seta 7-C with 2–7 branches (usually 2–4); seta 8-C branched; seta 13-C with 2–5 branches (rarely single); seta 7-I single; comb scales with a stout apical spine and short lateral denticles; seta 2-X with 3–6 branches. PUPAE: Cephalothorax with the ocular facets of the compound eye well developed; trumpet usually with the proximal 0.4 somewhat expanded; seta 2-IV–VI thin, inserted more cephalad than seta 4; seta 9-VIII usually double or triple.

****

Bionomics: Very little is known about the biology of most species of *Harbachius*. Larvae have been found in flood pools, residual pools in streambeds, ground pools in forest and once (*Ve. nobukonis*) in a fox hole. Larvae of *Ve. hamistylus* have been found in a *Pandanus* swamp as well as in residual stream pools. Females of *Ve. fragilis*, *Ve. hamistylus*, *Ve. nobukonis*, *Ve. stunga* and *Ve. yusafi* are known to attack and feed on humans in shaded areas, mainly in forest, during the daytime. *Verrallina yusafi* has been collected biting in both daytime and night. Adults of *Ve. ramalingami* were collected in a rubber plantation, but it is not known if the females will feed on humans. Species of *Harbachius* do not appear to be of medical importance.

****

Principal references: Reinert, 1974 (several species, as species of *Aedes* subgenus *Verrallina*, descriptions, keys, distributions, bionomics); Reinert, 1984 (Sri Lanka, several species, as a subgenus of *Aedes*, descriptions, keys, distributions, bionomics); Reinert, 1999f (new subgenus, subgenus and type species descriptions, keys); Reinert, 2001b (female genitalia); Reinert *et al.*, 2004, 2009 (morphology, phylogenetic relationships); Rattanarithikul *et al.*, 2010 (Thailand, keys, bionomics); Soghigian *et al.*, 2017 (phylogenetic relationships).

Neomacleaya Theobald, 1907

Neomacleaya, originally proposed as a genus by Theobald (1907), currently classified as a subgenus of *Verrallina*, is typified by *Verrallina indica* (Theobald, 1907), original combination: *Neomacleaya indica*. Subgenus abbreviation – *Nma*.

****

Systematics: *Neomacleaya* includes 52 extant species. Except for *Ve. atriisimilis* in Japan and *Ve. neomacrodixoa* and *Ve. panayensis* whose distributions extend south-eastward into the Australasian Region (the former reaches the Solomon Islands; both are recorded from Papua Island and Papua New Guinea), the species of *Neomacleaya* appear to be confined to the Oriental Region. *Neomacleaya* was recovered as the sister of *Verrallina sensu stricto* in the morphology-based phylogenetic analysis of Aedini conducted by Reinert *et al.* (2004), with the two subgenera in a sister relationship to *Harbachius*. The latter sister relationship was recovered in the more comprehensive analysis of Aedini conducted by Reinert *et al.* (2009) in which *Verrallina* was paraphyletic relative to *Neomacleaya*. Interestingly, the subgeneric relationships recovered in the study of Reinert *et al.* (2004) were also recovered in the maximum likelihood phylogeny of Soghigian *et al.* (2017) based on molecular markers.

****

Characteristics: The following combinations of characters distinguish species of *Neomacleaya* from the species of *Harbachius* and *Verrallina sensu stricto*. ADULTS: Eyes contiguous or nearly contiguous dorsally; mesepimeron with fine hair-like setae posterior and/or ventral to the scale-patch; hindungues of both females and males simple. FEMALE GENITALIA: Tergum VIII short, wider than long, nearly entirely covered with broad scales; sternum VIII with numerous thin setae on the apical lobes, largely covered with broad scales; postgenital lobe relatively narrow, with a deep median caudal indentation; upper vaginal lip with the median posterior area caudad of the spermathecal eminence developed into a large vertical shield; lower vaginal sclerite large, not contiguous with the lower vaginal lip, usually spiculate; insula ill-defined, small, with a few minute tuberculi; spermathecal eminence complex, with numerous well-developed spicules (less well developed in *Ve. indica*, *Ve. panayensis* and *Ve. rara*), without a cephalad pouch. MALE GENITALIA: Gonocoxite with a moderately broad basotergal apodeme; prosophallus consisting of a pair of sclerites separated apically, about as long as the phallus; phallus comprised of a pair of aedeagal sclerites joined mesally by a sternal bridge, each sclerite with a narrow projection basolaterally extending at approximately a right angle; opisthophallus short, caudal margin truncate or concave; paramere about as long as the phallus. LARVAE: Seta 1-A with 3–6 branches (rarely double); seta 7-C with 5–17 branches (usually 7–14); seta 8-C branched; seta 13-C short, with 2–6 branches (usually 3–5); seta 7-I long, double or triple; comb scales with a stout median apical spine and smaller lateral denticles; seta 2-X multi-branched (with 5 or more branches). PUPAE: Cephalothorax with the ocular facets of the compound eye moderately developed; trumpet usually expanded apically; seta 2-IV–VI usually thin, inserted more caudad than seta 4; seta 9-VIII usually single or double.

****

Bionomics: Virtually nothing is known about the biology of at least 14 species of *Neomacleaya*. The immature stages of the other species are found principally in small temporary bodies of freshwater in forests and plantations, including puddles/ground pools, road ruts, flood pools, residual pools in streambeds, seepage pools and animal hoof prints. *Verrallina atriisimilis*, *Ve. gibbosa* and *Ve. prioekanensis* have been found in forest swamps, *Ve. campylostylus* in a tin can, *Ve. clavata* in crab holes, *Ve. unca* in the axils of *Colocasia* and *Ve. vallistris* in a village well. Females have been collected feeding on cattle, horses and humans during the daylight hours, principally in forests and bamboo groves. The following species are known to feed on humans: *Ve. campylostylus*, *Ve. agrestis*, *Ve. atria*, *Ve. atriisimilis*, *Ve. cauta*, *Ve. clavata*, *Ve. cretata*, *Ve. johorensis*, *Ve. notabilis*, *Ve. panayensis*, *Ve. sabahensis*, *Ve. taeniata*, *Ve. unca*, *Ve. vallistris* and *Ve. varietas*. It seems likely that all species of *Neomacleaya* may attack humans who enter their realms. Species of *Neomacleaya* are not known to harbour or transmit pathogens of human diseases.

****

Principal references: Laffoon, 1946 (Philippines, as *Aedes* subgenus *Aedes*, less *Ve. dux*, *Ve. robertsi* and *Ve. umbrosus*, keys, descriptions, bionomics, distributions); Delfinado, 1967 (Southeast Asia, as a subgenus of *Aedes*, keys, descriptions, distributions, bionomics); Delfinado, 1968 (Southeast Asia, as a subgenus of *Aedes*, keys, descriptions, distributions, bionomics); Reinert, 1974 (several species, as species of *Aedes* subgenus *Verrallina*, descriptions, keys, distributions, bionomics); Reinert, 1984 (Sri Lanka, several species, as a subgenus of *Aedes*, classification, keys, descriptions, distributions, bionomics); Reinert, 1999c (keys, subgenus and type species descriptions, distribution); Reinert, 2001b (female genitalia); Reinert *et al.*, 2004, 2009 (morphology, phylogenetic relationships); Rattanarithikul *et al.*, 2010 (Thailand, keys, bionomics); Soghigian *et al.*, 2017 (phylogenetic relationships).

Verrallina Theobald, 1903 *sensu stricto*

Verrallina sensu stricto, the nominotypical subgenus of *Verrallina*, is typified by *Verrallina butleri* (Theobald, 1901c), original combination: *Aedes butleri*. Subgenus abbreviation – *Ver.*

****

Systematics: *Verrallina sensu stricto* includes 30 extant species. The species occur principally in the Australasian Region, with some minor extensions into the Oriental Region. Three species, *Ve. butleri*, *Ve. dux* and *Ve. lugubris*, are exclusively and widely distributed in the Oriental Region. *Verrallina iriomotensis* is only known from localities in the Ryukyu Islands of Japan. *Verrallina sensu stricto* was recovered in a paraphyletic relationship to *Neomacleaya* in the morphology-based phylogenetic analysis of Aedini conducted by Reinert *et al.* (2009), and the clade comprised of species of these groups was the sister of *Harbachius*. In agreement with the earlier study of Reinert *et al.* (2004), *Verrallina* + *Neomacleaya* was recovered as the sister of *Harbachius* in the maximum likelihood phylogeny of Soghigian *et al.* (2017) based on molecular markers.

****

Characteristics: The following combinations of characters distinguish species of *Verrallina sensu stricto* from species of *Harbachius* and *Neomacleaya*. ADULTS: Eyes contiguous dorsally; mesepimeron with fine hair-like setae posterior and/or ventral to the scale-patch; hindungues of both females and males simple (except the male of *Ve. prioekanensis* with a tooth on one unguis); FEMALE GENITALIA: Tergum VIII largely covered with broad scales; sternum VIII with numerous setae on the caudal lobes, largely with broad scales; postgenital lobe with a broadly rounded truncate or shallowly emarginate apex; upper vaginal lip not produced into a vertical shield; lower vaginal sclerite small, rather poorly developed (except the Butleri Series), separated from the lower vaginal lip; insula ill-defined, small, with a few minute tuberculi; spermathecal eminence consisting of a pair of comma-shaped plates, without spicules (except the Butleri Series) and a cephalad-produced pouch. MALE GENITALIA: Gonocoxite with a broad basotergal apodeme; prosophallus consisting of a pair of separated sclerites each developed apically into a series of overlapping leaf-like plates, more or less as long as the phallus (*Ve. lugubris* with a single leaf-like plate); phallus comprised of a pair of aedeagal sclerites joined near mid-length by a narrow sternal bridge, apices of the sclerites contiguous, angled outward giving the phallus an A-shaped appearance; opisthophallus with the caudal margin gently concave mesally; paramere noticeably shorter than the phallus. LARVAE: Seta 1-A with 3–9 branches (usually 3–6); seta 7-C with 5–14 branches; seta 8-C single (Carmenti Series) or branched (Butleri Series); seta 13-C usually single (with 2–4 branches in *Ve. butleri*, *Ve. iriomotensis* and *Ve. lugubris* of the Butleri Series); seta 7-I long, single; comb scales rounded apically, denticles usually uniformly developed (*Ve. dux* and *Ve. iriomotensis* with a longer median apical denticle); seta 2-X usually double or triple (with 4–8 branches in a few species). PUPAE: Compound eye of the cephalothorax with or without poorly developed ocular facets; seta 2-IV–VI (in most species) somewhat thicker and inserted more cephalad than seta 4; seta 9-VIII usually with 3–6 branches (except some species of the Butleri Series).

****

Bionomics: Species of *Verrallina sensu stricto* typically occur in groundwater habitats in forested areas. Common larval habitats include temporary ground pools, puddles, hoof prints, ditches, marshy depressions, swamp pools, road ruts, rock holes, ponds, pot holes in streambeds and wallows. In additional to some of these habitats, larvae of *Ve. butleri* occur in brackish tidal pools and mangrove swamps in coastal areas, larvae of *Ve. parasimilis* have been found in crayfish and crab holes in forest, and larvae of *Ve. lineata* have once been found in an old kettle. Females of at least 13 species readily feed on humans who enter their realms during the daytime. *Verrallina lineata* and *Ve. reesi* have been captured biting indoors and outdoors, *Ve. carmenti* sometimes enters houses in the evening and *Ve. lineata* and *Ve. funerea* are known to feed at night as well as during the day. *Dirofilaria* and Japanese encephalitis, Bebaru and Getah viruses have been isolated from *Ve. butleri* in Malaysia, but none of the other species of *Verrallina sensu stricto* are known or suspected of being of medical importance.

****

Principal references: King & Hoogstraal, 1947b (New Guinea, as *Aedes* subgenus *Aedes*, descriptions, keys); Belkin, 1962 (South Pacific, as a subgenus of *Aedes*, four species, keys, descriptions, bionomics, distributions); Huang, 1968a (Papuan Subregion, as a subgenus of *Aedes*, keys, descriptions, bionomics, distributions); Reinert, 1974 (several species, as species of *Aedes* subgenus *Verrallina*, descriptions, keys, distributions, bionomics); Reinert, 1984 (Sri Lanka, several species, as a subgenus of *Aedes*, classification, keys, descriptions, distributions, bionomics); Lee *et al.*, 1987a (Australasian Region, as species of *Aedes* subgenus *Verrallina*, taxonomy, keys, literature, distributions, bionomics); Reinert, 1999c (keys, subgenus and type species descriptions, distribution); Reinert, 2001b (female genitalia); Reinert *et al.*, 2004, 2009 (morphology, phylogenetic relationships); Rattanarithikul *et al.*, 2010 (Thailand, bionomics, keys); Soghigian *et al.*, 2017 (phylogenetic relationships).

Wyeomyia Theobald, 1901

Wyeomyia Theobald, 1901a, originally proposed and currently recognized as a genus, is typified by *Wyeomyia grayii* Theobald, 1901c. Genus abbreviation – *Wy.*

****

Systematics: *Wyeomyia* is a subordinate taxon of the subfamily Culicinae, tribe Sabethini. It includes 141 extant species. Twenty-seven species are without subgeneric placement; the remainder are divided between 17 subgenera: *Antunesmyia*, *Caenomyiella*, *Cruzmyia*, *Decamyia*, *Dendromyia*, *Dodecamyia*, *Exallomyia*, *Hystatomyia*, *Menolepis*, *Miamyia*, *Nunezia*, *Phoniomyia*, *Prosopolepis*, *Spilonympha*, *Triamyia*, *Wyeomyia* and *Zinzala*. The species principally occur in South and Central America, but a few species extend across the Caribbean and one extends into eastern North America. Based on morphological data, *Wyeomyia* was recovered as the sister of *Sabethes* in the phylogenetic analysis of Judd (1996), as the sister of *Limatus* + *Sabethes* in the analyses of Harbach & Kitching (1998) and Harbach & Peyton (2000), and was placed in an unresolved relationship with other New World sabethine genera when Harbach *et al.* (2007a) included the Old World *Kimia* in the data set of Harbach & Peyton (2000). A phylogenetic analysis of *Wyeomyia*, principally genus-group relationships, based on morphological and allozyme data (Motta *et al.*, 2007) provided evidence that *Wyeomyia* is not a monophyletic lineage because it includes *Onirion*. Contrary to the findings of Judd (1996), the results of Motta *et al.* showed that *Limatus* and *Phoniomyia* are separate monophyletic lineages outside of *Wyeomyia*; however, the authors pointed out that it was not possible to test the sister-group relationship of *Hystatomyia* and *Phoniomyia* that led Judd (1998) to formally reduce the latter to subgeneric status in *Wyeomyia*. Consequently, the placement of *Phoniomyia* in a sister relationship to *Wyeomyia* (including *Onirion*) may be due to the exclusion of *Hystatomyia* from the data set. Likewise, the placement of *Onirion* within *Wyeomyia* may have been influenced by the exclusion of morphological characters that distinguish the adults, larvae and pupae of *Onirion* from those life stages of *Wyeomyia*. Soghigian *et al.* (2023) included three species of *Wyeomyia sensu stricto* in their phylogenetic analysis of amino acid sequences of 709 genes, but the relationships of the species with each other and with other genus groups of Sabethini included in the analysis were not explicitly indicated.

****

Characteristics: *Wyeomyia* is difficult to characterize because it includes unnatural assemblages of species that require revision. The adults resemble *Limatus* and *Sabethes* more closely than the adults of other genus groups of Sabethini in the New World. They are distinguished from *Limatus* and *Sabethes* by having the scutum covered with scales ranging from a relatively dull bronzy colour with a slight metallic sheen (most species) to metallic gold. They also differ from the adults of *Limatus* in having prespiracular setae, and are distinguished from all species of *Sabethes* except *Sa. petrocchiae* by the presence of prealar setae. Those species that have the proboscis distinctly longer than the forefemur differ from similar species of other New World sabethines in having the upper calypter of the wing without setae. Some species with a short proboscis (about as long as or shorter than the forefemur) resemble certain species of *Shannoniana* and *Trichoprosopon*, but these can be distinguished by the absence of setae on the clypeus and the absence or presence of one to three setae on the upper calypter. *Wyeomyia* larvae are distinguished from the larvae of other New World sabethines by the following features: occipital foramen with dorsolateral slit-like extensions (distinction from *Johnbelkinia*, *Onirion*, *Shannoniana* and *Trichoprosopon*); seta 1-C not exceptionally close together (distinction from *Limatus*); maxillary brush present, rarely absent (distinction from *Johnbelkinia*, *Isostomyia*, *Shannoniana* and *Runchomyia*); maxillary palpus freely attached to the maxillary body (distinction from *Johnbelkinia* and *Sabethes*); seta 8-M present (distinction from *Trichoprosopon*).

****

Bionomics: *Wyeomyia* are principally forest mosquitoes. The larvae inhabit small collections of water in bromeliads and aroids, flower bracts, broken bamboo and bamboo stumps, tree holes, pitcher plants, and occasionally artificial and other containers. Adults are active during the day. They are usually found in damp forests near larval habitats. Various species are found at all elevations in forest canopy, but some seem to be restricted to ground level. Most of the species take blood meals and females readily feed on humans that enter their realm. *Wyeomyia* are not known to vector disease agents and have little if any economic importance; however, Ilheus and Venezuelan encephalitis viruses have been isolated from *Wy. medioalbipes* in Trinidad. Species of *Wyeomyia* engage in courtship displays that result in mating. Foster *et al.* (2024) found that two species of *Wyeomyia sensu stricto*, one species each of *Dendromyia* and *Triamyia* and a species of uncertain subgeneric placement all performed elaborate courtship behaviours and mated on vertical surfaces. The mating behaviour of the species differed in the positioning and movements of the antennae, proboscis, wings, legs, abdomen and genitalia.

****

Principal references: Lane, 1953 (Neotropical Region, taxonomy, keys, descriptions, distributions); Correa & Ramalho, 1956 (*Phoniomyia*, taxonomy, keys, descriptions, distributions); Forattini, 1965b (diagnosis, classification, medically important species, bionomics, distributions); Cova-Garcia *et al.*, 1966 (Venezuela, keys, descriptions, distributions, literature); Belkin *et al.*, 1970 (Jamaica, classification, keys, descriptions, bionomics, distributions); Clark-Gil & Darsie, 1983 (Guatemala, keys); Darsie, 1985 (Argentina, keys); Zavortink, 1986 (*Zinzala*, taxonomy, keys, descriptions, bionomics, distributions); Harbach & Peyton, 1990 (*Caenomyiella*, taxonomy, comparative morphology, type species, description, bionomics, distribution); Harbach & Peyton, 1992a (*Exallomyia*, taxonomy, descriptions, bionomics, distributions); Harbach & Peyton, 1993 (comparative morphology of larval maxillae); Judd, 1996 (morphology, phylogenetic relationships); Harbach & Kitching, 1998 (morphology, phylogenetic relationships); Harbach & Peyton, 2000 (phylogenetic relationships); Motta & Lourenço-de-Oliveira, 2000 (*Dendromyia*, keys, descriptions, bionomics, distributions), Motta & Lourenço-de-Oliveira, 2005 (*Spilonympha*, comparative morphology, descriptions, bionomics, distributions); Harbach *et al.*, 2007a (phylogenetic relationships); Motta *et al.*, 2007 (morphology, allozyme data, classification, phylogenetic relationships); Porter, 2014 (*Nunezia*, comparative morphology); Foster *et al.*, 2024 (courtship and mating behaviour).

Antunesmyia Lane & Cerqueira, 1942

Antunesmyia Lane & Cerqueira, 1942, originally proposed and currently recognized as a subgenus of *Wyeomyia*, is typified by *Wyeomyia alani* Lane & Cerqueira, 1957 [replacement name for *Wyeomyia rooti* Lane & Cerqueira, 1942, preoccupied by *Wyeomyia rooti* (del Ponte, 1939)]. Subgenus abbreviation – *Ant.*

****

Systematics: *Antunesmyia* includes three extant species which are only recorded from Brazil (*Wy. alani*, *Wy. flavifacies*), Colombia (*Wy. colombiana*, *Wy. flavifacies*) and Guyana (*Wy. flavifacies*). *Antunesmyia*, represented by *Wy. alani*, was recovered as the sister to *Miamyia* in the morphology-based phylogeny of *Wyeomyia* generated in the study of Motta *et al.* (2007).

****

Characteristics: The species of *Antunesmyia* are poorly known. The male, larva and pupa are known only for *Wy. alani*. The following is based principally on the morphology of that species. ADULTS: Proboscis short, expanded distally; scutum dark-scaled; mesopostnotum with scales; paratergite without scales; antepronotum dark-scaled dorsally, pale-scaled ventrally; lower mesokatepisternal setae extending above the dorsal margin of the mesomeron; upper calypter of the wing without marginal setae. MALE GENITALIA: Sternum VIII with posterolateral digitiform processes with a row of long setae on the inner margin and several thick peg-like setae on the narrowed apex; tergum IX lobes small, distinctly separated, each with a close-set row of thick tapered setae; gonocoxite elongate, distally attenuated, with a long tergal triad; gonostylus with a long narrow stem and well-developed distal setose lobes; aedeagus broad, width about 0.67 the length, apical tergal arms fused (forming an apical tergal bridge); proctiger with distinct apical teeth, cercal setae present. LARVAE: Seta 1-A single, flexible, inserted on the distal 0.33 of the antenna; setae 4–6-C single, widely separated, seta 5-C inserted far posterior to the level of seta 7-C, seta 6-C inserted anterior to 7-C approximately midway between setae 4- and 7-C; seta 3-Mx of the maxilla not borne in a lateral cup-like notch, seta 4-Mx forked, seta 6-Mx branched; seta 11-P,M,T single, 11-P,M piliform, 11-T spiniform; comb plate present, comb scales large, spine-like, borne in a single row on the posterior margin of the plate; siphon without a pecten, seta 1-S branched, inserted near mid-length of the siphon, seta la-S represented by 1 or 2 small posterolateral setae near the apex of the siphon, seta 2a-S with 2 small anterolateral setae near mid-length of the siphon; saddle with large spine-like spicules on the posterolateral margins; seta 4-X multi-branched, long, nearly as long as seta 1-X. PUPAE: Cephalothorax and abdomen without spots; trumpet short, broad, slightly expanded basally; seta 2-III–VII inserted mesal to seta 1; seta 6-VII inserted on the sternum ventrad of seta 9-VII; seta 9-VII,VIII strongly developed; paddle short, shorter than tergum VIII, acutely narrowed distally, triangular, apex spiculate.

****

Bionomics: The holotype female of *Wy. flavifacies* was reared from a larva taken from a bromeliad (Edwards, 1922d; Belkin, 1968b). Belkin *et al.* (1971) stated that larvae of *Wy. alani* 'probably inhabit bamboo', and Motta *et al.* (2007) mentioned that larvae of *Antunesmyia* and *Wyeomyia sensu stricto* 'inhabit bamboo-internode habitats'. Nothing is known about the bionomics of the adults.

****

Principal references: Lane, 1953 (in key to adult females, descriptions, distributions); Motta *et al.*, 2007 (morphology, phylogenetic relationships).

Caenomyiella Harbach & Peyton, 1990

Caenomyiella Harbach & Peyton, 1990, originally proposed and currently recognized as a subgenus of *Wyeomyia*, is typified by *Wyeomyia fernandezyepezi* (Cova Garcia, Sutil Oramas & Pulido F., 1974b), original combination: *Sabethes fernández-yepezi*. Subgenus abbreviation – *Cae.*

****

Systematics: *Caenomyiella* includes two extant species, the type species, known only from Panama and Venezuela, and an unnamed species in Colombia. The evolutionary affinities of *Caenomyiella* are unknown. It was not represented in the phylogenetic study of Sabethini conducted by Judd (1996) nor the phylogenetic study of *Wyeomyia* conducted by Motta *et al.* (2007). Harbach & Peyton (1990) noted that a number of morphological similarities suggest *Caenomyiella* is more closely related to *Decamyia*, *Wyeomyia* and *Zinzala* than to other genus groups of *Wyeomyia*.

****

Characteristics: Species of *Caenomyiella* exhibit the following principal anatomical features. ADULTS: Female unknown; male incompletely known, probably resembling the habitus of the species of *Decamyia* and *Wyeomyia sensu stricto*. Proboscis slightly shorter than the forefemur; mesopostnotal setae present. MALE GENITALIA: Tergum IX lobes small, with strongly developed setae; gonocoxite without long setae except for a long lateral seta and the tergal triad; gonostylus with a strongly developed lobe C, borne on the sternomesal margin near the base; aedeagus broad, width about 0.67 the length, submedian tergal arms weakly connected or unconnected, apical tergal arms fused (forming an apical tergal bridge); proctiger normal, cercal setae present. LARVAE: Seta 1-A single, inserted near the apex of the antenna; setae 4–6-C single, widely separated, seta 6-C inserted anterior to seta 7-C; seta 14-C strongly developed, significantly longer than seta 15-C, inserted anterior to setae 12,15-C; maxilla broad, with a strong apical tooth, seta 4-Mx single, shorter than the maxillary brush, hypostomal sclerite present, bearing seta 6-Mx; seta 1-P with few branches, as long as or slightly shorter than setae 2,3-P; seta 4-P relatively short, multi-branched; seta 11-P,M,T single, short, spine-like; seta 7-M weakly developed, branched, similar to seta 1-M; seta 4-T small, single or double; seta 8-T inserted laterally between seta 7-T and the sclerite bearing setae 9–12-T; seta 13-T multi-branched, not much longer than 0.5 the length of the thorax; seta 1-I,II small, developed like seta 4, seta 1-III–VI usually double, about 0.5 the length of the segment; seta 2-I–VII short, single, seta 2-III–VII inserted far mesad of seta 1 near the midline of the segment; seta 9-I–VI short, single; seta 13-II inserted far anterior to seta 7 and near the midline, seta 13-III–VI inserted on the level with or anterior to seta 7; comb plate absent, dorsal comb scales long, pointed, with a short fringe proximally at the sides, ventral scales short, broad, fringed distally; siphon with a single mid-posterior row of pecten spines on the distal 0.67, seta 1-S inserted near the base of the siphon, setae la,2a-S each in a more or less straight row; seta 4-X much shorter than setae 1–3-X, double. PUPAE: Cephalothorax and abdomen without spots; trumpet slightly flattened, not laterally expanded; seta 1-I strongly developed, with 3 or 4 primary stems and numerous distal branches; seta 2-II–VII inserted near the posterior margin of the segments, seta 2-II inserted between setae 1,3-II, seta 2-III–VII inserted mesal to seta 1; seta 3-II,III,VII inserted lateral to seta 1, seta 3-V,VI inserted mesad of seta 1; seta 6-II single, long, reaching well beyond the posterior margin of tergum III, seta 6-VII inserted on the sternum ventrad of seta 9-VII; paddle short, length less than 1.5 times the greatest width, extending slightly beyond the apex of the genital lobe in both sexes, apex fringed with long spicules that decrease rapidly and progressively on both the inner and outer margins.

****

Bionomics: The immature stages of species of *Caenomyiella* have only been found in bromeliads. Nothing is known about the bionomics of the adults. Species of *Caenomyiella* are not known or suspected of being vectors of pathogens of human diseases.

****

Principal reference: Harbach & Peyton, 1990 (new subgenus, type species, descriptions, comparative morphology, bionomics, distribution).

Cruzmyia Lane & Cerqueira, 1942

Cruzmyia Lane & Cerqueira, 1942, originally proposed and currently recognized as a subgenus of *Wyeomyia*, is typified by *Wyeomyia dyari* Lane & Cerqueira, 1942. Subgenus abbreviation – *Cru.*

****

Systematics: *Cruzmyia* includes four extant species, which are only recorded from Brazil (*Wy. dyari*, *Wy. kummi*, *Wy. mattinglyi*), Colombia (*Wy. kummi*, *Wy. mattinglyi*) and French Guiana (*Wy. forattinii*). The results of the morphology-based phylogenetic study of Motta *et al.* (2007) suggest that *Cruzmyia* is a basal lineage within *Wyeomyia*. Judd (1996) recovered *Cruzmyia* and *Limatus* as sister groups, also based on phylogenetic analysis of morphological data.

****

Characteristics: Species of *Cruzmyia* exhibit the following principal anatomical features. ADULTS: Proboscis longer than the forefemur; mesopostnotal setae present; paratergite without scales; lower mesokatepisternal setae not extending above the dorsal margin of the mesomeron; upper calypter of the wing without marginal setae. MALE GENITALIA: Tergum IX lobes weakly developed, widely separated, each with 3 or 4 stout apically blunt setae; gonocoxite without long setae except for a long lateral seta, tergal triad not apparent; gonostylus with a long narrow stem and rather small head comprising lobes M, AE and C, lobe C long and slender; aedeagus broadest in the distal 0.67, apical tergal arms not connected; paraproct with an apical tooth, cercal setae present. LARVAE: Seta 1-A single, inserted on the distal 0.25 of the antenna; setae 4–7-C single, seta 5-C inserted anterior to seta 7-C; seta 14-C strongly developed, significantly longer than seta 15-C, inserted anterior to seta 15-C; seta 3-Mx of the maxilla not borne in a lateral cup-like notch, seta 4-Mx single, truncate, seta 6-Mx branched; seta 1-P with few branches, inserted mesal to setae 2,3-P; seta 4-P relatively short, multi-branched; seta 11-P,M,T short, hair-like; seta 8-T inserted anterior to the sclerite bearing setae 9–12-T; seta 13-T multi-branched, nearly as long as the thorax; seta 1-I–VII branched, similarly developed; seta 13-I inserted anterior to the other ventral setae, 13-II,III inserted more or less on the level with seta 7, 13-IV–VI inserted distinctly posterior to seta 7; comb plate present, plate with a posterior row of strong spine-like scales; siphon without a pecten, seta 1-S inserted about 0.3 from the base of the siphon, seta la-S with relatively few elements in a straight posterolateral row beyond the level of seta 1-S, seta 2a-S in anterolateral and lateral rows with many and few setae, respectively; saddle without posterolateral spicules; seta 4-X much shorter than setae 1–3-X, multi-branched. PUPAE: Cephalothorax and abdomen with darkly pigmented areas; seta 1-I strongly developed, dendritic; seta 2-II inserted lateral to seta 1, 2-III–VII inserted anterior to setae 1 (at approximately the mid-lengths of terga IV–VI); seta 3-II inserted lateral and 3-III–VI inserted anterior to seta 1; seta 6-II single, long, longer than the following tergum, seta 6-VII inserted on the tergum posteromesal to seta 9; paddle long, much longer than tergum VIII, gradually narrowed to the apex, apex and the lateral margins with minute spicules.

****

Bionomics: Larvae of *Wy. dyari* and *Wy. forattinii* have been found in bromeliads (Clastrier, 1974; Heinemann & Belkin, 1978b, 1979) and larvae of the other two species of *Cruzmyia* probably also develop in bromeliads (Belkin *et al.*, 1971). Nothing is known about the bionomics of the adults.

****

Principal references: Lane, 1953 (in key to adult females, descriptions, distributions); Judd, 1996 (morphology, phylogenetic relationships); Motta & Lourenço-de-Oliveira, 2005 (comparative morphology); Motta *et al.*, 2007 (morphology, phylogenetic relationships).

Decamyia Dyar, 1919

Decamyia, originally proposed as a genus by Dyar (1919b), currently classified as a subgenus of *Wyeomyia*, is typified by *Wyeomyia onidus* Dyar & Knab, 1909 [junior synonym of *Wyeomyia ulocoma* (Theobald, 1903a)]. Subgenus abbreviation – *Dec.*

****

Systematics: *Decamyia* includes six extant species recorded from countries in northern South America and Central America southward from Guatemala, Honduras and the Caribbean. *Decamyia* was recovered as a monophyletic group in a sister relationship to a clade comprised of *Menolepis*, *Prosopolepis*, two species without subgeneric placement (*Wy. melanocephala* and *Wy. negrensis*) and *Onirion personatum* in the morphology-based phylogenetic analyses of Motta *et al.* (2007).

****

Characteristics: Species of *Decamyia* have not been studied in detail, particularly the immature stages. The following combinations of features should distinguish the four known species from species of the other genus groups of *Wyeomyia*. ADULTS: Proboscis short, 0.7–0.8 the length of the forefemur, expanded distally; scutum dark-scaled; paratergal scales present; antepronotum dark-scaled at least dorsally; lower mesokatepisternal setae extend above the dorsal margin of the mesomeron; upper calypter of the wing with marginal setae. MALE GENITALIA: Tergum IX lobes not developed, delineated by foliform setae (resembling fusiform or spatulate scales); gonocoxite with a long lateral seta and 1 (?) or 2 setae representing the tergal triad; gonostylus with a long stem, apical lobes not well developed; paraproct with 1 or 2 apical teeth, cercal setae present. LARVAE: Seta 1-A single, inserted beyond mid-length of the antenna; setae 4–6-C single or branched, widely separated, 6-C inserted anterior to seta 7-C; seta 14-C inserted on the level with seta 15-C; maxilla without an apical tooth; seta 11-P,M branched, 11-T single, spine-like; seta 4-T multi-branched; seta 13-T multi-branched, about as long as the thorax; seta 2-I,III branched, seta 2-II,IV–VII short, single, seta 2-III–VII inserted mesad of seta 1 but not near the midline of the segment, inserted noticeably anterior to setae 1 and 4; seta 9-I–VI branched; seta 13-II–VI inserted posterior to the other ventral setae, 13-III–VI inserted on the level with or anterior to seta 7; comb plate absent, comb scales in 2 of 3 rows; siphon with 2 posterolateral rows of pecten spines; saddle with posterolateral spines; seta 4-X noticeably shorter than setae 1–3-X, branched. PUPAE: Cephalothorax and abdomen without spots; seta 1-I strongly developed, with 3 or 4 primary stems and numerous apical branches; seta 2-II inserted between and on the same level or slightly anterior to setae 1 and 3, seta 2-III–VII inserted very near the posterior margins of the terga, inserted mesad of seta 1, seta 2-III inserted far mesad of seta 1; seta 6-II single, long, at least twice as long as seta 7-II, seta 6-VII inserted on the tergum posterior to seta 9-VII; paddle relatively long, longer than 1.5 times the greatest width, margins fringed entirely with long filamentous spicules.

****

Bionomics: The immature stages of species of *Decamyia* are predominantly found in the flower bracts and axils of *Heliconia*. They have also been found in bromeliads (*Wy. felicia*, *Wy. pseudopecten*, *Wy. ulocoma*); axils of *Aracea* (Araceae) and *Calathea* (Marantaceae) (*Wy. pseudopecten*); leaf axils of palm, fallen palm spathes, leaf axils of banana (*Wy. felicia*); and sections of split bamboo and tree holes (*Wy. ulocoma*) (Heinemann & Belkin, 1977a, 1978a, 1978b, 1978c, 1979; Heinemann *et al.*, 1980; Pecor *et al.*, 2000; Linton *et al.*, 2013). Little is known about the bionomics of the adults. Females of *Wy. ulocoma* have been taken on humans in Suriname (Panday, 1975) and *Wy. pseudopecten* has been captured on humans in Suriname and Peru (Panday, 1975; Pecor *et al.*, 2000, respectively).

****

Principal references: Harbach & Peyton, 1990 (comparative morphology); Motta *et al.*, 2007 (morphology, phylogenetic relationships).

Dendromyia Theobald, 1903

Dendromyia, originally proposed as a genus by Theobald (1903a), currently classified as a subgenus of *Wyeomyia*, is typified by *Wyeomyia luteoventralis* Theobald, 1901b. Subgenus abbreviation – *Den.*

****

Systematics: *Dendromyia* includes six extant species in tropical forests of countries in Central and South America between latitudes of 20° N and 20° S. An initial eletrophoretic analysis of allozyme data obtained from three species of *Dendromyia* and three species of other groups of *Wyeomyia* indicated that *Dendromyia* was a distinct taxon (Motta *et al.*, 1998). In a later study (Motta *et al.*, 2007), *Dendromyia* was recovered as a monophyletic clade based on morphological and allozyme data obtained from a larger number of species. Although the analysis of morphological data alone did not reveal close affinities with other genus groups of *Wyeomyia*, *Dendromyia* was recovered in a sister relationship with *Wy.* (*Triamyia*) *aporonoma* + (*Wy. melanocephala* + *Onirion personatum*) when the morphological data were combined with the allozyme data.

****

Characteristics: Species of *Dendromyia* differ principally from species of other genus groups of *Wyeomyia* in features of the male genitalia, larvae and pupae; members are morphologically distinct from but more similar to species of *Caenomyiella* than to species of the other genus groups. ADULTS: Clypeus with or without scales; mesopostnotum with a patch of numerous setae, no scales; paratergite without scales; lower mesokatepisternal setae extend above the dorsal margin of the mesomeron; upper calypter of the wing with or without marginal setae. MALE GENITALIA: Tergum IX lobes variable, indistinct or noticeably produced, with strongly developed setae; gonocoxite with a long lateral seta and a tergal triad; gonostylus without a stem, with 2 or 3 lobes arising at the base (lobe C, lobe M and lobe AE or AE/M), lobes variously developed; aedeagus as broad as long or longer than wide, broadest at mid-length, submedian tergal arms not joined; paraproct with 1 or 2 apical teeth, cercal setae present. LARVAE: Seta 1-A single, inserted near the apex of the antenna; setae 4–6-C single, 5-C inserted more or less level with or slightly anterior to seta 7-C, seta 6-C inserted anterolateral to seta 5-C on a line between setae 4- and 7-C; seta 14-C strongly developed, significantly longer than seta 15-C, inserted posterior to or on the level with setae 12,15-C; maxilla elongate, with a strongly developed apical tooth, seta 4-Mx single, stout, shorter than the maxillary brush, hypostomal sclerite ill-defined, seemingly fused to the maxillary body and maxillary palpus, seta 6-Mx small, single; seta 1-P usually single, inserted posterolateral to and longer than setae 2,3-P; seta 4-P relatively short, multi-branched; seta 11-P,M,T single, short, spine-like; seta 7-M small, branched, similar to seta 4-M; seta 4-T small, with 1–4 branches; seta 8-T inserted anterior to anterolaterad of the sclerite bearing setae 9–12-T; seta 13-T multi-branched, not much longer than 0.5 the length of the thorax, inserted near the lateral margin posterior to posteromesad of setae 9–12-T; seta 1-I,II small, usually similar in size to seta 2, seta 1-III–VI branched, 0.5–1.0 the length of the segment; seta 2-I–VII short, single, seta 2-III–VII inserted anteromesal to seta 1; seta 9-I–VI short, single; seta 13-I inserted near the anterior margin of the segment, 13-II–VI inserted slightly anterior to seta 7 on or near mid-length of the segment; comb plate absent, dorsal comb scales spine-like with a short fringe at the sides, ventral scales spine-like or squamiform with apical and lateral fringes; siphon with a single mid-posterior row of pecten spines beyond the level of seta 1-S, seta 1-S inserted some distance from the base of the siphon, only 1–4 seta la-S and normally 3 seta 2a-S present; seta 4-X much shorter than setae 1–3-X, multi-branched. PUPAE: Cephalothorax and abdomen without spots; trumpet gradually slightly broadened distally, narrowly funnel-like; seta 1-I strongly developed, dendritic, with 3 or 4 primary stems and usually numerous apical branches; setae 1,2-II–VII inserted near the posterior margin of the tergum, seta 2-I inserted lateral to seta 1, seta 2-II–VII inserted a greater distance mesal to seta 1; seta 3-II,III,VII inserted lateral to seta 1, seta 3-IV inserted anterior and seta 3-V,VI inserted mesal to seta 1; seta 6-II single, long, reaching well beyond the posterior margin of tergum III, seta 6-VII inserted on the sternum ventrad of seta 9-VII; paddle longer than tergum VIII, elongate triangular in shape, with a bluntly pointed apex fringed with short to long spicules that diminish in length and presence on both the inner and outer margins.

****

Bionomics: Little is known about the bionomics of *Dendromyia* species. Females of *Wy. luteoventralis*, *Wy. testei* and *Wy. ypsipola* have been collected on humans in forests during daylight hours. Immature stages have been found in collections of water in the axils of bromeliads and species of *Aracea*, *Calathea*, *Dieffenbachia*, *Heliconia* and *Montrichardia*. The larvae of *Wy. luteoventralis*, *Wy. testei* and *Wy. ypsipola*, and probably the other species of *Dendromyia*, aggressively attack and kill other larvae but do not eat them.

****

Principal references: Motta & Lourenço-de-Oliveira, 1995 (included species, type species, description, bionomics, distributions); Motta *et al.*, 1998 (allozymes, phylogenetic relationships); Motta & Lourenço-de-Oliveira, 2000 (keys, descriptions, bionomics, distributions); Motta & Lourenço-de-Oliveira, 2005 (comparative morphology); Motta *et al.*, 2007 (morphology, allozymes, phylogenetic relationships).

Dodecamyia Dyar, 1918

Dodecamyia, originally proposed as a genus by Dyar (1919b), currently classified as a subgenus of *Wyeomyia*, is typified by *Wyeomyia aphobema* Dyar, 1918c. Subgenus abbreviation – *Dod.*

****

Systematics: *Dodecamyia* is monobasic. It only includes the extant type species, which is known only from South America, with collections reported from Argentina, Bolivia, Brazil, Colombia, Ecuador, French Guiana, Guyana, Peru and Suriname. *Wyeomyia aphobema* was recovered as the sister to *Wy.* (*Nunezia*) *bicornis* + (*Phoniomyia* + an undescribed species of *Dendromyia*) in the study of Judd (1996) based on phylogenetic analyses of 59 morphological characters of larvae, pupae and adults. In contrast, the phylogenetic analyses of Motta *et al.* (2007) based on 88 characters from the three life stages indicated that *Dodecamyia* is more closely related to *Exallomyia* than to other genus groups of *Wyeomyia*. Some features of *Wy. aphobema*, *e.g.* the simple gonostylus of the male genitalia, indicate that *Dodecamyia* is a more primitive taxon of *Wyeomyia*, but the phylogenetic relationships of *Dodecamyia* must remain uncertain until most species of *Wyeomyia* are better known in all life stages.

****

Characteristics: It is likely that this currently monobasic subgenus is actually a group of closely related, morphologically similar species; however, the following combinations of characters should distinguish the group from the other genus groups of *Wyeomyia*. ADULTS: Proboscis short, slightly shorter than the forefemur, turned downward, expanded distally; scutum dark-scaled; mesopostnotum without scales and setae; paratergite bare; upper calypter of the wing without marginal setae. MALE GENITALIA: Tergum IX lobes not produced, each delineated by a row of 4 or 5 stout setae; gonocoxite without long setae, with a single stout mesal seta and prominent close-set row of 7–9 similar setae (homologous with the tergal triad); gonostylus attached to the apex of the gonocoxite, elongate, slender, simple, without expanded distal lobes; aedeagus broad, lateral plates each with a series of posterolateral peg-like projections proximal to a narrow distal extension (leaf-like in lateral view), apical tergal arms not fused; proctiger normal, paraproct with several apical teeth, cercal setae present. LARVAE: Seta 1-A single, inserted near the apex of the antenna; seta 14-C strongly developed, significantly longer than seta 15-C, inserted on level with or posterior to seta 15-C; maxilla broad, without an apical tooth, seta 4-Mx single, stout, brush-tipped, much longer than the maxillary brush, hypostomal sclerite not fused with the maxillary body or maxillary palpus; seta 4-P not fan-like; seta 11-P,M,T single, not spine-like; seta 13-III,IV inserted anterior to seta 12; comb plate absent, comb scales in several rows (patch); siphon with 2 posterolateral rows of pecten spines on the proximal 0.75, seta 1-S branched, inserted near the base of the siphon, setae 1a,2a-S numerous, not in distinctly straight rows; saddle without posterolateral spicules; seta 2-X single; seta 4-X much shorter than setae 1–3-X, multi-branched. PUPAE: Cephalothorax and abdomen without spots; seta 2-V inserted mesal to seta 1; seta 6-II single, long, reaching well beyond the posterior margin of tergum III, seta 6-VII inserted on the tergum posterior to seta 9; seta 9-VI hair-like; paddle relatively long, length 1.5 the length of tergum VIII, apex without spicules.

****

Bionomics: Larvae of *Wy. aphobema* inhabit epiphytic and terrestrial bromeliads (Heinemann & Belkin, 1978b; Peyton *et al.*, 1983; Linton *et al.*, 2013), and have been found at the bases of leaves of pineapple (Edwards, 1922d). Nothing is known about the bionomics of the adults.

****

Principal references: Judd, 1996 (morphology, phylogenetic relationships); Motta & Lourenço-de-Oliveira, 2005 (comparative morphology); Motta *et al.*, 2007 (morphology, phylogenetic relationships).

Exallomyia Harbach & Peyton, 1992

Exallomyia Harbach & Peyton, 1992a, originally proposed and currently recognized as a subgenus of *Wyeomyia*, is typified by *Wyeomyia tarsata* Lane & Cerqueira, 1942. Subgenus abbreviation – *Exm.*

****

Systematics: *Exallomyia* includes three extant species, which are known to occur in countries at latitudes roughly between 15° N and 15° S. *Wyeomyia arborea* is recorded from Panama and Nicaragua, *Wy. carrilloi* from Venezuela and *Wy. tarsata* from Bolivia, Brazil and Colombia. The phylogenetic analyses of Motta *et al.* (2007) based on morphological data indicated a closer relationship of *Exallomyia* with *Dodecamyia* than to other genus groups of *Wyeomyia*. However, as suggested by Harbach & Peyton (1992a), the phylogenetic relationships of *Exallomyia* cannot be known for certain until most species of *Wyeomyia* are thoroughly characterized in the immature stages and are known to represent natural groups of species.

****

Characteristics: *Exallomyia*, like the other genus groups of *Wyeomyia*, is diagnosed principally on morphological features of the male genitalia and larvae. ADULTS: Not differentiated much from the adults of the other genus groups of *Wyeomyia*; paratergal scales and mesopostnotal setae present; lower mesokatepisternal setae not extending above the dorsal margin of the mesomeron; upper calypter of the wing with 1 or 2 marginal setae. MALE GENITALIA: Tergum IX without lobes, with closely approximated rows of setae; gonocoxite with a prominent tergomesal group of setae, gonostylus with a flap-like lobe C; aedeagus narrow, distinctly longer than wide, apical tergal arms not joined; proctiger normal, cercal setae present. LARVAE: Seta 1-A multi-branched, inserted at or before mid-length of the antenna; seta 6-C inserted anterior to seta 7-C; seta 14-C very strongly developed, significantly longer than seta 15-C, inserted anterior to 15-C on the level with seta 12-C; maxilla broad, without an apical tooth, seta 4-Mx strongly developed, longer than the maxillary brush, split at the apex, hypostomal sclerite absent, maxillary palpus small, seta 6-Mx inserted on the basal margin of the maxillary body; seta 1-P multi-branched, much shorter and inserted well mesad and distinctly caudad of setae 2,3-P; seta 4-P long, double; seta 11-P,M,T well developed, with multiple branches; seta 7-M not strongly developed, single; seta 4-T developed like seta 1-T; seta 8-T inserted laterally between the sclerites bearing seta 7-T and setae 9–12-T; seta 13-T very strongly developed, longer than the length of the thorax; seta 1-I,II well developed, multi-branched, developed like seta 2-III–VII, seta 1-III–VI very long, single, borne on a basal plate; seta 2-I–VII well developed, with multiple branches, seta 2-III–VII inserted mesad and on the same level or slightly posterior to seta 1; seta 9-I–VI well developed, with multiple branches; setae 7,13-II inserted more or less on the same level, 13-III–VI inserted distinctly posterior to seta 7; comb plate absent, ventral comb scales small, fringed with spicules, becoming larger and progressively spine-like dorsally; siphon without a pecten, seta 1-S inserted near the base of the siphon, setae la,2a-S each in a more or less straight row; seta 4-X developed like setae 1–3-X, very long, single or double. PUPAE: Cephalothorax and abdomen without spots; trumpet slightly flattened, not laterally expanded, pinna shorter than the greatest width of the meatus; seta 1-I strongly developed, with numerous dendritic branches; seta 2-II–VII inserted near the posterior margins of the terga, 2-II,VII inserted lateral to seta 1, 2-III–VI inserted mesal to seta 1; seta 3-III inserted mesad of seta 1, 3-VII more or less in line with seta 1; seta 6-II variable, slightly shorter to longer than segment III, seta 6-VII inserted on the tergum caudomesad of seta 9-VII; paddle long, length about 1.5 times the width, apex fringed with minute spicules.

****

Bionomics: Very little is known about the bionomics of the three known species of *Exallomyia*. Larvae have been found in tree holes (*Wy. carrilloi*, *Wy. tarsata*) and bamboo internodes (*Wy. arborea*, *Wy. carrilloi*). Unconfirmed reports indicate that females of *Wy. tarsata* will feed on humans during the daytime. Females of *Wy. arborea* have been collected feeding on humans in forest canopy.

****

Principal references: Harbach & Peyton, 1992a (new subgenus, descriptions, bionomics, distributions); Motta & Lourenço-de-Oliveira, 2005 (comparative morphology); Motta *et al.*, 2007 (morphology, phylogenetic relationships).

Hystatomyia Dyar, 1919

Hystatomyia, originally proposed as a genus by Dyar (1919b), currently classified as a subgenus of *Wyeomyia*, is typified by *Wyeomyia circumcincta* Dyar & Knab, 1907c. Subgenus abbreviation – *Hys.*

****

Systematics: *Hystatomyia* includes nine extant species. The distributions of the species are poorly known. Only two of the species are known from more than one country. *Hystatomyia* (as *Dendromyia* 'species A group') was recovered as the sister of *Phoniomyia* in the phylogenetic analysis of combined larval, pupal and adult morphological data in the study of Sabethini conducted by Judd (1996). Later analyses (Judd, 1998) of the combined larval, pupal and adult data for clades most closely associated with *Phoniomyia* and *Hystatomyia* generated two topologies, one with the sister relationship of *Hystatomyia* + *Phoniomyia* and the other with *Hystatomyia* sister to a clade comprised of *Dodecamyia* + ((*Phoniomyia* + *Nunezia*) + (*Wy. albosquamata* + (*Wy. grayii* + *Onirion personatum*))). *Wyeomyia albosquamata* is not placed in a currently recognized genus group and *Wy. grayii* is a member of *Wyeomyia sensu stricto*. The former relationship was retained when the larval data were analysed alone, and the latter was recovered when only adult characters were analysed. *Hystatomyia* was not represented in the phylogenetic study of *Wyeomyia* conducted by Motta *et al.* (2007).

****

Characteristics: Species of *Hystatomyia* are readily distinguished from species of the other genus groups of *Wyeomyia* by the unusual development of the male genitalia and the pupal trumpet. ADULTS: Proboscis relatively stout, shorter than the forefemur, expanded distally; mesopostnotum with a cluster of setae; upper calypter of the wing without marginal setae. MALE GENITALIA: Tergum IX lobes more or less distinct, each with a row of stout setae; gonocoxite narrowed and extended well beyond the attachment of the gonostylus, lateral surface densely covered with scales, extended portion with long apical and tergomesal setae, broader portion proximal to the gonostylus with close-set mesal and tergomesal rows of long setae; gonostylus weakly developed, attached mesally near or proximal to mid-length of the gonocoxite, slender with apex slightly to distinctly expanded and with or without setae; aedeagus longer than broad, submedian tergal arms connected or fused apically, bowed outward subapically and fused at the apex (forming an apical tergal bridge), inner sternomesal margins of the lateral plates sternad of the apical tergal bridge minutely serrate or denticulate; proctiger without distinct apical teeth, cercal setae present. LARVAE: Seta 1-A branched, inserted on the distal 0.25 of the antenna; setae 4–6-C multi-branched, 5-C inserted more or less on the level with seta 7-C, 6-C inserted well anterior to seta 7-C; seta 14-C moderately developed, usually double, much longer than seta 15-C, inserted posterior to setae 12,15-C; maxilla longer than broad, without an apical tooth, seta 3-Mx borne in a cup-like lateral notch (as in species of *Phoniomyia*), seta 4-Mx single, stout, brush-tipped, as long as the maxillary brush, hypostomal sclerite reduced, more or less united with the maxillary body, bearing a single relatively long seta 6-Mx; seta 1-P multi-branched, inserted mesad of setae 2,3-P; seta 4-P long, double or triple; seta 11-P,M,T well developed, multi-branched; seta 7-M single, relatively long, much longer than seta 1-M; seta 13-T multi-branched, length varying from about 0.5 to nearly the length of the thorax; seta 1-I,II relatively large, multi-branched, similar to seta 1-III–VIII; seta 2-I usually inserted anterolateral and seta 2-II usually inserted anterior to seta 1, seta 2-III–VII inserted mesad of seta 1; seta 9-I–VI multi-branched; seta 13-II inserted anterior to the other ventral setae; seta 13-III–VI inserted on the level with or posterior to seta 7; comb plate absent, comb with numerous long fringed scales in 2 or 3 long rows; siphon with a single mid-posterior row of long distally minutely spiculate pecten spines, seta 1-S branched, usually inserted proximal to seta 1a-S, setae la,2a-S each in a straight row of numerous long single elements; seta 2-X with 2 or 3 branches; seta 4-X much shorter than setae 1–3-X, multi-branched. PUPAE: Cephalothorax and abdomen without spots; setae 1,5-CT very long, about as long as the cephalothorax; trumpet exceptionally long and slender, about as long as setae 1,5-CT; seta 1-I strongly developed, with 3 or 4 primary stems and numerous apical branches; seta 2-II inserted anterior to seta 1, seta 2-III–VII inserted mesal to seta 1; seta 3-II,III,VII inserted lateral to seta 1, seta 3-II,II single and very long, longer than the following 2 terga, seta 3-IV–VI inserted anterior to seta 1; seta 6-II single, very long, longer than the following 2 terga, seta 6-VII inserted on the tergum posteromesal to seta 9; seta 9-VII,VIII branched, fan-like; paddle short, length less than 1.5 times the length of tergum VIII, gently narrowed toward the apex, outer margin and the distal 0.5 of the inner margin with short spicules.

****

Bionomics: The immature stages of species of *Hystatomyia* develop in terrestrial and epiphytic bromeliads in humid tropical lowland forest environments (Porter, 2014). Nothing is known about the bionomics of the adults.

****

Principal references: Judd, 1996 (as *Dendromyia* 'species A group', morphology, phylogenetic relationships); Judd, 1998 (subgeneric status, morphology, taxonomy, phylogenetic relationships).

Menolepis Lutz, 1905

Menolepis, originally proposed as a genus by Lutz (1905), currently classified as a subgenus *Wyeomyia*, is typified by *Wyeomyia leucostigma* Lutz, 1904 (in Bourroul, 1904). Subgenus abbreviation – *Men.*

****

Systematics: *Menolepis* is monobasic. It only includes the extant type species, the distribution of which is poorly known. It is only recorded from localities in Argentina, Bolivia and Brazil. *Menolepis* was recovered as the sister to *Wy. negrensis* in the morphology-based phylogeny of *Wyeomyia* generated by Motta *et al.* (2007). A clade comprised of *Prosopolepis* + (*Onirion personatum* + *Wy. melanocephala*) was the sister of *Menolepis* + *Wy. negrensis*. *Wyeomyia melanocephala* and *Wy. negrensis* are currently without subgeneric placement.

****

Characteristics: *Menolepis* has not been studied in detail, but the following combinations of features should distinguish the only known species from species of the other genus groups of *Wyeomyia*. ADULTS: Proboscis short, expanded apically; scutum and scutellum dark-scaled; mesopostnotum with a patch of silvery scales anterior to the setae; paratergite without scales; antepronotum with iridescent blue scales; lower mesokatepisternal setae extending above the dorsal margin of the mesomeron; upper calypter of the wing without marginal setae. MALE GENITALIA: Tergum IX lobes small, well separated, each with 4 or 5 close-set apically curved setae; gonocoxite with a tergal triad; gonostylus short, stout, simple, mesal surface with numerous fine hair-like setae, apex with 3 short thick setae; aedeagus broad, widest at mid-length, submedian tergal arms weakly connected or unconnected, apical tergal arms fused (forming an apical tergal bridge); proctiger with 2 or more apical teeth, cercal setae present. LARVAE: Seta 1-A single, flexible, inserted on the apical 0.25 of the antenna; setae 4–6-C single, widely separated, 5-C inserted more or less on the level with seta 7-C; seta 3-Mx of the maxilla not borne in a lateral cup-like notch, seta 4-Mx single, acute, seta 6-Mx single; seta 11-P,M,T single, spiniform; comb plate absent, comb scales in 2 or 3 rows; siphon with 2 posterolateral rows of pecten spines, seta 1-S single, inserted about 0.25 from the base of the siphon, seta 1a-S absent, 5 or 6 seta 2a-S present, branched; saddle without posterolateral spicules; seta 4-X multi-branched, long, nearly as long as seta 1-X. PUPAE: Cephalothorax and abdomen without spots; trumpet not flattened or laterally expanded; seta 2-III–VII inserted mesal to seta 1; seta 6-VII inserted on the tergum posterior to seta 9-VII; paddle long, about twice as long as tergum VIII, narrowed and spiculate apically.

****

Bionomics: Larvae of *Wy. leucostigma* have only been found in the leaf axils of cattails (*Typha*) (Lutz, in Bourroul, 1904; Belkin *et al.*, 1971). Females have been collected on humans during all times of the day and occasionally at night (Lourenço-de-Oliveira & da Silva, 1985b). Nothing else is known about the bionomics of the species.

****

Principal references: Lane, 1953 (in key to adult females, *Wy. leucostigma*, description, distribution); Motta *et al.*, 2007 (morphology, phylogenetic relationships).

Miamyia Dyar, 1919

Miamyia, originally proposed as a genus by Dyar (1919b), currently classified as a subgenus of *Wyeomyia*, is typified by *Wyeomyia symmachus* Dyar & Knab, 1909 [junior synonym of *Wyeomyia hosautos* Dyar & Knab, 1907c]. Subgenus abbreviation – *Mia.*

****

Systematics: *Miamyia* includes eight extant species known to occur in locations from northern Argentina and Paraguay northward in Panama. *Miamyia* was found to be a monophyletic group in the morphology-based phylogenetic studies of Judd (1996) and Motta *et al.* (2007). Judd (1996) found the clade *Miamyia* + (*Wy. codiocampa* + *Wy. oblita*) in a sister relationship to a clade comprised of *Cruzmyia* + *Limatus*. In the study of Motta *et al.*, the *Miamyia* clade comprised of *Wy. sabethea* + (*Wy. oblita* + (*Wy. limai* + *Wy. lutzi*)) was sister to *Antunesmyia*, *i.e. Wy. alani.*

****

Characteristics: *Miamyia*, like the other genus groups of *Wyeomyia*, is diagnosed principally on morphological features of the male genitalia and larvae. ADULTS: Proboscis short, expanded distally; mesopostnotal setae present, scales absent; paratergite with or without scales; lower mesokatepisternal setae not extending above the dorsal margin of the mesomeron; upper calypter of the wing without marginal setae. MALE GENITALIA: Tergum VIII strongly produced posterolaterally; tergum IX lobes usually not developed (small in *Wy. oblita*), with a nearly continuous row of setae; gonocoxite with a tergal triad (2 setae in *Wy. sabethea*); gonostylus with a slender to stout stem, lobes M, AE and C developed; aedeagus longer than wide, apical tergal arms not fused (not forming an apical tergal bridge); proctiger expanded near mid-length, with a line of filamentous spicules extending and becoming longer from near the apex to the expansion, sometimes coalesced at the expansion to form a unique process, cercal setae present. LARVAE: Seta 1-A single, inserted near the apex of the antenna; setae 4–6-C single, widely separated, 5-C inserted posterior to seta 7-C, 6-C inserted anterior to seta 7-C; seta 14-C inserted anterior to seta 15-C; maxilla longer than broad, with a small apical tooth, seta 3-Mx of the maxilla not borne in a lateral cup-like notch, seta 4-Mx single, apically forked, hypostomal sclerite fused to the maxillary body, seta 6-Mx branched, often stellate; thorax and abdomen with numerous conspicuous stellate setae on the dorsal and ventral surfaces; seta 11-P,M,T single, hair-like; comb plate usually present, comb with a single row of large spine-like scales and often a variable number of much smaller scales anterior to their bases; siphon without a pecten, seta 1-S single or branched, inserted within or beyond the basal 0.33 of the siphon, setae la,2a-S single, each in a more or less straight row, 1a-S with 1–4 and 2a-S with 2–4 setae; saddle usually with conspicuous spicules on the posterior margin; seta 4-X branched, length variable, usually much shorter than setae 1–3-X. PUPAE: Cephalothorax and abdomen without spots; trumpet short, laterally expanded above the base; seta 1-I well developed, dendritic; seta 2-II–VII inserted anterior to the level of seta 1, seta 2-II inserted more or less midway between setae 1,3-II; seta 3-II,III inserted more or less directly lateral to seta 1; seta 6-II single, about as long as tergum III, seta 6-VII inserted on the sternum ventrad of seta 9-VII; paddle short, not longer than tergum VIII, more or less flask- or pear-shaped, apex more or less truncate and fringed with relatively long spicules.

****

Bionomics: Larvae of *Miamyia* are found in bamboo habitats (Belkin *et al.*, 1971; Heinemann & Belkin, 1978a, 1978c; Heinemann *et al.*, 1980; Linton *et al.*, 2013). Syntype males of *Wy. oblita* were reared from larvae found in the leaf axils of cattails (*Typha*) (Belkin *et al.*, 1971). Females of *Wy. codiocampa* have been collected feeding on humans in cacao plantations and forest areas in Trinidad (Heinemann *et al.*, 1980). Little else is known about the bionomics of the adults.

****

Principal references. Judd, 1996 (morphology, phylogenetic relationships); Motta *et al.*, 2007 (subgeneric status, morphology, phylogenetic relationships).

Nunezia Dyar, 1928

Nunezia Dyar, 1928, originally proposed and currently recognized as a subgenus of *Wyeomyia*, is typified by *Wyeomyia bicornis* (Root, 1928) (in Dyar, 1928), original combination: *Dendromyia bicornis*. Subgenus abbreviation – *Nuz.*

****

Systematics: *Nunezia* includes three extant species, which are only known from collections made in Argentina, Bolivia, Brazil, Peru and Venezuela. An undescribed species is known to occur in Panama (Heinemann & Belkin, 1978a). *Nunezia* was most closely associated with species of *Dodecamyia*, *Hystatomyia* and *Phoniomyia* in the morphology-based phylogenetic analyses of Sabethini conducted by Judd (1996, 1998). Porter (2014) pointed out that the immature stages of most species of these groups undergo development in bromeliads. Porter also noted that species of *Nunezia* and *Hystatomyia*, although larvae of both inhabit bromeliads, seem to have different altitudinal distributions, with species of the former occurring in humid tropical premontane and lower montane forests to elevations of at least 2700 m and the latter occurring principally in humid tropical lowland forests.

****

Characteristics: The following combinations of features should distinguish the known species of *Nunezia* from species of the other genus groups of *Wyeomyia*. ADULTS: Proboscis short, slightly shorter than the forefemur, curved downward; mesopostnotal setae present; paratergite without scales; lower mesokatepisternal setae not extending above the dorsal margin of the mesomeron; upper calypter of the wing without marginal setae. MALE GENITALIA: Tergum IX lobes weakly developed, distinctly separated, each with 3 short stout setae; gonocoxite elongate, tapered distally to the width of the base of the apical gonostylus, tergal triad represented by 2 long setae; gonostylus attached to the apex of the gonocoxite, long and slender, as long as or longer than the gonocoxite, with a relatively small head comprised principally of lobes A and M; aedeagus slightly longer than wide, submedian tergal arms weakly joined to form a bridge; apical tergal arms widely separated; paraproct with a denticulate apex, cercal setae absent. LARVAE: Seta 1-A branched, inserted near mid-length of the antenna; setae 4–6-C multi-branched, 4,6-C distinctly larger than 5-C, all 3 or only 4,6-C inserted anterior to seta 7-C; seta 14-C weakly developed, slightly longer than seta 15-C, inserted anterior to setae 12,15-C; maxilla longer than wide, without an apical tooth, seta 3-Mx absent, seta 4-Mx single, brush-tipped, as long as or longer than the maxillary brush, hypostomal sclerite fused with the maxillary palpus (incorrectly coded as separate from the maxillary palpus and maxillary body in the phylogenetic study of Judd, 1996), seta 6-Mx minute, single; seta 1-P with few branches, slightly shorter than setae 2,3-P; seta 4-P relatively short, multi-branched; seta 11-P,M,T branched, very small; seta 7-M weakly developed, branched; seta 13-T multi-branched, about 0.5 the length of the thorax; seta 1-I,II large, multi-branched, slightly smaller but otherwise similar to seta 1-III–VII; seta 2-I inserted anterolateral to seta 1, 2-II,III inserted anteromesal to seta 1, 2-IV–VII inserted far mesal and slightly anterior to seta 1; seta 13-I–VII inserted more or less on the level with seta 7; comb plate absent, comb scales numerous, in a patch; siphon with a long posterior row of filamentous pecten, seta 1-S well developed, long, multi-branched, inserted about 0.4 from the base of the siphon, 2 seta la-S inserted distal to seta 1-S on the distal 0.5 of the siphon, 5 mainly branched seta 2a-S in an anterolateral row extending from a point proximal to a point distal to the level of seta 1-S; seta 1-X noticeably shorter than setae 2,3-X, seta 2-X multi-branched; seta 4-X about 0.5 the length of seta 1-X. PUPAE: Cephalothorax and abdomen without spots; trumpet swollen in the mid-section, noticeably narrower basally and distally; seta 1-I strongly developed, dendritic; seta 2-II–VII inserted near the posterior margins of the terga, 2-II inserted between setae 1,3-II, 2-III–VII inserted mesal to seta 1; seta 3-II,III,VII inserted lateral to seta 1, 3-V,VI inserted in close association anterior to anteromesad of seta 1; seta 6-II single, very long, reaching well beyond the posterior margin of tergum III, seta 6-VII inserted on the tergum mesal and slightly posterior to seta 9-VII; paddle relatively short and broad, roughly ovoid, about as long as tergum VIII, outer part noticeably broader than the inner part, without marginal and apical spicules.

****

Bionomics: The immature stages of species of *Nunezia* principally develop in terrestrial and epiphytic bromeliads (Heinemann & Belkin, 1978a; Navarro & Liria, 2007; Parker *et al.*, 2012). Heinemann & Belkin (1978a) reported collecting larvae of an undescribed species (once) from the leaf axils of an aroid (*Dieffenbachia*).

****

Principal references: Lane, 1953 (keys to adult females, descriptions, distributions); Judd, 1996 (morphology, phylogenetic relationships); Judd, 1998 (morphology, phylogenetic relationships); Porter, 1914 (comparative morphology, bionomics).

Phoniomyia Theobald, 1903

Phoniomyia, originally proposed as a genus by Theobald (1903a), currently classified as a subgenus of *Wyeomyia*, is typified by *Wyeomyia longirostris* Theobald, 1901c. Subgenus abbreviation – *Pho.*

****

Systematics: *Phoniomyia* includes 22 extant species. All species of *Phoniomyia*, except *Wy. trinidadensis*, are known to occur in Brazil; one species, *Wy. trinidadensis*, is recorded from Trinidad and Tobago, and Venezuela. Only six of the species recorded from Brazil reportedly occur in other countries, including Argentina, Bolivia, French Guiana, Guyana, Paraguay, Suriname, and Trinidad and Tobago. *Phoniomyia* was sister to *Hystatomyia* (as *Dendromyia* 'species A group') in the cladistic analysis of combined larval, pupal and adult morphological data in the phylogenetic study of Sabethini conducted by Judd (1996). Later analyses of the combined larval, pupal and adult data of clades most closely associated with *Phoniomyia* and *Hystatomyia* (Judd, 1998) generated two topologies, one with the sister relationship of *Phoniomyia* + *Hystatomyia* and another with *Phoniomyia* + *Nunezia*. The former relationship was retained when the larval data were analysed alone, and the latter was recovered when only adult characters were analysed. In the absence of species of *Hystatomyia*, *Phoniomyia* was recovered, with very weak support, as a lineage outside *Wyeomyia* in the phylogenetic study of Motta *et al.* (2007).

****

Characteristics: The following combinations of characters distinguish species of *Phoniomyia* from those of the other genus groups of *Wyeomyia*. ADULTS: Proboscis unusually long and slender, much longer than the forefemur, curved downward; mesopostnotal setae present; paratergite without scales; lower mesokatepisternal setae not extending above the dorsal margin of the mesomeron; upper calypter of the wing without marginal setae. MALE GENITALIA: Tergum IX lobes weakly to strongly produced, setae short and stout or longer and narrower and often with apices bent laterad; gonocoxite rather long and relatively slender, tergal triad and basal mesal lobe absent; gonostylus either usually short and setose or divided proximally into 2 long arms, longer ventral arm with an elongate knife-like arm along one side; aedeagus elongate, apical tergal arms not fused to form an apical tergal bridge; proctiger with apical teeth, cercal setae present. LARVAE: Seta 1-A single or branched, inserted about 0.75 from the base of the antenna; setae 4–6-C usually multi-branched (may be single or double in *Wy. theobaldi*), 4-C not inserted close to the median labral plate, on level with the antennal prominence, 6-C inserted anterior to seta 7-C, 5-C usually inserted more or less on the level with seta 7-C (distinctly anterior to 7-C in *Wy. palmata*); seta 15-C inserted posterior to seta 12-C at the level of the mid-length of the labiogula; maxilla longer than wide, with a very small apical tooth, seta 3-Mx borne in a lateral cup-like notch (as in species of *Hystatomyia*), seta 4-Mx single, stout, brush-tipped, as long as or longer than the maxillary brush, seta 6-Mx single, arising from a distinct hypostomal sclerite or a sclerite unapparent with seta 6-Mx apparently arising from membrane; seta 1-P multi-branched, as long as or slightly shorter than seta 2-P; seta 4-P short, multi-branched; seta 11-P,M,T normal, not spine-like; seta 4-T relatively large, about as long as seta 1-M; seta 13-T multi-branched, about 0.5 the length of the thorax; comb plate absent, comb scales generally numerous, in a patch; siphon long and tapered, with 1 or 2 close-set posterior rows of pecten spines, seta 1-S branched, inserted about 0.25 from the base of the siphon, setae la,2a-S long and numerous; saddle without spicules on the posterior margin; seta 1-X with 1–3 branches, usually double; seta 4-X multi-branched, very short, shorter than 0.25 the length of seta 1-X. PUPAE: Cephalothorax and abdomen often with a variable pattern of pigmentation; trumpet slender and more or less cylindrical; seta 1-I strongly developed, dendritic; seta 2-II–VII inserted near the posterior margins of the terga, 2-II inserted between setae 1,3-II, 2-III–VII inserted mesal to seta 1; seta 3-II inserted lateral to seta 1, 3-IV–VI inserted anterolateral to seta 1; seta 6-II single, long, reaching well beyond the posterior margin of tergum III, seta 6-VII inserted on the tergum posterior to seta 9-VII; paddle length variable, as long as or distinctly longer than tergum VIII, generally ovoid, apex rounded or slightly narrowed, with or without short apical spicules.

****

Bionomics: The immature stages of species of *Phoniomyia* normally inhabit the axils of arboreal and terrestrial bromeliads, however larvae of *Wy. pilicauda* have been found in the leaf axils of a terrestrial aroid (*Xanthosoma*) (Heinemann & Belkin, 1979) and those of *Wy. trinidadensis* have been taken from the leaf axils of *Dracaena* in a domestic area (Heinemann *et al.*, 1980). The adults occur in damp forests and plantations in the immediate vicinity of larval habitats. They are principally active during the daylight hours, but some species remain active during diurnal periods and *Wy. lassalli* is active during the night (Guimarães *et al.*, 2000). Females are known to feed on humans (see, *e.g.* Heinemann *et al.*, 1980; Paterno & Marcondes, 2004) as well as other animals. The adults of various species have been collected in bird-baited traps at all elevations in forest canopy in Brazil. Species of *Phoniomyia* are of no medical or economic importance.

****

Principal references: Lane, 1953 (as a genus, keys, descriptions, distributions); Correa & Ramalho, 1956 (as a genus, taxonomy, keys, descriptions, distributions); Judd, 1996 (morphology, phylogenetic relationships); Judd, 1998 (subgeneric status, phylogenetic relationships); Motta *et al.*, 2007 (morphology, phylogenetic relationships); Porter, 2014 (comparative morphology).

Prosopolepis Lutz, 1905

Prosopolepis, originally proposed as a genus by Lutz (1905), currently classified as a subgenus of *Wyeomyia*, is typified by *Wyeomyia confusa* (Lutz, 1905), original combination: *Prosopolepis confusa*. Subgenus abbreviation – *Prl*.

****

Systematics: *Prosopolepis* is monobasic. It only includes the extant type species, which is only recorded from Brazil. The species seems to be confined to the Atlantic rain forests, mainly between latitudes of 8–30° S (the occurrence of the type species in Bolivia, Suriname and Venezuela listed in Wilkerson *et al.*, 2021 is doubtful). In an eletrophoretic analysis of allozyme data conducted by Motta *et al.* (1998), *Wy. confusa* was placed in a sister relationship to three species of *Dendromyia*. In a later phylogenetic analysis of morphological data (Motta *et al.*, 2007), it was recovered as the sister of *Wy. melanocephala* + *Onirion personatum*. The relationship was supported by three synapomorphies – larval seta 13-I single, seta 1-VIII branched and tergum VIII of the pupa bearing a posterolateral lobe – but the relationship was only weakly supported. The true phylogenetic relationships of *Prosopolepis* remain unknown.

****

Characteristics: The most distinctive features of *Prosopolepis* are found in the immature stages, particularly the larvae. ADULTS: Ornamentation and structural detail very similar to the adults of other species of *Wyeomyia*; distinctive characters include the clypeus largely covered with scales, translucent scales covering most of the proepisternum between the forecoxae, an elongate patch of scales extending from the laterotergite along the posterior edge of the metapostnotum, and presence of a few large, scale-like setae on the cerci of females. Proboscis shorter than the forefemur, expanded distally; mesopostnotal setae present, scales absent; paratergite without scales; lower mesokatepisternal setae not extending above the dorsal margin of the mesomeron; upper calypter of the wing without marginal setae. MALE GENITALIA: Tergum IX lobes small, distinctly separated, each with 3–5 short flattened setae; gonocoxite with 2 long tergomesal setae; gonostylus large, broad, lobe C long, slender, arising from the sternolateral edge of lobe M; aedeagus slightly longer than broad, broadest in the proximal 0.5, submedian tergal arms fused, forming a narrow median tergal bridge, apical tergal arms fused, forming a slightly produced apical tergal bridge; proctiger with broad basal sclerotization (tergum X), paraproct with a flattened fan-like process on the sternal margin at the base, cercal setae present. LARVAE: Seta 1-A single, inserted near the apex of the antenna; lateralia with denticles near the dorsolateral slit of the occipital foramen; the pair of seta 1-C close together; seta 5-C inserted very near seta 6-C and both inserted slightly anterior to seta 7-C; seta 14-C well developed, much longer than seta 15-C, inserted anterior to setae 12,15-C; maxilla highly modified for grasping, very long, slender, projecting far beyond the anterior margin of the head, seta 4-Mx present, spine-like, maxillary brush, other setae and laciniarastra absent, maxillary palpus highly reduced, borne (fused) dorsally at the base of the maxillary body; seta 1-P single, longer than setae 2,3-P; seta 4-P relatively short, multi-branched, fan-like; seta 11-P,M,T single, spine-like; seta 7-M weakly developed, branched, similar to seta 13-M; seta 8-T inserted immediately anterior to the sclerite bearing setae 9–12-T; seta 13-T strongly developed, multi-branched, about as long as the thorax; seta 1-I,II small, multi-branched, 1-III,IV small, single, 1-V,VI much longer, double or triple; seta 2-I–VII inserted mesad of seta 1; seta 9-I–VI short, single; seta 13-I,II inserted far anterior to seta 7, 13-I–VI single, inserted anteriorly at the level of seta 7; comb plate absent, comb with 12–27 spine-like scales in an uneven single or partially double row, some scales fused basally in pairs to form a forked composite; siphon with 3–9 pecten spines on either side of seta 1a-S on the distal 0.5 of the siphon, seta 2a-S comprised of frayed and branched setae; seta 4-X multi-branched, shorter than setae 1–3-X, about 0.6 the length of seta 1-X. PUPAE: Cephalothorax and abdomen unspotted; seta 1-CT sigmoid with a hooked tip; seta 2-II–VII near the posterior margin of the tergum, inserted mesad of seta 1; seta 3-VII inserted more or less on the level with seta 1; seta 6-II very long, extending well beyond the posterior margin of segment III, 6-VII inserted on the dorsal surface of the segment anterolateral to seta 9; seta 9-VIII strongly developed, considerably longer than the paddle; paddle short, only slightly longer than segment VIII, evenly tapered from the base, tip more or less pointed, surface spiculose, margins lined with spicules.

****

Bionomics: The immature stages of *Wy. confusa* have only been found in the leaf axils of *Heliconia* species and plants of the family Marantaceae, such as species of *Calathea*. Davis (1944) reported the finding of larvae of this species in bamboo internodes and the leaf axils of a cultivated aroid (*Arum esculentum*, Araceae), but this requires confirmation. The larvae exhibit aggressive and predacious behaviour. Females have been collected landing on humans in forest, more frequently at ground level than in the canopy. Females may also feed on humans and animals in cleared areas close to forest, but they rarely invade peridomestic environments (see references in Lourenço-de-Oliveira *et al.*, 1999).

****

Principal references: Lane, 1953 (*Wy. confusa*, as a species of subgenus *Dendromyia*, in keys, description, distribution); Motta *et al.*, 1998 (genetic relationships); Lourenço-de-Oliveira *et al.*, 1999 (subgeneric status, taxonomy, description, bionomics, distribution); Motta & Lourenço-de-Oliveira, 2005 (comparative morphology); Motta *et al.*, 2007 (morphology, phylogenetic relationships).

Spilonympha Motta & Lourenço-de-Oliveira, 2005

Spilonympha Motta & Lourenço-de-Oliveira, 2005, originally proposed and currently recognized as a subgenus of *Wyeomyia*, is typified by *Wyeomyia aningae* Motta & Lourenço-de-Oliveira, 2005. Subgenus abbreviation – *Spi.*

****

Systematics: *Spilonympha* includes seven extant species, all of which are known to occur in Brazil. Only two are recorded from other countries: *Wy. bourrouli* from French Guiana and Suriname, and *Wy. mystes* from Bolivia and French Guiana. The evolutionary relationships of *Spilonympha* are uncertain. It was recovered as a monophyletic group in the morphology-based phylogenetic study of Motta *et al.* (2007). The group shared a sister relationship with *Decamyia* and the '*flui* group' of species that are *incertae sedis* within *Wyeomyia*.

****

Characteristics: Species of *Spilonympha* differ from species of the other genus groups of *Wyeomyia* in features of the male genitalia, larvae and pupae. The species are morphologically distinct from but more similar to species of *Cruzmyia* than to species of the other groups. ADULTS: Similar to the adults of the other genus groups of *Wyeomyia* in overall habitus; proboscis shorter than the forefemur, thickened distally; mesopostnotal setae present, some scales also present; paratergite without scales; lower mesokatepisternal setae extend above the dorsal margin of the mesomeron; upper calypter of the wing without marginal setae. MALE GENITALIA: Tergum IX lobes small, each with a few strong setae; gonocoxite without long setae except for a long lateral seta and 2 tergomesal setae (= tergal triad); gonostylus with a conspicuous long stem, lobes A, C, E and M present at the apex; aedeagus longer than wide, broadest at mid-length, submedian tergal arms joined, apical tergal arms not joined; paraproct with 1 or 2 apical teeth, cercal setae present. LARVAE: Seta 1-A single, inserted near the apex of the antenna; setae 4–6-C single, widely separated, 6-C inserted anterior to seta 7-C; seta 14-C about 3 times as long as seta 15-C; seta 15-C inserted at most slightly posterior to seta 14-C, on the level with seta 12-C; maxilla short, with a strong apical tooth, seta 4-Mx single or bifid, relatively long, about as long as the maxillary brush, seta 6-Mx inserted on the margin of the maxillary body or on the hypostomal sclerite when present; seta 1-P inserted posterolateral to setae 2,3-P; seta 4-P relatively short, multi-branched; seta 11-P,M,T single, short, spine-like; seta 7-M relatively weakly developed, branched, about as long as seta 3-M; seta 8-T inserted laterally between seta 7-T and the sclerite bearing setae 9–12-T; seta 13-T multi-branched, shorter than 0.5 the length of the thorax; seta 1-I,II small, similar in size to seta 2, seta 1-IV–VII multi-branched, about 0.5 the length of the segment; seta 2-I–VII short, single, 2-III–VII inserted far mesad of seta 1, 2-III,IV inserted nearer to seta 1 than to the midline of the segment, 2-V–VII inserted approximately midway between seta 1 and the midline; seta 13-III,IV inserted posterior to seta 12, slightly posterior to the level of seta 7; comb plate present, comb scales spine-like, fringed at the sides, borne on the posterior margin of the sclerite; siphon with a distal posterior row of pecten spines, seta 1-S inserted near the base of the siphon, seta la-S in a more or less straight posterolateral row, seta 2a-S more numerous, inserted in a more or less double anterolateral row; seta 4-X much shorter than setae 1–3-X, multi-branched. PUPAE: Cephalothorax and abdomen darkly pigmented or with a pattern of dark pigmentation and interspersed pale spots; trumpet nearly cylindrical or expanded near mid-length; seta 1-I strongly developed, dendritic, seta 1-III,V,VI inserted lateral to setae 2 and 3; seta 2-II–VII inserted near the posterior margins of the terga, 2-II inserted lateral to seta 1, 2-III–VII inserted mesal to seta 1; seta 3-III,V,VI inserted mesal to seta 1, 3-IV inserted noticeably anterior to seta 1; seta 5-IV,V strongly developed, long, extending well beyond the posterior margin of the following tergum; seta 6-II as long as or longer than tergum III, 6-VII inserted on the tergum posterior to seta 9-VII; paddle long and broad, outer part much broader than the inner part, more than twice as broad, inner and outer margins serrate.

****

Bionomics: The immature stages of species of *Spilonympha* have been collected from the leaf axils of plants of the families Araceae, Asparagaceae and Bromeliaceae. Adults are active during the daytime and females are attracted to humans. Little else is known about the bionomics of species of *Spilonympha*. Species of the group are not known to be of medical importance.

****

Principal references: Motta & Lourenço-de-Oliveira, 2005 (new subgenus, taxonomy, descriptions, bionomics, distributions); Motta *et al.*, 2007 (morphology, phylogenetic relationships).

Triamyia Dyar, 1919

Triamyia, originally proposed as a genus by Dyar (1919b), currently classified as a subgenus of *Wyeomyia*, is typified by *Wyeomyia aporonoma* Dyar & Knab, 1906d. Subgenus abbreviation – *Tra.*

****

Systematics: *Triamyia* includes two extant species. The type species is widely distributed from Argentina northward into Central America as far as Mexico. *Wyeomyia staminifera* is only know from Brazil. *Triamyia* was recovered as a monophyletic clade in the phylogenetic study of Motta *et al.* (2007) based on morphological data. Although the analysis of morphological data alone did not reveal close affinities with other genus groups, *Wy. aporonoma*, in the absence of *Wy. staminifera*, was recovered in a sister relationship with *Wy. melanocephala* (subgenus uncertain) + *Onirion personatum* when the morphological data were combined with available allozyme data.

****

Characteristics: The following combinations of characters distinguish species of *Triamyia* from those of the other genus groups of *Wyeomyia*. ADULTS: Proboscis short, expanded distally; mesopostnotal setae present, scales absent; paratergite without scales; lower mesokatepisternal setae extend above the dorsal margin of the mesomeron; upper calypter of the wing without marginal setae. MALE GENITALIA: Tergum VIII without posterolateral lobes; tergum IX lobes weakly developed, with strongly developed setae; gonocoxite long, tapered and curved distally, with a tergal triad; gonostylus long, slender, curved, lobes not well developed; aedeagus longer than wide, apical tergal arms fused, forming an apical tergal bridge; proctiger normal, apex with 4 or 5 teeth, cercal setae present. LARVAE: Seta 1-A single, hair-like, inserted on the distal 0.5 of the antenna; setae 4–6-C single, widely separated, 5-C inserted posterior to seta 7-C, 6-C inserted anterior to seta 7-C; seta 14-C strongly developed, significantly longer than seta 15-C, inserted anterior to seta 15-C; seta 3-Mx of the maxilla not borne in a lateral cup-like notch, seta 4-Mx single, truncate, seta 6-Mx branched; seta 1-P with few branches, inserted mesal to and shorter than setae 2,3-P; seta 4-P long, multi-branched; seta 11-P,M,T single, short, hair-like; seta 7-M weakly developed, branched, similar to seta 1-M; seta 13-T strongly developed, multi-branched, about as long as the thorax; seta 1-I,II relatively large, much longer than seta 2-I,II, seta 1-III–VI usually double or triple, longer than the segment; seta 2-I–VII short, usually single, 2-II–VII inserted anteromesal to seta 1; seta 9-I–VI short, single or double; seta 13-I–VI branched, 13-I inserted anterior to the other ventral setae, 13-II–VI inserted mesad to seta 12; comb plate absent, comb scales in a single or partially double row, dorsal scales spine-like, ventral scales short, broad, fringed distally; siphon without a pecten, seta 1-S inserted on the proximal 0.33 of the siphon, setae la,2a-S each in a more or less straight row, each with 3 setae, la-S single or branched, 2a-S single; saddle without posterolateral spicules; seta 4-X branched, very short, about as long as the saddle. PUPAE: Cephalothorax and abdomen without spots; seta 1-I well developed, dendritic; seta 2-II inserted lateral to seta 1, 2-III–VII inserted mesal to seta 1; seta 3-II,III,VII inserted lateral to seta 1, 3-IV inserted anterolateral to seta 1, 3-V,VI inserted anteromesal to seta 1; seta 6-II single, about as long as tergum III, 6-VII inserted on the tergum posterior to seta 9-VII; paddle large, broad, about twice as long as tergum VIII, apex with minute spicules.

****

Bionomics: The immature stages of species of *Triamyia* are most often found in bamboo and tree-hole habitats. Larvae of *Wy aporonoma* have also been found in a hole in a fence post, a hollow in a banana stump, and in coconut and cacao shells (Dyar & Knab, 1906d; Belkin & Heinemann, 1976; Heinemann & Belkin, 1977b; Lourenço-de-Oliveira *et al.*, 1992). Females of *Wy. aporonoma* are known to land and feed on humans during the daytime (Belkin & Heinemann, 1976; Heinemann & Belkin, 1977b).

****

Principal reference: Motta *et al.*, 2007 (subgeneric status, morphology, phylogenetic relationships).

Wyeomyia Theobald, 1901 *sensu stricto*

Wyeomyia sensu stricto, the nominotypical subgenus of *Wyeomyia*, is typified by *Wyeomyia grayii* Theobald, 1901c. Subgenus abbreviation – *Wyo.*

****

Systematics: *Wyeomyia sensu stricto* includes 35 extant species, which occur mainly in areas from Brazil and Ecuador northward through Central America and the Caribbean Islands. One species (*Wy. hemisagnosta*) is recorded from as far south as Bolivia; two species, *Wy. vanduzeei* and *Wy. mitchellii*, are recorded from various Caribbean Islands, and Florida and Georgia in the southeastern USA. *Wyeomyia smithii* is the only temperate species of *Wyeomyia*, with a distribution in the northeastern USA from Illinois eastward to Delaware and southern Canada from Saskatchewan eastward to Newfoundland. The phylogenetic relationships of *Wyeomyia sensu stricto* are unknown. Belkin *et al.* (1970) recognized 'four more or less distinct phyletic lines' among species in Jamaica and noted that *Wy. nigritubus* 'is so distinct from the other species … that it should eventually be placed in a different subgenus'. Judd (1996) included only *Wy. arthrostigma* and *Wy. grayii* in her morphology-based phylogenetic analyses of Sabethini, and Motta *et al.* (2007) included only *Wy. arthrostigma* and *Wy. medioalbipes* in their phylogenetic analyses of *Wyeomyia*, also based on morphological data. The results of both studies suggest that *Wyeomyia sensu stricto* is not a monophyletic taxon. The two species included in the latter study shared a paraphyletic relationship with a clade comprised of species of *Miamyia*. Soghigian *et al.* (2023) included three species of *Wyeomyia sensu stricto* in their phylogenetic analysis of amino acid sequences of 709 genes, but the relationships of the species with each other and with other genus groups of Sabethini included in the analysis were not explicitly indicated.

****

Characteristics: *Wyeomyia sensu stricto* is difficult to characterize because most species are poorly known and probably do not comprise a monophyletic group; consequently, some of the characteristics listed here are unlikely to be shared by all species currently included in the group. ADULTS: Proboscis short, shorter than the forefemur, slightly enlarged distally; mesopostnotal setae present; paratergite with or without scales; lower mesokatepisternal setae not extending above the dorsal margin of the mesomeron; upper calypter of the wing without marginal setae. MALE GENITALIA: Tergum IX lobes weakly developed, each bearing 2 or more unmodified setae; gonocoxite elongate, lateral and ventral surfaces with numerous scales, basal mesal lobe and a tergal triad present; gonostylus with a distinct stem and a large head comprised of variously developed lobes, large median lobe M always present, lobe A indistinct to well developed, lobe E present, absent or not differentiated from lobes A or M, lobes B and C present or absent; aedeagus longer than wide, apical tergal arms not separate, not forming an apical tergal bridge; proctiger normal, paraproct with ill-defined apical teeth, cercal setae present, sometimes numerous and longer than usual. LARVAE: Seta 1-A single or branched, inserted on the distal 0.25 of the antenna, sometimes very near the apex; setae 4–6-C single or branched, 5-C inserted posterior more or less on level with or anterior to seta 7-C; seta 14-C moderately well developed, single or with relatively few branches, generally 2–4 times as long as seta 15-C (14-C very long with numerous branches in *Wy. nigritubus*), inserted more or less on the level with or slightly anterior to seta 15-C at approximately the mid-length of the labiogula; maxilla nearly as broad as long, without a prominent apical tooth, seta 3-Mx not borne in a lateral cup-like notch, seta 4-Mx single, simple, hypostomal sclerite fused (partially?) with the maxillary body; seta 1-P usually multi-branched, inserted mesad to posteromesad of setae 2,3-C approximately midway between these setae and the midline of the thorax; seta 4-P multi-branched, shorter to longer than seta 3-C; seta 4-T relatively large, multi-branched (except in *Wy. nigritubus*); seta 11-P,M,T branched (single in *Wy. nigritubus*), branches flexible; seta 13-T multi-branched, long, about 0.75 to full length of the thorax; insertions, positional relationships and development of setae 1,2,13-I–VII varied, seta 1 generally inserted more posterolaterad than usual, seta 2 generally inserted far anterior to anteromesad of seta 1; seta 9-I–VI with 2 or more branches; seta 13-I inserted far anterior to the other ventral setae but seta 9-I sometimes inserted more or less at the same level, 13-II–VI inserted posterior to the other setae (except in *Wy. nigritubus*); comb plate absent, comb a single long row of close-set spine-like scales; siphon without a pecten, seta 1-S inserted on the proximal 0.25 of the siphon, setae la,2a-S each a more or less straight row of many to numerous elements (few elements in *Wy. nigritubus*); saddle with or without posterolateral spicules; seta 4-X multi-branched, usually about 0.5 the length of seta 1-X (shorter in *Wy. nigritubus*). PUPAE: Cephalothorax and abdomen, especially the anterior terga, with a pattern of pigmentation (except *Wy. nigritubus*); trumpet short, laterally expanded; seta 1-I strongly developed, dendritic; seta 2-II–VII inserted anterior, anteromesal or mesal to seta 1; seta 3-V,VI inserted anterior to anteromesad of seta 1 (anterolateral in *Wy. nigritubus*); seta 6-II single, as long as to slightly longer than tergum III, 6-VII inserted on the tergum posteromesal to seta 9-VII (inserted on the sternum in *Wy. corona*); paddle variable, longer than wide, length usually more than 1.5 times the width, usually distinctly longer than tergum VIII, outer part distinctly wider than inner part, narrowed apically, apex and at least the distal margins minutely spiculate (margins fringed with long spicules in *Wy. mitchellii*).

****

Bionomics: The immature stages of species of *Wyeomyia sensu stricto* occupy a range of phytotelm habitats, but they are found primarily in terrestrial and arboreal bromeliads. Larvae of several species also have been found occasionally in leaf axils of aroids, flower bracts of *Heliconia*, bamboo sections and stumps, fallen cocoa pods, coconut shells and tree holes. Larvae of *Wy. arthrostigma* and *Wy. nigritubus* occur primarily in bamboo habitats (Belkin *et al.*, 1970; Heinemann & Belkin, 1977b, 1978a, 1978b, 1978c, 1979; Heinemann *et al.*, 1980). The immature stages of *Wy. smithii* develop in the leaves of the pitcher plant *Sarracenia purpurea* (see Carpenter & LaCasse, 1955). Females of several species are often encountered in forest during the daytime and are known to land and feed on humans, *e.g. Wy. hirsuta*, *Wy. mitchellii* and *Wy. vanduzeei*, as recorded by Carpenter & LaCasse (1955) and Belkin *et al.* (1970).

****

Principal references: Belkin *et al.*, 1970 (Jamaica, eight formally named and two unnamed species, keys, morphology, descriptions, bionomics, distributions); Judd, 1996 (morphology, phylogenetic relationships); Motta *et al.*, 2007 (morphology, phylogenetic relationships).

Zinzala Zavortink, 1986

Zinzala Zavortink, 1986, originally proposed and currently recognized as a subgenus of *Wyeomyia*, is typified by *Wyeomyia zinzala* Zavortink, 1986. Subgenus abbreviation – *Zin.*

****

Systematics: *Zinzala* includes two extant species, which are only known from the Gran Sabana region of southeastern Bolivar State, Venezuela. The species are probably restricted to the Guyana Highland where the host plants (*Heliamphora*) of the immature stages are endemic. *Zinzala* was recovered as the sister to *Wy.* (*Wyeomyia*) *arthrostigma* in the phylogenetic study of Sabethini conducted by Judd (1996) based on analysis of morphological data using successive approximations character weighting. *Zinzala* was not represented in the phylogenetic study of *Wyeomyia* conducted by Motta *et al.* (2007). Zavortink (1986a) noted that the affinities of the species were not obvious – resemblance of the adults to those of the other genus groups of *Wyeomyia* are superficial and the unusual male genitalia 'offer no clue as to the relationships of the subgenus'.

****

Characteristics: *Zinzala*, like the other genus groups of *Wyeomyia*, is diagnosed principally on morphological features of the male genitalia and larvae. ADULTS: The adults of *Zinzala* are not currently definitely distinguishable from the adults of the other genus groups of *Wyeomyia*, but they bear the following possible differential characters: interocular space broad, scutum with metallic purple, gold, bronze and blue scaling, mesepimeron without scales, abdominal sterna largely or completely without dark scales; proboscis shorter than the forefemur, gradually swollen distally; mesopostnotal setae present; paratergite without scales; lower mesokatepisternal setae not extended above the dorsal margin of the mesomeron; upper calypter of the wing without marginal setae. MALE GENITALIA: Tergum IX lobes produced; gonocoxite short, with numerous long setae; gonocoxite short, broad, with numerous long strong setae; gonostylus with a short stem, lobe C apparently not developed; aedeagus large, broad, complex, with an apical pair of fine divergent 'horns'; proctiger normal, cercal setae absent. LARVAE: Head elongate; seta 1-A branched, inserted near the apex of the antenna; setae 5,6-C usually single or double, 6-C inserted anterior to seta 7-C; seta 14-C very strongly developed, stellate, inserted slightly if at all anterior to seta 12-C; seta 15-C stellate but much smaller than seta 14-C, inserted far caudad on the labiogula; maxilla longer than broad, with a strong apical tooth, seta 4-Mx single, as long as or longer than the maxillary brush, hypostomal sclerite present, bearing a single or branched seta 6-Mx; setae 0-P, 1,8,11-P,M,T, 3-P, 14-P,M, 5-M, 4-T and 13-T strongly stellate with numerous branches, seta 13-T about as long as the thorax; seta 7-M weakly developed, single or branched; seta 8-T inserted anterior to setae 9–12-T; setae 1,2,5,9,13-I–VII, 11-I and 1-VIII strongly stellate with numerous branches; seta 13-I inserted far anterior to seta 7, 13-II–VI inserted slightly anterior to seta 7; comb plate absent, comb a single long row of close-set spine-like scales; siphon without a pecten, seta 1-S single or double, inserted near the base of the siphon, more or less in line with a straight row of single seta la-S, seta 2a-S also a straight row of single setae; saddle nearly complete, narrowly incomplete ventrally; seta 4-X stellate, much shorter than setae 1–3-X. PUPAE: Cephalothorax and abdomen without spots; trumpet slightly flattened, not laterally expanded; seta 13-CT often present; seta 1-I weakly developed, with 1–6 branches; seta 2-II–VII inserted near mid-length of the tergum lateral to the insertion of seta 1; seta 6-II single, about as long as seta 7, seta 6-VII inserted on the sternum ventrad of seta 9-VII; paddle short, length less than 1.5 times the greatest width, extending slightly beyond the apex of the genital lobe, apex fringed with minute spicules.

****

Bionomics: The immature stages of species of *Zinzala* develop in insectivorous pitcher plants of *Heliamphora*. Adult females are autogenous. Little else is known about the biology of the species.

****

Principal references: Zavortink, 1986 (new subgenus, descriptions, keys, distributions, bionomics); Harbach & Peyton, 1990 (comparative morphology); Judd, 1996 (morphology, phylogenetic relationships).

Zeugnomyia Leicester, 1908

Zeugnomyia Leicester, 1908, originally proposed and currently recognized as a genus, is typified by *Zeugnomyia gracilis* Leicester, 1908. Genus abbreviation – *Ze.*

****

Systematics: *Zeugnomyia* is a subordinate taxon of the subfamily Culicinae, tribe Aedini. It includes only four extant species, which occur in eastern areas of the Oriental Region, principally in the Philippine Islands. *Zeugnomyia* was recovered as the sister of *Belkinius*, and this pair was sister to *Udaya*, in the phylogeny of Aedini generated in the morphology-based phylogenetic study of Reinert *et al.* (2009). *Udaya* fell within a polytomy with *Belkinius*, *Bothaella*, *Catatassomyia* and *Zeugnomyia* in the phylogeny reconstructed by Wilkerson *et al.* (2015). The only species of *Zeugnomyia*, *Ze. gracilis*, included in the maximum likelihood phylogeny of Soghigian *et al.* (2017) based on molecular markers was recovered in a basal relationship to a clade comprised of *Armigeres* + *Eretmapodites*.

****

Characteristics: *Zeugnomyia* are rather small mosquitoes characterized by a broad vertical stripe of silver scales on the sides of the thorax. The head, scutum and abdominal segments are also ornamented with patches of flat silver scales. Major differential characters include the following: dorsocentral, prespiracular and postspiracular setae absent; 1 lower mesepimeral seta present; hindcoxa more or less in line with the base of the mesomeron; upper calypter of the wing with a fringe of setae, alula bare, cell R_2 distinctly shorter than vein R_{2+3}, anal vein (vein 1A) ends only slightly beyond the level of the origin of the mediocubital crossvein; pulvilli not evident. Larvae are similar to the larvae of other aedine genus groups in the Oriental Region. They exhibit the following characteristics: seta 7-C slender, delicate and single or double; hypostomal suture complete, reaching the posterior tentorial pit; comb scales in a single row (distinction from *Heizmannia*); pecten present (distinction from *Armigeres*); saddle incomplete, with strongly developed spines along the posterior margin (distinction from *Udaya*). The combined characteristics of seta 7-C, the comb scales and the saddle are needed to distinguish *Zeugnomyia* from many other aedine species.

****

Bionomics: Little is known about the bionomics of the species of *Zeugnomyia*. They appear to be restricted to shaded areas in forests. Larvae have been found in small collections of water in fallen leaves, axils of Anahaw palm, tree holes, cut bamboo and an artificial container (tin can). The larvae of *Ze. gracilis* reportedly prey on other mosquito larvae and the adult females of this species are known to feed on humans and other warm-blooded animals that enter their haunts. The species of *Zeugnomyia* are not known to vector pathogens of human diseases.

****

Principal references: Baisas & Feliciano, 1953 (Philippines, three species, descriptions); Delfinado, 1966a (Philippines, distribution, key to males, descriptions, bionomics); Reinert, 2000d (female genitalia); Reinert *et al.*, 2004, 2009 (morphology, phylogenetic relationships); Rattanarithikul *et al.*, 2010 (Thailand, bionomics, keys); Wilkerson *et al.*, 2015 (phylogenetic relationships); Soghigian *et al.*, 2017 (phylogenetic relationships).

Part V: Genus Groups and Informal Group Taxa

The practice of subdividing generic-level taxa into informal species groups is principally attributable to Edwards (1932a), who gave names to 'species groups which have been or might conceivably in the future be regarded as entitled to generic or subgeneric rank'. As more species were discovered and more fully described, significant changes were made to Edwards's generic, subgeneric and informal-group classifications. Many genera and subgenera were subdivided into hierarchical systems to reflect presumed evolutionary relationships. The formal and informal classification of the Culicidae recognized at the time of this writing is elaborated in this Part and tabulated in Part VI, but first some explanation about informal categories of classification.

The informal categories (ranks) of classification generally accepted by mosquito taxonomists include Sections, Series, Groups, Subgroups and Complexes. However, the application of these categories lacks uniformity, and most species-group taxa have not been examined on a worldwide basis to adequately ascertain the integrity of the informal groups, their relationships and the placement and relationships of the species. Consequently, the placement and presumed relationships of many species are questionable, and are subject to change when more information becomes available. However, in the face of uncertainty, the currently recognized informal groups provide the groundwork for phylogenetic studies aimed to establish a natural classification of the Culicidae.

Informal group taxa are given vernacular names, which are formed by placing the name of the category (rank) after the specific name of the group (Reid & Knight, 1961); thus, vernacular names consist of two words, a nominal genus- or species-group name (not always currently valid) followed by the name of the rank. Both words are printed in Roman type (neither is italicized) with the first letter capitalized, *e.g.* Pipiens Group and Dirus Complex. Informal group names formatted otherwise in the taxonomic literature, *e.g.* the 'series *Myzorhynchus*' and '*barbirostris* group' of Reid & Knight (1961) are restyled as the Myzorhynchus Series and Barbirostris Group, respectively.

The informal group taxa are based on anatomical similarities that are believed to, but may not, represent phylogenetically related assemblages of species. A number of genus-group taxa have not been divided into informal groups. These are mainly genus groups with relatively few species that are most likely to be monophyletic lineages.

In the listing below and the Tabulation in Part VI, subfamilies and genera are arranged alphabetically, with the nominotypical subgenus listed before the other subgenera of each genus that is subdivided into subgenera. This is done to place emphasis on the group that typifies the genus. In the Tabulation of formal and informal group taxa in Part VI, the authorities who first introduced or most recently redefined the informal taxonomic groups are indicated by literature citations. References for species complexes are those that include the first mention or treatment of all of the species currently recognized within the group. It is to be noted that the Tabulation includes numerous corrections and additions to the informal groupings listed in Wilkerson *et al.* (2021: 1099–1104). This is a comprehensive compilation of the internal classifications of the genus groups of the Culicidae.

Anophelinae Grassi, 1900

Anopheles Meigen, 1818

Anopheles Meigen, 1818 *sensu stricto*

Anopheles was established as a genus by Meigen (1818), with *An. maculipennis* Meigen, 1818 subsequently designated as the type species (International Commission on Zoological Nomenclature, 1959a: Opinion 547). Reid & Knight (1961) revised Edwards's (1932a) infrasubgeneric classification and divided the group into two Sections based on the shape of the pupal trumpet. The Laticorn Section was created for species with a wide funnel-shaped trumpet having the longest axis transverse to the stem, and the Angusticorn Section for species with a semi-tubular trumpet having the longest axis vertical more or less in line with the stem. These Sections were subsequently divided into Series, and the larger Series were divided into species Groups. The term 'section' was substituted for the 'group' category of Edwards to avoid confusion with 'species group' (which is equivalent to Group of current convention). The new classification retained the '*Anopheles* series' of Edwards, but the original Arribalzagia and Christya 'groups' were reduced to 'series'. Except for a few minor changes made at or below the 'group' level (Reid, 1968; Harrison, 1972; Vu *et al.*, 1991; Harbach, 2004), and the removal of *An. kyondawensis* to *Baimaia* by Harbach *et al.* (2005) and *An. implexus* and *An. okuensis* to *Christya* by Harbach & Kitching (2016), the internal classification of *Anopheles sensu stricto* has changed relatively little since it was introduced by Reid & Knight (1961).

Baimaia Harbach, Rattanarithikul & Harrison, 2005

Baimaia was proposed as a subgenus of *Anopheles*, with *An. kyondawensis* Abraham, 1947 as its type and only included species. Reid & Knight (1961) included *An. kyondawensis* in the Culiciformis Group (Anopheles Series, Angusticorn Section) based on features of the larva, which was the only life stage know at the time. Harrison & Scanlon (1975) considered the inclusion of the species in the Culiciformis Group to be tentative until the pupal and adult stages were known. Based on unique features of larvae and pupae and adults reared from larvae collected in Thailand, Harbach *et al.* (2005) concluded that *An. kyondawensis* belonged to a new subgenus, which they named in honour of Prof. Visut Baimai of Mahidol University in Bangkok.

Cellia Theobald, 1902

Cellia was originally established as a genus based on its inclusion in a tabular key of distinguishing anatomical characters (Theobald, 1902b), with *Anopheles pharoensis* Theobald, 1901b designated as the type species. Grjebine (1966), Reid (1968) and Gillies & de Meillon (1968) are largely responsible for the current internal classification of *Cellia*, although this system is based on Edwards (1932a). For the most part, these authors dealt with the *Anopheles* of Madagascar, Southeast Asia and Africa, respectively. Reid (1968) and Gillies & de Meillon (1968) independently substituted the term 'series' for the 'group' categories in Edwards's classification of *Cellia* (as *Myzomyia*). As Harrison (1980) noted, Gillies & de Meillon also introduced the term 'section' as a category below series level even though 'species group' was already in use for this taxon level and 'section' was in use as a category above the series level (Reid & Knight, 1961). Inasmuch as Gillies & de Meillon interchanged the terms section and group, *e.g.* Funestus Section and Funestus Group, Harbach (1994a) replaced their 'section' with 'group' to correspond with the convention followed by other authors. This action added uniformity to the internal classification of the genus *Anopheles*. Foster *et al.* (2017) proposed the elevation of *Cellia* to generic status based on phylogenetic analyses of mitochondrial protein-coding genes, but this action generally has not been accepted.

Christya Theobald, 1903

Christya was originally proposed as a distinct genus by Theobald (1903b), with *Christya implexa* Theobald, 1903b as its type and only originally included species. Edwards (1932a) included *Christya* as a synonym of *Anopheles* and followed Christophers (1915) in recognizing three main groups within *Anopheles sensu stricto*: 'Group A (*Anopheles*)', 'Group B (*Arribalzagia*)' and 'Group C (*Christya*)'. Reid & Knight (1961) proposed a new classification in which the term 'Group' was replaced by the word 'series'; thus, 'Group C' of Edwards became 'series *Christya*'. *Christya* remained in synonymy with *Anopheles* until it was elevated to subgeneric rank by Harbach & Kitching (2016) based on the recovery of *An. implexus* as the sister to all other anophelines in a phylogenetic analysis of morphological data.

Kerteszia Theobald, 1905

Kerteszia was originally proposed as a distinct genus by Theobald (1905f), with *Kerteszia boliviensis* Theobald, 1905f as its type and only originally included species. Edwards (1932a) followed Christophers (1924a) in treating *Kerteszia* as a synonym of *Nyssorhynchus*, but Dyar (1928) and Komp (1937b, 1942) recognized it as a distinct subgenus of *Anopheles*. Zavortink (1973) was the last reviser of *Kerteszia*, but he was uncertain whether it should remain a subgenus distinct from *Nyssorhynchus*. Peyton *et al.* (1992) conducted an in-depth comparative morphological analysis of the two taxa and discovered a 'preponderance of evidence for the subgeneric (perhaps generic) status of *Kerteszia*, and dispels the idea that the included species are merely specialized members of the subgenus *Nyssorhynchus*'. Nine of the 12 formally named species of *Kerteszia* are recognized as species complexes (Bourke *et al.*, 2023), each comprised of two or more undescribed species denoted by numbers: the Bellator (two species), Boliviensis (three species), Cruzii (three species), Homunculus (two species), Laneanus (two species), Neivai (eight species), Pholidotus (four species) and Rollai (five species) Complexes. Foster *et al.* (2017) proposed the elevation of *Kerteszia* to generic status based on phylogenetic analyses of mitochondrial protein-coding genes, but this action generally has not been accepted.

Lophopodomyia Antunes, 1937

Lophopodomyia was originally proposed as a subgenus of *Anopheles* by Antunes (1937), with *An. squamifemur* Antunes, 1937 as its type and only originally included species. Foster *et al.* (2017) proposed the elevation of *Lophopodomyia* to generic status based on phylogenetic analyses of mitochondrial protein-coding genes, but this action generally has not been accepted.

Nyssorhynchus Blanchard, 1902

Nyssorhynchus was originally established as a genus by Blanchard (1902), with *An. argyritarsis* Robineau-Desvoidy, 1827 designated as the type species. Although Christophers (1924a) was inclined to recognize *Nyssorhynchus* as a genus, he gave way to 'some practical objections to such a course' and treated it as a subgenus of *Anopheles*. The subgeneric status of *Nyssorhynchus* was adopted by Edwards (1930a, 1932a) and has continued to date despite the proposal of Foster *et al.* (2017) to return it to generic status based on phylogenetic analyses of mitochondrial protein-coding genes.

The internal classification of *Nyssorhynchus* has a complicated history. Christophers (1924a) included *Kerteszia* within *Nyssorhynchus*, and Edwards (1932a) followed this concept rather than the schemes proposed by either da Costa Lima (1928), who recognized *Kerteszia* and *Nyssorhynchus* as subgenera of *Nyssorhynchus*, or Dyar (1928), who treated these taxa as distinct subgenera of *Anopheles*. Edwards (1932a) divided *Nyssorhynchus* into three groups, 'Groups A (*Nyssorhynchus*)' and 'Group B

(*Myzorhynchella*)', which together equate to the current concept of *Nyssorhynchus*, and 'Group C (*Kerteszia*)', which equates to *Kerteszia*. *Kerteszia* was again treated as a subgenus by Komp (1937b, 1942), but its independent status remained in doubt (Zavortink, 1973) until Peyton *et al.* (1992) firmly established the uniqueness of this group. Based on phylogenetic analyses of mitochondrial protein-coding genes, Foster *et al.* (2017) proposed the elevation of *Nyssorhynchus* to generic status, which generally has not been accepted.

Edwards (1932a) subdivided his Group A (*Nyssorhynchus*) into the Argyritarsis, Tarsimaculatus and Rondoni Series. This system remained unchanged until Gabaldon (1940) referred to the Tarsimaculatus Series as the Albimanus Series and divided it into the Albimanus, Oswaldoi and Triannulatus Subseries. These categories were accepted by Gabaldon & Cova-Garcia (1952), who placed additional species in the Triannulatus and Oswaldoi Subseries.

Da Costa Lima (1928) split *Nyssorhynchus* into two groups, Groups A and B, which partially correspond to the contemporary Argyritarsis and Albimanus Sections introduced by Faran (1980). In another scheme, Galvão (1943) divided *Nyssorhynchus* into two series, the Argyritarsis and Tarsimaculatus Series, which correspond to Groups A and B of da Costa Lima (1928). Galvão further divided the Argyritarsis Series into two complexes (the *albitarsis* and *argyritarsis* complexes), and recognized a monobasic group (*albimanus*) and three complexes within the Tarsimaculatus Series (the *rondoni*, *tarsimaculatus* and *triannulatus* complexes). Leví Castillo (1949) divided *Nyssorhynchus* into the Albimanus and Argyritarsis Groups, and included three series within the former. These series, the Albimanus, Oswaldoi and Triannulatus Series, corresponded with the subseries of Gabaldon (1940), except that *An. galvaoi* and *An. nuneztovari* were shifted from the Oswaldoi Subseries to the Triannulatus Series.

The most recent comprehensive treatments of *Nyssorhynchus* include Faran (1980), Linthicum (1988) and Peyton *et al.* (1992). Faran divided *Nyssorhynchus* into the contemporary Albimanus and Argyritarsis Sections and recognized the '*Myzorhynchella* group'. The latter was originally described as a genus by Theobald (1907), recognized as a group (Group B) by Edwards (1932a), treated as a subgenus by Galvão (1941) and defined as a Section of coequal rank with the Albimanus and Argyritarsis Sections by Peyton *et al.* (1992). Faran (1980) and Linthicum (1988) revised the Albimanus and Argyritarsis Sections, respectively. Faran (1980) divided the Albimanus Section into the Albimanus (monobasic) and Oswaldoi Groups, separated the Oswaldoi Group into the Triannulatus (monobasic) and Oswaldoi Subgroups, and split the Oswaldoi Subgroup into the Oswaldoi and Strodei Complexes (later called the Oswaldoi and Strodei Series by Faran & Linthicum, 1981). Linthicum's (1988) classification of the Argyritarsis Section was first presented in Faran & Linthicum (1981). These authors divided the Argyritarsis Section into the Argyritarsis and Albitarsis Groups, subdivided the Argyritarsis Group into the Argyritarsis, Lanei (monobasic), Pictipennis (monobasic) and Darlingi (monobasic) Subgroups, and recognized the Albitarsis and Braziliensis (monobasic) Subgroups within the Albitarsis Group.

In view of the previous terms applied to different levels of classification within *Nyssorhynchus*, Harbach (1994a) recognized the 'groups' and 'subgroups' of Faran (1980) and Linthicum (1988) and the 'complexes' of Faran (1980) as Series, Groups and Subgroups, respectively, to bring the informal classification of *Nyssorhynchus* more in line with that established for *Anopheles sensu stricto* and *Cellia*.

Stethomyia Theobald, 1902

Stethomyia was originally proposed as a distinct genus by Theobald (1902b), with *Stethomyia boliviensis* Theobald, 1902b as its type and only originally included species. Christophers (1924a) treated *Stethomyia* as a synonym of *Anopheles*. It was elevated to subgeneric status by Edwards (1930a) and has retained that rank. Foster *et al.* (2017) proposed the elevation of *Stethomyia* to generic status based on phylogenetic analyses of mitochondrial protein-coding genes, but this action generally has not been accepted.

Bironella Theobald, 1905

Bironella Theobald, 1905 *sensu stricto*

Bironella was proposed as a distinct genus by Theobald (1905f), with *Bi. gracilis* Theobald, 1905f as its type and only originally included species. Christophers (1924a) treated *Bironella* as a subgenus of *Anopheles*. Edwards (1930a) considered it 'desirable to make the most of the [morphological] differences which exist, and to recognize *Bironella* as a genus distinct from *Anopheles*.' However, whether or not *Bironella* should retain generic status or be classified as a subgenus within *Anopheles* has not been satisfactorily resolved. Morphological and molecular phylogenetic studies have produced discordant results, making it impossible to objectively determine whether or not *Bironella* should be ranked as a genus or subgenus. Based on the complexity of the internal classification of *Anopheles sensu stricto*, which provides clear evidence of polyphyly, there should be little doubt that *Bironella* requires the rank of genus. Tenorio (1977) is the last reviser of the genus.

Brugella Edwards, 1930

Brugella was proposed as a subgenus of *Bironella* by Edwards (1930a), with *Anopheles travestitus* Brug, 1928b as its type and only originally included species.

Neobironella Tenorio, 1977

Neobironella was proposed as a subgenus of *Bironella* by Tenorio (1977), with *Bironella confusa* Bonne-Wepster, 1951 designated as the type species.

Chagasia Cruz, 1906

Chagasia was established as a genus by Cruz (1906), with *Ch. neivae* Cruz, 1906 [subjective synonym of *Ch. fajardi* (Lutz, 1904) (in Bourroul, 1904)] as its type and only originally included species. Christophers (1924a) stated that he was personally biased toward the recognition of *Chagasia* as a genus, but due to 'some practical objections to such a course' he treated it as a subgenus of *Anopheles*. Root (1927b) elucidated the main differences between the adults, larvae and pupae of *Chagasia* and those of *Anopheles* and treated *Chagasia* as a genus. Edwards (1930a), acknowledging the distinctions recorded by Root and other differences he had noted earlier (Edwards, 1926a), stated that '*Chagasia* appears to be amply distinct generically from *Anopheles*', and it has since been recognized as a genus separate from *Anopheles*. *Chagasia* is a small homogenous group of species that was last reviewed by Harbach & Howard (2009).

Culicinae Meigen, 1818

Aedeomyiini Theobald, 1901

Aedeomyia Theobald, 1901

Aedeomyia Theobald, 1901 *sensu stricto*

Aedeomyia was briefly defined as a genus by Theobald (1901a) and more fully described in his second volume of *A Monograph of the Culicidae* (Theobald, 1901c) based on *Ad. squamipennis* (Lynch Arribálzaga, 1878) and *Ad. venustipes* (Skuse, 1889). The former species was subsequently designated the type species of the genus by Brunetti (1914). *Aedeomyia* was last reviewed by Tyson (1970a).

Lepiothauma Enderlein, 1923

Lepiothauma was originally proposed as a genus by Enderlein (1923), with *Lepiothauma furfurea* Enderlein, 1923 as its type and only included species. Edwards (1925) considered the type species to be a synonym of *Ad. venustipes* (Skuse, 1889), where upon *Lepiothauma* became a synonym of *Aedeomyia*. Edwards (1929b) recognized *Ad. furfurea* as a species distinct from *An. venustipes*, but *Lepiothauma* remained a synonym of *Aedeomyia*, and was treated as such in the catalogues of Edwards (1932a) and Stone *et al.* (1959). Tyson (1970a) recognized *Lepiothauma* as a morphologically distinct taxon with subgeneric rank in *Aedeomyia*.

Aedini Neveu-Lemaire, 1902

Aedes Meigen, 1818

Aedes Meigen, 1818 *sensu stricto*

Aedes is one of two genera (the other being *Anopheles*) that were established by the pioneering German entomologist Johann Wilhelm Meigen (1764–1845). *Aedes sensu stricto* is typified by *Ae. cinereus* Wiedemann, 1818 (in Meigen, 1818), the only originally included species. Reinert *et al.* (2004) restricted the genus *Aedes* to the subgenus *Aedes*. Gornostaeva (2005) was the last reviser of *Aedes sensu stricto*. Half of the 12 known species of the subgenus are divided between the Cinereus and Esoensis Groups (Becker *et al.*, 2003).

Abraedes Zavortink, 1970

Abraedes was proposed as a subgenus of *Aedes* by Zavortink (1970c), with *Ae. papago* Zavortink, 1970c as its only originally included species.

Acartomyia Theobald, 1903

Acartomyia was originally proposed as a distinct genus by Theobald (1903a), with *Acartomyia zammitii* Theobald, 1903a as its type and only originally included species. *Acartomyia* was treated as a synonym of the genus *Ochlerotatus* by Edwards (1912b) and as a synonym of the genus *Aedes* by Howard *et al.* (1917) and Dyar (1923f). Edwards (1932a) accepted the earlier synonymy but treated *Ochlerotatus* as a subgenus of *Aedes*. *Ochlerotatus* was raised to generic rank by Reinert (2000c), and *Acartomyia* remained a synonym until it was recognized as a subgenus of *Ochlerotatus* by Reinert *et al.* (2008). *Acartomyia* was subsequently raised to generic status by Reinert *et al.* (2009) and reduced to its present status as a subgenus of *Aedes* by Wilkerson *et al.* (2015).

Aedimorphus Theobald, 1903

Aedimorphus was originally proposed as a distinct genus by Theobald (1903a), with *Uranotaenia domestica* Theobald, 1901c as the only included species. It was listed as a synonym of the genus *Ochlerotatus* by Edwards (1912b) and classified as 'Group

Aëdimorphus' in *Aedes* subgenus *Ochlerotatus* by Edwards (1917). *Aedimorphus* was listed as a synonym of *Aedes* by Howard *et al.* (1917) and Dyar (1923f). It seems that *Aedimorphus*, as a subgenus of *Aedes*, first appeared with the description of '*Aedes* (*Aedimorphus**) *lamborni*, sp. n.' by Edwards (1923a), the asterisk referring to an explanatory footnote: 'I have previously (following Dyar) used the name *Ecculex* for this subgenus, overlooking the fact that Theobald's *Aedimorphus* was published a year previously.' The following year, Edwards (1924b) included *Aedimorphus* in a key to the subgenera of *Aedes* in the Oriental Region. *Aedimorphus* was raised to generic status by Reinert *et al.* (2009) and returned to its present status as a subgenus of *Aedes* by Wilkerson *et al.* (2015). Edwards (1932a) divided the subgenus into eight groups, identified as Groups A–H, each identified with a specific name as well. Only two of his groups, 'Group C (*alboscutellatus*-group)' and 'Group G (*vexans*-group)', are still recognized (see the Tabulation in Part VI for the current internal classification of *Aedimorphus*).

Alanstonea Mattingly, 1960
Alanstonea was proposed as a subgenus of *Aedes* by Mattingly (1960), with *Scutomyia treubi* de Meijere, 1910 as the originally designated type species. The subgenus includes the two species that formed Edwards's (1932a) Group B of *Armigeres sensu stricto*. *Alanstonea* was raised to generic status by Reinert *et al.* (2004) and returned to its present status as a subgenus of *Aedes* by Wilkerson *et al.* (2015).

Albuginosus Reinert, 1987
Albuginosus was proposed as a subgenus of *Aedes* by Reinert (1987), with *Stegomyia marshallii* Theobald, 1901b as the originally designated type species. *Albuginosus* was raised to generic status by Reinert *et al.* (2004) and returned to its present status as a subgenus of *Aedes* by Wilkerson *et al.* (2015).

Ayurakitia Thurman, 1954
Ayurakitia was originally proposed as a distinct genus by Thurman (1954a), with *Ayurakitia griffithi* Thurman, 1954a as its type and only originally included species. Mattingly (1971a) reduced *Ayurakitia* to a subgenus of *Aedes* without explanation. It was restored to generic rank by Reinert (2000b) and returned to its present status as a subgenus of *Aedes* by Wilkerson *et al.* (2015).

Aztecaedes Zavortink, 1972
Aztecaedes was proposed as a subgenus of *Aedes* by Zavortink (1972), with *Ae.* (*Gualteria*) *ramirezi* Vargas & Downs, 1950 as the type and only included species. Reinert (2000c) treated *Aztecaedes* as a subgenus of *Ochlerotatus*, Reinert *et al.* (2004) raised it to generic status and it was returned to subgeneric status in *Aedes* by Wilkerson *et al.* (2015).

Belkinius Reinert, 1982
Belkinius was proposed as a subgenus of *Aedes* by Reinert (1982), with *Ae. aurotaeniatus* Edwards, 1922a as the type and only included species. *Belkinius* was raised to generic rank by Reinert *et al.* (2004) and returned to its original subgeneric status in *Aedes* by Wilkerson *et al.* (2015).

Bifidistylus Reinert, Harbach & Kitching, 2009
Bifidistylus was originally proposed as a genus by Reinert *et al.* (2009), with *Aedes* (*Aedimorphus*) *lamborni* Edwards, 1923a as the type species. Its current status as a subgenus of *Aedes* is accredited to Wilkerson *et al.* (2015).

Borichinda Harbach & Rattanarithikul, 2007
Borichinda was originally proposed as a genus by Harbach & Rattanarithikul, 2007 (in Harbach *et al.*, 2007b), with *Borichinda cavernicola* Rattanarithikul & Harbach, 2007 (in Harbach *et al.*, 2007b) as the type and only included species. *Borichinda* was reduced to its current rank as a subgenus of *Aedes* by Wilkerson *et al.* (2015).

Bothaella Reinert, 1973
Bothaella was proposed as a subgenus of *Aedes* by Reinert (1973f), with *Ae. helenae* Reinert, 1973f as the type species. *Bothaella* was raised to generic rank by Reinert *et al.* (2004) and returned to its original subgeneric status in *Aedes* by Wilkerson *et al.* (2015).

Bruceharrisonius Reinert, 2003
Bruceharrisonius was originally described as a subgenus of *Ochlerotatus* by Reinert (2003), with *Howardina greenii* Theobald, 1903a as the type species. *Bruceharrisonius* was raised to generic rank by Reinert *et al.* (2006) and reduced to subgeneric rank in *Aedes* by Wilkerson *et al.* (2015).

Cancraedes Edwards, 1929
Cancraedes was proposed as a subgenus of *Aedes* by Edwards (1929b), with *Ae. cancricomes* Edwards, 1922a as the type species. *Cancraedes* was raised to generic rank by Reinert *et al.* (2009) and returned to its original subgeneric status in *Aedes* by Wilkerson *et al.* (2015).

Catageiomyia Theobald, 1903

Catageiomyia was originally described as a distinct genus by Theobald (1903d), with *Catageiomyia senegalensis* Theobald, 1903d [subjective synonym of *Ae. irritans* (Theobald, 1901d)] as its type and only originally included species. *Catageiomyia* was listed as a synonym of *Aedes* by Howard *et al.* (1917) and Dyar (1923f). Edwards (1912b) listed *Catageiomyia senegalensis* as a synonym of *Ochlerotatus irritans*, and he later (Edwards, 1925) included *irritans* in his '*albocephalus* series' within 'Group IV. The *vexans* group' of *Aedes* subgenus *Aedimorphus*. This clearly indicates that Edwards considered *Catageiomyia* to be a synonym of *Aedimorphus*, which he listed as such in his catalogue of the Culicidae (Edwards, 1932a). *Catageiomyia* remained in synonymy with *Aedimorphus* until it was elevated to generic status by Reinert *et al.* (2009). It was downgraded to its present status as a subgenus of *Aedes* by Wilkerson *et al.* (2015).

Catatassomyia Dyar & Shannon, 1925

Catatassomyia was originally described as a distinct genus by Dyar & Shannon (1925a), with *Catatassomyia meronephada* Dyar & Shannon, 1925a as the type and only included species. Edwards (1929c) listed the species as a member of the subgenus *Stegomyia*, which implied the synonymy of *Catatassomyia* with *Stegomyia*. Edwards (1932a) explicitly listed *Catatassomyia* as a synonym of *Stegomyia* and placed *Ae. meronephada* in his 'Group B (*w-alba*-group).' Mattingly (1965) divided Group B into three subgroups, with *Ae. meronephada* being the sole member of 'Subgroup B2. (*Ae. meronephada* subgroup).' The species remained in subgenus *Stegomyia* until Huang (1978) transferred it to the subgenus *Diceromyia* based on morphological resemblance to species of that group. As a result of phylogenetic analyses of morphological data, Reinert *et al.* (2009) removed *Catatassomyia* from synonymy with *Diceromyia* and recognized it as a monobasic genus. It was reduced to its present status as a subgenus of *Aedes* by Wilkerson *et al.* (2015).

Christophersiomyia Barraud, 1923

Christophersiomyia was originally proposed as a distinct genus by Barraud (1923a), with *Stegomyia thomsoni* Theobald, 1905e as its originally designated type species. Barraud (1931b), without explanation, treated *Christophersiomyia* as a subgenus of *Aedes*. It subsequently remained a subgenus of *Aedes* until it was restored to its original generic rank by Reinert *et al.* (2004) based on phylogenetic analyses of morphological data. It was returned to its status as a subgenus of *Aedes* by Wilkerson *et al.* (2015).

Coetzeemyia Huang, Mathis & Wilkerson, 2010

Coetzeemyia was described as a subgenus of *Aedes* by Huang *et al.* (2010), with *Culicelsa fryeri* Theobald, 1912b as the type and only included species.

Collessius Reinert, Harbach & Kitching, 2006

Collessius was originally introduced as a genus by Reinert *et al.* (2006), with *Ochlerotatus macfarlanei* Edwards, 1914b as the originally designated type species. Reinert *et al.* divided the species of the genus between two informal groups recognized as Assemblages A and B. Two years later, the authors (Reinert *et al.*, 2008) formally recognized Assemblage A as the nominotypical subgenus *Collessius* and proposed the subgenus *Alloeomyia* for species of Assemblage B. Wilkerson *et al.* (2015) reduced *Collessius* to a subgenus of *Aedes*, and the two subgenera were concomitantly reduced to informal species groups, the Collessius and Alloeomyia Groups, respectively.

Cornetius Huang, 2005

Cornetius was proposed as a subgenus of *Aedes* by Huang (2005), with *Ae. cozi* Cornet, 1973 as the type and only included species. *Cornetius* was raised to genus level by Reinert *et al.* (2009) and returned to its original rank as a subgenus of *Aedes* by Wilkerson *et al.* (2015).

Dahliana Reinert, Harbach & Kitching, 2006

Dahliana was originally established as a genus by Reinert *et al.* (2006), with *Culex geniculatus* Olivier, 1791 designated as the type species. Its present status as a subgenus of *Aedes* was established by Wilkerson *et al.* (2015).

Danielsia Theobald, 1904

Danielsia was originally introduced as a genus by Theobald (1904c), with *Danielsia albotaeniata* Leicester, 1904 (in Theobald, 1904c) as its type and only originally included species. *Danielsia* was listed as a questionable synonym of the genus *Ochlerotatus* by Edwards (1912b) and as a definite synonym of 'Group *Finlaya*' of *Aedes* (*Ochlerotatus*) by Edwards (1917). Howard *et al.* (1917) and Dyar (1923f) listed *Danielsia*, as well as *Ochlerotatus*, as synonyms of *Aedes*. However, Dyar (1918b) had divided American species of the genus *Aedes* into groups based on features of the male genitalia, one of the groups being '*Finlaya* Theobald'. In a subsequent paper, Dyar (1920e) stated that the groups treated in his 1918 publication 'may properly be called subgenera'; thus, *Danielsia* became a synonym of the subgenus *Finlaya* of *Aedes*. Ironically, in the same year, Edwards (1920a) described and named a new southeastern Palaearctic species as *Ochlerotatus* (*Finlaya*) *echinus*. *Danielsia* remained a synonym of *Finlaya* until Reinert *et al.* (2008) resurrected it as a genus for three species previously classified as species of *Aedes* (*Finlaya*). *Danielsia* was relegated to its current status as a subgenus of *Aedes* by Wilkerson *et al.* (2015).

Dendroskusea Edwards, 1929

Dendroskusea was proposed as a subgenus of *Aedes* by Edwards (1929b), with *Culex micropterus* Giles, 1901a as the designated type species. Edwards (1932a) placed *Dendroskusea* in synonymy with the subgenus *Diceromyia*. He also included the African and Oriental species of the subgenus *Skusea*, and divided the composite *Diceromyia* into two groups, 'Group A (*furcifer*-group: *Diceromyia*)' and 'Group B (*micropterus*-group: *Dendroskusea*)'. *Dendroskusea* remained a synonym of *Diceromyia* until it was resurrected by Reinert *et al.* (2009) as a genus for five species known from India. Its present status as a subgenus of *Aedes* was established by Wilkerson *et al.* (2015).

Diceromyia Theobald, 1911

Diceromyia was originally established as a genus by Theobald (1911c), with *Diceromyia africana* Theobald, 1911c [subjective synonym of *Aedes furcifer* (Edwards, 1913b)] subsequently designated as the type species (Edwards, 1932a). As noted for *Dendroskusea* above, Edwards (1932a) divided *Diceromyia* into Groups A and B. The former, also denoted as the '*furcifer*-group: *Diceromyia*', included the four African species that were known at the time. Edwards (1936) added the African *Ae. taylori* to the group and described the adults and pupae of the five species in his compendium on the Afrotropical culicine mosquitoes (Edwards, 1941). That aside, the composite *Diceromyia* of Edwards (1932a) prevailed until the phylogenetic study of Reinert *et al.* (2009) revealed that it was not a natural group, and *Diceromyia* was restricted to the Afrotropical species (14 were known at the time). Huang & Rueda (2016) provided a pictorial key and proposed an informal classification for nine species of *Diceromyia* known to occur in sub-Saharan Africa (species known only from Madagascar were not included). They divided the species between two sections, designated Sections A and B; the former included three groups, the Adersi, Flavicollis and Furcifer Groups, and the later included two groups, the Fascipalpis and Zethus Groups. The following year, Brunhes *et al.* (2017) proposed the genus *Paulianius* for seven Madagascan species, three of them new, and included *Ae. zethus* de Meillon & Lavoipierre, 1944, the sole species of the Zethus Group created by Huang & Rueda (2016). This action limited *Diceromyia* to eight species which occur in countries of continental Africa south of the Sahara. *Diceromyia* was returned to its previous status as a subgenus of *Aedes* by Wilkerson *et al.* (2015).

Dobrotworskyius Reinert, Harbach & Kitching, 2006

Dobrotworskyius was originally proposed as a genus by Reinert *et al.* (2006) for seven Australian species, with *Aedes tubbutiensis* Dobrotworsky, 1959 as the designated type species. *Dobrotworskyius* was reduced to a subgenus of *Aedes* by Wilkerson *et al.* (2015).

Downsiomyia Vargas, 1950

Downsiomyia was established as a subgenus of *Aedes* by Vargas (1950), with *Stegomyia nivea* Ludlow, 1903 (in Theobald, 1903a) designated as the type species. Edwards (1932a) had previously placed the species in Group H of the subgenus *Finlaya*. Knight (1946) recognized *Ae. niveus* as a representative of a distinct section of Group H, the '*niveus* subgroup', which he seems to have introduced as a unit of informal classification. The subgroup was defined by Knight & Marks (1952) as 'Subgroup I, niveus' of Group H. Stone *et al.* (1959) listed *Downsiomyia* as a synonym of *Finlaya* in their world catalogue of the Culicidae, but there appears to be no previous record of the synonymy. All earlier works following Knight & Marks treated the type species as the nominotypical member of the Niveus Subgroup of the subgenus *Finlaya*. Reinert *et al.* (2004) removed *Downsiomyia* from synonymy with *Finlaya* and recognized it as a genus for the Niveus Group of species based on the results of a morphology-based phylogenetic analysis of Aedini. *Downsiomyia* was downgraded to its current subgeneric rank within *Aedes* by Wilkerson *et al.* (2015) based on reanalysis of the data set of Reinert *et al.* (2009). Reinert & Harbach (2006) provided a detailed description of *Downsiomyia* (as a genus) and divided it into the Alboniveus and Alboniveus Groups, which are retained as species groups of the subgenus *Downsiomyia*. The subgenus is equivalent to the Niveus Assemblage of Reinert (2002e), which he divided into two groups, the Alboniveus and Niveus Subassemblages, based on features of the female genitalia. The subassemblages are referred to herein as the Alboniveus and Niveus Groups, respectively.

Edwardsaedes Belkin, 1962

Edwardsaedes was established as a subgenus of *Aedes* by Belkin (1962), with *Culex imprimens* Walker, 1860 as the designated type species. Reinert *et al.* (2004) elevated *Edwardsaedes* to generic rank based on cladistic analysis of the Aedini determined by implied weighting of morphological data. Wilkerson *et al.* (2015) returned *Edwardsaedes* to its original rank as a subgenus of *Aedes* based on the results of an unweighted parsimony analyses of the more comprehensive data set of Reinert *et al.* (2009).

Elpeytonius Reinert, Harbach & Kitching, 2009

Elpeytonius was originally described as a genus by Reinert *et al.* (2009) for two species previously included in *Aedimorphus*, with *Ochlerotatus apicoannulatus* Edwards, 1912a designated as the type species. *Elpeytonius* was reduced to a subgenus of *Aedes* by Wilkerson *et al.* (2015).

Finlaya Theobald, 1903

Finlaya was originally proposed as a genus by Theobald (1903a), with *Culex kochi* Dönitz, 1901 subsequently designated as the type species by Blanchard (1905). *Finlaya* was listed as a synonym of the genus *Ochlerotatus* by Edwards (1912b) and as a synonym of 'Group *Finlaya*' of *Aedes* (*Ochlerotatus*) by Edwards (1917). Howard *et al.* (1917) listed both *Finlaya* and *Ochlerotatus* as synonyms of *Aedes*. Dyar (1918b) recognized '*Finlaya* Theobald' as a group of American species based on features of the male genitalia. In a subsequent paper, Dyar (1920e) stated that the groups treated in his 1918 publication 'may properly be called subgenera'; thus, *Finlaya* became a subgenus of *Aedes*. In the same year, Edwards (1920a) described and named the Palaearctic *Ochlerotatus echinus* as a new species of the subgenus *Finlaya*. The following year, Edwards (1921d) apparently agreed with the classification of Dyar (1920e) and recognized *Finlaya* as a subgenus of *Aedes* in *A revision of the mosquitoes of the Palaearctic Region*. In his review and catalogue of the Culicidae, Edwards (1932a) divided the subgenus into eight groups, designated as Groups A–H, with Group A identified as the '*kochi*-group: *Finlaya*, s. str.' for the species with spotted wings. This group is equivalent to the Kochi Assemblage of Reinert (2002e) based on morphological similarities of the female genitalia. Reinert (2000c) divided the traditional genus *Aedes* into the genera *Aedes* and *Ochlerotatus* and transferred *Finlaya* to *Ochlerotatus*. *Finlaya* remained a subgenus of *Ochlerotatus* until Reinert *et al.* (2004) restored it to genus rank for the Kochi Group of species. Wilkerson *et al.* (2015) re-established it as a subgenus of *Aedes*, restricted to the Kochi Group as it was by Reinert *et al.* (2004). Knight & Marks (1952) divided the Kochi Group into three subgroups, 'Subgroup I, **kochi** s. str.', 'Subgroup II, **lewelleni**' and 'Subgroup III, **gani**', based on characteristics of scaling and male genitalia. Bohart (1957) divided their Subgroup I into the 'Papuan-Australian subgroup I, *kochi*', which subsumed Subgroup II of Knight & Marks, and the 'Malayan subgroup II, *poicilius*'. As the Kochi Group is now equivalent in rank to the subgenus *Finlaya*, the three subgroups are recognized herein as species groups, the Gani, Kochi and Poicilius Groups. The Kochi Group includes the addition of the four species complexes recognized by Belkin (1962): the Bougainvillensis, Franclemonti, Knighti and Solomonis Complexes.

Fredwardsius Reinert, 2000

Fredwardsius was established as subgenus of *Aedes* by Reinert (2000a) for the distinctive *Culex vittatus* Bigot, 1861. Based on unique features of the adults, Edwards (1925, 1932a) recognized the species as the sole member of a distinct group of the subgenus *Stegomyia*, 'Group I. The *vittatus* group' and 'Group D (*A. vittatus* Big.)', respectively. Later authors who dealt with the species recognized its uniqueness and questioned its inclusion in *Stegomyia*. Huang (1977b) considered *Ae. vittatus* to bear a closer resemblance to species of *Aedimorphus* and transferred the species to that subgenus, where it remained until it was replaced into the monobasic subgenus *Fredwardsius*. As a result of morphology-based cladistic analyses of Aedini, Reinert *et al.* (2004) elevated *Fredwardsius* to generic status. *Edwardsaedes* was returned to its original rank as a subgenus of *Aedes* by Wilkerson *et al.* (2015) based on the results of reanalyses of the more comprehensive morphological data set of Reinert *et al.* (2009).

Georgecraigius Reinert, Harbach & Kitching, 2006

Georgecraigius, with *Culex atropalpus* Coquillett, 1902d the designated type species, was originally proposed as a genus by Reinert *et al.* (2006) for three species (*Ae. atropalpus*, *Ae. epactius* and *Ae. fluviatilis*) that were classified in the world catalogue of Stone *et al.* (1959) as species of *Aedes* (*Finlaya*), with *Ae. epactius* as a synonym of *Ae. atropalpus*. The three species (with *Ae. epactius* removed from synonymy) were transferred to *Aedes* (*Ochlerotatus*) by Zavortink (1972) and to *Ochlerotatus* (*Ochlerotatus*) by Reinert (2000c). The species were left without generic placement when the composition of *Ochlerotatus* was restricted by Reinert *et al.* (2004). Phylogenetic analysis of morphological data for taxa formerly placed in *Finlaya* led Reinert *et al.* (2006) to split *Georgecraigius* into two subgeneric groups, the nominotypical subgenus and the new subgenus *Horsfallius*. The genus was downgraded to subgeneric status in *Aedes* by Wilkerson *et al.* (2015), with concomitant recognition of the two subgenera as informal species groups, the Georgecraigius and Horsfallius Groups. The authors, however, were apparently unaware that the two groups were superseded by the Atropalpus and Fluviatilis Groups, respectively, which were established earlier by Zavortink (1972).

Geoskusea Edwards, 1929

Geoskusea was introduced by Edwards (1929b) as a subgenus of *Aedes*, with *Ae. fimbripes* Edwards, 1924b the designated type species. Reinert (2000c) returned *Ochlerotatus* to its original genus rank and recognized *Geoskusea* as a subgenus of that genus. *Geoskusea* remained a subgenus of *Ochlerotatus* until it was raised to generic status by Reinert *et al.* (2004), where it remained until it was returned to its original subgeneric rank in *Aedes* by Wilkerson *et al.* (2015). Belkin (1962) recognized three species groups for five species that occur on islands in the South Pacific, the Daggyi, Fimbripes and Longiforceps Groups. The other five species of *Geoskusea* are without group placement.

Gilesius Reinert, Harbach & Kitching, 2006

Gilesius was originally proposed as a genus by Reinert *et al.* (2006) for two species previously included in *Aedes* (*Finlaya*), *i.e. Ae. alius* Lien, 1968d and *Culex pulchriventer* Giles, 1901a, the latter being the designated type species. *Gilesius* was downgraded to subgeneric status in *Aedes* by Wilkerson *et al.* (2015).

Gymnometopa Coquillett, 1906

Gymnometopa was originally established as a genus by Coquillett (1906a), with *Stegomyia mediovittata* Coquillett, 1906a the designated type species. Coquillett also included a new species in the genus, *i.e. Gymnometopa albonotata*, which is now recognized as a species of *Aedes* (*Howardina*); thus, *Gymnometopa* is monobasic. Dyar & Knab (1906d), Howard *et al.* (1917) and Dyar (1920[illegible]) listed *Gymnometopa* as a synonym of *Aedes*. Bonne & Bonne-Wepster (1925) appear to be the first researchers to place the type species in the subgenus *Finlaya*, as did Edwards (1932a) and all later authors until Zavortink (1972) elevated *Gymnometopa* from synonymy with *Finlaya* as a monobasic subgenus of *Aedes*. *Gymnometopa* was raised to generic status by Reinert *et al.* (2004) based on the results of a phylogenetic analysis of morphological data for representatives of taxa included in Aedini. It was returned to subgeneric rank in *Aedes* by Wilkerson *et al.* (2015) based on the results of reanalyses of the more comprehensive morphological data set of Reinert *et al.* (2009).

Halaedes Belkin, 1962

Halaedes was proposed as a subgenus of *Aedes* by Belkin (1962), with *Culex australis* Erichson, 1842 designated the type species. The subgenus was established for species which Edwards (1932a) included in 'Group B (*Caenocephalus*)' of the subgenus *Pseudoskusea*. Edwards (1924b) had previously included a 'small group of species' in *Pseudoskusea* that were included in the genus *Caenocephalus* Taylor, 1914c, thus *Caenocephalus* became a synonym of *Pseudoskusea*. As *Caenocephalus* Taylor, 1914c was preoccupied by *Caenocephalus* van der Wulp, 1898, a genus of soldier flies (family Stratiomyiidae), Belkin effectively established *Halaedes* as a replacement name for *Caenocephalus*. *Halaedes* became a subgenus of *Ochlerotatus* when *Ochlerotatus* was returned to its original generic rank by Reinert (2000c). It was elevated to generic status by Reinert *et al.* (2004) based on the results of a morphology-based phylogenetic study of Aedini, and returned to its original rank as a subgenus of *Aedes* by Wilkerson *et al.* (2015).

Himalaius Reinert, Harbach & Kitching, 2006

Himalaius was originally established as a genus by Reinert *et al.* (2006), with *Finlaya gilli* Barraud, 1924d the designated type species. *Himalaius* was downgraded to subgeneric status in *Aedes* by Wilkerson *et al.* (2015).

Hopkinsius Reinert, Harbach & Kitching, 2008

Hopkinsius was originally introduced as a genus by Reinert *et al.* (2008) for seven species previously classified as species of *Aedes* (*Finlaya*). The authors designated *Ae.* (*Finlaya*) *ingrami* Edwards, 1930a as the type species and recognized two subgenera, the nominotypical subgenus and a new subgenus given the name *Yamada*. The genus was downgraded to subgeneric status in *Aedes* by Wilkerson *et al.* (2015), with the two subgenera relegated to the rank of informal species groups, the Hopkinsius and Yamada Groups.

Howardina Theobald, 1903

Howardina was originally proposed as a genus by Theobald (1903a), with *Culex walkeri* Theobald, 1901b subsequently designated as the type species by Dyar (1905a). Dyar & Knab (1906d), Howard *et al.* (1917) and Dyar (1923f) listed *Howardina* as a synonym of *Aedes*. Dyar (1918b) recognized '*Howardina* Theobald' as a group of American species based on features of the male genitalia. Two years later, Dyar (1920e) stated that the groups listed in his 1918 publication 'may properly be called subgenera', wherewith *Howardina* became a subgenus of *Aedes*, where it remained until it was returned to its original genus rank by Reinert *et al.* (2004). It was once again reduced to the rank of subgenus in *Aedes* by Wilkerson *et al.* (2015). Berlin (1969a), the last reviser of *Howardina*, divided the subgenus into three sections, the Fulvithorax, Sexlineatus and Walkeri Sections. The Sexlineatus Section included two species groups, the Aurivittatus and Eleanorae Groups, and the Walkeri Section included three groups, the Aurites, Ioliota and Walkeri Groups. Reinert (2002e) divided *Howardina* into Assemblages and Subassemblages based on morphological features of the female genitalia. Those informal categories were coequal for the most part with the Sections and Groups of Berlin (1969a), respectively. The Fulvithorax Section was divided into the Arborealis and Flavithorax Subassemblages; several species were removed from the Aurivittatus Group of the Sexlineatus Section and placed in the Inaequalis and Sexlineatus Subassemblages. However, species which Reinert did not examine, *e.g. Ae. guatemala* Berlin, 1969a, were not placed in an Assemblage or Subassemblage; therefore, the classification of Berlin (1969a) should be consulted for possible relationships of those species.

Huaedes Huang, 1968

Huaedes was established as a subgenus of *Aedes* by Huang (1968b), with the new species *Ae. wauensis* Huang, 1968b designated as the type species. *Huaedes* was raised to genus rank by Reinert *et al.* (2004), where it remained until it was returned to its original subgeneric rank in *Aedes* by Wilkerson *et al.* (2015).

Hulecoeteomyia Theobald, 1904

Hulecoeteomyia was originally introduced as a genus by Theobald (1904c), with *Hulecoeteomyia trilineata* Leicester, 1904 (in Theobald, 1904c) [subjective synonym of *Ae. chrysolineatus* (Theobald, 1907)] as its type and only originally included species. Edwards (1913c) treated *trilineata* as a species the genus *Stegomyia*, and Barraud (1923f, 1924a) and Borel (1928b) recognized it as a species of the genus *Finlaya*, perhaps without knowing that Edwards (1922b) had synonymized *trilineata* with *Ae.* (*Finlaya*) *chrysolineatus* in his synopsis of Oriental culicine mosquitoes. Edwards (1932a) listed *Hulecoeteomyia* as a synonym of *Aedes* subgenus *Finlaya* and divided the subgenus into eight species groups, with Group D denoted as the '*chrysolineata* group: *Hulecoeteomyia*'. Knight & Marks (1952) subdivided Group D, renamed the '*aureostriatus*-group: *Hulecoeteomyia*', into eight subgroups, with *Ae. chrysolineatus* being the nominotypical species of 'Subgroup I, *chrysolineatus*'. Knight (1968) provided a revisionary study of the

12 species of the group known at the time to occur in Southeast Asia. Three more recently described species and a species known from China have since been added to the group. Reinert (2002e) treated *Finlaya* as a subgenus of *Ochlerotatus* and divided it into seven Assemblages based on anatomical features of the female genitalia. *Hulecoeteomyia* was resurrected from synonymy with *Finlaya* and returned to its original generic status by Reinert *et al.* (2006) based on the results of a phylogenetic analysis of morphological data for representatives of taxa included in Aedini. It was returned to subgeneric rank in *Aedes* by Wilkerson *et al.* (2015) based on the results of reanalyses of the more comprehensive morphological data set of Reinert *et al.* (2009). *Hulecoeteomyia* is equivalent to the Chrysolineatus Assemblage of Reinert (2002e).

Indusius Edwards, 1934

Indusius was proposed as a subgenus of *Aedes* by Edwards, 1934 (in Barraud, 1934), with *Ae. pulverulentus* Edwards, 1922a as the type and only included species. *Indusius* was raised to genus level by Reinert *et al.* (2004) and returned to its original rank as a subgenus of *Aedes* by Wilkerson *et al.* (2015).

Isoaedes Reinert, 1979

Isoaedes was established as a subgenus of *Aedes* by Reinert (1979), with the new species *Ae. cavaticus* Reinert, 1979 as the type and only included species. *Isoaedes* was raised to the rank of genus by Reinert *et al.* (2004) and returned to its original subgeneric rank in *Aedes* by Wilkerson *et al.* (2015).

Jarnellius Reinert, Harbach & Kitching, 2006

Jarnellius was originally proposed as a genus by Reinert *et al.* (2006) for the five species which Arnell & Nielsen (1972) included in the Varipalpus Group of *Aedes* (*Ochlerotatus*). *Culex varipalpus* Coquillett, 1902d was designated as the type species, and the genus was divided into two subgeneric groups, the nominotypical subgenus and the new subgenus *Lewnielsenius*. The latter group was raised to generic rank by Reinert *et al.* (2009); however, *Jarnellius* and *Lewnielsenius* were both accorded subgeneric rank in *Aedes* by Wilkerson *et al.* (2015).

Jihlienius Reinert, Harbach & Kitching, 2006

Jihlienius was originally introduced as a genus by Reinert *et al.* (2006) for three Oriental species previously included as members of the subgenus *Finlaya*. The authors designated *Aedes chungi* Lien, 1968b as the type species. *Jihlienius* was relegated to subgeneric status in *Aedes* by Wilkerson *et al.* (2015).

Kenknightia Reinert, 1990

Kenknightia was established as a subgenus of *Aedes* by Reinert (1990b), with *Stegomyia dissimilis* Leicester, 1908 designated as the type species. Six of the 11 species included in the subgenus were described as new and five were transferred from the subgenus *Finlaya*. Reinert (2000c) transferred *Kenknightia* to *Ochlerotatus* when he returned *Ochlerotatus* to its original rank as a distinct genus. *Kenknightia* was raised to genus level by Reinert *et al.* (2004) and returned to its original subgeneric rank in *Aedes* by Wilkerson *et al.* (2015). *Kenknightia* includes two species groups, the Dissimilis and Luzonensis Groups, which were established concomitantly with the subgenus.

Kompia Aitkin, 1941

Kompia was described as a subgenus of *Aedes* by Aitkin (1941), with the new species *Ae. purpureipes* Aitken, 1941 as the type and only included species. *Kompia* was raised to genus level by Reinert *et al.* (2004) and returned to its original rank as a subgenus of *Aedes* by Wilkerson *et al.* (2015).

Leptosomatomyia Theobald, 1905

Leptosomatomyia was originally established as a genus by Theobald (1905f), with *Leptosomatomyia lateralis* Theobald, 1905f as the type and only included species [*lateralis* Theobald is the junior secondary homonym of *lateralis* Meigen, 1818, a synonym of *Ae. geniculatus* (Olivier, 1791), and *Ae. aurimargo* Edwards, 1922d is the available name of the type species]. *Leptosomatomyia* was first treated as a subgenus of *Aedes* by Edwards (1924b), the reason apparently being that he considered 67 of the 145 species known at the time to occur in the Australasian Region to 'belong to the genus *Aedes* in the broad sense'. *Leptosomatomyia* continued to be recognized as a monobasic subgenus of *Aedes* until it was returned to its original rank as a genus by Reinert *et al.* (2004) based on the phylogenetic relationships among Aedini generated in cladistic analyses of morphological data. It was returned to subgeneric rank in *Aedes* by Wilkerson *et al.* (2015) based on the results of reanalyses of the more comprehensive data set of Reinert *et al.* (2009).

Levua Stone & Bohart, 1944

Levua was introduced as a subgenus of *Aedes* by Stone & Bohart (1944), with the new species *Ae. suvae* Stone & Bohart, 1944 [subjective synonym of *Ae. geoskusea* Amos, 1944] as its type and only originally included species. Reinert (2000c) divided the traditional genus *Aedes* into the genera *Aedes* and *Ochlerotatus* and transferred *Levua* to *Ochlerotatus*. *Levua* was raised to genus level by Reinert *et al.* (2004) and returned to its original rank as a subgenus of *Aedes* by Wilkerson *et al.* (2015).

Lewnielsenius Reinert, Harbach & Kitching, 2006

Lewnielsenius was originally proposed as a subgenus of *Jarnellius* Reinert, Harbach & Kitching, 2006, with *Ae. muelleri* Dyar, 1920g as the type and only included species. The same authors raised *Lewnielsenius* to generic status 3 years later (Reinert *et al.*, 2009) based on a more comprehensive morphology-based phylogenetic study of Aedini than they had conducted previously. *Lewnielsenius* was transferred to subgeneric rank in *Aedes* by Wilkerson *et al.* (2015) based on the results of reanalyses of the data set of Reinert *et al.* (2009).

Lorrainea Belkin, 1962

Lorrainea was established as a subgenus of *Aedes* by Belkin (1962), with *Ae. dasyorrhus* King & Hoogstraal, 1946c designated the type species. *Lorrainea* was elevated to generic status by Reinert *et al.* (2004) based on the results of a morphology-based phylogenetic study of Aedini. It was returned to its original rank as a subgenus of *Aedes* by Wilkerson *et al.* (2015).

Luius Reinert, Harbach & Kitching, 2008

Luius was originally introduced as a genus by Reinert *et al.* (2008), with *Ae. fengi* Edwards, 1935a as the type and only included species. *Luius* was downgraded to subgeneric rank in *Aedes* by Wilkerson *et al.* (2015) based on the results of reanalyses of the comprehensive morphological data set of Reinert *et al.* (2009).

Macleaya Theobald, 1903

Macleaya was originally established as a genus by Theobald (1903e), with the new species *Macleaya tremula* Theobald, 1903e as its type and only originally included species. Edwards (1912b) listed *Macleaya* as a questionable synonym of the genus *Howardina* and it was formally recognized as a synonym of *Aedes* by Howard *et al.* (1917) and Dyar (1923f). Edwards (1924b) considered that *Macleaya* 'may be retained' as a subgenus of *Aedes* for the type species, 'which though apparently related to *Finlaya* shows several rather well-marked differences.' *Macleaya* became a subgenus of *Ochlerotatus* when *Ochlerotatus* was returned to its original generic rank by Reinert (2000c). Based on phylogenetic analyses of anatomical data, *Macleaya* was restored to its original rank as a genus by Reinert *et al.* (2006), with the concomitant recognition of *Chaetocruiomyia* Theobald, 1910c as a subgenus of the genus. The subgenus *Chaetocruiomyia* and the nominotypical subgenus were reduced to species groups, the Chaetocruiomyia and Macleaya Groups, when Wilkerson *et al.* (2015) restored *Macleaya* to its earlier status as a subgenus of *Aedes* based on the results of reanalyses of the more comprehensive anatomical data set of Reinert *et al.* (2009).

Molpemyia Theobald, 1910

Molpemyia was originally proposed as a genus by Theobald (1910c), with the new species *Molpemyia purpurea* Theobald, 1910c as the type and only originally included species. *Molpemyia* was formally recognized as a synonym of *Aedes* by Howard *et al.* (1917) and Dyar (1923f). Edwards (1912b) listed *Molpemyia* as a questionable synonym of the genus *Ochlerotatus*, but he later included the type species in a key to Australasian species of that genus (Edwards, 1912b). Edwards (1922d) recognized *Molpemyia* as a synonym of the subgenus *Finlaya* of *Aedes* by virtue of listing the type species as a species of that subgenus. Reinert (1993) reviewed the taxonomic history and morphology of four nominal species placed in various species groups of *Finlaya* by Edwards (1932a), Knight (1948b) and Knight & Marks (1952), including *Ae. auridorsum* Edwards, 1922d, *Ae. pecuniosus* Edwards, 1922d, *Calomyia priestleyi* Taylor, 1914c and *Molpemyia purpurea*, confirmed the synonymy of *Calomyia priestleyi* with *Molpemyia purpurea* and elevated *Molpemyia* from synonymy as a distinct subgenus of *Aedes* for the three species. Reinert (2000c) elevated *Ochlerotatus* from subgeneric status in *Aedes* to generic rank and transferred the subgenus *Molpemyia* to *Ochlerotatus*. *Molpemyia* was raised to genus level by Reinert *et al.* (2006) and returned to its original and current subgeneric rank in *Aedes* by Wilkerson *et al.* (2015).

Mucidus Theobald, 1901

Mucidus was originally briefly defined as a genus by Theobald (1901a) and more fully described in his first volume of *A Monograph of the Culicidae* (Theobald, 1901b). *Culex alternans* Westwood, 1835 was subsequently designated as the type species by Neveu-Lemaire (1902c). To understand the history of *Mucidus*, it is necessary to know that Theobald (1907) established the genus *Pardomyia*, with the new species *Pardomyia aurantia* as the type and only included species, and Leicester (1908) introduced the genus *Ekrinomyia* for the same species, which he named *Ekrinomyia aureostriata*. It appears that the three genera continued to be recognized as such until Edwards (1932a) placed *Pardomyia* and *Ekrinomyia* in synonymy with the subgenus *Mucidus* of *Aedes*, which he divided into two groups, 'Group A (*Mucidus*)' and 'Group B (*Pardomyia*)'. *Mucidus* remained a subgenus of *Aedes* until Reinert (2000c) split *Aedes* into the genera *Aedes* and *Ochlerotatus* and transferred the subgenus *Mucidus* to the latter genus. Based on morphological features of the female genitalia, Reinert (2002e) divided the subgenus into two groups, the Mucidus and Pardomyia Assemblages, which corresponded for the most part with Groups A and B of Edwards (1932a). *Mucidus* was raised to generic status by Reinert *et al.* (2004) based on the results of phylogenetic analyses of anatomical data, and 2 years later Reinert (2006b) resurrected *Pardomyia* from synonymy for recognition as a subgenus of *Mucidus*. Wilkerson *et al.* (2015) returned *Mucidus* to subgeneric status in *Aedes* based on the results of reanalyses of the more comprehensive anatomical data set of Reinert *et al.* (2009), with recognition of the two subgenera as informal species groups, the Mucidus and Pardomyia Groups.

Neomelaniconion Newstead, 1907

Neomelaniconion was originally established as a genus by Newstead, 1907 (in Newstead *et al.*, 1907), with *Neomelaniconion palpale* Newstead, 1907 as the type and only originally included species. The taxonomy of the group now known as *Neomelaniconion* dates back to its treatment as *Banksinella* Theobald, 1907 by Edwards (1932a) in his catalogue of the Culicidae, and in his book on the *Mosquitoes of the Ethiopian Region* (Edwards, 1941). *Banksinella* was in use until Stone (1957) pointed out that the name *Neomelaniconion* predated *Banksinella* by 22 days, the former published on 1 February 1907 and the latter on 23 February 1907. *Neomelaniconion* was subsequently recognized as a subgenus of *Aedes* until it was returned to its original rank as a genus by Reinert *et al.* (2004) based on the results of a morphology-based phylogenetic study of Aedini. Le Goff *et al.* (2007) accepted *Neomelaniconion* as a genus and proposed the creation of two species groups, named the Circumluteolus and Sylvaticum Groups. The name of the former group, however, is superseded by the Lineatopennis Group of Edwards (1941). The name of the second group was established for the newly described *Neomelaniconion sylvaticum* and related species. Based on the results of a reanalysis of the more comprehensive anatomical data set of Reinert *et al.* (2009), Wilkerson *et al.* (2015) returned *Neomelaniconion* to its former subgeneric rank in *Aedes*. With that action, *Neomelaniconion sylvaticum* became the junior secondary homonym of *sylvaticus* Meigen, 1818 [original combination: *Culex sylvaticus* = *Aedes* (*Ochlerotatus*) *communis* (de Geer, 1776)] and also *sylvaticus* Brunhes, 1983, currently a synonym of *Aedes* (*Paulianius*) *madagascarensis* van Someren, 1949a; consequently, *sylvaticum* Le Goff *et al.* (2007) was replaced with *hylaius* Harbach, 2018b, and the species is now known as *Aedes* (*Neomelaniconion*) *hylaius*. Ergo, the Sylvaticum Group of Le Goff *et al.*, 2007 is renamed here as the Hylaius Group.

Nyctomyia Harbach, 2013

Nyctomyia was proposed by Harbach (2013a) as a replacement for the genus name *Nyx* Harbach & Linton, 2013 (in Harbach *et al.*, 2013), which was preoccupied by *Nyx* Heppner, 1982, a genus in the lepidopteran family Choreutidae. *Nyctomyia* (as *Nyx* Harbach & Linton, 2013), was established for the new species *Nyx pholeocola* Linton & Harbach, 2013, the only originally included species. The generic rank of the taxon was based on phylogenetic relationships with other aedine taxa that were recovered when anatomical data for the type species were included in the data set of Reinert *et al.* (2009). *Nyctomyia* was relegated to its present subgeneric status in *Aedes* based on the results of reanalyses of the data set by Wilkerson *et al.* (2015).

Ochlerotatus Lynch Arribálzaga, 1891

Ochlerotatus was originally introduced as a genus based on its inclusion in a dichotomous key for distinguishing the genera of mosquitoes recognized at the time (Lynch Arribálzaga, 1891a), with *Ochlerotatus confirmatus* Lynch Arribálzaga, 1891b subsequently designated as the type species by Coquillett, 1910. *Ochlerotatus*, as currently defined, is a very large group with a complicated history. *Ochlerotatus* survived as a genus until Dyar & Knab (1906d) placed it in synonymy with *Aedes*. However, that action seems to have been overlooked by taxonomists working on the Old World fauna. Edwards (1917), for example, treated *Ochlerotatus* as a genus until he confusingly listed it as a subgenus of *Aedes* while at the same time listing species names in combination with the genus name *Ochlerotatus*. In his 1917 publication, Edwards stated that 'After further study of both larvae and adults of a large number of species of the *Aedes* group, I am now inclined to accept the view of Dyar and Knab that it would be most convenient to include almost all these forms within the single genus *Aedes*'. Edwards (1917) recognized five subgenera within *Aedes*, *Aedes sensu stricto*, *Armigeres*, *Ochlerotatus*, *Skusea* and *Stegomyia*, and divided *Ochlerotatus* into Groups *Diceromyia*, *Finlaya* and *Ochlerotatus sensu stricto*. Curiously, Edwards seemed to have been ambivalent about the status of *Ochlerotatus* as he indicated in a footnote in his first paper in a series of 'Mosquito Notes' (Edwards, 1920a) that 'The names *Stegomyia* and *Ochlerotatus* though used here as genera are to be understood in a subgeneric sense.' Dyar (1918b) separated the North American species of *Aedes* into five Groups: 'Group I (*Howardina* Theobald)', 'Group II (*Gualteria* Lutz)', 'Group III (*Taeniorhynchus* Arribálzaga)', 'Group IV (*Ochlerotatus* Arribálzaga)' and 'Group V (*Culicada* Felt)'. There was a brief hiatus in the history of *Ochlerotatus* when Dyar (1919c) misinterpreted *Heteronycha dolosa* Lynch Arribálzaga, 1891b as a species of *Aedes* and adopted *Heteronycha* to replace *Ochlerotatus* as a subgenus of *Aedes*. The following year, Dyar (1920e) restructured his 1918 informal classification of the North American species to include northern European species, with the recognition of 10 species groups named after the oldest American species: 'Group *curriei*', 'Group *impiger*', 'Group *innuitus*', 'Group *pullatus*', 'Group *punctor*', 'Group *scapularis*', 'Group *serratus*', 'Group *stimulans*', 'Group *thibaulti*' and 'Group *trichurus*'. In his review and catalogue of the Culicidae, Edwards (1932a) classified the known world fauna of the subgenus into eight groups, designated as Groups A–H, which subsumed the groups of Dyar (1920e). Undoubtedly, the groups recognized by Dyar and Edwards provided the foundation for the current informal classification, which developed and diversified as a result of taxonomic studies of *Ochlerotatus* in different geographical regions. The researchers who first proposed or who are accredited with the currently recognized infrasubgeneric groups of *Ochlerotatus* are indicated in the Tabulation in Part VI.

Orohylomyia Somboon & Harbach, 2023

Orohylomyia was established as a subgenus of *Aedes* by Somboon & Harbach. 2023 (in Somboon *et al.*, 2023b), with *Aedes yunnanensis* (Gaschen, 1934) as the type and only included species. The type species was originally described as a species of the subgenus *Finlaya* of *Aedes*. Knight & Marks (1952) classified it as a species of 'Group H (GENICULATUS-group; PROTOMACLEAYA)', without placement in one of seven subgroups recognized by the authors. Reinert *et al.* (2006), in their reclassification of Aedini, removed *Ae. yunnanensis* from *Finlaya* and recognized it as a species of *Hulecoeteomyia*. Somboon *et al.* (2023b) removed the species from *Hulecoeteomyia* and established *Orohylomyia* based on morphological distinctions and the results of phylogenetic analyses of mitochondrial *COI* gene sequences, in which the species was recovered as a distinct lineage among species representing 13 subgenera of *Aedes*.

Paraedes Edwards, 1934

Paraedes was originally proposed as a genus by Edwards, 1934 (in Barraud, 1934), with the new species *Paraedes barraudi* Edwards, 1934 designated the type species. The genus was reduced to subgeneric rank in *Aedes* by Mattingly (1958a), who divided the species known at the time into two groups, 'Group A (*Paraëdes* s. str.)' and 'Group B' for the new species *Ae. bonneae* Mattingly, 1958a. Mattingly noted that *Ae. bonneae* 'clearly deserves a separate species group' because it differed markedly from the other species of *Paraedes* 'in the ornamentation of the alula and scutellum and, in the female, of the vertex, as well as in details of the male terminalia'. Based only on anatomical features of the female genitalia, Reinert (2000e) divided the subgenus into two groups, the Barraudi and Ostentatio Assemblages, and included *Ae. bonneae* in the latter group. In view of the distinctions noted by Mattingly, the monobasic Group B is retained here as the Bonneae Group. *Paraedes* was raised to genus level by Reinert *et al.* (2004) and returned to its current subgeneric rank in *Aedes* by Wilkerson *et al.* (2015).

Patmarksia Reinert, Harbach & Kitching, 2006

Patmarksia was originally introduced as a genus by Reinert *et al.* (2006), with *Leucomyia australiensis* variety *papuensis* Taylor, 1914d designated as the type species. For the most part, *Patmarksia* includes the Australasian species that comprise 'Subgroup II (*papuensis*)' of 'Group F (*alboannulatus*)' of Knight & Marks (1952) and Lee *et al.* (1982), as well as the Papuensis Assemblage of Reinert (2002e), which were previously included in the subgenus *Finlaya*. Wilkerson *et al.* (2015) reduced *Patmarksia* to subgeneric status in *Aedes* based on the results of reanalyses of the anatomical data set of Reinert *et al.* (2009).

Paulianius Brunhes & Boussès, 2017

Paulianius was established as a genus by Brunhes & Boussès, 2017 (in Brunhes *et al.*, 2017) for seven Malagasy species (four previously included in *Diceromyia*) and one African species (previously included in *Diceromyia*), with *Ae. tiptoni* Grjebine, 1953a designated the type species. The authors arranged the species in three sections, the Coulangesi, Niveus and Tiptoni Sections. The Niveus Section was proposed to accommodate the African *Ae. niveus* de Meillon, 1943, which the authors considered to be a species of the genus *Diceromyia*. *Diceromyia nivea*, originally described as a species of *Aedes*, was recognized as the senior synonym of *Ae. zethus* de Meillon & Lavoipierre, 1944, which was introduced as a replacement for *niveus* de Meillon., *non* Ludlow, 1903 (in Theobald, 1903a) when it was classified as a species of *Aedes* (*Finlaya*) by Brug (1931c) and later workers. When *Paulianius* was relegated to subgeneric rank in *Aedes* by Harbach (2018b), *Ae. niveus* was returned to its earlier status as a synonym of *Ae. zethus*. As the 'sections' of Brunhes *et al.* (2017) are without subordinate groups, they are recognized here as the Coulangesi, Tiptoni and Zethus Groups, the last one being a replacement name for the Niveus Section.

Petermattinglyius Reinert, Harbach & Kitching, 2009

Petermattinglyius was originally introduced as a genus by Reinert *et al.* (2009) to accommodate five species that occur in the Oriental Region, with *Aedes iyengari* Edwards, 1923b being the designated type species. Two of the five species were originally placed in the subgenus *Skusea* by Edwards (1921a, 1923b) and later transferred to the subgenus *Dendroskusea* by Edwards (1929b). Edwards (1932a) synonymized *Dendroskusea* with *Diceromyia* and transferred the African and Oriental species of *Skusea* to the composite *Diceromyia*, which he divided into two groups, Group A for the African species and Group B for the Oriental species. Two of the five species which Edwards included in Group B are now species of *Petermattinglyius*; the other three are currently species of *Dendroskusea*. Reinert *et al.* (2009) recognized two subgenera, the nominotypical subgenus *Petermattinglyius* and the monobasic subgenus *Aglaonotus*. *Petermattinglyius* was relegated to subgeneric rank in *Aedes* by Wilkerson *et al.* (2015), with the two subgenera reduced to informal species groups, the Aglaonotus and Petermattinglyius Groups.

Phagomyia Theobald, 1905

Phagomyia was originally established as a distinct genus by Theobald (1905e), with *Culex gubernatoris* Giles, 1901a subsequently designated the type species by Edwards (1932a). *Phagomyia* was synonymized with the genus *Ochlerotatus* by Edwards (1912b) and considered a synonym of the genus *Aedes* by Howard *et al.* (1917). Edwards (1932a) treated *Phagomyia* as a synonym of *Finlaya* and included the type species among the Oriental species of 'Group B (*terrens-gubernatoris*-group: *Gualteria*)'. Knight & Marks (1952) omitted the name *gubernatoris* from Group B for conciseness and divided the group into Subgroups I–VII, with all but one of the Oriental species listed by Edwards contained in 'Subgroup III, *gubernatoris*'. *Phagomyia* was resurrected from synonymy with *Finlaya* by Reinert *et al.* (2006) to serve as a genus for Subgroup III of Knight & Marks, with the inclusion of two additional species. *Phagomyia* was reduced to subgeneric rank in *Aedes* by Wilkerson *et al.* (2015).

Polyleptiomyia Theobald, 1905

Polyleptiomyia was originally introduced as a distinct genus by Theobald (1905e), with *Stegomyia albocephala* Theobald, 1903c as the type and only originally included species. Edwards (1912b) treated *Polyleptiomyia* as a synonym of the genus *Ochlerotatus* and both Howard *et al.* (1917) and Dyar (1923f) considered it to be a synonym of the genus *Aedes*. Edwards (1917) recognized *Ochlerotatus* as a subgenus of *Aedes* and divided it into three groups: 'Group *Finlaya*', 'Group *Diceromyia*' and 'Group *Ochlerotatus*, s. str.' He further divided the last group into 'Group *Ochlerotatus*', 'Group *Ecculex*' and 'Group *Aëdimorphus*'. Edwards (1925) divided the subgenus *Aedimorphus* into Groups I–IV, the *furcifer*, *apicoannulatus*, *domesticus* and *vexans* groups, respectively (the last two equivalent to 'groups *Aëdimorphus* and *Ecculex*' outlined in his 1917 publication). Edwards further divided the *vexans* group into five series, the *albocephalus*, *lamborni*, *dentatus*, *vexans* and *ochraceus* series; thus, *Polyleptiomyia* became a synonym of the subgenus *Aedimorphus*. Reinert *et al.* (2009), as a result of a comprehensive morphology-based phylogenetic study of Aedini, resurrected *Polyleptiomyia* from synonymy with *Aedimorphus* for *albocephalus* and included *Ae. gandarai* da Cunha Ramos, Capela & Ribeiro, 1995 as a member of the genus. Wilkerson *et al.* (2015) relegated *Polyleptiomyia* to subgeneric status in *Aedes* based on the results of reanalyses of the data set of Reinert *et al.* (2009).

Protomacleaya Theobald, 1907

Protomacleaya was originally established as a genus by Theobald (1907), with *Culex triseriatus* Say, 1823 as the type and only originally included species. Edwards (1912b) treated *Protomacleaya* as a synonym of the genus *Ochlerotatus* and both Howard *et al.* (1917) and Dyar (1923f) considered it to be a synonym of the genus *Aedes*. Dyar (1928) listed *Protomacleaya* among genera which previously included species that had been moved to *Aedes* at one time or other, with the type species of *Protomacleaya* treated as a species of the subgenus *Finlaya*. Edwards (1932a) formally listed *Protomacleaya* as a synonym of *Finlaya*, which he divided into eight groups identified as Groups A–H. *Aedes triseriatus* was placed in Group H. As Edwards did not link a specific or a specific and generic name to Group H, as he did for most other groups, Knight & Marks (1952) added the names *geniculatus* and *Protomacleaya*. They further subdivided 'Group H (GENICULATUS-group: PROTOMACLEAYA)' into seven subgroups, with *Ae. triseriatus* placed in 'Subgroup VI, GENICULATUS (PROTOMACLEAYA, s. str.)'. Zavortink (1972) removed all New World *Aedes* that were included in the subgenus *Finlaya* and assigned them to other generic-level taxa. He resurrected *Protomacleaya* from synonymy with *Finlaya* for the Terrens Group treated by Schick (1970a, 1970b) and established the Kompi, Knabi, Triseriatus and Zoosophus Groups for 11 other species previously included in *Finlaya*. He also established *Gualteria* Lutz, 1904 (in Bourroul, 1904) as a synonym of *Protomacleaya*. Zavortink justified the recognition of *Protomacleaya* as a subgenus of *Aedes* by stating: 'Although the subgenus is quite diverse, is defined mainly on negative characters, and many of the included species depart from the "normal" in 1 or more characteristics of 1 or more stages, it is, I believe, natural because of the basically similar male genitalia of virtually all species.' To the contrary, the phylogenetic studies of Reinert *et al.* (2004, 2006, 2008, 2009) clearly showed that *Protomacleaya* is a polyphyletic assemblage, and therefore it was not possible to retain it as a valid genus-group entity. When Wilkerson *et al.* (2015) reanalysed the data set of Reinert *et al.* (2009), they inadvertently listed *Protomacleaya* as a synonym of the subgenus *Ochlerotatus*. This was corrected by Wilkerson & Linton (2015), who formally reinstated *Protomacleaya* as a subgenus of *Aedes*, rather than as a subgenus of *Ochlerotatus* where it was classified by Reinert (2000c) prior to the phylogenetic studies of Reinert *et al.* As shown in the Tabulation in Part VI, the current infrasubgeneric classification of *Protomacleaya* is comprised entirely of the informal groups recognized by Schick (1970a) and Zavortink (1972).

Pseudalbuginosus Huang & Rueda, 2015

Pseudalbuginosus was proposed as a subgenus of *Aedes* by Huang & Rueda (2015b), with *Aedes grjebinei* Hamon, Taufflieb & Maillot, 1957 as the type and only included species. *Aedes grjebinei* was original described as a species of the subgenus *Aedimorphus*. Reinert (1987) stated that 'there is no doubt this species should be included in *Albuginosus* based on the information provided in the original description, but it was not formally transferred to that subgenus.'

Pseudarmigeres Stone & Knight, 1956

Pseudarmigeres was originally proposed as a replacement name by Stone & Knight (1956) for the subgeneric name *Dunnius* Edwards, 1930a, which was preoccupied by *Dunnius* Distant, 1902. The type species, by original designation, being *Aedes* (*Armigeres*) *albomarginata* variety *dunni* Evans, 1928, was reclassified as a subspecies of *Ae. argenteoventralis* (Theobald, 1909) by White (1975) and synonymized with that species by Harbach & Wilkerson (2023). *Pseudarmigeres* was raised to genus level by Reinert *et al.* (2004) and returned to its original and current subgeneric rank in *Aedes* by Wilkerson *et al.* (2015).

Pseudoskusea Theobald, 1907

Pseudoskusea was originally established as a distinct genus by Theobald (1907), with *Skusea multiplex* Theobald, 1903a as the type and only originally included species. Edwards (1924b) stated that the name *Pseudoskusea* 'may be retained for a small group of [Australasian] species that resemble *Ochlerotatus*', and treated it as a subgenus of *Aedes*. His concept of *Pseudoskusea* included a 'small group of species' that were previously recognized as species of the genus *Caenocephalus* Taylor, 1914c. Edwards (1932a) clearly divided *Pseudoskusea* into two groups, denoted as 'Group A (*Pseudoskusea*)' and 'Group B (*Caenocephalus*)'. Dobrotworsky (1961a) remarked that the similarities of species of Group B with those of Group A 'hardly justify the inclusion of the species of Group B in the subgenus *Pseudoskusea*; they should probably be placed in a separate subgenus, but this step should be deferred pending a revision of the subgenus *Pseudoskusea*.' As noted above, Belkin (1962) established the subgenus *Halaedes* for Group B; thus, *Pseudoskusea* was limited to the species of Group A. *Pseudoskusea* remained a subgenus of *Aedes* until Reinert (2000c) restored *Ochlerotatus* to its original genus rank and transferred the subgenus *Pseudoskusea* to that genus. *Pseudoskusea* was raised to generic status by Reinert *et al.* (2006) based on the results of phylogenetic analyses of anatomical data, and then returned to its former subgeneric status in *Aedes* by Wilkerson *et al.* (2015) based on the results of reanalyses of the more comprehensive anatomical data set of Reinert *et al.* (2009).

Rampamyia Reinert, Harbach & Kitching, 2006

Rampamyia was originally introduced as a genus by Reinert *et al.* (2006), with *Culex notoscriptus* Skuse, 1889 designated the type species. The genus was established for three Australasian species that were included in 'Subgroup II, NOTOSCRIPTUS' of Knight & Marks (1952) of the subgenus *Finlaya*, a subgroup which they created for four species previously included in the undivided 'Group E (*mediovittatus*-group: *Gymnometopa*)' of Edwards (1932a). The fourth species originally included in the subgroup, *Ae. mallochi* Taylor, 1944, is currently without subgeneric placement in *Aedes*. It is noteworthy that Belkin (1962) suggested the Notoscriptus Subgroup 'probably should be recognized as a distinct group', which was confirmed by the results of the morphology-based phylogenetic studies of Reinert *et al.* (2006, 2008, 2009), who identified the group as the genus *Rampamyia*, and Wilkerson *et al.* (2015), who downgraded *Rampamyia* to its current status as a subgenus of *Aedes*.

Reinertia Somboon, Namgay & Harbach, 2021

Reinertia was established as a subgenus of *Aedes* by Somboon & Harbach, 2021 (in Somboon *et al.*, 2021a), with *Ae. suffusus* Edwards, 1922a as the type and only included species. The type species was originally described as a species of the subgenus *Finlaya* of *Aedes*. Edwards (1932a) included the species in Group H of the subgenus. Knight & Marks (1952) added specific and generic names to the Group, rendering it 'Group H (GENICULATUS-group; PROTOMACLEAYA)', and divided it into seven subgroups. Their 'Subgroup V, SUFFUSUS' was created as a monobasic group for *Ae. suffusus*. As a result of the morphology-based study of the phylogeny and classification of *Finlaya* conducted by Reinert *et al.* (2006), *Ae. suffusus* was excluded from *Finlaya* and recognized as a species without subgeneric placement (Wilkerson *et al.*, 2015: Appendix S1). Discovery of the previously unknown male of *Ae. suffusus* and phylogenetic analyses of the mitochondrial *COI* gene of species representing 38 subgenera of *Aedes* and six other genera of Aedini provided the evidence needed to establish *Reinertia* as a monobasic subgeneric lineage.

Rhinoskusea Edwards, 1929

Rhinoskusea was established as a subgenus of *Aedes* by Edwards (1929b), with *Ficalbia longirostris* Leicester, 1908 as the type and only originally included species. Prior to the creation of *Rhinoskusea*, the type species, in addition to being originally described as a species of *Ficalbia*, was also considered to be a species of the genus *Uranotaenia* by Taylor (1919, as the new species *Ur. hilli*) and the genus *Skusea* by Edwards (1917). *Rhinoskusea* remained a subgenus of *Aedes* until Reinert (2000c) transferred it to *Ochlerotatus* concomitant with its restoration as a distinct genus. *Rhinoskusea* was raised to generic status by Reinert *et al.* (2004) based on the results of morphology-based phylogenetic analyses of Aedini. It was later returned, along with *Ochlerotatus*, to its former subgeneric status, in *Aedes* by Wilkerson *et al.* (2015) based on the results of reanalyses of the more comprehensive anatomical data set of Reinert *et al.* (2009).

Rusticoidus Shevchenko & Prudkina, 1973

Rusticoidus was introduced as a subgenus of *Aedes* by Shevchenko & Prudkina (1973), with *Aedes refiki* Medschid, 1928 as the type and only originally (?) included species. The recognition of *Ae. refiki* as the only originally included species based on it being the only species described as a member of the subgenus is questionable because the authors subsequently indicated that *Rusticoidus* included all species of the '*rusticus* group'. Confusion caused by the failure of the authors to indicate which species should be included in *Rusticoidus* prompted Reinert (1999a) to review and clarify the taxonomic history of the '*rusticus* group' and establish which species should be included in the subgenus. Most of the species currently assigned to the subgenus were included in 'Group H (*rusticus*-group: *Feltianus*)' of Edwards (1932a) as valid species and synonyms of currently valid species. *Rusticoidus* was transferred from *Aedes* to subgeneric status in *Ochlerotatus* by Reinert (2000c) when he restored *Ochlerotatus* to its original generic rank. *Rusticoidus* was retained as a subgenus of *Ochlerotatus* in the phylogenetic studies of Reinert *et al.* (2004, 2006, 2008, 2009) and erroneously reduced to informal group status when Wilkerson *et al.* (2015) reanalysed the data set of Reinert *et al.* (2009). This was corrected by Wilkerson & Linton (2015), who formally reinstated *Rusticoidus* as a subgenus of *Aedes*.

Sallumia Reinert, Harbach & Kitching, 2008

Sallumia was originally established as a subgenus of *Ochlerotatus* by Reinert *et al.* (2008), with *Aedes hortator* Dyar & Knab, 1907a as the designated type species. *Sallumia* was raised to the rank of genus in a subsequent, more comprehensive phylogenetic study conducted by the same authors (Reinert *et al.*, 2009), and reduced to its current subgeneric rank in *Aedes* by Wilkerson *et al.* (2015) based on the results of reanalyses of their anatomical data set.

Scutomyia Theobald, 1904

Scutomyia was originally introduced as a distinct genus by Theobald (1904c), with the new species *Scutomyia albolineata* Theobald, 1904c as the type and only originally included species. Edwards (1912b) listed *Scutomyia* as a synonym of the genus *Howardina*. Howard *et al.* (1917) and Dyar (1923f) listed it as a synonym of the genus *Aedes*. Based on studies of the larval and adult stages of species of the '*Aedes* group', Edwards (1917) accepted 'the view of Dyar and Knab that it would be most convenient to include almost all of these forms within the single genus *Aedes*', but found it 'justifiable to retain as subgenera some of the groups which have hitherto been regarded as genera', including *Aedes*, *Armigeres*, *Ochlerotatus*, *Skusea* and *Stegomyia*. Edwards (1922b) and most other authors subsequently included the type species of *Scutomyia* as a member of the subgenus *Stegomyia*. Edwards (1932a) divided the subgenus *Stegomyia* into four groups: 'Group A (*Stegomyia* s. str., or *aegypti*-group)', 'Group B (*w-albus*-group)', 'Group C (*scutellaris*-group)' and 'Group D (*A. vittatus* Big.)', and placed *Ae. albolineatus* in Group C. Knight & Rozeboom (1946) removed *Ae. albolineatus* from Group C and placed it and related species in a new group appropriately named the Albolineatus Group, which, in keeping with the group names of Edwards (1932a), was later also denoted as 'Group E (*albolineatus*-group)' by Knight & Hurlbut (1949). *Scutomyia* was resurrected from synonymy with *Stegomyia* by Reinert (1985) to comprise the Albolineatus Group as a subgenus of *Aedes*, which he divided into two groups, the Laffooni and Hoogstraali Groups. *Scutomyia* was raised to generic status by Reinert *et al.* (2004) based on the results of morphology-based phylogenetic analyses of Aedini and was returned to its former subgeneric status in *Aedes* by Wilkerson *et al.* (2015) based on the results of reanalyses of the more comprehensive anatomical data set of Reinert *et al.* (2009).

Skusea Theobald, 1903

Skusea was originally described as a genus by Theobald (1903a), with *Aedes pembaensis* Theobald, 1901c subsequently designated the type species by Blanchard (1905). *Skusea* was synonymized with *Aedes* by Edwards (1913c) and classified as a subgenus of *Aedes* by Edwards (1917). Edwards (1932a) synonymized *Dendroskusea* with the subgenus *Diceromyia* and transferred the African (except *Ae. pembaensis*) and Oriental species of the subgenus *Skusea* to the composite subgenus, and recognized the subgenus *Skusea* to consist of the African *Ae. pembaensis* and three nominal Oriental species, *Ae. amesii* (Ludlow, 1903), *Ae. fumidus* Edwards, 1928a and *Ae. furvus* Edwards, 1928a, currently a synonym of the first species. In addition to the Oriental species included in *Skusea* by Edwards, Mattingly (1959) described the new species *Ae. celebicus* from the Indomalayan area and included it as a member of the subgenus. Belkin (1962) transferred the Oriental species to the subgenus *Lorrainea*, leaving only *Ae. pembaensis* in the subgenus *Skusea*. In addition to *Ae. pembaensis*, *Skusea* currently includes two species originally described as members of the subgenus, *Ae. lambrechti* van Someren, 1971 and *Ae. moucheti* Ravaonjanahary & Brunhes, 1977, and also *Stegomyia cartroni* Ventrillon, 1906b, which Brunhes (1977b) removed from synonymy with *Ae. pembaensis*. *Skusea* was raised to generic status by Reinert *et al.* (2004) based on the results of a morphology-based phylogenetic study of Aedini and returned to its former subgeneric status in *Aedes* by Wilkerson *et al.* (2015) based on reanalyses of the data set of Reinert *et al.* (2009).

Stegomyia Theobald, 1901

Stegomyia was originally established as a distinct genus by Theobald, 1901 (in Howard, 1901), with *Culex fasciatus* Fabricius, 1805 subsequently designated the type species by Neveu-Lemaire, 1902c. The name *Stegomyia fasciata* was in use by Edwards (1912b), who expressed a desire to retain it (Dyar, 1920h) even though he knew the specific name was preoccupied and *Culex aegypti* Linnaeus, 1762 was the next available name for the species. Dyar (1920f) was previously aware of this and had listed *Culex fasciatus* as a synonym of *Aedes aegypti*. However, the name of the species has had a long and complicated history that ranged from Hasselquist (1757) through much of the 20th century. Hasselquist gave the name *Culex aegypti* to a mosquito responsible for fierce attacks on humans in Egypt, but the name was never attributed to Hasselquist: it was accredited to Linnaeus even though the name *aegypti* never appeared in any of his publications. The remarkable history and controversy over the name of the species (see Clements & Harbach, 2018 for a detailed review) was settled by designation of a neotype specimen (Mattingly *et al.*, 1962) and validation of the specific name *aegypti* Linnaeus, 1762 by the International Commission on Zoological Nomenclature (1964). *Stegomyia* was returned to its original generic status by Reinert *et al.* (2004) based on the results of morphology-based phylogenetic analyses of Aedini, and was returned to its former subgeneric status in *Aedes* by Wilkerson *et al.* (2015) based on the results of reanalyses of the more comprehensive anatomical data set of Reinert *et al.* (2009).

Edwards (1925) arranged the Afrotropical *Stegomyia* in five groups: Group I, the *vittatus* group; Group II, the *africanus* group; Group III, the *argenteus* group; Group IV, the *apicoargenteus* group; and Group V, the *albopictus* group. Those groups, with the exception of the monobasic Group I, were abandoned when Edwards (1932a) divided the world fauna of the subgenus into four groups, designated Groups A, B, C and D, each with an associated scientific name, *i.e.* '*Stegomyia* s. str., or *aegypti*-group', '*w-alba*-group', '*scutellaris*-group' and '*A. vittatus* Big.', respectively. As noted above, *Ae. vittatus* (Bigot, 1861) was removed from *Stegomyia* and is now the sole member of the subgenus *Fredwardsius*. The other groups have since undergone significant reorganization and splitting into more homogeneous groups and subgroups, concomitant with the removal of some species to other currently recognized subgenera (*Catatassomyia* and *Scutomyia*) that Edwards had listed as synonyms of *Stegomyia*. The subgenus is currently divided into 16 species groups, two of which, the Scutellaris and W-albus Groups, are further divided into subgroups. Most of the groupings are attributable to the taxonomic studies of Yiau-Min Huang, which provide information on the history of the classification (Huang, 1972, 1977a, 1979b, 1988b, 1990, 1994, 2004).

Tanakaius Reinert, Harbach & Kitching, 2004

Tanakaius was originally introduced as a genus by Reinert *et al.* (2004) for two species of *Aedes* that were classified at the time as species of *Ochlerotatus* subgenus *Finlaya*, *i.e. Ae. togoi* (Theobald, 1907) and *Ae. savoryi* Bohart, 1957. The former species, originally described as *Culicelsa togoi* Theobald, 1907, is the designated type species. *Tanakaius* was reduced to its current status as a subgenus of *Aedes* by Wilkerson *et al.* (2015).

Tewarius Reinert, 2006

Tewarius was originally proposed as a new genus of Aedini by Reinert (2006a), with *Aedes agastyai* Tewari & Hiriyan, 1992b the designated type species. Reinert divided the species of the genus between two groups, the Agastyai and Nummatus Groups. Wilkerson *et al.* (2015) relegated *Tewarius* to its current subgeneric status in *Aedes* based on the results of phylogenetic analyses of the anatomical data set of Reinert *et al.* (2009).

Vansomerenis Reinert, Harbach & Kitching, 2006

Vansomerenis was originally established as a genus by Reinert *et al.* (2006) for three Afrotropical species, with *Ae. pulchrithorax* Edwards, 1939a the designated type species. *Vansomerenis* was reduced to subgeneric status in *Aedes* by Wilkerson *et al.* (2015) based on the results of phylogenetic reanalyses of the anatomical data set of Reinert *et al.* (2009).

Zavortinkius Reinert, 1999

Zavortinkius was introduced as a subgenus for 11 Afrotropical species of *Aedes* by Reinert (1999b), with *Stegomyia longipalpis* Grünberg, 1905 the designated type species. Reinert divided the species of *Zavortinkius* between three species groups, the Brygooi, Longipalpis and Monetus Groups. *Zavortinkius* was transferred from *Aedes* to the genus *Ochlerotatus* by Reinert (2000c) and was regarded as a subgenus of *Ochlerotatus* by Reinert (2002e), who treated the species groups as Assemblages with the same names based on morphological characters of the female genitalia. *Zavortinkius* was also treated as a subgenus of *Ochlerotatus* in the phylogenetic studies of Reinert *et al.* (2004, 2006, 2008, 2009). Based on the results of a reanalysis of the data set of Reinert *et al.* (2009), Wilkerson *et al.* (2015) returned *Zavortinkius* to its original subgeneric status in *Aedes*.

Armigeres Theobald, 1901

Armigeres Theobald, 1901 *sensu stricto*

Armigeres was briefly defined as a genus by Theobald (1901a) and more fully described in his first volume of *A Monograph of the Culicidae* (Theobald, 1901b). The genus is typified by *Culex obturbans* Walker, 1859, which was subsequently designated as the type species by Brunetti (1914). Edwards (1932a) divided *Armigeres sensu stricto* into Groups A, B and C. The monobasic Group A (*Ar. maximus* Edwards, 1922a) and Group C (less *mjobergi* Edwards, 1926b, now recognized as a species of the subgenus *Downsiomyia*) constitute *Armigeres sensu stricto*. The two species Edwards included in Group B now comprise *Aedes* subgenus *Alanstonea*. *Armigeres sensu stricto* has never undergone a comprehensive revision and is likely to include a number of distinctive groups. The study of six species of *Armigeres* in the Papuan Subregion conducted by Steffan (1968) is the only study since Edwards (1932a) to recognize species groups within *Armigeres sensu stricto*. Steffan recognized the Papuan *Armigeres* as belonging to Group C of Edwards and divided them into two 'subgroups' based on structures of adults and larvae. He established the '*malayi* group' for *Ar. breinli* (Taylor, 1914d) and *Ar. malayi* (Theobald, 1901c) and recognized a second 'rather heterogenous' unnamed group to include *Ar. denbesteni* Brug, 1925, *Ar. lacuum* Edwards, 1922d, *Ar. milnensis* Lee, 1944b and *Ar. papuensis* Peters, 1963a. Steffan wisely mentioned that 'clarification of the interrelationships of the members of this second group await a more detailed study of other members of this subgenus.'

Leicesteria Theobald, 1904

Leicesteria was originally introduced as a genus by Theobald (1904c), with *Leicesteria longipalpis* Leicester, 1904 (in Theobald, 1904c) as the only originally included species. When *Leicesteria* was first recognized as a subgenus of *Armigeres* is not immediately clear. In his second paper in a series of 'Mosquito Notes', Edwards (1921a, May 1921) listed *Leicesteria annulipalpis* (Theobald, 1910a) as a 'true' species of the genus. In a paper published later in the same year entitled 'H. Sauter's Formosan collections: Culicidae', Edwards (1921c, December 1921) listed *Armigeres* (*Leicesteria*) *annulitarsis* (Leicester, 1908) as one of the species 'collected by Herr Sauter'. In the following year, Edwards unmistakably treated *Leicesteria* as a subgenus of *Armigeres* in both parts of 'A Synopsis of Adult Oriental Culicine' mosquitoes (Edwards, 1922a, 1922b). Thurman (1959) resurrected the name *Leicesteriomyia* Brunetti, 1912 from synonymy with *Leicesteria* to serve as a third subgenus of *Armigeres* for *Ar. flavus* (Leicester, 1908). Macdonald (1960), the last significant study in the taxonomic history of *Leicesteria*, reinstated the synonymy of *Leicesteriomyia* and returned *Ar. flavus* to the subgenus *Leicesteria*. However, in view of noteworthy morphological and ecological differences, Macdonald suggested that *Ar. flavus* 'might reasonably be placed in a separate species group.'

Eretmapodites Theobald, 1901

Eretmapodites was first published as the name of a genus by Theobald (1901d), with *Er. quinquevittatus* Theobald, 1901d as the type and only originally included species. In this publication, Theobald described 25 of 26 species contained in a collection of mosquitoes made in West Africa, and stated that 'Thirteen of the new species are described in my Monograph of the Culicidae shortly to be published by the Trustees of the British Museum, and the remaining new ones here.' Ironically, *Er. quinquevittatus* was not one of the 'remaining' new species, and the genus and species names were both listed as having been published in volume 1 of the monograph (Theobald, 1901b). Consequently, the names *Eretmapodites* and *quinquevittatus* became available from the date of the earlier publication. *Eretmapodites quinquevittatus* was subsequently designated the type species of the genus by Brunetti (1914). *Eretmapodites* is the only genus of mosquitoes with species that only occur in the Afrotropical Region. Eight species of *Eretmapodites* were recognized by Edwards (1932a), one of which included two varieties and another a synonymous nominal species that are now recognized as distinct species. As a result of subsequent studies, taxonomic acts and the addition of new species, the genus has been divided into the six species groups listed with their authorship in the Tabulation in Part VI.

Haemagogus Williston, 1896

Haemagogus Williston, 1896 *sensu stricto*

Haemagogus was established as a new genus by Williston (1896), with *Hg. splendens* Williston, 1896 as the type and only originally included species. Dyar & Knab (1906d) listed *Haemagogus* as a synonym of *Aedes*, but they seemed to vacillate in a footnote by stating 'The genus *Haemagogus* may perhaps be recognized' based on a distinctive larval character ... 'If so, it will take *Howardina* and *Gymnometopa* as synonyms'. Arnell (1973) provided a detailed history of *Haemagogus* from its beginning with Williston (1896) to the removal of *Conopostegus* from synonymy with *Finlaya* and its classification as a subgenus of *Haemagogus* by Zavortink (1972). Arnell arranged the species of *Haemagogus sensu stricto* in three informal sections, the Albomaculatus, Splendens and Tropicalis Sections. Reinert (2002c) recognized two Assemblages based on features of the female genitalia, the Equinus and Splendens Assemblages, which correspond with the Albomaculatus and Splendens Sections, respectively.

Conopostegus Dyar, 1925

Conopostegus was originally proposed as a subgenus of *Aedes* by Dyar (1925d), with *Aedes leucocelaenus* Dyar & Shannon, 1924b as the type and only originally included species. *Conopostegus* continued to be recognized as a subgenus of *Aedes* until Edwards (1932a) placed it in synonymy with *Finlaya*. As noted above, Zavortink (1972) removed *Conopostegus* from synonymy and classified it as a subgenus of *Haemagogus*.

Heizmannia Ludlow, 1905

Heizmannia Ludlow, 1905 *sensu stricto*

Heizmannia was introduced as a distinct genus by Ludlow (1905a), with *Hz. scintillans* Ludlow, 1905a as the type and only originally included species. Theobald (1907) and Ludlow (1908b) later regarded the species as a member of the genus *Dendromyia*, which seems to be the reason why Howard *et al.* (1913) considered *Heizmannia* to be a distinct genus of Sabethini. In his synopsis of adult culicine mosquitoes, Edwards (1922b) stated that 'there can be no doubt that *H. scintillans* is closely related to the species of *Wyeomyia* and *Dendromyia* described from other parts of the Oriental Region by Theobald and Leicester' and transferred most of those species to the genus *Heizmannia*. He observed that morphological features indicated an affinity with the aedine group of genera, with particular mention of the 'closely related' genus *Haemagogus*. Mattingly (1957b) recognized two species groups within the genus, 'Group A (*Heizmannia* s. str.)' and 'Group B' for *Hz. achaetae* (Leicester, 1908), which was only provisionally included in the genus (see *Mattinglyia*).

Mattinglyia Lien, 1968

Mattinglyia was originally established as a distinct genus by Lien (1968c), with *Mattinglyia catesi* Lien, 1968c as the designated type species. *Mattinglyia* was downgraded to subgeneric status in *Heizmannia* by Mattingly (1970), in which he included *Hz. achaetae* (Leicester, 1908); thus, 'Group B' of Mattingly (1957b) was supplanted by *Mattinglyia*.

Opifex Hutton, 1902

Opifex Hutton, 1902 *sensu stricto*

Opifex was originally proposed as a genus of the family Tipulidae by Hutton (1902), with *Opifex fuscus* Hutton, 1902 as the only included species. Edwards (1921a) recognized the species as a remarkable mosquito and transferred the genus and species to the Culicidae. Miller (1922) considered the remarkable features of the adult and immature stages sufficient to justify the creation of the subfamily Opificinae to accommodate the genus and species. Edwards (1924b, 1932a) did not accept the new subfamily proposed by Miller and included *Opifex* in the '*Aedes* group' of genera, which comprises Aedini as currently defined (Belkin, 1962). *Opifex* remained a monobasic genus until Reinert *et al.* (2004) transferred *Nothoskusea* from subgeneric status in *Ochlerotatus* (see *Nothoskusea* below) to subgeneric status in *Opifex* based on a sister-group relationship recovered in a morphology-based phylogenetic analysis of the Aedini.

Nothoskusea Dumbleton, 1962

Nothoskusea was originally introduced as a subgenus of *Aedes* by Dumbleton (1962) to accommodate the new species *Aedes chathamicus* Dumbleton, 1962. *Nothoskusea* was transferred from *Aedes* to subgeneric status in *Ochlerotatus* by Reinert (2000c), concomitant with the restoration of *Ochlerotatus* from subgeneric rank in *Aedes* to its original generic rank. Based on two unique features of the antennae of males that supported a sister-group relationship between *Nothoskusea* and *Opifex* in a phylogenetic study of the Aedini, Reinert *et al.* (2004) transferred the subgenus *Nothoskusea* to the genus *Opifex*. Although *Nothoskusea* is currently retained as a subgenus of *Opifex*, phylogenetic analysis of New Zealand mosquito species based on rDNA ITS1 sequences (Cane *et al.*, 2020) showed that *Op. chathamicus* may be more closely related to species of *Aedes* than it is to *Op. fuscus*; thus, additional research is needed to establish the generic placement of the species.

Psorophora Robineau-Desvoidy, 1827

Psorophora Robineau-Desvoidy, 1827 *sensu stricto*

Psorophora was created as a new genus of mosquitoes by Robineau-Desvoidy (1827), with *Culex ciliatus* Fabricius, 1794 (originally incorrectly spelled '*ciliata*') subsequently designated the type species by implication (Theobald, 1901b). *Psorophora sensu stricto* includes 10 species. The group has not been revised or subdivided.

Grabhamia Theobald, 1903

Grabhamia was originally introduced as a genus by Theobald (1903a), with *Culex jamaicensis* Theobald, 1901b subsequently designated the type species by Felt (1904). *Grabhamia* was apparently first treated as a subgenus of *Psorophora* by Dyar (1923f). Prior to that, as indicted by Howard *et al.* (1917), *Grabhamia* was recognized as a 'section' of the subgenus *Janthinosoma*. *Grabhamia* includes 16 species. It has not been revised and only two informal groups have been mentioned, en passant, for five species that occur in the Caribbean Region (Belkin *et al.*, 1970). It seems likely that the two groups, the Cingulata and Confinnis Groups, include other species of *Grabhamia*.

Janthinosoma Lynch Arribálzaga, 1891

Janthinosoma was originally established as a genus by its inclusion in a dichotomous key of distinguishing characters (Lynch Arribálzaga, 1891a) and formally described in a second article published later in the same year Lynch Arribálzaga (1891b). *Culex discrucians* Walker, 1856 was subsequently designated the type species by Dyar (1905a). *Janthinosoma* was treated as a subgenus of *Psorophora* in the four volumes of *The Mosquitoes of North and Central America and the West Indies* (Howard *et al.*, 1913, 1915, 1917), but it included *Grabhamia* as a 'section' separate from *Janthinosoma* 'proper' (Howard *et al.*, 1917). *Janthinosoma* and *Grabhamia* were first treated as separate subgenera of *Psorophora* by Dyar (1923f). *Janthinosoma* includes 23 species. The group has not been revised or subdivided, and 'The taxonomy of the subgenus is in a very confused state' (Belkin *et al.*, 1970).

Udaya Thurman, 1954

Udaya was originally introduced as a subgenus of *Paraedes* by Thurman (1954b), with *Paraedes argyrurus* Edwards, 1934 (in Barraud, 1934) designated the type and only originally included species. Mattingly (1957b, 1958b) recognized *Udaya* as a distinct genus and downgraded *Paraedes* to subgeneric status in *Aedes*. *Udaya* now includes three species. Macdonald & Mattingly (1960) described and added *Ud. lucaris*, and *Ud. subsimilis* (Barraud, 1927c) was transferred to the genus from *Aedes* (*Finlaya*) by Reinert (1990a).

Verrallina Theobald, 1903

Verrallina Theobald, 1903 *sensu stricto*

Verrallina was described as a genus by Theobald (1903a), with *Aedes butleri* Theobald, 1901c subsequently designated the type species by Blanchard (1905). Edwards (1913c) synonymized *Verrallina* with *Aedes* and later (Edwards, 1917) transferred the species previously included in the genera *Aioretomyia* Leicester, 1908, *Neomacleaya* Theobald, 1907 and *Verrallina*, along with several additional species, to the subgenus *Aedes* of the genus *Aedes*. No changes were made to Edwards's concept of the subgenus *Aedes* until Belkin (1962) removed *Verrallina* from synonymy and established it as a separate subgenus of *Aedes*. Reinert (1974) revised the subgenus *Verrallina* and divided it into Sections A and B, and subdivided Section A into Series I, II and III. Twenty-five years later, Reinert (1999c) restored *Verrallina* to its original generic rank and recognized three subgenera: the nominotypical subgenus, *Neomacleaya*, which he transferred from subgeneric status in *Aedes*, and introduced the new subgenus *Harbachius*. He recognized two Series of species within *Verrallina sensu stricto*, the Butleri Series (newly established) and the Carmenti Series ('Series I, *carmenti* Series' of Reinert, 1974).

Harbachius Reinert, 1999

Harbachius was introduced as a subgenus of *Verrallina* by Reinert (1999c), with *Aedes yusafi* Barraud, 1931b the designated type species. *Harbachius* was established for the species which Reinert (1974) had included in Section B of *Verrallina* when it was classified as a subgenus of *Aedes*.

Neomacleaya Theobald, 1907

Neomacleaya was originally established as a genus by Theobald (1907), with *Neomacleaya indica* Theobald, 1907 as the type and only originally included species. Edwards (1913c) synonymized *Neomacleaya* with *Aedes* and later (Edwards, 1917) transferred the species previously included in the genera *Aioretomyia* Leicester, 1908, *Neomacleaya* and *Verrallina* Theobald (1903a), along with several other species, to the subgenus *Aedes* of the genus *Aedes*. Belkin (1962) removed *Verrallina* from synonymy and established it as a separate subgenus of *Aedes*. Five years later, Delfinado (1967) resurrected *Neomacleaya* from synonymy with *Verrallina* as a separate subgenus of *Aedes* to include mainly Oriental species that were previously included in *Aedes*, *Aioretomyia*, *Neomacleaya* and *Verrallina*. Reinert (1999c) transferred *Neomacleaya* from subgeneric status in *Aedes* to its current subgeneric status in *Verrallina*. *Neomacleaya* principally represents the species included in the Indicus Series of Reinert (1974).

Zeugnomyia Leicester, 1908

Zeugnomyia was introduced as a genus by Leicester (1908), with *Zeugnomyia gracilis* Leicester, 1908 as the type and only originally included species. *Zeugnomyia* has been recognized as a distinct genus of Aedini since it was established by Leicester. Three additional species were much later recognized as members of the genus: *Ze. lawtoni* Baisas, 1946b, *Ze. aguilari* Baisas & Feliciano, 1953 and *Ze. fajardoi* Baisas & Feliciano, 1953.

Culicini Linnaeus, 1758

Culex Linnaeus, 1758

Culex has a cosmopolitan distribution and includes more than 750 extant species divided among 28 subgenera. The current system of subgeneric classification is based primarily on external adult characters, especially features of the male genitalia. The species of the larger subgenera are arranged in informal classifications that variously include Sections, Series, Groups, Subgroups and Complexes. The informal infrasubgeneric categories are often based on superficial similarities that may not reflect natural relationships. In general, the larger the group, the less likely it is to be a monophyletic assemblage of species.

Culex Linnaeus, 1758 *sensu stricto*

Culex sensu stricto is a very large and difficult group taxonomically. The group has not been examined on a worldwide basis since Edwards (1932a), who divided it into two species groups, the Pipiens Group (cosmopolitan) and the Sitiens Group (Old World, principally Oriental Region). Both groups are highly complex and include species that do not readily fit into either group. Three additional species groups have been recognized subsequently, the Guiarti Group (Edwards, 1941) for several Afrotropical species, the Atriceps Group (Belkin, 1962) for three South Pacific species and the Duttoni Group (Harbach, 1988) for the unusual Afrotropical *Cx. duttoni* Theobald, 1901d and *Cx. watti* Edwards, 1920a. Heinemann & Belkin (1977a, and later publications) recognized two groups in the Neotropical Region, the Declarator and Inflictus Groups, but did not indicate which species they include. Strickman (1990) made reference to the Declarator Group, but he also did not mention which species comprise the group. The speciose Pipiens and Sitiens Groups are further divided into a number of subgroups and complexes (see the Tabulation in Part VI). The assignment of species to the informal groups is based principally on information contained in the works of Edwards (1932a, 1941), Belkin (1962), Forattini (1965a), Bram (1967a, 1967b), Sirivanakarn (1976), Tanaka *et al.* (1979) and Harbach (1988), but the inclusion of many species in groups and subgroups, especially New World species, is problematic. Edwards (1932a) and Sirivanakarn (1976) suggested that the Pipiens and Sitiens Groups possibly deserve to be ranked as separate subgenera. The phylogenetic study of Harbach *et al.* (2017a) based on morphological data seemed to support subgeneric status for the Sitiens Group, whereas the Pipiens Group definitely appeared to be polyphyletic. If the Sitiens Group were to be given subgeneric rank, it may need to include the subgenus *Oculeomyia*, which was previously classified as the Bitaeniorhynchus Subgroup of the Sitiens Group (Sirivanakarn, 1976). Harbach *et al.* (2017a) concluded that 'the internal classification of the subgenus based on intuitive assessment of morphological similarities and differences does not seem, in large part, to reflect natural (genealogical) relationships.'

Acalleomyia Leicester, 1908

Acalleomyia was originally proposed as a distinct genus by Leicester (1908), with *Acalleomyia obscurus* Leicester, 1908 as its type and only included species. Edwards (1913c) transferred the species to the genus *Micraedes* where it remained until Edwards (1922a) recognized *Acalleomyia* as a subgenus of *Culex*.

Acallyntrum Stone & Penn, 1948

Acallyntrum was proposed as a subgenus of *Culex* by Stone & Penn (1948), with the new species *Cx. perkinsi* Stone & Penn, 1948 as the type species. *Acallyntrum* currently includes eight species that comprise two species groups, the Bicki and Perkinsi Groups (Belkin, 1962).

Aedinus Lutz, 1904

Aedinus was originally proposed as a distinct genus by Lutz, 1904 (in Bourroul, 1904), but its authorship was not resolved until Belkin (1968b). *Aedinus* was classified as a subgenus of *Culex* by Edwards (1930a). Stone *et al.* (1959) included *Aedinus* (attributed to Bourroul, 1904), along with *Anoedioporpa*, *Micraedes* and *Tinolestes* (as synonyms), in the subgenus *Aedinus* based on the short maxillary palpus of males. Belkin (1968b) recognized that this character had evolved independently in several unrelated groups, and elevated *Aedinus*, as well as *Anoedioporpa*, *Micraedes* and *Tinolestes*, to subgeneric rank in *Culex*.

Afroculex Danilov, 1989

Afroculex was proposed as a subgenus of *Culex* by Danilov (1989), with *Pseudohowardina lineata* Theobald, 1912a as its type and the only included species. Edwards (1914b) transferred *Pseudohowardina lineata* to *Culex* and gave it the replacement name of *pulchrithorax* because *lineatus* was preoccupied by *Cx. lineatus* von Humboldt, 1819. The species was eventually placed in the subgenus *Neoculex*, but its taxonomic position was open to question (Edwards, 1941). Its provisional placement in *Neoculex* continued until Sirivanakarn (1971) transferred it to the subgenus *Maillotia*. Danilov (1989) realized that *Cx. lineatus* von Humboldt is actually a species of the genus *Psorophora*, as recorded by Knight & Stone (1977), reinstated *lineatus* (Theobald, 1912a) as the valid name of the species and transferred it to the new subgenus *Afroculex* based on unique features of the adults and male genitalia. The immature stages of *Cx. lineatus* remain unknown.

Allimanta Casal & García, 1968

Allimanta was proposed as a subgenus of *Culex* by Casal & García (1968a), with *Cx. tramazayguesi* Duret, 1954 as its type and only included species. *Culex tramazayguesi* was originally described as a species of the subgenus *Culex*.

Anoedioporpa Dyar, 1923

Anoedioporpa was proposed as a subgenus of *Culex* by Dyar (1923e), with *Cx. conservator* Dyar & Knab, 1906d as its designated type species. Most of the species currently included in *Anoedioporpa* were previously assigned to other groups that have been recognized as subgenera of *Culex*, including *Isostomyia* (by Dyar, 1918a; Edwards, 1932a), *Melanoconion* (by Dyar, 1925f, 1928) and *Tinolestes* (by Lane, 1953). Stone *et al.* (1959) included *Anoedioporpa*, along with *Tinolestes* and *Micraedes*, in the subgenus *Aedinus*, which at the time was attributed to Bourroul (1904). Belkin (1968b) recognized *Anoedioporpa* as a distinct taxon and restored it to subgeneric rank in *Culex*. Berlin & Belkin (1980) divided *Anoedioporpa* into two groups, the Conservator Group, which currently includes 11 species, and the monobasic Restrictor Group.

Barraudius Edwards, 1921

Barraudius was proposed as a subgenus of *Culex* by Edwards (1921d), with *Cx. pusillus* Macquart, 1850 as its designated type species. *Barraudius* currently includes four species that are considered to comprise a homogeneous group without subdivision.

Belkinomyia Adames & Galindo, 1973

Belkinomyia was proposed as a subgenus of *Culex* by Adames & Galindo (1973), with the new species *Cx. eldridgei* Adames & Galindo, 1973 as its type and only included species.

Carrollia Lutz, 1905

Carrollia was originally proposed as a distinct genus by Lutz (1905), with *Cx. iridescens* Lutz, 1905 as its type and only originally included species. Dyar (1918b) appears to have been the first person to regard *Carrollia* as a subgenus of *Culex*, followed by Edwards (1932a) and all later authors. The current internal division of *Carrollia* into two species groups, the Bihaicola and Iridescens Groups, the latter with two subgroups, the Urichii and Iridescens Subgroups, is attributable to Valencia (1973).

Culiciomyia Theobald, 1907

Culiciomyia was originally proposed as a distinct genus by Theobald (1907). Designation of the type species, *Culiciomyia inornata* Theobald, 1907 [subjective synonym of *Culex fragilis* Ludlow, 1903], is attributed to Edwards (1912b). *Culiciomyia* was reduced to subgeneric status in *Culex* by Edwards (1921d). Edwards (1932a) recognized two species groups in the subgenus: 'Group A (*fragilis*-group)', with species in the Oriental, Indomalayan and Australasian Regions, and 'Group B (*nebulosus*-group)', with species restricted to the Afrotropical Region. Three additional groups were recognized later for species in the Oriental Region, the Dispectus Group (Bram, 1969), Tricuspis Group (Harrison, 1987, as '*tricuspis* group') and the Shebbearei Group, which is the name proposed by Harbach (2011), based on nomenclatural priority, for the unnamed 'group or complex' of Sirivanakarn (1977c).

Eumelanomyia Theobald, 1909

Eumelanomyia was originally proposed as a distinct genus by Theobald (1909), with *Eumelanomyia inconspicuosa* Theobald, 1909 as its type and only included species. *Eumelanomyia* was described as being similar to *Culiciomyia* but the true identity of the type species was not known until Edwards (1922d) re-examined the specimens and identified them as members of the genus *Culex*. The name *inconspicuosa* was thus found to be preoccupied, and *Culex albiventris* was proposed for the name of the species. Theobald (1910c) recognized *Protomelanoconion* Theobald, 1909 as a distinct genus based on a misidentified species of *Culex*, which he named *Protomelanoconion fusca*. Edwards (1922d) also found this specific name to be preoccupied and proposed the replacement name *Culex horridus* for the species. In addition to these corrections, Edwards suggested that both *Eumelanomyia* and *Protomelanoconion* should be treated as separate subgenera of *Culex*. Eight years later, Edwards (1930a) established *Mochthogenes* as a subgenus of *Culex* with *Aedes malayi* Leicester, 1908 as the designed type species. No further changes were made to the taxonomy of these groups until Edwards (1932a) examined the classification of *Culex* on a worldwide basis. As a result, he retained *Mochthogenes* as a subgenus and included *Eumelanomyia*, *Protomelanoconion* and a number of other species within the subgenus *Neoculex*, which he divided into three groups: 'Group A (*Neoculex*, or *apicalis*-group)', 'Group B (*Eumelanomyia*, or *albiventris*-group)' and 'Group C (*Protomelanoconion*, or *uniformis*-group)'. In his later treatise on the Afrotropical Culicinae, Edwards (1941) recognized two additional groups, 'Group A (*C. pulchrithorax* Edw.)' and 'Group C (*rima* group)', for species previously included in his *apicalis*-group. No further changes were made to Edwards's classification until Sirivanakarn (1971) revised the classification of *Neoculex* to include nearly all species previously placed in *Mochthogenes* and a number of species previously included in *Neoculex*. Sirivanakarn synonymized *Protomelanoconion* and *Mochthogenes* with *Eumelanomyia* and recognized the subgenus *Maillotia* in addition to the subgenera *Eumelanomyia* and *Neoculex*. The current internal classification of *Eumelanomyia* was developed by Sirivanakarn (1971, 1972) (see the Tabulation in Part VI).

Kitzmilleria Danilov, 1989

Kitzmilleria was proposed as a subgenus of *Culex* by Danilov (1989), with *Cx. moucheti* Evans, 1923a as its type and only included species. *Culex moucheti* was originally placed in the Pipiens Group of *Culex sensu stricto*, and was considered a member of the Decens Series (Edwards, 1932a) until Danilov (1989) proposed the subgenus *Kitzmilleria* based on distinctive features of the adult, larval and pupal stages.

Lasioconops Theobald, 1903

Lasioconops was originally introduced as a genus by Theobald (1903c), with *Lasioconops poicilipes* Theobald, 1903a as the type and only included species. Edwards (1932a) relegated *Lasioconops* to synonymy with *Culex sensu stricto*, where it was retained until Tanaka (2004) resurrected *Oculeomyia* Theobald, 1907 from synonymy with *Culex* for species of the '*Bitaeniorhynchus* Subgroup' of Sirivanakarn (1976), which included *poicilipes* by virtue of equivalence of the group with the '*bitaeniorhynchus* series' of Edwards (1932a). Ironically, Tanaka noted that '*Culex poicilipes* may belong to a different group', apparently not realizing that Harbach (1988) had removed *Cx. poicilipes* from the Bitaeniorhynchus Subgroup and placed it provisionally in the Sitiens Subgroup ('*sitiens* subgroup' of Sirivanakarn, 1976; '*sitiens* series' of Edwards, 1932a). Unfortunately, this taxonomic act was overlooked and *Cx. poicilipes* was inadvertently classified as a species the subgenus *Oculeomyia* by later authors (*e.g.* Harbach, 2011, 2018b; Wilkerson *et al.*, 2021). For this reason, and because *Lasioconops* was published earlier than *Oculeomyia*, Berlov & Kuberskaya (2023) replaced *Oculeomyia* with *Lasioconops*. That action prompted Harbach (2023) to review the morphological distinctions which had previously led him (Harbach, 1988) to remove *Cx. poicilipes* from the Bitaeniorhynchus Subgroup. Based on significant morphological differences between the adult, larval and pupal stages of *Cx. poicilipes* and those life stages of the other species of the former Bitaeniorhynchus Group, *Lasioconops* was formally recognized as a distinct monobasic subgenus of *Culex*.

Lasiosiphon Kirkpatrick, 1925

Lasiosiphon was proposed as a subgenus of *Culex* by Kirkpatrick (1925a), with *Cx. pluvialis* Kirkpatrick, 1925a as its type and only included species. As *Cx. pluvialis* was preoccupied by *Cx. pluvialis* Barraud, 1924f, three replacement names were independently proposed: *Cx. adairi* Kirkpatrick, 1926, *kirkpatriki* Edwards, 1926c and *kirkpatricki* Stackelberg, 1927. The first name, *Cx. adairi*, has priority and is the accepted name of the species.

Lophoceraomyia Theobald, 1905

Lophoceraomyia was originally proposed as a distinct genus by Theobald (1905h), with *Lophoceraomyia uniformis* Theobald, 1905h as its type and only included species. *Lophoceraomyia* was reduced to a subgenus of *Culex* by Edwards (1917). Edwards (1932a) divided the subgenus (as *Lophoceratomyia*) into three groups: 'Group A (*minutissimus*-group)', 'Group B (*Lophoceratomyia* or *fraudatrix*-group)' and 'Group C. (*Cyathomyia* or *mammilifer*-group)'. Edwards later (in Barraud, 1934), amalgamated Groups A and B and subdivided Group C. Colless (1965), however, preferred to recognize only two major groups, the '*C. fraudatrix* Group' and the '*C. mammilifer* Group', with the latter group divided into two subgroups, the '*C. brevipalpus* Subgroup' and the '*C. mammilifer* Subgroup'. Sirivanakarn (1977a) modified the classifications of Edwards (1932a, 1934) and Colless (1965) to include three groups, the Fraudatrix, Mammilifer and Wilfredi Groups, based principally on structures of the antennae and male genitalia. The division of these groups into subgroups and complexes by Sirivanakarn (1968, 1977a) forms the backbone of the internal classification of *Lophoceraomyia* (see the Tabulation in Part VI).

Maillotia Theobald, 1907

Maillotia was originally proposed as a distinct genus by Theobald (1907), with *Maillotia pilifera* Theobald, 1907 [subjective synonym of *Culex hortensis* Ficalbi, 1889a] as its type and only included species. *Maillotia* was implicitly synonymized with *Culex* by Edwards (1911c) and placed in synonymy with the subgenus *Neoculex* by Edwards (1932a). Sirivanakarn (1971) removed *Maillotia* from synonymy to accommodate eight species divided between three groups, the '*hortensis* group' with three species, the monobasic '*pulchrithorax* group' for *Cx. pulchrithorax* Edwards, 1914b and the '*seyrigi* group' with four species. The subgenus currently includes only the Hortensis and Seyrigi Groups, and an unplaced species. The '*pulchrithorax* group' was eliminated when Danilov (1989) proposed *Afroculex* to accommodate *Cx. pulchrithorax*, the junior synonym of *Cx. lineatus* (Theobald, 1912a).

Melanoconion Theobald, 1903

Melanoconion was originally proposed as a distinct genus by Theobald (1903a). The type species, *Culex atratus* Theobald, 1901c, was subsequently designated by Dyar (1905a). Dyar & Knab (1906d) synonymized *Melanoconion* with *Culex* and proposed *Mochlostyrax* as a distinct genus with *Mochlostyrax caudelli* Dyar & Knab, 1906d as its type species. Howard *et al.* (1915) considered both *Melanoconion* and *Mochlostyrax* as synonyms of *Culex*, and 3 years later Dyar (1918b) recognized them as separate subgenera of *Culex*. In the same paper, Dyar also proposed *Choeroporpa* as a subgenus of *Culex*, with *Culex anips* Dyar, 1916a as its type species. *Choeroporpa* included most of the species that Dyar had previously placed in *Culex* or *Mochlostyrax*. In a second paper published in the same year, Dyar (1918c) proposed *Helcoporpa* as another subgenus of *Culex*, with *Culex menytes* Dyar, 1918c as its type species, currently a synonym of *Cx. spissipes* (Theobald, 1903a). Five years later, Dyar (1923e) reinstated *Gnophodeomyia* Theobald, 1905g as a subgenus of *Culex* and proposed *Anoedioporpa* as a replacement name for the subgenus *Isostomyia* Coquillett, 1906c. Dyar (1928) made significant changes to the classification of New World *Culex*. He recognized *Melanoconion* and *Mochlostyrax* as subgenera and reduced the other nominal generic-level groups to informal sections: *Choeroporpa*, *Helcoporpa* and the newly proposed *Dinoporpa* became sections of *Mochlostyrax*, and *Tinolestes* Coquillett, 1906a, *Gnophodeomyia* and *Anoedioporpa* became sections of *Melanoconion*, which also included *Culex americanus* (Neveu-Lemaire, 1902b), currently recognized as a *nomen dubium*, and *Cx. antillummagnorum* Dyar, 1928. Edwards (1932a), in his treatment of the Culicidae of the world, reinterpreted the taxonomy of *Melanoconion* and *Mochlostyrax*. He considered *Melanoconion* as a subgenus with *Gnophodeomyia*, *Asebeomyia* Aiken, 1911, *Tinolestes*, *Choeroporpa*, *Helcoporpa* and *Dinoporpa* as its synonyms; restricted the subgenus *Mochlostyrax* to include species included in the *Mochlostyrax* section of Dyar (1928); synonymized *Anoedioporpa* with the subgenus *Isostomyia*; and transferred *Cx. americanus* and *Cx. antillummagnorum* to the subgenus *Micraedes*. During the same year, Komp & Curry (1932) proposed *Upsiloporpa* as a new subgenus of *Culex*, with the new species *Cx. haynei* Komp & Curry, 1932 as its type and only included species. Komp (1935) found *Cx. haynei* to be conspecific with *Cx. menytes*; thus, *Upsiloporpa* became another synonym of *Melanoconion*. Except for the transfer of *Cx. ocellatus* Theobald, 1903a from *Microculex* Theobald, 1907 to *Melanoconion* by Lane & Whitman (1943), Edwards's classification remained unchanged until Rozeboom & Komp (1950a) treated *Melanoconion* and *Mochlostyrax* each as a 'Section' of the subgenus *Melanoconion*. Lane (1953) followed Rozeboom & Komp's classification but resurrected *Tinolestes* from synonymy with *Melanoconion* as a separate subgenus. A year later, Foote (1954) determined that *Mochlostyrax* was distinct based on larval morphology and considered it to be a subgenus separate from *Melanoconion*. Foote's separation of *Mochlostyrax* and *Melanoconion* prevailed until Belkin (1968b), Belkin *et al.* (1970) and Sirivanakarn (1983) considered *Melanoconion* and *Mochlostyrax* to form a single subgenus.

Dyar (1928) recognized four sections in the subgenus *Mochlostyrax*, the *Dinoporpa*, *Helcoporpa*, *Mochlostyrax* and *Choeroporpa* sections, and four sections in the subgenus *Melanoconion*, the *Tinolestes*, *Gnophodeomyia*, *Melanoconion* and *Anoedioporpa* sections. Edwards (1932a) recognized *Mochlostyrax*, without sections, and divided *Melanoconion* into Groups A, B and C based on external features of adults. Rozeboom & Komp (1950a) disagreed with Edwards's classification and largely adopted Dyar's (1928) scheme based chiefly on features of the male genitalia for their concept of *Melanoconion*, which included *Mochlostyrax* and excluded *Anoedioporpa*. Hence, Rozeboom & Komp divided the subgenus into seven sections, namely the *Choeroporpa*, *Dinoporpa*, *Gnophodeomyia*, *Helcoporpa*, *Melanoconion*, *Mochlostyrax* and *Tinolestes* sections. Nearly two decades later, Galindo (1969) established the Spissipes Group based on features of the male genitalia and larvae, and Duret (1969d) recognized the Ocellatus Group based on distinctive features of adults and male genitalia. Both groups were retained and redefined in the revised scheme of classification proposed by Sirivanakarn (1983).

Sirivanakarn (1983) distinguished three sections within *Melanoconion*, the Melanoconion, Ocellatus and Spissipes Sections, and divided the Melanoconion and Spissipes Sections into Groups and Subgroups based principally on structural differences of the male genitalia, characteristics of the scaling on the head and scutum of adults, and features of the larvae. Pecor *et al.* (1992) removed the Ocellatus Section from the subgenus, and it remains without subgeneric placement within *Culex*. Finally, Sallum & Forattini (1996) refined the Spissipes Section to include eight Groups and three Subgroups as opposed to the 13 groups recognized by Sirivanakarn (1983). Except for recognition of the Crybda, Taeniopus and Vomerifer Groups within the Spissipes Section by Sallum & Forattini, the internal classification of *Melanoconion* has changed little since it was established by Sirivanakarn (1983).

Micraedes Coquillett, 1906

Micraedes was originally proposed as a distinct genus by Coquillett (1906a), with the new species *Micraedes bisulcatus* Coquillett, 1906a as the type and only included species. Howard *et al.* (1915) synonymized *Micraedes* with *Culex* where it remained until Dyar (1918b) elevated it to subgeneric rank. Dyar (1928) synonymized *Micraedes bisulcatus* with *Cx.* (*Melanoconion*) *americanus* (Neveu-Lemaire, 1902b); thus, *Micraedes* became a synonym of *Melanoconion*. *Micraedes* remained in synonymy until Edwards (1932a) restored it to subgeneric rank. Lane (1953) synonymized it with *Tinolestes* and Stone *et al.* (1959) placed it, along with *Anoedioporpa* and *Tinolestes*, in synonymy with *Aedinus* Lutz, 1904 (incorrectly attributed to Bourroul, 1904). Berlin (1969b), following Foote (1954) and Belkin (1968b), once again treated *Micraedes* as a distinct subgenus of *Culex*. Berlin (1969b) revised *Micraedes* and divided it into three species groups, the Bisulcatus Group with three species, the Schicki Group with two species and the monobasic Erethyzonfer Group. Zavortink & O'Meara (1999) described *Cx. biscaynensis* as a new species of the Bisulcatus Group.

Microculex Theobald, 1907

Microculex was originally proposed as a distinct genus by Theobald (1907), with *Microculex argenteoumbrosus* Theobald, 1907 [subjective synonym of *Culex imitator* Theobald, 1903a] as the type and only included species. Brunetti (1914) regarded *Microculex* as a synonym of *Culex*, but it seems the synonymy was never recognized. *Microculex* has been treated as a subgenus of *Culex* since Dyar (1918b). Lane & Whitman (1951) recognized four groups (series) of species known to occur in Brazil, recast here as the Consolator, Imitator, Inimitabilis and Pleuristriatus Series, but no attempt has been made to develop a classification for all species of the subgenus.

Neoculex Dyar, 1905

Neoculex was originally proposed as a distinct genus by Dyar (1905a), with *Cx. territans* Walker, 1856 as its designated type species. *Neoculex* was regarded as a synonym of *Culex* by Brunetti (1914) and treated as a subgenus of *Culex* by Dyar (1918b). In his comprehensive treatment of *Culex*, Edwards (1932a) included *Maillotia*, *Eumelanomyia* and *Protomelanoconion* as synonyms of *Neoculex* and divided the subgenus into three groups: 'Group A (*Neoculex*, or *apicalis*-group)', 'Group B (*Eumelanomyia*, or *albiventris*-group)' and 'Group C (*Protomelanoconion*, or *uniformis*-group)'. In his later work on the Afrotropical Culicinae, Edwards (1941) retained Groups B and C and split Group A into three groups, 'Group A (*C. pulchrithorax* Edw.)', 'Group B (*Neoculex* s. str.; *C. salisburiensis* and related species)' and 'Group (*rima* group)'. His former Groups B and C became 'Group D' and 'Group E', respectively. King & Hoogstraal (1947c) followed this scheme and recognized a sixth group, 'group F', for the new species *Cx. pedicellus* King & Hoogstraal and *Cx. crassistylus* Brug, 1934 from New Guinea. As indicated by Mattingly & Marks (1955) and Belkin (1962), the groups recognized by Edwards (1932a, 1941) and King & Hoogstraal (1947b) give little idea of natural relationships because they are based on superficial characters that greatly overlap with characters exhibited by members of other genus-group taxa of *Culex*. This is obvious from his treatment of *Mochthogenes* as a subgenus separated from *Protomelanoconion* (*i.e. uniformis*-group) of *Neoculex* based on the relative length of the male maxillary palpi. As pointed out by Bram (1969), these groups are so similar in the larval stage that they should be included in the same subgenus. With this as background, Sirivanakarn (1971) proposed a reclassification of *Neoculex* based principally on structural differences observed in the male genitalia. Sirivanakarn removed *Eumelanomyia* and *Maillotia* from synonymy with *Neoculex*, established them as separate subgenera of *Culex* and synonymized *Mochthogenes* with *Eumelanomyia*. The downscaled concept of *Neoculex* that resulted from those actions included the recognition of three species groups, the Crassistylus, Pseudomelanoconia and Territans Groups.

Nicaromyia González Broche & Rodríguez R., 2001

Nicaromyia was proposed as a subgenus of *Culex* by González Broche & Rodríguez R. (2001), with *Culex nicaroensis* Duret, 1967a as its type and only included species. *Culex nicaroensis* was originally described as a species of the subgenus *Melanoconion*. Sallum & Forattini (1996) excluded the species from *Melanoconion* and it remained without subgeneric placement until *Nicaromyia* was proposed to accommodate it.

Oculeomyia Theobald, 1907

Oculeomyia was originally proposed as a distinct genus by Theobald (1907), with *Oculeomyia sarawaki* Theobald, 1907 [subjective synonym of *Culex infula* Theobald, 1901b] as the type and only included species. Brunetti (1914) considered *Oculeomyia* to be a genus of 'uncertain validity'. Edwards (1911c) synonymized *Oculeomyia sarawaki* with *Taeniorhynchus ager* Giles, 1901c and later (Edwards, 1913c) synonymized it with *Cx. bitaeniorhynchus* Giles, 1901a, thus relegating *Oculeomyia* to synonymy with *Culex*. *Oculeomyia* remained in synonymy with *Culex sensu stricto*, because *Cx. bitaeniorhynchus* was classified as a member of the Bitaeniorhynchus Series/Subgroup of the Sitiens Group (Edwards, 1932a, 1941; Belkin, 1962; Bram, 1967a; Sirivanakarn, 1976), until Tanaka (2004) resurrected it from synonymy and validated it as a subgenus to include *Cx. bitaeniorhynchus* and other species previously included in the Bitaeniorhynchus Subgroup.

Phalangomyia Dyar & Knab, 1914

Phalangomyia was originally proposed as a distinct genus by Dyar & Knab (1914), with *Phalangomyia debilis* Dyar & Knab, 1914 as the type and only originally included species. Without explanation, Dyar (1918c) treated *Phalangomyia* as a subgenus of *Culex*, with *Phalangomyia debilis* as the only included species. Dyar (1926) added *Cx. articularis* Philippi, 1865 (as *articulatus*) to the subgenus and later (Dyar, 1928) included *Cx. apicinus* Philippi, 1865 and *Cx. escomeli* Brèthes, 1920. Three years later, Martini (1931) described and named *Cx. alticola* as a fifth species of the subgenus. Edwards (1932a) was not aware of *Cx. alticola* when he formally recognized *Phalangomyia* as a synonym of *Culex sensu stricto*, concomitant with the synonymy of *Phalangomyia debilis* and *Cx. escomeli* with *Cx. apicinus*. *Culex alticola* continued to be recognized as a distinct species (Lane, 1939) until Lane (1951) synonymized it with *Cx. apicinus*. It remained in synonymy until Stone (1957) provided evidence 'to resurrect *alticola* as a valid species.' However, this action was reversed by Bram (1967b), who re-established the synonymy based on perceived variation in features of the male genitalia. As noted by Belkin (1968b), *alticola* may not be conspecific with *apicinus* as interpreted by Bram, which seems quite evident when the male gonocoxite illustrated for *Cx. apicinus* by Bram (1967b) is compared with the illustrations of Laurito & Almirón (2013) and González *et al.* (2023), who formally restored *Phalangomyia* to its earlier subgeneric rank. With the removal of *Cx. apicinus* from the Apicinus Subgroup (Harbach, 2011), there is no justification for retaining the group for the other included species (22), which are now classified as unplaced members of the Pipiens Group of *Culex sensu stricto*.

Phenacomyia Harbach & Peyton, 1992

Phenacomyia was proposed as a subgenus of *Culex* by Harbach & Peyton (1992b), with *Cx. corniger* Theobald, 1903a as its designated type species. Prior to the recognition of *Phenacomyia*, *Cx. corniger* and its two related species, *Cx. lactator* Dyar & Knab, 1906d and *Cx. airozai* Lane, 1945b, were included in *Culex sensu stricto*.

Phytotelmatomyia Rossi & Harbach, 2008

Phytotelmatomyia was proposed as a subgenus of *Culex* by Rossi & Harbach (2008), with *Cx. renatoi* Lane & Ramalho, 1960 as its designated type species. Prior to the recognition of *Phytotelmatomyia*, *Cx. renatoi* and its related species were included in *Culex sensu stricto*.

Sirivanakarnius Tanaka, 2004

Sirivanakarnius was proposed as a subgenus of *Culex* by Tanaka (2004), with *Cx. boninensis* Bohart, 1957 as its type and only included species. *Culex boninensis* was regarded as a member of the Sitiens Group of *Culex sensu stricto* until Tanaka (2004) established *Sirivanakarnius* based on distinct characters of the adults and male genitalia.

Tinolestes Coquillett, 1906

Tinolestes was originally proposed as a distinct genus by Coquillett (1906a), with the new species *Tinolestes latisquama* Coquillett, 1906a as its type and only originally included species. Howard *et al.* (1915) synonymized *Tinolestes* with *Culex* and Dyar (1918b) resurrected it to subgeneric rank. Dyar (1928) placed *Tinolestes latisquama* in *Melanoconion*, and as a consequence *Tinolestes* became a synonym of *Melanoconion*. Lane (1953) restored *Tinolestes* to subgeneric rank and synonymized *Micraedes* Coquillett, 1906a, *Isostomyia* Coquillett, 1906c and *Anoedioporpa* Dyar, 1923d with it. Stone *et al.* (1959) included *Tinolestes*, along with *Micraedes* and *Anoedioporpa*, in *Aedinus* Lutz, 1904 (incorrectly attributed to Bourroul, 1904) based on the short maxillary palpus of males, but Belkin (1968b) noted that this character occurs independently in several obviously unrelated groups and reinstated *Tinolestes* as a monobasic subgenus of *Culex*. Two species, *Cx. breviculus* Senevet & Abonnenc, 1939a and *Cx. cauchensis* Floch & Abonnenc, 1945a, were transferred from *Melanoconion* to *Tinolestes* by Sirivanakarn (1983).

Deinocerites Theobald, 1901

Deinocerites was briefly defined as a genus by Theobald (1901a) and more fully described in his second volume of *A Monograph of the Culicidae* (Theobald, 1901c), with *Deinocerites cancer* Theobald, 1901c as the type and only originally included species. In the third volume of *The Mosquitoes of North and Central America and the West Indies* (Howard *et al.*, 1915), *Deinocerites*, *Dinanamesus* Dyar & Knab, 1909 and *Dinomimetes* Knab, 1907b were recognized as a 'compact group' of genera termed 'the deinoceritines'. Dyar (1918b) reduced *Deinocerites* to subgeneric status in *Culex*, with *Dinanamesus* and *Dinomimetes* as synonyms. Seven years later, Dyar (1925d) and Bonne & Bonne-Wepster (1925) independently treated *Deinocerites* as a genus, with *Dinanamesus* and *Dinomimetes* retained as synonymous nominal genera. Dyar (1928) and Edwards (1932a) followed suit and recognized five and four species within the genus, respectively. Edwards (1932a) split the four species into three groups, 'Group A (*Deinocerites*)' with two species, 'Group B (*Dinomimetes*)' with one species and 'Group C (*Dinanamesus*)' also with one species. The genus remained limited to four species until Belkin & Hogue (1959) reviewed the genus and recognized 11 species, including the four species recognized by Edwards, two species previously considered to be synonyms of the type species and four new species. Belkin & Hogue recognized four species groups in the genus, identified as the Cancer, Dyari, Epitedeus and Spanius Groups. Adames (1971) revised *Deinocerites* and established a fifth group, the Pseudes Group, for three species which Belkin & Hogue had included in the Epitedeus Group.

Galindomyia Stone & Barreto, 1969

Galindomyia was established as a genus by Stone & Barreto (1969), with *Galindomyia leei* Stone & Barreto, 1969 as the type and only included species. The description of the genus and species was based on morphologically unusual adult males and females. The larva and pupa of the species were described later by Adames & Galindo (1975).

Lutzia Theobald, 1903

Lutzia Theobald, 1903 *sensu stricto*

Lutzia was originally proposed as a genus by Theobald (1903a), with *Culex bigoti* Bellardi, 1862 as the type and only originally included species. *Lutzia* continued to be recognized as a genus until Edwards (1932a) relegated it to subgeneric status in *Culex*. Edwards recognized two distinct groups within the subgenus, 'Group A (*Lutzia*)' for tropical South American species and 'Group B (*Jamesia*)' for species that occur in tropical areas of the Old World. Tanaka *et al.* (1979) followed previous workers in treating *Lutzia* as a subgenus of *Culex*, but asserted it was 'more reasonable to consider *Lutzia* as a genus' based on the distinctive morphological modifications of the included species. Tanaka (2003) accepted this and restored *Lutzia* to its original rank as a separate genus, and divided it into three subgenera: *Lutzia sensu stricto* for the South American species (Group A of Edwards), and the two new subgenera that follow.

Insulalutzia Tanaka, 2003

Insulalutzia was proposed as a subgenus of *Lutzia* by Tanaka (2003), with *Culex shinonagai* Tanaka, Mizusawa & Saugstad, 1979 as its type and only included species. The recognition of *Insulalutzia* is based on unique features of the adults and larva of the type species, which is endemic to the Bonin Islands and easily distinguished from species of the Old World 'Group B (*Jamesia*)' of Edwards (1932a), who treated *Lutzia* as a subgenus of *Culex*.

Metalutzia Tanaka, 2003

Metalutzia was proposed as a subgenus of *Lutzia* by Tanaka (2003), with *Lutzia vorax* Edwards, 1921d the designated type species. *Metalutzia* corresponds to 'Group B (*Jamesia*)' of Edwards (1932a), which was recognized for those species of *Lutzia* (as a subgenus of *Culex*) that were known at the time to occur in tropical areas of Africa, Asia and Australia. Tanaka (2003) considered *Jamesia* Christophers, 1906 as the subgeneric name of the group, but chose the new name *Metalutzia* because *Jamesia* Christophers is preoccupied by *Jamesia* Jekel, 1861, a genus of the longhorn beetle family Cerambycidae.

Culisetini Belkin, 1962

Culiseta Felt, 1904

Culiseta Felt, 1904 *sensu stricto*

Culiseta was introduced as a genus by Felt (1904), with *Culex absobrinus* Felt, 1904 [subjective synonym of *Culiseta impatiens* (Walker, 1848)] as the type and only originally included species. The preoccupied senior synonym *Theobaldia* Neveu-Lemaire, 1902a (*Theobaldia* Fischer, 1885 in Mollusca) was used extensively as the name of the genus principally for Old World species prior to the catalogue of Stone *et al.* (1959). In a revision of the mosquitoes of the Palaearctic Region, Edwards (1921d) divided *Theobaldia* into three subgenera, *Allotheobaldia* Brolemann, 1919, *Culicella* Felt, 1904 and *Theobaldia sensu stricto*. Edwards (1932a), apparently unaware that *Theobaldia* was preoccupied, arranged the species in five subgenera: *Allotheobaldia*, *Culicella*, *Climacura* Howard, Dyar & Knab, 1915, *Theobaldia sensu stricto* and *Theomyia* Edwards, 1930a. The genus was known as *Culiseta* in North America (Howard *et al.*, 1915) until Dyar (1928) unwittingly included it as a subgenus of *Culicella*, which included 'THEOBALDIA Neveu-Lemaire (not *Theobaldius* Nevill)' as a synonym. Dyar seemed to ignore the fact that *Theobaldia* was the senior synonym and dismissed it with the following ridiculous statement: 'The name *Theobaldia*, while not invalidated under the present rules of nomenclature, is too close to *Theobaldius* [Nevill, 1878] to be used. I think it sufficient to burden the memory with generic names, without adding thereto the necessity of learning terminations which are similar to the changing sexual terminations of specific names.' As noted above, *Theobaldia* disappeared from use with the publication of Stone *et al.* (1959), and obviously Edwards's subgenus *Theobaldia sensu stricto* gave way to *Culiseta sensu stricto*.

Allotheobaldia Brolemann, 1919

Allotheobaldia was originally introduced as a subgenus of *Theobaldia* Neveu-Lemaire, 1902a by Brolemann (1919), with *Culex longiareolatus* Macquart, 1838a as the type and only included species. Following global acceptance of *Culiseta* as the valid name of the genus (Stone *et al.*, 1959), Maslov (1964) removed *Allotheobaldia* from *Culiseta* and established it a separate genus. Dobrotworsky (1971) disagreed with this action and reinstated *Allotheobaldia* to its current status as a subgenus of *Culiseta*.

Austrotheobaldia Dobrotworsky, 1954

Austrotheobaldia was originally introduced as a subgenus of *Theobaldia* Neveu-Lemaire, 1902a by Dobrotworsky (1954), with *Chrysoconops littleri* Taylor, 1914c as the type and only included species. After *Culiseta* was universally accepted as the valid name of the genus (Stone *et al.*, 1959), Maslov (1964) removed *Austrotheobaldia* from *Culiseta* and established it a separate genus. In a review of the genus in Southeast Asia, Dobrotworsky (1971) disagreed with the generic status of *Austrotheobaldia* and returned it to its current status as a subgenus of *Culiseta*.

Climacura Howard, Dyar & Knab, 1915

Climacura was originally established as a subgenus of *Culex* by Howard *et al.* (1915), with *Culex melanurus* Coquillett, 1902b as the type and only originally included species. Dyar & Knab (1917), without explanation, removed *Climacura* from *Culex* and listed it as a separate genus, but they failed to recognize that *Climacura* is a feminine noun and incorrectly spelled the name of the type species as *Climacura melanurus*. Dyar must have been doubtful about the generic status because he reverted back to treating *Climacura* as a subgenus of *Culex* 6 years later (Dyar, 1923f). As was the case with *Culiseta sensu stricto*, Dyar (1928) treated *Climacura* as a subgenus of *Culicella*. Edwards (1932a) listed *Climacura* and four other genus-group taxa as subgenera of *Theobaldia* Neveu-Lemaire, 1902a, and those five subgenera became subgenera of *Culiseta* when *Culiseta* replaced *Theobaldia* as the valid name of the genus in the catalogue of Stone *et al.* (1959). *Climacura* currently includes six species with distributions in the Australasian, Oriental and Nearctic Regions. Edwards (1932a) included five Australian species in the subgenus *Climacura*, four of which were later placed in other subgenera (Dobrotworsky, 1954, 1960b, 1966), leaving only *Cs. melanura* and *Cs. tonnoiri* (Edwards, 1925) in *Climacura* until four additional species were described in later years.

Culicella Felt, 1904

Culicella was originally introduced as a distinct genus by Felt (1904), with *Culex dyari* Coquillett, 1902b [junior synonym of *Culiseta morsitans* (Theobald, 1901c)] as the originally designated type species. Howard *et al.* (1915) were 'unable to point out good structural characters for separating both sexes of the adults from *Culex*,' and treated *Culicella*, albeit 'reluctantly', as a subgenus of that genus. Probably because subgeneric status seemed uncertain, Dyar & Knab (1917) re-established *Climacura* as a separate genus. The vacillation between generic and subgeneric status ended with Dyar treating *Climacura* as a subgenus of *Culiseta* in later publications (Dyar, 1923f, 1928). Edwards (1921d) agreed with Brolemann (1919), who treated *Culex morsitans* Theobald, 1901c as a species of *Theobaldia* Neveu-Lemaire, 1902a, 'that *Culicella* is not generically separable from *Theobaldia*' and treated it as a subgenus of that genus. Dyar (1923f) followed Edwards in recognizing *Culicella* as a subgenus, but as a subgenus of *Culiseta* rather than *Theobaldia*. Without explanation, Dyar (1928) restored *Culicella* to generic status and included *Climacura* and *Culiseta* as subgenera. Edwards (1932a) listed *Culicella* as a subgenus of *Theobaldia*, and it became a subgenus of *Culiseta* when *Theobaldia* was listed as a synonym of *Culiseta* in the catalogue of Stone *et al.* (1959).

Neotheobaldia Dobrotworsky, 1958

Neotheobaldia was originally described as a subgenus of *Theobaldia* Neveu-Lemaire, 1902a by Dobrotworsky (1954), but it was not validated until Dobrotworsky (1958b) designated *Theobaldia hilli* Edwards, 1926a as the type species. *Neotheobaldia* was transferred to subgeneric rank in *Culiseta* by Stone *et al.* (1959).

Theomyia Edwards, 1930

Theomyia was originally introduced as a subgenus of *Theobaldia* Neveu-Lemaire, 1902a by Edwards (1930a), with *Leptosomatomyia fraseri* Edwards, 1914b as the type and only included species. *Theomyia* remained a subgenus of *Theobaldia* until Stone *et al.* (1959) showed that it was preoccupied by *Theobaldia* Fischer, 1885 in Mollusca and the next available name was *Culiseta* Felt, 1904. Maslov (1964) proposed the subtribe Culisetina to comprise the tribe Culisetini of Belkin (1962), with *Theomyia* included as a genus separate from *Culiseta*. Maslov's classification never gained acceptance, and Dobrotworsky (1971) returned *Theomyia* to its current status as a subgenus of *Culiseta*.

Ficalbiini Belkin, 1962

Ficalbia Theobald, 1903

Ficalbia was established as a genus by Theobald (1903a) for two species, *Uranotaenia minima* Theobald, 1901c from southern India and the newly recognized *Ficalbia simplex* from Sri Lanka. The former was subsequently, and fortunately, selected to be the type species by Brunetti (1914); the latter was recognized as a species of *Aedes* by Edwards (1928a). Edwards (1932a) classified *Ficalbia sensu stricto* as the nominotypical subgenus of the genus *Ficalbia*, which also included *Etorleptiomyia* Theobald, 1904b, *Ingramia* Edwards, 1912a and *Mimomyia* Theobald, 1903a as subgenera. Mattingly (1957a) added another subgenus, *i.e.* *Ravenalites* Doucet, 1957 (in Mattingly, 1957a). Surprisingly, Mattingly & Grjebine (1958) included *Ingramia* as a synonym of *Ficalbia sensu stricto*. Mattingly (1971a) treated *Mimomyia* as a separate genus with *Etorleptiomyia* and *Ravenalites* as subgenera.

Mimomyia Theobald, 1903

Mimomyia Theobald, 1903 *sensu stricto*

Mimomyia was established as a genus by Theobald (1903a), with *Mimomyia splendens* Theobald, 1903a as the type and only originally included species. *Mimomyia* appears to have been treated as a genus until it was recognized as a subgenus of *Ficalbia* by Edwards (1932a). While a wider treatment of *Mimomyia* was in press (Mattingly & Grjebine, 1958), Mattingly (1957a) introduced the Ludlowia Group (as 'Group B (*Ludlowia*)') and the Hybrida Group (as the '*F. hybrida* Group') for species of the subgenus in the Indomalayan area. Belkin (1962) subsequently referred to the latter group as the '*hybrida* complex'. In addition to the Ludlowia Group, which was expanded to include Afrotropical species, Mattingly & Grjebine (1958) established four groups for purely Afrotropical species of the subgenus: the monobasic Mimomyia Group ('Group A, *Mimomyia s. str.*') for the type species; the Boycia Group ('Group C, *Boycia*') for *Boycia mimomyiaformis* Newstead, 1907 (in Newstead *et al.*, 1907) and *Ficalbia parenti* de Meillon & Lavoipierre, 1944; the monobasic Megaculex Group ('Group D, *Megaculex*') for *Culex plumosus* Theobald, 1901b; and the monobasic 'Group E' for *Ficalbia flavopicta* Edwards, 1936. *Mimomyia* was reinstated as a separate genus by Mattingly (1971a), with *Etorleptiomyia* Theobald, 1904b and *Ravenalites* Doucet, 1957 (in Mattingly, 1957a) (= *Ingramia* Edwards, 1912a) as subgenera (see below).

Etorleptiomyia Theobald, 1904

Etorleptiomyia was originally established as a distinct genus by Theobald (1904b), with *Etorleptiomyia mediolineata* Theobald, 1904b as the only originally included species. Without explanation, Macfie & Ingram (1923) treated *mediolineata* as a species of the genus *Ficalbia*; thus, apparently prompting Edwards (1932a) to list *Etorleptiomyia* as a subgenus of *Ficalbia*. *Etorleptiomyia* remained a subgenus of *Ficalbia* until Mattingly (1971a) restored *Mimomyia* to generic status, which was previously treated as a subgenus of *Ficalbia* since Edwards (1932a), and transferred the subgenus *Etorleptiomyia* to the genus *Mimomyia*. The subgenus is currently divided into two species groups, which are modified versions of those groups that were initially introduced by Edwards (1932a) as 'Group A (*Etorleptiomyia s. str.*, ar [*sic*] *mediolineata*-group)' and 'Group B (*Dasymyia*, or *fusca*-group)'. Mattingly & Grjebine (1958) recognized Group A to include the three species originally included by Edwards (1932a) and added *Mi. xanthozona* (van Someren, 1948); Grjebine (1986) later added *Mi. bougainvillensis* (Belkin, 1962) and *Mi. solomonis* (Belkin, 1962). Edwards established Group B for a single nominal species, *Dasymyia fusca* Leicester, 1908, which Mattingly (1957a) later transferred to the subgenus *Ravenalites* Doucet, 1957 (in Mattingly, 1957a) when he restored *Mimomyia* to its original generic status. Mattingly apparently chose the name *Ravenalites* over *Dasymyia* knowing it was preoccupied by *Dasymyia* Egger, 1858 (see *Ingramia* below). Years later, White (1974) showed that the name *Ingramia* Edwards, 1912a had priority over *Ravenalites* Doucet, and its type species was *Dasymyia fusca*. Grjebine (1986) retained Group B as 'Group B (*Dixomyia*)' for the three species which he newly included in *Etorleptiomyia*: *Mi. bougainvillensis* (Belkin, 1962), *Mi. elegans* (Taylor, 1914c) and *Mi. solomonis* (Belkin, 1962).

Ingramia Edwards, 1912

Ingramia was originally introduced as a distinct genus by Edwards (1912b), with *Mimomyia malfeyti* Newstead, 1907 (in Newstead *et al.*, 1907) as the originally designated type species. However, as discovered by White (1974), *Mi. fusca* (Leicester, 1908) is the type species of *Ingramia*, deduced from the following chronicle of taxonomic history. To begin with, Leicester (1908) proposed the new genus *Dasymyia* with *fusca* as the haplotype, *i.e.* the only originally included species. Edwards (1912b) realized the genus name was preoccupied by *Dasymyia* Egger, 1858, a genus of Syrphidae, and therefore proposed the replacement name *Ingramia* for *Dasymyia* Leicester. It was, of course, inappropriate to change the type species of the genus. Mattingly (1957a) and Mattingly & Grjebine (1958), while working with species that at the time were considered to be species of *Ficalbia* Theobald, 1903a, classified *fusca* as a species of the subgenus *Ravenalites* Doucet, 1957 (in Mattingly, 1957a). *Ravenalites* and *Etorleptiomyia* were subsequently transferred from *Ficalbia* to *Mimomyia* by Mattingly (1971a), concomitant with the recognition of *Mimomyia* as a separate genus. As explained by White (1974), *Ingramia* is the senior primary objective synonym of *Dasymyia* Leicester, 1908 and the senior secondary subjective synonym of *Ravenalites* Doucet, 1957. Consequently, *Dasymyia fusca* Leicester, 1908 is the type species of *Ingramia*.

Mattingly & Grjebine (1958) divided *Ingramia* (as *Ravenalites*) into three species groups, 'Group A (*Ravenalites s. str.*)', 'Group B (*Ravenalites*)' and 'Group C'. These groups were later accepted by Grjebine (1986), with the recognition of two complexes in Group A, the Roubaudi and Spinosa Complexes, and notation of Group C as *Ingramia s. str.*, or Fusca Group.

Hodgesiini Belkin, 1962

Hodgesia Theobald, 1904

Hodgesia was introduced as a genus by Theobald (1904a), with *Hodgesia sanguinae* Theobald, 1904a as the type and only originally included species. *Hodgesia* has been recognized as a distinct genus since it was established by Theobald. The genus includes 11 currently recognized species, nine of which were known at the time of Edwards (1932a). Edwards found it difficult to place *Hodgesia* in the classification of the Culicidae – he noted that larvae were similar to those of *Ficalbia* but felt the genus should perhaps be placed near *Uranotaenia* based on features of the adults. The tenth species, *Ho. lampangensis*, was described by Thurman (1959) and the eleventh, *Ho. solomonis*, was added to the genus by Belkin (1962). Belkin, like Edwards, also noted some similarities of adults with those of *Uranotaenia*, and noted that larvae, in addition to *Ficalbia*, bore some resemblance to larvae of *Culiseta*. In view of a combination of unique characters exhibited by the adults and larvae, Belkin established Hodgesiini solely for the genus. *Hodgesia* has never been revised or subdivided into subordinate taxa.

Mansoniini Belkin, 1962

Mansonia Blanchard, 1901

Mansonia Blanchard, 1901 *sensu stricto*

Mansonia was proposed by Blanchard (1901) as a replacement name for the genus name *Panoplites* Theobald, 1900a, which was preoccupied by *Panoplites* Gould, 1854 (a family of humming birds). The type species of *Mansonia*, *Culex titillans* Walker, 1848, was selected by Neveu-Lemaire (1902b) and fixed by the International Commission on Zoological Nomenclature (1959c: Opinion 550). Prior to Edwards (1932a), some authors considered species of *Mansonia* to be species of *Taeniorhynchus* Lynch Arribálzaga, 1891a. The genus *Taeniorhynchus*, for which a type species was not specified, originally included three species, *Culex taeniorhynchus* Wiedemann, 1821, *Taeniorhynchus confinnis* Lynch Arribálzaga, 1891b and *Taeniorhynchus fasciolata* Lynch Arribálzaga, 1891b. According to Dyar & Knab (1910b) and Howard *et al.* (1915), the first listed species, *Cx. taeniorhynchus*, should have been considered to be the type of the genus, not only because it was the first treated, but also by absolute tautonymy. However, Theobald (1900a) considered that the specimens Lynch Arribálzaga had identified as *taeniorhynchus* must have been *Culex titillans*, which Lynch Arribálzaga included as a synonym of *taeniorhynchus*. This led Theobald to establish the genus *Panoplites*, with *titillans* as the type species, and recognize *Taeniorhynchus fasciolatus* as the type species of *Taeniorhynchus*. Howard *et al.* (1915) disagreed with Theobald's assessment and considered that the type species of *Taeniorhynchus* should be fixed as *Cx. taeniorhynchus*, an aedine mosquito, by tautonymy. As a consequence, Dyar & Knab (1910b), followed by Howard *et al.* (1915, 1917), relegated *Taeniorhynchus* to synonymy with *Aedes* Meigen, 1818. As noted, Blanchard (1901) discovered that *Panoplites* was preoccupied and replaced it with *Mansonia*. *Mansonia sensu stricto* has never been revised and the included species have only been treated at regional levels. With the exception of three species in Jamaica that are recognized as members of species complexes (Belkin *et al.*, 1970), there has been no attempt to established an internal classification of the group based on natural relationships. Two of the three species in Jamaica, *Ma. dyari* Belkin *et al.*, 1970 and *Ma. indubitans* Dyar & Shannon, 1925b, are recognized as representatives of the Indubitans Complex and the third species, the type species of *Mansonia*, is the nominotypical member of the Titillans Complex.

Mansonioides Theobald, 1907

Mansonioides was originally introduced as a distinct genus by Theobald (1907), with *Mansonioides septempunctata* Theobald, 1905f as the type and only originally included species; however, the type species was synonymized with *Mansonia annulifera* (Theobald, 1901c) (as *Mansonioides annuliferus*) by Edwards (1913c). *Mansonioides* continued to be recognized as a genus until Edwards (1922a, 1922b) decided to treat it, and also *Coquillettidia* Dyar, 1905a, as subgenera of *Taeniorhynchus* Lynch Arribálzaga, 1891a based on 'close similarity of the specialised larvae'. Dyar (1918d) used features of the male genitalia to recognize two subgenera for American species of *Mansonia*, the nominotypical subgenus and the subgenus *Coquillettidia*. Dyar (1925c), with reference to his earlier paper, stated 'I did not consider the oriental and African forms at the time, but if I had done so, all the species now classified as "*Taeniorhynchus*" would have been placed in *Coquillettidia*, and all those under *Mansonioides* in *Mansonia*.' *Mansonioides* was subsequently classified as a subgenus of *Mansonia* by Edwards (1932a) and all later authors. *Mansonioides* has never been revised and has not been subdivided into subordinate groups; however, Belkin (1962) considered it likely that *Ma. uniformis* (Theobald, 1901c) is a complex of species and described a new species, *Ma. melanesiensis*, as a member of the Uniformis Complex.

Coquillettidia Dyar, 1905

Coquillettidia Dyar, 1905 *sensu stricto*

Coquillettidia was originally introduced as a distinct genus by Dyar (1905a), with *Culex perturbans* Walker, 1856 as the originally designated type species. Despite being listed as a synonym of *Mansonia* by Howard *et al.* (1915), *Coquillettidia* continued to be recognized as a genus until Edwards (1922a, 1922b) decided to treat it as a subgenus of *Taeniorhynchus* Lynch Arribálzaga, 1891a. As shown by Stone & Knight in their application for the proposed suppression of *Taeniorhynchus* under the Plenary Powers of the International Commission on Zoological Nomenclature (1959c), the usage of *Taeniorhynchus* and *Mansonia* was mixed from the start of the 20th century to the advent of their application. Based on features of the male genitalia, Dyar (1918d) recognized *Coquillettidia* as a subgenus for American species of *Mansonia*. Some years later, Dyar (1925c) stated that species classified as

Taeniorhynchus should have been placed in the subgenus *Coquillettidia*, which was subsequently classified as a subgenus of *Mansonia* by Edwards (1932a) and all later authors until it was treated as a separate genus by Ronderos & Bachmann (1963a). However, not all authors agreed with Ronderos & Bachmann and continued to recognize *Coquillettidia* as a subgenus of *Mansonia* despite it being accepted as a genus by Stone (1967, 1970) and Knight & Stone (1977). Notable exceptions to its recognition as a separate genus include Mattingly (1971a), White (1980) and Service (1990). *Coquillettidia* has continued to be recognized as a distinct genus since the end of the 20th century. *Coquillettidia* has never been revised but three informal species groups have been established. These include the large Palaeotropical Crassipes (39 species) and Australasian Iracunda Groups (two species) of Belkin (1962), and the Holarctic Perturbans Group (three species) of Belkin (1968a).

Austromansonia Belkin, 1968

Austromansonia was proposed as a subgenus of *Coquillettidia* by Belkin (1968a), with *Taeniorhynchus* (*Coquillettidia*) *tenuipalpis* Edwards, 1924b as its type and only included species, known only from New Zealand. Belkin stated that '*Austromansonia* seems to be the most primitive and generalized phylad of *Coquillettidia* and of the entire tribe Mansoniini.'

Pseudotaeniorhynchus Theobald, 1911

Until very recently, this genus group was known as *Rhynchotaenia* Brèthes, 1910 until it was shown by Evenhuis & Harbach (2024) that the name was preoccupied by *Rhynchotaenia* Diesing, 1850, a genus of tapeworms (class Cestoda). *Rhynchotaenia* Brèthes, 1910 and *Pseudotaeniorhynchus* Theobald, 1911a, both originally established as genera, were treated as synonyms of *Mansonia* Blanchard, 1901 by Howard *et al.* (1915). The former was established as a subgenus of *Mansonia* by Dyar (1925c), with *Taeniorhynchus fasciolatus* Lynch Arribálzaga, 1891b subsequently identified as the type species by Edwards (1932a). The synonymy of *Pseudotaeniorhynchus*, for which *Taeniorhynchus fasciolatus* was also designated as the type species (Brunetti, 1914), was transferred from *Mansonia* to *Rhynchotaenia* by Edwards (1932a). *Rhynchotaenia* remained a subgenus of *Mansonia* until *Coquillettidia* was treated as a separate genus by Ronderos & Bachmann (1963a), with the recognition of *Rhynchotaenia* as one of its subgenera and *Pseudotaeniorhynchus* as its junior synonym. With the discovery that *Rhynchotaenia* Brèthes, 1910 is preoccupied, it was replaced by the next available name, *i.e. Pseudotaeniorhynchus* Theobald, 1911a (Evenhuis & Harbach, 2024). *Pseudotaeniorhynchus* currently includes 13 species that are considered to comprise a homogeneous group without subdivision.

Orthopodomyiini Belkin, Heinemann & Page, 1970

Orthopodomyia Theobald, 1904

Orthopodomyia was introduced as a genus by Theobald (1904c), with *Orthopodomyia albipes* Leicester, 1904 (in Theobald, 1904c) as the type and only originally included species. The history of the classification of *Orthopodomyia* was reviewed by Zavortink (1968) as an introduction to his revision of the genus. During the early part of the 20th century, species now recognized as members of *Orthopodomyia* were incorrectly placed in other genera, four of which became synonyms of *Orthopodomyia*. The internal classification of the genus began with Edwards (1932a), who divided the 11 species known at the time into two groups, 'Group A (*Orthopodomyia*)' and 'Group B (*Bancroftia*)'. Edwards (1939b) excluded *Or. phyllozoa* (Dyar & Knab, 1907c) from Group B but it was later returned to that group by Knight & Mattingly (1950) when they divided Group A into three subgroups based on the wing markings of adults: 'Subgroup I—*anopheloides*', 'Subgroup II—*arboricollis*' and 'Subgroup III—*fascipes*'. No further changes to the classification were made until Zavortink (1968) revised the genus from a world standpoint based on anatomical features of adults, male genitalia, larvae and pupae. Zavortink classified 24 species, including two newly named species and two unnamed species, into eight groups and divided them provisionally between four sections: Bancroftia Section, with two groups, the monobasic Albicosta Group and the Signifera Group (six species); Folicolae Section (monobasic); Orthopodomyia Section, with four groups, the Albipes Group (10 species), the monobasic Arboricollis Group, the Nkolbissonensis Group (two species) and the Vernoni Group (three species); and Thomasina Section (two species). The two largest groups were further divided into subgroups. Based on ornamentation of the adults, Zavortink recognized three subgroups within the Signifera Group, the Kummi Subgroup (two species), the monobasic Pulcripalpis Subgroup (misspelled as Pulchripalpis) and the Signifera Subgroup (three species). Zavortink explained that species of the Albipes Group, the largest and most diverse group, could be classified in different ways based on similarities exhibited by a particular life stage. He chose to divide the group into subgroups based on relationships shown by pupae. Except for the inclusion of *Or. flavicosta* Barraud, 1927a and *Or. flavithorax* Barraud, 1927a, the Albipes Group is otherwise equivalent to Subgroup I (Anopheloides Subgroup) of Knight & Mattingly (1950). Based solely on pupal morphology, Zavortink arranged the species of the Albipes Group into four subgroups, the Albipes Subgroup (monobasic), the Anopheloides Subgroup (four species), the Flavicosta Subgroup (four species) and the Wilsoni Subgroup (monobasic). Finally, Huang & Rueda (2013) added a fifth group to the Orthopodomyia Section, the monobasic Ototomoensis Group, to accommodate a new species which they named *Or. ototomoensis*.

Sabethini Blanchard, 1905

Sabethes Robineau-Desvoidy, 1827

Sabethes Robineau-Desvoidy, 1827 *sensu stricto*

Sabethes was established as a genus by Robineau-Desvoidy (1827), with *Sabethes locuples* Robineau-Desvoidy, 1827 [subjective synonym of *Sa. cyaneus* (Fabricius, 1805)] as the type and only originally included species. *Sabethes* was treated as a genus without subgenera until Lane & Cerqueira (1942) elevated *Sabethinus* Lutz, 1904 (in Bourroul, 1904) from synonymy with *Sabethoides* Theobald, 1903a and recognized both as subgenera of the genus. *Sabethes sensu stricto* has received little taxonomic attention and remains poorly known. Cerqueira (1961) divided the nine species known at the time and four new species into two groups based on the presence or absence of white markings on the legs: Group A, species with white markings (*Sa. amazonicus* Gordon & Evan, 1922, *Sa. belisarioi* Neiva, 1908, *Sa. bipartipes* Dyar & Knab, 1906a, *Sa. forattinii* Cerqueira, 1961, *Sa. shannoni* Cerqueira, 1961 and *Sa. tarsopus* Dyar & Knab, 1908) and Group B, species with entirely dark-scaled legs (*Sa. albiprivus* Theobald, 1903a, *Sa. batesi* Lane & Cerqueira, 1942, *Sa. cyaneus*, *Sa. lanei* Cerqueira, 1961, *Sa. purpureus* (Theobald, 1907), *Sa. quasicyaneus* Peryassú, 1922b and *Sa. spixi* Cerqueira, 1961). No changes were made to the classification of *Sabethes sensu stricto* until Harbach & Petersen (1992) separated a new species from *Sa. tarsopus*, *i.e. Sa. gymnothorax* Harbach & Petersen, 1992, and established a new Group, the Amazonicus Group, to include the new species, *Sa. amazonicus* and *Sa. forattinii*. Three species which have legs with white markings, *Sa. nitidus* Theobald, 1901c, *Sa. ortizi* Vargas & Diaz Nájera, 1961 and *Sa. paraitepuyensis* Anduze, 1941, are included herein in Group A and the dark-legged *Sa. schnusei* (Martini, 1931) is added to Group B. Taxonomic and phylogenetic studies based on all life stages and DNA sequence are needed before a natural classification of *Sabethes sensu stricto* can be achieved.

Davismyia Lane & Cerqueira, 1942

Davismyia was originally established as a subgenus of *Wyeomyia* Theobald, 1901a by Lane & Cerqueira (1942), with *Miamyia petrocchiae* Shannon & del Ponte, 1928 as the originally designated type species. The subgenus and its type species were transferred to the genus *Sabethes* Robineau-Desvoidy, 1827 by Harbach & Peyton (1991). Prior to the transfer, *Davismyia* included four nominal species. In addition to *Miamyia petrocchiae*, Lane & Cerqueira also included *Wyeomyia arborea* Galindo, Carpenter & Trapido, 1951 and *Dendromyia schnusei* Martini, 1931 in the subgenus. When *Wy. ininicola* Fauran & Pajot (1974) was described, the authors tentatively placed it in *Davismyia*. *Wyeomyia arborea* was subsequently placed in a new subgenus, *Exallomyia* Harbach & Peyton, 1992a, *Wy. ininicola* has remained in *Wyeomyia* without subgeneric placement and *Dendromyia schnusei* was later discovered to be a species of *Sabethes sensu stricto* (Harbach, 1992). *Davismyia* is a monobasic genus-level taxon as currently defined.

Peytonulus Harbach, 1991

Peytonulus was originally introduced as a subgenus of *Sabethes* Robineau-Desvoidy, 1827 by Harbach (1991a), with *Sabethinus aurescens* Lutz, 1905 as the originally designated type species. *Peytonulus* was established for seven species that were previously included in the subgenus *Sabethinus* Lutz, 1904 (in Bourroul, 1904). *Peytonulus* currently includes 13 species that are considered to comprise a homogeneous group without subdivision.

Sabethinus Lutz, 1904

Sabethinus was originally proposed as a distinct genus by Lutz, 1904 (in Bourroul, 1904), with *Sabethinus intermedius* Lutz, 1904 as the type and only originally included species. Dyar (1924b) united *Sabethinus* with *Sabethoides* Theobald, 1903a, and it was treated as a synonym of the later taxon until it was resurrected from synonymy by Lane & Cerqueira (1942), with *Sabethinus* and *Sabethoides* both recognized as subgenera of *Sabethes* Robineau-Desvoidy, 1827. With the transfer of seven species from *Sabethinus* to *Peytonulus* by Harbach (1991a), only two species, *Sa. intermedius* and *Sa. melanonymphe* Dyar, 1924b, remained in *Sabethinus* until Harbach (1994b) redefined the subgenus concomitant with the description of two new species, *Sa. idiogenes* and *Sa. xhyphydes*. *Sabethinus* was not the subject of further study until Harbach (2018a) described two new Central American species, *Sa. andreae* and *Sa. theresae*, and divided the subgenus into two groups based on correlated features of adults, larvae and pupae: Group A (Intermedius Group) (four South American species) and Group B (Andreae Group) (two Central American species).

Sabethoides Theobald, 1903

Sabethoides was originally established as a distinct genus by Theobald (1903a), with *Sabethoides confusus* Theobald, 1903a [subjective synonym of *Sabethes chloropterus* (von Humboldt, 1819)] as the type and only originally included species. As mentioned above, Dyar (1924b) united *Sabethinus* Lutz, 1904 with *Sabethoides* and it remained in synonymy until Lane & Cerqueira (1942) recognized it and *Sabethoides* as separate subgenera of *Sabethes* Robineau-Desvoidy, 1827. Lane & Cerqueira included three species in the subgenus *Sabethoides*: *Sa. chloropterus*, *Sa. glaucodaemon* (Dyar & Shannon, 1925b) and *Sa. imperfectus* (Bonne-Wepster & Bonne, 1920). The last two were originally described as species of the genus *Sabethoides* and the last one was placed in synonymy with *Sa. chloropterus* by Lane (1951). With the addition of *Sa. tridentatus* Cerqueira, 1961 and *Sa. conditus* Moses, Howard & Harbach, 2000, *Sabethoides* currently comprises a homogeneous group of four species.

Isostomyia Coquillett, 1906

Isostomyia was first published as the name of a genus by Coquillett (1906d), with *Aedes perturbans* Williston, 1896 as the type and only originally included species. Unfortunately, Coquillett (1906d) based *Isostomyia* on misidentified specimens of *Culex conservator* Dyar & Knab, 1906d (Howard *et al.*, 1915). Despite this error, Howard *et al.* rightly treated *Aedes perturbans* as the type species of *Isostomyia*. Bonne-Wepster & Bonne (1921) synonymized *Isostomyia* and four other nominal genus groups with *Goeldia* Theobald, 1903a. Three years later, Dyar & Shannon (1924) returned *Isostomyia* to its original status as a genus for species with short maxillary palpi in both sexes, and this was accepted by Dyar (1928). Edwards (1930a), however, did not agree that *Aedes perturbans* should be the type species of *Isostomyia* and proposed *Isogoeldia* as a subgenus of *Goeldia*, with *Aedes perturbans* as the type species. *Culex conservator* was therefore considered to be the type species of *Isostomyia*, which Edwards (1932a) classified as a subgenus of *Culex* Linnaeus, 1758. In the same publication, Edwards removed species with clypeal setae from *Goeldia* and placed them in the genus *Trichoprosopon* Theobald, 1901b, and recognized three subgenera for the remaining species of *Goeldia*: *Ctenogoeldia* Edwards, 1930a, *Goeldia* and *Isogoeldia*. Lane (1939) followed this classification but 3 years later Lane & Cerqueira (1942) subsumed *Goeldia* under *Trichoprosopon* and recognized *Isogoeldia* as one of seven subgenera. *Isostomyia* was included as a synonym of *Isogoeldia*, with *Aedes perturbans* as the type species. Lane (1953) followed the classification that he established with Cerqueira in 1942 except he replaced the subgenus *Hyloconops* Lutz, 1904 (in Bourroul, 1904) with *Runchomyia* Theobald, 1903a (misspelled as *Rhunchomyia*). Stone *et al.* (1959) retained the classification of Lane (1953) in their synoptic catalogue of the mosquitoes of the world but recognized *Isostomyia* in place of *Isogoeldia*. Zavortink (1979) provided the most up-to-date treatment of the mosquitoes which Lane included in the genus *Trichoprosopon*. His revised classification included four genera, *Johnbelkinia* gen. nov., *Trichoprosopon*, *Shannoniana* Lane & Cerqueira, 1942 and *Runchomyia*, with the last genus divided into three subgenera, *Ctenogoeldia*, *Isostomyia* and *Runchomyia sensu stricto*. *Isostomyia* remained a subgenus of *Runchomyia* until Harbach & Peyton (1993) returned it to generic rank based on the unique structure of the maxillae of larvae and distinctive features of the adult, larval and pupal stages. As currently defined, *Isostomyia* is a homogeneous group comprised of only four currently known species.

Johnbelkinia Zavortink, 1979

Johnbelkinia was introduced as a distinct genus by Zavortink (1979), with *Culex longipes* Fabricius, 1805 as the designated type species. The genus was established for the type species and two other species that were previously included in *Trichoprosopon* (*Runchomyia*) of previous authors. Morphological features of the adults, larvae and pupae clearly distinguish *Johnbelkinia* from other genus groups of Sabethini.

Kimia Vu & Harbach, 2007

Kimia was established as a genus by Vu & Harbach, 2007 (in Harbach *et al.*, 2007a), with *Topomyia decorabilis* Leicester, 1908 as the designated type species. The genus is a distinct group based on shared morphological features of the adult, larval and pupal stages of five species, all of which were previously placed in *Topomyia* (*Suaymyia*) of previous authors.

Limatus Theobald, 1901

Limatus was introduced as a distinct genus by Theobald (1901c), with *Limatus durhamii* Theobald, 1901c as the type and only originally included species. Howard *et al.* (1915) synonymized *Simondella* Laveran, 1902c with *Limatus* and listed the type species, *curvirostris* Laveran, 1902c, as a synonym of *Li. durhamii*; they included two other species in the genus, *Li. cacophrades* and *Li. methysticus*, both described and named by Dyar & Knab (1909). Dyar (1928) listed *Lemmamyia* Dyar, 1919b as a synonym of *Limatus* and included three species in the genus: *Li. asulleptus* (Theobald, 1903a), *Li. durhamii* and *Li. hoffmani* Root, 1927a. He listed the type species of *Lemmamyia*, *methysticus* Dyar & Knab, 1909, and *Lemmamyia pseudomethysticus* Bonne-Wepster & Bonne, 1920 as synonyms of *Li. asulleptus*, and recognized *Li. cacophrades* and *Dendromyia paraensis* Theobald, 1903a as synonyms of *Li. durhamii*. Edwards (1932a) recapitulated the classification of Dyar (1928) and recognized *Li. exhibitor* Shannon & del Ponte, 1928 as a synonym of *Li. durhamii*. Lane & Cerqueira (1942), in their revision of the Neotropical sabethines, recognized seven species of *Limatus*. In addition to the three species recognized by Dyar (1928) and two species described after Edwards (1932a), *Li. flavisetosus* de Oliveira Castro, 1935 and *Li. martiali* Senevet & Abonnenc, 1939b, they removed *paraensis* and *pseudomethysticus* from synonymy and described them as distinct species of the genus. With the later descriptions of *Li. andinus* and *Li. guayasi* by Levi-Castillo (1954), the genus, as currently known, includes nine species which have not been arranged into infrageneric groups.

Malaya Leicester, 1908

Malaya was introduced as a distinct genus by Leicester (1908), with *Malaya genurostris* Leicester, 1908 as the type and only originally included species. Of the 12 currently known species of *Malaya*, only the type species and *Ml. incomptas* Ramalingam & Pillai, 1972 were described in combination with this generic name. Nine of the other 10 species currently placed in *Malaya* were originally described as species of *Harpagomyia* de Meijere, 1909, and one was originally described as a species of *Grahamia* Theobald, 1909. With the exception of *Ml. incomptas*, all species of the genus were considered to be species of *Harpagomyia* until Stone & Knight (1957a) showed that the generic name *Malaya*, which was published earlier than *Harpagomyia*, was not preoccupied, as indicated by Edwards (1932a), by *Malaia* Heller, 1892 (Coleoptera: Scarabaeidae). *Malaya* has not been revised or subdivided.

Maorigoeldia Edwards, 1930

Maorigoeldia was originally established as a subgenus of *Rachionotomyia* Theobald, 1905h by Edwards (1930a), with *Culex argyropus* Walker, 1848 as the type and only originally included species. *Maorigoeldia* was raised to generic status by Belkin (1962), who stated that *Mg. argyropus* 'is a very distinctive sabethine, which shows strong affinities with *Tripteroides* but has so many peculiar features that it should be placed in a distinct genus.' The genus is monobasic.

Onirion Peyton & Harbach, 2000

Onirion was introduced as a distinct genus by Peyton & Harbach, 2000 (in Harbach & Peyton, 2000), with *Dendromyia personata* Lutz, 1904 (in Bourroul, 1904) as the designated type species. The genus includes seven species which share morphological features in the adult, larval and pupal stages. Three of the species were previously included in the genus *Wyeomyia* Theobald, 1901a and four were described as new. *Onirion* is divided into two groups based on the presence or absence of distal tergal setae on the gonocoxites of males, the Brucei and Personatum Groups. The former group includes three species in which distal tergal setae are absent, *On. brucei* (del Ponte & Cerqueira, 1938), *On. celatum* Peyton & Harbach, 2000 and *On. aenigma* Harbach, 2000 (in Harbach & Peyton, 2000), and the latter group includes four species in which the setae are present, *On. personatum* (Lutz, 1904), *On. sirivanakarni* (Duret, 1982), *On. imparis* Peyton & Harbach, 2000 and *On. regale* Peyton & Harbach, 2000.

Runchomyia Theobald, 1903

Runchomyia Theobald, 1903 *sensu stricto*

Runchomyia was established as a distinct genus by Theobald (1903a), with *Runchomyia frontosa* Theobald, 1903a as the type and only originally included species. Bonne-Wepster & Bonne (1921) synonymized *Runchomyia* with *Goeldia* Theobald, 1903a. In a reclassification of the New World sabethines, Lane & Cerqueira (1942) combined the genera *Trichoprosopon* Theobald, 1901b and *Goeldia* and recognized *Hyloconops* Lutz, 1904 (in Bourroul, 1904) as one of seven subgenera of *Trichoprosopon*. Two years later, Stone (1944) used the name *Runchomyia* in place of *Hyloconops*, and this was accepted by Lane (1953) in his treatise on the 'Neotropical Culicidae'. *Runchomyia* remained a subgenus of *Trichoprosopon* until Zavortink (1979) returned it to its original generic rank. As currently defined, *Runchomyia sensu stricto* includes only six described species.

Ctenogoeldia Edwards, 1930

Ctenogoeldia was introduced as a subgenus of *Goeldia* Theobald, 1903a by Edwards (1930a), with *Lesticocampa dicellaphora* Howard, Dyar & Knab, 1913 (= *Phoniomyia magna* Theobald, 1905f) as the originally designated type species. Lane & Cerqueira (1942) combined the genus *Goeldia* with *Trichoprosopon* Theobald, 1901b and recognized *Ctenogoeldia* as a subgenus of the composite genus *Trichoprosopon*. *Ctenogoeldia* remained a subgenus of *Trichoprosopon* until Zavortink (1979) reclassified the composite *Trichoprosopon* and included it as a subgenus of *Runchomyia* Theobald, 1903a. *Ctenogoeldia* currently includes only two described species.

Shannoniana Lane & Cerqueira, 1942

Shannoniana was introduced as a subgenus of *Trichoprosopon* Theobald, 1901b by Lane & Cerqueira (1942), with *Goeldia fluviatilis* Theobald, 1903a as the type and only originally included species. *Shannoniana* was elevated to generic rank when Zavortink (1979) reclassified the composite genus *Trichoprosopon* of Lane & Cerqueira (1942). The genus currently includes only four described species, but many others await formal description and more are likely to be discovered (Zavortink, 1979).

Topomyia Leicester, 1908

Topomyia Leicester, 1908 *sensu stricto*

Topomyia was originally established as a genus by Leicester (1908), with *Topomyia minor* Leicester, 1908 subsequently designated as the type species by Brunetti (1914). The genus *Pseudograhamia* Theobald, 1910a, which included only *P. aureoventer* Theobald, 1910a in India, was synonymized with *Topomyia* by Barraud (1929b). *Topomyia* was considered to be a homogeneous group until Thurman (1959) proposed the subgenus *Suaymyia* based on shared features of the male genitalia. *Topomyia sensu stricto* currently includes 44 formally recognized species that are not subdivided into groups.

Miyagiella Harbach, 2014

Miyagiella was established as a monobasic subgenus of *Topomyia* Leicester, 1908 by Harbach, 2014 (in Harbach & Culverwell, 2014) based on unique features of the genitalia of a single male mosquito captured in a light trap in Kampung Paradason of Sabah, Malaysia. *Miyagiella* is a clear indication that *Topomyia* is a diverse assemblage of taxa, the classification of which is very probably incomplete.

Suaymyia Thurman, 1959

Suaymyia was introduced as a subgenus of *Topomyia* Leicester, 1908 by Thurman (1959), with *Topomyia cristata* Thurman, 1959 as the originally designated type species. *Suaymyia* currently includes 21 species, but it formerly included two other species, *To. spathulirostris* Edwards, 1923b and *To. yanbarensis* Miyagi, 1976, which are considered to belong to a separate, as yet, unnamed lineage (Rattanarithikul *et al.*, 2007). This is an indication that *Suaymyia* is probably a heterogeneous assemblage that requires subdivision.

Trichoprosopon Theobald, 1901

Trichoprosopon was established as a genus by Theobald (1901b), with *Tr. nivipes* Theobald, 1901c [subjective synonym of *Tr. digitatum* (Rondani, 1848)] as its type and only originally included species. The taxonomic history of *Trichoprosopon* was reviewed by Zavortink (1979). In short, three genus-group names proposed shortly after the introduction of *Trichoprosopon*, *Joblotia* Blanchard, 1901, *Hyloconops* Lutz, 1904 (in Bourroul, 1904) and *Isogoeldia* Edwards, 1930a, feature prominently in the history of *Trichoprosopon*. Edwards (1932a) placed species with clypeal setae in *Trichoprosopon*, and species without these setae in *Goeldia*. Within the latter genus, he recognized the subgenera *Ctenogoeldia* Edwards, 1930a, *Goeldia* and *Isogoeldia*. Edwards's classification was accepted until Lane & Cerqueira (1942) united *Goeldia* with *Trichoprosopon* and *Goeldia* (preoccupied by *Goeldia* Keyserling, 1891, genus of Neotropical spiders) became a synonym of the newly introduced subgenus *Shannoniana* Lane & Cerqueira, 1942 by virtue of its type species, *fluviatilis* Theobald, 1903a, being established as the type species. Lane (1953), in the second volume of *Neotropical Culicidae*, and Stone *et al.* (1959), in *A Synoptic Catalog of the Mosquitoes of the World*, accepted the classification of Lane & Cerqueira, except Lane used *Runchomyia* Theobald, 1903a (misspelled as *Rhunchomyia*) in place of *Hyloconops* and Stone *et al.* recognized *Isostomyia* Coquillett, 1906c as the senior synonym of *Isogoeldia* (see *Isostomyia* above). The classification of *Trichoprosopon* accepted by Stone *et al.* was retained in the second edition of *A Synoptic Catalog of the Mosquitoes of the World* (Knight & Stone, 1977) and provisionally retained by Belkin (1968b) 'Pending a thorough revision of the genus'. However, Belkin found that the type species of *Hyloconops* (*pallidiventer* Lutz, 1905) did not belong in the subgenus *Runchomyia*, where it was placed by Lane (1953), and he transferred it to *Trichoprosopon sensu stricto*; thus, *Hyloconops* became a synonym of the nominotypical subgenus. Based on a study of the adults, larvae and pupae of 64 species, 45 recognized at the time as species of *Trichoprosopon* and 19 undescribed species, Zavortink (1979) recognized four generic-level taxa: *Johnbelkinia* (as new), *Shannoniana*, *Runchomyia* and *Trichoprosopon*. Except for *Runchomyia*, from which the subgenus *Isostomyia* was elevated to generic status by Harbach & Peyton (1993), the concepts of *Trichoprosopon* and the other two genera recognized by Zavortink have not changed. *Trichoprosopon*, as currently known, includes 17 formally recognized species. Zavortink (1981) divided the genus into four complexes, the Digitatum Complex (three species, two undescribed), Compressum Complex (four species, two undescribed), Lampropus Complex (two species) and the Pallidiventer Complex (six species, one undescribed). Three species were not included in a complex. More recently, Rivera-García *et al.* (2023) divided the genus into two groups, the Trichoprosopon and Hyloconops Groups, which 'could be proposed as subgenera' when the unknown immature stages of many species of the genus are known. Ironically, the two groups were previously treated as subgenera by Coher (1948) based on distinctive differences in structures of the female genitalia. Species of the Compressum, Digitatum and Lampropus Complexes of Zavortink (1981) are members of the Trichoprosopon Group, the species of the Pallidiventer Complex are members of the Hyloconops Group and the three species which Zavortink did not classify, as well as *Tr. mixtli* Rivera-García *et al.*, 2023, are not placed in a group or complex.

Tripteroides Giles, 1904

Tripteroides Giles, 1904 *sensu stricto*

Tripteroides was introduced by Giles (1904) as a tentative new genus for a single adult female mosquito which he initially characterized as a new species named *Runchomyia philippinensis*. After describing and illustrating the new species, and discussing peculiar structures of the antepronota, Giles stated 'I … await the receipt of further material before naming the species *Tripteroides philippinensis*, *sp. et gen. nov.* [*sic*].' Edwards (1932a) aptly asserted in a footnote to his treatment of *Tripteroides* that 'This must be regarded as a valid publication of the generic name *Tripteroides*, which takes precedence over *Runchomyia*. It is curious that both these names were introduced on account of misconceptions regarding thoracic structure.' Edwards recognized four subgenera, *Maorigoeldia* Edwards, 1930a, *Mimeteomyia* Theobald, 1910c, *Rachisoura* Theobald, 1910c and *Tripteroides sensu stricto*. He divided species of the last subgenus into two groups: 'Group A (*aranoides*-group: *Rachionotomyia*)' and 'Group B (*nitidiventer*-group [*sic*]: *Tripteroides*)'. Belkin (1950a) recognized the two groups (*nitidiventer*-group corrected to '*nitidoventer*-group') for species in the Solomon Islands. Belkin (1962) raised *Maorigoeldia* to generic status and recognized three subgenera: *Rachisoura*, *Rachionotomyia* Theobald, 1905h and *Tripteroides*. Belkin was aware of several distinct groups within the subgenus *Tripteroides*, and recognized two groups in the South Pacific, the monotypic Purpuratus Group in Fiji and the Bimaculipes Group for four species that are found in the Santa Cruz and Solomon Islands. Mattingly (1980, 1981) made extensive changes to the internal classification of *Tripteroides*. He redescribed the subgenus *Rachionotomyia* and included five species previously placed in *Tripteroides sensu stricto*; resurrected *Polylepidomyia* Theobald, 1905f and *Tricholeptomyia* Dyar & Shannon, 1925a from synonymy with *Rachionotomyia*; removed three species from *Rachionotomyia* and placed one in *Tricholeptomyia* and two in *Polylepidomyia*; transferred two species from *Rachionotomyia* to *Rachisoura*; and transferred one species from *Rachisoura* to *Tripteroides sensu stricto*. Mattingly primarily dealt with the internal classifications of *Polylepidomyia* and *Tricholeptomyia*. He redefined *Tripteroides sensu stricto*, and noted it was badly in need of further work. He included 48 species in a largely expanded version of the Nitidoventer Group founded by Edwards (1932a) and established the monobasic Mabinii Group for the distinctive *Tp. mabinii* Baisas & Ubaldo-Pagayon, 1953, but failed to recognize the two groups established by Belkin (1962). Lee (1946) accepted the two groups of Edwards (1932a) and recognized two unnamed groups for the Australasian species of Group B (Nitidoventer Group). Lee *et al.* (1989c) accepted the Bimaculipes and Purpuratus Groups of Belkin (1962) and recognized the two groups of Lee (1946) as the Littlechildi and Magnesianus Groups. *Tripteroides sensu stricto* is a large and heterogeneous taxon, which as currently interpreted, merely comprises six species groups, the Bimaculipes, Littlechildi, Mabinii, Magnesianus, Nitidoventer and Purpuratus Groups.

Polylepidomyia Theobald, 1905

Polylepidomyia was established as a distinct genus by Theobald (1905f), with *Polylepidomyia argenteiventris* Theobald, 1905f as the type and only originally included species. Edwards (1924b) placed *argenteiventris* in *Rachionotomyia* Theobald, 1905h, whereby *Polylepidomyia* became a synonym of that genus. Subsequently, Edwards (1932a) treated *Rachionotomyia* as a synonym of *Tripteroides* Giles, 1904; thus, *Polylepidomyia* also became a synonym of *Tripteroides*. Edwards's concept of *Tripteroides*, as well as that of Lee (1946) and Belkin (1950a), included four subgenera: *Maorigoeldia* Edwards, 1930a, *Mimeteomyia* Theobald, 1910c, *Rachisoura* Theobald, 1910c and *Tripteroides sensu stricto*. *Polylepidomyia* remained in synonymy with *Tripteroides* until it was raised to subgeneric rank by Belkin (1962). *Mimeteomyia* became a synonym of *Polylepidomyia* with the transfer of the type species, *Tp. apicotriangulatus* (Theobald, 1910c), as a synonym of *Tp. atripes* (Skuse, 1889), to *Polylepidomyia* by Mattingly (1980). Edwards (1932a) divided the subgenus *Mimeteomyia* into two groups: 'Group A (*atripes*-group)' and 'Group B (*caledonica* group)'. Lee (1946) recognized the two groups of Edwards and added a third group, the '*obscura*-group', for *Tp. obscurus* Brug, 1934 and *Tp. subobscurus* Lee, 1946, which were described post-Edwards (1932a). The current placement of species in the internal classification of *Polylepidomyia* is largely attributable to Belkin (1962), who ironically treated it as a synonym of *Rachionotomyia*. Belkin removed *Tp. argenteiventris* (Theobald, 1905f), *Tp. collessi* Lee, 1946 and *Tp. tasmaniensis* (Strickland, 1911) from the *caledonica*-group of Lee (1946), introduced the Argenteiventris Group to include the first two species and asserted that *Tp. tasmaniensis* 'should be placed in a separate monotypic group', which Lee *et al.* (1989c) appropriately termed the Tasmaniensis Group. Belkin (1962) also established the monobasic Floridensis Group for *Tp. floridensis* Belkin, 1950a, which in general appearance is suggestive of a species of *Mimeteomyia*, was not assigned to a subgenus at the time. Belkin provisionally accepted the Atripes Group as defined by Lee (1946), noting that the taxonomic status of *Tp. solomonis* (Edwards, 1924b) was uncertain. Pending further study, *Polylepidomyia* presently comprises five species groups, the Argenteiventris, Atripes, Caledonicus, Floridensis and Tasmaniensis Groups.

Rachionotomyia Theobald, 1905

Rachionotomyia was introduced as a genus by Theobald (1905h), with *Rachionotomyia ceylonensis* Theobald, 1905h as the type and only originally included species. *Rachionotomyia* continued to be recognized as a genus (Edwards, 1924b, 1927b, 1930a) until it was treated as a synonym of *Tripteroides* Giles, 1904 by Edwards (1932a). It remained a synonym (Brug, 1934; Lee, 1946) until it was raised to subgeneric status by Belkin (1962), to include *Polylepidomyia* Theobald, 1905f as one of its synonyms. Mattingly (1980, 1981) resurrected *Polylepidomyia* from synonymy as a valid subgenus of *Tripteroides* to include all of the Australasian species that had been included in *Rachionotomyia*. Additionally, Mattingly (1981) removed one species to *Tricholeptomyia* Dyar & Shannon, 1925a and added five species previously included in *Tripteroides sensu stricto*. He also temporarily transferred *Tp. rozeboomi* Baisas & Ubaldo-Pagayon, 1953 to *Rachionotomyia*, but noted that it would probably require 'a new subgenus when better known.' The current internal classification of *Rachionotomyia* is almost entirely attributable to Mattingly (1981). He adopted the name Aranoides Group for a revised version of 'Group A (*aranoides*-group)' of Edwards (1932a), and recognized three species of that group as nominotypical members of three subgroups, the Affinis, Aranoides and Nepenthis Subgroups. He established the Edwardsi Group for *Tp. edwardsi* (Barraud, 1929b), which was also included in Group A of Edwards, and provisionally placed *Tp. rozeboomi* in the group.

Rachisoura Theobald, 1910

Rachisoura was originally introduced as a distinct genus by Theobald (1910c), with *Rachisoura sylvestris* Theobald, 1910c as the type and only originally included species. Edwards (1924b) synonymized *Rachisoura* with *Rachionotomyia* Theobald, 1905h, stating 'Since the distinction between *Rachionotomyia* and *Rachisoura* rests almost entirely on the greater length of the proboscis in the former, and since, moreover, several species are more or less intermediate in this respect, I propose to unite the two genera under the earlier name.' Edwards (1930a) observed that '*Rachionotomyia* includes some species with long and some with short male palpi. For the former the names *Mimeteomyia* and *Rachisoura* are available and may be used with advantage for subgenera.' This apparently provided the motive for Edwards (1932a) to include *Mimeteomyia* Theobald, 1910c and *Rachisoura* as subgenera in a revised classification of *Tripteroides*. *Rachisoura* continued to be recognized as a subgenus of *Tripteroides* until the present (Lee, 1946; Belkin, 1950a; Baisas & Ubaldo-Pagayon, 1953; Mattingly, 1980, 1981; Lee *et al.*, 1989c). *Mimeteomyia*, on the other hand, has been treated as a synonym since its type species, *Mimeteomyia apicotriangulata* Theobald, 1910c, was first synonymized with *Tp. atripes* (Skuse, 1889) by Edwards (1924b). It subsequently became a synonym of *Tripteroides* when Edwards (1932a) treated *Rachionotomyia* as a synonym of *Tripteroides sensu stricto*. It once again became a synonym of *Rachionotomyia* when Belkin (1962) elevated this taxon to subgeneric status. Finally, *Mimeteomyia* became a synonym of *Polylepidomyia* with the transfer of the type species, as a synonym of *Tp. atripes*, to *Polylepidomyia* by Mattingly (1980). *Rachisoura* has never been the subject of a revisionary study. Lee (1946) recognized two species groups, the Filipes and Vanleeuweni Groups. As Belkin (1962) noted, *Rachisoura* is 'in a confused state nomenclaturally and taxonomically.' He only retained the Filipes Group of Lee for three South Pacific species and proclaimed 'It is evident that there are additional groups in the subgenus.'

Tricholeptomyia Dyar & Shannon, 1925

Tricholeptomyia was introduced as a genus by Dyar & Shannon (1925a), with *Wyeomyia nepenthicola* Banks, 1909 as the originally designated type species. Without explanation, Edwards (1929c) listed *Tricholeptomyia nepenthicola* as a subgenus and species of *Rachionotomyia* Theobald (1905h). The following year, Edwards (1930a) asserted that 'The Philippine *Wyeomyia nepenthicola*, Banks, belongs to this genus [*Rachionotomyia*] rather than to *Wyeomyia* ...', and subsequently (Edwards, 1932a) formally listed *Tricholeptomyia*, as well as *Rachionotomyia*, as synonyms of *Tripteroides* Giles, 1904. *Tricholeptomyia* remained a synonym of *Tripteroides* until Belkin (1962) raised *Rachionotomyia* to subgeneric status, with *Tricholeptomyia* as one of its synonyms. Nearly two decades later, Mattingly (1980, 1981) resurrected *Tricholeptomyia* from synonymy with *Rachionotomyia* and recognized it as a fourth subgenus of *Tripteroides* 'to include all the unornamented Philippines species (Group A of Baisas and Ubaldo-Pagayon, 1953) except *rozeboomi*', which he 'transferred temporarily to *Rachionotomyia*', and included *Tp. microcala* (Dyar, 1929c), which was previously recognized as a species of *Rachionotomyia*. It is noteworthy that Mattingly (1980) identified *Tricholeptomyia* as 'possibly the best characterized of the subgenera' and 'perhaps the best candidate for elevation to genus.' If considered a genus, as originally proposed, the two informal groups established by Mattingly (1981), the monobasic Delpilari Group and the Nepenthicola Group, would probably be recognized as subgenera, and it is likely that the Belkini (monobasic) and Nepenthicola Subgroups of the latter group would be elevated to full Group-level.

Wyeomyia Theobald, 1901

Wyeomyia was introduced as a genus by Theobald, first in an annotated list (Theobald, 1901a), then in a key to genera based on adults (Theobald, 1901b) and subsequently described (Theobald, 1901c) based on the adults of seven species, only four of which are currently classified as species of *Wyeomyia*. *Wyeomyia grayii* Theobald, 1901c was subsequently designated the type species by Neveu-Lemaire (1902c).

The taxonomic history of *Wyeomyia* is complicated, involving the following 27 nominal genus-group names: *Antunesmyia* Lane & Cerqueira, 1942; *Calladimyia* Dyar, 1919b; *Cleobonnea* Dyar, 1919b; *Cruzmyia* Lane & Cerqueira, 1942; *Davismyia* Lane & Cerqueira, 1942; *Decamyia* Dyar, 1919b; *Dendromyia* Theobald, 1903a; *Dinomyia* Dyar, 1919b; *Diphalangarpe* Dyar, 1919b; *Dodecamyia* Dyar, 1919b; *Dyarina* Bonne-Wepster & Bonne, 1921; *Eunicemyia* Dyar & Shannon, 1924b; *Heliconiamyia* Dyar, 1919b; *Hystatomyia* Dyar, 1919b; *Janicemyia* Dyar & Shannon, 1924b; *Melanolepis* Dyar, 1928; *Menolepis* Lutz, 1905; *Miamyia* Dyar, 1919b; *Nunezia* Dyar, 1928; *Pentemyia* Dyar, 1919b; *Phoniomyia* Theobald, 1903a; *Phyllozomyia* Dyar, 1924c; *Prosopolepis* Lutz, 1905; *Shropshirea* Dyar, 1922c; *Technicomyia* Dyar, 1925a; *Triamyia* Dyar, 1919b; and *Wyeomyia* Theobald, 1901a.

Dyar & Shannon (1924b) divided the New World sabethines (Sabethini) into two groups, the Sabethes and Wyeomyia Groups. They recognized the Wyeomyia Group as 'a large group of species of unusual difficulty to classify' and proposed a simplified classification consisting of 'four groups having generic rank on the basis of the pleural setae and wing scales.' The four groups (genera) were divided into subgenera based on characters of the male genitalia, as follows:

Genus *Menolepis*, represented by a single species, *Wy. leucostigma* Lutz, 1904 (in Bourroul, 1904).
Genus *Miamyia*, with four subgenera – *Cleobonnea*, *Dodecamyia*, *Miamyia* and *Shropshirea*.
Genus *Prosopolepis*, with nine subgenera – *Calladimyia*, *Decamyia*, *Dinomyia*, *Eunicemyia*, *Heliconiamyia*, *Hystatomyia*, *Janicemyia*, *Prosopolepis* and *Triamyia*.
Genus *Wyeomyia*, with four subgenera – *Dyarina*, *Pentemyia*, *Phyllozomyia* and *Wyeomyia*.

Dyar (1928) retained the classification of Dyar & Shannon, with a few notable changes. *Prosopolepis*, which was recognized as a genus by Dyar & Shannon, was replaced with the genus *Dendromyia*, which was not treated by Dyar & Shannon. *Prosopolepis* was retained as a questionable subgenus of *Dendromyia*, and *Melanolepis* and *Triamyia* (neither treated by Dyar & Shannon) were included as subgenera along with the other previously recognized subgenera. *Dodecamyia* was transferred from *Miamyia* to *Wyeomyia*, *Nunezia* (newly described) was included as a subgenus of *Wyeomyia*, the genus *Technicomyia* (previously valid) was included as a synonym of *Miamyia*, and the genera *Diphalangarpe* and *Phoniomyia* (also previously valid) were synonymized with *Wyeomyia*. Thus, Dyar's modified classification of the Wyeomyia Group consisted of four genera and 20 subgenera, as follows:

Genus *Dendromyia*, with 11 subgenera – *Calladimyia*, *Decamyia*, *Dendromyia*, *Dinomyia*, *Eunicemyia*, *Heliconiamyia*, *Hystatomyia*, *Janicemyia*, *Melanolepis*, *Prosopolepis*? and *Triamyia*.
Genus *Menolepis*, represented by *Wy. leucostigma*.
Genus *Miamyia*, with three subgenera – *Cleobonnea*, *Miamyia* and *Shropshirea*.
Genus *Wyeomyia*, with six subgenera – *Dodecamyia*, *Dyarina*, *Nunezia*, *Pentemyia*, *Phyllozomyia* and *Wyeomyia*.

Edwards (1932a) combined the Sabethes and Wyeomyia Groups of previous authors to form a single '*Sabethes* group', equivalent to the current concept of Sabethini, and merged the four genera recognized by Dyar (1928) into a single genus *Wyeomyia*. He retained *Dendromyia* and *Menolepis* as subgenera, revalidated *Phoniomyia* as a subgenus and treated *Technicomyia* (not treated by Dyar, 1928) as a synonym of *Wyeomyia*. The four subgenera and the genus-group names regarded as synonyms are as follows:

Subgenus *Dendromyia*, synonyms: *Calladimyia*, *Cleobonnea*, *Decamyia*, *Dinomyia*, *Eunicemyia*, *Heliconiamyia*, *Hystatomyia*, *Janicemyia*, *Melanolepis*, *Prosopolepis*, *Shropshirea* and *Triamyia*.
Subgenus *Menolepis*: monobasic.
Subgenus *Phoniomyia*, synonym: *Dyarina*.
Subgenus *Wyeomyia*, synonyms: *Diphalangarpe*, *Dodecamyia*, *Miamyia*, *Nunezia*, *Pentemyia*, *Phyllozomyia* and *Technicomyia*.

Lane & Cerqueira (1942), in their reclassification of the New World Sabethini, made some marked changes to Edwards's (1932a) classification of the genus. In brief, they retained subgenera *Dendromyia*, *Menolepis* and *Wyeomyia*, recognized *Nunezia* as a subgenus, described three new subgenera, *Antunesmyia*, *Cruzmyia* and *Davismyia*, and elevated *Phoniomyia* to its original generic rank. They subdivided the subgenus *Dendromyia* into three series, the Cleobonnea, Dendromyia and Prosopolepis Series, and divided the subgenus *Wyeomyia* into two series, the Pertinans and Serrata Series. This supraspecific classification of the genus was retained by Lane (1953), although he failed to explicitly mention the Serrata Series, and it was accepted by mosquito workers (Stone *et al.*, 1959; Knight & Stone, 1977) until gradual changes in the classification began with the introduction of the subgenus *Caenomyiella* Harbach & Peyton, 1990. Subsequent studies resulted in (1) the transfer of the subgenus *Davismyia* to the genus *Sabethes* by Harbach & Peyton (1991); (2) the introduction of the subgenus *Exallomyia* Harbach & Peyton, 1992a; (3) restriction of *Dendromyia* to a monophyletic group of seven species (Motta & Lourenço-de-Oliveira, 1995, 2000); (4) return of *Phoniomyia* and concomitant elevation of *Hystatomyia* to subgeneric status (Judd, 1998); (5) restriction and validation of *Prosopolepis* as a monobasic subgenus (Lourenço-de-Oliveira *et al.*, 1999); (6) removal of three species from *Dendromyia* to the new genus *Onirion* Peyton & Harbach, 2000; and (7) the introduction of the subgenus *Spilonympha* Motta & Lourenço-de-Oliveira, 2005.

Except for the discovery and description of a few new species, the 15 years between 1990 and 2005 were the last years since significant contributions were made to the supraspecific classification of *Wyeomyia*. At the present time (August 2024), the genus includes 141 species and 17 subgenera, but 27 species are without subgeneric placement.

Wyeomyia Theobald, 1901 *sensu stricto*

Wyeomyia sensu stricto was established as a genus in an annotated list of mosquito genera (Theobald, 1901a) and formally described later in the same year (Theobald, 1901c). The type species, *Wyeomyia grayii* Theobald, 1901c, was subsequently designated by Neveu-Lemaire (1902c). The species of *Wyeomyia sensu stricto* are generally poorly known and their relationships cannot be determined without a thorough revision. One study, however, has contributed to the internal classification of the group. Belkin *et al.* (1970) established four groups for the species of *Wyeomyia* that occur in Jamaica, the Caracula, Hirsuta, Pertinans and Vanduzeei Groups, and recognized three subgroups within the Hirsuta Group, the Hirsuta, Luna and Stellata Subgroups. How these groups and subgroups relate to the Pertinans and Serrata Series of Lane & Cerqueira (1942) and Lane (1953) is unknown.

Antunesmyia Lane & Cerqueira, 1942

Antunesmyia was introduced as a subgenus of *Wyeomyia* Theobald, 1901a by Lane & Cerqueira (1942), with *Wy. rooti* Lane & Cerqueira, 1942 as the originally designated type species, which was preoccupied by *Wy. rooti* (del Ponte, 1939) [original combination: *Dendromyia rooti*] and replaced with *Wy. alani* by Lane & Cerqueira, 1957. As currently defined, *Antunesmyia* is a homogeneous group comprised of only three currently known species.

Caenomyiella Harbach & Peyton, 1990

Caenomyiella was established as a subgenus of *Wyeomyia* Theobald, 1901a by Harbach & Peyton (1990), with *Sabethes fernandezyepezi* Cova Garcia, Sutil Oramas and Pulido F., 1974b as the originally designated type species. The authors included an undescribed species in the subgenus based on features of the larval and pupal stages: *Wyeomyia* sp. 69 from Colombia, exuviae on microscope slide no. CV600-1, deposited in the National Museum of Natural History, Smithsonian Institution, Washington, DC, USA. The adults of the two species could not be compared because the female of *Wy. fernandezyepezi* and both sexes of *Wyeomyia* sp. 69 are unknown.

Cruzmyia Lane & Cerqueira, 1942

Cruzmyia was introduced as a subgenus of *Wyeomyia* Theobald, 1901a by Lane & Cerqueira (1942), with *Wy. dyari* Lane & Cerqueira, 1942 as the originally designated type species. The subgenus also originally included *Wy. flavifacies* Edwards, 1922d and *Wy. kummi* Lane & Cerqueira, 1942. Lane (1953), without explanation, transferred *Wy. flavifacies* to *Antunesmyia* Lane & Cerqueira, 1942 and described *Wy. mattinglyi* as a new species of *Cruzmyia*. With the addition of *Wy. forattinii* Clastrier, 1974, *Cruzmyia* currently includes four species, which appear to comprise a monophyletic group (Motta *et al.*, 2007).

Decamyia Dyar, 1919

Decamyia was originally introduced as a distinct genus by Dyar (1919b), with *Wy. onidus* Dyar & Knab, 1909 [subjective synonym of *Wy. ulocoma* (Theobald, 1903a)] as the originally designated type species. The genus originally included the type species, *Wy. eloisa* Howard, Dyar & Knab, 1913 and *Wy. pseudopecten* Dyar & Knab, 1906a. *Decamyia* was treated as a subgenus of *Prosopolepis* Lutz, 1905 by Dyar & Shannon (1924b) to contain *Wy. eloisa*, *Wy. pseudopecten* and *Wy. ulocoma*. Four years later, Dyar (1928) recognized *Decamyia* as a subgenus of *Dendromyia* Theobald, 1903a, in which he included *Wy. eloisa*, *Wy. pseudopecten*, *Dendromyia felicia* Dyar & Núñez Tovar, 1927b and *Dendromyia ulocoma* Theobald, 1903a. [It is not clear whether *Wy. onidus* was first recognized as a synonym of *ulocoma* by Dyar & Shannon (1924b) or Dyar (1928).] *Decamyia* became a synonym of *Dendromyia* when Edwards (1932a) recognized this nominal genus group as a subgenus of *Wyeomyia* Theobald, 1901a. *Decamyia* remained a synonym of *Dendromyia* for the next 58 years (Lane & Cerqueira, 1942; Lane, 1953; Stone *et al.*, 1959; Knight & Stone, 1977) until it was raised to subgeneric rank in *Wyeomyia* by Harbach & Peyton (1990) 'to help stabilize the internal classification of the genus and to facilitate the comparison of *Decamyia*' with the newly recognized subgenus *Caenomyiella* Harbach & Peyton, 1990. *Wyeomyia rorotai* Senevet, Chabelard & Abonnenc, 1942, originally described as a species of the subgenus *Dendromyia*, was redescribed by Ribeiro *et al.* (2020) and transferred to *Decamyia*. With the addition of *Wy. anthica* Ribeiro, Pavan, da Silva, Galvão, Lourenço-de-Oliveira & Motta, 2021, *Decamyia*, as currently defined, includes six distinct species.

Dendromyia Theobald, 1903

Dendromyia was originally introduced as a distinct genus by Theobald (1903a), with *Wyeomyia luteoventralis* Theobald, 1901b subsequently designated as the type species by Blanchard (1905). Dyar & Shannon (1924b) recognized *Menolepis* Lutz, 1905, *Miamyia* Dyar, 1919b, *Prosopolepis* Lutz, 1905 and *Wyeomyia* Theobald, 1901a as a group of genera, the Wyeomyia Group. They divided the four genera into 15 subgenera based on features of the male genitalia. They did not mention *Dendromyia* except to say that they were unable to place *Wy. luteoventralis*, the type species, in any of the subgenera. Dyar (1928) recognized *Prosopolepis* as a synonym of the genus *Dendromyia* and retained *Menolepis*, *Miamyia* and *Wyeomyia* as separate genera. He divided the four genera into 17 subgenera, also based on features of the male genitalia. Edwards (1932a) disagreed with the recognition of numerous groups and merged them into a single genus, the genus *Wyeomyia*, comprised of four subgenera, the nominotypical subgenus, *Dendromyia*, *Menolepis* and *Phoniomyia* Theobald, 1903a, which he resurrected from synonymy with *Wyeomyia*. *Dendromyia* was a wastebasket taxon, a dumping ground for many unrelated species (Lane & Cerqueira, 1942; Lane, 1953; Judd, 1996), until it was restricted to six species, the type species, *Wy. complosa* (Dyar, 1928), *Wy. jocosa* (Dyar & Knab, 1908), *Wy. testei* Senevet & Abonnenc, 1939b, *Wy. trifurcata* Clastrier, 1973 and *Wy. ypsipola* Dyar, 1922c, based on shared morphological features of the adult, larval and pupal stages (Motta & Lourenço-de-Oliveira, 1995).

Dodecamyia Dyar, 1919

Dodecamyia was originally introduced as a distinct genus by Dyar (1919b), with *Wyeomyia aphobema* Dyar, 1918c as the type and only originally included species. Dyar & Shannon (1924b) treated *Dodecamyia* as a subgenus of *Miamyia* Dyar, 1919b and Dyar (1928) recognized it as a subgenus of *Wyeomyia* Theobald, 1901a. Edwards (1932a) incorporated the groups treated as separate genera by Dyar into four subgenera of the genus *Wyeomyia*, with *Dodecamyia* placed in synonymy with the nominotypical subgenus. It was retained as a synonym of *Wyeomyia* by later investigators until it was recognized and revalidated as a monobasic subgenus of *Wyeomyia* by Heinemann & Belkin (1978b) and Peyton *et al.* (1983), respectively. Subsequent morphological and molecular studies (Judd, 1996; Motta & Lourenço-de-Oliveira, 2005; Motta *et al.*, 2007) have corroborated the recognition of *Dodecamyia* as a distinct monobasic taxon.

Exallomyia Harbach & Peyton, 1992

Exallomyia was established as a distinct subgenus of *Wyeomyia* Theobald, 1901a by Harbach & Peyton (1992a), with *Wy. tarsata* Lane & Cerqueira, 1942 designated as the type species. *Exallomyia* is an easily distinguishable group comprised of three species, *Wy. arborea* Galindo, Carpenter & Trapido, 1951, *Wy. tarsata* and *Wy. carrilloi* (Sutil O. & Pulido F., 1978). *Wyeomyia arborea* was previously placed in the subgenus *Davismyia* Lane & Cerqueira, 1942 before *Davismyia* was transferred from *Wyeomyia* to

Sabethes by Harbach & Peyton (1991), *Wy. tarsata* was previously included in the subgenus *Dendromyia* Theobald, 1903a and *Sa. carrilloi* was originally described as a species of the nominate subgenus of *Sabethes* Robineau-Desvoidy, 1827.

Hystatomyia Dyar, 1919

Hystatomyia was introduced as a distinct genus by Dyar (1919b), with *Wyeomyia circumcincta* Dyar & Knab, 1907c as the originally designated type species. Without explanation, Dyar (1923e) included *Hystatomyia* as a subgenus of *Wyeomyia* Theobald, 1901a. Thereafter, Dyar & Shannon (1924b) treated *Hystatomyia* as a subgenus of the genus *Prosopolepis* Lutz, 1905, which they considered to be one of four genera of a composite Wyeomyia Group of New World genera of sabethine mosquitoes. Dyar & Shannon's concept of *Hystatomyia* included five species, *Wy. autocratica* Dyar & Knab, 1906d, *Wy. circumcincta*, *Wy. coenonus* Howard, Dyar & Knab, 1913, *Wy. intonca* Dyar & Knab, 1910a and *Wy. lamellata* (Bonne-Wepster & Bonne, 1920). Dyar (1928) retained the generic classification of Dyar & Shannon, but replaced *Prosopolepis* with *Dendromyia* Theobald, 1903a, which Dyar & Shannon did not include in their classification because they were not able to place the type species in one or other of the four genera they recognized. Consequently, Dyar (1928) included *Hystatomyia* as a subgenus of *Dendromyia*. Edwards (1932a) merged the four genera recognized by Dyar (1928) into a composite genus *Wyeomyia*, with *Dendromyia* included as one of four subgenera and *Hystatomyia* as one of its 12 synonyms. Lane & Cerqueira (1942) made major changes to the classification of Edwards (1932a), but retained *Hystatomyia* as a synonym of *Dendromyia*. This was followed by Lane (1953) and perpetuated until *Hystatomyia* was returned to subgeneric status by Judd (1998) based on phylogenetic analyses of morphological data (Judd, 1996, 1998). Judd resurrected *Hystatomyia* from synonymy to include seven species, the five species included in the subgenus by Dyar & Shannon (1924b) and two species which she transferred from *Phoniomyia* Theobald, 1903a, *Wy. lopezii* Cova Garcia, Sutil Oramas & Pulido F., 1974a and *Wy. esmeraldasi* (Levi-Castillo, 1955a). Two subsequently described species, *Wy. baltae* Porter, 2009 and *Wy. chocoensis* Porter & Wolff E., 2004, were added to the group. Judd (1996) recognized two distinct groups within *Hystatomyia* based on characteristics of the male genitalia, one represented by *Wy. autocratica* and the other by *Wy. circumcincta*, herein explicitly named the Autocratica and Circumcincta Groups, respectively. As noted by Porter & Wolff E. (2004) and Porter (2009), the Autocratica Group includes species found in the eastern foothills of the Andes and adjacent Amazon basin (*Wy. autocratica*, *Wy. baltae*, *Wy. lamellata*, *Wy. lopezii*) and the Circumcincta Group comprises species that occur in Meso-America and humid lowlands west of the Andes (*Wy. circumcincta*, *Wy. chocoensis*, *Wy. coenonus*, *Wy intonca*, *Wy esmeraldasi*).

Menolepis Lutz, 1905

Menolepis was originally established as a distinct genus by Lutz (1905), with *Wyeomyia leucostigma* Lutz (1904) (in Bourroul, 1904) as the type and only originally included species. *Menolepis* continued to be treated as a monobasic genus (Dyar & Shannon, 1924b; Dyar, 1928) until Edwards (1932a) combined the Sabethes and Wyeomyia Groups of previous authors and reduced the four genera recognized by Dyar (1928), *Dendromyia* Theobald, 1903a, *Menolepis*, *Miamyia* Dyar, 1919b and *Wyeomyia* Theobald, 1901a, to subgeneric rank in the genus *Wyeomyia*. *Menolepis* has continued to be recognized as a monobasic subgenus of *Wyeomyia* since Edwards (1932a). The results of a phylogenetic analysis of morphological data by Motta *et al.* (2007) indicate that *Menolepis* has no close relatives; thus, it remains a monobasic subgeneric taxon.

Miamyia Dyar, 1919

Miamyia was introduced as a genus by Dyar (1919b), with *Wyeomyia symmachus* Dyar & Knab, 1909 [subjective synonym of *Wy. hosautos* Dyar & Knab, 1907c] as the originally designated type species. Dyar & Shannon (1924b) recognized *Miamyia* as having four subgenera: *Cleobonnea* Dyar, 1919b, *Dodecamyia* Dyar, 1919b, *Miamyia* and *Shropshirea* Dyar, 1922c. Dyar (1928) transferred *Dodecamyia* to *Wyeomyia* Theobald, 1901a and retained *Miamyia* as a genus consisting of the subgenera *Cleobonnea*, *Miamyia* and *Shropshirea*. Edwards (1932a) disagreed with the classification of Dyar (1928) and combined his genus groups into a single genus *Wyeomyia*, consisting of the subgenera *Dendromyia* Theobald, 1903a, *Menolepis* Lutz, 1905, *Phoniomyia* Theobald, 1903a and *Wyeomyia*. He synonymized *Cleobonnea* and *Shropshirea* with *Dendromyia* and recognized *Miamyia* as a synonym of *Wyeomyia*. *Miamyia* remained a synonym of *Wyeomyia* until it was elevated to subgeneric status by Motta *et al.* (2007). As currently defined, *Miamyia* includes eight species that are not divided into groups.

Nunezia Dyar, 1928

Nunezia was established as a subgenus of *Wyeomyia* Theobald, 1901a by Dyar (1928) for *Dendromyia* (*Eunicemyia*) *bicornis* Root, 1928 (in Dyar, 1928), the designated type species. *Nunezia* currently only includes three species: *Wy. bicornis*, *Wy.* (*Dyarina*) *lateralis* Petrocchi, 1927 (in Shannon & del Ponte, 1927) and *Wy. paucartamboensis* Porter, 2014. *Wyeomyia bicornis* was placed in *Nunezia* by Lane & Cerqueira (1942) and *Wy. paucartamboensis* Porter, 2014 was expressly described as a species of the subgenus.

Phoniomyia Theobald, 1903

Phoniomyia was introduced as a distinct genus by Theobald (1903a), with *irostris* Theobald, 1901c subsequently designated as the type species by Dyar (1905a). *Phoniomyia* continued to be recognized as a genus until Dyar (1928) listed it as a synonym of *Wyeomyia* Theobald, 1901a without explanation (but see Dyar, 1924c). Six years later, Edwards (1932a) merged the four taxa recognized as genera by Dyar (1928) into a single genus *Wyeomyia* and resurrected *Phoniomyia* from synonymy to serve as one of its subgenera. A decade later, Lane & Cerqueira (1942) returned *Phoniomyia* to its original generic rank, and it continued to be recognized as a separate genus (Lane, 1953; Correa & Ramalho, 1956; Stone *et al.*, 1959; Knight & Stone, 1977) until Judd (1998) returned it to its former rank as a subgenus of *Wyeomyia* based on phylogenetic analyses of morphological data. [Note that Judd (1996, 1998) confused the taxonomic actions of Dyar (1928) and Edwards (1932a). She incorrectly attributed the synonymy of *Phoniomyia* with *Wyeomyia* to Edwards (1932a) (Judd, 1996; table 1) and credited its recognition as a subgenus of *Wyeomyia* to Dyar (1928), '... in keeping with the classification of Dyar (1928), *Phoniomyia* is reduced to a subgenus of *Wyeomyia*' (Judd, 1998).] *Phoniomyia* was recovered as a lineage outside *Wyeomyia* in the phylogenetic study of Motta *et al.* (2007), albeit with very weak support, but even so it would seem additional study is needed to definitively resolve its status as a subgenus or a separate genus. The internal classification of *Phoniomyia* has only been addressed by Lane & Cerqueira (1942), who divided it into three series of species based on the development of the gonostylus of the male genitalia: series I, gonostylus absent (actually present, highly reduced and unapparent, see Judd, 1998); series II, gonostylus simple; series III, gonostylus three-branched.

Prosopolepis Lutz, 1905

Prosopolepis was established as a genus by Lutz (1905), with *Prosopolepis confusa* Lutz, 1905 as the type and only originally included species. *Prosopolepis* continued to be recognized as a genus until it was included as a questionable subgenus of the genus *Dendromyia* Theobald, 1903a by Dyar (1928). Edwards (1932a) recognized *Dendromyia* as one of four subgenera of the genus *Wyeomyia* Theobald, 1901a and placed *Prosopolepis* in synonymy with *Dendromyia*. Lane & Cerqueira (1942) retained *Prosopolepis* as a synonym of the subgenus *Dendromyia* but used it as a vernacular name for a group of five species, Series Prosopolepis, which included *Dendromyia complosa* Dyar, 1928, *Prosopolepis confusa*, *Prosopolepis flui* Bonne-Wepster & Bonne, 1920, *Prosopolepis jocosa* Dyar & Knab, 1908 and *Wyeomyia kerri* del Ponte & Cerqueira, 1938 (all as species of *Wyeomyia*). Lane (1951) formally synonymized *Prosopolepis flui* with *Prosopolepis confusa*; thus, Lane (1953) recognized the Series Prosopolepis as a group of four species. *Prosopolepis* remained a synonym of *Dendromyia* until Motta & Lourenço-de-Oliveira (1995, 2000) redefined the subgenus to include only six species. All other species previously included in *Dendromyia*, including the type species of *Prosopolepis*, were left without subgeneric placement. Consequently, *Prosopolepis* became a synonym of *Wyeomyia*. Four years later, Lourenço-de-Oliveira *et al.* (1999) validated *Prosopolepis* as a monotypic subgenus of *Wyeomyia*, its current status, and elevated *Prosopolepis flui* from synonymy with *Prosopolepis confusa* and recognized *Wy. kerri* as its junior synonym.

Spilonympha Motta & Lourenço-de-Oliveira, 2005

Spilonympha was introduced as a distinct subgenus of *Wyeomyia* Theobald, 1901a by Motta & Lourenço-de-Oliveira (2005), with a new species, *Wyeomyia aningae* Motta & Lourenço-de-Oliveira, 2005, as the designated type species. The subgenus includes seven species, the type species and six other species that were excluded from *Dendromyia* by Motta & Lourenço-de-Oliveira (1995). The integrity of *Spilonympha* is supported by its recovery as a monophyletic lineage in the morphology-based phylogenetic study of Motta *et al.* (2007).

Triamyia Dyar, 1919

Triamyia was originally established as a distinct genus by Dyar (1919b), with *Wyeomyia aporonoma* Dyar & Knab, 1906d as the originally designated type species. *Triamyia* existed as a genus until Dyar (1928) reclassified it as a subgenus of *Dendromyia* Theobald, 1903a, one of four genera which Dyar included in his concept of the Wyeomyia Group of genera. Edwards (1932a) disagreed with the recognition of four genera and combined them to recognize a single genus, *Wyeomyia* Theobald, 1901a, consisting of four subgenera: *Dendromyia* Theobald, 1903a, *Menolepis* Lutz, 1905, *Phoniomyia* Theobald, 1903a and *Wyeomyia*. He placed *Triamyia* in synonymy with *Dendromyia*, where it remained until it was resurrected from synonymy and established as a subgenus of *Wyeomyia* by Motta *et al.* (2007) to include *Wy. aporonoma* and *Wy. staminifera* Lourenço-de-Oliveira, Motta & de Castro, 1992 based on phylogenetic analyses of morphological and allozyme data.

Zinzala Zavortink, 1986

Zinzala was introduced as a distinct subgenus of *Wyeomyia* Theobald, 1901a by Zavortink (1986a) for two newly recognized, 'remarkably different' species, *Wy. fishi* Zavortink, 1986 and *Wy. zinzala* Zavortink, 1986, the designated type species. In addition to being morphologically distinct, *Wy. fishi* and *Wy. zinzala* are the only Neotropical species whose immature stages are known to develop in pitcher plants.

Toxorhynchitini Lahille, 1904

Toxorhynchites Theobald, 1901

The generic name *Toxorhynchites* was first published by Howard (1901, 1 June), who used the name based on seeing 'proof sheets' of Theobald's (1901b) volume 1 of *A Monograph of the Culicidae*, published 23 November. Apparently unknown to Howard,

Theobald (1901a, 15 July) had briefly described the genus *Toxorhynchites* prior to describing it in more detail in his monograph for a single new species, *Tx. brevipalpis* Theobald, 1901b. In addition to complicating the authorship and availability of the name *Toxorhynchites*, Howard and Theobald, as well as later authors, also used *Megarhinus* Robineau-Desvoidy, 1827 for species now included in *Toxorhynchites*. In fact, since Howard had illustrated the female of *Megarhinus rutilus* Coquillett, 1896 as *Toxorhynchites rutilus*, some authors assumed that *Megarhinus rutilus* must be accepted as the type species of *Toxorhynchites*, in which case *Toxorhynchites* would become a synonym of *Megarhinus*. This issue was discussed by Edwards (1912b), who concluded that the name *Toxorhynchites* should be accepted as validly established by Howard (1901), apparently disregarding the nomenclatural priority of *Megarhinus* due to its recognition as a separate genus. Indeed, continued acceptance of *Megarhinus* as a valid genus eventually resulted in the recognition of *Toxorhynchites* as a junior synonym (Howard *et al.*, 1917; Edwards, 1932a). However, because of persisting uncertainty about the authorship and validity of *Toxorhynchites*, both *Megarhinus* and *Toxorhynchites* continued to be used by various authors. As revealed in a listing of genus-group names (Neave, 1940) and expounded by Stone (1948), the name *Megarhinus* Robineau-Desvoidy, 1827 is not an available genus-group name for mosquitoes because it is preoccupied by *Megarhinus* Rafinesque, 1820 (obsolete genus of paddlefish) and *Toxorhynchites* is the first available replacement name. The problem of authorship and validity of *Toxorhynchites* was definitively resolved when the International Commission on Zoological Nomenclature (1959b) ruled in favour of an application by Stone & Knight (1957b) to supress the name *Toxorhynchites* published by Howard (1901) and to validate *Toxorhynchites* Theobald, 1901a, with *Tx. brevipalpis* Theobald, 1901b as the type species.

Toxorhynchites Theobald, 1901 _sensu stricto_

Howard *et al.* (1917) recognized three groups of species within *Toxorhynchites* (as *Megarhinus*) based on differences in the maxillary palpus of females and geographic occurrence. Edwards (1932a) accepted these groups and identified them as 'Group A (*Megarhinus*)' (= *Lynchiella* Lahille, 1904), 'Group B (*Ankylorhynchus*)' and 'Group C (*Toxorhynchites*)'. The third division (Group C) included all Old World species of the genus. Edwards (1941) divided the Afrotropical species of this group into two groups, 'Group A (*brevipalpis* group)' and 'Group B (*lutescens* group)'. The first group (Brevipalpis Group) was further divided into two subgroups by Ribeiro (1991), the monobasic Brevipalpis Subgroup ('*brevipalpis* subgroup') and the Phytophagus Subgroup (misspelled as '*phytophygus* subgroup'), which currently includes eight species. The latter group (Group B) was later formally recognized as a distinct genus-group taxon, the subgenus *Afrorhynchus* Ribeiro, 1992a. The current division of the Old World *Toxorhynchites* into groups and subgroups is ascribable to the numerical phenetic, numerical phylogenetic and traditional taxonomic studies of Moss *et al.* (1979), Simon *et al.* (1983) and Steffan & Evenhuis (1985), respectively. As a result of those studies, the species of *Toxorhynchites* that occur in the Australasian, eastern Palaearctic and Oriental Regions are classified in seven groups, the Acaudatus (six species), Ater (three species), Christophi (nine species), Leicesteri (seven species), Magnificus (four species), Minimus (two species) and Splendens (six species) Groups. Two subgroups are recognized within the Acaudatus Group, the Acaudatus (five species) and monobasic Nepenthis Subgroups, and two are recognized within the Leicesteri Group, the Leicesteri (five species) and Manicatus (two species) Subgroups.

Afrorhynchus Ribeiro, 1992

Afrorhynchus was introduced as a distinct subgenus of *Toxorhynchites* Theobald, 1901a by Ribeiro (1992a), with *Megarhinus lutescens* Theobald, 1901b as the originally designated type species. The subgenus was established for the Afrotropical species previously included in 'Group B (*lutescens* group)' of Edwards (1941) – the updated '*lutescens* group' of Service (1990) and Ribeiro (1992b, 1992c). *Afrorhynchus* currently includes 19 species. Ribeiro (2004) recognized that six species endemic to Madagascar form a distinct group, which he named the Pauliani Group.

Ankylorhynchus Lutz, 1904

Ankylorhynchus was originally established as a genus by Lutz, 1904 (in Bourroul, 1904) for three nominal species: *Megarhinus purpureus* Theobald, 1901b, *Culex trichopygus* Wiedemann, 1828 and *Ankylorhynchus neglectus* Lutz, 1904 (in Bourroul, 1904), which was later synonymized with *trichopygus* by Dyar (1928). *Megarhinus purpureus* is regarded as the type species of *Ankylorhynchus* based on the recommendation of Stone (1961) that it 'should be considered' the type species, but its type status has not been formally established. Edwards (1932a), in agreement with Howard *et al.* (1917), divided *Toxorhynchites* (as *Megarhinus*) into three groups, and recognized *Ankylorhynchus* as an informal group taxon, 'Group B (*Ankylorhynchus*)', despite knowing that *Ankylorhynchus* had been treated as a subgenus of *Megarhinus* by Dyar (1928). Martini (1931), unbeknownst to Edwards, agreed with the subgeneric status of *Ankylorhynchus* when he described *Megarhinus hexacis* as a member of the subgenus. In fact, with one exception, *Ankylorhynchus* has been recognized as a subgenus in all works published on Neotropical toxorhynchitines after Edwards (1932a). The one exception is da Costa Lima *et al.* (1962), who treated *Ankylorhynchus* as a genus. As currently recognized, *Ankylorhynchus* includes four South American species.

Lynchiella Lahille, 1904

Lynchiella was established by Lahille (1904) as a replacement name for the genus *Megarhinus* Robineau-Desvoidy, 1827, preoccupied by *Megarhinus* Rafinesque, 1820 (obsolete genus of paddlefish), but its status as a replacement name was overlooked until it was brought to light by Neave (1940) and categorically recounted by Stone (1948). Prior to Stone (1948), *Lynchiella* was treated as a synonym of *Megarhinus*, apparently synonymized by Coquillett (1906d). Howard *et al.* (1917) recognized three divisions within the genus *Megarhinus* based on differences in the maxillary palpus of females and geographic areas. Edwards (1932a) accepted these divisions and recognized the '*Megarhinus* proper' division of Howard *et al.* as 'Group A (*Megarhinus*)' (= *Lynchiella*). With the invalidation of *Megarhinus*, *Toxorhynchites*, by virtue of being the first available replacement name, became the universally accepted name of the genus. As a consequence, *Lynchiella* was formally recognized as a subgenus of *Toxorhynchites* by Lane (1953). *Culex haemorrhoidalis* Fabricius, 1787, the type species of *Megarhinus* by virtue of being the only originally included species, was automatically fixed as the type species of *Lynchiella*. As currently defined, *Lynchiella* includes 20 species that are not divided into groups.

Uranotaeniini Lahille, 1904

Uranotaenia Lynch Arribálzaga, 1891

Uranotaenia Lynch Arribálzaga, 1904 *sensu stricto*

Uranotaenia was introduced as a genus by Lynch Arribálzaga in 1891, first in a key to the genera of mosquitoes recognized at the time (Lynch Arribálzaga, 1891a) and subsequently described (Lynch Arribálzaga, 1891b) based on the adult females of two newly described species, *Ur. nataliae* Lynch Arribálzaga, 1891b and *Ur. pulcherrima* Lynch Arribálzaga, 1891b. The latter species was subsequently designated the type species of *Uranotaenia* by Neveu-Lemaire (1902c). The taxonomic identity of *Uranotaenia* was obscured with the synonymy of *Pseudoficalbia* Theobald, 1912b by Edwards (1912b) until Edwards (1932a) recognized *Uranotaenia sensu stricto* and *Pseudoficalbia* as two informal groups of the genus, 'Group A (*Uranotaenia*)' and 'Group B (*Pseudoficalbia*)', respectively. However, *Uranotaenia sensu stricto* was not recognized as a separate genus-group taxon until *Pseudoficalbia* was removed from synonymy and afforded formal subgeneric rank by Peyton (1972). Coincidentally, most of the current internal classification of *Uranotaenia sensu stricto* is the result of taxonomic work completed after the recognition of *Pseudoficalbia* as a separate genus-group taxon. The current classification includes eight 'sections' and 11 'series', nine of which, the Oriental Campestris, Micans, Testacea, Trilineata and Unimaculiala Series (unpublished) and the Neotropical Leucoptera, Lowii, Pallidoventer and Pulcherrima Series (Galindo *et al.*, 1954), are not recognized as subordinate to sections. The other two 'series' are subordinate groups of the Anisocheleomyia Section (Belkin, 1962) comprised of species in the Australasian and Oriental Regions. The section also includes two 'series', the Ludlowae and Nivea Series (unpublished), the unplaced Argyrotarsis and Albescens Complexes (Belkin, 1962) and the unplaced Tibialis Group (King & Hoogstraal, 1947a), which includes the Novaguinensis Complex (Peters, 1963d; Lee *et al.*, 1989c). The monobasic Wysockii Section (Belkin, 1962) is the only other informal taxon that is currently recognized for an Australasian mosquito. The other six 'sections' are the major divisions of the Afrotropical *Uranotaenia sensu stricto*. These include the Alba, Caeruleocephala, Chorleyi and Hopkinsi Sections established by da Cunha Ramos (1993) and the Anopheloides and Dumonti Sections recognized by da Cunha Ramos & Brunhes (2004). Da Cunha Ramos (1993) divided the Alba Section into three species groups, the Alba, Balfouri and Bilineata Groups. She initially only recognized a single 'Group' within each of the other five sections, each having the same vernacular specific name as the section. In the subsequent collaborative study (da Cunha Ramos & Brunhes, 2004), a second group was distinguished for species of the Anopheloides, Caeruleocephala and Chorleyi Sections, the Neireti, Madagascarica and Hamoni Groups, respectively.

Pseudoficalbia Theobald, 1912

Pseudoficalbia was originally proposed as a distinct genus by Theobald (1912b), with the new species *Pseudoficalbia pandani* Theobald, 1912b subsequently designated as the type species by Howard *et al.* (1917). Edwards (1912b) synonymized *Pseudoficalbia* with *Uranotaenia* Arribálzaga (1891a), stating that in his 'opinion the name [*Pseudoficalbia*] should not be retained even as a subgenus, as there seems no important structural character to distinguish these species from the rest.' Edwards apparently was not certain of this and vacillated between subgeneric and informal group status. In a list of Afrotropical species (in Schwetz, 1927), Edwards treated *Pseudoficalbia* as a subgenus of *Uranotaenia*, then changed his mind (Edwards, 1932a) and recognized *Pseudoficalbia* as one of two informal groups within the genus, 'Group A (*Uranotaenia*)' and 'Group B (*Pseudoficalbia*)'. In his subsequent study of Afrotropical mosquitoes, Edwards (1941) replaced the two groups with four groups, designated groups A, B, C and D. *Pseudoficalbia* remained in synonymy until it was formally afforded subgeneric rank by Peyton (1972), who reviewed the taxonomic history of *Uranotaenia* and divided the subgenus *Pseudoficalbia* into Sections A and B, based principally on knowledge of species that occur in Southeast Asia. Peyton admitted he had considered raising *Pseudoficalbia* to generic rank based on the absence of an anapleural suture between the mesokatepisternum and the prealar area of adults and its presence in *Uranotaenia sensu stricto*. He also suggested that Section B could be accorded subgeneric status based on 'distinctive adult differences and, to a lesser extent, differences in other stages', but refrained from doing so. Section B encompassed Groups C and D of Edwards (1941). He divided Section B into two series, the Annulata and Maxima Series, for species in the Afrotropical and Oriental Regions, respectively, based on correlated anatomical features of the adults, male genitalia, larvae and pupae. Finally, he considered that the Afrotropical Group D of Edwards (1941) could be viewed as a separate subgenus, but elected instead to recognize it as the '*shillitonis* series' of Section A. By way of implication, Peyton's Section A embodied the species included in

Edwards's Groups A and B. Peyton (1977) further revised the *Pseudoficalbia* of Southeast Asia to include five groups (series) within Section A, correlated with the faunal regions defined by Belkin (1962), the Bicolor (Ethiopian, Indomalayan, Malagasy, North Australian, Oriental, Papuan and South Pacific), Bimaculata (Indomalayan, Oriental and Palaearctic), Harrisoni (Indomalayan), Maxima (Indomalayan, Nearctic, Oriental and Palaearctic) and Recondita Series (Indomalayan, North Australian, Oriental and Papuan). The internal classification of *Pseudoficalbia* established by Peyton (1972, 1977) served as the foundation for the revisionary study of the Afrotropical species completed by da Cunha Ramos (1993). Da Cunha Ramos divided Section A into three sections, the Mashonaensis, Nigripes and Shillitonis Sections, the last one equivalent to the '*shillitonis* series' of Peyton, and renamed Section B the Annulata Section. She divided the sections into subordinate groups based mainly on morphological characters of adults and male genitalia, and to a lesser extent on features of larvae and pupae. She recognized a single species group, the Mashonaensis Group, for the Mashonaensis Section, and a single group, Shillitonis Group with Subgroups A and B, for the Shillitonis Section, and divided the Nigripes Section into three species groups, the Fusca, Nigripes and Ornata Groups. In agreement with Peyton (1972), she implicitly recognized the Annulata Series of Section B, her Annulata Section, as the Annulata Group, with Subgroups A and B. The discovery and description of new species of Afrotropical *Pseudoficalbia* by da Cunha Ramos & Brunhes (2004) resulted in the refinement of the Annulata Series to include a second species group, the Lavieri Group, addition of the monobasic Bicincta Group to the Shillitonis Section and the introduction of a new Section, the Spinosa Section.

Part VI: Informal Group Taxa

Tabulation

ANOPHELINAE Grassi, 1900

Anopheles **Meigen, 1818** ***sensu stricto***

- Angusticorn Section (Reid & Knight, 1961)
  - Anopheles Series (Edwards, 1932a)
        - Claviger Complex (Coluzzi *et al.*, 1965)
    - Alongensis Group (Phan *et al.*, 1991)
    - Aitkenii Group (Reid & Knight, 1961)
    - Atratipes Group (Lee *et al.*, 1987b)
    - Culiciformis Group (Reid & Knight, 1961)
    - Lindesayi Group (Reid & Knight, 1961)
        - Baileyi Complex (Somboon *et al.*, 2020b)
        - Gigas Complex (redefined by Somboon *et al.*, 2020b)
      - Lindesayi Subgroup (Harrison *et al.*, 1991, as Lindesayi Complex)
        - Lindesayi Complex (Namgay *et al.*, 2020)
    - Maculipennis Group (Reid & Knight, 1961)
      - Maculipennis Subgroup (Linton, 2004)
      - Quadrimaculatus Subgroup (Linton, 2004)
      - Freeborni Subgroup (Linton, 2004)
    - Plumbeus Group (Reid & Knight, 1961)
    - Pseudopunctipennis Group (Reid & Knight, 1961)
    - Punctipennis Group (Reid & Knight, 1961)
        - Crucians Complex (Wilkerson *et al.*, 2004)
    - Stigmaticus Group (Reid & Knight, 1961)
  - Cycloleppteron Series (Edwards, 1932a) (monobasic)
  - Lophoscelomyia Series (Edwards, 1932a)
    - Asiaticus Group (Reid, 1968)
      - Asiaticus Subgroup (Rattanarithikul *et al.*, 2006b) (monobasic)
      - Interruptus Subgroup (Rattanarithikul *et al.*, 2006b) (monobasic)
- Laticorn Section (Reid & Knight, 1961)
  - Arribalzagia Series (Root, 1922)
  - Myzorhynchus Series (Edwards, 1932a)
    - Albotaeniatus Group (Reid & Knight, 1961)
    - Bancroftii Group (Reid & Knight, 1961)
    - Barbirostris Group (Reid & Knight, 1961)
      - Barbirostris Subgroup (Reid, 1968)
        - Barbirostris Complex (Satoto, 2001)
      - Vanus Subgroup (Reid, 1968)
    - Coustani Group (Reid & Knight, 1961)
    - Hyrcanus Group (Reid, 1953)
      - Lesteri Subgroup (Harrison, 1972)
      - Nigerrimus Subgroup (Harrison, 1972)
    - Umbrosus Group (Reid, 1950)
      - Baezai Subgroup (Rattanarithikul *et al.*, 2006b) (monobasic)
      - Letifer Subgroup (Reid, 1968)
      - Separatus Subgroup (Rattanarithikul *et al.*, 2006b) (monobasic)
      - Umbrosus Subgroup (Rattanarithikul *et al.*, 2006b) (monobasic)

***Baimaia* Harbach, Rattanarithikul & Harrison, 2005** (monobasic)

***Cellia* Theobald, 1902b**

- Cellia Series (Christophers, 1924a)
  - Squamosus Group (Grjebine, 1966)
- Myzomyia Series (Christophers, 1924a)
  - Demeilloni Group (Gillies & de Meillon, 1968)
  - Funestus Group (Garros *et al.*, 2005)
    - Aconitus Subgroup (Chen *et al.*, 2003)
    - Culicifacies Subgroup (Garros *et al.*, 2005)
    - Funestus Subgroup (Garros *et al.*, 2005)
    - Minimus Subgroup (Chen *et al.*, 2003)
      - Fluviatilis Complex (Sarala *et al.*, 1994)
      - Minimus Complex (Green *et al.*, 1990)
    - Rivulorum Subgroup (Garros *et al.*, 2005)
  - Marshallii Group (Gillies & de Meillon, 1968)
    - Marshallii Complex (Gillies & Coetzee, 1987)
  - Wellcomei Group (Gillies & de Meillon, 1968)
- Neocellia Series (Christophers, 1924a)
  - Annularis Group (Reid, 1968)
    - Annularis Complex (Atrie *et al.*, 1999)
    - Nivipes Complex (Green *et al.*, 1985; Harrison *et al.*, 1991)
  - Jamesii Group (Rattanarithikul *et al.*, 2006b)
  - Maculatus Group (Rattanarithikul & Green, 1987)
    - Maculatus Subgroup (Rattanarithikul *et al.*, 2006b)
    - Sawadwongporni Subgroup (Rattanarithikul *et al.*, 2006b)
- Neomyzomyia Series (Christophers, 1924a)
  - Annulipes Complex (Green, 1972)
  - Lungae Complex (Belkin, 1962)
  - Ardensis Group (Gillies & de Meillon, 1968)
    - Nili Complex (Gillies & de Meillon, 1968)
  - Kochi Group (Rattanarithikul *et al.*, 2006b) (monobasic)
  - Leucosphyrus Group (Reid, 1949)
    - Hackeri Subgroup (Sallum *et al.*, 2005a)
    - Leucosphyrus Subgroup (Peyton, 1990; Sallum *et al.*, 2005b)
      - Dirus Complex (Sallum *et al.*, 2005b)
      - Leucosphyrus Complex (Sallum *et al.*, 2005a)
    - Riparis Subgroup (Peyton, 1990)
  - Mascarensis Group (Harbach, 1994a) (monobasic)
  - Pauliani Group (Grjebine, 1966)
  - Punctulatus Group (Schmidt *et al.*, 2001)
    - Farauti Complex (Schmidt *et al.*, 2003)
  - Ranci Group (Grjebine, 1966)
    - Ranci Subgroup (Grjebine, 1966) (monobasic)
    - Roubaudi Subgroup (Grjebine, 1966)
  - Rhodesiensis Group (Gillies & de Meillon, 1968)
  - Smithii Group (Gillies & de Meillon, 1968)
  - Tessellatus Group (Rattanarithikul *et al.*, 2006b) (monobasic)
- Paramyzomyia Series (Christophers & Barraud, 1931)
  - Cinereus Group (Gillies & de Meillon, 1968)
  - Listeri Group (Gillies & de Meillon, 1968)
- Pyretophorus Series (Edwards, 1932a)
  - Gambiae Complex (White, 1985)
  - Subpictus Complex (Suguna *et al.*, 1994)
  - Sundaicus Complex (Sukowati *et al.*, 1999)

***Christya* Theobald, 1903b**

***Kerteszia* Theobald, 1905f**

- Bellator Complex (Bourke *et al.*, 2023)
- Boliviensis Complex (Bourke *et al.*, 2023)
- Cruzii Complex (Ramírez & Dessen, 2000; Bourke *et al.*, 2023)

Homunculus Complex (Bourke *et al.*, 2023)
Laneanus Complex (Bourke *et al.*, 2023)
Neivai Complex (Bourke *et al.*, 2023)
Pholidotus Complex (Bourke *et al.*, 2023)
Rollai Complex (Bourke *et al.*, 2023)
***Lophopodomyia* Antunes, 1937**
***Nyssorhynchus* Blanchard, 1902**
Albimanus Section (Leví Castillo, 1949)
Albimanus Series (Faran, 1980)
Oswaldoi Series (Faran, 1980)
Oswaldoi Group (Faran, 1980)
Oswaldoi Subgroup (Faran, 1980)
Konderi Complex (Ruiz-Lopez *et al.*, 2013)
Nuneztovari Complex (Mirabello & Conn, 2008; Foster *et al.*, 2013; Scarpassa *et al.*, 2016)
Oswaldoi Complex (Ruiz-Lopez *et al.*, 2013)
Strodei Subgroup (Faran, 1980)
Arthuri Complex (Bourke *et al.*, 2013)
Benarrochi Complex (Ruiz *et al.*, 2005)
Triannulatus Group (Faran, 1980)
Argyritarsis Section (Leví Castillo, 1949)
Albitarsis Series (Linthicum, 1988)
Albitarsis Group (Linthicum, 1988)
Albitarsis Complex (Wilkerson *et al.*, 1995)
Braziliensis Group (Linthicum, 1988) (monobasic)
Argyritarsis Series (Linthicum, 1988)
Argyritarsis Group (Linthicum, 1988)
Darlingi Group (Linthicum, 1988) (monobasic)
Lanei Group (Linthicum, 1988) (monobasic)
Pictipennis Group (Linthicum, 1988)
Myzorhynchella Section (Peyton *et al.*, 1992)
***Stethomyia* Theobald, 1902b**

Bironella* Theobald, 1905f *sensu stricto
***Brugella* Edwards, 1930a**
***Neobironella* Tenorio, 1977**

***Chagasia* Cruz, 1906**

CULICINAE Meigen, 1818

AEDEOMYIINI Theobald, 1901a
Aedeomyia* Theobald, 1901a *sensu stricto
***Lepiothauma* Enderlein, 1923** (monobasic)

AEDINI Neveu-Lemaire, 1902c
Aedes* Meigen, 1818 *sensu stricto
Cinereus Group (Becker *et al.*, 2003)
Esoensis Group (Becker *et al.*, 2003)
***Abraedes* Zavortink, 1970c** (monobasic)
***Acartomyia* Theobald, 1903a**
***Aedimorphus* Theobald, 1903a**
Domesticus Assemblage (Reinert, 2000e)
Domesticus Subassemblage (Reinert, 2000e)
Domesticus Group (Edwards, 1941)
Vexans Subassemblage (Reinert, 2000e)
Quasiunivittatus Group (Pao & Knight, 1970)
Vexans Group (Edwards, 1932a)

Mediolineatus Subassemblage (Reinert, 2000e)
Caecus Group (Reinert, 2000e)
Pampangensis Group (Reinert, 2000e)
Mediolineatus Group (Pao & Knight, 1970)
Taeniorhynchoides Group (Reinert, 2000e)
Punctifemoris Group (Reinert, 2000e) (monobasic)
Alboscutellatus Assemblage (Reinert, 2000e)
Alboscutellatus Group (Edwards, 1932a)
Abnormalis Group (Edwards, 1941)
***Alanstonea* Mattingly, 1960**
***Albuginosus* Reinert, 1987**
***Ayurakitia* Thurman, 1954a**
***Aztecaedes* Zavortink, 1972** (monobasic)
***Belkinius* Reinert, 1982** (monobasic)
***Bifidistylus* Reinert, Harbach & Kitching, 2009**
***Borichinda* Harbach & Rattanarithikul, 2007** (in Harbach *et al.*, 2007b) (monobasic)
***Bothaella* Reinert, 1973f**
***Bruceharrisonius* Reinert, 2003**
***Cancraedes* Edwards, 1929b**
***Catageiomyia* Theobald, 1903d**
***Catatassomyia* Dyar & Shannon, 1925a** (monobasic)
***Christophersiomyia* Barraud, 1923a**
***Coetzeemyia* Huang *et al.*, 2010** (monobasic)
***Collessius* Reinert, Harbach & Kitching, 2006**
Alloeomyia Group (Reinert *et al.*, 2008; Wilkerson *et al.*, 2015)
Collessius Group (Reinert *et al.*, 2008; Wilkerson *et al.*, 2015)
***Cornetius* Huang, 2005** (monobasic)
***Dahliana* Reinert, Harbach & Kitching, 2006**
***Danielsia* Theobald, 1904c**
***Dendroskusea* Edwards, 1929b**
***Diceromyia* Theobald, 1911c**
Section A (Huang & Rueda, 2016)
Adersi Group (Huang & Rueda, 2016)
Flavicollis Group (Huang & Rueda, 2016)
Furcifer Group (Huang, 1986a)
Section B (Huang & Rueda, 2016)
Fascipalpis Group (Huang & Rueda, 2016) (monobasic)
***Dobrotworskyius* Reinert, Harbach & Kitching, 2006**
***Downsiomyia* Vargas, 1950**
Alboniveus Group (Reinert & Harbach, 2006; Alboniveus Subassemblage of Reinert, 2002e)
Niveus Group (Reinert & Harbach, 2006; Niveus Subassemblage of Reinert, 2002e)
***Edwardsaedes* Belkin, 1962**
***Elpeytonius* Reinert, Harbach & Kitching, 2009**
***Finlaya* Theobald, 1903a**
Gani Group (Subgroup III, **gani** of Knight & Marks, 1952) (monobasic)
Kochi Group (Bohart, 1957, as Papuan-Australian subgroup I, *kochi*; Subgroup I, **kochi** *s. str. in* part of Knight & Marks, 1952; Subgroup II, **lewelleni** of Knight & Marks, 1952)
Bougainvillensis Complex (Belkin, 1962)
Franclemonti Complex (Belkin, 1962)
Knighti Complex (Belkin, 1962) + sp. 23
Solomonis Complex (Belkin, 1962)
Poicilius Group (Bohart, 1957, as Malayan subgroup II, *poicilius*)
***Fredwardsius* Reinert, 2000a** (monobasic)
***Georgecraigius* Reinert, Harbach & Kitching, 2006**
Atropalpus Group (Zavortink, 1972; supersedes the Georgecraigius Group of Wilkerson *et al.*, 2015)
Fluviatilis Group (Zavortink, 1972; supersedes the Horsfallius Group of Wilkerson *et al.*, 2015) (monobasic)
***Geoskusea* Edwards, 1929b**
Daggyi Group (Belkin, 1962)
Fimbripes Group (Belkin, 1962)
Longiforceps Group (Belkin, 1962) (monobasic)

- ***Gilesius* Reinert, Harbach & Kitching, 2006**
- ***Gymnometopa* Coquillett, 1906a** (monobasic)
- ***Halaedes* Belkin, 1962**
- ***Himalaius* Reinert, Harbach & Kitching, 2006**
- ***Hopkinsius* Reinert, Harbach & Kitching, 2008**
    - Hopkinsius Group (Reinert *et al.*, 2008; Wilkerson *et al.*, 2015)
    - Yamada Group (Reinert *et al.*, 2008; Wilkerson *et al.*, 2015)
- ***Howardina* Theobald, 1903a**
    - Fulvithorax Section (Berlin, 1969a) (Fulvithorax Assemblage of Reinert, 2002e)
        - Arborealis Subassemblage (Reinert, 2002e)
        - Flavithorax Subassemblage (Reinert, 2002e)
    - Sexlineatus Section (Berlin, 1969a) (Sexlineatus Assemblage of Reinert, 2002e)
        - Aurivittatus Group (Berlin, 1969a)
        - Eleanorae Group (Berlin, 1969a) (monobasic)
        - Inaequalis Subassemblage (Reinert, 2002e)
        - Sexlineatus Subassemblage (Reinert, 2002e)
    - Walkeri Section (Berlin, 1969a) (Walkeri Assemblage of Reinert, 2002e)
        - Aurites Group (Berlin, 1969a)
        - Ioliota Group (Berlin, 1969a)
        - Walkeri Group (Berlin, 1969a)
- ***Huaedes* Huang, 1968b**
- ***Hulecoeteomyia* Theobald, 1904c**
- ***Indusius* Edwards, 1934** (in Barraud, 1934) (monobasic)
- ***Isoaedes* Reinert, 1979** (monobasic)
- ***Jarnellius* Reinert, Harbach & Kitching, 2006** (Varipalpus Group of Arnell & Nielsen, 1972)
- ***Jihlienius* Reinert, Harbach & Kitching, 2006**
- ***Kenknightia* Reinert, 1990b**
    - Dissimilis Group (Reinert, 1990b)
    - Luzonensis Group (Reinert, 1990b)
- ***Kompia* Aitkin, 1941** (monobasic)
- ***Leptosomatomyia* Theobald, 1905f** (monobasic)
- ***Levua* Stone & Bohart, 1944** (monobasic)
- ***Lewnielsenius* Reinert, Harbach & Kitching, 2006** (monobasic)
- ***Lorrainea* Belkin, 1962**
- ***Luius* Reinert, Harbach & Kitching, 2008** (monobasic)
- ***Macleaya* Theobald, 1903e**
    - Chaetocruiomyia Group (Reinert *et al.*, 2006; Wilkerson *et al.*, 2015)
    - Macleaya Group (Reinert *et al.*, 2006; Wilkerson *et al.*, 2015)
- ***Molpemyia* Theobald, 1910c**
- ***Mucidus* Theobald, 1901a**
    - Mucidus Group (Edwards, 1932a; Tyson, 1970b; Wilkerson *et al.*, 2015)
    - Pardomyia Group (Edwards, 1932a; Tyson, 1970b; Wilkerson *et al.*, 2015)
- ***Neomelaniconion* Newstead, 1907** (in Newstead *et al.*, 1907)
    - Hylaius Group (Harbach, 2018b: replacement name for the Sylvaticum Group of Le Goff *et al.*, 2007)
    - Lineatopennis Group (Edwards, 1941: supersedes the Circumluteolus Group of Le Goff *et al.*, 2007)
- ***Nyctomyia* Harbach, 2013**
- ***Ochlerotatus* Lynch Arribálzaga, 1891a**
    - Albifasciatus Group (Edwards, 1932a)
    - Aculeatus Group (Aculeatus Section of Marks, 1957a; Gilesia Group of Wilkerson *et al.*, 2015)
    - Breedensis Group (Huang & Rueda, 2014)
    - Burpengaryensis Group (Burpengaryensis Section of Marks, 1957a)
    - Buvirilia Group (Reinert *et al.*, 2009; Wilkerson *et al.*, 2015) (monobasic)
    - Caballus Group (Huang & Rueda, 2014: supersedes the Juppius Group of Wilkerson *et al.*, 2015)
    - Cantans Group (Gutsevich *et al.*, 1974; Annulipes Group in part of Becker *et al.*, 2003)
    - Caspius Group (Gutsevich *et al.*, 1974, Palaearctic; Becker *et al.*, 2003, Europe; Huang & Rueda, 2014, Afrotropical)
    - Communis Group (Edwards, 1932a; Becker *et al.*, 2003, Europe)
        - Communis Aggregate (Ellis & Brust, 1973)
        - Leucomelas Subgroup (Danilov & Gornostaeva, 1987)
        - Punctor Subgroup (Knight, 1951a; Punctor Group in part of Dyar, 1920c; Punctor Group of Becker *et al.*, 2003)
    - Culicada Group (Reinert *et al.*, 2008; Wilkerson *et al.*, 2015) (monobasic)

Cunabulanus Group (Cunabulanus Section of Marks, 1957a)
Flavifrons Group (Flavifrons Section of Marks, 1957a; Pholeomyia Group of Wilkerson *et al.*, 2015)
Fulvus Group (Edwards, 1932a; Chrysoconops Group of Wilkerson *et al.*, 2015)
Impiger Group (Dyar, 1920c)
Perkinsi Group (Perkinsi Section of Dobrotworsky, 1960a; combined Perkinsi and Camptorhynchus Groups of Marks, 1957a)
Pullatus Group (Dyar, 1920c; supersedes the Intrudens Group of Becker *et al.*, 2003 and the Woodius Group of Wilkerson *et al.*, 2015)
Scapularis Group (Dyar, 1920c in part; Arnell, 1976; Ochlerotatus Group of Wilkerson *et al.*, 2015)
Bogotanus Subgroup (Arnell, 1976)
Crinifer Subgroup (Arnell, 1976)
Incomptus Subgroup (Arnell, 1976) (monobasic)
Infirmatus Subgroup (Arnell, 1976)
Obturbator Subgroup (Arnell, 1976)
Pectinatus Subgroup (Arnell, 1976) (monobasic)
Scapularis Subgroup (Arnell, 1976; Scapularis Complex of Belkin *et al.*, 1970)
Thelcter Subgroup (Arnell, 1976) (monobasic)
Tortilis Subgroup (Arnell, 1976; Tortilis Complex of Belkin *et al.*, 1970)
Trivittatus Subgroup (Arnell, 1976)
Serratus Group (Dyar, 1920c)
Stimulans Group (Dyar, 1920c; Annulipes Group in part of Edwards, 1932a and Becker *et al.*, 2003)
Increpitus Complex (Lanzaro & Eldridge, 1992)
Stricklandi Group (Stricklandi Section of Marks, 1957a; Lepidokeneon Group of Wilkerson *et al.*, 2015)
Taeniorhynchus Group (Edwards, 1932a, in part; Culicelsa Group of Wilkerson *et al.*, 2015)
Theobaldi Subgroup (Marks, 1949; Theobaldi Section of Marks, 1957a)
Vigilax Subgroup (Marks, 1949; Vigilax Section of Marks, 1957a; Empihals Group in part of Wilkerson *et al.*, 2015)
Thibaulti Group (Dyar, 1920c) (monobasic)

***Orohylomyia* Somboon & Harbach, 2023** (in Somboon *et al.*, 2023b)

***Paraedes* Edwards, 1934** (in Barraud, 1934)
Barraudi Assemblage (Reinert, 2000e)
Bonneae Group (replacement name for Group B of Mattingly, 1958a) (monobasic)
Ostentatio Assemblage (Reinert, 2000e, less Group B of Mattingly, 1958a)

***Patmarksia* Reinert, Harbach & Kitching, 2006**

***Paulianius* Brunhes & Boussès, 2017** (in Brunhes *et al.*, 2017)
Coulangesi Group (Coulangesi Section of Brunhes *et al.*, 2017)
Tiptoni Group (Tiptoni Section of Brunhes *et al.*, 2017)
Zethus Group (Niveus Section of Brunhes *et al.*, 2017: *zethus* is the senior synonym of *niveus*) (monobasic)

***Petermattinglyius* Reinert, Harbach & Kitching, 2009**
Aglaonotus Group (Wilkerson *et al.*, 2015) (monobasic)
Petermattinglyius Group (Wilkerson *et al.*, 2015)

***Phagomyia* Theobald, 1905e**

***Polyleptiomyia* Theobald, 1905e**

***Protomacleaya* Theobald, 1907**
Kompi Group (Zavortink, 1972)
Knabi Group (Zavortink, 1972) (monobasic)
Terrens Group (Schick, 1970a) (Subgroups I and II of the Terrens Group of Knight & Marks, 1952)
Aitkeni Subgroup (Schick, 1970a) (monobasic)
Alboapicus Subgroup (Schick, 1970a) (monobasic)
Bertrami Subgroup (Schick, 1970a) (monobasic)
Buenaventura Subgroup (Schick, 1970a) (monobasic)
Diazi Subgroup (Schick, 1970a) (monobasic)
Galindoi Subgroup (Schick, 1970a)
Heteropus Subgroup (Schick, 1970a)
Homoeopus Subgroup (Schick, 1970a)
Insolitus Subgroup (Schick, 1970a)
Metoecopus Subgroup (Schick, 1970a) (monobasic)
Podographicus Subgroup (Schick, 1970a) (monobasic)
Tehuantepec Subgroup (Schick, 1970a)
Terrens Subgroup (Schick, 1970a)

Thorntoni Subgroup (Schick, 1970a)
Triseriatus Group (Zavortink, 1972)
Zoosophus Group (Zavortink, 1972) (monobasic)
***Pseudalbuginosus* Huang & Rueda, 2015b** (monobasic)
***Pseudarmigeres* Stone & Knight, 1956**
***Pseudoskusea* Theobald, 1907**
***Rampamyia* Reinert, Harbach & Kitching, 2006**
***Reinertia* Somboon, Namgay & Harbach, 2021** (in Somboon *et al.*, 2021a) (monobasic)
***Rhinoskusea* Edwards, 1929b**
***Rusticoidus* Shevchenko & Prudkina, 1973**
***Sallumia* Reinert, Harbach & Kitching, 2008**
***Scutomyia* Theobald, 1904c**
Laffooni Group (Reinert, 1985)
Hoogstraali Group (Reinert, 1985)
***Skusea* Theobald, 1903a**
***Stegomyia* Theobald, 1901** (in Howard, 1901)
Aegypti Group (Belkin, 1962; Huang, 2004)
Africanus Group (Edwards, 1925; Huang, 1990)
Apicoargenteus Group (Huang, 2004)
Dendrophilus Group (Huang, 1994)
Edwardsi Group (Belkin, 1962)
Grantii Group (Huang, 2004) (monobasic)
Maehleri Group (Bohart, 1957)
Metallicus Group (Huang, 2004) (monobasic)
Pandani Group (Bohart, 1957)
Poweri Group (Huang, 2004)
Pseudonigeria Group (Huang, 1988b)
Scutellaris Group (Edwards, 1932a; Belkin, 1962; Huang, 1972)
Albopictus Subgroup (Huang, 1972)
Scutellaris Subgroup (Huang, 1972)
Pseudoscutellaris Subgroup/Complex (Belkin, 1962) (7 species)
Simpsoni Group (Huang, 2004) (Simpsoni Assemblage of Reinert, 2000e)
Simpsoni Complex (Huang, 1979b)
Unilineatus Group (Huang, 2004)
W-albus Group (Edwards, 1932a in part; Huang, 1977a)
Annandalei Subgroup (Huang, 1977a)
Desmotes Subgroup (Huang, 1977a) (monobasic)
Mediopunctatus Subgroup (Huang, 1977a)
W-albus Subgroup (Huang, 1977a)
Zoromorphus Group (Reinert *et al.*, 2009; Wilkerson *et al.*, 2015) (monobasic)
***Tanakaius* Reinert, Harbach & Kitching, 2004**
***Tewarius* Reinert, 2006a**
Agastyai Group (Reinert, 2006a)
Nummatus Group (Reinert, 2006a)
***Vansomerenis* Reinert, Harbach & Kitching, 2006**
***Zavortinkius* Reinert, 1999b**
Brygooi Group (Reinert, 1999b)
Longipalpis Group (Reinert, 1999b)
Monetus Group (Reinert, 1999b) (monobasic)

Armigeres* Theobald, 1901a *sensu stricto (Group A and C of Edwards, 1932a)
Malayi Group (Steffan, 1968)
***Leicesteria* Theobald, 1904c**

***Eretmapodites* Theobald, 1901d**
Chrysogaster Group (Edwards, 1941; Haddow, 1946)
Inornatus Group (Edwards, 1941; Haddow, 1946)
Leucopous Group (Rickenbach & Lombrici, 1975)
Oedipodeios Group (Haddow, 1946)

Plioleucus Group (Rickenbach & Eouzan, 1970)
Quinquevittatus Group (Haddow, 1946; Dracaenae Group of later authors)

Haemagogus **Williston, 1896** *sensu stricto*
Albomaculatus Section (Arnell, 1973; Equinus Assemblage of Reinert, 2002c)
Splendens Section (Arnell, 1973; Splendens Assemblage of Reinert, 2002c)
Tropicalis Section (Arnell, 1973) (monobasic)
Conopostegus **Dyar, 1925d**

Heizmannia **Ludlow, 1905a** *sensu stricto* (Group A of Mattingly, 1957b)
Mattinglyia **Lien, 1968c** (Group B of Mattingly, 1957b)

Opifex **Hutton, 1902** *sensu stricto* (monobasic)
Nothoskusea **Dumbleton, 1962** (monobasic)

Psorophora **Robineau-Desvoidy, 1827** *sensu stricto*
Grabhamia **Theobald, 1903a**
Cingulata Group (Belkin *et al.*, 1970)
Confinnis Group (Belkin *et al.*, 1970)
Confinnis Complex (Belkin *et al.*, 1970; Bickley, 1984; Lanzaro *et al.*, 2015)
Janthinosoma **Lynch Arribálzaga, 1891a**

Udaya **Thurman, 1954b**

Verrallina **Theobald, 1903a** *sensu stricto*
Butleri Series (Reinert, 1974 in part; Reinert, 1999c; Butleri Assemblage of Reinert, 2001b)
Carmenti Series (Reinert, 1974; Carmenti Assemblage of Reinert, 2001b)
Harbachius **Reinert, 1999c** (Section B of Reinert, 1974)
Neomacleaya **Theobald, 1907** (principally the Indicus Series of Reinert, 1974)

Zeugnomyia **Leicester, 1908**

CULICINI Linnaeus, 1758
Culex **Linnaeus, 1758** *sensu stricto*
Atriceps Group (Belkin, 1962)
Duttoni Group (Harbach, 1988)
Guiarti Group (Edwards, 1941)
Pipiens Group (Edwards, 1932a, in part)
Coronator Complex (Forattini, 2002)
Restuans Complex (Bram, 1967b, as *restuans–laticlasper–acharistus* complex)
Salinarius Complex (Bram, 1967b)
Decens Subgroup (Harbach, 1988)
Pipiens Subgroup (Sirivanakarn, 1976)
Simpsoni Subgroup (Harbach, 1988)
Tarsalis Subgroup (Edwards, 1932a, as *tarsalis* series)
Theileri Subgroup (Sirivanakarn, 1976)
Trifilatus Subgroup (Mattingly & Rageau, 1958)
Univittatus Subgroup (Sirivanakarn, 1976)
Sitiens Group (Edwards, 1932a, in part)
Barraudi Subgroup (Sirivanakarn, 1976)
Gelidus Subgroup (Sirivanakarn, 1976) (monobasic)
Mimeticus Subgroup (Mimeticus Series of Edwards, 1932a)
Mimeticus Complex (Sirivanakarn, 1976)
Mimulus Complex (Sirivanakarn, 1976)
Sitiens Subgroup (Bram, 1967a; Sitiens Series of Edwards, 1932a, in part)
Vishnui Subgroup (Bram, 1967a)
Acalleomyia **Leicester, 1908** (monobasic)

Acallyntrum **Stone & Penn, 1948**
Bicki Group (Belkin, 1962)
Perkinsi Group (Belkin, 1962)
Aedinus **Lutz, 1904** (in Bourroul, 1904)
Afroculex **Danilov, 1989** (monobasic)
Allimanta **Casal & García, 1968a** (monobasic)
Anoedioporpa **Dyar, 1923d**
Conservator Group (Berlin & Belkin, 1980)
Restrictor Group (Berlin & Belkin, 1980) (monobasic)
Barraudius **Edwards, 1921d**
Belkinomyia **Adames & Galindo, 1973** (monobasic)
Carrollia **Lutz, 1905**
Bihaicola Group (Valencia, 1973)
Iridescens Group (Valencia, 1973)
Urichii Subgroup (Valencia, 1973)
Iridescens Subgroup (Valencia, 1973)
Culiciomyia **Theobald, 1907**
Dispectus Group (Bram, 1969)
Fragilis Group (Edwards, 1932a) (Oriental and Australasian species)
Nebulosus Group (Edwards, 1932a) (Afrotropical species)
Shebbearei Group (Harbach, 2011) ("group or complex" of Sirivanakarn, 1977c)
Tricuspis Group (Harrison, 1987)
Eumelanomyia **Theobald, 1909**
Eumelanomyia Group (Sirivanakarn, 1971)
Mochthogenes Group (Sirivanakarn, 1971)
Bokorensis Subgroup (Sirivanakarn, 1972) (monobasic)
Femineus Subgroup (Sirivanakarn, 1971) (monobasic)
Foliatus Subgroup (Sirivanakarn, 1972)
Hinglungensis Subgroup (Sirivanakarn, 1971)
Inconspicuosus Subgroup (Sirivanakarn, 1971)
Iphis Subgroup (Sirivanakarn, 1972) (monobasic)
Khazani Subgroup (Sirivanakarn, 1972) (monobasic)
Malayi Subgroup (Sirivanakarn, 1971)
Okinawae Subgroup (Sirivanakarn, 1972)
Otachati Subgroup (Sirivanakarn, 1971)
Pluvialis Subgroup (Sirivanakarn, 1972)
Tenuipalpis Subgroup (Sirivanakarn, 1972)
Uncinatus Subgroup (Sirivanakarn, 1971) (monobasic)
Protomelanoconion Group (Sirivanakarn, 1971)
Rubinotus-rima Group (Sirivanakarn, 1971)
Rima Subgroup (Sirivanakarn, 1971)
Rubinotus Subgroup (Sirivanakarn, 1971)
Kitzmilleria **Danilov, 1989** (monobasic)
Lasioconops **Theobald, 1903c** (monobasic)
Lasiosiphon **Kirkpatrick, 1925a** (monobasic)
Lophoceraomyia **Theobald, 1905h**
Fraudatrix Group (Edwards, 1934, in Barraud, 1934)
Fraudatrix Subgroup (Sirivanakarn, 1977a) (the monobasic *ornatus* complex of Sirivanakarn, 1968 and the monobasic *alphus*, *cottlei*, *inculus* and *seniori* complexes of Sirivanakarn, 1977a are not recognized as complexes here)
Bergi Complex (Belkin, 1962)
Buxtoni Complex (Belkin, 1962)
Christiani Complex (Sirivanakarn, 1968)
Cinctellus Complex (Sirivanakarn, 1977a)
Fraudatrix Complex (Sirivanakarn, 1968)
Hilli Complex (Sirivanakarn, 1968)
Hurlbuti Complex (Belkin, 1962)
Marksae Complex (Sirivanakarn, 1968)
Petersi Complex (Sirivanakarn, 1968)
Pseudorubithoracis Complex (Sirivanakarn, 1968)

Quadripalpis Complex (Sirivanakarn, 1977a)
Rubithoracis Complex (Sirivanakarn, 1977a)
Solomonis Complex (Belkin, 1962)
Variatus Complex (Sirivanakarn, 1977a)
Minutissimus Subgroup (Sirivanakarn, 1977a)
Mammilifer Group (Edwards, 1932a)
Bolii Subgroup (Sirivanakarn, 1968) (monobasic)
Brevipalpus Subgroup (Colless, 1965) (the monobasic *curtipalpis*, *hewitti* and *jenseni* complexes of Sirivanakarn, 1977a are not recognized as complexes here)
Brevipalpus Complex (Sirivanakarn, 1977a)
Navalis Complex (Sirivanakarn, 1977a)
Digoelensis Subgroup (Sirivanakarn, 1968)
Mammilifer Subgroup (Colless, 1965) (the monobasic *singhbhumensis* complex of Natarajan & Rajavel, 2009 is not recognized here)
Flavicornis Complex (Sirivanakarn, 1977a)
Mammilifer Line (Sirivanakarn, 1977a) (the monobasic *impostor* complex of Sirivanakarn, 1977a is not recognized here)
Mammilifer Complex (Sirivanakarn, 1977a)
Traubi Complex (Sirivanakarn, 1977a)
Minor Line (Sirivanakarn, 1977a) (the monobasic *pholeter* complex of Sirivanakarn, 1977a is not recognized here)
Ganapathi Complex (Sirivanakarn, 1977a)
Minor Complex (Sirivanakarn, 1977a)
Peytoni Complex (Sirivanakarn, 1977a)
Wilfredi Group (Sirivanakarn, 1977a)
***Maillotia* Theobald, 1907**
Hortensis Group (Sirivanakarn, 1971)
Seyrigi Group (Sirivanakarn, 1971)
***Melanoconion* Theobald, 1903a**
Melanoconion Section (Sirivanakarn, 1983)
Atratus Group (Sirivanakarn, 1983)
Bastagarius Group (Sirivanakarn, 1983)
Bastagarius Subgroup (Sirivanakarn, 1983)
Iolambdis Subgroup (Sirivanakarn, 1983)
Conspirator Group (Sirivanakarn, 1983)
Distinguendus Group (Sirivanakarn, 1983)
Distinguendus Subgroup (Sirivanakarn, 1983)
Galindoi Subgroup (Sirivanakarn, 1983) (monobasic)
Putumayensis Subgroup (Sirivanakarn, 1983)
Rorotaensis Subgroup (Sirivanakarn, 1983) (monobasic)
Educator Group (Sirivanakarn, 1983)
Erraticus Group (Sirivanakarn, 1983)
Clarki Subgroup (Sirivanakarn, 1983) (monobasic)
Erraticus Subgroup (Sirivanakarn, 1983)
Psatharus Subgroup (Sirivanakarn, 1983) (monobasic)
Evansae Group (Sirivanakarn, 1983)
Inhibitator Group (Sirivanakarn, 1983)
Egcymon Subgroup (Sirivanakarn, 1983)
Inhibitator Subgroup (Sirivanakarn, 1983)
Mulrennani Subgroup (Sirivanakarn, 1983) (monobasic)
Intrincatus Group (Sirivanakarn, 1983)
Andricus Subgroup (Sirivanakarn, 1983) (monobasic)
Eastor Subgroup (Sirivanakarn, 1983) (monobasic)
Idottus Subgroup (Sirivanakarn, 1983)
Intrincatus Subgroup (Sirivanakarn, 1983)
Penai Subgroup (Sirivanakarn, 1983) (monobasic)
Tecmarsis Subgroup (Sirivanakarn, 1983) (monobasic)
Peccator Group (Sirivanakarn, 1983)
Pilosus Group (Sirivanakarn, 1983)
Caudelli Subgroup (Sirivanakarn, 1983)
Pilosus Subgroup (Sirivanakarn, 1983)

Saramaccensis Group (Sirivanakarn, 1983)
Trifidus Group (Sirivanakarn, 1983) (monobasic)
Spissipes Section (Sirivanakarn, 1983)
Crybda Group (Sallum & Forattini, 1996)
Paracrybda Subgroup (Sirivanakarn, 1983)
Pedroi Subgroup (Sirivanakarn, 1983)
Pereyrai Subgroup (Sirivanakarn, 1983) (monobasic)
Faurani Group (Sirivanakarn, 1983) (monobasic)
Jubifer Group (Sirivanakarn, 1983)
Lopesi Group (Sirivanakarn, 1983) (monobasic)
Ocossa Group (Sirivanakarn, 1983)
Spissipes Group (Sirivanakarn, 1983) (monobasic)
Taeniopus Group (Sallum & Forattini, 1996)
Vomerifer Group (Sallum & Forattini, 1996)
***Micraedes* Coquillett, 1906a**
Bisulcatus Group (Berlin, 1969b)
Schicki Group (Berlin, 1969b)
Erethyzonfer Group (Berlin, 1969b) (monobasic)
***Microculex* Theobald, 1907**
Consolator Series (Lane & Whitman, 1951)
Imitator Series (Lane & Whitman, 1951)
Inimitabilis Series (Lane & Whitman, 1951)
Pleuristriatus Series (Lane & Whitman, 1951)
***Neoculex* Dyar, 1905a**
Crassistylus Group (Sirivanakarn, 1971)
Pseudomelanoconia Group (Sirivanakarn, 1971)
Territans Group (Sirivanakarn, 1971)
***Nicaromyia* González Broche & Rodríguez R., 2001** (monobasic)
***Oculeomyia* Theobald, 1907**
Bitaeniorhynchus Complex (Sirivanakarn, 1976)
Geminus Complex (Sirivanakarn, 1976)
Sinensis Complex (Sirivanakarn, 1976)
***Phalangomyia* Dyar & Knab, 1914**
***Phenacomyia* Harbach & Peyton, 1992b**
***Phytotelmatomyia* Rossi & Harbach, 2008**
***Sirivanakarnius* Tanaka, 2004** (monobasic)
***Tinolestes* Coquillett, 1906a**

***Deinocerites* Theobald, 1901a**
Cancer Group (Belkin & Hogue, 1959)
Dyari Group (Belkin & Hogue, 1959)
Epitedeus Group (Belkin & Hogue, 1959 in part; Adames, 1971)
Spanius Group (Belkin & Hogue, 1959)
Pseudes Group (Adames, 1971)

***Galindomyia* Stone & Barreto, 1969** (monobasic)

Lutzia* Theobald, 1903a *sensu stricto (Group A, *Lutzia*, of Edwards, 1932a)
***Insulalutzia* Tanaka, 2003** (monobasic)
***Metalutzia* Tanaka, 2003** (Group B, *Jamesia*, of Edwards, 1932a)

CULISETINI Belkin, 1962
Culiseta* Felt, 1904 *sensu stricto
***Allotheobaldia* Brolemann, 1919** (monobasic)
***Austrotheobaldia* Dobrotworsky, 1954** (monobasic)
***Climacura* Howard, Dyar & Knab, 1915**
***Culicella* Felt, 1904**
***Neotheobaldia* Dobrotworsky, 1958b**
***Theomyia* Edwards, 1930a** (monobasic)

FICALBIINI Belkin, 1962
***Ficalbia* Theobald, 1903a**
Group A (Mattingly & Grjebine, 1958) Oriental Region Group (Grjebine, 1986)
Group B (Mattingly & Grjebine, 1958) Afrotropical Region Group (Grjebine, 1986)

Mimomyia* Theobald, 1903a *sensu stricto
Mimomyia Group (Group A) (Mattingly & Grjebine, 1958) (monobasic)
Hybrida Group (Mattingly, 1957a; *hybrida* complex of Belkin, 1962)
Ludlowia Group (Group B) (Mattingly, 1957a; Mattingly & Grjebine, 1958)
Boycia Group (Group C) (Mattingly & Grjebine, 1958)
Megaculex Group (Group D) (Mattingly & Grjebine, 1958) (monobasic)
Group E (Mattingly & Grjebine, 1958) (monobasic)
***Etorleptiomyia* Theobald, 1904b**
Group A (*Etorleptiomyia s. str.*) (Edwards, 1932a, in part; Mattingly & Grjebine, 1958; Grjebine, 1986)
Group B (*Dixomyia*) (Grjebine, 1986)
***Ingramia* Edwards, 1912a**
Group A (*Ravenalites s. str.*) (Mattingly & Grjebine, 1958; Grjebine, 1986)
Roubaudi Complex (Grjebine, 1986)
Spinosa Complex (Grjebine, 1986)
Group B (*Ravenalites*) (Mattingly & Grjebine, 1958; Grjebine, 1986)
Group C (Mattingly & Grjebine, 1958; *Ingramia s. str.*, Fusca Group of Grjebine, 1986)

HODGESIINI Belkin, 1962
***Hodgesia* Theobald, 1904a**

MANSONIINI Belkin, 1962
Mansonia* Blanchard, 1901 *sensu stricto
Indubitans Complex (Belkin *et al.*, 1970)
Titillans Complex (Belkin *et al.*, 1970)
***Mansonioides* Theobald, 1907**
Uniformis Complex (Belkin, 1962)

Coquillettidia* Dyar, 1905a *sensu stricto
Crassipes Group (Belkin, 1962) Palaeotropics
Iracunda Group (Belkin, 1962)
Perturbans Group (Belkin, 1968a) Holarctic
***Austromansonia* Belkin, 1968a** (monobasic)
***Pseudotaeniorhynchus* Theobald, 1911a**

ORTHOPODOMYIINI Belkin, Heinemann & Page, 1970
***Orthopodomyia* Theobald, 1904c**
Bancroftia Section (Zavortink, 1968; Bancroftia Group of Edwards, 1939b)
Albicosta Group (Zavortink, 1968) (monobasic)
Signifera Group (Zavortink, 1968)
Kummi Subgroup (Zavortink, 1968; includes unnamed species 5)
Pulcripalpis Subgroup (Zavortink, 1968) (monobasic)
Signifera Subgroup (Zavortink, 1968)
Folicolae Section (Zavortink, 1968) (monobasic)
Orthopodomyia Section (Zavortink, 1968; Group A, *Orthopodomyia*, of Edwards, 1932a less *Or. fascipes*)
Albipes Group (Zavortink, 1968; principally Subgroup I—*anopheloides* of Knight & Mattingly, 1950)
Albipes Subgroup (Zavortink, 1968) (monobasic)
Anopheloides Subgroup (Zavortink, 1968)
Flavicosta Subgroup (Zavortink, 1968)
Wilsoni Subgroup (Zavortink, 1968) (monobasic)
Arboricollis Group (Zavortink, 1968) (monobasic)
Nkolbissonensis Group (Zavortink, 1968)
Ototomoensis Group (Huang & Rueda, 2013) (monobasic)
Vernoni Group (Zavortink, 1968; Brunhes & Hervy, 1995)
Thomasina Section (Zavortink, 1968; Subgroup III—*fascipes* of Knight & Mattingly, 1950)

SABETHINI Blanchard, 1905
Sabethes **Robineau-Desvoidy, 1827** ***sensu stricto***
Amazonicus Group (Harbach & Petersen, 1992)
Group A (Cerqueira, 1961, less the species of the Amazonicus Group)
Group B (Cerqueira, 1961)
Davismyia **Lane & Cerqueira, 1942** (monobasic)
Peytonulus **Harbach, 1991a**
Sabethinus **Lutz, 1904** (in Bourroul, 1904)
Andreae Group (Harbach, 2018a)
Intermedius Group (Harbach, 2018a)
Sabethoides **Theobald, 1903a**

Isostomyia **Coquillett, 1906c**

Johnbelkinia **Zavortink, 1979**

Kimia **Vu & Harbach, 2007** (in Harbach *et al.*, 2007a)

Limatus **Theobald, 1901c**

Malaya **Leicester, 1908**

Maorigoeldia **Edwards, 1930a** (monobasic)

Onirion **Peyton & Harbach, 2000** (in Harbach & Peyton, 2000)
Brucei Group (Harbach & Peyton, 2000)
Personatum Group (Harbach & Peyton, 2000)

Runchomyia **Theobald, 1903a** ***sensu stricto***
Ctenogoeldia **Edwards, 1930a**

Shannoniana **Lane & Cerqueira, 1942**

Topomyia **Leicester, 1908** ***sensu stricto***
Miyagiella **Harbach, 2014** (in Harbach & Culverwell, 2014) (monobasic)
Suaymyia **Thurman, 1959**

Trichoprosopon **Theobald, 1901b,c**
Trichoprosopon Group (Rivera-García *et al.*, 2023; subgenus of Coher, 1948)
Digitatum Complex (Zavortink, 1981)
Compressum Complex (Zavortink, 1981)
Lampropus Complex (Zavortink, 1981)
Hyloconops Group (Rivera-García *et al.*, 2023; subgenus of Coher, 1948)
Pallidiventer Complex (Zavortink, 1981)

Tripteroides **Giles, 1904** ***sensu stricto***
Bimaculipes Group (Belkin, 1962)
Littlechildi Group (Lee *et al.*, 1989c)
Mabinii Group (Mattingly, 1981)
Magnesianus Group (Lee *et al.*, 1989c)
Nitidoventer Group (Edwards, 1932a partially; Mattingly, 1981 largely)
Purpuratus Group (Belkin, 1962) (monobasic)
Polylepidomyia **Theobald, 1905f**
Argenteiventris Group (Belkin, 1962)
Atripes Group (Edwards, 1932a; Lee, 1946)
Caledonicus Group (Lee, 1946; Belkin, 1962)
Floridensis Group (Belkin, 1962) (monobasic)
Tasmaniensis Group (Belkin, 1962; Lee *et al.*, 1989c) (monobasic)

***Rachionotomyia* Theobald, 1905h**
Aranoides Group (Edwards, 1932a in part; Mattingly, 1981)
Affinis Subgroup (Mattingly, 1981)
Aranoides Subgroup (Mattingly, 1981)
Nepenthis Subgroup (Mattingly, 1981)
Edwardsi Group (Mattingly, 1981)
***Rachisoura* Theobald, 1910c**
Filipes Group (Lee, 1946)
Vanleeuweni Group (Lee, 1946)
***Tricholeptomyia* Dyar & Shannon, 1925a**
Delpilari Group (Mattingly, 1981) (monobasic)
Nepenthicola Group (Mattingly, 1981)
Belkini Subgroup (Mattingly, 1981) (monobasic)
Nepenthicola Subgroup (Mattingly, 1981)

Wyeomyia* Theobald, 1901a *sensu stricto
Caracula Group (Belkin *et al.*, 1970)
Hirsuta Group (Belkin *et al.*, 1970)
Hirsuta Subgroup (Belkin *et al.*, 1970)
Luna Subgroup (Belkin *et al.*, 1970) (monobasic)
Stellata Subgroup (Belkin *et al.*, 1970)
Pertinans Group (Belkin *et al.*, 1970)
Vanduzeei Group (Belkin *et al.*, 1970) (monobasic)
***Antunesmyia* Lane & Cerqueira, 1942**
***Caenomyiella* Harbach & Peyton, 1990** (monobasic)
***Cruzmyia* Lane & Cerqueira, 1942**
***Decamyia* Dyar, 1919b**
***Dendromyia* Theobald, 1903a**
***Dodecamyia* Dyar, 1919b** (monobasic)
***Exallomyia* Harbach & Peyton, 1992a**
***Hystatomyia* Dyar, 1919b**
Autocratica Group (Judd, 1998; Porter, 2009)
Circumcincta Group (Judd, 1998; Porter, 2009)
***Menolepis* Lutz, 1905** (monobasic)
***Miamyia* Dyar, 1919b**
***Nunezia* Dyar, 1928**
***Phoniomyia* Theobald, 1903a**
First Series (Lane & Cerqueira, 1942; Lane, 1953)
Second Series (Lane & Cerqueira, 1942; Lane, 1953)
Third Series (Lane & Cerqueira, 1942; Lane, 1953)
***Prosopolepis* Lutz, 1905** (monobasic)
***Spilonympha* Motta & Lourenço-de-Oliveira, 2005**
***Triamyia* Dyar, 1919b**
***Zinzala* Zavortink, 1986**

TOXORHYNCHITINI Lahille, 1904
Toxorhynchites* Theobald, 1901a *sensu stricto
Acaudatus Group (Simon *et al.*, 1983 largely; Steffan & Evenhuis, 1985)
Acaudatus Subgroup (Simon *et al.*, 1983 less Ater Subgroup; Steffan & Evenhuis, 1985)
Nepenthis Subgroup (Steffan & Evenhuis, 1985)
Ater Group (Simon *et al.*, 1983 as Ater Subgroup; Steffan & Evenhuis, 1985)
Brevipalpis Group (Edwards, 1941)
Brevipalpis Subgroup (Ribeiro, 1991) (monobasic)
Phytophagus Subgroup (Ribeiro, 1991; misspelled as Phytophygus)
Christophi Group (Simon *et al.*, 1983 principally; Steffan & Evenhuis, 1985)
Leicesteri Group (Simon *et al.*, 1983 principally; Steffan & Evenhuis, 1985)
Leicesteri Subgroup (Steffan & Evenhuis, 1985)
Manicatus Subgroup (Steffan & Evenhuis, 1985)

Magnificus Group (Steffan & Evenhuis, 1985)
Minimus Group (Steffan & Evenhuis, 1985)
Splendens Group (Simon *et al.*, 1983 principally; Steffan & Evenhuis, 1985)
***Afrorhynchus* Ribeiro, 1992a**
Pauliani Group (Ribeiro, 2004)
***Ankylorhynchus* Lutz, 1904** (in Bourroul, 1904)
***Lynchiella* Lahille, 1904**

URANOTAENIINI Lahille, 1904
Uranotaenia* Lynch Arribálzaga, 1891a,b *sensu stricto
Alba Section (da Cunha Ramos, 1993)
Alba Group (da Cunha Ramos, 1993)
Balfouri Group (da Cunha Ramos, 1993)
Bilineata Group (da Cunha Ramos, 1993)
Anisocheleomyia Section (Belkin, 1962)
Argyrotarsis Complex (Belkin, 1962; Lee *et al.*, 1989c); Argyrotarsis Series (Peyton, unpublished, USNM specimen label data)
Albescens Complex (Belkin, 1962; *Lee et al.*, 1989c)
Tibialis Group (King & Hoogstraal, 1947a)
Novaguinensis Complex (Peters, 1963d; Lee *et al.*, 1989c)
Ludlowae Series (Peyton, unpublished, USNM specimen label data)
Lateralis Complex (Lee *et al.*, 1989c)
Nivea Series (Peyton, unpublished, USNM specimen label data)
Anopheloides Section (da Cunha Ramos & Brunhes, 2004)
Anopheloides Group (da Cunha Ramos & Brunhes, 2004)
Neireti Group (da Cunha Ramos & Brunhes, 2004)
Caeruleocephala Section (da Cunha Ramos, 1993)
Caeruleocephala Group (da Cunha Ramos, 1993)
Madagascarica Group (da Cunha Ramos & Brunhes, 2004)
Campestris Series (Peyton, unpublished, USNM specimen label data)
Chorleyi Section (da Cunha Ramos, 1993)
Chorleyi Group (da Cunha Ramos, 1993) (monobasic)
Hamoni Group (da Cunha Ramos & Brunhes, 2004)
Dumonti Section (da Cunha Ramos & Brunhes, 2004)
Dumonti Group (da Cunha Ramos & Brunhes, 2004)
Hopkinsi Section (da Cunha Ramos, 1993)
Hopkinsi Group (da Cunha Ramos, 1993)
Leucoptera Series (Galindo *et al.*, 1954)
Lowii Series (Galindo *et al.*, 1954)
Micans Series (Peyton, unpublished, USNM specimen label data)
Pallidoventer Series (Galindo *et al.*, 1954)
Pulcherrima Series (Galindo *et al.*, 1954)
Testacea Series (Peyton, unpublished, USNM specimen label data)
Trilineata Series (Peyton, unpublished, USNM specimen label data)
Unimaculiala Series (Peyton, unpublished, USNM specimen label data)
Wysockii Section (Belkin, 1962) (monobasic)
***Pseudoficalbia* Theobald, 1912b**
Annulata Section (da Cunha Ramos, 1993; Section B of Peyton, 1972)
Annulata Series (Peyton, 1972)
Annulata Group (da Cunha Ramos, 1993)
Subgroup A (da Cunha Ramos, 1993)
Subgroup B (da Cunha Ramos, 1993)
Lavieri Group (da Cunha Ramos & Brunhes, 2004)
Maxima Series (Peyton, 1972, 1977; Anhydor Series of Galindo *et al.*, 1954)
Bicolor Series (Peyton, 1977; Atra, Colocasiae & Painei Sections of Belkin, 1962)
Bimaculata Series (Peyton, 1977)
Harrisoni Series (Peyton, 1977) (monobasic)
Mashonaensis Section (da Cunha Ramos, 1993)

Mashonaensis Group (da Cunha Ramos, 1993)
Nigripes Section (da Cunha Ramos, 1993)
Fusca Group (da Cunha Ramos, 1993; undefined by Belkin, 1962: 159) (monobasic)
Nigripes Group (da Cunha Ramos, 1993)
Ornata Group (da Cunha Ramos, 1993; undefined by Belkin, 1962: 159)
Recondita Series (Peyton, 1977)
Shillitonis Section (da Cunha Ramos, 1993; Shillitonis Series of Peyton, 1972)
Shillitonis Group (da Cunha Ramos, 1993)
Subgroup A (da Cunha Ramos, 1993)
Subgroup B (da Cunha Ramos, 1993)
Bicincta Group (da Cunha Ramos & Brunhes, 2004) (monobasic)
Spinosa Section (da Cunha Ramos & Brunhes, 2004)
Spinosa Group (da Cunha Ramos & Brunhes, 2004)

Species and Their Classifications

The list below includes all species, except nomina dubia, that were recognized as valid at the time of this publication. The names are listed in alphabetical order, with two or more identical names listed alphabetically by author, and by date where identical names are by the same author. Nomina dubia can be found in the lexicon of Harbach (2018b) and the catalogue of Wilkerson *et al.* (2021).

Aa

abdita (Barraud, 1931b): *Verrallina* (*Harbachius*)
abdita Peyton, 1977: *Uranotaenia* (*Pseudoficalbia*), Recondita Series
abebela Dyar & Knab, 1908: *Wyeomyia* (*Wyeomyia*) (not placed in a Group)
aberrans Harrison & Scanlon, 1975: *Anopheles sensu stricto*, Angusticorn Section, Anopheles Series, Aitkenii Group
ablechra Dyar & Knab, 1908: *Wyeomyia* (*Wyeomyia*) (not placed in a Group)
ablusa Harbach, 2009 (in Harbach & Howard, 2009): *Chagasia*
abnormalis Lane, 1936: *Culex sensu stricto*, Pipiens Group (not placed in a Subgroup)
abnormalis (Theobald, 1909): *Aedes* (*Aedimorphus*), Alboscutellatus Assemblage, Abnormalis Group
abominator Dyar & Knab, 1909: *Culex* (*Melanoconion*), Melanoconion Section, Peccator Group
abonnenci Clastrier, 1970c: *Culex* (*Melanoconion*), Melanoconion Section, Inhibitator Group, Inhibitator Subgroup
aboriginis Dyar, 1917: *Aedes* (*Ochlerotatus*), Communis Group, Punctor Subgroup
abserratus (Felt & Young, 1904): *Aedes* (*Ochlerotatus*), Communis Group, Punctor Subgroup
abstrusa Peyton, 1977: *Uranotaenia* (*Pseudoficalbia*), Bicolor Series
abyssinicus Edwards, 1941: *Culex sensu stricto*, Pipiens Group (not placed in a Subgroup)
abyssinicus Ribeiro, 1991: *Toxorhynchites* (*Toxorhynchites*), Brevipalpis Group, Brevipalpis Subgroup
acaci Baisas, 1946a: *Anopheles sensu stricto*, Angusticorn Section, Anopheles Series, Aitkenii Group
acanthotorynus Komp, 1937a: *Anopheles* (*Stethomyia*)
acaudatus (Leicester, 1908): *Toxorhynchites* (*Toxorhynchites*), Acaudatus Group, Acaudatus Subgroup
accelerans Root, 1927c: *Culex* (*Aedinus*)
achaetae (Leicester, 1908): *Heizmannia* (*Mattinglyia*)
acharistus Root, 1927c: *Culex sensu stricto*, Pipiens Group, Restuans Complex (not placed in a Subgroup)
aconitus Dönitz, 1902: *Anopheles* (*Cellia*), Myzomyia Series, Funestus Group, Aconitus Subgroup
acrostichalis Edwards, 1941: *Culex* (*Eumelanomyia*), Eumelanomyia Group
aculeatus Colless, 1965: *Culex* (*Lophoceraomyia*), Fraudatrix Group, Fraudatrix Subgroup, Quadripalpis Complex
aculeatus (Theobald, 1903a): *Aedes* (*Ochlerotatus*), Aculeatus Group
acutipalus Colless, 1965: *Culex* (*Lophoceraomyia*), Mammilifer Group, Brevipalpus Subgroup, Brevipalpus Complex
acutisentis Arnell, 1973: *Haemagogus sensu stricto*, Albomaculatus Section
adairi Kirkpatrick, 1926: *Culex* (*Lasiosiphon*) (the Subgenus is monobasic)
adamesi Sirivanakarn & Galindo, 1980: *Culex* (*Melanoconion*), Spissipes Section, Crybda Group, Pedroi Subgroup
adami Ferrara & Eouzan, 1974: *Eretmapodites*, Oedipodeios Group
adami Geoffroy, 1971b: *Aedes* (*Catageiomyia*)
adami (Hamon & Mouchet, 1955): *Culex* (*Eumelanomyia*), Rubinotus-rima Group, Rima Subgroup
adamowiczi Szadziewski, Sontag & Szwedo, 2019: *Coquillettidia* (fossil species)
adelpha Dyar & Knab, 1906a: *Wyeomyia* (*Wyeomyia*) (not placed in a Group)
adenensis Edwards, 1941: *Aedes* (*Aedimorphus*), Domesticus Assemblage, Vexans Subassemblage, Vexans Group
adentata van den Assem, 1959: *Tripteroides* (*Rachisoura*), Vanleeuweni Group
adersi (Edwards, 1917): *Aedes* (*Diceromyia*), Section A, Adersi Group
adersianus Edwards, 1941: *Culex* (*Eumelanomyia*), Eumelanomyia Group
adusta (Laffoon, 1946): *Verrallina* (*Neomacleaya*)
aegypti (Linnaeus, 1762): *Aedes* (*Stegomyia*), Aegypti Group
aenea Thurman, 1959: *Topomyia* (*Topomyia*)
aenescens Edwards, 1941: *Culex sensu stricto*, Pipiens Group, Trifilatus Subgroup
aeneus (Edwards, 1921a): *Tripteroides* (*Tripteroides*), Nitidoventer Group
aeneus (Evans, 1926): *Toxorhynchites* (*Afrorhynchus*) (not placed in a Group)
aenigma Harbach, 2000 (in Harbach & Peyton, 2000): *Onirion*, Brucei Group
aenigmaticus Cerqueira & Costa, 1946: *Aedes* (*Ochlerotatus*), Serratus Group
aequatorianna Leví-Castillo, 1953b: *Uranotaenia* (*Uranotaenia*), Pulcherrima Series
aequatorianna Leví-Castillo, 1954: *Wyeomyia* (not placed in a Subgenus)
aerarius McIntosh, 1975: *Aedes* (*Aedimorphus*), Domesticus Assemblage, Vexans Subassemblage, Quasiunivittatus Group
aeritinctus Galindo & Trapido, 1967: *Haemagogus sensu stricto*, Splendens Section
aestivus Sirivanakarn, 1977a: *Culex* (*Lophoceraomyia*), Fraudatrix Group, Fraudatrix Subgroup, Quadripalpis Complex

affinis (Edwards, 1913c): *Tripteroides* (*Rachionotomyia*), Aranoides Group, Affinis Subgroup

africana Neveu-Lemaire, 1906: *Aedeomyia sensu stricto*

africana (Theobald, 1901d): *Mansonia* (*Mansonioides*)

africanus (Theobald, 1901d): *Aedes* (*Stegomyia*), Africanus Group

agastyai Tewari & Hiriyan, 1992b: *Aedes* (*Tewarius*), Agastyai Group

agranensis Singh & Prakash, 2008: *Lutzia* (*Metalutzia*)

agrestis (Barraud, 1931b): *Verrallina* (*Neomacleaya*)

agrihanensis Bohart, 1957: *Aedes* (*Stegomyia*), Pandani Group

aguilari Baisas & Feliciano, 1953: *Zeugnomyia*

ahomi Chowdhury, 1929: *Anopheles sensu stricto*, Laticorn Section, Myzorhynchus Series, Barbirostris Group, Vanus Subgroup

ainshamsi Gad, Harbach & Harrison, 2006: *Anopheles* (*Cellia*), Neocellia Series (not placed in a Group)

airozai Lane, 1945b: *Culex* (*Phenacomyia*)

airosai Lane & Cerqueira, 1942: *Wyeomyia* (*Spilonympha*)

aitkeni Schick, 1970a: *Aedes* (*Protomacleaya*), Terrens Group, Aitkeni Subgroup (the Subgroup is monobasic)

aitkenii James, 1903 (in Theobald, 1903a): *Anopheles sensu stricto*, Angusticorn Section, Anopheles Series, Aitkenii Group

akkeshiensis Tanaka, 1998: *Aedes* (*Ochlerotatus*), Communis Group, Punctor Subgroup

akritos Forattini & Sallum, 1995: *Culex* (*Melanoconion*), Spissipes Section, Taeniopus Group

alani Forattini, 1965a: *Culex sensu stricto*, Pipiens Group, Salinarius Complex (not placed in a Subgroup)

alani Lane & Cerqueira, 1957: *Wyeomyia* (*Antunesmyia*)

alaskaensis (Ludlow, 1906b): *Culiseta sensu stricto*

alba Baker, 1936: *Orthopodomyia*, Bancroftia Section, Signifera Group, Signifera Subgroup

alba Theobald, 1901c: *Uranotaenia* (*Uranotaenia*), Alba Section, Alba Group

albertianus Edwards, 1941: *Culex* (*Eumelanomyia*), Rubinotus-rima Group, Rima Subgroup

albertoi Unti, 1941: *Anopheles* (*Nyssorhynchus*), Albimanus Section, Oswaldoi Series, Oswaldoi Group, Strodei Subgroup

albescens Edwards, 1921d: *Aedes* (*Rusticoidus*)

albescens Taylor, 1914c: *Uranotaenia* (*Uranotaenia*), Anisocheleomyia Section, Albescens Complex

albicosta (Chagas, 1908) (in Peryassú, 1908): *Coquillettidia* (*Pseudotaeniorhynchus*)

albicosta (Edwards, 1913b): *Aedes* (*Neomelaniconion*), Lineatopennis Group

albicosta (Lutz, 1904) (in Bourroul, 1904): *Orthopodomyia*, Bancroftia Section, Albicosta Group (the Group is monobasic)

albifasciatus (Macquart, 1838b): *Aedes* (*Ochlerotatus*), Albifasciatus Group

albifera (Prado, 1931a): *Coquillettidia* (*Pseudotaeniorhynchus*)

albigenu (Lutz, 1908) (in Peryassú, 1908): *Psorophora* (*Janthinosoma*)

albilabris Edwards, 1925: *Aedes* (*Rampamyia*)

albimanus da Cunha Ramos & Brunhes, 2004: *Uranotaenia* (*Pseudoficalbia*), Annulata Section, Annulata Series, Lavieri Group

albimanus Wiedemann, 1820: *Anopheles* (*Nyssorhynchus*), Albimanus Section, Albimanus Series (the Series is monobasic)

albinensis Bonne-Wepster & Bonne, 1920: *Culex* (*Melanoconion*), Melanoconion Section, Inhibitator Group, Inhibitator Subgroup

albinervis Edwards, 1929b: *Culex* (*Oculeomyia*) (not placed in a Complex)

albineus Séguy, 1923: *Aedes* (*Ochlerotatus*), Caspius Group

albinotata da Cunha Ramos & Brunhes, 2004: *Uranotaenia* (*Pseudoficalbia*), Shillitonis Section, Shillitonis Group (not placed in a Subgroup)

albipes (Edwards, 1922a): *Toxorhynchites* (*Toxorhynchites*), Leicesteri Group, Manicatus Subgroup

albipes Leicester, 1904 (in Theobald, 1904c): *Orthopodomyia*, Orthopodomyia Section, Albipes Group, Albipes Subgroup (the Subgroup is monobasic)

albipes Lutz, 1904 (in Bourroul, 1904): *Culex* (*Microculex*), Pleuristriatus Series

albipes Peyton, 1977: *Uranotaenia* (*Pseudoficalbia*), Bimaculata Series

albipes (Theobald, 1907): *Psorophora* (*Janthinosoma*)

albiprivus Theobald, 1903a: *Sabethes* (*Sabethes*), Group B

albiradius (Le Goff, Boussès & Brunhes, 2007): *Aedes* (*Neomelaniconion*) (not placed in a Group)

albitarsis (Brug, 1939): *Toxorhynchites* (*Toxorhynchites*), Splendens Group

albitarsis Lynch Arribálzaga, 1878: *Anopheles* (*Nyssorhynchus*), Argyritarsis Section, Albitarsis Series, Albitarsis Group, Albitarsis Complex (the Complex also includes informally designated Species F, G, H and I, see Brochero *et al.*, 2007 and Ruiz-Lopez *et al.*, 2012)

albiventris Edwards, 1922d: *Culex* (*Eumelanomyia*), Eumelanomyia Group

alboabdominalis Theobald, 1910c: *Uranotaenia* (*Uranotaenia*), Alba Section, Alba Group

alboannulata (Theobald, 1905a): *Uranotaenia* (*Uranotaenia*) (not placed in a Section, Series, Group or Complex)

alboannulatus (Macquart, 1850): *Aedes* (*Dobrotworskyius*)

alboapicus Schick, 1970a: *Aedes* (*Protomacleaya*), Terrens Group, Alboapicus Subgroup (the Subgroup is monobasic)

albocephala da Cunha Ramos & Brunhes, 2004: *Uranotaenia* (*Uranotaenia*), Anopheloides Section, Anopheloides Group

albocephalus (Theobald, 1903c): *Aedes* (*Polyleptiomyia*)

albocinctus (Barraud, 1924d): *Aedes* (*Hopkinsius*), Yamada Group
albodorsalis Fontenille & Brunhes, 1985: *Aedes* (*Aedimorphus*), Domesticus Assemblage (not placed in a Subassemblage)
albolateralis (Theobald, 1908b): *Aedes* (*Downsiomyia*), Niveus Group
albolineatus (Theobald, 1904c): *Aedes* (*Scutomyia*), Laffooni Group
albomaculatus Theobald, 1903a: *Haemagogus sensu stricto*, Albomaculatus Section
albomarginatus (Newstead, 1907) (in Newstead *et al.*, 1907): *Aedes* (*Pseudarmigeres*)
alboniveus Barraud, 1934: *Aedes* (*Downsiomyia*), Alboniveus Group
albonotatus (Coquillett, 1906a): *Aedes* (*Howardina*), Sexlineatus Section, Inaequalis Subassemblage
albopictus (Skuse, 1895): *Aedes* (*Stegomyia*), Scutellaris Group, Albopictus Subgroup
alboscutellatus Lee, 1946: *Tripteroides* (*Tripteroides*), Magnesianus Group
alboscutellatus (Theobald, 1905f): *Aedes* (*Aedimorphus*), Alboscutellatus Assemblage, Alboscutellatus Group
albosquamata Bonne-Wepster & Bonne, 1919: *Wyeomyia* (not placed in a Subgenus)
albosternopleura Peters, 1963e: *Uranotaenia* (*Uranotaenia*) (not placed in a Section, Series, Group or Complex)
albotaeniatus (Leicester, 1904) (in Theobald, 1904c): *Aedes* (*Danielsia*)
albotaeniatus (Theobald, 1903a): *Anopheles sensu stricto*, Laticorn Section, Myzorhynchus Series, Albotaeniatus Group
albothorax (Theobald, 1907): *Aedes* (*Neomelaniconion*), Lineatopennis Group
alboventralis (Theobald, 1910c): *Aedes* (*Aedimorphus*), Alboscutellatus Assemblage, Abnormalis Group
alcasidi Huang, 1972: *Aedes* (*Stegomyia*), Scutellaris Group, Scutellaris Subgroup
alcocki Bonne-Wepster & Bonne, 1920: *Culex* (*Melanoconion*), Melanoconion Section, Distinguendus Group, Distinguendus Subgroup
alektorovi Stackelberg, 1943: *Aedes* (*Bruceharrisonius*)
alexandrei Gornostaeva, 2005: *Aedes sensu stricto* (not placed in a Group)
algeriensis Theobald, 1903a: *Anopheles sensu stricto*, Angusticorn Section, Anopheles Series (not placed in a Group)
aliciae Duret, 1953: *Culex* (*Melanoconion*), Melanoconion Section, Conspirator Group
alienus Colless, 1957a: *Culex sensu stricto*, Sitiens Group, Vishnui Subgroup
alinkios Sallum & Hutchings, 2003: *Culex* (*Melanoconion*), Melanoconion Section, Bastagarius Group, Bastagarius Subgroup
alis Theobald, 1903a: *Culex sensu stricto*, Sitiens Group, Sitiens Subgroup
alius Lien, 1968d: *Aedes* (*Gilesius*)
aliyusopi Miyagi & Toma, 2014 (in Miyagi *et al.*, 2014): *Topomyia* (*Topomyia*)
alkatirii Toma, Miyagi & Syafruddin, 1995: *Armigeres sensu stricto*
allostigma Howard, Dyar & Knab, 1915: *Lutzia sensu stricto*
allotecnon Kumm, Komp & Ruiz, 1940: *Aedes* (*Howardina*), Sexlineatus Section, Sexlineatus Subassemblage
alocasicola Marks, 1947a: *Aedes* (*Finlaya*), Kochi Group
alogistus Dyar, 1918c: *Culex* (*Melanoconion*), Melanoconion Section, Pilosus Group, Caudelli Subgroup
alongensis Venhuis, 1940: *Anopheles sensu stricto*, Angusticorn Section, Anopheles Series, Alongensis Group
alongi Galliard & Ngu, 1947: *Aedes* (*Bothaella*)
aloponotum Dyar, 1917: *Aedes* (*Ochlerotatus*), Stimulans Group
alorensis Bonne-Wepster & Brug, 1932: *Aedes* (*Stegomyia*), Scutellaris Group, Scutellaris Subgroup
alorensis Sirivanakarn, 1977a: *Culex* (*Lophoceraomyia*), Fraudatrix Group, Fraudatrix Subgroup, Minutissimus Subgroup
alphus Colless, 1965: *Culex* (*Lophoceraomyia*), Fraudatrix Group, Fraudatrix Subgroup (Sirivanakarn, 1977a unnecessarily recognized a monobasic '*alphus* complex' for this species in the absence of evidence that if is more than one species)
alternans (Westwood, 1835): *Aedes* (*Mucidus*), Mucidus Group
alticola (Bonne-Wepster, 1948): *Aedes* (subgenus uncertain)
alticola Peters, 1963: *Uranotaenia* (*Uranotaenia*), Anisocheleomyia Section, Tibialis Group, Novaguinensis Complex
altivallis Bonne-Wepster, 1948: *Tripteroides* (*Polylepidomyia*), ?Argenteiventris Group
amabilis Schick, 1970a: *Aedes* (*Protomacleaya*), Terrens Group, Heteropus Subgroup
amaltheus de Meillon & Lavoipierre, 1944: *Aedes* (*Stegomyia*), Dendrophilus Group
amamiensis Tanaka, Mizusawa & Saugstad, 1979: *Aedes* (*Hulecoeteomyia*)
amaniensis van Someren & Hamon, 1964: *Culex* (*Eumelanomyia*), Rubinotus-rima Group, Rima Subgroup
amateuri Ortega & Zavortink, 2019 (in Ortega-Morales *et al.*, 2019): *Aedes* (*Ochlerotatus*), Scapularis Group (not placed in a Subgroup)
amazonensis (Lutz, 1905): *Culex* (*Aedinus*)
amazonensis (Theobald, 1901c): *Mansonia* (*Mansonia*)
amazonica Cerqueira, 1960b: *Psorophora* (*Janthinosoma*)
amazonica Levi-Castillo, 1954: *Wyeomyia* (not placed in a Subgenus)
amazonicus Gordon & Evans, 1922: *Sabethes* (*Sabethes*), Amazonicus Group
ambodimanga da Cunha Ramos & Brunhes, 2004: *Uranotaenia* (*Pseudoficalbia*), Annulata Section, Annulata Series, Lavieri Group
amboinensis (Doleschall, 1857): *Toxorhynchites* (*Toxorhynchites*), Splendens Group

ambreensis Rodhain & Boutonnier, 1983a: *Aedes* (*Ochlerotatus*), Caspius Group
ambremontis (Brunhes & Boussès, 2017) (in Brunhes *et al.*, 2017): *Aedes* (*Paulianius*), Coulangesi Group
ambremontis Brunhes & Hervy, 1995: *Orthopodomyia*, Orthopodomyia Section, Vernoni Group
ameliae Casal, 1967: *Culex sensu stricto*, Pipiens Group (not placed in a Subgroup)
amesii (Ludlow, 1903): *Aedes* (*Lorrainea*)
amharicus Hunt, Wilkerson & Coetzee, 2013 (in Coetzee *et al.*, 2013): *Anopheles* (*Cellia*), Pyretophorus Series, Gambiae Complex
amictus Edwards, 1921a: *Anopheles* (*Cellia*), Neomyzomyia Series (not placed in a Group)
amiensis Peters, 1963e: *Uranotaenia* (*Uranotaenia*) (not placed in a Section, Series, Group or Complex)
amitis Komp, 1936d: *Culex* (*Melanoconion*), Melanoconion Section, Inhibitator Group, Inhibitator Subgroup
amurensis Maslov, 1964: *Culiseta* (*Culicella*)
ananae Knight & Laffoon, 1946: *Aedes* (*Finlaya*), Poicilius Group
anastasionis Dyar, 1921e: *Haemagogus sensu stricto*, Albomaculatus Section
anchietai Corrêa & Ramalho, 1968: *Anopheles sensu stricto*, Laticorn Section, Arribalzagia Series
andamanensis Barraud, 1934: *Orthopodomyia*, Orthopodomyia Section, Albipes Group, Anopheloides Subgroup
andamanensis (Edwards, 1922a): *Verrallina* (*Neomacleaya*)
andavakae Doucet, 1950: *Uranotaenia* (*Uranotaenia*), Anopheloides Section, Neireti Group
andersoni Edwards, 1914b: *Culex sensu stricto*, Pipiens Group (not placed in a Subgroup)
andersoni Edwards, 1926a: *Aedes* (*Ochlerotatus*), Cunabulanus Group
andinum Leví-Castillo, 1953c: *Trichoprosopon*, Hyloconops Group, Pallidiventer Complex
andinus Leví-Castillo, 1954: *Limatus*
andinus Osorno-Mesa, 1944: *Haemagogus sensu stricto*, Albomaculatus Section
andreae Doucet, 1962: *Uranotaenia* (*Pseudoficalbia*), Shillitonis Section, Shillitonis Group, Subgroup B
andreae Harbach, 2018a: *Sabethes* (*Sabethinus*), Andreae Group
andreanus Edwards, 1927a: *Culex* (*Eumelanomyia*), Rubinotus-rima Group, Rubinotus Subgroup
andrewsi Edwards, 1926a: *Aedes* (*Stegomyia*), Scutellaris Group, Scutellaris Subgroup
andricus Root, 1927c: *Culex* (*Melanoconion*), Melanoconion Section, Intrincatus Group, Andricus Subgroup (the Subgroup is monobasic)
anduzei Cerqueira & Lane, 1944: *Culex* (*Carrollia*), Iridescens Group, Urichii Subgroup
aneles Dyar & Ludlow, 1922: *Culex* (*Melanoconion*), Melanoconion Section, Educator Group
anggiensis Bonne-Wepster, 1937: *Aedes* (*Patmarksia*)
angkoris Klein, 1977b: *Topomyia* (*Topomyia*)
angolae Ribeiro & Ramos, 1974: *Aedes* (*Aedimorphus*), Domesticus Assemblage, Vexans Subassemblage, Vexans Group
angolensis da Cunha Ramos, 1985: *Uranotaenia* (*Uranotaenia*), Caeruleocephala Section, Caeruleocephala Group
angolensis da Cunha Ramos & Ribeiro, 1992: *Eretmapodites*, Oedipodeios Group
angolensis Ribeiro, 1992c: *Toxorhynchites* (*Afrorhynchus*) (not placed in a Group)
angularis Sá & Sallum, 2022 (in Sá *et al.*, 2022): *Culex* (*Melanoconion*), Melanoconion Section, Educator Group
angustus Edwards, 1935a: *Aedes* (*Stegomyia*), Poweri Group
angustiplatus Evenhuis & Steffan, 1986: *Toxorhynchites* (*Toxorhynchites*), Acaudatus Group, Acaudatus Subgroup
angustivittatus Dyar & Knab, 1907a: *Aedes* (*Ochlerotatus*), Scapularis Group, Trivittatus Subgroup
anhydor Dyar, 1907: *Uranotaenia* (*Pseudoficalbia*), Annulata Section, Maxima Series
aningae Motta & Lourenço-de-Oliveira, 2005: *Wyeomyia* (*Spilonympha*)
anips Dyar, 1916a: *Culex* (*Melanoconion*), Melanoconion Section, Peccator Group
ankaratrensis Brunhes & Hervy, 1995: *Orthopodomyia*, Orthopodomyia Section, Vernoni Group
annandalei Barraud, 1926: *Uranotaenia* (*Uranotaenia*), Testacea Series
annandalei Prashad, 1918: *Anopheles sensu stricto*, Angusticorn Section, Lophoscelomyia Series, Asiaticus Group
annandalei (Theobald, 1910b): *Aedes* (*Stegomyia*), W-albus Group, Annandalei Subgroup
annettii (Theobald, 1901d): *Coquillettidia* (*Coquillettidia*), Crassipes Group
annularis van der Wulp, 1884: *Anopheles* (*Cellia*), Neocellia Series, Annularis Group, Annularis Complex (*An. annularis* consists of two sibling species denoted as Species A and B (Atrie *et al.*, 1999); which is the nominotypical species is not known)
annulata Leicester, 1908: *Mansonia* (*Mansonioides*)
annulata (Schrank, 1776): *Culiseta sensu stricto*
annulata Theobald, 1901d: *Uranotaenia* (*Pseudoficalbia*), Annulata Section, Annulata Series, Annulata Group, Subgroup B
annulatus Haga, 1930: *Anopheles* (*Cellia*), Neomyzomyia Series (not placed in a Group)
annulifera (Theobald, 1901c): *Mansonia* (*Mansonioides*)
annulioris Theobald, 1901b: *Culex* (*Oculeomyia*) (not placed in a Complex)
annulipalpis Lynch Arribálzaga, 1878: *Anopheles sensu stricto*, Angusticorn Section (not placed in a Series or Group)
annulipalpis (Theobald, 1910a): *Armigeres* (*Leicesteria*)
annulipes (Meigen, 1830): *Aedes* (*Ochlerotatus*), Cantans Group

annulipes Walker, 1856: *Anopheles* (*Cellia*), Neomyzomyia Series, Annulipes Complex (not placed in a Group) (*An. annulipes* consists of 17 sibling species denoted as Species A–Q (Foley *et al.*, 2007); which is the nominotypical species is not known)
annulitarsis (Leicester, 1908): *Armigeres* (*Leicesteria*)
annulirostris Skuse, 1889: *Culex sensu stricto*, Sitiens Group, Sitiens Subgroup
annulirostris (Theobald, 1905h): *Aedes* (*Christophersiomyia*)
annuliventris (Blanchard, 1852): *Culex sensu stricto* (not placed in a Group)
annulus Theobald, 1901b: *Culex sensu stricto*, Sitiens Group, Vishnui Subgroup
anomalophyllus Komp, 1936e: *Anopheles* (*Nyssorhynchus*), Albimanus Section, Oswaldoi Series, Oswaldoi Group, Oswaldoi Subgroup
anopheloides Brunhes & Razafindrasolo, 1975: *Uranotaenia* (*Uranotaenia*), Anopheloides Section, Anopheloides Group
anopheloides (Giles, 1903) (in Thomson, 1903): *Orthopodomyia*, Orthopodomyia Section, Albipes Group, Anopheloides Subgroup
anoplicitus Forattini & Sallum, 1989a: *Culex* (*Melanoconion*), Melanoconion Section (not placed in a Group)
antalahaensis da Cunha Ramos & Brunhes, 2004: *Uranotaenia* (*Pseudoficalbia*), Spinosa Section, Spinosa Group
antennalis Bohart & Farner, 1944: *Tripteroides* (*Tripteroides*), Nitidoventer Group
antennalis Taylor, 1919: *Uranotaenia* (*Uranotaenia*), Anisocheleomyia Section, Tibialis Group
antennatus (Becker, 1903): *Culex sensu stricto*, Pipiens Group, Decens Subgroup
anthica Ribeiro, Pavan, da Silva, Galvão, Lourenço-de-Oliveira & Motta, 2021: *Wyeomyia* (*Decamyia*)
antillummagnorum Dyar, 1928: *Culex* (*Micraedes*), Bisulcatus Group
antipodea Dobrotworsky, 1962: *Culiseta* (*Climacura*)
antipodeus (Edwards, 1920a): *Aedes* (*Ochlerotatus*), Taeniorhynchus Group, Vigilax Subgroup
antiquus Borkent & Grimaldi, 2004: *Burmaculex* (fossil species)
antunesi Galvão & Franco do Amaral, 1940: *Anopheles* (*Nyssorhynchus*), Myzorhynchella Section
antunesi Lane & Guimarães, 1937: *Wyeomyia* (*Phoniomyia*), Second Series
antunesi Lane & Whitman, 1943: *Culex* (*Carrollia*), Iridescens Group, Iridescens Subgroup
aobae Belkin, 1962: *Aedes* (*Stegomyia*), Scutellaris Group, Scutellaris Subgroup
apeteticus Howard, Dyar & Knab, 1913: *Culex* (*Melanoconion*), Melanoconion Section, Educator Group
aphobema Dyar, 1918c: *Wyeomyia* (*Dodecamyia*) (the Subgenus is monobasic)
aphylactus Root, 1927c: *Culex* (*Microculex*), Inimitabilis Series
aphyllus Talaga, 2020 (in Talaga *et al.*, 2020b): *Culex* (*Melanoconion*), Melanoconion Section, Educator Group
apicalis Adams, 1903: *Culex* (*Neoculex*), Territans Group
apicalis Theobald, 1903a: *Uranotaenia* (*Uranotaenia*), Pulcherrima Series
apicimacula Dyar & Knab, 1906a: *Anopheles sensu stricto*, Laticorn Section, Arribalzagia Series
apicinus Philippi, 1865: *Culex* (*Phalangomyia*)
apicoannulatus (Edwards, 1912a): *Aedes* (*Elpeytonius*)
apicoargenteus (Theobald, 1909): *Aedes* (*Stegomyia*), Apicoargenteus Group
apicopilosus Cornel & Mayi, 2020 (in Cornel *et al.*, 2020): *Culex* (*Culiciomyia*), Nebulosus Group
apicosquamata da Cunha Ramos & Brunhes, 2004: *Uranotaenia* (*Pseudoficalbia*), Annulata Section, Annulata Series, Lavieri Group
apicotaeniata Theobald, 1909: *Uranotaenia* (*Pseudoficalbia*), Annulata Section, Annulata Series, Annulata Group, Subgroup B
apicotriangulatus (Theobald, 1910c): *Tripteroides* (*Polylepidomyia*), Atripes Group
apoci Marsh, 1933: *Anopheles* (*Cellia*), Myzomyia Series (not placed in a Group)
apoensis Baisas & Ubaldo-Pagayon, 1953: *Tripteroides* (*Tricholeptomyia*), Nepenthicola Group, Nepenthicola Subgroup
apoensis Bohart & Farner, 1944: *Armigeres sensu stricto*
aporonoma Dyar & Knab, 1906d: *Wyeomyia* (*Triamyia*)
approximata Peyton, 1977: *Uranotaenia* (*Pseudoficalbia*), Bicolor Series
apsarae Klein, 1977b: *Topomyia* (*Topomyia*)
aquarius Strickman, 1990: *Culex sensu stricto*, Pipiens Group (not placed in a Subgroup)
aquasalis Curry, 1932: *Anopheles* (*Nyssorhynchus*), Albimanus Section, Oswaldoi Series, Oswaldoi Group, Oswaldoi Subgroup
arabiensis (Patton, 1905): *Aedes* (*Aedimorphus*), Domesticus Assemblage, Vexans Subassemblage, Vexans Group
arabiensis Patton, 1905: *Anopheles* (*Cellia*), Pyretophorus Series, Gambiae Complex
aranoides (Theobald, 1901c): *Tripteroides* (*Rachionotomyia*), Aranoides Group, Aranoides Subgroup
arawak Berlin, 1969b: *Culex* (*Micraedes*), Bisulcatus Group
arbieeni Salem, 1938: *Culex* (*Maillotia*), Hortensis Group
arborea Galindo, Carpenter & Trapido, 1951: *Wyeomyia* (*Exallomyia*)
arborealis Bonne-Wepster & Bonne, 1920: *Aedes* (*Howardina*), Fulvithorax Section, Arborealis Subassemblage
arboricollis (de Charmoy, 1908): *Orthopodomyia*, Orthopodomyia Section, Arboricollis Group (the Group is monobasic)
arboricolus Galindo & Mendez, 1961: *Culex* (*Melanoconion*), Melanoconion Section, Pilosus Group, Caudelli Subgroup
arboricolus Knight & Rozeboom, 1946: *Aedes* (*Scutomyia*), Laffooni Group

arboricolus Zavortink, 1970b: *Anopheles sensu stricto*, Angusticorn Section, Anopheles Series, Plumbeus Group
archegus Dyar, 1929b: *Culex sensu stricto*, Pipiens Group, Salinarius Complex (not placed in a Subgroup)
ardensis (Theobald, 1905g): *Anopheles* (*Cellia*), Neomyzomyia Series, Ardensis Group
arenivaga Marks, 1968: *Culiseta* (subgenus uncertain)
argenteitarsis Brug, 1932: *Aedes* (*Patmarksia*)
argenteiventris (Theobald, 1905f): *Tripteroides* (*Polylepidomyia*), Argenteiventris Group
argenteolobatus (Gough, 1910): *Anopheles* (*Cellia*), Cellia Series (not placed in a Group)
argenteopunctatus (Theobald, 1901b): *Aedes* (*Catageiomyia*)
argenteopunctatus (Ventrillon, 1905): *Culex sensu stricto*, Pipiens Group (not placed in a Subgroup)
argenteorostris (Bonne-Wepster & Bonne, 1920): *Wyeomyia* (not placed in a Subgenus)
argenteoscutellatus Carter & Wijesundara, 1948: *Aedes* (*Aedimorphus*), Alboscutellatus Assemblage, Alboscutellatus Group
argenteoventralis Leicester, 1908: *Topomyia* (*Suaymyia*)
argenteoventralis (Theobald, 1909): *Aedes* (*Pseudarmigeres*)
argentinus Brèthes, 1912: *Anopheles sensu stricto*, Angusticorn Section, Anopheles Series, Pseudopunctipennis Group
argentipleura da Cunha Ramos & Brunhes, 2004: *Uranotaenia* (*Uranotaenia*), Caeruleocephala Section, Madagascarica Group
arguellesi Baisas, 1935c: *Uranotaenia* (*Uranotaenia*), Campestris Series
argyritarsis Robineau-Desvoidy, 1827: *Anopheles* (*Nyssorhynchus*), Argyritarsis Section, Argyritarsis Series, Argyritarsis Group
argyrites Dyar & Nuñez Tovar, 1927a: *Aedes* (*Howardina*), Walkeri Section, Walkeri Group
argyromeris Dyar & Ludlow, 1921: *Haemagogus sensu stricto*, Splendens Section
argyronotum Belkin, 1962: *Aedes* (*Patmarksia*)
argyropalpis Leicester, 1908: *Topomyia* (*Topomyia*)
argyropus (Swellengrebel, 1914): *Anopheles sensu stricto*, Laticorn Section, Myzorhynchus Series, Hyrcanus Group (not placed in a Subgroup)
argyropus (Walker, 1848): *Maorigoeldia* (the Genus is monobasic)
argyrotarsis Leicester, 1908: *Uranotaenia* (*Uranotaenia*), Anisocheleomyia Section, Argyrotarsis Complex
argyrothorax Bonne-Wepster & Bonne, 1920: *Aedes* (*Protomacleaya*), Terrens Group, Thorntoni Subgroup
argyrurus (Edwards, 1934) (in Barraud, 1934): *Udaya*
argyrurus Edwards, 1936: *Eretmapodites*, Inornatus Group
arizonensis Bohart, 1949: *Culex* (*Neoculex*), Territans Group
arribalzagae (Theobald, 1903a): *Coquillettidia* (*Pseudotaeniorhynchus*)
artemievi Gordeev, Zvantsov, Goriacheva, Shĕikevich & Ezhov, 2005: *Anopheles sensu stricto*, Angusticorn Section, Anopheles Series, Maculipennis Group, Maculipennis Subgroup
arthrostigma (Lutz, 1905): *Wyeomyia* (*Wyeomyia*) (not placed in a Group)
arthuri Unti, 1941: *Anopheles* (*Nyssorhynchus*), Albimanus Section, Oswaldoi Series, Oswaldoi Group, Strodei Subgroup, Arthuri Complex
articularis Philippi, 1865: *Culex sensu stricto*, Pipiens Group (not placed in a Subgroup)
arundinariae Kasper, 2020: *Aedes* (*Ochlerotatus*) (not placed in a Group)
aruni Sobti, 1968: *Anopheles* (*Cellia*), Myzomyia Series, Funestus Group, Funestus Subgroup
arvernensis (Piton, 1936): *Neoculicites* (fossil species)
ascidiicola de Meijere, 1910: *Uranotaenia* (*Pseudoficalbia*), Bimaculata Series
ashworthi Edwards, 1921a: *Aedes* (*Halaedes*)
asiaticus Edwards, 1926c: *Aedes* (*Ochlerotatus*), Caspius Group
asiaticus Leicester, 1903: *Anopheles sensu stricto*, Angusticorn Section, Lophoscelomyia Series, Asiaticus Group, Asiaticus Subgroup (the Subgroup is monobasic)
assamensis Bhattacharyya, Tewari, Prakash, Mohapatra & Mahanta, 2004b: *Verrallina* (*Neomacleaya*)
assamensis (Theobald, 1908b): *Aedes* (*Phagomyia*)
asteliae Belkin, 1968a: *Culex sensu stricto*, Pipiens Group, Trifilatus Subgroup
astridianus de Meillon, 1942: *Culex sensu stricto*, Pipiens Group (not placed in a Subgroup)
asulleptus (Theobald, 1903a): *Limatus*
atacamensis González & Sallum, 2010: *Anopheles* (*Nyssorhynchus*), Argyritarsis Section, Argyritarsis Series, Pictipennis Group
atactavittatus Arnell, 1976: *Aedes* (*Ochlerotatus*), Scapularis Group, Trivittatus Subgroup
ater (Daniels, 1908): *Toxorhynchites* (*Toxorhynchites*), Ater Group
atlantica (Edwards, 1932b): *Culiseta sensu stricto*
atlanticus Adames, 1971: *Deinocerites*, Spanius Group
atlanticus Dyar & Knab, 1906d: *Aedes* (*Ochlerotatus*), Serratus Group
atra (Lee, 1944b): *Culiseta* (*Culicella*)
atra Theobald, 1905f: *Uranotaenia* (*Pseudoficalbia*), Bicolor Series
atracus Colless, 1960: *Culex* (*Lophoceraomyia*), Fraudatrix Group, Fraudatrix Subgroup, Fraudatrix Complex
atrata Belkin, Heinemann & Page, 1970: *Wyeomyia* (*Wyeomyia*), Hirsuta Group, Hirsuta Subgroup
atratipes Skuse, 1889: *Anopheles sensu stricto*, Angusticorn Section, Anopheles Series, Atratipes Group

atratus Theobald, 1901c: *Culex* (*Melanoconion*), Melanoconion Section, Atratus Group
atria (Barraud, 1928): *Verrallina* (*Neomacleaya*)
atriceps Edwards, 1926a: *Culex sensu stricto*, Atriceps Group
atriisimilis (Tanaka & Mizusawa, 1973): *Verrallina* (*Neomacleaya*)
atripes (Skuse, 1889): *Tripteroides* (*Polylepidomyia*), Atripes Group
atritarsalis (Dobrotworsky, 1954): *Culiseta* (*Neotheobaldia*)
atropalpus (Coquillett, 1902d): *Aedes* (*Georgecraigius*), Atropalpus Group (Zavortink, 1972; supersedes the Georgecraigius Group of Wilkerson *et al.*, 2015)
atroparvus van Thiel, 1927: *Anopheles sensu stricto*, Angusticorn Section, Anopheles Series, Maculipennis Group, Maculipennis Subgroup
atropos Dyar & Knab, 1906b: *Anopheles sensu stricto*, Angusticorn Section, Anopheles Series, Maculipennis Group (not placed in a Subgroup)
aurantapex Edwards, 1914b: *Culex* (*Oculeomyia*) (not placed in a Complex)
auranticauda Lane, 1992: *Toxorhynchites* (*Toxorhynchites*), Christophi Group
aurantius (Theobald, 1907): *Aedes* (*Mucidus*), Pardomyia Group
aurata (Dobrotworsky, 1962): *Coquillettidia* (*Coquillettidia*), Crassipes Group
aurata (Doucet, 1951a): *Mimomyia* (*Ingramia*), Group B (*Ravenalites*)
auratus Grabham, 1906: *Aedes* (*Ochlerotatus*), Scapularis Group, Tortilis Subgroup
aurea Brug, 1932: *Heizmannia sensu stricto*
aurea (Edwards, 1915a): *Coquillettidia* (*Coquillettidia*), Crassipes Group
aurea (Leicester, 1908): *Mimomyia* (*Mimomyia*), Hybrida Group
aureoantennata Ferrara, 1973: *Orthopodomyia*, Orthopodomyia Section, Nkolbissonensis Group
aureochaeta (Leicester, 1908): *Heizmannia sensu stricto*
aureolineatus Berlin, 1969a: *Aedes* (*Howardina*), Fulvithorax Section, Arborealis Subassemblage
aureolineatus (Leicester, 1908): *Armigeres sensu stricto*
aureonotatus Duret & Barreto, 1956: *Culex* (*Melanoconion*), Melanoconion Section, Erraticus Group, Erraticus Subgroup
aureosquammata (Ludlow, 1909c): *Coquillettidia* (*Coquillettidia*), Crassipes Group
aureostriatus (Doleschall, 1857): *Aedes* (*Bruceharrisonius*)
aureoventer (Theobald, 1910a): *Topomyia* (*Topomyia*)
aurescens (Lutz, 1905): *Sabethes* (*Peytonulus*)
aureus Gutsevich, 1955: *Aedes* (*Neomelaniconion*), Lineatopennis Group
aureus Lane & Whitman, 1951: *Culex* (*Microculex*), Pleuristriatus Series
auriceps Brug, 1939: *Topomyia* (*Suaymyia*)
auridorsum Edwards, 1922d: *Aedes* (*Molpemyia*)
aurifer (Coquillett, 1903a): *Aedes* (*Ochlerotatus*), Pullatus Group
aurifluus (Edwards, 1921c): *Toxorhynchites* (*Toxorhynchites*), Christophi Group
aurimargo Edwards, 1922d: *Aedes* (*Leptosomatomyia*) (the Subgenus is monobasic)
auripes (Edwards, 1935b): *Toxorhynchites* (*Toxorhynchites*), Leicesteri Group, Manicatus Subgroup
aurirostris (Watson, 1910): *Anopheles* (*Cellia*), Neomyzomyia Series (not placed in a Group)
aurites (Theobald, 1901d): *Coquillettidia* (*Coquillettidia*), Crassipes Group
aurites (Theobald, 1907): *Aedes* (*Howardina*), Walkeri Section, Aurites Group
aurivittatus Cerqueira, 1943: *Aedes* (*Howardina*), Sexlineatus Section, Aurivittatus Group
auronitens Edwards, 1922a: *Aedes* (subgenus uncertain)
aurotaeniatus Edwards, 1922a: *Aedes* (*Belkinius*) (the Subgenus is monobasic)
aurovenatus Worth, 1960: *Aedes* (*Neomelaniconion*), Lineatopennis Group
austenii (Theobald, 1905b): *Anopheles* (*Cellia*), Myzomyia Series, Marshallii Group
australicus Dobrotworsky & Drummond, 1953: *Culex sensu stricto*, Pipiens Group, Pipiens Subgroup
australiensis (Theobald, 1910c): *Aedes* (subgenus uncertain)
australis (Erichson, 1842): *Aedes* (*Halaedes*)
autocratica Dyar & Knab, 1906d: *Wyeomyia* (*Hystatomyia*), Autocratica Group
auyantepuiensis Harbach & Navarro, 1996: *Anopheles* (*Kerteszia*) (not placed in a Complex)
ævianus de Meillon, 1943: *Culex* (*Maillotia*), Seyrigi Group
ævistylus Brug, 1939: *Aedes* (*Finlaya*), Poicilius Group
axillicolus Steffan, 1979: *Culex* (*Acallyntrum*), Perkinsi Group
axitiosus Kulasekera, Knight & Harbach, 1990: *Aedes* (*Downsiomyia*), Niveus Group
azaniae Bailly-Choumara, 1960: *Anopheles* (*Cellia*), Myzomyia Series (not placed in a Group)
azevedoi Ribeiro, 1969: *Anopheles* (*Cellia*), Paramyzomyia Series, Cinereus Group
aztecus Hoffmann, 1935: *Anopheles sensu stricto*, Angusticorn Section, Anopheles Series, Maculipennis Group (not placed in a Subgroup)

azureosquamata (Bonne-Wepster, 1948): *Verrallina sensu stricto*, Carmenti Series
azurini Basio, 1971: *Armigeres sensu stricto*
azurini Toma, Miyagi & Cabrera, 1984: *Culex* (*Culiciomyia*), Tricuspis Group
azymus Dyar & Knab, 1906b: *Culex* (*Microculex*), Pleuristriatus Series

Bb

babahoyensis Levi-Castillo, 1953d: *Culex* (*Carrollia*), Iridescens Group, Iridescens Subgroup
baezai Gater, 1933: *Anopheles sensu stricto*, Laticorn Section, Myzorhynchus Series, Umbrosus Group, Baezai Subgroup (the Subgroup is monobasic)
bahama Dyar & Knab, 1906a: *Wyeomyia* (*Wyeomyia*) (not placed in a Group)
bahamensis Berlin, 1969a: *Aedes* (*Howardina*), Sexlineatus Section, Inaequalis Subassemblage
bahamensis Dyar & Knab, 1906d: *Culex sensu stricto*, Pipiens Group, Tarsalis Subgroup
bahiensis Duret, 1969b: *Culex* (*Melanoconion*), Melanoconion Section, Distinguendus Group, Putumayensis Subgroup
bahri (Edwards, 1914b): *Culex* (*Culiciomyia*), Fragilis Group
baileyi Edwards, 1929b: *Anopheles sensu stricto*, Angusticorn Section, Anopheles Series, Lindesayi Group, Gigas Subgroup, Baileyi Complex (the Complex also includes informally designated Species D, see Somboon *et al.*, 2020b, 2022a)
bailyi Barraud, 1929a: *Hodgesia*
bailyi Barraud, 1934: *Culex* (*Culiciomyia*), Shebbearei Group
baimaii Sallum & Peyton, 2005 (in Sallum *et al.*, 2005b): *Anopheles* (*Cellia*), Neomyzomyia Series, Leucosphyrus Group, Leucosphyrus Subgroup, Dirus Complex
baisasi Colless, 1957b: *Anopheles* (*Cellia*), Neomyzomyia Series, Leucosphyrus Group, Leucosphyrus Subgroup (not placed in a Complex)
baisasi Knight & Hull, 1951: *Aedes* (*Geoskusea*) (not placed in a Group)
baisasi Sirivanakarn, 1972: *Culex* (*Eumelanomyia*), Mochthogenes Group, Hinglungensis Subgroup
baisasi Stone & Thurman, 1958: *Armigeres sensu stricto*
balabacensis Baisas, 1936: *Anopheles* (*Cellia*), Neomyzomyia Series, Leucosphyrus Group, Leucosphyrus Subgroup, Leucosphyrus Complex
balerensis Mendoza, 1947: *Anopheles sensu stricto*, Laticorn Section, Myzorhynchus Series, Albotaeniatus Group
balfouri Theobald, 1904b: *Uranotaenia* (*Uranotaenia*), Alba Section, Balfouri Group
baltae Porter, 2009: *Wyeomyia* (*Hystatomyia*), Autocratica Group
balteatus Macdonald, 1960: *Armigeres* (*Leicesteria*)
bambiotai Geoffroy, 1987: *Aedes* (*Aedimorphus*), Domesticus Assemblage, Domesticus Subassemblage, Domesticus Group
bamborum Rozeboom & Komp, 1948: *Culex* (*Anoedioporpa*), Conservator Group
bambusa (Yamada, 1917): *Tripteroides* (*Tripteroides*), Nitidoventer Group
bambusae Edwards, 1935a: *Aedes* (*Stegomyia*), Dendrophilus Group
bambusaihole Dong, Zhou & Dong, 1997a: *Topomyia* (*Suaymyia*)
bambusicola da Cunha Ramos & Brunhes, 2004: *Uranotaenia* (*Pseudoficalbia*), Shillitonis Section, Shillitonis Group (not placed in a Subgroup)
bambusicola (Lutz & Neiva, 1913): *Toxorhynchites* (*Lynchiella*)
bambusicolus Knight & Rozeboom, 1946: *Aedes* (*Scutomyia*), Hoogstraali Group
bambusicolus Komp, 1937b: *Anopheles* (*Kerteszia*) (not placed in a Complex)
bananea Wolfs, 1958: *Aedes* (*Diceromyia*), Section A, Adersi Group
bancoi Geoffroy, 1987: *Aedes* (*Aedimorphus*), Domesticus Assemblage, Domesticus Subassemblage, Domesticus Group
bancroftianus (Edwards, 1921a): *Aedes* (*Pseudoskusea*)
bancrofti (Taylor, 1914b): *Verrallina sensu stricto*, Butleri Series
bancroftii Giles, 1902: *Anopheles sensu stricto*, Laticorn Section, Myzorhynchus Series, Bancroftii Group
bandoengensis Brug, 1939: *Culex* (*Lophoceraomyia*), Mammilifer Group, Minor Line, Minor Complex
banksensis Maffi & Tenorio, 1977: *Culex sensu stricto*, Pipiens Group, Trifilatus Subgroup
banksi Edwards, 1922a: *Aedes* (*Collessius*), Alloeomyia Group
bannaensis Gong & Lu, 1995: *Topomyia* (*Suaymyia*)
baolini Gong, 1989b: *Topomyia* (*Topomyia*)
barberellus Evans, 1932: *Anopheles* (*Cellia*), Myzomyia Series (not placed in a Group)
barberi Coquillett, 1903b: *Anopheles sensu stricto*, Angusticorn Section, Anopheles Series, Plumbeus Group
barbipes Edwards, 1913b: *Toxorhynchites* (*Toxorhynchites*), Brevipalpis Group, Phytophagus Subgroup
barbirostris van der Wulp, 1884: *Anopheles sensu stricto*, Laticorn Section, Myzorhynchus Series, Barbirostris Group, Barbirostris Subgroup, Barbirostris Complex
barbiventris Brug, 1938: *Anopheles sensu stricto*, Laticorn Section, Myzorhynchus Series, Bancroftii Group

barbumbrosus Strickland & Chowdhury, 1927a: *Anopheles sensu stricto*, Laticorn Section, Myzorhynchus Series, Barbirostris Group, Vanus Subgroup
barbus Baisas, 1946c: *Topomyia* (*Topomyia*)
baresi Cerqueira, 1960a: *Haemagogus sensu stricto*, Albomaculatus Section
barianensis James, 1911 (in James & Liston, 1911): *Anopheles sensu stricto*, Angusticorn Section, Anopheles Series, Plumbeus Group
barnardi Edwards, 1924a: *Aedes* (*Hopkinsius*), Hopkinsius Group
barnesi Belkin, 1953a: *Uranotaenia* (*Uranotaenia*) (not placed in a Section, Series, Group or Complex)
barraudi Baisas & Ubaldo-Pagayon, 1953: *Tripteroides* (*Tricholeptomyia*), Nepenthicola Group, Nepenthicola Subgroup
barraudi Edwards, 1922a: *Culex sensu stricto*, Sitiens Group, Barraudi Subgroup
barraudi (Edwards, 1934) (in Barraud, 1934): *Aedes* (*Paraedes*), Barraudi Assemblage
barretoi Adames, 1971: *Deinocerites*, Dyari Group
barrinus Bram, 1967a: *Culex* (*Culiciomyia*), Fragilis Group
basilewskyi Leleup, 1957: *Anopheles* (*Cellia*), Myzomyia Series, Demeilloni Group
bastagarius Dyar & Knab, 1906b: *Culex* (*Melanoconion*), Melanoconion Section, Bastagarius Group, Bastagarius Subgroup
batesi Lane & Cerqueira, 1942: *Sabethes* (*Sabethes*), Group B
batesi Rozeboom & Komp, 1948: *Culex* (*Melanoconion*), Melanoconion Section, Evansae Group
bathana (Dyar, 1928) (in Curry, 1928): *Chagasia*
becki Belkin, 1962: *Aedes* (*Geoskusea*), Fimbripes Group
becki Belkin, 1962: *Culex* (*Lophoceraomyia*), Fraudatrix Group, Fraudatrix Subgroup, Solomonis Complex
bedfordi Edwards, 1936: *Aedes* (*Catageiomyia*)
behningi Martini, 1926: *Aedes* (*Ochlerotatus*), Cantans Group
bejaranoi Duret, 1953: *Culex* (*Melanoconion*), Melanoconion Section, Inhibitator Group, Inhibitator Subgroup
bejaranoi Martinez, Carcavallo & Prosen, 1960: *Aedes* (*Ochlerotatus*) (not placed in a Group)
bekkui Mogi, 1977: *Aedes* (*Edwardsaedes*)
beklemishevi Stegniy & Kabanova, 1976: *Anopheles sensu stricto*, Angusticorn Section, Anopheles Series, Maculipennis Group, Quadrimaculatus Subgroup
belemensis Duret & Damasceno, 1955: *Culex* (*Anoedioporpa*), Conservator Group
belenrae Rueda, 2005: *Anopheles sensu stricto*, Laticorn Section, Myzorhynchus Series, Hyrcanus Group (not placed in a Subgroup)
belisarioi Neiva, 1908: *Sabethes* (*Sabethes*), Group A
belkini Adames, 1971: *Deinocerites*, Spanius Group
belkini Baisas & Ubaldo-Pagayon, 1953: *Tripteroides* (*Tricholeptomyia*), Nepenthicola Group, Belkini Subgroup (the Subgroup is monobasic)
belkini Grjebine, 1979: *Uranotaenia* (*Pseudoficalbia*), Shillitonis Section, Shillitonis Group (not placed in a Subgroup)
belkini Stone & Penn, 1948: *Culex* (*Acallyntrum*), Perkinsi Group
bellator Dyar & Knab, 1906b: *Anopheles* (*Kerteszia*), Bellator Complex (*An. bellator* consists of two sibling species denoted as Species 1 (*bellator sensu stricto*) and 2) (Bourke *et al.*, 2023)
belleci (Le Goff, Boussès & Brunhes, 2007): *Aedes* (*Neomelaniconion*), Lineatopennis Group
benarrochi Gabaldon, Cova-Garcia & Lopez, 1941a: *Anopheles* (*Nyssorhynchus*), Albimanus Section, Oswaldoi Series, Oswaldoi Group, Strodei Subgroup, Benarrochi Complex (*An. benarrochi* consists of two sibling species, the nominotypical species and the other denoted as Species B) (Ruiz *et al.*, 2005)
bengalensis Barraud, 1934: *Culex* (*Lophoceraomyia*), Mammilifer Group, Minor Line, Minor Complex
bengalensis Puri, 1930: *Anopheles sensu stricto*, Angusticorn Section, Anopheles Series, Aitkenii Group
bengalensis Rosenberg & Evenhuis, 1985: *Toxorhynchites* (*Toxorhynchites*), Magnificus Group
benguetensis King, 1931: *Anopheles sensu stricto*, Angusticorn Section, Anopheles Series, Lindesayi Group, Lindesayi Subgroup, Lindesayi Complex
benoiti Wolfs, 1964 (in Benoit, 1964): *Uranotaenia* (*Uranotaenia*), Caeruleocephala Section, Caeruleocephala Group
bequaerti Dyar & Shannon, 1925b: *Culex* (*Melanoconion*), Melanoconion Section, Intrincatus Group, Intrincatus Subgroup
bequaerti Wolfs, 1947a: *Aedes* (*Neomelaniconion*), Lineatopennis Group
bergerardi Pajot & Geoffroy, 1971: *Aedes* (*Neomelaniconion*) (not placed in a Group)
berghei Vincke & Leleup, 1949: *Anopheles* (*Cellia*), Myzomyia Series, Marshallii Group
bergi Belkin, 1962: *Culex* (*Lophoceraomyia*), Fraudatrix Group, Fraudatrix Subgroup, Bergi Complex
bergrothi (Edwards, 1921b): *Culiseta sensu stricto*
berlandi Séguy, 1921: *Aedes* (*Ochlerotatus*), Caspius Group
berlini Schick, 1970a: *Aedes* (*Protomacleaya*), Terrens Group, Insolitus Subgroup
bernardi (Doucet, 1950): *Mimomyia* (*Ingramia*), Group B (Ravenalites)
bertii Cova García & Rausseo, 1964: *Uranotaenia* (*Uranotaenia*) (not placed in a Section, Series, Group or Complex)
bertrami Schick, 1970a: *Aedes* (*Protomacleaya*), Terrens Group, Bertrami Subgroup (the Subgroup is monobasic)

bervoetsi D'Haenens, 1961: *Anopheles* (*Cellia*), Myzomyia Series (not placed in a Group)
beta Séguy, 1924: *Culex sensu stricto* (not placed in a Group)
bevisi (Edwards, 1915a): *Aedes* (*Aedimorphus*), Domesticus Assemblage, Vexans Subassemblage, Quasiunivittatus Group
beytouti (Doucet, 1951b): *Mimomyia* (*Ingramia*), Group B (Ravenalites)
bhayungi Thurman & Thurman, 1958: *Armigeres sensu stricto*
bhutanensis Somboon & Harbach, 2020 (in Somboon *et al.*, 2020a): *Aedes* (*Hulecoeteomyia*)
bhutanensis Somboon, Namgay & Harbach, 2021 (in Somboon *et al.*, 2022a, first published online 17 November 2021): *Anopheles sensu stricto*, Angusticorn Section, Anopheles Series, Lindesayi Group, Gigas Subgroup, Baileyi Complex
bhutanensis Somboon, Namgay & Harbach, 2021 (in Somboon *et al.*, 2021c): *Culex sensu stricto*, Sitiens Group, Mimeticus Subgroup, Mimeticus Complex
bhutanensis Somboon, Namgay & Harbach, 2022c: *Uranotaenia* (*Pseudoficalbia*), Bimaculata Series
bibulus Dyar, 1920b: *Culex* (*Melanoconion*), Melanoconion Section, Educator Group
bicincta da Cunha Ramos & Brunhes, 2004: *Uranotaenia* (*Pseudoficalbia*), Shillitonis Section, Bicincta Group (the Group is monobasic)
bicki Stone & Penn, 1947: *Culex* (*Acallyntrum*), Bicki Group
bickleyi Forattini, 1965a: *Culex sensu stricto*, Pipiens Group (not placed in a Subgroup)
bickleyi Thurman, 1959: *Toxorhynchites* (*Toxorhynchites*), Christophi Group
bicolor Leicester, 1908: *Uranotaenia* (*Pseudoficalbia*), Bicolor Series
bicornis (Root, 1928) (in Dyar, 1928): *Wyeomyia* (*Nunezia*)
bicornutus (Theobald, 1910a): *Culex* (*Lophoceraomyia*), Mammilifer Group, Minor Line, Minor Complex
bicristatus Thurman & Winkler, 1950: *Aedes* (*Rusticoidus*)
bidens Dyar, 1922e: *Culex sensu stricto*, Pipiens Group, Tarsalis Subgroup
bidentata da Cunha Ramos & Brunhes, 2004: *Uranotaenia* (*Uranotaenia*), Chorleyi Section, Hamoni Group
bifasciata da Cunha Ramos & Brunhes, 2004: *Uranotaenia* (*Pseudoficalbia*), Annulata Section, Annulata Series, Lavieri Group
bifoliata (King & Hoogstraal, 1947b): *Verrallina sensu stricto*, Carmenti Series
bifoliolatus Duret & Barreto, 1956: *Culex* (*Melanoconion*), Melanoconion Section, Bastagarius Group, Iolambdis Subgroup
bifurcata Dong, Wang & Lu, 1995: *Topomyia* (*Topomyia*)
bigoti (Bellardi, 1862): *Lutzia sensu stricto*
bihaicolus Dyar & Núñez Tovar, 1928: *Culex* (*Carrollia*), Bihaicola Group
bihamatus Edwards, 1926a: *Culex sensu stricto*, Sitiens Group (not placed in a Subgroup)
bilineata Theobald, 1909: *Uranotaenia* (*Uranotaenia*), Alba Section, Bilineata Group
bilineatus Theobald, 1903a: *Culex sensu stricto*, Pipiens Group, Salinarius Complex (not placed in a Subgroup)
bimaculata Leicester, 1908: *Uranotaenia* (*Pseudoficalbia*), Bimaculata Series
bimaculatus (Coquillett, 1902c): *Aedes* (*Ochlerotatus*), Fulvus Group
bimaculiala Leicester, 1908: *Uranotaenia* (*Uranotaenia*), Micans Series
bimaculipes (Theobald, 1905f): *Tripteroides* (*Tripteroides*), Bimaculipes Group
binigrolineatus Knight & Rozeboom, 1945: *Culex* (*Acallyntrum*), Bicki Group
binotatus Belkin, 1950a: *Tripteroides* (*Tripteroides*), Bimaculipes Group
biocellatus (Taylor, 1914b): *Aedes* (subgenus uncertain)
bipartipes Dyar & Knab, 1906a: *Sabethes* (*Sabethes*), Group A
biscaynensis Zavortink & O'Meara, 1999: *Culex* (*Micraedes*), Bisulcatus Group
biskraensis Brunhes, 1999: *Aedes* (*Ochlerotatus*), Communis Group (not placed in a Subgroup)
bisquamatus Lee, 1946: *Tripteroides* (*Rachisoura*), Vanleeuweni Group
bisulcatus (Coquillett, 1906a): *Culex* (*Micraedes*), Bisulcatus Group
bitaeniorhynchus Giles, 1901a: *Culex* (*Oculeomyia*), Bitaeniorhynchus Complex
biunguiculatus (Harbach, 2014) (in Harbach & Taai, 2014): *Aedes* (*Nyctomyia*)
blacklocki Evans, 1925a: *Aedes* (*Stegomyia*), Apicoargenteus Group
bogotanus Arnell, 1976: *Aedes* (*Ochlerotatus*), Scapularis Group, Bogotanus Subgroup
boharti Brookman & Reeves, 1950: *Culex* (*Neoculex*), Territans Group
boharti Knight & Rozeboom, 1946: *Aedes* (*Scutomyia*), Laffooni Group
bokorensis Klein & Sirivanakarn, 1970: *Culex* (*Eumelanomyia*), Mochthogenes Group, Bokorensis Subgroup (the Subgroup is monobasic)
bolensis Edwards, 1936: *Aedes* (*Neomelaniconion*), Lineatopennis Group
bolii Sirivanakarn, 1968: *Culex* (*Lophoceraomyia*), Mammilifer Group, Bolii Subgroup (the Subgroup is monobasic)
boliviensis (Theobald, 1905f): *Anopheles* (*Kerteszia*), Boliviensis Complex (*An. boliviensis* consists of three sibling species denoted as Species 1, 2 and 3) (Bourke *et al.*, 2023); which is the nominotypical species is not known)
boneti Gil Collado, 1936: *Aedes* (*Bifidistylus*)
boninensis Bohart, 1957: *Culex* (*Sirivanakarnius*) (the Subgenus is monobasic)
bonneae Dyar & Knab, 1919: *Culex sensu stricto*, Pipiens Group (not placed in a Subgroup)

bonneae Edwards, 1930b: *Mansonia* (*Mansonioides*)
bonneae Mattingly, 1958a: *Aedes* (*Paraedes*), Ostentatio Assemblage
bonneae Root, 1927b: *Chagasia*
bonnei Dyar, 1921e: *Culex* (*Carrollia*), Iridescens Group, Iridescens Subgroup
bonnei (Lane & Cerqueira, 1942): *Wyeomyia* (*Phoniomyia*), First Series
bonneti Belkin, 1962: *Tripteroides* (*Tripteroides*), Bimaculipes Group
borneensis McArthur, 1949: *Anopheles sensu stricto*, Angusticorn Section, Anopheles Series, Aitkenii Group
boshelli Osorno-Mesa, 1944: *Haemagogus sensu stricto*, Splendens Section
bosseri Grjebine, 1979: *Uranotaenia* (*Pseudoficalbia*), Nigripes Section, Nigripes Group
bougainvillensis (Belkin, 1962): *Mimomyia* (*Etorleptiomyia*), Group B (Dixomyia)
bougainvillensis Marks, 1947a: *Aedes* (*Finlaya*), Kochi Group, Bougainvillensis Complex
bougainvillensis Steffan, 1979: *Culex* (*Acallyntrum*), Perkinsi Group
bourrouli (Lutz, 1905): *Wyeomyia* (*Spilonympha*)
boussesi da Cunha Ramos & Brunhes, 2004: *Uranotaenia* (*Pseudoficalbia*), Nigripes Section, Nigripes Group
brachiatus Hutchings & Sallum, 2008: *Culex* (*Melanoconion*), Melanoconion Section, Bastagarius Group, Bastagarius Subgroup
bradleyi King, 1939: *Anopheles sensu stricto*, Angusticorn Section, Anopheles Series, Punctipennis Group, Crucians Complex
brami Forattini, Rabello & Souza Lopes, 1967: *Culex sensu stricto*, Pipiens Group (not placed in a Subgroup)
braziliensis (Chagas, 1907): *Anopheles* (*Nyssorhynchus*), Argyritarsis Section, Albitarsis Series, Braziliensis Group (the Group is monobasic)
braziliensis Gordon & Evans, 1922: *Aedes* (*Protomacleaya*), Terrens Group, Terrens Subgroup
breedensis Muspratt, 1953: *Aedes* (*Ochlerotatus*), Breedensis Group
breinli (Taylor, 1914d): *Armigeres sensu stricto*, Malayi Group
brelandi Zavortink, 1972: *Aedes* (*Protomacleaya*), Triseriatus Group
brenguesi Brunhes & Ravaonjanahary, 1973: *Culex* (*Eumelanomyia*), Rubinotus-rima Group (not placed in a Subgroup)
brenguesi Rickenbach & Lombrici, 1975: *Eretmapodites*, Leucopous Group
brethesi Dyar, 1919c: *Culex sensu stricto*, Pipiens Group, Restuans Complex (not placed in a Subgroup)
breviculus Senevet & Abonnenc, 1939a: *Culex* (*Tinolestes*)
brevipalpis Brug, 1934: *Tripteroides* (*Tripteroides*), Bimaculipes Group
brevipalpis (Giles, 1902): *Culex* (*Eumelanomyia*), Protomelanoconion Group
brevipalpis Roper, 1914: *Anopheles sensu stricto*, Laticorn Section, Myzorhynchus Series, Umbrosus Group (not placed in a Subgroup)
brevipalpis Theobald, 1901b: *Toxorhynchites* (*Toxorhynchites*), Brevipalpis Group, Brevipalpis Subgroup
brevipalpus (Theobald, 1905f): *Culex* (*Lophoceraomyia*), Mammilifer Group, Brevipalpus Subgroup, Brevipalpus Complex
brevipes (da Costa Lima, 1931a): *Trichoprosopon*, Hyloconops Group, Pallidiventer Complex
brevirhynchus Brug, 1934: *Tripteroides* (*Rachisoura*), Vanleeuweni Group
brevirostris Reid, 1950: *Anopheles sensu stricto*, Laticorn Section, Myzorhynchus Series, Umbrosus Group (not placed in a Subgroup)
brevis Berlin, 1969a: *Aedes* (*Howardina*), Sexlineatus Section (not placed in a Group/Subassemblage)
brevis Edwards, 1941: *Eretmapodites*, Plioleucus Group
breviseta da Cunha Ramos & Brunhes, 2004: *Uranotaenia* (*Pseudoficalbia*), Annulata Section, Annulata Series, Lavieri Group
brevispinosus Bonne-Wepster & Bonne, 1920: *Culex sensu stricto*, Pipiens Group, Tarsalis Subgroup
brevitibia (Edwards, 1914c): *Aedes* (*Alanstonea*)
brevivittatus Berlin, 1969a: *Aedes* (*Howardina*), Sexlineatus Section, Sexlineatus Subassemblage
bricenoi Cova García, Pulido, Escalante de Ugueto, Amarista & Mora R., 1987c: *Uranotaenia* (*Uranotaenia*) (not placed in a Section, Series, Group or Complex)
briseis Dyar, 1925e: *Uranotaenia* (*Uranotaenia*), Lowii Series
britteni Marks & Hodgkin, 1958: *Aedes* (subgenus uncertain)
brohieri Edwards, 1929b: *Anopheles* (*Cellia*), Myzomyia Series, Marshallii Group
bromeliae (Theobald, 1911a): *Aedes* (*Stegomyia*), Simpsoni Group, Simpsoni Complex
brottesi Rickenbach, 1967: *Eretmapodites*, Chrysogaster Group
broussesi Edwards, 1929a: *Anopheles* (*Cellia*), Neocellia Series (not placed in a Group)
browni Komp, 1936d: *Culex* (*Anoedioporpa*), Conservator Group
browni Mattingly, 1955 (in Mattingly & Brown, 1955): *Uranotaenia* (*Pseudoficalbia*), Bicolor Series
brownscutumus Dong, Zhou & Dong, 1999: *Aedes* (*Bothaella*)
brucei (del Ponte & Cerqueira, 1938): *Onirion*, Brucei Group
brucei Service, 1960: *Anopheles* (*Cellia*), Myzomyia Series, Funestus Group, Rivulorum Subgroup
brucei Wilkerson, 2021 (in Wilkerson *et al.*, 2021): *Aedes* (*Paulianius*), Tiptoni Group
brumpti Doucet, 1951b: *Uranotaenia* (*Pseudoficalbia*), Nigripes Section, Nigripes Group
brumpti Gailiard, 1931a: *Culex sensu stricto* (not placed in a Group)

brumpti Hamon & Rickenbach, 1955a: *Anopheles* (*Cellia*), Cellia Series (not placed in a Group)
brunhesi Grjebine, 1979: *Uranotaenia* (*Pseudoficalbia*), Shillitonis Section, Shillitonis Group (not placed in a Subgroup)
brunhesi Reinert, 1999b: *Aedes* (*Zavortinkius*), Monetus Group
brunhesi Ribeiro, 2004: *Toxorhynchites* (*Afrorhynchus*), Pauliani Group
brunnipes (Theobald, 1910c): *Anopheles* (*Cellia*), Myzomyia Series (not placed in a Group)
brygooi Brunhes, 1971: *Aedes* (*Zavortinkius*), Brygooi Group
brygooi Grjebine, 1986: *Mimomyia* (*Ingramia*), Group A (*Ravenalites s. str.*), Spinosa Complex
buenaventura Schick, 1970a: *Aedes* (*Protomacleaya*), Terrens Group, Buenaventura Subgroup (the Subgroup is monobasic)
bukavuensis Wolfs, 1947b: *Culex sensu stricto*, Pipiens Group (not placed in a Subgroup)
bulkleyi Causey, 1937: *Anopheles sensu stricto*, Angusticorn Section, Lophoscelomyia Series (not placed in a Group)
burgeri Zavortink, 1972: *Aedes* (*Protomacleaya*), Kompi Group
burjaticus Kuchartshuk, 1973: *Aedes* (*Ochlerotatus*), Communis Group (not placed in a Subgroup)
burnetti Belkin, 1962: *Aedes* (*Finlaya*), Kochi Group
burnsi Basio & Reisen, 1971: *Aedes* (*Stegomyia*), Maehleri Group
burpengaryensis (Theobald, 1905g): *Aedes* (*Ochlerotatus*), Burpengaryensis Group
busckii (Coquillett, 1906d): *Aedes* (*Howardina*), Sexlineatus Section, Sexlineatus Subassemblage
bustamantei Galvão, 1955: *Anopheles sensu stricto*, Laticorn Section, Arribalzagia Series
butleri (Theobald, 1901c): *Verrallina sensu stricto*, Butleri Series
buxtoni Belkin, 1962: *Aedes* (*Patmarksia*)
buxtoni (Edwards, 1923b): *Coquillettidia* (*Coquillettidia*), Perturbans Group
buxtoni Edwards, 1926a: *Culex* (*Lophoceraomyia*), Fraudatrix Group, Fraudatrix Subgroup, Buxtoni Complex
buxtoni Service, 1958: *Anopheles* (*Cellia*), Neomyzomyia Series, Ardensis Group
bwamba van Someren, 1950c: *Aedes* (*Aedimorphus*), Alboscutellatus Assemblage, Abnormalis Group
bwambae White, 1985: *Anopheles* (*Cellia*), Pyretophorus Series, Gambiae Complex
bwambanus Edwards, 1941: *Culex sensu stricto*, Pipiens Group (not placed in a Subgroup)

Cc

caatingensis Andrade & la Corte, 2021: *Toxorhynchites* (*Lynchiella*)
caballus (Theobald, 1912a): *Aedes* (*Ochlerotatus*), Caballus Group
cabrerai Miyagi, Toma & Rivera, 1983: *Topomyia* (*Topomyia*)
cachani (Doucet, 1950): *Uranotaenia* (*Pseudoficalbia*), Nigripes Section, Nigripes Group
cacharanus (Barraud, 1923e): *Aedes* (*Phagomyia*)
cacozelus Marks, 1963b: *Aedes* (*Ochlerotatus*) (not placed in a Group)
caecus (Theobald, 1901b): *Aedes* (*Aedimorphus*), Domesticus Assemblage, Mediolineatus Subassemblage, Caecus Group
caeruleocephala Theobald, 1901d: *Uranotaenia* (*Uranotaenia*), Caeruleocephala Section, Caeruleocephala Group
caeruleocephalus (Leicester, 1908): *Tripteroides* (*Tripteroides*), Nitidoventer Group
caillardi Rickenbach, Ferrara & Eouzan, 1968: *Eretmapodites*, Oedipodeios Group
cairnsensis (Taylor, 1919): *Culex* (subgenus uncertain)
cairnsensis Taylor, 1919: *Hodgesia*
calabarensis Edwards, 1941: *Culex* (*Eumelanomyia*), Rubinotus-rima Group, Rima Subgroup
calabyi Marks, 1963a: *Aedes* (*Macleaya*), Chaetocruiomyia Group
calcariae Marks, 1957a: *Aedes* (*Ochlerotatus*), Flavifrons Group
calceatus Edwards, 1924 (in Haworth, 1924): *Aedes* (*Stegomyia*), Poweri Group
calderoni Wilkerson, 1991: *Anopheles sensu stricto*, Laticorn Section, Arribalzagia Series
caledonicus (Edwards, 1922d): *Tripteroides* (*Polylepidomyia*), Caledonicus Group
caliginosa Philip, 1931: *Uranotaenia* (*Uranotaenia*), Caeruleocephala Section, Caeruleocephala Group
caliginosus de Meillon, 1943: *Anopheles sensu stricto*, Laticorn Section, Myzorhynchus Series, Coustani Group
caliginosus (Graham, 1910): *Aedes* (*Aedimorphus*), Domesticus Assemblage, Vexans Subassemblage, Quasiunivittatus Group
calosomata Dyar & Knab, 1907c: *Uranotaenia* (*Uranotaenia*), Pallidoventer Series
calumnior Belkin, Heinemann & Page, 1970: *Aedes* (*Ochlerotatus*), Scapularis Group, Obturbator Subgroup (provisional placement)
calurus Edwards, 1935a: *Culex sensu stricto*, Pipiens Group (not placed in a Subgroup)
camaronis Ribeiro, 1991: *Toxorhynchites* (*Toxorhynchites*), Brevipalpis Group, Phytophagus Subgroup
cambournaci Hamon & Gandara, 1955: *Culex* (*Culiciomyia*), Nebulosus Group
cameronensis Edwards, 1929b: *Anopheles sensu stricto*, Angusticorn Section, Anopheles Series, Lindesayi Group, Lindesayi Subgroup, Lindesayi Complex
cameroni de Meillon & Evans, 1935: *Anopheles* (*Cellia*), Neomyzomyia Series, Rhodesiensis Group
campana Schick, 1970a: *Aedes* (*Protomacleaya*), Terrens Group, Galindoi Subgroup

campestris Dyar & Knab, 1907d: *Aedes* (*Ochlerotatus*), Caspius Group
campestris Leicester, 1908: *Uranotaenia* (*Uranotaenia*), Campestris Series
campestris Reid, 1962: *Anopheles sensu stricto*, Laticorn Section, Myzorhynchus Series, Barbirostris Group, Barbirostris Subgroup, Barbirostris Complex
campilunati Carter & Wijesundara, 1948: *Culex* (*Eumelanomyia*), Mochthogenes Group, Pluvialis Subgroup
camposi Dyar, 1925b: *Culex sensu stricto*, Pipiens Group, Coronator Complex (not placed in a Subgroup)
camptorhynchus (Thomson, 1869): *Aedes* (*Ochlerotatus*), Perkinsi Group
campylostylus (Laffoon, 1946): *Verrallina* (*Neomacleaya*)
canaanensis Lane & Whitman, 1943: *Culex* (*Anoedioporpa*), Conservator Group
canadensis (Theobald, 1901c): *Aedes* (*Ochlerotatus*), Culicada Group (the Group is monobasic)
cancer Theobald, 1901c: *Deinocerites*, Cancer Group
cancricomes Edwards, 1922a: *Aedes* (*Cancraedes*)
candelabrifer Brug, 1939: *Armigeres sensu stricto*
candidoscutellum Marks, 1947b: *Aedes* (subgenus uncertain)
canorii Floch & Abonnenc, 1945c: *Anopheles* (*Stethomyia*)
cantans (Meigen, 1818): *Aedes* (*Ochlerotatus*), Cantans Group
cantator (Coquillett, 1903a): *Aedes* (*Ochlerotatus*), Stimulans Group
capelai da Cunha Ramos, 1993: *Uranotaenia* (*Pseudoficalbia*), Shillitonis Section, Shillitonis Group, Subgroup B
capelai Ribeiro, 1992b: *Toxorhynchites* (*Afrorhynchus*) (not placed in a Group)
capensis de Meillon, 1935: *Culex* (*Maillotia*), Seyrigi Group
capensis Edwards, 1924a: *Aedes* (*Albuginosus*)
capricornii Lutz, 1904 (in Bourroul, 1904): *Haemagogus sensu stricto*, Albomaculatus Section
caracula Dyar & Núñez Tovar, 1928: *Wyeomyia* (*Wyeomyia*), Caracula Group
carcinicola da Cunha Ramos & Brunhes, 2004: *Uranotaenia* (*Pseudoficalbia*), Annulata Section, Annulata Series, Annulata Group, Subgroup A
carcinophilus Dyar & Knab, 1906d: *Culex* (*Melanoconion*), Melanoconion Section, Inhibitator Group, Inhibitator Subgroup
carcinoxenus de Oliveira Castro, 1932: *Culex sensu stricto*, Pipiens Group (not placed in a Subgroup)
caribeanus Galindo & Blanton, 1954: *Culex* (*Melanoconion*), Melanoconion Section, Atratus Group
carioca Lane & Whitman, 1951: *Culex* (*Microculex*), Imitator Series
carleti Brunhes & Ravaonjanahary, 1971: *Culex sensu stricto*, Pipiens Group (not placed in a Subgroup)
carmenti (Edwards, 1924b): *Verrallina sensu stricto*, Carmenti Series
carnevalei Brunhes, Le Goff & Geoffroy, 1999: *Anopheles* (*Cellia*), Neomyzomyia Series, Ardensis Group, Nili Complex
carolinensis Bohart & Ingram, 1946b: *Culex* (*Lophoceraomyia*), Fraudatrix Group, Fraudatrix Subgroup, Hilli Complex
caroni Adam, 1961: *Anopheles* (*Cellia*), Neomyzomyia Series, Smithii Group
carrilloi (Sutil O. & Pulido F., 1978): *Wyeomyia* (*Exallomyia*)
carteri Amerasinghe, 1993: *Heizmannia sensu stricto*
carteri Edwards, 1936: *Aedes* (*Neomelaniconion*) (not placed in a Group)
carteri Evans & de Meillon, 1933: *Anopheles* (*Cellia*), Myzomyia Series, Demeilloni Group
cartroni (Ventrillon, 1906b): *Aedes* (*Skusea*)
casali Schick, 1970b: *Aedes* (*Protomacleaya*), Terrens Group, Insolitus Subgroup
caspius (Pallas, 1771): *Aedes* (*Ochlerotatus*), Caspius Group
castaneus Sirivanakarn, 1973a: *Culex* (*Lophoceraomyia*), Fraudatrix Group, Fraudatrix Subgroup, Pseudorubithoracis Complex
castelli Hamon, 1957: *Culex sensu stricto*, Sitiens Group (not placed in a Subgroup)
castor de Meillon & Lavoipierre, 1944: *Culex* (*Eumelanomyia*), Mochthogenes Group, Inconspicuosus Subgroup
castrensis Edwards, 1922a: *Culex* (*Eumelanomyia*), Mochthogenes Group, Hinglungensis Subgroup
castroi Casal & García, 1967b: *Culex* (*Phytotelmatomyia*)
castroi Lane & Cerqueira, 1942: *Trichoprosopon*, Hyloconops Group, Pallidiventer Complex
cataphylla Dyar, 1916b: *Aedes* (*Ochlerotatus*), Communis Group, Leucomelas Subgroup
cataractarum Edwards, 1923b: *Culex* (*Eumelanomyia*), Mochthogenes Group, Hinglungensis Subgroup
catasticta Knab, 1909: *Aedeomyia sensu stricto*
catesi (Lien, 1968c): *Heizmannia* (*Mattinglyia*)
catharinensis (da Costa Lima, Guitton & Ferreira, 1962): *Toxorhynchites* (*Ankylorhynchus*)
cauchensis Floch & Abonnenc, 1945a: *Culex* (*Tinolestes*)
caudatus Clastrier, 1970c: *Culex* (*Melanoconion*), Melanoconion Section, Inhibitator Group, Egcymon Subgroup
caudelli (Dyar & Knab, 1906d): *Culex* (*Melanoconion*), Melanoconion Section, Pilosus Group, Caudelli Subgroup
cauta (Barraud, 1928): *Verrallina* (*Neomacleaya*)
cavalierii García & Casal, 1967: *Toxorhynchites* (*Lynchiella*)
cavaticus Reinert, 1979: *Aedes* (*Isoaedes*) (the Subgenus is monobasic)
cavernicola Mattingly, 1954b: *Uranotaenia* (*Pseudoficalbia*), Annulata Section, Annulata Series, Annulata Group, Subgroup A

cavernicola (Rattanarithikul & Harbach, 2007) (in Harbach *et al.*, 2007b): *Aedes* (*Borichinda*) (the Subgenus is monobasic)
cedecei Stone & Hair, 1968: *Culex* (*Melanoconion*), Spissipes Section, Taeniopus Group
celaenocephala Dyar & Knab, 1906a: *Wyeomyia* (*Wyeomyia*) (not placed in a Group)
celatum Peyton & Harbach, 2000 (in Harbach & Peyton, 2000): *Onirion*, Brucei Group
celebicus Mattingly, 1959: *Aedes* (*Lorrainea*)
celeste Dyar & Nuñez Tovar, 1927a: *Haemagogus sensu stricto*, Splendens Section
centropunctatus (Theobald, 1913): *Aedes* (*Aedimorphus*), Domesticus Assemblage, Vexans Subassemblage, Vexans Group
ceramensis Sirivanakarn & Kurihara, 1973: *Culex* (*Culiciomyia*), Fragilis Group
cerqueirai (Barreto & Coutinho, 1944): *Mansonia* (*Mansonia*)
cerqueirai Nascimento-Pereira, Neves, Lourenço-de-Oliveira & Motta, 2021 (in Nascimento-Pereira *et al.*, 2021): *Sabethes* (*Sabethes*), Group A
cerqueirai (Stone, 1944): *Runchomyia* (*Runchomyia*)
cerqueirai Valencia, 1973: *Culex* (*Carrollia*), Iridescens Group, Iridescens Subgroup
cesari del Ponte & Cerqueira, 1938: *Wyeomyia* (not placed in a Subgenus)
ceylonensis (Theobald, 1905h): *Tripteroides* (*Rachionotomyia*), Aranoides Group, Aranoides Subgroup
ceylonica (Edwards, 1917): *Verrallina* (*Neomacleaya*)
ceyx (von Heyden, 1870): *Neoculicites* (fossil species)
chaetoventralis (Theobald, 1910c): *Culex* (*Neoculex*), Pseudomelanoconia Group
chagasi (da Costa Lima, 1935a): *Mansonia* (*Mansonia*)
chaguanco Casal, García & Fernández, 1968: *Culex* (*Anoedioporpa*), Conservator Group
chaii Miyagi & Toma, 2012 (in Miyagi *et al.*, 2012): *Topomyia* (*Topomyia*)
chalcocephala Dyar & Knab, 1906a: *Wyeomyia* (not placed in a Subgenus)
chalcospilans Dyar, 1921b: *Haemagogus sensu stricto*, Splendens Section
chamberlaini Ludlow, 1904b: *Mimomyia* (*Mimomyia*), Hybrida Group
chamboni Cornet, 1968: *Aedes* (*Catageiomyia*)
champerico (Dyar & Knab, 1906a): *Psorophora* (*Janthinosoma*)
chandi Edwards, 1922a: *Heizmannia sensu stricto*
changuinolae Galindo & Blanton, 1954: *Culex* (*Melanoconion*), Melanoconion Section, Evansae Group
charmion Dyar, 1928: *Wyeomyia* (*Wyeomyia*) (not placed in a Group)
chathamicus (Dumbleton, 1962): *Opifex* (*Nothoskusea*) (the Subgenus is monobasic)
chaussieri Edwards, 1923a: *Aedes* (*Stegomyia*), Poweri Group
chauveti Brunhes & Rambelo, 1968: *Culex* (*Eumelanomyia*), Rubinotus-rima Group, Rima Subgroup
cheesmanae Mattingly & Marks, 1955: *Culex* (*Neoculex*), Pseudomelanoconia Group
chelli (Edwards, 1915a): *Aedes* (*Ochlerotatus*), Caballus Group
chemulpoensis Yamada, 1921: *Aedes* (*Stegomyia*), Pseudonigeria Group
chengi Lien, 1968c: *Heizmannia sensu stricto*
cheni Dong, Wang & Lu, 2003 (in Dong *et al.*, 2003b): *Culex* (*Culiciomyia*), Dispectus Group
cheni Lien, 1968a: *Tripteroides* (*Tripteroides*), Nitidoventer Group
cherrapunjiensis (Natarajan, Rajavel & Jambulingam, 2016): *Aedes* (*Hulecoeteomyia*)
chiangmaiensis Somboon & Harbach, 2019: *Lutzia* (*Metalutzia*)
chidesteri Dyar, 1921c: *Culex sensu stricto*, Pipiens Group (not placed in a Subgroup)
chionodes Belkin, 1962: *Aedes* (*Christophersiomyia*)
chionotum Zavortink, 1972: *Aedes* (*Protomacleaya*), Kompi Group
chiriquiensis Komp, 1936c: *Anopheles sensu stricto*, Angusticorn Section, Anopheles Series, Pseudopunctipennis Group
chitae Duret, 1967b: *Culex sensu stricto*, Pipiens Group, Tarsalis Subgroup
chloropterus (von Humboldt, 1819): *Sabethes* (*Sabethoides*)
chocoensis Porter & Wolff E., 2004: *Wyeomyia* (*Hystatomyia*), Circumcincta Group
chodukini Martini, 1929: *Anopheles sensu stricto*, Laticorn Section, Myzorhynchus Series, Hyrcanus Group (not placed in a Subgroup)
chorleyi Edwards, 1936: *Uranotaenia* (*Uranotaenia*), Chorleyi Section, Chorleyi Group (the Group is monobasic)
chorleyi Edwards, 1941: *Culex sensu stricto*, Pipiens Group (not placed in a Subgroup)
christiani Colless, 1960: *Culex* (*Lophoceraomyia*), Fraudatrix Group, Fraudatrix Subgroup, Christiani Complex
christophersi Baisas & Ubaldo-Pagayon, 1953: *Tripteroides* (*Tricholeptomyia*), Nepenthicola Group, Nepenthicola Subgroup
christophersi Barraud, 1926: *Uranotaenia* (*Uranotaenia*), Campestris Series
christophersi Edwards, 1922a: *Aedes* (*Bruceharrisonius*)
christophi (Portschinsky, 1884): *Toxorhynchites* (*Toxorhynchites*), Christophi Group
christyi (Newstead & Carter, 1911): *Anopheles* (*Cellia*), Pyretophorus Series
chryselatus Dyar & Knab, 1919: *Culex* (*Microculex*) (not placed in a Series)
chrysochlorus Arnell, 1973: *Haemagogus sensu stricto*, Albomaculatus Section

chrysogaster Graham, 1909: *Eretmapodites*, Chrysogaster Group
chrysogaster (Taylor, 1927): *Aedes* (*Mucidus*), Pardomyia Group
chrysolineatus (Theobald, 1907): *Aedes* (*Hulecoeteomyia*)
chrysonotum (Peryassú, 1922a): *Coquillettidia* (*Pseudotaeniorhynchus*)
chrysoscuta (Theobald, 1910c): *Aedes* (*Paraedes*), Ostentatio Assemblage
chrysosoma (Edwards, 1915a): *Coquillettidia* (*Coquillettidia*), Crassipes Group
chungi Lien, 1968b: *Aedes* (*Jihlienius*)
churchillensis Ellis & Brust, 1973: *Aedes* (*Ochlerotatus*), Communis Group, Communis Aggregate
ciliata (Fabricius, 1794): *Psorophora sensu stricto*
cilipes (Fabricius, 1805): *Psorophora sensu stricto*
cinctellus Edwards, 1922a: *Culex* (*Lophoceraomyia*), Fraudatrix Group, Fraudatrix Subgroup, Cinctellus Complex
cinctus (Newstead & Carter, 1910): *Anopheles* (*Cellia*), Neomyzomyia Series, Ardensis Group
cinerellus Edwards, 1922d: *Culex* (*Culiciomyia*), Nebulosus Group
cinereus Theobald, 1901b: *Anopheles* (*Cellia*), Paramyzomyia Series, Cinereus Group
cinereus Theobald, 1901c: *Culex* (*Culiciomyia*), Nebulosus Group
cinereus Wiedemann, 1818 (in Meigen, 1818): *Aedes sensu stricto*, Cinereus Group
cingulata (Fabricius, 1805): *Psorophora* (*Grabhamia*), Cingulata Group
cingulatus (Leicester, 1908): *Armigeres* (*Leicesteria*)
circumcincta Dyar & Knab, 1907c: *Wyeomyia* (*Hystatomyia*), Circumcincta Group
circumflava Cerqueira, 1943: *Psorophora* (*Janthinosoma*)
circumluteolus (Theobald, 1908a): *Aedes* (*Neomelaniconion*), Lineatopennis Group
circumtestacea (Theobald, 1908c): *Ficalbia*, Group B Afrotropical Region Group
civinskii Belkin, 1953a: *Uranotaenia* (*Uranotaenia*), Anisocheleomyia Section, Argyrotarsis Complex
claggi Bohart & Farner, 1944: *Tripteroides* (*Tripteroides*), Nitidoventer Group
clara Dyar & Shannon, 1925a: *Uranotaenia* (*Uranotaenia*), Anisocheleomyia Section, Ludlowae Series
clarki Evans, 1924: *Culex* (*Melanoconion*), Melanoconion Section, Erraticus Group, Clarki Subgroup (the Subgroup is monobasic)
clarki (Galindo, Carpenter & Trapido, 1953): *Haemagogus* (*Conopostegus*)
clasoleuca Dyar & Knab, 1908: *Wyeomyia* (not placed in a Subgenus)
clastrieri Casal & Garcia, 1968b: *Culex* (*Aedinus*)
clavata (Barraud, 1931b): *Verrallina* (*Neomacleaya*)
claviger (Meigen, 1804): *Anopheles sensu stricto*, Angusticorn Section, Anopheles Series, Claviger Complex (not placed in a Group)
clavipalpus (Theobald, 1908b): *Mimomyia* (*Mimomyia*), Hybrida Group
clelandi (Taylor, 1914c): *Aedes* (*Ochlerotatus*), Flavifrons Group
clintoni Taylor, 1946: *Aedes* (*Patmarksia*)
clivis Lanzaro & Eldridge, 1992: *Aedes* (*Ochlerotatus*), Stimulans Group, Increpitus Complex
clowi Rozeboom & Knight, 1946: *Anopheles* (*Cellia*), Neomyzomyia Series, Punctulatus Group
coatzacoalcos Dyar & Knab, 1906d: *Uranotaenia* (*Uranotaenia*), Pallidoventer Series
cockerelli (Edwards, 1923c): *Coquillettidia* (fossil species)
codiocampa Dyar & Knab, 1907c: *Wyeomyia* (*Miamyia*)
coenonus Howard, Dyar & Knab, 1913: *Wyeomyia* (*Hystatomyia*), Circumcincta Group
coerulescens Edwards, 1928a: *Culex* (*Lophoceraomyia*), Mammilifer Group, Brevipalpus Subgroup, Navalis Complex
coeruleus (Brug, 1934): *Toxorhynchites* (*Toxorhynchites*), Acaudatus Group, Acaudatus Subgroup
cogilli Edwards, 1922d: *Aedes* (*Phagomyia*)
coheni Belkin, 1950a: *Tripteroides* (*Polylepidomyia*), Argenteiventris Group
colledgei Marks, 1956: *Anopheles sensu stricto*, Angusticorn Section, Anopheles Series, Stigmaticus Group
collessi Grjebine, 1986: *Mimomyia* (*Ingramia*), Group B (Ravenalites)
collessi Lee, 1946: *Tripteroides* (*Polylepidomyia*), Argenteiventris Group
collessi Mattingly, 1958a: *Aedes* (*Paraedes*), Ostentatio Assemblage
collessi Reid, 1963: *Anopheles sensu stricto*, Laticorn Section, Myzorhynchus Series, Umbrosus Group, Letifer Subgroup
collessi Sirivanakarn, 1968: *Culex* (*Lophoceraomyia*), Fraudatrix Group, Fraudatrix Subgroup, Fraudatrix Complex
colocasiae Edwards, 1928a: *Uranotaenia* (*Pseudoficalbia*), Bicolor Series
colombiana Lane, 1945a: *Wyeomyia* (*Antunesmyia*)
colombianus Adames, 1971: *Deinocerites*, Epitedeus Group
columbiae (Dyar & Knab, 1906a): *Psorophora* (*Grabhamia*), Confinnis Group, Confinnis Complex
columnaris Sá & Hutchings, 2020 (in Sá *et al.*, 2020): *Culex* (*Melanoconion*), Melanoconion Section, Atratus Group
coluzzii Coetzee & Wilkerson, 2013 (in Coetzee *et al.*, 2013): *Anopheles* (*Cellia*), Pyretophorus Series, Gambiae Complex
coluzzii Rioux, Guilvard & Pasteur, 1998: *Aedes* (*Ochlerotatus*), Communis Group (not placed in a Subgroup)

comata (Barraud, 1931b): *Verrallina* (*Neomacleaya*)
comatus Senevet & Abonnenc, 1939a: *Culex* (*Melanoconion*), Melanoconion Section, Bastagarius Group, Bastagarius Subgroup
combesi Doucet, 1950: *Uranotaenia* (*Pseudoficalbia*), Nigripes Section, Nigripes Group
comitatus Arnell, 1976: *Aedes* (*Ochlerotatus*), Scapularis Group, Scapularis Subgroup
commevynensis Bonne-Wepster & Bonne, 1920: *Culex* (*Melanoconion*), Melanoconion Section, Atratus Group
comminutor Dyar, 1920b: *Culex* (*Melanoconion*), Melanoconion Section, Distinguendus Group, Distinguendus Subgroup
communis (de Geer, 1776): *Aedes* (*Ochlerotatus*), Communis Group, Communis Aggregate
communis (Leicester, 1908): *Heizmannia sensu stricto*
comorensis Brunhes, 1977a: *Culex sensu stricto*, Pipiens Group (not placed in a Subgroup)
comorensis Brunhes, 1977a: *Orthopodomyia*, Orthopodomyia Section, Vernoni Group
comorensis Brunhes, le Goff & Geoffroy, 1997: *Anopheles* (*Cellia*), Pyretophorus Series
comorensis da Cunha Ramos & Brunhes, 2004: *Uranotaenia* (*Pseudoficalbia*), Nigripes Section, Nigripes Group
comosa (Reinert, 1974): *Verrallina* (*Neomacleaya*)
complex (Theobald, 1910a): *Heizmannia sensu stricto*
complosa (Dyar, 1928): *Wyeomyia* (*Dendromyia*)
compressum Lutz, 1905: *Trichoprosopon*, Trichoprosopon Group, Compressum Complex
comptus Sá & Sallum (in Sá *et al.*, 2020): *Culex* (*Melanoconion*), Melanoconion Section, Atratus Group
conchobius Edwards, 1941: *Eretmapodites* (not placed in a Group)
concinnus Lee, 1946: *Tripteroides* (*Rachisoura*), Vanleeuweni Group
concolor Edwards, 1938 (in Evans, 1938): *Anopheles sensu stricto*, Angusticorn Section, Anopheles Series (not placed in a Group)
conditus Moses, Howard & Harbach, 2000: *Sabethes* (*Sabethoides*)
condolescens Dyar & Knab, 1907a: *Aedes* (*Ochlerotatus*), Scapularis Group, Infirmatus Subgroup
confinnis (Lynch Arribálzaga, 1891b): *Psorophora* (*Grabhamia*), Confinnis Group, Confinnis Complex
confundior Komp & Rozeboom, 1951: *Culex* (*Melanoconion*), Melanoconion Section, Bastagarius Group, Iolambdis Subgroup
confusa Bonne-Wepster, 1951: *Bironella* (*Neobironella*)
confusa (Lutz, 1905): *Wyeomyia* (*Prosopolepis*) (the Subgenus is monobasic)
confusa Peyton, 1977: *Uranotaenia* (*Pseudoficalbia*), Recondita Series
confusus Edwards, 1915b: *Armigeres sensu stricto*
confusus Evans & Leeson, 1935: *Anopheles* (*Cellia*), Myzomyia Series, Funestus Group, Funestus Subgroup
confusus Lee, 1946: *Tripteroides* (*Rachisoura*), Filipes Group
conjungens Edwards, 1914c: *Armigeres sensu stricto*
connali Edwards, 1912a: *Uranotaenia* (*Uranotaenia*), Alba Section, Bilineata Group
conradti Grünberg, 1907: *Toxorhynchites* (*Toxorhynchites*), Brevipalpis Group, Brevipalpis Subgroup
conservator Dyar & Knab, 1906d: *Culex* (*Anoedioporpa*), Conservator Group
consimilis Newstead, 1907 (in Newstead *et al.*, 1907): *Culex* (*Oculeomy*ia) (not placed in a Complex)
consolator Dyar & Knab, 1906b: *Culex* (*Microculex*), Consolator Series
consonensis (Reinert, 1973d): *Verrallina* (*Harbachius*)
conspirator Dyar & Knab, 1906d: *Culex* (*Melanoconion*), Melanoconion Section, Conspirator Group
contei Duret, 1968c: *Culex* (*Melanoconion*), Melanoconion Section, Inhibitator Group, Inhibitator Subgroup
contiguus Edwards, 1936: *Aedes* (*Stegomyia*), Poweri Group
continentalis Dobrotworsky, 1960a: *Aedes* (*Ochlerotatus*), Cunabulanus Group
contrastata da Cunha Ramos & Brunhes, 2004: *Uranotaenia* (*Pseudoficalbia*), Shillitonis Section, Shillitonis Group (not placed in a Subgroup)
cooki Belkin, 1962: *Aedes* (*Stegomyia*), Scutellaris Group, Scutellaris Subgroup, Pseudoscutellaris Subgroup/Complex
cooki Root, 1937: *Uranotaenia* (*Uranotaenia*), Pallidoventer Series
coonorensis Mattingly, 1981: *Tripteroides* (*Rachionotomyia*), Aranoides Group, Aranoides Subgroup
coppenamensis Bonne-Wepster & Bonne, 1920: *Culex* (*Melanoconion*), Melanoconion Section, Bastagarius Group, Bastagarius Subgroup
corbeti Hamon, 1962: *Eretmapodites*, Quinquevittatus Group
cordellieri Huang, 1986a: *Aedes* (*Diceromyia*), Section A, Furcifer Group
corentynensis Dyar, 1920b: *Culex* (*Melanoconion*), Melanoconion Section, Bastagarius Group, Iolambdis Subgroup
corethroides Theobald, 1907: *Anopheles sensu stricto*, Angusticorn Section, Anopheles Series, Stigmaticus Group
corneti Huang, 1986b: *Aedes* (*Stegomyia*), Africanus Group
corniger Theobald, 1903a: *Culex* (*Phenacomyia*)
cornuta da Cunha Ramos & Brunhes, 2004: *Uranotaenia* (*Pseudoficalbia*), Spinosa Section, Spinosa Group
cornutus Edwards, 1922a: *Culex* (*Oculeomyia*), Sinensis Complex
corona Belkin, Heinemann & Page, 1970: *Wyeomyia* (*Wyeomyia*), Hirsuta Group, Stellata Subgroup
coronator Dyar & Knab, 1906d: *Culex sensu stricto*, Pipiens Group, Coronator Complex (not placed in a Subgroup)
corrigani Dyar & Knab, 1907c: *Culex* (*Anoedioporpa*), Conservator Group

costai da Fonseca & da Silva Ramos, 1940: *Anopheles sensu stricto*, Laticorn Section, Arribalzagia Series
costaricensis Adames & Hogue, 1970: *Deinocerites*, Epitedeus Group
cottlei Sirivanakarn, 1968: *Culex* (*Lophoceraomyia*), Fraudatrix Group, Fraudatrix Subgroup (Sirivanakarn, 1977a unnecessarily recognized a monobasic '*cottlei* complex' for this species in the absence of evidence that it is more than one species)
coulangesi Rodhain & Boutonnier, 1983b: *Aedes* (*Paulianius*), Coulangesi Group
coursi Doucet, 1949a: *Culex* (*Maillotia*), Seyrigi Group
coustani Laveran, 1900: *Anopheles sensu stricto*, Laticorn Section, Myzorhynchus Series, Coustani Group
covagarciai Forattini, 1965a: *Culex sensu stricto*, Pipiens Group (not placed in a Subgroup)
covagarciai Sutil Oramas & Pulido F., 1974: *Wyeomyia* (not placed in a Subgenus)
covelli Barraud, 1929b: *Heizmannia sensu stricto*
cozi Cornet, 1973: *Aedes* (*Cornetius*) (the Subgenus is monobasic)
cozumelensis Díaz Nájera, 1966: *Aedes* (*Howardina*), Walkeri Section, Ioliota Group
cracens Sallum & Peyton, 2005 (in Sallum *et al.*, 2005b): *Anopheles* (*Cellia*), Neomyzomyia Series, Leucosphyrus Group, Leucosphyrus Subgroup, Dirus Complex
craggi (Barraud, 1923c): *Aedes* (*Stegomyia*), W-albus Group, Annandalei Subgroup
crassicomus Colless, 1965: *Culex* (*Lophoceraomyia*), Mammilifer Group, Minor Line, Minor Complex
crassiforceps Edwards, 1927a: *Aedes* (*Neomelaniconion*) (not placed in a Group)
crassipes (van der Wulp, 1881): *Coquillettidia* (*Coquillettidia*), Crassipes Group
crassistylus Brug, 1934: *Culex* (*Neoculex*), Crassistylus Group
crawfordi Reid, 1953: *Anopheles sensu stricto*, Laticorn Section, Myzorhynchus Series, Hyrcanus Group, Lesteri Subgroup
creole Anduze, 1949b: *Culex* (*Melanoconion*), Melanoconion Section, Bastagarius Group, Bastagarius Subgroup
cretata (Delfinado, 1967): *Verrallina* (*Neomacleaya*)
cretinus Edwards, 1921d: *Aedes* (*Stegomyia*), Unilineatus Group
crinicauda Edwards, 1921a: *Culex sensu stricto*, Sitiens Group (not placed in a Subgroup)
crinifer (Theobald, 1903a): *Aedes* (*Ochlerotatus*), Scapularis Group, Crinifer Subgroup
cristata (Theobald, 1904b): *Coquillettidia* (*Coquillettidia*), Crassipes Group
cristata Thurman, 1959: *Topomyia* (*Suaymyia*)
cristatus King & Baisas, 1936: *Anopheles* (*Cellia*), Neomyzomyia Series, Leucosphyrus Group, Riparis Subgroup
cristipalpis Service, 1977: *Anopheles* (*Cellia*), Cellia Series (not placed in a Group)
cristovaoi Duret, 1968a: *Culex* (*Melanoconion*), Melanoconion Section, Educator Group
croceus Knight & Laffoon, 1946: *Aedes* (*Finlaya*), Poicilius Group
crockeri Colless, 1955a: *Anopheles sensu stricto*, Angusticorn Section, Anopheles Series, Lindesayi Group, Gigas Subgroup, Gigas Complex
crossi Lien, 1967: *Aedes* (subgenus uncertain)
crowei Sirivanakarn, 1968: *Culex* (*Lophoceraomyia*), Fraudatrix Group, Fraudatrix Subgroup, Petersi Complex
crucians Wiedemann, 1828: *Anopheles sensu stricto*, Angusticorn Section, Anopheles Series, Punctipennis Group, Crucians Complex (*An. crucians* consists of five sibling species denoted as Species A, B, C, D and E (Wilkerson *et al.*, 2004); which is the nominotypical species is not known)
cruzii Dyar & Knab, 1908: *Anopheles* (*Kerteszia*), Cruzii Complex (*An. cruzii* consists of chromosomal species denoted as Species A, B and C (Ramírez & Dessen, 2000); however, later studies by de Carvalho-Pinto & Lourenço-de-Oliveira (2004) and Rona *et al.* (2009, 2010a, 2010b, 2013) suggest the complex consists of four or five species, but how these relate to the three chromosomal species and the three molecular genetic species (species 1, 2 and 3) of Bourke *et al.* (2023) is not known)
crybda Dyar, 1924f: *Culex* (*Melanoconion*), Spissipes Section, Crybda Group, Pedroi Subgroup
crypticus Coetzee, 1995: *Anopheles sensu stricto*, Laticorn Section, Myzorhynchus Series, Coustani Group
cubiculi Marks, 1989 (in Lee *et al.*, 1989b): *Culex* (*Lophoceraomyia*), Fraudatrix Group (not placed in a Subgroup)
cubitatus Colless, 1965: *Culex* (*Lophoceraomyia*), Fraudatrix Group, Fraudatrix Subgroup, Variatus Complex
cuccioi (Belkin, 1962): *Verrallina sensu stricto*, Carmenti Series
cucphuongensis Vu, Nguyen, Tran & Nguyen, 1991: *Anopheles sensu stricto*, Angusticorn Section, Anopheles Series, Alongensis Group
culicifacies Giles, 1901b: *Anopheles* (*Cellia*), Myzomyia Series, Funestus Group, Culicifacies Subgroup (*An. culicifacies* consists of five sibling species denoted as Species A, B, C, D and E (Kar *et al.*, 1999); which is the nominotypical species is not known)
culiciformis Cogill, 1903: *Anopheles sensu stricto*, Angusticorn Section, Anopheles Series, Culiciformis Group
culiciformis (Theobald, 1905f): *Aedes* (*Pseudoskusea*)
culicinus Edwards, 1922a: *Aedes* (*Aedimorphus*), Alboscutellatus Assemblage, Alboscutellatus Group
cumminsii (Theobald, 1903a): *Aedes* (*Aedimorphus*), Domesticus Assemblage, Vexans Subassemblage, Quasiunivittatus Group
cunabulanus Edwards, 1924b: *Aedes* (*Ochlerotatus*), Cunabulanus Group
curiche Adames, 1971: *Deinocerites*, Epitedeus Group
curtipalpis (Edwards, 1914c): *Culex* (*Lophoceraomyia*), Mammilifer Group, Brevipalpus Subgroup
curtipes Edwards, 1915b: *Aedes* (*Cancraedes*)

curvibrachius Angulo, 1993 (in Angulo & Olivares, 1993): *Culex sensu stricto*, Pipiens Group (not placed in a Subgroup)
cuttsi van den Assem, 1959: *Tripteroides* (*Rachisoura*), Vanleeuweni Group
cuyanus Duret, 1968c: *Culex sensu stricto*, Pipiens Group, Tarsalis Subgroup
cyanescens (Coquillett, 1902a): *Psorophora* (*Janthinosoma*)
cyaneus (Fabricius, 1805): *Sabethes* (*Sabethes*), Group B
cydippis de Meillon, 1931a: *Anopheles* (*Cellia*), Cellia Series, Squamosus Group
cylindricus Theobald, 1903a: *Culex* (*Lophoceraomyia*), Fraudatrix Group, Fraudatrix Subgroup, Minutissimus Subgroup
cyprioides Danilov & Stupin, 1982: *Aedes* (*Ochlerotatus*), Cantans Group
cyprius Ludlow, 1920b: *Aedes* (*Ochlerotatus*), Cantans Group
cyptopus Theobald, 1909: *Hodgesia*
cyrtolabis (Edwards, 1928a): *Verrallina* (*Neomacleaya*)

Dd

daciae Linton, Nicolescu & Harbach, 2004 (in Nicolescu *et al.*, 2004): *Anopheles sensu stricto*, Angusticorn Section, Anopheles Series, Maculipennis Group, Maculipennis Subgroup
daggyi Stone & Bohart, 1944: *Aedes* (*Geoskusea*), Daggyi Group
dahlae (Nielsen, 2009): *Aedes* (*Ochlerotatus*), Cantans Group
dahuricus Danilov, 1987: *Aedes sensu stricto* (not placed in a Group)
daisetsuzanus Tanaka, Mizusawa & Saugstad, 1979: *Aedes* (*Ochlerotatus*), Impiger Group
daitensis Miyagi & Toma, 1981: *Aedes* (*Stegomyia*), Scutellaris Group, Scutellaris Subgroup
daliensis (Taylor, 1916): *Aedes* (subgenus uncertain)
dalzieli (Theobald, 1910c): *Aedes* (*Aedimorphus*), Alboscutellatus Assemblage, Abnormalis Group
damascenoi Duret, 1969c: *Culex* (*Anoedioporpa*), Conservator Group
damasei Grjebine, 1979: *Uranotaenia* (*Pseudoficalbia*), Shillitonis Section, Shillitonis Group (not placed in a Subgroup)
damnatorum Scudder, 1890: *Culex* (fossil species)
damzeni (Szadziewski, 1998): *Eoaedes* (fossil species)
danaraji Ramalingam, 1988: *Topomyia* (*Topomyia*)
danaubento Mochtar & Walandouw, 1934: *Anopheles sensu stricto*, Angusticorn Section, Anopheles Series, Lindesayi Group, Gigas Subgroup, Gigas Complex
dancalicus Corradetti, 1939a: *Anopheles* (*Cellia*), Neocellia Series (not placed in a Group)
dandakaranyensis Natarajan, Rajavel & Jambulingam, 2017: *Uranotaenia* (*Pseudoficalbia*), Recondita Series
darjeelingensis Tyagi, Munirathinam, Krishnamoorthy, Baskaran, Govindaragan, Krishnamoorthi, Mariappan, Dhananjeyan & Venkatesh, 2015b: *Toxorhynchites* (*Toxorhynchites*) (not placed in a Group)
darlingi Root, 1926: *Anopheles* (*Nyssorhynchus*), Argyritarsis Section, Argyritarsis Series, Darlingi Group (the Group is monobasic)
daryi Schick, 1970a: *Aedes* (*Protomacleaya*), Terrens Group, Galindoi Subgroup
dasyorrhus King & Hoogstraal, 1946c: *Aedes* (*Lorrainea*)
daudi Coluzzi, 1958: *Anopheles* (*Cellia*), Pyretophorus Series
daumastocampa Dyar & Knab, 1908: *Culex* (*Microculex*) (not placed in a Series)
davidi Basio, 1971: *Aedes* (*Aedimorphus*), Domesticus Assemblage, Vexans Subassemblage, Vexans Group
davisi Kumm, 1933: *Culex* (*Microculex*), Pleuristriatus Series
davisi Lane, 1943: *Uranotaenia* (*Uranotaenia*), Pallidoventer Series
davisi (Lane & Cerqueira, 1942): *Wyeomyia* (*Phoniomyia*), Third Series
deanei (Lourenço de Oliveira, 1983): *Wyeomyia* (*Phoniomyia*), First Series
deaneorum Rosa-Freitas, 1989: *Anopheles* (*Nyssorhynchus*), Argyritarsis Section, Albitarsis Series, Albitarsis Group, Albitarsis Complex
deboeri Edwards, 1926a: *Aedes* (*Stegomyia*), Dendrophilus Group
deccanus (Barraud, 1923b): *Aedes* (*Phagomyia*)
decens Theobald, 1901d: *Culex sensu stricto*, Pipiens Group, Decens Subgroup
declarator Dyar & Knab, 1906d: *Culex sensu stricto*, Pipiens Group, Tarsalis Subgroup
decorabilis (Leicester, 1908): *Kimia*
decticus Howard, Dyar & Knab, 1917: *Aedes* (*Ochlerotatus*), Pullatus Group
deemingi Service, 1970a: *Anopheles* (*Cellia*), Neomyzomyia Series, Ardensis Group
deficiens Arnell, 1976: *Aedes* (*Ochlerotatus*), Scapularis Group, Bogotanus Subgroup
deguzmanae (Mattingly, 1957a): *Mimomyia* (*Ingramia*), Group C (*Ingramia s. str.*, Fusca Group)
dejesusi Baisas & Feliciano, 1953: *Topomyia* (*Topomyia*)
delfinadoae Sirivanakarn, 1973b: *Culex* (*Culiciomyia*), Tricuspis Group
delpilari Baisas & Ubaldo-Pagayon, 1953: *Tripteroides* (*Tricholeptomyia*), Delpilari Group (the Group is monobasic)

delpontei Duret, 1969a: *Culex* (*Melanoconion*), Spissipes Section, Crybda Group, Paracrybda Subgroup
delys Howard, Dyar & Knab, 1915: *Culex sensu stricto*, Pipiens Group (not placed in a Subgroup)
demeilloni Doucet, 1950: *Culex sensu stricto*, Pipiens Group (not placed in a Subgroup)
demeilloni Edwards, 1936: *Aedes* (*Stegomyia*), Dendrophilus Group
demeilloni Evans, 1933: *Anopheles* (*Cellia*), Myzomyia Series, Demeilloni Group
demeilloni Mattingly, 1970: *Heizmannia sensu stricto*
demeilloni Peyton & Rattanarithikul, 1970: *Uranotaenia* (*Pseudoficalbia*), Bicolor Series
demissus Colless, 1965: *Culex* (*Lophoceraomyia*), Mammilifer Group, Mammilifer Line, Mammilifer Complex
denbesteni Brug, 1925: *Armigeres sensu stricto*
denderensis Wolfs, 1949: *Aedes* (*Stegomyia*), Apicoargenteus Group
dendrophilus Edwards, 1921a: *Aedes* (*Stegomyia*), Dendrophilus Group
dentatus Barraud, 1927b: *Armigeres* (*Leicesteria*)
dentatus (Theobald, 1904b): *Aedes* (*Aedimorphus*), Domesticus Assemblage, Vexans Subassemblage, Quasiunivittatus Group
denticulatus Delfinado & Hodges, 1968: *Tripteroides* (*Tripteroides*), Nitidoventer Group
depereti (Meunier, 1915): *Neoculicites* (fossil species)
derivator Dyar & Knab, 1906d: *Culex* (*Neoculex*), Territans Group
derooki Brug, 1932: *Aedes* (*Patmarksia*)
derooki Soesilo & van Slooten, 1931a,b: *Bironella* (*Neobironella*)
deserticola Kirkpatrick, 1925a: *Culex* (*Maillotia*) (not placed in a Group)
deserticola Zavortink, 1970a: *Aedes* (*Jarnellius*)
desmotes (Giles, 1904): *Aedes* (*Stegomyia*), W-albus Group, Desmotes Subgroup (the Subgroup is monobasic)
detritus (Haliday, 1833): *Aedes* (*Ochlerotatus*), Communis Group (not placed in a Subgroup)
devemyi Hamon, 1955: *Uranotaenia* (*Pseudoficalbia*), Shillitonis Section, Shillitonis Group (not placed in a Subgroup)
diabolica (Lane & Forattini, 1952): *Wyeomyia* (*Phoniomyia*), Second Series
diagonalis Brug, 1934: *Uranotaenia* (*Pseudoficalbia*), Bicolor Series
dialloi Hamon & Brengues, 1965: *Aedes* (*Catageiomyia*)
diamphidius Peyton & Harbach, 1991: *Culex* (*Melanoconion*), Melanoconion Section, Bastagarius Group (not placed in a Subgroup)
diantaeus Howard, Dyar & Knab, 1913: *Aedes* (*Ochlerotatus*), Pullatus Group
diazi Schick, 1970a: *Aedes* (*Protomacleaya*), Terrens Group, Diazi Subgroup (the Subgroup is monobasic)
dibrugarhensis Bhattacharyya, Prakash, Mohapatra & Mahanta, 2004a: *Uranotaenia* (*Pseudoficalbia*), Recondita Series
diengensis Brug, 1931c: *Culex sensu stricto*, Sitiens Group, Mimeticus Subgroup, Mimeticus Complex
digitatum (Rondani, 1848): *Trichoprosopon*, Trichoprosopon Group, Digitatum Complex
digitatus (Edwards, 1914a): *Armigeres* (*Leicesteria*)
digoelensis Brug, 1932: *Culex* (*Lophoceraomyia*), Mammilifer Group, Digoelensis Subgroup
digoelensis Brug, 1934: *Tripteroides* (*Polylepidomyia*), Atripes Group
diluvialis Reinert, 1997 (in Reinert *et al.*, 1997): *Anopheles sensu stricto*, Angusticorn Section, Anopheles Series, Maculipennis Group, Quadrimaculatus Subgroup
dimidiata Cerqueira, 1943: *Psorophora* (*Grabhamia*) (not placed in a Group)
diplophyllum Dyar, 1929b: *Culex sensu stricto*, Pipiens Group (not placed in a Subgroup)
diraphati Peyton & Klein, 1970: *Uranotaenia* (*Uranotaenia*), Unimaculiala Series
dirus Peyton & Harrison, 1979: *Anopheles* (*Cellia*), Neomyzomyia Series, Leucosphyrus Group, Leucosphyrus Subgroup, Dirus Complex
discolor (Coquillett, 1903a): *Psorophora* (*Grabhamia*) (not placed in a Group)
discors Harbach, 2014 (in Harbach & Culverwell, 2014): *Topomyia* (*Miyagiella*) (the Subgenus is monobasic)
discrepans (Edwards, 1922a): *Heizmannia* (*Mattinglyia*)
discrucians (Walker, 1856): *Psorophora* (*Janthinosoma*)
dispar Rattanarithikul & Harbach, 1991: *Anopheles* (*Cellia*), Neocellia Series, Maculatus Group (not placed in a Subgroup)
dispectus Bram, 1966: *Culex* (*Culiciomyia*), Dispectus Group
dissidens Taai & Harbach, 2015: *Anopheles sensu stricto*, Laticorn Section, Myzorhynchus Series, Barbirostris Group, Barbirostris Subgroup, Barbirostris Complex
dissimilierodes Dong, Zhou & Dong, 2002a: *Aedes* (*Kenknightia*), Dissimilis Group
dissimilis (Leicester, 1908): *Aedes* (*Kenknightia*), Dissimilis Group
distigma (Edwards, 1925): *Tripteroides* (*Tripteroides*), Bimaculipes Group
distinctus (Newstead & Carter, 1911): *Anopheles* (*Cellia*), Myzomyia Series, Wellcomei Group
distinguendus Dyar, 1928: *Culex* (*Melanoconion*), Melanoconion Section, Distinguendus Group, Distinguendus Subgroup
ditaenionota Prado, 1931b: *Uranotaenia* (*Uranotaenia*), Pallidoventer Series
dives (Schiner, 1868): *Mansonia* (*Mansonioides*)
dmitryi Gornostaeva, 2005: *Aedes sensu stricto* (not placed in a Group)

dobodurus King & Hoogstraal, 1946b: *Aedes* (*Patmarksia*)
dobrotworskyi Marks, 1958: *Aedes* (*Finlaya*), Kochi Group
dofleini (Guenther, 1913): *Tripteroides* (*Rachionotomyia*), Aranoides Group, Affinis Subgroup
dohenyi Hogue, 1975: *Culex sensu stricto*, Pipiens Group (not placed in a Subgroup)
dolichocephalus (Leicester, 1908): *Armigeres* (*Leicesteria*)
dolichophyllus Clastrier, 1970b: *Culex* (*Melanoconion*), Melanoconion Section, Evansae Group
dolosus (Lynch Arribálzaga, 1891b): *Culex sensu stricto*, Pipiens Group, Salinarius Complex (not placed in a Subgroup)
domesticus (Theobald, 1901d): *Aedes* (*Aedimorphus*), Domesticus Assemblage, Domesticus Subassemblage, Domesticus Group
domicolus Edwards, 1916: *Anopheles* (*Cellia*), Myzomyia Series (not placed in a Group)
dominicanus Zavortink & Poinar, 2000: *Anopheles* (*Nyssorhynchus*) (fossil species)
domrey Macquart & Rahola, 2023 (in Macquart *et al.*, 2023): *Toxorhynchites* (*Toxorhynchites*), Acaudatus Group, Acaudatus Subgroup, Nepenthis Subgroup
donai da Cunha Ramos & Brunhes, 2004: *Uranotaenia* (*Pseudoficalbia*), Shillitonis Section, Shillitonis Group (not placed in a Subgroup)
donaldi Reid, 1962: *Anopheles sensu stricto*, Laticorn Section, Myzorhynchus Series, Barbirostris Group, Barbirostris Subgroup, Barbirostris Complex
doonii Wattal, Bhatia & Kalra, 1958: *Aedes* (*Bruceharrisonius*)
dorsalis (Meigen, 1830): *Aedes* (*Ochlerotatus*), Caspius Group
dorseyi Knight, 1946: *Aedes* (*Downsiomyia*), Niveus Group
douceti Adam & Hamon, 1959: *Eretmapodites*, Oedipodeios Group
douceti Grjebine, 1953a: *Uranotaenia* (*Pseudoficalbia*), Nigripes Section, Nigripes Group
douglasi Dobrotworsky, 1956: *Culex* (*Neoculex*), Pseudomelanoconia Group
downsi Bohart & Ingram, 1946a: *Aedes* (*Stegomyia*), Scutellaris Group, Albopictus Subgroup
downsi Lane, 1945a: *Wyeomyia* (*Wyeomyia*) (not placed in a Group)
dracaenae Edwards, 1916: *Eretmapodites*, Quinquevittatus Group
draconis Ingram & de Meillon, 1927: *Culex sensu stricto*, Pipiens Group (not placed in a Subgroup)
dravidicus Christophers, 1924c: *Anopheles* (*Cellia*), Neocellia Series, Maculatus Group, Maculatus Subgroup
druki Somboon, Namgay & Harbach, 2022 (in Somboon *et al.*, 2022b): *Anopheles sensu stricto*, Angusticorn Section, Anopheles Series, Lindesayi Group, Lindesayi Subgroup, Lindesayi Complex (the Complex also includes informally designated Species B, see Somboon *et al.*, 2022b)
drummondi (Dobrotworsky, 1960b): *Culiseta* (*Culicella*)
dthali Patton, 1905: *Anopheles* (*Cellia*), Myzomyia Series (not placed in a Group)
dualaensis Brunhes, Le Goff & Geoffroy, 1999: *Anopheles* (*Cellia*), Neomyzomyia Series (not placed in a Group)
dubitans Lane & Whitman, 1951: *Culex* (*Microculex*), Imitator Series
dubitans Leicester, 1908: *Topomyia* (*Topomyia*)
dufouri Hamon, 1953: *Aedes* (*Ochlerotatus*), Taeniorhynchus Group, Vigilax Subgroup
dugesi Dyar & Knab, 1906a: *Culiseta sensu stricto*
dulongensis Gong & Lu, 1995: *Topomyia* (*Suaymyia*)
dumbletoni Belkin, 1962: *Culex* (*Neoculex*), Pseudomelanoconia Group
dumonti Doucet, 1949b: *Uranotaenia* (*Uranotaenia*), Dumonti Section, Dumonti Group
dundo da Cunha Ramos, 1993: *Uranotaenia* (*Uranotaenia*), Hopkinsi Section, Hopkinsi Group
dundo da Cunha Ramos & Ribeiro, 1992: *Eretmapodites*, Oedipodeios Group
dundo Ribeiro, 1991: *Toxorhynchites* (*Toxorhynchites*), Brevipalpis Group, Phytophagus Subgroup
dunhami Causey, 1945: *Anopheles* (*Nyssorhynchus*), Albimanus Section, Oswaldoi Series, Oswaldoi Group, Oswaldoi Subgroup, Nuneztovari Complex
dunni Dyar, 1918c: *Culex* (*Melanoconion*), Melanoconion Section, Atratus Group
duplex Martini, 1926: *Aedes* (*Ochlerotatus*), Caspius Group
duplicator Dyar & Knab, 1909: *Culex sensu stricto*, Pipiens Group, Tarsalis Subgroup
dupreei (Coquillett, 1904a): *Aedes* (*Ochlerotatus*), Serratus Group
durbanensis (Theobald, 1903a): *Aedes* (*Aedimorphus*), Domesticus Assemblage, Vexans Subassemblage, Vexans Group
dureni Edwards, 1938 (in Evans, 1938): *Anopheles* (*Cellia*), Neomyzomyia Series, Ardensis Group
dureti Casal & García, 1968d: *Culex* (*Melanoconion*), Melanoconion Section, Bastagarius Group, Iolambdis Subgroup
durhami Edwards, 1917: *Armigeres sensu stricto*
durhami Sirivanakarn, 1968: *Culex* (*Lophoceraomyia*), Fraudatrix Group, Fraudatrix Subgroup, Solomonis Complex
durhamii Theobald, 1901c: *Limatus*
duttoni Theobald, 1901d: *Culex sensu stricto*, Duttoni Group
dux (Dyar & Shannon, 1925a): *Verrallina sensu stricto*, Butleri Series
dyari Belkin, Heinemann & Page, 1970: *Mansonia* (*Mansonia*), Indubitans Complex
dyari Belkin & Hogue, 1959: *Deinocerites*, Dyari Group

dyari Bohart & Farner, 1944: *Tripteroides* (*Tripteroides*), Nitidoventer Group
dyari Lane & Cerqueira, 1942: *Wyeomyia* (*Cruzmyia*)
dybasi Bohart, 1957: *Aedes* (*Stegomyia*), Scutellaris Group, Scutellaris Subgroup
dyi Baisas & Ubaldo-Pagayon, 1953: *Tripteroides* (*Tripteroides*), Nitidoventer Group
dyius Root, 1927c: *Culex* (*Melanoconion*), Melanoconion Section, Conspirator Group

Ee

ealaensis Huang, 2004: *Aedes* (*Stegomyia*), Apicoargenteus Group
earlei Vargas, 1943b: *Anopheles sensu stricto*, Angusticorn Section, Anopheles Series, Maculipennis Group, Freeborni Subgroup
eastor Dyar, 1920b: *Culex* (*Melanoconion*), Melanoconion Section, Intrincatus Group, Eastor Subgroup (the Subgroup is monobasic)
eatoni (Edwards, 1916): *Aedes* (subgenus uncertain)
ebogoensis Rickenbach & Ferrara, 1965: *Aedes* (*Aedimorphus*), Alboscutellatus Assemblage, Abnormalis Group
echinus (Edwards, 1920a): *Aedes* (*Dahliana*)
ecuadoriensis Berlin, 1969a: *Aedes* (*Howardina*), Sexlineatus Section, Sexlineatus Subassemblage
edgari Stone & Rosen, 1952: *Aedes* (*Ochlerotatus*), Buvirilia Group (the Group is monobasic)
eduardoi Casal & García, 1968e: *Culex sensu stricto*, Pipiens Group (not placed in a Subgroup)
educator Dyar & Knab, 1906d: *Culex* (*Melanoconion*), Melanoconion Section, Educator Group
edwardsi (Barraud, 1923a): *Aedes* (*Stegomyia*), Edwardsi Group
edwardsi Barraud, 1923d: *Culex sensu stricto*, Sitiens Group, Barraudi Subgroup
edwardsi (Barraud, 1924d): *Toxorhynchites* (*Toxorhynchites*), Christophi Group
edwardsi Barraud, 1926: *Uranotaenia* (*Uranotaenia*) (not placed in a Section, Series, Group or Complex)
edwardsi (Barraud, 1929b): *Tripteroides* (*Rachionotomyia*), Edwardsi Group
edwardsi (Lane & Cerqueira, 1942): *Wyeomyia* (*Phoniomyia*), Second Series
egcymon Dyar, 1923b: *Culex* (*Melanoconion*), Melanoconion Section, Inhibitator Group, Egcymon Subgroup
eidsvoldensis Mackerras, 1927: *Aedes* (*Ochlerotatus*), Taeniorhynchus Group, Theobaldi Subgroup (provisional placement)
eiseni Coquillett, 1902b: *Anopheles sensu stricto*, Angusticorn Section, Anopheles Series, Pseudopunctipennis Group
ejercitoi Baisas, 1935b: *Armigeres sensu stricto*
ejercitoi Mendoza, 1947: *Anopheles sensu stricto*, Laticorn Section, Myzorhynchus Series, Albotaeniatus Group
ekaterinae Gilka, Harbach & Perkosky, 2021: *Culex* (fossil species)
eknomios Forattini & Sallum, 1992: *Culex* (*Melanoconion*), Melanoconion Section, Educator Group
elchoensis (Taylor, 1929): *Aedes* (*Macleaya*), Chaetocruiomyia Group
eldridgei Adames & Galindo, 1973: *Culex* (*Belkinomyia*) (the Subgenus is monobasic)
eldridgei Reinert, 1973f: *Aedes* (*Bothaella*)
eleanorae Berlin, 1969a: *Aedes* (*Howardina*), Sexlineatus Section, Eleanorae Group (the Group is monobasic)
elegans Brug, 1934: *Tripteroides* (*Tripteroides*), Littlechildi Group
elegans (James, 1903) (in Theobald, 1903a): *Anopheles* (*Cellia*), Neomyzomyia Series, Leucosphyrus Group, Leucosphyrus Subgroup, Dirus Complex
elegans (Taylor, 1914c): *Mimomyia* (*Etorleptiomyia*), Group B (Dixomyia)
elephas Komp, 1936d: *Culex* (*Melanoconion*), Melanoconion Section, Inhibitator Group, Egcymon Subgroup
elevator Dyar & Knab, 1906d: *Culex* (*Melanoconion*), Melanoconion Section, Conspirator Group
elgonicus Edwards, 1941: *Culex sensu stricto*, Pipiens Group (not placed in a Subgroup)
ellinorae Edwards, 1941: *Aedes* (*Neomelaniconion*) (not placed in a Group)
ellinorae Ovazza, Hamon & Neri, 1956: *Culex* (*Oculeomyia*) (not placed in a Complex)
elongatus Rozeboom & Komp, 1950b: *Culex* (*Microculex*) (not placed in a Series)
elsiae (Barraud, 1923e): *Aedes* (*Collessius*), Collessius Group
embiensis (Huang, 1968a): *Verrallina sensu stricto*, Carmenti Series
embuensis Edwards, 1930a: *Aedes* (*Hopkinsius*), Hopkinsius Group
eminentia (Leicester, 1908): *Culex* (*Lophoceraomyia*), Mammilifer Group, Brevipalpus Subgroup, Brevipalpus Complex
engarensis Kanda & Oguma, 1978: *Anopheles sensu stricto*, Laticorn Section, Myzorhynchus Series, Hyrcanus Group (not placed in a Subgroup)
enigmatica Peyton, 1977: *Uranotaenia* (*Pseudoficalbia*), Recondita Series
ensiformis Bonne-Wepster & Bonne, 1920: *Culex* (*Melanoconion*), Melanoconion Section, Atratus Group
eouzani Brunhes, Le Goff & Bousses, 2003: *Anopheles* (*Cellia*), Neomyzomyia Series, Ardensis Group
eouzani Geoffroy, 1971a: *Culex* (*Culiciomyia*), Nebulosus Group
eouzani Rickenbach & Lombrici, 1974: *Eretmapodites*, Oedipodeios Group
epactius Dyar & Knab, 1908: *Aedes* (*Georgecraigius*), Atropalpus Group (Zavortink, 1972; supersedes the Georgecraigius Group of Wilkerson *et al.*, 2015)

epanastasis Dyar, 1922e: *Culex* (*Melanoconion*), Spissipes Section, Crybda Group, Pedroi Subgroup
epidesmus (Theobald, 1910a): *Culex* (*Oculeomyia*), Sinensis Complex
epiroticus Linton & Harbach, 2005 (in Linton *et al.*, 2005): *Anopheles* (*Cellia*), Pyretophorus Series, Sundaicus Complex
epitedeus (Knab, 1907b): *Deinocerites*, Epitedeus Group
equinoxialis Floch & Abonnenc, 1945b: *Culex* (*Melanoconion*), Melanoconion Section, Intrincatus Group, Intrincatus Subgroup
equinus Theobald, 1903g: *Haemagogus sensu stricto*, Albomaculatus Section
erepens Gillies, 1958: *Anopheles* (*Cellia*), Myzomyia Series, Wellcomei Group
erethyzonfer Galindo & Blanton, 1954: *Culex* (*Micraedes*), Erethyzonfer Group (the Group is monobasic)
erikae Szadziewski & Szadziewska, 1985: *Culex* (fossil species)
eritreae Lewis, 1942: *Aedes* (*Aedimorphus*), Alboscutellatus Assemblage, Abnormalis Group
erlindae Baisas & Ubaldo-Pagayon, 1953: *Tripteroides* (*Tripteroides*), Nitidoventer Group
ernanii Duret, 1968f: *Culex* (*Melanoconion*), Melanoconion Section, Inhibitator Group, Inhibitator Subgroup
ernsti Anduze, 1949a: *Culex* (*Melanoconion*), Melanoconion Section, Inhibitator Group, Inhibitator Subgroup
erraticus (Dyar & Knab, 1906d): *Culex* (*Melanoconion*), Melanoconion Section, Erraticus Group, Erraticus Subgroup
erythraeus Corradetti, 1939b: *Anopheles* (*Cellia*), Myzomyia Series (not placed in a Group)
erythrothorax Dyar, 1907: *Culex sensu stricto*, Pipiens Group (not placed in a Subgroup)
erythrurus (Edwards, 1941): *Toxorhynchites* (*Afrorhynchus*) (not placed in a Group)
eschirasi Galliard, 1931b: *Culex sensu stricto*, Pipiens Group (not placed in a Subgroup)
escodae Baisas & Ubaldo-Pagayon, 1953: *Tripteroides* (*Tripteroides*), Nitidoventer Group
esmeraldasi (Levi-Castillo, 1955a): *Wyeomyia* (*Hystatomyia*), Circumcincta Group
esoensis Yamada, 1921: *Aedes sensu stricto*, Esoensis Group
espini (Martini, 1914): *Isostomyia*
ethiopicus Gillies & Coetzee, 1987: *Anopheles* (*Cellia*), Myzomyia Series (not placed in a Group)
ethiopiensis Huang, 2004: *Aedes* (*Stegomyia*), Poweri Group
eucephalaeus Dyar, 1918c: *Aedes* (*Ochlerotatus*), Serratus Group
euedes Howard, Dyar & Knab, 1913: *Aedes* (*Ochlerotatus*), Cantans Group
euiris Dyar, 1922e: *Aedes* (*Ochlerotatus*) (not placed in a Group)
eukrines Bram & Rattanarithikul, 1967: *Culex* (*Lophoceraomyia*), Mammilifer Group, Minor Line, Peytoni Complex
euplocamus Dyar & Knab, 1906d: *Aedes* (*Ochlerotatus*), Scapularis Group, Infirmatus Subgroup
europaeus da Cunha Ramos, Ribeiro & Harrison, 2003: *Culex* (*Neoculex*), Territans Group
evandroi da Costa Lima, 1937: *Anopheles sensu stricto*, Laticorn Section, Arribalzagia Series
evansae Antunes, 1942 (in Lane & Cerqueira, 1942): *Trichoprosopon*, Trichoprosopon Group, Lampropus Complex
evansae (Brèthes, 1926): *Anopheles* (*Nyssorhynchus*), Albimanus Section, Oswaldoi Series, Oswaldoi Group, Oswaldoi Subgroup
evansae (Edwards, 1936): *Toxorhynchites* (*Toxorhynchites*), Brevipalpis Group, Phytophagus Subgroup
evansae Root, 1927c: *Culex* (*Melanoconion*), Melanoconion Section, Evansae Group
exallos Rocha, Lourenço-de-Oliveira & Motta, 2012: *Wyeomyia* (not placed in a Subgenus)
excrucians (Walker, 1856): *Aedes* (*Ochlerotatus*), Cantans Group
exedrus Root, 1927c: *Culex* (*Melanoconion*), Melanoconion Section, Atratus Group
exnebulis Bonne-Wepster, 1948: *Tripteroides* (*Rachisoura*), Filipes Group
explorator Marks, 1964a: *Aedes* (*Ochlerotatus*) (not placed in a Group)

Ff

fabricii Lane & Cerqueira, 1942: *Sabethes* (*Peytonulus*)
faini Leleup, 1952: *Anopheles* (*Cellia*), Neomyzomyia Series, Smithii Group
fairchildi Galindo & Blanton, 1954: *Culex* (*Melanoconion*), Melanoconion Section, Intrincatus Group, Idottus Subgroup
fajardi (Lutz, 1904) (in Bourroul, 1904): *Chagasia*
fajardoi Baisas & Feliciano, 1953: *Zeugnomyia*
falabreguesi Hamon, 1957: *Aedes* (*Catageiomyia*)
falcipes Banks, 1906: *Uranotaenia* (*Uranotaenia*), Testacea Series
farakoensis Hamon, 1955: *Culex sensu stricto*, Guiarti Group
farauti Laveran, 1902b: *Anopheles* (*Cellia*), Neomyzomyia Series, Punctulatus Group, Farauti Complex (the Complex also includes three unnamed species: *farauti* species 4 and 5 of Foley *et al.*, 1993 and *farauti* species 8 of Bower *et al.*, 2008)
farquharsoni (Edwards, 1922c): *Malaya*
fasciolata (Lynch Arribálzaga, 1891b): *Coquillettidia* (*Pseudotaeniorhynchus*)
fascipalpis (Edwards, 1912a): *Aedes* (*Diceromyia*), Section B, Fascipalpis Group (the Group is monobasic)
fascipes (Coquillett, 1906a): *Orthopodomyia*, Thomasina Section
fasyi Baisas, 1938: *Culex sensu stricto*, Sitiens Group, Mimeticus Subgroup, Mimeticus Complex
faurani Duret, 1968f: *Culex* (*Melanoconion*), Spissipes Section, Faurani Group (the Group is monobasic)

fausti Vargas, 1943a: *Anopheles sensu stricto*, Angusticorn Section, Anopheles Series, Plumbeus Group
feegradei Barraud, 1934: *Aedes* (*Phagomyia*)
felicia (Dyar & Núñez Tovar, 1927b): *Wyeomyia* (*Decamyia*)
felicitatis Bonne-Wepster, 1948: *Tripteroides* (*Rachisoura*), Filipes Group
femineus Edwards, 1926a: *Culex* (*Eumelanomyia*), Mochthogenes Group, Femineus Subgroup (the Subgroup is monobasic)
femorata (Edwards, 1936): *Mimomyia* (*Mimomyia*), Ludlowia Group
fengi Edwards, 1935a: *Aedes* (*Luius*) (the Subgenus is monobasic)
fergusoni (Taylor, 1914b): *Culex* (*Neoculex*), Pseudomelanoconia Group
ferinus Knight, 1947: *Aedes* (*Mucidus*), Mucidus Group
fernandezi Casal, García & Cavalieri, 1966: *Culex sensu stricto* (not placed in a Group)
fernandezyepezi (Cova Garcia, Sutil Oramas & Pulido F., 1974b): *Wyeomyia* (*Caenomyiella*) (the Subgenus is monobasic)
ferox (von Humboldt, 1819): *Psorophora* (*Janthinosoma*)
ferrarai Rickenbach & Eouzan, 1970: *Eretmapodites*, Plioleucus Group
ferreri Duret, 1968f: *Culex* (*Melanoconion*), Melanoconion Section, Intrincatus Group, Idottus Subgroup
fiebrigi Edwards, 1922d: *Psorophora* (*Janthinosoma*)
fijiensis (Belkin, 1962): *Coquillettidia* (*Coquillettidia*), Crassipes Group
fijiensis Marks, 1947a: *Aedes* (*Finlaya*), Kochi Group
filicis Ingram & de Meillon, 1927: *Aedes* (*Catageiomyia*)
filipes (Walker, 1861): *Tripteroides* (*Rachisoura*), Filipes Group
filipinae Manalang, 1930: *Anopheles* (*Cellia*), Myzomyia Series, Funestus Group, Aconitus Subgroup
fimbriata King & Hoogstraal, 1947a: *Uranotaenia* (*Uranotaenia*), Anisocheleomyia Section, Tibialis Group
fimbriatus Edwards, 1930a: *Armigeres sensu stricto*
fimbriforceps Edwards, 1935a: *Culex* (*Eumelanomyia*), Mochthogenes Group, Inconspicuosus Subgroup
fimbripes Edwards, 1924b: *Aedes* (*Geoskusea*), Fimbripes Group
finlayi Lane & Cerqueira, 1942: *Wyeomyia* (*Spilonympha*)
fishi Zavortink, 1986: *Wyeomyia* (*Zinzala*)
fitchii (Felt & Young, 1904): *Aedes* (*Ochlerotatus*), Cantans Group
flabellata (Lane & Cerqueira, 1942): *Wyeomyia* (*Phoniomyia*), Second Series
flabellifer Komp, 1936d: *Culex* (*Melanoconion*), Melanoconion Section, Inhibitator Group, Inhibitator Subgroup
flabelliger Bonne-Wepster, 1948: *Tripteroides* (*Rachisoura*), Filipes Group
flavens (King & Hoogstraal, 1946e): *Mimomyia* (*Mimomyia*), Ludlowia Group
flaveola (Coquillett, 1906a): *Mansonia* (*Mansonia*), Titillans Complex
flavescens (Müller, 1764): *Aedes* (*Ochlerotatus*), Cantans Group
flavicollis Edwards, 1928a: *Aedes* (*Diceromyia*), Section A, Flavicollis Group
flavicornis Barraud, 1924h: *Culex* (*Lophoceraomyia*), Mammilifer Group, Mammilifer Subgroup, Flavicornis Complex
flavicosta Barraud, 1927a: *Orthopodomyia*, Orthopodomyia Section, Albipes Group, Flavicosta Subgroup
flavicosta Edwards, 1911a: *Anopheles* (*Cellia*), Myzomyia Series (not placed in a Group)
flavifacies Edwards, 1922d: *Wyeomyia* (*Antunesmyia*)
flavifrons (Skuse, 1889): *Aedes* (*Ochlerotatus*), Flavifrons Group
flavimargo Edwards, 1941: *Aedes* (*Neomelaniconion*) (not placed in a Group)
flavipennis (Giles, 1904): *Aedes* (*Finlaya*), Poicilius Group
flavirostris (Ludlow, 1914a): *Anopheles* (*Cellia*), Myzomyia Series, Funestus Group, Minimus Subgroup
flavisetosus de Oliveira Castro, 1935: *Limatus*
flavithorax Barraud, 1927a: *Orthopodomyia*, Orthopodomyia Section, Albipes Group, Flavicosta Subgroup
flavocincta (Edwards, 1936): *Coquillettidia* (*Coquillettidia*), Crassipes Group
flavopicta (Edwards, 1936): *Mimomyia* (*Mimomyia*), Group E (the Group is monobasic)
flavopictus Yamada, 1921: *Aedes* (*Stegomyia*), Scutellaris Group, Albopictus Subgroup
flavus Gistl, 1831: *Culex* (fossil species)
flavus (Leicester, 1908): *Armigeres* (*Leicesteria*)
flochi Duret, 1969d: *Culex* (subgenus uncertain)
florestan Dyar, 1925a: *Wyeomyia* (*Wyeomyia*) (not placed in a Group)
floridensis Belkin, 1950a: *Tripteroides* (*Polylepidomyia*), Floridensis Group (the Group is monobasic)
flui (Bonne-Wepster & Bonne, 1920): *Wyeomyia* (not placed in a Subgenus)
fluminensis Root, 1927d: *Anopheles sensu stricto*, Laticorn Section, Arribalzagia Series
fluviatilis James, 1902: *Anopheles* (*Cellia*), Myzomyia Series, Funestus Group, Minimus Subgroup, Fluviatilis (*An. fluviatilis* consists of three sibling species denoted as Species S, T and U (Sarala *et al.*, 1994); which is the nominotypical species is not known)
fluviatilis (Lutz, 1904) (in Bourroul, 1904): *Aedes* (*Georgecraigius*), Horsfallius Group (the Group is monobasic)
fluviatilis (Theobald, 1903a): *Shannoniana*

foliaceus Lane, 1945b: *Culex sensu stricto*, Pipiens Group (not placed in a Subgroup)
foliafer Komp & Rozeboom, 1951: *Culex* (*Melanoconion*), Melanoconion Section, Pilosus Group, Caudelli Subgroup
foliatus Brug, 1931c: *Armigeres sensu stricto*
foliatus Brug, 1932: *Culex* (*Eumelanomyia*), Mochthogenes Group, Foliatus Subgroup
folicola Belkin, 1955: *Tripteroides* (*Polylepidomyia*), Caledonicus Group
foliformis (King & Hoogstraal, 1947b): *Verrallina sensu stricto*, Carmenti Series
fonsecai (Pinto, 1932): *Mansonia* (*Mansonia*)
fontenillei Brunhes & Hervy, 1995: *Orthopodomyia*, Orthopodomyia Section, Vernoni Group
fontenillei (Le Goff, Boussès & Brunhes, 2007): *Aedes* (*Neomelaniconion*), Hylaius Group
fontenillei Ribeiro, 2004: *Toxorhynchites* (*Afrorhynchus*), Pauliani Group
fontinalis Gillies & de Meillon, 1968: *Anopheles* (*Cellia*), Myzomyia Series (not placed in a Group)
forattinii Cerqueira, 1961: *Sabethes* (*Sabethes*), Amazonicus Group
forattinii Clastrier, 1974: *Wyeomyia* (*Cruzmyia*)
forattinii Wilkerson & Sallum, 1999: *Anopheles sensu stricto*, Laticorn Section, Arribalzagia Series
forceps Cerqueira, 1939: *Psorophora* (*Janthinosoma*)
forcipenis Lourenço-de-Oliveira & da Silva, 1985a: *Wyeomyia* (*Spilonympha*)
forcipulatus Edwards, 1936: *Eretmapodites*, Inornatus Group
formosensis Yamada, 1921: *Aedes* (*Hulecoeteomyia*)
formosus Ludlow, 1909a: *Anopheles sensu stricto*, Angusticorn Section, Anopheles Series, Lindesayi Group, Gigas Subgroup, Gigas Complex
formosus (Walker, 1848): *Aedes* (*Stegomyia*), Aegypti Group
fowleri (de Charmoy, 1908): *Aedes* (*Aedimorphus*), Domesticus Assemblage, Vexans Subassemblage, Vexans Group
fragilis Leicester, 1908: *Verrallina* (*Harbachius*)
fragilis Ludlow, 1903: *Culex* (*Culiciomyia*), Fragilis Group
fragilis (Theobald, 1903f): *Anopheles sensu stricto*, Angusticorn Section, Anopheles Series, Aitkenii Group
franciscanus McCracken, 1904: *Anopheles sensu stricto*, Angusticorn Section, Anopheles Series, Pseudopunctipennis Group
franciscoi Mattingly, 1959: *Aedes* (*Petermattinglyius*), Petermattinglyius Group
franciscoi Reid, 1962: *Anopheles sensu stricto*, Laticorn Section, Myzorhynchus Series, Barbirostris Group, Barbirostris Subgroup
franclemonti Belkin, 1962: *Aedes* (*Finlaya*), Kochi Group, Franclemonti Complex
fraseri (Edwards, 1912a): *Aedes* (*Stegomyia*), Apicoargenteus Group
fraseri Edwards, 1912a: *Uranotaenia* (*Uranotaenia*), Alba Section, Bilineata Group
fraseri (Edwards, 1914b): *Culiseta* (*Theomyia*) (the Subgenus is monobasic)
fraseri (Edwards, 1922c): *Malaya*
fraseri (Theobald, 1911a): *Coquillettidia* (*Coquillettidia*), Crassipes Group
fraudatrix (Theobald, 1905f): *Culex* (*Lophoceraomyia*), Fraudatrix Group, Fraudatrix Subgroup, Fraudatrix Complex
freeborni Aitken, 1939: *Anopheles sensu stricto*, Angusticorn Section, Anopheles Series, Maculipennis Group, Freeborni Subgroup
freetownensis Evans, 1925c: *Anopheles* (*Cellia*), Myzomyia Series, Demeilloni Group
frenchii (Theobald, 1901c): *Culiseta* (*Neotheobaldia*)
freycinetiae Laird, 1957: *Aedes* (*Finlaya*), Kochi Group
freyi Meng, 1957: *Anopheles sensu stricto*, Laticorn Section, Myzorhynchus Series, Barbirostris Group (not placed in a Subgroup)
frontosa Theobald, 1903a: *Runchomyia* (*Runchomyia*)
fryeri (Theobald, 1912b): *Aedes* (*Coetzeemyia*) (the Subgenus is monobasic)
fulgens da Cunha Ramos & Brunhes, 2004: *Uranotaenia* (*Pseudoficalbia*), Shillitonis Section, Shillitonis Group (not placed in a Subgroup)
fulgens (Edwards, 1917): *Aedes* (*Zavortinkius*), Longipalpis Group
fuliginosus Lee, 1946: *Tripteroides* (*Rachisoura*), Filipes Group
fulleri (Ludlow, 1909b): *Culex* (*Lophoceraomyia*), Fraudatrix Group, Fraudatrix Subgroup, Cinctellus Complex
fulvithorax (Lutz, 1904) (in Bourroul, 1904): *Aedes* (*Howardina*), Fulvithorax Section, Flavithorax Subassemblage
fulvus (Wiedemann, 1828): *Aedes* (*Ochlerotatus*), Fulvus Group
fumidus Edwards, 1928a: *Aedes* (*Lorrainea*)
fumipennis (Stephens, 1825): *Culiseta* (*Culicella*)
funerea (Leicester, 1908): *Heizmannia sensu stricto*
funerea (Theobald, 1903a): *Verrallina sensu stricto*, Carmenti Series
funestus Giles, 1900b: *Anopheles* (*Cellia*), Myzomyia Series, Funestus Group, Funestus Subgroup
funestus (Leicester, 1908): *Toxorhynchites* (*Toxorhynchites*), Magnificus Group
funiculus Dyar, 1920d: *Psorophora* (*Grabhamia*) (not placed in a Group)
furcifer (Edwards, 1913b): *Aedes* (*Diceromyia*), Section A, Furcifer Group
furfurea (Enderlein, 1923): *Aedeomyia* (*Lepiothauma*)
furlongi van Someren, 1954a: *Culex* (*Culiciomyia*), Nebulosus Group

fusca (Leicester, 1908): *Mimomyia* (*Ingramia*), Group C (*Ingramia s. str.*, Fusca Group)
fusca Theobald, 1907: *Uranotaenia* (*Pseudoficalbia*), Nigripes Section, Fusca Group (the Group is monobasic)
fuscana (Wiedemann, 1820): *Lutzia* (*Metalutzia*)
fuscatus Lane & Whitman, 1951: *Culex* (*Microculex*), Inimitabilis Series
fuscicinctus King & Hoogstraal, 1946a: *Culex* (*Culiciomyia*), Fragilis Group
fuscicolor van Someren, 1947b: *Anopheles sensu stricto*, Laticorn Section, Myzorhynchus Series, Coustani Group
fuscinervis (Edwards, 1914b): *Aedes* (*Neomelaniconion*) (not placed in a Group)
fuscipalpis Belkin, 1962: *Aedes* (*Finlaya*), Kochi Group, Solomonis Complex
fuscipes Edwards, 1922d: *Wyeomyia* (*Phoniomyia*), Second Series
fuscipleura Lee, 1946: *Tripteroides* (*Rachisoura*), Filipes Group
fuscitarsis Belkin, 1962: *Aedes* (*Finlaya*), Kochi Group, Solomonis Complex
fuscivenosus Leeson, 1930: *Anopheles* (*Cellia*), Myzomyia Series, Funestus Group, Rivulorum Subgroup
fuscocephala Theobald, 1907: *Culex sensu stricto*, Pipiens Group, Univittatus Subgroup
fuscopennata (Theobald, 1903a): *Coquillettidia* (*Coquillettidia*), Crassipes Group
fuscopteron (Theobald, 1911d): *Coquillettidia* (*Coquillettidia*), Crassipes Group
fuscus Hutton, 1902: *Opifex sensu stricto* (monobasic)
futunae Belkin, 1962: *Aedes* (*Stegomyia*), Zoromorphus Group (the Group is monobasic)

Gg

gabaldoni Cova García, Pulido F., Escalante de Ugueto & Mora R., 1987b: *Uranotaenia* (*Uranotaenia*) (not placed in a Section, Series, Group or Complex)
gabaldoni Vargas, 1941: *Anopheles sensu stricto*, Laticorn Section, Arribalzagia Series
gabonensis Rahola, Makanga & Paupy, 2014 (in Rahola *et al.*, 2014): *Anopheles* (*Cellia*), Myzomyia Series (not placed in a Group)
gabriel Schick, 1970a: *Aedes* (*Protomacleaya*), Terrens Group, Heteropus Subgroup
gaffigani Reinert, 1990b: *Aedes* (*Kenknightia*), Luzonensis Group
gagnei Evenhuis, 1989 (in Evenhuis & Gon, 1989): *Culex* (*Lophoceraomyia*), Fraudatrix Group, Fraudatrix Subgroup (Sirivanakarn, 1977a unnecessarily recognized a monobasic '*ornatus* complex' for this species in the absence of evidence that it is more than one species)
gahnicola Marks, 1947a: *Aedes* (*Finlaya*), Kochi Group
gairus Root, 1927c: *Culex* (*Microculex*), Pleuristriatus Series
galindoi Komp & Rozeboom, 1951: *Culex* (*Melanoconion*), Melanoconion Section, Distinguendus Group, Galindoi Subgroup (the Subgroup is monobasic)
galindoi Schick, 1970a: *Aedes* (*Protomacleaya*), Terrens Group, Galindoi Subgroup
galliardi Edwards, 1941: *Culex* (*Eumelanomyia*), Rubinotus-rima Group, Rima Subgroup
galloisi Yamada, 1921: *Aedes* (*Stegomyia*), Scutellaris Group, Albopictus Subgroup
galloisioides Liu & Lu, 1984: *Aedes* (*Stegomyia*), Scutellaris Group, Albopictus Subgroup
galvaoi Causey, Deane & Deane, 1943: *Anopheles* (*Nyssorhynchus*), Albimanus Section, Oswaldoi Series, Oswaldoi Group, Oswaldoi Subgroup
galvaoi (Correa & Ramalho, 1956): *Wyeomyia* (*Phoniomyia*), Third Series
galvaoi Duret, 1968b: *Culex* (*Melanoconion*), Melanoconion Section, Pilosus Group, Caudelli Subgroup
gambiae Giles, 1902: *Anopheles* (*Cellia*), Pyretophorus Series, Gambiae Complex
gameti Bailly-Choumara, 1966b: *Culex sensu stricto* (not placed in a Group)
gamma Séguy, 1924: *Culex* (*Neoculex*) (not placed in a Group)
ganapathi Colless, 1958: *Aedes* (*Downsiomyia*), Niveus Group
ganapathi Colless, 1965: *Culex* (*Lophoceraomyia*), Mammilifer Group, Minor Line, Ganapathi Complex
gandaensis Huang, 2004: *Aedes* (*Stegomyia*), Simpsoni Group
gandarai da Cunha Ramos, Capela & Ribeiro, 1995: *Aedes* (*Polyleptiomyia*)
gani Bonne-Wepster, 1940: *Aedes* (*Finlaya*), Gani Group (the Group is monobasic)
garciai Broche, 2000: *Culex sensu stricto*, Pipiens Group, Tarsalis Subgroup
garcesi Duret, 1968b: *Culex* (*Melanoconion*), Melanoconion Section, Pilosus Group, Caudelli Subgroup
gardnerii (Ludlow, 1905a): *Aedes* (*Stegomyia*), W-albus Group, W-albus Subgroup
garioui Bailly-Choumara & Rickenbach, 1966: *Culex* (*Eumelanomyia*), Eumelanomyia Group
garnhami Edwards, 1930a: *Anopheles* (*Cellia*), Myzomyia Series, Demeilloni Group
garnhami van Someren, 1948: *Uranotaenia* (*Pseudoficalbia*), Shillitonis Section, Shillitonis Group, Subgroup A
gaudeator Dyar & Knab, 1907c: *Culex* (*Microculex*) (not placed in a Series)
gaudians Dyar & Núñez Tovar, 1927b: *Wyeomyia* (*Wyeomyia*) (not placed in a Group)
gaufini Belkin, 1962: *Culex* (*Neoculex*), Pseudomelanoconia Group
gausapata Dyar & Núñez Tovar, 1927b: *Wyeomyia* (*Wyeomyia*) (not placed in a Group)

gedanica Szadziewski & Gilka, 2011: *Culiseta* (fossil species)
gediensis Edwards, 1941: *Culex sensu stricto*, Guiarti Group
gelidus Theobald, 1901c: *Culex sensu stricto*, Sitiens Group, Gelidus Subgroup (the Group is monobasic)
geminus Colless, 1955b: *Culex* (*Oculeomyia*), Geminus Complex
geminus Peus, 1970: *Aedes sensu stricto*, Cinereus Group
geniculata da Cunha Ramos & Brunhes, 2004: *Uranotaenia* (*Uranotaenia*), Caeruleocephala Section, Madagascarica Group
geniculatus (Olivier, 1791): *Aedes* (*Dahliana*)
genurostris Leicester, 1908: *Malaya*
geoffroyi Reinert, 1999b: *Aedes* (*Zavortinkius*), Longipalpis Group
geometrica Theobald, 1901c: *Uranotaenia* (*Uranotaenia*), Pulcherrima Series
geometricus Corrêa, 1944: *Anopheles sensu stricto*, Angusticorn Section, Anopheles Series, Pseudopunctipennis Group
georgianus King, 1939: *Anopheles sensu stricto*, Angusticorn Section, Anopheles Series, Punctipennis Group, Crucians Complex
geoskusea Amos, 1944: *Aedes* (*Levua*) (the Subgenus is monobasic)
gerbergi Belkin, 1977: *Toxorhynchites* (*Lynchiella*)
gerdae Slooff, 1963: *Uranotaenia* (*Uranotaenia*), Anisocheleomyia Section, Tibialis Group
germaini Geoffroy, 1974: *Culex* (*Eumelanomyia*), Mochthogenes Group, Inconspicuosus Subgroup
germaini Rickenbach & Eouzan, 1970: *Eretmapodites*, Plioleucus Group
gibbinsi Edwards, 1935a: *Aedes* (*Aedimorphus*), Domesticus Assemblage, Vexans Subassemblage, Quasiunivittatus Group
gibbinsi Evans, 1935: *Anopheles* (*Cellia*), Myzomyia Series, Marshallii Group
gibbosa (Delfinado, 1967): *Verrallina* (*Neomacleaya*)
gibbulus Delfinado, 1966a: *Culex* (*Lophoceraomyia*), Fraudatrix Group, Fraudatrix Subgroup, Rubithoracis Complex
giblini (Taylor, 1914d): *Coquillettidia* (*Coquillettidia*), Crassipes Group
gigantea Brug, 1931c: *Uranotaenia* (*Pseudoficalbia*), Bimaculata Series
giganteus Ventrillon, 1906a: *Culex* (*Oculeomyia*) (not placed in a Complex)
gigantulus (Dyar & Shannon, 1925a): *Toxorhynchites* (*Toxorhynchites*), Minimus Group
gigas Giles, 1901b: *Anopheles sensu stricto*, Angusticorn Section, Anopheles Series, Lindesayi Group, Gigas Subgroup, Gigas Complex
gilcolladoi Sánchez-Covisa Villa, Rodríguez Rodríguez & Guillén Llera, 1985: *Aedes* (*Dahliana*)
gilesi (Neiva, 1908) (in Peryassú, 1908): *Anopheles* (*Lophopodomyia*)
gilletti van Someren, 1949b: *Eretmapodites*, Chrysogaster Group
gilli (Barraud, 1924d): *Aedes* (*Himalaius*)
gilliesi Hamon & van Someren, 1961a: *Culex* (*Culiciomyia*), Nebulosus Group
gilliesi van Someren, 1962: *Aedes* (*Albuginosus*)
giveni Edwards, 1926a: *Armigeres sensu stricto*
glaphyroptera (Schiner, 1864): *Culiseta sensu stricto*
glaucodaemon (Dyar & Shannon, 1925b): *Sabethes* (*Sabethoides*)
globocoxitus Dobrotworsky, 1953: *Culex sensu stricto*, Pipiens Group, Pipiens Subgroup
glyptosalpinx Harbach, Peyton & Harrison, 1984: *Culex* (*Melanoconion*), Melanoconion Section, Intrincatus Group, Intrincatus Subgroup
gnomatos Sallum, Hutchings, Leila & Ferreira, 1997: *Culex* (*Melanoconion*), Spissipes Section, Vomerifer Group
goeldii Rozeboom & Gabaldon, 1941: *Anopheles* (*Nyssorhynchus*), Albimanus Section, Oswaldoi Series, Oswaldoi Group, Oswaldoi Subgroup, Nuneztovari Complex
gombakensis Mattingly, 1959: *Aedes* (*Christophersiomyia*)
gomezdelatorrei Leví-Castillo, 1955c: *Anopheles* (*Lophopodomyia*)
gonguoensis Gong & Lu, 1986: *Aedes* (*Jihlienius*)
gonzalezrinconesi Cova Garcia, Pulido F. & Escalante de Ugueto, 1977a: *Anopheles* (*Kerteszia*)
gorgasi Duret, 1971: *Sabethes* (*Peytonulus*)
gossi Bohart, 1957: *Culex* (*Lophoceraomyia*), Fraudatrix Group (not placed in a Subgroup)
gouldi Peyton & Klein, 1970: *Uranotaenia* (*Pseudoficalbia*), Bicolor Series
gouldi Reinert, 1972b: *Aedes* (*Aedimorphus*), Domesticus Assemblage, Mediolineatus Subassemblage, Pampangensis Group
grabhami Berlin, 1969a: *Aedes* (*Howardina*), Walkeri Section, Aurites Group
grabhamii Theobald, 1901b: *Anopheles sensu stricto*, Angusticorn Section, Cycloleppteron Series (the Series is monobasic)
gracicornis Sirivanakarn, 1977a: *Culex* (*Lophoceraomyia*), Fraudatrix Group, Fraudatrix Subgroup, Variatus Complex
gracilelineatus Bonne-Wepster, 1937: *Aedes* (subgenus uncertain)
gracilis Leicester, 1908: *Topomyia* (*Topomyia*)
gracilis Leicester, 1908: *Zeugnomyia*
gracilis Theobald, 1905f: *Bironella sensu stricto*
grahami Edwards, 1911b: *Eretmapodites*, Chrysogaster Group
grahamii (Theobald, 1909): *Aedes* (*Mucidus*), Mucidus Group

grahamii Theobald, 1910c: *Culex sensu stricto*, Guiarti Group
grandidieri (Blanchard, 1905): *Coquillettidia* (*Coquillettidia*), Crassipes Group
grandiosus (Williston, 1900): *Toxorhynchites* (*Lynchiella*)
grantii (Theobald, 1901b): *Aedes* (*Stegomyia*), Grantii Group (the Group is monobasic)
grassei da Cunha Ramos & Brunhes, 2004: *Uranotaenia* (*Uranotaenia*), Dumonti Section, Dumonti Group
grassei Doucet, 1951a: *Aedes* (*Paulianius*), Coulangesi Group
grassei Grjebine, 1953b: *Anopheles* (*Cellia*), Neomyzomyia Series, Pauliani Group
gravelyi (Edwards, 1921a): *Toxorhynchites* (*Toxorhynchites*), Leicesteri Group, Leicesteri Subgroup
grayii Theobald, 1901c: *Wyeomyia* (*Wyeomyia*) (not placed in a Group)
greeni Rattanarithikul & Harbach, 1991: *Anopheles* (*Cellia*), Neocellia Series, Maculatus Group (not placed in a Subgroup)
greenii (Theobald, 1903a): *Aedes* (*Bruceharrisonius*)
greenii (Theobald, 1905h): *Heizmannia sensu stricto*
grenieri Doucet, 1951b: *Uranotaenia* (*Pseudoficalbia*), Nigripes Section, Nigripes Group
grenieri Eouzan, 1969: *Culex* (*Culiciomyia*), Nebulosus Group
grenieri Grjebine, 1964: *Anopheles* (*Cellia*), Neomyzomyia Series, Pauliani Group
grenieri Hamon, Service, Adam & Taufflieb, 1961: *Aedes* (*Catageiomyia*)
grenieri Hamon & van Someren, 1961b: *Eretmapodites*, Oedipodeios Group
gressitti Bohart, 1957: *Aedes* (*Finlaya*), Kochi Group
gressitti Sirivanakarn, 1968: *Culex* (*Lophoceraomyia*), Fraudatrix Group, Fraudatrix Subgroup, Christiani Complex
griffithi (Thurman, 1954a): *Aedes* (*Ayurakitia*)
griveaudi Grjebine, 1961: *Anopheles* (*Cellia*), Neomyzomyia Series, Ranci Group (not placed in a Subgroup)
grjebinei (Brunhes, 1977b): *Mimomyia* (*Ingramia*), Group A (*Ravenalites s. str.*), Roubaudi Complex
grjebinei da Cunha Ramos & Brunhes, 2004: *Uranotaenia* (*Pseudoficalbia*), Spinosa Section, Spinosa Group
grjebinei Hamon, Taufflieb & Maillot, 1957: *Aedes* (*Pseudalbuginosus*) (the Subgenus is monobasic)
grjebinei Ribeiro, 2004: *Toxorhynchites* (*Afrorhynchus*), Pauliani Group
grossbecki Dyar & Knab, 1906d: *Aedes* (*Ochlerotatus*), Stimulans Group
guadeloupensis (Dyar & Knab, 1906c): *Toxorhynchites* (*Lynchiella*)
guamensis Farner & Bohart, 1944: *Aedes* (*Stegomyia*), Scutellaris Group, Scutellaris Subgroup
guarani Shannon, 1928 (in Dyar, 1928): *Anopheles* (*Nyssorhynchus*), Myzorhynchella Section
guarao Anduze & Capdevielle, 1949: *Anopheles sensu stricto*, Laticorn Section, Arribalzagia Series
guatemala Berlin, 1969a: *Aedes* (*Howardina*), Sexlineatus Section (not placed in a Group/Subassemblage)
guatemala Dyar & Knab, 1906a: *Wyeomyia* (*Wyeomyia*) (not placed in a Group)
guayasi Leví-Castillo, 1953b: *Culex sensu stricto*, Pipiens Group (not placed in a Subgroup)
guayasi Leví-Castillo, 1954: *Limatus*
gubernatoris (Giles, 1901a): *Aedes* (*Phagomyia*)
guedesi da Silva Mattos & Xavier, 1991: *Culex* (*Melanoconion*), Melanoconion Section, Saramaccensis Group
guerrero Berlin, 1969a: *Aedes* (*Howardina*), Sexlineatus Section (not placed in a Group/Subassemblage)
guerreroi Cova García, Sutil & Pulido, 1971: *Culex* (*Carrollia*), Bihaicola Group
guiarti Blanchard, 1905: *Culex sensu stricto*, Guiarti Group
guizhouensis Chen & Zhao, 1985: *Culex sensu stricto*, Pipiens Group, Trifilatus Subgroup
gurneri van Someren, 1946c: *Aedes* (*Pseudarmigeres*)
gurneyi (Belkin, 1962): *Mimomyia* (*Mimomyia*), Hybrida Group
gurneyi Stone & Bohart, 1944: *Aedes* (*Stegomyia*), Scutellaris Group, Scutellaris Subgroup
gutierrezi Duret, 1971: *Wyeomyia* (not placed in a Subgenus)
gutzevichi Dubitsky & Deshevykh, 1978: *Aedes* (*Ochlerotatus*), Caspius Group
guyanensis Clastrier, 1970a: *Culex* (*Aedinus*)
gyirongensis Ma, 1982: *Aedes* (*Ochlerotatus*), Communis Group, Punctor Subgroup
gymnothorax Harbach & Petersen, 1992: *Sabethes* (*Sabethes*), Amazonicus Group

Hh

habilitator Dyar & Knab, 1906d: *Culex sensu stricto*, Pipiens Group, Tarsalis Subgroup
hackeri Edwards, 1921a: *Anopheles* (*Cellia*), Neomyzomyia Series, Leucosphyrus Group, Hackeri Subgroup
hackeri Edwards, 1923b: *Culex* (*Eumelanomyia*), Mochthogenes Group, Tenuipalpis Subgroup
haddowi da Cunha Ramos & Brunhes, 2004: *Uranotaenia* (*Pseudoficalbia*), Annulata Section, Annulata Series, Lavieri Group
haddowi van Someren, 1949b: *Eretmapodites*, Chrysogaster Group
hadrognathus Harbach, 1995a: *Sabethes* (*Peytonulus*)
haemorrhoidalis (Fabricius, 1787): *Toxorhynchites* (*Lynchiella*)
hailarensis Xu & Luo, 1998: *Anopheles sensu stricto*, Laticorn Section, Myzorhynchus Series, Hyrcanus Group (not placed in a Subgroup)

hainanensis Chen, 1977: *Culex* (*Culiciomyia*), Dispectus Group
hakanssoni Knight & Hurlbut, 1949: *Aedes* (*Stegomyia*), Scutellaris Group, Scutellaris Subgroup
hakusanensis Yamaguti & Tamaboko, 1954: *Aedes* (*Ochlerotatus*), Communis Group, Punctor Subgroup
halifaxii (Theobald, 1903a): *Lutzia* (*Metalutzia*)
halophylus Silva do Nascimento & Lourenço-de-Oliveira, 2002: *Anopheles* (*Nyssorhynchus*), Albimanus Section, Oswaldoi Series, Triannulatus Group
hamistylus (Laffoon, 1946): *Verrallina* (*Harbachius*)
hamoni Adam, 1962: *Anopheles* (*Cellia*), Neomyzomyia Series, Smithii Group
hamoni Brunhes, Adam & Bailly-Choumara, 1967: *Culex* (*Eumelanomyia*), Mochthogenes Group, Inconspicuosus Subgroup
hamoni Grjebine, 1953a: *Uranotaenia* (*Uranotaenia*), Chorleyi Section, Hamoni Group
hamoni Grjebine, 1972: *Eretmapodites*, Oedipodeios Group
hancocki Edwards, 1929b: *Anopheles* (*Cellia*), Myzomyia Series, Marshallii Group
hancocki Edwards, 1930a: *Culex sensu stricto*, Pipiens Group (not placed in a Subgroup)
hancocki van Someren, 1962: *Aedes* (*Vansomerenis*)
hansfordi Huang, 1997: *Aedes* (*Stegomyia*), Dendrophilus Group
harbachi Nascimento-Pereira, Guimarães, Lourenço-de-Oliveira & Motta, 2021 (in Nascimento-Pereira *et al.*, 2021): *Sabethes* (*Peytonulus*)
harbachi Reinert, 1990b: *Aedes* (*Kenknightia*), Dissimilis Group
hardini Miyagi, Toma & Ramalingam, 1989: *Topomyia* (*Topomyia*)
hargreavesi (Edwards, 1920a): *Aedes* (*Ochlerotatus*), Caspius Group
hargreavesi Evans, 1927: *Anopheles* (*Cellia*), Myzomyia Series, Marshallii Group
harinasutai Knight, 1978a: *Aedes* (*Downsiomyia*), Niveus Group
harleyi Peters, 1955a: *Culex* (*Culiciomyia*), Nebulosus Group
harperi Evans, 1936: *Anopheles* (*Cellia*), Myzomyia Series, Marshallii Group
harperi Knight, 1948b: *Aedes* (*Danielsia*)
harperi van Someren, 1949b: *Eretmapodites*, Chrysogaster Group
harrisoni Harbach & Manguin, 2007 (in Harbach *et al.*, 2007c): *Anopheles* (*Cellia*), Myzomyia Series, Funestus Group, Minimus Subgroup, Minimus Complex
harrisoni Muspratt, 1953: *Aedes* (*Ochlerotatus*), Breedensis Group
harrisoni Peyton, 1977: *Uranotaenia* (*Pseudoficalbia*), Harrisoni Series (the Series is monobasic)
harrisoni Sirivanakarn, 1977c: *Culex* (*Culiciomyia*), Shebbearei Group
harrisonica (Reinert, 1974): *Verrallina* (*Neomacleaya*)
harveyi (Barraud, 1923e): *Aedes* (*Hulecoeteomyia*)
hastatus Dyar, 1922d: *Aedes* (*Ochlerotatus*), Serratus Group
hatorii Yamada, 1921: *Aedes* (*Collessius*), Collessius Group
haworthi Edwards, 1923d: *Aedes* (*Albuginosus*)
hayashii Yamada, 1917: *Culex* (*Eumelanomyia*), Mochthogenes Group, Tenuipalpis Subgroup
hebes Barraud, 1931b: *Uranotaenia* (*Uranotaenia*), Testacea Series
hebrardi da Cunha Ramos & Brunhes, 2004: *Uranotaenia* (*Uranotaenia*), Alba Section, Balfouri Group
hebrideus Edwards, 1926a: *Aedes* (*Stegomyia*), Scutellaris Group, Scutellaris Subgroup
hectoris Giaquinto Mira, 1931: *Anopheles sensu stricto*, Angusticorn Section, Anopheles Series, Pseudopunctipennis Group
hedys Root, 1927c: *Culex* (*Microculex*), Consolator Series
heiheensis Ma, 1981b: *Anopheles sensu stricto*, Laticorn Section, Myzorhynchus Series, Hyrcanus Group (not placed in a Subgroup)
heiseri Baisas, 1935c: *Uranotaenia* (*Uranotaenia*), Campestris Series
heischi van Someren, 1951: *Aedes* (*Stegomyia*), Dendrophilus Group
helenae Brunhes, Adam & Bailly-Choumara, 1967: *Culex* (*Eumelanomyia*), Mochthogenes Group, Inconspicuosus Subgroup
helenae Reinert, 1973f: *Aedes* (*Bothaella*)
helenae Ribeiro, 1992c: *Toxorhynchites* (*Afrorhynchus*) (not placed in a Group)
hemisagnosta Dyar & Knab, 1906d: *Wyeomyia* (*Wyeomyia*) (not placed in a Group)
hendersoni Cockerell, 1918: *Aedes* (*Protomacleaya*), Triseriatus Group
henrardi Edwards, 1935c: *Uranotaenia* (*Pseudoficalbia*), Shillitonis Section, Shillitonis Group (not placed in a Subgroup)
henriquei da Cunha Ramos, 1993: *Uranotaenia* (*Pseudoficalbia*), Nigripes Section, Nigripes Group
hensilli Farner, 1945: *Aedes* (*Stegomyia*), Scutellaris Group, Scutellaris Subgroup
hepperi Casal & Garcia, 1967c: *Culex* (*Phytotelmatomyia*)
hermanoi (Lane & Coutinho, 1940): *Coquillettidia* (*Pseudotaeniorhynchus*)
hermsi Barr & Guptavanij, 1989: *Anopheles sensu stricto*, Angusticorn Section, Anopheles Series, Maculipennis Group, Freeborni Subgroup
hernandoi Baisas & Feliciano, 1953: *Topomyia* (*Topomyia*)

herrerai Sutil Oramas, Pulido Florenzano & Amarista Meneses, 1987: *Culex* (*Melanoconion*), Melanoconion Section (not placed in a Group)
hervyi Brunhes, Le Goff & Geoffroy, 1999: *Anopheles* (*Cellia*), Neocellia Series (not placed in a Group)
hervyi da Cunha Ramos & Brunhes, 2004: *Uranotaenia* (*Pseudoficalbia*), Shillitonis Section, Shillitonis Group (not placed in a Subgroup)
hesperonotius Marks, 1959a: *Aedes* (*Ochlerotatus*) (not placed in a Group)
heteropus Dyar, 1921d: *Aedes* (*Protomacleaya*), Terrens Group, Heteropus Subgroup
heterospina Gong & Lu, 1986: *Heizmannia sensu stricto*
hewitti (Edwards, 1914b): *Culex* (*Lophoceraomyia*), Mammilifer Group, Brevipalpus Subgroup
hexacis (Martini, 1931): *Toxorhynchites* (*Ankylorhynchus*)
hexodontus Dyar, 1916b: *Aedes* (*Ochlerotatus*), Communis Group, Punctor Subgroup
hightoni van Someren, 1947c: *Eretmapodites*, Quinquevittatus Group
hilli Edwards, 1922d: *Culex* (*Lophoceraomyia*), Fraudatrix Group, Fraudatrix Subgroup, Hilli Complex
hilli (Edwards, 1926a): *Culiseta* (*Neotheobaldia*)
hilli Woodhill & Lee, 1944: *Anopheles* (*Cellia*), Neomyzomyia Series (not placed in a Group)
himalayensis Edwards, 1922a: *Heizmannia sensu stricto*
himalayensis Somboon, Namgay & Harbach, 2022 (in Somboon *et al.*, 2022b): *Anopheles sensu stricto*, Angusticorn Section, Anopheles Series, Lindesayi Group, Lindesayi Subgroup, Lindesayi Complex (the Complex also includes informally designated Species B, see Somboon *et al.*, 2022b)
hinesorum Schmidt, 2001 (in Schmidt *et al.*, 2001): *Anopheles* (*Cellia*), Neomyzomyia Series, Punctulatus Group, Farauti Complex
hinglungensis Chu, 1957: *Culex* (*Eumelanomyia*), Mochthogenes Group, Hinglungensis Subgroup
hirsuta Boussès & Brunhes, 2013: *Uranotaenia* (*Pseudoficalbia*), Annulata Section, Annulata Series, Lavieri Group
hirtusa Gong, 1989a: *Topomyia* (*Topomyia*)
hirsuta (Hill & Hill, 1946): *Wyeomyia* (*Wyeomyia*), Hirsuta Group, Hirsuta Subgroup
hirsutifemora Peters, 1964: *Uranotaenia* (*Pseudoficalbia*), Bicolor Series
hirsutus (Theobald, 1901b): *Aedes* (*Aedimorphus*), Domesticus Assemblage, Vexans Subassemblage, Vexans Group
hirtipalpis Sirivanakarn, 1977a: *Culex* (*Lophoceraomyia*), Wilfredi Group
hirtusa Gong, 1989a: *Topomyia* (*Topomyia*)
hispaniola (Theobald, 1903a): *Anopheles* (*Cellia*), Paramyzomyia Series, Cinereus Group
hispida (Delfinado, 1967): *Verrallina* (*Neomacleaya*)
hispida (Theobald, 1910c): *Mimomyia* (*Mimomyia*), Ludlowia Group
hodgkini Marks, 1959a: *Aedes* (*Ochlerotatus*), Perkinsi Group
hodgkini Reid, 1962: *Anopheles sensu stricto*, Laticorn Section, Myzorhynchus Series, Barbirostris Group, Barbirostris Subgroup
hodgkini (Wharton, 1962): *Coquillettidia* (*Coquillettidia*), Crassipes Group
hoffeinsorum (Szadziewski, 1998): *Aetheapnomyia* (fossil species)
hoffmani Root, 1927a: *Limatus*
hogsbackensis Huang, 2004: *Aedes* (*Stegomyia*), Poweri Group
hoguei Belkin, 1962: *Aedes* (*Stegomyia*), Scutellaris Group, Scutellaris Subgroup
hokkaidensis Tanaka, Mizusawa & Saugstad, 1979: *Aedes* (*Ochlerotatus*), Communis Group, Punctor Subgroup
hollandi Taylor, 1934: *Bironella* (*Brugella*)
hollandius King & Hoogstraal, 1946b: *Aedes* (*Patmarksia*)
hollingsheadi Belkin, 1962: *Aedes* (*Finlaya*), Kochi Group, Bougainvillensis Complex
holmbergii Lynch Arribálzaga, 1891a: *Psorophora sensu stricto*
holocinctus Edwards, 1941: *Aedes* (*Aedimorphus*), Domesticus Assemblage, Vexans Subassemblage, Quasiunivittatus Group
homoeopus Dyar, 1922c: *Aedes* (*Protomacleaya*), Terrens Group, Homoeopus Subgroup
homunculus Komp, 1937b: *Anopheles* (*Kerteszia*), Homunculus Complex (*An. homunculus* consists of two sibling species denoted as Species 1 (*homunculus sensu stricto*?) and 2) (Bourke *et al.*, 2023)
hongayi Galliard & Ngu, 1947: *Uranotaenia* (*Pseudoficalbia*), Recondita Series
hoogstraali Baisas, 1947: *Tripteroides* (*Tripteroides*), Nitidoventer Group
hoogstraali Knight & Rozeboom, 1946: *Aedes* (*Scutomyia*), Hoogstraali Group
hopkinsi Edwards, 1932b: *Culex sensu stricto*, Pipiens Group (not placed in a Subgroup)
hopkinsi Edwards, 1932b: *Uranotaenia* (*Uranotaenia*), Hopkinsi Section, Hopkinsi Group
hopkinsi Edwards, 1936: *Aedes* (*Catageiomyia*)
horotoi Taylor, 1972: *Aedes* (*Finlaya*), Kochi Group, Franclemonti Complex
horrescens Edwards, 1935a: *Aedes* (*Stegomyia*), Scutellaris Group, Scutellaris Subgroup, Pseudoscutellaris Subgroup/Complex
horrida (Dyar & Knab, 1908): *Psorophora* (*Janthinosoma*)
horridus Edwards, 1922d: *Culex* (*Eumelanomyia*), Protomelanoconion Group
hortator Dyar & Knab, 1907a: *Aedes* (*Sallumia*)

hortensis Ficalbi, 1889a: *Culex* (*Maillotia*), Hortensis Group
hosautos Dyar & Knab, 1907c: *Wyeomyia* (*Miamyia*)
houghtoni Feng, 1941: *Topomyia* (*Suaymyia*)
howardi Belkin & Hogue, 1959: *Deinocerites*, Pseudes Group
howardi Lane & Cerqueira, 1942: *Wyeomyia* (*Spilonympha*)
howardii Coquillett, 1901: *Psorophora sensu stricto*
huangae Meng, 1958: *Culex sensu stricto*, Pipiens Group (not placed in a Subgroup)
huangae Reinert, 1999b: *Aedes* (*Zavortinkius*), Longipalpis Group
huasteca Ortega, 2023 (in Ortega-Morales *et al*., 2023): *Shannoniana*
hughi Lambert & Coetzee, 1982: *Anopheles* (*Cellia*), Myzomyia Series, Marshallii Group, Marshallii Complex
hui Bohart, 1957: *Aedes* (*Finlaya*), Kochi Group
humboldti (Lane & Cerqueira, 1942): *Runchomyia* (*Runchomyia*)
humeralis Dyar & Knab, 1916: *Mansonia* (*Mansonia*)
humeralis Edwards, 1922d: *Aedes* (*Macleaya*), Chaetocruiomyia Group
hungaricus Mihályi, 1955: *Aedes* (*Ochlerotatus*), Communis Group (not placed in a Subgroup)
hunteri (Strickland, 1916): *Anopheles sensu stricto*, Laticorn Section, Myzorhynchus Series, Umbrosus Group (not placed in a Subgroup)
hurlbuti Belkin, 1962: *Culex* (*Lophoceraomyia*), Fraudatrix Group, Fraudatrix Subgroup, Hurlbuti Complex
hurlbuti Lien, 1967: *Aedes* (*Bruceharrisonius*)
husaini Qutubuddin, 1947: *Uranotaenia* (*Pseudoficalbia*), Recondita Series
hutchingsae Talaga, 2022 (in Talaga & Gendrin, 2022): *Culex* (*Melanoconion*), Melanoconion Section, Bastagarius Group, Bastagarius Subgroup
hutchinsoni Barraud, 1924e: *Culex sensu stricto*, Pipiens Group, Trifilatus Subgroup
hybrida (Leicester, 1908): *Mimomyia* (*Mimomyia*), Hybrida Group
hybridus Edwards, 1914c: *Armigeres sensu stricto*
hybridus (Leicester, 1908): *Tripteroides* (*Tripteroides*), Nitidoventer Group
hylaius Harbach, 2018b: *Aedes* (*Neomelaniconion*), Hylaius Group
hypoptes (Knab, 1907a): *Toxorhynchites* (*Lynchiella*)
hyrcanus (Pallas, 1771): *Anopheles sensu stricto*, Laticorn Section, Myzorhynchus Series, Hyrcanus Group (not placed in a Subgroup; *An. hyrcanus* comprises two sibling species, one informally denoted as *hyrcanus* $_{\mathrm{spIR}}$ by Djadid *et al*., 2009)
hystera Dyar & Knab, 1913: *Uranotaenia* (*Uranotaenia*), Leucoptera Series

Ii

ibiapabaensis (Sant'Ana & Sallum, 2023) (online 4 December 2023; see Sant'Ana & Sallum, 2024): *Anopheles* (*Nyssorhynchus*), Albimanus Section, Oswaldoi Series, Oswaldoi Group, Strodei Subgroup, Arthuri Complex
ibis Barraud, 1931b: *Aedes* (*Christophersiomyia*)
ichiromiyagii Toma & Higa, 2004: *Ficalbia*, Group A Oriental Region Group
idahoensis (Theobald, 1903a): *Aedes* (*Ochlerotatus*), Communis Group (not placed in a Subgroup)
idanus Schick, 1970a: *Aedes* (*Protomacleaya*), Terrens Group, Heteropus Subgroup
identicus Dyar & Knab, 1907c: *Sabethes* (*Peytonulus*)
idiogenes Harbach, 1994b: *Sabethes* (*Sabethinus*), Intermedius Group
idjenensis Brug, 1934: *Aedes* (*Downsiomyia*), Niveus Group
idottus Dyar, 1920b: *Culex* (*Melanoconion*), Melanoconion Section, Intrincatus Group, Idottus Subgroup
ignotus Harbach, 1995b: *Sabethes* (*Peytonulus*)
iguassuensis Barbosa, da Silva & Sallum, 2007: *Mansonia* (*Mansonia*)
ikelos Forattini & Sallum, 1995: *Culex* (*Melanoconion*), Spissipes Section, Taeniopus Group
imitata (Baisas, 1946c): *Kimia*
imitator (Leicester, 1908): *Aedes* (*Stegomyia*), W-albus Group, W-albus Subgroup
imitator Theobald, 1903a: *Culex* (*Microculex*), Imitator Series
imparis Peyton & Harbach, 2000 (in Harbach & Peyton, 2000): *Onirion*, Personatum Group
impatibilis (Walker, 1859): *Aedes* (*Scutomyia*), Hoogstraali Group
impatiens (Walker, 1848): *Culiseta sensu stricto*
imperfectus Dobrotworsky, 1962: *Aedes* (*Ochlerotatus*), Burpengaryensis Group
impiger (Walker, 1848): *Aedes* (*Ochlerotatus*), Impiger Group
implexus (Theobald, 1903b): *Anopheles* (Christya)
implicatus Vockeroth, 1954: *Aedes* (*Ochlerotatus*), Communis Group, Leucomelas Subgroup
impostor Schick, 1970a: *Aedes* (*Protomacleaya*), Terrens Group, Homoeopus Subgroup
impostor Sirivanakarn, 1977a: *Culex* (*Lophoceraomyia*), Mammilifer Group, Mammilifer Line

imprimens (Walker, 1860): *Aedes* (*Edwardsaedes*)
impudicus Ficalbi, 1890a: *Culex* (*Neoculex*), Territans Group
inadmirabilis Dyar, 1928: *Culex* (*Melanoconion*), Melanoconion Section, Educator Group
inaequalis (Grabham, 1907): *Aedes* (*Howardina*), Sexlineatus Section, Inaequalis Subassemblage
inatomii Kamimura & Wada, 1974: *Culex* (*Barraudius*)
incaudata Root, 1928 (in Dyar, 1928): *Wyeomyia* (*Phoniomyia*), Second Series
inchoatus Barraud, 1927b: *Armigeres* (*Leicesteria*)
incidens (Thomson, 1869): *Culiseta sensu stricto*
inclinata Thurman, 1959: *Topomyia* (*Topomyia*)
incognita Galindo, Blanton & Peyton, 1954: *Uranotaenia* (*Uranotaenia*), Pallidoventer Series
incognitus Baisas, 1938: *Culex sensu stricto*, Sitiens Group, Vishnui Subgroup
incognitus Brug, 1931b: *Anopheles* (*Cellia*), Neomyzomyia Series (not placed in a Group)
incomptas Ramalingam & Pillai, 1972: *Malaya*
incomptus Arnell, 1976: *Aedes* (*Ochlerotatus*), Scapularis Group, Incomptus Subgroup (the Subgroup is monobasic)
incomptus Bram & Rattanarithikul, 1967: *Culex* (*Lophoceraomyia*), Mammilifer Group, Minor Line, Minor Complex
inconspicua (Lee, 1937): *Culiseta* (*Culicella*)
inconspicuosus (Theobald, 1908a): *Culex* (*Eumelanomyia*), Mochthogenes Group, Inconspicuosus Subgroup
increpitus Dyar, 1916b: *Aedes* (*Ochlerotatus*), Stimulans Group, Increpitus Complex
inculus Colless, 1965: *Culex* (*Lophoceraomyia*), Fraudatrix Group, Fraudatrix Subgroup (Sirivanakarn, 1977a unnecessarily recognized a monobasic '*inculus* complex' for this species in the absence of evidence that it is more than one species)
indecorabilis Leicester, 1908: *Verrallina* (*Harbachius*)
indefinitus (Ludlow, 1904b): *Anopheles* (*Cellia*), Pyretophorus Series
indeterminatus Baisas & Ubaldo-Pagayon, 1953: *Tripteroides* (*Tripteroides*), Nitidoventer Group
indiana Edwards, 1930b: *Mansonia* (*Mansonioides*)
indica (Edwards, 1920a): *Culiseta sensu stricto*
indica (Theobald, 1905f): *Heizmannia sensu stricto*
indica (Theobald, 1907): *Verrallina* (*Neomacleaya*)
indicus (Barraud, 1929a): *Tripteroides* (*Tripteroides*), Nitidoventer Group
indicus Evenhuis & Steffan, 1986: *Toxorhynchites* (*Toxorhynchites*), Acaudatus Group, Acaudatus Subgroup
indonesiae Mattingly, 1958a: *Aedes* (*Cancraedes*)
indubitans Dyar & Shannon, 1925b: *Mansonia* (*Mansonia*), Indubitans Complex
inermis Colless, 1958: *Aedes* (*Downsiomyia*), Niveus Group
inexpectatus Bonne-Wepster, 1948: *Aedes* (*Ochlerotatus*), Taeniorhynchus Group, Vigilax Subgroup
inextricabilis Talaga, 2020 (in Talaga *et al.*, 2020b): *Culex* (*Melanoconion*), Melanoconion Section, Distinguendus Group, Putumayensis Subgroup
infantulus Edwards, 1922a: *Culex* (*Lophoceraomyia*), Fraudatrix Group, Fraudatrix Subgroup, Minutissimus Subgroup
infinis (Dyar & Knab, 1906d): *Psorophora* (*Grabhamia*), Cingulata Group
infirmatus Dyar & Knab, 1906d: *Aedes* (*Ochlerotatus*), Scapularis Group, Infirmatus Subgroup
inflictus Theobald, 1901c: *Culex sensu stricto*, Pipiens Group (not placed in a Subgroup)
infoliatus Bonne-Wepster & Bonne, 1920: *Culex* (*Carrollia*), Bihaicola Group
infula Theobald, 1901b: *Culex* (*Oculeomyia*), Bitaeniorhynchus Complex
ingrami Edwards, 1916: *Culex sensu stricto*, Guiarti Group
ingrami Edwards, 1930a: *Aedes* (*Hopkinsius*), Hopkinsius Group
inhibitator Dyar & Knab, 1906d: *Culex* (*Melanoconion*), Melanoconion Section, Inhibitator Group, Inhibitator Subgroup
inimitabilis Dyar & Knab, 1906d: *Culex* (*Microculex*), Inimitabilis Series
ininicola Fauran & Pajot, 1974: *Wyeomyia* (not placed in a Subgenus)
ininii Senevet & Abonnenc, 1938: *Anopheles* (*Nyssorhynchus*), Albimanus Section, Oswaldoi Series, Oswaldoi Group, Oswaldoi Subgroup
innovator Evans, 1924: *Culex* (*Melanoconion*), Melanoconion Section, Pilosus Group, Pilosus Subgroup
inornata (Williston, 1893): *Culiseta sensu stricto*
inornatus Newstead, 1907 (in Newstead *et al.*, 1907): *Eretmapodites*, Inornatus Group
inornatus (Theobald, 1905g): *Culex* (subgenus uncertain)
inornatus (Walker, 1865): *Toxorhynchites* (*Toxorhynchites*), Splendens Group
inquinatus Edwards, 1922a: *Aedes* (*Phagomyia*)
insequens Marks, 1989 (in Lee *et al.*, 1989b): *Culex* (*Lophoceraomyia*), Fraudatrix Group, Fraudatrix Subgroup, Fraudatrix Complex
insigniforceps Clastrier & Claustre, 1978a: *Culex* (*Carrollia*), Iridescens Group, Iridescens Subgroup
insignis (Carter, 1911): *Culex* (*Eumelanomyia*), Rubinotus-rima Group, Rima Subgroup
insolens Edwards, 1936: *Aedes* (*Catageiomyia*)

insolitus (Coquillett, 1906d): *Aedes* (*Protomacleaya*), Terrens Group, Insolitus Subgroup
insulaeflorum (Swellengrebel & Swellengrebel de Graaf, 1920): *Anopheles sensu stricto*, Angusticorn Section, Anopheles Series, Aitkenii Group
insularis Sirivanakarn, 1968: *Culex* (*Lophoceraomyia*), Fraudatrix Group, Fraudatrix Subgroup, Fraudatrix Complex
insularius (Dyar & Knab, 1906a): *Psorophora* (*Grabhamia*), Confinnis Group
interfor Dyar, 1928: *Culex sensu stricto*, Pipiens Group, Tarsalis Subgroup
intermedia Barraud, 1929a: *Mimomyia* (*Mimomyia*), Hybrida Group
intermediatus Baisas & Ubaldo-Pagayon, 1953: *Tripteroides* (*Tripteroides*), Nitidoventer Group
intermedius Danilov & Gornostaeva, 1987: *Aedes* (*Ochlerotatus*), Communis Group, Leucomelas Subgroup
intermedius Edwards, 1936: *Eretmapodites*, Chrysogaster Group
intermedius Lane & Whitman, 1951: *Culex* (*Microculex*), Pleuristriatus Series
intermedius (Lutz, 1904) (in Bourroul, 1904): *Sabethes* (*Sabethinus*), Intermedius Group
interrogator Dyar & Knab, 1906d: *Culex sensu stricto*, Pipiens Group, Tarsalis Subgroup
interruptus Puri, 1929a: *Anopheles sensu stricto*, Angusticorn Section, Lophoscelomyia Series, Asiaticus Group, Interruptus Subgroup (the Subgroup is monobasic)
interruptus Reinert, 1999b: *Aedes* (*Zavortinkius*), Brygooi Group
intonca Dyar & Knab, 1910a: *Wyeomyia* (*Hystatomyia*), Circumcincta Group
intonsus Galindo & Blanton, 1954: *Culex* (*Melanoconion*), Melanoconion Section, Bastagarius Group, Bastagarius Subgroup
intrincatus Brèthes, 1916: *Culex* (*Melanoconion*), Melanoconion Section, Intrincatus Group, Intrincatus Subgroup
introlatus Colless, 1957b: *Anopheles* (*Cellia*), Neomyzomyia Series, Leucosphyrus Group, Leucosphyrus Subgroup, Leucosphyrus Complex
intrudens Dyar, 1919a: *Aedes* (*Ochlerotatus*), Pullatus Group
inundatus Reinert, 1997 (in Reinert *et al.*, 1997): *Anopheles sensu stricto*, Angusticorn Section, Anopheles Series, Maculipennis Group, Quadrimaculatus Subgroup
invidiosus Theobald, 1901d: *Culex sensu stricto*, Pipiens Group, Decens Subgroup
invocator Pazos, 1908: *Culex* (*Melanoconion*), Melanoconion Section, Erraticus Group, Erraticus Subgroup
iolambdis Dyar, 1918b: *Culex* (*Melanoconion*), Melanoconion Section, Bastagarius Group, Iolambdis Subgroup
ioliota Dyar & Knab, 1913: *Aedes* (*Howardina*), Walkeri Section, Ioliota Group
iphis Barraud, 1924f: *Culex* (*Eumelanomyia*), Mochthogenes Group, Iphis Subgroup (the Subgroup is monobasic)
iracunda (Walker, 1848): *Coquillettidia* (*Coquillettidia*), Iracunda Group
irenicus Schmidt, 2003 (in Schmidt *et al.*, 2003): *Anopheles* (*Cellia*), Neomyzomyia Series, Punctulatus Group, Farauti Complex
irianensis Miyagi & Toma, 1997: *Topomyia* (*Topomyia*)
iriartei Cova García, Pulido F., Escalante de Ugueto & Mora R., 1987a: *Uranotaenia* (*Uranotaenia*) (not placed in a Section, Series, Group or Complex)
iridescens (Lutz, 1905): *Culex* (*Carrollia*), Iridescens Group, Iridescens Subgroup
iridicolor Dyar, 1921b: *Haemagogus sensu stricto*, Splendens Section
iriomotensis (Tanaka & Mizusawa, 1973): *Verrallina sensu stricto*, Butleri Series
irritans (Theobald, 1901d): *Aedes* (*Catageiomyia*)
isabelae Duret, 1968f: *Culex* (*Melanoconion*), Melanoconion Section, Inhibitator Group, Egcymon Subgroup
iwi Marks, 1955: *Aedes* (*Phagomyia*)
iyengari Edwards, 1923b: *Aedes* (*Petermattinglyius*), Petermattinglyius Group
iyengari Mattingly & Rageau, 1958: *Culex sensu stricto*, Pipiens Group, Trifilatus Subgroup

Jj

jacksoni Edwards, 1934 (in Barraud, 1934): *Culex sensu stricto*, Sitiens Group, Mimeticus Subgroup, Mimeticus Complex
jacksoni Edwards, 1935a: *Uranotaenia* (*Pseudoficalbia*), Recondita Series
jacksoni Mattingly, 1949: *Ficalbia*, Group A Oriental Region Group
jacobinae Serafim & Davis, 1933: *Aedes* (*Ochlerotatus*) (not placed in a Group)
jacobsoni (Edwards, 1930b): *Malaya*
jalisco Berlin, 1974: *Culex* (*Micraedes*)
jamaicensis (Theobald, 1901b): *Psorophora* (*Grabhamia*), Confinnis Group, Confinnis Complex
jamariensis (Sant'Ana & Sallum, 2022a): *Anopheles* (*Nyssorhynchus*), Albimanus Section, Oswaldoi Series, Oswaldoi Group, Oswaldoi Subgroup, Nuneztovari Complex
jambulingami (Natarajan, 2019): *Aedes* (*Paraedes*), Barraudi Assemblage
jamesi (Edwards, 1914b): *Aedes* (*Aedimorphus*), Alboscutellatus Assemblage, Alboscutellatus Group
jamesii Theobald, 1901b: *Anopheles* (*Cellia*), Neocellia Series, Jamesii Group
jamoti Hamon & Rickenbach, 1954: *Aedes* (*Neomelaniconion*) (not placed in a Group)
janconnae Wilkerson & Sallum, 2009 (in Motoki *et al.*, 2009): *Anopheles* (*Nyssorhynchus*), Argyritarsis Section, Albitarsis Series, Albitarsis Group, Albitarsis Complex (*An. janconnae* comprises two sibling species, one informally denoted as Lineage nr. *janconnae* by Gutiérrez *et al.*, 2010)

jani Rickenbach & Lombrici, 1975: *Eretmapodites*, Leucopous Group
janitor Theobald, 1903a: *Culex sensu stricto*, Pipiens Group, Tarsalis Subgroup
janthinomys Dyar, 1921b: *Haemagogus sensu stricto*, Albomaculatus Section
japonicus (Theobald, 1901b): *Aedes* (*Hulecoeteomyia*)
japonicus Yamada, 1918: *Anopheles sensu stricto*, Angusticorn Section, Anopheles Series, Lindesayi Group, Lindesayi Subgroup, Lindesayi Complex
javaensis Miyagi & Toma, 1995: *Topomyia* (*Topomyia*)
javanensis Bonne-Wepster, 1934: *Culex* (*Culiciomyia*), Shebbearei Group
jeansottei (Doucet, 1950): *Mimomyia* (*Ingramia*), Group B (Ravenalites)
jebudensis Froud, 1944: *Anopheles* (*Cellia*), Neomyzomyia Series, Smithii Group
jefferyi Sirivanakarn, 1977b: *Culex* (*Eumelanomyia*), Rubinotus-rima Group (not placed in a Subgroup)
jenningsi Dyar & Knab, 1907c: *Culex* (*Microculex*) (not placed in a Series)
jenseni (de Meijere, 1910): *Culex* (*Lophoceraomyia*), Mammilifer Group, Brevipalpus Subgroup
jeyporiensis James, 1902: *Anopheles* (*Cellia*), Myzomyia Series, Funestus Group (not placed in a Subgroup)
jinhongensis Dong, Dong & Zhou, 2003a: *Uranotaenia* (*Pseudoficalbia*), Recondita Series
jinjaensis Edwards, 1941: *Culex* (*Oculeomyia*) (not placed in a Complex)
jocasta Komp & Rozeboom, 1951: *Culex* (*Melanoconion*), Melanoconion Section, Conspirator Group
jocosa (Dyar & Knab, 1908): *Wyeomyia* (*Dendromyia*)
johni Cova Garcia, Pulido F. & Escalante de Ugueto, 1979: *Culex* (*Neoculex*) (not placed in a Group)
johnnyi Duret, 1968d: *Culex* (*Melanoconion*), Melanoconion Section, Evansae Group
johnsoni Galindo & Mendez, 1961: *Culex* (*Melanoconion*), Melanoconion Section, Intrincatus Group, Intrincatus Subgroup
johnsoni (Laffoon, 1946): *Verrallina* (*Neomacleaya*)
johnstonii (Grabham, 1905): *Psorophora* (*Janthinosoma*)
johorensis (Reinert, 1974): *Verrallina* (*Neomacleaya*)
joloensis (Ludlow, 1904a): *Armigeres sensu stricto*
jorgi Carpintero & Leguizamón, 2000: *Aedes* (*Ochlerotatus*), Fulvus Group
josephinae Marks, 1958: *Aedes* (*Finlaya*), Kochi Group
josephineae Baisas, 1935d: *Culex* (*Lophoceraomyia*), Fraudatrix Group, Fraudatrix Subgroup, Variatus Complex
josiahae Huang, 1988a: *Aedes* (*Stegomyia*), Simpsoni Group
joucouri da Cunha Ramos & Brunhes, 2004: *Uranotaenia* (*Uranotaenia*), Alba Section, Alba Group
joyoni Brunhes, 1977a: *Orthopodomyia*, Orthopodomyia Section, Vernoni Group
jubifer Komp & Brown, 1935: *Culex* (*Melanoconion*), Spissipes Section, Jubifer Group
judaicus Edwards, 1926c: *Culex* (*Neoculex*), Territans Group
judithae Zavortink, 1969: *Anopheles sensu stricto*, Angusticorn Section, Anopheles Series, Plumbeus Group
jugraensis (Leicester, 1908): *Aedes* (*Hulecoeteomyia*)
jugraensis (Leicester, 1908): *Armigeres sensu stricto*
juppi McIntosh, 1973: *Aedes* (*Ochlerotatus*), Caballus Group
juxtahirsuta Belkin, Heinemann & Page, 1970: *Wyeomyia* (*Wyeomyia*), Hirsuta Group, Hirsuta Subgroup
juxtamansonia (Chagas, 1907): *Coquillettidia* (*Pseudotaeniorhynchus*)

Kk

kabaenensis Brug, 1939: *Aedes* (*Geoskusea*) (not placed in a Group)
kabwachensis Edwards, 1941: *Aedes* (*Aedimorphus*), Alboscutellatus Assemblage, Abnormalis Group
kaimosi (van Someren, 1946b): *Toxorhynchites* (*Afrorhynchus*) (not placed in a Group)
kana Tanaka, Mizusawa & Saugstad, 1979: *Heizmannia sensu stricto*
kanarensis Edwards, 1934 (in Barraud, 1934): *Aedes* (*Dendroskusea*)
kanhsienensis Tung, 1955: *Heizmannia sensu stricto*
kanyamwerima van Someren, 1951: *Culex* (*Eumelanomyia*), Eumelanomyia Group
kapretwae Edwards, 1941: *Aedes* (*Albuginosus*)
karandalaensis (Wolfs, 1951): *Coquillettidia* (*Coquillettidia*), Crassipes Group
karooensis Muspratt, 1961: *Aedes* (*Aedimorphus*), Alboscutellatus Assemblage, Abnormalis Group
kartalae Brunhes, 1977a: *Culex sensu stricto*, Pipiens Group (not placed in a Subgroup)
karwari (Barraud, 1924a): *Aedes* (*Kenknightia*), Dissimilis Group
karwari (James, 1903) (in Theobald, 1903a): *Anopheles* (*Cellia*), Neocellia Series (not placed in a Group)
kasachstanicus Gutsevich, 1962: *Aedes* (*Ochlerotatus*), Communis Group (not placed in a Subgroup)
katherinensis Woodhill, 1949: *Aedes* (*Stegomyia*), Scutellaris Group, Scutellaris Subgroup
katibasensis Miyagi & Toma, 2012 (in Miyagi *et al.*, 2012): *Topomyia* (*Topomyia*)
kaviengensis Sirivanakarn, 1968: *Culex* (*Lophoceraomyia*), Fraudatrix Group, Fraudatrix Subgroup, Fraudatrix Complex
kazanjii Talaga, 2020 (in Talaga *et al.*, 2020b): *Culex* (*Melanoconion*), Melanoconion Section, Distinguendus Group (not placed in a Subgroup)

keefei King & Hoogstraal, 1946f: *Aedes* (subgenus uncertain)
kelabitensis Miyagi & Toma, 2010b: *Topomyia* (*Suaymyia*)
kempi (Edwards, 1921a): *Toxorhynchites* (*Toxorhynchites*), Leicesteri Group, Leicesteri Subgroup
keniensis Evans, 1931a: *Anopheles* (*Cellia*), Myzomyia Series, Demeilloni Group
keniensis van Someren, 1946a: *Aedes* (*Stegomyia*), Dendrophilus Group
kennethi Muspratt, 1956b: *Aedes* (*Albuginosus*)
kenyae van Someren, 1946c: *Aedes* (*Stegomyia*), Dendrophilus Group
kesseli Belkin, 1962: *Culex sensu stricto*, Atriceps Group
kesseli Huang & Hitchcock, 1980: *Aedes* (*Stegomyia*), Scutellaris Group, Scutellaris Subgroup
kesseli Ramalingam, 1987: *Armigeres sensu stricto*
khazani Edwards, 1922a: *Aedes* (*Phagomyia*)
khazani Edwards, 1922a: *Culex* (*Eumelanomyia*), Mochthogenes Group, Khazani Subgroup (the Subgroup is monobasic)
kiangsiensis Tung, 1955: *Aedes* (*Phagomyia*)
kilara van Someren, 1951: *Culex* (*Eumelanomyia*), Eumelanomyia Group
killertonis (Huang, 1968a): *Verrallina sensu stricto*, Carmenti Series
kinabaluensis Ramalingam, 1972: *Armigeres sensu stricto*
kinabaluensis Sirivanakarn, 1976: *Culex* (*Oculeomyia*), Geminus Complex
kingi Christophers, 1923: *Anopheles* (*Cellia*), Neomyzomyia Series, Ardensis Group
kingi Lee, 1946: *Tripteroides* (*Rachisoura*), Vanleeuweni Group
kingianus Edwards, 1922d: *Culex* (*Eumelanomyia*), Rubinotus-rima Group, Rubinotus Subgroup
kingii (Theobald, 1913): *Culex sensu stricto*, Pipiens Group (not placed in a Subgroup)
kiriensis Klein & Sirivanakarn, 1970: *Culex* (*Eumelanomyia*), Mochthogenes Group, Tenuipalpis Subgroup
kiriromi (Klein, 1969): *Mimomyia* (*Ingramia*), Group C (*Ingramia s. str.*, Fusca Group)
kishenehn Harbach & Greenwalt, 2012: *Culiseta* (fossil species)
kivuensis Edwards, 1941: *Aedes* (*Stegomyia*), Simpsoni Group
kleini Reinert, 1973f: *Aedes* (*Bothaella*)
kleini Rueda, 2005: *Anopheles sensu stricto*, Laticorn Section, Myzorhynchus Series, Hyrcanus Group (not placed in a Subgroup)
klossi (Edwards, 1921a): *Toxorhynchites* (*Toxorhynchites*), Leicesteri Group, Leicesteri Subgroup
knabi (Coquillett, 1906a): *Aedes* (*Protomacleaya*), Knabi Group (the Group is monobasic)
knabi Lane & Cerqueira, 1942: *Wyeomyia* (not placed in a Subgenus)
knighti Baisas & Ubaldo-Pagayon, 1953: *Tripteroides* (*Tripteroides*), Nitidoventer Group
knighti Stone & Bohart, 1944: *Aedes* (*Finlaya*), Kochi Group, Knighti Complex
kobayashii Nakata, 1956: *Aedes* (*Bruceharrisonius*)
kochi (Dönitz, 1901): *Aedes* (*Finlaya*), Kochi Group
kochi Dönitz, 1901: *Anopheles* (*Cellia*), Neomyzomyia Series, Kochi Group (the Group is monobasic)
kohkutensis Mattingly, 1958a: *Aedes* (*Cancraedes*)
kokhani Vythilingam, Jeffery & Harbach, 2007 (in Vythilingam *et al.*, 2007): *Anopheles* (*Cellia*), Neomyzomyia Series (not placed in a Group)
kolambuganensis Baisas, 1932: *Anopheles* (*Cellia*), Neomyzomyia Series (not placed in a Group)
koli Peyton & Klein, 1970: *Uranotaenia* (*Pseudoficalbia*), Recondita Series
koliensis Owen, 1945: *Anopheles* (*Cellia*), Neomyzomyia Series, Punctulatus Group
kompi Edwards, 1930a: *Anopheles* (*Stethomyia*)
kompi Valencia, 1973: *Culex* (*Carrollia*), Iridescens Group, Iridescens Subgroup
kompi Vargas & Downs, 1950: *Aedes* (*Protomacleaya*), Kompi Group
komponga (Klein, 1973): *Verrallina* (*Neomacleaya*)
konderi Galvão & Damasceno, 1942a: *Anopheles* (*Nyssorhynchus*), Albimanus Section, Oswaldoi Series, Oswaldoi Group, Oswaldoi Subgroup, Konderi Complex
koreicoides Sasa, Kano & Hayashi, 1950: *Aedes* (subgenus uncertain)
koreicus (Edwards, 1917): *Aedes* (*Hulecoeteomyia*)
koreicus Yamada & Watanabe, 1918: *Anopheles sensu stricto*, Laticorn Section, Myzorhynchus Series, Barbirostris Group (not placed in a Subgroup)
kosiensis Coetzee, Segerman & Hunt, 1987: *Anopheles* (*Cellia*), Myzomyia Series, Marshallii Group, Marshallii Complex
kowiroensis Sirivanakarn, 1968: *Culex* (*Lophoceraomyia*), Fraudatrix Group, Fraudatrix Subgroup, Marksae Complex
kraussi Grjebine, 1953a: *Uranotaenia* (*Pseudoficalbia*), Nigripes Section, Nigripes Group
krombeini Huang, 1975: *Aedes* (*Stegomyia*), Scutellaris Group, Scutellaris Subgroup
krymmontanus Alekseev, 1989: *Aedes* (*Rusticoidus*)
kuchingensis Edwards, 1915b: *Armigeres sensu stricto*
kuhnsi King & Hoogstraal, 1955: *Culex* (*Lophoceraomyia*), Mammilifer Group, Minor Line, Minor Complex
kukenan Anduze, 1942: *Culex* (*Microculex*) (not placed in a Series)

kummi Edwards, 1930a: *Aedes* (*Pseudarmigeres*)
kummi Edwards, 1939b: *Orthopodomyia*, Bancroftia Section, Signifera Group, Kummi Subgroup
kummi Komp & Rozeboom, 1951: *Culex* (*Melanoconion*), Melanoconion Section, Inhibitator Group, Inhibitator Subgroup
kummi Lane & Cerqueira, 1942: *Wyeomyia* (*Cruzmyia*)
kusaiensis Bohart, 1957: *Culex* (*Lophoceraomyia*), Fraudatrix Group (not placed in a Subgroup)
kweiyangensis Yao & Wu, 1944: *Anopheles sensu stricto*, Laticorn Section, Myzorhynchus Series, Hyrcanus Group (not placed in a Subgroup)
kyondawensis Abraham, 1947: *Anopheles* (*Baimaia*) (the Subgenus is monobasic)
kyotoensis Yamaguti & LaCasse, 1952: *Culex* (*Culiciomyia*), Shebbearei Group

Ll

labesba Howard, Dyar & Knab, 1913: *Wyeomyia* (*Wyeomyia*) (not placed in a Group)
labranchiae Falleroni, 1926: *Anopheles sensu stricto*, Angusticorn Section, Anopheles Series, Maculipennis Group, Maculipennis Subgroup
lacani Grjebine, 1953a: *Anopheles* (*Cellia*), Neomyzomyia Series, Ranci Group, Roubaudi Subgroup
lacani Rickenbach & Eouzan, 1970: *Eretmapodites*, Plioleucus Group
lacertosus Komp & Rozeboom, 1951: *Culex* (*Melanoconion*), Melanoconion Section, Pilosus Group, Caudelli Subgroup
lactator Dyar & Knab, 1906d: *Culex* (*Phenacomyia*)
lacteus Knight, 1946: *Aedes* (*Downsiomyia*), Niveus Group
lacustris (Edwards, 1935a): *Mimomyia* (*Mimomyia*), Ludlowia Group
lacuum Edwards, 1922d: *Armigeres sensu stricto*
laffooni Baisas & Ubaldo-Pagayon, 1953: *Tripteroides* (*Tripteroides*), Nitidoventer Group
laffooni Belkin, 1962: *Culex* (*Lophoceraomyia*), Fraudatrix Group, Fraudatrix Subgroup, Bergi Complex
laffooni Knight & Rozeboom, 1946: *Aedes* (*Scutomyia*), Laffooni Group
laffosseae da Cunha Ramos & Brunhes, 2004: *Uranotaenia* (*Pseudoficalbia*), Nigripes Section, Nigripes Group
laguna Arnell & Nielsen, 1972: *Aedes* (*Jarnellius*)
lahillei Bachmann & Casal, 1962: *Culex sensu stricto*, Pipiens Group (not placed in a Subgroup)
lairdi Belkin, 1962: *Culex* (*Lophoceraomyia*), Fraudatrix Group, Fraudatrix Subgroup, Buxtoni Complex
lakei Sirivanakarn, 1968: *Culex* (*Lophoceraomyia*), Fraudatrix Group, Fraudatrix Subgroup, Hilli Complex
lamborni Edwards, 1923a: *Aedes* (*Bifidistylus*)
lambrechti van Someren, 1971: *Aedes* (*Skusea*)
lamellata (Bonne-Wepster & Bonne, 1920): *Wyeomyia* (*Hystatomyia*), Autocratica Group
lamelliferus Bohart & Ingram, 1946b: *Aedes* (*Lorrainea*)
lampangensis Sirivanakarn, 1973b: *Culex* (*Culiciomyia*), Fragilis Group
lampangensis Thurman, 1959: *Hodgesia*
lampropus (Howard, Dyar & Knab, 1913): *Trichoprosopon*, Trichoprosopon Group, Lampropus Complex
laneanus Corrêa & Cerqueira, 1944: *Anopheles* (*Kerteszia*), Laneanus Complex (*An. laneanus* consists of two sibling species denoted as Species 1 and 2 (*laneanus sensu stricto*)) (Bourke *et al.*, 2023)
lanei (Antunes, 1937): *Trichoprosopon* (not placed in a Group or Complex)
lanei Cerqueira, 1961: *Sabethes* (*Sabethes*), Group B
lanei Galvão & Franco do Amaral, 1938: *Anopheles* (*Nyssorhynchus*), Argyritarsis Section, Argyritarsis Series, Lanei Group (the Group is monobasic)
lanei Martinez & Prosen, 1953a: *Uranotaenia* (*Uranotaenia*) (not placed in a Section, Series, Group or Complex)
lanei Oliveira Coutinho & Forattini, 1962: *Culex* (*Microculex*), Pleuristriatus Series
lanei Shannon & Cerqueira, 1943: *Psorophora* (*Janthinosoma*)
langata van Someren, 1946c: *Aedes* (*Stegomyia*), Poweri Group
laniger (Wiedemann, 1820): *Aedes* (*Mucidus*), Mucidus Group
lanyuensis Lien, 1968a: *Orthopodomyia*, Orthopodomyia Section, Albipes Group (not placed in a Subgroup)
lanzaroi Cornel & Mayi, 2020 (in Cornel *et al.*, 2020): *Culex* (*Culiciomyia*), Nebulosus Group
laoagensis Knight, 1946: *Aedes* (*Downsiomyia*), Niveus Group
laoensis Toma & Miyagi, 2003: *Armigeres sensu stricto*
laplantei (Hamon, Adam & Mouchet, 1955): *Culex* (*Eumelanomyia*), Rubinotus-rima Group, Rima Subgroup
lasaensis Meng, 1962: *Aedes* (*Ochlerotatus*), Communis Group, Punctor Subgroup
lasiopalpis Sirivanakarn, 1977a: *Culex* (*Lophoceraomyia*), Mammilifer Group, Mammilifer Subgroup, Flavicornis Complex
lassalli (Bonne-Wepster & Bonne, 1921): *Wyeomyia* (*Phoniomyia*), Third Series
latens Sallum & Peyton, 2005 (in Sallum *et al.*, 2005b): *Anopheles* (*Cellia*), Neomyzomyia Series, Leucosphyrus Group, Leucosphyrus Subgroup, Leucosphyrus Complex
lateralis Ludlow, 1905c: *Uranotaenia* (*Uranotaenia*), Anisocheleomyia Section, Ludlowae Series, Lateralis Complex

lateralis Petrocchi, 1927 (in Shannon & del Ponte, 1927): *Wyeomyia* (*Nunezia*)
laticinctus Edwards, 1913a: *Culex sensu stricto*, Pipiens Group, Theileri Subgroup
laticlasper Galindo & Blanton, 1954: *Culex sensu stricto*, Pipiens Group, Restuans Complex (not placed in a Subgroup)
latifoliatus Delfinado, 1966a: *Culex* (*Eumelanomyia*), Mochthogenes Group, Foliatus Subgroup
latipennis (Delfinado, 1967): *Verrallina* (*Neomacleaya*)
latispinus Gong & Ji, 1989: *Tripteroides* (*Tripteroides*), Nitidoventer Group
latisquama (Coquillett, 1906a): *Culex* (*Tinolestes*)
latisquama (Edwards, 1927b): *Tripteroides* (*Rachisoura*), Filipes Group
latus Dobrotworsky, 1956: *Culex* (*Neoculex*), Pseudomelanoconia Group
laureli Baisas, 1935d: *Culex* (*Eumelanomyia*), Mochthogenes Group, Malayi Subgroup
lauriei (Carter, 1920): *Aedes* (subgenus uncertain)
lavatae Stone & Bohart, 1944: *Culex* (*Lophoceraomyia*), Mammilifer Group, Mammilifer Line, Traubi Complex
lavieri Doucet, 1950: *Uranotaenia* (*Pseudoficalbia*), Annulata Section, Annulata Series, Lavieri Group
lawtoni Baisas, 1946b: *Zeugnomyia*
leberi Boreham, 1970: *Mansonia* (*Mansonia*)
lebiedi da Cunha Ramos & Brunhes, 2004: *Uranotaenia* (*Uranotaenia*), Dumonti Section, Dumonti Group
ledgeri Huang, 1981: *Aedes* (*Stegomyia*), Poweri Group
leei Berlin, 1969a: *Aedes* (*Howardina*), Sexlineatus Section, Sexlineatus Subassemblage
leei King & Hoogstraal, 1955: *Culex* (*Lophoceraomyia*), Fraudatrix Group, Fraudatrix Subgroup, Marksae Complex
leei Peters, 1959: *Tripteroides* (*Rachisoura*), Filipes Group
leei Stone & Barreto, 1969: *Galindomyia* (the Genus is monobasic)
leei (Wharton, 1947): *Malaya*
leesoni Edwards, 1932b: *Aedes* (*Aedimorphus*), Alboscutellatus Assemblage, Abnormalis Group
leesoni Evans, 1931b: *Anopheles* (*Cellia*), Myzomyia Series, Funestus Group, Minimus Subgroup
legoffi da Cunha Ramos & Brunhes, 2004: *Uranotaenia* (*Pseudoficalbia*), Spinosa Section, Spinosa Group
lehcharlesi Miyagi & Toma, 2008: *Topomyia* (*Suaymyia*)
leiboensis Chu, 1981: *Uranotaenia* (*Pseudoficalbia*), Bimaculata Series
leicesteri (Edwards, 1917): *Verrallina* (*Neomacleaya*)
leicesteri Theobald, 1904c: *Toxorhynchites* (*Toxorhynchites*), Leicesteri Group, Leicesteri Subgroup
leilae (King & Hoogstraal, 1947b): *Verrallina sensu stricto*, Carmenti Series
lemniscata Harbach & Greenwalt, 2012: *Culiseta* (fossil species)
lemuriae Ribeiro, 2004: *Toxorhynchites* (*Afrorhynchus*), Pauliani Group
leonardi Belkin, 1962: *Culex* (*Neoculex*), Crassistylus Group
leonis Colless, 1958: *Aedes* (*Downsiomyia*), Niveus Group
lepchanus (Barraud, 1923b): *Aedes* (*Danielsia*)
lepidocoxitus Dong, Zhou & Dong, 1995b: *Armigeres* (*Leicesteria*)
lepidonotus (Edwards, 1920a): *Aedes* (*Rusticoidus*)
lepidotus Zavortink, 1973: *Anopheles* (*Kerteszia*) (not placed in a Complex)
lepidus Cerqueira & Paraense, 1945: *Aedes* (*Ochlerotatus*) (not placed in a Group)
leptolabis Edwards, 1936: *Aedes* (*Aedimorphus*), Domesticus Assemblage, Domesticus Subassemblage, Domesticus Group
lerozeboomi Reinert, 1990b: *Aedes* (*Kenknightia*), Luzonensis Group
lesteri Baisas & Hu, 1936: *Anopheles sensu stricto*, Laticorn Section, Myzorhynchus Series, Hyrcanus Group, Lesteri Subgroup
letabensis Lambert & Coetzee, 1982: *Anopheles* (*Cellia*), Myzomyia Series, Marshallii Group, Marshallii Complex
letifer Sandosham, 1944: *Anopheles sensu stricto*, Laticorn Section, Myzorhynchus Series, Umbrosus Group, Letifer Subgroup
leucarthrius (Speiser, 1909): *Aedes* (*Aedimorphus*), Domesticus Assemblage, Vexans Subassemblage, Quasiunivittatus Group
leucocelaenus (Dyar & Shannon, 1924b): *Haemagogus* (*Conopostegus*)
leucocnemis Martini, 1931: *Psorophora* (*Grabhamia*) (not placed in a Group)
leucomelas (Meigen, 1804): *Aedes* (*Ochlerotatus*), Communis Group, Leucomelas Subgroup
leucomeres (Giles, 1904): *Aedes* (*Kenknightia*), Luzonensis Group
leucophoebus (Galindo, Carpenter & Trapido, 1953): *Haemagogus* (*Conopostegus*)
leucopous Graham, 1909: *Eretmapodites*, Leucopous Group
leucoptera (Theobald, 1907): *Uranotaenia* (*Uranotaenia*), Leucoptera Series
leucopus (Dyar & Knab, 1906a): *Johnbelkinia*
leucosphyrus Dönitz, 1901: *Anopheles* (*Cellia*), Neomyzomyia Series, Leucosphyrus Group, Leucosphyrus Subgroup, Leucosphyrus Complex
leucostigma Lutz, 1904 (in Bourroul, 1904): *Wyeomyia* (*Menolepis*) (the Subgenus is monobasic)
leucotaeniatus (Komp, 1938): *Haemagogus* (*Conopostegus*)
leucotarsis Lane, 1936: *Wyeomyia* (*Wyeomyia*), Hirsuta Group, Hirsuta Subgroup
leucotarsis Thurman, 1959: *Topomyia* (*Suaymyia*)

levicastilloi Grjebine, 1986: *Mimomyia* (*Ingramia*), Group A (*Ravenalites s. str.*)
levicastilloi Lane, 1945b: *Culex sensu stricto*, Pipiens Group (not placed in a Subgroup)
levicastilloi Levi Castillo, 1944: *Anopheles sensu stricto*, Angusticorn Section, Anopheles Series, Pseudopunctipennis Group
lewelleni Starkey & Webb, 1946: *Aedes* (*Finlaya*), Kochi Group
lewisi Ludlow, 1920a: *Anopheles sensu stricto*, Angusticorn Section, Anopheles Series, Maculipennis Group (not placed in a Subgroup)
lewisi Ribeiro, 1991: *Toxorhynchites* (*Toxorhynchites*), Brevipalpis Group, Phytophagus Subgroup
lewnielseni Ortega & Zavortink, 2019 (in Ortega-Morales *et al.*, 2019): *Aedes* (*Protomacleaya*), Kompi Group
liangshanensis Kang, Tan, Cao, Cheng, Yang & Huang, 1984: *Anopheles sensu stricto*, Laticorn Section, Myzorhynchus Series, Hyrcanus Group (not placed in a Subgroup)
liberiensis Peters, 1955a: *Culex* (*Culiciomyia*), Nebulosus Group
lii Wu, 1936: *Heizmannia sensu stricto*
lilii (Theobald, 1910c): *Aedes* (*Stegomyia*), Simpsoni Group, Simpsoni Complex
limacifer Komp, 1936d: *Culex* (*Melanoconion*), Melanoconion Section, Bastagarius Group, Iolambdis Subgroup
limai Lane & Cerqueira, 1942: *Wyeomyia* (*Miamyia*)
limosus King, 1932: *Anopheles* (*Cellia*), Pyretophorus Series
lindesayi Giles, 1900a: *Anopheles sensu stricto*, Angusticorn Section, Anopheles Series, Lindesayi Group, Lindesayi Subgroup, Lindesayi Complex (the Complex also includes informally designated Species B, see Somboon *et al.*, 2022b)
lindsayi Thurman, 1959: *Topomyia* (*Topomyia*)
linealis (Skuse, 1889): *Coquillettidia* (*Coquillettidia*), Iracunda Group
lineata (Taylor, 1914d): *Verrallina sensu stricto*, Carmenti Series
lineata (von Humboldt, 1819): *Psorophora sensu stricto*
lineatopennis (Ludlow, 1905a): *Aedes* (*Neomelaniconion*), Lineatopennis Group
lineatus (Theobald, 1912a): *Culex* (*Afroculex*) (the Subgenus is monobasic)
linesi Marks, 1964a: *Aedes* (*Ochlerotatus*) (not placed in a Group)
lipovskyi Belkin, 1950a: *Tripteroides* (*Tripteroides*), Bimaculipes Group
listeri de Meillon, 1931b: *Anopheles* (*Cellia*), Paramyzomyia Series, Listeri Group
litoralis Bohart, 1946: *Culex sensu stricto*, Sitiens Group, Sitiens Subgroup
litoralis King, 1932: *Anopheles* (*Cellia*), Pyretophorus Series
litorea (Shute, 1928): *Culiseta* (*Culicella*)
litoreus Colless, 1958: *Aedes* (*Downsiomyia*), Niveus Group
littlechildi (Edwards, 1930b): *Tripteroides* (*Tripteroides*), Littlechildi Group
littlechildi Taylor, 1933: *Aedes* (*Macleaya*), Macleaya Group
littleri (Taylor, 1914c): *Culiseta* (*Austrotheobaldia*) (the Subgenus is monobasic)
litwakae Harbach, 1985: *Culex sensu stricto*, Pipiens Group, Decens Subgroup
litwakae Reinert, 1990b: *Aedes* (*Kenknightia*), Luzonensis Group
lloreti Gil Collado, 1936: *Anopheles* (*Cellia*), Myzomyia Series, Demeilloni Group
lokojensis Service, 1959b: *Aedes* (*Catageiomyia*)
longiareolata (Macquart, 1838a): *Culiseta* (*Allotheobaldia*) (the Subgenus is monobasic)
longicornis Grjebine, 1986: *Mimomyia* (*Ingramia*), Group A (*Ravenalites s. str.*), Spinosa Complex
longicornis Sirivanakarn, 1976: *Culex* (*Oculeomyia*), Bitaeniorhynchus Complex
longifilamentus Su & Zhang, 1988: *Aedes* (*Ochlerotatus*), Caspius Group
longiforceps Edwards, 1929 (in Paine & Edwards, 1929): *Aedes* (*Geoskusea*), Longiforceps Group (the Group is monobasic)
longipalpatus Lee, 1946: *Tripteroides* (*Rachisoura*), Filipes Group
longipalpis Dong, Zhou & Dong, 1997b: *Tripteroides* (*Rachionotomyia*), Edwardsi Group
longipalpis (Grünberg, 1905): *Aedes* (*Zavortinkius*), Longipalpis Group
longipalpis (Leicester, 1904) (in Theobald, 1904c): *Armigeres* (*Leicesteria*)
longipalpis (Theobald, 1903a): *Anopheles* (*Cellia*), Myzomyia Series, Funestus Group, Funestus Subgroup (Type C of Koekemoer *et al.*, 2009; their Type A is placed in the Minimus Subgroup)
longipalpus Randolph & O'Neill, 1944 (in Division of Medical Entomology, Bureau of Laboratories, 1944): *Psorophora* (*Janthinosoma*)
longipes (Fabricius, 1805): *Johnbelkinia*
longirostris Brug, 1928a: *Anopheles* (*Cellia*), Neomyzomyia Series (not placed in a Group)
longirostris (Leicester, 1908): *Aedes* (*Rhinoskusea*)
longirostris Leicester, 1908: *Uranotaenia* (*Uranotaenia*), Micans Series
longirostris Theobald, 1901c: *Wyeomyia* (*Phoniomyia*), First Series
longiseta Edwards, 1936: *Aedes* (*Aedimorphus*), Domesticus Assemblage, Domesticus Subassemblage, Domesticus Group
longisetosa Gong, 1994: *Topomyia* (*Topomyia*)
longisetosus Sá & Sallum, 2020 (in Sá *et al.*, 2020): *Culex* (*Melanoconion*), Melanoconion Section, Atratus Group

longisiphonus Dong, Zhou & Dong, 2001: *Tripteroides* (*Tripteroides*), Nitidoventer Group
longistriatus Sá & Hutchings, 2022 (in Sá *et al.*, 2022): *Culex* (*Melanoconion*), Melanoconion Section, Educator Group
longistylus Sá & Sallum, 2020 (in Sá *et al.*, 2020): *Culex* (*Melanoconion*), Melanoconion Section, Atratus Group
longitubus da Cunha Ramos & Brunhes, 2004: *Uranotaenia* (*Pseudoficalbia*), Annulata Section, Annulata Series, Lavieri Group
longitubus Somboon, Namgay & Harbach, 2021 (in Somboon *et al.*, 2021b): *Culex sensu stricto*, Sitiens Group, Mimeticus Subgroup, Complex uncertain
lopesi (Correa & Ramalho, 1956): *Wyeomyia* (*Phoniomyia*), First Series
lopesi Sirivanakarn & Jakob, 1979: *Culex* (*Melanoconion*), Spissipes Section, Lopesi Group (the Group is monobasic)
lopezii Cova Garcia, Sutil Oramas & Pulido F., 1974a: *Wyeomyia* (*Hystatomyia*), Autocratica Group
lophoventralis (Theobald, 1910a): *Aedes* (*Phagomyia*)
lorengaui Peters, 1963b: *Tripteroides* (*Tripteroides*), Bimaculipes Group
lorraineae Berlin, 1969a: *Aedes* (*Howardina*), Sexlineatus Section (not placed in a Group/Subassemblage)
lottei Hamon & Brengues, 1965: *Aedes* (*Catageiomyia*)
loturus Dyar, 1925e: *Culex* (*Melanoconion*), Melanoconion Section, Atratus Group
lounibosi Gillies & Coetzee, 1987: *Anopheles* (*Cellia*), Neomyzomyia Series, Rhodesiensis Group
lousthei Boussès & Brunhes, 2013: *Uranotaenia* (*Pseudoficalbia*), Spinosa Section, Spinosa Group
lovettae Evans, 1934: *Anopheles* (*Cellia*), Neomyzomyia Series, Smithii Group
lowii Theobald, 1901c: *Uranotaenia* (*Uranotaenia*), Lowii Series
lowisii (Theobald, 1910c): *Aedes* (*Aedimorphus*), Alboscutellatus Assemblage, Alboscutellatus Group
lucackermanni Talaga, 2022 (in Talaga & Gendrin, 2022): *Culex* (*Melanoconion*), Spissipes Section, Vomerifer Group
lucaris Colless, 1965: *Culex* (*Lophoceraomyia*), Mammilifer Group, Brevipalpus Subgroup, Brevipalpus Complex
lucaris Macdonald & Mattingly, 1960: *Udaya*
lucianus Muspratt, 1959: *Aedes* (*Mucidus*), Mucidus Group
lucifer (Howard, Dyar & Knab, 1913): *Haemagogus sensu stricto*, Splendens Section
lucifugus Komp, 1936d: *Culex* (*Melanoconion*), Melanoconion Section, Conspirator Group
lucyae van Someren, 1954a: *Uranotaenia* (*Pseudoficalbia*), Annulata Section, Annulata Series, Annulata Group, Subgroup A
ludlowae (Blanchard, 1905): *Aedes* (*Ochlerotatus*), Taeniorhynchus Group, Vigilax Subgroup
ludlowae Brunetti, 1920: *Ficalbia*, Group A Oriental Region Group
ludlowae Dyar & Shannon, 1925a: *Uranotaenia* (*Uranotaenia*), Anisocheleomyia Section, Ludlowae Series
ludlowae (Theobald, 1903a): *Anopheles* (*Cellia*), Pyretophorus Series
lugubris (Barraud, 1928): *Verrallina sensu stricto*, Butleri Series
lui Gong & Li, 1999: *Heizmannia* (*Mattinglyia*)
lui Lien, 1968a: *Uranotaenia* (*Pseudoficalbia*), Bimaculata Series
luna Belkin, Heinemann & Page, 1970: *Wyeomyia* (*Wyeomyia*), Hirsuta Group, Luna Subgroup (the Subgroup is monobasic)
lunata (Theobald, 1901c): *Isostomyia*
lunda da Cunha Ramos, 1993: *Uranotaenia* (*Pseudoficalbia*), Nigripes Section, Ornata Group
lungae Belkin & Schlosser, 1944: *Anopheles* (*Cellia*), Neomyzomyia Series, Lungae Complex (not placed in a Group)
lunulatus King & Hoogstraal, 1946c: *Aedes* (*Geoskusea*) (not placed in a Group)
luridus McIntosh, 1971: *Aedes* (*Neomelaniconion*), Lineatopennis Group
lutea (Belkin, 1962): *Coquillettidia* (*Coquillettidia*), Crassipes Group
luteifemur Edwards, 1926a: *Aedes* (*Ochlerotatus*), Perkinsi Group
luteocephalus (Newstead, 1907) (in Newstead *et al.*, 1907): *Aedes* (*Stegomyia*), Africanus Group
luteola Edwards, 1934 (in Barraud, 1934): *Uranotaenia* (*Pseudoficalbia*), Recondita Series
luteolateralis (Theobald, 1901c): *Aedes* (*Neomelaniconion*), Lineatopennis Group
luteopleurus (Theobald, 1903a): *Culex* (*Anoedioporpa*), Conservator Group
luteostriatus Robinson, 1950: *Aedes* (*Vansomerenis*)
luteoventralis Theobald, 1901b: *Wyeomyia* (*Dendromyia*)
lutescens Leicester, 1908: *Uranotaenia* (*Pseudoficalbia*), Bicolor Series
lutescens (Theobald, 1901b): *Toxorhynchites* (*Afrorhynchus*) (not placed in a Group)
luteus (Ludlow, 1905a): *Aedes* (*Finlaya*), Poicilius Group
lutzi (da Costa Lima, 1930): *Wyeomyia* (*Miamyia*)
lutzii Cruz, 1901: *Anopheles* (*Nyssorhynchus*), Myzorhynchella Section
lutzii (Theobald, 1901b): *Psorophora* (*Janthinosoma*)
luxodens Hall, Howard & Harbach, 1999: *Sabethes* (*Peytonulus*)
luzonensis (Ludlow, 1905a): *Mimomyia* (*Etorleptiomyia*), Group A (*Etorleptiomyia s. str.*)
luzonensis Rozeboom, 1946: *Aedes* (*Kenknightia*), Luzonensis Group
luzonensis Sirivanakarn, 1976: *Culex* (*Oculeomyia*), Bitaeniorhynchus Complex
lygrus Root, 1927c: *Culex sensu stricto*, Pipiens Group, Tarsalis Subgroup
lynchi (Shannon, 1931): *Coquillettidia* (*Pseudotaeniorhynchus*)

Mm

mabinii Baisas & Ubaldo-Pagayon, 1953: *Tripteroides* (*Tripteroides*), Mabinii Group
macaensis Ribeiro, 1997: *Toxorhynchites* (*Toxorhynchites*), Splendens Group
macarthuri Colless, 1956: *Anopheles* (*Cellia*), Neomyzomyia Series, Leucosphyrus Group, Riparis Subgroup
macdonaldi Colless, 1965: *Culex* (*Lophoceraomyia*), Fraudatrix Group, Fraudatrix Subgroup, Variatus Complex
macdonaldi Mattingly, 1957b: *Heizmannia sensu stricto*
macdougalli Edwards, 1922a: *Aedes* (*Collessius*), Collessius Group
macfarlanei (Edwards, 1914b): *Aedes* (*Collessius*), Collessius Group
macfarlanei Edwards, 1914c: *Uranotaenia* (*Uranotaenia*), Campestris Series
macfiei Edwards, 1923a: *Culex* (*Culiciomyia*), Nebulosus Group
machadoi da Cunha Ramos, 1986: *Uranotaenia* (*Uranotaenia*), Caeruleocephala Section, Caeruleocephala Group
machadoi da Silva Mattos, da Silveira Guedes & Xavier, 1978: *Culex* (*Phytotelmatomyia*)
machardyi Edwards, 1930b: *Anopheles* (*Cellia*), Neomyzomyia Series, Ardensis Group
macintoshi Marks, 1959a: *Aedes* (*Ochlerotatus*), Perkinsi Group
mackerrasi Taylor, 1927: *Aedes* (*Patmarksia*)
macmahoni Evans, 1936: *Anopheles* (*Cellia*), Myzomyia Series, Demeilloni Group
macmillani Marks, 1964b: *Aedes* (*Macleaya*), Chaetocruiomyia Group
macrodixoa (Dyar & Shannon, 1925a): *Verrallina* (*Neomacleaya*)
macrophyllus Edwards & Gibbins, 1939: *Culex sensu stricto*, Pipiens Group (not placed in a Subgroup)
macrostylus Sirivanakarn & Ramalingam, 1976: *Culex* (*Eumelanomyia*), Mochthogenes Group, Tenuipalpis Subgroup
maculatus Theobald, 1901b: *Anopheles* (*Cellia*), Neocellia Series, Maculatus Group, Maculatus Subgroup
maculipalpis Giles, 1902: *Anopheles* (*Cellia*), Neocellia Series (not placed in a Group)
maculipennis Meigen, 1818: *Anopheles sensu stricto*, Angusticorn Section, Anopheles Series, Maculipennis Group, Maculipennis Subgroup
maculipennis (Theobald, 1911a): *Coquillettidia* (*Coquillettidia*), Crassipes Group
maculipes (Theobald, 1903a): *Anopheles sensu stricto*, Laticorn Section, Arribalzagia Series
maculipleura Leicester, 1908: *Uranotaenia* (*Pseudoficalbia*), Bimaculata Series
madagascarensis da Cunha Ramos & Brunhes, 2004: *Uranotaenia* (*Pseudoficalbia*), Shillitonis Section, Shillitonis Group (not placed in a Subgroup)
madagascarensis Ribeiro, 2004: *Toxorhynchites* (*Afrorhynchus*), Pauliani Group
madagascarensis van Someren, 1949a: *Aedes* (*Paulianius*), Coulangesi Group
madagascarica Brunhes, Boussès & da Cunha Ramos, 2011: *Aedeomyia sensu stricto*
madagascarica da Cunha Ramos & Brunhes, 2004: *Uranotaenia* (*Uranotaenia*), Caeruleocephala Section, Madagascarica Group
madininensis Senevet, 1936: *Culex* (*Melanoconion*), Melanoconion Section, Conspirator Group
madrensis Baisas, 1946a: *Orthopodomyia*, Orthopodomyia Section, Albipes Group, Flavicosta Subgroup
maehleri Bohart, 1957: *Aedes* (*Stegomyia*), Maehleri Group
maffii Taylor & Tenorio, 1974: *Aedes* (*Finlaya*), Kochi Group, Bougainvillensis Complex
magna (Theobald, 1905f): *Runchomyia* (*Ctenogoeldia*)
magnesianus (Edwards, 1924b): *Tripteroides* (*Tripteroides*), Magnesianus Group
magnificus (Leicester, 1908): *Toxorhynchites* (*Toxorhynchites*), Magnificus Group
magnus (Theobald, 1901c): *Deinocerites*, Cancer Group
magnus (Theobald, 1908b): *Armigeres* (*Leicesteria*)
mahaffyi van Someren, 1949b: *Eretmapodites*, Chrysogaster Group
mahantai Bhattacharyya, Prakash, Mohapatra & Sarma, 2009: *Armigeres sensu stricto*
maiae Edwards, 1917: *Armigeres sensu stricto*
maikalensis Natarajan, Rajavel & Jambulingam, 2017: *Uranotaenia* (*Pseudoficalbia*), Recondita Series
majidi Young & Majid, 1928: *Anopheles* (*Cellia*), Myzomyia Series (not placed in a Group)
major Edwards, 1935a: *Culex* (*Oculeomyia*) (not placed in a Complex)
malariager Poinar, 2005: *Culex* (fossil species)
malayensis Colless, 1962: *Aedes* (*Stegomyia*), Scutellaris Group, Scutellaris Subgroup
malayensis Sirivanakarn, 1972: *Culex* (*Eumelanomyia*), Rubinotus-rima Group, Rubinotus Subgroup
malayi Delfinado & Hodges, 1968: *Tripteroides* (*Tripteroides*), Nitidoventer Group
malayi (Leicester, 1908): *Culex* (*Eumelanomyia*), Mochthogenes Group, Malayi Subgroup
malayi Leicester, 1908: *Hodgesia*
malayi (Theobald, 1901c): *Armigeres sensu stricto*, Malayi Group
malaysiensis Ramalingam & Banu, 1987: *Topomyia* (*Topomyia*)
malaysisimilis Miyagi & Toma, 2023 (in Miyagi *et al.*, 2023): *Topomyia* (*Topomyia*)
malefactor Dyar & Knab, 1907c: *Anopheles sensu stricto*, Laticorn Section, Arribalzagia Series
malfeyti (Newstead, 1907) (in Newstead *et al.*, 1907): *Ficalbia*, Group B Afrotropical Region Group

maliensis Bailly-Choumara & Adam, 1959: *Anopheles* (*Cellia*), Neomyzomyia Series, Ardensis Group
malikuli Huang, 1973: *Aedes* (*Stegomyia*), W-albus Group, Mediopunctatus Subgroup
mallochi Taylor, 1944: *Aedes* (subgenus uncertain)
malvari Baisas & Ubaldo-Pagayon, 1953: *Tripteroides* (*Tripteroides*), Nitidoventer Group
mammilifer (Leicester, 1908): *Culex* (*Lophoceraomyia*), Mammilifer Group, Mammilifer Line, Mammilifer Complex
mamoedjoensis Mattingly, 1958a: *Aedes* (*Cancraedes*)
manakaraensis da Cunha Ramos & Brunhes, 2004: *Uranotaenia* (*Pseudoficalbia*), Annulata Section, Annulata Series, Lavieri Group
manalangi Baisas, 1935b: *Armigeres sensu stricto*
manalangi Mendoza, 1940: *Anopheles sensu stricto*, Laticorn Section, Myzorhynchus Series, Barbirostris Group, Vanus Subgroup
mangyanus (Banks, 1906): *Anopheles* (*Cellia*), Myzomyia Series, Funestus Group, Aconitus Subgroup
manhi (Harbach & Cook, 2010) (in Cook *et al.*, 2010): *Aedes* (*Bothaella*)
manicatus (Edwards, 1921c): *Toxorhynchites* (*Toxorhynchites*), Leicesteri Group, Manicatus Subgroup
manopi Thurman, 1959: *Toxorhynchites* (*Toxorhynchites*), Christophi Group
mansouri Qutubuddin, 1959: *Aedes* (*Aedimorphus*), Domesticus Assemblage, Vexans Subassemblage, Quasiunivittatus Group
manusensis Sirivanakarn, 1975: *Culex* (*Eumelanomyia*), Mochthogenes Group, Hinglungensis Subgroup
maplei Knight & Hurlbut, 1949: *Culex* (*Culiciomyia*), Fragilis Group
maracayensis Evans, 1923b: *Culex sensu stricto*, Pipiens Group, Tarsalis Subgroup
marajoara Galvão & Damasceno, 1942b: *Anopheles* (*Nyssorhynchus*), Argyritarsis Section, Albitarsis Series, Albitarsis Group, Albitarsis Complex (*An. marajoara* includes two lineages, Lineages 1 and 2, see McKeon *et al.*, 2010)
marceli (Mattingly, 1953b): *Malaya*
marcelleae Adam & Hamon, 1959: *Eretmapodites*, Oedipodeios Group
marchettei Garcia, Jeffery & Rudnick, 1969: *Culiseta* (*Climacura*)
margarsen (Dyar & Shannon, 1925a): *Verrallina* (*Neomacleaya*)
margina Gong & Lu, 1995: *Topomyia* (*Topomyia*)
mariae (Bourroul, 1904): *Toxorhynchites* (*Lynchiella*)
mariae (Sergent & Sergent, 1903): *Aedes* (*Acartomyia*)
marianae Bohart & Ingram, 1946b: *Culex sensu stricto*, Sitiens Group, Sitiens Subgroup
marinkellei Berlin, 1969a: *Aedes* (*Howardina*), Sexlineatus Section, Sexlineatus Subassemblage
marksae Dobrotworsky, 1965: *Tripteroides* (*Polylepidomyia*), Argenteiventris Group
marksae Grjebine, 1986: *Mimomyia* (*Ingramia*), Group B (Ravenalites)
marksae King & Hoogstraal, 1955: *Culex* (*Lophoceraomyia*), Fraudatrix Group, Fraudatrix Subgroup, Marksae Complex
marquesensis Stone & Rosen, 1953: *Culex sensu stricto*, Atriceps Group
marshallensis Stone & Bohart, 1944: *Aedes* (*Stegomyia*), Scutellaris Group, Scutellaris Subgroup
marshallii (Theobald, 1901b): *Aedes* (*Albuginosus*)
marshallii (Theobald, 1903a): *Anopheles* (*Cellia*), Myzomyia Series, Marshallii Group, Marshallii Complex
marteri Senevet & Prunnelle, 1927: *Anopheles sensu stricto*, Angusticorn Section, Anopheles Series (not placed in a Group)
martiali Senevet & Abonnenc, 1939b: *Limatus*
martinei (Doucet, 1951a): *Mimomyia* (*Ingramia*) Group A (*Ravenalites s. str.*)
martineti Senevet, 1937: *Aedes* (*Ochlerotatus*) (not placed in a Group)
martinezi Berlin, 1969a: *Aedes* (*Howardina*), Sexlineatus Section, Sexlineatus Subassemblage
martinezi Casal & García, 1968c: *Culex* (*Melanoconion*), Melanoconion Section, Conspirator Group
martinii Medschid, 1930: *Culex* (*Neoculex*), Territans Group
martinii (Statz, 1944): *Coquillettidia* (fossil species)
martinius Shingarev, 1926: *Anopheles sensu stricto*, Angusticorn Section, Anopheles Series, Maculipennis Group, Maculipennis Subgroup
mascarensis de Meillon, 1947: *Anopheles* (*Cellia*), Neomyzomyia Series, Mascarensis Group (the Group is monobasic)
mascarensis MacGregor, 1924: *Aedes* (*Stegomyia*), Aegypti Group
masculinus Mattingly, 1958a: *Aedes* (*Cancraedes*)
mashonaensis Theobald, 1901c: *Uranotaenia* (*Pseudoficalbia*), Mashonaensis Section, Mashonaensis Group
masoalensis Fontenille & Brunhes, 1985 *Aedes* (*Aedimorphus*), Domesticus Assemblage (not placed in a Subassemblage)
masseyi Edwards, 1923a: *Aedes* (*Stegomyia*), Dendrophilus Group
mathesoni Belkin, 1950a: *Tripteroides* (*Rachisoura*), Filipes Group
mathesoni Belkin & Heinemann, 1975: *Psorophora* (*Janthinosoma*)
mathesoni Belkin & Hogue, 1959: *Deinocerites*, Spanius Group
mathesoni Lane, 1943: *Uranotaenia* (*Uranotaenia*), Pallidoventer Series
mathioti Fontenille & Brunhes, 1985: *Aedes* (*Aedimorphus*), Domesticus Assemblage (not placed in a Subassemblage)
mattinglyi Baisas & Ubaldo-Pagayon, 1953: *Tripteroides* (*Tripteroides*), Nitidoventer Group
mattinglyi Grjebine, 1986: *Mimomyia* (*Ingramia*), Group B (Ravenalites)

mattinglyi Hamon & Rickenbach, 1954: *Aedes* (*Aedimorphus*), Alboscutellatus Assemblage, Abnormalis Group
mattinglyi Hamon & van Someren, 1961a: *Eretmapodites*, Chrysogaster Group
mattinglyi Knight, 1953: *Culex sensu stricto*, Pipiens Group, Theileri Subgroup
mattinglyi Lane, 1953: *Wyeomyia* (*Cruzmyia*)
mattinglyi Qutubuddin, 1951: *Uranotaenia* (*Pseudoficalbia*), Recondita Series
mattinglyi Thurman, 1959: *Heizmannia sensu stricto*
mattinglyorum Huang, 1994: *Aedes* (*Stegomyia*), Dendrophilus Group
mattogrossensis Lutz & Neiva, 1911: *Anopheles sensu stricto*, Laticorn Section, Arribalzagia Series
mauesensis Lane, 1945b: *Culex sensu stricto* (not placed in a Group)
maverlius Reinert, 1997 (in Reinert *et al.*, 1997): *Anopheles sensu stricto*, Angusticorn Section, Anopheles Series, Maculipennis Group, Quadrimaculatus Subgroup
maxgermaini Huang, 1990: *Aedes* (*Stegomyia*), Africanus Group
maxi Dyar, 1928: *Culex sensu stricto*, Pipiens Group, Tarsalis Subgroup
maxima Leicester, 1908: *Uranotaenia* (*Pseudoficalbia*), Annulata Section, Maxima Series
maximalepido Dong, Zhou & Dong, 2005: *Heizmannia sensu stricto*
maximus Edwards, 1922a: *Armigeres sensu stricto*
maxinocca Dyar, 1920b: *Culex* (*Melanoconion*), Melanoconion Section, Distinguendus Group, Distinguendus Subgroup
mayeri Edwards, 1912a: *Uranotaenia* (*Uranotaenia*), Alba Section, Alba Group
mayottensis Brunhes, 1977b: *Uranotaenia* (*Uranotaenia*), Dumonti Section, Dumonti Group
mccormicki (Belkin, 1962): *Verrallina sensu stricto*, Carmenti Series
mcdonaldi Belkin, 1962: *Aedes* (*Ochlerotatus*), Aculeatus Group
mcdonaldi Belkin & Hogue, 1959: *Deinocerites*, Pseudes Group
mcintoshi Huang, 1985: *Aedes* (*Neomelaniconion*), Lineatopennis Group
medialis Brug, 1932: *Aedes* (*Huaedes*)
medialis Harbach, 2018b: *Anopheles sensu stricto*, Laticorn Section, Arribalzagia Series
medioalbipes Lutz, 1904 (in Bourroul, 1904): *Wyeomyia* (*Wyeomyia*) (not placed in a Group)
mediolineata (Theobald, 1904b): *Mimomyia* (*Etorleptiomyia*), Group A (*Etorleptiomyia s. str.*)
mediolineatus (Theobald, 1901c): *Aedes* (*Aedimorphus*), Domesticus Assemblage, Mediolineatus Subassemblage, Mediolineatus Group
mediopunctatus (Lutz, 1903) (in Theobald, 1903a): *Anopheles sensu stricto*, Laticorn Section, Arribalzagia Series
mediopunctatus (Theobald, 1905h): *Aedes* (*Stegomyia*), W-albus Group, Mediopunctatus Subgroup
mediovittatus Coquillett, 1906d: *Aedes* (*Gymnometopa*) (the Subgenus is monobasic)
medleri Knight & Laffoon, 1946: *Aedes* (*Finlaya*), Poicilius Group
mefouensis Ferrara, 1974: *Aedes* (*Diceromyia*), Section A, Flavicollis Group
megafolius Chen & Dong, 1992: *Culex* (*Eumelanomyia*), Mochthogenes Group, Tenuipalpis Subgroup
megaloba Luh, Chao & Xu, 1974: *Culiseta sensu stricto*
megaonychus Yang, Li & Chen, 1993: *Culex* (*Culiciomyia*), Shebbearei Group
melanesiensis Belkin, 1955: *Tripteroides* (*Polylepidomyia*), Caledonicus Group
melanesiensis Belkin, 1962: *Mansonia* (*Mansonioides*), Uniformis Complex
melanimon Dyar, 1924e: *Aedes* (*Ochlerotatus*), Caspius Group
melanocephala Dyar & Knab, 1906a: *Wyeomyia* (not placed in a Subgenus)
melanonymphe Dyar, 1924b: *Sabethes* (*Sabethinus*), Intermedius Group
melanoon Hackett, 1934: *Anopheles sensu stricto*, Angusticorn Section, Anopheles Series, Maculipennis Group, Maculipennis Subgroup
melanophylum Dyar & Knab, 1907c: *Deinocerites*, Cancer Group
melanopous Graham, 1909: *Eretmapodites*, Inornatus Group
melanopterus (Giles, 1904): *Aedes* (*Phagomyia*)
melanopus Dyar, 1919b: *Wyeomyia* (*Wyeomyia*) (not placed in a Group)
melanota Cerqueira, 1943: *Psorophora* (*Janthinosoma*)
melanura (Coquillett, 1902b): *Culiseta* (*Climacura*)
melas (Theobald, 1903a): *Anopheles* (*Cellia*), Pyretophorus Series, Gambiae Complex
memorans (Bonne-Wepster, 1930): *Coquillettidia* (*Coquillettidia*), Crassipes Group
mendacis (Daniels, 1908): *Tripteroides* (*Tripteroides*), Nitidoventer Group
mendiolai Baisas, 1935c: *Uranotaenia* (*Uranotaenia*), Campestris Series
mengi Chen, Wang & Zhao, 1989: *Uranotaenia* (*Pseudoficalbia*), Recondita Series
mengi Dong, Wang & Lu, 1990: *Topomyia* (*Suaymyia*)
menglaensis Dong, Zhou & Dong, 2002b: *Armigeres* (*Leicesteria*)
menglangensis Ma, 1981a: *Anopheles sensu stricto*, Angusticorn Section, Anopheles Series, Lindesayi Group, Lindesayi Subgroup
menglianensis Lu & Gong, 1986: *Heizmannia sensu stricto*

menglianeroides Dong, Dong & Zhou, 2003 (in Dong *et al.*, 2003b): *Heizmannia sensu stricto*
menoni Mattingly, 1958b: *Aedes* (*Paraedes*), Barraudi Assemblage
meprai Martínez & Prosen, 1953b: *Aedes* (*Ochlerotatus*), Scapularis Group, Trivittatus Subgroup
meraukensis Venhuis, 1932: *Anopheles* (*Cellia*), Neomyzomyia Series (not placed in a Group)
mercurator Dyar, 1920a: *Aedes* (*Ochlerotatus*), Stimulans Group
meronephada (Dyar & Shannon, 1925a): *Aedes* (*Catatassomyia*) (the Subgenus is monobasic)
merus Dönitz, 1902: *Anopheles* (*Cellia*), Pyretophorus Series, Gambiae Complex
mesodentatus Komp & Kumm, 1938: *Haemagogus sensu stricto*, Albomaculatus Section
mesodenticulatus Galindo & Mendez, 1961: *Culex* (*Melanoconion*), Melanoconion Section, Inhibitator Group, Inhibitator Subgroup
mesostictus Harbach, 2018b: *Aedes* (*Aedimorphus*), Domesticus Assemblage, Vexans Subassemblage, Quasiunivittatus Group
messeae Falleroni, 1926: *Anopheles sensu stricto*, Angusticorn Section, Anopheles Series, Maculipennis Group, Maculipennis Subgroup
metallica (Theobald, 1901d): *Coquillettidia* (*Coquillettidia*), Crassipes Group
metallicus (Edwards, 1912a): *Aedes* (*Stegomyia*), Metallicus Group (the Group is monobasic)
metallicus Leicester, 1904 (in Theobald, 1904c): *Toxorhynchites* (*Toxorhynchites*), Leicesteri Group, Leicesteri Subgroup
metatarsata Edwards, 1914b: *Uranotaenia* (*Uranotaenia*), Campestris Series
metempsytus Dyar, 1921e: *Culex* (*Carrollia*), Bihaicola Group
metoecopus Dyar, 1925b: *Aedes* (*Protomacleaya*), Terrens Group, Metoecopus Subgroup (the Subgroup is monobasic)
mexicana (Bellardi, 1859): *Psorophora* (*Janthinosoma*)
mexicanus Zavortink & Poinar, 2008: *Toxorhynchites* (*Toxorhynchites*) (fossil species)
miaolingensis Chen, 1982: *Culex* (*Eumelanomyia*), Mochthogenes Group, Okinawae Subgroup
micans Leicester, 1908: *Uranotaenia* (*Uranotaenia*), Micans Series
michaelikati van Someren, 1946c: *Aedes* (*Pseudarmigeres*)
mickevichae Huang, 1988b: *Aedes* (*Stegomyia*), Pseudonigeria Group
micolo Ribeiro, da Cunha Ramos, Capela & Pires, 1998: *Culex* (*Eumelanomyia*), Mochthogenes Group, Inconspicuosus Subgroup
microannulata (Theobald, 1911a): *Coquillettidia* (*Coquillettidia*), Crassipes Group
microcala (Dyar, 1929c): *Tripteroides* (*Tricholeptomyia*), Nepenthicola Group, Nepenthicola Subgroup
microlepis (Edwards, 1927b): *Tripteroides* (*Polylepidomyia*), Argenteiventris Group
micromelas Edwards, 1934: *Uranotaenia* (*Pseudoficalbia*), Mashonaensis Section, Mashonaensis Group
microphyllus Root, 1927c: *Culex* (*Microculex*), Inimitabilis Series
micropterus (Giles, 1901a): *Aedes* (*Dendroskusea*)
microstictus Edwards, 1936: *Aedes* (*Catageiomyia*)
mijanae Brunhes, Adam & Bailly-Choumara, 1967: *Culex* (*Eumelanomyia*), Mochthogenes Group, Inconspicuosus Subgroup
mikrokopion Knight & Harrison, 1988: *Aedes* (*Downsiomyia*), Niveus Group
millecampsi Lips, 1960: *Anopheles* (*Cellia*), Neomyzomyia Series, Ardensis Group
milleri Dyar, 1922e: *Aedes* (*Ochlerotatus*) (not placed in a Group)
millironi Belkin, 1962: *Culex* (*Neoculex*), Pseudomelanoconia Group
milloti Doucet, 1949b: *Culex* (*Culiciomyia*), Nebulosus Group
milloti Doucet, 1951a: *Orthopodomyia*, Orthopodomyia Section, Vernoni Group
milloti Grjebine, 1986: *Mimomyia* (*Ingramia*), Group A (*Ravenalites s. str.*), Roubaudi Complex
milloti Grjebine & Lacan, 1953 (in Grjebine, 1953a): *Anopheles* (*Cellia*), Neomyzomyia Series, Pauliani Group
milnensis (King & Hoogstraal, 1947b): *Verrallina sensu stricto*, Carmenti Series
milnensis Lee, 1944b: *Armigeres sensu stricto*
milsoni (Taylor, 1915): *Aedes* (*Dobrotworskyius*)
milwardi Xavier & da Silva Mattos, 1972: *Culex* (*Melanoconion*), Melanoconion Section, Intrincatus Group, Intrincatus Subgroup
mimeticus Noè, 1899: *Culex sensu stricto*, Sitiens Group, Mimeticus Subgroup, Mimeticus Complex
mimomyiaformis (Newstead, 1907) (in Newstead *et al.*, 1907): *Mimomyia* (*Mimomyia*), Boycia Group
mimuloides Barraud, 1924c: *Culex sensu stricto*, Sitiens Group, Mimeticus Subgroup, Mimeticus Complex
mimulus Edwards, 1915b: *Culex sensu stricto*, Sitiens Group, Mimeticus Subgroup, Mimulus Complex
minima (Theobald, 1901c): *Ficalbia*, Group A Oriental Region Group
minimus Theobald, 1901b: *Anopheles* (*Cellia*), Myzomyia Series, Funestus Group, Minimus Subgroup, Minimus Complex
minimus (Theobald, 1905h): *Toxorhynchites* (*Toxorhynchites*), Minimus Group
minjensis Sirivanakarn, 1968: *Culex* (*Lophoceraomyia*), Fraudatrix Group, Fraudatrix Subgroup, Christiani Complex
minnesotae Barr, 1957: *Culiseta* (*Culicella*)
minor da Costa Lima, 1929: *Anopheles sensu stricto*, Laticorn Section, Arribalzagia Series
minor (Leicester, 1908): *Culex* (*Lophoceraomyia*), Mammilifer Group, Minor Line, Minor Complex
minor Leicester, 1908: *Topomyia* (*Topomyia*)

minutus Poinar, Zavortink, Pike & Johnston, 2000: *Paleoculicis* (fossil species)
minutus (Theobald, 1901b): *Aedes* (*Catageiomyia*)
minutissimus (Theobald, 1907): *Culex* (*Lophoceraomyia*), Fraudatrix Group, Fraudatrix Subgroup, Minutissimus Subgroup
miraculosus Bonne-Wepster, 1937: *Culex sensu stricto*, Pipiens Group, Trifilatus Subgroup
mirans Sallum & Peyton, 2005 (in Sallum *et al.*, 2005b): *Anopheles* (*Cellia*), Neomyzomyia Series, Leucosphyrus Group, Hackeri Subgroup
mirificus Edwards, 1913b: *Culex sensu stricto*, Pipiens Group (not placed in a Subgroup)
misionensis Duret, 1953: *Culex* (*Melanoconion*), Melanoconion Section, Intrincatus Group, Intrincatus Subgroup
mitchellae (Dyar, 1905b): *Aedes* (*Ochlerotatus*), Taeniorhynchus Group (not placed in a Subgroup)
mitchellii (Theobald, 1905d): *Wyeomyia* (*Wyeomyia*), Pertinans Group
mixtli Rivera-García, Mendez-Andrade & Ibáñez-Bernal, 2023: *Trichoprosopon* (not placed in a Group or Complex)
mixtus Edwards, 1936: *Aedes* (*Catageiomyia*)
miyagii (Toma & Mogi, 2003): *Kimia*
miyagii Mogi & Toma, 1999: *Culex* (*Acallyntrum*), Bicki Group
miyarai Tanaka, Mizusawa & Saugstad, 1979: *Aedes* (*Stegomyia*), Scutellaris Group, Albopictus Subgroup
mjobergi (Edwards, 1926b): *Aedes* (*Downsiomyia*), Niveus Group
moctezuma (Dyar & Knab, 1906c): *Toxorhynchites* (*Lynchiella*)
modesta (King & Hoogstraal, 1946e): *Mimomyia* (*Mimomyia*), Ludlowia Group
modesta Leicester, 1908: *Uranotaenia* (*Pseudoficalbia*), Bimaculata Series
modestus Ficalbi, 1890b: *Culex* (*Barraudius*)
moerbista (Dyar & Knab, 1919): *Wyeomyia* (not placed in a Subgenus)
moghulensis Christophers, 1924c: *Anopheles* (*Cellia*), Neocellia Series (not placed in a Group)
mogii Miyagi, Toma & Higa, 2004: *Mimomyia* (*Ingramia*), Group A (*Ravenalites s. str.*)
mogilasium (Dyar & Knab, 1907c): *Trichoprosopon*, Trichoprosopon Group, Compressum Complex
mohani Knight, 1969: *Aedes* (*Downsiomyia*), Niveus Group
mohani Sirivanakarn, 1977d: *Culex* (*Eumelanomyia*), Mochthogenes Group (not placed in a Subgroup)
mollis Dyar & Knab, 1906b: *Culex sensu stricto*, Pipiens Group (not placed in a Subgroup)
moloiensis (Taylor, 1929): *Aedes* (*Macleaya*), Chaetocruiomyia Group
monetifer (Dyar, 1920f): *Tripteroides* (*Tripteroides*), Nitidoventer Group
monetus Edwards, 1935a: *Aedes* (*Zavortinkius*), Monetus Group (the Group is monobasic)
mongiro van Someren, 1951: *Culex* (*Culiciomyia*), Nebulosus Group
monocellatus Marks, 1948: *Aedes* (subgenus uncertain)
monotrichus Edwards, 1936: *Aedes* (*Neomelaniconion*) (not placed in a Group)
montana Ingram & de Meillon, 1927: *Uranotaenia* (*Pseudoficalbia*), Annulata Section, Annulata Series, Annulata Group, Subgroup B
montanus Brug, 1939: *Aedes* (*Rampamyia*)
montanus Stanton & Hacker, 1917: *Anopheles sensu stricto*, Laticorn Section, Myzorhynchus Series, Albotaeniatus Group
montchadskyi Dubitsky, 1968: *Aedes* (*Ochlerotatus*), Communis Group (not placed in a Subgroup)
monticola Belkin & McDonald, 1957: *Aedes* (*Jarnellius*)
monticola Ribeiro, Lourenço-de-Oliveira & Motta, 2022 (in Ribeiro *et al.*, 2022): *Wyeomyia* (*Decamyia*)
monticola Somboon, Namgay & Harbach, 2021 (in Somboon *et al.*, 2022a, first published online 17 November 2021): *Anopheles sensu stricto*, Angusticorn Section, Anopheles Series, Lindesayi Group, Gigas Subgroup, Baileyi Complex
moralesi (Dyar & Knab, 1919): *Shannoniana*
moramangae da Cunha Ramos & Brunhes, 2004: *Uranotaenia* (*Uranotaenia*), Chorleyi Section, Hamoni Group
moresbyensis Peters, 1963e: *Uranotaenia* (*Uranotaenia*) (not placed in a Section, Series, Group or Complex)
morsitans (Theobald, 1901c): *Culiseta* (*Culicella*)
mortiauxi da Cunha Ramos & Ribeiro, 1990: *Eretmapodites*, Leucopous Group
mortiauxi Edwards, 1938 (in Evans, 1938): *Anopheles* (*Cellia*), Myzomyia Series, Marshallii Group
moucheti Evans, 1923a: *Culex* (*Kitzmilleria*) (the Subgenus is monobasic)
moucheti Evans, 1925b: *Anopheles* (*Cellia*), Myzomyia Series (not placed in a Group)
moucheti (Hamon & Adam, 1955): *Malaya*
moucheti Ravaonjanahary & Brunhes, 1977: *Aedes* (*Skusea*)
moufiedi Peyton, 1977: *Uranotaenia* (*Pseudoficalbia*), Bimaculata Series
moultoni Edwards, 1914c: *Armigeres sensu stricto*
moultoni Edwards, 1914c: *Uranotaenia* (*Pseudoficalbia*), Bimaculata Series
mousinhoi de Meillon & de Carvalho Pereira, 1940: *Anopheles* (*Cellia*), Myzomyia Series, Marshallii Group
mpusiensis Huang, 2004: *Aedes* (*Stegomyia*), Poweri Group
mubiensis Luh & Shih, 1958: *Aedes sensu stricto* (not placed in a Group)
mucidus (Karsch, 1887): *Aedes* (*Mucidus*), Mucidus Group

muehlensi Petrocchi, 1927 (in Shannon & del Ponte, 1927): *Wyeomyia* (*Phoniomyia*), Third Series
muelleri Dyar, 1920g: *Aedes* (*Lewnielsenius*) (the Subgenus is monobasic)
mulrennani Basham, 1948: *Culex* (*Melanoconion*), Melanoconion Section, Inhibitator Group, Mulrennani Subgroup (the Subgroup is monobasic)
multicinctus Edwards, 1930a: *Anopheles* (*Cellia*), Neomyzomyia Series, Ardensis Group
multicolor Cambouliu, 1902: *Anopheles* (*Cellia*), Paramyzomyia Series, Listeri Group
multifolium (King & Hoogstraal, 1947b): *Verrallina sensu stricto*, Carmenti Series
multiplex (Theobald, 1903a): *Aedes* (*Pseudoskusea*)
mundulus Grünberg, 1905: *Culex* (*Eumelanomyia*), Rubinotus-rima Group, Rima Subgroup
muroafcete Huang, 1997: *Aedes* (*Stegomyia*), Dendrophilus Group
murphyi Gillies & de Meillon, 1968: *Anopheles* (*Cellia*), Cellia Series (not placed in a Group)
murrelli Lien, 1968d: *Culex sensu stricto*, Sitiens Group, Mimeticus Subgroup, Mimulus Complex
muruae Sirivanakarn, 1968: *Culex* (*Lophoceraomyia*), Fraudatrix Group, Fraudatrix Subgroup, Marksae Complex
murudensis Miyagi, Toma & Okazawa, 2021 (in Miyagi *et al.*, 2021a): *Topomyia sensu stricto*
musarum Edwards, 1932b: *Culex sensu stricto*, Pipiens Group (not placed in a Subgroup)
musarum Edwards, 1936: *Uranotaenia* (*Pseudoficalbia*), Nigripes Section, Ornata Group
muspratti Hamon & Lambrecht, 1959: *Culex* (*Culiciomyia*), Nebulosus Group
mutator Dyar & Knab, 1906d: *Culex* (*Melanoconion*), Melanoconion Section, Intrincatus Group, Intrincatus Subgroup
mutilus Edwards, 1936: *Aedes* (*Catageiomyia*)
mystes Dyar, 1924a: *Wyeomyia* (*Spilonympha*)
mzooi van Someren, 1962: *Aedes* (*Zavortinkius*), Longipalpis Group

Nn

nailoni King & Hoogstraal, 1946a: *Culex* (*Culiciomyia*), Fragilis Group
nairobiensis (van Someren, 1946b): *Toxorhynchites* (*Afrorhynchus*) (not placed in a Group)
nakuruensis Mattingly, 1951: *Culex sensu stricto*, Pipiens Group (not placed in a Subgroup)
namibiensis Coetzee, 1984: *Anopheles sensu stricto*, Laticorn Section, Myzorhynchus Series, Coustani Group
natalensis Edwards, 1930a: *Aedes* (*Pseudarmigeres*)
natalensis (Hill & Haydon, 1907): *Anopheles* (*Cellia*), Neomyzomyia Series, Ardensis Group
nataliae Belkin, 1945: *Anopheles* (*Cellia*), Neomyzomyia Series, Lungae Complex (not placed in a Group)
nataliae Gornostaeva, 2005: *Aedes sensu stricto* (not placed in a Group)
nataliae Lynch Arribálzaga, 1891b: *Uranotaenia* (*Uranotaenia*), Leucoptera Series
natronius Edwards, 1932b: *Aedes* (*Aedimorphus*), Domesticus Assemblage, Vexans Subassemblage, Vexans Group
navalis Edwards, 1926a: *Culex* (*Lophoceraomyia*), Mammilifer Group, Brevipalpus Subgroup, Navalis Complex
neavei Theobald, 1906: *Culex sensu stricto*, Pipiens Group, Univittatus Subgroup
nebulosus Arnell, 1973: *Haemagogus sensu stricto*, Albomaculatus Section
nebulosus Theobald, 1901d: *Culex* (*Culiciomyia*), Nebulosus Group
neghmei Mann, 1950: *Anopheles sensu stricto*, Angusticorn Section, Anopheles Series, Pseudopunctipennis Group
neglectus Lutz, 1904 (in Bourroul, 1904): *Culex* (*Microculex*), Inimitabilis Series
negrensis Gordon & Evans, 1922: *Wyeomyia* (not placed in a Subgenus)
neireti Edwards, 1920b: *Uranotaenia* (*Uranotaenia*), Anopheloides Section, Neireti Group
neivai (Lane & Coutinho, 1940): *Coquillettidia* (*Pseudotaeniorhynchus*)
neivai Howard, Dyar & Knab, 1913: *Anopheles* (*Kerteszia*), Neivai Complex (*An. neivai* consists of eight sibling species denoted as Species 1 (*neivai sensu stricto*) and 2–8) (Bourke *et al.*, 2023)
nemophilous Peyton & Ramalingam, 1988: *Anopheles* (*Cellia*), Neomyzomyia Series, Leucosphyrus Group, Leucosphyrus Subgroup, Dirus Complex
nemorosa (Gong, 1996): *Kimia*
neoafricanus Cornet, Valade & Dieng, 1978: *Aedes* (*Stegomyia*), Africanus Group
neogalloisi Chen & Chen, 2000: *Aedes* (*Stegomyia*), Scutellaris Group, Albopictus Subgroup
neogeorgianus Belkin, 1962: *Aedes* (*Finlaya*), Kochi Group, Bougainvillensis Complex
neomacrodixoa (King & Hoogstraal, 1947b): *Verrallina* (*Neomacleaya*)
neomaculipalpus Curry, 1931: *Anopheles sensu stricto*, Laticorn Section, Arribalzagia Series
neopandani Bohart, 1957: *Aedes* (*Stegomyia*), Pandani Group
neotibialis King & Hoogstraal, 1947a: *Uranotaenia* (*Uranotaenia*), Anisocheleomyia Section, Tibialis Group
nepenthes (Theobald, 1912b): *Uranotaenia* (*Pseudoficalbia*), Bicolor Series
nepenthicola (Banks, 1909): *Tripteroides* (*Tricholeptomyia*), Nepenthicola Group, Nepenthicola Subgroup
nepenthicola Miyagi & Toma, 2007: *Topomyia* (*Suaymyia*)
nepenthicola Steffan & Evenhuis, 1982: *Toxorhynchites* (*Toxorhynchites*), Splendens Group

nepenthis (Dyar & Shannon, 1925a): *Toxorhynchites* (*Toxorhynchites*), Acaudatus Group, Nepenthis Subgroup
nepenthis (Edwards, 1915b): *Tripteroides* (*Rachionotomyia*), Aranoides Group, Nepenthis Subgroup
nepenthisimilis Mattingly, 1981: *Tripteroides* (*Rachionotomyia*), Aranoides Group, Nepenthis Subgroup
nevadensis Chapman & Barr, 1964: *Aedes* (*Ochlerotatus*), Communis Group, Communis Aggregate
ngong van Someren, 1950a: *Aedes* (*Albuginosus*)
nicaroensis Duret, 1967a: *Culex* (*Nicaromyia*) (the Subgenus is monobasic)
nicceriensis Bonne-Wepster & Bonne, 1920: *Culex* (*Melanoconion*), Melanoconion Section, Distinguendus Group, Distinguendus Subgroup
nicksoni Miyagi & Toma, 2012 (in Miyagi *et al.*, 2012): *Topomyia* (*Topomyia*)
nicoyae Adames & Hogue, 1970: *Deinocerites*, Dyari Group
niger (Leicester, 1908): *Culex* (*Lophoceraomyia*), Fraudatrix Group, Fraudatrix Subgroup, Rubithoracis Complex
nigeriae Edwards, 1930a: *Hodgesia*
nigeriensis Evans, 1931a: *Anopheles* (*Cellia*), Myzomyia Series (not placed in a Group)
nigerrimus Giles, 1900a: *Anopheles sensu stricto*, Laticorn Section, Myzorhynchus Series, Hyrcanus Group, Nigerrimus Subgroup
nigeriensis Ribeiro, 2005a: *Toxorhynchites* (*Afrorhynchus*) (not placed in a Group)
nigerrimus (Theobald, 1913): *Aedes* (*Mucidus*), Mucidus Group
nigra Leicester, 1908: *Topomyia* (*Topomyia*)
nigra (Theobald, 1901d): *Ficalbia*, Group B Afrotropical Region Group
nigrescens (Edwards, 1929) (in Paine & Edwards, 1929): *Aedes* (*Mucidus*), Pardomyia Group
nigricans (Coquillett, 1904c): *Coquillettidia* (*Pseudotaeniorhynchus*)
nigricephala Clastrier & Claustre, 1978b: *Wyeomyia* (not placed in a Subgenus)
nigricephala da Cunha Ramos & Brunhes, 2004: *Uranotaenia* (*Pseudoficalbia*), Nigripes Section, Nigripes Group
nigricephalus (Theobald, 1901d): *Aedes* (*Aedimorphus*), Alboscutellatus Assemblage, Abnormalis Group
nigrimacula Lane & Whitman, 1943: *Culex* (subgenus uncertain)
nigrinus (Eckstein, 1918): *Aedes* (*Ochlerotatus*), Communis Group (not placed in a Subgroup)
nigripalpus Theobald, 1901c: *Culex sensu stricto*, Pipiens Group (not placed in a Subgroup)
nigripes (Edwards, 1935b): *Toxorhynchites* (*Toxorhynchites*), Ater Group
nigripes (Theobald, 1905i): *Uranotaenia* (*Pseudoficalbia*), Nigripes Section, Nigripes Group
nigripes (Zetterstedt, 1838): *Aedes* (*Ochlerotatus*), Communis Group, Punctor Subgroup
nigripleura da Cunha Ramos & Brunhes, 2004: *Uranotaenia* (*Pseudoficalbia*), Spinosa Section, Spinosa Group
nigritarsis (Chagas, 1907): *Anopheles* (*Nyssorhynchus*), Myzorhynchella Section
nigritarsis (Wolfs, 1958): *Coquillettidia* (*Coquillettidia*), Crassipes Group
nigrithorax (Macquart, 1847): *Aedes* (*Ochlerotatus*), Burpengaryensis Group
nigrithorax (Theobald, 1910c): *Coquillettidia* (*Coquillettidia*), Crassipes Group
nigritubus Galindo, Carpenter & Trapido, 1951: *Wyeomyia* (*Wyeomyia*), Caracula Group
nigrocanus Martini, 1927: *Aedes* (*Ochlerotatus*) (not placed in a Group)
nigrochracea (Bonne-Wepster, 1930): *Coquillettidia* (*Coquillettidia*), Crassipes Group
nigromaculata Edwards, 1941: *Uranotaenia* (*Pseudoficalbia*), Mashonaensis Section, Mashonaensis Group
nigromaculis (Ludlow, 1906a): *Aedes* (*Ochlerotatus*), Taeniorhynchus Group (not placed in a Subgroup)
nigropterus (Le Goff, Boussès & Brunhes, 2007): *Aedes* (*Neomelaniconion*), Lineatopennis Group
nigropunctatus Edwards, 1926a: *Culex* (*Culiciomyia*), Fragilis Group
nigrorhynchus Brug, 1931c: *Aedes* (*Hulecoeteomyia*)
nigrosignata (Edwards, 1917): *Coquillettidia* (*Coquillettidia*), Crassipes Group
nigrostriatus (Barraud, 1927c): *Aedes* (*Aedimorphus*), Domesticus Assemblage, Mediolineatus Subassemblage, Mediolineatus Group
nigrotarsis (Ludlow, 1908a): *Verrallina* (*Neomacleaya*)
nilgiricus Christophers, 1924b: *Anopheles sensu stricto*, Angusticorn Section, Anopheles Series, Lindesayi Group, Lindesayi Subgroup
nilgiricus Edwards, 1916: *Culex sensu stricto*, Pipiens Group (not placed in a Subgroup)
nili (Theobald, 1904b): *Anopheles* (*Cellia*), Neomyzomyia Series, Ardensis Group, Nili Complex
nimbus (Theobald, 1902b): *Anopheles* (*Stethomyia*)
nimpe Nguyen, Tran & Harbach, 2000 (in Nguyen *et al.*, 2000): *Anopheles sensu stricto*, Laticorn Section, Myzorhynchus Series, Hyrcanus Group (not placed in a Subgroup)
ninagongoensis Edwards, 1928b: *Culex sensu stricto*, Pipiens Group (not placed in a Subgroup)
ningheensis Lei, 1989b: *Aedes* (*Collessius*), Collessius Group
niphadopsis Dyar & Knab, 1918: *Aedes* (*Ochlerotatus*), Impiger Group
nipponii (Theobald, 1907): *Aedes* (*Aedimorphus*), Domesticus Assemblage, Vexans Subassemblage, Vexans Group
nippononiveus Sasa & Nakahashi, 1952: *Aedes* (*Downsiomyia*), Niveus Group
nipponica LaCasse & Yamaguti, 1950: *Culiseta* (*Culicella*)

nipponicus LaCasse & Yamaguti, 1948: *Aedes* (*Downsiomyia*), Niveus Group
nishikawai Tanaka, Mizusawa & Saugstad, 1979: *Aedes* (*Downsiomyia*), Niveus Group
nissanensis Lee, 1946: *Tripteroides* (*Tripteroides*), Bimaculipes Group
nitens (Cerqueira, 1943): *Coquillettidia* (*Pseudotaeniorhynchus*)
nitidoventer (Giles, 1904): *Tripteroides* (*Tripteroides*), Nitidoventer Group
nitidus Harrison, Scanlon & Reid, 1973: *Anopheles sensu stricto*, Laticorn Section, Myzorhynchus Series, Hyrcanus Group, Nigerrimus Subgroup
nitidus Theobald, 1901c: *Sabethes* (*Sabethes*), Group A
nivalis Edwards, 1926a: *Aedes* (*Ochlerotatus*), Perkinsi Group
nivea Leicester, 1908: *Uranotaenia* (*Uranotaenia*), Anisocheleomyia Section, Nivea Series
niveitaeniata (Theobald, 1907): *Culiseta sensu stricto*
niveoides Barraud, 1934: *Aedes* (*Downsiomyia*), Niveus Group
niveoscutum Zavortink, 1972: *Aedes* (*Protomacleaya*), Kompi Group
niveus (Ludlow, 1903) (in Theobald, 1903a): *Aedes* (*Downsiomyia*), Niveus Group
nivipes (Theobald, 1903f): *Anopheles* (*Cellia*), Neocellia Series, Annularis Group, Nivipes Complex (*An. nivipes* consists of two cytogenetic species in Thailand)
nivipes (Theobald, 1905a): *Uranotaenia* (*Uranotaenia*) (not placed in a Complex, Group or Series)
nivipleura Leicester, 1908: *Uranotaenia* (*Pseudoficalbia*), Bimaculata Series
nivipous Theobald, 1912a: *Uranotaenia* (*Pseudoficalbia*), Annulata Section, Annulata Series, Annulata Group, Subgroup B
njombiensis Huang, 1997: *Aedes* (*Stegomyia*), Dendrophilus Group
njombiensis Peters, 1955b: *Anopheles* (*Cellia*), Myzomyia Series, Marshallii Group
nkolbissonensis Rickenbach & Hamon, 1966: *Orthopodomyia*, Orthopodomyia Section, Nkolbissonensis Group
nobukonis (Yamada, 1932): *Verrallina* (*Harbachius*)
nocticola Peyton, 1977: *Uranotaenia* (*Pseudoficalbia*), Recondita Series
nocturnus (Theobald, 1903a): *Aedes* (*Aedimorphus*), Domesticus Assemblage, Vexans Subassemblage, Vexans Group
noei Mann, 1950: *Anopheles sensu stricto*, Angusticorn Section, Anopheles Series, Pseudopunctipennis Group
noniae Reid, 1963: *Anopheles sensu stricto*, Angusticorn Section, Lophoscelomyia Series, Asiaticus Group
normanensis (Taylor, 1915): *Aedes* (*Ochlerotatus*), Taeniorhynchus Group, Theobaldi Subgroup
notabilis (Delfinado, 1967): *Verrallina* (*Neomacleaya*)
notanandai Rattanarithikul & Green, 1987: *Anopheles* (*Cellia*), Neocellia Series, Maculatus Group, Sawadwongporni Subgroup
notleyi van Someren, 1949a: *Anopheles* (*Cellia*), Neomyzomyia Series, Ranci Group, Roubaudi Subgroup
notoscriptus (Skuse, 1889): *Aedes* (*Rampamyia*)
novaezealandiae Pillai, 1966: *Culiseta* (*Climacura*)
novaguinensis Peters, 1963d: *Uranotaenia* (*Uranotaenia*), Anisocheleomyia Section, Tibialis Group, Novaguinensis Complex
novaguinensis Venhuis, 1933: *Anopheles* (*Cellia*), Neomyzomyia Series (not placed in a Group)
novalbitarsis King & Hoogstraal, 1946d: *Aedes* (*Patmarksia*)
novalbopictus Barraud, 1931a: *Aedes* (*Stegomyia*), Scutellaris Group, Albopictus Subgroup
novobscura Barraud, 1934: *Uranotaenia* (*Pseudoficalbia*), Bimaculata Series
novochracea (Barraud, 1927d): *Coquillettidia* (*Coquillettidia*), Crassipes Group
novohanoverae Peters, 1963b: *Tripteroides* (*Tripteroides*), Bimaculipes Group
novoniveus Barraud, 193): *Aedes* (*Downsiomyia*), Niveus Group
nubicola (Laffoon, 1946): *Verrallina* (*Neomacleaya*)
nubilus (Theobald, 1903a): *Aedes* (*Ochlerotatus*), Serratus Group
nummatus Edwards, 1923b: *Aedes* (*Tewarius*), Nummatus Group
nuneztovari Gabaldon, 1940: *Anopheles* (*Nyssorhynchus*), Albimanus Section, Oswaldoi Series, Oswaldoi Group, Oswaldoi Subgroup, Nuneztovari Complex (*An. nuneztovari* includes five species denoted as Lineages 1–5 by Mirabello & Conn, 2008 and Scarpassa *et al.*, 2016; which is the nominotypical species is not known)
nyasae Edwards, 1930a: *Aedes* (*Hopkinsius*), Hopkinsius Group
nyounae Hamon & Adam, 1959a: *Aedes* (*Catageiomyia*)

Oo

oakleyi Stone, 1939: *Aedes* (*Aedimorphus*), Alboscutellatus Assemblage, Alboscutellatus Group
oblita (Lutz, 1904) (in Bourroul, 1904): *Wyeomyia* (*Miamyia*)
obscura Edwards, 1915b: *Uranotaenia* (*Pseudoficalbia*), Bicolor Series
obscura Tenorio, 1975: *Bironella* (*Brugella*)
obscurum Lane & Cerqueira, 1942: *Trichoprosopon*, Trichoprosopon Group, Compressum Complex
obscurus Brug, 1934: *Tripteroides* (*Rachisoura*), Vanleeuweni Group
obscurus (Grünberg, 1905): *Anopheles sensu stricto*, Laticorn Section, Myzorhynchus Series (not placed in a Group)

obscurus (Leicester, 1908): *Culex* (*Acalleomyia*) (the Subgenus is monobasic)
obsoleta (Huang, 1968a): *Verrallina sensu stricto*, Carmenti Series
obturbans (Walker, 1859): *Armigeres sensu stricto*
obturbator Dyar & Knab, 1907a: *Aedes* (*Ochlerotatus*), Scapularis Group, Obturbator Subgroup
occidentalis Dyar & Knab, 1906b: *Anopheles sensu stricto*, Angusticorn Section, Anopheles Series, Maculipennis Group, Freeborni Subgroup
occidentalis (Skuse, 1889): *Aedes* (*Dobrotworskyius*)
occidentayunnana (Gong & Lu, 1991): *Heizmannia* (*Mattinglyia*)
occulta Bonne-Wepster & Bonne, 1919: *Wyeomyia* (not placed in a Subgenus)
oceanicus Belkin, 1962: *Aedes* (*Finlaya*), Kochi Group
ocellatus Theobald, 1903a: *Culex* (subgenus uncertain)
ochracea (Theobald, 1903a): *Coquillettidia* (*Coquillettidia*), Crassipes Group
ochraceus (Theobald, 1901c): *Aedes* (*Aedimorphus*), Domesticus Assemblage, Mediolineatus Subassemblage, Mediolineatus Group
ochripes (Macquart, 1850): *Psorophora sensu stricto*
ochroptera (Peus, 1935): *Culiseta* (*Culicella*)
ocossa Dyar & Knab, 1919: *Culex* (*Melanoconion*), Spissipes Section, Ocossa Group
oedipodeios Graham, 1909: *Eretmapodites*, Oedipodeios Group
oedipus Root, 1927c: *Culex* (*Melanoconion*), Melanoconion Section, Inhibitator Group, Inhibitator Subgroup
ohamai Tanaka, Mizusawa & Saugstad, 1975a: *Uranotaenia* (*Pseudoficalbia*), Recondita Series
oiketorakras Osorno-Mesa, 1947: *Anopheles* (*Lophopodomyia*)
okinawae Bohart, 1953: *Culex* (*Eumelanomyia*), Mochthogenes Group, Okinawae Subgroup
okinawanus Bohart, 1946: *Aedes* (*Bruceharrisonius*)
okinawensis Toma, Miyagi & Tanaka, 1990: *Toxorhynchites* (*Toxorhynchites*), Christophi Group
okuensis Brunhes, Le Goff & Geoffroy, 1997: *Anopheles* (*Christya*)
oligopistus Dyar, 1918a: *Aedes* (*Ochlerotatus*), Serratus Group
olimpioi Xavier, da Silva & da Silva Mattos, 1970a: *Culex* (*Melanoconion*), Melanoconion Section, Conspirator Group
omani Belkin, 1962: *Culex sensu stricto*, Sitiens Group (not placed in a Subgroup)
omissus (Edwards, 1914b): *Armigeres* (*Leicesteria*)
omorii Lien, 1968b: *Aedes* (*Downsiomyia*), Niveus Group
omorii Sakakibara, 1959: *Anopheles sensu stricto*, Angusticorn Section, Anopheles Series, Plumbeus Group
opok Corbet & van Someren, 1962: *Aedes* (*Stegomyia*), Africanus Group
orbitae Edwards, 1922a,b: *Aedes* (*Aedimorphus*), Domesticus Assemblage, Mediolineatus Subassemblage, Caecus Group
orbostiensis Dobrotworsky, 1958a: *Culex* (*Lophoceraomyia*), Fraudatrix Group (not placed in a Subgroup)
oreios Bangs & Harbach, 2014 (in Bangs *et al.*, 2015, first published online 23 December 2014): *Anopheles* (*Cellia*), Neomyzomyia Series, Punctulatus Group, Farauti Complex
oreophilus (Edwards, 1916): *Aedes* (subgenus uncertain)
oresbius Harbach & Rattanarithikul, 1988: *Culex* (*Eumelanomyia*), Mochthogenes Group, Tenuipalpis Subgroup
orfilai Duret, 1953: *Culex* (*Melanoconion*), Melanoconion Section, Inhibitator Group, Inhibitator Subgroup
organaboensis Talaga & Duchemin, 2023: *Culex* (*Melanoconion*), Melanoconion Section, Distinguendus Group, Distinguendus Subgroup
orientalis Barraud, 1926: *Uranotaenia* (*Uranotaenia*), Anisocheleomyia Section, Nivea Series
orientalis Edwards, 1921d: *Culex sensu stricto*, Sitiens Group, Mimeticus Subgroup, Mimulus Complex
originator Gordon & Evans, 1922: *Culex* (*Anoedioporpa*), Conservator Group
ornata Theobald, 1909: *Uranotaenia* (*Pseudoficalbia*), Nigripes Section, Ornata Group
ornatothoracis Theobald, 1909: *Culex sensu stricto*, Pipiens Group (not placed in a Subgroup)
orstom Brunhes, Adam & Bailly-Choumara, 1967: *Culex* (*Eumelanomyia*), Mochthogenes
orthodoxa Dyar, 1921c: *Uranotaenia* (*Uranotaenia*), Lowii Series
ortizi Vargas & Diaz Nájera, 1961: *Sabethes* (*Sabethes*), Group A
oryzalimnetes Wilkerson & Motoki, 2009 (in Motoki *et al.*, 2009): *Anopheles* (*Nyssorhynchus*), Argyritarsis Section, Albitarsis Series, Albitarsis Group, Albitarsis Complex
osornoi Berlin, 1969a: *Aedes* (*Howardina*), Sexlineatus Section, Sexlineatus Subassemblage
ostentatio (Leicester, 1908): *Aedes* (*Paraedes*), Ostentatio Assemblage
oswaldoi (Peryassú, 1922b): *Anopheles* (*Nyssorhynchus*), Albimanus Section, Oswaldoi Series, Oswaldoi Group, Oswaldoi Subgroup, Oswaldoi Complex (*An. oswaldoi* consists of three sibling species, the nominotypical species and two denoted as Species A and B) (Ruiz *et al.*, 2005, 2010; Ruiz-Lopez *et al.*, 2013)
otachati Klein & Sirivanakarn, 1970: *Culex* (*Eumelanomyia*), Mochthogenes Group, Otachati Subgroup
oteizai Perez Vigueras, 1956: *Uranotaenia* (*Uranotaenia*) (not placed in a Section, Series, Group or Complex)
ototomo da Cunha Ramos, 1993: *Uranotaenia* (*Pseudoficalbia*), Shillitonis Section, Shillitonis Group (not placed in a Subgroup)

ototomoensis Huang & Rueda, 2013: *Orthopodomyia*, Orthopodomyia Section, Ototomoensis Group (the Group is monobasic)
otwayensis (Dobrotworsky, 1960b): *Culiseta* (*Culicella*)
ousqua Dyar, 1918b: *Culex sensu stricto*, Pipiens Group, Coronator Complex (not placed in a Subgroup)
ovazzai Hamon & Adam, 1959b: *Aedes* (*Aedimorphus*), Domesticus Assemblage, Domesticus Subassemblage, Domesticus Group
ovengensis Awono-Ambene, Kengne, Simard, Antonio-Nkonkjio & Fontenille, 2004: *Anopheles* (*Cellia*), Neomyzomyia Series, Ardensis Group, Nili Complex
oweni Belkin, 1962: *Culex* (*Lophoceraomyia*), Fraudatrix Group, Fraudatrix Subgroup, Bergi Complex

Pp

pachyurus Edwards, 1936: *Aedes* (*Aedimorphus*), Domesticus Assemblage, Vexans Subassemblage, Quasiunivittatus Group
pacificus Edwards, 1916: *Culex sensu stricto*, Pipiens Group, Trifilatus Subgroup
pagei (Ludlow, 1911): *Aedes* (*Paraedes*), Ostentatio Assemblage
pahangi (Delfinado, 1968): *Verrallina* (*Harbachius*)
painei Edwards, 1935a: *Uranotaenia* (*Pseudoficalbia*), Bicolor Series
pairoji Sirivanakarn, 1977a: *Culex* (*Lophoceraomyia*), Fraudatrix Group, Fraudatrix Subgroup, Variatus Complex
pajoti da Cunha Ramos & Ribeiro, 1982: *Culex sensu stricto*, Guiarti Group
palaciosi Duret, 1968d: *Culex* (*Melanoconion*), Melanoconion Section, Pilosus Group, Caudelli Subgroup
palauensis Bohart, 1957: *Aedes* (*Stegomyia*), Scutellaris Group, Scutellaris Subgroup
palawanicus Mattingly, 1958a: *Aedes* (*Cancraedes*)
pallens Coquillett, 1898: *Culex sensu stricto*, Pipiens Group, Pipiens Subgroup
pallens Ross, 1943: *Aedes* (*Ochlerotatus*), Fulvus Group
pallescens Edwards, 1922d: *Psorophora sensu stricto*
pallida Edwards, 1925: *Mimomyia* (*Mimomyia*), Ludlowia Group
pallidiceps (Theobald, 1905f): *Culex* (*Acallyntrum*), Perkinsi Group
pallidipleura da Cunha Ramos & Brunhes, 2004: *Uranotaenia* (*Pseudoficalbia*), Shillitonis Section, Shillitonis Group (not placed in a Subgroup)
pallidiventer (Lutz, 1905): *Trichoprosopon*, Hyloconops Group, Pallidiventer Complex
pallidocephala Theobald, 1908c: *Uranotaenia* (*Uranotaenia*), Caeruleocephala Section, Caeruleocephala Group
pallidostriatus (Theobald, 1907): *Aedes* (*Aedimorphus*), Domesticus Assemblage, Mediolineatus Subassemblage, Mediolineatus Group
pallidothorax Dong, Dong & Wu, 2008: *Tripteroides* (*Rachionotomyia*), Aranoides Group, Aranoides Subgroup
pallidothorax Theobald, 1905g: *Culex* (*Culiciomyia*), Fragilis Group
pallidoventer Theobald, 1903a: *Uranotaenia* (*Uranotaenia*), Pallidoventer Series
pallidoventer (Theobald, 1907): *Wyeomyia* (*Phoniomyia*), Third Series
pallidus Lee, 1946: *Tripteroides* (*Rachisoura*), Vanleeuweni Group
pallidus Theobald, 1901b: *Anopheles* (*Cellia*), Neocellia Series, Annularis Group, Annularis Complex
pallirostris Edwards, 1922a: *Aedes* (*Hulecoeteomyia*)
pallithorax Dong, Zhou & Dong, 2004: *Armigeres sensu stricto*
palmarum Edwards, 1924b: *Aedes* (*Patmarksia*)
palmata (Lane & Cerqueira, 1942): *Wyeomyia* (*Phoniomyia*), Third Series
palmatus (Rodenwaldt, 1926): *Anopheles sensu stricto*, Angusticorn Section, Anopheles Series, Aitkenii Group
palmeirimi de Meillon & Rebêlo, 1941: *Uranotaenia* (*Uranotaenia*), Hopkinsi Section, Hopkinsi Group
palpalis (Newstead, 1907) (in Newstead *et al.*, 1907): *Aedes* (*Neomelaniconion*) (not placed in a Group)
palpalis Taylor, 1912: *Culex sensu stricto*, Sitiens Group, Sitiens Subgroup
paltrinierii Shidrawi & Gillies, 1988: *Anopheles* (*Cellia*), Neocellia Series (not placed in a Group)
paludis Theobald, 1900b: *Anopheles sensu stricto*, Laticorn Section, Myzorhynchus Series, Coustani Group
paludosa Galindo, Blanton & Peyton, 1954: *Uranotaenia* (*Uranotaenia*), Lowii Series
pampanai Büttiker & Beales, 1959: *Anopheles* (*Cellia*), Myzomyia Series, Funestus Group, Aconitus Subgroup
pampangensis (Ludlow, 1905a): *Aedes* (*Aedimorphus*), Domesticus Assemblage, Mediolineatus Subassemblage, Pampangensis Group
pampithes (Dyar & Núñez Tovar, 1928) (in Dyar, 1928): *Wyeomyia* (not placed in a Subgenus)
panamensis Adames, 1971: *Deinocerites*, Epitedeus Group
panarchys Dyar, 1921b: *Haemagogus sensu stricto*, Albomaculatus Section
panayensis (Ludlow, 1914c): *Verrallina* (*Neomacleaya*)
pandani Brunhes, 1969: *Culex* (*Culiciomyia*), Nebulosus Group
pandani Stone, 1939: *Aedes* (*Stegomyia*), Pandani Group
pandani (Theobald, 1912b): *Uranotaenia* (*Pseudoficalbia*), Bicolor Series
panocossa Dyar, 1923c: *Culex* (*Melanoconion*), Spissipes Section, Ocossa Group

papago Zavortink, 1970c: *Aedes* (*Abraedes*) (the Subgenus is monobasic)
papuae (Swellengrebel & Swellengrebel de Graaf, 1920): *Bironella* (*Neobironella*)
papuensis Dobrotworsky, 1957: *Anopheles sensu stricto*, Angusticorn Section, Anopheles Series, Stigmaticus Group
papuensis Marks, 1960: *Topomyia* (*Suaymyia*)
papuensis Peters, 1963a: *Armigeres sensu stricto*
papuensis (Taylor, 1914d): *Aedes* (*Patmarksia*)
papuensis (Taylor, 1914d): *Culex* (*Culiciomyia*), Fragilis Group
papuensis (Taylor, 1914d): *Mansonia* (*Mansonioides*)
papuensis Zavortink, 1968: *Orthopodomyia*, Orthopodomyia Section, Albipes Group, Anopheloides Subgroup
paracrybda Komp, 1936d: *Culex* (*Melanoconion*), Spissipes Section, Crybda Group, Paracrybda Subgroup
paraculeatus Sirivanakarn, 1977a: *Culex* (*Lophoceraomyia*), Fraudatrix Group, Fraudatrix Subgroup, Quadripalpis Complex
paradissimilis Rozeboom, 1946: *Aedes* (*Kenknightia*), Luzonensis Group
paraensis (Theobald, 1903a): *Limatus*
paraitepuyensis Anduze, 1941: *Sabethes* (*Sabethes*), Group A
paralateralis Peters, 1964: *Uranotaenia* (*Uranotaenia*), Anisocheleomyia Section, Ludlowae Series, Lateralis Complex
paramaxi Duret, 1968b: *Culex sensu stricto*, Pipiens Group, Tarsalis Subgroup
paranensis (Brèthes, 1910): *Isostomyia*
parangensis (Ludlow, 1914b): *Anopheles* (*Cellia*), Pyretophorus Series
paranovaguinensis Peters, 1963d: *Uranotaenia* (*Uranotaenia*), Anisocheleomyia Section, Tibialis Group, Novaguinensis Complex
parapunctipennis Martini, 1932: *Anopheles sensu stricto*, Angusticorn Section, Anopheles Series, Pseudopunctipennis Group
parasimilis (King & Hoogstraal, 1947b): *Verrallina sensu stricto*, Carmenti Series
parensis Gillies, 1962: *Anopheles* (*Cellia*), Myzomyia Series, Funestus Group, Funestus Subgroup
parenti (de Meillon & Lavoipierre, 1944): *Mimomyia* (*Mimomyia*), Boycia Group
particeps (Adams, 1903): *Culiseta sensu stricto*
parvipluma Edwards, 1941: *Eretmapodites*, Oedipodeios Group
parvus (Chagas, 1907): *Anopheles* (*Nyssorhynchus*), Myzorhynchella Section
patersoni Shannon & del Ponte, 1928: *Aedes* (*Ochlerotatus*), Scapularis Group, Infirmatus Subgroup
patriciae Mattingly, 1954a: *Aedes* (*Stegomyia*), Scutellaris Group, Albopictus Subgroup
patriciae Peyton, 1977: *Uranotaenia* (*Pseudoficalbia*), Bimaculata Series
pattoni Christophers, 1926: *Anopheles* (*Cellia*), Neocellia Series (not placed in a Group)
paucartamboensis Porter, 2014: *Wyeomyia* (*Nunezia*)
pauliani Doucet, 1949b: *Uranotaenia* (*Pseudoficalbia*), Nigripes Section, Nigripes Group
pauliani (Doucet, 1951b): *Toxorhynchites* (*Afrorhynchus*), Pauliani Group
pauliani Grjebine, 1950: *Eretmapodites*, Chrysogaster Group
pauliani Grjebine, 1953a: *Aedeomyia sensu stricto*
pauliani Grjebine, 1953a: *Anopheles* (*Cellia*), Neomyzomyia Series, Pauliani Group
paulli Paterson & Shannon, 1927: *Psorophora* (*Grabhamia*) (not placed in a Group)
paullusi Stone & Farner, 1945: *Aedes* (*Stegomyia*), Scutellaris Group, Scutellaris Subgroup
pavlovskyi Casal & García, 1967a: *Culex* (*Melanoconion*), Melanoconion Section, Inhibitator Group, Inhibitator Subgroup
peccator Dyar & Knab, 1909: *Culex* (*Melanoconion*), Melanoconion Section, Peccator Group
pecori Reinert, 1990b: *Aedes* (*Kenknightia*), Dissimilis Group
pectinatus Arnell, 1976: *Aedes* (*Ochlerotatus*), Scapularis Group, Pectinatus Subgroup (the Subgroup is monobasic)
pectinatus (Edwards, 1914a): *Armigeres* (*Leicesteria*)
pecuniosus Edwards, 1922d: *Aedes* (*Molpemyia*)
pedicellus King & Hoogstraal, 1947c: *Culex* (*Neoculex*), Crassistylus Group
peditaeniatus (Leicester, 1908): *Anopheles sensu stricto*, Laticorn Section, Myzorhynchus Series, Hyrcanus Group, Lesteri Subgroup
pedroi Sirivanakarn & Belkin, 1980: *Culex* (*Melanoconion*), Spissipes Section, Crybda Group, Pedroi Subgroup
peipingensis Feng, 1938: *Aedes* (subgenus uncertain)
pembaensis Theobald, 1901c: *Aedes* (*Skusea*)
penai Sirivanakarn, 1979: *Culex* (*Melanoconion*), Melanoconion Section, Intrincatus Group, Penai Subgroup (the Subgroup is monobasic)
pendleburyi (Edwards, 1930a): *Toxorhynchites* (*Toxorhynchites*), Splendens Group
pendulus (Edwards, 1914a): *Armigeres* (*Leicesteria*)
penghuensis Lien, 1968b: *Aedes* (*Cancraedes*)
penicillatus Edwards, 1941: *Eretmapodites*, Inornatus Group
pennai Antunes & Lane, 1938: *Aedes* (*Ochlerotatus*), Fulvus Group
perexiguus Theobald, 1903a: *Culex sensu stricto*, Pipiens Group, Univittatus Subgroup

pereyrai Duret, 1967a: *Culex* (*Melanoconion*), Spissipes Section, Crybda Group, Pereyrai Subgroup (the Subgroup is monobasic)
perfidiosus Edwards, 1914b: *Culex sensu stricto*, Pipiens Group (not placed in a Subgroup)
perfuscus Edwards, 1914b: *Culex sensu stricto*, Pipiens Group (not placed in a Subgroup)
peringueyi Edwards, 1924a: *Culex* (*Maillotia*), Seyrigi Group
periskelatus (Giles, 1902): *Aedes* (*Dendroskusea*)
perkinsi Marks, 1949: *Aedes* (*Ochlerotatus*), Perkinsi Group
perkinsi Stone & Penn, 1948: *Culex* (*Acallyntrum*), Perkinsi Group
pernotatus Farner & Bohart, 1944: *Aedes* (*Stegomyia*), Scutellaris Group, Scutellaris Subgroup
perplexens (Edwards, 1932b): *Mimomyia* (*Mimomyia*), Ludlowia Group
perplexens Ludlow, 1907: *Anopheles sensu stricto*, Angusticorn Section, Anopheles Series, Punctipennis Group
perplexus (Leicester, 1908): *Aedes* (*Stegomyia*), W-albus Group, Mediopunctatus Subgroup
perplexus Leicester, 1908: *Culex sensu stricto*, Sitiens Group, Vishnui Subgroup
perplexus Peters, 1963c: *Tripteroides* (*Polylepidomyia*), ?Argenteiventris Group
perryi Belkin, 1962: *Aedes* (*Geoskusea*), Daggyi Group
perryi Belkin, 1962: *Culex* (*Lophoceraomyia*), Fraudatrix Group, Fraudatrix Subgroup, Hurlbuti Complex
persiensis Linton, Sedaghat & Harbach, 2003 (in Sedaghat *et al.*, 2003): *Anopheles sensu stricto*, Angusticorn Section, Anopheles Series, Maculipennis Group, Maculipennis Subgroup
persimilis Mattingly, 1970: *Heizmannia sensu stricto*
personatum (Lutz, 1904) (in Bourroul, 1904): *Onirion*, Personatum Group
pertinans (Williston, 1896): *Wyeomyia* (*Wyeomyia*), Pertinans Group
pertinax Grabham, 1906: *Aedes* (*Ochlerotatus*), Serratus Group
perturbans (Walker, 1856): *Coquillettidia* (*Coquillettidia*), Perturbans Group
perturbans (Williston, 1896): *Isostomyia*
perventor Cerqueira & Costa, 1946: *Aedes* (*Sallumia*)
pervigilans Bergroth, 1889: *Culex sensu stricto*, Pipiens Group, Trifilatus Subgroup
peryassui Dyar & Knab, 1908: *Anopheles sensu stricto*, Laticorn Section, Arribalzagia Series
pessoai (Barreto & Coutinho, 1944): *Mansonia* (*Mansonia*)
petersi Colless, 1960: *Culex* (*Lophoceraomyia*), Fraudatrix Group, Fraudatrix Subgroup, Petersi Complex
petragnani del Vecchio, 1939: *Anopheles sensu stricto*, Angusticorn Section, Anopheles Series, Claviger Complex (not placed in a Group)
petrocchiae (Shannon & del Ponte, 1928): *Sabethes* (*Davismyia*) (the Subgenus is monobasic)
petroelephantus (Wijesundara, 1951): *Verrallina* (*Neomacleaya*)
pexus Colless, 1958: *Aedes* (*Downsiomyia*), Niveus Group
peytoni Bram & Rattanarithikul, 1967: *Culex* (*Lophoceraomyia*), Mammilifer Group, Minor Line, Peytoni Complex
peytoni Kulasekera, Harrison & Amerasinghe, 1989: *Anopheles sensu stricto*, Angusticorn Section, Anopheles Series, Aitkenii Group
peytoni Leguizamón & Carpintero, 2004: *Orthopodomyia*, Bancroftia Section (not placed in a Group)
peytoni Reinert, 1972a: *Aedes* (*Ayurakitia*)
phaecasiatus Marks, 1964a: *Aedes* (*Ochlerotatus*) (not placed in a Group)
phaeonotus Arnell, 1976: *Aedes* (*Ochlerotatus*), Scapularis Group, Scapularis Subgroup
phangngae Sirivanakarn, 1972: *Culex* (*Eumelanomyia*), Protomelanoconion Group
pharoensis Theobald, 1901b: *Anopheles* (*Cellia*), Cellia Series (not placed in a Group)
philipi Edwards, 1929b: *Culex sensu stricto*, Pipiens Group (not placed in a Subgroup)
philippinensis (Delfinado, 1968): *Verrallina* (*Neomacleaya*)
philippinensis Ludlow, 1902b: *Anopheles* (*Cellia*), Neocellia Series, Annularis Group, Annularis Complex
philippinensis Sirivanakarn, 1976: *Culex sensu stricto*, Sitiens Group, Vishnui Subgroup
phillipi van Someren, 1949a: *Aedes* (*Zavortinkius*), Brygooi Group
philonuxia Philip, 1931: *Uranotaenia* (*Uranotaenia*), Caeruleocephala Section, Caeruleocephala Group
phlabistus Dyar, 1920b: *Culex* (*Melanoconion*), Melanoconion Section, Distinguendus Group, Putumayensis Subgroup
phlogistus Dyar, 1920b: *Culex* (*Melanoconion*), Melanoconion Section, Inhibitator Group, Inhibitator Subgroup
phnoma (Klein, 1973): *Verrallina* (*Neomacleaya*)
phoeniciae Coluzzi & Sabatini, 1968: *Aedes* (*Acartomyia*)
pholeocola (Linton & Harbach, 2013) (in Harbach *et al.*, 2013): *Aedes* (*Nyctomyia*)
pholeter Bram & Rattanarithikul, 1967: *Culex* (*Lophoceraomyia*), Mammilifer Group, Minor Line
pholidotus Zavortink, 1973: *Anopheles* (*Kerteszia*), Pholidotus Complex (*An. pholidotus* consists of four sibling species denoted as Species 1–4 (Bourke *et al.*, 2023); which is the nominotypical species is not known)
phroso Howard, Dyar & Knab, 1915: *Wyeomyia* (not placed in a Subgenus)
phyllados Hutchings & Sallum, 2008: *Culex* (*Melanoconion*), Melanoconion Section, Bastagarius Group, Bastagarius Subgroup

phyllolabis Edwards, 1929b: *Aedes* (*Catageiomyia*)
phyllozoa (Dyar & Knab, 1907c): *Orthopodomyia*, Folicolae Section
phytophagus Theobald, 1909: *Toxorhynchites* (*Toxorhynchites*), Brevipalpis Group, Phytophagus Subgroup
pia (Le Goff & Robert, 2013) (in Le Goff *et al.*, 2013): *Aedes* (*Stegomyia*), Aegypti Group
pictipennis (Philippi, 1865): *Anopheles* (*Nyssorhynchus*), Argyritarsis Section, Argyritarsis Series, Pictipennis Group
pifanoi Anduze, 1949a: *Culex* (*Melanoconion*), Melanoconion Section, Intrincatus Group, Intrincatus Subgroup
pifanoi Cova Garcia, Pulido F. & Escalante de Ugueto, 1981: *Uranotaenia* (*Uranotaenia*) (not placed in a Section, Series, Group or Complex)
pilicauda Root, 1928 (in Dyar, 1928): *Wyeomyia* (*Phoniomyia*), Second Series
pilifemoralis Wang & Feng, 1964: *Culex* (*Lophoceraomyia*), Wilfredi Group
pilinotum Harrison & Scanlon, 1974: *Anopheles sensu stricto*, Angusticorn Section, Anopheles Series, Aitkenii Group
pilipes (Macquart, 1834): *Psorophora sensu stricto*
pillaii Mattingly, 1958a: *Aedes* (*Rhinoskusea*)
pilosa Brug, 1931c: *Topomyia* (*Topomyia*)
pilosa da Cunha Ramos & Brunhes, 2004: *Uranotaenia* (*Pseudoficalbia*), Spinosa Section, Spinosa Group
pilosa Duret, 1971: *Psorophora* (*Janthinosoma*)
pilosus (Dyar & Knab, 1906d): *Culex* (*Melanoconion*), Melanoconion Section, Pilosus Group, Pilosus Subgroup
pilosus Lee, 1946: *Tripteroides* (*Rachisoura*), Vanleeuweni Group
pinarocampa Dyar & Knab, 1908: *Culex sensu stricto*, Pipiens Group, Tarsalis Subgroup
pingpaensis Chang, 1965: *Aedes* (*Edwardsaedes*)
pinjaurensis Barraud, 1932: *Anopheles sensu stricto*, Angusticorn Section, Anopheles Series, Aitkenii Group
pionips Dyar, 1919a: *Aedes* (*Ochlerotatus*), Impiger Group
pipersalatus (Giles, 1902): *Aedes* (*Aedimorphus*), Domesticus Assemblage, Mediolineatus Subassemblage, Taeniorhynchoides Group
pipiens Linnaeus, 1758: *Culex sensu stricto*, Pipiens Group, Pipiens Subgroup
pipkini (Bohart, 1957): *Verrallina sensu stricto*, Carmenti Series
plagosus Marks, 1959b: *Aedes* (subgenus uncertain)
platylepidus Knight & Hull, 1951: *Aedes* (*Scutomyia*), Hoogstraali Group
pleccau Koidzumi, 1924: *Anopheles sensu stricto*, Angusticorn Section, Anopheles Series, Lindesayi Group, Lindesayi Subgroup, Lindesayi Complex
plectoporpe Root, 1927c: *Culex* (*Melanoconion*), Melanoconion Section, Inhibitator Group, Inhibitator Subgroup
pleuristriatus Theobald, 1903a: *Culex* (*Microculex*), Pleuristriatus Series
plicatus Olivares, 1993 (in Angulo & Olivares, 1993): *Culex sensu stricto*, Pipiens Group (not placed in a Subgroup)
plioleucus Edwards, 1941: *Eretmapodites*, Plioleucus Group
plumbeus Stephens, 1828: *Anopheles sensu stricto*, Angusticorn Section, Anopheles Series, Plumbeus Group
plumiferus King & Hoogstraal, 1946f: *Aedes* (*Phagomyia*)
plumiger Bonne-Wepster, 1948: *Tripteroides* (*Rachisoura*), Vanleeuweni Group
plumosa (Theobald, 1901b): *Mimomyia* (*Mimomyia*), Megaculex Group (the Group is monobasic)
plumosus (Brug, 1931c): *Tripteroides* (*Tripteroides*), Nitidoventer Group
pluvialis Barraud, 1924f: *Culex* (*Eumelanomyia*), Mochthogenes Group, Pluvialis Subgroup
podographicus Dyar & Knab, 1906b: *Aedes* (*Protomacleaya*), Terrens Group, Podographicus Subgroup (the Subgroup is monobasic)
pogonurus Edwards, 1936: *Aedes* (*Neomelaniconion*) (not placed in a Group)
poicilipes (Theobald, 1903a): *Culex* (*Lasioconops*) (the Subgenus is monobasic)
poicilius (Theobald, 1903a): *Aedes* (*Finlaya*), Poicilius Group
pollicaris Reid, 1962: *Anopheles sensu stricto*, Laticorn Section, Myzorhynchus Series, Barbirostris Group, Barbirostris Subgroup
pollinctor (Graham, 1910): *Aedes* (*Zavortinkius*), Longipalpis Group
polynesiensis Marks, 1951: *Aedes* (*Stegomyia*), Scutellaris Group, Scutellaris Subgroup, Pseudoscutellaris Subgroup/Complex
ponmeki Miyagi & Toma, 2001: *Tripteroides* (*Rachionotomyia*), Edwardsi Group
portesi Senevet & Abonnenc, 1941: *Culex* (*Melanoconion*), Spissipes Section, Vomerifer Group
portonovoensis Tewari & Hiriyan, 1992a: *Aedes* (*Rhinoskusea*)
portoricensis (von Röder, 1885): *Toxorhynchites* (*Lynchiella*)
postspiraculosus Dobrotworsky, 1961a: *Aedes* (*Pseudoskusea*)
postspiraculosus Lee, 1944b: *Culex* (*Neoculex*), Pseudomelanoconia Group
powderi Zavortink, 1970b: *Anopheles sensu stricto*, Angusticorn Section, Anopheles Series, Plumbeus Group
powelli Lee, 1944a: *Anopheles sensu stricto*, Angusticorn Section, Anopheles Series, Stigmaticus Group
powelli (Ludlow, 1909c): *Tripteroides* (*Tripteroides*), Nitidoventer Group
poweri (Theobald, 1905g): *Aedes* (*Stegomyia*), Poweri Group

prachongae Rattanarithikul & Harrison, 2017 (in Harbach *et al.*, 2017b): *Anopheles sensu stricto*, Angusticorn Section, Anopheles Series, Lindesayi Group, Gigas Subgroup, Gigas Complex
prajimi Peyton & Rattanarithikul, 1970: *Uranotaenia* (*Uranotaenia*), Testacea Series
pretoriensis (Theobald, 1903a): *Anopheles* (*Cellia*), Neocellia Series (not placed in a Group)
principensis da Cunha Ramos, 1993: *Uranotaenia* (*Pseudoficalbia*), Shillitonis Section, Shillitonis Group, Subgroup A
prioekanensis (Brug, 1931a): *Verrallina* (*Neomacleaya*)
pristinus Nagaki & Sallum, 2010 (in Nagaki *et al.*, 2010): *Anopheles* (*Nyssorhynchus*), Myzorhynchella Section
procax (Skuse, 1889): *Aedes* (*Ochlerotatus*), Taeniorhynchus Group, Vigilax Subgroup
productus Edwards, 1941: *Eretmapodites*, Leucopous Group
productus Senevet & Abonnenc, 1939a: *Culex* (*Melanoconion*), Melanoconion Section, Distinguendus Group, Distinguendus Subgroup
prominens (Barraud, 1923c): *Aedes* (*Phagomyia*)
propinqua Mattingly, 1970: *Heizmannia sensu stricto*
propinqua Peyton, 1977: *Uranotaenia* (*Pseudoficalbia*), Bimaculata Series
propinquus Colless, 1955b: *Culex sensu stricto*, Sitiens Group, Mimeticus Subgroup, Mimulus Complex
prosecutor Séguy, 1927: *Culex sensu stricto*, Pipiens Group (not placed in a Subgroup)
protolepis Cockerell, 1916: *Aedes* (*Ochlerotatus*) (fossil species)
protorhinus Cockerell, 1916: *Culex* (fossil species)
protuberans (Delfinado, 1967): *Verrallina* (*Neomacleaya*)
provocans (Walker, 1848): *Aedes* (*Rusticoidus*)
proxima Mattingly, 1970: *Heizmannia sensu stricto*
proximus (Edwards, 1915b): *Tripteroides* (*Tripteroides*), Nitidoventer Group
pruina Theobald, 1901d: *Culex sensu stricto*, Pipiens Group (not placed in a Subgroup)
psatharus Dyar, 1920e: *Culex* (*Melanoconion*), Melanoconion Section, Erraticus Group, Psatharus Subgroup (the Subgroup is monobasic)
psectropus Edwards, 1930a: *Hodgesia*
pseudalbolineatus Brug, 1939: *Aedes* (*Scutomyia*), Laffooni Group
pseudalbopictus (Borel, 1928): *Aedes* (*Stegomyia*), Scutellaris Group, Albopictus Subgroup
pseudes Dyar & Knab, 1909: *Deinocerites*, Pseudes Group
pseudoafricanus Chwatt, 1949: *Aedes* (*Stegomyia*), Africanus Group
pseudoalbimanus da Cunha Ramos & Brunhes, 2004: *Uranotaenia* (*Pseudoficalbia*), Annulata Section, Annulata Series, Lavieri Group
pseudoalbipes Duret, 1971: *Psorophora* (*Janthinosoma*)
pseudoandreanus Bailly-Choumara, 1966a: *Culex* (*Eumelanomyia*), Rubinotus-rima Group, Rubinotus Subgroup
pseudoauriceps Miyagi & Toma, 2010a: *Topomyia* (*Suaymyia*)
pseudobarbirostris Ludlow, 1902a: *Anopheles sensu stricto*, Laticorn Section, Myzorhynchus Series, Bancroftii Group
pseudobarbus Baisas, 1946c: *Topomyia* (*Topomyia*)
pseudocinereus Theobald, 1901c: *Culex* (*Culiciomyia*), Nebulosus Group
pseudoconopas (Theobald, 1910c): *Coquillettidia* (*Coquillettidia*), Crassipes Group
pseudodiurna (Theobald, 1910a): *Verrallina* (*Neomacleaya*)
pseudodominicii Komp, 1936b: *Aedes* (*Howardina*), Sexlineatus Section, Sexlineatus Subassemblage
pseudohenrardi Peters, 1955c: *Uranotaenia* (*Pseudoficalbia*), Shillitonis Section, Shillitonis Group (not placed in a Subgroup)
pseudojamesi Strickland & Chowdhury, 1927b: *Anopheles* (*Cellia*), Neocellia Series, Jamesii Group
pseudoleucotarsis Thurman, 1959: *Topomyia* (*Suaymyia*)
pseudomaculipes (Chagas, 1908) (in Peryassú, 1908): *Anopheles sensu stricto*, Laticorn Section, Arribalzagia Series
pseudomaculipleura Peyton & Rattanarithikul, 1970: *Uranotaenia* (*Pseudoficalbia*), Bimaculata Series
pseudomediofasciata (Theobald, 1910c): *Verrallina* (*Neomacleaya*)
pseudomelanoconia Theobald, 1907: *Culex* (*Neoculex*), Pseudomelanoconia Group
pseudomelanota Barata & Cotrim, 1971: *Psorophora* (*Janthinosoma*)
pseudomethysticus (Bonne-Wepster & Bonne, 1920): *Limatus*
pseudonigeria (Theobald, 1910c): *Aedes* (*Stegomyia*), Pseudonigeria Group
pseudoniveus (Theobald, 1905f): *Aedes* (*Downsiomyia*), Niveus Group
pseudonormanensis Marks, 1949: *Aedes* (*Ochlerotatus*), Taeniorhynchus Group, Theobaldi Subgroup
pseudonummatus Reinert, 1973e: *Aedes* (*Tewarius*), Nummatus Group
pseudopecten Dyar & Knab, 1906a: *Wyeomyia* (*Decamyia*)
pseudopictus Grassi, 1899 (in Grassi *et al.*, 1899): *Anopheles sensu stricto*, Laticorn Section, Myzorhynchus Series, Hyrcanus Group (not placed in a Subgroup)
pseudopruina van Someren, 1951: *Culex sensu stricto*, Pipiens Group (not placed in a Subgroup)

pseudopunctipennis Theobald, 1901c: *Anopheles sensu stricto*, Angusticorn Section, Anopheles Series, Pseudopunctipennis Group
pseudornatus Colless, 1960: *Culex* (*Lophoceraomyia*), Fraudatrix Group, Fraudatrix Subgroup, Christiani Complex
pseudorobusta Pajot & Fauran, 1975: *Wyeomyia* (*Wyeomyia*) (not placed in a Group)
pseudorubithoracis Sirivanakarn, 1968: *Culex* (*Lophoceraomyia*), Fraudatrix Group, Fraudatrix Subgroup, Pseudorubithoracis Complex
pseudoscutellaris (Theobald, 1910b): *Aedes* (*Stegomyia*), Scutellaris Group, Scutellaris Subgroup, Pseudoscutellaris Subgroup/Complex
pseudoshillitonis da Cunha Ramos & Brunhes, 2004: *Uranotaenia* (*Pseudoficalbia*), Shillitonis Section, Shillitonis Group (not placed in a Subgroup)
pseudosinensis Baisas, 1935a: *Anopheles sensu stricto*, Laticorn Section, Myzorhynchus Series, Hyrcanus Group, Nigerrimus Subgroup
pseudosinensis Colless, 1955b: *Culex* (*Oculeomyia*), Bitaeniorhynchus Complex
pseudostigmaticus Dobrotworsky, 1957: *Anopheles sensu stricto*, Angusticorn Section, Anopheles Series, Stigmaticus Group
pseudostigmatosoma Strickman, 1990: *Culex sensu stricto*, Pipiens Group, Tarsalis Subgroup
pseudostricklandi Natarajan, Rajavel & Jambulingam, 2018: *Uranotaenia* (*Pseudoficalbia*), Recondita Series
pseudosubaequalis Cornel & Mayi, 2020 (in Cornel *et al.*, 2020): *Culex* (*Culiciomyia*), Nebulosus Group
pseudosundaicus Tyagi, Hiriyan, Tewari, Ayanar, Samuel, Arunachalam, Paramasivan, Krishnamoorthy, Dhananjeyan, Leo & Rajendran, 2009: *Anopheles* (*Cellia*), Pyretophorus Series
pseudotaeniatus (Giles, 1901a): *Aedes* (*Collessius*), Alloeomyia Group
pseudotarsalis van Someren, 1946a: *Aedes* (*Catageiomyia*)
pseudotibiamaculatus Galvão & Barretto, 1941: *Anopheles* (*Lophopodomyia*)
pseudotitillans (Theobald, 1901c): *Mansonia* (*Mansonia*)
pseudovarietas (Reinert, 1974): *Verrallina* (*Neomacleaya*)
pseudovishnui Colless, 1957a: *Culex sensu stricto*, Sitiens Group, Vishnui Subgroup
pseudowillmori (Theobald, 1910c): *Anopheles* (*Cellia*), Neocellia Series, Maculatus Group (not placed in a Subgroup)
pubescens Edwards, 1925: *Aedes* (*Aedimorphus*), Domesticus Assemblage, Vexans Subassemblage, Quasiunivittatus Group
puehensis Miyagi, Toma & Okazawa, 2022 (in Miyagi *et al.*, 2022): *Topomyia* (*Suaymyia*)
pujutensis Colless, 1948: *Anopheles* (*Cellia*), Neomyzomyia Series, Leucosphyrus Group, Hackeri Subgroup
pulcherrima Lynch Arribálzaga, 1891b: *Uranotaenia* (*Uranotaenia*), Pulcherrima Series
pulcherrimus Theobald, 1902a: *Anopheles* (*Cellia*), Neocellia Series (not placed in a Group)
pulchrithorax Edwards, 1939a: *Aedes* (*Vansomerenis*)
pulchriventer (Giles, 1901a): *Aedes* (*Gilesius*)
pulcripalpis (Rondani, 1872): *Orthopodomyia*, Bancroftia Section, Signifera Group, Pulcripalpis Subgroup (the Subgroup is monobasic)
pulcritarsis (Rondani, 1872): *Aedes* (*Ochlerotatus*), Caspius Group
pulidoi Cova Garcia & Sutil Oramas, 1974: *Culex* (*Microculex*) (not placed in a Series)
pullatus (Coquillett, 1904c): *Aedes* (*Ochlerotatus*), Pullatus Group
pullus Theobald, 1905f: *Culex* (*Culiciomyia*), Fragilis Group
pullus Yamada, 1937: *Anopheles sensu stricto*, Laticorn Section, Myzorhynchus Series, Hyrcanus Group (not placed in a Subgroup)
pulverulentus Edwards, 1922a: *Aedes* (*Indusius*) (the Subgenus is monobasic)
punctifemoris (Ludlow, 1921): *Aedes* (*Aedimorphus*), Domesticus Assemblage, Mediolineatus Subassemblage, Punctifemoris Group (the Group is monobasic)
punctimacula Dyar & Knab, 1906a: *Anopheles sensu stricto*, Laticorn Section, Arribalzagia Series
punctipennis (Say, 1823): *Anopheles sensu stricto*, Angusticorn Section, Anopheles Series, Punctipennis Group
punctipes Edwards, 1921a: *Aedes* (*Petermattinglyius*), Petermattinglyius Group
punctiscapularis Floch & Abonnenc, 1946b: *Culex* (subgenus uncertain)
punctocostalis (Theobald, 1909): *Aedes* (*Neomelaniconion*) (not placed in a Group)
punctodes Dyar, 1922a: *Aedes* (*Ochlerotatus*), Communis Group, Punctor Subgroup
punctolateralis (Theobald, 1903e): *Tripteroides* (*Polylepidomyia*), Atripes Group
punctor (Kirby, 1837): *Aedes* (*Ochlerotatus*), Communis Group, Punctor Subgroup
punctothoracis (Theobald, 1909): *Aedes* (*Catageiomyia*)
punctulatus Dönitz, 1901: *Anopheles* (*Cellia*), Neomyzomyia Series, Punctulatus Group (the Group includes an unnamed species near *punctulatus*, see Foley *et al.*, 1995)
purpuraceus Brug, 1932: *Aedes* (*Ochlerotatus*), Aculeatus Group
purpuratus (Edwards, 1921a): *Tripteroides* (*Tripteroides*), Purpuratus Group (the Group is monobasic)
purpureifemur Marks, 1959a: *Aedes* (*Ochlerotatus*) (not placed in a Group)
purpureipes Aitken, 1941: *Aedes* (*Kompia*) (the Subgenus is monobasic)

purpureus (Theobald, 1901b): *Toxorhynchites* (*Ankylorhynchus*)
purpureus (Theobald, 1907): *Sabethes* (*Sabethes*), Group B
purpureus (Theobald, 1910c): *Aedes* (*Molpemyia*)
purpuriventris Edwards, 1926a: *Aedes* (*Ochlerotatus*), Flavifrons Group
pursati Laveran, 1902a: *Anopheles sensu stricto*, Laticorn Section, Myzorhynchus Series, Hyrcanus Group, Nigerrimus Subgroup
pusillus (da Costa Lima, 1931b): *Toxorhynchites* (*Lynchiella*)
pusillus Macquart, 1850: *Culex* (*Barraudius*)
putumayensis Matheson, 1934: *Culex* (*Melanoconion*), Melanoconion Section, Distinguendus Group, Putumayensis Subgroup
pygmaea Theobald, 1901c: *Uranotaenia* (*Uranotaenia*) (not placed in a Section, Series, Group or Complex)
pygmaea (Theobald, 1903a): *Psorophora* (*Grabhamia*), Confinnis Group
pylei Baisas, 1946a: *Uranotaenia* (*Pseudoficalbia*), Recondita Series

Qq

quadriannulatus (Theobald, 1911b): *Anopheles* (*Cellia*), Pyretophorus Series, Gambiae Complex
quadrifoliatus Komp, 1936d: *Culex* (*Melanoconion*), Melanoconion Section, Intrincatus Group, Intrincatus Subgroup
quadrifolium (Brug, 1934): *Verrallina sensu stricto*, Carmenti Series
quadrimaculata Edwards, 1929 (in Paine & Edwards, 1929): *Uranotaenia* (*Pseudoficalbia*), Bicolor Series
quadrimaculatus Say, 1824 (in Keating, 1824): *Anopheles sensu stricto*, Angusticorn Section, Anopheles Series, Maculipennis Group, Quadrimaculatus Subgroup
quadripalpis (Edwards, 1914b): *Culex* (*Lophoceraomyia*), Fraudatrix Group, Fraudatrix Subgroup, Quadripalpis Complex
quadripunctis (Ludlow, 1910) (in Theobald, 1910c): *Aedes* (*Mucidus*), Pardomyia Group
quadrispinata (King & Hoogstraal, 1947b): *Verrallina sensu stricto*, Carmenti Series
quadrivittatus (Coquillett, 1902d): *Aedes* (*Howardina*), Sexlineatus Section, Sexlineatus Subassemblage
quasicyaneus Peryassú, 1922b: *Sabethes* (*Sabethes*), Group B
quasiferinus Mattingly, 1961: *Aedes* (*Mucidus*), Mucidus Group
quasiferox (Leicester, 1908): *Toxorhynchites* (*Toxorhynchites*), Christophi Group
quasiguiarti Theobald, 1910c: *Culex sensu stricto*, Pipiens Group (not placed in a Subgroup)
quasihibridus Galindo & Blanton, 1954: *Culex* (*Melanoconion*), Melanoconion Section, Bastagarius Group, Iolambdis Subgroup
quasilongirostris (Theobald, 1907): *Wyeomyia* (*Phoniomyia*), First Series
quasimodesta Peyton, 1977: *Uranotaenia* (*Pseudoficalbia*), Bimaculata Series
quasioriginator Duret, 1972: *Culex* (*Anoedioporpa*), Conservator Group
quasiornatus (Taylor, 1915): *Tripteroides* (*Tripteroides*), Bimaculipes Group
quasirubithorax (Theobald, 1910c): *Aedes* (subgenus uncertain)
quasirusticus Torres Cañamares, 1951: *Aedes* (*Rusticoidus*)
quasisanguinae Leicester, 1908: *Hodgesia*
quasiscutellaris Farner & Bohart, 1944: *Aedes* (*Stegomyia*), Scutellaris Group, Scutellaris Subgroup
quasiunivittatus (Theobald, 1901c): *Aedes* (*Aedimorphus*), Domesticus Assemblage, Vexans Subassemblage, Quasiunivittatus Group
quettensis Mattingly, 1955: *Culex* (*Maillotia*), Hortensis Group
qui Dong, Dong & Zhou, 2003 (in Dong *et al.*, 2003b): *Uranotaenia* (*Pseudoficalbia*), Bimaculata Series
quinquefasciatus Say, 1823: *Culex sensu stricto*, Pipiens Group, Pipiens Subgroup
quinquelineatus Edwards, 1922d: *Aedes* (*Rampamyia*)
quinquemaculata Bonne-Wepster, 1934: *Uranotaenia* (*Pseudoficalbia*), Bimaculata Series
quinquevittatus Theobald, 1901d: *Eretmapodites*, Quinquevittatus Group
quintetti Brunhes, Adam & Bailly-Choumara, 1967: *Culex* (*Eumelanomyia*), Mochthogenes Group, Inconspicuosus Subgroup
quitensis Levi-Castillo, 1953a: *Culex sensu stricto* (not placed in a Group)

Rr

rabanicolus Floch & Abonnenc, 1946a: *Culex* (*Melanoconion*), Melanoconion Section, Intrincatus Group, Intrincatus Subgroup
rabelloi Forattini & Sallum, 1987a: *Culex* (*Melanoconion*), Melanoconion Section, Inhibitator Group, Inhibitator Subgroup
rachoui Duret, 1968b: *Culex* (*Melanoconion*), Melanoconion Section, Educator Group
rachoui Galvão, 1952: *Anopheles sensu stricto*, Laticorn Section, Arribalzagia Series
rachoui Xavier, da Silva Mattos & da Silva, 1970b: *Uranotaenia* (*Uranotaenia*), Pulcherrima Series
radama de Meillon, 1943: *Anopheles* (*Cellia*), Neomyzomyia Series, Pauliani Group
rageaui (Hamon & Rickenbach, 1955b): *Culex* (*Eumelanomyia*), Protomelanoconion Group
rageaui Mattingly & Adam, 1954: *Anopheles* (*Cellia*), Neomyzomyia Series, Smithii Group
raghavanii Rahman, Chowdhury & Kalra, 1969: *Culex* (*Lophoceraomyia*), Mammilifer Group, Mammilifer Subgroup, Flavicornis Complex

rajah Tsukamoto, 1989: *Culex* (*Culiciomyia*), Shebbearei Group
rajah Tsukamoto, 1989: *Toxorhynchites* (*Toxorhynchites*) (not placed in a Group)
rajaneeae Sirivanakarn, 1968: *Culex* (*Lophoceraomyia*), Fraudatrix Group, Fraudatrix Subgroup, Fraudatrix Complex
rajagopalani Natarajan, Eapen & Jambulingam, 2020: *Heizmannia sensu stricto*
rajaonariveloi Brunhes & Hervy, 1995: *Orthopodomyia*, Orthopodomyia Section, Vernoni Group
rajaveli (Natarajan, Gopalakrishnan, Kumar & Kumar, 2022): *Aedes* (*Downsiomyia*), Alboniveus Group
ramachandrai Reuben, 1967: *Aedes* (*Dendroskusea*)
ramakrishnii Wattal & Kalra, 1965: *Culex* (*Culiciomyia*), Fragilis Group
ramalai Grjebine, 1986: *Mimomyia* (*Ingramia*), Group A (*Ravenalites s. str.*), Spinosa Complex
ramalingami Evenhuis & Steffan, 1986: *Toxorhynchites* (*Toxorhynchites*), Acaudatus Group, Acaudatus Subgroup
ramalingami (Reinert, 1974): *Verrallina* (*Harbachius*)
ramalingami Sirivanakarn, 1973b: *Culex* (*Culiciomyia*), Fragilis Group
rami (Barraud, 1928): *Verrallina* (*Neomacleaya*)
ramirezi Vargas & Downs, 1950: *Aedes* (*Aztecaedes*) (the Subgenus is monobasic)
ramosa da Cunha Ramos, 1993: *Uranotaenia* (*Pseudoficalbia*), Nigripes Section, Ornata Group
rampae Harbach & Somboon, 2011 (in Somboon *et al.*, 2011): *Anopheles* (*Cellia*), Neocellia Series, Maculatus Group, Sawadwongporni Subgroup
rampae Peyton & Klein, 1970: *Uranotaenia* (*Uranotaenia*), Campestris Series
ranci Grjebine, 1953a: *Anopheles* (*Cellia*), Neomyzomyia Series, Ranci Group, Ranci Subgroup (the Subgroup is monobasic)
rangeli Gabaldon, Cova-Garcia & Lopez, 1940: *Anopheles* (*Nyssorhynchus*), Albimanus Section, Oswaldoi Series, Oswaldoi Group, Oswaldoi Subgroup, Oswaldoi Complex
rapax (Dyar & Knab, 1906a): *Runchomyia* (*Runchomyia*)
rara (Delfinado, 1968): *Verrallina* (*Neomacleaya*)
raris (Leicester, 1908): *Toxorhynchites* (*Toxorhynchites*), Magnificus Group
ratcliffei Marks, 1959a: *Aedes* (*Ochlerotatus*), Perkinsi Group
rausseoi Cova Garcia, Sutil O. & Pulido F., 1972: *Culex* (*Carrollia*), Bihaicola Group
ravaonjanaharyi Brunhes & Hervy, 1995: *Orthopodomyia*, Orthopodomyia Section, Vernoni Group
ravenalaphila Boussès & Brunhes, 2013: *Uranotaenia* (*Pseudoficalbia*), Annulata Section, Annulata Series, Lavieri Group
ravenalicola da Cunha Ramos & Brunhes, 2004: *Uranotaenia* (*Pseudoficalbia*), Nigripes Section, Nigripes Group
ravissei Rickenbach & Eouzan, 1970: *Eretmapodites*, Plioleucus Group
raymondi del Ponte, Castro & Garcia, 1951: *Aedes* (*Ochlerotatus*), Scapularis Group, Infirmatus Subgroup
reali Hamon & Adam, 1959a: *Aedes* (*Catageiomyia*)
recens Sallum & Peyton, 2005 (in Sallum *et al.*, 2005b): *Anopheles* (*Cellia*), Neomyzomyia Series, Leucosphyrus Group, Hackeri Subgroup
recondita Edwards, 1922d: *Uranotaenia* (*Pseudoficalbia*), Recondita Series
reducens Lane & Whitman, 1951: *Culex* (*Microculex*), Consolator Series
reesi (King & Hoogstraal, 1947b): *Verrallina sensu stricto*, Carmenti Series
reevesi Wirth, 1948: *Culex* (*Neoculex*), Territans Group
refiki Medschid, 1928: *Aedes* (*Rusticoidus*)
refutans Alcock, 1913: *Anopheles sensu stricto*, Angusticorn Section, Anopheles Series, Lindesayi Group, Gigas Subgroup, Gigas Complex
regale Peyton & Harbach, 2000 (in Harbach & Peyton, 2000): *Onirion*, Personatum Group
regalis Dyar & Knab, 1906b: *Haemagogus sensu stricto*, Splendens Section
reginae Edwards, 1922a: *Aedes* (*Dendroskusea*)
reidi Colless, 1965: *Culex* (*Lophoceraomyia*), Fraudatrix Group, Fraudatrix Subgroup, Quadripalpis Complex
reidi Harrison, 1973: *Anopheles sensu stricto*, Laticorn Section, Myzorhynchus Series, Barbirostris Group, Vanus Subgroup
reidi Mattingly, 1957b: *Heizmannia sensu stricto*
reinerti Peyton, 1977: *Uranotaenia* (*Pseudoficalbia*), Bimaculata Series
reinerti Rattanarithikul & Harrison, 1988: *Aedes* (*Hulecoeteomyia*)
reiseni Basio, 1971: *Tripteroides* (*Tripteroides*), Nitidoventer Group
rejector Dyar & Knab, 1906d: *Culex* (*Microculex*) (not placed in a Series)
rempeli Vockeroth, 1954: *Aedes* (*Ochlerotatus*), Communis Group, Punctor Subgroup
renatoi Lane & Ramalho, 1960: *Culex* (*Phytotelmatomyia*)
rennellensis Taylor & Maffi, 1991: *Anopheles* (*Cellia*), Neomyzomyia Series, Punctulatus Group
restrictor Dyar & Knab, 1906d: *Culex* (*Anoedioporpa*), Restrictor Group (the Group is monobasic)
restuans Theobald, 1901c: *Culex sensu stricto*, Pipiens Group, Restuans Complex (not placed in a Subgroup)
retrosus Lane & Whitman, 1951: *Culex* (*Microculex*), Imitator Series
reubenae Tewari & Hiriyan, 1992b: *Aedes* (*Tewarius*), Agastyai Group
reunionensis Brunhes & Hervy, 1995: *Orthopodomyia*, Orthopodomyia Section, Vernoni Group

reversa (Lane & Cerqueira, 1942): *Runchomyia sensu stricto*
reyi Baisas, 1935c: *Uranotaenia* (*Uranotaenia*), Anisocheleomyia Section, Ludlowae Series
rhodesiensis Theobald, 1901b: *Anopheles* (*Cellia*), Neomyzomyia Series, Rhodesiensis Group
rhungkiangensis Chang & Chang, 1974: *Aedes* (*Stegomyia*), W-albus Group, Mediopunctatus Subgroup
rhyacophilus da Costa Lima, 1933: *Aedes* (*Ochlerotatus*), Scapularis Group (not placed in a Subgroup)
ribeirensis Forattini & Sallum, 1985: *Culex* (*Melanoconion*), Spissipes Section, Crybda Group, Pedroi Subgroup
richardgarciai Jeffery, Oothuman & Rudnick, 1987: *Culex* (*Eumelanomyia*), Mochthogenes Group, Otachati Subgroup
richei Klein, 1970: *Culex* (*Eumelanomyia*), Mochthogenes Group, Tenuipalpis Subgroup
richeti Brunhes & Venard, 1966: *Culex* (*Barraudius*)
richiardii (Ficalbi, 1889b): *Coquillettidia* (*Coquillettidia*), Perturbans Group
rickenbachi da Cunha Ramos, 1993: *Uranotaenia* (*Pseudoficalbia*), Shillitonis Section, Shillitonis Group (not placed in a Subgroup)
rickenbachi Ferrara & Eouzan, 1974: *Eretmapodites*, Oedipodeios Group
rickenbachi Hamon & Adam, 1959b: *Aedes* (*Aedimorphus*), Alboscutellatus Assemblage, Abnormalis Group
rickenbachi Ribeiro, 1991: *Toxorhynchites* (*Toxorhynchites*), Brevipalpis Group, Phytophagus Subgroup
rima Theobald, 1901d: *Culex* (*Eumelanomyia*), Rubinotus-rima Group, Rima Subgroup
riojanus Duret, 1968e: *Culex sensu stricto*, Pipiens Group (not placed in a Subgroup)
riparioides Su & Zhang, 1987: *Aedes* (*Ochlerotatus*), Stimulans Group
riparius Dyar & Knab, 1907d: *Aedes* (*Ochlerotatus*), Cantans Group
riparis King & Baisas, 1936: *Anopheles* (*Cellia*), Neomyzomyia Series, Leucosphyrus Group, Riparis Subgroup
rivadeneirai Leví Castillo: *Anopheles sensu stricto*, Angusticorn Section, Anopheles Series, Pseudopunctipennis Group
riverai Duret, 1970: *Uranotaenia* (*Uranotaenia*), Lowii Series
riverai Miyagi, Toma & Tsukamoto, 1983: *Tripteroides* (*Tripteroides*), Mabinii Group
riversi Bohart & Ingram, 1946a: *Aedes* (*Stegomyia*), Scutellaris Group, Scutellaris Subgroup
rivulorum Leeson, 1935: *Anopheles* (*Cellia*), Myzomyia Series, Funestus Group, Rivulorum Subgroup
rizali (Banks, 1906): *Aedes* (*Hulecoeteomyia*)
rizzoi (de Deus Palma & Galvão, 1969): *Toxorhynchites* (*Lynchiella*)
roai Belkin, 1962: *Aedes* (subgenus uncertain)
roberti da Cunha Ramos & Brunhes, 2004: *Uranotaenia* (*Uranotaenia*), Chorleyi Section, Hamoni Group
robertsi (Laffoon, 1946): *Verrallina* (*Harbachius*)
robinsoni Belkin, 1962: *Aedes* (*Stegomyia*), Edwardsi Group
robusta Senevet & Abonnenc, 1939b: *Wyeomyia* (*Wyeomyia*) (not placed in a Group)
rochei (Doucet, 1951b): *Coquillettidia* (*Coquillettidia*), Crassipes Group
rodhaini (Brunhes & Boussès, 2017) (in Brunhes *et al.*, 2017): *Aedes* (*Paulianius*), Coulangesi Group
rodhaini Brunhes & Hervy, 1995: *Orthopodomyia*, Orthopodomyia Section, Vernoni Group
rodhaini Leleup & Lips, 1950: *Anopheles* (*Cellia*), Neomyzomyia Series, Rhodesiensis Group
rodhaini Ribeiro, 1991: *Toxorhynchites* (*Toxorhynchites*), Brevipalpis Group, Phytophagus Subgroup
rollai Cova Garcia, Pulido F. & Escalante de Ugueto, 1977b: *Anopheles* (*Kerteszia*), Rollai Complex (*An. rollai* consists of five sibling species denoted as Species 1–5 (Bourke *et al.*, 2023); which is the nominotypical species is not known)
romeroi Surcouf & Gonzalez-Rincones, 1912: *Culex* (subgenus uncertain)
ronderosi de Linero, 1967: *Culex* (*Melanoconion*), Melanoconion Section, Intrincatus Group, Idottus Subgroup
rondoni (Neiva & Pinto, 1922a): *Anopheles* (*Nyssorhynchus*), Albimanus Section, Oswaldoi Series, Oswaldoi Group, Strodei Subgroup
rondoniensis (Sant'Ana & Sallum, 2022b): *Anopheles* (*Nyssorhynchus*), Albimanus Section, Oswaldoi Series, Oswaldoi Group, Strodei Subgroup, Arthuri Complex
rooti (del Ponte, 1939): *Wyeomyia* (not placed in a Subgenus)
rooti Rozeboom, 1935: *Culex* (*Melanoconion*), Melanoconion Section, Pilosus Group, Pilosus Subgroup
roperi Reid, 1950: *Anopheles sensu stricto*, Laticorn Section, Myzorhynchus Series, Umbrosus Group, Letifer Subgroup
rorotaensis Floch & Abonnenc, 1946a: *Culex* (*Melanoconion*), Melanoconion Section, Distinguendus Group, Rorotaensis Subgroup (the Subgroup is monobasic)
rorotai Senevet, Chabelard & Abonnenc, 1942: *Wyeomyia* (*Decamyia*)
roseni Belkin, 1962: *Culex sensu stricto*, Sitiens Group (not placed in a Subgroup)
roslihashimi Miyagi & Toma, 2005: *Topomyia* (*Suaymyia*)
rossi Delfinado, 1966b: *Uranotaenia* (*Pseudoficalbia*), Recondita Series
rossicus Dolbeškin, Gorickaja & Mitrofanova, 1930: *Aedes sensu stricto*, Esoensis Group
rotanus Bohart & Ingram, 1946b: *Aedes* (*Stegomyia*), Pandani Group
rotoruae Belkin, 1968a: *Culex sensu stricto*, Pipiens Group, Trifilatus Subgroup
rottensis Statz, 1944: *Anopheles*? (fossil species)
rotumae Belkin, 1962: *Aedes* (*Stegomyia*), Scutellaris Group, Scutellaris Subgroup, Pseudoscutellaris Subgroup/Complex

rotumanus (Edwards, 1929b): *Tripteroides* (*Polylepidomyia*), Caledonicus Group
roubaudi (Doucet, 1950): *Mimomyia* (*Ingramia*), Group A (*Ravenalites s. str.*), Roubaudi Complex
roubaudi Grjebine, 1953a: *Anopheles* (*Cellia*), Neomyzomyia Series, Ranci Group, Roubaudi Subgroup
roucouyana (Bonne-Wepster & Bonne, 1920): *Wyeomyia* (not placed in a Subgenus)
roxasi Baisas & Ubaldo-Pagayon, 1953: *Tripteroides* (*Tricholeptomyia*), Nepenthicola Group, Nepenthicola Subgroup
rozeboomi Baisas & Ubaldo-Pagayon, 1953: *Tripteroides* (*Rachionotomyia*), Edwardsi Group
rozeboomi Causey, Deane & Deane, 1944: *Chagasia*
ruarinus Edwards, 1940: *Anopheles* (*Cellia*), Neomyzomyia Series, Rhodesiensis Group
rubensis Sasa & Takahashi, 1948: *Culex* (*Neoculex*), Territans Group
rubiginosus Belkin, 1962: *Aedes* (subgenus uncertain)
rubinotus Theobald, 1906: *Culex* (*Eumelanomyia*), Rubinotus-rima Group, Rubinotus Subgroup
rubithoracis (Leicester, 1908): *Culex* (*Lophoceraomyia*), Fraudatrix Group, Fraudatrix Subgroup, Rubithoracis Complex
rubithoracis Leicester, 1908: *Topomyia* (*Topomyia*)
rubrithorax (Macquart, 1850): *Aedes* (*Dobrotworskyius*)
rufipes (Gough, 1910): *Anopheles* (*Cellia*), Neocellia Series (not placed in a Group)
ruiliensis Dong, Zhou & Wang, 1997 (in Dong, Zhou, Wang & Wang, 1997): *Heizmannia sensu stricto*
rupestris Dobrotworsky, 1959: *Aedes* (*Dobrotworskyius*)
rupicolus Lewis, 1937: *Anopheles* (*Cellia*), Neomyzomyia Series, Rhodesiensis Group
rusticus (Rossi, 1790): *Aedes* (*Rusticoidus*)
ruthae Peters, 1958: *Culex* (*Culiciomyia*), Nebulosus Group
rutherfordi Edwards, 1922a: *Uranotaenia* (*Uranotaenia*), Trilineata Series
rutilus (Coquillett, 1896): *Toxorhynchites* (*Lynchiella*)
ruwenzori Haddow & van Someren, 1950: *Aedes* (*Stegomyia*), Africanus Group
ruwenzori (van Someren, 1948): *Toxorhynchites* (*Afrorhynchus*) (not placed in a Group)
ryukyensis Bohart, 1946: *Culex* (*Culiciomyia*), Fragilis Group
ryukyuana Tanaka, Mizusawa & Saugstad, 1979: *Uranotaenia* (*Pseudoficalbia*), Bimaculata Series
ryukyuanus Tanaka, Mizusawa & Saugstad, 1979: *Culex* (*Eumelanomyia*), Mochthogenes Group, Tenuipalpis Subgroup

Ss

sabahensis Ramalingam & Ramakrishna, 1988: *Topomyia* (*Topomyia*)
sabahensis (Reinert, 1974): *Verrallina* (*Neomacleaya*)
sabethea Lane & Cerqueira, 1942: *Wyeomyia* (*Miamyia*)
sacchettae Sirivanakarn & Jakob, 1982: *Culex* (*Melanoconion*), Spissipes Section, Vomerifer Group
sacharovi Favre, 1903: *Anopheles sensu stricto*, Angusticorn Section, Anopheles Series, Maculipennis Group, Maculipennis Subgroup
saeungae Taai & Harbach, 2015: *Anopheles sensu stricto*, Laticorn Section, Myzorhynchus Series, Barbirostris Group, Barbirostris Subgroup, Barbirostris Complex
saeva Dyar & Knab, 1906a: *Psorophora sensu stricto*
sagax (Skuse, 1889): *Aedes* (*Ochlerotatus*), Burpengaryensis Group
saimedres Huang, 1988b: *Aedes* (*Stegomyia*), Pseudonigeria Group
saipanensis Stone, 1945: *Aedes* (*Stegomyia*), Pandani Group
salauni Rickenbach, Ferrara & Eouzan, 1968: *Eretmapodites*, Oedipodeios Group
salbaii Maffi & Coluzzi, 1958: *Anopheles* (*Cellia*), Neocellia Series (not placed in a Group)
salinarius Coquillett, 1904b: *Culex sensu stricto*, Pipiens Group, Salinarius Complex (not placed in a Subgroup)
salisburiensis Theobald, 1901c: *Culex* (*Maillotia*), Seyrigi Group
sallumae González & Reyes, 2017 (in González *et al.*, 2017): *Aedes* (*Ochlerotatus*), Albifasciatus Group
sallumae Talaga, 2022 (in Talaga & Gendrin, 2022): *Culex* (*Melanoconion*), Spissipes Section, Vomerifer Group
saltanensis Dyar, 1928: *Culex sensu stricto*, Pipiens Group, Tarsalis Subgroup
samarensis Rozeboom, 1951: *Anopheles sensu stricto*, Laticorn Section, Myzorhynchus Series, Umbrosus Group (not placed in a Subgroup)
samoaensis Stone, 1966: *Coquillettidia* (*Coquillettidia*), Crassipes Group
samoaensis (Theobald, 1914): *Culex* (*Oculeomyia*) (not placed in a Complex)
samoanus (Grünberg, 1913): *Aedes* (*Finlaya*), Kochi Group, Solomonis Complex
sampaioi da Costa Lima, 1935b: *Orthopodomyia*, Thomasina Section
sampi Huang, 2004: *Aedes* (*Stegomyia*), Simpsoni Group
sanctielii Senevet & Abonnenc, 1938: *Anopheles* (*Nyssorhynchus*), Albimanus Section, Oswaldoi Series, Oswaldoi Group, Oswaldoi Subgroup, Oswaldoi Complex
sandrae Berlin, 1969b: *Culex* (*Micraedes*), Schicki Group

sandrae Zavortink, 1972: *Aedes* (*Protomacleaya*), Kompi Group
sangengluoensis Wang, 1984: *Culex* (*Lophoceraomyia*), Fraudatrix Group, Fraudatrix Subgroup, Rubithoracis Complex
sanguinae Theobald, 1904a: *Hodgesia*
santamarinai Broche, 2000: *Psorophora* (*Grabhamia*), Confinnis Group
saotomensis Lien, Lin, Lin & Tseng, 2008: *Coquillettidia* (*Coquillettidia*), Crassipes Group
saperoi Bohart & Ingram, 1946a: *Anopheles sensu stricto*, Laticorn Section, Myzorhynchus Series, Albotaeniatus Group
saperoi Knight, 1946: *Aedes* (*Downsiomyia*), Alboniveus Group
sapiens Marks, 1964a: *Aedes* (*Ochlerotatus*), Burpengaryensis Group
sapphirina (Osten Sacken, 1868): *Uranotaenia* (*Uranotaenia*), Pulcherrima Series
saramaccensis Bonne-Wepster & Bonne, 1920: *Culex* (*Melanoconion*), Melanoconion Section, Saramaccensis Group
sarawakensis Miyagi, Toma & Okazawa, 2021 (in Miyagi *et al.*, 2021b): *Topomyia* (*Topomyia*)
sardinerae Fox, 1953: *Culex* (*Melanoconion*), Melanoconion Section, Intrincatus Group, Idottus Subgroup
sarpangensis Somboon, Namgay & Harbach, 2024 (in Somboon *et al.*, 2024): *Anopheles sensu stricto*, Laticorn Section, Myzorhynchus Series, Barbirostris Group, Barbirostris Subgroup, Barbirostris Complex
sasai Kano, Nitahara & Awaya, 1954: *Culex* (*Culiciomyia*), Shebbearei Group
sasai Tanaka, Mizusawa & Saugstad, 1975b: *Aedes sensu stricto*, Cinereus Group
satpuraensis Natarajan, Rajavel & Jambulingam, 2017: *Uranotaenia* (*Pseudoficalbia*), Recondita Series
saungi Colless, 1955a: *Anopheles* (*Cellia*), Neomyzomyia Series (not placed in a Group)
savoryi Bohart, 1957: *Aedes* (*Tanakaius*)
sawadwongporni Rattanarithikul & Green, 1987: *Anopheles* (*Cellia*), Neocellia Series, Maculatus Group, Sawadwongporni Subgroup
sawyeri Causey, Deane, Deane & Sampaio, 1943: *Anopheles* (*Nyssorhynchus*), Argyritarsis Section, Argyritarsis Series, Argyritarsis Group
saxicola (Leicester, 1908): *Aedes* (*Hulecoeteomyia*)
scanloni Bram, 1967a: *Culex* (*Culiciomyia*), Fragilis Group
scanloni Mattingly, 1970: *Heizmannia sensu stricto*
scanloni Reinert, 1970a: *Aedes* (*Petermattinglyius*), Petermattinglyius Group
scanloni Sallum & Peyton, 2005 (in Sallum *et al.*, 2005b): *Anopheles* (*Cellia*), Neomyzomyia Series, Leucosphyrus Group, Leucosphyrus Subgroup, Dirus Complex
scapularis (Rondani, 1848): *Aedes* (*Ochlerotatus*), Scapularis Group, Scapularis Subgroup
scatophagoides (Theobald, 1901b): *Aedes* (*Mucidus*), Mucidus Group
schedocyclia (Dyar & Knab, 1908): *Shannoniana*
scheuberi Carpintero & Leguizamón, 2004: *Culex sensu stricto* (not placed in a Group)
schicki Berlin, 1969b: *Culex* (*Micraedes*), Schicki Group
schicki Zavortink, 1972: *Aedes* (*Protomacleaya*), Kompi Group
schilfgaardei Sirivanakarn, 1968: *Culex* (*Lophoceraomyia*), Fraudatrix Group, Fraudatrix Subgroup, Fraudatrix Complex
schizopinax Dyar, 1929a: *Aedes* (*Ochlerotatus*), Communis Group, Punctor Subgroup
schlosseri Belkin, 1962: *Aedes* (*Finlaya*), Kochi Group, Bougainvillensis Complex
schnusei (Martini, 1931): *Sabethes* (*Sabethes*), Group B
schoutedeni (Wolfs, 1948a): *Coquillettidia* (*Coquillettidia*), Crassipes Group
schroederi Schick, 1970a: *Aedes* (*Protomacleaya*), Terrens Group, Tehuantepec Subgroup
schtakelbergi Shingarev, 1928: *Aedes* (*Ochlerotatus*), Communis Group (not placed in a Subgroup)
schueffneri Stanton, 1915: *Anopheles* (*Cellia*), Neocellia Series, Annularis Group, Annularis Complex
schwetzi Edwards, 1926a: *Aedes* (*Stegomyia*), Apicoargenteus Group
schwetzi Edwards, 1929b: *Culex sensu stricto*, Guiarti Group
schwetzi Evans, 1934: *Anopheles* (*Cellia*), Myzomyia Series (not placed in a Group)
scimitar Branch & Seabrook, 1959: *Culex sensu stricto*, Pipiens Group (not placed in a Subgroup)
scintillans Ludlow, 1905a: *Heizmannia sensu stricto*
scotinomus (Dyar & Knab, 1907c): *Wyeomyia* (*Wyeomyia*) (not placed in a Group)
scottii Theobald, 1912b: *Culex sensu stricto*, Pipiens Group (not placed in a Subgroup)
scutellalbum Boshell-Manrique, 1939: *Aedes* (*Ochlerotatus*) (not placed in a Group)
scutellaris (Walker, 1859): *Aedes* (*Stegomyia*), Scutellaris Group, Scutellaris Subgroup
scutoscriptus Bohart & Ingram, 1946b: *Aedes* (*Stegomyia*), Scutellaris Group, Scutellaris Subgroup
scutostriata da Cunha Ramos & Brunhes, 2004: *Uranotaenia* (*Pseudoficalbia*), Annulata Section, Annulata Series, Lavieri Group
seampi Huang, 1974: *Aedes* (*Stegomyia*), Edwardsi Group
seatoi Huang, 1969: *Aedes* (*Stegomyia*), Scutellaris Group, Albopictus Subgroup
sechani Brunhes & Boussès, 2009: *Culex sensu stricto*, Sitiens Group (not placed in a Subgroup)
seculata (Menon, 1950): *Verrallina* (*Neomacleaya*)
secundus Bonne-Wepster & Bonne, 1920: *Culex* (*Carrollia*), Iridescens Group, Iridescens Subgroup

secutor Theobald, 1901c: *Culex sensu stricto*, Pipiens Group, Tarsalis Subgroup
sedaensis Lei, 1989a: *Aedes* (*Ochlerotatus*) (not placed in a Group)
sedlacekae Sirivanakarn, 1968: *Culex* (*Lophoceraomyia*), Fraudatrix Group, Fraudatrix Subgroup, Pseudorubithoracis Complex
segermanae Huang, 1997: *Aedes* (*Stegomyia*), Dendrophilus Group
selai Klein & Sirivanakarn, 1970: *Culex* (*Eumelanomyia*), Mochthogenes Group, Pluvialis Subgroup
selangorensis Sirivanakarn, 1976: *Culex* (*Oculeomyia*), Bitaeniorhynchus Complex
seldeslachtsi Wolfs, 1947c: *Culex sensu stricto*, Pipiens Group (not placed in a Subgroup)
sembeli Toma & Miyagi, 2002: *Armigeres sensu stricto*, Malayi Group
semibrunneus Edwards, 1927a: *Culex* (*Culiciomyia*), Nebulosus Group
semisimplicipes Edwards, 1914b: *Eretmapodites*, Chrysogaster Group
semlikiensis van Someren, 1950b: *Aedes* (*Aedimorphus*), Alboscutellatus Assemblage, Abnormalis Group
seniori Barraud, 1934: *Culex* (*Lophoceraomyia*), Fraudatrix Group, Fraudatrix Subgroup (Sirivanakarn, 1977a unnecessarily recognized a monobasic '*seniori* complex' for this species in the absence of evidence that it is more than one species)
sentania (King & Hoogstraal, 1947b): *Verrallina sensu stricto*, Carmenti Series
senyavinensis Knight & Hurlbut, 1949: *Aedes* (*Aedimorphus*), Alboscutellatus Assemblage, Alboscutellatus Group
seoulensis Yamada, 1921: *Aedes* (*Hopkinsius*), Yamada Group
separatus (Leicester, 1908): *Anopheles sensu stricto*, Laticorn Section, Myzorhynchus Series, Umbrosus Group, Separatus Subgroup (the Subgroup is monobasic)
separatus (Lynch Arribálzaga, 1891b): *Toxorhynchites* (*Lynchiella*)
septempunctata Theobald, 1905f: *Mansonia* (*Mansonioides*)
septemstriatus Dyar & Knab, 1907a: *Aedes* (*Howardina*), Fulvithorax Section, Flavithorax Subassemblage
septentrionalis (Dyar & Knab, 1906c): *Toxorhynchites* (*Lynchiella*)
serafini (Szadziewski, 1998): *Aedes* (*Ochlerotatus*) (fossil species)
seretsei Abdulla-Khan, Coetzee & Hunt, 1998: *Anopheles* (*Cellia*), Paramyzomyia Series, Listeri Group
sergentii (Theobald, 1907): *Anopheles* (*Cellia*), Myzomyia Series, Demeilloni Group
sergievi Danilov, Markovich & Proskuryakova, 1978: *Aedes* (*Ochlerotatus*) (not placed in a Group)
serrata (Lutz, 1905): *Wyeomyia* (*Miamyia*)
serratimarge Root, 1927c: *Culex* (*Melanoconion*), Melanoconion Section, Inhibitator Group, Egcymon Subgroup
serratoria (Dyar & Núñez Tovar, 1927b): *Wyeomyia* (not placed in a Subgenus)
serratus (Barraud, 1929a): *Tripteroides* (*Rachionotomyia*), Aranoides Group, Aranoides Subgroup
serratus (Theobald, 1901c): *Aedes* (*Ochlerotatus*), Serratus Group
seticoxitus Luh & Li, 1981: *Armigeres sensu stricto*
setifer Delfinado, 1966a: *Armigeres sensu stricto*
setosa King & Hoogstraal, 1947a: *Uranotaenia* (*Uranotaenia*), Anisocheleomyia Section, Tibialis Group
sexaueri Belkin, 1953a: *Uranotaenia* (*Uranotaenia*) (not placed in a Complex, Group or Series)
sexlineatus (Theobald, 1901b): *Aedes* (*Howardina*), Sexlineatus Section, Sexlineatus Subassemblage
seydeli Edwards, 1929b: *Anopheles* (*Cellia*), Myzomyia Series, Marshallii Group
seyrigi Edwards, 1941: *Culex* (*Maillotia*), Seyrigi Group
shanahani Sirivanakarn, 1968: *Culex* (*Lophoceraomyia*), Fraudatrix Group, Fraudatrix Subgroup, Petersi Complex
shannoni Davis, 1931: *Anopheles sensu stricto*, Laticorn Section, Arribalzagia Series
shannoni (Lane & Antunes, 1937): *Coquillettidia* (*Pseudotaeniorhynchus*)
shannoni (Lane & Cerqueira, 1942): *Sabethes* (*Peytonulus*)
shannoni Vargas & Downs, 1950: *Aedes* (*Ochlerotatus*), Stimulans Group
shebbearei Barraud, 1924g: *Culex* (*Culiciomyia*), Shebbearei Group
shehzadae Qutubuddin, 1972: *Aedes* (*Downsiomyia*), Niveus Group
sherki Knight, 1948a: *Aedes* (*Hulecoeteomyia*)
shillitonis Edwards, 1932b: *Uranotaenia* (*Pseudoficalbia*), Shillitonis Section, Shillitonis Group, Subgroup A
shinonagai (Tanaka, Mizusawa & Saugstad, 1979): *Lutzia* (*Insulalutzia*) (the Subgenus is monobasic)
shintienensis Tsai & Lien, 1950: *Aedes* (*Hulecoeteomyia*)
shoae Hamon & Ovazza, 1954: *Culex sensu stricto*, Pipiens Group (not placed in a Subgroup)
shopei Forattini & Toda, 1966: *Culex* (*Microculex*) (not placed in a Series)
shortti (Barraud, 1923e): *Aedes* (*Collessius*), Collessius Group
siamensis Zavortink, 1968: *Orthopodomyia*, Orthopodomyia Section, Albipes Group, Flavicosta Subgroup
sibiricus Danilov & Filippova, 1978: *Aedes* (*Stegomyia*), Scutellaris Group, Albopictus Subgroup
sierrensis (Ludlow, 1905b): *Aedes* (*Jarnellius*)
signata da Cunha Ramos, 1993: *Uranotaenia* (*Pseudoficalbia*), Nigripes Section, Ornata Group
signifera (Coquillett, 1896): *Orthopodomyia*, Bancroftia Section, Signifera Group, Signifera Subgroup
signipennis (Coquillett, 1904c): *Psorophora* (*Grabhamia*) (not placed in a Group)
silvai Duret, 1968f: *Culex* (*Melanoconion*), Melanoconion Section, Intrincatus Group, Intrincatus Subgroup

silvestris Dobrotworsky, 1961b: *Aedes* (*Ochlerotatus*), Perkinsi Group
silvestris Ingram & de Meillon, 1927: *Eretmapodites* (not placed in a Group)
simanini Gutsevich, 1966: *Aedes* (*Ochlerotatus*), Communis Group (not placed in a Subgroup)
simile Lane & Cerqueira, 1942: *Trichoprosopon*, Hyloconops Group, Pallidiventer Complex
similis (Leicester, 1908): *Tripteroides* (*Tripteroides*), Nitidoventer Group
similis (Theobald, 1910c): *Verrallina sensu stricto*, Carmenti Series
similissimus Strickland & Chowdhury, 1927a: *Anopheles sensu stricto*, Laticorn Section, Myzorhynchus Series, Umbrosus Group (not placed in a Subgroup)
simlensis Edwards, 1922a: *Aedes* (*Himalaius*)
simlensis (James, 1911) (in James & Liston, 1911): *Anopheles sensu stricto*, Angusticorn Section, Anopheles Series, Lindesayi Group, Gigas Subgroup, Baileyi Complex
simmondsi Tenorio, 1977: *Bironella sensu stricto*
simmsi (Dyar & Knab, 1908): *Wyeomyia* (*Wyeomyia*) (not placed in a Group)
simpla (King & Hoogstraal, 1947b): *Verrallina sensu stricto*, Carmenti Series
simplex Brug, 1934: *Tripteroides* (*Rachisoura*), Vanleeuweni Group
simplex (Theobald, 1903a): *Aedes* (*Cancraedes*)
simplicicornis Edwards, 1930a: *Culex* (*Eumelanomyia*), Rubinotus-rima Group, Rubinotus Subgroup
simpliciforceps Edwards, 1941: *Culex* (*Eumelanomyia*), Mochthogenes Group, Inconspicuosus Subgroup
simpsoni (Theobald, 1905c): *Aedes* (*Stegomyia*), Simpsoni Group, Simpsoni Complex
simpsoni Theobald, 1905g: *Culex sensu stricto*, Pipiens Group, Simpsoni Subgroup
simulans (Newstead & Carter, 1911): *Aedes* (*Elpeytonius*)
simulator Dyar & Knab, 1906d: *Culex* (*Melanoconion*), Spissipes Section, Jubifer Group
simulatus Baisas & Ubaldo-Pagayon, 1953: *Tripteroides* (*Tripteroides*), Nitidoventer Group
sinaiticus Kirkpatrick, 1925a: *Culex sensu stricto*, Pipiens Group, Simpsoni Subgroup
sinensis Chow, 1950: *Aedes* (*Downsiomyia*), Niveus Group
sinensis Theobald, 1903a: *Culex* (*Oculeomyia*), Sinensis Complex
sinensis Wiedemann, 1828: *Anopheles sensu stricto*, Laticorn Section, Myzorhynchus Series, Hyrcanus Group (not placed in a Subgroup)
sineroides Yamada, 1924: *Anopheles sensu stricto*, Laticorn Section, Myzorhynchus Series, Hyrcanus Group (not placed in a Subgroup)
singhbhumensis Natarajan & Rajavel, 2009: *Culex* (*Lophoceraomyia*), Mammilifer Group, Mammilifer Subgroup
singuawaensis Sirivanakarn, 1969: *Culex* (*Lophoceraomyia*), Mammilifer Group, Digoelensis Subgroup
singularis (Leicester, 1908): *Verrallina* (*Neomacleaya*)
sinkiangensis Hsiao, 1977: *Aedes* (*Ochlerotatus*), Caspius Group
sintoni (Barraud, 1924b): *Aedes* (subgenus uncertain)
sintoni Puri, 1929b: *Anopheles sensu stricto*, Angusticorn Section, Anopheles Series, Culiciformis Group
sintonoides Ho, 1938: *Anopheles sensu stricto*, Angusticorn Section, Anopheles Series, Culiciformis Group
siphanulatus Lourenço-de-Oliveira & da Silva, 1987: *Culex* (*Microculex*) (not placed in a Series)
sirivanakarni (Duret, 1982): *Onirion*, Personatum Group
sitiens Wiedemann, 1828: *Culex sensu stricto*, Sitiens Group, Sitiens Subgroup
smaragdinus Reinert, 1997 (in Reinert *et al.*, 1997): *Anopheles sensu stricto*, Angusticorn Section, Anopheles Series, Maculipennis Group, Quadrimaculatus Subgroup
smithburni van Someren, 1950b: *Aedes* (*Catageiomyia*)
smithii (Coquillett, 1901): *Wyeomyia* (*Wyeomyia*) (not placed in a Group)
smithii Theobald, 1905b: *Anopheles* (*Cellia*), Neomyzomyia Series, Smithii Group
soaresi Lane & Cerqueira, 1942: *Trichoprosopon* (not placed in a Group or Complex)
socialis Theobald, 1901c: *Uranotaenia* (*Uranotaenia*), Pulcherrima Series
sohni (Reinert, 1974): *Verrallina* (*Neomacleaya*)
soleatus Edwards, 1924 (in Haworth, 1924): *Aedes* (*Stegomyia*), Apicoargenteus Group
solitarius Bonne-Wepster, 1938: *Culex sensu stricto*, Sitiens Group, Mimeticus Subgroup, Mimeticus Complex
sollicitans (Walker, 1856): *Aedes* (*Ochlerotatus*), Taeniorhynchus Group (not placed in a Subgroup)
solomonis Belkin, 1953a: *Uranotaenia* (*Uranotaenia*), Anisocheleomyia Section, Albescens Complex
solomonis (Belkin, 1962): *Mimomyia* (*Etorleptiomyia*), Group B (Dixomyia)
solomonis Belkin, 1962: *Hodgesia*
solomonis Belkin, Knight & Rozeboom, 1945: *Anopheles* (*Cellia*), Neomyzomyia Series, Lungae Complex (not placed in a Group)
solomonis (Edwards, 1924b): *Tripteroides* (*Polylepidomyia*), Atripes Group
solomonis Edwards, 1929 (in Paine & Edwards, 1929): *Culex* (*Lophoceraomyia*), Fraudatrix Group, Fraudatrix Subgroup, Solomonis Complex
solomonis Stone & Bohart, 1944: *Aedes* (*Finlaya*), Kochi Group

solomonis (Wharton, 1947): *Malaya*
solstitialis (Lutz, 1904) (in Bourroul, 1904): *Toxorhynchites* (*Lynchiella*)
somalicus Rivola & Holstein, 1957: *Anopheles* (*Cellia*), Neomyzomyia Series, Ardensis Group, Nili Complex
somblooni Peyton & Klein, 1970: *Uranotaenia* (*Uranotaenia*), Testacea Series
soperi Antunes & Lane, 1937: *Culex* (*Carrollia*), Iridescens Group, Iridescens Subgroup
soperi Lane & Cerqueira, 1942: *Sabethes* (*Peytonulus*)
soperi Leví-Castillo, 1955b: *Haemagogus sensu stricto*, Albomaculatus Section
sororcula Dyar & Knab, 1906a: *Wyeomyia* (*Wyeomyia*) (not placed in a Group)
sorsogonensis Bañez & Jueco, 1966: *Aedes* (*Finlaya*), Poicilius Group
spanius (Dyar & Knab, 1909): *Deinocerites*, Spanius Group
spathifurca (Edwards, 1915b): *Culex* (*Culiciomyia*), Fragilis Group
spathulatus Forattini & Sallum, 1987b: *Culex* (*Melanoconion*), Melanoconion Section, Inhibitator Group (not placed in a Subgroup)
spathulirostris Edwards, 1923b: *Topomyia* (not placed in a Subgenus)
speciosus (Skuse, 1889): *Toxorhynchites* (*Toxorhynchites*), Splendens Group
spegazzinii Brèthes, 1912: *Haemagogus sensu stricto*, Albomaculatus Section
spencerii (Theobald, 1901c): *Aedes* (*Ochlerotatus*), Communis Group (not placed in a Subgroup)
spermatheca (Wijesundara, 1951): *Verrallina* (*Neomacleaya*)
sphinx Howard, Dyar & Knab, 1913: *Culex sensu stricto*, Pipiens Group (not placed in a Subgroup)
spiculosa Peyton & Rattanarithikul, 1970: *Uranotaenia* (*Pseudoficalbia*), Recondita Series
spiculostylus Chen, 1989: *Culex* (*Culiciomyia*), Shebbearei Group
spiculosus Bram & Rattanarithikul, 1967: *Culex* (*Lophoceraomyia*), Mammilifer Group, Minor Line, Ganapathi Complex
spilotus Marks, 1963b: *Aedes* (*Ochlerotatus*), Stricklandi Group
spinifer Sá & Sallum, 2020 (in Sá *et al.*, 2020): *Culex* (*Melanoconion*), Melanoconion Section, Atratus Group
spiniformis Sá & Hutchings, 2022 (in Sá *et al.*, 2022): *Culex* (*Melanoconion*), Melanoconion Section, Educator Group
spinitubus da Cunha Ramos & Brunhes, 2004: *Uranotaenia* (*Pseudoficalbia*), Annulata Section, Annulata Series, Lavieri Group
spinophallus Zhou, Zhu & Lu, 1999: *Topomyia* (*Suaymyia*)
spinosa da Cunha Ramos & Brunhes, 2004: *Uranotaenia* (*Pseudoficalbia*), Spinosa Section, Spinosa Group
spinosa (Doucet, 1951a): *Mimomyia* (*Ingramia*), Group A (*Ravenalites s. str.*), Spinosa Complex
spinosipes Edwards, 1922d: *Aedes* (*Macleaya*), Chaetocruiomyia Group
spinosus Berlin, 1969a: *Aedes* (*Howardina*), Walkeri Section, Walkeri Group
spinosus Lutz, 1905: *Culex sensu stricto*, Pipiens Group, Salinarius Complex (not placed in a Subgroup)
spiraculata da Cunha Ramos & Brunhes, 2004: *Uranotaenia* (*Pseudoficalbia*), Shillitonis Section, Shillitonis Group (not placed in a Subgroup)
spissipes (Theobald, 1903a): *Culex* (*Melanoconion*), Spissipes Section, Spissipes Group
spixi Cerqueira, 1961: *Sabethes* (*Sabethes*), Group B
splendens (de Meijere, 1909): *Malaya*
splendens Lee, 1946: *Tripteroides* (*Tripteroides*), Littlechildi Group
splendens Theobald, 1903a: *Mimomyia* (*Mimomyia*), Mimomyia Group (the Group is monobasic)
splendens (Wiedemann, 1819): *Toxorhynchites* (*Toxorhynchites*), Splendens Group
splendens Williston, 1896: *Haemagogus sensu stricto*, Splendens Section
splendida Bonne-Wepster & Bonne, 1919: *Wyeomyia* (*Phoniomyia*), Second Series
splendidus Koidzumi, 1920: *Anopheles* (*Cellia*), Neocellia Series, Jamesii Group
spoliata Edwards, 1923b: *Hodgesia*
squamifemur Antunes, 1937: *Anopheles* (*Lophopodomyia*)
squamiger (Coquillett, 1902c): *Aedes* (*Ochlerotatus*), Stimulans Group
squamipennis (Lynch Arribálzaga, 1878): *Aedeomyia sensu stricto*
squamosus (Taylor, 1914c): *Culex* (*Oculeomyia*) (not placed in a Complex)
squamosus Theobald, 1901b: *Anopheles* (*Cellia*), Cellia Series, Squamosus Group
srilankensis Peyton, 1974: *Uranotaenia* (*Pseudoficalbia*), Recondita Series
srilankensis (Reinert, 1977): *Verrallina* (*Harbachius*)
staminifera Lourenço-de-Oliveira, Motta & de Castro, 1992: *Wyeomyia* (*Triamyia*)
standfasti Peters, 1959: *Tripteroides* (*Polylepidomyia*), ?Argenteiventris Group
stanleyi Peters, 1963a: *Aedes* (subgenus uncertain)
starckeae Stone & Knight, 1958: *Culex* (*Oculeomyia*) (not placed in a Complex)
steffani Sirivanakarn, 1968: *Culex* (*Lophoceraomyia*), Fraudatrix Group, Fraudatrix Subgroup, Petersi Complex
stellata Belkin, Heinemann & Page, 1970: *Wyeomyia* (*Wyeomyia*), Hirsuta Group, Stellata Subgroup
stellata Grjebine, 1986: *Mimomyia* (*Ingramia*), Group A (*Ravenalites s. str.*)
stellatus van Someren, 1947a: *Culex* (*Eumelanomyia*), Protomelanoconion Group

stenei Thompson, 1956: *Aedes* (*Howardina*), Walkeri Section, Aurites Group
stenoetrus (Theobald, 1907): *Aedes* (*Aedimorphus*), Domesticus Assemblage, Vexans Subassemblage, Vexans Group
stenolepis Dyar & Knab, 1908: *Culex sensu stricto*, Pipiens Group, Tarsalis Subgroup
stenoscutus (Edwards, 1912a): *Aedes* (*Catageiomyia*)
stenostylus (Cornel, Kowo & Mayi, 2023) (in Kowo *et al.*, 2023): *Aedes* (*Aedimorphus*), Domesticus Assemblage, Domesticus Subassemblage, Domesticus Group
stephensi Liston, 1901: *Anopheles* (*Cellia*), Neocellia Series (not placed in a Group), potentially a species complex (Firooziyan *et al.*, 2018)
stevensoni (Barraud, 1923b): *Aedes* (*Phagomyia*)
sticticus (Meigen, 1838): *Aedes* (*Ochlerotatus*), Communis Group (not placed in a Subgroup)
stigmaticus Edwards, 1922d: *Aedes* (*Ochlerotatus*), Fulvus Group
stigmaticus Skuse, 1889: *Anopheles sensu stricto*, Angusticorn Section, Anopheles Series, Stigmaticus Group
stigmatosoma Dyar, 1907: *Culex sensu stricto*, Pipiens Group, Tarsalis Subgroup
stimulans (Walker, 1848): *Aedes* (*Ochlerotatus*), Stimulans Group
stokesi Evans, 1929: *Aedes* (*Albuginosus*)
stonei Belkin, 1950a: *Tripteroides* (*Rachisoura*), Filipes Group
stonei Knight & Laffoon, 1946: *Aedes* (*Finlaya*), Poicilius Group
stonei Lane & Whitman, 1943: *Culex* (*Microculex*) (not placed in a Series)
stonei Vargas, 1956: *Psorophora sensu stricto*
stonei Vargas & Martínez Palacios, 1953: *Wyeomyia* (*Wyeomyia*) (not placed in a Group)
stoneorum Marks, 1977: *Aedes* (*Macleaya*), Macleaya Group
stookesi Colless, 1955a: *Anopheles* (*Cellia*), Neomyzomyia Series (not placed in a Group)
stramineus Dubitzky, 1970: *Aedes* (*Ochlerotatus*), Caspius Group
strelitziae Muspratt, 1950: *Aedes* (*Stegomyia*), Simpsoni Group
striatipes Edwards, 1941: *Culex sensu stricto*, Pipiens Group (not placed in a Subgroup)
striatus Sant'Ana & Sallum, 2016 (online 31 December 2016; see Sant'Ana & Sallum, 2017): *Anopheles* (*Nyssorhynchus*), Albimanus Section, Oswaldoi Series, Oswaldoi Group, Strodei Subgroup
stricklandi Barraud, 1926: *Uranotaenia* (*Pseudoficalbia*), Recondita Series
stricklandi (Edwards, 1912b): *Aedes* (*Ochlerotatus*), Stricklandi Group
stricklandi Reid, 1965: *Anopheles sensu stricto*, Angusticorn Section, Anopheles Series, Aitkenii Group
strodei Root, 1926: *Anopheles* (*Nyssorhynchus*), Albimanus Section, Oswaldoi Series, Oswaldoi Group, Strodei Subgroup
stunga (Klein, 1973): *Verrallina* (*Harbachius*)
suarezi Cova García & Sutil Oramas, 1976b: *Mansonia* (*Mansonia*)
subaequalis Edwards, 1941: *Culex* (*Culiciomyia*), Nebulosus Group
subalbatus (Coquillett, 1898): *Armigeres sensu stricto*
subalbirostris Klein & Marks, 1960: *Aedes* (*Ochlerotatus*), Perkinsi Group
subalbitarsis King & Hoogstraal, 1946d: *Aedes* (*Patmarksia*)
subalbopictus Barraud, 1931a: *Aedes* (*Stegomyia*), Scutellaris Group, Albopictus Subgroup
subargenteus Edwards, 1925: *Aedes* (*Stegomyia*), Simpsoni Group
subauridorsum Marks, 1948: *Aedes* (subgenus uncertain)
subbasalis Dobrotworsky, 1962: *Aedes* (*Dobrotworskyius*)
subcomplosa (del Ponte, 1939): *Wyeomyia* (not placed in a Subgenus)
subdentatus Edwards, 1936: *Aedes* (*Aedimorphus*), Domesticus Assemblage, Vexans Subassemblage, Quasiunivittatus Group
subdiversus Martini, 1926: *Aedes* (*Rusticoidus*)
submarginalis Sirivanakarn, 1973a: *Culex* (*Lophoceraomyia*), Fraudatrix Group, Fraudatrix Subgroup, Fraudatrix Complex
subniveus Edwards, 1922a: *Aedes* (*Downsiomyia*), Niveus Group
subnormalis Martini, 1920: *Uranotaenia* (*Uranotaenia*), Campestris Series
subnudipennis (Edwards, 1927b): *Tripteroides* (*Rachisoura*), Vanleeuweni Group
subochrea (Edwards, 1921) (in Wesenberg-Lund, 1921): *Culiseta sensu stricto*
subpictus Grassi, 1899 (in Grassi *et al.*, 1899): *Anopheles* (*Cellia*), Pyretophorus Series, Subpictus Complex (*An. subpictus* consists of four sibling species denoted as Species A, B, C and D (Suguna *et al.*, 1994); which is the nominotypical species is not known)
subrima Edwards, 1941: *Culex* (*Eumelanomyia*), Rubinotus-rima Group, Rima Subgroup
subsalisburiensis Hervé & Geoffroy, 1974: *Culex* (*Maillotia*), Seyrigi Group
subsimilis (Barraud, 1927c): *Udaya*
subsimplicipes Edwards, 1914b: *Eretmapodites*, Chrysogaster Group
subtibioclada King & Hoogstraal, 1947a: *Uranotaenia* (*Uranotaenia*), Anisocheleomyia Section, Tibialis Group
suchariti (Miyagi & Toma, 1989): *Kimia*
sudanensis (Theobald, 1908c): *Aedes* (*Mucidus*), Mucidus Group

suffusus Edwards, 1922a: *Aedes* (*Reinertia*) (the Subgenus is monobasic)
sulawesi Waktoedi Koesoemawinangoen, 1954: *Anopheles* (*Cellia*), Neomyzomyia Series, Leucosphyrus Group, Hackeri Subgroup
sullivanae Baisas & Ubaldo-Pagayon, 1953: *Tripteroides* (*Tripteroides*), Nitidoventer Group
sumatranus Brug, 1931a: *Culex* (*Lophoceraomyia*), Mammilifer Group, Brevipalpus Subgroup
sumatranus (Brug, 1939): *Toxorhynchites* (*Toxorhynchites*), Ater Group
sumatranus Swellengrebel & Rodenwaldt, 1932: *Anopheles sensu stricto*, Angusticorn Section, Anopheles Series, Lindesayi Group, Gigas Subgroup, Gigas Complex
sumethi Peyton & Rattanarithikul, 1970: *Uranotaenia* (*Pseudoficalbia*), Recondita Series
sumidero Schick, 1970a: *Aedes* (*Protomacleaya*), Terrens Group, Heteropus Subgroup
summorosus Dyar, 1920f: *Culex sensu stricto*, Sitiens Group, Vishnui Subgroup
sundaicus (Rodenwaldt, 1925): *Anopheles* (*Cellia*), Pyretophorus Series, Sundaicus Complex (the Complex also includes informally designated Species B, C, D and E, see Dusfour *et al.*, 2007)
sunthorni Thurman, 1959: *Toxorhynchites* (*Toxorhynchites*), Christophi Group
sunyaniensis Edwards, 1941: *Culex* (*Eumelanomyia*), Rubinotus-rima Group, Rima Subgroup
superbus (Dyar & Knab, 1906c): *Toxorhynchites* (*Lynchiella*)
superpictus Grassi, 1899: *Anopheles* (*Cellia*), Neocellia Series (not placed in a Group) (*An. superpictus* consists of two sibling species denoted as Species A and B (Oshaghi *et al.*, 2007, 2008); which is the nominotypical species is not known)
surcoufi (Theobald, 1912) (in Surcouf, 1912): *Aedes* (*Ochlerotatus*), Stimulans Group
surinamensis Bruijning, 1959: *Wyeomyia* (not placed in a Subgenus)
surinamensis Dyar, 1918c: *Culex sensu stricto*, Pipiens Group, Tarsalis Subgroup
sursumptor Dyar, 1924d: *Culex* (*Melanoconion*), Melanoconion Section, Intrincatus Group, Intrincatus Subgroup
sutili Cova Garcia & Pulido F., 1974: *Culex* (*Microculex*) (not placed in a Series)
suzannae Boussès, 2022 (in Boussès *et al.*, 2022): *Aedes* (*Aedimorphus*), Domesticus Assemblage, Vexans Subassemblage, Vexans Group
svastii Thurman, 1959: *Topomyia* (*Topomyia*)
swahilicus Gillies, 1964: *Anopheles* (*Cellia*), Cellia Series (not placed in a Group)
sylvanensis (Dobrotworsky, 1960b): *Culiseta* (*Culicella*)
sylvatica Lu, Dong & Wang, 1986: *Topomyia* (*Topomyia*)
sylvestris (Theobald, 1910c): *Tripteroides* (*Rachisoura*), Filipes Group
symbletos Sallum & Hutchings, 2003: *Culex* (*Melanoconion*), Melanoconion Section, Inhibitator Group, Inhibitator Subgroup
symesi Edwards, 1928a: *Anopheles sensu stricto*, Laticorn Section, Myzorhynchus Series, Coustani Group
synchytus Arnell, 1976: *Aedes* (*Ochlerotatus*), Scapularis Group, Crinifer Subgroup
syntheta Dyar & Shannon, 1924a: *Uranotaenia* (*Pseudoficalbia*), Annulata Section, Maxima Series
syntheticus Barraud, 1928: *Aedes* (*Aedimorphus*), Domesticus Assemblage, Vexans Subassemblage, Vexans Group
szemaonensis Wang & Feng, 1964: *Culex* (*Lophoceraomyia*), Mammilifer Group (not placed in a Subgroup)

Tt

tabu Ramalingam & Belkin, 1965: *Aedes* (*Stegomyia*), Scutellaris Group, Scutellaris Subgroup
tadei Saraiva & Scarpassa, 2021: *Anopheles* (*Nyssorhynchus*), Albimanus Section, Oswaldoi Series, Oswaldoi Group, Oswaldoi Subgroup, Konderi Complex
taeniarostris (Theobald, 1909): *Aedes* (*Neomelaniconion*), Lineatopennis Group
taeniarostris (Theobald, 1911a): *Malaya*
taeniata (Leicester, 1908): *Verrallina* (*Neomacleaya*)
taeniopus Dyar & Knab, 1907b: *Culex* (*Melanoconion*), Spissipes Section, Taeniopus Group
taeniorhynchoides (Christophers, 1911): *Aedes* (*Aedimorphus*), Domesticus Assemblage, Mediolineatus Subassemblage, Taeniorhynchoides Group
taeniorhynchus (Wiedemann, 1821): *Aedes* (*Ochlerotatus*), Taeniorhynchus Group (not placed in a Subgroup)
tahoensis Dyar, 1916b: *Aedes* (*Ochlerotatus*), Impiger Group
taiwanensis Lien, 1968c: *Heizmannia sensu stricto*
taiwanica Lien, Lin & Weng, 1999: *Culiseta* (*Climacura*)
taiwanus Lien, 1968d: *Aedes* (*Bruceharrisonius*)
takasagoensis Morishita, 1946: *Anopheles* (*Cellia*), Neomyzomyia Series, Leucosphyrus Group, Leucosphyrus Subgroup, Dirus Complex (*An. takasagoensis* comprises two sibling species, one informally denoted as aff. *takasagoensis* by Takano *et al.*, 2010)
tamsi Edwards, 1934: *Culex sensu stricto*, Pipiens Group, Trifilatus Subgroup
tanakai Miyagi & Toma, 2013: *Uranotaenia* (*Pseudoficalbia*), Bicolor Series
tanzaniae da Cunha Ramos, 1993: *Uranotaenia* (*Pseudoficalbia*), Shillitonis Section, Shillitonis Group (not placed in a Subgroup)
tarsalis Coquillett, 1896: *Culex sensu stricto*, Pipiens Group, Tarsalis Subgroup

tarsalis Delfinado & Hodges, 1968: *Tripteroides* (*Tripteroides*), Nitidoventer Group
tarsalis (Newstead, 1907) (in Newstead *et al.*, 1907): *Aedes* (*Catageiomyia*)
tarsata Lane & Cerqueira, 1942: *Wyeomyia* (*Exallomyia*)
tarsopus Dyar & Knab, 1908: *Sabethes* (*Sabethes*), Group A
tasmaniensis Dobrotworsky, 1966: *Anopheles sensu stricto*, Angusticorn Section, Anopheles Series, Atratipes Group
tasmaniensis (Strickland, 1911): *Tripteroides* (*Polylepidomyia*), Tasmaniensis Group (the Group is monobasic)
tatoi Casal & García, 1971: *Culex sensu stricto*, Pipiens Group (not placed in a Subgroup)
taufflιebi Geoffroy & Hervé, 1976: *Culex* (*Eumelanomyia*), Eumelanomyia Group
tauffliebi Rickenbach & Ferrara, 1965: *Aedes* (*Aedimorphus*), Alboscutellatus Assemblage, Abnormalis Group
taurepana Anduze, 1941: *Wyeomyia* (not placed in a Subgenus)
taylori Edwards, 1936: *Aedes* (*Diceromyia*), Section A, Furcifer Group
tchekedii de Meillon & Leeson, 1940: *Anopheles* (*Cellia*), Myzomyia Series (not placed in a Group)
tecmarsis Dyar, 1918c: *Culex* (*Melanoconion*), Melanoconion Section, Intrincatus Group, Tecmarsis Subgroup (the Subgroup is monobasic)
teesdalei van Someren, 1954a: *Aedes* (*Albuginosus*)
tehuantepec Schick, 1970a: *Aedes* (*Protomacleaya*), Terrens Group, Tehuantepec Subgroup
telesilla de Meillon & Lavoipierre, 1945 (in de Meillon *et al.*, 1945): *Culex sensu stricto*, Pipiens Group (not placed in a Subgroup)
telmatophila Galindo, Blanton & Peyton, 1954: *Uranotaenia* (*Uranotaenia*), Lowii Series
tenagius van Someren, 1954b: *Culex sensu stricto*, Pipiens Group, Theileri Subgroup
tenax (de Meijere, 1910): *Tripteroides* (*Rachionotomyia*), Aranoides Group, Aranoides Subgroup
tendeiroi da Cunha Ramos, Ribeiro & de Barros Machado, 1992: *Eretmapodites*, Plioleucus Group
tenebrosus Dönitz, 1902: *Anopheles sensu stricto*, Laticorn Section, Myzorhynchus Series, Coustani Group
tengchongensis Dong, Wang & Zhou, 2002: *Heizmannia sensu stricto*
tenuipalpis Barraud, 1924f: *Culex* (*Eumelanomyia*), Mochthogenes Group, Tenuipalpis Subgroup
tenuipalpis (Edwards, 1924b): *Coquillettidia* (*Austromansonia*) (the Subgenus is monobasic)
tenuis Edwards, 1922a: *Topomyia* (*Topomyia*)
terebor Dyar, 1920b: *Culex* (*Melanoconion*), Melanoconion Section, Conspirator Group
termi Thurman, 1955: *Culex* (*Culiciomyia*), Fragilis Group
terrens (Walker, 1856): *Aedes* (*Protomacleaya*), Terrens Group, Terrens Subgroup
territans Walker, 1856: *Culex* (*Neoculex*), Territans Group
terzii Edwards, 1941: *Culex sensu stricto*, Pipiens Group (not placed in a Subgroup)
tessellatus Theobald, 1901b: *Anopheles* (*Cellia*), Neomyzomyia Series, Tessellatus Group (the Group is monobasic)
testacea Theobald, 1905f: *Uranotaenia* (*Uranotaenia*), Testacea Series
testei Senevet & Abonnenc, 1939b: *Wyeomyia* (*Dendromyia*)
thailandensis Reinert, 1976a: *Aedes* (*Paraedes*), Ostentatio Assemblage
thalassius Theobald, 1903a: *Culex sensu stricto*, Sitiens Group (not placed in a Subgroup)
theileri Edwards, 1912c: *Anopheles* (*Cellia*), Myzomyia Series, Wellcomei Group
theileri Theobald, 1903a: *Culex sensu stricto*, Pipiens Group, Theileri Subgroup
thelcter Dyar, 1918c: *Aedes* (*Ochlerotatus*), Scapularis Group, Thelcter Subgroup (the Subgroup is monobasic)
thelmae Mattingly, 1970: *Heizmannia sensu stricto*
theobaldi Barraud, 1934: *Armigeres sensu stricto*
theobaldi (Dyar & Knab, 1906c): *Toxorhynchites* (*Lynchiella*)
theobaldi Giles, 1901b: *Anopheles* (*Cellia*), Neocellia Series (not placed in a Group)
theobaldi (Lane & Cerqueira, 1942): *Runchomyia* (*Runchomyia*)
theobaldi (Lane & Cerqueira, 1942): *Wyeomyia* (*Phoniomyia*), Second Series
theobaldi (Lutz, 1904) (in Bourroul, 1904): *Culex* (*Melanoconion*), Melanoconion Section, Educator Group
theobaldi (Taylor, 1914a): *Aedes* (*Ochlerotatus*), Taeniorhynchus Group, Theobaldi Subgroup
theresae Harbach, 2018a: *Sabethes* (*Sabethinus*), Andreae Group
thibaulti Dyar & Knab, 1910a: *Aedes* (*Ochlerotatus*), Thibaulti Group (the Group is monobasic)
thimphuensis Somboon, Namgay & Harbach, 2022 (in Somboon *et al.*, 2022b): *Anopheles sensu stricto*, Angusticorn Section, Anopheles Series, Lindesayi Group, Lindesayi Subgroup, Lindesayi Complex (the Complex also includes informally designated Species B, see Somboon *et al.*, 2022b)
thomasi Shannon, 1933: *Anopheles* (*Stethomyia*)
thomsoni (Theobald, 1905e): *Aedes* (*Christophersiomyia*)
thorntoni Dyar & Knab, 1907a: *Aedes* (*Protomacleaya*), Terrens Group, Thorntoni Subgroup
thriambus Dyar, 1921a: *Culex sensu stricto*, Pipiens Group, Tarsalis Subgroup
thurmanae Mattingly, 1958a: *Aedes* (*Cancraedes*)
thurmanorum Bram, 1967a: *Culex* (*Culiciomyia*), Fragilis Group
tianpingensis Chen, 1981: *Culex sensu stricto*, Sitiens Group, Mimeticus Subgroup, Complex uncertain

tibialis Taylor, 1919: *Uranotaenia* (*Uranotaenia*), Anisocheleomyia Section, Tibialis Group
tibiamaculatus (Neiva, 1906): *Anopheles sensu stricto*, Angusticorn Section, Anopheles Series, Pseudopunctipennis Group
tibioclada King & Hoogstraal, 1947a: *Uranotaenia* (*Uranotaenia*), Anisocheleomyia Section, Tibialis Group
tigertti Scanlon & Peyton, 1967: *Anopheles sensu stricto*, Angusticorn Section, Anopheles Series, Aitkenii Group
tigripes (de Grandpré & de Charmoy, 1901): *Lutzia* (*Metalutzia*)
timorensis (Miyagi, Toma & Lien, 2004): *Aedes* (*Geoskusea*) (not placed in a Group)
tiptoni Grjebine, 1953a: *Aedes* (*Paulianius*), Tiptoni Group
tipuliformis Leicester, 1908: *Topomyia* (*Topomyia*)
titillans (Walker, 1848): *Mansonia* (*Mansonia*), Titillans Complex
tityae Slooff, 1961: *Tripteroides* (*Rachisoura*), Filipes Group
toffaletii Baisas & Ubaldo-Pagayon, 1953: *Tripteroides* (*Tripteroides*), Nitidoventer Group
togoi (Theobald, 1907): *Aedes* (*Tanakaius*)
toltecum (Dyar & Knab, 1906a): *Psorophora* (*Grabhamia*) (not placed in a Group)
tongae Edwards, 1926a: *Aedes* (*Stegomyia*), Scutellaris Group, Scutellaris Subgroup, Pseudoscutellaris Subgroup/Complex
tonkinensis Galliard & Ngu, 1947: *Aedes* (*Collessius*), Alloeomyia Group
tonkingi (Gebert, 1948): *Aedes* (*Mucidus*), Mucidus Group
tonnoiri (Edwards, 1925): *Culiseta* (*Climacura*)
tonsus Edwards, 1924b: *Aedes* (*Geoskusea*) (not placed in a Group)
tonsus Edwards, 1941: *Eretmapodites* (not placed in a Group)
torakala Stoker & Waktoedi Koesoemawinangoen, 1949: *Anopheles* (*Cellia*), Pyretophorus Series
tormentor Dyar & Knab, 1906c: *Aedes* (*Ochlerotatus*), Serratus Group
toroensis Edwards & Gibbins, 1939: *Culex sensu stricto*, Pipiens Group (not placed in a Subgroup)
torokinae Belkin, 1950a: *Tripteroides* (*Rachisoura*), Filipes Group
torosa (Delfinado, 1967): *Verrallina* (*Neomacleaya*)
torrentium Martini, 1925: *Culex sensu stricto*, Pipiens Group, Trifilatus Subgroup
torresiensis Schmidt, 2001 (in Schmidt *et al.*, 2001): *Anopheles* (*Cellia*), Neomyzomyia Series, Punctulatus Group, Farauti Complex
tortilis (Theobald, 1903g): *Aedes* (*Ochlerotatus*), Scapularis Group, Tortilis Subgroup
totonaci Lassmann, 1951: *Psorophora* (*Janthinosoma*)
tournieri Senevet & Abonnenc, 1939a: *Culex* (*Melanoconion*), Melanoconion Section, Bastagarius Group, Bastagarius Subgroup
toviiensis Klein, Rivière & Séchan, 1984: *Culex sensu stricto*, Sitiens Group (not placed in a Subgroup)
towadensis (Matsumura, 1916): *Toxorhynchites* (*Toxorhynchites*), Christophi Group
townsendi Stone, 1944: *Trichoprosopon*, Trichoprosopon Group, Digitatum Complex
toxopeusi Bonne-Wepster, 1948: *Aedes* (subgenus uncertain)
tramazayguesi Duret, 1954: *Culex* (*Allimanta*) (the Subgenus is monobasic)
trapidoi Galindo, Blanton & Peyton, 1954: *Uranotaenia* (*Uranotaenia*), Pallidoventer Series
traubi Colless, 1965: *Culex* (*Lophoceraomyia*), Mammilifer Group, Mammilifer Line, Traubi Complex
traubi Macdonald, 1960: *Armigeres* (*Leicesteria*)
travestita (Brug, 1928b): *Bironella* (*Brugella*)
tremulus (Theobald, 1903e): *Aedes* (*Macleaya*), Macleaya Group
treubi (de Meijere, 1910): *Aedes* (*Alanstonea*)
triannulatus (Neiva & Pinto, 1922b): *Anopheles* (*Nyssorhynchus*), Albimanus Section, Oswaldoi Series, Triannulatus Group (*An. triannulatus* consists of two sibling species, the nominotypical species and the other denoted as Species C by Silva-do-Nascimento & Lourenço-de-Oliveira, 2007)
tricholabis Edwards, 1941: *Aedes* (*Aedimorphus*), Alboscutellatus Assemblage, Abnormalis Group
trichopygus (Wiedemann, 1828): *Toxorhynchites* (*Ankylorhynchus*)
trichorostris (Theobald, 1909): *Malaya*
trichorryes Dyar & Knab, 1907c: *Trichoprosopon*, Trichoprosopon Group, Digitatum Complex
tricolor da Cunha Ramos & Brunhes, 2004: *Uranotaenia* (*Pseudoficalbia*), Spinosa Section, Spinosa Group
tricuspis Edwards, 1930a: *Culex* (*Culiciomyia*), Tricuspis Group
tridentata da Cunha Ramos & Brunhes, 2004: *Uranotaenia* (*Pseudoficalbia*), Shillitonis Section, Shillitonis Group (not placed in a Subgroup)
tridentatus Cerqueira, 1961: *Sabethes* (*Sabethoides*)
trifida Edwards, 1922a: *Topomyia* (*Topomyia*)
trifidus Dyar, 1921c: *Culex* (*Melanoconion*), Melanoconion Section, Trifidus Group (the Group is monobasic)
trifilatus Edwards, 1914b: *Culex sensu stricto*, Pipiens Group, Trifilatus Subgroup
trifoliatus Edwards, 1914b: *Culex sensu stricto*, Pipiens Group (not placed in a Subgroup)
trifurcata Clastrier, 1973: *Wyeomyia* (*Dendromyia*)
trigeminatus Clastrier, 1970c: *Culex* (*Melanoconion*), Melanoconion Section, Atratus Group

trilineata Leicester, 1908: *Uranotaenia* (*Uranotaenia*), Trilineata Series
trilobulatus Duret & Barreto, 1956: *Culex* (*Melanoconion*), Melanoconion Section, Intrincatus Group, Intrincatus Subgroup
trimaculatus (Theobald, 1905f): *Aedes* (*Aedimorphus*), Domesticus Assemblage, Mediolineatus Subassemblage, Mediolineatus Group
trinidadensis Theobald, 1901c: *Wyeomyia* (*Phoniomyia*), Third Series
trinkae Faran, 1979: *Anopheles* (*Nyssorhynchus*), Albimanus Section, Oswaldoi Series, Oswaldoi Group, Oswaldoi Subgroup, Oswaldoi Complex
tripartita (Bonne-Wepster & Bonne, 1921): *Wyeomyia* (*Phoniomyia*)
tripunctata (Theobald, 1908b): *Heizmannia* (*Mattinglyia*)
triseriatus (Say, 1823): *Aedes* (*Protomacleaya*), Triseriatus Group
trisetosus Fauran, 1961: *Culex* (*Melanoconion*), Melanoconion Section, Intrincatus Group, Intrincatus Subgroup
trispinata (King & Hoogstraal, 1947b): *Verrallina sensu stricto*, Carmenti Series
tritaeniorhynchus Giles, 1901a: *Culex sensu stricto*, Sitiens Group, Vishnui Subgroup
trivittatus (Coquillett, 1902b): *Aedes* (*Ochlerotatus*), Scapularis Group, Trivittatus Subgroup
tropicalis Cerqueira & Antunes, 1938: *Haemagogus sensu stricto*, Tropicalis Section (the Section is monobasic)
trukensis Bohart, 1957: *Aedes* (*Aedimorphus*), Alboscutellatus Assemblage, Abnormalis Group
tsaratananae Doucet, 1950: *Uranotaenia* (*Pseudoficalbia*), Nigripes Section, Nigripes Group
tsengi Lien, 1968d: *Culex sensu stricto*, Sitiens Group, Mimeticus Subgroup, Mimeticus Complex
tsiliensis King & Hoogstraal, 1946f: *Aedes* (subgenus uncertain)
tubbutiensis Dobrotworsky, 1959: *Aedes* (*Dobrotworskyius*)
tuberis Bohart, 1946: *Culex* (*Lophoceraomyia*), Mammilifer Group, Minor Line, Minor Complex
tulagiensis Edwards, 1926a: *Aedes* (*Stegomyia*), Edwardsi Group
tulliae (Taylor, 1929): *Aedes* (*Macleaya*), Chaetocruiomyia Group
tumetarsalis Chen & Zhang, 1988: *Topomyia* (*Suaymyia*)
turkhudi Liston, 1901: *Anopheles* (*Cellia*), Paramyzomyia Series, Cinereus Group
turneri Marks, 1963b: *Aedes* (*Ochlerotatus*), Stricklandi Group
tutuilae Ramalingam & Belkin, 1965: *Ades* (*Finlaya*), Kochi Group, Solomonis Complex
tyagii Krishnamoorthy, Munirathinam, Dhananjeyan, Hiriyan, Mariappan, Samuel & Venkatesh, 2013: *Toxorhynchites* (*Toxorhynchites*) (not placed in a Group)
typhlosomata Dyar & Knab, 1907c: *Uranotaenia* (*Uranotaenia*), Pallidoventer Series

Uu

ugandae da Cunha Ramos, 1993: *Uranotaenia* (*Pseudoficalbia*), Shillitonis Section, Shillitonis Group (not placed in a Subgroup)
ugandae Evans, 1934: *Anopheles* (*Cellia*), Myzomyia Series, Wellcomei Group
ugandae van Someren, 1967: *Culex sensu stricto*, Pipiens Group (not placed in a Subgroup)
uichancoi Baisas & Ubaldo-Pagayon, 1953: *Tripteroides* (*Tripteroides*), Nitidoventer Group
ulocoma (Theobald, 1903a): *Wyeomyia* (*Decamyia*)
ulopus (Dyar & Knab, 1906a): *Johnbelkinia*
umbripes Edwards, 1941: *Culex sensu stricto*, Pipiens Group (not placed in a Subgroup)
umbrosus (Theobald, 1903a): *Anopheles sensu stricto*, Laticorn Section, Myzorhynchus Series, Umbrosus Group, Umbrosus Subgroup (the Subgroup is monobasic)
unalom Maquart & Hide, 2024 (in Hide *et al.*, 2024): *Aedes* (*Stegomyia*), Aegypti Group
unca (Theobald, 1901c): *Verrallina* (*Neomacleaya*)
uncinatus Delfinado, 1966a: *Culex* (*Eumelanomyia*), Mochthogenes Group, Uncinatus Subgroup (the Subgroup is monobasic)
undosus (Coquillett, 1906a): *Sabethes* (*Peytonulus*)
undulata del Ponte & Cerqueira, 1938: *Wyeomyia* (not placed in a Subgenus)
unguiculata Edwards, 1913a: *Uranotaenia* (*Pseudoficalbia*), Annulata Section, Maxima Series
ungujae White, 1975: *Anopheles* (*Cellia*), Myzomyia Series, Wellcomei Group
unicinctus Edwards, 1922a: *Aedes* (*Jihlienius*)
unicornis Root, 1928 (in Dyar, 1928): *Culex* (*Melanoconion*), Melanoconion Section, Pilosus Group, Pilosus Subgroup
unidentatus McIntosh, 1971: *Aedes* (*Neomelaniconion*), Lineatopennis Group
uniformis (Theobald, 1901c): *Mansonia* (*Mansonioides*), Uniformis Complex
uniformis (Theobald, 1904b): *Ficalbia*, Group B Afrotropical Region Group
uniformis (Theobald, 1905h): *Culex* (*Lophoceraomyia*), Mammilifer Group, Mammilifer Line, Traubi Complex
uniformis (Theobald, 1910a): *Verrallina* (*Harbachius*)
unilineatus (Theobald, 1906): *Aedes* (*Stegomyia*), Unilineatus Group
unimaculiala Leicester, 1908: *Uranotaenia* (*Uranotaenia*), Unimaculiala Series
unispinosa Thurman, 1959: *Topomyia* (*Topomyia*)

univittatus Theobald, 1901c: *Culex sensu stricto*, Pipiens Group, Univittatus Subgroup
untii (Sant'Ana & Sallum, 2023) (online 4 December 2023; see Sant'Ana & Sallum, 2024): *Anopheles* (*Nyssorhynchus*), Albimanus Section, Oswaldoi Series, Oswaldoi Group, Strodei Subgroup, Arthuri Complex
upatensis Anduze & Hecht, 1943: *Aedes* (*Ochlerotatus*) (not placed in a Group)
upolensis Marks, 1957b: *Aedes* (*Stegomyia*), Scutellaris Group, Scutellaris Subgroup, Pseudoscutellaris Subgroup/Complex
urichii (Coquillett, 1906d): *Culex* (*Carrollia*), Iridescens Group, Urichii Subgroup
usambara Mattingly, 1953c: *Aedes* (*Stegomyia*), Poweri Group
usquatissimus Dyar, 1922b: *Culex sensu stricto*, Pipiens Group, Coronator Complex (not placed in a Subgroup)
usquatus Dyar, 1918c: *Culex sensu stricto*, Pipiens Group, Coronator Complex (not placed in a Subgroup)

Vv

vagans Wiedemann, 1828: *Culex sensu stricto*, Pipiens Group, Trifilatus Subgroup
vagus Dönitz, 1902: *Anopheles* (*Cellia*), Pyretophorus Series
valeryi Gornostaeva, 2005: *Aedes sensu stricto* (not placed in a Group)
vallistris (Barraud, 1928): *Verrallina* (*Neomacleaya*)
vanapa (Huang, 1968a): *Verrallina sensu stricto*, Carmenti Series
vanderwulpi Townson & Harbach, 2013 (in Townson *et al.*, 2013): *Anopheles sensu stricto*, Laticorn Section, Myzorhynchus Series, Barbirostris Group, Barbirostris Subgroup, Barbirostris Complex
vanduzeei Dyar & Knab, 1906a: *Wyeomyia* (*Wyeomyia*), Vanduzeei Group (the Group is monobasic)
vaneedeni Gillies & Coetzee, 1987: *Anopheles* (*Cellia*), Myzomyia Series, Funestus Group, Funestus Subgroup
vanemdeni Martini, 1931: *Aedes* (*Howardina*), Sexlineatus Section, Aurivittatus Group
vanhoofi Wanson & Lebied, 1945: *Anopheles* (*Cellia*), Neomyzomyia Series, Smithii Group
vanleeuweni (Edwards, 1927b): *Tripteroides* (*Rachisoura*), Vanleeuweni Group
vanoyei (Wolfs, 1948b): *Coquillettidia* (*Coquillettidia*), Crassipes Group
vansomerenae Grjebine, 1986: *Mimomyia* (*Ingramia*), Group B (Ravenalites)
vansomerenae Mattingly, 1955 (in Mattingly & Brown, 1955): *Aedes* (*Ochlerotatus*), Taeniorhynchus Group, Vigilax Subgroup
vansomereni Edwards, 1926a: *Culex sensu stricto*, Pipiens Group (not placed in a Subgroup)
vansomereni Hamon, 1962: *Eretmapodites*, Chrysogaster Group
vanus Colless, 1958: *Aedes* (*Downsiomyia*), Niveus Group
vanus Walker, 1859: *Anopheles sensu stricto*, Laticorn Section, Myzorhynchus Series, Barbirostris Group, Vanus Subgroup
vargasi Gabaldon, Cova-Garcia & Lopez, 1941b: *Anopheles* (*Lophopodomyia*)
vargasi Schick, 1970a: *Aedes* (*Protomacleaya*), Terrens Group, Heteropus Subgroup
variabilis (Huang, 1968a): *Verrallina sensu stricto*, Carmenti Series
variatus (Leicester, 1908): *Culex* (*Lophoceraomyia*), Fraudatrix Group, Fraudatrix Subgroup, Variatus Complex
variegata (Dobrotworsky, 1962): *Coquillettidia* (*Coquillettidia*), Crassipes Group
variepictus King & Hoogstraal, 1946c: *Aedes* (*Huaedes*)
varietas (Leicester, 1908): *Verrallina* (*Neomacleaya*)
varinervis Edwards, 1922d: *Psorophora* (*Grabhamia*) (not placed in a Group)
varipalpus (Coquillett, 1902d): *Aedes* (*Jarnellius*)
varipes (Coquillett, 1904a): *Psorophora* (*Janthinosoma*)
varivestita (Statz, 1944): *Coquillettidia* (fossil species)
varuae Belkin, 1962: *Aedes* (*Stegomyia*), Scutellaris Group, Scutellaris Subgroup
varuna Iyengar, 1924: *Anopheles* (*Cellia*), Myzomyia Series, Funestus Group, Aconitus Subgroup
vattieri Geoffroy, 1971c: *Culex* (*Eumelanomyia*), Rubinotus-rima Group, Rima Subgroup
vaxus Dyar, 1920b: *Culex* (*Melanoconion*), Melanoconion Section, Educator Group
vectensis Edwards, 1923c: *Culex* (fossil species)
veeniae McIntosh, 1975: *Aedes* (*Catageiomyia*)
venezuelensis (Theobald, 1912) (in Surcouf, 1912): *Coquillettidia* (*Pseudotaeniorhynchus*)
ventrilloni Edwards, 1920b: *Culex sensu stricto*, Pipiens Group (not placed in a Subgroup)
ventrovittis Dyar, 1916b: *Aedes* (*Ochlerotatus*), Communis Group (not placed in a Subgroup)
venustipes (Skuse, 1889): *Aedeomyia sensu stricto*
vernoni van Someren, 1949a: *Orthopodomyia*, Orthopodomyia Section, Vernoni Group
vernus Gillies & de Meillon, 1968: *Anopheles* (*Cellia*), Neomyzomyia Series, Ardensis Group
versabilis Sirivanakarn, 1968: *Culex* (*Lophoceraomyia*), Fraudatrix Group, Fraudatrix Subgroup, Marksae Complex
versicolor (Barraud, 1924i): *Aedes* (subgenus uncertain)
versicolor (Edwards, 1913b): *Coquillettidia* (*Coquillettidia*), Crassipes Group
veruslanei Vargas, 1979: *Anopheles sensu stricto*, Laticorn Section, Arribalzagia Series
verutus Harbach, 1987: *Culex sensu stricto*, Guiarti Group

vestitipennis Dyar & Knab, 1906a: *Anopheles sensu stricto*, Laticorn Section, Arribalzagia Series
vexans (Meigen, 1830): *Aedes* (*Aedimorphus*), Domesticus Assemblage, Vexans Subassemblage, Vexans Group
vexillatus Edwards, 1941: *Culex sensu stricto*, Pipiens Group, Decens Subgroup
vexillifer Komp, 1936d: *Culex* (*Melanoconion*), Melanoconion Section, Pilosus Group, Caudelli Subgroup
vicinalis de Meillon & Lavoipierre, 1944: *Culex sensu stricto*, Pipiens Group, Decens Subgroup
vicarius Lien, 1968d: *Aedes* (*Collessius*), Collessius Group
vicinus (Edwards, 1914c): *Tripteroides* (*Tripteroides*), Nitidoventer Group
vicinus (Taylor, 1916): *Culex sensu stricto*, Sitiens Group (not placed in a Subgroup)
victoriensis (Dobrotworsky, 1954): *Culiseta* (*Culicella*)
vietnamensis Nguyen, Tran & Nguyen, 1993: *Anopheles sensu stricto*, Laticorn Section, Myzorhynchus Series, Hyrcanus Group, Lesteri Subgroup
vigilax (Skuse, 1889): *Aedes* (*Ochlerotatus*), Taeniorhynchus Group, Vigilax Subgroup
vijayae Ramalingam, 1975: *Topomyia* (*Topomyia*)
vimoli Thurman & Thurman, 1958: *Armigeres* (*Leicesteria*)
vinckei de Meillon, 1942: *Anopheles* (*Cellia*), Neomyzomyia Series, Ardensis Group
vinckei Hamon, Holstein & Rivola, 1957: *Culex* (*Eumelanomyia*), Eumelanomyia Group
vinsoni Mattingly, 1953a: *Aedes* (*Stegomyia*) (not place in a Group)
violaceus (Wiedemann, 1820): *Toxorhynchites* (*Lynchiella*)
viridibasis (Edwards, 1935a): *Toxorhynchites* (*Afrorhynchus*) (not placed in a Group)
viridis Barraud, 1929b: *Heizmannia sensu stricto*
viridiventer Giles, 1901a: *Culex* (*Culiciomyia*), Fragilis Group
virilis Leicester, 1908: *Verrallina* (*Neomacleaya*)
vishnui Theobald, 1901b: *Culex sensu stricto*, Sitiens Group, Vishnui Subgroup
vittatus (Bigot, 1861): *Aedes* (*Fredwardsius*)
vittiger (Skuse, 1889): *Aedes* (*Ochlerotatus*), Burpengaryensis Group
voltaensis (Danilov, 1982): *Coquillettidia* (*Coquillettidia*), Crassipes Group
vomerifer Komp, 1932: *Culex* (*Melanoconion*), Spissipes Section, Vomerifer Group
vonplesseni (Dyar & Knab, 1906a): *Trichoprosopon* (not placed in a Group or Complex)
vorax Edwards, 1921d: *Lutzia* (*Metalutzia*)

Ww

wadai Tanaka, Mizusawa & Saugstad, 1979: *Aedes* (*Stegomyia*), Pseudonigeria Group
wahlbergi (Edwards, 1936): *Coquillettidia* (*Coquillettidia*), Crassipes Group
w-albus (Theobald, 1905f): *Aedes* (*Stegomyia*), W-albus Group, W-albus Subgroup
walcotti (Lane & Cerqueira, 1942): *Runchomyia* (*Ctenogoeldia*)
walkeri (Theobald, 1901b): *Aedes* (*Howardina*), Walkeri Section, Walkeri Group
walkeri Theobald, 1901b: *Anopheles sensu stricto*, Angusticorn Section, Anopheles Series, Maculipennis Group (not placed in a Subgroup)
wallacei Edwards, 1926a: *Aedes* (*Finlaya*), Kochi Group
walravensi Edwards, 1930a: *Anopheles* (*Cellia*), Myzomyia Series (not placed in a Group)
walukasi Belkin, 1962: *Culex* (*Lophoceraomyia*), Fraudatrix Group, Fraudatrix Subgroup, Solomonis Complex
wamanguae Sirivanakarn, 1968: *Culex* (*Lophoceraomyia*), Fraudatrix Group, Fraudatrix Subgroup, Marksae Complex
wannonii Cova Garcia & Sutil O., 1976a: *Culex* (*Carrollia*), Iridescens Group, Iridescens Subgroup
wansoni Edwards, 1941: *Eretmapodites*, Oedipodeios Group
wansoni Wolfs, 1945: *Culex* (*Eumelanomyia*), Rubinotus-rima Group, Rima Subgroup
wanxianensis Lei & Li, 1990: *Orthopodomyia*, Orthopodomyia Section (not placed in a Group)
wardangensis Brust, Ballard, Driver, Hartley, Galway & Curran, 1998: *Aedes* (*Halaedes*)
wardi Reinert, 1976b: *Aedes* (*Rhinoskusea*)
wardi Sirivanakarn, 1977a: *Culex* (*Lophoceraomyia*), Mammilifer Group, Mammilifer Line, Mammilifer Complex
washinoi Lanzaro & Eldridge, 1992: *Aedes* (*Ochlerotatus*), Stimulans Group, Increpitus Complex
wasselli Marks, 1947b: *Aedes* (subgenus uncertain)
watasei Yamada, 1921: *Aedes* (*Phagomyia*)
watsonii (Leicester, 1908): *Anopheles* (*Cellia*), Neomyzomyia Series (not placed in a Group)
watteni Lien, 1968b: *Aedes* (*Downsiomyia*), Niveus Group
wattensis Taylor, 1929: *Aedes* (*Macleaya*), Chaetocruiomyia Group
watti Edwards, 1920a: *Culex sensu stricto*, Duttoni Group
wauensis Huang, 1968b: *Aedes* (*Huaedes*)
waverleyi (Grabham, 1907): *Orthopodomyia*, Bancroftia Section, Signifera Group, Signifera Subgroup

weindorferi (Edwards, 1926a): *Culiseta* (*Culicella*)
wejchoochotei Taai & Harbach, 2015: *Anopheles sensu stricto*, Laticorn Section, Myzorhynchus Series, Barbirostris Group, Barbirostris Subgroup, Barbirostris Complex
wellcomei Theobald, 1904b: *Anopheles* (*Cellia*), Myzomyia Series, Wellcomei Group
wellingtonianus Alcock, 1912: *Anopheles sensu stricto*, Angusticorn Section, Anopheles Series, Lindesayi Group, Lindesayi Subgroup
wellmanii (Theobald, 1905b): *Aedes* (*Hopkinsius*), Hopkinsius Group
wendyae Service, 1959a: *Aedes* (*Catageiomyia*)
wepsterae Komp & Rozeboom, 1951: *Culex* (*Melanoconion*), Melanoconion Section, Inhibitator Group, Inhibitator Subgroup
werneri Baisas & Ubaldo-Pagayon, 1953: *Tripteroides* (*Tricholeptomyia*), Nepenthicola Group, Nepenthicola Subgroup
weschei Edwards, 1935a: *Culex sensu stricto*, Guiarti Group
whartoni Colless, 1965: *Culex* (*Lophoceraomyia*), Fraudatrix Group, Fraudatrix Subgroup, Variatus Complex
whartoni Mattingly, 1965: *Aedes* (*Petermattinglyius*), Aglaonotus Group (the Group is monobasic)
whartoni Reid, 1963: *Anopheles sensu stricto*, Laticorn Section, Myzorhynchus Series, Umbrosus Group, Letifer Subgroup
whitei Barraud, 1923d: *Culex sensu stricto*, Sitiens Group, Vishnui Subgroup
whitmani Lane & Cerqueira, 1942: *Sabethes* (*Peytonulus*)
whitmorei Dunn, 1918: *Aedes* (*Howardina*), Sexlineatus Section, Sexlineatus Subassemblage
whitmorei (Giles, 1904): *Culex sensu stricto*, Sitiens Group, Sitiens Subgroup
whittingtoni Belkin, 1962: *Culex sensu stricto*, Sitiens Group (not placed in a Subgroup)
wigglesworthi Edwards, 1941: *Aedes* (*Aedimorphus*), Alboscutellatus Assemblage, Alboscutellatus Group
wigglesworthi Edwards, 1941: *Culex* (*Eumelanomyia*), Rubinotus-rima Group, Rima Subgroup
wilfredi Colless, 1965: *Culex* (*Lophoceraomyia*), Wilfredi Group
wilkersoni Reinert, 1990b: *Aedes* (*Kenknightia*), Dissimilis Group
willmori (James, 1903) (in Theobald, 1903a): *Anopheles* (*Cellia*), Neocellia Series, Maculatus Group (not placed in a Subgroup)
wilsoni (Barreto & Coutinho, 1944): *Mansonia* (*Mansonia*)
wilsoni Evans, 1934: *Anopheles* (*Cellia*), Neomyzomyia Series, Smithii Group
wilsoni Lane & Whitman, 1943: *Culex* (*Carrollia*), Iridescens Group, Iridescens Subgroup
wilsoni Macdonald, 1958: *Orthopodomyia*, Orthopodomyia Section, Albipes Group, Wilsoni Subgroup (the Subgroup is monobasic)
winchesteri Cockerell, 1919: *Culex* (fossil species)
winkleri Belkin, 1962: *Culex* (*Lophoceraomyia*), Fraudatrix Group, Fraudatrix Subgroup, Bergi Complex
winter Dong, Wu & Mao, 2006: *Topomyia* (*Topomyia*)
wolfsi Ribeiro, 2005a: *Toxorhynchites* (*Afrorhynchus*) (not placed in a Group)
woodi Edwards, 1922d: *Aedes* (*Stegomyia*), Simpsoni Group
worontzowi Pessoa & Galvão, 1936: *Culex* (*Microculex*), Consolator Series
wui Lin, Tseng & Lien, 2008: *Culiseta* (subgenus uncertain)
wysockii Belkin, 1953a: *Uranotaenia* (*Uranotaenia*), Wysockii Section (the Section is monobasic)

Xx

xanthogaster (Edwards, 1924b): *Coquillettidia* (*Coquillettidia*), Crassipes Group
xanthomelaena Edwards, 1925: *Uranotaenia* (*Pseudoficalbia*), Bimaculata Series
xanthozona (van Someren, 1948): *Mimomyia* (*Etorleptiomyia*), Group A (*Etorleptiomyia s. str.*)
xelajuensis de Leon, 1938: *Anopheles sensu stricto*, Angusticorn Section, Anopheles Series, Plumbeus Group
xenismus Harbach, 1995b: *Sabethes* (*Peytonulus*)
xenophobus Ronderos, 1965: *Culex* (*Microculex*), Pleuristriatus Series
xhyphydes Harbach, 1994b: *Sabethes* (*Sabethinus*), Intermedius Group
xui Dong, Zhou, Dong & Mao, 2007: *Anopheles sensu stricto*, Laticorn Section, Myzorhynchus Series, Hyrcanus Group (not placed in a Subgroup)

Yy

yaeyamae Bohart, 1956: *Toxorhynchites* (*Toxorhynchites*), Leicesteri Group, Manicatus Subgroup
yaeyamaensis Somboon & Harbach, 2010 (in Somboon *et al.*, 2010): *Anopheles* (*Cellia*), Myzomyia Series, Funestus Group, Minimus Subgroup, Minimus Complex
yaeyamana Tanaka, Mizusawa & Saugstad, 1975a: *Uranotaenia* (*Pseudoficalbia*), Recondita Series
yaeyamensis Tanaka, Mizusawa & Saugstad, 1979: *Aedes* (*Hulecoeteomyia*)
yaeyamensis Tanaka, Mizusawa & Saugstad, 1979: *Tripteroides* (*Tripteroides*), Nitidoventer Group
yamadai (Ôuchi, 1939): *Toxorhynchites* (*Toxorhynchites*), Leicesteri Group, Manicatus Subgroup
yamadai Sasa, Kano & Takahasi, 1950: *Aedes sensu stricto*, Esoensis Group

yanbarensis Miyagi, 1976: *Topomyia* (not placed in a Subgenus)
yanbareroides Dong & Miyagi, 1995: *Topomyia* (*Suaymyia*)
yangambiensis de Meillon & Lavoipierre, 1944: *Aedes* (*Catageiomyia*)
yaoi Tung, 1955: *Culex* (*Culiciomyia*), Fragilis Group
ybarmis Dyar, 1920b: *Culex* (*Melanoconion*), Melanoconion Section, Intrincatus Group, Intrincatus Subgroup
yeageri Baisas, 1935d: *Culex* (*Eumelanomyia*), Mochthogenes Group, Malayi Subgroup
yerburyi (Edwards, 1917): *Verrallina* (*Neomacleaya*)
yojoae Strickman, 1990: *Culex sensu stricto*, Pipiens Group (not placed in a Subgroup)
yongi Miyagi, Toma & Ramalingam, 1991: *Topomyia* (*Topomyia*)
yovani van Someren, 1951: *Uranotaenia* (*Pseudoficalbia*), Nigripes Section, Ornata Group
ypsipola Dyar, 1922c: *Wyeomyia* (*Dendromyia*)
yunnanensis Dong, Dong & Wu, 2004: *Uranotaenia* (*Uranotaenia*), Testacea Series
yunnanensis Dong, Zhou & Dong, 1995a: *Armigeres sensu stricto*
yunnanensis (Gaschen, 1934): *Aedes* (*Orohylomyia*) (the Subgenus is monobasic)
yusafi (Barraud, 1931b): *Verrallina* (*Harbachius*)
yvonneae Edwards, 1941: *Aedes* (*Catageiomyia*)

Zz

zabanicus Talaga & Duchemin, 2023: *Culex* (*Melanoconion*), Melanoconion Section, Bastagarius Group, Iolambdis Subgroup
zairensis Ribeiro, 2005a: *Toxorhynchites* (*Afrorhynchus*) (not placed in a Group)
zammitii (Theobald, 1903a): *Aedes* (*Acartomyia*)
zavortinki Schick, 1970a: *Aedes* (*Protomacleaya*), Terrens Group, Terrens Subgroup
zeteki Dyar, 1918c: *Culex* (*Melanoconion*), Melanoconion Section, Atratus Group
zethus de Meillon & Lavoipierre, 1944: *Aedes* (*Paulianius*), Zethus Group (the Group is monobasic)
zhangi Gong, 1991: *Topomyia* (*Topomyia*)
ziemanni Grünberg, 1902: *Anopheles sensu stricto*, Laticorn Section, Myzorhynchus Series, Coustani Group
zinzala Zavortink, 1986: *Wyeomyia* (*Zinzala*)
zombaensis Theobald, 1901c: *Culex sensu stricto*, Pipiens Group (not placed in a Subgroup)
zoosophus Dyar & Knab, 1918: *Aedes* (*Protomacleaya*), Zoosophus Group (the Group is monobasic)

Section 3: Anatomical Terminology

DOI: 10.1079/9781800628014.0003

Anatomical Terminology

Adults

This section contains the principal anatomical terminology applied to features of the external skeleton of adult mosquitoes that is used in taxonomic works. It differs in two ways from the section on adult morphology in the anatomical glossary of Harbach & Knight (1980). It includes some new terms and updated information, *e.g.* the terminology applied to wing spots (Wilkerson & Peyton, 1990), and excludes terms that are not normally used in the mosquito taxonomic literature. For synonymous anatomical terms, see Harbach & Knight (1980, 1982).

abdomen (Figs 1, 10a) (pl. abdomens, abdomina; adj. abdominal). The third, posterior division (tagma) of the body, consisting of 10 apparent segments.

abdominal segment (I, II, *etc.*) (Figs 1, 10a). One of the annular subdivisions of the abdomen; labelled I (most anterior) to X (most caudal).

accessory setae (AsS) (Fig. 12b). In male *Anopheles*, usually a pair of prominent setae located near the mesal margin of the dorsal (prerotation sense) surface of the gonocoxite; dorsal to the internal seta and distal to the parabasal setae.

acrostichal area (AA) (Fig. 5a). The median longitudinal area of the scutum from the anterior promontory to the prescutellar space; bearing the acrostichal setae.

acrostichal scales (AcSc) (Fig. 6a). Scales usually occurring in one or two lines on the acrostichal area of the scutum; when two rows are present they are borne on either side of the acrostichal setae (unless absent); sometimes merging posteriorly with the prescutellar scales.

acrostichal setae (AcS) (Fig. 7a). Setae occurring in a median longitudinal row on the acrostichal area of the scutum; absent in some genera and/or species.

adult (Fig. 1). A fully developed, sexually mature animal; the final stage in the mosquito life cycle.

aedeagal sclerite (AeS) (Fig. 12d). In males of certain Culicidae, *e.g. Verrallina*, a supporting sclerite on either side of the phallus; in most male mosquitoes, one of a pair of lateral sclerotizations of the aedeagus.

aedeagus (Ae) (Figs 12b,d, 13b,d) (pl. aedeagi; adj. aedeagal). In most male mosquitoes, the central body of the phallosome serving as the intromittent organ; in certain male mosquitoes, *e.g. Verrallina*, the distal part of the phallus.

alula (Al) (Fig. 8a,d) (pl. alulae). A weakly differentiated lobe of the posterior margin of the wing bounded proximally by the upper calypter, distally by the axillary incision and anteriorly by the base of the anal vein.

anal vein (A) (Fig. 8a). A principal longitudinal vein between the cubitus and the alula; only one anal vein (1A) is normally developed with scales; vein 2A is usually developed with scales in *Culex univittatus* (see Harbach, 1988: fig. 2D).

anepisternal cleft (AEC) (Fig. 5b). The membranous space separating the anterior and posterior mesanepisterna and extending anteroventrally between the anterior mesanepisternum and the mesokatepisternum; postspiracular scales sometimes extend onto the membrane of the anepisternal cleft.

anepisternum (pl. anepisterna; adj. anepisternal). The upper or dorsal area of the episternum, when the latter is divided by a longitudinal mesanapleural suture.

antealar area (AnA) (Fig. 5a,b). The lateral area of the scutum at the level of the paratergite; bearing the antealar scales and setae.

antealar scales (AnSc) (Fig. 6a). Scales occurring on the antealar area of the scutum. A distinct line of antealar scales may be continuous anteriorly with the lateral scutal fossal scales and posteriorly with a lateral line of supraalar scales.

antealar setae (AnS) (Fig. 7b). Setae occurring in a longitudinal row on the antealar area of the scutum immediately above the paratergite; sometimes more or less continuous with the supraalar setae posteriorly.

antenna (A) (Fig. 1) (pl. antennae; adj. antennal). One of the paired anterior appendicular sensory organs of the head, consisting of a narrow ring-like basal segment (scape), a second globular and cup-shaped segment (pedicel) and a third segment (flagellum) comprised of a series of 13 flagellomeres.

antepronotal scales (ApSc) (Fig. 6b). Scales occurring in a group on the antepronotum; sometimes forming a tuft.

antepronotal setae (ApS) (Fig. 7b). Setae occurring on the antepronotum.

antepronotum (Ap) (Figs 1, 5b) (pl. antepronota; adj. antepronotal). The anterior division of the pronotum; visible as a lateral setose lobe just above the proepisternum.

anterior dorsocentral area (ADA). The part of the dorsocentral area cephalad of about the level of the antealar area; with setae and scales.
anterior dorsocentral scales (ADSc) (Fig. 6a). Scales occurring in a more or less distinct line located cephalad of the posterior scutal fossal scales on the anterior dorsocentral area of the scutum.
anterior dorsocentral setae (ADS) (Fig. 7a). Setae occurring on the anterior dorsocentral area; occurring in one or more rows.
anterior hard palate (AHP) (Fig. 4d,e). The irregular plate-like anterior area of the clypeopalatum; bearing one or more palatal setae.
anterior mesanepisternum (AMas) (pl. anterior mesanepisterna). The area of the mesopleuron consisting of the prespiracular area anterior to the spiracle, the hypostigmal area immediately below the spiracle, the subspiracular area ventral to the hypostigmal area and the postspiracular area posterior to the hypostigmal and subspiracular areas.
anterior mesepimeral scales (AMSc) (Fig. 6b). Scales occurring in a group on the cephalodorsal quarter of the mesanepimeron.
anterior promontory (AnP) (Fig. 5a,b). The transverse marginal area of the mesonotum at the anterior end of the acrostichal and dorsocentral areas, more or less projecting cephalad over the cervix; often bearing setae; the area traditionally referred to as the anterior promontory includes the antedorsocentral area of Reinert (1999b).
anterior scutal fossal setae (ASFS) (Fig. 7b). Setae occurring in a group on the anterolateral margin of the scutal fossa above the posterior margin of the antepronotum.
anterior tentorial pit (ATP) (Fig. 3a). An external depression in the cranium at the base of each anterior tentorial arm, occurring where the epistomal suture meets the antennal socket.
apicodorsal lobe (ADL) (Fig. 10c). In male mosquitoes, the most distal area of the dorsal (prerotation sense) surface of the gonocoxite; an apical dorsomesal lobe.
apodeme. Any ridge or ingrowth of the cuticle forming the endoskeleton; usually marked externally by a suture or pit.
apodeme of gonocoxite (AG) (Fig. 13a). In male mosquitoes, an apodematous continuation of the proximal mesal dorsal (prerotation sense) corner of the gonocoxite; fused or articulated with the basal piece.
arculus (Ar) (Fig. 8a). A crossvein between the radius and cubitus near the base of the wing.
axillary incision (AI) (Fig. 8a,d). A notch on the posterior margin of the wing between the alula and distal part.
basal dorsomesal lobe (BDL) (Fig. 13a). In male mosquitoes, a basal lobe on the mesal margin of the dorsal (prerotation sense) surface of the gonocoxite.
basal lateral arm (BLA) (Fig. 12c). In *Culex* males, a laterally directed basal process of the paraproct; may bend in caudal and/or ventral directions.
basal mesal lobe (BML) (Fig.13c). In male mosquitoes, a small lobe or sclerite located basomesally on the gonocoxite; usually connected ventrally with its mate; homologous with the claspette.
basal spur (BS) (Fig. 8a). A longitudinal proximal extension of the radial sector or radius-four-plus-five when these are sharply bent in the form of a crossvein near their points of origin.
calypter (Cl) (Fig. 8a). Either of two lobes of the posterior margin of the wing proximad of the alula. See **lower calypter** and **upper calypter**.
capitellum (Ca) (Fig. 8b) (pl. capitella). The expanded, knob-like distal part of the halter.
cell. An area of wing membrane delimited by veins or by veins and the wing margin; named after the vein immediately anterior to it, *e.g.* cell M_1 is the area of wing membrane posterior to vein M_1, or after the posterior element if the vein is formed by the 'fusion' of two vein branches, *e.g.* cell M_2 is the area of wing membrane posterior to vein M_{1+2}. The area of wing membrane posterior to media-one-plus-two and media-two.
cercal sclerite (CSc) (Fig.12c). In male mosquitoes, the more or less distinctly sclerotized area between the paraprocts on the dorsal (prerotation sense) surface of the proctiger or one of a pair of sclerites in the same area.
cercal setae (Cse) (Figs 12c, 13a,b). Setae borne on a cercus. All setae on the proctiger of male mosquitoes are considered to be from the cercus and are thus cercal setae; usually one to a few pairs is present distally.
cercus (Ce) (pl. cerci; adj. cercal). One of the pair of appendages of abdominal segment XI; in females (Figs 1, 10a,b, 11a–d), a pair of more or less conspicuous lobes posterior to tergum IX; in males (Figs 12c, 13b), largely membranous, not sharply delimited from adjacent parts of the proctiger and with the paraproct representing the main area of sclerotization.
cervical sclerite (CS) A sclerite situated laterally on the cervix; articulates with the head anteriorly and the proepisternum posteriorly.
cervix (Cv) (Figs 1, 2b,c, 13a,b) (pl. cervices, cervixes; adj. cervical). The membranous tube connecting the thorax to the head, laterally supported by the cervical sclerites.
cibarial armature (CA). In some female mosquitoes, a series of specialized spicules, cibarial teeth, borne on a transverse ridge, cibarial crest, and a group of spiculate ridges, cibarial ridges, lying ventrally at the posterior margin of the cibarium.
cibarial crest (CC). The transverse ridge located ventrally at the posterior margin of the cibarium that supports the cibarial teeth.
cibarial dome (CD) (Fig. 4e). A hemispherical, spiculate structure projecting from the clypeopalatum; at the posterior margin of the cibarium, presumably the posterior part of the posterior hard palate.

cibarial ridge (CiR). In female mosquitoes of some genera, one of a series of short spiculate ridges lying just posterior to the cibarial crest and forming part of the cibarial armature.

cibarial setae (CiS). Various types of setae borne within the cibarium; in groups designated by location as dorsal, palatal and ventral setae. See **dorsal setae, palatal setae** and **ventral setae**.

cibarial teeth (CT) (Fig. 4d,e). In some female mosquitoes, a series of specialized spicules borne on the cibarial crest located ventrally at the posterior margin of the cibarium; commonly of two kinds called cones and rods; when a single type is present they have more or less the character of cones. See **cone** and **rod**.

cibarium (Cib) (Fig. 4d,e) (pl. cibaria; adj. cibarial). The preoral food cavity between the base of the hypopharynx and the under surface of the clypeus, clypeopalatum.

claspette (Cl) Figs 10c, 13a). In male mosquitoes, a descriptive term applied to a variably shaped lobe arising mesally at the base of the gonocoxite; homologous with the basal mesal lobe. In anophelines, the claspette is often divided into a dorsal lobe and a ventral. The dorsal lobe in some species is more or less divided so that in addition to the ventral lobe there is a middle lobe and a dorsal lobe. In species of the subgenus *Nyssorhynchus*, the ventral lobes are fused (median lobe). The latter usually bears a small basal lobule on either side that may be homologous with the middle lobe.

claspette filament (CF) (Fig. 13a). In male mosquitoes, a specialized apical seta of the claspette; usually simple or foliform.

claspette stem (CSt) (Fig. 13a). In male mosquitoes, the basal part of the claspette; bearing the claspette filament apically.

claw. See **unguis**.

clypeopalatum (Cp) (pl. clypeopalata; adj. clypeopalatal). The part of the palatum formed by the under surface of the clypeus, the roof of the cibarium.

clypeus (Clp) (Figs 1, 2a,b, 3a) (pl. clypei; adj. clypeal). The conspicuous and projecting facial area delimited above and below by sutures.

compound eye (CE) (Figs 1–3, 15a). A light-perceptive organ consisting of an aggregation of optic elements (ommatidia); convex, kidney-shaped and occupying much of the anterolateral surface of the head.

cone (Cn) (Fig. 8d). One of the specialized spicules comprising the cibarial teeth.

costa (C) (Fig. 8a) (adj. costal). The usual first principal longitudinal vein of the wing; typically extending around the wing apex from its articulation with the thorax; sometimes absent posteriorly and apically; sometimes somewhat removed from the margin.

coxa (Cx) (Figs 1, 5b) (pl. coxae; adj. coxal). The basal segment of the leg articulating with the sternum and/or pleuron; referred to as fore- (Cx-I), mid- (Cx-II) or hindcoxa (Cx-III) as appropriate.

coxal setae (CxS) (Fig. 6b). Any setae occurring on a coxa; may be described as inner, outer, anterior, posterior, upper and lower fore-, mid- and hindcoxal setae as appropriate.

crossvein. A short, usually transverse vein between two longitudinal veins; the usual crossveins are the humeral crossvein between the costa and subcosta, the radiomedial crossvein between the costa and subcosta, the radiomedial crossvein between the radius and media and the mediocubital crossvein between the media and cubitus.

cubitus (Cu). The usual fifth longitudinal vein of the wing; usually with two main branches, cubitus anterior and cubitus posterior.

cubitus anterior (CuA) (Fig. 8a). The anterior branch of the cubitus, unbranched and bearing scales.

cubitus posterior (CuP) (Fig. 8a). The posterior branch of the cubitus; weakly developed, unbranched and without scales; closely parallels the cubitus anterior.

distal sclerite (DS) (Fig. 4b,c). The labellar sclerite marking the distal segment of the labellum; sometimes partially or completely fused with the proximal sclerite.

dorsal aedeagal bridge (DAB) (Figs 12d, 13b,d). In male mosquitoes, the transverse sclerotization connecting the aedeagal sclerites nearest the anus.

dorsal arm (DOA) (Fig. 12d). In *Culex* males, the dorsal process (in normal resting position) of the outer division of the lateral plate.

dorsal setae (DSe) (Fig. 4d,e). The cibarial setae borne lateral and/or posterior to the anterior hard palate of the clypeopalatum; the dorsal setae usually consist of two peg-like setae located immediately adjacent to the posterolateral margin of the anterior hard palate, the more anterior of these is borne within a pit, and a linear series of one to six small simple setae borne lateral to the former.

dorsocentral area (DA) (Fig. 5a,b). A longitudinal area of the scutum on each side of the acrostichal area extending from the antedorsocentral area to the prescutellar area; indefinitely divided into anterior and posterior dorsocentral areas at the level of the antealar area; bearing the dorsocentral scales and setae. The dorsocentral area was traditionally defined as the linear area between the anterior promontory and the prescutellar area. Reinert (1999b) restricted the anterior promontory to the marginal area of the scutum anterior to the acrostichal area and defined the marginal area anterior to the dorsocentral area as the antedorsocentral area.

dorsocentral scales (DSc). Scales borne on the dorsocentral area of the scutum; may occur in more or less distinct anterior, posterior, inner and/or outer longitudinal rows. See **anterior dorsocentral scales** and **posterior dorsocentral scales**.

dorsocentral setae (DS). Setae occurring on the dorsocentral area; divided into anterior and posterior rows near the median posterior corner of the scutal fossa at about the level of the antealar area. See **anterior dorsocentral setae** and **posterior dorsocentral setae**.

empodium (Em) (Fig. 9b–d) (pl. empodia). The distal median prolongation of the unguitractor plate.

epimeron (pl. epimera; adj. epimeral). The posterior area of a pleuron; separated from the anterior area (episternum) by the pleural suture. See **mesepimeron** and **metepimeron**.

episternum (pl. episterna; adj. episternal). The anterior area of a pleuron; separated from the posterior area (epimeron) by the pleural suture. See **mesepisternum** and **metepisternum**.

erect scales (ESc) (Fig. 3c,d). The usually numerous straight, rigid scales standing up on the dorsum (vertex and occiput) of the head; sometimes confined to a transverse line at the back of the head (occiput). See *forked* under **scale** in the **Vestiture** section.

femur (Fe) (Fig. 1) (pl. femora; adj. femoral). The usual third segment of the leg; generally the largest leg segment; referred to as fore- (Fe-I), mid- (Fe-II) or hindfemur (Fe-III) as appropriate.

flagellar whorl (FW) (Figs 1, 2b). A ring of long curved setae borne at the base of each of flagellomeres 2–13 of the antenna; a weak whorl occurs at the base of flagellomere 1 in males.

flagellomere (Flm) (Fig. 1). An individual unit of the antennal flagellum.

flagellum (Fl) (Fig. 1) (pl. flagella; adj. flagellar). The third segment of the antenna, comprised of 13 flagellomeres.

forecoxa (Cx-I) (Fig. 5b) (pl. forecoxae; adj. forecoxal). The first (basal) segment of a foreleg, articulating with a process of the proepisternum. See **coxa**.

forefemur (Fe-I) (Fig. 9a) (pl. forefemora). The third segment of a foreleg. See **femur**.

forefemur length. The length of the forefemur measured from its point of articulation with the foretrochanter to its apex along the ventral margin (the anteroventral margin when in the resting position).

forefemur/proboscis ratio (f/p). The ratio of the forefemur length to the proboscis length.

foreleg (Figs 1, 9a). One of the pair of legs of the prothorax.

foretarsus (Ta-I) (Fig. 9a,b) (pl. foretarsi; adj. foretarsal). The fifth segment of a foreleg, consisting of five tarsomeres numbered 1–5 from proximal to distal. See **tarsus**.

foretibia (Ti-I) (Fig. 9a) (pl. foretibiae; adj. foretibial). The fourth segment of a foreleg. See **tibia**.

foretrochanter (Tr-I) (Fig. 1). The second segment of a foreleg. See **trochanter**.

foreunguis (U-I) (Figs 1, 9a–c) (pl. foreungues). The anterior or posterior claw of one of the forelegs. See **unguis**.

fringe scales (FS) (Fig. 8c). Scales that usually occur in four rows along the posterior edge of the wing. See **primary fringe scales, secondary fringe scales, tertiary fringe scales** and **wing fringe**.

frons (Fr) (Figs 2a, 3a). The facial area of the cranium lying between the postfrontal sutures dorsally and the epistomal suture ventrally; largely obscured by the convergent antennal sockets; comprised of a narrow ventral prefrons between the convergent antennal sockets and a broader dorsal postfrons bearing the antennal pulsating organ.

frontal tuft (FT) (Fig. 3b). A specialized name applied to a group of erect elongate simple setae and fusiform scales arising from the interocular space and the immediately adjacent area of the vertex; applied in particular to anophelines.

gena (Ge) (Figs 2c, 3e) (pl. genae; adj. genal). The anterolateral area of the cranium extending ventroposteriorly from the anterior tentorial pit behind the clypeus and united ventrally with its mate between the compound eyes.

genitalia (G) (Figs 10–13) (pl. genitalia, no singular form). In general, the abdominal structures involved in reproduction, *i.e.* abdominal segments IX (the genital segment) and X (parts of the proctiger may take part in the copulatory process); in some species, abdominal segment VIII may be sexually differentiated from the preceding ones in one or more details and would in such cases be treated as a part of the genitalia. The complex of anal and genital parts borne caudad of abdominal segment VII in males rotate 180° of a few hours after emergence, and this has caused much confusion in the morphological and taxonomic literature. To solve this, the terms 'tergal' and 'sternal' have sometimes been used. This usage, however, is equally confusing because many parts of the genitalia are neither tergal nor sternal in origin. Consequently, the terms 'dorsal' and 'ventral' are used here in the prerotation sense, *i.e.* with their true morphological connotation.

gonocoxal apodeme (GA) (Figs 10c, 12d, 13a,b,d). In male mosquitoes, one of the pair of basal supporting sclerotizations of the phallosome; lying within abdominal segment IX and often extending anteriorly into segment VIII; articulated with the paramere, sometimes articulated with the base of the proctiger and often connected with the apodeme of gonocoxite.

gonocoxite (Gc) (Figs 10c, 12a–c, 13a,c). The coxite of the gonocoxopodite. In male mosquitoes, the largest unit of the genitalia; the dorsal and ventral surfaces (prerotation sense) with various lobes that are described with appropriate combinations of positional adjectives without intent to imply strict homologies; bearing the gonostylus apically.

gonocoxopodite (Gp). The coxopodite, *i.e.* the basal segment or coxa, of an appendage of the genital segment, abdominal segment IX, consisting of the gonocoxite and gonostylus.

gonostylar claw (GC) (Figs 10c, 12a–c, 13a). The differentiated, more or less spiniform seta(e) at or near the apex of the gonostylus in some male mosquitoes; frequently flattened or variously modified and markedly different from other setae; not strictly homologous between all species.

gonostylus (Gs) (pl. gonostyli; adj. gonostylar). The stylus of the gonocoxopodite; movable and attached at or near the apex of the gonocoxite.

halter (Hl) (Figs 1, 5b, 6b, 7b, 8b) (pl. halteres). The modified, reduced, drumstick-shaped metathoracic wing consisting of a thick base, scabellum, a slender stem-like part, pedicel, and an expanded distal part, capitellum. Compare **metathoracic wing** in the **Pupa** section.

head (Figs 1–3). The anterior section (tagma) of the body bearing the compound eyes, antennae and mouthparts; separated from the thorax by the cervix.

hindcoxa (Cx-III) (Fig. 5b) (pl. hindcoxae; adj. hindcoxal). The first (basal) segment of a hindleg. See **coxa**.

hindfemur (Fe-III) (Fig. 9a) (pl. hindfemora; adj. hindfemoral). The third segment of a hindleg. See **femur**.

hindleg (Figs 1, 9a). One of the pair of legs of the metathorax.

hindtarsus (Ta-III) (Fig. 9a) (pl. hindtarsi; adj. hindtarsal). The fifth segment of a hindleg. See **tarsus**.

hindtibia (Ti-III) (Fig. 9a) (pl. hindtibiae; adj. hindtibial). The fourth segment of a hindleg. See **tibia**.

hindtrochanter (Tr-III) (Fig. 1). The second segment of a hindleg. See **trochanter**.

hindunguis (U-III) (Fig. 9a,b) (pl. hindungues). The anterior or posterior claw of one of the hindlegs. See **unguis**.

humeral crossvein (h) (Fig. 8a,d,e). The crossvein extending between the costa and subcosta near the base of the wing.

hypostigmal area (HyA) (Fig. 5b). The weakly sclerotized or membranous area of the anterior mesanepisternum immediately below the mesothoracic spiracle; its boundary with the subspiracular area is usually poorly defined; scales sometimes occur on this area.

hypostigmal scales (HySc) (Fig. 6b). Scales occurring in a small cluster on the hypostigmal area of the anterior mesanepisternum.

inner division (ID) (Fig. 12d). In *Culex* males, the sternal and most mesal part of the lateral plate of the phallosome in the normal resting position.

insula (I) (Fig. 11a–c,e) (pl. insulae; adj. insular). In some female mosquitoes, a median, usually enlarged area of the lower vaginal lip; often oriented differently than adjacent parts of the lip and connected with them by weakly sclerotized areas; spiculate and usually with setae and tuberculi.

insular setae (IsS) (Fig. 11c,e). Setae borne on the insula of the female genitalia.

interlobar space (ILS) (Fig. 12c). In male mosquitoes, the area between the tergum IX lobes.

internal seta (InS) (Fig. 12a,b) (pl. internal setae). A strong subapical seta arising mesally on the gonocoxite of *Anopheles* males.

interocular setae (ISe) (Fig. 2a). The ocular setae arising on the interocular space; often somewhat longer than other ocular setae.

interocular space (IS) (Figs 2a, 3a,c,d). The narrow part of the vertex between the compound eyes and above the postfrontal sutures.

katepimeron (pl. katepimera; adj. katepimeral). The lower or ventral area of the epimeron, when the latter is divided by a longitudinal suture. See **mesokatepimeron**.

katepisternum (pl. katepisterna; adj. katepisternal). The lower or ventral area of an episternum, separated from the upper or dorsal area (anepisternum) by a suture or fissure. See **mesokatepisternum**.

knee. The joint between the femur and tibia.

knee spot (KS). A group of pale scales occurring at the terminations of the femur, tibia and first tarsomere; normally applied to a band of pale scales occurring at the apex of the femur.

labellar basal sclerite (LaBS) (Fig. 4a–c). One of several small sclerites occurring ventrally in the membranous articulation between the prementum and the labella of the labium.

labellar mesal sclerite (LMs). One of two narrow sclerotized strips borne on the inner surface of each labellum.

labellar scales (LS) (Fig. 4c). Scales borne on the labellum; absent in anophelines, present on the proximal sclerite of the labellum in certain species and groups of culicine mosquitoes; covering the labella of *Toxorhynchites*.

labellar sclerite (LaS). A cuticular plate of the labellum; two are present in most mosquitoes; partially or completely fused in certain taxa; however, the labella of toxorhynchitine and anopheline mosquitoes consist of a single sclerite, but in *Chagasia* and subgenus *Anopheles* of *Anopheles* (Fig. 4a) the sclerite is divided dorsally.

labellum (La) (Figs 1, 4a–c) (pl. labella; adj. labellar). One lobe of a partially bilobed structure situated at the apex of the labium; between the lobes on the midline is the ligula; each labellum possesses several labellar sclerites which possibly represent labial palpal segments.

labial basal setae (LBS) (Fig. 2b). A row or cluster of setae occurring ventrally near the base of the labium.

labium (Lb) (pl. labia; adj. labial). The elongate stout part of the proboscis consisting of the prementum, a pair of short labella articulated with the distal end of the prementum and a short ligula between the labella.

lateral anterior promontory scales (APSL). Scales usually forming a tuft on the anterolateral corner of the scutum in anopheline mosquitoes.

lateral anterior promontory setae (LAPS). Setae occurring in a small cluster on the anterolateral comer of the anterior promontory at or near the anterior end of the dorsocentral setae.

lateral arm (LA) (Fig. 12d). In *Culex* males, the variously developed lateral process (in normal resting position) of the outer division of the lateral plate of the phallosome.

lateral flange (LF) (Fig. 4d). The lateral expansion of the posterior part of the cibarium.

lateral plate (LP) (Fig. 12c,d). In *Culex* males, one of the pair of variously developed lateral sclerotizations of the phallosome located between the aedeagus and proctiger; often consisting of inner and outer divisions which may also be subdivided; homologous with the opisthophallic sclerite.

lateral prescutellar scales (SSPL) (Fig. 6a). The prescutellar scales occurring in a line on each lateral margin of the prescutellar area of the scutum; difficult to separate anteriorly from the dorsocentral scales.

lateral scutal fossal scales (SFSL) (Fig. 6a). Scales occurring in a more or less distinct line on the lateral margin of the scutal fossa above the ante- and postpronotum; may be continuous posteriorly with the antealar scales.

lateral scutal fossal setae (LSFS) (Fig. 7a). Setae occurring in a small group on the lateral margin of the scutal fossa above the postpronotum.

lateral scutellar scales (LSSc) (Fig. 6a). Scales occurring in a small group on each lateral lobe of the scutellum.

lateral scutellar setae (LSS) (Fig. 7a). Setae occurring in a small group on each lateral lobe of the scutellum.

laterotergite (Lt) (Fig. 1). The lateral part of abdominal tergum I when more or less distinctly separated from the median dorsal part; usually not visible from above.

leaflet of aedeagus (LAe) (Fig. 12a). In male anopheline mosquitoes, one of the elongate leaflet-like structures usually occurring at the apex of the aedeagus; homologous with the ventral arms in *Culex*.

Leg (Figs 1, 9a). One of the paired appendages ordinarily used in locomotion and support, composed of the coxa, trochanter, femur, tibia, tarsus and posttarsus. The coxal surfaces are described as though the coxa is in the normal resting position, the surfaces being the outer, inner, anterior and posterior; the surfaces of the other leg segments are described as though the leg is extended horizontally perpendicular to the longitudinal axis of the body, the surfaces being the dorsal, ventral anterior and posterior; individual legs and leg segments are denoted by prefixing fore-, mid- or hind- as appropriate, thus foreleg, midfemur, hindtrochanter, *etc.*

ligula (Lg) (Fig. 4a–c) (pl. ligulae; adj. ligular). The sharply pointed lobe occurring distally on the midline of the labium between the labella.

lower calypter (LC). The proximal calypter of the dipteran wing; typically situated below the upper calypter. See **calypter** and **upper calypter**.

lower mesepimeral scales (LMSc) (Fig. 6b). Scales occurring in a group on the ventroanterior quarter of the mesanepimeron.

lower mesepimeral setae (MeSL) (Fig. 7b). Setae occurring in groups on the anterior, middle and/or posterior area of the mesanepimeron below the level of the metathoracic spiracle.

lower mesokatepisternal scales (MScL) (Fig. 6b). Scales occurring in a group located immediately in front of the lower mesokatepisternal setae.

lower mesokatepisternal setae (MkSL) (Fig. 7b). Setae occurring in a more or less vertical line along the posterior margin of the mesokatepisternum at about the level of the mesotrochantin; sometimes continuous with the upper mesokatepisternal setae.

lower ocular setae (LOcS). The ocular setae occurring along the lateral margin of the compound eye. See **ocular setae**.

lower postpronotal scales (LPSc) (Fig. 6b). Scales occurring in a group on the ventral part of the postpronotum.

lower prealar scales (LPrSc) (Fig. 6b). Scales occurring in a group on the posterior mesanepisternum below the prealar setae.

lower proepisternal scales (PScL). The proepisternal scales occurring in a group mesad of the forecoxa and below the upper proepisternal scales; sometimes continuous dorsolaterally with the upper proepisternal scales. *Synonym*: Prosternal scaling.

lower proepisternal setae (PeSL). The proepisternal setae occurring in a group mesad of the forecoxa and below the upper proepisternal setae. See **proepisternal scales** and **upper proepisternal scales**.

lower vaginal lip (LVL) (Fig. 11a–c). In female mosquitoes, the sclerotized and pigmented rim of the floor of the vagina; articulated with the upper vaginal lip; its median area is joined to the insula; usually spiculate.

lower vaginal sclerite (LVS) (Fig. 11b). In female mosquitoes, a pigmented sclerite of the vaginal floor other than the lower vaginal lip; usually absent; well developed in certain aedine species, *e.g.* those of subgenera *Neomacleaya* and *Verrallina* of the genus *Verrallina*.

maxillary palpus (MPlp) (Figs 1, 2b,c, 3a, 4e) (pl. maxillary palpi; adj. maxillary palpal). The appendage of the maxilla, inserted immediately below the clypeus and laterad to the base of the proboscis; varies widely in form according to genus and sex; primitively consisting of five palpomeres.

media (M) (Fig. 8a) (adj. medial). The usual fourth principal longitudinal vein of the wing; branches into media-one-plus-two and media-three-plus-four with the former branching into media-one and media-two.

median anterior promontory scales (APSM). Scales usually forming a tuft at the middle of the anterior margin of the anterior promontory in anopheline mosquitoes. See **median anterior promontory setae**.

median anterior promontory setae (MAPS) (Fig. 7a). Setae occurring in a small cluster anteriorly on the midline of the anterior promontory.

median prescutellar scales (SSPM) (Fig. 6a). Scales occurring in a transverse line on the anterior margin and/or in a median longitudinal line on the prescutellar area of the scutum.

median scutal fossal scales (SFSM) (Fig. 6a). Scales occurring in a group on the central area of the scutal fossa.

median scutal fossal setae (MSFS) (Fig. 7a). Setae occurring in a group near the centre of the scutal fossa.

median scutellar scales (MSSc). The scales occurring in a group on the middle lobe of the scutellum.

median scutellar setae (MSS). The setae occurring in a group on the middle lobe of the scutellum.

media-one (M_1) (Fig. 8a). The anterior branch of media-one-plus-two which reaches the margin of the wing near the apex. See **media**.

media-one-plus-two (M_{1+2}) (Fig. 8a). The anterior branch of the media (media posterior) which splits into media-one and media-two near the wing apex. See **media**.

media-three-plus-four (M_{3+4}) (Fig. 8a). The posterior branch of the media (media posterior); with a short basal crossvein-like segment which traditionally has been considered to be the mediocubital crossvein; the longer distal part has been considered to be cubitus-one. See **media**.

media-two (M_2) (Fig. 8a). The posterior branch of media-one-plus-two. See **media**.

mediocubital crossvein (mcu) (Fig. 8a). The crossvein extending between the media and cubitus, connecting media-three-plus-four and the cubitus anterior.

meron (m) (pl. mera; adj. meral). The lateral postarticular area of the coxal base, more associated with the pleuron than the coxa; present on the meso- and metathorax, mesomeron and metameron, respectively.

mesal membrane (MM) (Fig. 12c). In some male mosquitoes, the unsclerotized mesal surface of the gonocoxite.

mesanapleural suture (MAnS). The suture occurring between the mesanepisternum and the mesokatepisternum; in some mosquitoes, *e.g.* species of the nominotypical subgenus of *Uranotaenia*, weakly developed between the posterior mesanepisternum and the mesokatepisternum.

mesanepimeron (Mam) (Fig. 5b) (pl. mesanepimera; adj. mesanepimeral). The large upper area of the mesepimeron, bearing scales and setae referred to simply as mesepimeral elements because of the insignificant nature of the mesokatepimeron.

mesanepisternum (Mas) (pl. mesanepisterna; adj. mesanepisternal). The upper area of the mesepisternum separated from the mesokatepisternum by the mesanapleural suture.

mesepimeral scales (MeSc). Scales borne on the mesepimeron; anterior, posterior, upper and lower mesepimeral scales occur on the mesanepimeron.

mesepimeral setae (MeS). Setae borne on the mesepimeron, occurring on the mesanepimeron, usually in a group inserted dorsally, but when, as in a few species, a ventral group is also present the latter are termed the lower mesepimeral setae and the former the upper mesepimeral setae.

mesepimeron (Mm) (Fig. 1) (pl. mesepimera; adj. mesepimeral). The area of the mesopleuron posterior to the mesopleural suture, divided horizontally into a large mesanepimeron and an insignificant mesokatepimeron located immediately below the former; scales and setae that occur on the mesanepimeron are referred to simply as mesepimeral because of the reduced nature of the mesokatepimeron.

mesepisternum (Ms) (Fig. 1) (pl. mesepisterna; adj. mesepisternal). The area of the mesopleuron anterior to the mesopleural suture.

mesokatepimeron (Mkm) (Fig. 5b) (pl. mesokatepimera; adj. mesokatepimeral). The lower division of the mesepimeron, an insignificant strip of cuticle projecting over the dorsal edge of the mesomeron.

mesokatepisternal scales (MkSc). Scales borne on the mesokatepisternum, occurring in two groups, the upper mesokatepisternal scales and lower mesokatepisternal scales.

mesokatepisternal setae (MkS). Setae borne on the mesokatepisternum, occurring in two groups, the upper mesokatepisternal setae and lower mesokatepisternal setae, along the posterior margin.

mesokatepisternum (Mks) (Fig. 5b) (pl. mesokatepisterna; adj. mesokatepisternal). The lower area of the mesepisternum, separated from the anterior mesanepisternum by the anepisternal cleft; usually united with the posterior mesanepisternum but sometimes separated from it by the mesanapleural suture; united ventrally with its mate of the opposite side.

mesomeron (Msm) (Figs 1, 5b) (pl. mesomera; adj. mesomeral). The meron of the mesothorax located slightly above and immediately behind the midcoxa below the mesepimeron.

mesonotum (Mn) (pl. mesonota; adj. mesonotal). The notum (tergum) of the mesothorax comprised mainly of the scutum and scutellum.

mesopleural suture (ms) (Fig. 5b). The external groove (pleural suture) extending from the base of the wing to the midcoxal articulation; separating the mesepisternum from the mesepimeron.

mesopleuron (Mp) (pl. mesopleura; adj. mesopleural). The pleuron of the mesothorax, comprised principally of the mesepisternum and mesepimeron.

mesopostnotal scales (MpSc) (Fig. 6b). Scales occurring in a small cluster on the median posterior area of the mesopostnotum.

mesopostnotal setae (MpnS) (Fig. 7b). Setae occurring in a small group on the median posterior area of the mesopostnotum.

mesopostnotum (Mpn) (Figs 1, 5a,b) (pl. mesopostnota; adj. mesopostnotal). The well-developed postnotum of the mesothorax.

mesothoracic spiracle (MS) (Figs 1, 5b). The spiracle of the mesothorax, located just below the scutal margin behind the postpronotum and prespiracular area.

mesothorax (M) (pl. mesothoraces, mesothoraxes; adj. mesothoracic). The second or middle segment and largest part of the thorax, especially dorsally, that bears the wings.
mesotrochantin (Mtr) (Fig. 5b). The precoxal sclerite of the mesopleuron, the triangular sclerite lying between the mesokatepisternum and the mesomeron just above the midcoxa.
metameral scales (MeSc) (Fig. 6b). Scales occurring in a small cluster on the metameron.
metameron (Mem) (Figs 1, 5b) (pl. metamera; adj. metameral). The meron of the metathorax represented by a vertically narrow sclerite above the hindcoxa.
metapleural suture (mts) (Fig. 5b). The external groove (pleural suture) separating the metepisternum from the metepimeron; an irregular groove extending downward from near the base of the halter.
metapleuron (Mtp) (pl. metapleura; adj. metapleural). The pleuron of the metathorax, represented by only a reduced metepisternum and metepimeron.
metathoracic spiracle (MtS) (Figs 1, 5b). The spiracle of the metathorax, located between the upper area of the mesepimeron and the halter.
metepimeron (Mtm) (Fig. 5b) (pl. metepimera; adj. metepimeral). The area of the metapleuron, reduced to a small triangular area which is sometimes poorly separated from the metepisternum and appears to be continuous dorsally with the metanotum.
metepisternal scales (MtSc) (Fig. 6b). Scales occurring in a small group on the metepisternum just below the metathoracic spiracle.
metepisternum (Mts) (Figs 1, 5b) (pl. metepisterna; adj. metepisternal). The anterior area of the metapleuron, the area immediately behind and below the metathoracic spiracle.
metapostnotum (Mtpn) (Figs 1, 5b) (pl. metapostnota; adj. metapostnotal). The postnotum of the metathorax, a narrow plate behind the intersegmental cleft which extends ventrally to the metameron of each side; apparently reflected upon and united with the anterior part of abdominal segment I.
metathorax (T) (pl. metathoraces, metathoraxes; adj. metathoracic). The third, posterior segment of the thorax; reduced, appearing as part of the abdomen and bearing the halteres.
midcoxa (Cx-II) (Fig. 5b) (pl. midcoxae; adj. midcoxal). The first (basal) segment of a midleg, articulating with the mesotrochantin laterally and a process of the mesosternum ventrally. See **coxa**.
midfemur (Fe-II) (Fig. 9a) (pl. midfemora; adj. midfemoral). The third segment of a midleg. See **femur**.
midleg (Fig. 9a). One of the pair of legs of the mesothorax.
midtarsus (Ta-II) (Fig. 9a,b) (pl. midtarsi; adj. midtarsal). The fifth segment of a midleg. See **tarsus**.
midtibia (Ti-II) (Fig. 9a) (pl. midtibiae; adj. midtibial). The fourth segment of a midleg. See **tibia**.
midtrochanter (Tr-II) (Fig. 1). The second segment of a midleg. See **trochanter**.
midunguis (U-II) (Fig. 9a,b) (pl. midungues). The anterior or posterior claw of one of the midlegs. See **unguis**.
ninth tergal lobe. See **tergum IX lobe**.
notum (N) (pl. nota; adj. notal). The tergum of a thoracic segment comprising the primary dorsal segmental sclerotization. See **tergum**.
occiput (Occ) (Fig. 2a,b) (pl. occipita, occiputs; adj. occipital). The posterior dorsal part of the cranium separated from the postocciput by the postoccipital suture, its boundaries with the vertex and postgenae are not delimited; usually with erect scales.
ocular line (OL) (Fig. 2a,b). The area along the anterior border of the vertex between the margin of the compound eye and the row of ocular setae.
ocular scales (OS) (Fig. 2a). Scales occurring on the ocular line of the head.
ocular setae (OcS) (Figs 1b, 3b,f). Setae occurring in a line on the head near the posterior margin of each compound eye; one or more pairs are usually differentiated as interocular setae; the ocular setae along the dorsal margin of the compound eye are the upper ocular setae and those along the lateral margin are the lower ocular setae.
ocular suture (OSc). A line of inflection encircling the compound eye.
opisthophallic sclerite (OPS) (Fig. 12c,d). In some culicid males, a supporting sclerite on each side of the opisthophallus; in *Verrallina*, fused with its mate to form a dorsal (prerotation sense) transverse bridge between the gonocoxal apodemes; homologous with the lateral plate of *Culex*.
opisthophallus (OP) (pl. opisthophalli; adj. opisthophallic). In some culicid males, a transverse lobiform element of the phallosome located between the phallus and proctiger; supported on either side by a sclerite, opisthophallic sclerite; not studied or recognized in most mosquitoes but is well developed in *Culex*, *Maorigoeldia* and *Verrallina*.
outer division (OD) (Fig. 12d). In *Culex* males, the dorsal and lateral part of the lateral plate of the phallosome in normal resting position; variously developed.
palatal setae (PSe) (Fig. 4d,e). Four small peg-like cibarial setae borne on the anterior hard palate.
palpifer (Pl) (Figs 2b,c, 3b). A small lobe laterad of the clypeus at the base of the maxillary palpus; anatomically a lobe of the maxillary stipes to which the maxillary palpus articulates.
palpomere (Plp). An individual segment of the maxillary palpus.
parabasal lobe (PBL) (Fig. 12b). In male anopheline mosquitoes, a variably developed lobe located dorsobasally (prerotation sense) on the gonocoxite; bearing one or more strongly developed setae, parabasal setae.

parabasal setae (PBS) (Fig. 12a,b). In male anopheline mosquitoes, the distinctive setae usually borne on a lobe, the parabasal lobe, located dorsobasally (prerotation sense) on the gonocoxite.

paramere (Par) (Fig. 12b,c). In male mosquitoes, a lateral sclerite of the phallosome interposed between and articulating with the gonocoxal apodeme and the aedeagus (aedeagal sclerite); the paramere may have one or more dorsal arms.

paranotal suture (PnS) (Fig. 5b). The strongly developed suture separating the paratergite from the scutum; often appearing to be continuous posteriorly with the supraalar groove.

paraproct (Ppr) (Figs 12b,c, 13a,b). In male mosquitoes, a paired sclerite occurring laterally on the proctiger; articulated basally with tergum X; apparently a part of the cercus.

paraproct crown (PpC) (Fig. 12c). In male *Culex* mosquitoes, a collection of spicules borne at or near the apex of the paraproct.

paratergal scales (PsSc) (Fig. 6b). Scales occurring in a group on the paratergite.

paratergite (Pa) (Fig. 5b). A narrow lateral part of the mesonotum just before the wing root; separated from the scutum by the paranotal suture. *Synonyms*: Lateral lobe, paranotal fold, prescutum.

pedicel [of antenna] (Pe) (Figs 1, 2a,b, 3a,b,e). The second segment of the antenna, usually enlarged and appearing as the basal segment; often bearing setae and/or scales.

pedicel [of halter] (Pc) (Fig. 8b). The slender, stalk-like part of the halter bearing the capitellum distally.

pediment (Pd). The basal part of a cone; bearing the filament anteriorly; the posterior part, crest, extends posteriorly and lies between the bases of adjacent rods.

phallosome (PH) (adj. phallosomal). In male mosquitoes, the complex of structures surrounding the gonopore between the proctiger, gonocoxites and sternum IX; presumably consisting primitively of the gonocoxal apodemes, parameres, prosophallus, phallus and opisthophallus, *e.g.* in *Verrallina*; usually comprised of the gonocoxal apodemes, parameres and aedeagus, but including the lateral plates in *Culex*; the homologies of the various components are unknown.

phallus (Ph) (pl. phalli). In some culicid males, *e.g. Verrallina*, the median lobe of the phallosome located between the prosophallus and opisthophallus; supported by a sclerite, aedeagal sclerite, on either side that connects with the base of the prosophallic sclerite; bearing the aedeagus distally.

pleural membrane (PMe) (Figs 1, 10c). The membrane occurring between the tergum and sternum of a body segment.

pleural suture (pls). The external groove separating the episternum and epimeron; referred to as pro-, meso- or metapleural suture as appropriate.

pleuron (P) (pl. pleura; adj. pleural). The sclerotization of the pleural or lateral area of a body segment; in the thorax, termed pro-, meso- or metapleuron as appropriate.

plume scales (PlSc) (Fig. 8a). A set of erect and usually long linear wing scales which, together with the wing vein to which they attach, somewhat resemble a plume feather.

posterior dorsocentral area (PDA). The part of the dorsocentral area caudad of a point near the median corner of the scutal fossa at about the level of the antealar area; with scales and setae.

posterior dorsocentral scales (PDSc) (Fig. 6a). Scales occurring in a more or less distinct longitudinal row usually located laterad of the posterior dorsocentral setae on the posterior dorsocentral area of the scutum.

posterior dorsocentral setae (PDS) (Fig. 7a). Dorsocentral setae occurring on the posterior dorsocentral area; occurring in one or more rows that merge posteriorly with the prescutellar setae.

posterior hard palate (PHP) (Fig. 4d,e). The large plate-like posterior area of the clypeopalatum; bearing ventral setae near the base of each lateral flange.

posterior medial scutal setae (PMSS) (Fig. 7a). Setae occurring in a V-shaped group on the midline of the scutum at the posterior end of the row of acrostichal setae and anterior to the prescutellar area.

posterior mesanepisternum (PMas) (Fig. 5b) (pl. posterior mesanepisterna; adj. posterior mesanepisternal). The vertically narrow posterior part of the mesanepisternum; separated from the anterior mesanepisternum by the anepisternal cleft, usually united with the mesokatepisternum but sometimes separated from it by a weakly developed mesanapleural suture.

posterior mesepimeral scales (PMSc) (Fig. 6b). Scales occurring in a group on the ventroposterior quarter of the mesanepimeron.

posterior scutal fossal scales (SFSP) (Fig. 6a). Scales occurring in a more or less distinct line extending diagonally from the scutal angle to the dorsocentral scales along the posterior margin of the scutal fossa.

posterior scutal fossal setae (PSFS) (Fig. 7a). Setae usually occurring in a short row on the posterior margin of the scutal fossa between the scutal angle and the dorsocentral area.

posterior tentorial pit (PTP) (Fig. 3e). An external depression in the cranium at the base of each posterior tentorial arm, occurring at the lower end of the postoccipital suture.

postfrontal suture (pfs) (Fig.3a). One of the bilateral pair of sutures separating the interocular space from the frons; short and confluent laterally with the ocular suture.

postgena (PG) (Figs 2b,c, 3b,f) (pl. postgenae; adj. postgenal). The lateral and ventral parts of the cranium behind the compound eye, not differentiated from the gena, occiput and vertex.

postgenal setae (PgS) (Figs 2b,c, 3b,e,f). Setae occurring in a group on the postgena at the posteroventral area of the cranium.

postgenital lobe (PGL) (Figs 1, 10a,b, 11a–d). In female mosquitoes, the weakly to moderately sclerotized median caudal lobe below the anus and above the upper vaginal lip; setose and with or without an apical median indentation.

postnotum (Pn) (pl. postnota; adj. postnotal). The phragma-bearing plate in the dorsum of a pterothoracic segment; derived from the anterior part of the following notum. See **mesopostnotum** and **metapostnotum**.

postoccipital suture (POS). A line of inflection extending dorsad from each posterior tentorial pit and separating the postocciput from more anterior areas of the cranium.

postocciput (POc) (Fig. 2a) (pl. postoccipita, postocciputs; adj. postoccipital). The posterior area of the cranium separated from the more anterior areas by the postoccipital suture.

postprocoxal membrane (PM) (Fig. 5b). The membrane between the forecoxa and the mesokatepisternum; sometimes bearing scales.

postprocoxal scales (PSc) (Fig. 6b). Scales occurring in a small group on the postprocoxal membrane.

postpronotal scales (PpSc). Scales borne on the postpronotum; occurring in upper and lower groups, the upper and lower postpronotal scales.

postpronotal setae (PpS) (Fig. 7b). Setae occurring in an arcuate line on the upper posterior margin of the postpronotum.

postpronotum (Ppn) (Figs 1, 5b) (pl. postpronota; adj. postpronotal). The posterior division of the pronotum generally lying between the antepronotum and the scutum cephalad of the scutal angle, usually not clearly demarcated ventrally from the proepimeron.

postspiracular area (PA) (Figs 1, 5b). The sclerotized area of the anterior mesanepisternum lying posterior to the mesothoracic spiracle; connected or continuous ventrally with the subspiracular area; scales and setae are borne on this area.

postspiracular scales (PoSc) (Fig. 6b). Scales occurring in a group on the postspiracular area of the anterior mesanepisternum; sometimes extending posteriorly onto the membrane of the anepisternal cleft.

postspiracular setae (PS) (Fig. 7b). Setae occurring in a group on the postspiracular area of the anterior mesanepisternum.

posttarsus (Pta) (Fig. 9c) (pl. posttarsi; adj. posttarsal). The terminal segment of the leg distal to the tarsus.

prealar knob (PK) (Figs 1, 5b). The convex protuberance borne on the upper part of the posterior mesanepisternum anterior to the wing; with scales and setae.

prealar scales (PrSc). Scales borne anterior to the wing on the posterior mesanepisternum; an upper group occurs among and above the prealar setae and a lower group sometimes occurs below them. See **upper prealar scales**.

prealar setae (PaS) (Fig. 7b). Setae occurring in a group on the prealar knob of the posterior mesanepisternum.

prearculus (PAr). A more or less distinct crossvein or transverse furrow between the radius and subcosta in line with the arculus.

prementum (Prm) (Figs 1, 2b,c, 3e,f, 4a–c) (adj. premental). The stipital part of the labium, forming the elongate stout ventral part of the proboscis; bearing the ligula and labella distally.

prescutal suture (PrS) (Fig. 7a). The largely obsolescent suture extending caudomesad from the posterior edge of the scutal angle at the posterior margin of the scutal fossa.

prescutellar area (PrA) (Figs 5a, 7a). The median posterior area of the scutum between the acrostichal area and the scutellum; frequently devoid of scales and setae.

prescutellar scales (SScP). Scales occurring on the prescutellar area of the scutum; usually occurring in a curved line on the lateral and anterior margins; those extending posteriorly from the dorsocentral scales on the lateral margins may be referred to as lateral prescutellar scales; those occurring on the anterior margin and sometimes forming a median longitudinal row may be termed median prescutellar scales. See **lateral prescutellar scales** and **median prescutellar scales**.

prescutellar setae (PrsS) (Fig. 7a). Setae usually occurring in several rows on the anterior and/or lateral margins of the prescutellar area; often difficult to separate anteriorly from the dorsocentral setae.

prespiracular area (PsA) (Fig. 5b). The small area of the anterior mesanepisternum just anterior to the mesothoracic spiracle; set off from the postpronotum by a strong ridge; sometimes bearing scales and setae.

prespiracular scales (PsSc) (Fig. 6b). Scales occurring in a small group on the prespiracular area of the anterior mesanepisternum.

prespiracular setae (PsS) (Fig. 7b). Setae occurring on the prespiracular area of the anterior mesanepisternum.

primary fringe scales (PFS) (Fig. 8c). The long prominent fusiform scales occurring in a row along the posterior edge of the wing. See **fringe scales, secondary fringe scales. tertiary fringe scales** and **wing fringe**.

proboscis (P) (Fig. 1) (pl. proboscises, proboscides; adj. probascidial). The greatly elongate sheath-like labium and the enclosed fascicle, which includes the labrum, two mandibles, two laciniae and the hypopharynx.

proboscis length. The length of the extended proboscis measured from the labial basal setae to the apex of the labella.

proctiger (Pr). The anal lobe of the male genitalia; comprised of structures posterior to abdominal segment IX, including abdominal segment X and the cerci.

proepimeral scales (PmSc) (Fig. 6b). Scales occurring in a group on the proepimeron.

proepimeron (Pm) (Fig. 5b) (pl. proepimera; adj. proepimeral). The posterior area of the propleuron, usually not clearly demarcated from the lower margin of the postpronotum but sometimes separated from it by a weak suture.

proepisternal scales (PeSc). Scales borne on the proepisternum, occurring in upper and lower groups, upper proepisternal scales and lower proepisternal scales, which may be continuous. See **lower proepisternal scales** and **upper proepisternal scales**.

proepisternal setae (PeS). Setae borne on the proepisternum, occurring in two groups, one above and one mesad of the forecoxa. See **lower proepisternal setae** and **upper proepisternal setae**.
proepisternum (Ps) (Figs 1, 5b) (pl. proepisterna; adj. proepisternal). The anterior area of the propleuron, extending from the antepronotum to the base of the forecoxa where it expands anteroventrally around the coxa to meet its mate of the opposite side.
pronotum (PrN) (Fig. 1) (pl. pronota; adj. pronotal). The notum of the prothorax, transversely divided on each side into an antepronotum and a postpronotum.
propleuron (Pp) (pl. propleura; adj. propleural). The pleuron of the prothorax, consisting of a small proepimeron and a proepisternum.
prosophallic sclerite (POS). In some male mosquitoes, *e.g. Verrallina*, one of a pair of sclerites which embody the prosophallus.
prosophallus (PO) (pl. prosophalli; adj. prosophallic). In some culicid males, a ventral (prerotation sense) lobe-like element of the phallosome located between the bases of the gonocoxites; embodied by a pair of sclerites, prosophallic sclerites; known only in *Verrallina*; possibly homologous with the claspettes.
prothorax (P) (pl. prothoraces, prothoraxes; adj. prothoracic). The first or anterior segment of the thorax; comparatively small, especially dorsally.
proximal sclerite (PS) (Fig. 4a–c). The labellar sclerite marking the proximal segment of the labellum; sometimes partially or completely fused with the distal sclerite.
pulvillus (Pv) (Fig. 9d) (pl. pulvilli). One of two pad-like or divided lobes on the posttarsus; one arising below the base of each unguis.
radial sector (R_s) (Fig. 8a). The stem of radius-one-plus-two and radius-three-plus-four; splitting from the stem of the radius with radius-one.
radiomedial crossvein (rm) (Fig. 8a). The crossvein extending between the radius and media, connecting radius-four-plus-five and media-one-plus-two.
radius (R) (Fig. 8a). The usual third principal longitudinal vein of the wing, four-branched, the stem branching into radius-one and the radial sector, the radial sector branching into radius-two-plus-three and radius-four-plus-five and radius-two-plus-three branching into radius-two and radius-three.
radius-four-plus-five (R_{4+5}) (Fig. 8a). The posterior branch of the radius; with a short basal crossvein-like segment between the point of separation from radius-two-plus-three and the radiomedial crossvein.
radius-one (R_1) (Fig. 8a). The anterior branch of the radius terminating anteriorly at the apex of the wing.
radius-three (R_3) (Fig. 8a). The posterior branch of radius-two-plus-three extending to the wing apex.
radius-two (R_2) (Fig. 8a). The anterior branch of radius-two-plus-three represented by a short vein near the wing apex.
radius-two-plus-three (R_{2+3}) (Fig. 8a). The stem of radius-one and radius-two; branching from the radial sector with radius-three-plus-four.
remigial setae (ReS). Setae occurring on the posterior margin of the radius proximal to the arculus (the remigium).
remigium (Re) (Fig. 8a,d) (pl. remiges; adj. remigial). The part of the radius basad of the arculus.
rod (Rd) (Fig. 4d). One of the specialized spicules comprising the cibarial teeth; having circular, oval or elliptical origins between the crests of the cones; each rod has a tapered or bulbous base and usually has a simple tapered, rod-like termination; bearing small barb-like processes laterally.
scabellum (Sl) (Fig. 8b). The bulbous basal part of the halter supporting the pedicel.
scape (Sc) (Fig. 2a). The first or basal segment of the antenna; greatly reduced and hidden by the pedicel.
scutal angle (ScA) (Figs 5a, 7b). The more or less distinct angular projection of the scutal margin just anterior to the prescutal suture, or just in front of the prespiracular area when the prescutal suture is not discernible.
scutal fossa (SF) (Fig. 5a,b) (pl. scutal fossae; adj. scutal fossal). The somewhat depressed anterolateral area of the scutum extending caudad to the level of the scutal angle and mesad before the prescutal suture to the dorsocentral area.
scutal fossal scales (SFSc). Scales borne on the scutal fossa; often occurring in three more or less distinct groups, the lateral scutal fossal scales, posterior scutal fossal scales and median scutal fossal scales.
scutal fossal setae (SFS). Setae borne on the scutal fossa; often occurring in four distinct groups, the anterior scutal fossal setae, lateral scutal fossal setae, posterior scutal fossal setae and median scutal fossal setae.
scutellar scales (ScSc). Scales borne on the scutellum; frequently separated into distinct median and lateral groups, the median scutellar scales and lateral scutellar scales.
scutellar setae (ScS). Setae occurring on the scutellum, usually confined to the caudal margin; occurring in a median and a pair of lateral groups in the Culicinae (except *Toxorhynchites*), the median scutellar setae and lateral scutellar setae, respectively.
scutellum (Stm) (Figs 1, 5a) (pl. scutella; adj. scutellar). The posterior area of the mesonotum separated from the scutum by a suture, a transverse lobe that is trilobed in the Culicinae, except *Toxorhynchites*.
scutum (Scu) (Figs 1, 5a,b) (pl. scuta; adj. scutal). The principal dorsal area of the thorax belonging to the mesonotum.
secondary fringe scales (SFS) (Fig. 8c). The short fusiform scales (shorter than primary fringe scales) occurring in a row along the ventral surface of the wing edge. See **fringe scales, primary fringe scales, tertiary fringe scales** and **wing fringe**.
setae *a–h* (*a, h, etc.*) (Fig. 12c). In many male Culicini, the setae borne on the subapical lobe of the gonocoxite; lettered from proximal to distal.

spermatheca (Spt) (pl. spermathecae; adj. spermathecal). The sperm storage organ; one or three may be present, each consisting of a spermathecal capsule at the end of a spermathecal duct.
spermathecal capsule (SCa) (Figs 10b, 11a). Any one of the one to three enlarged sperm storage reservoirs often differentiated at the inner end or ends of the spermathecal ducts; typically spherical and with darkly pigmented cuticle. See **spermatheca**.
spiracle (S). The opening into a trachea (primary tracheal orifice). See **mesothoracic spiracle** and **metathoracic spiracle**.
squame scales (SS) (Fig. 8a). Decumbent scales, usually only applied to the generally short broad squame scales on the wing veins.
sternum (S) (Figs 1, 10a) (pl. sterna; adj. sternal; adv. sternad, sternally). The ventral sclerotization of a body segment, applied mainly to sterna of abdominal segments, labelled S-I (most anterior) to S-IX (most caudal).
sternum VIII (S-VIII) (Figs 1, 10a–c). The sternum of abdominal segment VIII; variable in shape; dorsal in males post-rotation (see **genitalia**).
sternum IX (S-IX). The sternum of abdominal segment IX; dorsal in males post-rotation (see **genitalia**), a small sclerite continuous laterally with tergum IX.
subapical lobe (SL) (Fig. 12c). In *Culex* males, a mesal or dorsomesal (prerotation sense) lobe located near or distal to the middle of the gonocoxite; bearing setae *a–h*; divided into proximal and distal lobes in some species or groups.
subcosta (Sc) (Fig. 8a) (pl. subcostae; adj. subcostal). The usual second principal longitudinal vein of the wing.
subcostal setae (SuS). Setae occurring in a group on the base of the subcosta on the ventral surface of the wing.
subspiracular area (SA) (Fig. 5b). The rather indefinite area of the anterior mesanepisternum lying below the hypostigmal area between the proepimeron and the postspiracular area, usually continuous with or connected to the postspiracular area ventrally; scales and rarely setae may be on this area.
subspiracular scales (SSc) (Fig. 6b). Scales occurring in a group on the subspiracular area of the anterior mesanepisternum.
subspiracular setae (SuS). Setae which rarely occur on the subspiracular area of the anterior mesanepisternum.
supraalar area (SA) (Fig. 5a,b). The lateral area of the scutum just above and in front of the wing; a line of setae in this area may extend anterior to the posterior end of the paratergite mesad to the antealar setae; a line of scales in this area is sometimes continuous with an antealar line.
supraalar groove (SaG) (Fig. 5b). The furrow lateral to the scutum above the wing base; often appearing to be continuous anteriorly with the paranotal suture.
supraalar scales (SaSc) (Fig. 6a). Scales occurring on the supraalar area of the scutum; a distinct lateral line of supraalar scales is often present otherwise they are difficult to separate from the antealar scales.
supraalar setae (SaS) (Fig. 7a,b). Setae occurring in a cluster on the supraalar area of the scutum; a row of supraalar setae may extend cephalad above the antealar area or be more or less continuous with the caudal end of a row of antealar setae.
suture. Applied either to a seam produced by the union of two areas of sclerotization or to a groove produced incidental to the formation of a ridge on the inner surface of the cuticle.
tagma (pl. tagmata). One of the three morphologically distinct sections of the body, the head, thorax and abdomen.
tarsomere ($Ta\text{-}I_1$, $Ta\text{-}II_4$, *etc.*) (Fig. 9). An individual subsegment of a tarsus, five tarsomeres comprise each tarsus; referred to as the first through the fifth tarsomeres of the appropriate tarsus and denoted by adding numerical subscripts (1–5) to the abbreviation of the tarsus.
tarsus (Ta) (Figs 1, 9a) (pl. tarsi; adj. tarsal). The usual fifth leg segment located distad to the tibia, consisting of five tarsomeres, numbered 1–5 from proximal to distal; referred to a fore- (Ta-I), mid- (Ta-II) or hindtarsus (Ta-III) as appropriate.
tergal triad (tt). Three very long close-set setae inserted on the tergal surface of the gonocoxite of certain sabethine males, mainly species of *Wyeomyia*; homologous in part and referred to as **tergomesal setae** (see) when only two setae are present, mainly in species of *Sabethes*; sometimes represented by a single seta.
tergomesal setae (ts) (Fig. 13c). One, two or three usually very long setae inserted on the tergal surface of the gonocoxite of certain male sabethine mosquitoes, mainly species of *Sabethes* (see Harbach, 194b) and *Wyeomyia* (**tergal triad** of Belkin *et al.*, 1970).
tergum (T) (Figs 1, 10a) (pl. terga; adj. tergal; adv. tergad, tergally). The dorsal sclerotization of a body segment; called notum in the thorax; abdominal terga are labelled T-I (most anterior) to T-IX (most caudal).
tergum VIII (T-VIII) (Figs 1, 10a–c). The tergum of abdominal segment VIII; in females, variable in shape; usually more or less trapezoidal; setose apically; scales present or absent; in males, ventral in post-rotation position.
tergum VIII comb (TC-VIII). In mosquitoes of the gnus *Mansonia*, a row of spiniform setae along and near the posterior margin of tergum VIII.
tergum IX (T-IX) (Fig. 10c). The tergum of abdominal segment IX; in females, a small plate situated between tergum VIII and the cerci (tergum X if present) dorsally; variable in shape and pigmentation; setae present or absent; in males, ventral in post-rotation position; usually markedly bilobed.
tergum IX lobe (TL-IX) (Figs 12b,c, 13a). In male mosquitoes, one of the caudally directed, bilaterally paired projections of abdominal tergum IX; often bearing setae distally.
tergum X (T-X) (Figs 12c, 13a). The tergum of abdominal segment X; in males, usually developed as a pair of sclerites laterally at the base of the proctiger; often articulating with tergum IX and the gonocoxal apodeme; sometimes confluent with the basal

part of the paraproct sclerotization; in females, a small sclerite situated dorsally between tergum IX and the cerci in some species, particularly anophelines.

tertiary fringe scales (TFS) (Fig. 8c). The small linear scales set in two rows on the wing margin, on the dorsal surface just inside the row of primary fringe scales and the other on the ventral surface just inside the row of secondary fringe scales; poorly developed or absent in males. See **fringe scales, primary fringe scales, tertiary fringe scales** and **wing fringe**.

thorax (Figs 1, 5–7) (pl. thoraces, thoraxes; adj. thoracic). The second or intermediate division (tagma) of the body; bearing the legs and wings; comprised of the pro-, meso- and metathoracic segments.

tibia (Ti) (Figs 1, 9a) (pl. tibiae; adj. tibial). The usual fourth segment of the leg distal to the femur; referred to as fore- (Ti-I), mid- (Ti-II) or hindtibia (Ti-III) as appropriate.

trochanter (Tr) (Fig. 1). The short second segment of the leg between the coxa and femur; referred to as fore- (Tr-I), mid- (Tr-II) or hindtrochanter (Tr-III) as appropriate.

unguis (U) (Figs 1, 9a–d) (pl. ungues). An anterior or posterior claw of the posttarsus; both are usually present; referred to as fore- (U-I), mid- (U-II) or hindungues (U-III) as appropriate.

unguitractor plate (Un) (Fig. 9c). The ventral sclerite of the posttarsus, produced apically as an empodium.

upper calypter (UC) (Fig. 8d). The distal calypter of the wing; typically situated above the lower calypter. See **calypter** and **lower calypter**.

upper mesepimeral scales (UMSc) (Fig. 6b). Scales occurring in a group on the dorsoposterior quarter of the mesanepimeron; occupying the same area as the upper mesepimeral setae.

upper mesepimeral setae (MeSU) (Fig. 7b). Setae occurring in a group on the upper posterior area of the mesanepimeron before the metathoracic spiracle.

upper mesokatepisternal scales (MScU) (Fig. 6b). Scales occurring in a group located immediately below the upper mesokatepisternal setae.

upper mesokatepisternal setae (MkSU) (Fig.7b). Setae occurring in more or less a horizontal line on the upper part of the mesokatepisternum.

upper ocular setae (UOcS). The ocular setae occurring along the dorsal margin of the compound eye. See **ocular setae**.

upper postpronotal scales (UPSc) (Fig. 6b). Scales usually occurring in a large group on the dorsal part of the postpronotum.

upper prealar scales (UPrSc) (Fig. 6b). Scales occurring principally on the prealar knob among and above the prealar setae.

upper proepisternal scales (PScU) (Fig. 6b). Scales occurring in a group on the proepisternum above the base of the forecoxa; sometimes continuous with the lower proepisternal scales mesoventrally.

upper proepisternal setae (PeSU) (Fig. 7b). The proepisternal setae occurring in a group above the forecoxa.

upper vaginal lip (UVL) (Fig. 11b,d). In female mosquitoes, the sclerotized and pigmented rim of the roof of the vagina; articulating at each end with the lower vaginal lip; the caudal margin is joined to the basoventral margin of the postgenital lobe; usually spiculate.

upper vaginal sclerite (UVS). In female mosquitoes, a pigmented sclerite of the vaginal roof other than the upper vaginal lip; usually attached to the inner median cephalic margin of the upper vaginal lip; may be continuous with the lip and distinguishable by position only; non-spiculate.

vein. One of the tubular thickenings of a wing often containing blood, tracheae and nerves. The principal longitudinal veins of mosquitoes are the anal vein, costa, cubitus, media, radius and subcosta.

vein 1A. See **anal vein**.

vein CuA. See **cubitus anterior**.

vein CuP. See **cubitus posterior**.

vein M. See **media**.

vein M_1 (M_1). See **media-one**.

vein M_2 (M_2). See **media-two**.

vein M_{1+2} (M_{1+2}). See **media-one-plus-two**.

vein M_{3+4} (M_{3+4}). See **media-three-plus-four**.

vein R (R). See **radius**.

vein R_s (R_s). See **radius-sector**.

vein R_1 (R_1). See **radius-one**.

vein R_2 (R_2). See **radius-two**.

vein R_{2+3} (R_{2+3}). See **radius-two-plus-three**.

vein R_3 (R_3). See **radius-three**.

vein R_{4+5} (R_{4+5}). See **radius-four-plus-five**.

vein Sc (Sc). See **subcosta**.

ventral aedeagal bridge (VAB) (Figs 12d, 13b). In male mosquitoes, the transverse sclerotization connecting the aedeagal sclerites furthest from the anus.

ventral arm (VA) (Fig. 12d). In *Culex* males, the lateral pointed process (in normal resting position) of the inner division of the lateral plate of the phallosome; homologous with the leaflets of aedeagus in *Anopheles*.

ventral setae (VSe) (Fig. 4d). Four small peg-like cibarial setae borne ventrally at the posterior margin of the cibarium.
vertex (V) (Figs 2a,b, 3b) (pl. vertices, vertexes). The dorsal surface of the head behind the compound eyes and the interocular space; its boundaries with the occiput, genae and postgenae are not delimited.
wing (W) (Figs 1, 8a,d). One of the paired organs of flight occurring on the mesothoracic segment. Compare **mesothoracic wing** in the **Pupa** section.
wing fringe (WF) (Fig. 8a,c). Four rows of scales borne on the posterior margin of the wing extending from just beyond the alula to the apex: a row of primary fringe scales, one of secondary fringe scales and two of tertiary fringe scales. See **fringe scales**.
wing membrane (WM). The thin transparent or translucent tissue of the wing between the veins and/or margins; usually with microtrichia.
wing spots (Fig. 8d,e) (the names of the spots are listed after this entry). Dark and pale spots on the wings correspond to areas with dark and pale scales on the veins. The spots on the costa and anterior veins are usually regular and distinct and are identified using the standardized nomenclature (Wilkerson & Peyton, 1990) based on morphological reference points. The names are those which have been used for the wing spots of anophelines and species of *Finlaya*, but there should be little difficulty in applying them to the wing spots of species of *Aedeomyia*, *Culex* (Mimeticus Group), *Orthopodomyia*, *Psorophora* and *Uranotaenia*. Dark and pale spots occurring on the stem and branches of posterior veins can usually be designated by the adjectives basal, median, preapical or apical preceded by the abbreviation of the vein, *e.g.* M_1 median pale spot. Pale spots of the wing fringe are termed 'pale fringe spots' and are named after the vein at the end of which they occur, *e.g.* 1A pale fringe spot.

1A pale fringe spot (1APFS) (Fig. 8d). The pale fringe spot located at the outer end of the anal vein. See **pale fringe spot** below.
accessory preapical dark spot (APD). In most species of the Arribálzaga Series of *Anopheles* subgenus *Anopheles*, a dark spot that divides the preapical pale spot; not found in other anophelines.
accessory sector dark spot (ASD). In the Arribálzaga Series of *Anopheles* subgenus *Anopheles*, the dark spot that divides the sector pale spot into two pale areas.
accessory sector pale spot (ASP). The group of pale scales sometimes occurring on the radius at or just distad of the splitting of radius-one and the radial sector and usually separated by a dark spot from the sector pale but sometimes fused with it, sometimes also present on the costa and subcosta.
apical dark spot (AD) (Fig. 8d,e). The group of dark scales on the costa and radius-one distad of the preapical pale spot and proximad of the apical pale spot.
apical pale spot (AP) (Fig. 8d,e). The area of pale scales on the costa at the end of radius-one (pale scales are also usually, but not always, present on the end of radius-one); absent when the apical dark spot is at radius-one.
basal dark spot (BD) (Fig. 8d). The area of dark scales at or proximal to the end of radius-one and distal to the preapical pale spot.
basal pale fringe spot (BPFS) (Fig. 8d). A row of pale fringe scales occurring near the middle of the posterior margin of cell 1A.
basal pale spot (BP) (Fig. 8d). The group of pale scales occurring at the extreme base of the costa.
humeral dark spot (HD) (Fig. 8d). The group of dark scales occurring on the costa between the humeral pale spot and presector pale spot.
humeral pale spot (HP) (Fig. 8d). The group of pale scales occurring on the costa at or slightly distal to the humeral crossvein.
pale fringe spot (PFS) (Fig. 8d). A row of pale fringe scales named after the vein at the end of which it occurs, *e.g.* 1A pale fringe spot; an exception is the basal pale fringe spot.
postsubcostal dark spot (POSCD). In species of the Arribálzaga Series of *Anopheles* subgenus *Anopheles* and genus *Orthopodomyia*, the dark wing spot of the subcostal area located distal to the subcostal dark spot.
postsubcostal pale spot (POSCP). In species of the Arribálzaga Series of *Anopheles* subgenus *Anopheles* and genus *Orthopodomyia*, a pale wing spot of the subcostal area located distal to the subcostal dark spot.
preapical dark spot (PD) (Fig. 8d,e). The group of dark scales occurring beyond the subcostal pale spot; frequently interrupted by a small pale spot on radius-one.
preapical pale spot (PP) (Fig. 8d,e). The area of pale scales on the costa and radius-one proximal to the end of radius-one, between the preapical dark and apical dark spots.
prehumeral dark spot (PHD) (Fig. 8d). The group of dark scales occurring on the costa just before the humeral crossvein.
prehumeral pale spot (PHP) (Fig. 8d). The group of pale scales occurring on the costa before the humeral crossvein and usually separated from it by the prehumeral dark spot.
presector dark spot (PSD) (Fig. 8d,e). The group of dark scales occurring between the presector pale and sector pale spots; when the presector pale is absent, this spot forms part of the subbasal dark spot.
presector pale spot (PSP) (Fig. 8d,e). The group of pale scales often interrupting the subbasal dark spot to give rise to the humeral dark and presector dark spots.
presubcostal dark spot (PRSCD). In species of the Arribálzaga Series of *Anopheles* subgenus *Anopheles* and genus *Orthopodomyia*, the dark wing spot of the subcostal area located basal to the subcostal dark spot.

presubcostal pale spot (PRSCP). In species of the Arribálzaga Series of *Anopheles* subgenus *Anopheles*, a pale wing spot of the subcostal area located proximal to the subcostal dark spot.

sector dark spot (SD) (Fig. 8d,e). A usually large group of dark scales occurring on the costa near the middle of the wing beyond the sector pale spot; sometimes interrupted by an accessory sector pale spot.

sector pale spot (SP) (Fig. 8d,e). The group of pale scales occurring on the costa, subcosta and radius between the presector dark and median dark spots.

subapical dark spot (SAD). The dark spot distal to the subcostal pale spot when the preapical pale spot is absent, the combined preapical and apical dark spots.

subbasal dark spot (SBD) (Fig. 8e). The large group of dark scales distad of the humeral pale spot; sometimes including the prehumeral dark spot or even the basal dark spot; often interrupted by a pale spot, the presector pale spot, to give rise to the humeral dark and presector dark spots.

subcostal area (SCA). In species of the Arribálzaga Series of *Anopheles* subgenus *Anopheles*, the area of the costal region of the wing occupied by the subcostal dark spot and the presubcostal and postsubcostal dark and pale spots.

subcostal dark spot (SCD). The dark wing spot located at the end of the subcosta in species of the Arribálzaga Series of *Anopheles* subgenus *Anopheles*. Spots basal to the subcostal dark spot include the presubcostal pale spot and presubcostal dark spot; those distal to it are the postsubcostal pale spot and postsubcostal dark spot. The area occupied by these spots is termed the subcostal area.

subcostal pale spot (SCP) (Fig, 8d,e). The group of pale scales occurring on the costa and radius-one at and beyond the end of the subcosta between the sector dark and preapical dark spots. This definition is suitable for species of *Anopheles* subgenera *Anopheles* (except the Arribálzaga Series), *Cellia*, *Nyssorhynchus* and *Lophopodomyia*. It does not always apply in species of the subgenus *Kerteszia* because the subcosta joins the costa in the sector dark spot or at its distal end.

Eggs

The eggs of mosquitoes bear the general morphological characters of many insect eggs and consist, before fertilization, of the ovum and the remains of the nurse cells surrounded by a covering, the chorion, formed from the follicular epithelium and consisting of an outer layer, the outer chorion, and an inner layer, the inner chorion, which together form the eggshell. Structural features of the outer chorion provide useful taxonomic information, but, unfortunately, the eggs of most mosquito species are unknown.

anterior end. The end of the egg bearing the micropyle, adjacent to the head of the embryo.

cell. An area of the inner or outer chorion bounded by elements of the inner or outer chorionic reticulum, respectively; differing in shape and relief in different areas of the surface. When referring to the inner and outer chorion or the inner and outer chorionic sculpture, the term 'cell' may be used for brevity. See **inner chorionic cell** and **outer chorionic cell**.

cellule (Ce) (Fig. 14e). In some culicine eggs, a small impressed or elevated subdivision of an inner chorionic cell bounded by minute ridges or shallow grooves, the cellule wall.

cellule wall (CeW) (Fig. 14e). In some culicine eggs, a small ridge or shallow groove completely or incompletely enclosing a cellule of an inner chorionic cell.

chorion (C) (adj. chorionic). The outer covering or shell of the egg secreted by the follicular epithelium and consisting of one or more layers, usually consisting of two layers, the inner and outer chorions, but a delicate middle layer, lamellate membrane, occurs in some species.

corolla (Co) (Fig. 14g,h). In many mosquito eggs, notably those of the genera *Culex*, *Culiseta* and *Toxorhynchites*, a delicate frill-like collar surrounding the micropyle; often readily detachable and frequently lost; associated with a micropylar collar in some *Culex* eggs.

deck (De) (Fig. 14c). In many anopheline eggs, the tuberculate outer chorion of the ventral surface that normally is not covered by water and is completely or incompletely enclosed by the frill; sometimes divided so that in effect there are two decks, one anterior and one posterior. In the eggs of species of *Hodgesia*, *Mimomyia*, *Orthopodomyia* and *Deinocerites*, that part of the ventral surface enclosed by the frill, or the flange in the case of *Orthopodomyia*.

deck tubercle (DeT) (Fig. 14d). In anopheline eggs, one of the outer chorionic tubercles that ornament the deck; a variable number of large deck tubercles with notched or convoluted margins located at each end of the deck are referred to as lobed tubercles.

dehiscence. The splitting of the chorion of the egg in the emergence of the larva. Four modes of dehiscence occur in mosquito eggs: (1) *longitudinal* – the splitting of the chorion in the direction of the long axis of the egg; (2) *oblique* – the splitting of the chorion in a plane forming an acute angle to the longitudinal axis of the egg; (3) *spiral* – the transverse splitting of an egg in constantly changing planes (helical); (4) *transverse* – the splitting of the chorion of an egg in a plane perpendicular to the longitudinal axis. Adjectives used to qualify the types of dehiscence include apical (= capsular), subapical, equatorial, complete and incomplete.

denticle (D). In the eggs of *Mansonia* species, one of the small tooth-like projections of the inner chorion located within the micropylar collar.

dorsal surface. The surface of the egg opposite the dorsum of the embryo. In anopheline and sometimes other horizontally oriented eggs, notably of species of Aedini, the lower surface, *i.e.* the surface in contact with the substratum.

egg. The first life stage. Mosquito eggs are deposited so that they rest either horizontally or vertically. This poses a serious problem when trying to describe the orientation of an egg, *e.g.* whereas the terms 'upper surface' and 'lower surface' are applicable to the horizontally oriented egg, they are inapplicable to the egg oriented vertically. Since, however, the micropyle is always opposite the head of the developing embryo, the orientation of the embryo be used as reference for describing the orientation of the deposited egg. See **anterior end**, **dorsal surface**, **posterior end** and **ventral surface**.

egg batch. The group of eggs deposited by a female mosquito during a single oviposition. The floating egg batch consisting of eggs that are laterally attached to one another in species of *Coquillettidia*, *Culex*, *Culiseta*, *Trichoprosopon* and *Uranotaenia* is termed an egg raft.

egg cap (EC). The usually small anterior lid-like part of the chorion that may completely or incompletely separate from the posterior part following oblique, spiral or transverse dehiscence.

egg raft. In the culicine genera *Coquillettidia*, *Culex*, *Culiseta*, *Trichoprosopon* and *Uranotaenia*, the floating mass (egg batch) of laterally cohering, vertically oriented eggs.

flange (Fg). In *Orthopodomyia* eggs, one of the flattened longitudinal ribs borne on each side of the egg; bearing a series of holes, *lacunae*, traversed by little bars, *transverse trabeculae*.

float (F) (Fig. 14a,c). In anopheline eggs, a narrow longitudinal part of the outer chorion containing a hollow space and bearing a series of transverse ridges on its outer wall; multiple floats occur in *Chagasia* eggs otherwise two is the usual number, one on each side and at varying distances from the ventral midline.

float ridge (FR) (Fig. 14a,c). In anopheline eggs, one of the raised sections of a float; appearing membranous in slide preparations as compared to the grooves on each side.

frill (Fr) (Fig. 14c,d). In many anopheline eggs and eggs of the genera *Hodgesia* and *Mimomyia*, a narrow, usually ribbed, ridge-like formation of the outer chorion located on or near the ventral surface and commonly enclosing an area termed the 'deck'. In anopheline eggs, the frill may (1) completely or incompletely enclose an undivided deck, (2) in essence be two frills, one surrounding the anterior part and the other the posterior part of a divided deck or (3) have the form of a collar around the anterior or posterior end of the egg. In *Hodgesia* eggs, a short frill encloses a narrow deck on the ventral surface at the posterior end. In *Mimomyia* eggs, the frill completely encloses an undivided deck.

horn (H). In the eggs of at least some *Mansonia* species, one of the elongate pointed projections arising at the anterior end. The relationship of the horns to the micropylar collar is presently unknown.

inner chorion (IC). The innermost of usually two layers of chorion secreted by the follicular epithelium; with a smooth outer surface in anophelines, *Toxorhynchites* and many other culicines; sculptured in most aedines.

inner chorionic cell (ICC) (Fig. 14e). In many mosquito eggs, mainly those of aedine species, an area of the inner chorion bounded by elements of the inner chorionic reticulum; differing in shape in different areas of the surface.

inner chorionic reticulum (ICR) (Fig. 14e). In many mosquito eggs, particularly those of aedine species, a feature of the inner chorionic sculpture consisting of an interconnected system of ridges which separates adjacent inner chorionic cells; corresponding exactly with the boundaries of the follicular epithelial cells that secrete the chorion.

inner chorionic sculpture (ICS) (Fig. 14e). In many mosquito eggs, primarily eggs of aedine species, the pattern of impressed and raised markings of the inner chorion; usually consisting of cells bounded by a ridge-like reticulum.

lobed tubercle (LoT) (Fig. 14b,d). In anopheline eggs, one of the variable number of large deck tubercles with notched margins that are usually located at the anterior and posterior ends of the deck.

micropylar collar (MiC) (Fig.14b,f). In many mosquito eggs, the prominent, usually thickened and darkened annular wall or ridge encircling the micropylar disc.

micropylar disc (MiD) (Fig. 14b). In many mosquito eggs, a thin, thick or raised area of outer chorion surrounding the micropyle; often encircled by a micropylar collar.

micropylar mound (MiM). In certain *Culex* and *Psorophora* eggs, perhaps others, a dome or lobe of inner chorion lying directly under the micropylar disc; pierced at its centre by the micropyle.

micropylar plug (MiP). In many mosquito eggs, a piece of inner chorion that obstructs or fills in the micropyle.

micropyle (Mi) (Fig. 14b,h). A minute openings in the chorion through which spermatozoa enter. In mosquito eggs, a single pore located at the anterior end that is often filled with a micropylar plug and surrounded by a micropylar disc.

outer chorion (OC). The outermost of usually two layers of chorion secreted by the follicular epithelium; always sculptured and usually formed into a corolla and/or micropylar collar around the micropyle at the anterior end of an egg.

outer chorionic cell (OCC) (Fig. 14f,g). In many mosquito eggs, an area of the outer chorion bounded by elements of the outer chorionic reticulum; differing in shape and relief in different areas of the surface.

outer chorionic reticulum (OCR) (Fig. 14f,g). In many mosquito eggs, a feature of the outer chorionic sculpture consisting of an interconnected system of ridges and/or tubercles, or an interrelated chain of unconnected tubercles, which separates adjacent outer chorionic cells; corresponding exactly with the boundaries of the follicular epithelial cells that secrete the chorion.

outer chorionic sculpture (OCS). The pattern of impressed and raised markings of the outer chorion; often consisting of cells bounded by a reticulum.
outer chorionic tubercle (OCT) (Fig. 14b,f,g). A small rounded or irregularly shaped prominence of the outer chorion.
posterior end. The extremity of the egg opposite the end bearing the micropyle; opposed to the hindmost part of the embryo.
ventral surface. The surface of the egg opposite the venter of the embryo. In anopheline and sometimes other horizontally oriented eggs, notably those of species of Aedini, the upper surface.

Larvae

Mosquito larvae have a greater number of anatomical characters amenable to taxonomic study than any other life stage. They possess many secondary adaptive features, but they also tend to show phyletic relationships better than the adults. This section defines the head, thoracic and abdominal structures of mosquito larvae, excluding sensory setae and other cuticular projections, which are of primary importance in taxonomy. The complement of setae (innervated sensory organs) that arise from various body parts of mosquito larvae is treated in the **Larval Chaetotaxy** section, and the assortment of setal types and non-innervated projections of the cuticle are defined in the **Vestiture** section.

abdomen (Ab) (Fig. 15a) (pl. abdomens, abdomina; adj. abdominal). The third, posterior division (tagma) of the body, consisting of nine apparent segments.
abdominal segment (I, II, *etc.*) (Figs 15a, 18a,c, 19a,b, 20a–d). One of the annular subdivisions of the abdomen; labelled I (most anterior) to IX (most caudal).
abdominal puncture (p) (Fig. 19a, 23a). The empty alveolus that occurs medially on the dorsolateral area of abdominal segments III–V (occurring anomalously on segment II); associated with setae 4-, 5- and 6-III–V.
abdominal segment VIII (VIII) (Figs 15a, 18c, 19b, 20a–d). The apparent eighth segment of the abdomen; structurally comprising embryonic abdominal segments VIII and IX, bearing a siphon and/or spiracular apparatus dorsally.
abdominal segment X (X) (Figs 15a, 18c, 19b, 20a–d). The ultimate or apparent ninth abdominal segment; largely comprising embryonic abdominal segment X but probably including parts of the true ninth and eleventh segments and the telson.
accessory tergal plate (ATP) (Fig. 18a,c). In some anopheline larvae, one of the small median or submedian dorsal sclerites located posterior to the tergal plates of abdominal segments I–VII. See **tergal plate**, **median accessory tergal plate** and **submedian accessory tergal plate**.
anal papilla (APP) (Figs 15a, 18c, 19b, 10a–d) (pl. anal papillae). One of usually four soft elongate processes borne posteriorly around the anus of abdominal segment X; presumably representing the modified cerci of embryonic abdominal segment XI.
antenna (A) (Figs 15a, 16a–d) (pl. antennae; adj. antennal). One of the paired anterior appendicular sensory organs of the head, inserted anterolaterally, consisting of a narrow ring-like scape, which is often united with a large distal part, the fused pedicel and flagellum.
anterior median process (AMPc) (Figs 20a, 21a). In anopheline larvae, an elevated dome- or knob-like membranous area situated posteriorly on the midline of the anterior spiracular lobe of the spiracular apparatus; in some species, produced into an elongate process, *e.g.* in *Chagasia* species.
anterior spiracular lobe (ASL) (Fig. 21a,b). The anterior, unpaired flap-like projection of the spiracular apparatus, bearing setae 3- (usually a puncture), 4- and 5-S on its posterolateral margins; partly homologous with the saw of Mansoniini larvae.
anterolateral spiracular lobe (LSL) (Fig. 21a,b). One of the anterior of two bilateral pairs of flap-like projections of the spiracular apparatus; weakly developed (indistinct) in Mansoniini.
apical tooth (AT). In the majority of sabethine mosquitoes, a distal process of the maxillary body ranging in development from a small protuberance to a prominent tooth- or claw-like structure.
boss (B). In certain mosquito larvae, a more or less sclerotized elevated area without a grid at the base of the ventral brush. See **grid**.
boss setae (BS). The ventral brush setae borne on a boss. See **grid**.
cervix (Cv) (Fig. 15a) (pl. cervices, cervixes; adj. cervical). The membranous tube connecting the thorax to the head at the margins of the occipital foramen, bearing setae 18- and 19-C laterally and a small sclerite, the ventromedian cervical sclerite, on its ventral midline.
clypeolabral suture (ClS) (Fig. 16d). The transverse suture separating the labrum and clypeus, sometimes difficult to trace mesally because of the curvature of the head.
collar (Col) (Fig. 16a–f). The modified, usually heavily pigmented, posterior rim of the cranium, including much of the postocciput, the back edge of the gula and a band just in front of the main part of the postoccipital suture.
comb (C) (Figs 15a, 19b, 20b,c). In most culicine and first-instar anopheline larvae, a row or patch of specialized spicules borne on each side of abdominal segment VIII; sometimes arising from a sclerite, the comb plate.
comb plate (CP) (Fig. 20c). In certain culicine larvae, a lateral sclerite on abdominal segment VIII that bears the comb scales; sometimes the comb plates of opposite sides are joined dorsally; not homologous with the tergal plate of anopheline larvae; also not homologous with the large setal support plate that commonly bears setae 2–5-VIII in *Toxorhynchites* larvae.

comb scale (CS) (Fig. 19b). In most culicine and first-instar anopheline larvae, one of the specialized spicules forming the comb; of variable form but usually scale- or spine-like; often bearing a fringe of spinules or denticles.

coronal ecdysial line (CEL) (Fig. 16e). The median unpaired posterior part of the epicranial ecdysial line; considered absent if the paired frontal ecdysial lines reach the posterior cranial margin separately; sometimes absent, sometimes short and ending posteriorly in the coronal gap in late instars. See **ecdysial line**.

coronal gap (CG) (Fig. 16a–d). The narrow or broad emargination in the posterodorsal edge of the cranium to which the epicranial ecdysial lines extend, present in late instars; sometimes regarded as part of the coronal ecdysial line.

cranium (C). (pl. craniums, crania; adj. cranial). The sclerotized skull-like part of the head.

cratal setae (CrS). The ventral brush setae borne on a grid.

dorsal apotome (DAp) (Fig. 16a,c,d,f). The cranial area bounded laterally by the frontal ecdysial lines and apically by an imaginary line between the most apical parts of the frontal ecdysial lines; in mosquito larvae, perhaps homologous with the dorsal apotome of mosquito pupae.

dorsal brush (DB) (Figs 15a, 18c, 19b). Setae 2- and 3-X; arising in a group from the dorsocaudal angle of abdominal segment X.

dorsomentum (Dm) (Fig. 16b, 23b). In some mosquito larvae, the upper and more distal of the two subdivisions of the mentum.

ecdysial line (EL). A preformed line of weakness of the cuticle involved in opening or folding at ecdysis.

flagellum (Fl) (pl. flagella; adj. flagellar). The third segment of the antenna, usually unrecognizably fused with the pedicel, and often the scape; possibly comprised of three flagellomeres, one basad of seta l-A, flagellomere 1, one between seta 1-A and seta 2-A and seta 3-A, flagellomere 2 and in Mansoniini larvae, one distal to setae 2- and 3-A, flagellomere 3; seta 5-A may represent additional distal flagellomeres.

fringing setae (FS) (Fig. 20a). In *Chagasia* larvae, the setae bordering the outer margins of the posterolateral spiracular lobes of the spiracular apparatus; vestiges of the fringing setae appear as a corrugated or undulated membrane in many mosquito larvae.

frontal ecdysial line (FEL) (Fig. 16a,c,e,f). One of the arms of the dorsal ecdysial line separating the dorsal apotome and lateralia, normally ending in the lateral part of dorsal apotome about opposite the antennal socket. See **ecdysial line**.

galea (Gl) (Fig. 17c) (pl. galeae; adj. galeal). The outer endite lobe of the maxilla, either united with part of the stipes to form the galeastipes or united with the mesal part of the stipes and the lacinia to form the maxillary body; in anopheline larvae, a small triangular area partly separated from the main part of the maxillary body by a fissure.

galeastipes (GSt) (Fig. 17d) (pl. galeastipites; adj. galeastipital). Recognizable in many mosquito larvae, mostly culicines, as the lateral half of the maxillary body; representing the fused stipes and galea which are partly separated in anophelines; usually bearing the maxillary brush and setae 1-, 3-, 4- and 5-Mx, the maxillary palpus is articulated laterally at its base; the dorsal and ventral sutures, when present, separate it from the laciniastipes. See **ventral maxillary suture**.

grid (G) (Fig. 19c). The network of sclerotized ridges that bear the cratal setae of the ventral brush; with transverse grid bars at the base of individual setae; sometimes with lateral grid bars that may be confluent with the ventroposterior margin of the saddle. See **boss**.

Head (Fig. 15a). The anterior section (tagma) of the body bearing the antennae and mouthparts; separated from the thorax by the cervix.

hypocranial ecdysial line (HEL) (Fig. 16b,d). Any ecdysial line extending anteriad from the posteroventral margin of the cranium; in some instars of some mosquito larvae, such a line occurs medially. See **ecdysial line**.

hypopharynx (Hy) (Fig. 17e,f). (pl. hypopharynges, hypopharyxes; adj. hypopharyngeal). The median postoral appendage of the head; separated from the prelabium by a narrow space but united with it laterally to form the labiohypopharynx.

hypostomal sclerite (HSc) (Fig. 17c,d). A small, independent plate bearing seta 6-Mx at the base of the maxilla; often fused with the maxillary palpus, partially attached to the margin of the lateralia or completely united with the lateralia.

hypostomal suture (HyS) (Fig. 16b,d). The faint or undeveloped line of weakness borne at the margin of the labiogula; forming the border between the labiogula and lateralia; often faint or absent posteriorly.

labial palpus (LP) (Fig. 17e,f) (pl. labial palpi). One of a pair of vestigial labial palpi that are represented by a membranous area on each side of the ligula or prementoligular teeth; usually bearing four or five peg-like setae.

labiogula (LG) (Fig. 16b,d) (pl. labiobulae; adj. labiogular). The ventral area of the head enclosed by the hypostomal sutures.

labiohypopharynx (Lh) (Fig. 17e,f) (pl. labiohypopharynges, labiohypopharynxes; adj. labiohypopharyngeal). The protruding surface located between the mouth and the dorsomentum, comprising the prelabium and hypopharynx; forming the ventroposterior wall of the preoral cavity.

lacinia (Lc) (pl. laciniae; adj. lacinial). The inner lobe of the maxilla, either united with part of the stipes to form the laciniastipes or united with the mesal part of the stipes and the galea to form the maxillary body.

laciniarastrum (LR) (Fig. 17c,d) (pl. laciniarastra). One of two or three rows of variously developed spicules, laciniarastra 1–3, borne on the mesodorsal surface of the laciniastipes, or the maxillary body when the limits of the laciniastipes are indistinct; the rows are numbered from most mesal to most lateral and identified as LR_1, LR_2 and LR_3, respectively.

laciniastipes (LSt) (Fig. 17d) (pl. laciniastipites; adj. laciniastipital). Recognizable in many larvae, mostly culicines, as the mesal half of the maxillary body; representing the fused stipes and lacinia which cannot be separately distinguished; bearing the laciniarastra and seta 2-Mx mesodorsally; the dorsal and ventral maxillary sutures, when present, serve as boundaries between the laciniastipes and galeastipes.

larva (Fig. 15a) (pl. larvae; adj. larval). The second stage in the life cycle of the mosquito larva, which develops through four stadia, *i.e.* four instars.

lateral grid bar (LGB) (Fig. 19c). One of the more or less strongly developed longitudinal sclerotizations forming the lateral margins of the grid; sometimes confluent with the ventroposterior margin of the saddle.

lateral palatal brush (LPB). One of the brushes borne anterolaterally on the front of the head, consisting of a few to hundreds of filaments which bring food into the mouth.

lateral sclerite (LP). In *Orthopodomyia* larvae, a small transverse lateral plate located cephalad of the saddle and perhaps belonging to embryonic abdominal segment IX; the lateral plates of opposite sides may join to form a complete ring; probably not homologous with the saddle acus.

lateral premental teeth (LPT) (Fig. 17e,f). The bilaterally paired, usually heavily sclerotized and thickened teeth or toothed processes that flank the labial palpi on the mid-ventral region of the labiohypopharynx; in some culicines, occurring in three interconnected groups, lateral premental teeth 1–3, identified as LPT_1, LPT_2 and LPT_3 (Fig. 17f), respectively.

lateralia (Lat) (Fig. 16a–f) (pl. lateralia). The lateral and ventral areas of the cranium lateral to the frontal ecdysial lines dorsally and the hypostomal sutures ventrally.

ligula (Lg) (Fig. 17e,f). (pl. ligulae; adj. ligular). The separate or fused proximal lobes of the labium; in anopheline and some culicine larvae, an elevated, usually cusped tongue-like structure located between the labial palpi.

mandible (Mn) (Fig. 17a,b) (adj. mandibular). One of the first pair of gnathal appendages implanted obliquely on the underside of the head; typically flattened lobes with their mesal ends produced into strongly sclerotized teeth and a spiculate lobe.

mandibular brush (MnB) (Fig. 17a,b). A group of prominent, curved, acute spicules linearly arranged and varying in position from the dorsoanterior (in most species) to the dorsomesal margin (in predacious species) of the mandible; the number of spicules is reduced in predacious species; the most lateral spicules may be flattened and plumose in anophelines.

mandibular lobe (MnL) (Fig. 17a,b). A lightly sclerotized, rounded protuberance on the mesal margin of the mandible just posterior to the mandibular teeth; bearing as many as five groups of spicules; highly reduced or absent in predacious species.

mandibular spiculose area brush (MSAB) (Fig. 16a). In anopheline larvae, a linearly arranged dorsolateral group of variously developed filaments.

mandibular sweeper (MnS) (Fig. 17a,b). A linear group of filaments arising along the dorsal basal margin of the mandible; often divided into two groups, adoral and oral groups.

mandibular teeth (MnT) (Fig. 17a,b). A cluster of heavily sclerotized variously developed projections occurring on the mesodistal margin of the mandible.

maxilla (Mx) (Fig. 17c,d) (pl. maxillae; adj. maxillary). One of the second pair of gnathal appendages borne between the mandibles and the transverse margin of the lateralia; typically flattened lobe-like appendages consisting of the variously developed and/or fused hypostomal sclerite, stipes, galea and maxillary palpus.

maxillary body (MxBo) (Fig. 17c,d). The principal part of the maxilla consisting of the fused lacinia, galea and mesal part of the stipes; dorsal and ventral maxillary sutures demarcate the laciniastipes and galeastipes in a number of culicine larvae (see Fig. 17d).

maxillary brush (MxB) (Fig. 17c,d). A collection of variously developed spicules arising distally from the maxillary body, or the galeastipes when distinguishable (see Fig. 17d).

maxillary bundle (MxBn). In certain sabethine larvae, a flexible bundle of partially fused, bound or ensheathed spicules (individual elements are clearly evident internally or at least apically) arising distally from the maxillary body; sometimes accompanied by a number of independent basal spicules; homologous with the maxillary brush.

maxillary claw (MxC). In certain sabethine larvae, a solid claw-like structure borne at the apex of the maxillary body; representing completely fused and unrecognizable elements of the maxillary brush; sometimes accompanied by a number of independent basal spicules; homologous with the maxillary brush.

maxillary palpus (MPlp) (Fig. 17c,d) (pl. maxillary palpi; adj. maxillary palpal). The lateral cylindrical or fusiform lobe of the maxilla that is articulated or fused basally with the maxillary body; part of the stipes has been incorporated into the maxillary palpus in anophelines and many culicines.

median accessory tergal plate (MATP) (Fig. 18c). In some anopheline larvae, a small dorsal sclerite located immediately posterior to the tergal plate on the midline of abdominal segments I–VII. See **accessory tergal plate**.

median labral plate (MLP) (Fig. 16a,d). The dorsal sclerite of the labrum; separated from the dorsal apotome by the clypeolabral suture; indistinguishably fused with the dorsal apotome in certain taxa.

median plate (MdP) (Fig. 21a). In anopheline larvae, a large sclerite located centrally on the dorsal surface of the spiracular apparatus which receives the attachment of the muscles which close the apparatus.

merostipital sclerite (mSS) (Fig. 17d). In many culicine larvae, a small band of cuticle located laterally below the base of the maxillary palpus; articulated with the base of the mandible.

mesothorax (M) (Figs 15a, 18a, 19a) (pl. mesothoraces, mesothoraxes; adj. mesothoracic). The second or middle segment of the thorax; fused with the pro- and metathorax; with a series of 14 pairs of circumferentially arranged setae.

metathorax (T) (Figs 15a, 18a, 19a) (pl. metathoraces, metathoraxes; adj. metathoracic). The third, posterior segment of the thorax; fused with the mesothorax; with a series of 13 pairs of circumferentially arranged setae.

notal plate (NP). In certain taxa, one of a number of small, usually paired sclerites occurring on the dorsal surface of the thorax; their occurrence, size, number and location are variable.

occipital foramen (OF) (Fig. 16b,d,e,f) (pl. occipital foramina). The posterior opening of the head into the cervix.

pecten (Pt) (Figs 15a, 18c, 19b, 20a–c). In culicine larvae, a comb-like row of variously developed spicules, pecten spines, borne posterolaterally on the proximal part of the siphon; in anopheline larvae, borne on the posterior margin of the pecten plate.

pecten plate (PP) (Fig. 18c). In anopheline larvae, a more or less lateral sclerite bearing the pecten along its posterior margin.

pecten spine (PS) (Fig. 19b). One of the specialized spicules comprising the pecten; of various forms but usually spine-like with a denticulate margin.

pleural setal group (psg) (Figs 18a,b, 19a). The four (three in early instars of some species) setae borne on a common, usually partially sclerotized, ventrolateral tubercle of each thoracic segment.

postoccipital suture (POS). A line of inflection separating the postocciput from more anterior areas of the cranium; faint and sometimes only partly developed; interrupted dorsally by the coronal gap in late instars.

postocciput (POc) (Fig. 16c) (pl. postoccipita, postocciputs; adj. postoccipital). The posterior area of the cranium at the dorsal and lateral margins of the occipital foramen; separated from the more anterior areas of the cranium by the postoccipital suture; the narrow posterior part of the collar; interrupted dorsally by the coronal gap; ventrally continuous with the gula.

posterior median process (PMPc) (Fig. 21b). In certain culicine larvae, a filamentous projection arising from the posterior margin of the spiracular apodeme; protruding caudad between the posterolateral spiracular lobes of the spiracular apparatus.

posterolateral spiracular lobe (PSL) (Fig. 21a–c). One of the paired posterior flap-like projections of the spiracular apparatus; highly developed in larvae of Mansoniini.

preboss setae (PS). The ventral brush setae borne anterior to a boss.

precratal setae (PcrS) (Fig. 20b). The ventral brush setae borne anterior to a grid.

prelabium (pLb) (Fig. 17e,f) (pl. prelabia; adj. prelabial). The distal part of the labium comprising the mentum, ligula and labial palpi.

premental mala (PM) (Fig. 17e) (pl. premental malae). In anopheline and certain culicine larvae, one of a pair of denticulate tritural surfaces borne on each dorsolateral surface of the labiohypopharynx adjacent to the lateral premental teeth.

prothorax (P) (Figs 15a, 18a, 19a) (pl. prothoraces, prothoraxes; adj. prothoracic). The first or anterior segment of the thorax; fused with the mesothorax; with a series of 15 pairs of circumferentially arranged setae.

saddle (Sa) (Figs 15a, 18c, 19b, 20a–d). A large sclerite usually covering most of the dorsal and lateral surfaces of abdominal segment X; sometimes continuous ventrally to form a girdle.

saddle acus (SaA) (Fig. 20b). In some mosquito larvae, a small sclerite often attached anterolaterally to the base of the saddle.

saw (SAW) (Fig. 21c). In Mansoniini larvae, the anterior serrated plate of the spiracular apparatus; at least partly homologous with the anterior spiracular lobe of other mosquito larvae.

scape (Sc). The first or basal segment of the antenna; greatly reduced and often unrecognizably fused with the pedicel and flagellum; sometimes detached.

setal support plate (SSP) (Figs 18b, 19a). A small sclerite from which one or more setae project; it may or may not cover an elevation or tubercle.

setal support plate spine (SSPS) (Fig. 18b). A spine-like process of a setal support plate, occurring mainly on the thorax.

siphon (S) (Figs 15a, 19b, 20a–d, 21c). The dorsally located, elongate sclerotized tube (incompletely developed in early instars) of abdominal segment VIII; bearing the spiracular apparatus at its apex; structurally comprising parts of embryonic abdominal segments VIII and IX; at least partly homologous with the pecten plate of anopheline larvae.

siphon acus (SA) (Figs 19b, 20b). In some mosquito larvae, a small sclerite often attached posterolaterally to the base of the siphon; in *Mansonia* larvae, sometimes forming a complete ring.

siphon index. In siphon-bearing larvae, the ratio of the siphon length to the siphon width.

siphon length. In siphon-bearing larvae, the dorsal (anterior) length of the siphon measured in a straight line from base to apex.

siphon width. In siphon-bearing larvae, the width at the mid-length of the siphon measured at a right angle to the longitudinal axis.

spiracle (s) (Fig. 21a,b). The opening into a trachea; the spiracles of the meso- and metathorax and abdominal segments I–VII are rudimentary and non-functional; the postabdominal spiracles of abdominal segment VIII, enclosed by the spiracular apparatus, are functional.

spiracular apodeme (SAd) (Figs 19b, 20b,d, 21b). In culicine larvae, the usually hollow funnel-shaped ingrowth of the dorsal surface of the spiracular apparatus located between but largely posterior to the spiracles; receiving the muscles responsible for folding up the spiracular apparatus and hence closing the spiracles; partly homologous with the median plate of anopheline larvae.

spiracular apparatus (SAp) (Figs 15a, 18c, 19a,20a–d, 21a–e) (pl. apparatuses). The five-lobed valvular structure (including the spiracular apodeme or equivalent) encompassing the spiracles; structurally comprising parts of embryonic abdominal segments VIII and IX; borne at the apex of a siphon in culicine larvae; highly modified in *Mansonia* and *Coquillettidia* larvae for piercing plant tissues.

spiracular filament (SF). In some *Uranotaenia* larvae, a short or elongate filamentous process with an enlarged base borne externally on the posterolateral margin of each postabdominal spiracle.

spiracular lobe (SL). One of five rounded flap-like projections of the spiracular apparatus; highly modified in Mansoniini larvae. See **anterior spiracular lobe, anterolateral spiracular lobe** and **posterolateral spiracular lobe**.

submedian accessory tergal plate (SATP). In some anopheline larvae, one of usually a pair of small dorsal sclerites borne on each side of the midline immediately posterior to the median accessory tergal plates of abdominal segments I–VII. See **tergal plate** and **median accessory tergal plate**.

tagma (pl. tagmata). One of the three morphologically distinct sections of the body, the head, thorax and abdomen.

tergal plate (TP) (Figs 18a,c, 20a). In anopheline larvae, a small sclerite occurring anteriorly on the dorsal midline of abdominal segments I–VIII; in *Orthopodomyia* larvae, generally larger sclerites located on the dorsal surface of abdominal segments VI–VIII and VII–VIII, respectively; the tergal plates of *Orthopodomyia* may not be strictly homologous with those of anopheline larvae. See **accessory tergal plate** and **median accessory tergal plate**.

thorax (Fig. 15a) (pl. thoraces, thoraxes; adj. thoracic). The second or intermediate division (tagma) of the body; comprised of the fused pro-, meso- and metathoracic segments, which are distinguished by series of bilaterally paired, circumferentially arranged setae.

transverse grid bar (TGB) (Fig. 19c). One of the transverse sclerotizations of a grid supporting the bases of individual cratal setae; sometimes joined laterally by lateral grid bars. *Synonym*: Transverse bar.

ventral brush (VB) (Figs 15a, 18c, 19b). Seta 4-X, usually a more or less linear series of irregularly paired setae borne posteroventrally on the midline of abdominal segment X; often divided into two groups, a posterior group of cratal or boss setae borne on a grid or boss, respectively, and an anterior group of precratal or preboss setae borne cephalad of the grid or boss, respectively.

ventral maxillary suture (VMxS) (Fig. 17d). In some culicine larvae, a groove or furrow on the ventral surface of the maxillary body between the galeastipes and laciniastipes; often appearing only as a line of slightly more heavily sclerotized cuticle.

Larval Chaetotaxy

Chaetotaxy is defined as the arrangement (pattern and positions), branching and characteristics of setae (singular seta), the innervated sensory organs. Setal definitions all begin with the word **seta** and are arranged numerically for each body area. The body area treatments begin with the antenna and proceed posteriorly to abdominal segment X. It should be noted that serially homologous setae are denoted with the singular 'seta', *e.g.* seta 13-M,T (seta 13 on the meso- and metathoracic segments) and seta 1-I–IV (seta 1 on abdominal segments I, II, III and IV). Because of the considerable variation that occurs throughout the Culicidae in vertical and horizontal orientation of the living larva and its terminal parts, it is impractical to describe the positional relationships of the setae in terms of their true morphological orientations. Instead, all descriptions of positional relationships are based on a simple horizontal orientation of all sections of the body. Fourth-instar larvae may bear as many as 223 bilateral pairs of setae. All setae are potentially of taxonomic significance. They are distributed and defined as follows.

Antenna

Six setae are borne on the antenna, individually numbered by their approximate position from base to apex and bearing the hyphenated suffix 'A'.

seta 1-A (1-A) (Figs 22b, 23b). The dorsal seta borne proximal to the other antennal setae, except in *Toxorhynchites* in which it arises slightly distal to setae 2- and 3-A.

seta 2-A (2-A) (Figs 22b, 23b). The subapical seta on the mesoventral surface between the base of seta 1-A and the antennal apex, except in *Toxorhynchites* as noted for seta 1-A; sometimes as in Mansoniini arising well before the apex, or arising apically as in Anophelinae; in the Anophelinae and some Uranotaeniini, flattened and blade-like; arising dorsoventral in some taxa due to an apparent antennal rotation.

seta 3-A (3-A) (Figs 22b, 23b). The subapical seta borne on the lateroventral surface of the antenna; paired with seta 2-A; in some taxa it may occur with seta 2-A on the mesal surface where it is then ventral to 2-A; flattened and blade-like in Anophelinae and some Uranotaeniini.

seta 4-A (4-A) (Figs 22b, 23b). The seta borne on the upper median margin of the antennal apex; seemingly lateral in Anophelinae but actually occurring in the same relative position, the distal part of the antenna having rotated on its axis sufficiently to bring the ventral surface mesad and even somewhat dorsad.

seta 5-A (5-A) (Fig. 23b). A peg-like seta borne at the centre of the antennal apex, slightly preapical in some culicines; distinctive in being opaque basally and hyaline distally, the two parts being sharply demarcated.

seta 6-A (6-A) (Fig. 23b). The seta borne on the dorsolateral margin of the antennal apex; not as sharply tapered as setae 2-, 3- and 4-A; usually with a transverse transparent part near its base.

Cranium

Twenty pairs of setae (numbered 0–19) are known to occur on the cranium (two pairs borne on the cervix are included as cranial setae). These are designated by numerals beginning anterodorsally and preceding posteriorly and ventrally, each numeral bearing the hyphenated suffix 'C'.

seta 0-C (0-C) (Fig. 22b). The lateral seta on the median labral plate, lateral to seta 1-C, and the most anterolateral seta of the cranium; minute, usually ventral in position; apparently absent in some taxa.

seta 1-C (1-C) (Figs 22b, 23b). The mesal seta on the median labral plate, mesal to seta 0-C, and the most anteromesal seta of the cranium; usually single, often spiniform.

seta 2-C (2-C) (Fig. 22b). The most anteromesal seta in the series of setae (maximally consisting of seven pairs, seta 2- to 8-C) occurring on the dorsal apotome, just posterior to the median labral plate; interpreted as being absent when only one of the seven pairs of setae of this series is missing; usually well developed in anophelines; absent in most culicines.

seta 3-C (3-C) (Figs 22b, 23b). The most anterolateral seta in the series of setae (maximally consisting of seven pairs, seta 2- to 8-C) occurring on the dorsal apotome, just posterior to the median labral plate; well developed in anophelines; in culicines, sometimes arising from the underside of the head; in Sabethini and *Toxorhynchites*, associated with seta 0-C on the median labral plate; interpreted as being absent along with seta 2-C when two of the seven pairs of this series are missing.

seta 4-C (4-C) (Figs 22b, 23b). Generally the most anterior and/or anteromesal seta in the series of setae (maximally consisting of seven pairs, seta 2- to 8-C) occurring on the dorsal apotome; in *Toxorhynchites*, occurring on the median labral plate where it is interpreted as being the most mesal seta; occasionally lateral to seta 5-C, *e.g.* in *Huaedes*.

seta 5-C (5-C) (Figs 22b, 23b). The most posteromesal seta in the series of setae (maximally consisting of seven pairs, seta 2- to 8-C) occurring on the dorsal apotome.

seta 6-C (6-C) (Figs 22b, 23b). The seta located anterior and/or lateral to seta 5-C in the series of setae (maximally consisting of seven pairs, seta 2- to 8-C) occurring on the dorsal apotome.

seta 7-C (7-C) (Figs 22b, 23b). The most anterolateral seta in the series of setae (maximally consisting of seven pairs, seta 2- to 8-C) occurring on the dorsal apotome; the dorsal seta nearest to the base of the antenna.

seta 8-C (8-C) (Figs 22b, 23b). Usually the most posterior (always the most posterolateral) seta in the series of setae (maximally consisting of seven pairs, seta 2- to 8-C) occurring on the dorsal apotome; in *Toxorhynchites*, indistinguishably aligned with setae 5-, 6- and 7-C, interpreted as being between seta 6-C and seta 7-C.

seta 9-C (9-C) (Figs 22b, 23b). One of six setae occurring on the lateralia; specifically, the seta situated immediately lateral to the frontal ecdysial line, usually more or less opposite seta 8-C; its alveolus is visible from the dorsal aspect.

seta 10-C (10-C) (Figs 22b, 23b). One of six setae occurring on the lateralia; specifically, the seta situated lateroventral to seta 9-C; its alveolus is usually visible from the dorsal aspect; homologous with pupal seta 1-CT.

seta 11-C (11-C) (Figs 22b, 23b). One of six setae occurring on the lateralia; specifically, the seta located dorsolaterally near the base of the antenna; may sometimes arise sufficiently dorsad so that its alveolus is visible from above.

seta 12-C (12-C) (Figs 22b, 23b). One of six setae occurring on the lateralia; specifically, the seta (with seta 13-C) borne posterior and ventral to seta 11-C; usually situated posterior to seta 13-C, when linearly arranged with 13-C it is interpreted as being the more mesal seta; homologous with pupal seta 2-CT.

seta 13-C (13-C) (Figs 22b, 23b). One of six setae occurring on the lateralia; specifically, the seta (with seta 12-C) borne posterior to seta 11-C; usually anterior to seta 12-C, when linearly arranged with 12-C it is interpreted as being the more lateral seta; homologous with pupal seta 3-CT.

seta 14-C (14-C) (Figs 22b, 23b). One of six setae occurring on the lateralia; specifically, the seta situated immediately posterior to the base of the maxilla; the most mesoventral seta of the lateralia.

seta 15-C (15-C) (Figs 22b, 23b). The seta situated on the labiogula, usually anteriorly; the only seta arising on this cranial area.

seta 16-C (16-C). The more distant from the postocciput of two small setae borne on the lateralia in certain *Culex* larvae; if only one of these setae (16- and 17-C) is present, it is interpreted as being 16-C.

seta 17-C (17-C). The closer to the postocciput of two small setae (16- and 17-C) borne on the lateralia in certain *Culex* larvae; if only one of the two setae is present, 17-C is regarded as being absent.

seta 18-C (18-C). The more lateral of two small setae (18- and 19-C) borne ventrolaterally on the cervix at the level of the ventromedian cervical sclerite; when only one of the two setae is present, it is interpreted as being 18-C.

seta 19-C (19-C). The more ventral or anterior of two small setae (18- and 19-C) borne ventrolaterally on the cervix at the level of the ventromedian cervical sclerite; when only one of these is present, 19-C is regarded as being absent.

Mouth region

A total of 29 setae (or multiple setae) are known to occur on the areas and appendages about the mouth. These are numbered individually on each area or appendage, each number being followed by a hyphen and the abbreviation for the appropriate area or appendage. Three pairs occur on the labropalatum (Lp), one pair at the margin of the mouth (Mo), four pairs on the mandible (Mn), 15 pairs on the maxilla (Mx) and six pairs on the labiohypopharynx (Lh). Except for setae of the maxilla, the other setae of the mouth region, unfortunately, are not normally mentioned in taxonomic works, and for this reason only setae of the maxilla are listed here. See Harbach & Knight (1980) for descriptions of the other setae that occur in the region of the mouth.

seta 1-Mx (1-Mx). A usually paired peg-like seta borne on the dorsal surface of the maxillary body, or galeastipes when distinguishable; usually located near the dorsal maxillary suture but sometimes borne apically (extremely so in *Uranotaenia* larvae).

seta 2-Mx (2-Mx). A seta borne on the mesodorsal surface of the maxillary body, or the laciniastipes when distinguishable; usually located between laciniarastra 2 and 3.

seta 3-Mx (3-Mx) (Fig. 17c). A seta borne dorsally on the maxillary body, or the galeastipes when distinguishable; occurring near the distolateral margin in most culicines.

seta 4-Mx (4-Mx) (Fig. 17c,d). A seta borne ventrally on the maxillary body, or the galeastipes when distinguishable; occurring near the distolateral margin in most culicines.

seta 5-Mx (5-Mx) (Fig. 17c). A seta occurring distomesally on the maxillary body, or the galeastipes when distinguishable, at the base of the maxillary brush; almost always located at the apex of the dorsal maxillary suture.

seta 6-Mx (6-Mx) (Figs 17c,d, 22b, 23b). A prominent seta borne on the hypostomal sclerite or the anterior margin of the lateralia at the base of the maxilla.

seta 7-Mx (7-Mx) (Fig. 17c). In anopheline and *Mimomyia* larvae, a seta borne laterally on the maxillary palpus; prominent and dendritic in anophelines.

setae 8–15-Mx, maxillary pulpal sensoria (MS) (Fig. 17c,d). A series of setae (maximum of seven, setae 8–15-Mx) located at the apex of the maxillary palpus. Note: the homologies of these setae are somewhat tentative. They are presently based on similarities between anopheline and *Uranotaenia* larvae. If the maxillary palpus of certain anophelines is rotated so that the dorsal pair of setae (8- and 9-Mx) lie laterally, the positions of the structures more or less correspond to the positions of those in *Uranotaenia* species. An additional twisting of the palpus and a fusion of the centrally located setae, which appears to have taken place in *Uranotaenia*, would account for the observed positions of the structures. The setae of *Uranotaenia* larvae are easily homologized with those of most other culicines. See Harbach & night (1980) for information concerning the positions and recognition of the individual setae.

Thorax

A maximum of 15 pairs of setae are borne on the prothorax, 14 pairs on the mesothorax and 13 pairs on the metathorax. Each of these setae is individually designated first by a numeral showing its actual, or in some cases presumed, primitive position in the 'basic pattern' in which seta 1 arises closest to the dorsal longitudinal midline with the positions of successively higher numbered setae following progressively and circumferentially toward the ventral longitudinal midline, and second by a hyphen and a capital letter P, M or T indicating a position on the prothorax, mesothorax or metathorax, respectively. The presumptive segmental homologies between all of the setae of the thorax (except for seta 0-P) were established by Belkin (1950b, 1952, 1953b, 1954b, 1960, 1962).

Prothorax

seta 0-P (0-P) (Fig. 23a). A dorsal prothoracic seta borne posterior to a line between setae 3- and 4-P; serial homologues of this seta are not known to occur on succeeding segments. For probable ontogenetic homologies, see pupal setae 6- and 7-CT.

seta 1-P (1-P) (Figs 22a, 23a). The seta borne nearest the dorsal midline of the prothorax; the positions of setae 1-, 2- and 3-P are usually close-set in a series, 1-P is interpreted as having the most posterior and/or mesal position of the three. For probable ontogenetic homologies, see pupal setae 4- and 5-CT.

seta 2-P (2-P) (Figs 22a, 23a). The dorsal seta located in the second position in the circumferential series of prothoracic setae; in some taxa, seta 2-P may be posterior to seta 3-P. For probable ontogenetic homologies, see pupal setae 4- and 5-CT.

seta 3-P (3-P) (Figs 22a, 23a). The dorsal seta located in the third position in the circumferential series of prothoracic setae; nearly always closely associated with setae 1- and 2-P; may be partly homologous with pupal seta 4-CT. For probable ontogenetic homologies, see pupal setae 3–6-CT.

seta 4-P (4-P) (Figs 22a, 23a). The dorsal seta usually arising fourth from the dorsal midline of the prothorax. For probable ontogenetic homologies, see pupal setae 4–7-CT.

seta 5-P (5-P) (Figs 22a, 23a). The dorsolateral seta arising fifth from the dorsal midline of the prothorax; if occurring more or less longitudinally aligned with either seta 5- or 6-M, it is interpreted as being more anterior. For probable ontogenetic homologies, see pupal seta 6- and 7-CT.

seta 6-P (6-P) (Figs 22a, 23a). The dorsolateral seta arising sixth from the dorsal midline in the circumferential series of prothoracic setae; if more or less longitudinally aligned with either seta 5- or 7-M, it is interpreted as being the most posterior.

seta 7-P (7-P) (Figs 22a, 23a). The dorsolateral seta arising seventh from the dorsal midline in the circumferential series of prothoracic setae, however if seta 6- and 7-P are nearly equidistant from the dorsal midline, the more anterior seta is interpreted as seta 7-P even if slightly more mesal.

seta 8-P (8-P) (Figs 22a, 23a). The lateroventral seta arising eighth from the dorsal midline in the circumferential series of prothoracic setae.

seta 9-P (9-P) (Figs 22a, 23a). The dorsal (when viewed laterally) or external (when viewed ventrally) anterior seta in the pleural seta group (setae 9–12-P); elongate and usually branched. See **pleural seta group** in the **Larva** section.

seta 10-P (10-P) (Figs 22a, 23a). The ventral (when viewed laterally) or mesal (when viewed ventrally) seta in the pleural seta group (setae 9–12-P); usually the longest seta in the group. See **pleural seta group** in the **Larva** section.

seta 11-P (11-P) (Figs 22a, 23a). The dorsal (when viewed laterally) or external (when viewed ventrally) posterior seta in the pleural seta group (setae 9–12-P); usually reduced in size. See **pleural seta group** in the **Larva** section.

seta 12-P (12-P) (Figs 22a, 23a). The ventral (when viewed laterally) or mesal (when viewed ventrally) posterior seta in the pleural setal group (setae 9–12-P). See **pleural seta group** in the **Larva** section.

seta 13-P (13-P) (Fig. 22a). The lateroventral or ventral prothoracic seta borne anterior (in anophelines) or mesal (in many sabethines) to the tubercle bearing setae 9- to 12-P; absent in most culicines.

seta 14-P (14-P) (Figs 22a, 23a). The seta located closest to the ventral midline on the prothorax; sometimes aligned longitudinally with seta 13-P.

Mesothorax

seta 1-M (1-M) (Figs 22a, 23a). The seta borne nearest the dorsal midline of the mesothorax; probably homologous with pupa seta 8-CT.

seta 2-M (2-M) (Figs 22a, 23a). The dorsal seta located in the second position in the circumferential series of mesothoracic setae; in some taxa, seta 2-M may be posterior to seta 3-M.

seta 3-M (3-M) (Figs 22a, 23a). The dorsal seta located in the third position in the circumferential series of mesothoracic setae; nearly always closely associated with setae 1- and 2-M; may be anterior to and even widely separated from setae 1- and 2-M in some taxa; in cases where setae 3- and 4-M become more or less longitudinally aligned, the anterior seta is interpreted as being seta 3-M.

seta 4-M (4-M) (Figs 22a, 23a). The dorsal seta usually arising fourth from the dorsal midline on the mesothorax; in some anophelines it is posterior or posterolateral to seta 5-M and thus fifth from the dorsal midline; in cases where setae 3- and 4-M become more or less longitudinally aligned, the posterior seta is interpreted as being seta 4-M; possibly homologous with pupal seta 9-CT.

seta 5-M (5-M) (Figs 22a, 23a). The dorsolateral seta arising fifth from the dorsal midline on the mesothorax; if occurring more or less longitudinally aligned with either seta 5-P,T or seta 6-P,T, it is interpreted as being more anterior; possibly homologous with pupal seta 9-CT.

seta 6-M (6-M) (Figs 22a, 23a). The dorsolateral seta arising sixth from the dorsal midline in the circumferential series of mesothoracic setae; if more or less longitudinally aligned with either seta 5-P,T or seta 7-P,T, it is interpreted as being the most posterior.

seta 7-M (7-M) (Figs 22a, 23a). The dorsolateral seta arising seventh from the dorsal midline in the circumferential series of mesothoracic setae, however if setae 6- and 7-M are nearly equidistant from the dorsal midline, the more anterior seta is interpreted as seta 7-M even if slightly more mesal.

seta 8-M (8-M) (Figs 22a, 23a). The lateroventral seta arising eighth from the dorsal midline in the circumferential series of mesothoracic setae.

seta 9-M (9-M) (Figs 22a, 23a). The dorsal (when viewed laterally) or external (when viewed ventrally) anterior seta in the pleural seta group (setae 9–12-M); elongate and usually branched. See **pleural seta group** in the **Larva** section.

seta 10-M (10-M) (Figs 22a, 23a). The ventral (when viewed laterally) or mesal (when viewed ventrally) seta in the pleural seta group (setae 9–12-M); usually the longest seta in the group. See **pleural seta group** in the **Larva** section.

seta 11-M (11-M) (Figs 22a, 23a). The dorsal (when viewed laterally) or external (when viewed ventrally) posterior seta in the pleural seta group (setae 9–12-M); usually reduced in size. See **pleural seta group** in the **Larva** section.

seta 12-M (12-M) (Figs 22a, 23a). The ventral (when viewed laterally) or mesal (when viewed ventrally) posterior seta in the pleural setal group (setae 9–12-M). See **pleural seta group** in the **Larva** section.

seta 13-M (13-M) (Figs 22a, 23a). The lateroventral or ventral mesothoracic seta associated with the mesothoracic setal group (setae 9–12-M).

seta 14-M (14-M) (Figs 22a, 23a). The seta located closest to the ventral midline on the mesothorax.

Metathorax

seta 1-T (1-T) (Figs 22a, 23a). The seta borne nearest the dorsal midline of the metathorax; homologous with pupal seta 10-CT.

seta 2-T (2-T) (Figs 22a, 23a). The dorsal seta located in the second position in the circumferential series of metathoracic setae; in some taxa, seta 2-T may be posterior to seta 3-T; homologous with pupal seta 11-CT.

seta 3-T (3-T) (Figs 22a, 23a). The dorsal seta located in the third position in the circumferential series of metathoracic setae; nearly always closely associated with setae 1- and 2-T; may be anterior to and even widely separated from these setae in some taxa; in cases where setae 3- and 4-T become more or less longitudinally aligned, the anterior seta is interpreted as being seta 3-T; homologous with pupal seta 12-CT.

seta 4-T (4-T) (Figs 22a, 23a). The dorsal seta usually arising fourth from the dorsal midline on the metathorax; in cases where setae 3- and 4-M become more or less longitudinally aligned, the posterior seta is interpreted as being seta 4-T; in *Toxorhynchites*, seta 4-T is closely associated with seta 3-T in a position immediately anterior to setae 1- and 2-T.

seta 5-T (5-T) (Figs 22a, 23a). The dorsolateral seta arising fifth from the dorsal midline on the metathorax; if occurring more or less longitudinally aligned with either seta 5- or 6-M, it is interpreted as being more anterior.

seta 6-T (6-T) (Figs 22a, 23a). The dorsolateral seta arising sixth from the dorsal midline in the circumferential series of metathoracic setae; if more or less longitudinally aligned with either seta 5- or 7-M, it is interpreted as being the most posterior.

seta 7-T (7-T) (Figs 22a, 23a). The dorsolateral seta arising seventh from the dorsal midline in the circumferential series of metathoracic setae, however if seta 6- and 7-T are nearly equidistant from the dorsal midline, the more anterior seta is interpreted as seta 7-T even if slightly more mesal.

seta 8-T (8-T) (Figs 22a, 23a). The lateroventral seta arising eighth from the dorsal midline in the circumferential series of metathoracic setae.

seta 9-T (9-T) (Figs 22a, 23a). The dorsal (when viewed laterally) or external (when viewed ventrally) anterior seta in the pleural seta group (setae 9–12-T); elongate and usually branched. See **pleural seta group** in the **Larva** section.

seta 10-T (10-T) (Figs 22a, 23a). The ventral (when viewed laterally) or mesal (when viewed ventrally) seta in the pleural seta group (setae 9–12-T); usually the longest seta in the group. See **pleural seta group** in the **Larva** section.

seta 11-T (11-T) (Figs 22a, 23a). The dorsal (when viewed laterally) or external (when viewed ventrally) posterior seta in the pleural seta group (setae 9–12-T); usually reduced in size. See **pleural seta group** in the **Larva** section.

seta 12-T (12-T) (Figs 22a, 23a). The ventral (when viewed laterally) or mesal (when viewed ventrally) posterior seta in the pleural setal group (setae 9–12-T). See **pleural seta group** in the **Larva** section.

seta 13-T (13-T) (Figs 22a, 23a). The metathoracic seta located closest to the ventral midline.

Abdomen

A maximum of 12 pairs of setae are borne on abdominal segment I, 15 pairs on abdominal segments II through VII, seven pairs on abdominal segment VIII, 13 pairs on the siphon (pecten plate in anophelines) and spiracular apparatus (derived from embryonic abdominal segments VIII and IX) and five pairs on abdominal segment X. The terminology for these setae consists of an Arabic number for each indicating its presumed primitive sequential relationship in which seta 1 is closest to the dorsal longitudinal midline with the higher numbered setae following progressively around the side to the ventral longitudinal midline, connected by a hyphen with a capitalized Roman numeral indicating the abdominal segment on which the seta occurs. The presumptive segmental homologies between all of the setae of the first seven abdominal segments were established by Belkin (1950b, 1952, 1953b, 1954b, 1960, 1962).

Abdominal segments I–VII

A maximum of 12 pairs of setae are borne on abdominal segment I (setae 0, 8 and 14 are presumed to be absent); a maximum of 15 pairs of setae are borne on each of abdominal segments II through VII. The sequential Arabic numerals applied to the setae of abdominal segments I through VII are believed to indicate serial homologies down the row of segments, ontogenetic homologies between the larval instars, as well as with the pupa, and phylogenetic homologies throughout the Culicidae. Typically on these abdominal segments, setae 0–5 are dorsal, setae 6–9 lateral and setae 10–14 ventral. Serial homologies for setae 0–5 are most clearly apparent on abdominal segment I and for setae 6–14 on abdominal segment II. On these segments, the setae designated are typically present in a consecutive numerical sequence or in departures from this that can be more or less readily determined. In determining serial homologies, similarities in setal branching and development should be noted, as well as relative position. Difficult setae can sometimes be named through a process of elimination.

seta 0-II–VII (0-II, 0-III, *etc.*) (Figs 22a,c, 23a,c). The most anterior of the dorsal setae on abdominal segments II–VII; inconspicuous except in some sabethines where it may be prominent and stellate; absent from abdominal segment I.

seta 1-I–VII (1-I, 1-II, *etc.*) (Figs 22a,c, 23a,c). The dorsal seta nearest both the posterior margin and the longitudinal midline of abdominal segments I–VII; seta 3-VI and sometimes seta 2-VI may be nearer in some taxa (in such cases these setae can be identified through a process of elimination); usually well developed, may be stellate or palmate in some taxa.

seta 2-I–VII (2-I, 2-II, *etc.*) (Figs 22a,c, 23a,c). The dorsal seta nearest (except for the minute seta 0) to both the anterior margin and the longitudinal midline of abdominal segments I–VII; occasionally so atypically placed as to be identifiable only through a process of elimination, *e.g.* in species of *Aedes* subgenus *Alanstonea.*

seta 3-I–VII (3-I, 3-II, *etc.*) (Figs 22a,c, 23a,c). The dorsal seta sequentially in line lateral to seta 2 on abdominal segment I, variable in position on succeeding segments but determinable based on branching, development and relative position, usually mesal to seta 4 on abdominal segments I, VI and VII, even with or lateral to seta 4 on abdominal segments II–V where it is associated with seta 5; on abdominal segment VI it arises in the vicinity of seta 1 in some taxa.

seta 4-I–VII (4-I, 4-II, *etc.*). The dorsal seta sequentially in line lateral to seta 3 on abdominal segment I, variable in position on succeeding segments, usually lateral to seta 3 on abdominal segments I, VI and VII, on line with or mesal to seta 3 on abdominal segments II–V, usually located anterior and mesal to seta 5 (more often lateral on VII); stellate in some taxa.

seta 5-I–VII (5-I, 5-II, *etc.*) (Figs 22a,c, 23a,c). The dorsal seta sequentially in line lateral to seta 4 on abdominal segment I, its position relative to seta 4 is variable on succeeding segments but it is identifiable because of its very characteristic dorsal position on the posterolateral angle of the segment, usually proximal and dorsal to seta 6.

seta 6-I–VII (6-I, 6-II, *etc.*) (Figs 22a,c, 23a,c). The lateral seta borne regularly on the prominent lateral bulge about midway between the dorsal and ventral longitudinal midlines on abdominal segments I–VII; well developed on abdominal segments I and II, commonly somewhat less well developed on III–VI, still smaller on VII.

seta 7-I–VII (7-I, 7-II, *etc.*) (Figs 22a,c, 23a,c). The lateral seta either directly ventral or slightly posteroventral to seta 6 on abdominal segments I–VII; when reduced, as is often the case on abdominal segments III–VII, it can be identified by its position relative to seta 6; stellate in some taxa.

seta 8-II–VII (8-I, 8-II, *etc.*) (Figs 22a,c, 23a,c). The lateral seta nearest the anterior margin of abdominal segments II–VII; absent from abdominal segment I.

seta 9-I–VII (9-I, 9-II, *etc.*) (Figs 22a,c, 23a,c). The lateral seta associated anteriorly, posteriorly or ventrally with seta 7 on abdominal segments I–VII; stellate in some taxa.

seta 10-I–VII (10-I, 10-II, *etc.*) (Figs 22a,c, 23a,c). The seta with setae 11, 12 and 13 on abdominal segments I–VII comprising a transverse median or somewhat posterior group of ventral setae so diversely positioned and developed down the line of segments as frequently to obscure the recognition of serial homologies. Theoretically, setae 10 and 11 can be identified by tracing their sensory neurons to their homologues in the developing pupa. This is possible because setae 12 and 13 are absent (rarely one may be anomalously present) in the pupa. In the larva, not only are setae 12 and 13 routinely present (12 is absent on abdominal segment I in some taxa), but they arise in such a variety of relationships to each other and to setae 10 and 11 as to make positive identifications very difficult. If the pupa of the species being examined is available, setae 11 and 12 can in most cases be recognized through cross comparisons. In many taxa, abdominal segment II is helpful because setae 10 through 13 are identifiably arranged sequentially from laterad to mesad. From this segment one can proceed anteriorly and posteriorly, identifying setae by noting similarities in development and relative position. Seta 13 is the nearest of the group to the ventral longitudinal midline on abdominal segment II, and in many taxa it is also the most developed (sometimes even stellate) of the four grouped ventral setae on the more posterior segments.

seta 11-I–VII (11-I, 11-II, *etc.*) (Figs 22a,c, 23a,c). The seta with setae 10, 12 and 13 on abdominal segment I comprising a transverse median or somewhat posterior group of ventral setae so diversely positioned and developed down the line of segments as frequently to obscure the recognition of serial homologies; usually shorter than seta 10; its relationship with seta 10 regarding appearance and relative position closely approximates that found on the pupa; stellate in some sabethines. See **seta 10-I–VII** for explanation.

seta 12-I–VII (12-I, 12-II, *etc.*) (Figs 22a,c, 23a,c). The seta with setae 10, 11 and 13 on segments I–VII comprising a transverse median or somewhat posterior group of ventral setae so diversely positioned and developed down the line of segments as frequently to obscure the recognition of serial homologies; interpreted as being absent when only three in the group of setae 10–13 are present on abdominal segment I. See **seta 10-I–VII** for explanation.

seta 13-I–VII (13-I, 13-II, *etc.*) (Figs 22a,c, 23a,c). The seta with setae 10, 11 and 12 on segments I–VII comprising a transverse median or somewhat posterior group of ventral setae; seta 13 is the nearest of the group to the ventral longitudinal midline of abdominal segment II; in many taxa, it is the most developed (sometimes even stellate) of the four grouped ventral setae on at least the more posterior segments. See **seta 10-I–VII** for explanation.

seta 14-II–VII (14-II, 14-III, *etc.*). The ventral seta of abdominal segments II–VII located close to the longitudinal midline far forward on the intersegmental membrane; minute except in some sabethines; absent from abdominal segment I and usually absent from segment II; an apparent homologue is present on abdominal segment VIII.

Abdominal segment VIII

Of the seven pairs of setae that occur on this segment, the minute setae 0 and 14 are treated as homologues of those similarly designated on abdominal segments II–VII. The numerical designations applied from dorsal to ventral to the other five pairs of setae are arbitrary and do not imply homologies with setae of the segments anterior and posterior to segment VIII.

seta 0-VIII (0-VIII) (Figs 22c, 23c). The minute seta borne anteriorly on the dorsal surface of abdominal segment VIII in the same relative position as the corresponding seta on the preceding segments.

seta 1-VIII (1-VIII) (Figs 22c, 23c). Usually the most dorsal seta borne on abdominal segment VIII; often almost directly anterior to seta 2-VIII; typically branched.

seta 2-VIII (2-VIII) (Figs 22c, 23c). The second seta in the set of setae 1–5-VIII; located anteroventral to seta 1-VIII in anophelines; located ventral, posteroventral or posterior to seta 1-VIII in culicines, often unbranched in these taxa; homologous with pupal seta 4-VIII.

seta 3-VIII (3-VIII) (Figs 22c, 23c). The middle seta in the set of setae 1–5-VIII; usually conspicuously branched and typically the most posterior in the set.

seta 4-VIII (4-VIII) (Figs 22c, 23c). The seta located ventral to seta 3-VIII in the set of setae 1–5-VIII; usually branched.

seta 5-VIII (5-VIII) (Figs 22c, 23c). The most ventral of the conspicuous setae of abdominal segment VIII; usually branched.

seta 14-VIII (14-VIII) (Figs 22c, 23c). The minute ventral seta of abdominal segment VIII located anteriorly on the intersegmental membrane in the same relative position as the corresponding seta on the preceding segments.

Siphon and spiracular apparatus

The siphon (pecten plate in anophelines) and spiracular apparatus possess 13 pairs of setae of uniform occurrence. Additional setae occur in sabethines but since they do not occur in other taxa they are regarded as being adventitious and are therefore not included in the standard setal nomenclature. The setae on these structures are numbered from proximal to distal. No homologies are implied with setae borne elsewhere on the abdomen. The terminology consists of an Arabic number for each seta connected by a hyphen with the capital letter 'S'. The structures referred to above are comprised of parts of embryonic abdominal segments VIII and IX.

seta 1-S (1-S) (Figs 21c, 22c, 23c). In anopheline larvae, the seta closely associated with and usually located posterior to the pecten below the posterolateral spiracular lobe; in other mosquito larvae, borne posterolaterally on the siphon; in species where more than one seta is borne in a similar position on the siphon, they are designated as seta 1a-S, seta 1b-S, *etc.* (Fig. 23c) proceeding distad from the most proximal seta.

seta 2-S (2-S) (Figs 21c, 22c, 23c). In anopheline larvae, the seta borne on the pecten plate; in other mosquito larvae, borne on the dorsal preapical area of the siphon or on the membrane just beyond its apex (in Mansoniini and some Sabethini).

seta 2a-S (2a-S). In certain culicine larvae, *e.g. Aedes* (*Aedes*), *Aedes* (*Rusticoidus*) and *Culex* (*Maillotia*), one or more setae borne anterolaterally on the siphon.

seta 3-S (3-S) (Fig. 21d,e). The minute anterior of the three close-set setae borne laterally on the unpaired anterior spiracular lobe; usually represented by an empty alveolus (see **puncture** in the **Vestiture** section).

seta 4-S (4-S) (Fig. 21d,e). The minute middle of the three close-set setae borne laterally on the unpaired anterior spiracular lobe.

seta 5-S (5-S) (Fig. 21d,e). The minute posterior of the three close-set setae borne laterally on the unpaired anterior spiracular lobe.

seta 6-S (6-S) (Fig. 21c,d,e). The proximal seta of the paired anterolateral spiracular lobe.

seta 7-S (7-S) (Fig. 21d,e). The distal seta of the paired anterolateral spiracular lobe.

seta 8-S (8-S) (Fig. 21c,d,e). The proximal seta borne on the outer surface (when the postabdominal spiracles are closed) of the paired posterolateral spiracular lobe.

seta 9-S (9-S) (Fig. 21d,e). The distal seta borne preapically on the surface (when the postabdominal spiracles are closed) of the paired posterolateral spiracular lobe; in Mansoniini larvae, a small seta located near the outer spiracular teeth.

seta 10-S (10-S) (Fig. 21d,e). The seta borne basally on the posterior margin of the inner surface (when the postabdominal spiracles are closed) of the paired posterolateral spiracular lobe.

seta 11-S (11-S) (Fig. 21d,e). The more proximal of the two setae borne apically on the inner surface (when the postabdominal spiracles are closed) of the paired posterolateral spiracular lobe.

seta 12-S (12-S) (Fig. 21d,e). The more distal of the two setae borne apically on the inner surface (when the postabdominal spiracles are closed) of the paired posterolateral spiracular lobe; with an expanded base and a proximal kink in *Uranotaenia* species of the subgenus *Uranotaenia*.

seta 13-S (13-S) (Fig. 21d,e). The seta borne medially on the anterior margin of the inner surface (when the postabdominal spiracles are closed) of the paired posterolateral spiracular lobe; minute and perhaps vestigial in species of *Uranotaenia* (compare **seta 12-S**).

Abdominal segment X

Four pairs of setae are of uniform occurrence on this segment. No homologies with other abdominal setae are implied by the numbers assigned to these setae.

seta 1-X (1-X) (Figs 22c, 23c). The lateral seta of abdominal segment X borne on the saddle, usually near the posterior margin; arising from the integument adjacent to the saddle in some species.

seta 2-X (2-X) (Figs 22c, 23c). The seta located closest to the dorsal longitudinal midline of abdominal segment X and slightly posterior to the saddle; along with seta 3-X, referred to as the dorsal brush.
seta 3-X (3-X) (Figs 22c, 23c). The seta borne on abdominal segment X immediately ventral to seta 2-X; along with seta 2-X, referred to as the dorsal brush.
seta 4-X (4-X) (Figs 22c, 23c). Any seta of the ventral brush, the variable number of paired or unpaired setae arising from the ventral longitudinal midline of abdominal segment X; the most anterior seta is designated as seta 4a-X, the next posterior one as 4b-X, *etc.*; represented by a single pair of setae in all Sabethini, except species of subgenus *Sabethinus* of *Sabethes*, which have a pair of small auxiliary setae 4-X.
seta x-X, seta y-X and **seta z-X** (x-X, y-X, z-X). The paired accessory setae located on the side of abdominal segment X on the membranous part or on the saddle, seta x-X being the most proximal, seta z-X the most posterior and seta y-X between them; one or more pairs are present in some Mansoniini and *Uranotaenia*.

Pupae

This section contains specific terminology applied to the structures that comprise the external skeleton of mosquito pupae. Extensive anatomical changes take place within the pupa, and the adult emerges by a splitting of the dorsal surface of the cephalothorax. The pupal–adult apolysis occurs shortly after the larval–pupal ecdysis. Hence, there is a very short exposed pupal period followed by a very long pharate adult phase. Therefore, most of the conventional 'pupal stage' is an adult, not a pupa, enclosed by the remains of the pupal cuticle. Consequently, much of the terminology applied to pupal structures is the same as that applied to adult mosquitoes.

abdomen (Figs 15b, 15a,b) (pl. abdomens, abdomina; adj. abdominal). The third, posterior division (tagma) of the body, consisting of 10 apparent segments.
abdominal puncture (p) (Figs 15b, 25a,b). The empty alveolus that occurs medially on the dorsolateral area of abdominal segments III to V (occurring anomalously on segment II); associated with setae 4-, 5- and 6-III–V.
abdominal segment (I, II, *etc.*) (Figs 15b, 25,a,b). One of the annular subdivisions of the abdomen; labelled I (most anterior) to X (most caudal).
abdominal segment IX (IX) (Figs 15b, 25a,b). The ninth annular subdivision of the abdomen, represented dorsally chiefly by the median caudal lobe; sternum IX may be present but is usually reduced and indistinct.
angusticorn trumpet (AT) (Fig. 24d). A trumpet having the longest axis vertical more or less in line with the stem; narrowly funnel-shaped when closed; with a split (meatal cleft) down one side which allows it to open widely at the surface of the water. Compare with **laticorn trumpet**.
antenna (A) (Figs 15b, 24a,b). One of the paired anterior appendicular organs of the head; curved upward and backward over the side of the thorax beneath the lower edge of the wing.
buttress (Bu) (Fig. 25a,b) (pl. buttresses). The thickened, sclerotized, basolateral part of the paddle.
cephalothorax (CT) (Figs 15, 24a,b) (pl. cephlothoraxes, cephalothoraces; adj. cephalothoracic). The combined head and thorax. This term is entirely descriptive and is not meant to indicate that the head and thorax are united.
cercus (Ce) (Figs 15c, 25a) (pl. cerci; adj. cercal). One of the pair of appendages of abdominal segment XI; in female pupae, one of the more or less distinct lateral lobes of the proctiger; sometimes bearing a seta, seta 1-XI.
compound eye (CE) (Fig. 15b). A light-perceptive organ consisting of an aggregation of optic elements (ommatidia) generally located on each side of the head; present in a developing state; perhaps functional.
dorsal apotome (DAp) (Fig. 24a,b). A dorsal cranial area perhaps homologous with the dorsal apotome of larvae.
genital lobe (GL) (Figs 15b,c, 25a,b). The rounded projection lying ventral to the proctiger; in females, usually small, inconspicuous and spiculate; in males, large and partially bilobed, representing fused gonocoxopodites. See **gonocoxopodite** in the **Adult** section.
head. The anterior section (tagma) of the body bearing the compound eyes, antennae and mouthparts; separated from the thorax by the cervix, the cervix is reduced and the head abuts the thorax, the combined head and thorax being referred to as the cephalothorax.
inner margin (Fig. 25a,b). The inner or mesal margin of the paddle from the apex of the midrib, or its projection, to the base.
inner part (Fig. 25a,b). The part of the paddle mesad of the midrib.
laticorn trumpet (LT) (Fig. 24c,e). A trumpet having the longest axis transverse to the stem; often with a secondary cleft in the pinna opposite the meatal cleft; sometimes less wide-mouthed but with a more or less elaborate lobe, the tragus, on the rim of the pinna. Compare with **angusticorn trumpet**.
lateralia (Lat) (Fig. 15b). The principally lateral area of the head, bearing setae 1–3-CT. See **lateralia** in the **Larvae** section.
leg (Fig. 15b). One of the non-functional paired appendages of the thorax; folded in loops against the sides of the thorax; comprised of the usual segments. See **leg** in the **Adult** section.
maxillary palpus (MPlp) (Figs 15b, 24a,b). The appendage of the maxilla folded along the side of the proboscis. See **maxillary palpus** in the **Adult** section.

meatal cleft (MC) (Fig. 24d). In some pupae, a more or less distinct slit or line extending into the meatus from the spiracular opening; allowing the spiracular opening to enlarge at the surface of the water.

meatus (Mea) (Fig. 24g). The part of the trumpet from the base to an imaginary line drawn more or less perpendicular to the longitudinal axis at the most proximal margin of the spiracular opening; extent expressed as a fraction of the trumpet length.

median caudal lobe (MCL) (Figs 15b,c, 25a,b). The transverse lobe borne caudal to tergum VIII and dorsal to the proctiger; representing the dorsum of abdominal segment IX and at least a small part of abdominal segment VIII.

median keel (MK) (Figs 15b, 24a,b). The crest-like median longitudinal ridge of the scutum; bearing the dorsal ecdysial line along its midline.

mesothoracic wing (MW) (Fig. 15b). See **wing** in the **Adult** section; present but non-functional in the pupal stage; extended downward on the sides of the thorax; with little or no indication of venation.

metanotum (Mtn) (Figs 15b, 25a,b). The notum (tergum) of the metathorax; an ill-defined sclerite appearing as part of the abdomen; bearing the metathoracic wings laterally.

metathoracic wing (MtW) (Figs 15b, 25a,b). See **halter** and **wing** in the **Adult** section); present but non-functional in the pupal stage; lying behind the mesothoracic wing with the apex directed posteriorly along the side of abdominal segment I.

midrib (Mr) (Figs 15b, 25a,b). The thickened median vein-like supporting structure of the paddle.

outer margin (Fig. 25a,b). The outer or lateral margin of the paddle from the apex of the midrib, or its projection, to the base.

outer part (Fig. 25a,b). The part of the paddle laterad of the midrib.

paddle (Pa) (Figs 15b, 25a,b). The movable or immovable paired appendage of abdominal segment IX; appearing to project from the caudolateral part of abdominal segment VIII. The paddle has inner and outer parts and inner and outer margins, respectively.

paddle fringe (PF). The serration, aciculae and/or filaments occurring along the inner and outer margins of the paddle.

paddle index. The ratio of the paddle length to the paddle width.

paddle length. The maximum length of the paddle from the basal articulation to the most distal point measured along the midrib or its projection.

paddle marginal serration (Se) (Fig. 25a). The serration of the paddle fringe, usually confined to the outer margin.

paddle marginal spicules (Sp) (Fig. 25a,b). The aciculae or filaments of the paddle fringe, usually confined to the outer margin.

paddle width. The maximum width of the paddle measured along a line perpendicular to the midrib.

pinna (Pi) (Fig. 24f,g). The part of the trumpet from the apex to an imaginary line drawn more or less perpendicular to the longitudinal axis at the most proximal margin of the opening, the part distal to the meatus; extent expressed as a fraction of the trumpet length.

proboscis (P) (Fig. 15b). The greatly elongate bundle of closely adherent oral appendages curled beneath the cephalothorax. See **proboscis** in the **Adult** section.

proctiger (Pr) (Fig. 15c). The poorly defined segmentation behind abdominal segment IX. Compare **proctiger** in the **Adult** section.

pronotum. The ill-defined dorsal area of the cephalothorax bearing setae 4–7-CT.

pupa (Fig. 15b) (pl. pupae; adj. pupal). The third or penultimate stage in the life cycle of a mosquito.

refractile border (RB). The part of the outer margin of the paddle bearing the paddle marginal serration; refractile to light.

refractile index. In mosquito pupae, the ratio of the paddle length to the length of the refractile border.

reticulate area (RA) (Fig. 24f,g). The part of the trumpet meatus where the surface has an interconnected system of shallow grooves which separate raised areas of the cuticula.

secondary cleft (SeC) (Fig. 24c). In many pupae with a laticorn trumpet, a small slit extending into the pinna from the margin of the opening opposite the meatal cleft.

scutum (Scu) (Figs 15b, 24a,b) (pl. scuta; adj. scutal). The principal dorsal area of the mesothorax; an imperfectly delimited area between the bases of the mesothoracic wings, bearing setae 8- and 9-CT.

sternum (S) (pl. sterna; adj. sternal; adv. sternad, sternally). The ventral sclerotization of an abdominal segment.

sternum IX (S-IX). The sternum of abdominal segment IX; may be present as a reduced and indistinct transverse sclerite at the base of the genital lobe.

tagma (pl. tagmata). One of the three morphologically distinct sections of the body, the head, thorax and abdomen; the head and thorax are united for form the cephalothorax.

tergum (Te) (pl. terga; adj. tergal; adv. tergad, tergally). The dorsal sclerotization of an abdominal segment.

thorax (pl. thoraces, thoraxes; adj. thoracic). The second or intermediate division (tagma) of the body; bearing the legs and wings; the head abuts the thorax, and the combined head and thorax are referred to as the cephalothorax.

tracheoid area (TA) (Fig. 24f,g). The proximal part of the trumpet meatus when it shows more or less distinct transverse striations on the external surface; extent expressed as a fraction of its greatest length to the trumpet length.

tragus (Tg) (Fig. 24e). A more or less elaborate lobe sometimes occurring on the rim of the pinna of a laticorn trumpet.

trumpet (Tr) (Figs 15b, 24a–g). One of the paired, usually movable, dorsolateral appendages of the cephalothorax containing the mesothoracic spiracle; apex cut off obliquely to form the pinna, the remainder being the meatus; the surface sometimes with a proximal tracheoid area and a distal reticulate area or entirely reticulate. See **angusticorn trumpet** and **laticorn trumpet**.

trumpet index. The ratio of the trumpet length to the trumpet width.
trumpet length. The maximum length of the trumpet measured along a straight line from the point of articulation at the base to the extreme apex.
trumpet width. The width of the trumpet measured at one-half the trumpet length; measured along a line perpendicular to the line of projection for the trumpet length.

Pupal Chaetotaxy

This section deals with the chaetotaxy, the arrangement (pattern and positions), branching and characteristics, of the innervated sensory structures, setae (singular seta), of mosquito pupae. The components of the pupal chaetotaxy are homologous phylogenetically, ontogenetically and in part serially. The discovery (Belkin, 1960, 1962) that the developing pupal setae are connected by sensory neurons to their homologues in the fourth-instar larva established beyond doubt the ontogenetic homology of the setae. However, the fate of many larval setae during the ontogenetic transition from the larval stage is unknown. The setal definitions presented below are based on the interpretations of Belkin (1962) and many later publications. Borkent (2012) proposed a different setal nomenclature based on comparative study of the families of Culicomorpha, but it is not accepted as a replacement for the traditional nomenclature used by culicid taxonomists. The setal definitions all begin with the word **seta** and are arranged numerically for each body area. One hundred and nineteen (119) pairs of setae are found on mosquito pupae. All setae are potentially of taxonomic significance. The setae are distributed and defined as follows.

Cephalothorax

Fifteen pairs of setae occur on the cephalothorax, three on the lateralia of the head, four on the pronotum, two on the scutum of the mesothorax and usually three, sometimes four, five or six, on the metathorax (metanotum). The setae of this area are designated with a number connected by a hyphen to the suffix 'CT'.

seta 1-CT (1-CT) (Fig. 24a,b). The dorsal of the three head setae borne on the lateralia; homologous with larval seta 10-C; in slide-mounted exuviae, the lateralia fold in such a manner that this seta becomes the most mesal and caudal whereas seta 3-CT becomes the most lateral and anterior.
seta 2-CT (2-CT) (Fig. 24a,b). The median of the three head setae borne on the lateralia; homologous with larval seta 12-C.
seta 3-CT (3-CT) (Fig. 24a,b). The ventral of the three head setae borne on the lateralia; homologous with larval seta 13-C.
seta 4-CT (4-CT) (Fig. 24a,b). The seta associated with seta 5-CT to form the more anterior setal group of the pronotum; usually interpreted as being the more mesal and/or posterior of the two; with seta 5-CT, probably homologous with either larval setae 1- and 2-P, setae 1- and 3-P, or setae 2- and 3-P, respectively.
seta 5-CT (5-CT) (Fig. 24a,b). The seta associated with seta 4-CT to form the more anterior setal group of the pronotum; usually interpreted as being the more lateral and/or anterior of the two; with seta 4-CT, probably homologous with either larval setae 1- and 2-P, setae 1- and 3-P, or setae 2- and 3-P, respectively.
seta 6-CT (6-CT) (Fig. 24a,b). The seta associated with seta 7-CT to form the more posterior setal group of the pronotum; usually interpreted as being the more anterior and/or lateral of the two; with seta 7-CT, probably homologous with either larval setae 3- and 4-P, setae 4- and 0-P, or setae 4- and 5-P, respectively.
seta 7-CT (7-CT) (Fig. 24a,b). The seta associated with seta 6-CT to form the more posterior setal group of the pronotum; usually interpreted as being the more posterior and/or mesal of the two; with seta 6-CT, probably homologous with either larval setae 3- and 4-P, setae 4- and 0-P, or setae 4- and 5-P, respectively.
seta 8-CT (8-CT) (Fig. 24a,b). The seta associated with seta 9-CT on the scutum of the mesothorax mesad of the trumpet; interpreted as being either the more anterior and/or mesal of the two; probably homologous with larval seta 1-M based on position and the degree of development.
seta 9-CT (9-CT) (Fig. 24a,b). The seta associated with seta 8-CT on the scutum of the mesothorax mesad of the trumpet; interpreted as being either the more posterior and/or lateral of the two; homology not clear, possibly homologous with either larval seta 4-M or seta 5-M.
seta 10-CT (10-CT) (Fig. 24a,b). The seta associated with setae 11- and 12-CT to form a group anterolaterally on the metanotum; interpreted as being the most mesal of the group; homologous with larval seta 1-T.
seta 11-CT (11-CT) (Fig. 24a,b). The seta associated with setae 10- and 12-CT to form a group lateroanteriorly on the metanotum; interpreted as being the intermediate seta of the group; homologous with larval seta 2-T.
seta 12-CT (12-CT) (Fig. 24a,b). The seta associated with setae 10- and 11-CT to form a group lateroanteriorly on the metanotum; interpreted as being the most lateral of the group; homologous with larval seta 3-T.
seta 13-CT (13-CT). An evanescent metanotal seta, sometimes represented by an alveolus, borne slightly caudolaterad to seta 12-CT in pupae of *Bironella* species, some species of *Aedes* subgenera *Paraedes* and *Stegomyia*, and the genus *Uranotaenia*.

seta 14-CT (14-CT). A sporadic seta borne at the centre of the metathoracic wing far laterad to seta 12-CT in pupae of species of *Aedes* subgenus *Stegomyia*; perhaps other taxa as well.

seta 15-CT (15-CT). A sporadic seta borne on the posterior margin of the metanotum slightly laterad to the level of seta 12-CT in pupae of species of *Aedes* subgenus *Stegomyia*; perhaps other taxa as well.

Abdomen

A total of 104 pairs of setae are found on the pupal abdomen, including the paddles. The numbers applied to the abdominal setae indicate homologies with larval abdominal setae, unless otherwise noted. Individual setae, as in the case of larval abdominal setae, are denoted by a number followed with a hyphen and a Roman numeral to indicate the abdominal segment on which it occurs. As with larval setae, serially homologous setae are referred to using the singular 'seta', *e.g.* seta 5-I–IV (seta 5 on abdominal segments I, II, III and IV). Paddle setae are denoted by a number followed by a hyphen and the abbreviation 'Pa'.

seta 0-II–VIII (0-II, 0-III, *etc.*) (Fig. 25a,b). The most anterior of the dorsal setae on abdominal segments II–VIII; minute except in some Anophelinae and Sabethini; absent from abdominal segment I.

seta 1-I (1-I) (Fig. 25a,b). The dorsal seta nearest both the posterior margin and the longitudinal midline of abdominal segment I; usually inserted mesad of a large membranous area; typically a large multi-branched dendritic seta; simple or sparsely branched in species with non-surface resting pupae, *e.g.* some Mansoniini pupae.

seta 1-II–VII (1-II, 1-III, *etc.*) (Fig. 25a,b). The mesodorsal seta nearest the posterior margin of abdominal segments II–VII; if seta 2 is equally near or nearer the posterior margin, it is also usually nearer the longitudinal midline and typically shorter than seta 1; if seta 3 is equally near or nearer the posterior margin, it is also nearer the lateral margin, but mesad of seta 1 in *Anopheles*.

seta 1-IX (1-IX) (Fig. 25a,b). A seta sometimes present dorsally near the lateral margin of the median caudal lobe; arbitrarily named.

seta 1-XI (1-XI). A lateral seta of the cercus; found only in *Toxorhynchites*.

seta 2-I (2-I) (Fig. 25a,b). One of the two (usually noticeably paired with seta 3-I) dorsal setae occurring nearest both the anterior margin and the longitudinal midline of abdominal segment I; the shorter of the two except in pupae of Anophelinae, Aedeomyiini, some Mansoniini and *Uranotaenia* (*Uranotaenia*).

seta 2-II–VII (2-II, 2-III, *etc.*) (Fig. 25a,b). The mesodorsal seta on abdominal segments II–VII associated with setae 1 and 3 and occurring anywhere along an anterior arc extending from lateral to mesad of seta 1; sometimes distinguishable from seta 3 on one or more segments only because of a serial resemblance down the line of segments; nearly always the shortest of the three mesodorsal setae. Seta 2 is absent from abdominal segment VIII.

seta 3-I (3-I) (Fig. 25a,b). One of the two (usually noticeably paired with seta 2-I) dorsal setae occurring nearest both the anterior margin and the longitudinal midline of abdominal segment I; in pupae of most genera, interpreted as being the more posterior and/or lateral of the two; usually also the longer.

seta 3-II–VII (3-II, 3-III, *etc.*) (Fig. 25a,b). The mesodorsal seta variously associated either with setae 1 and 2 or 4 and 5 or standing alone on abdominal segments II–VII; sometimes exhibiting sufficient lack of serial resemblance in form and position on one or more segments to be difficult to name; usually determinable by process of elimination; at least on segments IV and V, believed to be homologous with larval seta 4. Seta 3 is absent from abdominal segment VIII.

seta 4-I (4-I) (Fig. 25a,b). The dorsoanterior seta grouped with seta 5 on abdominal segment I; usually posterior to seta 5 but if at the same level it is regarded as being the mesal one of the pair.

seta 4-II–VIII (4-II, 4-III, *etc.*) (Fig. 25a,b). The dorsoposterior seta that along with seta 5 arises mesal to setal groups 1-2-3 and 6-9 on abdominal segments II–VII; in some cases, seta 4 and/or 5 are distinguishably grouped with one or more of the 1-2-3 group; seta 4 is usually shorter than seta 5; on segment II, seta 4 is relatively stable in position with seta 5 occurring on any side of it; on segments III–V, seta 4 is generally nearer the puncture than seta 5; on segment III, seta 4 is usually anterior and lateral to seta 5; on segments IV–VI, seta 4 is anterior to seta 5; on segment VII, seta 4 is nearly always more anterior and lateral than seta 5; on VIII, arbitrarily applied to the inner posterior seta.

seta 5-I (5-I) (Fig. 25a,b). The dorsoanterior seta grouped with seta 4 on abdominal segment I; usually anterior to seta 4 but if at the same level it is regarded as being the lateral one of the pair.

seta 5-II–VII (5-II, 5-III, *etc.*) (Fig. 25a,b). The dorsoposterior seta that along with seta 4 arises mesal to setal groups 1-2-3 and 6-9 on abdominal segments II–VII; in some cases, seta 5 and/or 4 are distinguishably grouped with one or more of the 1-2-3 group; seta 5 is usually longer than seta 4; on segment II, seta 5 occurs on any side of it; on segments III–V, seta 5 is generally farther from the puncture than seta 4; on segment III, seta 5 is usually posterior and mesal to seta 4; on segments IV–VI, seta 5 is posterior to seta 4; on segment VII, seta 5 is nearly always more posterior and mesal than seta 4; absent from abdominal segment VIII, except in species of *Aedes* subgenus *Kenknightia*.

seta 5-VIII (5-VIII). The minute dorsoposterior seta borne caudomesad of seta 9 and laterad of seta 4 on abdominal segment VIII; arbitrarily named; known only in species of *Aedes* subgenus *Kenknightia*.

seta 6-I (6-I) (Fig. 25a,b). The seta interpreted as being the most anterior (and often the most mesal) in the dorsolateral setal group 6-7-9 on abdominal segment I.

seta 6-II–VII (6-II, 6-III, *etc.*) (Fig. 25a,b). The dorsoposterior seta generally inserted lateral to setae 4 and 5 and mesal to seta 9 on abdominal segments II–VII. Seta 6 is absent from abdominal segment VIII.
seta 7-I (7-I) (Fig. 25a,b). The seta closely associated with setae 6 and 9 on abdominal segment I; posterior to seta 6.
seta 7-II (7-II) (Fig. 25a,b). The seta usually dorsally associated with setae 6 and 9 on abdominal segment II; identifiable by comparison with seta 7 on abdominal segment I.
seta 7-III–VII (7-III, 7-IV, *etc.*) (Fig. 25a,b). The seta lateroventrally associated with seta 8 on abdominal segments III–VII; interpreted as being the more posterior of the two. Seta 7 is absent from abdominal segment VIII.
seta 8-II–VII (8-II, 8-III, *etc.*) (Fig. 25a,b). The ventral seta nearest both the lateral and the anterior margins of abdominal segments II–VII; associated with seta 7; often absent on segment II. Seta 8 is absent from abdominal segments I and VIII.
seta 9-I (9-I) (Fig. 25a,b). The lateral seta closely associated with setae 6 and 7 on abdominal segment I; the shortest (except in some Anophelinae) and generally the most posterior of the three setae.
seta 9-II–VIII (9-II, 9-III, *etc.*) (Fig. 25a,b). In flattened exuviae, the seta inserted on or near the lateral margin of abdominal segments II–VIII, generally lateral to seta 6; in most taxa, small to minute on segments II–VI and elongate on segments VII and VIII.
seta 10-II–VIII (10-II, 10-III, *etc.*) (Fig. 25a,b). The ventroposterior seta associated with seta 11 in a position mesal to setal group 7-8 on abdominal segments II–VII; usually longer than seta 11. Seta 10 occurs on abdominal segment VIII in *Aedes* (*Stegomyia*) *futunae*.
seta 11-I (11-I) (Fig. 25b). A small ventral seta on the membranous ventral surface of abdominal segment I; borne posterior or lateral to seta 13 when both are present; may be absent or represented by an alveolus. See **seta 13-I**.
seta 11-II–VII (11-II, 11-III, *etc.*) (Fig. 25a,b). The ventroposterior seta associated with seta 10 in a position mesal to setal group 7-8 on abdominal segments II–VII; usually shorter than seta 10; occurring in any relationship to seta 10 except immediately posterior. Seta 11 is absent from abdominal segment VIII.
seta 12-V,VI (12-V, 12-VI). A transitory ventral seta of abdominal segments V and VI reported for *Anopheles freeborni* by Belkin (1953b, as hair 9); regarded as being homologous with larval seta 12.
seta 13-I (13-I) (Fig. 25b). A small ventral seta present on the membranous ventral surface of abdominal segment I; when seta 11 is also present, seta 13 is the one nearer both the longitudinal midline and the anterior margin of the segment; may be absent or represented by an alveolus.
seta 13-II–VII (13-II, 13-III, *etc.*) (Fig. 25b). In pupae of some sabethine genera, a ventral puncture present approximately midway between the longitudinal midline and the lateral margin and between the anterior and posterior margins of abdominal segments II–VII; sometimes with a minute seta as an anomaly.
seta 14-III–VIII (14-III, 14-IV, *etc.*) (Fig. 25a,b). The minute ventral seta near the anterior margin at the longitudinal midline of abdominal segments III–VIII.
seta 1-Pa (1-Pa) (Fig. 25a,b). The terminal or outer seta of the paddle, arbitrarily named.
seta 2-Pa (2-Pa) (Fig. 25a,b). The pre-terminal or inner seta of the paddle; often absent; arbitrarily named.

Vestiture

The vestiture (general surface covering) of mosquitoes consists of two types of cuticular projections: **setae** (singular **seta**), which are articulated processes arising from a basal alveolus, and **spicules**, which are non-articulated continuous outgrowths of the cuticle. Flattened setae with longitudinal ridges that arise from minute alveoli are known as **scales**. Ordinary setae are innervated sensory organs (mechanical receptors) borne in various locations on all body areas of adults, larvae and pupae. Scales form the principal body covering (ornamentation and colouration) of adult mosquitoes.

The terminology applicable to the vestiture of mosquitoes consists of names applied to the fundamental types of cuticular projections and descriptors (adjectival names) that characterize their common structural forms. The latter are listed as italicized adjectival names under each of the fundamental types of cuticular projections.

acicula (Fig. 28c) (pl. aciculae). A small, slender, rigid, needle- or thorn-like spicule. See *aciculate* under **seta**.
aculea (Fig. 27d,e,g,m,n) (pl. aculeae). One of the microtrichium-like spicules comprising the tomentum that covers the cuticle (except the wing membrane); aculeae usually form a dense covering in adults but are normally sparse or absent in larvae and pupae.
alveolus (a) (Fig. 27b) (pl. alveoli). A cup-like depression (socket) bearing a seta or scale (absent in the figure); its floor is formed by the membrane that supports the seta or scale; its wall is formed by a ring-like swelling of the cuticula (collar). An alveolus-like structure not bearing a seta or scale is termed a **puncture** (see).
blade. An elongate, flattened, usually stiff spicule; may lie in a single plane or be wavy, curved and/or twisted; sometimes movable. The more common types of blades are named by the adjectives listed and defined below.
bipectunculate (Fig. 28e) – With two rows of acicula-like or small tooth-like processes.
branch-tipped – With a few short branches arising apically or subapically.
divided (Fig. 26z4) – With long branches of relatively equal diameter arising below the distal third.

incised (Figs 26z7, 28d) – With apical or subapical notches; the notches are usually rounded as are the tooth-like processes between them.

pectunculate (Figs 26z9, 28e) – With a row of acicula- or tooth-like processes arising along one side.

simple (Fig. 28d) – Without lateral or apical processes. See Note for *simple* under **seta**.

cuticular projection. Any elongate process jutting from the outer surface of the cuticle. There are two basic types of cuticular projections: setae (see **seta**), which are articulated processes arising from a basal alveolus, and spicules (see **spicule**), which are non-articulated, continuous processes of the cuticle.

dendrite (Fig. 26z5). A spicule branched to resemble a tree in form; having a basal stem bearing irregular or dichotomous branches which may be repeatedly branched in turn.

denticle (Fig. 28f). A small tooth-like process or spicule.

echinoid (Fig. 26z6). A spicule consisting of numerous spine-like rays projecting at various angles from a tubercle-like base.

filament. A long, slender, flexible spicule that gradually tapers to a point or is of equal or nearly equal diameter throughout. In adult females with cibarial teeth, the tapered terminal part of a cone borne upon the pediment; bearing short lateral processes. The more common types of filaments occurring in mosquitoes are named by the adjectives listed and defined below.

barbed (Fig. 26y) – With minute, short, pointed processes projecting obliquely from the surface.

branch-tipped (Fig. 26z1) – With a few subdivisions arising apically; the subdivisions may be short and thick or long and slender.

comb-tipped (Fig. 26z2) – With a short row of small rigid processes located at the side near the apex.

divided (Fig. 28g) – With long branches of equal diameter usually arising within the basal two-thirds.

pectinate (Fig. 26x) – With long or short branches arising at regular intervals along one side.

simple (Fig. 26x) – Without lateral or distal processes, usually sharply pointed. See Note for *simple* under **seta**.

microtrichium (Fig. 27a,f,h–l) (pl. microtrichia). One of the minute, slender, tapered, flexible spicules closely covering the wing membrane; characteristically bent so that they extend more or less parallel to the longitudinal veins with their apices pointed toward the wing margin.

pedicel (p) (Fig. 27g,i). The slender basal stalk of a scale which supports the squame; with or without longitudinal ridges.

puncture (p) (Fig. 28a). An alveolus-like depression not bearing a seta or scale. See **alveolus**.

rachis (r) (Fig. 26l–n,q–t,v) (pl. rachises, rachides). The central shaft or main stem of a seta, the axial structure that bears the branches.

scale. A modified seta comprised of a slender basal stalk, the pedicel, and an expanded and/or flattened distal part, the squame; all scales have longitudinal ridges and arise from alveoli that are usually minute. Two main types of scales are recognized by cross-section of the squame. If the squame is round or elliptical, the scale is **piliform**; if it is thin and flat, the scale is **lamellar**. Many forms of lamellar scales exist but only one type of piliform scale. The piliform and the more common types of lamellar scales are listed and defined below. It should be noted that piliform scales tend to grade into the lamellar form and that the types of lamellar scales grade into one another. For this reason, scales are divided into a number of extreme types with broad definitions so that the transitional forms may be classified more easily.

asymmetrical (Fig. 27a) – A lamellar scale that is obviously unevenly developed on opposite sides of a plane which is parallel to the plane of the squame.

falcate (Fig. 27b) – A sickle-shaped lamellar scale; the squame narrow and curved with a sharp or narrowly rounded apex.

forked (Fig. 27g) – A lamellar scale that is cuneate or gently flared distally; with a long thick pedicel that gradually merges into the squame; distal margin of squame cut off squarely or notched.

fusiform (Fig. 27f) – Widest at the middle and tapered toward the ends; apex sharply pointed or narrowly rounded; ranging from short and broad to long and slender.

linear (Fig. 27h) – A strap-like lamellar scale; much longer than broad with parallel sides; always truncate.

piliform (Fig. 27n) – Any scale of which the squame is circular or elliptical in cross-section; normally narrow, curved and pointed; seta-like.

spatulate (Fig. 27i–m) – A lamellar scale that is very broad distally, attenuate at base; top rounded or truncate, rarely emarginate (heart-shaped); the squame may be parallel-sided, obovate or accrescent.

twisted (Fig. 27c) – A curled lamellar scale with a spiral or winding form; the squame is usually accrescent.

serration (Fig. 28h). A linear series of recurved denticle-like processes. See **paddle marginal serration** in the **Pupa** section.

seta (pl. setae; adj. setal). A cuticular projection that arises from a basal alveolus. Flattened setae with longitudinal ridges are known as scales (see **scale**). Setae may be **single** (unbranched) or **branched**. Single setae may be simple (with smooth surface) or bear short lateral processes that are not to be considered as branches. Single setae may have longitudinal ridges but unlike scales they are broadest at the base and taper distally. Branched setae have branches radiating from the base, originating as divisions of the main stem or arising along the length of the main stem (rachis). The descriptors used to name single setae include aciculate, barbed, brush-tipped, foliform, lanceolate, pectunculate, peg-like, simple, spiniform and spinulate; those used to name branched setae include dendritic, fan-like, forked, palmate, pectinate, plumose and stellate. The names of single setae are used to describe the branches of branched setae. The types of setae are listed and defined below.

aciculate (Fig. 26i) – Furnished with slender needle-like processes along the stem; the processes are somewhat flexible but often appear to be rigid.

barbed (Fig. 26c) – With minute, short, heavy, pointed processes projecting obliquely from the surface.
brush-tipped (Fig. 26j) – With numerous moderately long, slender processes arising apically.
dendritic (Fig. 26q) – Branched to resemble a tree in form; having a stem-like part bearing irregular or dichotomous branches which may be repeatedly forked or branched in turn.
fan-like (Fig. 26o,r–t) – With branches spreading out in a single plane from a short stem; in the case of some ventral brush setae the branches successively arise on one side of the stem; branches may be simple, barbed, aciculate or dendritic.
foliform (Figs 26h, 28b) – Flattened and leaf-like; similar to some scales but lacking longitudinal ridges, *e.g.* seta *g* borne on the subapical lobe of the gonocoxite of many *Culex* males.
forked (Fig. 26l,m) – With a few branches arising beyond the basal third of the main stem.
lanceolate (Fig. 26e) – Oblong or spear-shaped; tapering distally to a point.
palmate (Fig. 26u) – With flattened, movable, usually horizontal branches radiating from a common point on a short stem.
pectinate (Fig. 26n) – With long branches arising at regular intervals from one side of the main stem; a few small, sparsely arranged branches may arise from the opposite side. See *pectinate* (Fig. 26z3) under **filament**.
pectunculate (Fig. 26k) – With a row of short rigid processes resembling the teeth of a comb.
peg-like (Fig. 28b) – A small cylindrical, usually blunt-tipped simple seta.
plumose (Fig. 26v) – With numerous usually regularly arranged branches arising on either side of the main stem (rachis); the branches on either side may be directly opposite one another or alternate.
simple (Fig. 26a,b) – Without lateral or apical processes. Note: the descriptor single refers to an unbranched seta, which may be simple or bear lateral spicules. The setae shown in Fig. 26a,b are simple single setae.
spiniform (Fig. 26g) – Thick, spine-like and usually not markedly attenuate or sharply pointed.
spinulate (Fig. 26d) – Beset with small spine-like processes.
stellate (Fig. 26p,u) – Tufted (multiradiate); with numerous stiff branches projecting at various angles from a single base; branches may be simple, barbed or aciculate. The term 'stellate' refers to a condition where the rays (branches) project in all directions from a single point in the same plane. A better term would be 'multiradiate' but owing to the long use of the term 'stellate' in mosquito taxonomy it has been retained.

setiform (Fig. 28i). A spicule that resembles a single seta; usually stiff with a diameter greater than a filament; may be simple or bear short lateral or apical processes.

spicule. A non-articulated cuticular projection, directly continuous with the cuticle. Spicules are named by nouns. The common types that occur in mosquitoes include acicula, aculea, blade, dendrite, denticle, echinoid, filament, microtrichia, serration, setiform, spine, spinule, spur, teeth and villi. Spurs and some filaments and blades are movable, *i.e.* they have a ring of non-sclerotized cuticle at the base.

spine (Fig. 28j). A very large, sturdy, immovable spicule; with a sharp or narrowly rounded tip.

spinule (Fig. 28k). A minute spine-like spicule; always stiff; commonly occurring in short rows on the saddle and siphon of larvae, and in the pharynx of *Toxorhynchites* larvae.

spur. A movable spine-like spicule; with a ring of unsclerotized cuticle at the base.

squame (s) (Fig. 27g,i). The expanded and/or flattened distal part of a scale supported by the pedicel; invariably with longitudinal ridges.

tomentum (pl. tomenta). A covering of aculeae on the body and its appendages, except the wings. See **aculea**.

tooth (Fig. 28h) (pl. teeth). A very stout heavy process or spicule with a blunt apex.

vestiture. The general surface covering comprised of cuticular projections, *i.e.* setae, scales and spicules.

villus (Fig. 28l) (pl. villi). A minute, slender, flexible, filament-like spicule; usually with a blunt tip.

Figures

Figure 1. Adult female mosquito (diagrammatic), lateral aspect (left side) (modified from Belkin, 1962).

Abbreviations

Ant, antenna
Ap, antepronotum
Ce, cercus
CE, compound eye
Clp, clypeus
Cv, cervix
Cx, coxa
Fe, femur
Fl, flagellum
Flm$_{11}$, flagellomere 11
FW, flagellar whorl
Hl, halter
La, labellum
Lt, laterotergite
Mam, mesanepimeron
Mem, metameron
MPlp, maxillary palpus
Mks, mesokatepisternum
Mpn, mesopostnotum
MS, mesothoracic spiracle
Msm, mesomeron
Mtpn, metapostnotum
Mts, metepisternum
MtS, metathoracic spiracle
P, proboscis
Pa, paratergite
PA, postspiracular area
Pe, pedicel
PGL, postgenital lobe
PK, prealar knob
PMe, pleural membrane
Ppn, postpronotum
Prm, prementum
PrN, pronotum
Ps, proepisternum
S, sternum
Scu, scutum
Stm, scutellum
T, tergum
Ta, tarsus
Ti, tibia
Tr, trochanter
U, ungues
W, wing
1–5, tarsomeres
I–VIII, abdominal segment

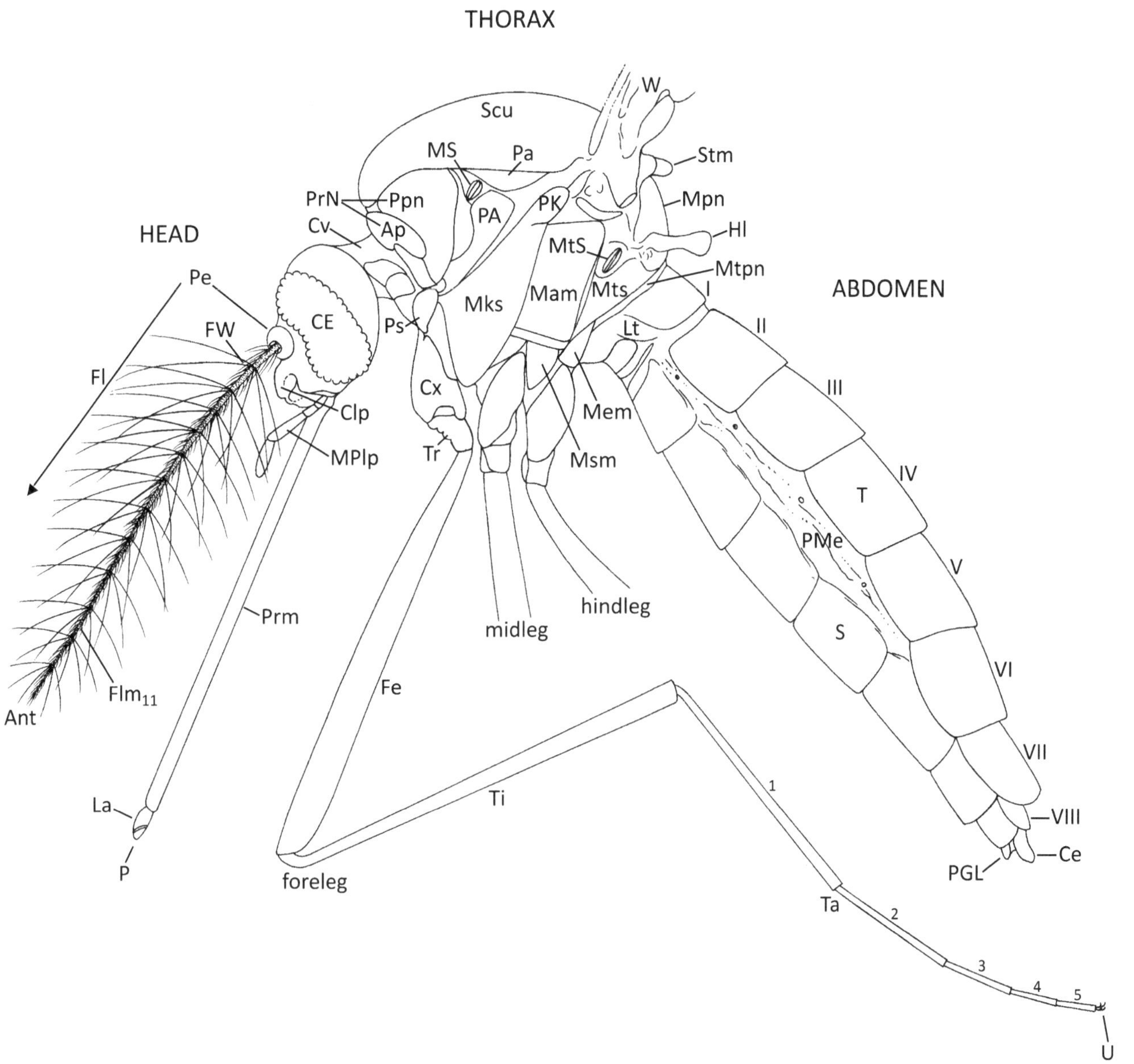
THORAX
HEAD
ABDOMEN
W
Scu
Stm
MS
Pa
Mpn
PrN
Ppn
PA
PK
Hl
Cv
Ap
MtS
Mtpn
Pe
Mam
Mts
I
Mks
II
FW
CE
Ps
Lt
Fl
III
Cx
Clp
Mem
Tr
MPlp
Msm
IV
T
PMe
V
hindleg
Prm
midleg
S
VI
Fe
Flm11
Ant
VII
1
Ti
La
VIII
Ce
PGL
P
foreleg
Ta
2
3
4
5
U

Figure 2. Head of adults of *Aedes* (*Ochlerotatus*) *grossbecki* Dyar & Knab (modified from Harbach & Knight, 1980).

a) Female, anterodorsal aspect.
b) Male, lateral aspect (left side).
c) Female, ventral aspect.

Abbreviations

CE, compound eye
Clp, clypeus
Cv, cervix
Fr, frons
FW, flagellar whorl
Ge, gena
IS, interocular space
ISe, interocular seta
LBS, labial basal seta
MPlp, maxillary palpus
Occ, occiput
OcS, ocular seta
OL, ocular line
OS, ocular scale
Pe, pedicel
PG, postgena
PgS, postgenal seta
Pl, palpifer
POc, postocciput
Prm, prementum
Sc, scape
V, vertex

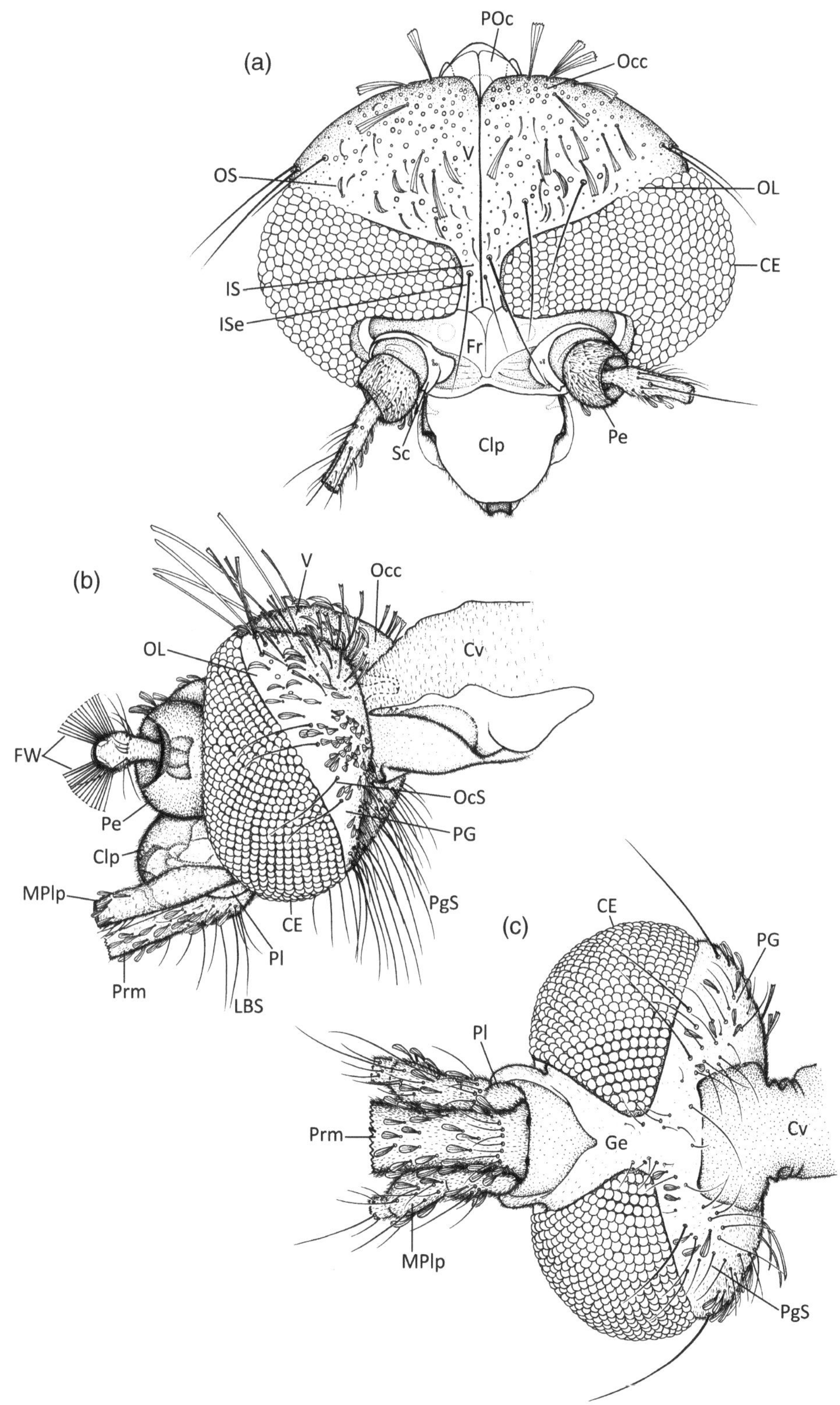
(a)
POc
Occ
V
OS
OL
CE
IS
ISe
Fr
Pe
Sc
Clp
(b)
V
Occ
OL
Cv
FW
OcS
Pe
PG
Clp
MPlp
PgS
CE
Pl
Prm
LBS
(c)
CE
PG
Pl
Prm
Ge
Cv
MPlp
PgS

Figure 3. Head of adults of (a) *Aedes* (*Ochlerotatus*) *tormentor* Dyar & Knab, (b,f) *Anopheles* (*Anopheles*) *quadrimaculatus* Say, (c) *Aedes* (*Ochlerotatus*) *caspius* (Pallas) and (d) *Sabethes* (*Sabethoides*) *chloropterus* (von Humboldt) (a, b, e and f modified from Harbach & Knight, 1980; c and d modified from Harbach & Kitching, 1998).

a) Female, anterior aspect.
b) Female, lateral aspect (left side).
c) Female, dorsal aspect.
d) Female, anterodorsal aspect.
e) Male, ventral aspect.
f) Male, posteroventral aspect.

Abbreviations

AS, antennal socket
ATP, anterior tentorial pit
CE, compound eye
Clp, clypeus
ClS, clypeolabral suture
CoF, corneal facet
Cv, cervix
Esc, erect scales
Fr, frons
FT, frontal tuft
Ge, gena
ig, interantennal groove
is, interantennal suture
IS, interocular space
MPlp, maxillary palpus
OcS, ocular seta
OL, ocular line
OSc, ocular suture
OSl, ocular sclerite
Pe, pedicel
Pfs, postfrontal suture
PG, postgena
PgS, postgenal seta
Pl, palpifer
Prm, prementum
PTP, posterior tentorial pit
SG, subgenal suture
V, vertex

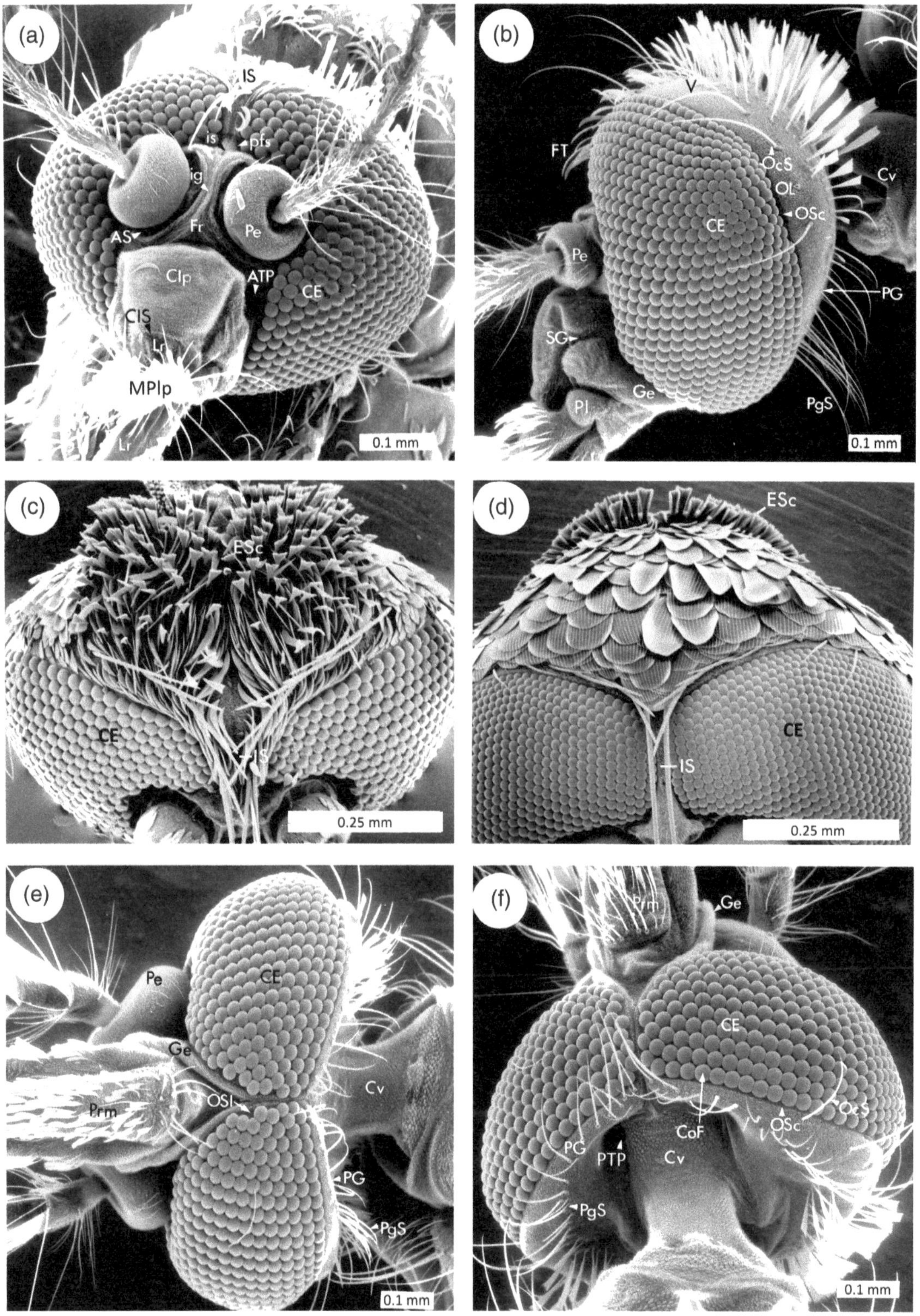
(a)
IS
is
pfs
ig
AS
Fr
Pe
Clp
ATP
CE
CIS
Lr
MPlp
Lr
0.1 mm
(b)
V
FT
OcS
OL
Cv
OSc
CE
Pe
PG
SG
Ge
Pl
PgS
0.1 mm
(c)
ESc
CE
IS
0.25 mm
(d)
ESc
CE
IS
0.25 mm
(e)
Pe
CE
Ge
Cv
Prm
OSl
PG
PgS
0.1 mm
(f)
Prm
Ge
CE
CoF
OcS
OSc
PG
PTP
Cv
PgS
0.1 mm

Figure 4. Distal part of prementum and labella (a–c) and cibaria (d,e) of female mosquitoes (a and c modified from Harbach & Kitching, 1998; b modified from Harbach & Knight, 1980; c modified from Christophers, 1933; d modified from Sirivanakarn, 1978).

a–c) Dorsal aspects of labella of (a) *Anopheles* (*Cellia*) *leucosphyrus* Dönitz, (b) *Aedes* (*Ochlerotatus*) *grossbecki* Dyar & Knab and (c) *Tripteroides* (*Polylepidomyia*) *melanesiensis* Belkin.

d,e) Dorsal aspects of cibaria (generalized) of anopheline and *Culex* females, respectively.

Abbreviations

AHP, anterior hard palate
Cib, cibarium
Cn, cone
CT, cibarial teeth
DS, distal sclerite
DSe, dorsal seta
LaBS, labellar basal sclerite
LF, lateral flange
Lg, ligula
LS, labellar scales
MPlp, maxillary palpus
PHP, posterior hard palate
Prm, prementum
PS, proximal sclerite
PSe, palatal seta
Rd, rod
VSe, ventral seta

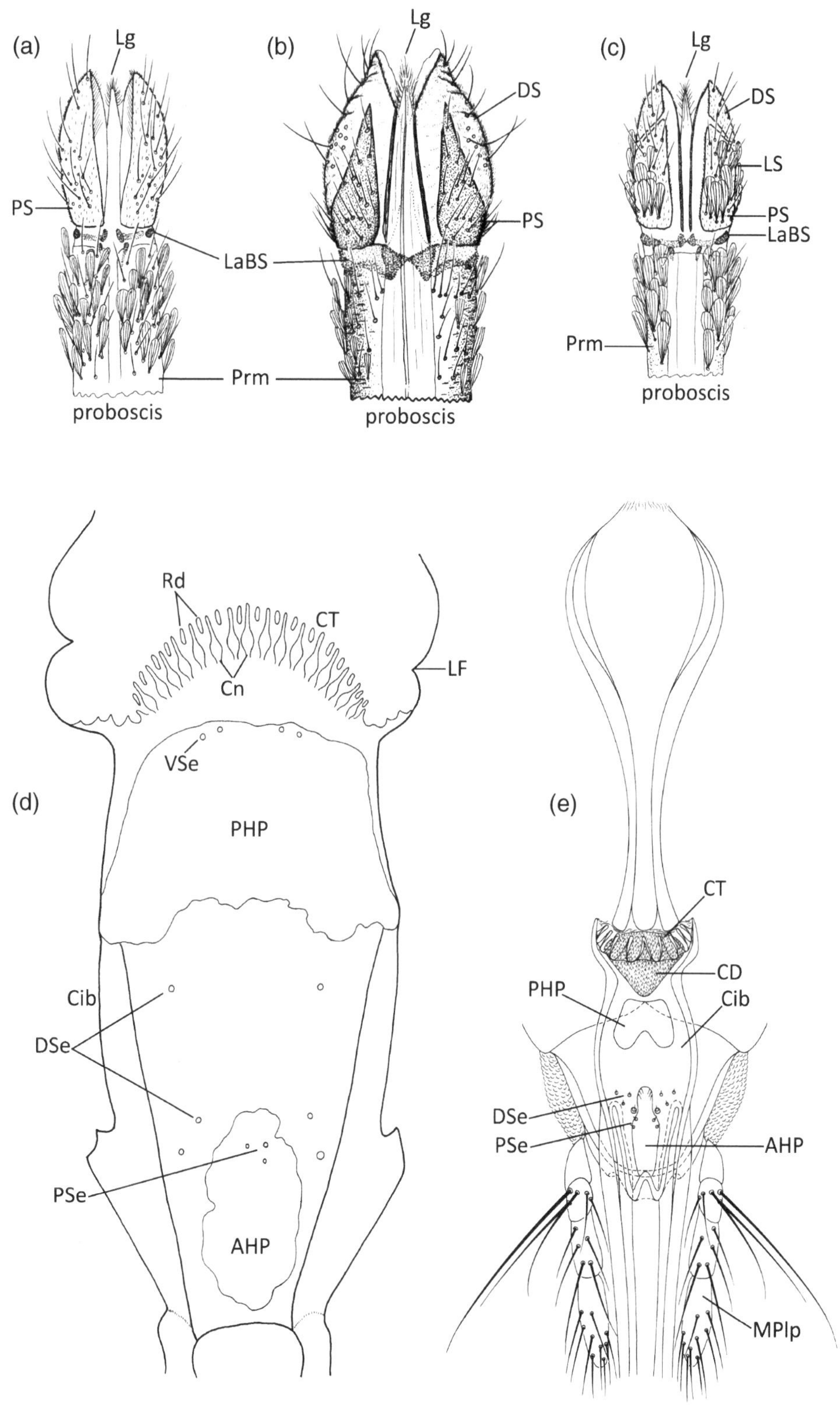
(a)
Lg
PS
LaBS
Prm
proboscis
(b)
Lg
DS
PS
LaBS
Prm
proboscis
(c)
Lg
DS
LS
PS
LaBS
Prm
proboscis
(d)
Rd
CT
Cn
LF
VSe
PHP
Cib
DSe
PSe
AHP
(e)
CT
CD
PHP
Cib
DSe
PSe
AHP
MPlp

Figure 5. Thorax of male of *Aedes* (*Ochlerotatus*) *grossbecki* Dyar & Knab (modified from Harbach & Knight, 1980).

a) Dorsal aspect.
b) Lateral aspect (left side).

Abbreviations
AA, acrostichal area
AEC, anepisternal cleft
AnA, antealar area
AnP, anterior promontory
Ap, antepronotum
Cx-I, forecoxa
Cx-II, midcoxa
Cx-III, hindcoxa
DA, dorsocentral area
Hl, halter
HyA, hypostigmal area
Mam, mesanepimeron
Mem, metameron
Mkm, mesokatepimeron
Mks, mesokatepisternum
Mpn, mesopostnotum
ms, mesopleural suture
MS, mesothoracic spiracle
Msm, mesomeron
Mtm, metepimeron
Mtpn, metapostnotum
Mtr, mesotrochantin
mts, metapleural suture
Mts, metepisternum
MtS, metathoracic spiracle
Pa, paratergite
PA, postspiracular area
PK, prealar knob
Pm, proepimeron
PM, postprocoxal membrane
PMas, posterior mesanepisternum
PnS, paranotal suture
Ppn, postpronotum
PrA, prescutellar area
Ps, proepisternum
PsA, prespiracular area
SA, subspiracular area
SaA, supraalar area
SaG, supraalar groove
ScA, scutal angle
Scu, scutum
SF, scutal fossa
Stm, scutellum

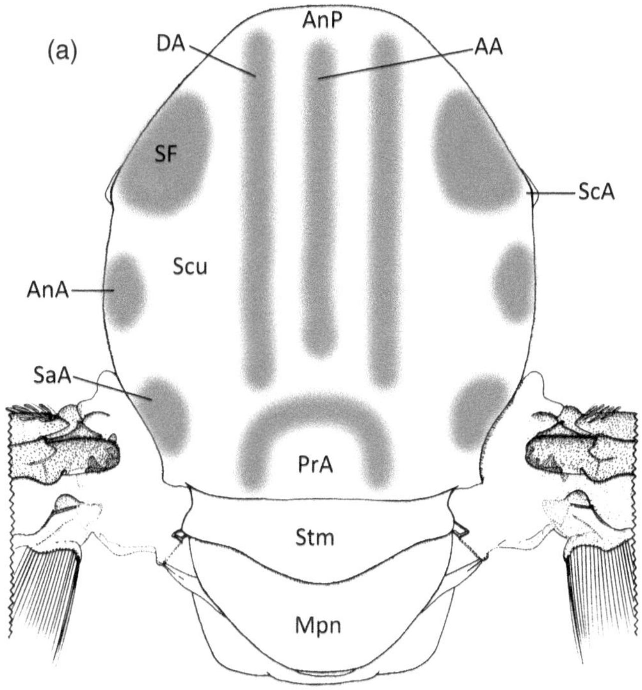
(a)
AnP
DA
AA
SF
ScA
Scu
AnA
SaA
PrA
Stm
Mpn

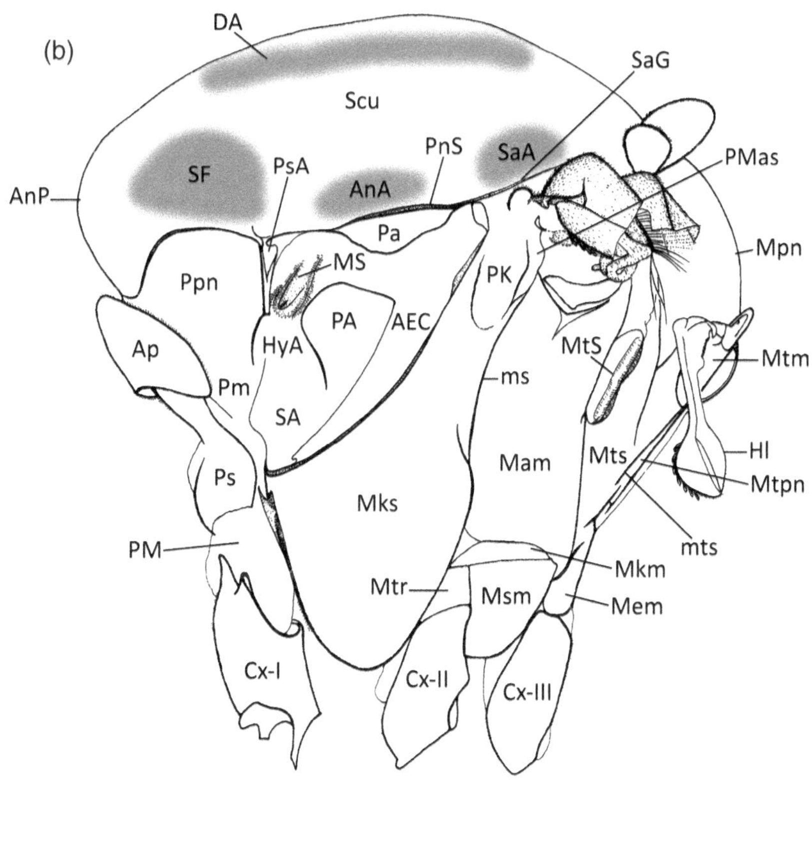
(b)
DA
Scu
SaG
AnP
SF
PsA
AnA
PnS
SaA
PMas
Pa
Mpn
MS
Ppn
PK
Ap
HyA
PA
AEC
MtS
Mtm
Pm
ms
SA
Hl
Ps
Mks
Mam
Mts
Mtpn
mts
PM
Mkm
Mtr
Msm
Mem
Cx-I
Cx-II
Cx-III

Figure 6. Patches of scales that occur on the thorax (diagrammatic) of adult mosquitoes (modified from Belkin, 1962).

a) Dorsal aspect.
b) Lateral aspect (left side).

Abbreviations

AcSc, acrostichal scales
ADSc, anterior dorsocentral scales
AMSc, anterior mesepimeral scales
AnSc, antealar scales
ApSc, antepronotal scales
HySc, hypostigmal scales
Hl, halter
LMSc, lower mesepimeral scales
LPrSc, lower prealar scales
LPSc, lower postpronotal scales
LSSc, lateral scutellar scales
MeSc, metameral scales
MpSc, mesopostnotal scales
MScL, lower mesokatepisternal scales
MScU, upper mesokatepisternal scales
MSSc, median scutellar scales
MtSc, metepisternal scales
PaSc, paratergal scales
PDSc, posterior dorsocentral scales
PmSc, proepimeral scales
PMSc, posterior mesepimeral scales
PoSc, postspiracular scales
PSc, postprocoxal scales
PScU, upper proepisternal scales
PsSc, prespiracular scales
SaSc, supraalar scales
SFSL, lateral scutal fossal scales
SFSM, median scutal fossal scales
SFSP, posterior scutal fossal scales
SSc, subspiracular scales
SSPL, lateral prescutellar scales
SSPM, median prescutellar scales
UMSc, upper mesepimeral scales
UPrSc, upper prealar scales
UPSc, upper postpronotal scales

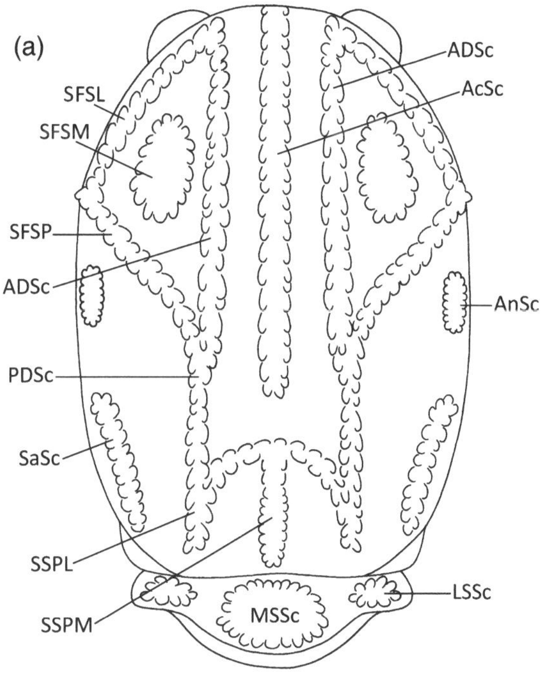
(a)
ADSc
AcSc
SFSL
SFSM
SFSP
ADSc
AnSc
PDSc
SaSc
SSPL
LSSc
MSSc
SSPM

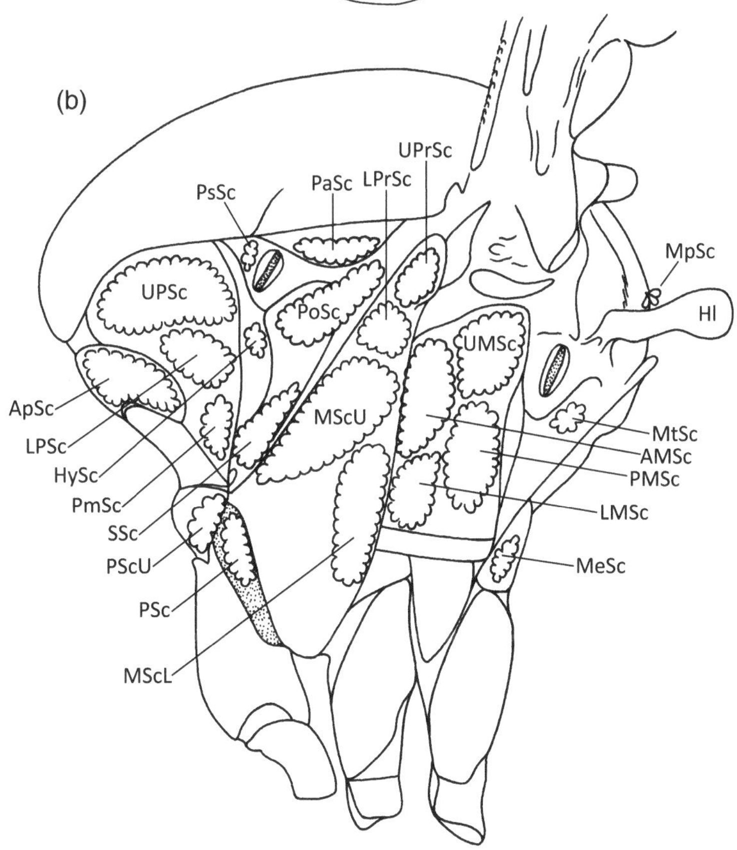
(b)
UPrSc
PsSc
PaSc
LPrSc
MpSc
UPSc
PoSc
Hl
UMSc
ApSc
MScU
LPSc
HySc
MtSc
AMSc
PMSc
PmSc
SSc
LMSc
PScU
MeSc
PSc
MScL

Figure 7. Groups of setae that occur on the thorax (diagrammatic) of adult mosquitoes (modified from Belkin, 1962).

a) Dorsal aspect.
b) Lateral aspect (left side).

Abbreviations

AcS, acrostichal setae
ADS, anterior dorsocentral setae
AnA, antealar area
AnS, antealar setae
ApS, antepronotal setae
ASFS, anterior scutal fossal setae
CxS, coxal setae
Hl, halter
LSFS, lateral scutal fossal setae
LSS, lateral scutellar setae
MAPS, median anterior promontory setae
MeSL, lower mesepimeral setae
MeSU, upper mesepimeral setae
MkSL, lower mesokatepisternal setae
MkSU, upper mesokatepisternal setae
MpnS, mesopostnotal setae
MSFS, median scutal fossal setae
MSS, median scutellar setae
Pa, paratergite
PaS, prealar setae
PDS, posterior dorsocentral setae
PeSU, upper proepisternal setae
PMSS, posterior medial scutal setae
PpS, postpronotal setae
PrA, prescutellar area
PrsS, prescutellar setae
PrS, prescutal suture
PS, postspiracular setae
PSFS, posterior scutal fossal setae
PsS, prespiracular setae
SaS, supraalar setae
ScA, scutal angle

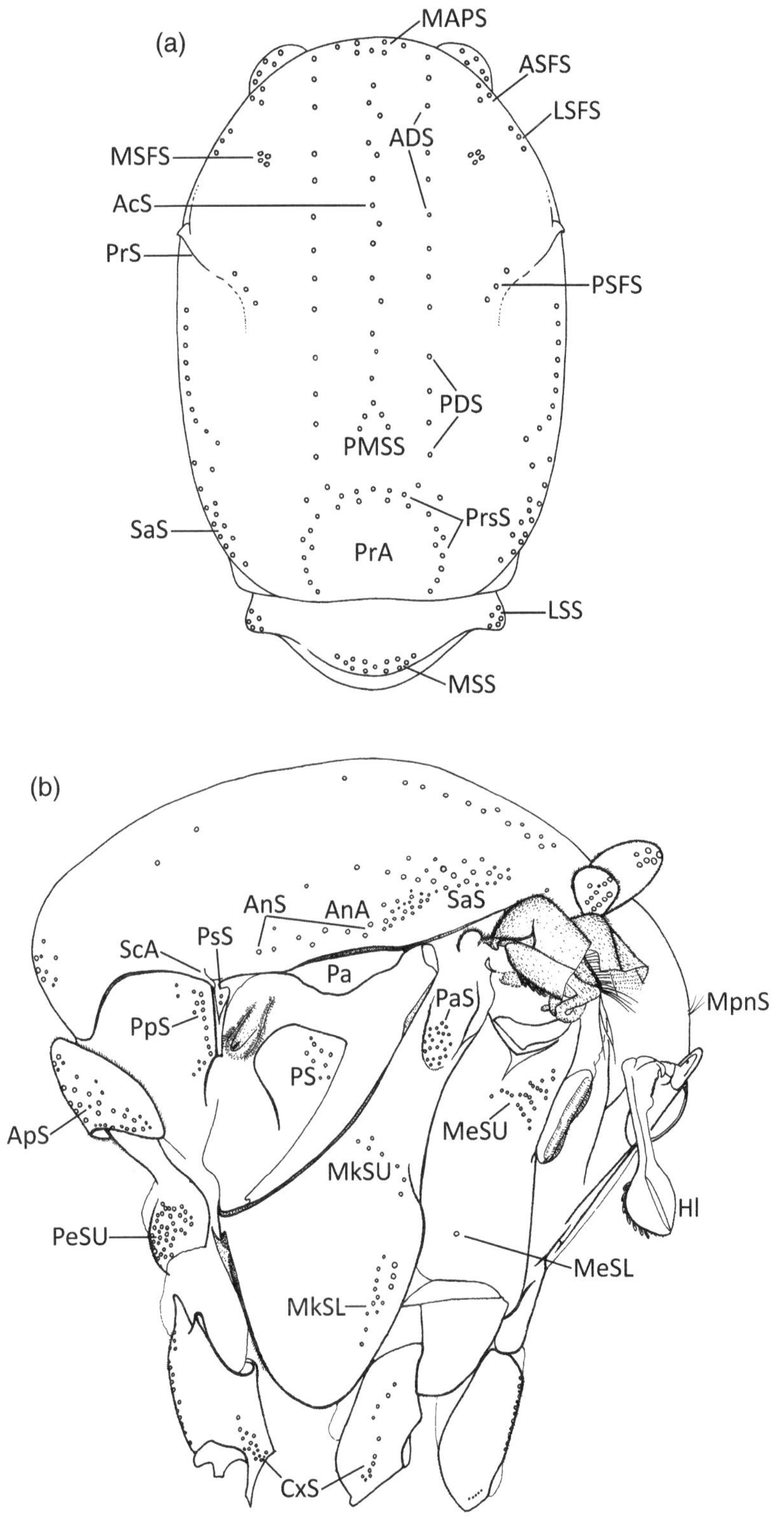
(a)
MAPS
ASFS
LSFS
MSFS
ADS
AcS
PrS
PSFS
PDS
PMSS
PrsS
SaS
PrA
LSS
MSS
(b)
AnS
AnA
SaS
ScA
PsS
Pa
PaS
MpnS
PpS
PS
ApS
MeSU
MkSU
Hl
PeSU
MeSL
MkSL
CxS

Figure 8. Wing and halter of mosquitoes (a–c modified from Harbach & Knight, 1980; d modified from Harrison, 1980; e modified from Glick, 1992 and Azari-Hamidian & Harbach, 2009).

a) Right wing (dorsal aspect) of *Uranotaenia* (*Uranotaenia*) *lowii* Theobald.
b) Left halter (lateral aspect) of *Aedes* (*Ochlerotatus*) *grossbecki* Dyar & Knab.
c) Area of fringe scales of the left wing (ventral aspect) of *Aedes* (*Ochlerotatus*) *sticticus* (Meigen).
d) Right wing (hypothetical, dorsal aspect) of *Anopheles* subgenus *Cellia*.
e) Costal wing spots (dorsal aspects) of *Anopheles* (*Cellia*) *demeilloni* Evans (above) and *An.* (*Cel.*) *fluviatilis* James (below).

Abbreviations

AD, apical dark spot
AI, axillary incision
Al, alula
AP, apical pale spot
Ar, arculus
ASP, accessory sector pale spot
BD, basal dark spot
BP, basal pale spot
BPFS, basal pale fringe spot
BS, basal spur
C, costa
Ca, capitellum
CuA, cubitus anterior
CuP, cubitus posterior
FS, fringe scales
H, humeral crossvein
HD, humeral dark spot
HP, numeral pale spot
M, media
M_1, media-one
M_{1+2}, media-one-plus-two
M_2, media-two
M_{3+4}, media-three-plus-four
Mcu, mediocubital crossvein
Pc, pedicel
PD, preapical dark spot
PFS, primary fringe scales
PHD, prehumeral dark spot
PHP, prehumeral pale spot
PlSc, plume scales
PP, preapical pale spot
PSD, presector dark spot
PSP, presector pale spot
R, radius
R_1, radius-one
R_2, radius-two
R_{2+3}, radius-two-plus-three
R_3, radius-three
R_{4+5}, radius-four-plus-five
Rs, radial sector
Re, remigium
Rm, radiomedial crossvein
SBD, subbasal dark spot
Sc. subcosta
SCP, subcostal pale spot
SD, sector dark spot
SP, sector pale spot
SFS, secondary fringe scales
Sl, scabellum
SS, squame scales
TFS, tertiary fringe scales
UC, upper calypter
WF, wing fringe
1A, anal vein
1APFS, 1A pale fringe spot

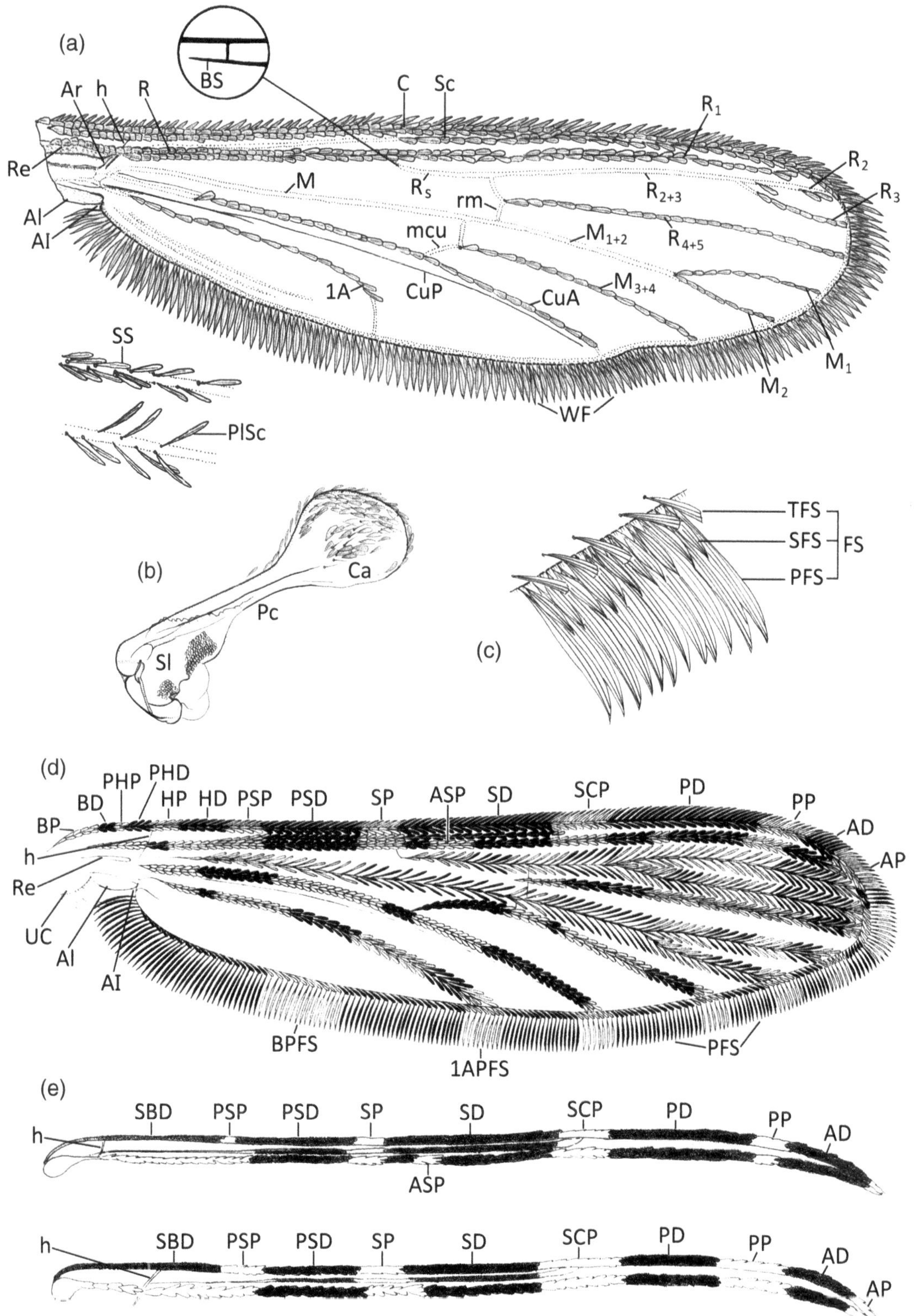
(a)
BS
Ar h R
C Sc
R1
Re
R2
R3
M
Rs
R2+3
rm
Al
AI
mcu
M1+2
R4+5
1A
CuP
CuA
M3+4
SS
M1
M2
WF
PlSc
TFS
SFS
FS
PFS
(b)
Ca
Pc
Sl
(c)
(d)
PHP PHD
BD HP HD PSP PSD SP ASP SD SCP PD
BP
PP
AD
h
AP
Re
UC
Al
AI
BPFS
1APFS
PFS
(e)
SBD PSP PSD SP SD SCP PD PP AD
h
ASP
h
SBD PSP PSD SP SD SCP PD PP AD
AP

Figure 9. Legs of adult mosquitoes (a original; b and c modified from Harbach & Knight, 1980; d modified from Edwards, 1941 and Azari-Hamidian & Harbach, 2009).

a) Fore-, mid- and hindlegs (lateral aspects) of a female of *Anopheles* (*Cellia*) *gambiae* Giles.

b,c) Terminal parts of legs of a male of *Aedes* (*Ochlerotatus*) *grossbecki* Dyar & Knab. (b) Lateral aspects of fore-, mid- and hindlegs; (c) ventral aspect of foreleg.

d) Ventral aspect of the posttarsus of a foreleg of an *Aedes* female.

Abbreviations

Em, empodium
Fe-I, forefemur
Fe-II, midfemur
Fe-III, hindfemur
Pta, posttarsus
Pv, pulvillus
Ta-I, foretarsus
Ta-II, midtarsus
Ta-III, hindtarsus
Ti-I, foretibia
Ti-II, midtibia
Ti-III, hindtibia
U, unguis
U-I, foreunguis
U-II, midunguis
U-III, hindunguis
1–5, tarsomeres

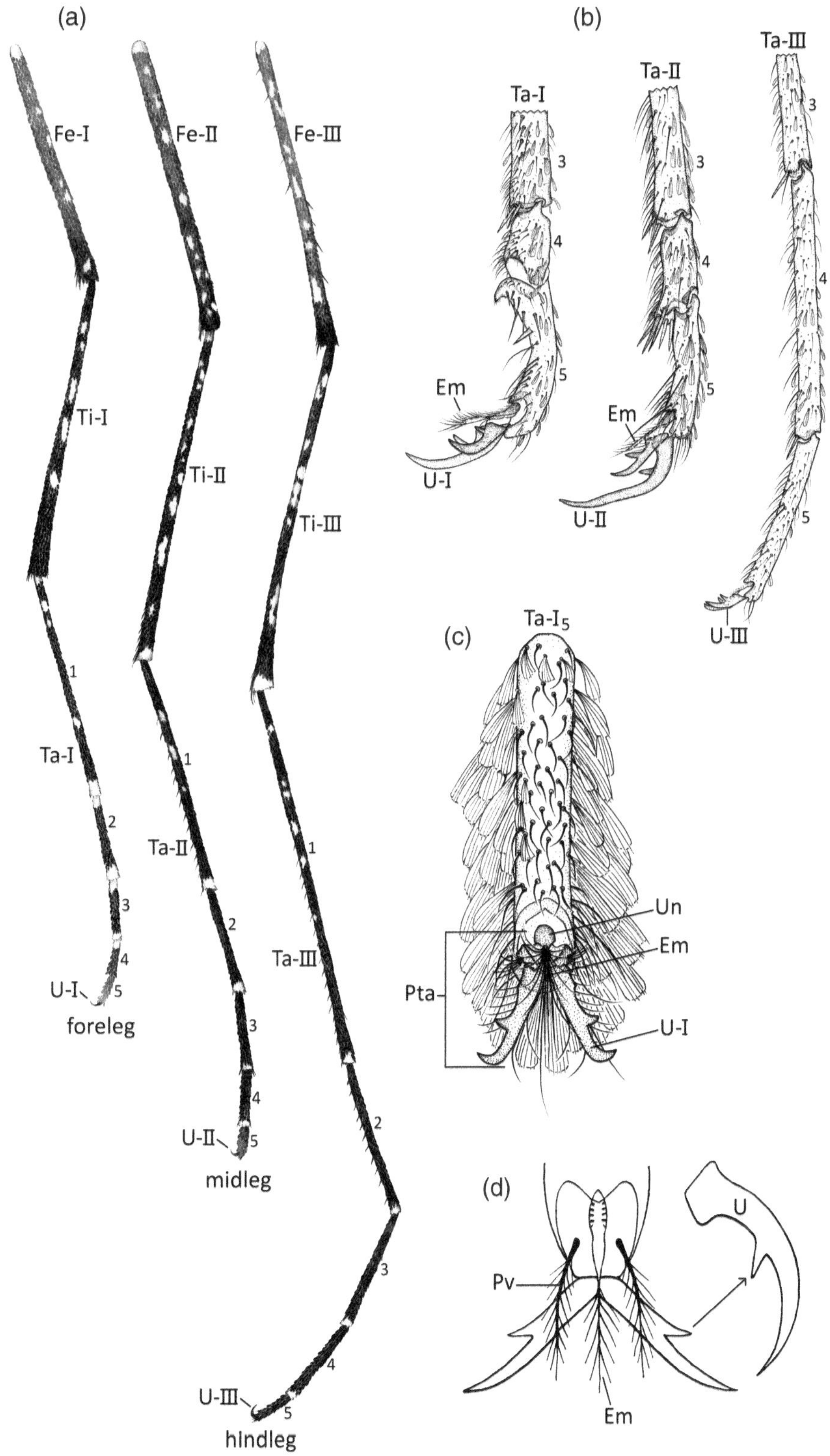
(a)
Fe-I
Fe-II
Fe-III
Ti-I
Ti-II
Ti-III
Ta-I
Ta-II
Ta-III
U-I
U-II
U-III
foreleg
midleg
hindleg
(b)
Ta-I
Ta-II
Ta-III
Em
U-I
U-II
U-III
(c)
Ta-I5
Un
Em
Pta
U-I
(d)
Pv
U
Em

Figure 10. Abdomen of adults of *Aedes* (*Ochlerotatus*) *grossbecki* Dyar & Knab (modified from Harbach & Knight, 1980).

a) Female, lateral aspect (left side) of the entire abdomen.
b) Female, lateral aspect (left side) of terminal abdominal segments and genitalia (scales of segment VII and cerci not shown).
c) Male, lateral aspect (left side) of terminal abdominal segments and genitalia (segment VIII and genitalia in post-rotation position).

Abbreviations
ADL, apicodorsal lobe
Ce, cercus
Cl, claspette
GA, gonocoxal apodeme
Gc, gonocoxite
GC, gonostylar claw
Gs, gonostylus
PGL, postgenital lobe
PMe, pleural membrane
SCa, spermathecal capsule
S-VII, sternum VII
S-VIII, sternum VIII
S-IX, sternum IX
T-VII, tergum VII
T-VIII, tergum VIII
T-IX, tergum IX
I–VIII, abdominal segments

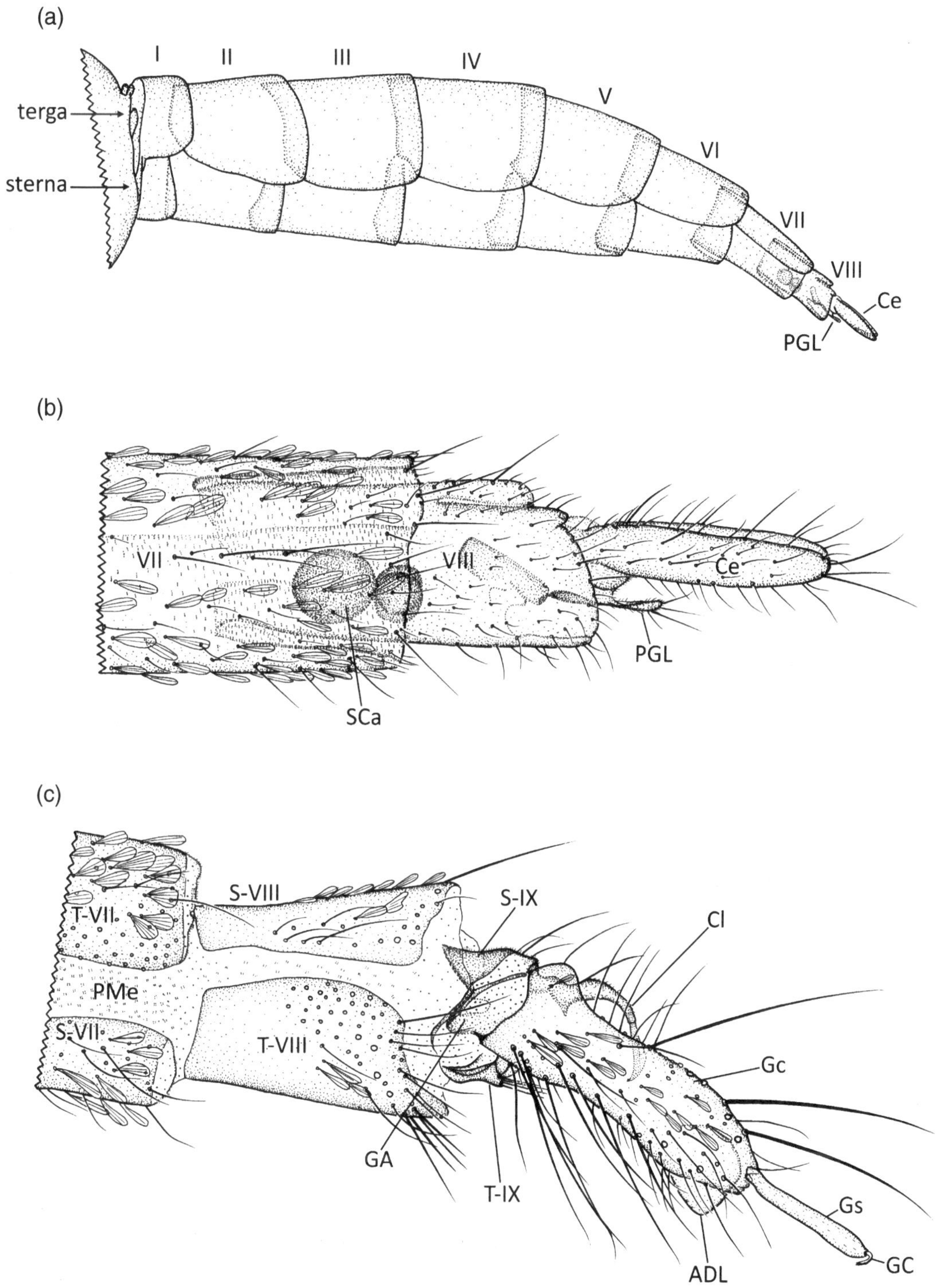
(a)
I
II
III
IV
V
VI
VII
VIII
terga
sterna
Ce
PGL
(b)
VII
VIII
Ce
PGL
SCa
(c)
S-VIII
S-IX
Cl
T-VII
PMe
S-VII
T-VIII
Gc
GA
T-IX
Gs
ADL
GC

Figure 11. Female genitalia of (a) *Culex* (*Neoculex*) *territans* Walker, (b) *Aedes* (*Borichinda*) *cavernicola* (Rattanarithikul & Harbach), (c) *Aedes* (*Ochlerotatus*) *ludlowae* (Blanchard) and (d,e) *Sabethes* (*Sabethinus*) *xhyphydes* Harbach (a and c modified from Harbach & Knight, 1980; b modified from Harbach *et al.*, 2007b, c modified from Harbach, 1994b).

a) Lateral aspect (left side) of terminal abdominal segments and genitalia.
b) Ventrocaudal aspect of genitalia.
c) Ventral aspect of genitalia.
d) Ventral aspect of dorsal parts of genitalia.
e) Caudal aspect of insula.

Abbreviations

Ce, cercus
I, insula
IsS, insular setae
LVL, lower vaginal lip
PGL, postgenital love
SCa, spermathecal capsule
UVL, upper vaginal lip
UVS, upper vaginal sclerite

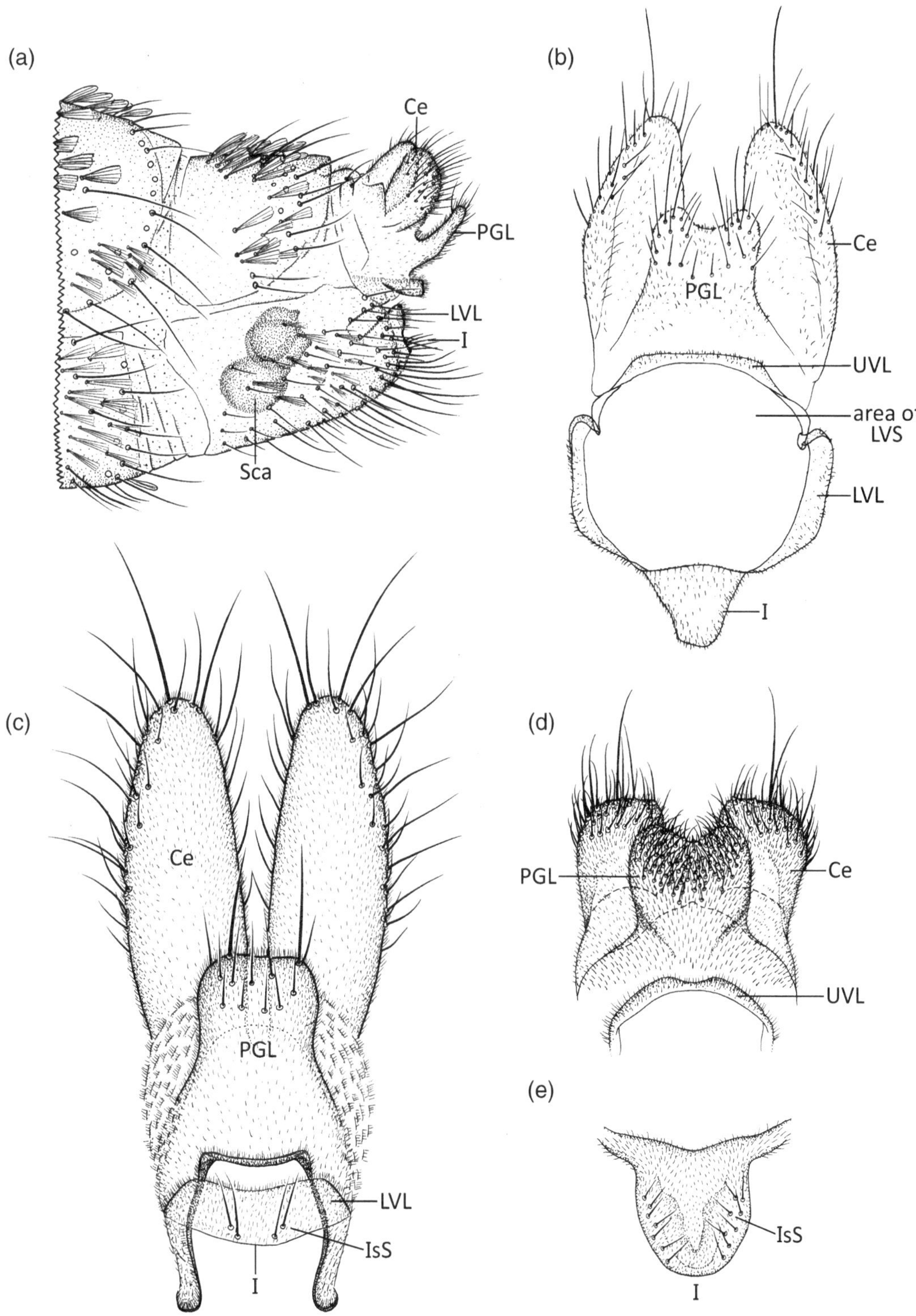
(a)
Ce
PGL
LVL
I
Sca
(b)
PGL
Ce
UVL
area of LVS
LVL
I
(c)
Ce
PGL
LVL
IsS
I
(d)
PGL
Ce
UVL
(e)
IsS
I

Figure 12. Male genitalia of (a) *Anopheles* (*Anopheles*) *barbirostris* van der Wulp, (b) *An.* (*Nyssorhynchus*) *trinkae* Faran and (c,d) *Culex* (*Culex*) *pipiens* Linnaeus (a modified from Harrison & Scanlon, 1975; b–d modified from Harbach & Knight, 1980).

a) Dorsal (tergal, post-rotation) aspect of genitalia (segment IX omitted).
b) Dorsal (tergal) aspect of genitalia (right gonocoxopodite, post-rotation position, omitted).
c) Dorsal (tergal, post-rotation) aspect of intact genitalia.
d) Dorsal (tergal, post-rotation) aspect of phallosome.

Abbreviations
AeS, aedeagal sclerite
Ae, aedeagus
a–h, setae *a–h*
AsS, accessory setae
BLA, basal lateral arm
Ce, cercus
CSc, cercal sclerite
CSe, cercal setae
DAB, dorsal aedeagal bridge
DOA, dorsal arm
GA, gonocoxal apodeme
Gc, gonocoxite
GC, gonostylar claw
Gs, gonostylus
ID, inner division
ILS, interlobar space
InS, internal seta
LA, lateral arm
LAe, leaflets of aedeagus
LP, lateral plate
MM, mesal membrane
OD, outer division
OPS, opisthophallic sclerite
Par, paramere
PBL, parabasal lobe
PBS, parabasal setae
PpC, paraproct crown
Ppr, paraproct
Pr, proctiger
SL, subapical lobe
T-X, tergum X
TL-IX, tergum IX lobe
VA, ventral arm
VAB, ventral aedeagal bridge

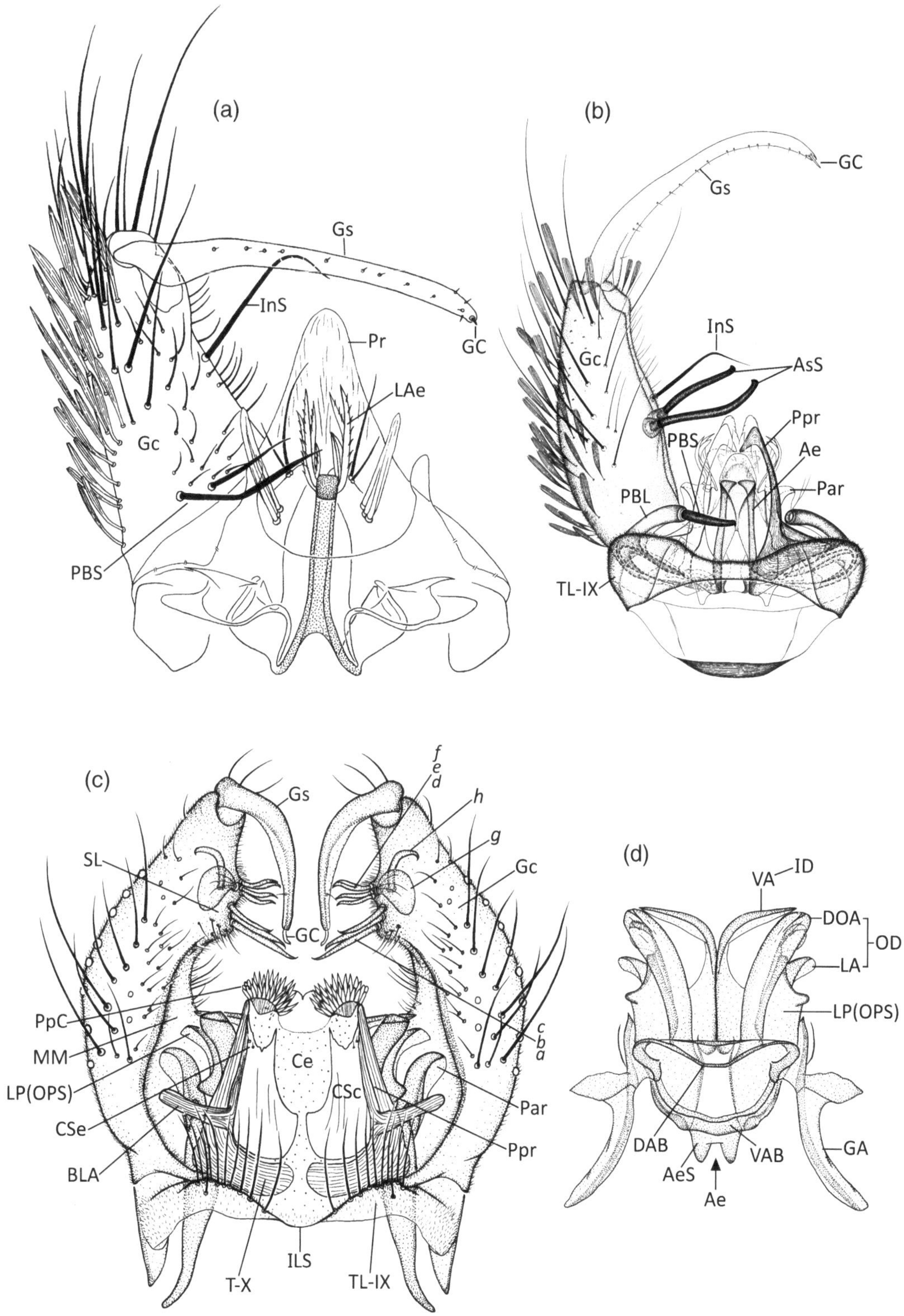
(a)
Gs
InS
GC
Pr
LAe
Gc
PBS
(b)
GC
Gs
InS
AsS
Gc
Ppr
PBS
Ae
Par
PBL
TL-IX
(c)
f
e
d
Gs
h
g
SL
Gc
GC
PpC
MM
LP(OPS)
CSe
BLA
Ce
CSc
c
b
a
Par
Ppr
ILS
T-X
TL-IX
(d)
VA
ID
DOA
OD
LA
LP(OPS)
DAB
VAB
AeS
Ae
GA

Figure 13. Male genitalia of (a) *Aedes* (*Downsiomyia*) *dorseyi* Knight, (b) *Ae.* (*Ochlerotatus*) *stimulans* (Walker) and (c,d) *Sabethes* (*Peytonulus*) *aurescens* (Lutz) (a modified from Harbach & Knight, 1980; b modified from Matheson, 1929; c and d modified from Hall *et al.*, 1999).

a) Dorsal (tergal, post-rotation) aspect of intact genitalia.
b) Dorsal (tergal, post-rotation) aspect of phallosome and proctiger.
c) Mesal aspect of gonocoxopodite.
d) Dorsal (tergal, post-rotation) aspect of phallosome.

Abbreviations
Ae, aedeagus
BDL, basal dorsomesal lobe
BML, basal mesal lobe
Ce, cercus
DF, claspette filament
Cl, claspette
CSe, cercal setae
CSt, claspette stem
DAB, dorsal aedeagal bridge
GA, gonocoxal apodeme
Gc, gonocoxite
GC, gonostylar claw
Gs, gonostylus
Ppr, paraproct
VAB, ventral aedeagal bridge
T-X, tergum X
TL-IX, tergum IX lobe

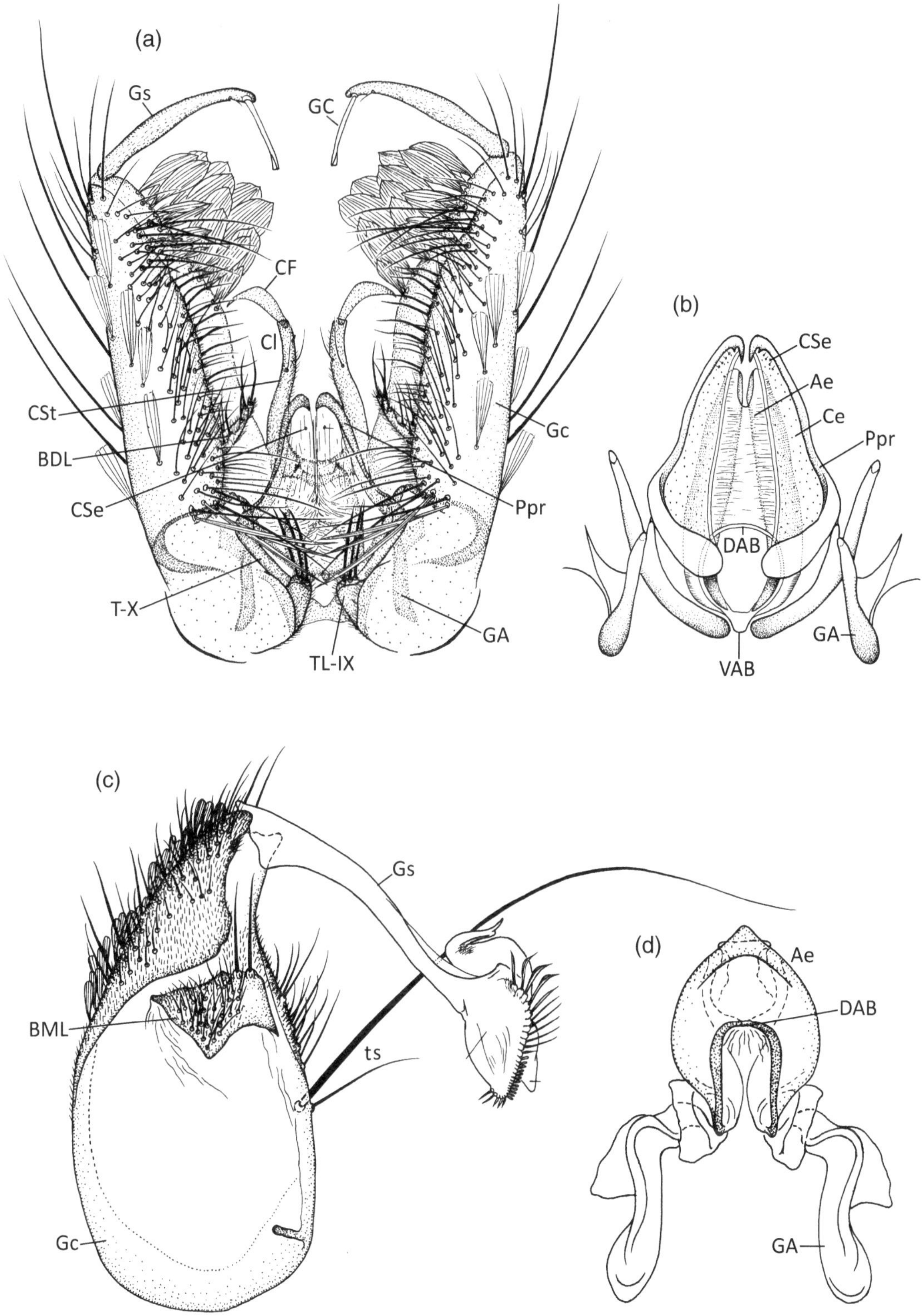
(a)
Gs
GC
CF
Cl
CSt
BDL
CSe
Gc
Ppr
T-X
GA
TL-IX
(b)
CSe
Ae
Ce
Ppr
DAB
GA
VAB
(c)
Gs
BML
ts
Gc
(d)
Ae
DAB
GA

Figure 14. Mosquito eggs (modified from Harbach & Knight, 1980).

a–d) Egg of *Anopheles* (*Cellia*) *aconitus* Dönitz. (a) Lateral (left aspect (anterior end at right); (b) anterior end; (c) ventral surface; (d) ventral view of anterior end.

e,f) Egg of *Aedes* (*Stegomyia*) *aegypti* (Linnaeus). (e) Lateral (left) aspect of inner chorion (anterior end above), with enlarged area of inner chorionic sculpture; (f) ventral surface of outer chorion (anterior end above), with enlarged area of outer chorionic sculpture.

g,h) Egg of *Toxorhynchites* (*Toxorhynchites*) *brevipalpus* Theobald. (g) Lateral aspect (anterior end above), with enlarged area of outer chorionic sculpture (note absence of outer chorionic reticulum).

Abbreviations

Ce, cellule
CeW, cellule wall
Co, corolla
De, deck
DeT, deck tubercle
F, float
Fr, frill
FR, float ridge
ICC, inner chorionic cell
ICR, inner chorionic reticulum
ICS, inner chorionic sculpture
LoT, lobed tubercle
Mi, micropyle
MiC, micropylar collar
MiD, micropylar disc
OCC, outer chorionic cell
OCR, outer chorionic reticulum
OCT, outer chorionic tubercle

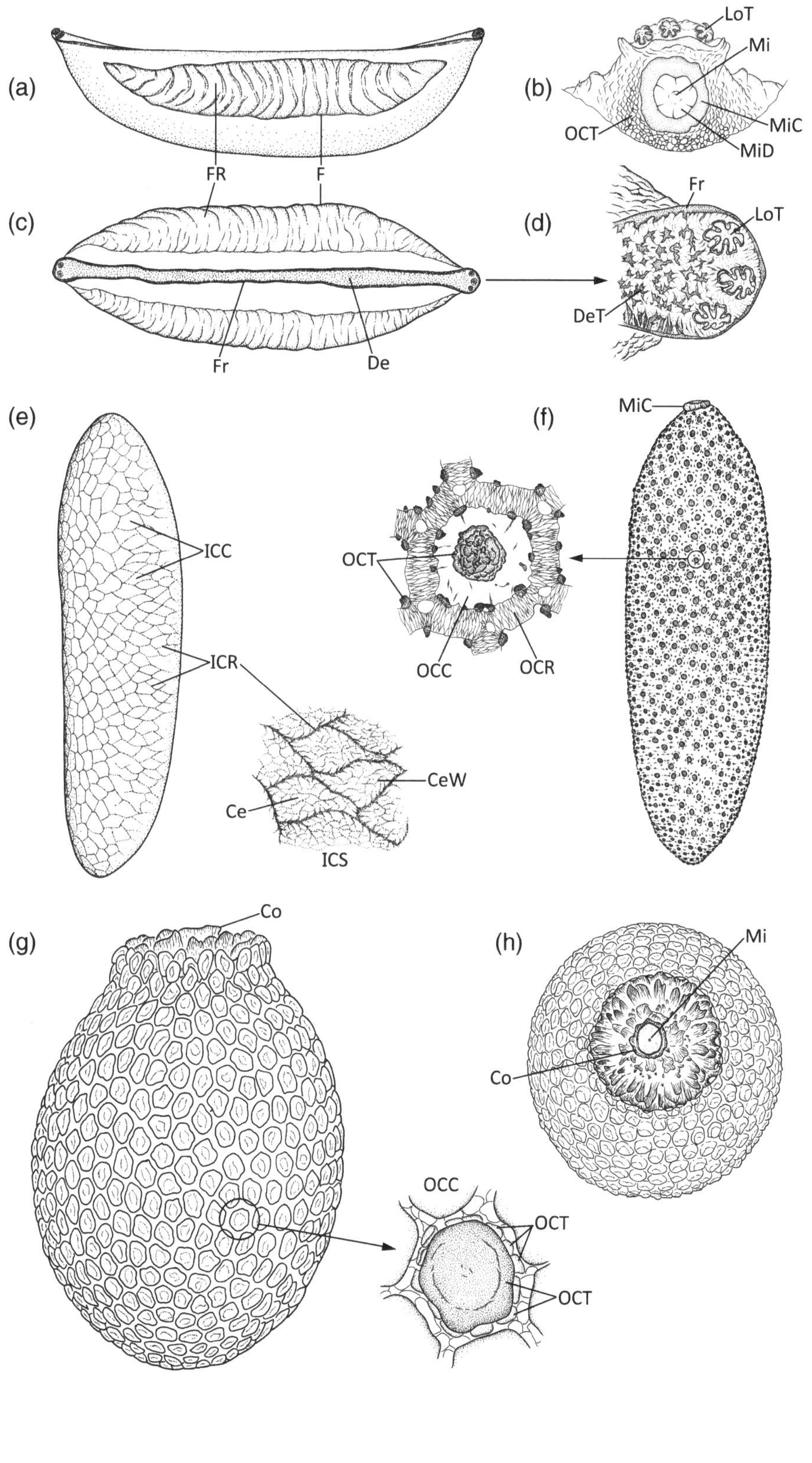
(a)
FR
F
(b)
LoT
Mi
MiC
MiD
OCT
(c)
Fr
De
(d)
Fr
LoT
DeT
(e)
ICC
ICR
CeW
Ce
ICS
(f)
MiC
OCT
OCC
OCR
(g)
Co
OCC
OCT
OCT
(h)
Mi
Co

Figure 15. Mosquito larva and pupa (modified from Harbach & Knight, 1980).

a) Larva (left side) of *Aedes* (*Ochlerotatus*) *pallens* Ross.
b) Pupa (male, left side) of *Ae.* (*Ochlerotatus*) *canadensis* (Theobald).
c) Pupa (female), terminal abdominal structures (diagrammatic).

Abbreviations

A, antenna
APP, anal papillae
CE, compound eye
Cv, cervix
DB, dorsal brush
GL, genital lobe
Lat, lateralia
M, mesothorax
MCL, median caudal lobe
MK, median keel
MPlp, maxillary palpus
Mr, midrib
Mtn, metanotum
MtW, metathoracic wing
MW, mesothoracic wing
P, prothorax (larva); proboscis (pupa)
Pa, paddle
Pt, pecten
S, siphon
Sa, saddle
SAp, spiracular apparatus
Scu, scutum
T, metathorax
Tr, trumpet
VB, ventral brush
I–X, abdominal segments

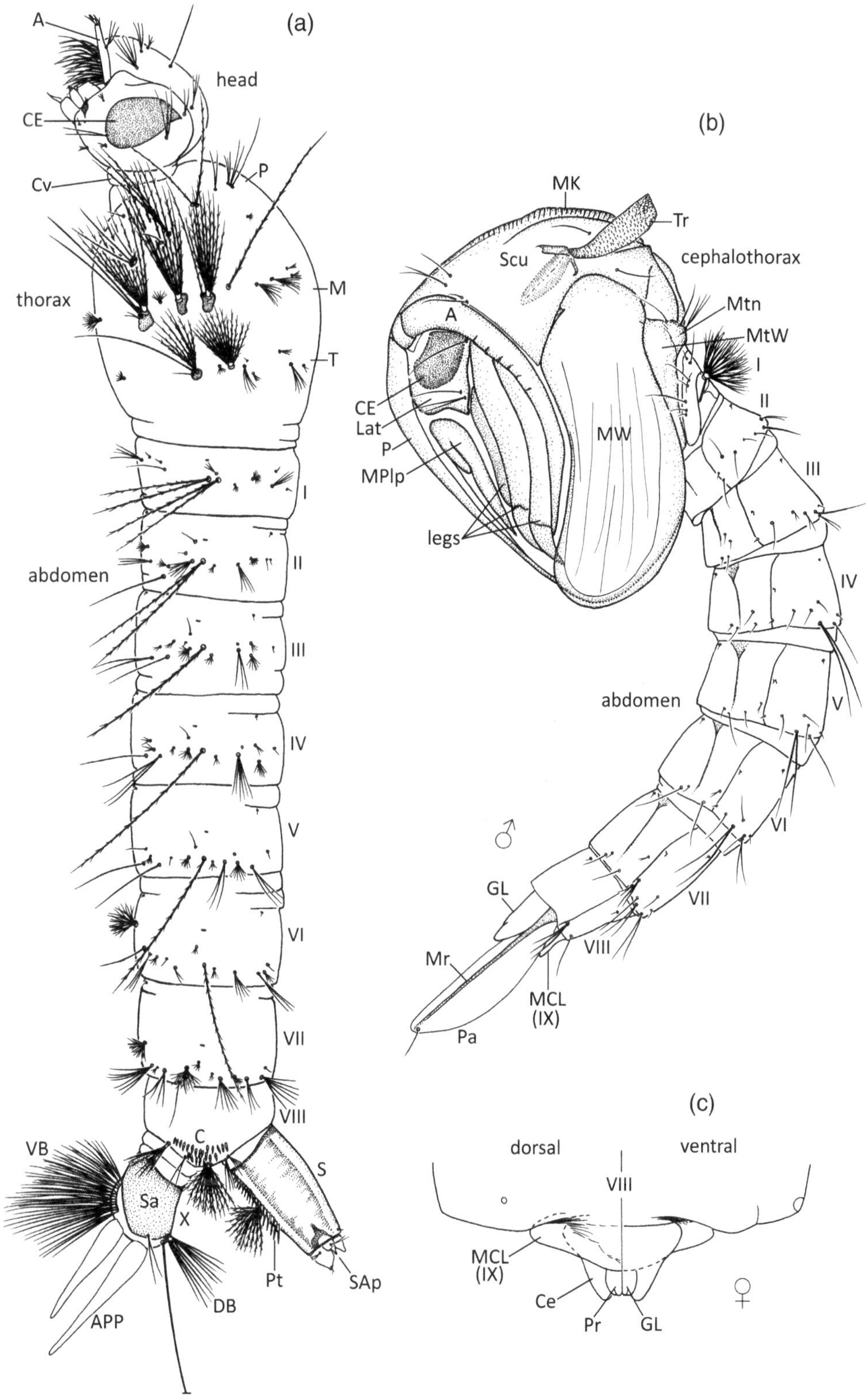
(a)
A
head
CE
Cv
P
M
thorax
T
I
abdomen
II
III
IV
V
VI
VII
VIII
C
VB
Sa
X
S
Pt
SAp
DB
APP
(b)
MK
Tr
Scu
cephalothorax
Mtn
MtW
A
I
CE
Lat
P
MPlp
MW
II
III
legs
IV
abdomen
V
♂
VI
GL
VII
Mr
VIII
MCL
(IX)
Pa
(c)
dorsal
ventral
VIII
MCL
(IX)
Ce
Pr
GL
♀

Figure 16. Heads of larvae of (a,b) *Anopheles* (*Anopheles*) *quadrimaculatus* Say, (c,d) *Aedes* (*Ochlerotatus*) *pallens* Ross and *Culex* (*Neoculex*) *martinii* Medschid, respectively, and (f) *Sabethes* (*Peytonulus*) *undosus* (Coquillett) (a–d modified from Harbach & Knight, 1980; e and f modified from Harbach & Kitching, 1998).

a,b) Dorsal and ventral aspects, respectively, with alveoli shown but setae omitted and many filaments of the lateral palatal brushes removed.
c,d) Dorsal and ventral aspects, respectively, with alveoli shown but setae (except 1-C) omitted; c with oral brushes and mouthparts removed.
e,f) Posterior (caudal) aspects, with setae and alveoli omitted.

Abbreviations

A, antenna
CG, coronal gap
ClS, clypeolabral suture
Col, collar
DAp, dorsal apotome
Dm, dorsomentum
FEL, frontal ecdysial line
HEL, hypocranial ecdysial line
HyS, hypostomal suture
Lat, lateralia
LG, labiogula
LPB, lateral palatal brush
MLP, median labral plate
OF, occipital foramen
POc, postocciput

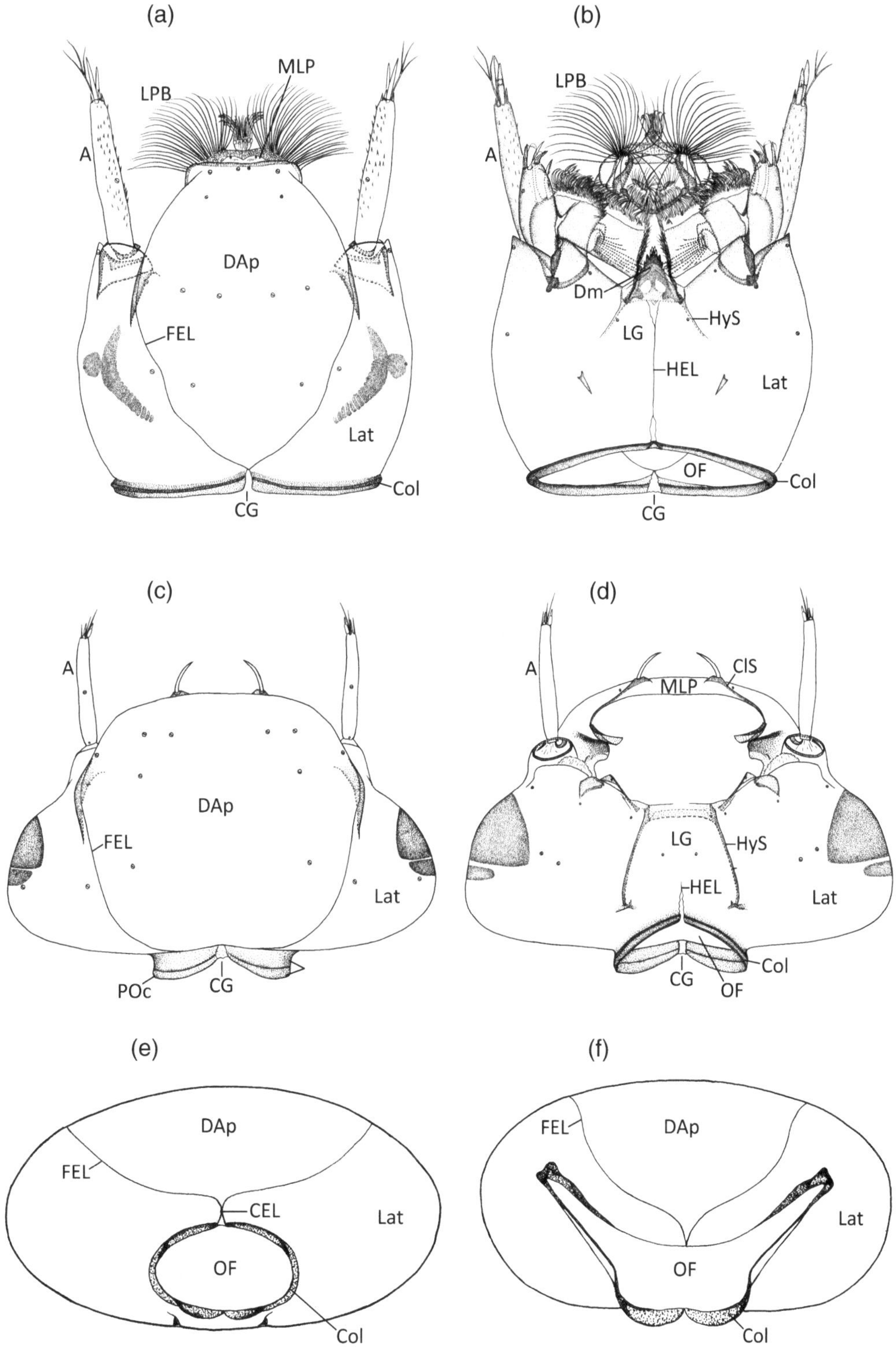
(a)
MLP
LPB
A
DAp
FEL
Lat
Col
CG
(b)
LPB
A
Dm
HyS
LG
HEL
Lat
OF
Col
CG
(c)
A
DAp
FEL
Lat
POc
CG
(d)
A
CIS
MLP
LG
HyS
HEL
Lat
Col
CG
OF
(e)
DAp
FEL
CEL
Lat
OF
Col
(f)
FEL
DAp
Lat
OF
Col

Figure 17. Mouthparts of mosquito larvae (a–c, e and f modified from Harbach & Knight, 1980; d modified from Knight & Harbach, 1977).

a) Mandible (dorsal aspect) of *Anopheles* (*Anopheles*) *crucians* Wiedemann.
b) Mandible (dorsal aspect) of *Culex* (*Culex*) *quinquefasciatus* Say.
c) Maxilla (ventral aspect) of *An.* (*Ano.*) *crucians*.
d) Maxilla (ventral aspect) of *Opifex* (*Opifex*) *fuscus* Hutton.
e) Labiohypopharynx (anterior aspect) of *An.* (*Nyssorhynchus*) *albimanus* Wiedemann.
f) Labiohypopharynx (anterior aspect) of *Aedes* (*Ochlerotatus*) *canadensis* (Theobald).

Abbreviations

Gl, galea
GSt, galeastipes
HSc, hypostomal sclerite
Hy, hypopharynx
Lg, ligula
LP, labial palpus
LPT, lateral premental teeth
LR_1, laciniarastrum 1
LPT_1, lateral premental teeth 1
LPT_2, lateral premental teeth 2
LPT_3, lateral premental teeth 3
LSt, laciniastipes
MnB, mandibular brush
MnL, mandibular lobe
MnS, mandibular sweeper
MnT, mandibular teeth
MPlp, maxillary palpus
MS, maxillary palpal sensoria
mSS, merostipital sclerite
MSAB, mandibular spiculose area brush
MxB, maxillary brush
MxBo, maxillary body
pLb, prelabium
PM, premental mala
VMxS, ventral maxillary suture
3–6, setae 1- to 6-Mx

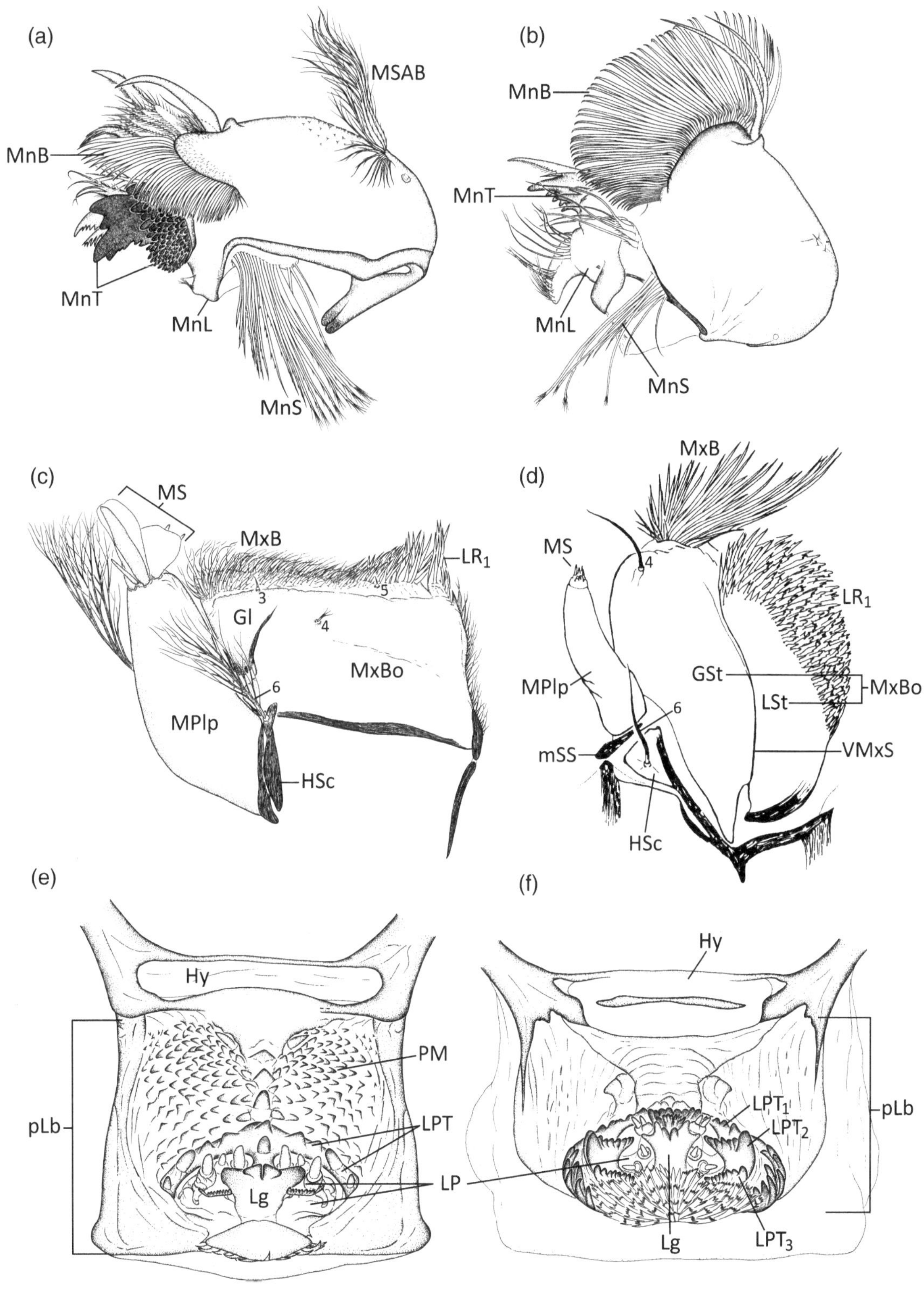
(a)
MSAB
MnB
MnT
MnL
MnS
(b)
MnB
MnT
MnL
MnS
(c)
MS
MxB
LR1
Gl
MxBo
MPlp
HSc
(d)
MxB
MS
LR1
MPlp
GSt
LSt
MxBo
VMxS
mSS
HSc
(e)
Hy
PM
pLb
LPT
LP
Lg
(f)
Hy
LPT1
LPT2
pLb
LP
Lg
LPT3

Figure 18. Anopheline larva (a and c modified from Nguyen *et al.*, 2000; b modified from Reid, 1968).

a) Thorax and abdominal segments I–VI (dorsal on left; ventral on right).
b) Pleural setal group and basal support (generalized).
c) Abdominal segments VII, VIII and X (lateral, left side).

Abbreviations

APP, anal papillae
ATP, accessory tergal plate
DB, dorsal brush
M, mesothorax
MATP, median accessory tergal plate
P, prothorax
psg, pleural setal group
PP, pecten plate
Pt, pecten
Sa, saddle
SAp, spiracular apparatus
SSP, setal support plate
SSPS, setal support plate spine
T, metathorax
TP, tergal plate
VB, ventral brush
I–VIII, X, abdominal segments

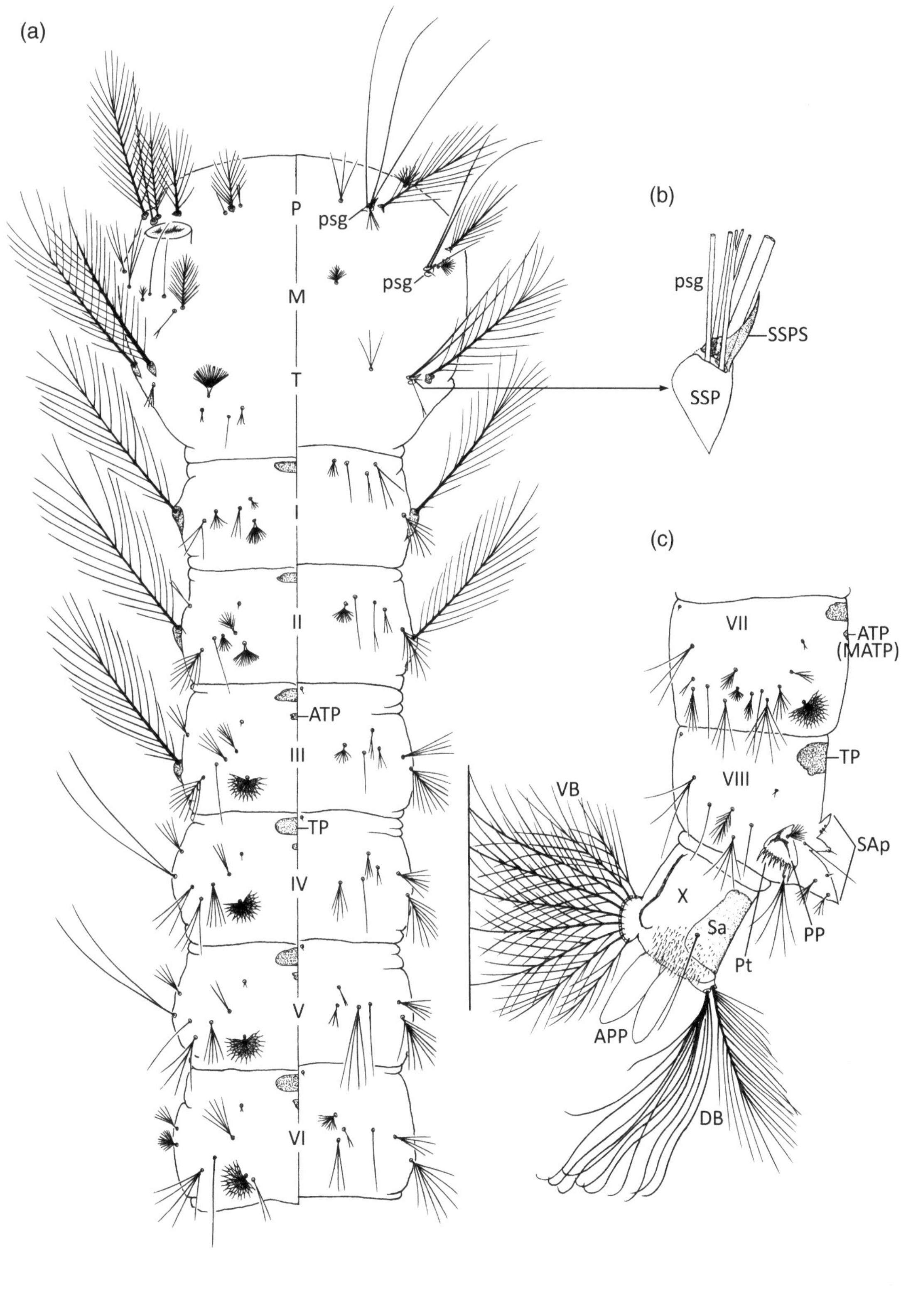
(a)
P
psg
psg
M
T
I
II
ATP
III
TP
IV
V
VI
(b)
psg
SSPS
SSP
(c)
VII
ATP
(MATP)
TP
VIII
VB
SAp
X
Sa
PP
Pt
APP
DB

Figure 19. Culicine larva (a and b modified from Harbach *et al.*, 2013; c modified from Harbach & Knight, 1980).

a) Thorax and abdominal segments I–VI (dorsal on left; ventral on right).
b) Abdominal segments VII, VIII and X (lateral, left side).
c) Grid (generalized, lateral, left side).

Abbreviations

APP, anal papillae
C, comb
CS, comb scale
DB, dorsal brush
LGB, lateral grid bar
M, mesothorax
P, prothorax
PS, pecten spine
psg, pleural setal group
Pt, pecten
S, siphon
Sa, saddle
SA, siphon acus
SAd, spiracular apodeme
SAp, spiracular apparatus
SSP, setal support plate
T, metathorax
TGB, transverse grid bar
VB, ventral brush
I–VIII, X, abdominal segments

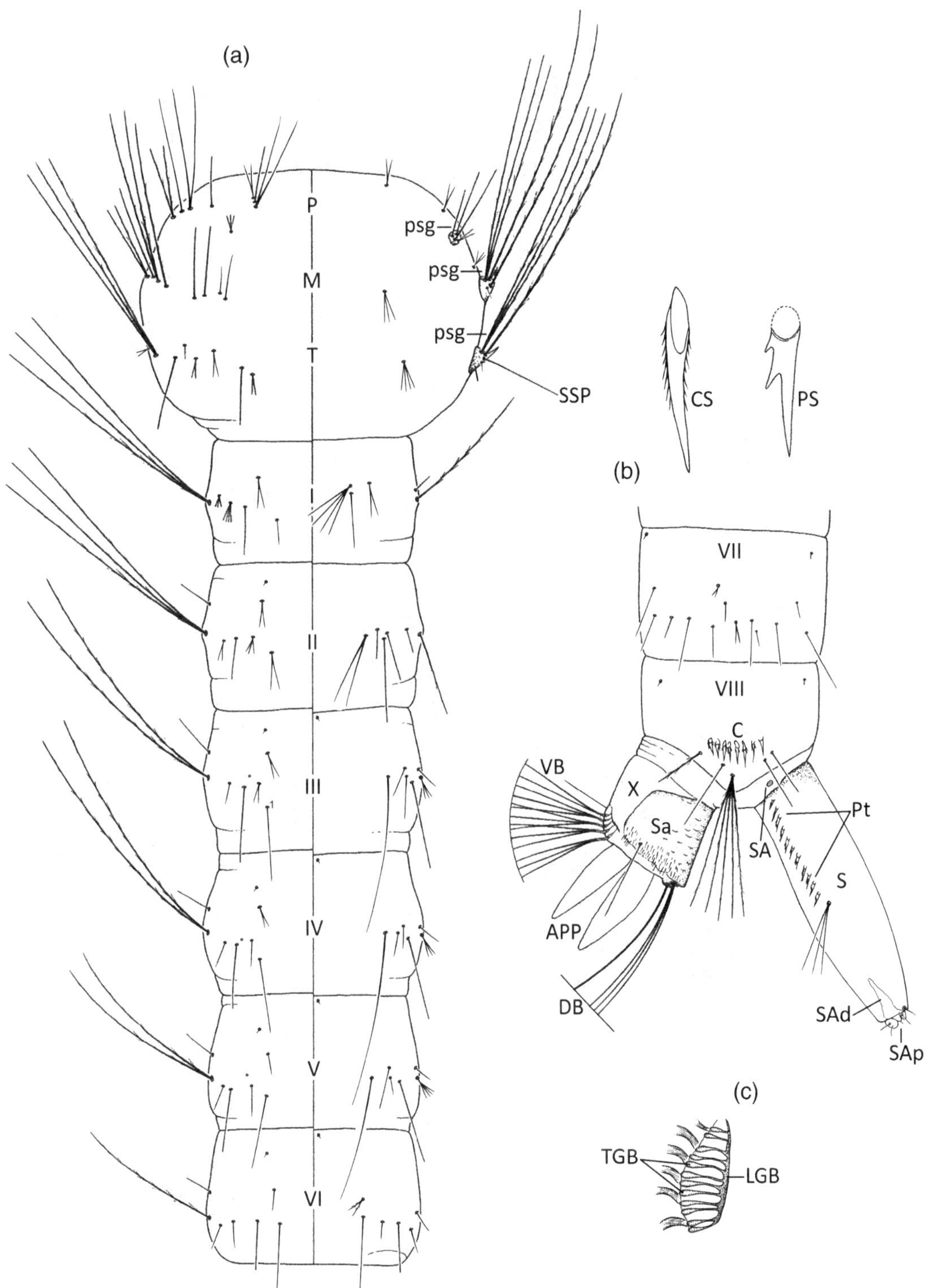
(a)
P
M
T
psg
psg
psg
SSP
I
II
III
IV
V
VI
CS
PS
(b)
VII
VIII
C
VB
X
Sa
Pt
SA
S
APP
DB
SAd
SAp
(c)
TGB
LGB

Figure 20. Terminal abdominal segments (lateral, left side) of mosquito larvae (a modified from Harbach & Howard, 2009; b and d modified from Harbach & Knight, 1980; c original).

a) *Chagasia bonneae* Root.
b) *Culiseta* (*Culiseta*) *inornata* (Williston).
c) *Sabethes* (*Sabethes*) *bipartipes* Dyar & Knab.
d) *Toxorhynchites* (*Toxorhynchites*) *brevipalpis* Theobald.

Abbreviations

AMPc, anterior median process
APP, anal papillae
C, comb
CP, comb plate
FS, fringing setae
PcrS, precratal setae
PP, pecten plate
S, siphon
Sa, saddle
SA, siphon acus
SaA, saddle acus
SAd, spiracular apodeme
SAp, spiracular apparatus
Pt, pecten
TP, tergal plate
VII, VIII, X, abdominal segments

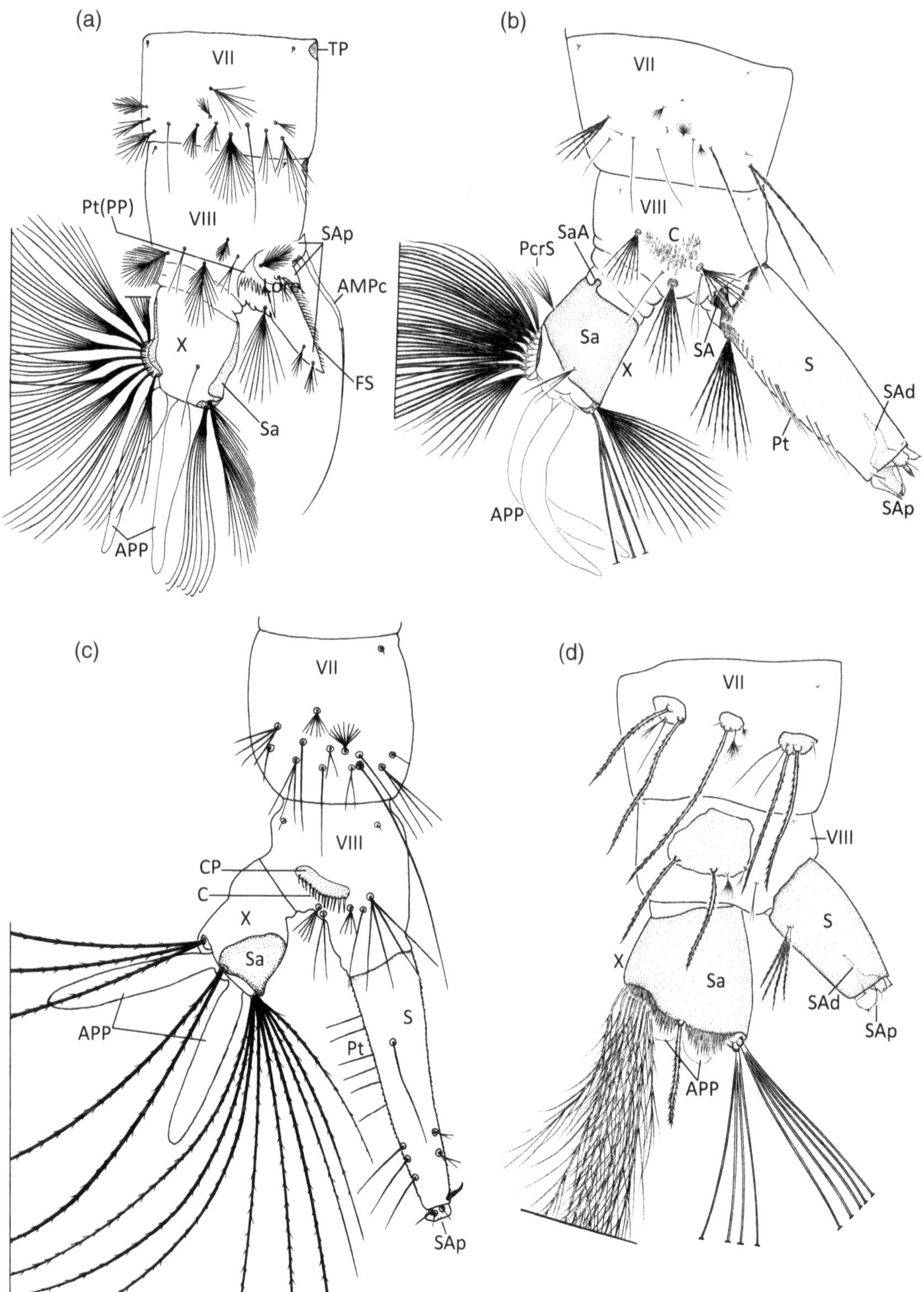
(a)
VII
TP
VIII
Pt(PP)
SAp
pre
AMPc
X
FS
Sa
APP
(b)
VII
VIII
C
SaA
PcrS
Sa
X
SA
S
SAd
Pt
SAp
APP
(c)
VII
VIII
CP
C
X
Sa
APP
S
Pt
SAp
(d)
VII
VIII
S
X
Sa
SAd
SAp
APP

Figure 21. Spiracular apparatuses of mosquito larvae (modified from Harbach & Knight, 1980) (a, b, d and e, dorsal aspects; c, lateral (right side) of siphon).

a) *Anopheles* (*Anopheles*) *crucians* Wiedemann.
b) *Culex* (*Culex*) *nigripalpus* Theobald.
c) *Mansonia* (*Mansonioides*) *uniformis* (Theobald).
d) Anopheline (generalized).
e) Culicine (generalized).

Abbreviations
AMPc, anterior median process
ASL, anterior spiracular lobe
LSL, anterolateral spiracular lobe
MdP, median plate
PMPc, posterior median process
PSL, posterolateral spiracular lobe
S, siphon (c); spiracle (a, b)
SAd, spiracular apodeme
SAp, spiracular apparatus
SAW, saw
1–13, setae of siphon and spiracular apparatus

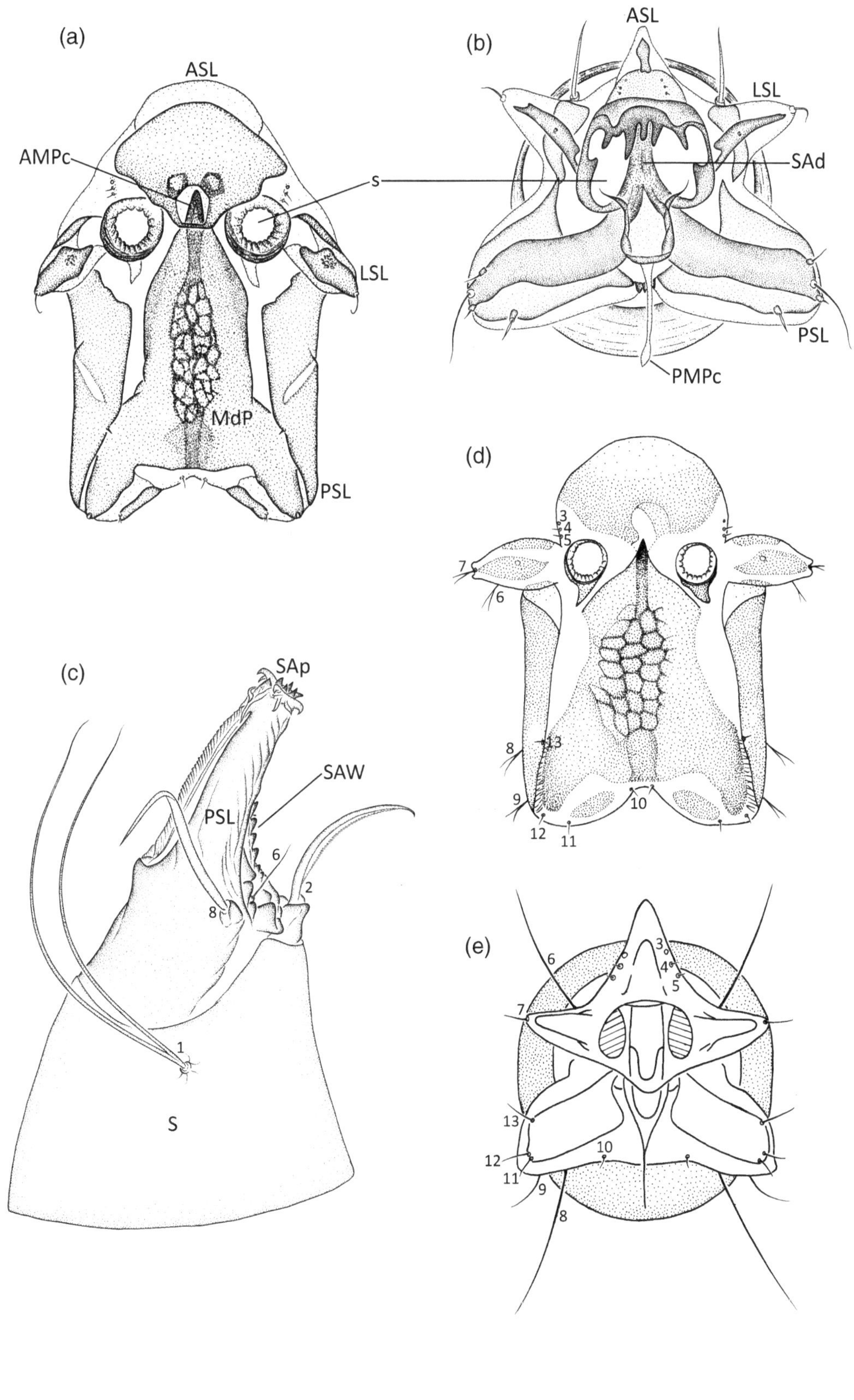
(a)
ASL
AMPc
s
LSL
MdP
PSL
(b)
ASL
LSL
SAd
PSL
PMPc
(c)
SAp
SAW
PSL
6
8
2
1
S
(d)
3
4
5
7
6
8
13
9
10
12
11
(e)
6
3
4
5
7
13
12
10
11
9
8

Figure 22. Chaetotaxy of anopheline larvae (original, based on *Anopheles* (*Cellia*) *subpictus* Grassi *sensu lato*).

a) Thorax and abdominal segments I–VI (dorsal on left; ventral on right).
b) Head (dorsal on left; ventral on right).
c) Abdominal segments VII, VIII and X (lateral, left side).

Abbreviations

A, antenna
C, cranium
M, mesothorax
P, prothorax
S, spiracular apparatus
T, metathorax
I–VIII, X, abdominal segments
0–14, setae of specified areas

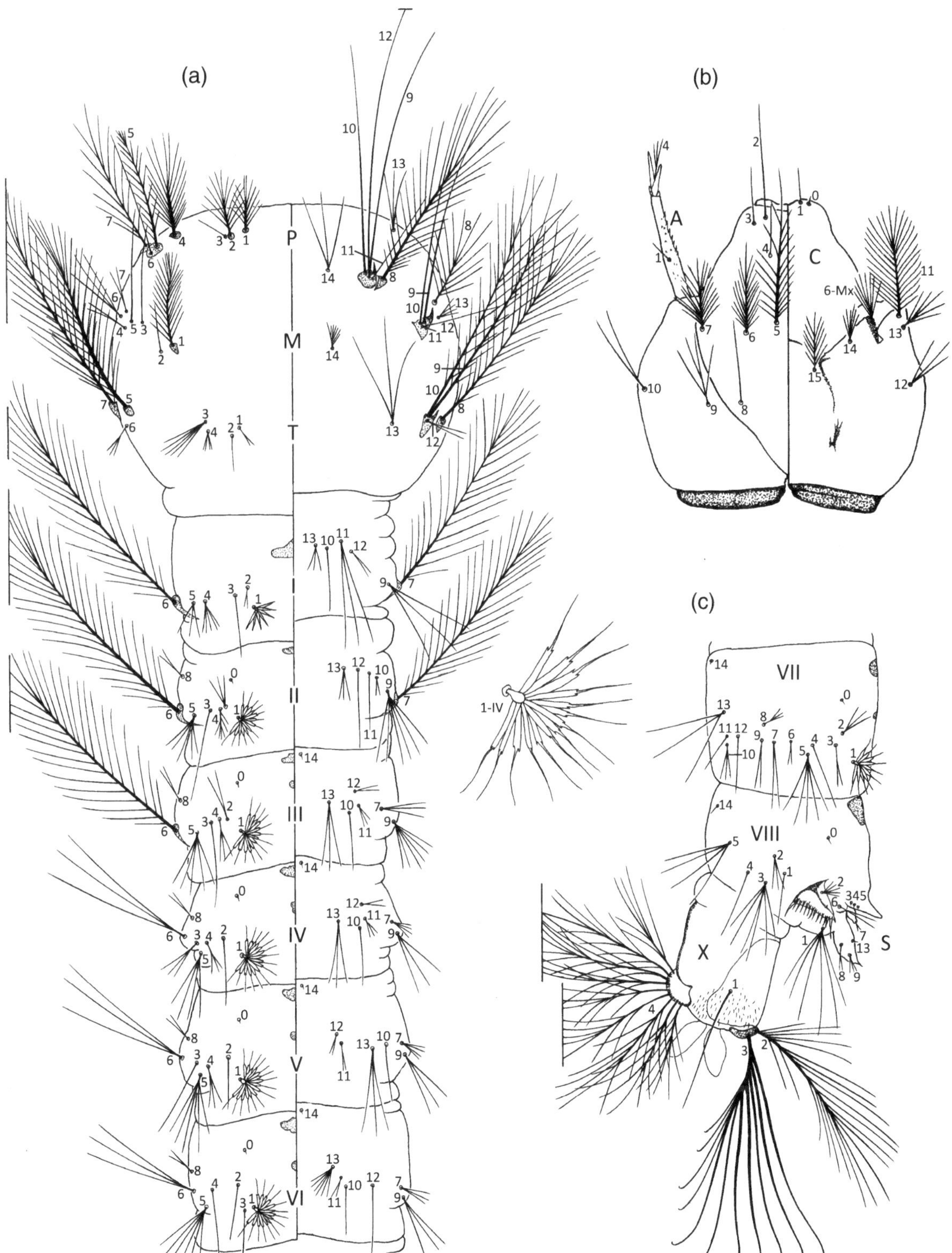
(a)
P
M
T
I
II
III
IV
V
VI
1-IV
(b)
A
C
6-Mx
(c)
VII
VIII
X
S

Figure 23. Chaetotaxy of culicine larvae (based on *Culex* (*Culex*) *pipiens* Linnaeus, modified from Harbach *et al.*, 1985).

a) Thorax and abdominal segments I–VI (dorsal on left; ventral on right).
b) Head (dorsal on left; ventral on right).
c) Abdominal segments VII, VIII and X (lateral, left side).

Abbreviations

A, antenna
C, cranium
Dm, dorsomentum
M, mesothorax
p, puncture
P, prothorax
S, siphon and spiracular apparatus
T, metathorax
I–VIII, X, abdominal segments
0–14, setae of specified areas
6-Mx, seta 6 of maxilla

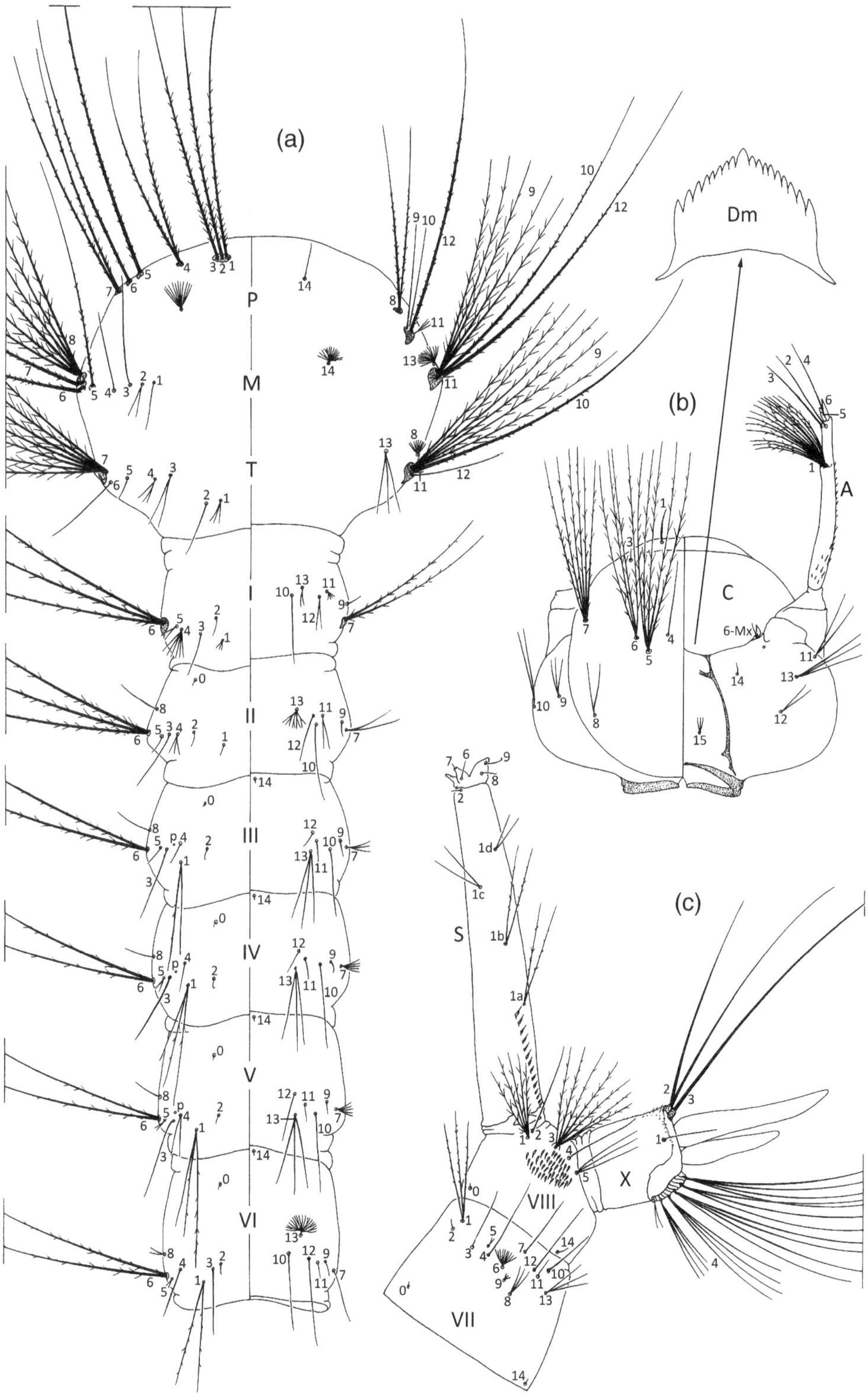
(a)
P
M
T
I
II
III
IV
V
VI
(b)
Dm
A
C
6-Mx
(c)
S
1a
1b
1c
1d
VII
VIII
X

Figure 24. Cephalothorax and trumpets of mosquito pupae (a–e and g modified from Harbach & Knight, 1980; f modified from Belkin *et al.*, 1970).

a) Anopheline (generalized), cephalothorax of pupal exuviae.
b) Culicine (generalized), cephalothorax of pupal exuviae.
c) Laticorn trumpet of *Anopheles* (*Anopheles*) *barbirostris* van der Wulp.
d) Angusticorn trumpet of *An.* (*Cellia*) *maculatus* Theobald.
e) Laticorn trumpet of *An.* (*Ano.*) *letifer* Sandosham.
f) Trumpet of *Mansonia* (*Mansonia*) *dyari* Belkin, Heinemann & Page.
g) Trumpet (diagrammatic).

Abbreviations

A, antenna
CT, cephalothorax
DAp, dorsal apotome
MC, meatal cleft
Mea, meatus
MK, median keel
MPlp, maxillary palpus
Pi, pinna
RA, reticulate area
Scu, scutum
SeC, secondary cleft
TA, tracheoid area
Tg, tragus
Tr, trumpet
1–9, setae of cephalothorax

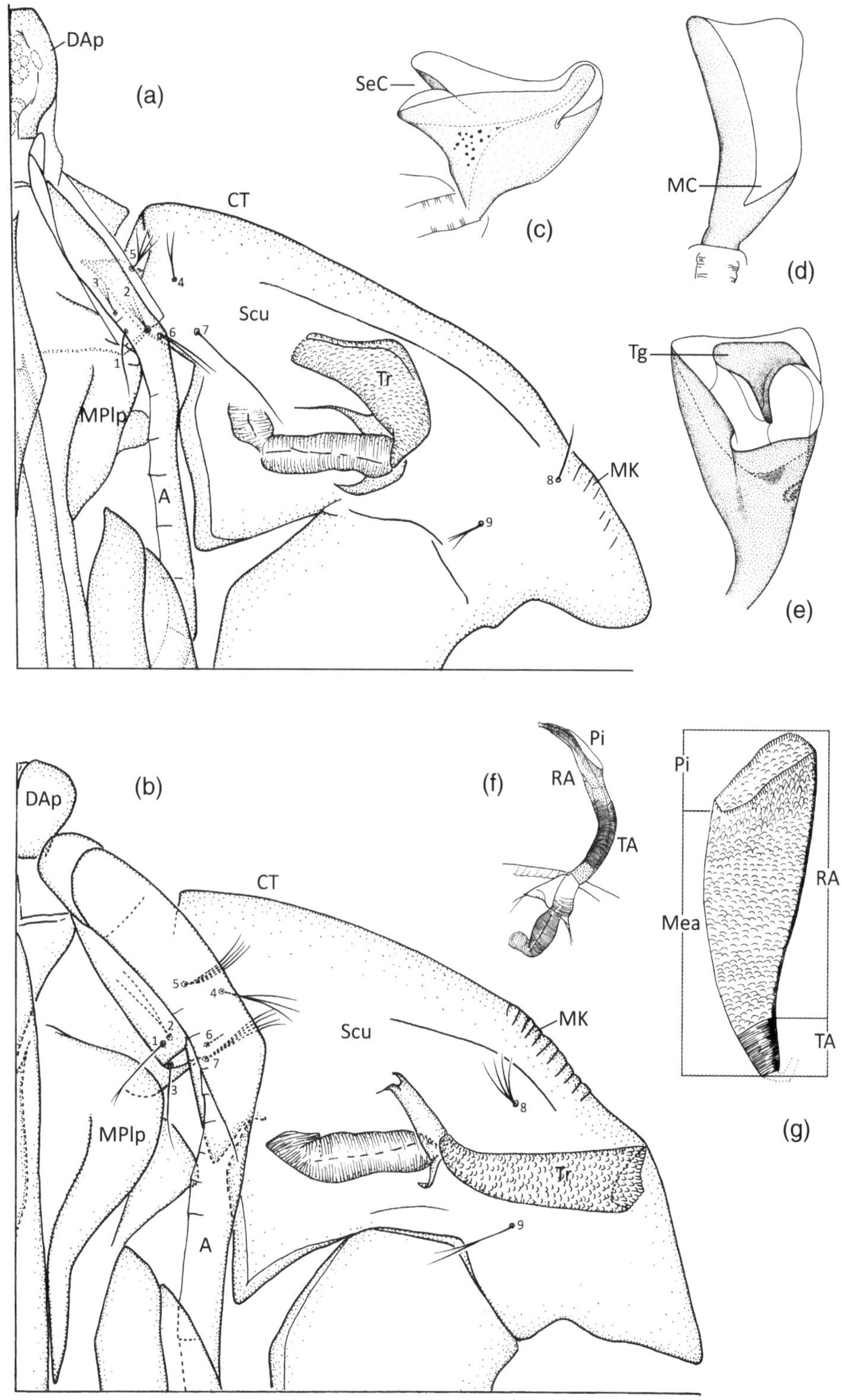
DAp
(a)
SeC
MC
(c)
(d)
CT
Scu
Tr
MPlp
A
MK
Tg
(e)
Pi
RA
TA
(f)
Mea
(b)
(g)

Figure 25. Chaetotaxy of mosquito pupae (modified from Harbach & Knight, 1980).

a) Anopheline (composite) (dorsal on left; ventral on right).
b) Culicine (composite) (dorsal on left; ventral on right).

Abbreviations

Bu, buttress
Ce, cercus
GL, genital lobe
MCL, median caudal lobe
Mr, midrib
Mtn, metanotum
MtW, metathoracic wing
p, puncture
Pa, paddle
Se, paddle marginal serration
Sp, paddle margin spicules
I–IX, abdominal segments
0–14, setae of specified areas

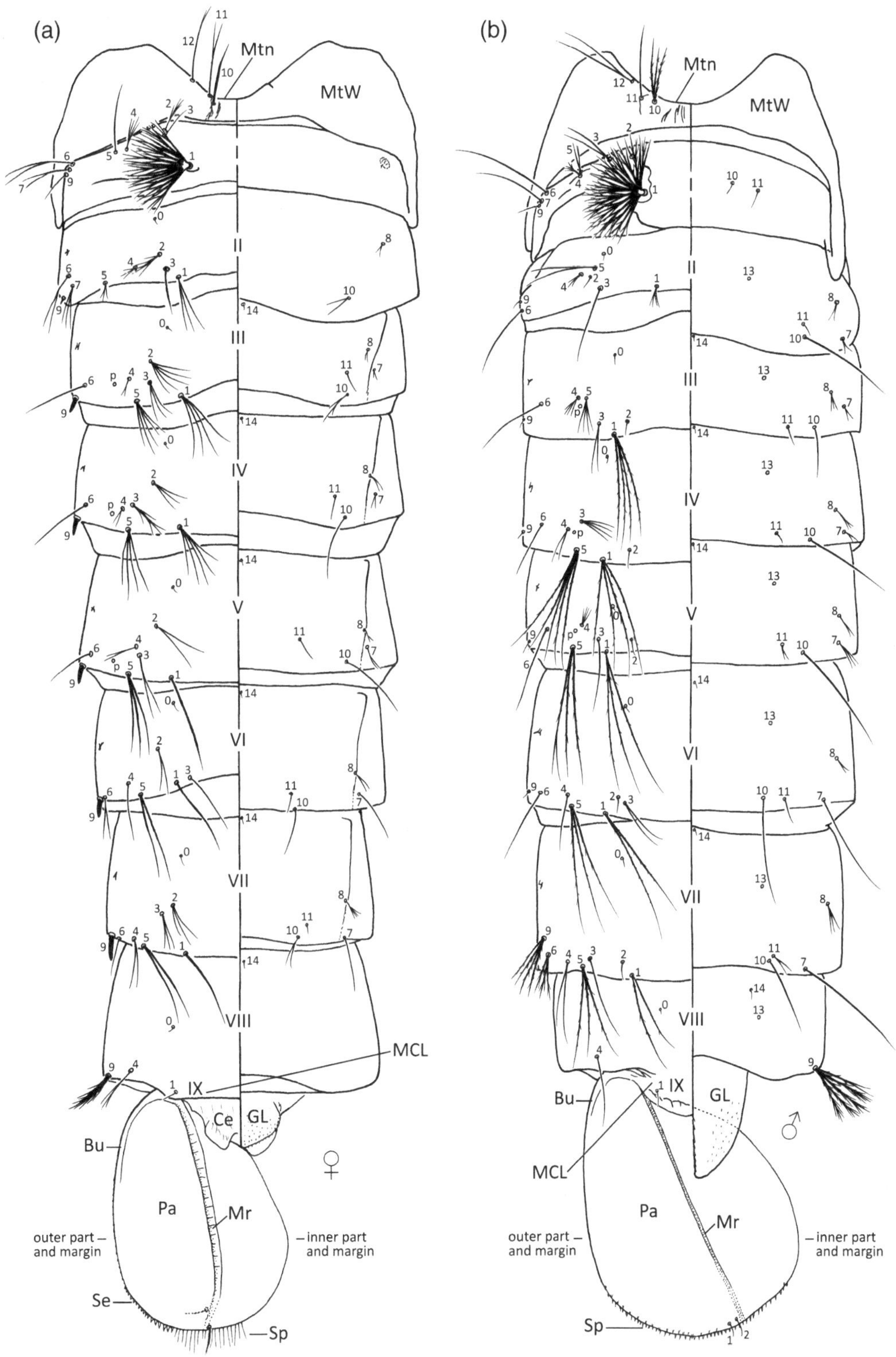
(a)
(b)
Mtn
MtW
II
III
IV
V
VI
VII
VIII
IX
MCL
Ce
GL
Bu
Pa
Mr
outer part and margin
inner part and margin
Se
Sp
♀
♂

Figure 26. Types of setae and spicules that constitute the vestiture of mosquitoes, principally larvae and pupae (modified from Harbach & Knight, 1980).

Setae

Single setae

a,b) Simple setae.
c) Barbed seta.
d) Spinulate seta.
e,f) Lanceolate setae.
g) Spiniform seta.
h) Foliform seta.
i) Aciculate seta.
j) Brush-tipped seta.
k) Pectunculate seta.

Branched setae

l,m) Forked setae.
n) Pectinate seta.
o) Fan-like seta.
p) Stellate seta with aciculate branches.
q) Dendritic seta with aciculate branches.
r) Fan-like seta with aciculate-dendritic branches.
s) Fan-like seta with dendritic branches.
t) Fan-like seta with aciculate branches.
u) Palmate seta.
v) Plumose seta.
w) Stellate seta with simple branches.

Spicules

Filaments

x) Simple filament.
y) Barbed filament.
z1) Branch-tipped filament.
z2) Comb-tipped filament.
z3) Pectinate filament.

Blades

z4) Divided blade.
z5) Incised blade.
z6,z7) Pectunculate blades.

Others

z8) Dendrite.
z9) Echinoid.

Abbreviations

r, rachis

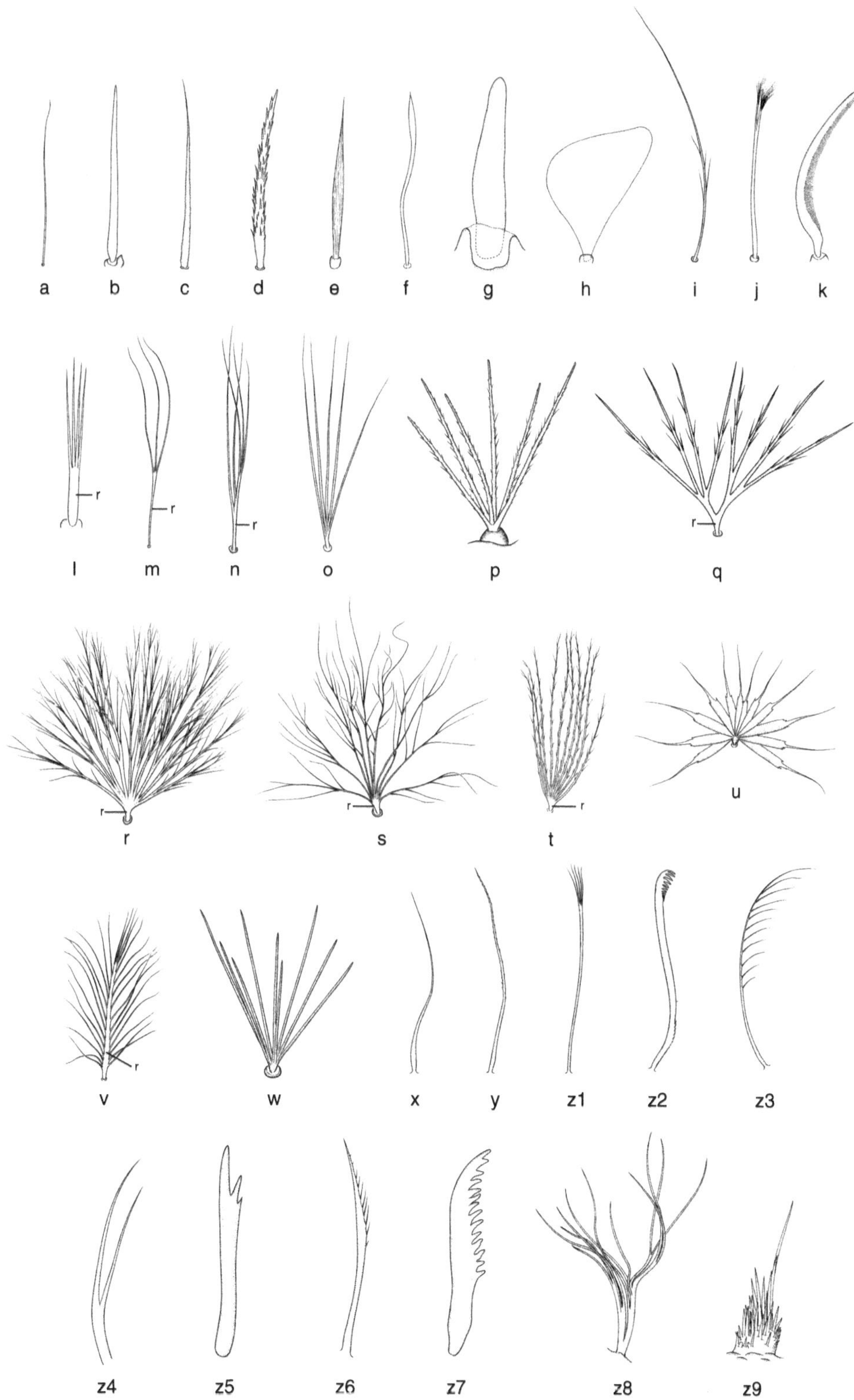
a
b
c
d
e
f
g
h
i
j
k
l
m
n
o
p
q
r
s
t
u
v
w
x
y
z1
z2
z3
z4
z5
z6
z7
z8
z9

Figure 27. Types of scales that constitute the vestiture of adult mosquitoes (modified from Harbach & Knight, 1980).

a) Asymmetrical scale.
b) Falcate scales.
c) Twisted scales.
d–f) Fusiform scales.
g) Forked scales.
h) Linear scales.
i–m) Spatulate scales.
n) Piliform scales.

Abbreviations
a, alveolus (seta removed)
p, pedicel
s, squame

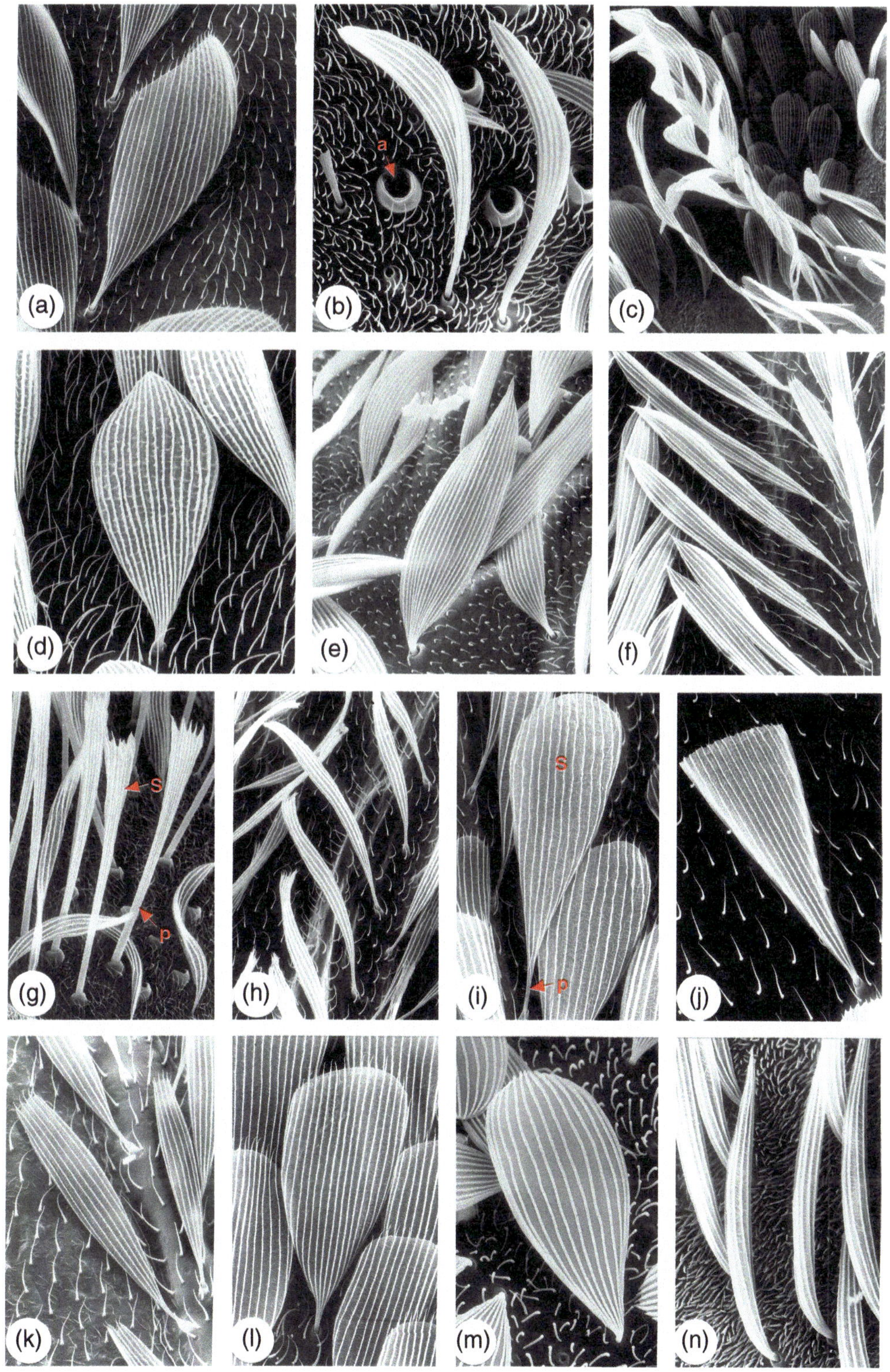
(a)
(b)
a
(c)
(d)
(e)
(f)
(g)
S
p
(h)
(i)
S
p
(j)
(k)
(l)
(m)
(n)

Figure 28. Some types of cuticular projections that constitute the vestiture of mosquito larvae (modified from Harbach & Knight, 1980).

a) Puncture.
b) Peg-like and foliform setae.
c) Aciculae (on caudal margin of saddle).
d) Simple and incised (back rows) blades (front and back rows, respectively).
e) Pectunculate and bipectunculate blades (lower and upper arrows, respectively.
f) Denticles.
g) Divided filaments.
h) Teeth and serration (upper and lower arrows, respectively).
i) Setiforms.
j) Spines (of pecten).
k) Spinules.
l) Villi.

Abbreviations
p, puncture
9-S, setae 9-S (at apex of spiracular apparatus of *Mansonia* larva)

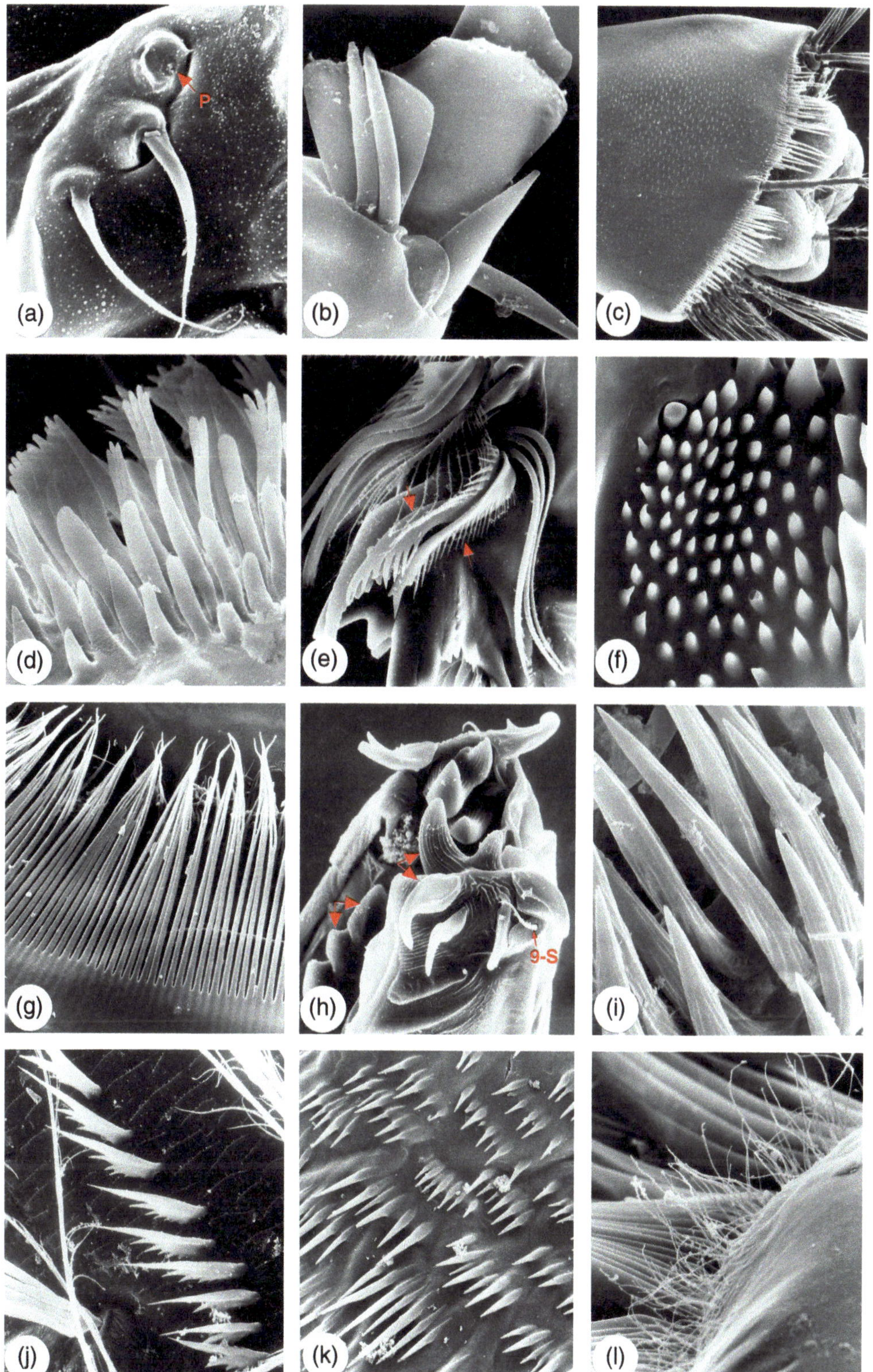
(a)
P
(b)
(c)
(d)
(e)
(f)
(g)
(h)
9-S
(i)
(j)
(k)
(l)

Literature Cited

The bibliography presented here includes all cited references and those attributable to the authorship of scientific names. The references are listed by author and date, alphabetized by the surname name of the single or first author of each publication. Articles by single authors are followed by articles with two authors, and those in turn are followed by articles with three or more authors. Multiple works by the same author or authors are listed in chronological order beginning with the earliest. For multiple works with the same authorship and year of publication, lower case letters are added to the date in alphabetical order, with the sequence of letters from 'a' onward corresponding in most cases with the issue date of the publication. For publications where the surnames of the first and one or more of the following authors are the same, the articles are listed in alphabetical order by the surname of the first co-author that is different than the authorship of the preceding reference, regardless of the number of authors. In cases where this results in year-letter dates that are not consecutive or are out of sequence, the letter-dates are placed in boldface to make it easier to locate such publications in the list. For citations in the text where the full authorship is not necessary, publications are identified by the surname of the first author followed by the Latin phrase *et alia* (abbreviated *et al.*), meaning 'and others'.

The names of authors and the titles of their published articles are replicated here, *i.e.* they are written exactly as they were published, including compound names, diacritical marks, special symbols, improper italicization and the spelling of mosquitoes as 'mosquitos'. The word '*sic*' is only used to indicate obviously misspelled words. Although it is not a misspelling, the symbol v that was used in publications of the 17th and 18th centuries to represent the Latin consonantal u, is replaced with the letter u in applicable references. Also, the symbol æ is replaced with ae. Titles of articles published in capital letters are printed here in lower case letters (except for the first letter of proper names that must be capitalized, *e.g.* genus names and German nouns), and genus and species names that were not italicized in published works are italicized here. The titles of books are italicized, the names of journals are written in full and italicized, and Arabic numerals are used for the volume numbers of journals that were published as Roman numerals. Subdivisions of journal volumes (issues), if applicable and known, are enclosed in parentheses after volume numbers.

Abdulla-Khan, R., Coetzee, M. & Hunt, R.H. (1998) Description of *Anopheles* (*Cellia*) *seretsei* sp. nov. from Kasane, Botswana. *Journal of the American Mosquito Control Association*, 14 (3), 248–252.

Abercrombie, J. (1977) Medical entomology studies – IX. The subgenus *Christophersiomyia* of the genus *Aedes* (Diptera: Culicidae). *Contributions of the American Entomological Institute*, 14 (2), 1–72.

Abraham, O.K. (1947) *Anopheles* (*Anopheles*) *kyondawensis* n. sp., from Lower Burma. *Medical Journal of Malaya*, 1 (3), 173–176.

Adam, J.-P. (1961) *Anopheles caroni* n. sp. (*Diptera, Culicidae*) anophèle cavernicole nouveau de la République du Congo. *Bulletin de la Société de Pathologie exotique*, 54 (4), 714–717.

Adam, J.-P. (1962) Un anophèle cavernicole nouveau de la République du Congo (Brazzaville): *Anopheles* (*Neomyzomyia*) *hamoni* n. sp. (*Diptera-Culicidae*). *Bulletin de la Société de Pathologie exotique*, 55 (1), 153–165.

Adam, J.P. & Hamon, J. (1959) Description de deux nouvelles sous-espèces d'*Eretmapodites œdipodius* Graham, capturées en Afrique Occidentale. *Bulletin de la Société de Pathologie exotique* (for 1958), 51 (4), 524–527.

Adames, A.J. (1971) Mosquito studies (Diptera, Culicidae) XXIV. A revision of the crabhole mosquitoes of the genus *Deinocerites*. *Contributions of the American Entomological Institute*, 7 (2), 1–154.

Adames, A.J. & Galindo, P. (1973) Mosquito studies (Diptera, Culicidae) XXX. A new subgenus and species of *Culex* from Colombia. *Contributions of the American Entomological Institute*, 9 (3), 55–61.

Adames, A.J. & Galindo, P. (1975) Description of the immature stages of *Galindomyia leei* Stone and Barreto, 1969. *Mosquito Systematics*, 7 (2), 132–136.

Adames, A.J. & Hogue, C.L. (1970) Mosquito studies (Diptera, Culicidae) XVII. Two new species of *Deinocerites* from Costa Rica. *Contributions of the American Entomological Institute* (for 1969), 5 (1), 9–20.

Adams, C.F. (1903) Dipterological contributions. *Kansas University Science Bulletin*, 2 (2), 21–47.

Adugna, F., Wale, M. & Nibret, E. (2021) Review of *Anopheles* mosquito species, abundance, and distribution in Ethiopia. *Journal of Tropical Medicine*, 2021, 6726622. DOI: 10.1155/2021/6726622

Ahumada, M.L., Orjuela, L.I., Pareja, P.X., Conde, M., Cabarcas, D.M., Cubillos, E.F.G., Lopez, J.A., Beier, J.C., Herrera, S. & Quiñones, M.L. (2016) Spatial distributions of *Anopheles* species in relation to malaria incidence at 70 localities in the highly endemic Northwest and South Pacific coast regions of Colombia. *Malaria Journal*, 15, 407. DOI: 10.1186/s12936-016-1421-4

Aiken, J. (1911) A synoptical view of the mosquitoes of British Guiana. *Timehri*, 1 (2), 187–204.

Aitken, T.H.G. (1939) The *Anopheles maculipennis* complex of western America (Diptera, Culicidae). *Pan-Pacific Entomologist*, 15 (4), 191–192.

Aitken, T.H.G. (1941) A new American subgenus and species of *Aëdes* (Diptera, Culicidae). *Pan-Pacific Entomologist*, 17 (2), 81–84.

Aitken, T.H.G. (1960) A survey of Trinidadian arthropods for natural viral infections (August, 1953 to December, 1958). *Mosquito News*, 20 (1), 1–10.

Alcock, A. (1912) Description of a new species of *Anopheles* from the Malay Peninsula. *Journal of the London School of Tropical Medicine*, 2 (1), 1–2.

Alcock, A. (1913) Synopsis of the anopheline mosquitos of Africa and of the Oriental Region. *Journal of the London School of Tropical Medicine*, 2 (3), 153–166.

Alekseev, E.V. (1989) Bloodsucking mosquito *Aedes* (*Rusticoidus*) *krymmontanus* Alekseev, sp. n., a relic of the entomofauna of the Crimea [in Russian]. *Parazitologiia*, 23 (2), 173–178.

Alencar, J., Ferreira de Mello, C., Marcondes, C.B., Guimarães, A.É., Toma, H.K., Queiroz Bastos, A., Olsson Freitas Silva, S. & Lisboa Machado, S. (2021) Natural infection and vertical transmission of Zika virus in sylvatic mosquitoes *Aedes Albopictus* [*sic*] and *Haemagogus Leucocelaenus* [*sic*] from Rio de Janeiro, Brazil. *Tropical Medicine and Infectious Disease*, 6 (2), 99. DOI: 10.3390/tropicalmed6020099

Amerasinghe, F.P. (1993) A new species of *Heizmannia* from Sri Lanka (Diptera: Culicidae). *Mosquito Systematics*, 25 (2), 110–115.

Amos, D.W. (1944) *Mosquito Control Training Manual*. Suva, Fiji.

Andrade, D.C. & la Corte, R. (2021) *Toxorhynchites* (*Lynchiella*) *caatingensis* sp. nov. (Diptera: Culicidae) from the semi-arid region of Brazil. *Zootaxa*, 5072 (2), 130–144. DOI: 10.11646/zootaxa.5072.2.2

Anduze, P.J. (1941) Primer informe sobre entomologia medica del Estado Bolivar (Venezuela). Serie 1. La fauna culicidiana. Descripción de tres especies nuevas (*Diptera, Culicidae*). *Revista de Sanidad y Asistencia Social*, 6 (6), 812–836.

Anduze, P.J. (1942) II. La fauna culicidiana de Venezuela. Descripcion de una especie nueva (Diptera: Culicidae). *Revista de Sanidad y Asistencia Social*, 7 (4), 557–560.

Anduze, P.J. (1949a) Fauna culicina de Venezuela. Descripcion de cinco nuevas especies. *Boletín de Entomologia Venezolana* (for 1948), 7 (1–2), 60–66.

Anduze, P.J. (1949b) Fauna culicina de Venezuela. Descripcion de una especie nueva. *Boletín de Entomologia Venezolana* (for 1948), 7 (3–4), 149–150.

Anduze, P.J. & Capdevielle, A. (1949) *Anopheles* (*Anopheles*) *guarao* sp. nov. *Boletín de Entomologia Venezolana*, 8 (3–4), 119–124.

Anduze, P.J. & Hecht, O. (1943) *Aedes* (*Finlaya*) *upatensis* sp. nov. *Boletín de Entomologia Venezolana*, 2 (4), 185–188.

Angulo, A.O. & Olivares, T.S. (1993) Catalogo de los culicidos de Chile (Diptera Culicidae), y dos especies nuevas de *Culex* (*Culex*) Linnaeus. *Boletin Sociedad Biologica Concepción, Chile*, 64, 21–26.

Antunes, P.C.A. (1937) A new *Anopheles* and a new *Goeldia* from Colombia (Dipt. Culic.). *Bulletin of Entomological Research*, 28 (1), 69–73.

Antunes, P.C.A. & Lane, J. (1937) Um novo *Culex*, *Culex* (*Carrollia*) *soperi*, encontrado em São Paulo (*Diptera, Culicidae*). *Revista de Biologia e Hygiene*, 8 (1), 21–23.

Antunes, P.C.A. & Lane, J. (1938) Um novo *Aedes*, *Aedes* (*Ochlerotatus*) *pennai*, encontrado em São Paulo. *Revista do Museu Paulista*, 23, 605–614, 3 pls.

Arnell, J.H. (1973) Mosquito studies (Diptera, Culicidae) XXXII. A revision of the genus *Haemagogus*. *Contributions of the American Entomological Institute*, 10 (2), ii + 1–174.

Arnell, J.H. (1976) Mosquito studies (Diptera, Culicidae) XXXIII. A revision of the Scapularis Group of *Aedes* (*Ochlerotatus*). *Contributions of the American Entomological Institute*, 13 (3), 1–144.

Arnell, J.H. & Nielsen, L.T. (1972) Mosquito studies (Diptera, Culicidae) XXVII. The Varipalpus Group of *Aedes* (*Ochlerotatus*). *Contributions of the American Entomological Institute*, 8 (2), 1–48.

Atrie, B., Subbarao, S.K., Pillai, M.K.K., Rao, S.R.V. & Sharma, V.P. (1999) Population cytogenetic evidence for sibling species in *Anopheles annularis* (Diptera: Culicidae). *Annals of the Entomological Society of America*, 92 (2), 243–249. DOI: 10.1093/aesa/92.2.243

Awono-Ambene, H.P., Kengne, P., Simard, F., Antonio-Nkondjio, C. & Fontenille, D. (2004) Description and bionomics of *Anopheles* (*Cellia*) *ovengensis* (Diptera: Culicidae), a new malaria vector species of the *Anopheles nili* group from South Cameroon. *Journal of Medical Entomology*, 41 (4), 561–568. DOI: 10.1603/0022-2585-41.4.561

Azari-Hamidian, S. & Harbach, R.E. (2009) Keys to the adult females and fourth-instar larvae of the mosquitoes of Iran (Diptera: Culicidae). *Zootaxa*, 2078, 1–33. DOI: 10.11646/zootaxa.2078.1.1

Azari-Hamidian, S., Abai, M.-R. & Norouzi, B. (2020) *Mansonia uniformis* (Diptera: Culicidae), a genus and species new to southwestern Asia, with a review of its medical and veterinary importance. *Zootaxa*, 4772 (2), 385–395. DOI: 10.11646/zootaxa.4772.2.10

Azari-Hamidian, S., Norouzi, B. & Noorallahi, A. (2017) *Orthopodomyia pulcripalpis* (Diptera: Culicidae), a genus and species new to the Iranian mosquito fauna, with a review of bionomical information. *Zootaxa*, 4299 (1), 141–145. DOI: 10.11646/zootaxa.4299.1.11

Bachmann, A.O. & Casal, O.H. (1962) *Culex* (*Culex*) *lahillei* una nueva especie de la Argentina (Diptera: Culicidae). *Physis*, 23 (65), 267–270.

Bailly-Choumara, H. (1960) Une espèce nouvelle d'anophèle du golfe d'Aden, *Anopheles* (*Myzomyia*) *azaniae* (Diptera, Culicidae), notes morphologiques, systématiques et écologiques. *Bulletin de la Société de Pathologie exotique*, 53 (3), 531–542.

Bailly-Choumara, H. (1966a) Contribution a l'étude des *Culex* (*Neoculex*) (*Diptera: Culicidae*) de la région Éthiopienne. Description de *Culex* (*Neoculex*) *pseudoandreanus* sp. n. *Bulletin de la Société de Pathologie exotique* (for 1965), 58 (4), 660–664.

Bailly-Choumara, H. (1966b) *Culex (Culex) gameti (Diptera, Culicidae)*, moustique nouveau du Cameroun, description du male. *Bulletin de la Société de Pathologie exotique*, 59 (1), 148–150.

Bailly-Choumara, H. & Adam, J.-P. (1959) *Anopheles maliensis* n. sp. (Diptera Culicidae), une espèce nouvelle d'anophélinés du Fouta-Djalon (République de Guinée). *Compte Rendu de l'Académie des Sciences*, 248 (26), 3742–3743.

Bailly-Choumara, H. & Rickenbach, A. (1966) Contribution a l'étude des *Culex* (*Diptera, Culicidae*) de la région Éthiopienne. Description de *Culex* (*Neoculex*) *garioui* sp. n., moustique nouveau du Cameroun. *Bulletin de la Société de Pathologie exotique*, 59 (1), 144–148.

Baisas, F.E. (1932) The morphology of some Philippine *Anopheles*. *Transactions of the Eighth Congress of the Far East Association of Tropical Medicine*, 1930, 252–257.

Baisas, F.E. (1935a) Notes on Philippine mosquitoes, V. The pupal characters of anophelines under the Myzorhynchus Series and group Neocellia, with further comments on the larvae and adults of *sinensis*. *Monthly Bulletin of the Bureau of Health (Manila)*, 15 (9), 291–339.

Baisas, F.E. (1935b) Notes on Philippine mosquitoes, I. The *Armigeres* group. *Philippine Journal of Science*, 56 (4), 485–497, 4 pls.

Baisas, F.E. (1935c) Notes on Philippine mosquitoes, II. *Uranotaenia* group. *Philippine Journal of Science*, 57 (1), 63–80, 4 pls.

Baisas, F.E. (1935d) Notes on Philippine mosquitoes, III. Genus *Culex*: groups *Lophoceratomyia*, *Mochthogenes*, and *Neoculex*. *Philippine Journal of Science*, 57 (2), 167–179, 4 pls.

Baisas, F.E. (1936) Notes on Philippine mosquitoes, IV. The pupal and certain adult characters of some rare species of *Anopheles*. *Philippine Journal of Science*, 59 (1), 65–84, 15 pls.

Baisas, F.E. (1938) Notes on Philippine mosquitoes, VII. A–*Culex* (*Culex*) with banded proboscis and tarsi. B–*Anopheles*: the pupae of three rare species; the *leucosphyrus*-subgroup. *Monthly Bulletin of the Bureau of Health (Manila)*, 18 (5), 175–232, 18 pls.

Baisas, F.E. (1946a) Notes on Philippine mosquitoes, VIII. Species found in the jungles of Llavac. *Monthly Bulletin of the Bureau of Health (Manila)*, 22 (1), 27–49.

Baisas, F.E. (1946b) Notes on Philippine mosquitoes, IX. A new species of *Zeugnomyia*. *Monthly Bulletin of the Bureau of Health (Manila)*, 22 (2), 27–35.

Baisas, F.E. (1946c) Notes on Philippine mosquitoes, XII. *Topomyia*. *Monthly Bulletin of the Bureau of Health (Manila)*, 22 (4), 31–47.

Baisas, F.E. (1947) Notes on Philippine mosquitoes, XI. A new species of *Tripteroides*. *Fieldiana, Zoology*, 31 (15), 121–124.

Baisas, F.E. & Feliciano, P. (1953) Notes on Philippine mosquitoes, XIII. Four new species of *Zeugnomyia* and *Topomyia*. *Fieldiana, Zoology*, 33 (3), 161–179.

Baisas, F.E. & Hu, S.M.K. (1936) *Anopheles hyrcanus* var. *sinensis* of the Philippines and certain parts of China, with some comments on *Anopheles hyrcanus* var. *nigerrimus* of the Philippines. *Monthly Bulletin of the Bureau of Health (Manila)*, 16 (6), 205–242, 6 pls.

Baisas, F.E. & Ubaldo-Pagayon, A. (1953) Notes on Philippine mosquitoes, XVI. Genus *Tripteroides*. Monograph, Institute of Science and Technology, Manila, Philippines.

Baisas, F.E., Bañez, L.F.L. & Leuenberger, N. (1962) Notes on Philippine mosquitoes, XXII. The axil-breeding species. *Philippine Journal of Science* (for 1960), 89 (2), 183–199.

Baker, F.C. (1936) A new species of *Orthopodomyia*, *O. alba* sp. n. (Diptera, Culicidae). *Proceedings of the Entomological Society of Washington*, 38 (1), 1–7.

Bañez, L.F.L. & Jueco, N.L. (1966) Adult and larva of *Aedes* (*Finlaya*) *sorsogonensis* Bañez, 1963. Comparison with *Aedes* (*Finlaya*) *ananae* Knight and Laffoon, 1946 and *Aedes* (*Finlaya*) *poecilus* [*sic*] (Theobald), 1903. *Acta Medica Philippina*, 2 (4), 194–200.

Bangs, M.J., Taai, K., Howard, T.M., Cook, S. & Harbach, R.E. (2015) The mosquito *Anopheles* (*Cellia*) *oreios* sp. n., formerly species 6 of the Australasian *Anopheles farauti* complex, and a critical review of its biology and relation to disease. *Medical and Veterinary Entomology*, 29 (1), 68–81. DOI: 10.1111/mve.12092

Banks, C.S. (1906) A list of Philippine Culicidae with descriptions of some new species. *Philippine Journal of Science*, 1 (9), 977–1005.

Banks, C.S. (1909) Four new Culicidae from the Philippines. *Philippine Journal of Science*, 4 (6), 545–551.

Barata, J.M.S. & Cotrim, M.D. (1971) Uma nova espécie de *Psorophora* (Diptera, Culicidae). *Revista Brasileira de Entomologia*, 15 (6), 35–40.

Barbosa, A.A., da Silva, M.A.N. & Sallum, M.A.M. (2007) *Mansonia* (*Mansonia*) *iguassuensis* sp. nov. (Diptera: Culicidae) from Brasil. *Zootaxa*, 1527, 45–52. DOI: 10.11646/zootaxa.1527.1.5

Barr, A.R. (1957) A new species of *Culiseta* (Diptera: Culicidae) from North America. *Proceedings of the Entomological Society of Washington*, 59 (4), 163–167.

Barr, A.R. & Barr, S. (1969) Mosquito studies (Diptera, Culicidae) XIII. Pupae of the genus *Psorophora* in North America and Puerto Rico. *Contributions of the American Entomological Institute*, 4 (4), 1–20.

Barr, A.R. & Guptavanij, P. (1989) *Anopheles hermsi* N. SP., an unrecognized American species of the *Anopheles maculipennis* group (Diptera: Culicidae). *Mosquito Systematics* (for 1988), 20 (3), 352–356.

Barraud, P.J. (1923a) A revision of the culicine mosquitoes of India. Part I. The genera *Stegomyia*, Theo. and *Christophersiomyia* gen. n. adult stage. *Indian Journal of Medical Research*, 10 (3), 772–788, 17 pls.

Barraud, P.J. (1923b) A revision of the culicine mosquitoes of India. Part III. Notes on certain Indian species of the genus *Finlaya*, Theo. and descriptions of new species. *Indian Journal of Medical Research*, 11 (1), 214–219.

Barraud, P.J. (1923c) A revision of the culicine mosquitoes of India. Part V. Further notes on the genera *Stegomyia*, Theo. and *Finlaya*, Theo. with descriptions of new species. *Indian Journal of Medical Research*, 11 (1), 224–228, 3 pls.

Barraud, P.J. (1923d) Two new species of *Culex* (Diptera, Culicidae) from Assam. *Indian Journal of Medical Research*, 11 (2), 507–509.

Barraud, P.J. (1923e) Some new culicine mosquitos found in India, and a note on *Finlaya assamensis* (Theo.). *Bulletin of Entomological Research*, 13 (4), 405–408.

Barraud, P.J. (1923f) A revision of the culicine mosquitoes of India. Part VII. The larvae of some species of *Stegomyia* (Theo.). The larvae of some species of *Finlaya* (Theo.). The larva of *Christophersiomyia thomsoni* (Theo.). The larva of *Mimomyia chamberlaini* (Ludl.). The larva of *Aedomyia catasticta* (Knab.). *Indian Journal of Medical Research*, 11 (2), 495–505, 7 pls.

Barraud, P.J. (1924a) A revision of the culicine mosquitoes of India. Part VIII. Further descriptions of Indian species of *Finlaya* Theo. *Indian Journal of Medical Research*, 11 (3), 845–865, 10 pls.

Barraud, P.J. (1924b) A new mosquito from Kashmir. *Indian Journal of Medical Research*, 11 (3), 967–968.

Barraud, P.J. (1924c) A revision of the culicine mosquitoes of India. Part XI. Some Indian species of *Culex* L. *Indian Journal of Medical Research*, 11 (4), 979–998, 4 pls.

Barraud, P.J. (1924d) Four new mosquitoes from the western Himalayas. *Indian Journal of Medical Research*, 11 (4), 999–1006.

Barraud, P.J. (1924e) A revision of the culicine mosquitoes of India. Part XII. Further descriptions of Indian species of *Culex* L. including two new species. *Indian Journal of Medical Research*, 11 (4), 1259–1274, 3 pls.

Barraud, P.J. (1924f) A revision of the culicine mosquitoes of India. Part XIII. Further descriptions of Indian species of *Culex* L. including three new species. *Indian Journal of Medical Research*, 11 (4), 1275–1282, 2 pls.

Barraud, P.J. (1924g) A revision of the culicine mosquitoes of India. Part XIV. The Indian species of the subgenus *Culiciomyia* (Theo.) Edw., including one new species. *Indian Journal of Medical Research*, 12 (1), 15–22, 1 pl.

Barraud, P.J. (1924h) A revision of the culicine mosquitoes of India. XV. The Indian species of the subgenus *Lophoceratomyia* (Theo.) Edw., including two new species. *Indian Journal of Medical Research*, 12 (1), 39–46, 2 pls.

Barraud, P.J. (1924i) A new mosquito from Kashmir and the North-west Frontier Province. *Indian Journal of Medical Research*, 12 (1), 73–74.

Barraud, P.J. (1926) A revision of the culicine mosquitoes of India. Part XVIII. The Indian species of *Uranotaenia* and *Harpagomyia*, with descriptions of five new species. *Indian Journal of Medical Research*, 14 (2), 331–350, 1 pl.

Barraud, P.J. (1927a) A revision of the culicine mosquitoes of India. Part XIX. The Indian species of *Aëdomyia* [*sic*] and *Orthopodomyia* with descriptions of two new species. *Indian Journal of Medical Research*, 14 (3), 523–532, 1 pl.

Barraud, P.J. (1927b) A revision of the culicine mosquitoes of India. XX. The Indian species of *Armigeres* (including *Leicesteria*), with descriptions of two new species. *Indian Journal of Medical Research*, 14 (3), 533–548, 2 pls.

Barraud, P.J. (1927c) A revision of the culicine mosquitoes of India. Part XXI. Descriptions of new species of *Aedimorphus* and *Finlaya*, and notes on *Stegomyia albolineata* (Theo.). *Indian Journal of Medical Research*, 14 (3), 549–554.

Barraud, P.J. (1927d) A revision of the culicine mosquitoes of India. Part XXII. The Indian species of the genus *Taeniorhynchus* (including *Mansonioides*), with a description of one new species. *Indian Journal of Medical Research*, 14 (3), 555–563, 2 pls.

Barraud, P.J. (1928) A revision of the culicine mosquitoes of India. Part XXIV. The Indian species of the subgenera *Skusea* and *Aedes*, with descriptions of eight new species, and remarks on a new method for identifying the females of the subgenus *Aedes*. *Indian Journal of Medical Research*, 16 (2), 357–375, 8 pls.

Barraud, P.J. (1929a) A revision of the culicine mosquitoes of India. Part XXV. The genera *Mucidus*, *Mimomyia*, *Ficalbia*, *Rachionotomyia*, and *Hodgesia*. *Indian Journal of Medical Research*, 16 (4), 1052–1063, 1 pl.

Barraud, P.J. (1929b) A revision of the culicine mosquitoes of India. Part XXVI. The genera *Heizmannia*, *Haemagogus*, *Topomyia*, and *Megarhinus*. *Indian Journal of Medical Research*, 17 (1), 259–280, 2 pls.

Barraud, P.J. (1931a) Notes on some Indian mosquitoes of the subgenus *Stegomyia*, with descriptions of new species. *Indian Journal of Medical Research*, 19 (1), 221–227, 1 pl.

Barraud, P.J. (1931b) Descriptions of eight new species of Indian culicine mosquitoes. *Indian Journal of Medical Research*, 19 (2), 609–616, 1 pl.

Barraud, P.J. (1932) *Anopheles aitkeni* James var. n. *pinjaurensis*. *Records of the Malaria Survey of India*, 3 (2), 353–355.

Barraud, P.J. (1934) *The fauna of British India, including Ceylon and Burma. Diptera. Volume V. Family Culicidae. Tribes Megarhinini and Culicini.* Taylor & Francis, London, UK.

Barraud, P.J. & Covell, G. (1928) The morphology of the buccal cavity in anopheline and culicine mosquitoes. *Indian Journal of Medical Research*, 15, 671–679, 8 pls.

Barreto, M.P. & Coutinho, J.O. (1944) Sôbre o gênero *Taeniorhynchus* Arribálzaga, 1891, com a descrição de três novas espécies do subgênero *Taeniorhynchus* (*Diptera, Culicidae*). *Arquivos de Higiene e Saúde Pública*, 9 (21), 53–85.

Basham, E.H. (1948) *Culex* (*Melanoconion*) *mulrennani*, a new species from Florida (Diptera: Culicidae). *Annals of the Entomological Society of America*, 41 (1), 1–7.

Basio, R.G. (1971) On Philippine mosquitoes III. Some new species from Mt. Makiling and its vicinity in Luzon (Diptera: Culicidae). *Philippine Entomologist*, 2 (1), 51–57.

Basio, R.G. & Reisen, W.K. (1971) On some mosquitoes of Guam, Marianas Islands (Diptera: Culicidae). *Philippine Entomologist*, 2 (1), 57–61.

Bates, M. (1949) *The Natural History of Mosquitoes*. The Macmillan Company, New York, NY, USA.

Becker, N., Petrić, D., Zgomba, M., Boase, C., Dahl, C., Lane, J. & Kaiser, A. (2003) *Mosquitoes and their Control*. Kluwer Academic/Plenum Publishers, New York, NY, USA. DOI: 10.1007/978-1-4757-5897-9

Becker, N., Petrić, D., Zgomba, M., Boase, C., Madon, M., Dahl, C. & Kaiser, A. (2010) *Mosquitoes and their Control*. 2nd edn. Springer-Verlag, Berlin, Germany. DOI: 0.1007/978-3-540-92874-4

Becker, N., Petrić, D., Zgomba, M., Boase, C., Madon, M., Dahl, C. & Kaiser, A. (2020) *Mosquitoes: Identification, Ecology and Control*. 3rd edn. Springer Nature, Cham, Switzerland. DOI: 10.1007/978-3-030-11623-1

Becker, T. (1903) Ägyptische Dipteren. *Mitteilungen aus dem zoologischen Museum in Berlin*, 2 (3), 67–195, 4 pls.

Beebe, N.W., Russell, T., Burkot, T.R. & Cooper, R.D. (2015) *Anopheles punctulatus* group: evolution, distribution, and control. *Annual Review of Entomology*, 60, 335–350. DOI: 10.1146/annurev-ento-010814-021206

Belkin, J.N. (1945) *Anopheles nataliae*, a new species from Guadalcanal. *Journal of Parasitology*, 31 (5), 315–318.

Belkin, J.N. (1950a) Mosquitoes of the genus *Tripteroides* in the Solomon Islands. *Proceedings of the United States National Museum*, 100 (3262), 201–274.

Belkin, J.N. (1950b) A revised nomenclature for the chaetotaxy of the mosquito larva (Diptera: Culicidae). *American Midland Naturalist*, 44 (3), 678–698.

Belkin, J.N. (1952) The homology of the chaetotaxy of immature mosquitoes and a revised nomenclature for the chaetotaxy of the pupa (Diptera, Culicidae). *Proceedings of the Entomological Society of Washington*, 54 (3), 115–130.

Belkin, J.N. (1953a) Mosquitoes of the genus *Uranotaenia* in the Solomon Islands (Diptera: Culicidae). *Pacific Science*, 7 (3), 312–391.

Belkin, J.N. (1953b) Corrected interpretations of some elements of the abdominal chaetotaxy of the mosquito larva and pupa (Diptera, Culicidae). *Proceedings of the Entomological Society of Washington*, 55 (6), 318–324.

Belkin, J.N. (1954a) The dorsal hairless setal ring of mosquito pupae (Diptera: Culicidae). *Pan-Pacific Entomologist*, 30 (3), 227–230.

Belkin, J.N. (1954b) Remarks on the larval chaetotaxy of the subgenus *Melanoconion* (Diptera: Culicidae). *Entomological News*, 45 (2), 29–32.

Belkin, J.N. (1955) The *Tripteroides caledonica* complex of mosquitoes in Melanesia (Diptera: Culicidae). *Pacific Science*, 9 (2), 221–246.

Belkin, J.N. (1960) Innervation as a criterion of homology of the elements of the larval and pupal chaetotaxy of mosquitoes (Diptera: Culicidae). *Proceedings of the Entomological Society of Washington*, 62 (3), 197.

Belkin, J.N. (1962) *The Mosquitoes of the South Pacific (Diptera, Culicidae)*. 2 vols. University of California Press, Berkeley, CA, USA.

Belkin, J.N. (1968a) Mosquito studies (Diptera, Culicidae) VII. The Culicidae of New Zealand. *Contributions of the American Entomological Institute*, 3 (1), 1–182.

Belkin, J.N. (1968b) Mosquito studies (Diptera, Culicidae). IX. The type specimens of New World mosquitoes in European museums. *Contributions of the American Entomological Institute*, 3 (4), 1–69.

Belkin, J.N. (1977) *Toxorhynchites* (*Lyn.*) *gerbergi*, a new species from the southern Lesser Antilles. *Mosquito Systematics*, 9 (3), 329–332.

Belkin, J.N. & Heinemann, S.J. (1975) *Psorophora (Janthinosoma) mathesoni*, sp. nov. for "*varipes*" of the southeastern U.S.A. *Mosquito Systematics*, 7 (4), 363–366.

Belkin, J.N. & Heinemann, S.J. (1976) Collection records of the project "Mosquitoes of Middle America" 6. Southern Lesser Antilles: Barbados (BAR), Dominica (DOM), Grenada (GR, GRR), St. Lucia (LU), St. Vincent (VT). *Mosquito Systematics*, 8 (3), 237–297.

Belkin, J.N. & Hogue, C.L. (1959) A review of the crabhole mosquitoes of the genus *Deinocerites* (Diptera, Culicidae). *University of California Publications in Entomology*, 14 (6), 411–458.

Belkin, J.N. & McDonald, W.A. (1957) A new species of *Aedes* (*Ochlerotatus*) from tree holes in southern Arizona and a discussion of the *varipalpus* complex (Diptera: Culicidae). *Annals of the Entomological Society of America*, 50 (2), 179–191.

Belkin, J.N. & Schlosser, R.J. (1944) A new species of *Anopheles* from the Solomon Islands. *Journal of the Washington Academy of Sciences*, 34 (8), 268–273.

Belkin, J.N., Heinemann, S.J. & Page, W.A. (1970) The Culicidae of Jamaica (Mosquito Studies. XXI). *Contributions of the American Entomological Institute*, 6 (1), 1–458.

Belkin, J.N., Knight, K.L. & Rozeboom, L.E. (1945) Anopheline mosquitoes of the Solomon Islands and New Hebrides. *Journal of Parasitology*, 31 (4), 241–265.

Belkin, J.N., Schick, R.X. & Heinemann, S.J. (1971) Mosquito studies (Diptera, Culicidae) XXV. Mosquitoes originally described from Brazil. *Contributions of the American Entomological Institute*, 7 (5), 1–64.

Bellardi, L. (1859) *Saggio di ditterologia messicana. Parte I.* Torino, Italy.

Bellardi, L. (1862) *Saggio di ditterologia messicana. Appendice.* Torino, Italy.

Benoit, P.L.G. (1964) Diptera – Culicidae. Observations biologiques et écologiques, pp. 297–320. Mission de Zoologie médicale au Maniema (Congo, Léopoldville) (P.L.G. Benoit, 1959): résultats scientifiques. Musee Royal de l'Afrique centrale — Tervuren, Belgique. *Annales du Musée Royal de l'Afrique Centrale (Serie in 8vo Sciences Zoologiques)*, 132, 69–71.

Bergroth, E. (1889) Zwei neue Diptera. *Wiener entomologische Zeitung*, 8 (9), 295–298.

Berlin, O.G.W. (1969a) Mosquito studies (Diptera, Culicidae) XII. A revision of the Neotropical subgenus *Howardina* of *Aedes*. *Contributions of the American Entomological Institute*, 4 (2), 1–190.

Berlin, O.G.W. (1969b) Mosquito studies (Diptera, Culicidae) XVIII. The subgenus *Micraedes* of *Culex*. *Contributions of the American Entomological Institute*, 5 (1), 21–63.

Berlin, O.G.W. (1974) A new species of bromeliad breeding *Culex* (*Micraedes*) from Mexico (Diptera, Culicidae). *Mosquito Systematics*, 6 (4), 273–278.

Berlin, O.G.W. & Belkin, J.N. (1980) Mosquito studies (Diptera, Culicidae) XXXVI. Subgenera *Aedinus*, *Tinolestes* and *Anoedioporpa* of *Culex*. *Contributions of the American Entomological Institute*, 17 (2), ii, 1–104.

Berlov, O.E. & Kuberskaya, O.V. (2023) Additions and corrections to the Catalog of blood-sucking mosquitoes of the world (Insecta: Diptera, Culicidae) by Wilkerson *et al.* (2021) [in Russian]. *Amurian Zoological Journal*, 15 (1), 110–118.

Besansky, N.J. & Fahey, G.T. (1997) Utility of the *white* gene in estimating phylogenetic relationships among mosquitoes (Diptera: Culicidae). *Molecular Biology and Evolution*, 14 (4), 442–454. DOI: 10.1093/oxfordjournals.molbev.a025780

Besansky, N.J., Powell, J.R., Caccone, A., Hamm, D.M., Scott, J.A. & Collins, F.H. (1994) Molecular phylogeny of the *Anopheles gambiae* complex suggests genetic introgression between principal malaria vectors. *Proceedings of the National Academy of Sciences*, 91 (15), 6885–6888. DOI: 10.1073/pnas.91.15.6885

Bhat, H.R. (1975) A survey of haematophagous arthropods in Western Himalayas, Sikkim and Hill Districts of West Bengal: records of mosquitoes collected from Himalayan region of West Bengal and Sikkim with ecological notes. *Indian Journal of Medical Research*, 63 (2), 232–241.

Bhattacharyya, D.R., Prakash, A., Mohapatra, P.K. & Mahanta, J. (**2004a**) *Uranotaenia dibrugarhensis*, a new species in subgenus *Pseudoficalbia* from Assam, India. *Journal of the American Mosquito Control Association*, 20 (1), 1–5.

Bhattacharyya, D.R., Prakash, A., Mohapatra, P.K. & Sarma, D.K. (2009) *Armigeres* (*Armigeres*) *mahantai*, a new mosquito species from India. *Journal of the American Mosquito Control Association*, 25 (1), 1–5. DOI: 10.2987/08-5785.1

Bhattacharyya, D.R., Rajavel, A.R., Natarajan, R., Mohapatra, P.K., Jambulingam, P., Mahanta, J. & Prakash, A. (2014) Faunal richness and the checklist of Indian mosquitos (Diptera: Culicidae). *Check List*, 10 (6), 1342–1358. DOI: 10.15560/10.6.1342

Bhattacharyya, D.R., Tewari, S.C., Prakash, A., Mohapatra, P.K. & Mahanta, J. (**2004b**) *Verrallina* (*Neomacleaya*) *assamensis*, a new species from Assam, Indian. *Journal of the American Mosquito Control Association*, 20 (2), 115–120.

Bick, G.H. (1951) The ecology of the mosquito larvae of New Guinea. *Pacific Science*, 5 (4), 392–431.

Bickley, W.E. (1984) The *Psorophora confinnis* complex. *Mosquito News*, 36 (2), 376.

Bigot, J. (1861) Trois diptères nouveaux de la Corse. *Annales de la Société Entomologique de France*, 1, 227–229.

Blanchard, E. (1852) Orden IX. Dipteros, pp. 327–468. In: Gay, C. (ed.), *Historia fisica y politica de Chile. Zoologia*. Volume 7. Paris, France.

Blanchard, R. (1901) Observations sur quelques moustiques. *Comptes Rendus Hebdomadaires des Séances et Mémoires de la Société de Biologie*, 53, 1045–1046.

Blanchard, R. (1902) Nouvelle note sur les moustiques. *Comptes Rendus Hebdomadaires des Séances et Mémoires de la Société de Biologie*, 54, 793–795.

Blanchard, R. (1905) *Les moustiques, histoire naturelle et médicale*. F.R. de Rudeval, Imprimeur-Éditeur, Paris, France.

Bohart, R.M. (1946) New species of mosquitoes from the Marianas and Okinawa (Diptera, Culicidae). *Proceedings of the Biological Society of Washington*, 59, 39–46.

Bohart, R.M. (1949) The subgenus *Neoculex* in America north of Mexico (Diptera, Culicidae). *Annals of the Entomological Society of America* (for 1948), 41 (3), 330–345. DOI: 10.1093/aesa/41.3.330

Bohart, R.M. (1953) A new species of *Culex* and notes on other species of mosquitoes from Okinawa (Diptera, Culicidae). *Proceedings of the Entomological Society of Washington*, 55 (4), 183–188.

Bohart, R.M. (1957) Insects of Micronesia, Diptera: Culicidae. *Insects of Micronesia* (for 1956), 12 (1), 1–85.

Bohart, R.M. & Farner, D.S. (1944) New culicine mosquitoes from the Philippine Islands (Diptera, Culicidae). *Proceedings of the Biological Society of Washington*, 57, 69–74.

Bohart, R.M. & Ingram, R.L. (1946a) Four new species of mosquitoes from Okinawa (Diptera: Culicidae). *Journal of the Washington Academy of Sciences*, 36 (2), 46–52.

Bohart, R.M. & Ingram, R.L. (1946b) *Mosquitoes of Okinawa and islands in the central Pacific*. Navmed 1055. Bureau of Medicine and Surgery, Navy Department, Washington, DC, USA.

Bonne, C. & Bonne-Wepster, J. (1925) Mosquitoes of Surinam: a study on neotropical mosquitoes. Mededeeling/Koloniaal Instituut te Amsterdam; no. 21. Afdeeling Tropische Hygiene no. 13. Het Instituut Druk de Bussy, Amsterdam, Netherlands.

Bonne-Wepster, J. (1930) The genus *Taeniorhynchus* Arribalzaga in the Dutch East Indies. Part II. Subgenus *Coquillettidia* Dyar. *Mededeelingen van den Dienst Volksgezondheid in Nederlandsch-Indië*, 19 (3), 387–399.

Bonne-Wepster, J. (1934) New mosquitos (Dipt.) from the Netherlands Indies. *Stylops*, 3 (12), 272–276.

Bonne-Wepster, J. (1937) Notes on mosquitoes from the Netherlands Indies. A new *Culex* and two new *Finlaya*'s from New-Guinea. *Mededeelingen van den Dienst Volksgezondheid in Nederlandsch-Indië*, 26 (1–2), 92–98.

Bonne-Wepster, J. (1938) Notes on mosquitoes from the Netherlands Indies. A new *Culex* from New Guinea: *Culex* (*Culex*) *solitarius* nov. sp. Description of male terminalia of *Culex* (*C*) *miraculosus* B.-W. *Mededeelingen van den Dienst Volksgezondheid in Ned.-Indië*, 27 (3), 394–396.

Bonne-Wepster, J. (1940) Notes on mosquitoes from the Netherlands Indies: a new *Finlaya* from New-Guinea. *Mededeelingen van den Dienst der Volksgezondheid in Nederlandsch-Indië*, 29 (3–4), 158–159, 2 pls.

Bonne-Wepster, J. (1948) Results of the third Archbold Expedition 1938–1939. *Diptera Culicidae*. Notes on the mosquitoes collected by the Neth. Indian — American expedition to Central and North New Guinea. *Treubia*, 19 (2), 305–322.

Bonne-Wepster, J. (1951) Notes on Oriental and Australasian mosquitoes of the Malay Archipelago. *Documenta Neerlandica et Indonesica de Morbis Tropicis*, 3 (1), 67–74.

Bonne-Wepster, J. & Bonne, C. (1919) Four new South American mosquitoes (*Diptera*, *Culicidae*). *Insecutor Inscitiae Menstruus*, 7 (7–9), 105–113.

Bonne-Wepster, J. & Bonne, C. (1920) Diagnoses of new mosquitoes from Surinam, with a note on synonymy *(Diptera, Culicidae)*. *Insecutor Inscitiae Menstruus* (for 1919), 7 (10–12), 165–180.

Bonne-Wepster, J. & Bonne, C. (1921) Notes on South American mosquitoes in the British Museum (*Diptera*, *Culicidae*). *Insecutor Inscitiae Menstruus*, 9 (1–3), 1–26.

Bonne-Wepster, J. & Brug, S.L. (1932) The subgenus *Stegomyia* in Netherland India. *Geneeskundig Tijdschrift voor Nederlandsch-Indië*, 72, 39–119.

Boreham, M.M. (1970) *Mansonia leberi*, a new mosquito from the Panama Canal Zone, with notes on its biology (Diptera: Culicidae). *Journal of Medical Entomology*, 7 (3), 383–390. DOI: 10.1093/jmedent/7.3.383

Boreham, P.F.L., Chandler, J.A. & Highton, R.B. (1975) Studies on the feeding patterns of mosquitoes of the genera *Ficalbia*, *Mimomyia* and *Uranotaenia* in the Kisumu area of Kenya. *Bulletin of Entomological Research*, 65 (1), 69–74. DOI: 10.1017/S0007485300005770

Borel, E. (1928) Les moustiques de la Cochinchine et du Sud-Annam (*Suite*). *Archives des Instituts Pasteur d'Indochine*, 7, 75–106, 8 pls.

Borel, E. (1928b) Les moustiques de la Cochinchine et du Sud-Annam. *Archives des Instituts Pasteur d'Indochine*, 8, 41–76, 12 pls.

Borkent, A. (2012) The pupae of Culicomorpha—morphology and a new phylogenetic tree. *Zootaxa*, 3396, 1–98. DOI: 10.11646/zootaxa.3396.1.1

Borkent, A. (2018) The state of phylogenetic analysis: narrow visions and simple answers—examples from the Diptera (flies). *Zootaxa*, 4374 (1), 107–143. DOI: 10.11646/zootaxa.4374.1.7

Borkent, A. & Grimaldi, D.A. (2004) The earliest fossil mosquito (Diptera: Culicidae), in mid-Cretaceous amber. *Annals of the Entomological Society of America*, 97 (5), 882–888. DOI: 10.1603/0013-8746(2004)097[0882:TEFMDC]2.0.CO;2

Borkent, A. & Grimaldi, D.A. (2016) The Cretaceous fossil *Burmaculex antiquus* confirmed as the earliest known lineage of mosquito (Diptera: Culicidae). *Zootaxa*, 4079 (4), 457–466. DOI: 10.11646/ZOOTAXA.4079.4.5

Boshell-Manrique, J. (1939) A new species of *Aedes* from Colombia (Dipt. Culicidae). *Revista de Entomologia*, 10 (2), 309–312.

Boshell-Manrique, J. & Osorno-Mesa, E. (1944) Observations on the epidemiology of the jungle yellow fever in Santander and Boyaca, Colombia, September, 1941, to April, 1942. *American Journal of Hygiene*, 40 (2), 170–181. DOI: 10.1093/oxfordjournals.aje.a118984

Bourke, B.P., Foster, P.G., Bergo, E.S., Calado, D.C. & Sallum, M.A.M. (2010) Phylogenetic relationships among species of *Anopheles* (*Nyssorhynchus*) Diptera, Culicidae) based on nuclear and mitochondrial gene sequences. *Acta Tropica*, 114 (2), 88–96. DOI: 10.1016/j.actatropica.2010.01.009

Bourke, B.P., Nagaki, S.S., Bergo, E.S., da Cruz Cardoso, J. & Sallum, M.A.M. (2011) Molecular phylogeny of the Myzorhynchella Section of *Anopheles* (*Nyssorhynchus*) (Diptera: Culicidae): genetic support for recently described and resurrected species. *Memórias do Instituto Oswaldo Cruz*, 106 (6), 705–715. DOI: 10.1590/S0074-02762011000600010

Bourke, B.P., Oliveira, T.P., Suesdek, L., Bergo, E.S. & Sallum, M.A.M. (2013) A multi-locus approach to barcoding in the *Anopheles strodei* subgroup (Diptera: Culicidae). *Parasites & Vectors*, 6, 111. DOI: 10.1186/1756-3305-6-111

Bourke, B.P., Wilkerson, R.C., Ruiz-Lopez, F., Justi, S.A., Pecor, D.B., Quinones, M.L., Navarro, J.-C., Ormaza, J.A., Ormaza, J.A., Jr, González, R., Flores-Mendoza, C., Castro, F., Escovar, J. & Linton, Y.-M. (2023) High levels of diversity in *Anopheles* subgenus *Kerteszia* revealed by species delimitation analyses. *Genes*, 14 (2), 344. DOI: 10.3390/genes14020344

Bourroul, C. (1904) *Mosquitos do Brasil*. These de Doutoramento, Faculdade de Medicina da Bahia. Officina Typographica de João Baptista de Oliveira Costa, Bahia. [Consisting of Prefacio (pp. i–viii), Novas especies (pp. 1–6, 30–32), Quadro das especies encontradas na Bahia (4 pp.) and Proposições (18 pp.) by Bourroul; remainder and bulk of work by Adolpho Lutz. Continuous pagination used. See Belkin, 1968b: 50, 62 for details].

Boussès, P. & Brunhes, J. (2013) Trois nouveaux *Uranotaenia* Lynch Arribálzaga de Madagascar (Diptera, Culicidae). *Bulletin de la Société Entomologique de France*, 118 (3), 399–410. DOI: 10.3406/bsef.2013.3124

Boussès, P., Le Goff, G., Barnabé, C., Brenguesa, C. & Robert, V. (2022) Description d'un nouveau moustique d'Aldabra (Seychelles), *Aedes* (*Aedimorphus*) *suzannae* n. sp., et du stade larvaire d'*Aedes* (*Coetzeemyia*) *fryeri* (Theobald, 1912) (Diptera: Culicidae). *Annales de la Société entomologique de France*, 58 (1), 63–80. DOI: 10.1080/00379271.2022.2036639

Bower, J.E., Dowton, M., Cooper, R.D. & Beebe, N.W. (2008) Intraspecific concerted evolution of the rDNA ITS1 in *Anopheles farauti* sensu stricto (Diptera: Culicidae) reveal recent patterns of population structure. *Journal of Molecular Evolution*, 67 (4), 397–411. DOI: 10.1007/s00239-008-9161-x

Braima, K.A., Muslimin, M., Ghazali, A.-R.M., Wan-Nor, F., Wilson, J.J., Jeffery, J. & Abdul-Aziz, N.M. (2017) Feeding behavior of *Mimomyia (Etorleptiomyia) luzonensis* (Ludlow, 1905) (Diptera, Culicidae) in Peninsular Malaysia. *Acta Tropica*, 171, 138–140. DOI: 10.1016/j.actatropica.2017.03.025

Bram, R.A. (1966) *Culex* (*Thaiomyia*) *dispectus*, a new subgenus and species from Thailand (Diptera: Culicidae). *Proceedings of the Entomological Society of Washington*, 68 (2), 73–77.

Bram, R.A. (1967a) Contributions to the mosquito fauna of Southeast Asia. II. The genus *Culex* in Thailand (Diptera: Culicidae). *Contributions of the American Entomological Institute*, 2 (1), 1–296.

Bram, R.A. (1967b) Classification of *Culex* subgenus *Culex* in the New World (Diptera: Culicidae). *Proceedings of the United States National Museum*, 120 (3557), 1–122.

Bram, R.A. (1968) A redescription of *Culex* (*Acalleomyia*) *obscurus* (Leicester). *Proceedings of the Entomological Society of Washington*, 70 (1), 52–57.

Bram, R.A. (1969) Relationships of adult and larval anatomy in the supraspecific classification of the genus *Culex* in Southeast Asia (Diptera: Culicidae). *Mosquito Systematics Newsletter*, 1 (2), 9–12.

Bram, R.A. & Rattanarithikul, M. (1967) Six new species of the *Culex* (*Lophoceraomyia*) *mammilifer* group from Thailand (Diptera: Culicidae). *Proceedings of the Entomological Society of Washington*, 69 (1), 1–17.

Branch, N. & Seabrook, E.L. (1959) *Culex* (*Culex*) *scimitar*, a new species of mosquito from the Bahama Islands (Diptera, Culicidae). *Proceedings of the Entomological Society of Washington*, 61 (5), 216–218.

Brèthes, J. (1910) Dípteros nuevos ó poco conocidos de Sud-América. *Anales del Museo Nacional de Buenos Aires*, 20, 469–484.

Brèthes, J. (1912) Los mosquitos de la República Argentina. *Boletín de Instituto Entomológia Patagonia Vegetal*, 1, 1–48.

Brèthes, J. (1916) Algunas notas sobre mosquitos argentinos su relación con las enfermedades palúdicas, etc. y descripción de tres especies nuevas. *Anales del Museo Nacional de Buenos Aires*, 28, 193–218.

Brèthes, J. (1920) Description d'un nouveau moustique du Pérou. *Revista Chilena de Historia Natural*, 24, 41–43.

Brèthes, J. (1926) Description provisoire de deux especes nouvelles d'Anophelinae argentins. *Prensa Medica Argentina*, 13 (3), 106–107.

Broche, R.G. (2000) Dos nuevas especies de mosquitos de los géneros *Psorophora* (*Grabhamia*) y *Culex* (*Culex*) (Diptera, Culicidae), de Cuba. *Revista Brasileira de Entomologia*, 44 (1/2), 27–33.

Brochero, H.H.L., Li, C. & Wilkerson, R.C. (2007) A newly recognized species in the *Anopheles* (*Nyssorhynchus*) *albitarsis* complex (Diptera: Culicidae) from Puerto Carreño, Colombia. *American Journal of Tropical Medicine and Hygiene*, 76 (6), 1113–1117. DOI: 10.4269/ajtmh.2007.76.1113

Brolemann, H.-W. (1919) Sur quelques *Culex* des Pyrénées II ([1]). Campagne 1918. *Annales de la Société Entomologique de France*, 88 (1–2), 65–103.

Bromley-Schnur, H.J. (2021) *The Mosquitoes of Israel and Surrounding Areas: An Illustrated Guide*. Wageningen Academic Publishers, Wageningen, Netherlands. DOI: 10.3920/978-90-8686-913-8

Brookman, B. & Reeves, W.C. (1950) A new name for a California mosquito (Diptera, Culicidae). *Pan-Pacific Entomologist*, 26 (4), 159–160.

Brug, S.L. (1925) A new *Armigeres* (Diptera, Culicidae) from Ceram (Moluccas). *Bulletin of Entomological Research*, 15 (4), 345–347. DOI: 10.1017/S0007485300046320

Brug, S.L. (1928a) *Anopheles* (*Neomyzomyia*) *longirostris* n. sp. (Diptera, Culicidae). *Geneeskundig Tijdschrift voor Nederlandsch-Indië*, 68, 278–279, 1 pl.

Brug, S.L. (1928b) Aanteekeningen omtrent Muskieten (IV). *Geneeskundig Tijdschrift voor Nederlandsch-Indië*, 68, 921–927, 3 pls.

Brug, S.L. (1931a) New Culicidae from Sumatra. *Tijdschrift voor Entomologie*, 74, 245–250.

Brug, S.L. (1931b) *Anopheles incognitus* n. sp. *Geneeskundig Tijdschrift voor Nederlandsch-Indië*, 71 (2), 136–137.

Brug, S.L. (1931c) Culiciden der Deutschen Limnologischen Sunda-Expedition. *Archiv für Hydrobiologie Suppl.-Bd 9, Tropische Binnengewässer*, 2, 1–42.

Brug, S.L. (1932) Notes on Dutch East Indian mosquitos. *Bulletin of Entomological Research*, 23 (1), 73–83. DOI: 10.1017/S0007485300002583

Brug, S.L. (1934) Notes on Dutch East Indian mosquitos. *Bulletin of Entomological Research*, 25 (4), 501–519. DOI: 10.1017/S0007485300012864

Brug, S.L. (1938) Waarnemingen bij anophelinen. *Geneeskundig Tijdschrift voor Nederlandsch-Indië*, 78, 520–523.

Brug, S.L. (1939) Notes on Dutch East-Indian mosquitoes. *Tijdschrift voor Entomologie*, 82, 91–113.

Bruijning, C.F.A. (1959) Notes on *Wyeomyia* mosquitoes of Suriname, with a description of *Wyeomyia surinamensis* sp.n. *Studies on the Fauna of Suriname and other Guyanas*, 3 (8), 99–146.

Brunetti, E. (1912) Annotated catalogue of Oriental Culicidae – Supplement. *Records of the Indian Museum*, 4 (10), 403–517.

Brunetti, E. (1914) Critical review of "general" in Culicidae. *Records of the Indian Museum*, 10 (1), 15–73.

Brunetti, E. (1920) Catalogue of Oriental and South Asiatic Nemocera. *Records of the Indian Museum*, 17, preface + 300 pp.

Brunhes, J. (1969) Culicidés de Madagascar: IV. Description des imagos, de la nymphe et de la larve de *Culex (Culiciomyia) pandani* sp. n. *Cahiers ORSTOM Série Entomologie médicale et Parasitologie*, 7 (3), 175–180.

Brunhes, J. (1971) Culicidés de Madagascar. V. Quelques *Aedes* (sous-genre *Finlaya*) de Madagascar. *Cahiers ORSTOM Série Entomologie médicale et Parasitologie*, 9 (4), 335–349.

Brunhes, J. (1977a) Les moustiques de l'archipel des Comores I. – Inventaire, répartition et description de quatre espèces ou sous-espèces nouvelles. *Cahiers ORSTOM Série Entomologie médicale et Parasitologie*, 15 (2), 131–152.

Brunhes, J. (1977b) Les moustiques de l'archipel des Comores. II. – Description de quatre espèces nouvelles ou peu connues. Répartition des membres du sous-genre *Skusea* dans l'océan Indien accidental. Affinités de la faune culicidienne des Comores. *Cahiers ORSTOM Série Entomologie médicale et Parasitologie*, 15 (2), 153–170.

Brunhes, J. (1983) Culicidés de Madagascar IX. Compléments à la description d'*Aedes (Diceromyia) grassei* Doucet, 1951 et description d'*Aedes (Diceromyia) sylvaticus* n.sp. (Diptères, Nématocères). *Cahiers ORSTOM Série Entomologie médicale et Parasitologie* (for 1982), 20 (4), 313–318.

Brunhes, J. (1999) Culicidae du Maghreb. Description d'*Aedes (Ochlerotatus) biskraensis* n. sp. d'Algérie (Diptera, Nematocera). *Bulletin de la Société Entomologique de France*, 104 (1), 25–30. DOI: 10.3406/bsef.1999.16537

Brunhes, J. & Boussès, P. (2009) Description de *Culex* (*Culex*) *sechani* n. sp. de Tahiti (îles de la Société (Diptera, Culicidae). *Bulletin de la Société Entomologique de France*, 114 (1), 21–28. DOI: 10.3406/bsef.2009.2758

Brunhes, J. & Hervy, J.-P. (1995) *Faune de Madagascar. 85. Insectes diptères Culicidae Culicinae genre* Orthopodomyia *de la sous-région malgache et de la région afrotropicale*. Muséum nationale d'Histoire naturelle, Paris, France.

Brunhes, J. & Rambelo, J. (1968) Contribution a l'étude des culicides de Madagascar. Description des adultes, nymphe et larve de *Culex (Neoculex) chauveti* sp. n. *Cahiers ORSTOM Entomologie médicale*, 6 (1), 113–118.

Brunhes, J. & Ravaonjanahary, C. (1971) Culicides de Madagascar: VI. Description de *Culex (Culex) carleti* sp. n. *Cahiers ORSTOM Série Entomologie médicale et Parasitologie*, 9 (2), 177–182.

Brunhes, J. & Ravaonjanahary, C. (1973) Culicides de Madagascar. VII. Description des imagos, de la nymphe et de la larve de *Culex (Eumelanomyia) brenguesi* sp. n. *Cahiers ORSTOM Série Entomologie médicale et Parasitologie*, 11 (3), 169–173.

Brunhes, J. & Razafindrasolo, E. (1975) Culicidés de Madagascar VIII. Description des imagos, de la nymphe et de la larve d'*Uranotaenia (Uranotaenia) anopheloides* sp. n. *Cahiers ORSTOM Série Entomologie médicale et Parasitologie*, 13 (4), 223–227.

Brunhes, J. & Venard, P. (1966) Contribution a l'étude des culicidés de la région éthiopienne. Description de *Culex (Barraudius) richeti* n. sp. Notes complémentaires sur *Culex (B) pusillus* Macquart et *Culex (B.) modestus* Ficalbi. *Cahiers ORSTOM Série Entomologie médicale et Parasitologie*, 4 (7), 75–84.

Brunhes, J., Adam, J.P. & Bailly-Choumara, H. (1967) Contribution a l'etude des *Culex* de la region Ethiopienne appartenant au sous-genre *Mochthogenes* (Diptères, Culicidés), avec description des males de cinq nouvelles especes. *Cahiers ORSTOM Série Entomologie médicale*, 5 (1), 43–52.

Brunhes, J., Boussès, P. & da Cunha Ramos, H. (2011) Les *Aedeomyia* Theobald, 1901, des régions afro-tropicale et malgache (Diptera, Culicidae). *Bulletin de la Société Entomologique de France*, 116 (1), 99–128. DOI: 10.3406/bsef.2011.2911

Brunhes, J., Boussès, P., Tantely, M.L. & Kengne, P. (2017) Un nouveau genre de Culicidae (Diptera), *Paulianius* n. gen., avec la description de trois nouvelles espèces malgaches. *Annales de la Société Entomologique de France*, 53 (5), 344–373. DOI: 10.1080/00379271.2017.1365627

Brunhes, J., Le Goff, G. & Bousses, P. (2003) Anophèles afrotropicaux. – V. Description du mâle et des stades pré-imaginaux d'*An. deemingi* Service, 1970 et description d'*An. eouzani* n. sp. (Diptera: Culicidae). *Annales de la Société Entomologique de France*, 39 (2), 179–185. DOI: 10.1080/00379271.2003.10697372

Brunhes, J., Le Goff, G. & Geoffroy, B. (1997) Anophèles afro-tropicaux. I. – Descriptions d'espèces nouvelles et changements de statuts taxonomiques (Diptera: Culicidae). *Annales de la Société Entomologique de France*, 33 (2), 173–183. DOI: 10.1080/21686351.1997.12277748

Brunhes, J., Le Goff, G. & Geoffroy, B. (1999) Afro-tropical anopheline mosquitoes. III. Description of three new species: *Anopheles carnevalei* sp. nov., *An. hervyi* sp. nov., and *An. dualaensis* sp. nov., and resurrection of *An. rageaui* Mattingly and Adam. *Journal of the American Mosquito Control Association*, 15 (4), 552–558.

Brust, R.A. (1997) Ecology of *Aedes australis* (Erichson) 1842 complex (Diptera: Culicidae) in Australia and New Zealand. *Australian Journal of Entomology*, 36 (3), 287–292. DOI: 10.1111/j.1440-6055.1997.tb01471.x

Brust, R.A. & Mahon, R.J. (1997) Resurrection of *Aedes (Halaedes) ashworthi* Edwards 1921: morphological characterisation and separation from its sibling *Aedes (Halaedes) australis* (Erichson) 1842 (Diptera: Culicidae). *Australian Journal of Entomology*, 36 (2), 129–136. DOI: 10.1111/j.1440-6055.1997.tb01445.x

Brust, R.A., Ballard, J.W.O., Driver, F., Hartley, D.M., Galway, N.J. & Curran, J. (1998) Molecular systematics, morphological analysis, and hybrid crossing identify a third taxon, *Aedes* (*Halaedes*) *wardangensis* sp. nov., of the *Aedes* (*Halaedes*) *australis* species-group (Diptera: Culicidae). *Canadian Journal of Zoology*, 76 (7), 1236–1246. DOI: 10.1139/z98-051

Bugher, J.C., Boshell-Manrique, J., Roca-García, M. & Osorno-Mesa, E. (1944) Epidemiology of jungle yellow fever in eastern Colombia. *American Journal of Hygiene*, 39 (1), 16–51. DOI: 10.1093/oxfordjournals.aje.a118895

Burke, D.S. & Leake, J.C. (1988) Japanese encephalitis. In: Monath, T.P. (ed.), *The Arboviruses: Epidemiology and Ecology*. Vol. 3. CRC Press, Boca Raton, FL, USA, pp. 63–92.

Büttiker, W. & Beales, P. (1959) *Anopheles pampanae*, a new species of mosquito from Cambodia. *Acta Tropica*, 16 (1), 63–69.

Byrd, B.D., Harrison, B.A., Zavortink, T.J. & Wesson, D.M. (2012) Sequence, secondary structure, and phylogenetic analyses of the ribosomal internal transcribed spaces 2 (ITS2) in members of the North American Signifera Group of *Orthopodomyia* (Diptera: Culicidae). *Journal of Medical Entomology*, 49 (6), 1189–1197. DOI: 10.1603/ME10243

Cagampang-Ramos, A. & Darsie, R.F. (1970) Illustrated keys to the *Anopheles* mosquitoes of the Philippine Islands. USAF Fifth Epidemiological Flight, PACAF Technical Report 70-1.

Cambouliu, M. (1902) Contribution à l'étude des *Anopheles* de l'isthme de Suez. *Compte Rendu de l'Académie des Sciences*, 135 (17), 704–706.

Camp, J.V., Bakonyi, T., Soltész, Z., Zechmeister, T. & Nowotny, N. (2018) *Uranotaenia unguiculata* Edwards, 1913 are attracted to sound, feed on amphibians, and are infected with multiple viruses. *Parasites & Vectors*, 11, 456. DOI: 10.1186/s13071-018-3030-2

Campos, R.E. (2013) The aquatic communities inhabiting internodes of two sympatric bamboos in Argentinean subtropical forest. *Journal of Insect Science*, 13, 93. DOI: 10.1673/031.013.9301

Campos, R.E. & Zavortink, T.J. (2010) Description of the larva and pupa and redescription of the adults of *Isostomyia paranensis* (Brèthes) (Diptera: Culicidae). *Zootaxa*, 2689, 27–36. DOI: 10.11646/zootaxa.2689.1.3

Cane, R.P., Li, D., Turbitt, E. & Chambers, G.K. (2020) Molecular phylogenetic analysis of New Zealand mosquito species. *New Zealand Journal of Zoology*, 47 (4), 324–349. DOI: 10.1080/03014223.2020.1770305

Carpenter, S.J. & LaCasse, W.J. (1955) *Mosquitoes of North America (North of Mexico)*. University of California Press, Berkeley, CA, UK.

Carpintero, D.J. & Leguizamón, M.N. (2000) Una neuva especie de mosquito en la la [*sic*] Provincia de Buenos Aires *Aëdes* (*Ochlerotatus*) *jorgi*. Nueva especies. *Mariposas del Mundo*, 5, 20–23.

Carpintero, D.J. & Leguizamón, M.N. (2004) Description of a new *Culex* (*Culex*) species (Diptera: Culicidae) from La Pampa Province, Argentina, and a preliminary list of the mosquitoes of the province. *Studia dipterologica*, 11 (2), 501–503.

Carter, H.F. (1911) A new mosquito from Uganda. *Bulletin of Entomological Research*, 2 (1), 37–38. DOI: 10.1017/S0007485300001164

Carter, H.F. (1920) Descriptions of the adult, larval, and pupal stages of a new mosquito from Lord Howe Island, S. Pacific. *Proceedings of the general meetings for scientific business of the Zoological Society of London*, 1920, 623–628.

Carter, H.F. (1950) The genus *Taeniorhynchus* Lynch Arribalzaga (Diptera, Culicidae) with special reference to the bionomics and relation to disease of the species occurring in Ceylon. *Ceylon Journal of Science (Biological Sciences)*, 25 (1), 1–26, 2 pls.

Carter, H.F. & Wijesundara, D.P. (1948) Notes on some Ceylon culicine mosquitoes. *Ceylon Journal of Science, Section B: Zoology and Geology*, 23 (3), 135–151.

Casal, O.H. (1967) *Culex (Culex) ameliae*, una nueva especie de la Republic Argentina *(Diptera, Culicidae)*. *Neotropica*, 13 (40), 29–30.

Casal, O.H. & Garcia, M. (1967a) *Culex (Melanoconion) pavlovskyi*, nueva especie de la Republica Argentina (Diptera, Culicidae). *Neotropica*, 13 (40), 19–20.

Casal, O.H. & García, M. (1967) *Culex (Culex) castroi*, nueva especie de la República Argentina *(Diptera, Culicidae)*. *Physis*, 26 (73), 451–457.

Casal, O.H. & Garcia, M. (1967c) *Culex (Culex) hepperi*, nueva especie del Delta Bonaerense del Rio Parana *(Diptera, Culicidae)*. *Physis*, 27 (74), 87–94.

Casal, O.H. & García, M. (1968a) *Culex* (*Allimanta*), nuevo subgenero para *Culex* (*Culex*) *tramazayguesi* Duret, 1954 (Diptera, Culicidae). *Physis*, 27 (75), 329–335.

Casal, O.H. & Garcia, M. (1968) *Culex (Eubonnea) clastrieri*, una nueva especie del Brasil (Diptera, Culicidae). *Neotropica*, 14 (45), 119–120.

Casal, O.H. & García, M. (1968c) *Culex (Melanoconion) martinezi*, una nueva especie de la Republica Argentina *(Diptera, Culicidae)*. *Physis*, 27 (75), 455–459.

Casal, O.H. & García, M. (1968d) *Culex (Melanoconion) dureti*, una nueva especie Argentino-Paraguaya (*Diptera, Culicidae)*. *Physis*, 28 (76), 123–125.

Casal, O.H. & García, M. (1968e) *Culex* (*Culex*) *eduardoi*, una nueva especie de la Republica Argentina (Diptera, Culicidae). *Revista de la Sociedad Entomólogica Argentina*, 30 (1–4), 105–111.

Casal, O.H. & García, M. (1971) *Culex (Culex) tatoi*, una nueva especie de la Republica Argentina *(Diptera, Culicidae)*. *Physis*, 30 (81), 631–635.

Casal, O.H., García, M. & Cavalieri, F. (1966) *Culex (Culex) fernandezi*, una nueva especie bromelicola de la *Republica Argentina (Diptera, Culicidae)*. *Physis*, 26 (71), 185–192.

Casal, O.H., García, M. & Fernández, H.I. (1968) El subgenero *Culex (Aedinus)* Bourroul, 1904, nuevo para la entomofauna Argentina, con la descripcion de una nueva especie *(Diptera, Culicidae)*. *Physis*, 28 (76), 217–218.

Causey, O.R. (1937) New anopheline and culicine mosquitoes from Siam. *Journal of Parasitology*, 23 (5), 543–545.

Causey, O.R. (1945) Description of *Anopheles* (*Nyssorhynchus*) *dunhami*, a new species from the Upper Amazon Basin. *Journal of the National Malaria Society*, 4 (3), 231–234.

Causey, O.R., Deane, L.M. & Deane, M.P. (1943) Descrição de um novo anofelino da parte alta do Vale do Amazonas "*Anopheles* (*Nyssorhynchus*) *galvãoi*" n. sp. *Revista Paulista de Medicina*, 23, 293–296.

Causey, O.R., Deane, L.M. & Deane, M.P. (1944) An illustrated key to the eggs of thirty species of Brazilian anophelines, with several new descriptions. *American Journal of Hygiene*, 39 (1), 1–7. DOI: 10.1093/oxfordjournals.aje.a118894

Causey, O.R., Deane, L.M., Deane, M.P. & Sampaio, M.M. (1943) *Anopheles* (*Nyssorhynchus*) *sawyeri*, a new anopheline mosquito from Ceara, Brazil. *Annals of the Entomological Society of America*, 36 (1), 11–20. DOI: 10.1093/aesa/36.1.11

Cerqueira, N. (1939) Sobre *Psorophora lutzi* Theobald, *Psorophora albipes* Theobald e *Psorophora forceps* n. sp. (Dipt. Culicidae). *Revista de Entomologia*, 10 (1), 78–85.

Cerqueira, N.L. (1943) Algumas espécies novas da Bolívia, e referência a três espécies do *Haemagogus* (Diptera, Culicidae). *Memórias do Instituto Oswaldo Cruz*, 39 (1), 1–14, 4 pls. DOI: 10.1590/S0074-02761943000400001

Cerqueira, N.L. (1960a) Uma nova espécie de *Haemagogus* do Amazonas (Diptera Culicidae). *Boletim de Museu Paraense Emilio Goeldi*, 25, 1–5, 2 pls.

Cerqueira, N.L. (1960b) Sôbre uma n. sp. de *Janthinosoma* da Série Lutzi e a descrição do macho de *Psorophora* (*J.*) *circumflava* Cerqueira, 1943 (Diptera Culicidae). *Boletim de Museu Paraense Emilio Goeldi*, 26, 1–7, 3 pls.

Cerqueira, N.L. (1961) Cinco novos sabetinos da Amazônia (Diptera Culicidae). *Revista Brasileira de Entomologia*, 10, 37–52.

Cerqueira, N. & Antunes, P.C.A. (1938) *Haemagogus tropicalis*, a new species from Pará, Brazil (Diptera, Culicidae). *Proceedings of the Entomological Society of Washington*, 40 (1), 1–9.

Cerqueira, N.L. & Costa, A.F. (1946) Duas novas espécies de *Aëdes (Ochlerotatus) (Diptera Culicidae)*. *Livro de homenagem a R. F. d'Almeida*, 11, 133–140.

Cerqueira, N.L. & Lane, J. (1944) Sôbre algumas espécies de *Culex* (*Carrollia*) (Diptera, Culicidae). *Revista de Entomologia*, 15 (1–2), 216–220.

Cerqueira, N.L. & Paraense, W.L. (1945) Uma nova espécie de *Aedes*, transmissora de *Plasmodium gallinaceum* (Diptera Culicidae). *Memórias do Instituto Oswaldo Cruz*, 42 (1), 11–15. DOI: 10.1590/S0074-02761945000100002

Chagas, C. (1907) *Novas especies de culicidios brazileiros*. Besnard Freres, Rio de Janeiro, Brazil.

Chamberlain, R.W. (1981) Epidemiology of arthropod-borne togaviruses: the role of arthropods as hosts and vectors and of vertebrate hosts in natural transmission cycles. In: Schlesinger, R.W. (ed.), *The Togaviruses: Biology, Structure, Replication*. Academic Press, New York, NY, USA, pp. 175–228.

Chang, H. (1965) Description of a mosquito *Aëdes* (*Edwardsaedes*) *pingpaensis*, sp. nov. [in Chinese]. *Acta Parasitologica Sinica*, 2 (4), 397–402.

Chang, P.-X. & Chang, S.-M. (1974) Two new species of Culicini from Kweichow [in Chinese]. *Acta Entomologica Sinica*, 17 (3), 347–352.

Chapman, H.C. & Barr, A.R. (1964) *Aedes communis nevadensis*, a new subspecies of mosquito from western North America (Diptera: Culicidae). *Mosquito News*, 24 (4), 439–447.

Chen, B., Butlin, R.K. & Harbach, R.E. (2003) Molecular phylogenetics of the Oriental members of the Myzomyia Series of *Anopheles* subgenus *Cellia* (Diptera: Culicidae) inferred from nuclear and mitochondrial DNA sequences. *Systematic Entomology*, 28 (1), 57–69. DOI: 10.1046/j.1365-3113.2003.00200.x

Chen, D.-H., He, S.-L., Fu, W.B., Yan, Z.-T., Hu, Y.-J., Yuan, H., Wang, M.-B. & Chen, B. (2024) Mitogenome-based phylogeny of mosquitoes (Diptera: Culicidae). *Insect Science*, 31 (2), 599–612. DOI: 10.1111/1744-7917.13251

Chen, H. & Zhao, H. (1985) A new species of the genus *Culex* from China (Diptera: Culicidae) [in Chinese]. *Entomotaxonomia*, 7 (4), 265–269.

Chen, H.-B. (1977) A new species of *Culex* (Diptera: Culicidae) [in Chinese]. *Acta Entomologica Sinica*, 20 (1), 95–98.

Chen, H.-B. (1981) A new species of *Culex* (Diptera: Culicidae) [in Chinese]. *Acta Entomologica Sinica*, 24 (4), 426–429.

Chen, H.-B. (1982) A new species of *Culex* from Guizhou, *Culex* (*Eumelanomyia*) *miaolingensis*, sp. nov. [in Chinese]. *Acta Zootaxonomica Sinica*, 7 (3), 312–316.

Chen, H.-B. (1989) A new species of *Culex* (*Culiciomyia*) from Yunnan (Diptera: Culicidae) [in Chinese]. *Acta Zootaxonomica Sinica*, 14 (1), 86–90.

Chen, H.-B. & Dong, X.-S. (1992) A new species of *Culex* (*Eumelanomyia*) from China (Diptera: Culicidae) [in Chinese]. *Acta Zootaxonomica Sinica*, 17 (1), 78–83.

Chen, H.B. & Zhang, P.-X. (1988) *Topomyia* (*Suaymyia*) *tumetarsalis*, a new species from China (Diptera: Culicidae) [in Chinese]. *Acta Entomologica Sinica*, 31 (2), 213–218.

Chen, H.-B., Wang, S.-H. & Zhao, H. (1989) A new species of *Uranotaenia* breeding in crab holes from Guizhou Province of China (Diptera: Culicidae) [in Chinese]. *Acta Zootaxonomica Sinica*, 14 (2), 224–228.

Chen, H.-L. & Chen, H.-B. (2000) Description of a new species of the *Aedes* from Henan, China (Diptera: Culicidae) [in Chinese]. *Acta Zootaxonomica Sinica*, 25 (3), 341–344.

Chow, C.Y. (1950) Collection of culicine mosquitoes (Diptera, Culicidae) in Taiwan (Formosa), China, with description of a new species. *Quarterly Journal of the Taiwan Museum*, 3 (4), 281–287.

Chowdhury, K.L. (1929) A new variety of protanopheline *A. barbirostris* van der Wulp, var. *ahomi*, found in Upper Assam. *Indian Journal of Medical Research*, 16 (4), 986–988.

Christophers, S.R. (1906) On the importance of larval characters in the classification of mosquitoes. *Scientific memoirs by officers of the medical and sanitary departments of the government of India*, 25, 1–18 pp., 3 pls.

Christophers, S.R. (1911) [Notes on mosquitoes.] II.—A new anopheline. *Paludism*, 2, 64–68, 1 pl.

Christophers, S.R. (1915) The male genitalia of *Anopheles*. *Indian Journal of Medical Research*, 3 (2), 371–394, 6 pls.

Christophers, S.R. (1922) The development and structure of the terminal abdominal segments and hypopygium of the mosquito, with observations on the homologies of the terminal segments of the larva. *Indian Journal of Medical Research*, 10 (2), 530–572 + 4 pls.

Christophers, S.R. (1923) A new East-African anopheline (*A. kingi*) related to *A. natalensis*, Hill and Hayden, and *A. watsonii*, Leicester. *Indian Journal of Medical Research*, 10 (4), 1008–1019, 2 pls.

Christophers, S.R. (1924a) Provisional list and reference catalogue of the Anophelini. *Indian Medical Research Memoirs*, 3, 1–105.

Christophers, S.R. (1924b) Some Himalayan and peninsular varieties of Indian species of *Anopheles*. *Indian Journal of Medical Research*, 12 (1), 11–13.

Christophers, S.R. (1924c) Some further varieties of Indian species of *Anopheles* with notes on the species *A. pallidus* Theobald and *A. philippinensis* Ludlow. *Indian Journal of Medical Research*, 12 (2), 295–301.

Christophers, S.R. (1926) *A.* (*Myzomyia*) *pattoni*, a new *Anopheles* from Shan-tung, North China; with notes on some other species of *Anopheles* from the same locality. *Indian Journal of Medical Research*, 13 (4), 871 –877, 1 pl.

Christophers, S.R. (1933) *The fauna of British India, including Ceylon and Burma. Diptera. Vol.* IV. *Family Culicidae. Tribe Anophelini.* Taylor & Francis, London, UK.

Christophers, S.R. & Barraud, P.J. (1931) The eggs of Indian *Anopheles*, with descriptions of the hitherto undescribed eggs of a number of species. *Records of the Malaria Survey of India*, 2 (1), 161–192, 5 pls.

Chu, F.-I. (1957) Collection of megarhine and culicine mosquitoes from Hainan Island, South China, with description of a new species [in Chinese]. *Acta Zoologica Sinica*, 9 (2), 145–164, 3 pls.

Chu, F.-I. (1981) A new species of Chinese *Uranotaenia* (Diptera: Culicidae) [in Chinese]. *Entomotaxonomia*, 3 (4), 273–275.

Chu, H., Li, C., Guo, X., Zhang, H., Luo, P., Wu, Z., Wang, G. & Zhao, T. (2018) The phylogenetic relationships of known mosquito (Diptera: Culicidae) mitogenomes. *Mitochondrial DNA Part A: DNA Mapping, Sequencing, and Analysis*, 29 (1), 31–35. DOI: 10.1080/24701394.2016.1233533

Chwatt, L.J. (1949) *Aedes* (*Stegomyia*) *pseudoafricanus* sp. nov.: a new species of *Aedes* from the coast of Nigeria (British West Africa). *Nature*, 163 (4151), 808–809. DOI: 10.1038/163808a0

Clark-Gil, S. & Darsie, R.F., Jr (1983) The mosquitoes of Guatemala, their identification, distribution and bionomics, with keys to adult females and larvae. *Mosquito Systematics*, 15 (3), 151–284.

Clastrier, J. (1970a) *Culex* (*Eubonnea*) *guyanensis n.sp.* (*Diptera, Culicidae*) nouveau moustique de la Guyane française. *Annales de Parasitologie*, 45 (1), 115–118.

Clastrier, J. (1970b) *Culex (Melanoconion) dolichophyllus* n. sp. et *Culex (Microculex) stonei* Lane et Whitman, 1943 *(Diptera, Culicidae)* en Guyane française. *Annales de Parasitologie Humaine et Comparée*, 45 (5), 857–861. DOI: 10.1051/parasite/1970456857

Clastrier, J. (1970c) Quatre nouveau *Melanoconion* (*Diptera, Culicidae*) de la Guyane française. *Annales de Parasitologie*, 45 (4), 463–476. DOI: 10.1051/parasite/1970454463

Clastrier, J. (1973) *Wyeomyia* (*Dendromyia*) *trifurcata* n. sp. nouveau moustique de la Guyane française (Diptera, Culicidae). *Nouvelle Revue d'Entomologie*, 3 (1), 39–47.

Clastrier, J. (1974) *Wyeomyia (Dendromyia) bourrouli* (Lutz, 1905) et *Wyeomyia (Cruzmyia) forattinii* n.sp. en Guyane française *(Diptera, Culicidae). Cahiers ORSTOM Série Entomologie médicale et Parasitologie*, 12 (3), 153–163.

Clastrier, J. & Claustre, J. (1978a) *Culex (Carrollia) insigniforceps* n. sp. de la Guyane française (Diptera, Culicidae). *Nouvelle Revue d'Entomologie*, 8 (3), 359–362.

Clastrier, J. & Claustre, J. (1978b) *Limatus martiali* Senevet et Abonnenc, 1939 et *Wyeomyia nigricephala* n. sp. (Diptera, Culicidae) de la Guyane française. *Annales de Parasitologie*, 53 (5), 539–546. DOI: 10.1051/parasite/1978535539

Clements, A.N. (1992) *The Biology of Mosquitoes. Volume* 1. *Development, Nutrition and Reproduction*. Chapman & Hall, London, UK.

Clements, A.N. (1999) *The Biology of Mosquitoes. Volume* 2. *Sensory Reception and Behaviour*. CABI, Wallingford, UK.

Clements, A.N. (2012) *The Biology of Mosquitoes. Volume* 3. *Transmission of Viruses and Interactions with Bacteria*. CABI, Wallingford, UK.

Clements, A.N. & Harbach, R.E. (2018) Controversies over the scientific name of the principal mosquito vector of yellow fever virus – expediency versus validity. *Journal of Vector Ecology*, 43 (1), 1–14. DOI: 10.1111/jvec.12277

Cockerell, T.D.A. (1916) British fossil insects. *Proceedings of the United States National Museum*, 49 (2119), 469–499, 6 pls.

Cockerell, T.D.A. (1918) The mosquitoes of Colorado. *Journal of Economic Entomology*, 11 (2), 195–200.

Cockerell, T.D.A. (1919) The oldest mosquitoes. *Nature*, 103 (2577), 43–46.

Coetzee, M. (1984) A new species of *Anopheles* (*Anopheles*) from Namibia (Diptera: Culicidae). *Systematic Entomology*, 9 (1), 1–8. DOI: 10.1111/j.1365-3113.1984.tb00498.x

Coetzee, M. (1995) *Anopheles crypticus*, new species from South Africa is distinguished from *Anopheles coustani* (Diptera: Culicidae). *Mosquito Systematics* (for 1994), 26 (3), 125–131.

Coetzee, M. (2017) Culicidae (mosquitoes). In: Kirk-Spriggs, A.H. & Sinclair, B.J. (eds), *Manual of Afrotropical Diptera. Volume* 2. *Nematocerous Diptera and lower Brachycera*. South African National Biodiversity Institute, Pretoria, South Africa, pp. 277–692.

Coetzee, M. (2020) Key to the females of Afrotropical *Anopheles* mosquitoes (Diptera: Culicidae). *Malaria Journal*, 19, 70. DOI: 10.1186/s12936-020-3144-9

Coetzee, M. (2022) Literature review of the systematics, biology and role in malaria transmission of species in the Afrotropical *Anopheles* subgenus *Anopheles* (Diptera: Culicidae). *Zootaxa*, 5133 (2), 182–200. DOI: 10.11646/zootaxa.5133.2.2

Coetzee, M., Hunt, R.H., Wilkerson, R., della Torre, A., Coulibaly, M.B. & Besansky, N.J. (2013) *Anopheles coluzzii* and *Anopheles amharicus*, new members of the *Anopheles gambiae* complex. *Zootaxa*, 3619 (3), 246–274. DOI: 10.11646/zootaxa.3619.3.2

Coetzee, M., Segerman, J. & Hunt, R.H. (1987) Description of a new species *Anopheles* (*Cellia*) *kosiensis* (Diptera: Culicidae) from Zululand, South Africa. *Systematic Entomology*, 12 (1), 23–28. DOI: 10.1111/j.1365-3113.1987.tb00543.x

Cogill, H. (1903) The *Anopheles* of Karwar (North Kanara). *Journal of the Bombay Natural History Society*, 15 (2), 327–336, 1 pl.

Coher, E.I. (1948) A study of the female genitalia of Culicidae: with particular reference to characters of generic value. *Entomologica Americana*, 28 (3), 75–112.

Colless, D.H. (1948) The anopheline mosquitoes of north-west Borneo. *Proceedings of the Linnean Society of New South Wales*, 73 (3–4), 71–119.

Colless, D.H. (1955a) New anopheline mosquitoes from North Borneo. *Sarawak Museum Journal*, 6 (5), 331–342.

Colless, D.H. (1955b) Notes on the culicine mosquitoes of Singapore. I.–Three new species of *Culex* (Diptera, Culicidae) and a redescription of *Culex hutchinsoni* Barraud. *Annals of Tropical Medicine and Parasitology*, 49 (3), 311–319. DOI: 10.1080/00034983.1955.11685680

Colless, D.H. (1956) The *Anopheles leucosphyrus* group. *Transactions of the Royal Entomological Society of London*, 108 (3), 37–116. DOI: 10.1111/j.1365-2311.1956.tb02266.x

Colless, D.H. (1957a) Notes on the culicine mosquitoes of Singapore II.–The *Culex vishnui* group (Diptera, Culicidae), with descriptions of two new species. *Annals of Tropical Medicine and Parasitology* (for 1956), 51 (1), 87–101. DOI: 10.1080/00034983.1957.11685798

Colless, D.H. (1957b) Further notes on the systematics of the *Anopheles leucosphyrus* group (Diptera: Culicidae). *Proceedings of the Royal Entomological Society of London Series B Taxonomy*, 26 (7–8), 131–139. DOI: 10.1111/j.1365-3113.1957.tb00392.x

Colless, D.H. (1958) Notes on the culicine mosquitoes of Singapore IV.–The *Aëdes niveus* subgroup (Diptera, Culicidae): introduction and description of five new species and of one new subspecies. *Annals of Tropical Medicine and Parasitology*, 52 (4), 468–483. DOI: 10.1080/00034983.1958.11685885

Colless, D.H. (1960) Some species of *Culex* (*Lophoceraomyia*) from New Guinea and adjacent islands, with descriptions of four new species and notes on the male of *Culex fraudatrix* Theobald (Diptera: Culicidae). *Proceedings of the Linnean Society of New South Wales* (for 1959), 84 (3), 382–390.

Colless, D.H. (1962) Notes on the taxonomy of the *Aëdes scutellaris* group, and new records of *A. paullusi* and *A. albopictus* (Diptera: Culicidae). *Proceedings of the Linnean Society of New South Wales*, 87 (3), 312–315.

Colless, D.H. (1965) The genus *Culex*, subgenus *Lophoceraomyia* in Malaya (Diptera: Culicidae). *Journal of Medical Entomology*, 2 (3), 261–307. DOI: 10.1093/jmedent/2.3.261

Colless, D.H. (1979) Two little-used characters in the mosquito wing (Diptera: Culicidae). *Mosquito Systematics*, 11 (2), 127–129.

Colless, D.H. & McAlpine, D.K. (1970) Diptera (flies). In: M.J. Mackerras (ed.), *The Insects of Australia*. CSIRO/Melbourne University Press, Melbourne, Australia, pp. 656–740.

Collins, L.E. & Blackwell, A. (2000) The biology of *Toxorhynchites* mosquitoes and their potential as bicontrol agents. *Biocontrol News and Information*, 21 (4), 105N–116N.

Collucci, E. & Sallum, M.A.M. (2003) Phylogenetic analysis of the subgenus *Kerteszia* of *Anopheles* (Diptera: Culicidae: Anophelinae) based on morphological characters. *Insect Systematics & Evolution*, 34 (4), 361–372. DOI: 10.1163/187631203X00018

Collucci, E. & Sallum, M.A.M. (2007) Cladistic analysis of the subgenus *Anopheles* (*Anopheles*) Meigen (Diptera: Culicidae) based on morphological characters. *Memórias do Instituto Oswaldo Cruz*, 102 (3), 277–291. DOI: 10.1590/S0074-02762007005000026

Coluzzi, M. (1958) Descrizione di una nouva specie rinvenuta in Somalia. *Revista di Malariologia*, 37 (1–3), 69–71.

Coluzzi, M. & Sabatini, A. (1968) Divergenze morfologiche e barriere di sterilita' nel complesso *Aedes mariae* (*Diptera, Culicidae*). *Rivista de Parassitologia*, 29 (1), 49–70, 5 pls.

Coluzzi, M., Bianchi-Bullini, A.P. & Bullini, L. (1976) Speciation in the *mariae* complex of the genus *Aedes* (Diptera, Culicidae). *Atti - Associazione genetica italiana*, 21, 221–223.

Coluzzi, M., Sacca, G. & Feliciangeli, D. (1965) Il complesso *A. claviger* nella sottoregione mediterranea. *Cahiers ORSTOM Série Entomologie médicale et Parasitologie*, 3, 97–102.

Cook, S., Diallo, M., Sall, A.A., Cooper, A. & Holmes, E.C. (2005) Mitochondrial markers for molecular identification of *Aedes* mosquitoes (Diptera: Culicidae) involved in transmission of arboviral disease in West Africa. *Journal of Medical Entomology*, 42 (1), 19–28. DOI: 10.1093/jmedent/42.1.19

Cook, S., Lien, N.G., McAlister, E. & Harbach, R.E. (2010) *Bothaella manhi*, a new species of tribe Aedini (Diptera: Culicidae) from the Cuc Phuong National Park of Vietnam based on morphology and DNA sequence. *Zootaxa*, 2661, 33–46.

Coquillett, D.W. (1896) New Culicidae from North America. *Canadian Entomologist*, 28 (1), 43–44. DOI: 10.4039/Ent2843-2

Coquillett, D.W. (1898) Report on a collection of Japanese Diptera, presented to the U. S. National Museum by the Imperial University of Tokyo. *Proceedings of the United Stated National Museum*, 21 (1146), 301–340.

Coquillett, D.W. (1901) Three new species of Culicidae. *Canadian Entomologist*, 33 (9), 258–260. DOI: 10.4039/Ent33258-9

Coquillett, D.W. (1902a) New orthorrhaphous Diptera from Mexico and Texas. *Journal of the New York Entomological Society*, 10 (3), 136–141.

Coquillett, D.W. (1902b) New forms of Culicidae from North America. *Journal of the New York Entomological Society*, 10 (3), 191–194.

Coquillett, D.W. (1902c) New Diptera from North America. *Proceedings of the United States National Museum*, 25 (1280), 83–126. DOI: 10.5479/si.00963801.25-1280.83

Coquillett, D.W. (1902d) Three new species of *Culex*. *Canadian Entomologist*, 34 (11), 292–293. DOI: 10.4039/Ent34292-11

Coquillett, D.W. (1903a) Four new species of *Culex*. *Canadian Entomologist*, 35 (9), 255–257. DOI: 10.4039/Ent35255-9

Coquillett, D.W. (1903b) A new *Anopheles* with unspotted wings. *Canadian Entomologist*, 35 (11), 310. DOI: 10.4039/Ent35310-11

Coquillett, D.W. (1904a) Several new Diptera from North America. *Canadian Entomologist*, 36 (1), 10–12. DOI: 10.4039/Ent3610-1

Coquillett, D.W. (1904b) Notes on *Culex nigritulus*. *Entomological News and Proceedings of the Entomological Section of the Academy of Natural Sciences of Philadelphia*, 15 (2), 73–74.

Coquillett, D.W. (1904c) New North American Diptera. *Proceedings of the Entomological Society of Washington*, 6 (3), 166–192.

Coquillett, D.W. (1906a) New Culicidae from the West Indies and Central America. *Proceedings of the Entomological Society of Washington* (for 1905), 7 (4), 182–186.

Coquillett, D.W. (1906b) On the breaking-up of the old genus *Culex*. *Science*, 23 (582), 312–314. DOI: 10.1126/science.23.582.312

Coquillett, D.W. (1906c) A classification of the mosquitoes of North and Middle America. *U.S. Department of Agriculture, Bureau of Entomology, Technical Series*, 11, 1–31.

Coquillett, D.W. (1906d) Five new Culicidae from the West Indies. *Canadian Entomologist*, 38 (2), 60–62. DOI: 10.4039/Ent3860-2

Coquillett, D.W. (1910) Type-species of the North American genera of Diptera. *Proceedings of the United States Nation Museum*, 37 (1719), 499–647. DOI: 10.5479/si.00963801.37-1719.499

Corbet, P.S. & van Someren, E.C.C. (1962) *Aëdes (Stegomyia) opok* sp. nov., a new species of mosquito from Uganda. *Annals of Tropical Medicine and Parasitology*, 56 (1), 73–77. DOI: 10.1080/00034983.1962.11686093

Cornel, A.J., Mayi, M.P.A., Kowo, C., Foncha, D., Andongma, E., Anong, D.N., Elad, M., Djomo, C., Tchuinkam, T., Brisco, K.K. & Sehgal, R. (2020). New species of *Culex* (*Culiciomyia*) (Diptera: Culicidae) from Talangaye Forest in Cameroon and descriptions and identification keys for males of the Afrotropical species of the subgenus. *Zootaxa*, 4858 (4), 451–506. DOI: 10.11646/zootaxa.4858.4.1

Corner, L.C., Robertson, A.K., Hayles, L.B. & Iversen, J.O. (1980) Cache Valley virus: experimental infection in *Culiseta inornata*. *Canadian Journal of Microbiology*, 26 (3), 287–290. DOI: 10.1139/m80-04

Cornet, M. (1968) *Aedes (Aedimorphus) chamboni* n. sp., une nouvelle espèce de *Culicidae* du Sénégal. *Bulletin de la Société de Pathologie exotique* (for 1967), 60 (4), 398–402.

Cornet, M. (1973) *Aedes (Stegomyia) cozi* n. sp., une nouvelle espèce de Culicidae au Sénégal. *Cahiers ORSTOM Série Entomologie médicale et Parasitologie*, 11 (3), 175–180.

Cornet, M., Valade, M. & Dieng, P.Y. (1978) *Aedes (Stegomyia) neoafricanus* une nouvelle espèce de moustique capturée au Sénégal Oriental *(Diptera: Culicidae)*. *Cahiers ORSTOM Série Entomologie médicale et Parasitologie*, 16 (3), 227–230.

Corradetti, A. (1939a) Una nuova specie di *Anopheles* Rinvenuta in Dancalia: *Anopheles (Neocellia) dancalicus* n. sp. Nota preliminare. *Rivista di Parassitologia*, 3 (3), 277–278.

Corradetti, A. (1939b) Su alcune larve di *Anopheles* rinvenutea Ghinda (Eritrea) e descrizione di una nuova larva: *Anopheles (Myzomyia) erythraeus* n.sp. *Rivista di Parassitologia*, 3 (4), 287–292.

Corrêa, R.R. (1944) Deserição de uma nova subespécie do subgénero "*Anopheles*" (Diptera, Culicidae). *Anals Paulistas de Medicina e Cirurgia*, 47 (5), 407–408.

Corrêa, R.R. & Cerqueira, F.M.C. (1944) Descrição de *Anopheles (Kerteszia) laneanus*, nova espécie de anofelino de Campos do Jordão *(Diptera, Culicidae)*. *Arquivos de Higiene e Saúde Pública*, 9, 111–117.

Correa, R.R. & Ramalho, G.R. (1956) Revisão de *Phoniomyia* Theobald, 1903 (Diptera, Culicidae, Sabethini). *Folia Clinica et Biologica*, 25 (1/6), 1–176.

Corrêa, R.R. & Ramalho, G.R. (1968) *Anopheles* (*Arribalzagia*) *anchietai* nova espécie de anofelino de São Paula, Brasil *(Diptera Culicidae)*. *Revista Brasileira de Malariologia e Doenças Tropicais*, 20 (8–9), 115–132.

Cotrim, M. das D. & Galati, E.A.B. (1977) Revisão da série *pleuristriatus* do subgênero *Microculex* Theobald, 1907 (Diptera, Culicidae). *Revista Brasileira de Entomologia*, 20 (3/4), 169–205.

Coutinho, J.O. & Forattini, O.P. (1962) Notas sôbre Culicidae (Diptera). 4ª — Contribuição ao estudo do subgêero *Microculex*, com a descrição duma nova espécie. *Studia Entomologica*, 5 (1–4), 209–234.

Cova-Garcia, P. (1961) *Notas sobre los anofelinos de Venezuela y su identificación*. Segunda edicion. Editora Grafos, C.A., Caracas, Venezuela.

Cova Garcia, P. & Pulido F., J. (1974) Nueva especie de *Culex* (*Microculex*) (Diptera, Culicidae): *Culex* (*Microculex*) *sutili*. *Boletin de la Direccion de Malariologia y Saneamiento Ambiental*, 14 (1–2), 27–28.

Cova García, P. & Rausseo, J.A. (1964) *Uranotaenia bertii*, n. sp. *Revista Venezolana de Sanidad y Asistencia Social*, 29 (2), 221–229.

Cova Garcia, P. & Sutil Oramas, E. (1974) *Culex* (*Microculex*) *pulidoi* (Diptera, Culicidae). *Boletin de la Direccion de Malariologia y Saneamiento Ambiental*, 14 (1–2), 29–30.

Cova Garcia, P. & Sutil O., E. (1976) *Culex (Carrollia) wannonii* (Diptera, Culicidae), una nueva especie de Venezuela. *Boletin de la Direccion de Malariologia y Saneamiento Ambiental*, 16 (4), 354–359.

Cova Garcia, P. & Sutil Oramas, E. (1976) Nueva especie para Venezuela *Mansonia* (*Mansonia*) *suarezi* (Diptera, Culicidae). *Boletin de la Direccion de Malariologia y Saneamiento Ambiental*, 16 (4), 360–362.

Cova Garcia, P., Pulido F., J. & Escalante de Ugueto, C. (1977a) *Anopheles (Kerteszia) gonzalezrinconesi* n. sp. (Diptera, Culicidae) de Venezuela. *Boletin de la Direccion de Malariologia y Saneamiento Ambiental*, 17 (2), 140–149.

Cova Garcia, P., Pulido F., J. & Escalante de Ugueto, C. (1977b) *Anopheles* (*Kerteszia*) *rollai,* nomen novum *Anopheles* (*Kerteszia*) *hilli* Cova García, 1976 (non *Anopheles hilli* Woodhill & Lee 1944). *Boletin de la Direccion de Malariologia y Saneamiento Ambiental*, 17 (3), 241.

Cova Garcia, P., Pulido F., J. & Escalante de Ugueto, C. (1979) *Culex (Neoculex) johni* (Diptera: Culicidae) nueva especie de Venezuela. *Boletin de la Direccion de Malariologia y Saneamiento Ambiental*, 19 (1), 21–28.

Cova Garcia, P., Pulido F., J. & Escalante de Ugueto, C. (1981) *Uranotaenia pifanoi* (Diptera, Culicidae) nueva especie de Venezuela. *Boletin de la Direccion de Malariologia y Saneamiento Ambiental*, 21 (3–4), 210–218.

Cova García, P., Pulido F., J., Escalante de Ugueto, C. & Mora F., J.D. (1987a) *Uranotaenia iriartei* (Diptera, Culicidae), nueva especie de Venezuela. *Boletin de la Direccion de Malariologia y Saneamiento Ambiental*, 27 (1–4), 45–52.

Cova García, P., Pulido F., J., Escalante de Ugueto, C. & Mora R., J.D. (1987b) *Uranotaenia gabaldoni* (Diptera, Culicidae), nueva especie de Venezuela. *Boletin de la Direccion de Malariologia y Saneamiento Ambiental*, 27 (1–4), 94–104.

Cova García, P., Pulido, J., Escalante de Ugueto, C., Amarista, J. & Mora, J.D. (1987c) *Uranotaenia briceñoi* (Diptera, Culicidae), nueva especie de Venezuela. *Boletin de la Direccion de Malariologia y Saneamiento Ambiental*, 27 (1–4), 65–73.

Cova García, P., Sutil, E. & Pulido, J. (1971) Descripcion de nueva especie de *Culex*, del subgenero *Carrollia* (Diptera, Culicidae). *Boletín de la Sociedad Venezolana de Ciencias Naturales*, 29, 203–210.

Cova Garcia, P., Sutil O., E. & Pulido F., J. (1972) Neuva especie de *Culex* (*carrollia* [*sic*]) *rausseoi*. *Boletin de la Direccion de Malariologia y Saneamiento Ambiental*, 12, 207–209.

Cova Garcia, P., Sutil Oramas, E. & Pulido F., J. (1974a) *Wyeomyia lopezii* n. sp. (Diptera, Culicidae). *Boletin de la Direccion de Malariologia y Saneamiento Ambiental*, 14 (1–2), 21–22.

Cova Garcia, P., Sutil Oramas, E. & Pulido F., J. (1974b) *Sabethes fernández-yepezi* n. sp. *Boletin de la Direccion de Malariologia y Saneamiento Ambiental*, 14 (1–2), 23–24.

Cova Garcia, P., Sutil, E. & Rausseo, J.A. (1966) *Mosquitos (Culicinos) de Venezuela*. *Volumes* I & II. Ministerio de Sanidad y Asistencia Social, Caracas, Venezuela.

Crabtree, M.B., Kading, R.C., Mutebi, J.-P., Lutwama, J.J. & Miller, B.R. (2013) Identification of host blood from engorged mosquitoes collected in western Uganda using cytochrome oxidase I gene sequences. *Journal of Wildlife Diseases*, 49 (3), 611–626. DOI: 10.7589/2012-08-213

Cranston, P.S., Ramsdale, C.D., Snow, K.R. & White, G.B. (1967) Keys to the adults, male hypopygia, fourth-instar larvae and pupae of the British mosquitoes (Culicidae) with notes on their ecology and medical importance. *Freshwater Biological Association Scientific Publication*, 48, 1–152.

Cruz, O.G. (1901) Contribuição para o estudo dos culicidios do Rio de Janeiro. *Brazil-Medico*, 15 (43), 423–426.

Cruz, O.G. (1906) Um novo genero da sub-familia «Anophelina». *Brazil-Medico*, 20 (20), 199–200.

Curry, D.P. (1928) A new anopheline mosquito, *Anopheles* (*Chagasia*) *bathanus* Dyar, discovered in the Canal Zone. *American Journal of Tropical Medicine*, 8 (3), 243–248. DOI: 10.4269/AJTMH.1928.S1-8.243

Curry, D.P. (1931) *Anopheles* (*Anopheles*) *neomaculipalpus*. A new species of the *Arribalzagaia* group of *Anopheles* from Panama. *American Journal of Hygiene*, 13 (2), 643–647. DOI: 10.1093/oxfordjournals.aje.a117142

Curry, D.P. (1932) Some observations on the *Nyssorhynchus* group of the *Anopheles* (*Culicidae*) of Panama. *American Journal of Hygiene*, 15 (2), 566–572. DOI: 10.1093/oxfordjournals.aje.a117832

da Costa Lima, A. (1928) Sobre algumas anophelinas encontradas no Brasil. *Supplemento das Memórias do Instituto Oswaldo Cruz*, 21 (Suppl. 3), 91–113. DOI: 10.1590/S0074-02761928000500005

da Costa Lima, A. (1929) Sobre um novo *Anopheles* do Brasil. *Brasil-Medico*, 43 (2), 1100–1101.

da Costa Lima, A. (1930) Sobre especies do genero *Miamyia*, subgenero *Miamyia* (Diptera: Culicidae). *Memórias do Instituto Oswaldo Cruz*, 24 (2), 73–78, 3 pls. DOI: 10.1590/S0074-02761930000700004

da Costa Lima, A. (1931a) Nota sobre sabethineos do grupo *Joblotia* (Diptera: Culicidae). *Memórias do Instituto Oswaldo Cruz*, 25 (1), 65–71, 9 pls. DOI: 10.1590/S0074-02761931000100004

da Costa Lima, A. (1931b) Sobre as especies de *Megarhinus* do Brasil (Diptera: Culicidae). *Memórias do Instituto Oswaldo Cruz*, 25 (4), 307–315, 2 pls. DOI: 10.1590/S0074-02761931000400002

da Costa Lima, A. (1933) Sobre um novo *Aëdes* (*Ochlerotatus*) do Brasil. *Memórias do Instituto Oswaldo Cruz*, 27 (4), 403–406, 1 pl. DOI: 10.1590/S0074-02761933000400002

da Costa Lima, A. (1935a) Especies de *Taeniorhynchus* (*Taeniorhynchus*) (Diptera: Culicidae). *Memórias do Instituto Oswaldo Cruz*, 30 (3), 453–469, 10 pls. DOI: 10.1590/S0074-02761935000900008

da Costa Lima, A. (1935b) Sobre as especies de *Orthopodomyia* encontradasno Brasil (Diptera: Culicidae). *Revista Medico-Cirurgica do Brasil*, 43 (6), 175–179, 2 pls.

da Costa Lima, A. (1937) Um novo *Anopheles* da Baixada Fluminense (*Diptera: Culicidae*). *Revista Medico-Cirurgica do Brasil*, 45 (1), 5–7.

da Costa Lima, A., Guitton, N. & Ferreira, O. (1962) Comentários relativos às espécies da tribo Toxorhynchitini (Megarhinini) com a desçricão de uma espécie nova de *Lynchiella* (Diptera, Culicidae). *Memórias do Instituto Oswaldo Cruz*, 60 (2), 225–251, 4 pls. DOI: 10.1590/S0074-02761962000200008

da Cunha Ramos, H. (1985) Research on the mosquitoes of Angola (*Diptera: Culicidae*) XVII – Description of *Uranotaenia (Uranotaenia) angolensis* sp. nov. *Boletim da Sociedade Portuguesa de Entomologia*, 1 (Supl. l), 73–82.

da Cunha Ramos, H. (1986) Research on the mosquitoes of Angola (Diptera: Culicidae). XVIII – Description of *Uranotaenia* (*Uranotaenia*) *machadoi* sp. nov. *Mosquito Systematics*, 18 (1), 110–117.

da Cunha Ramos, H. (1993) *Revisão do género* Uranotaenia *(Diptera, Culicidae) na região afrotropical*. Ministério do Planeamento e da Administração do Território, Secretaria de Estado da Ciêcia e Technologia, Instituto de Investigação Científica Tropical, Lisbon, Portugal.

da Cunha Ramos, H. & Brunhes, J. (2004) *Insecta Diptera Culicidae* Uranotaenia. Faune de Madagascar 91. Institut de recherche pour le développement (Montpellier); Centre de cooperation internationale en recherché agronomique pour le développement (Paris); Muséum national d'histoire naturelle (Paris), France.

da Cunha Ramos, H. & Ribeiro, H. (1975) Research on the mosquitoes of Angola *(Diptera: Culicidae)* [*sic*] IX—Genera *Toxorhynchites* Theo., 1901, *Malaya* Leicester, 1908, *Ficalbia* Theo [*sic*], 1903, *Coquillettidia* Dyar, 1905, *Mansonia* Blanchard, 1901, *Uranotaenia* L. Arribálzaga, 1891, *Aedeomyia* Theo., 1901, *Eretmapodites* Theo., 1901 and *Culiseta* Felt, 1904. *Anais do Instituto de Higiene e Medicina Tropical*, 2 (1/4), 111–141.

da Cunha Ramos, H. & Ribeiro, H. (1982) Research on the mosquitoes of Angola (Diptera: Culicidae) XII – Description of *Culex (Culex) pajoti* sp. nov. *Mosquito Systematics* (for 1981), 13 (2), 176–190.

da Cunha Ramos, H. & Ribeiro, H. (1990) Research on the mosquitoes of Angola (Diptera, Culicidae). XX – *Eretmapodites mortiauxi*, a new species of the *leucopus* group. *Boletim da Sociedade Portuguesa de Entomologia*, 4 (115), 138–144.

da Cunha Ramos, H. & Ribeiro, H. (1992) Research on the mosquitoes of Angola *(Diptera, Culicidae)* XXI – Description of *Eretmapodites angolensis* sp. nov. and *Eretmapodites dundo* sp. nov., of the «*oedipodeios* group». *Garcia de Orta, Série de Zoologia* (for 1990), 17 (1–2), 31–35.

da Cunha Ramos, H., Capela, R.A. & Ribeiro, H. (1995) Description of *Aedes* (*Aedimorphus*) *gandarai*, a new species from the island of São Tomé, with keys to all known species of the Tarsalis Group (Diptera: Culicidae). *Mosquito Systematics* (for 1994), 26 (3), 97–104.

da Cunha Ramos, H., Ribeiro, H. & de Barros Machado, A. (1992) Research on the mosquitoes of Angola *(Diptera, Culicidae)*. XIX – *Eretmapodites tendeiroi*, a new species of the «*plioleucus* group». *Garcia de Orta, Série de Zoologia* (for 1989), 16 (1–2), 133–136.

da Cunha Ramos, H., Ribeiro, H. & Harrison, B.A. (2003) A new European mosquito species: *Culex (Neoculex) europaeus* (Diptera: Culicidae). *European Mosquito Bulletin*, 15, 6–11.

da Fonseca, F. & da Silva Ramos, A. (1940) Novo subgenero e novas especies de anofelinas neotropicas (*Diptera. Culicidae*). (Nota previa). *Memorias do Instituto Butantan* (for 1939), 13, 383–387, 2 pls.

Daniels, C.W. (1908) Notes on the mosquitoes on the river and coast districts of the eastern side of the Peninsula. *Studies from the Institute for Medical Research, Federated Malay States*, 3 (3), 262–266.

Danilov, V.N. (1982) Mosquito fauna (Diptera: Culicidae) of the Ethiopian zoogeographical region 2. A new species of *Mansonia* (*Coquillettidia*) and a key for adults of the genus *Mansonia* [in Russian]. *Meditsinskaya Parazitologiya i Parazitarnye Bolezni*, 51 (4), 26–30.

Danilov, V.N. (1985a) Mosquitoes (Diptera, Culicidae) of Afghanistan. Communication I. Identification table of the females [in Russian]. *Meditsinskaya Paraziologiya i Parazitarnye Bolezni*, 1985 (2), 67–72.

Danilov, V.N. (1985b) Mosquitoes of Afghanistan. Communication 2. A key to fourth-stage larvae [in Russian]. *Meditsinskaya Paraziologiya i Parazitarnye Bolezni*, 198 (4), 51–55.

Danilov, V.N. (1987) Mosquitoes of the subgenus *Aedes* (Diptera, Culicidae) of the USSR fauna. II. *Aedes dahuricus* sp. n. [in Russian]. *Vestnik Zoologii*, 1987 (4), 35–41.

Danilov, V.N. (1989) Two new Afrotropical subgenera of the mosquito genus *Culex* (Diptera, Culicidae) [in Russian]. *Entomologischeskoe Obozrenie*, 68 (4), 790–797.

Danilov, V.N. & Filippova, V.V. (1978) A new species of mosquito, *Aedes* (*Stegomyia*) *sibiricus* sp. n. (Culicidae) [in Russian]. *Parazitologiia*, 12 (2), 170–176.

Danilov, V.N. & Gornostaeva, R.M. (1987) A new species of mosquito, *Aedes* (*Ochlerotatus*) *intermedius* sp. n. (Diptera, Culicidae) [in Russian]. *Parazitologiia*, 21 (5), 612–619.

Danilov, V.N. & Stupin, V.I. (1982) A new species of the mosquito genus *Aedes* Mg. (Diptera, Culicidae) from southern Siberia and Priamuria [in Russian]. *Entomologischeskoe Obozrenie*, 61 (4), 824–829.

Danilov, V.N., Markovich, N.Ya. & Proskuryakova, A.M. (1978) A new species of mosquito, *Aedes* (*Ochlerotatus*) *sergievi* sp. n. [in Russian]. *Parazitologiia*, 12 (1), 44–49.

Darsie, R.F., Jr (1985) Mosquitoes of Argentina. Part I. Keys for identification of adult females and fourth stage larvae in English and Spanish (Diptera, Culicidae). *Mosquito Systematics*, 17 (3), 153–253.

Darsie, R.F., Jr (2000) Description of the pupae of five species in the subgenus *Armigeres*, genus *Armigeres* Theobald, with a key to species of the known pupae of the subgenus (Diptera: Culicidae). *Proceedings of the Entomological Society of Washington*, 102 (1), 108–119.

Darsie, R.F., Jr & Pradhan, S.P. (1990) The mosquitoes of Nepal: their identification, distribution and biology. *Mosquito Systematics*, 22 (2), 69–130.

Darsie, R.F., Jr & Ward, R.A. (1981) Identification and geographical distribution of the mosquitoes of North America, north of Mexico. *Mosquito Systematics Supplement*, 1, 1–313.

Darsie, R.F., Jr & Ward, R.A. (2005) *Identification and Geographical Distribution of the Mosquitoes of North America, North of Mexico*. University Press of Florida, Gainesville, FL, USA.

Darsie, R.F., Jr, Courtney, G.W. & Pradhan, S.P. (1993) Notes on the mosquitoes of Nepal: III. Additional new records in 1992 (Diptera: Culicidae). *Mosquito Systematics*, 25 (3), 186–191.

Darsie, R.F., Jr, Courtney, G.W. & Pradhan, S.P. (1996) Notes on the mosquitoes of Nepal IV. Results of the 1994 collecting in the midwestern region, including new country records and voucher confirmation (Diptera, Culicidae). *Journal of the American Mosquito Control Association*, 12 (1), 130–134.

Das, B.P., Rajagopal, R. & Akiyama, J. (1990) Pictorial key to the species of Indian anopheline mosquitoes. *Zoology (New Delhi)*, 2 (3), 131–162.

da Silva, A.F., Machado, L.C., de Paula, M.B., da Silva Pessoa Vieira, C.J., de Morais Bronzoni, R.V., de Melo Santos, M.A.V. & Wallau, G.L. (2020a) Culicidae evolutionary history focusing on the Culicinae subfamily based on mitochondrial phylogenomics. *Scientific Reports*, 10 (1), 18823. DOI: 10.1038/s41598-020-74883-3

da Silva, F.S., Cruz, A.C.R., de Almeida Medeiros, D.B., da Silva, S.P., Nunes, M.R.T., Martins, L.C., Chiang, J.O., da Silva Lemos, P., Cunha, G.M., de Araujo, R.F., de Oliveira Monteiro, H.A. & Neto, J.P.H. (2020b) Mitochondrial genome sequencing and phylogeny of *Haemagogus albomaculatus*, *Haemagogus leucocelaenus*, *Haemagogus spegazzinii*, and *Haemagogus tropicalis* (Diptera: Culicidae). *Scientific Reports*, 10, 16948. DOI: 10.1038/s41598-020-73790-x

da Silva e Silva, L.H., da Silva, F.S., de Almeida Medeiros, D.B., Cruz, A.C.R., da Silva, S.P., de Oliveira Aragao, A., Dias, D.D., Sena do Nascimento, B.L., Júnior, J.W.R., Vieira, D.B.R., de Oliveira Monteiro, H.A. & Neto, J.P.N. (2022) Description of the mitogenome and phylogeny of *Aedes* spp. (Diptera: Culicidae) from the Amazon region. *Acta Tropica*, 232, 106500. DOI: 10.1016/j.actatropica.2022.106500

da Silva Mattos, S. & Xavier, S.H. (1991) *Culex (Melanoconion) guedesi* uma nova espécie do Brasil (Diptera, Culicidae). *Revista Brasileira de Biologia*, 51 (1), 193–196.

da Silva Mattos, S., de Silveira Guedes, A. & Xavier, S.H. (1978) Uma espécie nova de *Culex (Culex)* que se cria em umbelífera (Diptera, Culicidae). *Revista Brasileira de Biologia*, 38 (1), 31–35.

Davis, N.C. (1931) A new anopheline mosquito from Pará, Brazil. *American Journal of Hygiene*, 13 (1), 345–348, 2 pl. DOI: 10.1093/oxfordjournals.aje.a117122

Davis, D.E. (1944) Larval habits of some Brazilian mosquitoes. *Revista de Entomologia*, 15, 221–235.

Davis, N.C. & Shannon, R.C. (1931) Further attempts to transmit yellow fever with mosquitoes of South America. *American Journal of Hygiene*, 14 (3), 715–722. DOI: 10.1093/oxfordjournals.aje.a117801

Debenham, M.L., Hicks, M.M. & Griffiths, M. (1989) *The Culicidae of the Australasian Region*. Vol. 12. Summary of taxonomic changes. Revised alphabetical list of species. Supplementary bibliography. Errata and addenda. Geographic guide to species. Synopsis of disease relationships. Indexes. Monograph Series, Entomology Monograph No. 2. Australian Government Publishing Service, Canberra, Australia.

de Carvalho-Pinto, C.J. & Lourenço-de-Oliveira, R. (2004) Isoenzimatic analysis of four *Anopheles* (*Kerteszia*) *cruzii* (Diptera: Culicidae) populations of Brazil. *Memórias do Instituto Oswaldo Cruz*, 99 (5), 471–475. DOI: 10.1590/S0074-02762004000500002

de Charmoy, d'E. (1908) On three new species of *Culex* collected during the anti-malarial campaign in Mauritius in 1908. *Annals of Tropical Medicine and Parasitology*, 2 (3), 257–264. DOI: 10.1080/00034983.1908.11684887 de Deus Palma, J. & Galvão, A.B. (1969) *Lynchiella rizzoi* sp. n., toxorrinquitíneo bromelícola do Estado de Goiás (Diptera, Culicidae). *Revista Goiana de Medicina*, 15 (1,2), 3–13.

de Geer, C. (1776) *Memoires pour servir a l'histoire des insectes*. Tome sixieme. De l'imprimerie de Pierre Hesselberg, Stockholm, Sweden.

de Grandpré, A.D. & de Charmoy, D. d'E. (1901) *Les moustiques: Anatomie et biologie. Contribution à l'étude des culicidées et principalement des genres* Culex *et* Anopheles *de leur rôle dans la propagation de la malaria et de la filariose et des moyens de s'en préserver*. Ile Maurice, Port Louis (for 1900), Mauritius.

de Leon, J.R. (1938) El anophelismo de altura en Guatemala. *Boletin de la Direccion General de Sanidad Publica (Guatemala)*, 9, 411–424.

Delfinado, M.D. (1966a) The culicine mosquitoes of the Philippines, tribe Culicini (Diptera, Culicidae). *Memoirs of the American Entomological Institute*, 7, 1–252.

Delfinado, M.D. (1966b) Two undescribed species of *Uranotaenia* from the Philippines (Diptera: Culicidae). *Journal of Medical Entomology*, 3 (1), 36–37. DOI: 10.1093/jmedent/3.1.36

Delfinado, M.D. (1967) Contributions to the mosquito fauna of Southeast Asia. I. The genus *Aedes*, subgenus *Neomacleaya* Theobald in Thailand. *Contributions of the American Entomological Institute*, 1 (8), 1–56.

Delfinado, M.D. (1968) Contributions to the mosquito fauna of Southeast Asia. III. The genus *Aedes*, subgenus *Neomacleaya* Theobald in Southeast Asia. *Contributions of the American Entomological Institute*, 2 (4), 1–74.

Delfinado, M.D. & Hodges, E.R. (1968) Three new species of the genus *Tripteroides*, subgenus *Tripteroides* Giles (Diptera: Culicidae). *Proceedings of the Entomological Society of Washington*, 70 (4), 361–375.

de Linero, M.Q. (1967) Nueva especie de "*Culex* (*Melanoconion*)". *Revista Brasileira de Biologia*, 27 (3), 289–291.

del Ponte, E. (1939) Identificación de «Sabethini» (Dip. Culicidae) por medio de tarjetas perforadas. *Physis*, 17, 535–541.

del Ponte, E. & Cerqueira, N. (1938) Alguns sabethineos do Brasil (Diptera, Culicidae). *Revista de Entomologia*, 8 (3–4), 225–237.

del Ponte, E., Castro, M.P. & Garcia, M. (1951) Clave para las especies de *Psorophora* y *Aëdes* de la Argentina y comarcas vecinas. Diagnosis de *Aëdes* (*O.*) *raymondi* n. sp. (Diptera, Culicidæ). *Anales de la Sociedad Cientifica Agentina*, 151 (5), 228–243.

del Vecchio, G. (1939) Sulle varietà di *A. clæviger* (*bifurcatus*). Nota I. *Rivista di Parassitologia*, 3 (1), 27–37, 1 pl.

Demari-Silva, B., Vesgueiro, F.T., Sallum, M.A.M. & Marrelli, M.T. (2011) Taxonomic and phylogenetic relationships between species of the genus *Culex* (Diptera: Culicidae) from Brazil inferred from the cytochrome *c* oxidase I mitochondrial gene. *Journal of Medical Entomology*, 48 (2), 272–279. DOI: 10.1603/ME09293

Demari-Silva, B., Foster, P.G., de Oliveira, M.P., Bergo, E.S., Sanabani, S.S., Pessôa, R. & Sallum, M.A.M. (2015) Mitochondrial genomes and comparative analyses of *Culex camposi*, *Culex coronator*, *Culex usquatus* and *Culex usquatissimus* (Diptera:Culicidae [*sic*]), members of the coronator group [*sic*]. *BMC Genomics*, 16, 831. DOI: 10.1186/s12864-015-1951-0

de Meijere, J.C.H. (1909) Drei myrmecophile Dipteren aus Java. *Tijdschrift voor Entomologie*, 52, 165–174, 1 pl.

de Meijere, J.C.H. (1910) *Nepenthes*-Tiere. I. Systematik. *Annales du Jardin Botanique de Buitenzorg, Supplément*, 3 (2), 917–940, 4 pls.

de Meillon, B. (1931a) Illustrated keys to the full-grown larvae and adults of South African anopheline mosquitos. *Publications of the South African Institute for Medical Research*, 4 (28), 275–375.

de Meillon, B. (1931b) A new South African anopheline. *Journal of the Medical Association of South Africa*, 5, 482–483.

de Meillon, B. (1935) Entomological studies. Studies on insects of medical importance in South Africa – Part II. *Publications of the South African Institute for Medical Research*, 6 (35), 323–365.

de Meillon, B. (1942) New Nematocera from the Ethiopian Region. *Journal of the Entomological Society of Southern Africa*, 5, 87–98.

de Meillon, B. (1943) New records, and new species of Nematocera (Diptera) from the Ethiopian region. *Journal of the Entomological Society of Southern Africa*, 6, 90–113.

de Meillon, B. (1947) New records and species of biting insects from the Ethiopian region. II. *Journal of the Entomological Society of Southern Africa*, 10 (1), 110–124.

de Meillon, B. & de Carvalho Pereira, M. (1940) Notes on some anophelines (Dipt. Culicidae) from Portuguese East Africa. *Moçambique Documentário Trimestral*, 23, 69–83, 1 photo, 7 pls, map.

de Meillon, B. & Evans, A.M. (1935) Two new anophelines from South Africa. *Annals of Tropical Medicine and Parasitology*, 29 (1), 91–94.

de Meillon, B. & Lavoipierre, M. (1944) New records and species of biting insects from the Ethiopian region. *Journal of the Entomological Society of Southern Africa*, 7, 38–67.

de Meillon, B. & Leeson, H.S. (1940) Notes on Ethiopian Anophelini. *Bulletin of Entomological Research*, 31 (1), 61–67. DOI: 10.1017/S0007485300004818

de Meillon, B. & Rebêlo, A. (1941) Culicini (Diptera, Nematocera) from the Colony of Mocambique. *Moçambique Documentário Trimestral*, 27, 69–77.

de Meillon, B., Parent, M. & Black, L.O'C. (1945) Descriptions of new larvae and pupae of Ethiopian Culicini. *Bulletin of Entomological Research*, 36 (1), 85–101. DOI: 10.1017/S0007485300023932

de Oliveira Castro, G.M. (1932) Estudo sobre uma especie de *Culex*, que se cria em buracos de guayamú. (Dipt., Culicidae). *Revista de Entomologia*, 2 (1), 97–105.

de Oliveira Castro, G.M. (1935) Estudo sobre uma especie de "*Limatus*" que se cria em buracos de pau (Dipt., Culicidae). *Revista do Departamento Nacional da Produção Animal*, 2, 143–155.

Deus, S. (2009) *Phylogeny of selected species of subgenus* Culex *mosquitoes (Diptera: Culicidae) from the U.S., Puerto Rico, and Guatemala, based on nucleotide sequences from three genes*. PhD Thesis, Department of Microbiology, Immunology and Pathology, Colorado State University, Fort Collins, CO, USA.

D'Haenens, G. (1961) Description d'un Anophele [*sic*] nouveau du Congo (Kwango) *A. moucheti bervoetsi* n. subsp. *Bulletin et Annales de la Société Royale d'Entomologie de Belgique*, 97 (7–8), 188–200.

Diallo, M., Ba, Y., Sall, A.A., Diop, O.M., Ndione, J.A., Mondo, M., Girault, L. & Mathiot, C. (2003) Amplification of the sylvatic cycle of dengue virus type 2, Senegal, 1999–2000: entomologic findings and epidemiologic considerations. *Emerging Infectious Diseases*, 9 (3), 362–367. DOI: 10.3201/eid0903.020219

Diallo, M., Thonnon, J., Traore-Lamizana, M. & Fontenille, D. (1999) Vectors of Chikungunya virus in Senegal: current data and transmission cycles. *American Journal of Tropical Medicine and Hygiene*, 60 (2), 281–286. DOI: 10.4269/ajtmh.1999.60.281

Díaz Nájera, A. (1966) *Aedes* de Mexico. Descripcion de una nueva especie del subgenero *Howardina* (Diptera: Culicidae). *Revista de Investigacion en Salud Pública*, 26 (4), 331–344.

Diesing, C.M. (1850) *Systema helminthum*. Vol. I. C. Gerold's Sohn, Vindobonae [Vienna].

Distant, W.L. (1902) Rhynchota.—Vol. I. (Heteroptera). In: Blanford, W.T. (ed.), *The fauna of British India, including Ceylon and Burma*. Taylor & Francis, London, UK.

Division of Medical Entomology, Bureau of Laboratories (1944) *The Mosquitoes of Texas*. Texas State Health Department. Austin, TX, USA, 100 pp. [Authorship of an included new species attributed to N.M. Randolph and K. O'Neill, Preface, p. 3].

Dixit, J., Srivastava, H., Sharma, M., Das, M.K., Singh, O.P., Raghavendra, K., Nanda, N., Dash, A.P., Saksena, D.N. & Das, A. (2010) Phylogenetic inference of Indian malaria vectors from multilocus DNA sequences. *Infection, Genetics and Evolution*, 10 (6), 755–763. DOI: 10.1016/j.meegid.2010.04.008

Djadid, N.D., Jazayeri, H., Gholizadeh, S., Pashaeirad, S. & Zakeri, S. (2009) First record of a new member of *Anopheles* Hyrcanus Group from Iran: molecular identification, diagnosis, phylogeny, status of kdr resistance and *Plasmodium* infection. *Journal of Medical Entomology*, 46 (5), 1084–1093. DOI: 10.1603/033.046.0515

Dobrotworsky, N.V. (1953) The *Culex pipiens* group in south-eastern Australia. I. *Proceedings of the Linnean Society of New South Wales*, 77 (5–6), 357–360.

Dobrotworsky, N.V. (1954) The genus *Theobaldia* (Diptera, Culicidae) in Victoria. *Proceedings of the Linnean Society of New South Wales*, 79 (3–4), 65–78.

Dobrotworsky, N.V. (1956) Notes on Australian mosquitoes (Diptera, Culicidae). I. Some species of the subgenus *Neoculex*. *Proceedings of the Linnean Society of New South Wales*, 81 (2), 105–114.

Dobrotworsky, N.V. (1957) Notes on Australian mosquitoes (Diptera, Culicidae). II. Notes on *Anopheles stigmaticus* Skuse and description of new species of *Anopheles* from Australia and New Guinea. *Proceedings of the Linnean Society of New South Wales*, 82 (2), 180–188.

Dobrotworsky, N.V. (1958a) Notes on Australian mosquitoes (Diptera, Culicidae). III. The subgenus *Lophoceraomyia* in Victoria. *Proceedings of the Linnean Society of New South Wales* (for 1957), 82 (3), 317–321.

Dobrotworsky, N.V. (1958b) Designation of the type species of the subgenus *Neotheobaldia* Dobrotworsky (genus *Theobaldia* Neveu-Lemaire, 1902) (Diptera, Culicidae). *Proceedings of the Entomological Society of Washington*, 60 (4), 186.

Dobrotworsky, N.V. (1959) Notes on Australian mosquitoes (*Diptera, Culicidae*). IV. *Aëdes alboannulatus* complex in Victoria. *Proceedings of the Linnean Society of New South Wales*, 84 (1), 131–145.

Dobrotworsky, N.V. (1960a) The subgenus *Ochlerotatus* in the Australian Region (Diptera: Culicidae). III. Review of the Victorian species of *perkinsi* and *cunabulanus* sections with descriptions of two new species. *Proceedings of the Linnean Society of New South Wales*, 85 (1), 53–74.

Dobrotworsky, N.V. (1960b) The genus *Theobaldia* (Diptera, Culicidae) in Victoria. II. *Proceedings of the Linnean Society of New South Wales*, 85 (2), 240–247.

Dobrotworsky, N.V. (1961a) Notes on Australian mosquitoes (Diptera, Culicidae). V. Subgenus *Pseudoskusea* in Victoria. *Proceedings of the Linnean Society of New South Wales* (for 1960), 85 (3), 257–267.

Dobrotworsky, N.V. (1961b) A new name for *Aedes waterhousei* Dobrotworsky (Diptera: Culicidae). *Proceedings of the Entomological Society of Washington* (for 1960), 62 (4), 248.

Dobrotworsky, N.V. (1962) Notes on Australian mosquitoes (Diptera, Culicidae). VI. Five new Victorian species and a description of the larva of *Aëdes milsoni* (Taylor). *Proceedings of the Linnean Society of New South Wales*, 87 (3), 291–302.

Dobrotworsky, N.V. (1965) *The Mosquitoes of Victoria (Diptera, Culicidae)*. Melbourne University Press, Melbourne, Australia.

Dobrotworsky, N.V. (1966) Mosquitoes of Tasmania and Bass Straits Islands. *Proceedings of the Linnean Society of New South Wales*, 91 (2), 121–146.

Dobrotworsky, N.V. (1971) Contributions to the mosquito fauna of Southeast Asia. X. The genus *Culiseta* Felt in Southeast Asia. *Contributions of the American Entomological Institute*, 7 (3), 38–61.

Dobrotworsky, N.V. & Drummond, F.H. (1953) The *Culex pipiens* group in south-eastern Australia. II. *Proceedings of the Linnean Society of New South Wales*, 78 (3–4), 131–146.

Dolbeškin, B.I., Gorickaya, V.V. & Mitrofanova, Yu.G. (1930) Description of a new species of the genus *Aedes* (n. sp.) from eastern Europe [in Russian]. *Magazine of Parasitology Zoological Museum of Sciences of the USSR*, 1, 253–260, 2 pls.

Doleschall, C.L. (1857) Tweede bijdrage tot de kennis der dipterologische fauna van Nederlandsch Indië. *Natuurkundig Tijdschrift voor Nederlandsch Indië*, 14, 377–418, 11 pls.

Donald, C.L., Siriyasatien, P. & Kohl, A. (2020) *Toxorhynchites* species: a review of current knowledge. *Insects*, 11, 747. DOI: 10.3390/insects11110747

do Nascimento, B.L.S., da Silva, F.S., Nunes-Neto, J.P., de Almeida Medeiros, D.B., Cruz, A.C.R., da Silva, S.P., da Silva e Silva, L.H., de Oliveira Monteiro, H.A., Dias, D.D., Vieira, D.B.R., Rosa, J.W., Jr, Brandão, R.C.F., Chiang, J.O., Martins, L.C. & da Costa Vasconcelos, P.F. (2021) First description of the mitogenome and phylogeny of Culicinae species from the Amazon region. *Genes*, 12, 1983. DOI: 10.3390/genes12121983

Dong, X. & Miyagi, I. (1995) A new species description of subgenus *Suaymyia* of genus *Topomyia* (Diptera: Culicidae) [in Chinese]. *Zoological Research*, 16 (4), 343–347.

Dong, X.-S., Dong, L.-M. & Wu, C. (2004) A new species of the genus *Uranotaenia* from China (Diptera, Culicidae) [in Chinese]. *Acta Zootaxonomica Sinica*, 29 (3), 566–568.

Dong, X.-S., Dong, L.-M. & Wu, C. (2008) A new species of the genus *Tripteroides* (Diptera: Culicidae) from China [in Chinese]. *Acta Zootaxonomica Sinica*, 33 (1), 187–190.

Dong, X.-S., Dong, L.-M. & Zhou, H.-N. (2003a) A new species of genus *Uranotaenia* (Diptera: Culicidae) [in Chinese]. *Entomotaxonomia*, 25 (3), 204–208.

Dong, X.-S., Dong, L.-Y.-M., Zhou, H.-N., Wang, X.-Z. & Lu, Y.-R. (2003b) Three new species of mosquitoes from Yunnan, China (Diptera: Culicidae) [in Chinese]. *Acta Entomologica Sinica*, 46 (4), 519–532.

Dong, X.-S., Wang, X.-Z. & Lu, B.-L. (1990) A new species of *Topomyia* from Yunnan Province, China (Diptera: Culicidae) [in Chinese]. *Acta Zootaxonomica Sinica*, 15 (1), 64–66.

Dong, X.-S., Wang, X.-Z. & Lu, B.-L. (1995) A new species of *Topomyia* from Yunnan Province, China (Diptera: Culicidae) [in Chinese]. *Acta Zootaxonomica Sinica*, 20 (1), 381–186.

Dong, X.-S., Wang, X.-Z. & Zhou, H.-N. (2002) A new species of genus *Heizmannia* (Diptera: Culicidae) [in Chinese]. *Entomotaxonomia*, 24 (2), 131–133.

Dong, X.-S., Wu, C. & Mao, X.-H. (2006) A new species of the genus *Topomyia* (Diptera: Culicidae) [in Chinese]. *Entomotaxonomia*, 28 (3), 217–221.

Dong, X., Zhou, H. & Dong, L. (1995a) Studies on genus *Armigeres* of Yunnan with description of a new species (Diptera: Culicidae) [in Chinese]. *Zoological Research*, 16 (2), 95–104.

Dong, X., Zhou, H. & Dong, L. (1995b) A new species of the genus *Armigeres* (Diptera: Culicidae) [in Chinese]. *Entomotaxonomia*, 17 (4), 281–286.

Dong, X., Zhou, H. & Dong, L. (1997a) Study on *Topomyia* of Yunnan with one new species and one new record (Diptera: Culicidae) [in Chinese]. *Zoological Research*, 18 (3), 247–251.

Dong, X.-S., Zhou, H.N. & Dong, L.M. (1997b) A new species of subgenus *Rachionotomyia* of the genus *Tripteroides* (Diptera: Culicidae) [in Chinese]. *Acta Zootaxonomica Sinica*, 22 (2), 197–200.

Dong, X., Zhou, H. & Dong, L. (1999) A new species and a new Chinese record of the subgenus *Bothaella* of genus *Aedes* (Diptera: Culicidae) [in Chinese]. *Entomotaxonomia*, 21 (2), 133–137.

Dong, X.-S., Zhou, H.-N. & Dong, L.M. (2001) A new species of *Tripteroides* (s.str.) from Yunnan China (Diptera: Culicidae) [in Chinese]. *Acta Zootaxonomica Sinica*, 26 (2), 221–224.

Dong, X.-S., Zhou, H.-N. & Dong, L.-M. (2002a) One new species and two new records of the subgenus *Kenknightia* of the genus *Aedes* in Yunnan (Diptera: Culicidae) [in Chinese]. *Acta Entomologica Sinica*, 45 (4), 516 –521.

Dong, X., Zhou, H. & Dong, L. (2002b) A new species of *Armigeres* from Yunnan, China (Diptera: Culicidae) [in Chinese]. *Acta Parasitologica et Medica Entomologica*, 9 (3), 155–159.

Dong, X.-S., Zhou, H.N. & Dong, L.M. (2004) A new species of the genus *Armigeres* (Diptera: Culicidae) [in Chinese]. *Entomotaxonomia*, 26 (2), 123–125.

Dong, X.-S., Zhou, H.-N. & Dong, L.-M. (2005) A new species of the genus *Heizmannia* (Diptera: Culicidae) from China. *Entomotaxonomia*, 27 (3), 209–211.

Dong, X.-S., Zhou, H.-N., Dong, L.-M. & Mao, X.-H. (2007) A new species of genus *Anopheles* (Diptera: Culicidae) [in Chinese]. *Entomotaxonomia*, 29 (1), 37–43.

Dong, X., Zhou, H., Wang, P. & Wang, X. (1997) A new species of the subgenus *Heizmannia* (Diptera: Culicidae) [in Chinese]. *Zoological Research*, 18 (1), 27–31.

Dönitz, W. (1901) Nachrichten aus dem Berliner Entomologischen Verein. *Insektenbörse*, 18 (5), 36–38.

Dönitz, W. (1902) Beiträge zur Kenntniss der *Anopheles*. *Zeitschrift für Hygiene und Infektionskrankheiten*, 41, 15–88, 2 pls.

Doucet, J. (1949a) Étude des Culicidae (Diptera) du lac Alaotra. *Mémoires de l'Institut Scientifique de Madagascar Série A*, 3 (2), 121–145.

Doucet, J. (1949b) Recherches sur les Culicides de Madagascar. *Mémoires de l'Institut Scientifique de Madagascar*, 3 (3), 325–332.

Doucet, J. (1950) Les culicines de Madagascar [Dipt.]. *Mémoires de l'Institut Scientifique de Madagascar Série A*, 4 (1), 39–65.

Doucet, J. (1951a) Les moustiques de la region de Perinet. *Mémoires de l'Institut Scientifique de Madagascar Série A*, 6 (1), 63–82.

Doucet, J. (1951b) Etude des culicides de la region de Vangaindrano (Diptera). *Mémoires de l'Institut Scientifique de Madagascar Série A*, 6 (1), 83–114.

Doucet, J. (1962) Moustiques forestiers de la République de Côte-d'Ivoire. III. – Notes biologiques et systématiques sur *Uranotaenia andreae* sp. n. et *Culex (Neoculex) vinckei* Hamon, Holstein et Rivola, 1957. *Bulletin de la Société de Pathologie exotique* (for 1961), 54 (5), 1156–1164.

Dubitzky, A.M. (1968) A new species *Aedes montchadskyi* sp. n. from the basin of the Ili River [in Russian]. *Parazitologiia*, 2 (3), 218–224.

Dubitzky, A.M. (1970) A new species of mosquito *Aedes stramineus* sp. n. (Diptera, Culicidae) [in Russian]. *Parazitologiia*, 4 (5), 408–413.

Dubitzky, A.M. & Deshevykh, N.D. (1978) A new species of mosquito *Aedes (Ochlerotatus) gutzevichi* sp.n. from central Asia (Culicidae) [in Russian]. *Parazitologiia*, 12 (2), 177–182.

DuBose, W.P. & Curtin, T.J. (1965) Identification keys to the adults and larval mosquitoes of the Mediterranean area. *Journal of Medical Entomology*, 1 (4), 349–355. DOI: 10.1093/jmedent/1.4.349

Dumbleton, L.J. (1962) A new species and new sub-genus of *Aedes* (Diptera: Culicidae) from New Zealand. *New Zealand Journal of Science*, 5 (1), 17–27.

Dunn, L.H. (1918) A new mosquito (*Aedes whitmorei)* from Colombia. *Proceedings of the Entomological Society of Washington*, 20 (6), 128–130.

Duret, J.P. (1953) Las especies argentinas de *Culex (Melanoconion) (Diptera-Culicidae)*. *Revista de la Sociedad Entomólogica Argentina*, 16 (3), 67–76.

Duret, J.P. (1954) Nuevo mosquito argentino (Diptera, Culicidae). *Neotropica*, 1 (2), 23–26.

Duret, J.P. (1967a) Dos especies nuevas de *Culex (Melanoconion)* (Diptera - Culicidae). *Neotropica*, 13 (41), 77–84.

Duret, J.P. (1967b) *Culex (Culex) chitae* una nueva especie de Colombia (Diptera, Culicidae). *Neotropica*, 13 (42), 113–116.

Duret, J.P. (1968a) *Culex (Melanoconion) kerri* y *C. (M.) cristovaoi* dos nuevas especies del Brasil. (Diptera, Culicidae). *Neotropica*, 14 (43), 11–16.

Duret, J.P. (1968b) Cuatro especies nuevas de *Culex* (Diptera, Culicidae). *Neotropica*, 14 (44), 57–64.

Duret, J.P. (1968c) *Culex (Melanoconion) contei* una nueva especie del Brasil y Panama (Diptera, Culicidae). *Neotropica*, 14 (45), 121–124.

Duret, J.P. (1968d) Dos especies nuevas de *Culex* del Brasil *(Diptera, Culicidae)*. *Physis*, 28 (76), 15–19.

Duret, J.P. (1968e) Dos especies nuevas de *Culex (Culex)* de la Argentina *(Diptera, Culicidae)*. *Physis*, 27 (75), 323–328.

Duret, J.P. (1968f) Cinco especies nuevas de *Culex (Melanoconion)* (Diptera, Culicidae). *Revista de la Sociedad Entomológica Argentina*, 30 (1–4), 69–81.

Duret, J.P. (1969a) Contribucion al conocimiento de los *Culex* del Paraguay (Diptera-Culicidae). *Revista de la Sociedad Entomólogica Argentina*, 31 (1–4), 3–13.

Duret, J.P. (1969b) *Culex (Melanoconion) bahiense* una especie nueva del Brasil (Diptera, Culicidae). *Neotropica*, 15 (46), 40–42.

Duret, J.P. (1969c) *Culex (Anoedioporpa) damascenoi* una nueva especie del Brasil (Diptera - Culicidae). *Neotropica*, 15 (48), 143–145.

Duret, J.P. (1969d) Estudio sobre *Culex (Melanoconion) ocellatus* y sus especies afines *(Diptera. Culicidae)*. *Physis*, 29 (78), 113–129.

Duret, J.P. (1970) *Uranotaenia riverai* una nueva especie de Colombia *(Diptera, Culicidae)*. *Revista de la Sociedad Entomólogica Argentina*, 32 (1–4), 43–46.

Duret, J.P. (1971) Cinco especies nuevas de culicidos neotropicales (Diptera-Culicidae). *Neotropica*, 17 (52), 15–28.

Duret, J.P. (1972) *Culex (Anoedioporpa) quasioriginator*, una nueva especie del Brasil. *Neotropica*, 18 (55), 1–5.

Duret, J.P. (1982) *Wyeomyia (Dendromyia) sirivanakarni* una especie nueva de Panama (Diptera, Cuicidae). *Revista de la Sociedad Entomólogica Argentina*, 41 (1–4), 167–170.

Duret, J.P. & Barreto, M.P. (1956) Notas sôbre culícidas do Estado de São Paulo, Brasil, com descrições de três novas espécies de *Culex (Diptera, Culicidae)*. *Revista Brasileira de Entomologia*, 5, 81–99.

Duret, J.P. & Damasceno, R.G. (1955) Notas sobre *Culex (Tinolestes)* de Pará, Brasil (Diptera Culicidae). *Revista Ecuatoriana de Entomologia y Parasitologia*, 2 (3–4), 393–409, 3 pls.

Dusfour, I., Michaux, J.R., Harbach, R.E. & Manguin, S. (2007) Speciation and phylogeography of the Southeast Asian *Anopheles sundaicus* complex. *Infection, Genetics and Evolution*, 7 (4), 484–493. DOI: 10.1016/j.meegid.2007.02.003

Dyar, H.G. (1905a) Remarks on genitalic genera in the Culicidae. *Proceedings of the Entomological Society of Washington*, 7 (1), 42–49.

Dyar, H.G. (1905b) A new mosquito. *Journal of the New York Entomological Society*, 13 (2), 74.

Dyar, H.G. (1906) On the classification of the Culicidae. *Proceedings of the Entomological Society of Washington*, 7 (4), 188–191.

Dyar, H.G. (1907) Report on the mosquitoes of the coast region of California, with descriptions of new species. *Proceedings of the United States National Museum*, 32 (1516), 121–129. DOI: 10.5479/SI.00963801.32-1516.121

Dyar, H.G. (1916a) Mosquitoes at San Diego, California. *Insecutor Inscitiae Menstruus*, 4 (4–6), 46–51.

Dyar, H.G. (1916b) New *Aedes* from the mountains of California *(Diptera, Culicidae)*. *Insecutor Inscitiae Menstruus*, 4 (7–9), 80–90.

Dyar, H.G. (1917) The mosquitoes of the Pacific Northwest *(Diptera, Culicidae). Insecutor Inscitiae Menstruus*, 5 (7–9), 97–102, 1 pl.

Dyar, H.G. (1918a) The male genitalia of *Aedes* as indicative of natural affinities *(Diptera, Culicidae). Insecutor Inscitiae Menstruus*, 6 (4–6), 71–86.

Dyar, H.G. (1918b) A revision of the American species of *Culex* on the male genitalia *(Diptera, Culicidae). Insecutor Inscitiae Menstruus*, 6 (4–6), 86–111, 2 pls.

Dyar, H.G. (1918c) New American mosquitoes (*Diptera, Culicidae). Insecutor Inscitiae Menstruus*, 6 (7–9), 120–129.

Dyar, H.G. (1918d) A note on the American species of *Mansonia (Diptera, Culicidae). Insecutor Inscitiae Menstruus*, 6 (4–6), 112–115.

Dyar, H.G. (1919a) Westward extension of the Canadian mosquito fauna (*Diptera, Culicidae). Insecutor Inscitiae Menstruus*, 7 (1–3), 11–39.

Dyar, H.G. (1919b) A revision of the American Sabethini of the *Sabethes* group by the male genitalia *(Diptera, Culicidae). Insecutor Inscitiae Menstruus*, 7 (7–9), 114–142, 1 pl.

Dyar, H.G. (1919c) A note on Argentine mosquitoes (*Diptera, Culicidae). Insecutor Inscitiae Menstruus*, 7 (4–6), 85–89.

Dyar, H.G. (1920a) The mosquitoes of British Columbia and Yukon Territory, Canada *(Diptera, Culicidae). Insecutor Inscitiae Menstruus*, 8 (1–3), 1–27.

Dyar, H.G. (1920b) The species of *Choeroporpa*, a subgenus of *Culex (Diptera, Culicidae). Insecutor Inscitiae Menstruus*, 8 (4–6), 54–81.

Dyar, H.G. (1920c) The classification of American *Aedes (Diptera, Culicidae). Insecutor Inscitiae Menstruus*, 8 (7–9), 103–126.

Dyar, H.G. (1920d) The *Grabhamia* group of *Psorophora (Diptera, Culicidae). Insecutor Inscitiae Menstruus*, 8 (7–9), 140–141.

Dyar, H.G. (1920e) A new *Culex* from Panama *(Diptera, Culicidae). Insecutor Inscitiae Menstruus*, 8 (10–12), 173–174.

Dyar, H.G. (1920f) A collection of mosquitoes from the Philippine Islands *(Diptera, Culicidae). Insecutor Inscitiae Menstruus*, 8 (10–12), 175–186.

Dyar, H.G. (1920g) A new mosquito from Mexico *(Diptera, Culicidae). Insecutor Inscitiae Menstruus*, 8 (4–6), 81–82.

Dyar, H.G. (1920h) The earliest name of the yellow fever mosquito *(Diptera: Culicidae). Insecutor Inscitiae Menstruus*, 8 (10–12), 204.

Dyar, H.G. (1921a) Ring-legged *Culex* in Texas *(Diptera, Culicidae). Insecutor Inscitiae Menstruus*, 9 (1–3), 32–34.

Dyar, H.G. (1921b) The genus *Haemagogus* Williston *(Diptera, Culicidae). Insecutor Inscitiae Menstruus*, 9 (7–9), 101–114.

Dyar, H.G. (1921c) Illustrations of certain mosquitoes *(Diptera, Culicidae). Insecutor Inscitiae Menstruus*, 9 (7–9), 114–118, 1 pl.

Dyar, H.G. (1921d) The species of *Finlaya* allied to *terrens* Walker *(Diptera, Culicidae). Insecutor Inscitiae Menstruus*, 9 (10–12), 151–153.

Dyar, H.G. (1921e) New mosquitoes from Costa Rica *(Diptera, Culicidae). Insecutor Inscitiae Menstruus*, 9 (10–12), 154–155.

Dyar, H.G. (1922a) New mosquitoes from Alaska *(Diptera, Culicidae). Insecutor Inscitiae Menstruus*, 10 (1–3), 1–3.

Dyar, H.G. (1922b) Note on the male genitalia of *Culex coronator* and allied forms *(Diptera, Culicidae). Insecutor Inscitiae Menstruus*, 10 (1–3), 18–19, 1 pl.

Dyar, H.G. (1922c) Mosquito notes *(Diptera, Culicidae). Insecutor Inscitiae Menstruus*, 10 (4–6), 92–99.

Dyar, H.G. (1922d) The American *Aedes* of the *serratus* group (*Diptera, Culicidae). Insecutor Inscitiae Menstruus*, 10 (7–9), 157–166.

Dyar, H.G. (1922e) Notes on tropical American mosquitoes (*Diptera, Culicidae). Insecutor Inscitiae Menstruus*, 10 (10–12), 188–196, 1 pl.

Dyar, H.G. (1923a) The mosquitoes of the Yellowstone National Park (*Diptera, Culicidae). Insecutor Inscitiae Menstruus*, 11 (1–3), 36–46.

Dyar, H.G. (1923b) Mosquito notes (*Diptera, Culicidae). Insecutor Inscitiae Menstruus*, 11 (4–6), 64–72, 3 pls.

Dyar, H.G. (1923c) Notes on American *Culex (Diptera, Culicidae). Insecutor Inscitiae Menstruus*, 11 (7–9), 118–121.

Dyar, H.G. (1923d) On some of the American subgenera of *Culex (Diptera, Culicidae). Insecutor Inscitiae Menstruus,* 11 (10–12), 187–190, 2 pls.

Dyar, H.G. (1923e) The mosquitoes of Panama (*Diptera, Culicidae). Insecutor Inscitiae Menstruus*, 11 (10–12), 167–187.

Dyar, H.G. (1923f) The mosquitoes of the United States. *Proceedings of the United States National Museum*, 62 (2447), 1–119.

Dyar, H.G. (1924a) A new sabethid from Brazil (*Diptera, Culicidae). Insecutor Inscitiae Menstruus*, 12 (7–9), 92.

Dyar, H.G. (1924b) A note on *Sabethes* Robineau-Desvoidy (*Diptera, Culicidae). Insecutor Inscitiae Menstruus*, 12 (7–9), 97–100.

Dyar, H.G. (1924c) *Phoniomyia* and *Dendromyia* Theobald (*Diptera, Culicidae). Insecutor Inscitiae Menstruus*, 12, 107–113.

Dyar, H.G. (1924d) Some new mosquitoes from Colombia (*Diptera, Culicidae). Insecutor Inscitiae Menstruus*, 12, 119–124.

Dyar, H.G. (1924e) Two new mosquitoes from California (*Diptera, Culicidae). Insecutor Inscitiae Menstruus*, 12, 125–127.

Dyar, H.G. (1924f) Some new mosquitoes from Colombia—III (*Diptera, Culicidae). Insecutor Inscitiae Menstruus*, 12, 183–186.

Dyar, H.G. (1925a) A new sabethid from Panama *(Diptera, Culicidae). Insecutor Inscitiae Menstruus*, 13, 20–21.

Dyar, H.G. (1925b) Some mosquitoes from Ecuador *(Diptera, Culicidae). Insecutor Inscitiae Menstruus*, 13, 27–31.

Dyar, H.G. (1925c) The subgenera of *Mansonia* Blanchard *(Diptera, Culicidae). Insecutor Inscitiae Menstruus*, 13 (1–3), 40–43, 1 pl.

Dyar, H.G. (1925d) The mosquitoes of Panama (*Diptera, Culicidae). Insecutor Inscitiae Menstruus*, 13 (7–9), 101–195.

Dyar, H.G. (1925e) Some mosquitoes from Venezuela (*Diptera, Culicidae). Insecutor Inscitiae Menstruus*, 13, 213–216.

Dyar, H.G. (1926) A note on *Phalangomyia* Dyar & Knab *(Diptera, Culicidae). Insecutor Inscitiae Menstruus*, 14 (1–3), 41–43.

Dyar, H.G. (1928) *The mosquitoes of the Americas*. Publication no. 387. Carnegie Institution of Washington, Washington, DC, USA.

Dyar, H.G. (1929a) A new species of mosquito from Montana, with annotated list of the species known from the state. *Proceedings of the United States National Museum*, 75 (2794), 1–8.

Dyar, H.G. (1929b) Remarks on the subgenus *Phalangomyia* of *Culex* Linn. (Diptera, Culicidae). *American Journal of Hygiene*, 9 (2), 509–511. DOI: 10.1093/oxfordjournals.aje.a121665

Dyar, H.G. (1929c) A new mosquito from the Philippine Islands. *Proceedings of the Entomological Society of Washington*, 31 (3), 61–62.

Dyar, H.G. & Knab, F. (1906a) Diagnoses of new species of mosquitoes. *Proceedings of the Biological Society of Washington*, 19, 133–141.

Dyar, H.G. & Knab, F. (1906b) Notes on some American mosquitoes with descriptions of new species. *Proceedings of the Biological Society of Washington*, 19, 159–172.

Dyar, H.G. & Knab, F. (1906c) The species of mosquitoes in the genus *Megarhinus*. *Smithsonian Miscellaneous Collections*, 48 (3), 241–258.

Dyar, H.G. & Knab, F. (1906d) The larvae of Culicidae classified as independent organisms. *Journal of the New York Entomological Society*, 14 (4), 169–230, 13 pls.

Dyar, H.G. & Knab, F. (1907a) Descriptions of some American mosquitoes. *Journal of the New York Entomological Society*, 15 (1), 9–13.

Dyar, H.G. & Knab, F. (1907b) New American mosquitoes. *Journal of the New York Entomological Society*, 15 (2), 100–101.

Dyar, H.G. & Knab, F. (1907c) Descriptions of new mosquitoes from the Panama Canal Zone. *Journal of the New York Entomological Society*, 15 (4), 197–212.

Dyar, H.G. & Knab, F. (1907d) Descriptions of three new North American mosquitoes. *Journal of the New York Entomological Society*, 15 (4), 213–214.

Dyar, H.G. & Knab, F. (1908) Descriptions of some new mosquitoes from tropical America. *Proceedings of the United States National Museum*, 35 (1632), 53–70. DOI: 10.5479/si.00963801.35-1632.53

Dyar, H.G. & Knab, F. (1909) Descriptions of some new species and a new genus of American mosquitoes. *Smithsonian Miscellaneous Collections*, 52 (1822), 253–266.

Dyar, H.G. & Knab, F. (1910a) Description of three new American mosquitoes. *Proceedings of the Entomological Society of Washington* (for 1909), 11 (1), 173–174.

Dyar, H.G. & Knab, F. (1910b) The genus *Mansonia*. *Entomological News and Proceedings of the Entomological Section of the Academy of Natural Sciences of Philadelphia*, 21 (6), 259–264.

Dyar, H.G. & Knab, F. (1913) Three new Neotropical mosquitoes. *Insecutor Inscitiae Menstruus*, 1 (6), 76–78.

Dyar, H.G. & Knab, F. (1914) New mosquitoes from Peru *(Diptera, Culicidae)*. *Insecutor Inscitiae Menstruus*, 2 (4), 58–62.

Dyar, H.G. & Knab, F. (1916) Eggs and oviposition in certain species of *Mansonia (Diptera; Culicidae)*. *Insecutor Inscitiae Menstruus*, 4 (4–6), 61–68.

Dyar, H.G. & Knab, F. (1917) The genus *Culex* in the United States *(Diptera, Culicidae)*. *Insecutor Inscitiae Menstruus*, 5 (10–12), 170–183.

Dyar, H.G. & Knab, F. (1918) New American mosquitoes *(Diptera, Culicidae)*. *Insecutor Inscitiae Menstruus* (for 1917), 5 (10–12), 165–169.

Dyar, H.G. & Knab, F. (1919) New species of tropical American mosquitoes *(Diptera, Culicidae)*. *Insecutor Inscitiae Menstruus*, 7 (1–3), 1–9.

Dyar, H.G. & Ludlow, C.S. (1921) A note on two Panama mosquitos *(Diptera, Culicidae)*. *Military Surgeon*, 48, 677–680.

Dyar, H.G. & Ludlow, C.S. (1922) Notes on American mosquitoes *(Diptera, Culicidae)*. *Military Surgeon*, 50, 61–64.

Dyar, H.G. & Nuñez Tovar, M. (1927) Notes on biting flies from Venezuela *(Diptera, Culicidae, Psychodidae)*. *Insecutor Inscitiae Menstruus* (for 1926), 14, 152–155, 190–191.

Dyar, H.G. & Núñez Tovar, M. (1927) *Notas sobre nuevos dipteros hematofágos de Venezuela (Culicidae)*. Maracay, Venezuela.

Dyar, H.G. & Núñez Tovar, M. (1928) Descriptions of new species of mosquitoes from Venezuela. *American Journal of Hygiene*, 8 (1), 89–92. DOI: 10.1093/oxfordjournals.aje.a120992

Dyar, H.G. & Shannon, R.C. (1924a) The American species of *Uranotaenia (Diptera, Culicidae)*. *Insecutor Inscitiae Menstruus*, 12 (10–12), 187–192.

Dyar, H.G. & Shannon, R.C. (1924b) The subfamilies, tribes, and genera of American Culicidae. *Journal of the Washington Academy of Sciences*, 14 (20), 472–486.

Dyar, H.G. & Shannon, R.C. (1925a) The types of Philippine mosquitos described by Ludlow and other notes on the fauna *(Diptera, Culicidae)*. *Insecutor Inscitiae Menstruus*, 13 (4–6), 66–89.

Dyar, H.G. & Shannon, R.C. (1925b) New mosquitoes from Brazil (Diptera, Culicidae). *Journal of the Washington Academy of Sciences*, 15 (2), 39–41.

Ebejer, M.J. & Gatt, P. (2021) A checklist of the Diptera (Insecta) of the Maltese Islands. *Zootaxa*, 5018 (1), 1–75. DOI: 10.11646/zootaxa.5018.1.1

Eckstein, F. (1918) Zur Systematik der einheimischen Stechmücken. *Zentralblatt fur Bakteriologie, Abt. 1, Originale*, 82 (2), 57–68.

Edwards, F.W. (1911a) Some new West African species of *Anopheles (sensu lato)*, with notes on nomenclature. *Bulletin of Entomological Research*, 2 (2), 141–143. DOI: 10.1017/S0007485300001334

Edwards, F.W. (1911b) On the culicid genus *Eretmapodites*, Theobald. *Annals and Magazine of Natural History*, 8 (43), 67–73. DOI: 10.1080/00222931108693001

Edwards, F.W. (1911c) The African species of *Culex* and allied genera. *Bulletin of Entomological Research*, 2 (3), 241–268. DOI: 10.1017/S0007485300001450

Edwards, F.W. (1912a) A synopsis of the species of African Culicidae, other than *Anopheles*. *Bulletin of Entomological Research*, 3 (1), 1–53. DOI: 10.1017/S0007485300001656

Edwards, F.W. (1912b) A key to the Australasian species of *Ochlerotatus* (Culicidae). *Annals and Magazine of Natural History*, 9 (53), 521–527.

Edwards, F.W. (1912c) A key for determining the African species of *Anopheles (sensu lato)*. *Bulletin of Entomological Research*, 3 (3), 241–250. DOI: 10.1017/S0007485300001917

Edwards, F.W. (1913a) Tipulidae and Culicidae from the Lake of Tiberias and Damascus. *Journal and Proceedings of the Asiatic Society of Bengal*, 9, 47–51.

Edwards, F.W. (1913b) Further notes on African Culicidae. *Bulletin of Entomological Research*, 4 (1), 47–59. DOI: 10.1017/S0007485300016941

Edwards, F.W. (1913c) New synonymy in Oriental Culicidae. *Bulletin of Entomological Research*, 4 (3), 221–242. DOI: 10.1017/S0007485300043133

Edwards, F.W. (1914a) On the Oriental culicid genus *Leicesteria*, Theobald. *Bulletin of Entomological Research*, 4 (4), 255–263. DOI: 10.1017/S0007485300043182

Edwards, F.W. (1914b) New species of Culicidae in the British Museum, with notes on the genitalia of some African *Culex*. *Bulletin of Entomological Research*, 5 (1), 63–81. DOI: 10.1017/S0007485300034568

Edwards, F.W. (1914c) New Culicidae from Borneo and Hong Kong. *Bulletin of Entomological Research*, 5 (2), 125–128. DOI: 10.1017/S0007485300033976

Edwards, F.W. (1915a) New and little-known East African Culicidae. *Bulletin of Entomological Research*, 5 (4), 273–281. DOI: 10.1017/S0007485300030303

Edwards, F.W. (1915b) Diagnoses of new Bornean Culicidae. *Bulletin of Entomological Research*, 5 (4), 283–285. DOI: 10.1017/S0007485300030315

Edwards, F.W. (1916) Eight new mosquitos in the British Museum collection. *Bulletin of Entomological Research*, 6 (4), 357–364. DOI: 10.1017/S0007485300043637

Edwards, F.W. (1917) Notes on Culicidae, with descriptions of new species. *Bulletin of Entomological Research*, 7 (3), 201–229. DOI: 10.1017/S0007485300017557

Edwards, F.W. (1920a) Mosquito notes. *Bulletin of Entomological Research*, 10 (2), 129–137. DOI: 10.1017/S0007485300043923

Edwards, F.W. (1920b) Notes on the mosquitos of Madagascar, Mauritius and Réunion. *Bulletin of Entomological Research*, 11 (2), 133–138. DOI: 10.1017/S0007485300044539

Edwards, F.W. (1921a) Mosquito notes.–II. *Bulletin of Entomological Research*, 12 (1), 69–80. DOI: 10.1017/S0007485300044874

Edwards, F.W. (1921b) A synonymic list of the mosquitoes hitherto recorded from Sweden, with keys for determining the genera and species. *Entomologisk Tidskrift*, 42 (1), 46–52.

Edwards, F.W. (1921c) H. Sauter's Formosan collections: Culicidae. *Annals and Magazine of Natural History*, 8 (48), 629–632. DOI: 10.1080/00222932108632632

Edwards, F.W. (1921d) A revision of the mosquitos of the Palaearctic Region. *Bulletin of Entomological Research*, 12 (3), 263–351. DOI: 10.1017/S0007485300040207

Edwards, F.W. (1922a) A synopsis of adult Oriental culicine (including megarhinine and sabethine) mosquitoes. Part I. *Indian Journal of Medical Research*, 10 (1), 249–293, 3 pls.

Edwards, F.W. (1922b) A synopsis of adult Oriental culicine (including megarhinine and sabethine) mosquitoes. Part II. *Indian Journal of Medical Research*, 10 (2), 430–475, 3 pls.

Edwards, F.W. (1922c) A revision of the genus *Harpagomyia* de Meij. (Diptera, Culicidae). *Transactions of the Royal Entomological Society of London*, 1921 (3–4), 496–503, 1 pl.

Edwards, F.W. (1922d) Mosquito notes.–III. *Bulletin of Entomological Research*, 13 (1), 75–102. DOI: 10.1017/S0007485300045259

Edwards, F.W. (1923a) Four new African mosquitos. *Bulletin of Entomological Research* (for 1922), 13 (4), 397–399. DOI: 10.1017/S0007485300045466

Edwards, F.W. (1923b) Mosquito notes.–IV. *Bulletin of Entomological Research*, 14 (1), 1–9. DOI: 10.1017/S0007485300028157

Edwards, F.W. (1923c) Oligocene mosquitoes in the British Museum; with a summary of our present knowledge concerning fossil Culicidae. *Quarterly Journal of the Geological Society of London*, 79 (2), 139–155, 1 pl. DOI: 10.1144/GSL.JGS.1923.079.01-04.10

Edwards, F.W. (1923d) Mosquitoes reared by Dr. W. E. Haworth from coconut palms in East Africa. *Transactions of the Royal Society of Tropical Medicine and Hygiene*, 16 (8), 498–501. DOI: 10.1016/S0035-9203(23)90354-8

Edwards, F.W. (1924a) Some mosquitoes from Ovamboland, S.W. Africa, and from the Cape Province. *Annals of the South African Museum*, 19 (1), 159–163

Edwards, F.W. (1924b) A synopsis of the adult mosquitos of the Australasian Region. *Bulletin of Entomological Research*, 14 (4), 351–401. DOI: 10.1017/S0007485300045624

Edwards, F.W. (1925) Mosquito notes.–V. *Bulletin of Entomological Research*, 15 (3), 257–270. DOI: 10.1017/S0007485300046204

Edwards, F.W. (1926a) Mosquito notes.–VI. *Bulletin of Entomological Research*, 17 (2), 101–131. DOI: 10.1017/S0007485300019143

Edwards, F.W. (1926b) Diptera Nematocera from the mountains of Borneo. *Sarawak Museum Journal*, 3 (8–11), 243–278, 2 pls.

Edwards, F.W. (1926c) Una revisione delle zanzare delle regioni paleartiche. *Rivista di Malariologia*, 5 (3–6), 1–152.

Edwards, F.W. (1927a) Four new mosquitoes from the Belgian Congo. *Revue de Zoologie Africaine*, 15 (3), 351–355.

Edwards, F.W. (1927b) New mosquitoes of the genus *Rachionotomyia* from New Guinea. *Nova Guinea: Résultats des Expéditions Scientifiques Néerlandaises à la Nouvelle Guinée*, 15 (3), 352–356.

Edwards, F.W. (1928a) Mosquito notes.–VII. *Bulletin of Entomological Research*, 18 (3), 267–284. DOI: 10.1017/S0007485300019994

Edwards, F.W. (1928b) A new mosquito from the Kivu volcanoes, Belgian Congo. *Revue de zoologie et de botanique Africaines*, 16 (3), 324–325.

Edwards, F.W. (1929a) A new *Anopheles* from the central Sahara. *Archives de l'Institut Pasteur d'Algérie*, 7 (1), 82–87.

Edwards, F.W. (1929b) Mosquito notes.–VIII. *Bulletin of Entomological Research*, 20 (3), 321–343. DOI: 10.1017/S0007485300021283

Edwards, F.W. (1929c) Philippine nematocerous Diptera II. *Notulae Entomologicae*, 9, 1–14.

Edwards, F.W. (1930a) Mosquito notes.–IX. *Bulletin of Entomological Research*, 21 (3), 287–306. DOI: 10.1017/S0007485300021805

Edwards, F.W. (1930b) Mosquito notes.–X. *Bulletin of Entomological Research*, 21 (4), 541–545. DOI: 10.1017/S000748530002486X

Edwards, F.W. (1932a) *Genera Insectorum. Diptera, Fam. Culicidae*. Fascicle 194. V. Verteneuil & L. Desmet, Bruxelles [Brussels], Belgium, 258 pp., 5 pls.

Edwards, F.W. (1932b) Mosquito notes.–XI. *Bulletin of Entomological Research*, 23 (4), 559–562. DOI: 10.1017/S0007485300004338

Edwards, F.W. (1934) The Percy Sladen and Godman Trusts expedition to the islands in the Gulf of Guinea, October 1932–March 1933.–II. Diptera Nematocera. *Annals and Magazine of Natural History*, 14 (81), 321–336. DOI: 10.1080/00222933408654901

Edwards, F.W. (1935a) Mosquito notes.–XII. *Bulletin of Entomological Research*, 26 (2), 127–136. DOI: 10.1017/S0007485300038116

Edwards, F.W. (1935b) Results of the Oxford University expedition to Borneo, 1932. Two new species of *Megarhinus* (Diptera, Culicidae). *Annals and Magazine of Natural History*, 15 (89), 579–581. DOI: 10.1080/00222933508655002

Edwards, F.W. (1935c) A new *Uranotaenia* from the Belgian Congo (Diptera, Culicidae). *Revue de zoologie et de botanique Africaines*, 27 (1), 96.

Edwards, F.W. (1936) New African culicine mosquitoes (Diptera, Culicidae). *Proceedings of the Royal Entomological Society of London, Series B: Taxonomy*, 5 (3), 49–55. DOI: 10.1111/j.1365-3113.1936.tb00593.x

Edwards, F.W. (1939a) A new East African *Aëdes* (Dipt., Culicidae). *Proceedings of the Royal Entomological Society of London, Series B: Taxonomy*, 8 (2), 17. DOI: 10.1111/j.1365-3113.1939.tb00484.x

Edwards, F.W. (1939b) A new species of *Orthopodomyia* (Diptera, Culicidae). *Proceedings of the Royal Entomological Society of London, Series B: Taxonomy*, 8 (6), 121–123. DOI: 10.1111/j.1365-3113.1939.tb00505.x

Edwards, F.W. (1940) A new plain-winged *Anopheles* from Rhodesia. *Annals of Tropical Medicine and Parasitology*, 34 (2), 93–96. DOI: 10.1080/00034983.1940.11685087

Edwards, F.W. (1941) *Mosquitoes of the Ethiopian Region III. – Culicine adults and pupae.* Printed by order of the Trustees, British Museum (Natural History), London, UK.

Edwards, F.W. & Gibbins, E.G. (1939) *Mosquitoes. Ruwenzori Expedition 1934-5.* Vol. 1, no. 3, pp. 29–33, 1 pl. British Museum (Natural History), London, UK.

Egger, J. (1858) Dipterologische Beiträge. *Verhandlungen der Zoologisch-Botanischen Gesellschaft in Wien*, 8 (Abhanglungen), 701–716.

Ellis, R.A. & Brust, R.A. (1973) Sibling species delimitation in the *Aedes communis* (Degeer [*sic*]) aggregate (Diptera: Culicidae). *Canadian Journal of Zoology*, 51 (9), 915–959. DOI: 10.1139/z73-138

Enderlein, G. (1923) Zur Kenntnis afrikanischer und orientalischer Culiciden. *Wiener Entomologische Zeitung*, 40 (1–4), 25–29. DOI: 10.5962/bhl.part.2580

Eouzan, J.-P. (1969) Description du mâle de *Culex (Culiciomyia) grenieri* sp. n. (Diptera, Culicidae). *Cahiers ORSTOM Série Entomologie médicale et Parasitologie*, 7 (3), 163–166.

Erichson, W.F. (1842) Beitrag zur Insecten-Fauna von Vandiemensland, mit besonderer Berücksichtigung der geographischen verbreitung der Insecten, vom Herausgeber. *Archiv für Naturgeschichte*, 8 (1), 83–287, 2 pls. DOI: 10.5962/bhl.part.21656

Evans, A.M. (1923a) A new species and a new variety of *Culex* from the Belgian Congo. *Annals of Tropical Medicine and Parasitology*, 17 (1), 89–91. DOI: 10.1080/00034983.1923.11684352

Evans, A.M. (1923b) Notes on Culicidae in Venezuela, with descriptions of new species. Part III. *Annals of Tropical Medicine and Parasitology*, 17 (1), 101–111. DOI: 10.1080/00034983.1923.11684354

Evans, A.M. (1924) Descriptions of new mosquitos from South America. *Annals of Tropical Medicine and Parasitology*, 18 (3), 363–375. DOI: 10.1080/00034983.1924.11684421

Evans, A.M. (1925a) Notes on Culicidae collected in Sierra Leone, with descriptions of a new species and a new variety. *Annals of Tropical Medicine and Parasitology*, 19 (1), 119–126, 1 pl. DOI: 10.1080/00034983.1925.11684445

Evans, A.M. (1925b) A new variety of *Anopheles marshalli* Theobald from the Belgian Congo. *Annals of Tropical Medicine and Parasitology*, 19 (2), 211–213. DOI: 10.1080/00034983.1925.11684451

Evans, A.M. (1925c) A new variety of *Anopheles marshallii*, from Sierra Leone. *Annals of Tropical Medicine and Parasitology*, 19 (4), 461–462. DOI: 10.1080/00034983.1925.11684473

Evans, A.M. (1926) Notes on Freetown mosquitos, with descriptions of new and little-known species. *Annals of Tropical Medicine and Parasitology*, 20 (1), 97–108, 1 pl. DOI: 10.1080/00034983.1926.11684481

Evans, A.M. (1927) A short illustrated guide to the anophelines of tropical and South America. *Memoirs of the Liverpool School of Tropical Medicine*, 3, 1–79.

Evans, A.M. (1928) A new variety of *Armigeres* from Lagos, with descriptions of the larval and pupal stages. *Annals of Tropical Medicine and Parasitology*, 22 (1), 39–42. DOI: 10.1080/00034983.1928.11684559

Evans, A.M. (1929) *Aëdes (Aëdimorphus) apicoannulatus* Edwards and yellow fever: a correction. *Annals of Tropical Medicine and Parasitology*, 23 (4), 521–522. DOI: 10.1080/00034983.1929.11684620

Evans, A.M. (1931a) Notes on African anophelines. *Annals of Tropical Medicine and Parasitology*, 25 (1), 129–143. DOI: 10.1080/00034983.1931.11684674

Evans, A.M. (1931b) A new subspecies of *Anopheles funestus* Giles, from southern Rhodesia. *Annals of Tropical Medicine and Parasitology*, 25 (3–4), 545–549. DOI: 10.1080/00034983.1931.11684699

Evans, A.M. (1932) Notes on African mosquitoes. *Annals of Tropical Medicine and Parasitology*, 26 (1), 85–108. DOI: 10.1080/00034983.1932.11684706

Evans, A.M. (1933) *Anopheles demeilloni* sp. n., a new name for *Anopheles transvaalensis* of authors. *Annals of Tropical Medicine and Parasitology*, 27 (2), 265–269. DOI: 10.1080/00034983.1933.11684758

Evans, A.M. (1934) Further notes on African anophelines, with a description of a new group of *Myzomyia*. *Annals of Tropical Medicine and Parasitology*, 28 (4), 549–570. DOI: 10.1080/00034983.1934.11684821

Evans, A.M. (1935) Notes on anophelines. I.–Description of *Anopheles marshalli* [*sic*] var. *gibbinsi* from Uganda. II.–The characters of *A. maculipennis* var. *messeae* in Wirral, England. *Annals of Tropical Medicine and Parasitology*, 29 (4), 469–473. DOI: 10.1080/00034983.1935.11684868

Evans, A.M. (1936) Two new anopheline mosquitoes from Kenya. *Annals of Tropical Medicine and Parasitology*, 30 (4), 533–540. DOI: 10.1080/00034983.1936.11684958

Evans, A.M. (1938) *Mosquitoes of the Ethiopian Region. II. – Anophelini adults and early stages*. British Museum (Natural History), London, UK.

Evans, A.M. & de Meillon, B. (1933) Notes on *Anopheles demeilloni (transvaalensis)* and *Anopheles garnhami* in South Africa, with descriptions of new varieties of these species. *Annals of Tropical Medicine and Parasitology*, 27 (2), 271–282. DOI: 10.1080/00034983.1933.11684759

Evans, A.M. & Leeson, H.S. (1935) The *funestus* series of *Anopheles* in Southern Rhodesia, with description of a new variety. *Annals of Tropical Medicine and Parasitology*, 29 (1), 33–47. DOI: 10.1080/00034983.1935.11684829

Evenhuis, N.L. (1994) *Catalogue of the Fossil Flies of the World (Insecta: Diptera)*. Backhuys Publishers, Leiden, Netherlands. DOI: 10.1093/aesa/89.2.307

Evenhuis, N.L. (2018) Nomenclatural studies toward a World List of Diptera genus-group names. Part VI: Daniel William Coquillett. *Zootaxa*, 4381 (1), 1–95. DOI: 10.11646/zootaxa.4381.1.1

Evenhuis, N.L. & Gon III, S.M. (1989) Family Culicidae. In: Evenhuis, N.L. (ed.), *Catalog of the Diptera of the Australasian and Oceanian Regions*. Bishop Museum Special Publication 86, Bishop Museum Press and E.J. Brill, Honolulu, HI, USA, pp. 191–218.

Evenhuis, N.L. & Harbach, R.E. (2024) *Rhynchotaenia* Brèthes, 1910 (preoccupied): transfer of its species to *Pseudotaeniorhynchus* Theobald, 1911 (Diptera: Culicidae). In: Evenhuis, N.L. & Pape, T. (eds), Systema Dipterorum Nomenclatural Notes. IV. *Bishop Museum Occasional Papers*, 157, 29–31.

Evenhuis, N.L. & Steffan, W.A. (1986) Classification of the subgenus *Toxorhynchites* (Diptera: Culicidae) II. Revision of the *Toxorhynchites acaudatus* group. *Journal of Medical Entomology*, 23 (5), 538–574. DOI: 10.1093/jmedent/23.5.538

Fabricius, J.C. (1787) *Mantissa Insectorum, sistens eorum species nuper detectas: adjectis synonymis, observationibus, descriptionibus, emendationibus*. Vol. 2. Christ. Gottl. Proft, Hafniæ [Copenhagen], Denmark. DOI: 10.5962/bhl.title.36471

Fabricius, J.C. (1794) *Entomologia systematica emendata et aucta: secundum classes, ordines, genera, species, adjectis synonimis, locis, observationibus, descriptionibus*. Volume 4. C.G. Proft, fil. & soc., Hafniae [Copenhagen], Denmark. DOI: 10.5962/bhl.title.125869

Fabricius, J.C. (1805) *Systema antliatorum secundum ordines, genera, species, adiectis synonymis, locis, observationibus, descriptionibus*. C. Reichard, Brunsvigae [Brunswick], Germany. DOI: 10.5962/bhl.title.15806

Falleroni, D. (1926) Fauna anofelica italiana e suo "habitat" (paludi, risaie, canali). Metodi di lotta contro la malaria. *Rivista di Malariologia*, 5 (5–6), 553–593.

Faran, M.E. (1979) *Anopheles (Nyssorhynchus) trinkae*, a new species in the Albimanus Section (Diptera: Culicidae). *Mosquito Systematics*, 11 (1), 26–39.

Faran, M.E. (1980) Mosquito studies (Diptera, Culicidae) XXXIV. A revision of the Albimanus Section of the subgenus *Nyssorhynchus* of *Anopheles*. *Contributions of the American Entomological Institute*, 15 (7), 1–215.

Faran, M.E. & Linthicum, K.J. (1981) A handbook of the Amazonian species *of Anopheles (Nyssorhynchus)* (Diptera: Culicidae). *Mosquito Systematics*, 13 (1), 1–81.

Farner, D.S. (1945) A new species of *Aedes* from the Caroline Islands. (Diptera, Culicidae). *Proceedings of the Biological Society of Washington*, 58, 59–61.

Farner, D.S. & Bohart, R.M. (1944) Three new species of Australasian *Aedes* (Diptera, Culicidae). *Proceedings of the Biological Society of Washington*, 57, 117–122.

Fauran, P. (1961) Notes sur les moustiques de la Guyane. *Archives de l'Institut Pasteur de la Guyane Française l'Inini*, 22 (464), 1–15.

Fauran, P. & Pajot, F.X. (1974) *Wyeomyia ininicola* n. sp. *(Diptera, Culicidae)*, nouveau moustique de la Guyane française. *Cahiers ORSTOM Série Entomologie médicale et Parasitologie*, 12 (2), 145–148.

Favre, V.V. (1903) *An Attempt in the Study of Malaria in Russia* [in Russian]. Kharkov, Russia.

Felt, E.P. (1904) Mosquitos or Culicidae of New York State. *New York State Museum Bulletin*, 79 (323), 241–391, 57 pls, 391a–391f, 393–400.

Felt, E.P. & Young, D.B. (1904) Importance of isolated rearings from culicid larvae. *Science*, 20, 312–313. DOI: 10.1126/science.20.505.312

Feng, L.-C. (1938) The tree hole species of mosquitoes of Peiping, China. *Chinese Medical Journal, Supplement*, 2, 503–525, 5 pls.

Feng, L.C. (1941) A new mosquito, *Topomyia houghtoni* sp. nov. from western Yunnan, China. *Bulletin of the Fan Memorial Institute of Biology, Peiping (Zoology Series)*, 10, 243–254.

Ferrara, L. (1973) Une nouvelle espèce de moustique *(Diptera Culicidae)* capturée au Cameroun. *Orthopodomyia aureoantennata* n. sp. *Cahiers ORSTOM Série Entomologie médicale et Parasitologie*, 11 (3), 199–203.

Ferrara, L. (1974) Description d'un nouvel *Aedes* du sous genre *Diceromyia (Diptera, Culicidae)* découvert au Cameroun *Aedes (Diceromyia) mefouensis* n. sp. *Cahiers ORSTOM Série Entomologie médicale et Parasitologie*, 12 (1), 3–5.

Ferrara, L. & Eouzan, J.-P. (1974) Description de deux nouveaux *Eretmapodites*, du groupe *oedipodius* Graham, 1909 *(Diptera, Culicidae)* découverts au Cameroun. *Cahiers ORSTOM Série Entomologie médicale et Parasitologie*, 12 (4), 259–262.

Ferreira de Freitas, L. & Bartholomay, L.C. (2021) The taxonomic history of *Ochlerotatus* Lynch Arribálzaga, 1891 (Diptera: Culicidae). *Insects*, 12, 452. DOI: 10.3390/insects12050452

Ficalbi, E. (1889a) Notizie preventive sulle zanzare italiane. I[a] Nota preventiva. Alcune generalità. – Descrizione di una specie nuova (*Culex hortensis.*). *Bullettino della Società Entomologica Italiana*, 21, 20–30.

Ficalbi, E. (1889b) Notizie preventive sulle zanzare italiane. II[a] Nota preventiva (1). Descrizione di una specie nuova. *Bullettino della Società Entomologica Italiana*, 21, 50–53.

Ficalbi, E. (1890a) Notizie preventive sulle zanzare italiane. VII[a] Nota preventiva (1). Descrizione di una specie nuova. *Bullettino della Società Entomologica Italiana*, 22, 81–84.

Ficalbi, E. (1890b) Notizie preventive sulle zanzare italiane. IV. Nota preventiva. Descrizione di una specie nuova. Zanara di colorito modesto *Culex modestus*, n. sp. *Bullettino della Società Entomologica Italiana* (for 1889), 21, 93–94.

Firooziyan, S., Djadid, N.D. & Gholizadeh, S. (2018) Speculation on the possibility for introducing *Anopheles stephensi* as a species complex: preliminary evidence based on odorant binding protein 1 intron I sequence. *Malaria Journal*, 17, 366. DOI: 10.1186/s12936-018-2523-y

Fischer, P.H. (1885) Fascicule VIII, pp. 689–784, *Manuel de conchyliologie et de paleontologie conchyliologique ou histoire naturelle des mollusques vivants et fossiles suivi d'un appendice sur les brachiopodes par D. P. Oehlert*. Librairie F. Savy, Paris. [Issued in 11 parts 1880–1887, with their publication dates given on the verso of the half-title]. DOI: 10.5962/bhl.title.13213

Floch, H. & Abonnenc, E. (1945a) Description de *Culex* nouveaux de la Guyane française. *Culex (Melanoconion) cauchensis* n. sp. *Culex (Melanoconion) cayennensis* n. sp. *Publications de Institut Pasteur de la Guyane et du territoire de l'Inini*, 112, 1–6.

Floch, H. & Abonnenc, E. (1945b) Description de *Culex* nouveaux de la Guyane Française (II). *Culex (Melanoconion) cavernicolus* n. sp. *Culex (Melanoconion) equinoxialis* n. sp. *Publications de Institut Pasteur de la Guyane et du territoire de l'Inini*, 114, 1–5.

Floch, H. & Abonnenc, E. (1945c) Description d'une nouvelle espèce d'anophèle: *A.* (*Stethomyia) canorii*. *Publications de Institut Pasteur de la Guyane et du territoire de l'Inini*, 116, 1– 3.

Floch, H. & Abonnenc, E. (1946a) Description de *Culex* nouveaux de la Guyane française (III). *Culex (Melanoconion) rabanicolus* n. sp. *Culex (Melanoconion) rorotaensis* n. sp. *Publications de Institut Pasteur de la Guyane et du territoire de l'Inini*, 120, 1–4.

Floch, H. & Abonnenc, E. (1946b) Description de *Culex* nouveaux de la Guyane Française (IV). *Culex (Melanoconion) punctiscapularis* n. sp. *Publications de Institut Pasteur de la Guyane et du territoire de l'Inini*, 122, 1–3.

Foley, D.H., Bryan, J.H., Yeates, D. & Saul, A. (1998) Evolution and systematics of *Anopheles*: insights from a molecular phylogeny of Australasian mosquitoes. *Molecular Phylogenetics and Evolution*, 9 (2), 262–275. DOI: 10.1006/mpev.1997.0457

Foley, D.H., Cooper, R.D. & Bryan, J.H.(1995) A new species within the *Anopheles punctulatus* complex in Western Province, Papua New Guinea. *Journal of the American Mosquito Control Association*, 11 (1), 122–127.

Foley, D.H., Paru, R., Dagoro, H. & Bryan, J.H. (1993) Allozyme analysis reveals six species within the *An. punctulatus* complex of mosquitoes in Papua New Guinea. *Medical and Veterinary Entomology*, 7 (1), 37–48. DOI: 10.1111/j.1365-2915.1993.tb00649.x

Foley, D.H., Wilkerson, R.C., Cooper, R.D., Volovsek, M.E. & Bryan, J.H. (2007) A molecular phylogeny of *Anopheles annulipes* (Diptera: Culicidae) sensu lato: the most species-rich anopheline complex. *Molecular Phylogenetics and Evolution*, 43 (1), 283–297. DOI: 10.1016/j.ympev.2006.10.008

Foote, R.H. (1954) The larvae and pupae of the mosquitoes belonging to the *Culex* subgenera *Melanoconion* and *Mochlostyrax*. *US Department of Agriculture Technical Bulletin*, 1091, [i], 126 pp.

Fontenille, D. & Brunhes, J. (1985) Trois nouveaux culicidés de Madagascar: *Aedes (Aedimorphus) albodorsalis* n. sp., *Aedes (Aedimorphus) masoalensis* n. sp. et *Aedes (Aedimorphus) mathioti* n. sp. *Cahiers ORSTOM Série Entomologie médicale et Parasitologie* (for 1984), 22 (2), 151–155.

Forattini, O.P. (1962) *Entomologia Médica. Vol. 1. Parte geral, Diptera, Anophelini*. Faculdade de Higiene e Saúde Pública, Departamento de Parasitologia, Universidade de São Paulo, São Paulo, Brazil.

Forattini, O.P. (1965a) *Entomologia Médica. Vol. 2. Culicini:* Culex, Aedes *e* Psorophora. Universidade de São Paulo, São Paulo, Brazil.

Forattini, O.P. (1965b) *Entomologia Médica. Vol.* 3. *Culicini:* Haemagogus, Mansonia, Culiseta *Sabethini. Toxorhynchitini. Arboviruses. Filariose bancroftiana. Genética*. Universidade de São Paulo, São Paulo, Brazil.

Forattini, O.P. (2002) *Culicidologia Médica. Vol. 2. Identificação, Biologia, Epidemiologia*. Universidade de São Paulo, São Paulo, Brazil.

Forattini, O.P. & Sallum, M.A.M. (1985) A new species of *Culex (Melanoconion)* from southern Brazil (Diptera: Culicidae). *Revista de Saúde Pública*, 19 (2), 171–182. DOI: 10.1590/S0034-89101985000200007

Forattini, O.P. & Sallum, M.A.M. (1987a) Studies on some species of *Culex (Melanoconion)*, with the description of a new one from southern Brazil (Diptera: Culicidae). *Revista de Saúde Pública*, 21 (2), 123–156. DOI: 10.1590/S0034-89101987000200006

Forattini, O.P. & Sallum, M.A.M. (1987b) A new species of *Culex (Melanoconion)* from inland southern Brazil (Diptera: Culicidae). *Mosquito Systematics*, 19 (2), 167–172.

Forattini, O.P. & Sallum, M.A.M. (1989) A new species of *Culex (Melanoconion)* from Atlantic tropical system of southern Brazil (Diptera: Culicidae). *Memórias do Instituto Oswaldo Cruz*, 84 (1), 1–4. DOI: 10.1590/S0074-02761989000100001

Forattini, O.P. & Sallum, M.A.M. (1992) A new species of *Culex (Melanoconion)* from the Amazonian region (Diptera: Culicidae). *Memórias do Instituto Oswaldo Cruz*, 87 (2), 265–274. DOI: 10.1590/S0074-02761992000200015

Forattini, O.P. & Sallum, M.A.M. (1995) Two new species of the Spissipes Section of *Culex (Melanoconion)* (Diptera: Culicidae). *Mosquito Systematics*, 27 (2), 125–142.

Forattini, O.P. & Toda, A. (1966) Notas sôbre Culicidae (Diptera). 11. Algumas espécies amazônicas de *Microculex*. *Studia Entomologica*, 9 (1–4), 501–514.

Forattini, O.P., Rabello, E.X. & Souza Lopes, O. (1967) Notas sôbre *Culicidae (Diptera)*. 12- Nova espécie do subgênero *Culex*. *Revista Brasileira de Entomologia*, 12, 12–16, 1 pl.

Foster, P.G., Bergo, E.S., Bourke, B.P., Oliveira, T.M.P., Nagaki, S.S., Sant'Ana, D.C. & Sallum, M.A.M. (2013) Phylogenetic analysis and DNA-based species confirmation in *Anopheles (Nyssorhynchus)*. *PLoS ONE*, 8 (2), e54063. DOI: 10.1371/journal.pone.0054063

Foster, P.G., de Oliveira, T.M.P., Bergo, E.S., Conn, J.E., Sant'Ana, D.C., Nagaki, S.S., Nihei, S., Lamas, C.E., González, C., Moreira, C.C. & Sallum, M.A.M. (2017) Phylogeny of Anophelinae using mitochondrial protein coding genes. *Royal Society Open Science*, 4, 170758. DOI: 10.1098/rsos.170758

Foster, W.A., Mynhardt, G., McHugh, M. & Zavortink, T.J. (2024) Courtship diversity and similarity in the mosquito genus *Wyeomyia* (Culicidae, Sabethini). *Journal of Vector Ecology*, 49 (1), 105–113. DOI: 10.52707/1081-1710-49.1.105

Fox, I. (1953) Notes on Puerto Rican mosquitoes including a new species of *Culex*. *Journal of Parasitology*, 39 (2), 178–182. DOI: 10.2307/3274114

Francy, D.B., Jaenson, T.G.T., Lundström, J.O., Schildt, E.-B., Espmark, Å., Henriksson, B. & Niklasson, B. (1989) Ecologic studies of mosquitoes and birds as hosts of Ockelbo virus in Sweden and isolation of Inkoo and Batai viruses from mosquitoes. *American Journal of Tropical Medicine and Hygiene*, 41 (3), 355–363. DOI: 10.4269/ajtmh.1989.41.355

Freier, J.E. & Beier, J.C. (1984) Oral and transovarial transmission of La Crosse virus by *Aedes atropalpus*. *American Journal of Tropical Medicine and Hygiene*, 33 (4), 708–714. DOI: 10.4269/ajtmh.1984.33.708

Freitas, L.A., Russo, C.A.M., Voioch, C.M., Mutaquiha, O.C.F., Marques, L.P. & Schrago, C.G. (2015) Diversification of the genus *Anopheles* and a Neotropical clade from the late Cretaceous. *PLoS ONE*, 10 (8), e0134462. DOI: 10.1371/journal.pone.0134462

Froud, M.D. (1944) *Anopheles jebudensis* sp. nov., a new anopheline mosquito from southern Nigeria. *Annals of Tropical Medicine and Parasitology*, 38 (1), 73–77.

Gabaldon, A. (1940) Estudios sobre anofelinos. Serie I. 1. Descripción de *Anopheles (Nyssorhynchus) nuñez-tovari* n. sp. y consideraciones sobre una sub-division del grupo *Nyssorhynchus* (Diptera, Culicidae). *Publicaciones del División de Malariología*, 5, 3–7.

Gabaldon, A. & Cova-Garcia, P. (1952) Zoogeografía de los anofelinos en Venezuela IV Su posición en la región Neotrópica y observaciones sobre las especies de esta región. *Revista Venezolana Sanidad y Asistencia Social*, 17, 171–209, 12 pls.

Gabaldon, A., Cova-Garcia, P. & Lopez, J.A. (1940) Estudios sobre anofelinos. Serie I. 2. *Anopheles (Nyssorhynchus) rangeli*, una nueva especie de la sub-serie *oswaldoi* (Diptera, Culicidae) de amplia distribucion en Venezuela. *Publicaciones de la Division de Malariologia*, 5, 9–23.

Gabaldon, A., Cova-Garcia, P. & Lopez, J.A. (1941a) Estudios sobre anofelinos. Serie II. 1. *Anopheles (Nyssorhynchus) benarrochi*, una especie de la subserie *triannulatus*. *Publicaciones de la Division de Malariologia*, 7, 3–24, 2 pls.

Gabaldon, A., Cova-Garcia, P. & Lopez, J.A. (1941b) Estudias sobre anofelines. Serie II. 2. *Anopheles (Arthuromyia) vargasi*, n. sp. y redescripcion del subgenero *Arthuromyia* Galvao, 1941. *Publicaciones de la Division de Malariologia*, 7, 25–56, 3 pls.

Gad, A., Harbach, R.E. & Harrison, B.A. (2006) *Anopheles (Cellia) ainshamsi*, n. sp. (Diptera: Culicidae), a saltwater species from the Red Sea coast of Egypt. *Proceedings of the Entomological Society of Washington*, 108 (2), 366–380.

Gaffigan, T.V. & Ward, R.A. (1985) Index to the second supplement to "A Catalog of the Mosquitoes of the World", with corrections and additions (Diptera: Culicidae). *Mosquito Systematics*, 17 (1), 52–63.

Galindo, P. (1958) Bionomics of *Sabethes chloropterus* Humboldt, a vector of sylvan yellow fever in Middle America. *American Journal of Tropical Medicine and Hygiene*, 7 (4), 429–440. DOI: 10.4269/ajtmh.1958.7.429

Galindo, P. (1969) Notes on the systematics of *Culex (Melanoconion) Taeniopus* [*sic*] Dyar and Knab and related species, gathered during arbovirus investigations in Panama. *Mosquito Systematics Newsletter*, 1 (4), 82–89.

Galindo, P. & Blanton, F.S. (1954) Nine new species of Neotropical *Culex*, eight from Panama and one from Honduras (Diptera, Culicidae). *Annals of the Entomological Society of America*, 47 (2), 231–247. DOI: 10.1093/aesa/47.2.231

Galindo, P. & Mendez, E. (1961) Descriptions of four new species of *Culex* from Panama (Diptera: Culicidae). *Annals of the Entomological Society of America*, 54 (1), 1–4. DOI: 10.1093/aesa/54.1.1

Galindo, P. & Trapido, H. (1967) Description of *Haemagogus aeritinctus*, a new species from British Honduras, with a note on the validity of *Haemagogus lucifer* (H. D. and K.) (Diptera: Culicidae). *Proceedings of the Entomological Society of Washington*, 69 (2), 103–111.

Galindo, P., Blanton, F.S. & Peyton, E.L. (1954) A revision of the *Uranotaenia* of Panama with notes on other American species of the genus (Diptera, Culicidae). *Annals of the Entomological Society of America*, 47 (1), 107–177. DOI: 10.1093/aesa/47.1.107

Galindo, P., Carpenter, S.J. & Trapido, H. (1951) Descriptions of two new species of *Wyeomyia* and the male of *Sabethes tarsopus* Dyar and Knab (Diptera, Culicidae). *Proceedings of the Entomological Society of Washington*, 53 (2), 86–96.

Galindo, P., Carpenter, S.J. & Trapido, H. (1953) The taxonomic status of the *Aedes leucocelaenus* complex with descriptions of two new forms (Diptera, Culicidae). *Annals of the Entomological Society of America* (for 1952), 45 (4), 529–542. DOI: 10.1093/aesa/45.4.529

Galliard, H. (1931a) *Culex brumpti*, n. sp. Moustique nouveau trouvé en Corse. *Annales de Parasitologie Humaine et Comparée*, 9 (2), 134–139. DOI: 10.1051/parasite/1931092134

Galliard, H. (1931b) Culicides du Gabon. I. – Culicinés, avec la description d'une espéce et de deux variétés nouvelles. *Annales de Parasitologie Humaine et Comparée*, 9 (3), 225–232. DOI: 10.1051/parasite/1931093225

Galliard, H. & Ngu, D.-V. (1947) Culicinés du Tonkin. I. Liste des espèces et leur répartition. *Annales de Parasitologie Humaine et Comparée*, 22 (1–2), 75–80. DOI: 10.1051/parasite/1947221075

Galvão, A.B. (1952) *Anopheles (Anopheles) rachoui*, uma nova espécie do grupo *Arribalzagia*. *Revista Brasileira de Malariologia e Doenças Tropicais*, 4 (1), 69–74.

Galvão, A.B. (1955) *Anopheles (Arribalzagia) bustamantei*, uma nova espécie *(Diptera, Culicidae)*. *Revista Brasileira de Malariologia e Doenças Tropicais*, 7, 255–258.

Galvão, A.L.A. (1941) Contribuição ao conhecimento das especies de *Myzorhynchella (Diptera, Culicidae)*. *Arquivos de Zoologia*, 2 (18), 505–576, 13 pls. DOI: 10.11606/issn.2176-7793.19412505-576

Galvão, A.L.A. (1943) Chaves para a determinação das espécies do subgênero *Nyssorhynchus* do Brasil. *Arquivos de Higiene Saúde Pública* 8 (19), 141–162.

Galvão, A.L.A. & Barretto, M.P. (1941) Sôbre um novo anofelino de São Paulo, *Anopheles (Anopheles) pseudotibiamaculatus* n. sp. *(Diptera, Culicidae)*. *Revista de Biologia e Hygiene*, 11 (2), 65–70.

Galvão, A.L.A. & Damasceno, R.G. (1942a) *Anopheles (Nyssorhynchus) konderi* nova espécie de *Anopheles* do vale do Amazonas e considerações sôbre as espécies do complexo *tarsimaculatus* (Diptera, Culicidae). *Folia Clinica et Biologica*, 14 (5–6), 115–135.

Galvão, A.L.A. & Damasceno, R.G. (1942b) Sôbre um novo anofelino da Ilha de Marajó, *Anopheles (N.) Marajoara* [*sic*] N. Sp. *Anais Paulistas de Medicina e Cirurgia*, 44 (5), 424–427.

Galvão, A.L.A. & Franco do Amaral, A.D. (1938) Sobre um novo anofelino de Campos do Jordão, Estado de São Paulo, *Anopheles (Nyssorhynchus) lanei* n. sp. *(Diptera, Culicidae)*. *Revista de Biologia e Hygiene*, 9 (1), 8–16, 4 pls.

Galvão, A.L.A. & Franco do Amaral, A.D. (1940) Estudos sôbre os anofelinos do grupo *Myzorhynchella* com a descrição de uma espécie nova, *Anopheles (Nyssorhynchus) antunesi* n. sp. (Dipt. Culicidae). *Folia Clinica et Biologica*, 12 (5), 150–160.

García, M. & Casal, O.H. (1967) *Toxorhynchites (Lynchiella) cavalierii*, nueva especie bromelícola de la República Argentina *(Diptera, Culicidae)*. *Physis*, 26 (73), 435–441.

Garcia, R., Jeffery, J. & Rudnick, A. (1969) *Culiseta marchettei*, a new species of the subgenus *Climacura* Howard, Dyar, and Knab from Malaysia, with notes on its biology (Diptera: Culicidae). *Journal of Medical Entomology*, 6 (3), 251–256. DOI: 10.1093/jmedent/6.3.251

Gardner, C.F., Nielsen, L.T. & Knight, K.L. (1973) Morphology of the mouthparts of larval *Aedes communis* (DeGeer): (Diptera: Culicidae). *Mosquito Systematics*, 5 (2), 163–182.

Garros, C., Harbach, R.E. & Manguin, S. (2005) Morphological assessment and molecular phylogenetics of the Funestus and Minimus Groups of *Anopheles (Cellia)*. *Journal of Medical Entomology*, 42 (4), 522–536. DOI: 10.1093/jmedent/42.4.522

Gaschen, H. (1934) Description d'un nouvel *Aedes* du Yunnan *Finlaya yunnanensis* sp. n. *Archives des Instituts Pasteur d'Indochine*, 19, 331–335, 3 pls.

Gater, B.A.R. (1933) On *Anopheles baezai* n.sp., from the Malay Peninsula. *Bulletin of the Raffles Museum*, 8, 162–168.

Gebert, S. (1948) Notes on a new species of *Aëdes*, sub-genus *Mucidus* (Diptera: Culicidae) found in Mauritius. *Entomologist*, 81, 96–99.

Geoffroy, B. (1971a) Description du mâle de *Culex (Culiciomyia) eouzani* sp. n. *(Diptera Culicidae)* nouveau moustique de la République Centrafricaine. *Cahiers ORSTOM Série Entomologie médicale et Parasitologie*, 9 (3), 279–283.

Geoffroy, B. (1971b) Description d'un moustique nouveau de la République Centrafricaine *Aedes (Aedimorphus) adami* sp. n. *Cahiers ORSTOM Série Entomologie médicale et Parasitologie*, 9 (3), 273–277.

Geoffroy, B. (1971c) Description d'un nouveau moustique de la République Centrafricaine: *Culex (Neoculex) vattieri* sp. n. *Cahiers ORSTOM Série Entomologie médicale et Parasitologie*, 9 (3), 265–268.

Geoffroy, B. (1974) Description du mâle de *Culex (Mochthogenes) germaini* n. sp. *(Diptera, Culicidae)* moustique nouveau de République Centrafricaine. *Cahiers ORSTOM Série Entomologie médicale et Parasitologie*, 12 (2), 93–96.

Geoffroy, B. (1987) The *Aedes (Aedimorphus)* Domesticus Group (Diptera, Culicidae). I. New species, descriptions of *Aedes bambiotai* and *Aedes bancoi*. *Mosquito Systematics*, 19 (1), 100–110.

Geoffroy, B. & Hervé, J.-P. (1976) Description d'un nouveau moustique de République Centrafricaine, *Culex (Eumelanomyia) taufﬂiebi* n. sp. *Cahiers ORSTOM Série Entomologie médicale et Parasitologie*, 14 (2), 151–154.

Germain, M., Francy, D.B., Ferrara, L., Sanyang, Y., Monath, T.P., Adam, C. & Salaun, J.-J. (1980) Yellow fever in the Gambia, 1978–1979: a complementary entomological survey done in October 1979. *Cahiers ORSTOM Série Entomologie médicale et Parasitologie*, 18 (1), 3–12.

Ghassemi-Khademi, T., Oshaghi, M.A., Vatandoost, H., Madjdzadeh, S.M. & Gorouhi, M.A. (2021) Utility of complete mitochondrial genomes in phylogenetic classification of the species of *Anopheles* (Culicidae: Anophelinae). *Journal of Arthropod-Borne Diseases*, 15 (1), 1–20. DOI: 10.18502/jad.v15i1.6483

Giaquinto Mira, M. (1931) Una nueva especie de *Anopheles* en Guatemala: *Anopheles hectoris* mihi. *Boletín de la Dirección General de Salubridad, Guatemala*, 1 (20/25), 606–615.

Gil Collado, J. (1936) Culicidos de la Isla de Fernando Poo recogidos por la expedición J. Gil-F. Bonet. *EOS - Revista Espanola de Entomologia*, 11 (4), 311–329.

Giles, G.M. (1900a) *A handbook of the gnats or mosquitoes, giving the anatomy and life history of the Culicidae*. John Bale, Sons & Danielsson, London, UK.

Giles, G.M. (1900b) Addendum I. Description of two species of *Anopheles* from West Africa. *Memoirs of the Liverpool School of Tropical Medicine*, 2, 49–51.

Giles, G.M. (1901a) A plea for the collective investigation of Indian Culicidae, with suggestions as to moot points for enquiry, and a prodromus of species known to the author. *Journal of the Bombay Natural History Society*, 13 (4), 592–610, 2 pls.

Giles, G.M. (1901b) Descriptions of four new species of *Anopheles* from India. *Entomologist's Monthly Magazine*, 37, 196–198.

Giles, G.M. (1901c) Six new species of Culicidae from India. *Entomologist*, 34 (458), 192–197.

Giles, G.M. (1902) *A handbook of the gnats or mosquitoes giving the anatomy and life history of the Culicidae together with descriptions of all species noticed up to the present date*. 2nd edn. John Bale, Sons & Danielsson, London, UK.

Giles, G.M. (1904) Notes on some collections of mosquitoes, &c., received from the Philippine Islands and Angola; with some incidental remarks upon classification. *Journal of Tropical Medicine*, 7, 365–369.

Giłka, W., Harbach, R.E. & Perkovsky, E.E. (2021) Mosquitoes (Culicidae: Diptera) in Eocene amber from the Rovno region, Ukraine. *Zootaxa*, 5016 (2), 257–270. DOI: 10.11646/zootaxa.5016.2.6

Gillett, J.D. (1971) *Mosquitos*. Weidenfeld & Nicolson, London, UK.

Gillies, M.T. (1958) Notes on the biology of a new subspecies of *Anopheles wellcomei* (Diptera: Culicidae) from East Africa, and on the distribution of related forms. *Proceedings of the Royal Entomological Society of London, Series A: General Entomology*, 33 (1–3), 9–14. DOI: 10.1111/j.1365-3032.1958.tb00386.x

Gillies, M.T. (1962) A new species of the *Anopheles funestus* complex (Diptera: Culicidae) from East Africa. *Proceedings of the Royal Entomological Society of London, Series B: Taxonomy*, 31 (7–8), 81–86. DOI: 10.1111/j.1365-3113.1962.tb01190.x

Gillies, M.T. (1964) A new *Anopheles* (Diptera: Culicidae) from the East African coastal lowlands. *Proceedings of the Royal Entomological Society of London, Series B: Taxonomy*, 33 (3–4), 67–70. DOI: 10.1111/j.1365-3113.1964.tb01614.x

Gillies, M.T. & Coetzee, M. (1987) A supplement to the Anophelinae of Africa south of the Sahara (Afrotropical Region). *Publications of the South African Institute for Medical Research*, 55, 1–143.

Gillies, M.T. & de Meillon, B. (1968) The Anophelinae of Africa south of the Sahara (Ethiopian Zoogeographical Region). *Publications of the South African Institute for Medical Research*, 54, 1–343.

Gistl, J.N.F.X. (1831) Kerfe in Copal eingeschlossen. *Isis (Oken's)*, 24, 247–248.

Glick, J.I. (1992) Illustrated key to the female *Anopheles* of southwestern Asia and Egypt (Diptera: Culicidae). *Mosquito Systematics*, 24 (2), 125–153.

Gómez, G.F., Bickersmith, S.A., González, R., Conn, J.E. & Correa, M.M. (2015) Molecular taxonomy provides new insights into *Anopheles* species of the Neotropical Arribalzagia Series. *PLoS ONE*, 10, 137. DOI: 10.1371/journal.pone.0119488

Gong, Z.-D. (1989a) A new species of genus *Topomyia* from Yunnan, China (Diptera: Culicidae) [in Chinese]. *Acta Entomologica Sinica*, 32 (1), 90–91.

Gong, Z.-D. (1989b) A new species of the genus *Topomyia* from Yunnan Province, China (Diptera: Culicidae) [in Chinese]. *Acta Zootaxonomica Sinica*, 14 (2), 220–223.

Gong, Z.-D. (1991) A new species of the subgenus *Topomyia* of genus *Topomyia* from Yunnan, China (Diptera: Culicidae) [in Chinese and English]. *Acta Zootaxonomica Sinica*, 16 (2), 228–231.

Gong, Z.-D. (1994) A new species of the genus *Topomyia* from West Yunnan, China (Diptera: Culicidae) [in Chinese]. *Acta Zootaxonomica Sinica*, 19 (2), 210–213.

Gong, Z.-D. (1996) A new species of *Topomyia* from the bamboo stumps in Yunnan, China (Diptera: Culicidae) [in Chinese]. *Acta Zootaxonomica Sinica*, 21 (3), 362–365.

Gong, Z.-D. & Ji, S.-H. (1989) A new species of *Tripteroides* from Yunnan (Diptera: Culicidae) [in Chinese]. *Acta Entomologica Sinica*, 32 (4), 482–484.

Gong, Z.D. & Li, Z.-H. (1999) A new species of *Heizmannia* from Yunnan, China (Diptera: Culicidae) [in Chinese]. *Acta Zootaxonomica Sinica*, 24 (1), 96–99.

Gong, Z. & Lu, B. (1986) A new *Aedes* of subgenus *Finlaya* from Yunnan, China [in Chinese]. *Zoological Research*, 7 (1), 1–4.

Gong, Z. & Lu, B. (1991) A new subgenus and species of the *Aedes* (Diptera: Culicidae) from Yunnan, China [in Chinese]. *Entomotaxonomia*, 13 (1), 55–59.

Gong, Z. & Lu, B. (1995) Studies on the genus *Topomyia* of China (Diptera: Culicidae) [in Chinese]. *Acta Entomologica Sinica*, 38 (4), 475–485.

González Broche, R. (2006) *Culicidos de Cuba (Diptera: Culicidae)*. Editorial Cientifico-Tecnica, Habana, Cuba.

González Broche, R. & Rodríguez Rodríguez, R. (2001) Nuevo subgénero de *Culex*, descripción de la pupa y de la larva y redescripción de la hembra de *Culex nicaroensis* Duret, (Díptera: Culicidae). *Avicennia*, 14, 65–74.

González, C.R. & Sallum, M.A.M. (2010) *Anopheles (Nyssorhynchus) atacamensis* (Diptera: Culicidae), a new species from northern Chile. *Memórias do Instituto Oswaldo Cruz*, 105 (1), 13–24. DOI: 10.1590/S0074-02762010000100002

González, C.R., Jercic, M.I., Reyes, C., Mejias, G., Plavetic, C. & Parra, A. (2008) A pictorial key to the genera of Culicidae (Diptera) from Chile of medical importance [in Spanish]. *Acta Entomológica Chilena*, 32 (1–2), 35–42.

González, C.R., Reyes, C., Rada-Chaparro, V. & Saldarriaga-Córdoba, M. (2017) A new species of *Aedes* Meigen subgenus *Ochlerotatus* Lynch Arribálzaga (Diptera: Culicidae) from the coastal wetlands of the desert in northern Chile: morphological and molecular identification. *Zootaxa*, 4273 (1), 31–49. DOI: 10.11646/zootaxa.4273.1.3

González, C.R., Reyes-Valenzuela, C., Rossi, G.C. & Laurito, M. (2023) Revalidation of *Phalangomyia* Dyar & Knab as a subgenus of *Culex* L. (Diptera: Culicidae) based on morphological and molecular evidence. *Zootaxa*, 5256 (6), 544–564. DOI: 10.11646/zootaxa.5256.6.2

Gordeev, M.I., Zvantsov, A.B., Goriacheva, I.I., Sheĭkevich, E.V. & Ezhov, M.N. (2005) Description of the new species *Anopheles artemievi* sp.n. (Diptera, Culicidae) [in Russian]. *Meditsinskaia Parazitologiia i Parazitarnye Bolezni*, 2005 (2), 4–5.

Gordon, R.M. & Evans, A.M. (1922) Mosquitoes collected in the Manáos Region of the Amazon. *Annals of Tropical Medicine and Parasitology*, 16 (3), 315–338, 1 pl. DOI: 10.1080/00034983.1922.11684325

Gornostaeva, R.M. (2003) The status of taxa of subgenus *Aedes* (Diptera, Culicidae, *Aedes*): *esoensis* Yamada, *rossicus* Dolbeskin, Gorickaja & Mitrofanova, *cinereus* Meigen, *geminus* Peus. I. Overview. *European Mosquito Bulletin*, 16, 22–26.

Gornostaeva, R.M. (2005) To the revision of the mosquitoes of the subgenus *Aedes* (Diptera, Culicidae) from the Palaearctic [in Russian]. *Parazitologiia*, 39 (6), 457–507.

Gough, L.H. (1910) On a collection of *Anopheles* made at Onderstepoort in the autumn of 1909. *Report of the Government Veterinary Bacteriologist / Union of South Africa*, 1908–09, 115–121.

Gould, D.J., Bailey, C.L. & Vongpradist, S. (1982) Implication of forest mosquitoes in transmission of *Wuchereria bancrofti* in Thailand. *Mosquito News*, 42 (4), 560–564.

Gould, J. (1854) *A monograph of the Trochilidae, or family of humming-birds.* Vol. 2. Taylor & Francis, London, UK.

Grabham, M. (1905) Notes on some Jamaican Culicidae. *Canadian Entomologist*, 37 (12), 401–411. DOI: 10.4039/Ent37401-12

Grabham, M. (1906) Four new Culicidae from Jamaica, West Indies. *Canadian Entomologist*, 38 (11), 311–320.

Grabham, M. (1907) Notes on some new mosquitoes from Jamaica, West Indies. *Canadian Entomologist*, 39 (1), 25–26. DOI: 10.4039/Ent3925-1

Graham, W.M. (1909) Four new species of the genus *Eretmapodites* (Theobald) from Ashanti. *Entomologist*, 42 (551, 553), 86–89, 157–159.

Graham, W.M. (1910) On new species of West-African *Culicidae*. *Annals and Magazine of Natural History, including Zoology, Botany, and Geology* (Eighth Series), 5, 264–273. DOI: 10.1080/00222931008692764

Grassi, B. (1899) Ancora sulla malaria. *Rendiconti delle seduti delle Reale Accademia dei Lincei. Classe de scienze fisiche, matematische e naturali*, 8, 559–561.

Grassi, B., Bignami, A. & Bastianelli, G. (1899) Resoconto degli studi fatti sulla malaria durante il mese di gennaio. *Reale Accademia dei Lincei. Rendiconti. Memorie della classe de scienze fisiche, matematische e naturali*, 8, 100–104.

Grassi, G.B. (1900) Studi di uno zoologo sulla marlaria. Memoria del Socio Battista Grassi. *Reale Accademia dei Lincei. Rendiconti. Classe de scienze fisiche, matematische e naturali. Memoirs*, (5) 3, 299–516, 5 pls.

Green, C.A. (1972) The *Anopheles annulipes* complex of species. Abstract. 14th International Congress of Entomology, Canberra, Australia, August 22–30, 286.

Green, C.A., Cass, R.F., Munstermann, L.E. & Baimai, V. (1990) Population-genetic evidence for two species in *Anopheles minimus* in Thailand. *Medical and Veterinary Entomology*, 4 (1), 25–34. DOI: 10.1111/j.1365-2915.1990.tb00256.x

Green, C.A., Harrison, B.A., Klein, T.A. & Baimai, V. (1985) Cladistic analysis of polytene chromosome rearrangements in anopheline mosquitoes, subgenus *Cellia*, series *Neocellia*. *Canadian Journal of Genetics and Cytology*, 27 (2), 123–133. DOI: 10.1139/g85-020

Grjebine, A. (1950) Un nouveau moustique forestier de la basse Cote d'Ivoire *Eretmapodites pauliani* n. sp. *Bulletin de la Société de Pathologie exotique*, 43 (1–2), 45–50.

Grjebine, A. (1953a) Observations sur les nematoceres vulnerants de Madagascar, regions de Majunga et de la Mandraka. *Mémoires de l'Institut Scientifique de Madagascar Séries E*, 4, 443–502.

Grjebine, A. (1953b) Une nouvelle espece d'anophele de Madagascar [Dipt. Culicidae]. *Le Naturaliste Malgache*, 5 (2), 203–209.

Grjebine, A. (1961) Un nouvel anophèle de Madagascar. *Bulletin de la Société de Pathologie exotique* (for 1960), 53 (6), 939–942.

Grjebine, A. (1964) *Anopheles (Neomyzomyia) grenieri* n. sp. *(Diptera, Culicidae)*, nouvel anophèle de Madagascar. *Bulletin de la Société de Pathologie exotique*, 57 (1), 38–42.

Grjebine, A. (1966) *Faune de Madagascar. XXII. Insecies Diptéres Culicidae Anophelinae*. Centre National de la Recherche Scientifique, Office de la Recherche Scientifique et Technique Outre-Mer, Paris, France.

Grjebine, A. (1972) Une nouvelle espèce de moustique du Congo-Brazzaville: *Eretmapodites hamoni* n. sp. (Dipt. Culicidae). *Bulletin de la Société Entomologique de France*, 77 (9–10), 308–313. DOI: 10.3406/bsef.1972.21287

Grjebine, A. (1979) Les moustiques des *Nepenthes* de Madagascar. Espèces nouvelles du genre *Uranotaenia*. *(Diptera, Culicidae)*. *Annales de la Société Entomologique de France*, 15 (1), 53–74. DOI: 10.1080/21686351.1979.12278194

Grjebine, A. (1986) *Faune de Madagascar. 68. Insectes Diptères Culicidae Culicinae Ficalbiini*. Muséum nationale d'Histoire naturelle, Paris, France.

Grünberg, K. (1902) Ein neuer *Anopheles* aus Westafrika, *Anopheles Ziemanni* [*sic*] nov. spec. *Zoologischen Anzeiger*, 25 (677), 550–551.

Grünberg, K. (1905) Zur Kenntnis der Culicidenfauna von Kamerun und Togo. *Zoologischen Anzeiger*, 29 (12), 377–390.

Grünberg, K. (1907) Ein neuer *Toxorhynchites* aus Kamerun. (Dipt.). *Deutsche Entomologische Zeitschrift*, 1907, 405–406.

Grünberg, K. (1913) Zur Kenntnis der Culiciden-Fauna von Samoa. *Entomologische Rundschau*, 30 (22), 130–131.

Guenther, K. (1913) Die lebenden Bewohner der Kannen der insektenfressenden Pflanze *Nepenthes destillatoria* auf Ceylon. *Zeitschrift für wissenschaftliche Insektenbiologie*, 9 (6/7), 198–207.

Guimarães, A.É. Gentile, C., Lopes, C.M. & de Mello, R.P. (2000) Ecology of mosquitoes (Diptera: Culicidae) in areas of Serra do Mar State Park, State of São Paulo, Brazil. III - Daily biting rhythms and lunar cycle influence. *Memórias do Instituto Oswaldo Cruz*, 95 (6), 753–760. DOI: 10.1590/S0074-02762000000600002

Gunathilaka, N. (2017) Illustrated key to the adult female *Anopheles* (Diptera: Culicidae) mosquitoes of Sri Lanka. *Applied Entomology and Zoology*, 52 (1), 69–77. DOI: 10.1007/s13355-016-0455-y

Gunathilaka, N., Fernando, T., Hapugoda, M., Abeyewickreme, W. & Wickremasinghe, R. (2014) Revised morphological identification key to the larval anopheline (Diptera: Culicidae) of Sri Lanka. *Asian Pacific Journal of Tropical Biomedicine*, 4 (Suppl. 1), S222–S227. DOI: 10.12980/APJTB.4.2014C941

Guo, J., Yan, Z.-T., Fu, W.-B., Yuan, H., Li, X.-D. & Chen, B. (2021) Complete mitogenomes of *Anopheles peditaeniatus* and *Anopheles nitidus* and phylogenetic relationships within the genus *Anopheles* inferred from mitogenomes. *Parasites & Vectors*, 14, 452. DOI: 10.1186/s13071-021-04963-4

Gutiérrez, L.A., Orrego, L.M., Gómez, G.F., López, A., Luckhart, S., Conn, J.E. & Correa, M.M. (2010) A new mtDNA COI gene lineage closely related to *Anopheles janconnae* of the Albitarsis complex in the Caribbean region of Colombia. *Memórias do Instituto Oswaldo Cruz*, 105 (8), 1019–1025. DOI: 10.1590/S0074-02762010000800011

Gutsevich, A.V. (1955) New and little known forms of mosquitoes (Diptera, Culicidae) [in Russian]. *Trudy Zoologicheskogo Instituta Akademii Nauk*, 18, 320–324.

Gutsevich, A.V. (1962) A new species of the genus *Aedes* (Diptera, Culicidae) from Kazakhstan [in Russian]. *Entomologicheskoe Obozrenie*, 41 (4), 886–888.

Gutsevich, A.V. (1966) A new mosquito species from Uzbekistan, *Aedes (Ochlerotatus) simanini* (Culicidae) [in Russian]. *Zoologicheskii Zhurnal*, 45 (3), 457–459.

Gutsevich, A.V., Monchadskii, A.S. & Shtakel'berg, A.A. (1974) *Fauna of the U.S.S.R. Diptera. Volume 3, No. 4. Mosquitoes Family* Culicidae [*sic*]. Keter Publishing House Jerusalem, Jerusalem, Israel.

Hackett, L.W. (1934) The present status of our knowledge of the sub-species of *Anopheles maculipennis*. *Transactions of the Royal Society of Tropical Medicine and Hygiene*, 28 (2), 109–128. DOI: 10.1016/S0035-9203(34)90135-8

Haddow, A.J. (1946) The mosquitoes of Bwamba County, Uganda. IV.—Studies on the genus *Eretmapodites*, Theobald. *Bulletin of Entomological Research*, 37 (1), 57–82. DOI: 10.1017/S0007485300021994

Haddow, A.J. & Ssenkubuge, Y. (1963) Studies on the biting habits of East African mosquitos in the genera *Uranotaenia*, *Ficalbia* and *Hodgesia*. *Bulletin of Entomological Research*, 53 (4), 639–652. DOI: 10.1017/S0007485300048380

Haddow, A.J. & van Someren, E.C.C. (1950) A new species of *Stegomyia* Theobald from the Ruwenzori Range, Uganda. *Annals of Tropical Medicine and Parasitology*, 44 (3), 281–284. DOI: 10.1080/00034983.1950.11685450

Haga, J. (1930) Tabellen voor determinatie der in Nederländsch-Oost-Indië voorkomende Anophelinen. *Geneeskundig Tijdschrift voor Nederlandsch-Indië* (for 1929), 70, 363–382.

Haliday, A.H. (1833) Catalogue of Diptera occurring about Holywood in Downshire. *Entomological Magazine*, 1, 147–180.

Hall, C.R., Howard, T.M. & Harbach, R.E. (1999) *Sabethes (Peytonulus) luxodens*, a new species of Sabethini (Diptera: Culicidae) from Ecuador. *Memórias do Instituto Oswaldo Cruz*, 94 (3), 329–338. DOI: 10.1590/S0074-02761999000300009

Hamon, J. (1953) Une nouvelle espece d'*Aedes* de l'ile de la Reunion (Diptere Culicidae). *Naturaliste Malgache*, 5, 35–41.

Hamon, J. (1955) Contribution a l'étude des culicides d'Afrique Occidentale. Description de *Uranotaenia devemyi* sp. n., *Culex grahami* var. *farakoensis* var. n. et de la larve de *Harpagomyia trichorostris* Theobald. *Bulletin de la Société de Pathologie exotique* (for 1954), 47 (5), 672–678.

Hamon, J. (1957) Deux moustiques nouveaux de Cote-d'Ivoire: *Aedes (Aedimorphus) falabreguesi* sp. n., *Culex (Culex) castelli* sp. n. *Bulletin de la Société de Pathologie exotique*, 50 (5), 685–691.

Hamon, J. (1961) Les *Eretmapodites* du groupe *oedipodius* Graham. 1. Descriptions complémentaires et position systématique des six sous-espèces actuellement cannues. *Bulletin de la Société de Pathologie exotique*, 54 (4), 892–906.

Hamon, J. (1962) Description de deux nouveaux *Eretmapodites (Diptera, Culicidae)* capturés en Uganda: *E. vansomereni* sp. n. et *E. corbeti* sp. n. *Bulletin de la Société de Pathologie exotique*, 55 (2), 253–259.

Hamon, J. & Adam, J.P. (1955) Contribution a l'étude des culicidés (Diptères) d'Afrique. Description de *Harpagomyia moucheti* sp. n. *Bulletin de la Société de Pathologie exotique*, 48 (6), 882–885.

Hamon, J. & Adam, J.-P. (1959a) Description de deux nouveaux *Aedes* appartenant au groupe *tarsalis* Newstead en basse Cote-d'Ivoire (A. O. F.): *A. (Aedimorphus) reali* sp. n. et *A. (Aedimorphus) nyounae* sp. n. *Bulletin de la Société de Pathologie exotique* (for 1958), 51 (1), 530–534.

Hamon, J. & Adam, J.P. (1959b) Description de deux nouveaux *Aedes* de Cote-d'Ivoire appartenant au sous-genre *Aedimorphus: A. ovazzai* sp. n. et *A. rickenbachi* sp. n. *Bulletin de la Société de Pathologie exotique*, 52 (2), 147–154.

Hamon, J. & Brengues, J. (1965) Observations sur les *Aedes (Aedimorphus)* d'Afrique avec description de deux nouvelles espèces: *Ae. lottei* n. sp. et *Ae. dialloi* n. sp. *Bulletin de la Société de Pathologie exotique*, 58 (1), 101–108.

Hamon, J. & Gandara, A. (1955) Contribution a l'étude des culicidés (Diptères) de la région Éthiopienne. Description de *Culex (Culiciomyia) cambournaci* n. sp. *Bulletin de la Société de Pathologie exotique*, 48 (6), 866–872.

Hamon, J. & Lambrecht, F.L. (1959) Description de *Culex (Culiciomyia) muspratti* n. sp. accompagnée de quelques notes sur les moustiques de la Forêt d'Irangi, Kivu, Congo Belge. *Bulletin de la Société de Pathologie exotique*, 52 (5), 582–587.

Hamon, J. & Mouchet, J. (1955) Contribution a l'étude des *Neoculex* (Diptères: Culicidés) de la région éthiopienne 2. — Description de *Neoculex adami* sp. n. *Bulletin de la Société de Pathologie exotique*, 48 (6), 860–862.

Hamon, J. & Ovazza, M. (1954) Une nouvelle espèce de moustique (*Culex shoae* n. sp.) vivant sur les bananiers du haut plateau d'Éthiopie. *Bulletin de la Société de Pathologie exotique*, 47 (3), 416–421.

Hamon, J. & Rickenbach, A. (1954) Contribution a l'étude des culicidés d'Afrique occidentale. Description d'*Aedes (Aedimorphus) mattinglyi* sp. n., *Aedes (Banksinella) jamoti* sp. n. Notes complémentaires sur *Aedes (Aedimorphus) stokesi* Evans, *Aedes (Banksinella) bolensis* Edwards. *Bulletin de la Société de Pathologie exotique,* 47 (6), 930–941.

Hamon, J. & Rickenbach, A. (1955a) Contribution a l'étude des culicidés d'Afrique occidentale. Description d'*Anopheles brumpti* sp. n. *Bulletin de la Société de Pathologie exotique*, 48 (3), 342–344.

Hamon, J. & Rickenbach, A. (1955b) Contribution a l'etude des *Néoculex* (Diptères: Culicidés) de la Région éthiopienne. 1. — Corrections de quelques descriptions de terminalia mâles données par Edwards, avec étude d'une nouvelle variété. *Bulletin de la Société de Pathologie exotique*, 48 (6), 848–859.

Hamon, J. & van Someren, E.C.C. (1961a) Description de deux nouveaux moustiques du Tanganyika: *Culex (Culiciomyia) gilliesi* sp. n. et *Eretmapodites mattinglyi* sp. n. *Bulletin de la Société de Pathologie exotique*, 54 (4), 724–731.

Hamon, J. & van Someren, E.C.C. (1961b) Les *Eretmapodites* du groupe *oedipodius* Graham. II. Description de *E. grenieri* sp. n. et clé de détermination des espèces et sous-espèces du groupe *oedipodius*. *Bulletin de la Société de Pathologie exotique*, 54 (4), 907–913.

Hamon, J., Adam, J.P. & Mouchet, J. (1955) Contribution a l'étude des *Neoculex* (Diptères: Culicidés) de la région éthiopienne. 3. — Description de *Neoculex laplantei* sp. n. *Bulletin de la Société de Pathologie exotique*, 48 (6), 862–865.

Hamon, J., Holstein, M. & Rivola, E. (1957a) Description d'un nouveau moustique du Congo Belge: *Culex (Neoculex) vinckei* sp. n. *Bulletin de la Société de Pathologie exotique*, 50 (5), 681–685.

Hamon, J., Service, M.W., Adam, J.P. & Taufflieb, R. (1961) Observations sur les *Aedes (Aedimorphus)* du groupe *tarsalis* Newstead avec description d'une nouvelle espèce: *Aedes grenieri* n. sp. *Bulletin de la Société de Pathologie exotique*, 54 (2), 375–388.

Hamon, J., Taufflieb, R. & Maillot, L. (1957b) Description d'*Aedes (Aedimorphus) grjebinei* sp. n. accompagnée de quelques données nouvelles sur les moustiques d'A. E. F. *Bulletin de la Société de Pathologie exotique*, 50 (5), 692–696.

Hancock, R.G., Boyd, T., MacFadden, S., Sowders, A., Foster, W.A. & Lounibos, L.P. (2022) Mosquitoes eating mosquitoes: how *Toxorhynchites amboinensis*, *Psorophora ciliata*, and *Sabethes cyaneus* (Diptera: Culicidae) capture prey. *Annals of the Entomological Society of America*, 115 (6), 2022, 461–471. DOI: 10.1093/aesa/saac017

Hancock, R.G., Foster, W.A. & Yee, W.L. (1990) Courtship behavior of the mosquito *Sabethes cyaneus* (Diptera: Culicidae). *Journal of Insect Behavior*, 3 (3), 401–416. DOI: 10.1007/BF01052117

Hao, Y.-J., Zou, Y.-L., Ding, Y.-R., Xu, W.-Y., Yan, Z.-T., Li, X.-D., Fu, W.-B., Li, T.-J. & Chen, B. (2017) Complete mitochondrial genomes of *Anopheles stephensi* and *An. dirus* and comparative evolutionary mitochondriomics of 50 mosquitoes. *Scientific Reports*, 7 (1), 7666. DOI: 10.1038/s41598-017-07977-0

Harbach, R.E. (1977) Comparative and functional morphology of the mandibles of some fourth stage mosquito larvae (Diptera: Culicidae). *Zoomorphologie*, 87 (3), 217–236. DOI: 10.1007/BF00995821

Harbach, R.E. (1978) Comparative structure of the labiohypopharynx of fourth stage mosquito larvae (Diptera: Culicidae), with comments on larval morphology, evolution and feeding habits. *Mosquito Systematics*, 10 (3), 301–333.

Harbach, R.E. (1985) A new species, *Culex (Culex) litwakae* (Diptera: Culicidae), from the coastal region of Kenya. *Mosquito Systematics*, 17 (3), 254–265.

Harbach, R.E. (1987) *Culex verutus*, a new species of the subgenus *Culex* (Diptera: Culicidae) from Sierra Leone. *Mosquito Systematics* (for 1986), 18 (3–4), 216–229.

Harbach, R.E. (1988) The mosquitoes of the subgenus *Culex* in southwestern Asia and Egypt (Diptera: Culicidae). *Contributions of the American Entomological Institute*, 24 (1), vi, 1–240.

Harbach, R.E. (1991a) A new subgenus of the genus *Sabethes* (Diptera: Culicidae). *Mosquito Systematics*, 23 (1), 1–9.

Harbach, R.E. (1991b) Ontogeny of the larval stage of *Sabethes chloropterus*, with special reference to setal development and phylogenetic implications for the family Culicidae (Diptera). *Mosquito Systematics*, 23 (1), 10–24.

Harbach, R.E. (1992) Neotype designation, generic realignment and description of *Dendromyia schnusei* Martini (Diptera: Culicidae). *Mosquito Systematics* (for 1991), 23 (3), 175–181.

Harbach, R.E. (1994a) Review of the internal classification of the genus *Anopheles* (Diptera: Culicidae): the foundation for comparative systematics and phylogenetic research. *Bulletin of Entomological Research*, 84 (6), 331–342. DOI: 10.1017/S0007485300032454

Harbach, R.E. (1994b) The subgenus *Sabethinus* of *Sabethes* (Diptera: Culicidae). *Systematic Entomology*, 19 (3), 207–234. DOI: 10.1111/j.1365-3113.1994.tb00588.x

Harbach, R.E. (1995a) A new *Sabethes* of the subgenus *Peytonulus* (Diptera: Culicidae) with an unusual fourth-instar larva. *Entomologica scandinavica*, 26 (1), 87–96.

Harbach, R.E. (1995b) Two new species of the subgenus *Peytonulus* of *Sabethes* (Diptera: Culicidae) from Colombia. *Memórias do Instituto Oswaldo Cruz*, 90 (5), 583–587. DOI: 10.1590/S0074-02761995000500007

Harbach, R.E. (2004) The classification of genus *Anopheles* (Diptera: Culicidae): a working hypothesis of phylogenetic relationships. *Bulletin of Entomological Research*, 94 (6), 537–553. DOI: 10.1079/BER2004321

Harbach, R.E. (2007) The Culicidae (Diptera): a review of taxonomy, classification and phylogeny. *Zootaxa*, 1668, 591–688. DOI: 10.11646/zootaxa.1668.1.28

Harbach, R.E. (2011) Classification within the cosmopolitan genus *Culex* (Diptera: Culicidae): the foundation for molecular systematics and phylogenetic research. *Acta Tropica*, 120 (1–2), 1–14. DOI: 10.1016/j.actatropica.2011.06.005

Harbach, R.E. (2013a) *Nyctomyia*, a replacement name for the preoccupied genus *Nyx* Harbach & Linton, 2013 (Diptera: Culicidae). *Zootaxa*, 3691 (1), 199. DOI: 10.11646/zootaxa.3691.1.9

Harbach, R.E. (2013b) The phylogeny and classification of *Anopheles*. In: Manguin, S. (ed.), Anopheles *Mosquitoes – New Insights into Malaria Vectors*. InTech, Rijeka, Croatia, pp. 3–55. DOI: 10.5772/54695

Harbach, R.E. (2018a) Two new species of the subgenus *Sabethinus* of *Sabethes* (Diptera: Culicidae) from Costa Rica, first confirmation of members of the subgenus in Central America. *Zootaxa*, 4429 (2), 269–280. DOI: 10.11646/zootaxa.4429.2.3

Harbach, R.E. (2018b) *Culicipedia: Species-group, genus-group and family-group names in Culicidae (Diptera)*. CABI, Wallingford, UK. DOI: 10.1079/9781786399052.0000

Harbach, R.E. (2023) Recognition of *Lasioconops* Theobald, 1903 and *Oculeomyia* Theobald, 1907 as separate subgenera of the genus *Culex* Linnaeus, 1758 (Diptera: Culicidae). *Zootaxa*, 5319 (4), 595–599. DOI: 10.11646/zootaxa.5319.4.10

Harbach, R.E. & Culverwell, C.L. (2014) A new subgenus and species of *Topomyia* (Diptera: Culicidae: Sabethini) based on a remarkable male mosquito from Sabah, Malaysia. *Zootaxa*, 3794 (4), 575–580. DOI: 10.11646/zootaxa.3794.4.8

Harbach, R.E. & Greenwalt, D. (2012) Two Eocene species of *Culiseta* (Diptera: Culicidae) from the Kishenehn Formation in Montana. *Zootaxa*, 3530, 25–34. DOI: 10.11646/zootaxa.3530.1.2

Harbach, R.E. & Howard, T.M. (1997) Metathoracic sensilla in mosquitoes (Culicidae) and related Diptera, and their potential importance in taxonomy and behavioural physiology. *Bulletin of Entomological Research*, 87 (1), 101–104. DOI: 10.1017/S0007485300036427

Harbach, R.E. & Howard, T.M. (2007) Corrections in the status and rank of names used to denote varietal forms of mosquitoes (Diptera: Culicidae). *Zootaxa*, 1542, 35–48. DOI: 10.11646/zootaxa.1542.1.3

Harbach, R.E. & Howard, T.M. (2009) Review of the genus *Chagasia* (Diptera: Culicidae: Anophelinae). *Zootaxa*, 2210, 1–25. DOI: 10.11646/zootaxa.2210.1.1

Harbach, R.E. & Kitching, I.J. (1998) Phylogeny and classification of the Culicidae (Diptera). *Systematic Entomology*, 23 (4), 327–370. DOI: 10.1046/j.1365-3113.1998.00072.x

Harbach, R.E. & Kitching, I.J. (2005) Reconsideration of anopheline phylogeny (Diptera: Culicidae: Anophelinae) based on morphological data. *Systematics and Biodiversity*, 3 (4), 345–374. DOI: 10.1017/S147720000500174X

Harbach, R.E. & Kitching, I.J. (2016) The phylogeny of Anophelinae revisited: inferences about the origin and classification of *Anopheles* (Diptera: Culicidae). *Zoologica Scripta*, 45 (1), 34–47. DOI: 10.1111/zsc.12137

Harbach, R.E. & Knight, K.L. (1977) A mosquito taxonomic glossary. XI. The larval maxilla. *Mosquito Systematics*, 9 (2), 128–175.

Harbach, R.E. & Knight, K.L. (1980) *Taxonomists' Glossary of Mosquito Anatomy*. Plexus Publishing, Marlton, NJ, USA.

Harbach, R.E. & Knight, K.L. (1982) Corrections and additions to *Taxonomists' Glossary of Mosquito Anatomy*. *Mosquito Systematics* (for 1981), 13 (2), 201–217.

Harbach, R.E. & Navarro, J.C. (1996) A new species of *Anopheles* subgenus *Kerteszia* (Diptera: Culicidae) from Venezuela. *Entomologica scandinavica*, 27 (2), 207–216. DOI: 10.1163/187631296X00052

Harbach, R.E. & Petersen, J.L. (1992) Two species previously confused under the concept of *Sabethes tarsopus* in Central America (Diptera: Culicidae). *Mosquito Systematics*, 24 (2), 102–123.

Harbach, R.E. & Peyton, E.L. (1990) A new subgenus in *Wyeomyia* (Diptera: Culicidae), with the reclassification and redescription of the type species, *Sabethes fernandezyepezi*. *Mosquito Systematics*, 22 (1), 15–23.

Harbach, R.E. & Peyton, E.L. (1991) Transfer of the subgenus *Davismyia* from *Wyeomyia* to *Sabethes* and description of the type species, *Miamyia petrocchiae* (Diptera: Culicidae). *Mosquito Systematics* (for 1990), 22 (3), 149–159.

Harbach, R.E. & Peyton, E.L. (1992a) A new subgenus of *Wyeomyia* (Diptera: Culicidae), with the reclassification and redescription of *Wyeomyia (Davismyia) arborea*, *Wyeomyia (Dendromyia) tarsata* and *Sabethes (Sabethes) carrilloi*. *Mosquito Systematics* (for 1991), 23 (2), 92–109.

Harbach, R.E. & Peyton, E.L. (1992b) A new subgenus of *Culex* in the Neotropical Region (Diptera: Culicidae). *Mosquito Systematics*, 24 (3), 242–252.

Harbach, R.E. & Peyton, E.L. (1993) Morphology and evolution of the larval maxilla and its importance in the classification of the Sabethini (Diptera: Culicidae). *Mosquito Systematics*, 25 (1), 1–16.

Harbach, R.E. & Peyton, E.L. (2000) Systematics of *Onirion*, a new genus of Sabethini (Diptera: Culicidae) from the Neotropical Region. *Bulletin of the Natural History Museum Entomology Series*, 69 (2), 115–169.

Harbach, R.E. & Rattanarithikul, R. (1988) A new species of the subgenus *Eumelanomyia* of *Culex* (Diptera: Culicidae) from Thailand. *Mosquito Systematics*, 20 (1), 69–76.

Harbach, R.E. & Sandlant, G.R. (1997) *CABIKEY Mosquito Genera of the World*. Windows CD-ROM. CABI, Wallingford, UK.

Harbach, R.E. & Snow, K.R. (2003) Definitions of types. *European Mosquito Bulletin*, 16, 13.

Harbach, R.E. & Taai, K. (2014) *Nyctomyia biunguiculata*, a new cavernicolous species of tribe Aedini (Diptera: Culicidae) from southern Thailand. *Zootaxa*, 3895 (3), 427–432. DOI: 10.11646/zootaxa.3895.3.7

Harbach, R.E. & Wilkerson, R.C. (2023) The insupportable validity of mosquito subspecies (Diptera: Culicidae) and their exclusion from culicid classification. *Zootaxa*, 5303 (1), 1–184. DOI: 10.11646/zootaxa.5303.1.1

Harbach, R.E., Culverwell, C.L. & Kitching, I.J. (**2017a**) Phylogeny of the nominotypical subgenus of *Culex* (Diptera: Culicidae): insights from analyses of anatomical data into interspecific relationships and species groups in an unresolved tree. *Systematics and Biodiversity*, 15 (4), 296–306. DOI: 10.1080/14772000.2016.1252439

Harbach, R.E., Dahl, C. & White, G.B. (1985) *Culex (Culex) pipiens* Linnaeus (Diptera: Culicidae): concepts, type designations, and description. *Proceedings of the Entomological Society of Washington*, 87 (1), 1–24.

Harbach, R.E., Garros, C., Nguyen, D.M. & Manguin, S. (**2007c**) Formal taxonomy of species C of the *Anopheles minimus* sibling species complex (Diptera: Culicidae). *Zootaxa*, 1654, 41–54. DOI: 10.11646/zootaxa.1654.1.3

Harbach, R.E., Kitching, I.J., Culverwell, C.L., Dubois, J. & Linton, Y.-M. (2012) Phylogeny of mosquitoes of tribe Culicini (Diptera: Culicidae) based on morphological diversity. *Zoologica Scripta*, 41 (5), 499–514. DOI: 10.1111/j.1463-6409.2012.00546.x

Harbach, R.E., Kitching, I.J., Culverwell, C.L., Howard, T.M. & Linton, Y.-M. (2013) *Nyx pholeocola*, a new genus and cavernicolous species of tribe Aedini (Diptera: Culicidae) from southern Thailand based on morphological and molecular data. *Zootaxa*, 3683 (2), 159–177. DOI: 10.11646/zootaxa.3683.2.5

Harbach, R.E., Peyton, E.L. & Harrison, B.A. (1984) A new species of *Culex (Melanoconion)* from southern South America (Diptera: Culicidae). *Mosquito Systematics*, 16 (2), 185–200.

Harbach, R.E., Rattanarithikul, R. & Harrison, B.A. (2005) *Baimaia*, a new subgenus for *Anopheles kyondawensis* Abraham, a unique crabhole-breeding anopheline in southeastern Asia. *Proceedings of the Entomological Society of Washington*, 107 (4), 750–761.

Harbach, R.E., Rattanarithikul, R. & Harrison, B.A. (**2017b**) *Anopheles prachongae*, a new species of the Gigas Complex of subgenus *Anopheles* (Diptera: Culicidae) in Thailand, contrasted with known forms of the complex. *Zootaxa*, 4236 (1), 65–81. DOI: 10.11646/zootaxa.4236.1.3

Harbach, R.E., Rattanarithikul, R., Howard, T.M., Linton, Y.-M. & Kitching, I.J. (**2007b**) Systematics of a new genus and cavernicolous species of the mosquito tribe Aedini (Diptera: Culicidae) from Thailand. *Proceedings of the Entomological Society of Washington*, 109 (2), 469–488.

Harbach, R.E., Vu, D.H. & Kitching, I.J. (**2007a**) Systematics of *Kimia*, a new genus of Sabethini (Diptera: Culicidae) in the Oriental Region. *Proceedings of the Entomological Society of Washington*, 109 (1), 102–120.

Harrison, B.A. (1972) A new interpretation of affinities within the *Anopheles hyrcanus* complex of Southeast Asia. *Mosquito Systematics*, 4 (3), 73–83.

Harrison, B.A. (1973) *Anopheles (An.) reidi*, a new species of the *barbirostris* species complex from Sri Lanka (Diptera: Culicidae). *Proceedings of the Entomological Society of Washington*, 75 (3), 365–371.

Harrison, B.A. (1980) Medical entomology studies – XIII. The Myzomyia Series of *Anopheles (Cellia)* in Thailand, with emphasis on intra-interspecific variations (Diptera: Culicidae). *Contributions of the American Entomological Institute*, 17 (4), iv, 1–195.

Harrison, B.A. (1987) *Culex* subgenus *Thaiomyia* Bram, a synonym of *Culex* subgenus *Culiciomyia* Theobald (Diptera: Culicidae). *Mosquito Systematics*, 19 (1), 111–116.

Harrison, B.A. & Bickley, W.E. (1990) Occurrence of mesopostnotal setae and scales in the family Culicidae. *Mosquito Systematics*, 22 (2), 131–143.

Harrison, B.A. & Peyton, E.L. (1984) The value of the pupal stage to anopheline taxonomy, with notes on anomalous setae (Diptera: Culicidae). *Mosquito Systematics*, 16 (2), 201–210.

Harrison, B.A. & Scanlon, J.E. (1974) *Anopheles (An.) pilinotum*, a new species name in the *aitkenii* complex for *An. insulaeflorum* from the Philippines and eastern Indonesia (Diptera: Culicidae). *Mosquito Systematics*, 6 (1), 32–40.

Harrison, B.A. & Scanlon, J.E. (1975) Medical entomology studies – II. The subgenus *Anopheles* in Thailand (Diptera: Culicidae). *Contributions of the American Entomological Institute*, 12 (1), iv, 1–307.

Harrison, B.A., Rattanarithikul, R., Peyton, E.L. & Mongkolpanya, K. (1991) Taxonomic changes, revised occurrence records and notes on the Culicidae of Thailand and neighboring countries. *Mosquito Systematics* (for 1990), 22 (3), 196–227.

Harrison, B.A., Ruiz-Lopez, F., Falero, G.C., Savage, H.M., Pecor, J.E. & Wilkerson, R.C. (2012) *Anopheles (Kerteszia) lepidotus* (Diptera: Culicidae), not the malaria vector we thought it was: revised male and female morphology; larva, pupa, and male genitalia characters; and molecular verification. *Zootaxa*, 3218, 1–17.

Harrison, B.A., Scanlon, J.E. & Reid, J.A. (1973) A new synonymy and new species name in the Southeast Asian *Anopheles hyrcanus* complex. *Mosquito Systematics*, 5 (4), 263–268.

Harrison, B.A., Varnado, W., Whitt, P.B. & Goddard, J. (2008) New diagnostic characters for females of *Psorophora (Janthinosoma)* species in the United States, with notes on *Psorophora mexicana* (Bellardi) (Diptera: Culicidae). *Journal of Vector Ecology*, 33 (2), 232–237. DOI: 10.3376/1081-1710-33.2.232

Hasselquist, F. (1757) *Frederic Hasselquists … Iter Palæstinum, eller Resa til Heliga Landet, förrättad ifrån År 1749 Til 1752, med beskrifningar, rön, anmårkningar, öfver de mårkvårdigaste Naturalier … utgifven af Carl Linnæus*. Trykt på Lars Salvii, Stockholm, Sweden. DOI: 10.5962/bhl.title.112563

Haworth, W.E. (1924) Mosquitoes and coconut palms. A mosquito survey of palm trees in East Africa and the problems resulting therefrom. *Transactions of the Royal Society of Tropical Medicine and Hygiene*, 18 (4), 162–198. DOI: 10.1016/S0035-9203(24)90847-9

Heinemann, S.J. & Belkin, J.N. (1977a) Collection records of the project "Mosquitoes of Middle America" 7. Costa Rica (CR). *Mosquito Systematics*, 9 (2), 237–287.

Heinemann, S.J. & Belkin, J.N. (1977b) Collection records of the project "Mosquitoes of Middle America" 8. Central America: Belize (BH), Guatemala (GUA), El Salvador (SAL), Honduras (HON), Nicaragua (NI, NIC). *Mosquito Systematics*, 9 (4), 403–454.

Heinemann, S.J. & Belkin, J.N. (1978a) Collection records of the project "Mosquitoes of Middle America" 10. Panama, including Canal Zone (PA, GG). *Mosquito Systematics*, 10 (2), 119–196.

Heinemann, S.J. & Belkin, J.N. (1978b) Collection records of the project "Mosquitoes of Middle America" 11. Venezuela (VZ); Guianas: French Guiana (FG, FGC), Guyana (GUY), Surinam (SUR). *Mosquito Systematics*, 10 (3), 365–459.

Heinemann, S.J. & Belkin, J.N. (1978c) Collection records of the project "Mosquitoes of Middle America" 12. Colombia (COA, COB, COL, COM). *Mosquito Systematics*, 10 (4), 493–539.

Heinemann, S.J. & Belkin, J.N. (1979) Collection records of the project "Mosquitoes of Middle America" 13. South America: Brazil (BRA, BRAP, BRB), Ecuador (ECU), Peru (PER), Chile (CH). *Mosquito Systematics*, 11 (2), 61–118.

Heinemann, S.J., Aitken, T.H.G. & Belkin, J.N. (1980) Collection records of the project "Mosquitoes of Middle America" 14. Trinidad and Tobago (TR, TRM, TOB). *Mosquito Systematics*, 12 (2), 179–284.

Heller, K.M. (1892) Die mit der Ruteliden-Gattung *Sinyhala* Burm. verwandten Gattungen und Arten. *Deutsche Entomologische Zeitschrift*, 1891 (2), 289–306.

Heppner, J.B. (1982) Millieriinae, a new subfamily of Choreutidae, with new taxa from Chile and the United States (Lepidoptera: Sesioidea). *Smithsonian Contributions to Zoology*, 370, iii, 27 pp., 63 figs, 1 table.

Hervé, J.-P. & Geoffroy, B. (1974) Description d'un nouveau moustique de République Centrafricaine *Culex (Neoculex) subsalisburiensis* n.sp. *Cahiers ORSTOM Série Entomologie médicale et Parasitologie*, 12 (3), 191–193.

Hide, M., Maquart, P.-O., Yean, S., Suor, K., Heng, K., Sen, S., Yim, C., Leng, S., Rakotonirina, A., Fontenille, D. & Boyer, S. (2024) *Aedes unalom* sp. nov. in Cambodia, a new *Stegomyia* species close to *Aedes albopictus* (Diptera, Culicidae). *Journal of Asia-Pacific Entomology*, 27 (2), 102233. DOI: 10.1016/j.aspen.2024.102233

Hill, E. & Haydon, L.G. (1907) A contribution to the study of the characteristics of larvae of species of *Anophelina* in South Africa. *Annals of the Natal Government Museum*, 1 (2), 111–157, 12 pls.

Hill, R.B. & Hill, C.M. (1946) *Phoniomyia hirsuta*, a new sabethine from Jamaica (Diptera, Culicidae). *Proceedings of the Entomological Society of Washington*, 48 (2), 39–41.

Ho, C. (1938) A new tree-hole breeding *Anopheles* from the island of Hainan. *Annals of Tropical Medicine and Parasitology*, 32 (3), 279–285. DOI: 10.1080/00034983.1938.11685034

Hoffmann, C.C. (1935) La formacion de razas en los *Anopheles* mexicanos. I. *A. maculipennis* y *A. quadrimaculatus* y una raza nueva del *maculipennis*. *Anales del Instituto de Biología, Universidad nacional Autónoma de México*, 6, 3–22.

Hogue, C.L. (1975) A new species of bromeliad-breeding *Culex (Culex)* from Cocos Island. *Mosquito Systematics*, 7 (4), 357–362.

Hopkins, G.H.E. (1952) *Mosquitoes of the Ethiopian Region I. – Larval bionomics of mosquitoes and taxonomy of culicine larvae*. 2nd edn. Printed by order of the Trustees, British Museum (Natural History), London, UK.

Horsfall, W.R. (1972) *Mosquitoes—Their Bionomics and Relation to Disease*. Hafner, New York, NY, USA.

Horsfall, W.R., Miles, R.C. & Sokatch, J.T. (1952) Eggs of floodwater mosquitoes. I. Species of *Psorophora* (Diptera: Culicidae). *Annals of the Entomological Society of America*, 45 (4), 618–624. DOI: 10.1093/aesa/45.4.618

Howard, L.O. (1901) *Mosquitoes: how they live; how they carry disease; how they are classified; how they may be destroyed*. McClure, Phillips & Co., New York, NY, USA.

Howard, L.O., Dyar, H.G. & Knab, F. (1913) *The mosquitoes of North and Central America and the West Indies*. Volumes 1 and 2. Carnegie Institution of Washington Publication No. 159 [for 1912]. The Lord Baltimore Press, Baltimore, MD, USA.

Howard, L.O., Dyar, H.G. & Knab, F. (1915) *The mosquitoes of North and Central America and the West Indies. Volume three. Systematic description (in two parts). Part I*. Publication No. 159. Carnegie Institution of Washington, Washington, DC, USA.

Howard, L.O., Dyar, H.G. & Knab, F. (1917) *The mosquitoes of North and Central America and the West Indies. Volume four. Systematic description (in two parts). Part II*. Publication No. 159. Carnegie Institution of Washington, Washington, DC, USA.

Hsiao, C.-E. (1977) A new species of *Aedes* (Diptera: Culicidae) [in Chinese]. *Acta Entomologica Sinica*, 20 (4), 451–454.

Huang, Y.-M. (1968a) *Aedes (Verrallina)* of the Papuan Subregion (Diptera: Culicidae). *Pacific Insects Monograph*, 17, 1–73.

Huang, Y.-M. (1968b) A new subgenus of *Aedes* (Diptera, Culicidae) with illustrated key to the subgenera of the Papuan Subregion (Diptera: Culicidae). *Journal of Medical Entomology*, 5 (2), 169–188. DOI: 10.1093/jmedent/5.2.169

Huang, Y.-M. (1969) A new species of *Aedes (Stegomyia)* from Thailand (Diptera: Culicidae). *Proceedings of the Entomological Society of Washington*, 71 (2), 234–239.

Huang, Y.-M. (1972) Contributions to the mosquito fauna of Southeast Asia. XIV. The subgenus *Stegomyia* of *Aedes* in Southeast Asia. I – The *scutellaris* group of species. *Contributions of the American Entomological Institute*, 9 (1), 1–109.

Huang, Y.-M. (1973) A new species of *Aedes (Stegomyia)* from Thailand and notes on the Mediopunctatus Subgroup (Diptera: Culicidae). *Proceedings of the Entomological Society of Washington*, 75 (2), 224–232.

Huang, Y.-M. (1974) A new species of *Aedes (Stegomyia)* from the Andaman Islands (Diptera: Culicidae). *Mosquito Systematics*, 6 (2), 137–141.

Huang, Y.-M. (1975) A new species of *Aedes (Stegomyia)* from Sri Lanka (Ceylon) (Diptera: Culicidae). *Mosquito Systematics*, 7 (4), 345–356.

Huang, Y.-M. (1977a) Medical entomology studies – VII. The subgenus *Stegomyia* of *Aedes* in Southeast Asia. II – The *edwardsi* group of species. III – The *w-albus* group of species. (Diptera: Culicidae). *Contributions of the American Entomological Institute*, 14 (1), 1–111.

Huang, Y.-M. (1977b) Medical entomology studies – VIII. Notes on the taxonomic status of *Aedes vittatus* (Diptera: Culicidae). *Contributions of the American Entomological Institute*, 14 (1), 113–132.

Huang, Y.-M. (1978) Redescription and subgeneric position of *Aedes meronephada* (Dyar and Shannon) with notes on the subgenus *Diceromyia* (Diptera: Culicidae). *Mosquito Systematics*, 10 (3), 334–350.

Huang, Y.-M. (1979a) Medical entomology studies – XI. The subgenus *Stegomyia* of *Aedes* in the Oriental region with keys to the species (Diptera: Culicidae). *Contributions of the American Entomological Institute*, 15 (6), 1–79.

Huang, Y.-M. (1979b) *Aedes (Stegomyia) simpsoni* complex in the Ethiopian Region with lectotype designation for *simpsoni* (Theobald) (Diptera: Culicidae). *Mosquito Systematics*, 11 (3), 221–234.

Huang, Y.-M. (1981) A redescription of *Aedes (Stegomyia) calceatus* Edwards and description of a new Afrotropical species, *Aedes (Stegomyia) ledgeri* (Diptera: Culicidae). *Mosquito Systematics*, 13 (1), 92–113.

Huang, Y.-M. (1985) A new African species of *Aedes* (Diptera: Culicidae). *Mosquito Systematics*, 17 (2), 108–120.

Huang, Y.-M. (1986a) Notes on the *Aedes (Diceromyia) furcifer* group, with a description of a new species (Diptera: Culicidae). *Proceedings of the Entomological Society of Washington*, 88 (4), 634–649.

Huang, Y.-M. (1986b) *Aedes (Stegomyia) corneti*, a new species of the *africanus* subgroup (Diptera: Culicidae). *Proceedings of the Entomological Society of Washington*, 88 (4), 764–776.

Huang, Y.-M. (1988a) *Aedes (Stegomyia) josiahae*, a new species of the *simpsoni* subgroup (Diptera: Culicidae). *Proceedings of the Entomological Society of Washington*, 90 (2), 155–163.

Huang, Y.-M. (1988b) The *Aedes (Stegomyia) pseudonigeria* group with emphasis on the species from the Afrotropical Region (Diptera: Culicidae). *Mosquito Systematics*, 20 (1), 1–20.

Huang, Y.-M. (1990) The subgenus *Stegomyia* of *Aedes* in the Afrotropical Region I. The *africanus* group of species (Diptera: Culicidae). *Contributions of the American Entomological Institute*, 26 (1), 1–90.

Huang, Y.-M. (1994) *Aedes (Stegomyia) mattinglyorum*, a new species of the *dendrophilus* group (Diptera: Culicidae). *Proceedings of the Entomological Society of Washington*, 96 (1), 11–21.

Huang, Y.-M. (1997) The subgenus *Stegomyia* of *Aedes* in the Afrotropical Region. II. The *dendrophilus* group of species (Diptera: Culicidae). *Contributions of the American Entomological Institute*, 29 (4), 1–104.

Huang, Y.-M. (2001) A pictorial key for the identification of the subfamilies of Culicidae, genera of Culicinae, and subgenera of *Aedes* mosquitoes of the Afrotropical Region (Diptera: Culicidae). *Proceedings of the Entomological Society of Washington*, 103 (1), 1–53.

Huang, Y.-M. (2002) A pictorial key to the mosquito genera of the world, including subgenera of *Aedes* and *Ochlerotatus* (Diptera: Culicidae). *Insecta Koreana*, 19 (1), 1–130.

Huang, Y.-M. (2004) The subgenus *Stegomyia* of *Aedes* in the Afrotropical Region with keys to the species (Diptera: Culicidae). *Zootaxa*, 700, 1–120. DOI: 10.11646/zootaxa.700.1.1

Huang, Y.-M. (2005) *Cornetius*, a new subgenus of *Aedes*, and a redescription of *Aedes (Cornetius) cozi* Cornet (Diptera: Culicidae). *Proceedings of the Entomological Society of Washington*, 107 (3), 517–529.

Huang, Y.-M. & Hitchcock, J.C. (1980) Medical entomology studies – XII. A revision of the *Aedes scutellaris* group of Tonga (Diptera: Culicidae). *Contributions of the American Entomological Institute*, 17 (3), 1–107.

Huang, Y.-M. & Rueda, L.M. (1998) Description of a paralectotype female of *Aedes (Finlaya) niveus* (Ludlow) (Diptera: Culicidae). *Proceedings of the Entomological Society of Washington*, 100 (4), 824–827.

Huang, Y.-M. & Rueda, L.M. (2013) A new species of the genus *Orthopodomyia* from Ototomo, Cameroon, with notes on other *Orthopodomyia* group-species. *Journal of the American Mosquito Control Association*, 29 (3), 270–274.

Huang, Y.-M. & Rueda, L.M. (2014) A pictorial key to the species of *Aedes (Ochlerotatus* and *Coetzeemyia)* in the Afrotropical Region (Diptera: Culicidae). *Zootaxa*, 3754 (5), 592–600. DOI: 10.11646/ZOOTAXA.3754.5.5

Huang, Y.-M. & Rueda, L.M. (2015a) Pictorial keys to the species of *Albuginosus* and *Aedimorphus* (Grjebinei and Apicoannulatus Groups) of the genus *Aedes* Meigen in the Afrotropical Region (Diptera: Culicidae). *Zootaxa*, 3925 (1), 25–36. DOI: 10.11646/zootaxa.3925.1.2

Huang, Y.-M. & Rueda, L.M. (2015b) *Pseudalbuginosus*, a new subgenus of *Aedes*, and a redescription of *Aedes (Pseudalbuginosus) grjebinei* Hamon, Taufflieb, and Maillot (Diptera: Culicidae). *Proceedings of the Entomological Society of Washington*, 117 (3), 381–388.

Huang, Y.-M. & Rueda, L.M. (2015c) A pictorial key to the species of the *Aedes (Zavortinkius)* in the Afrotropical Region (Diptera: Culicidae). *Zootaxa*, 4027 (4), 593–599. DOI: 10.11646/zootaxa.4027.4.9

Huang, Y.-M. & Rueda, L.M. (2016) A pictorial key to the sections, groups, and species of the *Aedes (Diceromyia)* in the Afrotropical Region (Diptera: Culicidae). *Zootaxa*, 4079 (2), 281–290. DOI: 10.11646/zootaxa.4079.2.9

Huang, Y.-M. & Rueda, L.M. (2017) Pictorial keys to the sections, groups, and species of the *Aedes (Finlaya)* in the Afrotropical Region (Diptera: Culicidae). *Zootaxa*, 4221 (1), 131–141. DOI: 10.11646/zootaxa.4221.1.7

Huang, Y.-M., Mathis, W.N. & Wilkerson, R.C. (2010) *Coetzeemyia*, a new subgenus of *Aedes*, and a redescription of the holotype female of *Aedes (Coetzeemyia) fryeri* (Theobald) (Diptera: Culicidae). *Zootaxa*, 2638, 1–24. DOI: 10.11646/zootaxa.2638.1.1

Huang, Y.-M., Mathis, W.N. & Wilkerson, R.C. (2012) Subgeneric position of *Aedes dufouri* Hamon with notes on the subgenus *Levua* Stone and Bohart (Diptera: Culicidae). *Proceedings of the Entomological Society of Washington*, 114 (2), 194–204.

Hudson, J.E. & Abul-Hab, J. (1987) Key to the species of adult female culicine (Diptera, Culicidae) mosquitoes of Iraq. *Bulletin of Endemic Diseases*, 28 (1–4), 53–59.

Hutchings, R.S.G. & Sallum, M.A.M. (2008) Two new species of *Culex* subgenus *Melanoconion* (Diptera: Culicidae) from the Amazon forest. *Zootaxa*, 1920, 41–50. DOI: 10.11646/zootaxa.1920.1.3

Hutton, F.W. (1902) Additions to the *Diptera* fauna of New Zealand. *Transactions and Proceedings of the New Zealand Institute* (for 1901), 34, 179–196.

Ibrahim, I.K., Al-Samarae, T.Y.M., Zaini, M.A. & Kassal, S. (1983) Identification key for Iraqi culcine [*sic*] mosquitoe [*sic*] larvae (Culicinae-Diptera). *Bulletin of Endemic Diseases*, 22/23, 89–113.

Ingram, A. & de Meillon, B. (1927) A mosquito survey of certain parts of South Africa, with special reference to the carriers of malaria and their control. (Part I). *Publications of the South African Institute for Medical Research*, 4 (4), 1–81, 15 pls.

International Commission on Zoological Nomenclature (1959a) Opinion 547. Designation under the plenary powers of a type-species in harmony with accustomed usage for the nominal genus '*Anopheles*' Meigen, 1818 (Class Insecta, Order Diptera). *Opinions and declarations rendered by the International Commission on Zoological Nomenclature*, 20 (14), 153–164.

International Commission on Zoological Nomenclature (1959b) Opinion 548. Validation under the plenary powers of the generic name "*Toxorhynchites*" Theobald (Class Insecta, Order Diptera), as published in 1901 in the "Journal of Tropical Medicine". *Opinions and declarations rendered by the International Commission on Zoological Nomenclature*, 20 (14), 167–174.

International Commission on Zoological Nomenclature (1959c) Opinion 550. Suppression under the Plenary Powers of the generic name *Taeniorhynchus* Lynch-Arribálzaga, 1891, for the purpose of protecting the generic name *Mansonia* Blanchard, 1901 (Class Insecta, Order Diptera) and matters relating to other names in the same class and in the Class Aves incidental thereto. *Opinions and declarations rendered by the International Commission on Zoological Nomenclature*, 20 (17), 185–198.

International Commission on Zoological Nomenclature (1964) Opinion 711. *Culex aegypti* Linnaeus, 1762 (Insecta, Diptera): validated and interpreted under the plenary powers. *Bulletin of Zoological Nomenclature*, 21 (4), 246–248.

Irish, S.R., Kyalo, D., Snow, R.W. & Coetzee, M. (2020) Updated list of *Anopheles* species (Diptera: Culicidae) by country in the Afrotropical Region and associated islands. *Zootaxa*, 4747 (3), 401–449. DOI: 10.11646/zootaxa.4747.3.1

Isoe, J. (2000) *Comparative analysis of the vitellogenin genes of the Culicidae*. PhD Thesis, Faculty of the Interdisciplinary Program in Insect Science, University of Arizona, Tucson, AZ, USA.

Iyengar, M.O.T. (1924) On the Indian anophelines of the *funestus* group and the description of a new species (Diptera, Culicidae). *Indian Journal of Medical Research*, 12, 23–29, 1 pl.

Iyengar, M.O.T. & Menon, M.A.U. (1948) Notes on *Harpagomyia genurostris* Leicester (Dipt. Culicidae). *Proceedings of the Royal Entomological Society of London, Series A, General Entomology*, 23 (4–6), 39–43. DOI: 10.1111/j.1365-3032.1948.tb00612.x

James, S.P. (1902) *Malaria in India*. Scientific Memoirs by Officers of the Medical and Sanitary Department of the Government of India. (New Series) No. 2. Office of the Superintendent of Government Printing, Calcutta, India.

James, S.P. & Liston, W.G. (1911) *A monograph of the anopheline mosquitoes of India*. 2nd edn. Thacker, Spink and Co., Calcutta, India.

Jeffery, J., Oothuman, P. & Rudnick, A. (1987) *Culex (Eumelanomyia) richardgarciai*, a new species of mosquito from peninsular Malaysia (Diptera: Culicidae). *Mosquito Systematics*, 19 (2), 139–142.

Jeffery, J., Rohela, M., Muslimin, M., Abdul Aziz, S.M.N., Jamaiah, I., Kumar, S., Tan, T.C., Lim, Y.A.L., Nissapatorn, V. & Abdul-Aziz, N.M. (2012) *Illustrated keys: some mosquitoes of Peninsula Malaysia*. University of Malaya Press, Kuala Lumur, Malaysia.

Jekel, H. (1861) Observations suggérées par les Notes de M. Chevrolat sur les Cérambycides de M. Thomson. *Journal of Entomology*, 1 (4), 255–262.

Judd, D.D. (1996) Review of the systematics and phylogenetic relationships of the Sabethini (Diptera: Culicidae). *Systematic Entomology*, 21 (2), 129–150. DOI: 10.1046/j.1365-3113.1996.d01-1.x

Judd, D.D. (1998) Review of a bromeliad-ovipositing lineage in *Wyeomyia* and the resurrection of *Hystatomyia* (Diptera: Culicidae). *Annals of the Entomological Society of America*, 91 (5), 572–589. DOI: 10.1093/aesa/91.5.572

Jupp, P.G. (1996) *Mosquitoes of southern Africa: Culicinae and Toxorhynchitinae*. Ekogilde Publishers, Hartebeespoort, South Africa.

Juthayothin, T. (2004) *Molecular phylogenetic study of culicine mosquitoes using the mitochondrial cytochrome oxidase I gene and the relationships with mosquito-borne flaviviruses*. MSc Thesis, Faculty of Graduate Studies, Mahidol University, Bangkok, Thailand.

Kamimura, K. & Wada, Y. (1974) A new subspecies of *Culex (Barraudius) modestus* from Japan (Diptera, Culicidae). *Japanese Journal of Sanitary Zoology*, 25 (1), 13–20. DOI: 10.7601/mez.25.13

Kanda, T. & Oguma, Y. (1978) *Anopheles engarensis*, a new species related to *sinensis* from Hokkaido Island, Japan. *Mosquito Systematics*, 10 (1), 45–52.

Kang, W., Tan, J., Cao, C., Cheng, H., Yang, Y. & Huang, R. (1984) *Anopheles liangshanensis*, a new species related to *A. kweiyangensis* and *A. sinensis* from Sichuan, China (Diptera: Culicidae) [in Chinese]. *Sichuan Journal of Zoology*, 3 (2), 11–14.

Kano, R., Nitahara, M. & Awaya, J. (1954) Description of a new mosquito, *Culex (Culiciomyia) sasai* n. sp., collected in the southwestern part of Japan (Culicidae, Diptera) [in Japanese]. *Japanese Journal of Sanitary Zoology*, 5 (1–2), 14–20. DOI: 10.7601/mez.5.14

Kar, I., Subbarao, S.K., Eapen, A., Ravendaran, J., Satyanarayana, T.S., Raghavendra, K., Nanda, N. & Sharma, V.P. (1999) Evidence for a new malaria vector species, species E, within the *Anopheles culicifacies* complex (Diptera: Culicidae). *Journal of Medical Entomology*, 36 (5), 595–600. DOI: 10.1093/jmedent/36.5.595

Karsch, F. (1887) Dipterologisches von der Delagoabei. *Entomologische Nachrichten*, 13 (2), 22–26.

Karimian, F., Oshaghi, M.A., Sedaghat, M.M., Waterhouse, R.M., Vatandoost, H., Hanafi-Bojd, A.A., Ravasan, N.M. & Chavshin, A.R. (2014) Phylogenetic analysis of the Oriental-Palearctic-Afrotropical members of *Anopheles* (Culicidae: Diptera) based on nuclear rDNA and mitochondrial DNA characteristics. *Japanese Journal of Infectious Diseases*, 67 (5), 361–367. DOI: 10.7883/yoken.67.361

Kasper, J. (2020) Description of a new species of mosquito, *Aedes (Ochlerotatus) arundinariae* (Diptera: Culicidae) from the Chatham Islands, New Zealand. *New Zealand Entomologist*, 43 (2), 93–102. DOI: 10.1080/00779962.2020.1846988

Keating, W.H. (main author). (1824) *Narrative of an expedition to the source of St. Peter's River, Lake Winnepeek, Lake of the Woods, &c. &c. performed in the year1823, by order of the Hon. J. C. Calhoun, Secretary of War, under the command of Stephen H. Long, Major U.S. T.E.: compiled from the notes of Major Long, Messrs. Say, Keating, & Colhoun*. Volume II. H. C. Carey & I. Lea, Philadelphia, PA, USA.

Keyserling, E. (1891) Die Spinnen Insektuto. *Konchuaikokai*, 26, 14–18.

King, W.V. (1931) The Philippine varieties of *Anopheles gigas* and *Anopheles lindesayi*. *Philippine Journal of Science*, 46 (4), 751–757, 2 pls.

King, W.V. (1932) The Philippine *Anopheles* of the *rossi-ludlowi* group. *Philippine Journal of Science*, 47 (3), 305–342, 4 pls.

King, W.V. (1939) Varieties of *Anopheles crucians* Wied. *American Journal of Tropical Medicine*, 19 (5), 461–471. DOI: 10.4269/ajtmh.1939.s1-19.461

King, W.V. & Baisas, F.E. (1936) A new species and a new variety of Philippine *Anopheles* related to *Anopheles leucosphyrus* Dönitz. *Proceedings of the Entomological Society of Washington*, 38 (5), 79–89.

King, W.V. & Hoogstraal, H. (1946a) The New Guinea species of *Culex (Culiciomyia)*, with descriptions of two new species. *Proceedings of the Biological Society of Washington*, 59, 143–153.

King, W.V. & Hoogstraal, H. (1946b) Two new species of *Aedes (Finlaya)* from New Guinea (Diptera: Culicidae). *Proceedings of the Entomological Society of Washington*, 48 (2), 37–38.

King, W.V. & Hoogstraal, H. (1946c) Three new species of *Aedes* from Netherlands New Guinea. *Proceedings of the Entomological Society of Washington*, 48 (4), 95–106.

King, W.V. & Hoogstraal, H. (1946d) Species of *Aedes (Finlaya)* of the *papuensis* group in the Australasian Region (Diptera, Culicidae). *Proceedings of the Entomological Society of Washington*, 48 (6), 135–157.

King, W.V. & Hoogstraal, H. (1946e) Two new species of mosquitoes of the genus *Ficalbia* from Netherlands New Guinea. *Proceedings of the Entomological Society of Washington*, 48 (7), 186–190.

King, W.V. & Hoogstraal, H. (1946f) Descriptions of three new species of mosquitoes of the genus *Aedes*, subgenus *Finlaya*, from New Guinea. *Journal of the Washington Academy of Sciences*, 36 (9), 305–313.

King, W.V. & Hoogstraal, H. (1947a) New species of New Guinea *Uranotaenia* of the *tibialis* group (Diptera: Culicidae). *Annals of the Entomological Society of America* (for 1946), 39 (4), 585–596. DOI: 10.1093/aesa/39.4.585

King, W.V. & Hoogstraal, H. (1947b) New Guinea species of mosquitoes of the genus *Aedes*, subgenus *Aedes*. *Journal of the Washington Academy of Sciences*, 37 (4), 113–134.

King, W.V. & Hoogstraal, H. (1947c) Two new species of *Culex (Neoculex)* from New Guinea (Diptera, Culicidae). *Proceedings of the Entomological Society of Washington*, 49 (2), 65–69.

King, W.V. & Hoogstraal, H. (1955) Three new species of New Guinea *Culex*, subgenus *Lophoceraomyia*, with notes on other species (Diptera, Culicidae). *Proceedings of the Entomological Society of Washington*, 57 (1), 1–11.

Kirby, W. (1837) Fauna Boreali-Americana. Part the fourth and last. The Insects. In: Richardson, J. (ed.), *Fauna Boreali-Americana; or the Zoology of the northern parts of British America: containing descriptions of the objects of natural history collected on the late Northern Land Expeditions, under command of Captain Sir John Franklin, R.N.* Josiah Fletcher, Norwich, UK.

Kirchgatter, K., Guimarães, L.O., Trujillano, H.H.Y., Arias, F.R., Cáceres, A.G., Duarte, A.M.R.C., Malafronte, R.S., Tubaki, R.M. & Sallum, M.A.M. (2020) Phylogeny of *Anopheles (Kerteszia)* (Diptera: Culicidae) using mitochondrial genes. *Insects*, 11, 324. DOI: 10.3390/insects11050324

Kirkpatrick, T.W. (1925a) Preliminary notes on the determination of Egyptian mosquitos, with descriptions of three species new to science. *Bulletin Société Entomologique d'Égypte* (for 1924), 8, 362–391.

Kirkpatrick, T.W. (1925b) *The mosquitoes of Egypt*. Egyptian Government. Anti-Malaria Commission. Government Press, Cairo, Egypt.

Kirkpatrick, T.W. (1926) *Culex (Lasiosiphon) Adairi* [*sic*], Nom. Nov. *Bulletin Société Entomologique d'Égypte*, 10, 112.

Kitching, I.J., Culverwell, L.C. & Harbach, R.E. (2015) The phylogenetic conundrum of *Lutzia* (Diptera: Culicidae: Culicini): a cautionary account of conflict and support. *Insect Systematics and Evolution*, 46 (3), 269–290. DOI: 10.1163/1876312X-45032119

Klein, J.M. (1969) A new species of *Ficalbia (Ravenalites)* from Cambodia (Diptera: Culicidae). *Proceedings of the Entomological Society of Washington*, 71 (2), 215–217.

Klein, J.M. (1970) A new species of *Culex (Neoculex)* from Cambodia (Diptera: Culicidae). *Proceedings of the Entomological Society of Washington*, 72 (4), 504–506.

Klein, J.M. (1973) Contributions to the mosquito fauna of Southeast Asia. XVII. The Cambodian *Aedes (Neomacleaya)* species with some new descriptions (Diptera: Culicidae). *Contributions of the American Entomological Institute*, 10 (1), 1–21.

Klein, J.-M. (1977a) La faune des moustiques du Cambodge *I. Anophelinae – (Diptera, Culicidae). Cahiers ORSTOM Série Entomologie médicale et Parasitology*, 15 (2), 107–122.

Klein, J.-M. (1977b) Deux nouvelles espèces de *Topomyia* du Cambodge *(Diptera, Culicidae). Cahiers ORSTOM Série Entomologie médicale et Parasitologie*, 15 (2), 123–129.

Klein, J.-M. & Marks, E.N. (1960) Australian mosquitoes described by Macquart. I. Species in the Paris Museum, *Aëdes (Finlaya) alboannulatus* (Macquart), *Aëdes (Finlaya) rubrithorax* (Macquart), *Aëdes (Ochlerotatus) albirostris* (Macquart). New synonymy and a new species from New Zealand. *Proceedings of the Linnean Society of New South Wales*, 85 (1), 107–116.

Klein, J.M. & Sirivanakarn, S. (1970) Four new species of *Culex*, subgenus *Mochthogenes* from Southeast Asia (Diptera: Culicidae). *Proceedings of the Entomological Society of Washington* (for 1969), 71 (4), 582–592.

Klein, J.-M., Rivière, F. & Séchan, Y. (1984) Une nouvelle espèce de moustique (Diptera: Culicidae) des îles Marquises (Polynésie Française): *Culex (Culex) toviiensis* n. sp. Description de la larve et de la nymphe. *Cahiers ORSTOM Série Entomologie médicale et Parasitologie* (for 1983), 21 (2), 71–76.

Knab, F. (1907a) A new species of *Megarhinus*. *Canadian Entomologist*, 39 (2), 50–51. DOI: 10.4039/Ent3950-2

Knab, F. (1907b) A new genus and species of sabethid mosquito. *Journal of the New York Entomological Society*, 15 (3), 120–121.

Knab, F. (1909) Two new Philippine Culicidae. *Entomological News*, 20 (9), 386–388.

Knight, K.L. (1946) The *Aedes (Finlaya) niveus* subgroup of Oriental mosquitoes. *Journal of the Washington Academy of Sciences*, 36 (8), 270–280.

Knight, K.L. (1947) The *Aedes (Mucidus)* mosquitoes of the Pacific (Diptera: Culicidae). *Journal of the Washington Academy of Sciences*, 37 (9), 315–325.

Knight, K.L. (1948a) The *Aedes (Finlaya) chrysolineatus* group of mosquitoes (Diptera: Culicidae). *Annals of the Entomological Society of America* (for 1947), 40 (4), 624–649. DOI: 10.1093/aesa/40.4.624

Knight, K.L. (1948b) The *Aedes (Finlaya) albotaeniatus* group of mosquitoes (Diptera, Culicidae). *Proceedings of the Entomological Society of Washington*, 50 (1), 1–8.

Knight, K.L. (1951) The *Aedes (Ochlerotatus) punctor* subgroup in North America (Diptera, Culicidae). *Annals of the Entomological Society of America*, 44 (1), 87–99. DOI: 10.1093/aesa/44.1.87

Knight, K.L. (1953) Two new species of mosquitoes from the Yemen (Diptera: Culicidae). *Journal of the Washington Academy of Sciences*, 43 (10), 320–325.

Knight, K.L. (1968) Contributions to the mosquito fauna of Southeast Asia. – IV. Species of the subgroup *chrysolineatus* of group D, genus *Aedes*, subgenus *Finlaya* Theobald. *Contributions of the American Entomological Institute*, 2 (5), 1–45.

Knight, K.L. (1969) A new species of the genus *Aedes*, subgenus *Finlaya* Theobald, from India (Diptera: Culicidae). *Journal of the Kansas Entomological Society*, 42 (4), 382–386.

Knight, K.L. (1971a) Comparative anatomy of the mandible of the fourth instar mosquito larva (Diptera: Culicidae). *Journal of Medical Entomology*, 8 (2), 189–205. DOI: 10.1093/jmedent/8.2.189

Knight, K.L. (1971b) A mosquito taxonomic glossary. VII. The pupa. *Mosquito Systematics Newsletter*, 3 (2), 42–65.

Knight, K.L. (1972) History of mosquito systematics. Part I. Eighteen century. *Mosquito Systematics*, 4 (1), 10–15.

Knight, K.L. (1978a) A new *Aedes (Finlaya)* mosquito from Thailand. *Mosquito Systematics*, 10 (1), 106–116.

Knight, K.L. (1978b) *Supplement to a catalog of the mosquitoes of the world (Diptera: Culicidae)*. The Thomas Say Foundation, Supplement to Volume VI. Entomological Society of America, College Park, MD, USA.

Knight, K.L. & Harbach, R.E. (1977) Maxillae of fourth stage mosquito larvae (Diptera: Culicidae). *Mosquito Systematics*, 9 (4), 455–477.

Knight, K.L. & Harrison, B.A. (1988) A new *Aedes (Finlaya)* of the Niveus-Subgroup (Diptera: Culicidae). *Mosquito Systematics* (for 1987), 19 (3), 212–236.

Knight, K.L. & Hull, W.B. (1951) Three new species of *Aedes* from the Philippines (Diptera, Culicidae). *Pacific Science*, 5 (2), 197–203.

Knight, K.L. & Hull, W.B. (1952) The *Aedes* mosquitoes of the Philippine Islands. II. Subgenera *Skusea, Christophersiomyia, Geoskusea, Rhinoskusea*, and *Stegomyia* (Diptera, Culicidae). *Pacific Science*, 6 (2), 157–189.

Knight, K.L. & Hurlbut, H.S. (1949) The mosquitoes of Ponape Island, eastern Carolines. *Journal of the Washington Academy of Sciences*, 39 (1), 20–34.

Knight, K.L. & Laffoon, J.L. (1946) The Oriental species of the *Aedes (Finlaya) kochi* group (Diptera: Culicidae). *Transactions of the American Entomological Society*, 72 (4), 203–225, 3 pls.

Knight, K.L. & Marks, E.N. (1952) An annotated checklist of the mosquitoes of the subgenus *Finlaya*, genus *Aedes*. *Proceedings of the United States National Museum*, 101 (3288), 513–574.

Knight, K.L. & Mattingly, P.F. (1950) The *Orthopodomyia anopheloides* subgroup of mosquitoes (Diptera, Culicidae). *Proceedings of the Entomological Society of Washington*, 52 (1), 1–20.

Knight, K.L. & Rozeboom, L. (1945) A new species of *Culex* from New Guinea (Diptera, Culicidae). *Proceedings of the Biological Society of Washington*, 47 (9), 289–295.

Knight, K.L. & Rozeboom, L.E. (1946) The *Aedes (Stegomyia) albolineatus* group (Diptera, Culicidae). *Proceedings of the Biological Society of Washington*, 59, 83–98.

Knight, K.L. & Stone, A. (1977) *A catalog of the mosquitoes of the world (Diptera: Culicidae)*. 2nd edn. The Thomas Say Foundation. Volume VI. Entomological Society of America, College Park, MD, USA.

Koekemoer, L.L., Misiani, E.A., Hunt, R.H., Kent, R.J., Norris, D.E. & Coetzee, M. (2009) Cryptic species within *Anopheles longipalpis* from southern Africa and phylogenetic comparison with members of the *An. funestus* group. *Bulletin of Entomological Research*, 99 (1), 41–49. DOI: 10.1017/S0007485308006123

Koidzumi, M. (1920) Preventive medical study of mosquitoes in Formosa [in Japanese]. *Daiwan Kenkyujo Hokoku [Report of the Formosa Government]*, 8, 1–158.

Koidzumi, M. (1924) The anophelines of Formosa. In: Hoops, A.S. & Scharff, J.W. (eds), *Far Eastern Association of Tropical Medicine—Transactions of the Fifth Biennial Congress held at Singapore, 1924*. John Bale, Sons & Danielson, London, UK, pp. 96–101.

Komp, W.H.W. (1932) A new *Culex, Culex vomerifer,* from Panama (Dipt., Culicidæ). *Psyche*, 39 (3), 79–82. DOI: 10.1155/1932/97349

Komp, W.H.W. (1935) Notes on the validity of the types of the species in the subgenera *Mochlostyrax* and *Melanoconion* in the U. S. National Museum. (Diptera, Culicidae.). *Proceedings of the Entomological Society of Washington*, 37 (1), 1–11.

Komp, W.H.W. (1936a) An annotated list of the mosquitoes found in the vicinity of an endemic focus of yellow fever in the Republic of Colombia. *Proceedings of the Entomological Society of Washington*, 38 (4), 57–70.

Komp, W.H.W. (1936b) The male and larva of *Aedes dominicii* Rangel & Romero Sierra, and the male of *Aedes pseudodominicii* sp. nov., representatives of a new subgenus (*Soperia)* of the genus *Aedes*, from Colombia. *Proceedings of the Entomological Society of Washington*, 38 (4), 71–75.

Komp, W.H.W. (1936c) *Anopheles (Anopheles) chiriquiensis*, a new species of *Anopheles* from Panama (Diptera, Culicidae). *Proceedings of the Entomological Society of Washington*, 38 (7), 156–160.

Komp, W.H.W. (1936d) Description of nine new species of *Culex*, seven from Panama and two from Venezuela (Diptera, Culicidae). *Annals of the Entomological Society of America*, 29 (2), 319–334. DOI: 10.1093/aesa/29.2.319

Komp, W.H.W. (1936e) *Anopheles (Nyssorhynchus) anomalophyllus*, a new species of *Anopheles* from Panama and Costa Rica (Dipt., Culicidae). *Proceedings of the Entomological Society of Washington*, 38 (7), 160–164.

Komp, W.H.W. (1937a) *Anopheles acanthotorynus*, a new species of the subgenus *Stethomyia* from Peru. (Diptera, Culicidae.). *Annals of the Entomological Society of America*, 30 (2), 358–360. DOI: 10.1093/aesa/30.2.358

Komp, W.H.W. (1937b) The species of the subgenus *Kerteszia* of *Anopheles* (Diptera, Culicidae). *Annals of the Entomological Society of America*, 30 (3), 492–529. DOI: 10.1079/BER2004321

Komp, W.H.W. (1938) *Aedes leucotaeniatus*, a new species of *Aedes* allied to *A. leucocelaenus* D. & S.; and descriptions of the male and larva of *A. leucocelaenus* D. & S. (Diptera, Culicidae). *Proceedings of the Entomological Society of Washington*, 40 (8), 260–266.

Komp, W.H.W. (1942) The anopheline mosquitoes of the Caribbean Region. *National Institute of Health Bulletin*, 179, 1–195.

Komp, W.H.W. & Brown, C.G. (1935) *Culex jubifer*, a new species of *Culex* from Panama (Diptera: Culicidae). *Annals of the Entomological Society of America*, 28 (2), 254–255. DOI: 10.1093/aesa/28.2.254

Komp, W.H.W. & Curry, D.P. (1932) A new *Culex* from Panama (Dipt., Culicidae). *Psyche*, 39 (3), 82–84. DOI: 10.1155/1932/54725

Komp, W.H.W. & Kumm, H.W. (1938) A new species of *Haemagogus, mesodentatus*, from Costa Rica, and a description of the larva of *Haemagogus anastasionis* Dyar (Diptera, Culicidae). *Proceedings of the Entomological Society of Washington*, 40 (9), 253–259.

Komp, W.H.W. & Rozeboom, L.E. (1951) Descriptions of eight new species of *Culex* subgenus *Melanoconion* (Diptera, Culicidae). *Proceedings of the Entomological Society of Washington*, 53 (3), 121–137.

Kowo, C., Mayi, M.P.A., de Almeida, A.P.G., Foncha, D., Elad, M., Andongma, E., Djomo, C., Fru-Cho, J., Anong, D.N., Sehgal, R. & Cornel, A.J. (2023) Descriptions of a new *Aedes* species and subspecies of the subgenus *Aedimorphus*, from southwest Cameroon and updated key for the species of the "Domesticus group". *African Entomology*, 31, e15181. DOI: 10.17159/2254-8854/2023/a15181

Krishnamoorthy, R., Munirathinam, A., Dhananjeyan, K.J., Hiriyan, J., Mariappan, T., Samuel, P.P. & Venkatesh, A. (2013) Description of a new species, *Toxorhynchites (Toxorhynchites) tyagii* (Diptera: Culicidae), from Nilgiri hills, Western Ghats, southern India. *Zootaxa*, 3701 (4), 447–459. DOI: 10.11646/zootaxa.3701.4.4

Krzywinski, J., Wilkerson, R.C. & Besansky, N. (2001a) Evolution of mitochondrial and ribosomal gene sequences in Anophelinae (Diptera: Culicidae): implications for phylogeny reconstruction. *Molecular Phylogenetics and Evolution*, 18 (3), 479–487. DOI: 10.1006/mpev.2000.0894

Krzywinski, J., Wilkerson, R.C. & Besansky, N. (2001b) Toward understanding Anophelinae (Diptera, Culicidae) phylogeny: insights from nuclear single copy genes and the weight of evidence. *Systematic Biology*, 50 (4), 540–556. DOI: 10.1080/10635150119931

Kuchartshuk, L.P. (1973) New species of mosquito from the genus *Aedes* (Diptera, Culicidae) [in Russian]. In: *New and Little-known Species of Siberian Fauna. Part 7* [in Russian]. Academy of Sciences, USSR, Siberian Branch, Institute of Biology, Publishing House 'Nauka', Novosibirsk, Russia, pp. 98–101.

Kulasekera, V.L., Harrison, B.A. & Amerasinghe, F.P. (1989) *Anopheles (Anopheles) peytoni* new species, the "*An. insulaeflorum*" auct. from Sri Lanka (Diptera: Culicidae). *Mosquito Systematics* (for 1988), 20 (3), 302–316.

Kulasekera, V., Knight, K.L. & Harbach, R.E. (1990) *Aedes (Finlaya) axitiosus,* new species of the Niveus Subgroup (Diptera: Culicidae) from East Malaysia. *Mosquito Systematics*, 22 (1), 26–33.

Kumm, H.W. (1933) Mosquitos breeding in bromeliads, at Bahia, Brazil. *Bulletin of Entomological Research*, 24 (4), 561–573, 2 pls. DOI: 10.1017/S0007485300035513

Kumm, H.W., Komp, W.H.W. & Ruiz, H. (1940) The mosquitoes of Costa Rica. *American Journal of Tropical Medicine*, 20 (3), 385–422. DOI: 10.4269/ajtmh.1940.s1-20.385

La Casse, W.J. & Yamaguti, S. (1948) *Mosquito fauna of Japan and Korea*. Office of the Surgeon, Headquarters I Corps, APO 301, U.S. Army, Japan.

La Casse, W.J. & Yamaguti, S. (1950) *Mosquito fauna of Japan and Korea*. Office of the Surgeon, Headquarters 8th Army, APO 343, Kyoto, Honshu, Japan.

Laffoon, J. (1946) The Philippine mosquitoes of the genus *Aedes*, subgenus *Aedes*. *Journal of the Washington Academy of Sciences*, 36 (7), 228–245.

Lahille, F. (1904) Notes sur la classification des moustiques. *Actas y Trabajos del Segundo Congreso Médico Latino-Americano (Buenos Aires)*, 2, 71–96.

Laird, M. (1957) A new mosquito from Fiji, *Aedes (Finlaya) freycinetiae* n.sp. (Diptera: Culicidae). *Pacific Science*, 11 (3), 342–351.

Laird, M. (1988) *The Natural History of Larval Mosquito Habitats*. Academic Press, London, UK.

Lambert, D.M. & Coetzee, M. (1982) A dual genetical and taxonomic approach to the resolution of the mosquito taxon, *Anopheles (Cellia) marshallii* (Culicidae). *Systematic Entomology*, 7 (3), 321–332. DOI: 10.1111/j.1365-3113.1982.tb00449.x

Lane, C.J. (1992) *Toxorhynchites auranticauda* sp.n., a new Indonesian mosquito and potential biocontrol agent. *Medical and Veterinary Entomology*, 6 (3), 301–305. DOI: 10.1111/j.1365-2915.1992.tb00621.x

Lane, J. (1936) Notas sobre culicideos de Matto Grasso. *Revista do Museu Paulista*, 20, 173–206, 4 pls.

Lane, J. (1939) Catálogo dos mosquitos neotrópicos. *Boletim Biológico Série Monográfica*, 1, xi, 218 pp.

Lane, J. (1943) Sôbre o gênero *Uranotaenia* (Diptera, Culicidae, Culicini). *Revista de Entomologia*, 14 (1–2), 137–161.

Lane, J. (1945a) Os sabetineos da América. (Addenda e corrigenda). *Revista de Entomologia*, 16 (1–2), 132–157.

Lane, J. (1945b) Quatro espécies novas de *Culex* da Região Neotrópica. *Revista de Entomologia*, 16 (1–2), 204–209.

Lane, J. (1951) Synonymy of Neotropical Culicidae (Diptera). *Proceedings of the Entomological Society of Washington*, 53 (6), 333–336.

Lane, J. (1953) *Neotropical Culicidae*. Volumes I and II. University of São Paulo, São Paulo, 1112 pp.

Lane, J. & Antunes, P.C.A. (1937) Nota sobre o genero *Mansonia*, sub-genero *Rhynchotaenia*, com a descripção de uma nova especie *(Diptera, Culicidae)*. *Revista do Museu Paulista*, 23, 227–232.

Lane, J. & Causey, O.R. (1955) Additional data on Sabethini (Diptera, Culicidae). *Proceedings of the Entomological Society of Washington*, 57 (1), 11–17.

Lane, J. & Cerqueira, N.L. (1942) Os sabetíneos da América (Diptera, Culicidae). *Arquivos de Zoologia do Estudo de São Paulo*, 3 (9), 473–849.

Lane, J. & Cerqueira, N.L. (1957) The validity and change of name of two species of *Wyeomyia* (Diptera, Culicidae). *Proceedings of the Entomological Society of Washington*, 59 (5), 244.

Lane, J. & Coutinho, J.O. (1940) *Mansonia* subg. *Rhynchotaenia*: Descrição de duas espécies novaes dados sobre o subgênero (Dipt. Culicidae). *Revista de Entomologia*, 11 (1–2), 589–597.

Lane, J. & Forattini, O.P. (1952) New data on Sabethini (Diptera, Culicidae). *Proceedings of the Entomological Society of Washington*, 54 (5), 254–257.

Lane, J. & Guimarães, L.R. (1937) Notas sobre culicideos de Campos do Jordão. *Annaes Paulistas de Medicina e Cirugia*, 33, 211–217.

Lane, J. & Ramalho, G.R. (1960) A new neotropical *Culex*. *Revista Brasileira de Entomologia*, 9, 173–176.

Lane, J. & Whitman, L. (1943) Novas espécies de *Culex* do Brasil (Diptera, Culicidae). *Revista de Entomologia*, 14 (3), 389–408.

Lane, J. & Whitman, L. (1951) The subgenus "*Microculex*" in Brazil (Diptera, Culicidae). *Revista Brasileira de Biologia*, 11 (3), 341–366.

Lanzaro, G.C. & Eldridge, B.F. (1992) A classical and population genetic description of two new sibling species of *Aedes (Ochlerotatus) increpitus* Dyar. *Mosquito Systematics*, 24 (2), 85–101.

Lanzaro, G.C., Collier, T.C. & Lee, Y. (2015) Defining genetic, taxonomic, and geographic boundaries among species of the *Psorophora confinnis* (Diptera: Culicidae) complex in North and South America. *Journal of Medical Entomology*, 52 (5), 907–917. DOI: 10.1093/jme/tjv084

Lassmann, G.W. (1951) *Psorophora (Janthinosoma) totonaci*, a new species from Mexico (Diptera, Culicidae). *Proceedings of the Entomological Society of Washington*, 53 (5), 284–287.

Laurito, M. & Almirón, W.R. (2013) Phylogenetic relationships of *Culex (Culex)* species (Diptera, Culicidae) from Argentina based on morphological characters. *Zootaxa*, 3652 (1), 117–155. DOI: 10.11646/zootaxa.3652.1.4

Laurito, M., de Oliveira, T.M.P., Almirón, W.R. & Sallum, M.A.M. (2013) COI barcode versus morphological identification of *Culex (Culex)* (Diptera: Culicidae) species: a case study using samples from Argentina and Brazil. *Memóriasdo Instituto Oswaldo Cruz*, 108 (Suppl.1), 110–122. DOI: 10.1590/0074-0276130457

Laveran, [A.] (1900) Sur un *Anopheles* provenant de Madagascar. *Comptes Rendus Hebdomadaires des Séances et Mémoires de la Société de Biologie*, 52, 109–110.

Laveran, A. (1902a) Sur des culicides du Cambodge. *Comptes Rendus Hebdomadaires des Séances et Mémoires de la Société de Biologie*, 54, 906–908.

Laveran, A. (1902b) Sur des culicides de Nouvelles-Hébrides. *Comptes Rendus Hebdomadaires des Séances et Mémoires de la Société de Biologie*, 54, 908–910.

Laveran, [A.] (1902c) [Note]. In: Simond, P.L., Description d'un moustique dont le male possede un trompe faucille. *Comptes Rendus Hebdomadaires des Séances et Mémoires de la Société de Biologie*, 54, 1158–1160.

Lee, D.J. (1937) Notes on Australian mosquitoes (Diptera, Culicidae). Part IV. The genus *Theobaldia*, with description of a new species. *Proceedings of the Linnean Society of New South Wales*, 62 (5–6), 294–298.

Lee, D.J. (1944a) A new species of the genus *Anopheles* from northern Australia (Diptera, Culicidae). *Proceedings of the Linnean Society of New South Wales*, 69 (1–2), 21–25.

Lee, D.J. (1944b) Notes on Australian mosquitoes (Diptera, Culicidae). Part V. The genus *Armigeres* and new species of *Armigeres, Theobaldia* and *Culex*. *Proceedings of the Linnean Society of New South Wales*, 69 (5–6), 215–225.

Lee, D.J. (1946) Notes on Australian mosquitoes (Diptera, Culicidae). Part VI. The genus *Tripteroides* in the Australasian Region. *Proceedings of the Linnean Society of New South Wales*, 70 (5–6), 219–275, 2 pls.

Lee, D.J., Hicks, M.M., Debenham, M.L., Griffiths, M., Bryan, J.H. & Marks, E.N. **(1988b)** *The Culicidae of the Australasian Region. Volume 9. Nomenclature, synonymy, literature, distribution, biology and relation to disease. Genus* Culex (*Subgenera* Lutzia, Neoculex, *Subgenus undecided). Genera* Culiseta, Ficalbia, Heizmannia, Hodgesia, Malaya, Mansonia. Monograph Series, Entomology Monograph No. 2. Australian Government Publishing Service, Canberra, Australia.

Lee, D.J., Hicks, M.M., Debenham, M.L., Griffiths, M., Bryan, J.H. & Marks, E.N. **(1988c)** *The Culicidae of the Australasian Region. Volume 10. Nomenclature, synonymy, literature, distribution, biology and relation to disease. Genera* Maorigoeldia, Mimomyia, Opifex, Orthopodomyia, Topomyia, Toxorhynchites. Monograph Series, Entomology Monograph No. 2. Australian Government Publishing Service, Canberra, Australia.

Lee, D.J., Hicks, M.M., Debenham, M.L., Griffiths, M., Marks, E.N., Bryan, J.H. & Russell, R.C. **(1989a)** *The Culicidae of the Australasian Region. Volume 7. Nomenclature, synonymy, literature, distribution, biology and relation to disease. Genus* Culex. *Subgenera* Acallyntrum, Culex. Monograph Series, Entomology Monograph No. 2. Australian Government Publishing Service, Canberra, Australia.

Lee, D.J., Hicks, M.M., Debenham, M.L., Griffiths, M., Marks, E.N. & Bryan, J.H. **(1989b)** *The Culicidae of the Australasian Region. Volume 8. Nomenclature, synonymy, literature, distribution, biology and relation to disease. Genus* Culex. *Subgenera* Culiciomyia, Eumelanomyia, Lophoceraomyia. Monograph Series, Entomology Monograph No. 2. Australian Government Publishing Service, Canberra, Australia.

Lee, D.J., Hicks, M.M., Griffiths, M., Debenham, M.L., Bryan, J.H., Russell, R.C., Geary, M. & Marks, E.N. **(1987a)** *The Culicidae of the Australasian Region. Volume 4. Nomenclature, synonymy, literature, distribution, biology and relation to disease. Genus* Aedes. *Subgenera* Scutomyia, Stegomyia, Verrallina. Monograph Series, Entomology Monograph No. 2. Australian Government Publishing Service, Canberra, Australia.

Lee, D.J., Hicks, M.M., Griffiths, M., Debenham, M.L., Bryan, J.H., Russell, R.C., Geary, M. & Marks, E.N. **(1987b)** *The Culicidae of the Australasian Region. Volume 5. Nomenclature, synonymy, literature, distribution, biology and relation to disease. Genus* Anopheles. *Subgenera* Anopheles, Cellia. Monograph Series, Entomology Monograph No. 2. Australian Government Publishing Service, Canberra, Australia.

Lee, D.J., Hicks, M.M., Griffiths, M., Debenham, M.L., Bryan, J.H., Russell, R.C., Geary, M. & Marks, E.N. **(1988a)** *The Culicidae of the Australasian Region. Volume 6. Nomenclature, synonymy, literature, distribution, biology and relation to disease. Genera* Armigeres, Bironella *and* Coquillettidia. Monograph Series, Entomology Monograph No. 2. Australian Government Publishing Service, Canberra, Australia.

Lee, D.J., Hicks, M.M., Griffiths, M., Debenham, M.L., Marks, E.N., Bryan, J.H. & Russell, R.C. **(1989c)** *The Culicidae of the Australasian Region. Volume 11. Nomenclature, synonymy, literature, distribution, biology and relation to disease. Genera* Tripteroides, Uranotaenia, Wyeomyia, Zeugnomyia. Monograph Series, Entomology Monograph No. 2. Australian Government Publishing Service, Canberra, Australia.

Lee, D.J., Hicks, M.M., Griffiths, M., Russell, R.C. & Marks, E.N. (**1980**) *The Culicidae of the Australasian Region. Volume 1.* Monograph Series, Entomology Monograph No. 2. Australian Government Publishing Service, Canberra, Australia.

Lee, D.J., Hicks, M.M., Griffiths, M., Russell, R.C. & Marks, E.N. (**1982**) *The Culicidae of the Australasian Region. Volume 2. Nomenclature, synonymy, literature, distribution, biology and relation to disease. Genus* Aedeomyia. *Genus* Aedes *(subgenera [*Aedes*]* Aedimorphus, Chaetocruiomyia, Christophersiomyia, Edwardsaedes *and* Finlaya). Monograph Series, Entomology Monograph No. 2. Australian Government Publishing Service, Canberra, Australia.

Lee, D.J., Hicks, M.M., Griffiths, M., Russell, R.C. & Marks, E.N. (**1984**) *The Culicidae of the Australasian Region. Volume 3. Nomenclature, synonymy, literature, distribution, biology and relation to disease. Genus* Aedes. *Subgenera* Geoskusea, Halaedes, Huaedes, Leptosomatomyia, Levua, Lorrainea, Macleaya, Mucidus, Neomelaniconion, Nothoskusea, Ochlerotatus, Paraedes, Pseudoskusea, Rhinoskusea. Monograph Series, Entomology Monograph No. 2. Australian Government Publishing Service, Canberra, Australia.

Leeson, H.S. (1930) A new variety of *Anopheles funestus*, Giles, from southern Rhodesia. *Bulletin of Entomological Research*, 21 (2), 125–126. DOI: 10.1017/S0007485300021623

Leeson, H.S. (1935) Another anopheline of the *funestus* series from Southern Rhodesia. *Annals of Tropical Medicine and Parasitology*, 29 (1), 69–71.

Le Goff, G., Boussès, P. & Brunhes, J. (2007) Révision des *Neomelaniconion* Newstead (Diptera: Culicidae) de Madagascar: espèces présentes et description de cinq nouvelles espèces. *Bulletin de la Société Entomologique de France*, 43 (2), 187–200. DOI: 10.1080/00379271.2007.10697510

Le Goff, G., Brengues, C. & Robert, V. (2013) *Stegomyia* mosquitoes in Mayotte, taxonomic study and description of *Stegomyia pia* n. sp. *Parasite*, 20, 31. DOI: 10.1051/parasite/2013030

Leguizamón, M.N. & Carpintero, D.J. (2004) Description of a new species of *Orthopodomyia* (Diptera: Culicidae) from the Neotropical region. *Studia dipterologia*, 11 (1), 203–205.

Lei, X. (1989a) A new species of *Aedes (Ochlerotatus)* (Diptera: Culicidae) [in Chinese]. *Sichuan Journal of Zoology*, 8 (1), 6–8.

Lei, X. (1989b) *The mosquito fauna of Sichuan, China* [in Chinese]. Chengdu University of Science and Technology Press, Chengdu, China.

Lei, X. & Li, K. (1990) A new species of *Orthopodomyia* in Sichuan (Diptera, Culicidae) [in Chinese]. *Sichuan Journal of Zoology*, 9 (1), 8–10.

Leicester, G.F. (1903) A breeding place of certain forest mosquitoes in Malaya. *Journal of Tropical Medicine*, 6, 291–292.

Leicester, G.F. (1908) The Culicidae of Malaya. *Studies from the Institute for Medical Research, Federated Malay States*, 3 (3), 18–261.

Leleup, N. (1952) Un Anophèle cavernicole nouveau du Kibali-Ituri: *Anopheles Faini* [*sic*] n. sp. *Revue de zoologie et de botanique Africaines*, 46 (1–2), 151–158.

Leleup, N. (1957) *Anopheles garnhami* sbsp. *basilewskyi*, sous-espece nouvelle de tres haute altitude (2.900 M.). *Folia Scientifica Africae Centralis*, 3 (3), 69.

Leleup, N. & Lips, M. (1950) Un Anophèle cavernicole nouveau du Katanga: *Anopheles Rodhaini* [*sic*] n. sp. *Revue de zoologie et de botanique Africaines*, 43 (4), 303–308.

Lemine, A.M.M., Lemrabott, M.A.O., Ebou, M.H., Lekweiry, K.M., Salem, M.S.O.A., Brahim, K.O., Moukah, M.O., Bouraya, I.N.O., Brengues, C., Trape, J.-F., Basco, L., Bogreau, H., Simard, F., Faye, O. & Boukhary, A.O.M.S. (2017) Mosquitoes (Diptera: Culicidae) in Mauritania: a review of their biodiversity, distribution and medical importance. *Parasites & Vectors*, 10, 35. DOI: 10.1186/s13071-017-1978-y

Levi Castillo, R. (1944) *El complejo* "Pseudopunctipennis" *en el Ecuador (Diptera - Culicidae)*. Imprenta de la Universidad, Guayaquil, Ecuador.

Leví Castillo, R. (1945) *Los anofelinos de la Republica del Ecuador*. Tomo primero. Artes Graficas Senefelder, Guayaquil, Ecuador.

Leví Castillo, R. (1949) *Atlas de los anofelinos Sudamericanos*. Sociedad Filantrópica de Guayas, Guayaquil, Ecuador.

Levi-Castillo, R. (1953a) Observations on the subgenus *Phalangomyia* of the genus *Culex* in Ecuador with description of a new species (Diptera: Culicidae). *Pacific Science*, 7 (2), 187–192.

Leví-Castillo, R. (1953b) Dos especies nuevas de mosquitos ecuatorianos (Diptera-culicidae). *Revista Ecuatoriana de Entomología y Parasitología*, 1 (1), 14–18, 2 pls.

Leví-Castillo, R. (1953c) Dos especies nuevas de mosquitos de la sierra ecuatoriana (Diptera-culicidae). *Revista Ecuatoriana de Entomología y Parasitología*, 1 (4), 63–70, 5 pls.

Levi-Castillo, R. (1953d) Una nueva especie de *Culex* procedente de la Provincia de Los Ríos, Ecuador: *Culex (Carrollia) Babahoyersis* N. Sp. (Diptera-Culicidae). *Revista Ecuatoriana de Entomología y Parasitología*, 1 (3), 91–94, 2 pls.

Leví-Castillo, R. (1954) Cuatro especies nuevas de Sabethini del Ecuador (Diptera–Culicidae). *Revista Ecuatoriana de Entomologia y Parasitología*, 2 (1–2), 247–255, 5 pls.

Levi-Castillo, R. (1955) *Phoniomyia esmeraldasi*, a new mosquito from Ecuador. *Revista Ecuatoriana de Entomología y Parasitología*, 2 (3–4), 389–391, 1 pl.

Leví-Castillo, R. (1955b) *Haemagogus soperi* n. sp. mosquito trasmisor de fiebre amarilla selvática en el Ecuador (Diptera-Culicidae). *Revista Ecuatoriana de Entomología y Parasitología*, 2 (3–4), 479–484, 2 pls.

Leví-Castillo, R. (1955c) Un nuevo anofelino de altura del Ecuador: *Anopheles gomezdelatorrei* n. sp. *Revista Ecuatoriana de Entomología y Parasitología*, 2 (3–4), 509–514, 3 pls.

Lewis, D.J. (1937) A new species of *Anopheles* from the Anglo-Egyptian Sudan. *Proceedings of the Royal Entomological Society of London, Series B: Taxonomy*, 6 (9), 181–183, 1 pl. DOI: 10.1111/j.1365-3113.1937.tb01685.x

Lewis, D.J. (1942) A new species of *Aëdes* (Dipt., Culicidae) from Eritrea. *Proceedings of the Royal Entomological Society of London, Series B: Taxonomy*, 11 (11), 165–166. DOI: 10.1111/j.1365-3113.1942.tb00717.x

Lewis, D.J. (1943) Mosquitoes in relation to yellow fever in the Nuba Mountains, Anglo-Egyptian Sudan. *Annals of Tropical Medicine and Parasitology*, 37 (1), 65–76, 3 pls.

Lien, J.C. (1967) New species of mosquitoes from Taiwan (Diptera: Culicidae). Part I. Two new species of *Aedes (Finlaya)*. *Tropical Medicine*, 9 (4), 177–185.

Lien, J.C. (1968a) New species of mosquitoes from Taiwan (Diptera: Culicidae). Part II. New species of *Tripteroides, Orthopodomyia, Culiseta* and *Uranotaenia*. *Tropical Medicine*, 10 (1), 1–20.

Lien, J.C. (1968b) New species of mosquitoes from Taiwan (Diptera: Culicidae). Part III. Five new species of *Aedes*. *Tropical Medicine*, 10 (2), 95–115.

Lien, J.C. (1968c) New species of mosquitoes from Taiwan (Diptera: Culicidae). Part IV. *Mattinglyia catesi*, n. gen., n. sp. and four new species of genus *Heizmannia* Ludlow. *Tropical Medicine*, 10 (3), 127–153.

Lien, J.C. (1968d) New species of mosquitoes from Taiwan (Diptera: Culicidae). Part V. Three new subspecies of *Aedes* and seven new species of *Culex*. *Tropical Medicine*, 10 (4), 217–262.

Lien, J.C., Kosman, L., Partono, F., Joesoef, A., Kosin, E. & Cross, J.H. (1975) A brief survey of mosquitoes in North Sumatra, Indonesia. *Journal of Medical Entomology*, 12 (2), 233–239. DOI: 10.1093/jmedent/12.2.233

Lien, J.-C., Lin, C.-C., Lin, M.-C. & Tseng, L.F. (2008) A new species of the genus *Coquillettidia* (Diptera, Culicidae) from the Democratic Republic Sao [*sic*] Tome and Principe. *Journal of the National Taiwan Museum*, 61 (4), 55–60.

Lien, R., Lin, C. & Weng, M. (1999) *Culiseta taiwanica*, a new species of the subgenus *Climacura* Felt (Diptera: Culicidae). *Republic of China Journal of Parasitology*, 12 (1/2), 1–9.

Lin, C.-C., Tseng, L.-F. & Lien, J.-C. (2008) A new species of the genus *Culiseta* (Diptera, Culicidae) from the Democratic Republic of Sao [*sic*] Tome and Principe. *Journal of the National Taiwan Museum*, 61 (2), 89–96.

Linnaeus, C. (1758) *Systema naturae per regna tria naturae, secundum classes, ordines, genera, species, cum characteribus, differentiis, synonymis, locis.* Tomus 1, Editio Decima. Impensis Direct. Laurentii Salvii, Holmiae [Stockholm]. DOI: 10.5962/bhl.title.542

Linnaeus, C. (1762) Zweyter Theil, enthalt Beschreibungen verschiedener wichtiger Naturalien, pp. 267–606. In: Hasselquist, D.F. *Reise nach Palästina in den Jahren von 1749 bis 1752. Auf Befehl Ihro Majestat der Königinn von Schweden.* Johann Christian Koppe, Rostock, Germany.

Linthicum, K.J. (1988) A revision of the Argyritarsis Section of the subgenus *Nyssorhynchus* of *Anopheles* (Diptera: Culicidae). *Mosquito Systematics*, 20 (2), 98–271.

Linton, Y. (2004) Systematics of the Holarctic *maculipennis* complex. The 70th Annual Meeting of the American Mosquito Control Association, February 22–26, 2004, Savannah, GA, USA.

Linton, Y.-M., Dusfour, I., Howard, T.M., Ruiz, F., Manh, N.D., Dinh, T.H., Sochanta, T., Coosemans, M. & Harbach, R.E. (2005) *Anopheles (Cellia) epiroticus* (Diptera: Culicidae), a new malaria vector species in the Southeast Asian Sundaicus Complex. *Bulletin of Entomological Research*, 95 (4), 329–339. DOI: 10.1079/BER2005364

Linton, Y.-M., Pecor, J.E., Porter, C.H., Mitchell, L.B., Garzón-Moreno, A., Foley, D.H., Pecor, D.B. & Wilkerson, R.C. (2013) Mosquitoes of eastern Amazonian Ecuador: biodiversity, bionomics and barcodes. *Memórias do Instituto Oswaldo Cruz*, 108 (Suppl. 1), 100–109. DOI: 10.1590/0074-0276130440

Lips, M. (1960) *Anopheles* du Congo Belge. 2. Quelques especes des galeries forestieres. *Rivista di Parassitologia*, 21 (1), 39–64.

Liria, J. & Navarro, J.-C. (2014) Phylogenetic relationships in the genus *Psorophora* Robineau-Desvoidy (Diptera: Culicidae), based on morphological characters. *Entomology and Applied Science Letters*, 1 (4), 22–50.

Liston, W.G. (1901) A year's experience of the habits of *Anopheles* in Ellichpur. The description of the species of *Anopheles* found in Ellichpur during the year. *Indian Medical Gazette*, 36, 361–366, 441–443, 1 pl.

Liu, S.Z. & Lu, B.-L. (1984) A new *Aedes* from China (Diptera: Culicidae) [in Chinese]. *Acta Zootaxonomica Sinica*, 9 (1), 76–79.

Lorenz, C., Alves, J.M.P., Foster, P.G., Suesdek, L. & Sallum, M.A.M. (2021) Phylogeny and temporal diversification of mosquitoes (Diptera: Culicidae) with an emphasis on the Neotropical fauna. *Systematic Entomology*, 46 (4), 798–811. DOI: 10.1111/syen.12489

Lorenz, C., Patané, J.S.L. & Suesdek, L. (2015) Morphogenetic characterisation, date of divergence, and evolutionary relationships of malaria vectors *Anopheles cruzii* and *Anopheles homunculus*. *Infection, Genetics and Evolution*, 35, 144–152. DOI: 10.1016/j.meegid.2015.08.011

Lounibos, L.P., Rey, J.R. & Frank, J.H. (eds) (1985) *Ecology of Mosquitoes: Proceedings of a Workshop*. Florida Medical Entomology Laboratory, Vero Beach, FL, USA.

Lourenço de Oliveira, R. (1983) Sobre uma nova espécie de *Phoniomyia* (Diptera: Culicidae) do Rio de Janeiro, Brasil. *Memórias do Instituto Oswaldo Cruz*, 78 (4), 501–505. DOI: 10.1590/S0074-02761983000400015

Lourenço-de-Oliveira, R. & da Silva, T.F. (1985a) *Wyeomyia forcipenis*, new species of mosquito (Diptera: Culicidae) from Rio de Janeiro, Brazil. *Memórias do Instituto Oswaldo Cruz*, 80 (3), 321–326. DOI: 10.1590/S0074-02761985000300008

Lourenço-de-Oliveira, R. & da Silva, T.F. (1985b) Alguns aspectos da ecologia dos mosquitos (Diptera: Culicidae) de uma área de planície (Granjas Calábria), em Jacarepaguá, Rio de Janeiro. III. Preférencia horária das fêmeas para o hematofagismo. *Memórias do Instituto Oswaldo Cruz*, 80 (2), 195–201. DOI: 10.1590/S0074-02761985000200001

Lourenço-de-Oliveira, R. & da Silva, T.F. (1987) *Culex siphanulatus*, a new species of mosquito from the coast of Rio de Janeiro State, Brazil (Diptera: Culicidae). *Memórias do Instituto Oswaldo Cruz*, 82 (1), 101–110. DOI: 10.1590/S0074-02761987000100017

Lourenço-de-Oliveira, R., Harbach, R.E., Castro, M.G., Motta, M.A. & Peyton, E.L. (1999) *Wyeomyia (Prosopolepis) confusa* (Lutz): subgeneric validation, species description, and recognition of *Wyeomyia flui* (Bonne-Wepster and Bonne) as the senior synonym of *Wyeomyia kerri* del Ponte and Cerqueira. *Journal of the American Mosquito Control Association*, 15 (2), 200–212.

Lourenço-de-Oliveira, R., Motta, M.A. & de Castro, M.G. (1992) *Wyeomyia staminifera*, a new species of mosquito from Brazil (Diptera: Culicidae). *Memórias do Instituto Oswaldo Cruz*, 87 (1), 115–121. DOI: 10.1590/S0074-02761992000100018

Lu, B.-L. & Gong, Z.-D. (1986) Studies on the genus *Heizmannia* of China with descriptions of two new species (Diptera: Culicidae) [in Chinese]. *Acta Entomologica Sinica*, 29 (2), 181–189.

Lu, B. & Ji, S. (1997) Subgenus *Finlaya* Theobald, 1903. In: Lu, B., Li, B., Ji, S., Chen, H., Meng, Q., Su, L., Qu, F., Gong, Z. & Zhang, Z. (eds), *Diptera: Culicidae 1. Fauna Sinica, Insecta* [in Chinese]. Volume 8. Science Press, Beijing, China, pp. 99–173 and figs 18–60.

Lu, B.L. & Li, B.S. (1982) Identification of Chinese mosquitoes. In: Lu, B.S. (ed.), *Identification Handbook for Medically Important Animals in China* [in Chinese]. People's Health Publication Company, Beijing, China, pp. 1–159.

Lu, B.-L. & Su, L. (1987) *A Handbook for the Identification of Chinese Aedine Mosquitoes* [in Chinese]. Science Press, Beijing, China.

Lu, B.-L., Dong, X.-S. & Wang, X.-Z. (1986) A new species of subgenus *Topomyia* of genus *Topomyia* (Diptera: Culicidae) [in Chinese]. *Acta Zootaxonomica Sinica*, 11 (4), 406–408.

Lu, B., Li, B., Ji, S., Chen, H., Meng, Q., Su, L., Qu, F., Gong, Z. & Zhang, Z. (eds) (1997a) *Fauna Sinica, Insecta Vol. 8, Diptera: Culicidae 1*. Science Press, Beijing, China.

Lu, B., Li, B., Ji, S., Chen, H., Meng, Q., Su, L., Qu, F., Gong, Z. & Zhang, Z. (eds) (1997b) *Fauna Sinica, Insecta Vol. 9, Diptera: Culicidae II*. Science Press, Beijing, China.

Ludlow, C.S. (1902a) Two Philippine mosquitoes. *Journal of the New York Entomological Society*, 10 (3), 127–131.

Ludlow, C.S. (1902b) Description of a new *Anopheles*. *Journal of the American Medical Association*, 39 (8), 426–427.

Ludlow, C.S. (1903) Some Philippine mosquitoes. *Journal of the New York Entomological Society*, 11 (3), 137–144.

Ludlow, C.S. (1904a) Mosquito notes. *Canadian Entomologist*, 36 (8), 233–236. DOI: 10.4039/Ent36233-8

Ludlow, C.S. (1904b) Mosquito notes.—No. 2. *Canadian Entomologist*, 36 (10), 297–301. DOI: 10.4039/Ent36297-10

Ludlow, C.S. (1905a) Mosquito notes.—No. 3. *Canadian Entomologist*, 37 (3), 94–102, 129–135. DOI: 10.4039/Ent3794-3

Ludlow, C.S. (1905b) A new North American *Taeniorhynchus*. *Canadian Entomologist*, 37 (6), 231–232. DOI: 10.4039/Ent37231-6

Ludlow, C.S. (1905c) Mosquito notes—No. 4. *Canadian Entomologist*, 37 (11), 385–388. DOI: 10.4039/Ent37385-11
Ludlow, C.S. (1906a) A new American mosquito. *George Washington University Bulletin*, 5, 83–84.
Ludlow, C.S. (1906b) An Alaskan mosquito. *Canadian Entomologist*, 38 (10), 326–328. DOI: 10.4039/Ent38326-10
Ludlow, C.S. (1907) Mosquito notes.—No. 5.—Continued. *Canadian Entomologist*, 39 (4, 8, 12), 129–131, 266–268, 413–414. DOI: 10.4039/Ent39413-12
Ludlow, C.S. (1908a) Mosquito notes.—No. 6. *Canadian Entomologist*, 40 (1, 2, 9), 32–34, 50–52, 331–332. DOI: 10.4039/Ent4032-1
Ludlow, C.S. (1908b) *The mosquitos of the Philippine Islands; the distribution of certain species, and their occurrence in relation to the incidence of certain diseases.* PhD Thesis, The George Washington University, Washington, DC, USA.
Ludlow, C.S. (1909a) Mosquito comment. *Canadian Entomologist*, 41 (1), 21–24. DOI: 10.4039/Ent4121-1
Ludlow, C.S. (1909b) New Philippine mosquitoes. *Canadian Entomologist*, 41 (3), 97. DOI: 10.4039/Ent4197-3
Ludlow, C.S. (1909c) Mosquito observations. *Canadian Entomologist*, 41 (7), 233–235. DOI: 10.4039/Ent41233-7
Ludlow, C.S. (1911) The Philippine mosquitoes. *Psyche*, 18 (4), 125–133. DOI: 10.1155/1911/704241
Ludlow, C.S. (1914a) Philippine mosquitoes. *Psyche*, 21 (1), 30–32. DOI: 10.1155/1914/24037
Ludlow, C.S. (1914b) A new anopheline. *Psyche*, 21 (4), 129–130. DOI: 10.1155/1914/83576
Ludlow, C.S. (1914c) A new aëdine. *Psyche*, 21 (5), 159–160. DOI: 10.1155/1914/14720
Ludlow, C.S. (1920a) Siberian *Anopheles*. *Psyche*, 27 (4), 74–78. DOI: 10.1155/1920/76829
Ludlow, C.S. (1920b) New Siberian Culicidae *(Diptera)*. *Insecutor Inscitiae Menstruus* (for 1919), 7 (10–12), 151–161.
Ludlow, C.S. (1921) A new Philippine mosquito (*Diptera, Culicidae*). *Military Surgeon*, 49, 690–691.
Luh, P.-L. & Li, B.S. (1981) A new *Armigeres* from China (Diptera: Culicidae) [in Chinese]. *Acta Zootaxonomica Sinica*, 6 (2), 188–190.
Luh, P.-L. & Shih, J.-M. (1958) A new species of *Aedes* and a new subspecies of *Armigeres* from Kwangsi (Diptera: Culicidae) [in Chinese]. *Junshi Yixue Zazhi [Journal of Military Medicine]*, 1 (3), 222–226.
Luh, P.-L., Chao, K. & Xu, J.-M. (1974) A new *Culiseta* from China (Diptera: Culicidae) [in Chinese]. *Acta Entomologica Sinica*, 17 (4), 458–462.
Lutz, A. (1905) Novas especies de mosquitos do Brasil. *Imprensa Medica*, 13 (2, 3, 4, 5, 6, 7, 9, 11, 14, 15, 16, 18), 26–29, 48–52, 65–70, 81–84, 101–104, 125–128, 169–172, 202–204, 269–271, 287–290, 311–314, 347–350.
Lutz, A. & Neiva, A. (1911) Notas dipterolojicas. Dipterologische Mitteilungen. *Memórias do Instituto Oswaldo Cruz*, 3 (2), 295–300. DOI: 10.1590/S0074-02761911000200005
Lutz, A. & Neiva, A. (1913) Contribuições para a biolojia das megarininas com descrições de duas especies novas. *Memórias do Instituto Oswaldo Cruz*, 5 (2), 129–141. DOI: 10.1590/S0074-02761913000200003
Lynch Arribálzaga, F. (1878) Descripción de tres nuevos Culicidae de Buenos Aires. *El Naturalista Argentino*, 1 (5), 149–152.
Lynch Arribálzaga, F. (1891a) Dipterologia Argentina. *Revista del Museo de La Plata*, 1, 345–377.
Lynch Arribálzaga, F. (1891b) Dipterologia Argentina. *Revista del Museo de La Plata*, 2, 131–174, 5 pls.
Ma, S. (1981a) A new species of *Anopheles (Anopheles)* from China [in Chinese]. *Entomotaxonomia*, 3 (4), 277–279.
Ma, S.-F. (1981b) Studies on the *Anopheles (A.) sinensis* group of mosquitoes in China, including four new sibling species [in Chinese]. *Sinozoologia*, 1 (1), 59–70, 4 pls.
Ma, S.-F. (1982) Diptera: Culicidae [in Chinese]. In: Qingzang Gaoyuan Zonghe Kexue Kaochadui (corporate author), *Xizang kunchong* [*Insects of Xizang (Tibet)*], Volume 2. Qingzang gaoyuan kexue kaocha congshu [Series of the Scientific Expedition to Qinghai-Xizang Plateau]. Science Press, Beijing, China, pp. 157–163.
Macdonald, W.W. (1957) Malaysian parasites — XVI. An interim review of the non-anopheline mosquitoes of Malaya. In: Andy, J.R. (ed.), *Malaysian Parasites XVI–XXXIV*. [Series] Studies of the Institute for Medical Research of the Federation of Malaya, no. 28. Institute for Medical Research, Kuala Lumpur, Malaysia, pp. 1–34.
Macdonald, W.W. (1958) Description of a new species of *Orthopodomyia* Theobald from Malaya, with a note on the genus in that country (Diptera: Culicidae). *Proceedings of the Royal Entomological Society of London, Series B: Taxonomy*, 27 (7–8), 121–126. DOI: 10.1111/j.1365-3113.1958.tb00424.x
Macdonald, W.W. (1960) Malaysian parasites. XXXVIII. On the systematics and ecology of *Armigeres* subgenus *Leicesteria* (Diptera, Culicidae). *Studies from the Institute for Medical Research, Federated Malay States Malaya*, 29, 110–153.
Macdonald, W.W. & Mattingly, P.F. (1960) A new species of *Udaya* from Malaya and a description of the early stages of *U. argyrurus* (Edwards, 1934) (Diptera: Culicidae). *Proceedings of the Royal Entomological Society of London, Series B: Taxonomy*, 29 (1–2), 22–28. DOI: 10.1111/j.1365-3113.1960.tb01124.x
Macdonald, W.W. & Traub, R. (1960) Malaysian parasites XXXVII an introduction to the ecology of the mosquitoes of the lowland dipterocarp forest of Selangor, Malaya. *Studies from the Institute for Medical Research, Federation of Malaya*, 29, 79–109.
Macdonald, W.W., Smith, C.E.G. & Webb, H.E. (1965) Arbovirus infections in Sarawak: observations on the mosquitoes. *Journal of Medical Entomology*, 1 (4), 335–347. DOI: 10.1093/jmedent/1.4.335
Macfie, J.W.S. (1920) The chaetotaxy of the pupa of *Stegomyia fasciata*. *Bulletin of Entomological Research*, 10 (2), 161–172. DOI: 10.1017/S0007485300043960
Macfie, J.W.S. & Ingram, A. (1923) The early stages of West African mosquitos.—VI. *Bulletin of Entomological Research*, 10 (4), 409–442. DOI: 10.1017/S000748530004551X
MacGregor, M.E. (1924) *Aëdes (Stegomyia) mascarensis*, MacGregor: a new mosquito from Mauritius. *Bulletin of Entomological Research*, 14 (4), 409–412. DOI: 10.1017/S0007485300045648
Mackerras, I.M. (1927) Notes on Australian mosquitoes (Diptera, Culicidae). Part ii. The zoogeography of the subgenus *Ochlerotatus*, with notes on the species. *Proceedings of the Linnean Society of New South Wales*, 52 (3), 284–298.
Mackerras, I.M. (1946) Transmission of dengue fever by *Aedes (Stegomyia) scutellaris* Walk. in New Guinea. *Transactions of the Royal Society of Tropical Medicine and Hygiene*, 40 (3), 295–312. DOI: 10.1016/0035-9203(46)90070-3
Macquart, J. (1834) *Histoire naturelle des Insectes. Diptères. Tome premier.* Librairie Encyclopédique de Roret, Paris, France. DOI: 10.5962/bhl.title.14274
Macquart, J. (1838a) Diptères exotiques nouveaux ou peu connus. *Mémoires de la Société Royale des Sciences, de l'Agriculture et des Arts, de Lille*, 2, 9–225, 25 pls. DOI: 10.5962/bhl.title.15792
Macquart, J. (1838b) *Diptères exotiques nouveaux ou peu connus.* Tome premier. – 1.re Partie. Libraire Encyclopédique de Roret, Paris, France.
Macquart, J. (1847) *Diptères exotiques nouveaux ou peu connus.* 2.e Supplément. Pp. 5–104. Paris, France [published separately].
Macquart, J. (1850) Diptères exotiques nouveaux ou peu connus. 4.e Supplément. *Mémoires de la Société des Sciences, de l'Agriculture et des Arts de Lille*, Année 1849, 309–479, 14 pls. DOI: 10.5962/bhl.title.15792

Macquart, P.-O., Rahola, N., Chhuoy, K., Heng, K., Chhum, M., Suor, K. & Boyer, S. (2023) Description of a new species of *Toxorhynchites* (Diptera: Culicidae) from *Nepenthes* pitchers in Cambodia. *Journal of Asia-Pacific Entomology*, 26 (2), 102064. DOI: 10.1016/j.aspen.2023.102064

Maffi, M. & Coluzzi, M. (1958) *Anopheles (Neocellia) salbaii* n. sp. Morfologia ed ambiente. *Rivista di Malariologia*, 37 (1–3), 45–55, 3 pls.

Maffi, M. & Tenorio, J.A. (1977) *Culex (Culex) banksensis*, a new species of mosquito (Diptera: Culicidae) from the Banks Islands, New Hebrides. *Pacific Insects*, 17 (4), 503–509.

Maguire, T. (1970) The laboratory transmission of Coxsackie A6 virus by mosquitoes. *Journal of Hygiene*, 68, 625–630. DOI: 10.1017/S0022172400042558

Mahgoub, M.M., Colucci, M.E. & Odone, A. (2020) An updated checklist of mosquitoes (Diptera: Culicidae) of Sudan: taxonomy, vectorial importance and pictorial keys. *International Journal of Mosquito Research*, 7 (3), 9–18.

Manalang, C. (1930) Morphology and classification of the Philippine variety of *Anopheles aconitus* Donitz, 1902, and *Anopheles minimus*, 1901. *Philippine Journal of Science*, 43 (2), 247–261, 1 pl.

Mallampalli, V.L. (1995) *Phylogenetic relationships among taxa in the genus* Culex, *and the phylogenetic relationships of vectors of New World alphaviruses in the subgenus* Melanoconion *(Diptera: Culicidae)*. PhD Thesis, Department of Entomology, University of Maryland, College Park, MD, USA.

Mann F., G. (1950) Dos nuevas sub-especies del *Anopheles pseudopunctipennis* Th. 1901. *Biológica*, 8–11, 33–42.

Maquart, P.-O., Fontenille, D., Rahola, N., Yean, S. & Boyer, S. (2021) Checklist of the mosquito fauna (Diptera, Culicidae) of Cambodia. *Parasite*, 28, 60. DOI: 10.1051/parasite/2021056

Marcondes, C.B. & Alencar, J. (2010) Revisão de mosquitos *Haemagogus* Williston (Diptera: Culicidae) do Brasil. *Revista Biomédica*, 21 (3), 221–238.

Marks, E.N. (1947a) Studies of Queensland mosquitoes. 1. The *Aëdes (Finlaya) kochi* group with descriptions of new species from Queensland, Bougainville and Fiji. *Papers of the Department of Biology, University of Queensland*, 2 (5), 1–66.

Marks, E.N. (1947b) Studies of Queensland mosquitoes. 2. New species of *Aëdes* (subgenus *Finlaya)*. *Papers of the Department of Biology, University of Queensland*, 2 (6), 1–10.

Marks, E.N. (1948) Studies of Queensland mosquitoes. 3. The *Aëdes (Finlaya) australiensis* group. *Papers of the Department of Biology, University of Queensland*, 2 (8), 1–42.

Marks, E.N. (1949) Studies of Queensland mosquitoes. 4. Some species of *Aëdes* (subgenus *Ochlerotatus)*. *Papers of the Department of Biology, University of Queensland*, 2 (11), 1–41.

Marks, E.N. (1951) The vector of filariasis in Polynesia: a change in nomenclature. *Annals of Tropical Medicine and Parasitology*, 45 (2), 137–140. DOI: 10.1080/00034983.1951.11685481

Marks, E.N. (1955) Studies of Queensland mosquitoes. Part V—Some species of *Aedes* (subgenus *Finlaya)*. *Papers of the Department of Entomology, University of Queensland*, 1 (2), 11–29.

Marks, E.N. (1956) A new species of *Anopheles* from Queensland and notes on related species (Diptera: Culicidae). *Proceedings of the Royal Society of Queensland*, 67 (6), 41–52.

Marks, E.N. (1957a) The subgenus *Ochlerotatus* in the Australian Region (Diptera: Culicidae). I. Notes on classification, with the description of a new species. *Papers of the Department of Entomology, University of Queensland*, 1 (5), 71–83.

Marks, E.N. (1957b) Some mosquitoes from Western Samoa, with a description of a new species of *Aëdes (Stegomyia)* (Diptera, Culicidae). *Annals of Tropical Medicine and Parasitology*, 51 (1), 50–57. DOI: 10.1080/00034983.1957.11685794

Marks, E.N. (1958) New species and records of the *Aedes kochi* group from eastern Australia (Diptera: Culicidae). *Proceedings of the Royal Society of Queensland*, 69 (6), 57–74.

Marks, E.N. (1959a) The subgenus *Ochlerotatus* in the Australian region (Diptera: Culicidae). II. Five new species from Western Australia. *Papers of the Department of Entomology, University of Queensland*, 1 (8), 111–134.

Marks, E.N. (1959b) On two mosquitoes of the subgenus *Finlaya* (Diptera: Culicidae). *Proceedings of the Royal Society of Queensland*, 70 (4), 21–25.

Marks, E.N. (1960) Taxonomy and biology of some Papuan Culicidae. I. Genus *Topomyia*. *Pacific Insects*, 2 (1), 91–98.

Marks, E.N. (1963a) A revision of subgenus *Chaetocruiomyia* Theobald (Diptera: Culicidae). *Papers of the Department of Entomology, University of Queensland*, 1 (13), 189–211.

Marks, E.N. (1963b) The subgenus *Ochlerotatus* in the Australian region (Diptera: Culicidae) V. The Stricklandi Section. *Journal of the Entomological Society of Queensland*, 2 (1), 31–47.

Marks, E.N. (1964a) The subgenus *Ochlerotatus* in the Australasian Region (Diptera: Culicidae). VII. Four new species. *Papers of the Department of Entomology, University of Queensland*, 2 (3), 59–71.

Marks, E.N. (1964b) Notes on the subgenus *Chaetocruiomyia* Theobald (Diptera: Culicidae). *Proceedings of the Linnean Society of New South Wales*, 89 (1), 131–147.

Marks, E.N. (1968) Northern records of the genus *Culiseta* Felt in Australia with the description of a new species (Diptera: Culicidae). *Journal of the Australian Entomological Society*, 7 (1), 43–56. DOI: 10.1111/j.1440-6055.1968.tb00700.x

Marks, E.N. (1971) A metapleural generic character in *Topomyia* Leicester. *Mosquito Systematics Newsletter*, 3 (1), 4–6.

Marks, E.N. (1974) Notes on some taxonomic characters of Culicidae. *Mosquito Systematics*, 6 (3), 211–213.

Marks, E.N. (1977) *Aedes (Macleaya) stoneorum*, a new species from Queensland (Diptera: Culicidae). *Proceedings of the Entomological Society of Washington*, 79 (1), 33–37.

Marks, E.N. & Hodgkin, E.P. (1958) A new species of *Aedes (Finlaya)* from northern Australia (Diptera, Culicidae). *Proceedings of the Linnean Society of New South Wales*, 83 (1), 33–39.

Marsh, F. (1933) A new species of *Anopheles (Myzomyia* group) from South-west Persia. *Stylops: Journal of Taxonomic Entomology*, 2 (9), 193–197, 1 pl.

Marshall, J.F. & Staley, J. (1935) Generic and subgeneric differences in the mouth-parts of male mosquitos. *Bulletin of Entomological Research*, 26 (4), 531–532. DOI: 10.1017/S0007485300036865

Martinez, A. & Prosen, A.F. (1953) Una especie nueva de *Uranotaenia* (Dipt. Culic.) de Argentina. *Boletín de la Sociedad Entomológica Argentina*, 2, 4.

Martínez, A. & Prosen, A.F. (1953) Una nueva especie de mosquito de la Argentina. *Publicaciones de la Misión de Estudios de Patología Regional Argentina, Jujuy, Universidad de Buenos Aires*, 1, 1–2.

Martínez, A., Carcavallo, R.U. & Prosen, A.F. (1960) Una nueva especie de Culicidae de Bolivia (Díptera). *Revista de la Cátedra de Microbiología y Parasitología*, 28 (89), 25–29.

Martínez, A., Carcavallo, R.U. & Prosen, A.F. (1961) El genero *Haemagogus* Williston, 1896, en la Argentina (*Diptera – Culicidae). Anales del Instituto de Medicina Regional* (for 1960), 5 (2), 63–86.

Martini, E. (1914) Some new American mosquitoes. *Insecutor Inscitiae Menstruus*, 2 (5), 65–76, 1 pl.

Martini, E. (1920) Ueber mecklenburgische Culicinen. *Sitzungsberichte und Abhandlungen der Naturforschenden Gesellschaft zu Rostock*, 7, 203–209, 1 pl.

Martini, E. (1923) Über einige für das System bedeutungsvolle Merkmale der Stechmücken. *Zoologische Jahrbücher. Abteilung für Systematik, Geographie und Biologie der Tiere*, 46 (6), 517–590.

Martini, E. (1925) Zwei bemerkenswerte Culiciden von einem eigenartigen Biotop. *Internationale Revue der gesamte Hydrobiologie und Hydrographie*, 12 (5/6), 333–337.

Martini, E. (1926) Über die Stechmücken der Umgebung von Saratow. *Raboty Volzhskoi Biologicheskoi Stantsii [Arbeiten der Biologischen Wolga-Station]*, 8, 189–228.

Martini, E. (1927) Über zwei neue Stechmückenarten aus Anatolien. *Archiv für Schiffs- und Tropenhygiene*, 31, 386–390.

Martini, E. (1929) 11. u. 12. Culicidae. In: Lindner, E. (ed.), *Die Fliegen der palaerktischen Region. Band III*. E. Schweizerbart'sche Verlagsbuchhandlung (Erwin Nägele), Stuttgart, Germany, pp. 1–144.

Martini, E. (1931) Ueber einige südamerikanische Culiciden. *Revista de Entomologia Rio de Janeiro*, 1 (2), 199–219.

Martini, E. (1932) Dos nuevos mosquitos *Anopheles*, procedentes del Estado de Chiapas, México. *Revista Mexicana de Biología*, 12, 99–102.

Maslov, A.V. (1964) On the systematics of bloodsucking mosquitoes of the group *Culiseta* (Diptera, Culicidae) [in Russian]. *Entomologicheskoe Obozrenie*, 43 (1), 193–217.

Maslov, A.V. (1967) Krovososushchie komarȳ podtribȳ Culisetina (Diptera, Culicidae) mirovoĭ faunȳ [Bloodsucking mosquitoes of the subtribe Culisetina (Diptera, Culicidae) of the world]. Académie Impériale des Sciences, Musée Zoologique, St Petersburg. *Opredeliteli po faune SSSR*, Vȳp. 93, 182 pp., 1 pl.

Maslov, A.V. (1989) *Blood-sucking Mosquitoes of the Subtribe Culisetina (Diptera, Culicidae) in world fauna*. Smithsonian Institution Libraries and the National Science Foundation, Washington, DC, USA.

Matheson, R. (1929) *A Handbook of the Mosquitoes of North America — their structure — how they live — how they carry disease — how they may be studied — how they may be controlled — how they may be identified*. Charles C. Thomas, Springfield, IL, USA.

Matheson, R. (1934) Notes on mosquitoes from South America, with a description of a new species (Diptera, Culicidae). *Proceedings of the Entomological Society of Washington*, 36 (5), 120–122.

Matsumura, S. (1916) *Thousand Insects of Japan* [Shin Nihon senchu zukai]. *Additamenta* [Shin Nihon senchu zukai]. Vol. II. Keisei-sha, Tokyo, Japan.

Mattingly, P.F. (1949) *Ficalbia (Ficalbia) jacksoni* sp. n., a new species of mosquito (Diptera, Culicidae) from Hong-Kong. *Proceedings of the Royal Entomological Society of London, Series B: Taxonomy*, 18 (1–2), 9–11. DOI: 10.1111/j.1365-3113.1949.tb01407.x

Mattingly, P.F. (1951) *Culex (Culex) nakuruensis* sp. n. (Diptera, Culicidae), a remarkable new mosquito from the East African Highlands. *Proceedings of the Royal Entomological Society of London, Series B: Taxonomy*, 20 (3–4), 44–46. DOI: 10.1111/j.1365-3113.1951.tb01016.x

Mattingly, P.F. (1953a) The sub-genus *Stegomyia* (Diptera: Culicidae) in the Ethiopian Region. II. Distribution of species confined to the East and South African sub-region. *Bulletin of the British Museum (Natural History) Entomology*, 3 (1), 1–65. DOI: 10.5962/bhl.part.1046

Mattingly, P.F. (1953b) Notes on the Culicini of the Katanga (Diptera, Culicidae). Part I. Taxonomy. *Revue de Zoologie et de Botanique Africaines*, 47 (3–4), 311–343.

Mattingly, P.F. (1953c) New records and a new species of the subgenus *Stegomyia* (Diptera, Culicidae) from the Ethiopian Region. *Annals of Tropical Medicine and Parasitology*, 47 (3), 294–298. DOI: 10.1080/00034983.1953.11685571

Mattingly, P.F. (1954a) Notes on the subgenus *Stegomyia* (Diptera, Culicidae), with a description of a new species. *Annals of Tropical Medicine and Parasitology*, 48 (3), 259–270. DOI: 10.1080/00034983.1954.11685623

Mattingly, P.F. (1954b) Notes on Ethiopian *Uranotaenia* (Diptera: Culicidae) with a description of a new species. *Proceedings of the Royal Entomological Society of London, Series B: Taxonomy*, 23 (9–10), 167–171. DOI: 10.1111/j.1365-3113.1954.tb00444.x

Mattingly, P.-F. (1955) Le sous-genre *Neoculex (Diptera, Culicidae)* dans la sous-région mediterraneenne. I. – Espèce, sous-espèce et synonymie nouvelles. *Annales de Parasitologie Humaine et Comparée*, 30 (4), 374–388. DOI: 10.1051/parasite/1955304374

Mattingly, P.F. (1957a) *The Culicine Mosquitoes of the Indomalayan Area. Part I: Genus* Ficalbia *Theobald*. British Museum (Natural History), London, UK.

Mattingly, P.F. (1957b) *The Culicine Mosquitoes of the Indomalayan Area. Part II: Genus* Heizmannia *Ludlow*. British Museum (Natural History), London, UK.

Mattingly, P.F. (1958a) *The Culicine Mosquitoes of the Indomalayan Area. Part III. Genus* Aëdes *Meigen, subgenera* Paraëdes *Edwards,* Rhinoskusea *Edwards and* Cancraëdes *Edwards*. British Museum (Natural History), London, UK.

Mattingly, P.F. (1958b) A revision of *Paraëdes* Edwards and *Cancraëdes* Edwards (Diptera: Culicidae). *Proceedings of the Royal Entomological Society of London, Series B: Taxonomy*, 27 (5–6), 76–83. DOI: 10.1111/j.1365-3113.1958.tb00416.x

Mattingly, P.F. (1959) *The Culicine Mosquitoes of the Indomalayan Area. Part IV. Genus* Aëdes *Meigen, subgenera* Skusea *Theobald,* Diceromyia *Theobald,* Geoskusea *Edwards and* Christophersiomyia *Barraud*. British Museum (Natural History), London, UK.

Mattingly, P.F. (1960) *Aëdes* subgenus *Alanstonea* subgen. nov. (Diptera: Culicidae). *Proceedings of the Royal Entomological Society of London, Series B: Taxonomy*, 29 (11–12), 170–171. DOI: 10.1111/j.1365-3113.1960.tb01132.x

Mattingly, P.F. (1961) *The Culicine Mosquitoes of the Indomalayan Area. Part V. Genus* Aëdes *Meigen, subgenera* Mucidus *Theobald,* Ochlerotatus *Lynch Arribalzaga and* Neomelaniconion *Newstead*. British Museum (Natural History), London, UK.

Mattingly, P.F. (1965) *The Culicine Mosquitoes of the Indomalayan Area. Part VI. Genus* Aedes *Meigen, subgenus* Stegomyia *Theobald (Groups A, B and D)*. British Museum (Natural History), London, UK.

Mattingly, P.F. (1969) *The Biology of Mosquito-borne Disease*. George Allen & Unwin, London, UK.

Mattingly, P.F. (1970) Contributions to the mosquito fauna of Southeast Asia. – VI. The genus *Heizmannia* Ludlow in Southeast Asia. *Contributions of the American Entomological Institute*, 5 (7), 1–104.

Mattingly, P.F. (1971a) Contributions to the mosquito fauna of Southeast Asia. – XII. Illustrated keys to the genera of mosquitoes (Diptera, Culicidae). *Contributions of the American Entomological Institute*, 7 (4), 1–84.

Mattingly, P.F. (1971b) Mosquito eggs XIV. Genus *Armigeres* Theobald (continued) and *Aedes* subgenus *Alanstonea* Mattingly. *Mosquito Systematics Newsletter*, 3 (3), 130–137.

Mattingly, P.F. (1980) An interim reclassification of the genus *Tripteroides* with particular reference to the Australasian subgenera. *Mosquito Systematics*, 12 (2), 164–171.

Mattingly, P.F. (1981) Medical entomology studies – XIV. The subgenera *Rachionotomyia*, *Tricholeptomyia* and *Tripteroides* (Mabinii Group) of genus *Tripteroides* in the Oriental region (Diptera: Culicidae). *Contributions of the American Entomological Institute*, 17 (5), ii, 1–147.

Mattingly, P.F. & Adam, J.P. (1954) A new species of cave-dwelling anopheline from the French Cameroons. *Annals of Tropical Medicine and Parasitology*, 48 (1), 55–57. DOI: 10.1080/00034983.1954.11685598

Mattingly, P.F. & Brown, E.S. (1955) The mosquitos (Diptera: Culicidae) of the Seychelles. *Bulletin of Entomological Research*, 46 (1), 69–110, 3 pls. DOI: 10.1017/S0007485300030777

Mattingly, P.F. & Grjebine, A. (1958) Revision du genre *Ficalbia* Theobald et discussion de la position systematique des *Ravenalites* Doucet (Diptera, Culicidae). *Mémoires de l'Institut Scientifique de Madagascar Série E*, 9, 259–290.\

Mattingly, P.F. & Knight, K.L. (1956) The mosquitoes of Arabia I. *Bulletin of the British Museum (Natural History) Entomology*, 4 (3), 91–141.

Mattingly, P.F. & Marks, E.N. (1955) Some Australian mosquitoes (Diptera, Culicidae) of the subgenera *Pseudoskusea* and *Neoculex*. *Proceedings of the Linnean Society of New South Wales*, 80 (2), 163–176.

Mattingly, P.F. & Rageau, J. (1958) *Culex (Culex) iyengari* n. sp., a new species of mosquito (Diptera, Culicidae) from the South Pacific. *Pacific Science*, 12 (3), 241–250.

Mattingly, P.F., Stone, A. & Knight, K.L. (1962) *Culex aegypti* Linnaeus, 1762 (Insect, Diptera); proposed validation and interpretation under the plenary powers of the species so named. *Bulletin of Zoological Nomenclature*, 19 (4), 208–219.

McArthur, J. (1949) A new variety of *Anopheles aitkeni* from Borneo. *Bulletin of Entomological Research*, 40 (1), 49–52. DOI: 10.1017/S0007485300022690

McCracken, I. (1904) *Anopheles* in California, with description of a new species. *Entomological News*, 15 (1), 9–14, 1 pl.

McIntosh, B.M. (1971) The aedine subgenus *Neomelanoconion* Newstead (Culicidae, Diptera) in southern Africa with descriptions of two new species. *Journal of the Entomological Society of Southern Africa*, 34 (2), 319–333.

McIntosh, B.M. (1973) A taxonomic re-assessment of *Aedes (Ochlerotatus) caballus* (Theobald) (Diptera: Culicidae) including a description of a new species of *Ochlerotatus*. *Journal of the Entomological Society of Southern Africa*, 36 (2), 261–269.

McIntosh, B.M. (1975) A taxonomic revision of certain *Aedes* species (Diptera: Culicidae) of the subgenus *Aedimorphus* in southern Africa. *Journal of the Entomological Society of Southern Africa*, 38 (2), 251–287.

McIntosh, B.M., Jupp, P.G. & de Sousa, J. (1972) Further isolations of arboviruses from mosquitoes collected in Tongaland, South Africa, 1960–1968. *Journal of Medical Entomology*, 9 (2), 155–159. DOI: 10.1093/jmedent/9.2.155

McIntosh, B.M., Weinbren, M.P., Worth, C.B. & Kokernot, R.H. (1962) Isolation of viruses from mosquitoes collected at Lumbo, Mozambique: III. Isolation of Spondweni virus from *Aedes (Ochlerotatus) fryeri* (Theobald) and/or *Aedes (Aedimorphus) fowleri* (D'Emmerez de Charmoy). *American Journal of Tropical Medicine and Hygiene*, 11 (5), 685–686.

McIntosh, B.M., Worth, C.B. & Kokernot, R.H. (1961) Isolation of Semliki Forest virus from *Aedes (Aedimorphus) argenteopunctatus* (Theobald) collected in Portuguese East Africa. *Transactions of the Royal Society of Tropical Medicine and Hygiene*, 55 (2), 192–198. DOI: 10.1016/0035-9203(61)90025-6

McKeon, S., Lehr, M.A., Wilkerson, R.C., Ruiz, J.F., Sallum, M.A., Lima, J.B.P., Povoa, M.M. & Conn, J.E. (2010) Lineage divergence detected in the malaria vector *Anopheles marajoara* (Diptera: Culicidae) in Amazonian Brazil. *Malaria Journal*, 9, 271. DOI: 10.1186/1475-2875-9-271

Medschid. (1928) Über *Aëdes lepidonotus* Edw. und *Aëdes Refiki* n. sp. *Archiv für Schiffs- und Tropen-Hygiene*, 32, 306–315.

Medschid. (1930) *Culex martinii* n. sp. *Archiv für Schiffs- und Tropen-Hygiene*, 34, 364–369.

Meigen, J.W. (1804) *Klassifikazion und Beschreibung der europäischen Zweiflügeligen Insekten. (Diptera Linn.)*. Erster Band, ester Abtheilung, mit VIII Kupfertafeln. Karl Reichard, Braunschweig [Brunswick], Germany.

Meigen, J.W. (1818) *Systematische Beschreibung der bekannten europäischen zweiflügeligen Insekten*. Volume 1. Aachen, Germany. DOI: 10.5962/bhl.title.12464

Meigen, J.W. (1830) *Systematische Beschreibung der bekannten europäischen zweiflügeligen Insekten*. Volume 6. Hamm, Germany. DOI: 10.5962/bhl.title.12464

Meigen, J.W. (1838) *Systematische Beschreibung der bekannten europaeischen zweifluegeligen Insekten*. Volume 7. Hamm, Germany.

Meinert, F. (1886) De eucephale myggelarver. Sur les larves eucéphales de diptères. Leurs moeurs et leurs metamorphoses. *Videnskabernes selskabs skrifter. 6te Række. Naturvidenskabelig og mathematisk afdeling, Series 3*, 4, [2 pp.], 373–493, 8 pls.

Melero-Alcíbar, R.M. (2004) The pupae of Spanish *Ochlerotatus* I. *Ochlerotatus quasirusticus* (Diptera: Culicidae). *European Mosquito Bulletin*, 17, 20–22.

Melero-Alcíbar, R.M. (2005) The pupae of Spanish *Ochlerotatus* II. *Ochlerotatus rusticus* (Rossi) (Diptera: Culicidae). *European Mosquito Bulletin*, 19, 14–16.

Mendoza, J.B. (1940) New *Anopheles* of the *Myzorhynchus* series. *Monthly Bulletin of the Bureau of Health (Manila)*, 20 (11), 353–369.

Mendoza, J.B. (1947) Two more Philippine *Anopheles* in the *Myzorhynchus* series. *Monthly Bulletin of the Bureau of Health (Manila)*, 23 (3), 171–184.

Meng, C.-H. (1957) *The Taxonomy, Biology and Control of Chinese Anopheline Mosquitoes*. People's Hygiene Press, Peking, China.

Meng, C.H. (1958) A new species of *Culex* [in Chinese]. *Acta Entomologica Sinica*, 8 (4), 351–354.

Meng, C.H. (1962) A new species of *Aedes* (Diptera, Culicidae). *Aedes (Ochlerotatus) lasaensis* sp. nov. [in Chinese]. *Acta Entomologica Sinica*, 11 (2), 159–163.

Menon, M.A.U. (1950) The male of a new species of the subgenus *Aëdes* Meigen, 1818 (Diptera: Culicidae). *Proceedings of the Royal Entomological Society of London, Series B: Taxonomy*, 19 (9–10), 139–141. DOI: 10.1111/j.1365-3113.1950.tb00949.x

Merabti, B., Boumaza, M., Ouakid, M.L., Carvajal, T.M. & Harbach, R.E. (2021) An updated checklist of the mosquitoes (Diptera: Culicidae) present in Algeria, with assessments of doubtful records and problematic species. *Zootaxa*, 5027 (4), 515–545. DOI: 10.11646/ZOOTAXA.5027.4.3

Meunier, F. (1915) Nouvelles recherches sur quelques insectes des plâtrières d'Aix en Provence. *Verhandelingen der Koninklijke Akademie van Wetenschappen le Amsterdam (Tweede Sectie)*, 18 (5), 1–17, 5 pls.

Mihályi, F. (1955) *Aëdes hungaricus* n. sp. (Culicidae, Diptera). *Annales Historico-Naturales Musei Nationalis Hungarici*, 47 (series nova 6), 343–345.

Miller, B.R., Crabtree, M.B. & Savage, H.M. (1996) Phylogeny of fourteen *Culex* mosquito species, including the *Culex pipiens* complex, inferred from the internal transcribed spacers of ribosomal DNA. *Insect Molecular Biology*, 5 (2), 93–107. DOI: 10.1111/j.1365-2583.1996.tb00044.x

Miller, D. (1922) A remarkable mosquito, *Opifex fuscus*, Hutton. *Bulletin of Entomological Research*, 13 (1), 115–126. DOI: 10.1017/S0007485300045284

Minkeu, F.N. & Vernick, K.D. (2018) A systematic review of the natural virome of *Anopheles* mosquitoes. *Viruses*, 10, 222. DOI: h10.3390/v10050222

Mirabello, L. & Conn, J.E. (2008) Population analysis using the nuclear *white* gene detects Pliocene/Pleistocene lineage divergence within *Anopheles nuneztovari* in South America. *Medical and Veterinary Entomology*, 22 (2), 109–119. DOI: 10.1111/j.1365-2915.2008.00731.x

Mitchell, A., Sperling, F.A.H. & Hickey, D.A. (2002) Higher-level phylogeny of mosquitoes (Diptera: Culicidae): mtDNA data support a derived placement for *Toxorhynchites*. *Insect Systematics and Evolution*, 33 (2), 163–174. DOI: 10.1163/187631202X00118

Mitchell, C.J. & Darsie, R.F., Jr (1985) Mosquitoes of Argentina Part II. Geographic distribution and bibliography (Diptera, Culicidae). *Mosquito Systematics*, 17 (4), 279–360.

Miyagi, I. (1976) Description of a new species of the genus *Topomyia* Leicester from the Ryukyu Islands, Japan (Diptera: Culicidae). *Tropical Medicine*, 17 (4), 201–210.

Miyagi, I. & Toma, T. (1981) A new species of *Aedes (Stegomyia)* from Daito Islands, Ryukyu, Japan (Diptera: Culicidae). *Mosquito Systematics* (for 1980), 12 (4), 428–440.

Miyagi, I. & Toma, T. (1989) A new species of *Topomyia (Suaymyia) suchariti* from Thailand (Diptera: Culicidae). *Mosquito Systematics*, 21 (1), 16–24.

Miyagi, I. & Toma, T. (1995) *Topomyia javaensis*: a new Indonesian species in the subgenus *Topomyia* (Diptera: Culicidae). *Mosquito Systematics*, 27 (3), 179–186.

Miyagi, I. & Toma, T. (1997) Description of *Topomyia irianensis* n. sp. and new records of *To. papuensis* from Maluku and Irian Jaya, Indonesia (Diptera: Culicidae). *Journal of the American Mosquito Control Association*, 13 (2), 134–139.

Miyagi, I. & Toma, T. (2001) *Tripteroides (Rachionotomyia) ponmeki* (Diptera: Culicidae): a new species from Khammouane Province, Lao PDR. *Medical Entomology and Zoology*, 52 (4), 269–277. DOI: 10.7601/mez.52.269

Miyagi, I. & Toma, T. (2005) *Topomyia roslihashimi*, a new species of the subgenus *Suaymyia* (Diptera: Culicidae) from Gombak, Peninsular Malaysia. *Medical Entomology and Zoology*, 56 (4), 275–282. DOI: 10.7601/mez.56.275

Miyagi, I. & Toma, T. (2007) A new mosquito of the genus *Topomyia* (Diptera, Culicidae) from a *Nepenthes* pitcher plant in a Bario highland of Sarawak, Malaysia. *Medical Entomology and Zoology*, 58 (3), 167–174. DOI: 10.7601/mez.58.167

Miyagi, I. & Toma, T. (2008) Description of a new species *Topomyia (Suaymyia) lehcharlesi* (Diptera: Culicidae) from Sarawak, Malaysia. *Medical Entomology and Zoology*, 59 (3), 163–170. DOI: 10.7601/mez.59.163

Miyagi, I. & Toma, T. (2010a) Descriptions of *Topomyia auriceps* Brug, 1939 and *Topomyia pseudoauriceps*, n. sp. from Sarawak, Malaysia (Diptera: Culicidae). *Medical Entomology and Zoology*, 61 (1), 27–38. DOI: 10.7601/mez.61.27

Miyagi, I. & Toma, T. (2010b) *Topomyia (Suaymyia) kelabitense* (Diptera: Culicidae), a new species from Sarawak, Malaysia. *Medical Entomology and Zoology*, 61 (4), 353–361. DOI: 10.7601/mez.61.353

Miyagi, I. & Toma, T. (2013) *Uranotaenia (Pseudoficalbia) tanakai* (Diptera, Culicidae), a new species from forest swamps, Iriomote Is., the Ryukyu Archipelago, Japan. *Medical Entomology and Zoology*, 64 (3), 167–174. DOI: 10.7601/mez.64.167

Miyagi, I., Toma, T. & Cabrera, B.D. (1983a) Synonyms of *Topomyia rausai* with *Topomyia apsarae*, with description of the larva (Diptera: Culicidae). *Mosquito Systematics*, 15 (2), 111–116.

Miyagi, I., Toma, T. & Higa, Y. (2004a) A new species of *Mimomyia (Ingramia)* from Indonesia (Diptera: Culicidae). *Medical Entomology and Zoology*, 55 (1), 11–20. DOI: 10.7601/mez.55.11

Miyagi, I., Toma, T. & Lien, J.C. (2004b) *Ochlerotatus (Geoskusea) timorensis* (Culicidae: Diptera), a new species from crab-holes, West Timor, Indonesia. *Medical Entomology and Zoology*, 55 (2), 107–114. DOI: 10.7601/mez.55.107

Miyagi, I., Toma, T., Okazawa, T. & Wong, S.F. (2021a) Description of a new species of *Topomyia* (Diptera: Culicidae) from Mount Murud, Sarwak, Malaysia. *Medical Entomology and Zoology*, 72 (2), 59–65. DOI: 10.7601/mez.72.59

Miyagi, I., Toma, T., Okazawa, T. & Wong, S.F. (2021b) Description of a new species *Topomyia (Topomyia) sarawakensis* (Diptera: Culicidae) from Lanjak-Entimau Wildlife Sanctuary, Sarawak, Malaysia. *Medical Entomology and Zoology*, 72 (3), 189–197. DOI: 10.7601/mez.72.189

Miyagi, I., Toma, T., Okazawa, T. & Wong, S.F. (2022) Description of a new mosquito species, *Topomyia (Suaymyia) puehensis*, (Diptera: Culicidae) from Sarawak, Malaysia. *Medical Entomology and Zoology*, 73 (1), 7–14. DOI: 10.7601/mez.73.7

Miyagi, I., Toma, T., Okazawa, T. & Wong, S.F. (2023) A new species of the genus *Topomyia* Leicester (Diptera: Culicidae) from Sarawak, Malaysia. *Medical Entomology and Zoology*, 74 (4), 139–145. DOI: 10.7601/mez.74.139

Miyagi, I., Toma, T., Okazawa, T., Wong, S.F., Leh, M.U. & Yong, H.S. (2012) Three new phytotelma [*sic*] mosquitoes of the genus *Topomyia* (Diptera: Culicidae) from Katibas, Lanjak-Entimau, Sarawak, Malaysia. *Journal of Science and Technology in the Tropics*, 8 (2), 97–117.

Miyagi, I., Toma, T., Okazawa, T., Wong, S.F., Leh, M.U. & Yong, H.S. (2014) Description of a new species of *Topomyia (Topomyia) aliyusopi* (Diptera: Culicidae) from Kelabit highlands, Sarawak, Malaysia. *Journal of Science and Technology in the Tropics*, 10 (1), 27–38.

Miyagi, I., Toma, T. & Ramalingam, S. (1989) *Topomyia (Topomyia) hardini*, a new species from Sarawak Malaysia (Diptera: Culicidae). *Tropical Biomedicine*, 6, 91–98.

Miyagi, I., Toma, T. & Ramalingam, S. (1991) *Topomyia (Topomyia) yongi*, a new species of mosquito from Peninsular Malaysia (Diptera: Culicidae). *Mosquito Systematics* (for 1990), 22 (3), 185–191.

Miyagi, I., Toma, T. & Rivera, D.G. (1983b) *Topomyia (Topomyia) cabrerai*, a new species from the Philippines (Diptera: Culicidae). *Mosquito Systematics*, 15 (1), 1–4.

Miyagi, I., Toma, T. & Tsukamoto, M. (1983c) Description of a new species, *Tripteroides (Tripteroides) riverai* from the Philippines (Diptera: Culicidae). *Mosquito Systematics*, 15 (1), 5–17.

Mochtar, R. & Walandouw, E.K. (1934) Een variëteit van *A. gigas*, (Uit Sumatra's Westkust). *Geneeskundig Tijdschrift voor Nederlandsch-Indië*, 74, 932–935.

Mogi, M. (1977) A new species of *Aedes* (Diptera: Culicidae) from Japan. *Tropical Medicine*, 19 (2), 129–140.

Mogi, M. & Chan, K.L. (1996) Predatory habits of dipteran larvae inhabiting *Nepenthes* pitchers. *Raffles Bulletin of Zoology*, 44 (1), 233–245.

Mogi, M. & Toma, T. (1999) *Culex (Acallyntrum) miyagii* (Diptera: Culicidae): new species from Seram Island, Indonesia, with keys to the species of the subgenus. *Journal of Medical Entomology*, 36 (4), 473–480. DOI: 10.1093/jmedent/36.4.473

Mogi, M. & Yong, H.S. (1992) Aquatic arthropod communities in *Nepenthes* pitchers: the role of niche differentiation, aggregation, predation and competition in community organization. *Oecologia*, 90 (2), 172–184. DOI: 10.1007/BF00317174

Mohamed, A.H., Ali, A.M., Harbach, R.E., Reeves, R.G., Ibrahim, K.M., AhmedMohmed, M.A. & Azrag, R.S. (2017) *Aedes* mosquitoes in the Repulic of the Sudan, with dichotomous keys for the adult and larval stages. *Journal of Natural History*, 51 (9–10), 513–529. DOI: 10.1080/00222933.2017.1285069

Mohanty, A., Swain, S., Kar, S.K. & Hazra, R.K. (2009) Analysis of the phylogenetic relationship of *Anopheles* species, subgenus *Cellia* (Diptera: Culicidae) and using it to define the relationship of morphologically similar species. *Infection, Genetics and Evolution*, 9 (6), 1204–1224. DOI: 10.1016/j.meegid.2009.06.021

Morishita, K. (1946) A new variety of Formosan *Anopheles – Anopheles leucosphyrus* var. *takasagoensis* var. nov. [in Japanese]. *Journal of the Formosan Medical Association*, 45 (2), 21, 65–70.

Moses, D.A., Howard, T.M. & Harbach, R.E. (2000) A new species of the subgenus *Sabethoides* of *Sabethes* (Diptera: Culicidae) from Venezuela and Brazil. *Proceedings of the Entomological Society of Washington*, 102 (4), 991–1002.

Moss, W.W., Steffan, W.A., Evenhuis, N.L. & Manning, D.L. (1979) The genus *Toxorhynchites* (Diptera, Culicidae); analysis of *T. splendens* and allies using techniques of numerical taxonomy. *Mosquito Systematics*, 11 (4), 258–274.

Motoki, M.T., Linton, Y.-M., Conn, J.E., Ruiz-Lopez, F. & Wilkerson, R.C. (2021) Phylogenetic network of mitochondrial *COI* gene sequences distinguishes 10 taxa within the neotropical Albitarsis Group (Diptera: Culicidae), confirming the separate species status of *Anopheles albitarsis* H (Diptera: Culicidae) and revealing a novel lineage, *Anopheles albitarsis* J. *Journal of Medical Entomology*, 58 (2), 599–607. DOI: 10.1093/jme/tjaa211

Motoki, M.T., Wilkerson, R.C. & Sallum, M.A.M. (2009) The *Anopheles albitarsis* complex with the recognition of *Anopheles oryzalimnetes* Wilkerson and Motoki, n. sp. and *Anopheles janconnae* Wilkerson and Sallum, n. sp. (Diptera: Culicidae). *Memórias do Instituto Oswaldo Cruz*, 104 (6), 823–850. DOI: 10.1590/S0074-02762009000600004

Motta, M.A. & Lourenço-de-Oliveira, R. (1995) *Wyeomyia luteoventralis* Theobald, the type species of the subgenus *Dendromyia* Theobald (Diptera: Culicidae). *Memórias do Instituto Oswaldo Cruz*, 90 (3), 375–385. DOI: 10.1590/S0074-02761995000300012

Motta, M.A. & Lourenço-de-Oliveira, R. (2000) The subgenus *Dendromyia* Theobald: a review with redescriptions of four species (Diptera: Culicidae). *Memórias do Instituto Oswaldo Cruz*, 95 (5), 649–683. DOI: 10.1590/S0074-02762000000500011

Motta, M.A. & Lourenço-de-Oliveira, R. (2005) *Spilonympha*, a new subgenus of *Wyeomyia* (Diptera: Culicidae) and description of a new species *Wyeomyia aningae*. *Annals of the Entomological Society of America*, 98 (6), 838–852. DOI: 10.1603/0013-8746(2005)098[0838:SANSOW]2.0.CO;2

Motta, M.A., Lourenço-de-Oliveira, R., Monteiro, F.A. & Barros, L.R. (1998) Preliminary evaluation of the genetic relatedness of three species of the subgenus *Dendromyia* Theobald and other species of the genus *Wyeomyia* Theobald (Diptera: Culicidae). *Memórias do Instituto Oswaldo Cruz*, 93 (2), 189–194. DOI: 10.1590/S0074-02761998000200010

Motta, M.A., Lourenço-de-Oliveira, R. & Sallum, M.A.M. (2007) Phylogeny of genus *Wyeomyia* (Diptera: Culicidae) inferred from morphological and allozyme data. *Canadian Entomologist*, 139 (5), 591–627. DOI: 10.4039/n06-088

Müller, O.F. (1764) *Fauna insectorum Fridrichsdalina, sive methodica descriptio insectorum agri Fridrichsdalensis, cum characteribus genericis et specificis, nominibus trivialibus, locis natalibus, iconibus allegatis, novisque pluribus speciebus additis.* In officina libraria I.F. Gleditschii, Hafniae et Lipsiae (Copenhagen, Denmark and Leipzig, Germany).

Munawar, K., Saleh, A., Afzal, M., Qasim, M., Khan, K.A., Zafar, M.I. & Khater, E.I. (2020) Moledular characterization and phylogenetic analysis of anopheline (Anophelinae: Culicidae) mosquitoes of the Oriental and Afrotropical zoogeographic zones in Saudi Arabia. *Acta Tropica*, 207, 105494. DOI: 10.1016/j.actatropica.2020.105494

Muñoz-Gamba, A.S., Laiton-Donato, K., Perdomo-Balaguera, E., Castro, L.R., Usme-Ciro, J.A. & Parra-Henao, G. (2021) Molecular characterization of mosquitoes (Diptera: Culicidae) from the Colombian rainforest. *Revista do Instituto de Medicina Tropical de São Paulo*, 63, e24. DOI: 10.1590/S1678-9946202163024

Muspratt, J. (1950) Notes on *Aëdes* (Diptera, Culicidae) from Natal, with a description of a new species of the subgenus *Stegomyia*. *Journal of the Entomological Society of Southern Africa*, 13 (1), 73–79.

Muspratt, J. (1951) Research on South African Culicini (Diptera: Culicidae). VI. Two new sub-species and notes on bionomics and distribution. *Journal of the Entomological Society of Southern Africa*, 24 (1), 92–103.

Muspratt, J. (1953) Research on South African Culicini (Diptera Culicidae) 1. Descriptions of two new *Aëdes (Ochlerotatus)* Lynch Arribalzaga. *Journal of the Entomological Society of Southern Africa*, 16 (1), 51–58.

Muspratt, J. (1955) Research on South African Culicini (Diptera, Culicidae). III. – A check-list of the species and their distribution, with notes on taxonomy, bionomics and identification. *Journal of the Entomological Society of Southern Africa*, 18 (2), 149–207.

Muspratt, J. (1956a) The *Stegomyia* mosquitoes of South Africa and some neighbouring territories. Including chapters on the mosquito-borne virus diseases of the Ethiopian zoo-geographical region of Africa. *Memoirs of the Entomological Society of Southern Africa*, 4, 1–138.

Muspratt, J. (1956b) Research on South African Culicini (Diptera, Culicidae). IV. – Additional distribution records, taxonomic descriptions and miscellaneous notes. *Journal of the Entomological Society of Southern Africa*, 19 (2), 37–46.

Muspratt, J. (1959) Research on South African Culicini (Diptera: Culicidae). V. A new species of and notes on the sub-genus *Mucidus*, with other observations and records. *Journal of the Entomological Society of Southern Africa*, 22 (1), 64–77.

Muspratt, J. (1961) Research on South African Culicini (Diptera: Culicidae). VI. Two new sub-species and notes on bionomics and distribution. *Journal of the Entomological Society of Southern Africa*, 24 (1), 92–103.

Nagaki, S.S., de Albuquerque Motta, M. & Sallum, M.A.M. (2010) Redescription of *Anopheles (Nyssorhynchus) antunesi* Galvão & Amaral and description of a new species of the Myzorhynchella Section (Diptera: Culicidae) from Serra da Mantiqueira, Brazil. *Memórias do Instituto Oswaldo Cruz*, 105 (3), 278–285. DOI: 10.1590/S0074-02762010000300006

Nagpal, B.N. & Sharma, V.P. (1995) *Indian Anophelines*. Science Publishers, Lebanon, NH, USA.

Nakata, G. (1956) A new species of mosquito, *Aëdes (Finlaya) kobayashii* n. sp. from Kyoto, Japan [in Japanese]. *Japanese Journal of Sanitary Zoology*, 7 (2), 135.

Namgay, R., Drukpa, T., Wangdi, T., Pemo, D., Harbach, R.E. & Somboon, P. (2018) A checklist of the *Anopheles* mosquito species (Diptera: Culicidae) in Bhutan. *Acta Tropica*, 188, 206–212. DOI: 10.1016/j.actatropica.2018.09.006

Namgay, R., Pemo, D., Wangdi, T., Phanitchakun, T., Harbach, R.E. & Somboon, P. (2020) Molecular and morphological evidence for sibling species within *Anopheles (Anopheles) lindesayi* Giles (Diptera: Culicidae) in Bhutan. *Acta Tropica*, 207, 105455. DOI: 10.1016/j.actatropica.2020.105455

Nascimento-Pereira, A.C., Neves, M.S.A.S., Guimarães, A.E., de Albuquerque Motta, M. & Lourenço-de-Oliveira, R. (2021) *Wyeomyia shannoni* Lane & Cerqueira, 1942, a taxonomic puzzle (Diptera: Culicidae): synonymy, genus transfer, homonymy, and description of a new species of *Sabethes* Robineau-Desvoidy, 1827. *Zootaxa*, 5082 (3), 259–277. DOI: 10.11646/zootaxa.5082.3.4

Natarajan, R. (2019) Description of a new species of the genus *Paraedes* (Diptera: Culicidae) in Assam, India. *Zootaxa*, 4691 (3), 286–292. DOI: 10.11646/zootaxa.4691.3.9

Natarajan, R. & Rajavel, A.R. (2009) Description of a new species of *Culex (Lophoceraomyia)* from Orissa, India. *Journal of the American Mosquito Control Association*, 25 (4), 403–408. DOI: 10.2987/09-5918.1

Natarajan, R., Eapen, A. & Jambulingam, P. (2020) *Heizmannia rajagopalani* n. sp. (Diptera: Culicidae) in Kerala, India, a species previously misidentified as *Hz. metallica* (Leicester). *Zootaxa*, 4722 (5), 472–478. DOI: 10.11646/zootaxa.4722.5.5

Natarajan, R., Gopalakrishnan, S., Kumar, P. & Kumar, A. (2022) A new species of the genus *Downsiomyia* Vargas (Diptera: Culicidae) from Himachal Pradesh, India. *Zootaxa*, 5205 (6), 575–584. DOI: 10.11646/zootaxa.5205.6.5

Natarajan, R., Rajavel, A.R. & Jambulingam, P. (2016) Description of a new species of the genus *Hulecoeteomyia* (Diptera: Culicidae) from Meghalaya, India. *Zootaxa*, 4137 (3), 330–338. DOI: 10.11646/zootaxa.4137.3.2

Natarajan, R., Rajavel, A.R. & Jambulingam, P. (2017) Descriptions of three new species of *Uranotaenia (Pseudoficalbia)* (Diptera: Culicidae) from India. *Zootaxa*, 4227 (2), 251–265. DOI: 10.11646/zootaxa.4227.2.6

Natarajan, R., Rajavel, A.R. & Jambulingam, P. (2018) *Uranotaenia pseudostricklandi*, a new species in subgenus *Pseudoficalbia* (Diptera: Culicidae) from Kerala, India. *Zootaxa*, 4429 (1), 181–188. DOI: 10.11646/ ootaxa.4429.1.11

Navarro, J.-C. & Liria, J. (2000) Phylogenetic relationships among eighteen Neotropical Culicini species. *Journal of the American Mosquito Control Association*, 16 (2), 75–85.

Navarro, J.-C. & Liria, J. (2007) *Wyeomyia trujilloi* Pulido y Sutil, 1981, nuevo sinónimo de *Wyeomyia bicornis* (Root, 1928) (Culicidae: Sabethini), con redescripción de la pupa y parte de la larva. *Boletín de Malariología y Salud Ambiental*, 47 (1), 89–102.

Navarro, J.-C. & Weaver, S.C. (2004) Molecular phylogeny of the Vomerifer and Pedroi Groups in the Spissipes Section of the subgenus *Culex (Melanoconion)*. *Journal of Medical Entomology*, 41 (4), 575–581. DOI: 10.1603/0022-2585-41.4.575

Ndiaye, E.H., Boukhary, A.O.M.S., Diallo, M., Diallo, D., Labbo, R., Boussès, P., Le Goff, G. & Robert, V. (2021) Moustiques, distribution et richesse spécifique dans huit pays d'Afrique: Cap-Vert, Mauritanie, Sénégal, Gambie, Mali, Burkina Faso, Niger et Tchad. *Médecine Tropicale et Santé Internationale*, 1 (2), 1–25. DOI: 10.48327/mtsibulletin.2021.109

Neafsey, D.E., Waterhouse, R.M., Abai, M.R., Aganezov, S.S., Alekseyev, M.A., Allen, J.E., Amon, J., Arcà, B., Arensburger, P., Artemov, G., Assour, L.A., Basseri, H., Berlin, A., Birren, B.W., Blandin, S.A., Brockman, A.I., Burkot, T.R., Burt, A., Chan, C.S., Chauve, C., Chiu, J.C., Christensen, M., Costantini, C., Davidson, V.L.M., Deligianni, E., Dottorini, T., Dritsou, V., Gabriel, S.B., Guelbeogo, W.M., Hall, A.B., Han, M.V., Hlaing, T., Hughes, D.S.T., Jenkins, A.M., Jiang, X., Jungreis, I., Kakani, E.G., Kamali, M., Kemppainen, P., Kennedy, R.C., Kirmitzoglou, I.K., Koekemoer, L.L., Laban, N., Langridge, N., Lawniczak, M.K.N., Lirakis, M. Lobo, N.F., Lowy, E., MacCallum, R.M., Mao, C., Maslen, G., Mbogo, C., McCarthy, J., Michel, K., Mitchell, S.N., Moore, W., Murphy, K.A., Naumenko, A.N., Nolan, T., Novoa, E.M., O'Loughlin, S., Oringanje, C., Oshaghi, M.A., Pakpour, N., Papathanos, P.A., Peery, A.N., Povelones, M., Prakash, A., Price, D.P., Rajaraman, A., Reimer, L.J., Rinker, D.C., Rokas, A., Russell, T.L., Sagnon, N., Sharakhova, M.V., Shea, T., Simão, F.A., Simard, F., Slotman, M.A., Somboon, P., Stegniy, V., Struchiner, C.J., Thomas, G.W.C., Tojo, M., Topalis, P., Tubio, J.M.C., Unger, M.F., Vontas, J., Walton, C., Wilding, C.S., Willis, J.H., Wu, Y.-C., Yan, G., Zdobnov, E.M., Zhou, X., Catteruccia, F., Christophides, G.K., Collins, F.H., Cornman, R.S., Crisanti, A., Donnelly, M.J., Emrich, S.J., Fontaine, M.C., Gelbart, W., Hahn, M.W., Hansen, I.A., Howell, P.I., Kafatos, F.C., Kellis, M., Lawson, D., Louis, C., Luckhart, S., Muskavitch, M.A.T., Ribeiro, J.M., Riehle, M.A., Sharakhov, I.V., Tu, Z., Zwiebel, L.J.& Besansky, N.J. (2015) Highly evolvable malaria vectors: the genomes of 16 *Anopheles* mosquitoes. *Science*, 347 (6217), 1258522. DOI: 10.1126/science.1258522

Neave, S.A. (ed.) (1940) *Nomenclator zoologicus*. Volume 3. M–P. Zoological Society of London, London, UK.

Neiva, A. (1906) Uma nova especie de anophelina brazileira. *Brazil-Medico*, 20 (28), 288–289.

Neiva, A. (1908) Contribuição ao estudo dos dipteros brazileiros. Uma nova especie de *Sabethes*. *Brazil-Medico*, 22 (36), 351–352.

Neiva, A. & Pinto, C. (1922a) Contribuição para o conhecimento das anophelinas do estado de Motto Grosso com a descripção de uma nova especie. *Brazil-Medico*, 36 (2), 321–322.

Neiva, A. & Pinto, C. (1922b) Considerações sobre o genero "*Cellia*" Theobald, com a deseripção de uma nova especie. *Brazil-Medico*, 36 (2), 355–357.

Netherlands, E.C., Cook, C.A., Du Preez, L.H., Vanhove, M.P.M., Brendonck, L. & Smit, N.J. (2020) An overview of the Dactylosomatidae (Apicomplexa: Adeleorina: Dactylosomatidae), with the description of *Dactylosoma kermiti* n. sp. parasitising *Ptychadena anchietae* and *Sclerophrys gutturalis* from South Africa. *International Journal for Parasitology: Parasites and Wildlife*, 11, 246–260. DOI: 10.1016/j.ijppaw.2019.12.006

Neves, M.S.A.S., Motta, M.A., Maciel-de-Freitas, R., Xavier, A.S.da, Lourenço-de-Oliveira, R. & Silva-do-Nascimento, T.F. (2024) Illustrated identification key to females of the genus *Sabethes* Robineau-Desvoidy recorded from Brazil (Diptera: Culicidae), in dichotomous and interactive formats, including an updated list of species and new records for the states. *Zootaxa*, 5406 (2), 253–287. DOI: 10.11646/zootaxa.5406.2.3

Neveu-Lemaire, M. (1902a) Sur la classification des Culicides. *Comptes Rendus Hebdomadaires des Séances et Mémoires de la Société de Biologie*, 54, 1329–1332.

Neveu-Lemaire, M. (1902b) Description de quelques moustiques de la Guyane. *Archives de Parasitologie*, 6 (1), 5–25.

Neveu-Lemaire, M. (1902c) Classification de la famille des Culicidae. *Mémoires de la Société Zoologique de France*, 15, 195–227.

Neveu-Lemaire, M. (1906) Étude des culicides africains. *Archives de Parasitologie*, 10 (2), 238–288.

Nevill, G. (1878) *Hand list of Mollusca in the Indian Museum, Calcutta. Part I. Gastropoda. Pulmonata and Prosobranchia-Neurobranchia*. Office of the Superintendent of Government Printing, Calcutta, India.

Newstead, R. & Carter, H.F. (1910) Descriptions of a new genus and three new species of anopheline mosquitos. *Annals of Tropical Medicine and Parasitology*, 4 (3), 377–383, 2 pls. DOI: 10.1080/00034983.1910.11685726

Newstead, R. & Carter, H.F. (1911) On some new species of African mosquitos (Culicidae). *Annals of Tropical Medicine and Parasitology*, 5 (2), 233–244, 1 pl. DOI: 10.1080/00034983.1911.11686353

Newstead, R., Dutton, J.E. & Todd, J.L. (1907) Insects and other Arthropoda collected in the Congo Free State. *Annals of Tropical Medicine and Parasitology*, 1 (1–5), 3–113. DOI: 10.1080/00034983.1907.11719252

Nguyen, D.M., Tran, D.H., Harbach, R.E., Elphick, J. & Linton, Y.-M. (2000) A new species of the Hyrcanus Group of *Anopheles*, subgenus *Anopheles*, a secondary vector of malaria in coastal areas of southern Vietnam. *Journal of the American Mosquito Control Association*, 16 (3), 189–198.

Nguyen, D.M., Tran, D.H. & Nguyen, T.H. (1993) *Anopheles (Anopheles) vietnamensis*: a new species of the *Anopheles hyrcanus* group from Vietnam (Diptera: Culicidae). *Mosquito Systematics*, 25 (3), 215–221.

Nicolescu, G., Linton, Y.-M., Vladimirescu, A., Howard, T.M. & Harbach, R.E. (2004) Mosquitoes of the *Anopheles maculipennis* group (Diptera: Culicidae) in Romania, with the discovery and formal recognition of a new species based on molecular and morphological evidence. *Bulletin of Entomological Research*, 95 (6), 525–535. DOI: 10.1079/BER2004330

Nielsen, L.T. (2009) *Ochlerotatus dahlae*, a new mosquito species from Utah. *Journal of the American Mosquito Control Association*, 25 (2), 123–125. DOI: 10.2987/08-5751.1

Njabo, K.Y., Cornel, A.J., Sehgal, R.N.M., Loiseau, C., Buermann, W., Harrigan, R.J., Pollinger, J., Valkiūnas, G. & Smith, T.B. (2009) *Coquillettidia* (Culicidae, Diptera) mosquitoes are natural vectors of avian malaria in Africa. *Malaria Journal*, 8, 193. DOI: 10.1186/1475-2875-8-193

Noè, G. (1899) Contribuzione allo studio dei Culicidi. *Bollettino della Società Entomologica Italiana*, 31, 235–262.

Norris, L.C. & Norris, D.E. (2015) Phylogeny of anopheline (Diptera: Culicidae) species in southern Africa, based on nuclear and mitochondrial genes. *Journal of Vector Ecology*, 40 (1), 16–27. DOI: 10.1111/jvec.12128

Nugroho, S.S. & Mujiyono, M. (2021) Updated mosquito taxonomic information and photographic identification key to mosquito genera (Diptera: Culicidae) in Indonesia [in Indonesian]. *Jurnal Entomologi Indonesia*, 18 (1) 55–73.

Nugroho, S.S., Ayuningrum, F.D., Setiyaningsih, R. & Astuti, U.N.W. (2020) A revised checklist of mosquitoes genus *Coquillettidia* Dyar, 905 (Diptera: Culicidae) from Indonesia with key to species. *Biodiversitas*, 21 (12), 5772–5777. DOI: 10.13057/biodiv/d211238

Nuttall, G.H.F. & Shipley, A.E. (1901) Studies in relation to malaria. II. The structure and biology of *Anopheles (Anopheles maculipennis)*. The egg and larva. *Journal of Hygiene*, 1 (1), 45–77. DOI: 10.1017/S002217240000005X

Nye, E.R. & Lien, J.C. (1960) Laboratory transmission of Semliki Forest virus by *Aedes togoi* Theo. *Transactions of the Royal Society of Tropical Medicine and Hygiene*, 54 (3), 263–264. DOI: 10.1016/0035-9203(60)90072-3

Olivier, M. (1791) *Encyclopedique méthodique. Histoire naturelle. Insectes.* Volume 6. Panckoucke, Paris, France.

Oliveira Coutinho, J. & Forattini, O.P. (1962) Notas sôbre Culicidae (Diptera). 4ª — Contribuição ao estudo do subgênero *Microculex* com a descriçao duma nova espécie. *Studia Entomologica*, 5 (1–4), 209–234.

Oliveira, R.M.P., Foster, P.G., Bergo, E.S., Nagaki, S.S., Sanabani, S.S., Marinotti, O., Marinotti, P.N. & Sallum, M.A.M. (2016) Mitochondrial genomes of *Anopheles (Kerteszia)* (Diptera: Culicidae) from the Atlantic Forest, Brazil. *Journal of Medical Entomology*, 53 (4), 790–797. DOI: 10.1093/jme/tjw001

Omondi, D., Masiga, D.K., Ajamma, Y.U., Fielding, B.C., Njoroge, L. & Villinger, J. (2015) Unraveling [*sic*] host-vector-arbovirus interactions by two-gene high resolution melting mosquito bloodmeal [*sic*] analysis in a Kenyan wildlife-livestock interface. *PLoS ONE*, 10 (7), e0134375. DOI: 10.1371/journal.pone.0134375

Oo, T.T., Kaiser, A. & Becker, N. (2006) Illustrated keys to the anopheline mosquitoes of Myanmar. *Journal of Vector Ecology*, 31 (1), 9–16. DOI: 10.3376/1081-1710(2006)31[9:ikttam]2.0.co;2

Oo, T.T., Storch, V. & Becker, N. (2004) Review of the anopheline mosquitoes of Myanmar. *Journal of Vector Ecology*, 29 (1), 21–40.

Ortega-Morales, A.I., Hernández-Triana, L.M., Garza-Hernández, J.A., Ramírez-Huicochea, C.M., Martínez-Gaona, A.J., Quijano-Barraza, J.M., González-Acosta, C. & Correa-Morales, F. (2024) The mosquitoes of Morelos, Mexico: DNA barcodes, distribution, ecology and the resurrection of the name *Culiseta dugesi* Dyar and Knab (Diptera: Culicidae). *Diversity*, 16, 261. DOI: 10.3390/d16050261

Ortega-Morales, A.I., Hernández-Triana, L.M. & Siller-Rodríguez, Q.K. (2023) The mosquitoes of Querétaro, Mexico: distribution, ecology, and discovery of *Shannoniana huasteca* n. sp. (Diptera: Culicidae). *Diversity*, 15, 697. DOI: 10.3390/d15060697

Ortega-Morales, A.I., Zavortink, T.J., Garza-Hernández, J.A., Siller-Rodríguez, Q.K. & Fernández-Salas, I. (2019) The mosquitoes (Diptera: Culicidae) of Nuevo León, Mexico, with descriptions of two new species. *PLoS ONE*, 14 (8), e0217694. DOI: 10.1371/journal.pone.0217694

Oshaghi, M.A., Shemshad, Kh., Yaghobi-Ershadi, M.R., Pedram, M., Vatandoost, H., Abaie, M.R., Akbarzadeh, K. & Mohtarami, F. (2007) Genetic structure of the malaria vector *Anopheles superpictus* in Iran using mitochondrial cytochrome oxidase (COI and COII) and morphologic markers: a new species complex? *Acta Tropica*, 101 (3), 241–248. DOI: 10.1016/j.actatropica.2007.02.006

Oshaghi, M.A., Yaghobi-Ershadi, M.R., Shemshad, Kh., Pedram, M. & Amani, H. (2008) The *Anopheles superpictus* complex: introduction of a new malaria vector complex in Iran. *Bulletin de la Société de Pathologie exotique*, 101 (5), 429–434.

Osorno-Mesa, E. (1944) Two new species of *Haemagogus* from Colombia, *H. andinus* and *H. boshelli* (Diptera: Culicidae). *Proceedings of the Entomological Society of Washington*, 46 (7), 165–175.

Osorno-Mesa, E. (1947) Una nueva especie de *Anopheles* de Bogota, Colombia. *Caldasia*, 4 (20), 431–446.

Osten Sacken, [C.]R. (1868) Description of a new species of Culicidae. *Transactions of the American Entomological Society*, 2, 47–48. DOI: 10.2307/25076193

Ôuchi, Y. (1939) A new species of *Megarhinus* mosquito from Amami-Oshiina, the southern Japan. *Journal of the Shanghai Science Institute, Section III*, 4 (21), 223–225, 1 pl.

Ovazza, M., Hamon, J. & Neri, P. (1956) Contribution a l'étude des diptères vulnérants de l'empire d'Éthiopie. I. Culicidae. *Bulletin de la Société de Pathologie exotique*, 49 (1), 151–182.

Owen, W.B. (1945) A new anopheline from the Solomon Islands with notes on its biology. *Journal of Parasitology*, 31 (4), 236–240. DOI: 10.2307/3273000

Paine, R.W. & Edwards, F.W. (1929) Mosquitos from the Solomon Islands. *Bulletin of Entomological Research*, 20 (3), 303–316, 2 pls. DOI: 10.1017/S0007485300021271

Pajot, F.-X. & Fauran, P. (1975) *Wyeomyia (Wyeomyia) pseudorobusta* n. sp. *(Diptera, Culicidae)*, nouveau moustique découvert en Guyane française. *Cahiers ORSTOM Série Entomologie médicale et Parasitologie*, 13 (3), 125–129.

Pajot, F.-X. & Geoffroy, B. (1971) *Aedes (Neomelanoconion) bergerardi* sp. n. une nouvelle espèce de *Culicidae* de la République Centrafricaine. *Cahiers ORSTOM Série Entomologie médicale et Parasitologie*, 9 (3), 269–272.

Pallas, P.S. (1771) *Reise durch versehiedene Provinzen des Russischen Reichs.* Vol. 1. Kayserliche Academie der Wissenschaften, St Petersburg, Russia.

Panday, R.S. (1975) Mosquito identification studies in a savanna forest in Surinam. *Mosquito News*, 35 (2), 141–146.

Pao, B. & Knight, K.L. (1970) Morphology of the fourth stage larval mouthparts of *Aedes (Aedimorphus) vexans* (Diptera: Culicidae). *Journal of the Georgia Entomological Society*, 5 (3), 115–137.

Parker, D.M., Zavortink, T.J., Billo, T.J., Valdez, U. & Edwards, J.S. (2012) Mosquitoes and other arthropod macro fauna associated with tank bromeliads in a Peruvian cloud forest. *Journal of the American Mosquito Control Association*, 29 (1), 45–46. DOI: 10.2987/11-6189.1

Paronyan, L., Babayan, L., Manucharyan, A., Manukyan, D., Vardanyan, H., Andrasyan, G.M., Schaffner, F. & Robert, V. (2020) The mosquitoes of Armenia: review of knowledge and results of a filed survey with first report of *Aedes albopictus*. *Parasite*, 27, 42. DOI: 10.1051/parasite/2020039

Paterno, U. & Marcondes, C.B. (2004) Mosquitoes antropofílicos de atividade matutina em Mata Atlântica, Florianópolis, SC. *Revista* de *Saúde Pública*, 38 (1), 133–135. DOI: 10.1590/S0034-89102004000100019

Paterson, G.C. & Shannon, R.C. (1927) Mosquitos de Embarcación (Salta) con notas sobre la zona biológica del Chaco (Chaco Life Zone). *Boletín del Instituto de Clínica Quirúrgica*, 21–25, 1–10.

Patton, W.S. (1905) The culicid fauna of the Aden Hinterland, their haunts and habits. *Journal of the Bombay Natural History Society*, 16 (4), 623–637, 5 pls.

Pazos, J.H. (1908) Descripcion de nuevas especies de mosquitos de Cuba. *Anales de la Academia de Ciencias de la Habana*, 45, 417–432, 4 pls.

Pecor, J.E., Jones, J., Turell, M.J., Fernandez, R., Carbajal, F., O'Guinn, M., Sardalis, M., Watts, D., Zyzak, M., Calampn, C. & Klein, T.A. (2000) Annotated checklist of the mosquito species encountered during arboviral studies in Iquitos, Peru (Diptera: Culicidae). *Journal of the American Mosquito Control Association*, 16 (3), 210–218.

Pecor, J.E., Mallampalli, V.L., Harbach, R.E. & Peyton, E.L. (1992) Catalog and illustrated review of the subgenus *Melanoconion* of *Culex* (Diptera: Culicidae). *Contributions of the American Entomological Institute*, 27 (2), v, 1–228.

Penn, G.H. (1947) The larva of *Aedes (Leptosomatomyia) aurimargo* (Diptera, Culicidae). *Proceedings of the Entomological Society of Washington*, 49 (4), 103–105.

Penn, G.H. (1949) The pupae of the mosquitoes of New Guinea. *Pacific Science*, 3 (1), 3–85.

Perez Vigueras, I. (1956) *Los ixodidos y culicidos de Cuba. Su historia natural y medica*. Imprenta Universidad de la Habana, La Habana, Cuba.

Peryassú, A.G. (1908) *Os culicideos do Brazil.* Instituto de Manguinhos, Rio de Janeiro, Brazil.

Peryassú, A. (1922a) Uma nova especie do Culicineo Brasileiro. *Folha medica*, 3 (15), 117–118.

Peryassú, A. (1922b) Duas novas especies de mosquitos do Brasil. *Folha medica*, 3 (23), 179.

Pessoa, S.B. & Galvão, A.A. (1936) Notes sobre algumas especies de mosquitos que se criam em bromelias (Diptera, Culicidae). *Revista de Biologia e Hygiene*, 6 (2), 79–90.

Peters, W. (1955a) The mosquitoes of Liberia (Diptera: Culicidae). *Proceedings of the Royal Entomological Society of London, Series B: Taxonomy*, 24 (5–6), 81–90. DOI: 10.1111/j.1365-3113.1955.tb01480.x

Peters, W. (1955b) New Anophelini from the southern highlands of Tanganyika and notes on some other members of the genus *Anopheles* Meigen (Diptera: Culicidae). *Proceedings of the Royal Entomological Society of London, Series B: Taxonomy*, 24 (5–6), 95–103. DOI: 10.1111/j.1365-3113.1955.tb01482.x

Peters, W. (1955c) The mosquitoes of Liberia (Diptera: Culicidae). *Proceedings of the Royal Entomological Society of London, Series B: Taxonomy*, 24 (7–8), 117–128. DOI: 10.1111/j.1365-3113.1955.tb01487.x

Peters, W. (1958) Mosquitoes of New Guinea (Diptera: Culicidae). Part 1. *Proceedings of the Royal Entomological Society of London, Series B: Taxonomy*, 27 (5–6), 89–92. DOI: 10.1111/j.1365-3113.1958.tb00418.x

Peters, W. (1959) Mosquitoes of New Guinea (Diptera: Culicidae). Part II.—Notes on *Tripteroides* Giles, 1904. *Proceedings of the Royal Entomological Society of London, Series B: Taxonomy*, 28 (9–10), 135–154. DOI: 10.1111/j.1365-3113.1959.tb00150.x

Peters, W. (1963a) Mosquitoes of New Guinea (Diptera: Culicidae). IV.–Two new species, genera *Aedes* Meigen and *Armigeres* Theobald. *Proceedings of the Royal Entomological Society of London, Series B: Taxonomy*, 32 (1–2), 1–11. DOI: 10.1111/j.1365-3113.1963.tb01203.x

Peters, W. (1963b) Mosquitoes of New Guinea (Diptera: Culicidae). Part V.—New species and previously undescribed pupae of the genus *Tripteroides* Giles. *Proceedings of the Royal Entomological Society of London, Series B: Taxonomy*, 32 (5–6), 63–71. DOI: 10.1111/j.1365-3113.1963.tb01662.x

Peters, W. (1963c) Mosquitoes of New Guinea (Diptera: Culicidae). VI.—The genus *Tripteroides* Giles: a new species of the subgenus *Polylepidomyia* Theobald, notes on the taxonomy and distribution and a revised key to described pupae. *Proceedings of the Royal Entomological Society of London, Series B: Taxonomy*, 32 (7–8), 89–100. DOI: 10.1111/j.1365-3113.1963.tb01667.x

Peters, W. (1963d) Mosquitoes of New Guinea (Diptera: Culicidae). VII.—A new species complex of *Uranotaenia* Lynch Arribálzaga, 1891. *Proceedings of the Royal Entomological Society of London, Series B: Taxonomy*, 32 (9–10), 135–144. DOI: 10.1111/j.1365-3113.1963.tb01595.x

Peters, W. (1963e) Mosquitoes of New Guinea (Diptera: Culicidae). Part VIII.—New species of the genus *Uranotaenia* Lynch Arribálzaga, 1891. *Proceedings of the Royal Entomological Society of London, Series B: Taxonomy*, 32 (11–12), 203–209. DOI: 10.1111/j.1365-3113.1963.tb01593.x

Peters, W. (1964) Mosquitoes of New Guinea (Diptera: Culicidae). Part IX.—Further new species of *Uranotaenia* Lynch Arribálzaga and notes on the genus. *Proceedings of the Royal Entomological Society of London, Series B: Taxonomy*, 33 (1–2), 19–29. DOI: 10.1111/j.1365-3113.1964.tb01221.x

Peus, F. (1935) *Theobaldia* (Subg. *Culicella) ochroptera* sp. n., eine bisher unbekannte Stechmücke. *Markische Tierwelt: Zeitschrift für die faunistische Erforschung der Kurmark*, 1, 113–121.

Peus, F. (1970) Bemerkenswerte Mücken am Tegeler Fliess. *Berliner Naturschutzblätter*, Special issue (May), 18–26.

Peyton, E.L. (1972) A subgeneric classification of the genus *Uranotaenia* Lynch Arribalzaga, with a historical review and notes on other categories. *Mosquito Systematics*, 4 (2), 16–40.

Peyton, E.L. (1974) *Uranotaenia srilankensis*, a new species of the subgenus *Pseudoficalbia* from Sri Lanka (Diptera: Culicidae). *Mosquito Systematics*, 6 (3), 222–227.

Peyton, E.L. (1977) Medical entomology studies – X. A revision of the subgenus *Pseudoficalbia* of the genus *Uranotaenia* in Southeast Asia (Diptera: Culicidae). *Contributions of the American Entomological Institute*, 14 (3), iv, 1–273.

Peyton, E.L. (1990) A new classification for the Leucosphyrus Group of *Anopheles (Cellia)*. *Mosquito Systematics* (for 1989), 21 (3), 197–205.

Peyton, E.L. & Harbach, R.E. (1991) A new species of *Culex (Melanoconion)* encountered during arbovirus surveillance in Mexico (Diptera: Culicidae). *Mosquito Systematics*, 23 (3), 199–204.

Peyton, E.L. & Harrison, B.A. (1979) *Anopheles (Cellia) dirus*, a new species of the Leucosphyrus Group from Thailand (Diptera: Culicidae). *Mosquito Systematics*, 11 (1), 40–52.

Peyton, E.L. & Klein, J.M. (1970) Five new species of *Uranotaenia* from Southeast Asia (Diptera: Culicidae). *Proceedings of the Entomological Society of Washington*, 72 (2), 243–251.

Peyton, E.L. & Ramalingam, S. (1988) *Anopheles (Cellia) nemophilous*, a new species of the Leucosphyrus Group from Peninsular Malaysia and Thailand (Diptera: Culicidae). *Mosquito Systematics*, 20 (2), 272–299.

Peyton, E.L. & Rattanarithikul, R. (1970) Five additional new species of *Uranotaenia* from Southeast Asia (Diptera: Culicidae). *Proceedings of the Entomological Society of Washington*, 72 (3), 403–413.

Peyton, E.L., Roberts, D.R., Pinheiro, F.P., Vargas, R. & Balderama, F. (1983) Mosquito collections from a remote unstudied area of southeastern Bolivia. *Mosquito Systematics*, 15 (2), 61–89.

Peyton, E.L., Wilkerson, R.C. & Harbach, R.E. (1992) Comparative analysis of the subgenera *Kerteszia* and *Nyssorhynchus* of *Anopheles* (Diptera: Culicidae). *Mosquito Systematics*, 24 (1), 51–69.

Phan, V.T., Manh, N.D., Hinh, T.D. & Vien, N.T. (1991) *Anopheles (Anopheles) cucphuongensis*: a new species from Vietnam (Diptera: Culicidae). *Mosquito Systematics* (for 1990), 22 (3), 145–148.

Philip, C.B. (1929) Preliminary report of further tests with yellow fever transmission by mosquitoes other than *Aedes aegypti*. *American Journal of Tropical Medicine and Hygiene*, 9 (4), 267–269.

Philip, C.B. (1931) Two new species of *Uranotaenia* (Culicidae) from Nigeria, with notes on the genus in the Ethiopian Region. *Bulletin of Entomological Research*, 22 (2), 183–193. DOI: 10.1017/S0007485300035124

Philippi, R.A. (1865) Aufzählung der chilenische Dipteren. *Verhandlungen der kaiserlich-königlichen zoologisch-botanischen Gesellschaft in Wien*, 15, 595–782, pls 23–29. DOI: 10.5962/bhl.title.9295

Pillai, J.S. (1966) *Culiseta novaezealandiae*, a new species of the subgenus *Climacura* Felt (Diptera: Culicidae: Culisetini), with notes on its ecology and development. *Transactions of the Royal Society of New Zealand, Zoology*, 8 (11), 125–133, 1 pl.

Pinto, C. (1932) Alguns mosquitos do Brasil e do oriente do Bolivia (Diptera. Culicidae). *Revista medico-cirurgica do Brasil*, 40 (10), 285–309.

Piton, L. (1936) Addition a la faune entomologique des cinérites du Lac Chambon (Puy-de-Dome). *Revue Scientifique du Bourbonmais*, 1936, 16–24.

Podėnas, S. (1999) New species of fossil mosquitoes (Diptera, Culicidae) from the Baltic amber. *Acta Zoologica Lituanica*, 9 (3), 111–114. DOI: 10.1080/13921657.1999.10512303

Poinar, G., Jr (2005) *Culex malariager*, n. sp. (Diptera: Culicidae) from Dominican amber: the first fossil mosquito vector of *Plasmodium*. *Proceedings of the Entomological Society of Washington*, 107 (3), 548–553.

Poinar, G., Zavortink, T.J. & Brown, A. (2019) *Priscoculex burmanicus* n. gen. et sp. (Diptera: Culicidae: Anophelinae) from mid-Cretaceous Myanmar amber. *Historical Biology*, 32 (9), 1–6. DOI: 10.1080/08912963.2019.1570185

Poinar, G.O., Jr, Zavortink, T.J., Pike, T. & Johnston, P.A. (2000) *Paleoculicis minutus* (Diptera: Culicidae) n. gen., n. sp., from Cretaceous Canadian amber, with a summary of described fossil mosquitoes. *Acta Geologica Hispanica*, 35 (1–2), 119–128.

Poole-Smith, B.K., Hemme, R.R., Delorey, M., Felix, G., Gonzalez, A.L., Amador, M., Hunsperger, E.A. & Barrera, R. (2015) Comparison of vector competence of *Aedes mediovittatus* and *Aedes aegypti* for dengue virus: implications for dengue control in the Caribbean. *PLoS Neglected Tropical Diseases*, 9 (2), e0003462. DOI: 10.1371/journal.pntd.0003462

Porter, C.H. (2009) *Wyeomyia (Hystatomyia) baltae*, a new species of Sabethini (Diptera: Culicidae) from Peru. *Proceedings of the Entomological Society of Washington*, 111 (4), 807–825.

Porter, C.H. (2014) *Wyeomyia (Nunezia) paucartamboensis*, a new species of Sabethini (Diptera: Culicidae) from the Peruvian Andes with a diagnosis of the subgenus *Nunezia*. *Proceedings of the Entomological Society of Washington*, 116 (3), 311–338.

Porter, C.H. & Wolff E., M.I. (2004) A new species of *Wyeomyia (Hystatomyia)* (Diptera: Culicidae) from Colombia and a redescription of *Wy. (Hystatomyia) intonca* Dyar & Knab. *Zootaxa*, 477, 1–31. DOI: 10.11646/zotaxa.477.1.1

Portschinsky, J. (1884) Diptera europaea et asiatica nova aut minus cognita (Cum notis biologicis). *Trudy Russkago entomologicheskago obshchestva. Horae Societatis entomologicae rossicae, variis semonibus in Russia usitatis editae*, 18, 122–134.

Prado, A. (1931a) Contribuições ao conhecimento dos culicideos de São Paulo I. Notas sobre *Mansonia albifera* Prado e sobre o macho de *Mansonia albicosta* (Chagas). *Memorias do Instituto Butantan*, 6, 193–198.

Prado, A. (1931b) Contribuições ao conhecimento dos culicideos de São Paulo IV. Uma nova especie de *Uranotaenia* (Diptera, Culicidae). *Memorias do Instituto Butantan*, 6, 209–213.

Prashad, B. (1918) XVI. The description and life-history of a new species of *Anopheles* that breeds in holes in trees. *Records of the Indian Museum*, 15 (3), 123–127, 1 pl.

Puri, I.M. (1929a) Description of the male, female, egg and larva of *Anopheles annandalei* var. *interruptus* nov. var., with corrections for the previous descriptions of the type species. *Indian Journal of Medical Research*, 17 (2), 385–395, 1 pl.

Puri, I.M. (1929b) A new tree-hole breeding *Anopheles* from South India—*Anopheles sintoni* sp. nov.—and a revised description of the larva of *A. culiciformis* Cogill. *Indian Journal of Medical Research*, 17 (2), 397–404, 1 pl.

Puri, I.M. (1930) A note on two species of Indian anopheline mosquitoes—*A. insulaeflorum* Swellengrebel and *A. aitkenii* James, with its variety *bengalensis* nov. var. *Indian Journal of Medical Research*, 17 (3), 953–956.

Qutubuddin, M. (1947) A new species of *Uranotaenia*, Culicidae-Diptera from Hyderabad (Deccan). *Indian Journal of Entomology* (for 1946), 8 (1), 117–118.

Qutubuddin, M. (1951) *Uranotaenia mattinglyi* sp. n., a new species of mosquito (Diptera, Culicidae) from Hyderabad (Deccan), India. *Proceedings of the Royal Entomological Society of London, Series B: Taxonomy*, 20 (9–10), 107–109. DOI: 10.1111/j.1365-3113.1951.tb01030.x

Qutubuddin, M. (1959) A new species of *Aëdimorphus* (Diptera: Culicidae) from Sudan Republic. *Annals and Magazine of Natural History*, 2 (13), 21–23. DOI: 10.1080/00222935908651009

Qutubuddin, M. (1972) A new species of *Finlaya* (Diptera: Culicidae) from Papua. *Pakistan Journal of Health*, 22 (1–4), 119–120.

Rafinesque, C.S. (1820) Fishes of the River Ohio. *Western Review and Miscellaneous Magazine*, 3 (4), 244–252.

Rahman, S.J., Choudhury, D.S. & Kalra, N.L. (1969) Description of *Culex (Lophoceraomyia) raghavanii* sp. n. (Diptera: Culicidae) from Nilgiri Hills, South India. *Bulletin of the Indian Society for Malaria and Other Communicable Diseases* (for 1968), 5 (1–2), 28–34.

Rahola, N., Makanga, B., Yangari, P., Jiolle, D., Fontenille, D., Renaud, F., Ollomo, B., Ayala, D., Prugnolle, F. & Paupy, C. (2014) Description of *Anopheles gabonensis*, a new species potentially involved in rodent malaria transmission in Gabon, Central Africa. *Infection, Genetics and Evolution*, 28, 628–634. DOI: 10.1016/j.meegid.2014.05.012

Ramalingam, S. (1972) A new species of *Armigeres* from Sabah, Borneo (Diptera: Culicidae). *Proceedings of the Entomological Society of Washington*, 74 (4), 459–467.

Ramalingam, S. (1975) A new species of *Topomyia* from peninsular Malaysia (Diptera: Culicidae). *Mosquito Systematics*, 7 (3), 185–192.

Ramalingam, S. (1987) On the restriction of *Armigeres durhami* Edwards and the description of *Armigeres kesseli* n. sp. (Diptera: Culicidae). *Tropical Biomedicine*, 4 (1), 55–65.

Ramalingam, S. (1988) Description of a new species of *Topomyia* from peninsular Malaysia (Diptera: Culicidae). *Mosquito Systematics* (for 1987), 19 (3), 245–250.

Ramalingam, S. & Banu, Q. (1987) Studies on the genus *Topomyia*: 2. Description of a new species from Sabah, Malaysia (Diptera: Culicidae). *Tropical Biomedicine*, 4 (2), 119–124.

Ramalingam, S. & Belkin, J.N. (1965) Mosquito studies (Diptera, Culicidae). III. Two new species of *Aedes* from Tonga and Samoa. *Contributions of the American Entomological Institute*, 1 (4), 1–10.

Ramalingam, S. & Pillai, A.G. (1972) A new species of *Malaya* from West Malaysia (Diptera: Culicidae). *Proceedings of the Entomological Society of Washington*, 74 (4), 451–459.

Ramalingam, S. & Ramakrishna [*sic*], K. (1988) Studies on the genus *Topomyia*: 1. A new species from Sabah, Malaysia. *Mosquito Systematics*, 20 (1), 33–40.

Ramalingam, S. & Ramakrishnan, K. (1971) Redescription of *Aedes (Alanstonea) brevitibia* (Edwards) from Brunei, Borneo (Diptera: Culicidae). *Proceedings of the Entomological Society of Washington*, 73 (2), 231–238.

Ramírez, C.C.L. & Dessen, E.M.B. (2000) Chromosome differentiated populations of *Anopheles cruzii*: evidence for a third sibling species. *Genetica*, 108 (1), 73–80. DOI: 10.1023/A:1004020904877

Rao, R. (1984) *The anophelines of India*. 2nd edn. Malaria Research Centre, Indian Council of Medical Research, New Delhi, India.

Rattanarithikul, R. & Green, C.A. (1987) Formal recognition of the species of the *Anopheles maculatus* group (Diptera: Culicidae) occurring in Thailand, including the descriptions of two new species and a preliminary key to females. *Mosquito Systematics* (for 1986), 18 (3,4), 246–278.

Rattanarithikul, R. & Harbach, R.E. (1991) *Anopheles maculatus* (Diptera: Culicidae) from the type locality of Hong Kong and two new species of the Maculatus Complex from the Philippines. *Mosquito Systematics* (for 1990), 22 (3), 160–183.

Rattanarithikul, R. & Harrison, B.A. (1988) *Aedes (Finlaya) reinerti*, a new species from northern Thailand related to *Aedes (Finlaya) formosensis* Yamada (Diptera: Culicidae). *Mosquito Systematics*, 20 (1), 77–96.

Rattanarithikul, R. & Panthusiri, P. (1994) Illustrated keys to the medically important mosquitos [*sic*] of Thailand. *Southeast Asian Journal of Tropical Medicine and Public Health*, 25 (Suppl. 1), 1–66.

Rattanarithikul, R., Harbach, R.E., Harrison, B.A., Panthusiri, P. & Coleman, R.E. (**2007**) Illustrated keys to the mosquitoes of Thailand. V. Genera *Orthopodomyia, Kimia, Malaya, Topomyia, Tripteroides*, and *Toxorhynchites*. *Southeast Asian Journal of Tropical Medicine and Public Health*, 38 (Suppl. 2), 1–65.

Rattanarithikul, R., Harbach, R.E., Harrison, B.A., Panthusiri, P., Coleman, R.E. & Richardson, J.H. (**2010**) Illustrated keys to the mosquitoes of Thailand VI. Tribe Aedini. *Southeast Asian Journal of Tropical Medicine and Public Health*, 41 (Suppl. 1), 1–225.

Rattanarithikul, R., Harrison, B.A., Panthusiri, P., Peyton, E.L. & Coleman, R.E. (**2006a**) Illustrated keys to the mosquitoes of Thailand. III. Genera *Aedeomyia, Ficalbia, Mimomyia, Hodgesia, Coquillettidia, Mansonia*, and *Uranotaenia*. *Southeast Asian Journal of Tropical Medicine and Public Health*, 37 (Suppl. 1), 1–85.

Rattanarithikul, R., Harrison, B.A., Harbach, R.E., Panthusiri, P. & Coleman, R.E. (**2006b**) Illustrated keys to the mosquitoes of Thailand. IV. *Anopheles*. *Southeast Asian Journal of Tropical Medicine and Public Health*, 37 (Suppl. 2), 1–128.

Rattanarithikul, R., Harrison, B.A., Panthusiri, P. & Coleman, R.E. (**2005a**) Illustrated keys to the mosquitoes of Thailand. I. Background; geographic distribution; lists of genera, subgenera, and species; and a key to the genera. *Southeast Asian Journal of Tropical Medicine and Public Health*, 36 (Suppl. 1), 1–80.

Rattanarithikul, R., Harbach, R.E., Harrison, B.A., Panthusiri, P., Jones, J.W. & Coleman, R.E. (**2005b**) Illustrated keys to the mosquitoes of Thailand. II. Genera *Culex* and *Lutzia*. *Southeast Asian Journal of Tropical Medicine and Public Health*, 36 (Suppl. 2), 1–97.

Ravaonjanahary, C. & Bruhnes [*sic*], J. (1977) Un nouvel *Aedes* du genre *Skusea* découvert à Madagascar: *Aedes (Skusea) moucheti* sp. n. *Cahiers ORSTOM Série Entomologie médicale et Parasitologie*, 15 (3), 213–215.

Reeves, L.E., Holderman, C.J., Blosser, E.M., Gillett-Kaufman, J.L., Kawahara, A.Y., Kaufman, P.E. & Burkett-Cadena, N.D. (2018) Identification of *Uranotaenia sapphirina* as a specialist of annelids broadens known mosquito host use patterns. *Communications Biology*, 1 (1), 192. DOI: 10.1038/s42003-018-0096-5

Reid, J.A. (1949) A preliminary account of the forms of *Anopheles leucosphyrus* Dönitz (Diptera: Culicidae). *Proceedings of the Royal Entomological Society of London Series B Taxonomy*, 18 (3–4), 42–53. DOI: 10.1111/j.1365-3113.1949.tb01420.x

Reid, J.A. (1950) The *Anopheles umbrosus* group (Diptera: Culicidae). Part 1: systematics, with descriptions of two new species. *Transactions of the Royal Entomological Society of London*, 101 (9), 281–318.

Reid, J.A. (1953) The *Anopheles hyrcanus* group in south-east Asia (Diptera: Culicidae). *Bulletin of Entomological Research*, 44 (1), 5–76. DOI: 10.1017/S0007485300022938

Reid, J.A. (1962) The *Anopheles barbirostris* group (Diptera, Culicidae). *Bulletin of Entomological Research*, 53 (1), 1–57. DOI: 10.1017/S0007485300047945

Reid, J.A. (1963) Notes on anopheline mosquitoes from Malaya, with descriptions of three new species. *Annals of Tropical Medicine and Parasitology*, 57 (1), 97–116. DOI: 10.1080/00034983.1963.11686166

Reid, J.A. (1965) A revision of the *Anopheles aitkenii* group in Malaya and Borneo. *Annals of Tropical Medicine and Parasitology*, 59 (1), 106–125. DOI: 10.1080/00034983.1965.11686289

Reid, J.A. (1968) *Anopheline mosquitoes of Malaya and Borneo*. Studies from the Institute for Medical Research Malaya, No. 31. Government of Malaysia, Kuala Lumpur, Malaysia.

Reid, J.A. & Knight, K.L. (1961) Classification within the subgenus *Anopheles* (Diptera, Culicidae). *Annals of Tropical Medicine and Parasitology*, 55 (4), 474–488. DOI: 10.1080/00034983.1961.11686077

Reidenbach, K.R., Cook, S., Bertone, M.A., Harbach, R.E., Wiegmann, B.M. & Besansky, N.J. (2009) Phylogenetic analysis and temporal diversification of mosquitoes (Diptera: Culicidae) based on nuclear genes and morphology. *BMC Evolutionary Biology*, 9, 298. DOI: 10.1186/1471-2148-9-298

Reinert, J.F. (1970a) Contributions to the mosquito fauna of Southeast Asia. V. Genus *Aedes*, subgenus *Diceromyia* Theobald in Southeast Asia. *Contributions of the American Entomological Institute*, 5 (4), 1–43.

Reinert, J.F. (1970b) The zoogeography of *Aedes (Diceromyia)* Theobald (Diptera: Culicidae). *Journal of the Entomological Society of South Africa*, 33 (1), 129–141.

Reinert, J.F. (1972a) Contributions to the mosquito fauna of Southeast Asia. – XV. Genus *Aedes* Meigen, subgenus *Ayurakitia* Thurman. *Contributions of the American Entomological Institute*, 9 (2), 1–42.

Reinert, J.F. (1972b) *Aedes gouldi*, a new species of the subgenus *Aedimorphus* Theobald from West Pakistan (Diptera: Culicidae). *Proceedings of the Entomological Society of Washington*, 74 (4), 357–362.

Reinert, J.F. (1973a) Notes on the genus *Heizmannia* and the inclusion in subgenus *Mattinglyia* of two species previously placed in *Aedes* subgenus uncertain (Diptera: Culicidae). *Mosquito Systematics*, 5 (1), 14–23.

Reinert, J.F. (1973b) *Aedes wainwrighti* Baisas, a synonym of *Aedes (Stegomyia) meronephada* (Dyar and Shannon), with notes on the subgenus *Stegomyia* Theobald (Diptera: Culicidae). *Mosquito Systematics*, 5 (1), 27–30.

Reinert, J.F. (1973c) Contributions to the mosquito fauna of Southeast Asia. – XVI. Genus *Aedes* Meigen, subgenus *Aedimorphus* Theobald in Southeast Asia. *Contributions of the American Entomological Institute*, 9 (5), 1–218.

Reinert, J.F. (1973d) *Aedes consonensis*, a new species of the subgenus *Neomacleaya* from South Vietnam (Diptera: Culicidae). *Mosquito Systematics*, 5 (3), 252–262.

Reinert, J.F. (1973e) Contributions to the mosquito fauna of Southeast Asia. – XVIII. A reconsideration of *Diceromyia* Theobald with the inclusion of *Aedes nummatus* Edwards and *Aedes pseudonummatus* new species (Diptera: Culicidae). *Contributions of the American Entomological Institute*, 10 (1), 22–40.

Reinert, J.F. (1973f) Contributions to the mosquito fauna of Southeast Asia. – XIX. *Bothaella*, a new subgenus of *Aedes* Meigen. *Contributions of the American Entomological Institute*, 10 (3), 1–51.

Reinert, J.F. (1974) Medical entomology studies – I. A new interpretation of the subgenus *Verrallina* of the genus *Aedes* (Diptera: Culicidae). *Contributions of the American Entomological Institute*, 11 (1), 1–249.

Reinert, J.F. (1976a) A new man-biting species of *Aedes (Paraedes)* from Southeast Asia (Diptera: Culicidae). *Mosquito Systematics*, 8 (3), 319–331.

Reinert, J.F. (1976b) Medical entomology studies – V. The subgenus *Rhinoskusea* of the genus *Aedes* (Diptera: Culicidae). *Contributions of the American Entomological Institute*, 13 (2), 1–60.

Reinert, J.F. (1976c) Medical entomology studies – IV. The subgenera *Indusius* and *Edwardsaedes* of the genus *Aedes* (Diptera: Culicidae). *Contributions of the American Entomological Institute*, 13 (1), 1–45.

Reinert, J.F. (1977) Description of a new species of *Aedes (Verrallina)* from Sri Lanka (Diptera: Culicidae). *Mosquito Systematics*, 9 (3), 366–371.

Reinert, J.F. (1979) A description of *Isoaedes*, a new subgenus of *Aedes* Meigen, and its type-species, *Ae. (Isa.) cavaticus* new species (Diptera: Culicidae). *Mosquito Systematics*, 11 (2), 144–162.

Reinert, J.F. (1981a) A literature review of pupal seta 13-CT in the family Culicidae (Diptera) and a report of its occurrence in *Aedes aurotaeniatus* Edwards. *Mosquito Systematics* (for 1980), 12 (4), 441–444.

Reinert, J.F. (1981b) Medical entomology studies – XV. A revision of the subgenus *Paraedes* of the genus *Aedes* (Diptera: Culicidae). *Contributions of the American Entomological Institute*, 18 (4), 1–91.

Reinert, J.F. (1982) *Belkinius*, a new subgenus of *Aedes* and a description of *Ae. (Blk.) aurotaeniatus* Edwards (Diptera: Culicidae). *Mosquito Systematics*, 14 (2), 101–121.

Reinert, J.F. (1984) Medical entomology studies – XVI. A review of the species of the subgenus *Verrallina*, genus *Aedes*, from Sri Lanka and a revised description of the subgenus (Diptera: Culicidae). *Mosquito Systematics*, 16 (1), 1–130.

Reinert, J.F. (1985) A description of *Scutomyia*, a subgenus resurrected for the Albolineatus Group of the genus *Aedes* (Diptera: Culicidae). *Mosquito Systematics*, 17 (2), 121–131.

Reinert, J.F. (1987) *Albuginosus*, a new subgenus of *Aedes* Meigen (Diptera: Culicidae) described from the Afrotropical region. *Mosquito Systematics* (for 1986), 18 (3,4), 307–326.

Reinert, J.F. (1990a) *Udaya subsimilis*, a new generic combination, a redescription of the holotype, and notes on the genus *Udaya* (Diptera: Culicidae). *Mosquito Systematics* (for 1989), 21 (3), 206–215.

Reinert, J.F. (1990b) Medical entomology studies – XVII. Biosystematics of *Kenknightia*, a new subgenus of the mosquito genus *Aedes* from the Oriental Region (Diptera: Culicidae). *Contributions of the American Entomological Institute*, 26 (2), 1–119.

Reinert, J.F. (1993) Redescription of *Molpemyia*, and its revalidation as a subgenus of *Aedes* (Diptera: Culicidae). *Mosquito Systematics*, 25 (1), 41–63.

Reinert, J.F. (1999a) The subgenus *Rusticoidus* of genus *Aedes* (Diptera: Culicidae) in Europe and Asia. *European Mosquito Bulletin*, 4, 1–7.

Reinert, J.F. (1999b) Descriptions of *Zavortinkius*, a new subgenus of *Aedes*, and the eleven included species from the Afrotropical Region (Diptera: Culicidae). *Contributions of the American Entomological Institute*, 31 (2), 1–105.

Reinert, J.F. (1999c) Restoration of *Verrallina* to generic rank in tribe Aedini (Diptera: Culicidae) and descriptions of the genus and three included subgenera. *Contributions of the American Entomological Institute*, 31 (3), 1–83.

Reinert, J.F. (1999d) The dorsal apotome of pupae and fourth-instar larvae of Culicidae (Diptera), a structure of phylogenetic significance. *Journal of the American Mosquito Control Association*, 15 (1), 77–83.

Reinert, J.F. (2000a) Description of *Fredwardsius*, a new subgenus of *Aedes* (Diptera: Culicidae). *European Mosquito Bulletin*, 6, 1–7.

Reinert, J.F. (2000b) Restoration of *Ayurakitia* to generic rank in tribe Aedini and a revised definition of the genus. *Journal of the American Mosquito Control Association*, 16 (2), 57–65.

Reinert, J.F. (2000c) New classification for the composite genus *Aedes* (Diptera: Culicidae: Aedini), elevation of subgenus *Ochlerotatus* to generic rank, reclassification of the other subgenera, and notes on certain subgenera and species. *Journal of the American Mosquito Control Association*, 16 (3), 175–188.

Reinert, J.F. (2000d) Comparative anatomy of the female genitalia of genera and subgenera in tribe Aedini (Diptera: Culicidae). Part I. Introduction, preparation techniques, and anatomical terminology. Part II. Genus *Psorophora* Robineau-Desvoidy. Part III. Genus *Udaya* Thurman. Part IV. Genus *Zeugnomyia*. *Contributions of the American Entomological Institute*, 32 (2), 1–52.

Reinert, J.F. (2000e) Comparative anatomy of the female genitalia of genera and subgenera in tribe Aedini (Diptera: Culicidae). Part V. Genus *Aedes* Meigen. *Contributions of the American Entomological Institute*, 32 (3), 1–102.

Reinert, J.F. (2001a) Comparative anatomy of the female genitalia of genera and subgenera in tribe Aedini (Diptera: Culicidae). Part VII. Genus *Opifex* Hutton. *Contributions of the American Entomological Institute*, 32 (4), 9–15.

Reinert, J.F. (2001b) Comparative anatomy of the female genitalia of genera and subgenera in tribe Aedini (Diptera: Culicidae). Part VIII. Genus *Verrallina* Theobald. *Contributions of the American Entomological Institute*, 32 (4), 17–39.

Reinert, J.F. (2002a) A comparative analysis of the ventral brush and its attachment to segment X of fourth-stage larvae in tribe Aedini (Diptera: Culicidae). *Journal of the American Mosquito Control Association*, 18 (1), 16–251.

Reinert, J.F. (2002b) Comparative anatomy of the female genitalia of genera and subgenera in tribe Aedini (Diptera: Culicidae). Part X. Genus *Heizmannia* Ludlow. *Contributions of the American Entomological Institute*, 32 (5), 1–15.

Reinert, J.F. (2002c) Comparative anatomy of the female genitalia of genera and subgenera in tribe Aedini (Diptera: Culicidae). Part XI. Genus *Haemagogus* Williston. *Contributions of the American Entomological Institute*, 32 (5), 17–30.

Reinert, J.F. (2002d) Comparative anatomy of the female genitalia of genera and subgenera in tribe Aedini (Diptera: Culicidae). Part XII. Genus *Armigeres* Theobald. *Contributions of the American Entomological Institute*, 32 (5), 31–46.

Reinert, J.R. (2002e) Comparative anatomy of the female genitalia of genera and subgenera in tribe Aedini (Diptera: Culicidae). Part XIII. Genus *Ochlerotatus* Lynch Arribalzaga. *Contributions of the American Entomological Institute*, 33 (1), 1–111.

Reinert, J.F. (2003) Description of *Bruceharrisonius*, a new subgenus of *Ochlerotatus*, and a redescription of its type species *Oc. (Brh.) greenii*. *Journal of the American Mosquito Control Association*, 19 (4), 309–322.

Reinert, J.F. (2006a) *Tewarius*, a new genus of Aedini (Diptera: Culicidae). *Proceedings of the Entomological Society of Washington*, 108 (3), 639–645.

Reinert, J.F. (2006b) Description of subgenus *Pardomyia* Theobald and comparison with the nominotypical subgenus of *Mucidus* Theobald (Diptera: Culicidae: Aedini). *Journal of the American Mosquito Control Association*, 22 (4), 586–600. DOI: 10.2987/8756-971X(2006)22[586:DOSPTA]2.0.CO;2

Reinert, J.F. (2007) Development and taxonomic value of dorsal tertiary fringe scales on the wings of adult mosquitoes (Diptera: Culicidae). *Proceedings of the Entomological Society of Washington*, 109 (4), 948–956.

Reinert, J.F. (2008a) Comparative anatomy of the female genitalia of generic-level taxa in tribe Aedini (Diptera: Culicidae). Part XV. Genus *Georgecraigius* Reinert, Harbach and Kitching. *Contributions of the American Entomological Institute*, 35 (2), 1–10.

Reinert, J.F. (2008b) Comparative anatomy of the female genitalia of generic-level taxa in tribe Aedini (Diptera: Culicidae). Part XVI. Genus *Phagomyia* Theobald. *Contributions of the American Entomological Institute*, 35 (2), 11–19.

Reinert, J.F. (2008c) Comparative anatomy of the female genitalia of generic-level taxa in tribe Aedini (Diptera: Culicidae). Part XVII. Genus *Dahliana* Reinert, Harbach and Kitching. *Contributions of the American Entomological Institute*, 35 (2), 21–29.

Reinert, J.F. (2008d) Comparative anatomy of the female genitalia of generic-level taxa in tribe Aedini (Diptera: Culicidae). Part XVIII. Genus *Hulecoeteomyia* Theobald. *Contributions of the American Entomological Institute*, 35 (2), 31–39.

Reinert, J.F. (2008e) Comparative anatomy of the female genitalia of generic-level taxa in tribe Aedini (Diptera: Culicidae). Part XIX. Genus *Danielsia* Theobald. *Contributions of the American Entomological Institute*, 35 (3), 1–9.

Reinert, J.F. (2008f) Comparative anatomy of the female genitalia of generic-level taxa in tribe Aedini (Diptera: Culicidae). Part XX. Genus *Rampamyia* Reinert, Harbach and Kitching. *Contributions of the American Entomological Institute*, 35 (3), 11–18.

Reinert, J.F. (2008g) Comparative anatomy of the female genitalia of generic-level taxa in tribe Aedini (Diptera: Culicidae). Part XXI. Genus *Patmarksia* Reinert, Harbach and Kitching. *Contributions of the American Entomological Institute*, 35 (3), 19–27.

Reinert, J.F. (2008h) Comparative anatomy of the female genitalia of generic-level taxa in tribe Aedini (Diptera: Culicidae). Part XXII. Genus *Downsiomyia* Vargas. *Contributions of the American Entomological Institute*, 35 (3), 29–38.

Reinert, J.F. (2008i) Comparative anatomy of the female genitalia of generic-level taxa in tribe Aedini (Diptera: Culicidae). Part XXIII. Genus *Tanakaius* Reinert, Harbach and Kitching. *Contributions of the American Entomological Institute*, 35 (4), 1–9.

Reinert, J.F. (2008j) Comparative anatomy of the female genitalia of generic-level taxa in tribe Aedini (Diptera: Culicidae). Part XXIV. Genus *Vansomerenis* Reinert, Harbach and Kitching. *Contributions of the American Entomological Institute*, 35 (4), 11–19.

Reinert, J.F. (2008k) Comparative anatomy of the female genitalia of generic-level taxa in tribe Aedini (Diptera: Culicidae). Part XXV. Genus *Dobrotworskyius* Reinert, Harbach and Kitching. *Contributions of the American Entomological Institute*, 35 (4), 21–30.

Reinert, J.F. (2008l) Comparative anatomy of the female genitalia of generic-level taxa in tribe Aedini (Diptera: Culicidae). Part XXVI. Genus *Collessius* Reinert, Harbach and Kitching. *Contributions of the American Entomological Institute*, 35 (4), 31–41.

Reinert, J.F. (2009a) Descriptions of *Danielsia* Theobald, a genus recently resurrected from synonymy (Diptera: Culicidae: Aedini), and its type species *Dn. albotaeniata* Leicester. *Zootaxa*, 1976, 25–38. DOI: 10.11646/zootaxa.1976.1.2

Reinert, J.F. (2009b) Comparative anatomy of the female genitalia of generic-level taxa in tribe Aedini (Diptera: Culicidae). Part XXVII. Genus *Hopkinsius* Reinert, Harbach and Kitching. *Contributions of the American Entomological Institute*, 36 (1), 1–11.

Reinert, J.F. (2009c) Comparative anatomy of the female genitalia of generic-level taxa in tribe Aedini (Diptera: Culicidae). Part XXVIII. Genus *Luius* Reinert, Harbach and Kitching. *Contributions of the American Entomological Institute*, 36 (1), 13–21.

Reinert, J.F. (2009d) Comparative anatomy of the female genitalia of generic-level taxa in tribe Aedini (Diptera: Culicidae). Part XXIX. Genus *Jihlienius* Reinert, Harbach and Kitching. *Contributions of the American Entomological Institute*, 36 (1), 23–32.

Reinert, J.F. (2009e) Comparative anatomy of the female genitalia of generic-level taxa in tribe Aedini (Diptera: Culicidae). Part XXX. Genus *Gilesius* Reinert, Harbach and Kitching. *Contributions of the American Entomological Institute*, 36 (1), 33–42.

Reinert, J.F. (2010a) Comparative anatomy of the female genitalia of generic-level taxa in tribe Aedini (Diptera: Culicidae). Part XXXI. Genus *Sallumia* Reinert, Harbach and Kitching. *Contributions of the American Entomological Institute*, 36 (2), 1–10.

Reinert, J.F. (2010b) Comparative anatomy of the female genitalia of generic-level taxa in tribe Aedini (Diptera: Culicidae). Part XXXII. Genus *Jarnellius* Reinert, Harbach and Kitching. *Contributions of the American Entomological Institute*, 36 (2), 11–20.

Reinert, J.F. (2010c) Comparative anatomy of the female genitalia of generic-level taxa in t0ribe Aedini (Diptera: Culicidae). Part XXXIII. Genus *Lewnielsenius* Reinert, Harbach and Kitching. *Contributions of the American Entomological Institute*, 36 (2), 21–30.

Reinert, J.F. (2010d) Comparative anatomy of the female genitalia of generic-level taxa in tribe Aedini (Diptera: Culicidae). Part XXXIV. Genus *Catageiomyia* Theobald. *Contributions of the American Entomological Institute*, 36 (2), 31–42.

Reinert, J.F. (2010e) Comparative anatomy of the female genitalia of generic-level taxa in tribe Aedini (Diptera: Culicidae). Part XXXV. Genus *Elpeytonius* Reinert, Harbach and Kitching. *Contributions of the American Entomological Institute*, 36 (3), 1–11.

Reinert, J.F. (2010f) Comparative anatomy of the female genitalia of generic-level taxa in tribe Aedini (Diptera: Culicidae). Part XXXVI. Genus *Polyleptiomyia* Theobald. *Contributions of the American Entomological Institute*, 36 (3), 13–22.

Reinert, J.F. (2010g) Comparative anatomy of the female genitalia of generic-level taxa in tribe Aedini (Diptera: Culicidae). Part XXXVII. Genus *Bifidistylus* Reinert, Harbach and Kitching. *Contributions of the American Entomological Institute*, 36 (3), 23–34.

Reinert, J.F. (2010h) Comparative anatomy of the female genitalia of generic-level taxa in tribe Aedini (Diptera: Culicidae). Part XXXVIII. Genus *Petermattinglyius* Reinert, Harbach and Kitching. *Contributions of the American Entomological Institute*, 36 (3), 35–48.

Reinert, J.F. & Harbach, R.E. (2005) Descriptions of *Finlaya* Theobald, a genus in tribe Aedini (Diptera: Culicidae), and its type species *Fl. kochi* Dönitz. *Journal of the American Mosquito Control Association*, 21 (2), 154–170. DOI: 10.2987/8756-971X(2005)21[154:DOFTAG]2.0.CO;2

Reinert, J.F. & Harbach, R.E. (2006) Descriptions of genus *Downsiomyia* Vargas (Diptera: Culicidae: Aedini) and its type species *Do. nivea* (Ludlow). *Zootaxa*, 1196, 33–61. DOI: 10.11646/zootaxa.1196.1.2

Reinert, J.F., Harbach, R.E. & Kitching, I.J. (2004) Phylogeny and classification of Aedini (Diptera: Culicidae) based on morphological characters of all life stages. *Zoological Journal of the Linnean Society*, 142 (3), 289–368. DOI: 10.1111/j.1096-3642.2004.00144.x

Reinert, J.F., Harbach, R.E. & Kitching, I.J. (2006) Phylogeny and classification of *Finlaya* and allied taxa (Diptera: Culicidae: Aedini) based on morphological data from all life stages. *Zoological Journal of the Linnean Society*, 148 (1), 1–101. DOI: 10.1111/j.1096-3642.2006.00254.x

Reinert, J.F., Harbach, R.E. & Kitching, I.J. (2008) Phylogeny and classification of *Ochlerotatus* and allied taxa (Diptera: Culicidae: Aedini) based on morphological data from all life stages. *Zoological Journal of the Linnean Society*, 153 (1), 29–114. DOI: 10.1111/j.1096-3642.2008.00382.x

Reinert, J.F., Harbach, R.E. & Kitching, I.J. (2009) Phylogeny and classification of tribe Aedini (Diptera: Culicidae). *Zoological Journal of the Linnean Society*, 157 (4), 700–794. DOI: 10.1111/j.1096-3642.2009.00570.x

Reinert, J.F., Kaiser, P.E. & Seawright, J.A. (1997) Analysis of the *Anopheles (Anopheles) quadrimaculatus* complex of sibling species (Diptera: Culicidae) using morphological, cytological, molecular, genetic, biochemical, and ecological techniques in an integrated approach. *Journal of the American Mosquito Control Association*, 13 (Suppl.), 1–102.

Remington, C.L. (1945) Feeding habits of *Uranotaenia lowii* Theobald (Diptera: Culicidae). *Entomological News*, 56 (2), 32–37, (3), 64–68.

Reuben, R. (1967) *Aedes (Diceromyia) ramachandrai*, n. sp. (Diptera: Culicidae) from southern India. *Journal of Medical Entomology*, 4 (2), 234–236. DOI: 10.1093/jmedent/4.2.234

Reuben, R., Tewari, S.C., Hiriyan, J. & Akiyama, J. (1994) Illustrated keys to species of *Culex (Culex)* associated with Japanese encephalitis in Southeast Asia (Diptera: Culicidae). *Mosquito Systematics*, 26 (2), 75–96.

Ribeiro, H. (1969) Research on the mosquitoes of Angola. IV – Description of *Anopheles (Cellia) azevedoi* sp. nov. *(Diptera, Culicidae)*. *Anais da Escola Nacional de Saúde Pública e de Medicina Tropical*, 3 (1/4), 113–123.

Ribeiro, H. (1991) Research on the mosquito subfamily *Toxorhynchitinae (Diptera: Culicidae)*. I—The Afrotropical group *brevipalpis* (Adults) (1). *Arquivos do Museu Bocage*, 2 (3), 31–62.

Ribeiro, H. (1992a) Research on the mosquito subfamily Toxorhynchitinae (Diptera: Culicidae). II. Description of the Afrotropical subgenus *Afrorhynchus* subg. nov. *Mosquito Systematics* (for 1991), 23 (3), 195–198.

Ribeiro, H. (1992b) On the terrestrial and riverine fauna of the Democratic Republic of São Tomé and Príncipe. VII – Description of *Toxorhynchites capelai* sp. nov. *(Diptera, Culicidae)*. *Garcia de Orta, Série de Zoologia*, 19 (1–2), 135–128.

Ribeiro, H. (1992c) Research on the mosquito subfamily Toxorhynchitinae (Diptera: Culicidae). III – Description of two new species from Angola. *Boletim da Sociedade Portuguesa de Entomologia*, 2 (Supl. 3), 187–196.

Ribeiro, H. (1997) A new species of *Toxorhynchites* (Diptera: Culicidae) from Macau (China). *Journal of the American Mosquito Control Association*, 13 (3), 213–217.

Ribeiro, H. (2004) Les *Toxorhynchites* Theobald de Madagascar (Diptera: Culicidae). *Annales de la Société Entomologique de France*, 40 (3–4), 243–257. DOI: 10.1080/00379271.2004.10697423

Ribeiro, H. (2005a) Three new species and one new subspecies of *Toxorhynchites* (Diptera: Culicidae) of the Afrotropical Region. *Journal of Medical Entomology*, 42 (3), 224–232. DOI: 10.1093/jmedent/42.3.224

Ribeiro, H. (2005b) Systematics and identification of Afrotropical Toxorhynchitinae (Diptera: Culicidae). *Journal of Medical Entomology*, 42 (6), 953–958. DOI: 10.1093/jmedent/42.6.953

Ribeiro, H. & Ramos, H.C. (1974) Research on the mosquitoes of Angola. VII – Redescription of the larva of *Aedes durbanensis durbanensis* (Theo., 1903) and description of *Aedes durbanensis angolae* subsp. nov. *Mosquito Systematics*, 6 (2), 93–98.

Ribeiro, H. & da Cunha Ramos, H. (1980) Research on the mosquitoes of Angola (Diptera: Culicidae). X. – The genus *Culex* L., 1758. Check-list with new records, keys to females and larvae, distribution, and taxonomic and bioecological notes. *Junta de Investigações Ultlramar Estudos, Ensaios et Documentos*, 134, 1–175.

Ribeiro, H., da Cunha Ramos, H., Capela, R.A. & Pires, C.A. (1998) Os mosquitos (Diptera: Culicidae) da Ilha de São Tomé. *Garcia de Orta, Série de Zoologia*, 22 (1–2), 1–20.

Ribeiro, H., da Cunha Ramos, H., Pires, C.A. & Antunes Capela, R. (1988) An annotated checklist of the mosquitoes of continental Portugal *(Diptera Culicidae)*. Actas, III Congreso Ibérico Entomologia, 1988 December 1988, Universidad de Granada, Granada, Spain, pp. 233–253.

Ribeiro, P.S., Galvão, C., Talaga, S., Carinci, R., Pavan, M.G., Lourenço-de-Oliveira, R. & Motta, M.A. (2020) Redescription and placement of *Wyeomyia rorotai* Senevet, Chabelard & Abonnenc (Diptera: Culicidae) in the subgenus *Decamyia* based on morphological and molecular analyses. *Zootaxa*, 4830 (2), 291–309. DOI: 10.11646/zootaxa.4830.2.4

Ribeiro, P.S., Motta, M.A., Seiblitz, G.C., Pereira, G.R., Galvão, C., Pecor, D.B. & Lourenço-de-Oliveira, R. (2022) Redescription of *Wyeomyia (Decamyia) felicia* (Dyar & Núñez Tovar, 1927) and description of a new species of the subgenus from Trinidad (Diptera: Culicidae). *Zootaxa*, 5175 (1), 101–125. DOI: 10.11646/zootaxa.5175.1.5

Ribeiro, P.S., Pavan, M.G., da Silva, M.B., Galvão, C., Lourenço-de-Oliveira, R. & Motta, M.A. (2021) A new species of *Wyeomyia* (Diptera: Culicidae) from *Heliconia* flower bracts in northern South America. *Zootaxa*, 4999 (6), 534–552. DOI: 10.11646/zootaxa.4999.6.2

Rickenbach, A. (1967) Description d'un nouvel *Eretmapodites (Diptera, Culicidae)* capturé au Cameroun. *Bulletin de la Société de Pathologie exotique* (for 1966), 59 (3), 374–376.

Rickenbach, A. & Eouzan, J.-P. (1970) Description do quatre *Eretmapodites* nouveaux du groupe *plioleucus* Edwards 1941, capturés au Cameroun (Diptera, Culicidae). *Cahiers ORSTOM Série Entomologie médicale et Parasitologie*, 8 (2), 131–135.

Rickenbach, A. & Ferrara, L. (1965) Description de deux nouveaux *Aedes* du Cameroun appartenant au sous-genre *Aedimorphus* (Diptera, Culicidae). *Bulletin de la Société de Pathologie exotique*, 58 (1), 24–29.

Rickenbach, A. & Hamon, J. (1966) Description d'une nouvelle espèce d'*Orthopodomyia (Diptera, Culicidae)* capturée au Cameroun: *O. nkolbissonensis* n. sp. *Bulletin de la Société de Pathologie exotique* (for 1965), 58 (6), 1112–1117.

Rickenbach, A. & Lombrici, G. (1974) Un nouvel *Eretmapodites* Theobald capturé au Cameroun: *E. eouzani* n. sp. [Dipt. Culicidae]. *Bulletin de la Société Entomologique de France*, 79 (7–8), 218–222. DOI: 10.3406/bsef.1974.21410

Rickenbach, A. & Lombrici, G. (1975) Les *Eretmapodites* du groupe *leucopus* Graham, 1909. Position systématique des deux sous-espèces actuellement connues et description de deux espèces nouvelles [Dipt. Culicidae]. *Bulletin de la Société Entomologique de France*, 80 (1–2), 39–42. DOI: 10.3406/bsef.1975.21434

Rickenbach, A., Ferrara, L. & Eouzan, J.-P. (1968) Deux *Eretmapodites* nouveaux du Cameroun, appartenant au groupe *oedipodius* Graham, 1909 [Dipt. Culicidae]. *Bulletin de la Société Entomologique de France* (for 1967), 72 (10), 297–300. DOI: 10.3406/bsef.1968.21028

Rioux, J.A. & Quézel, P. (1957) Présence de *Culex (Lasiosiphon) adairi* Kirkpatrick 1924 au Tibesti. *Bulletin de la Société d'Histoire Naturelle de l'Afrique du Nord*, 48 (7–8), 461–470.

Rioux, J.A., Guilvard, E. & Pasteur, N. (1998) Description d'*Aedes (Ochlerotatus) coluzzii* n. sp. (Diptera, Culicidae), espèce jumelle A du complexe *detritus*. *Parassitologia*, 40 (3), 353–360.

Rivera-García, K.D., Mendez-Andrade, A. & Ibáñez-Bernal, S. (2023) *Trichoprosopon mixtli* sp. nov., a new sabethine species (Diptera: Culicidae) from a Mexican cloud forest, with an assessment of the genus and keys for the identification of known species. *Zootaxa*, 5254 (1), 94–116. DOI: 10.11646/zootaxa.5254.1.5

Rivera-García, K.D., Rísquez-Pérez, A. & Ibáñez-Bernal, S. (2019) Description of the pupa and additional characters of the fourth-instar larva, female, and male genitalia of *Uranotaenia (Uranotaenia) coatzacoalcos* Dyar & Knab, with keys for the identification of Mexican species of *Uranotaenia* (Diptera: Culicidae). *Zootaxa*, 4608 (2), 247–260. DOI: 10.11646/zootaxa.4608.2.3

Rivola, E. & Holstein, M.H. (1957) Note sur une variété d'*Anopheles nili* Theo. *Bulletin de la Société de Pathologie exotique*, 50 (3), 382–387.

Robert, V., Lhuillier, M., Meunier, D., Sarthou, J.L., Monteny, N., Digoutte, J.P., Cornet, M., Germain, M. & Cordellier, M. (1993) Virus amaril, dengue 2 et autres arbovirus isolés de moustiques, au Burkina Faso, de 1983 a 1986: Considérations entomologiques et épidémiologiques. *Bulletin de la Société de Pathologie exotique*, 86 (2), 90–100.

Robert, V., Ndiaye, E.H., Rahola, N., Le Goff, G., Boussès, P., Diallo, D., Le Goff, V., Mariamé, L. & Diallo, M. (2022) Clés dichotomiques illustrées d'identification des femelles et des larves de moustiques (Diptera: Culicidae) du Burkina Faso, Cap-Vert, Gambie, Mali, Mauritanie, Niger, Sénégal et Tchad. Institut de Recherche pour le Développement (IRD), Montpellier, France.

Roberts, D.R., Peyton, E.L., Pinheiro, F.P., Balderama, F. & Vargas, R. (1984) Associations of arbovirus vectors with gallery forests and domestic environments in southeastern Bolivia. *Bulletin of the Pan American Health Organization*, 18 (4), 337–350.

Robineau-Desvoidy, J.-B. (1827) Essai sur la tribu des culicides. *Mémoires de la Société d'Histoire Naturelle de Paris*, 3, 390–413, 1 pl.

Robinson, G.G. (1950) A new species of *Aedes (Finlaya)* from Northern Rhodesia. *Journal of the Entomological Society of Southern Africa*, 13 (1), 80–82.

Rocha, G.P., Lourenço-de-Oliveira, R. & Motta, M. de A. (2012) *Wyeomyia exallos*, a new species of sylvatic mosquito (Diptera: Culicidae) from Brazil. *Memórias do Instituto Oswaldo Cruz*, 107 (7), 928–934. DOI: 10.1590/S0074-02762012000700015

Roche, J.C., Cordellier, R., Hervy, J.P., Digoutte, J.P. & Monteny, N. (1983) Isolement de 96 souches de virus dengue 2 à partir de moustiques capturés en Cote-d'Ivoire et Haute-Volta. *Annales de l'Institut Pasteur/Virologie*, 134 (2), 233–244. DOI: 10.1016/S0769-2617(83)80062-8

Rodenwaldt, E. (1925) Entomologische notities. III. *Geneeskundig Tijdschrift voor Nederlandsch-Indië*, 65, 173–201, 3 pls.

Rodenwaldt, E. (1926) Entomologische notities. IV. *Geneeskundig Tijdschrift voor Nederlandsch-Indië*, 66 (6), 787–799.

Rodhain, F. & Boutonnier, A. (1983a) Description d'un nouvel *Aedes* du sous-genre *Ochlerotatus* (Diptera: Culicidae) de Madagascar: *Aedes ambreensis* nova species et considérations générales sur les femelles de sous-genre *Ochlerotatus* dans la région afro-tropicale. *Bulletin de la Société de Pathologie exotique*, 76 (5 Pt 2), 825–833.

Rodhain, F. & Boutonnier, A. (1983b) *Aedes*, subgenus *Diceromyia* (Diptera: Culicidae). *Mosquito Systematics*, 15 (4), 330–336.

Rona, L.D.P., Carvalho-Pinto, C.J., Gentile, C., Grisard, E.C. & Peixoto, A.A. (2009) Assessing the molecular divergence between *Anopheles (Kerteszia) cruzii* populations from Brazil using the *timeless* gene: further evidence of a species complex. *Malaria Journal*, 8, 60. DOI: 10.1186/1475-2875-8-60

Rona, L.D.P., Caarvalho-Pinto, C.J. & Peixtoto, A.A. (2010a) Molecular evidence for the occurrence of a new sibling species within the *Anopheles (Kerteszia) cruzii* complex in south-east Brazil. *Malaria Journal*, 9, 33. DOI: 10.1186/1475-2875-9-33

Rona, L.D.P., Carvalho-Pinto, C.J., Mazzoni, C.J. & Peixoto, A.A. (2010b) Estimation of divergence time between two sibling species of the *Anopheles (Kerteszia) cruzii* complex using a multilocus approach. *BMC Evolutionary Biology*, 10, 91. DOI: 10.1186/1471-2148-10-91

Rona, L.D.P., Carvalho-Pinto, D.J. & Peixoto, A.A. (2013) Evidence for the occurrence of two sympatric sibling species within the *Anopheles (Kerteszia) cruzii* complex in southeast Brazil and the detection of asymmetric introgression between them using a multilocus analysis. *BMC Evolutionary Biology*, 13, 207. DOI: 10.1186/1471-2148-13-207

Rondani, C. (1848) Esme di varie specie d'insetti ditteri brasiliani. *Studi entomologici (Turino)*, 1, 63–112, 3 pls.

Rondani, C. (1872) Sulle specie Italiane del genere *Culex* Lin. *Bollettino della Societa Entomologica Italiana*, 4, 29–31.

Ronderos, R.A. (1965) Notas sobre *Culicidae* Venezolanos *(Diptera)*. *Physis*, 25 (70), 249–253.

Ronderos, R.A. & Bachmann, A.O. (1963a) A proposito del complejo *Mansonia (Diptera-Culicidae)*. *Revista de la Sociedad Entomológica Argentina* (for 1962), 25, 43–51.

Ronderos, R.A. & Bachmann, A.O. (1963b) *Mansoniini* [*sic*] neotropicales. I *(Diptera-Culicidae)*. *Revista de la Sociedad Entomológica Argentina*, 26, 57–65.

Root, F.M. (1922) The classification of American *Anopheles* mosquitoes. *American Journal of Hygiene*, 2 (3), 321–322.

Root, F.M. (1926) Studies on Brazilian mosquitoes. I. The anophelines of the *Nyssorhynchus* group. *American Journal of Hygiene*, 6 (5), 684–717, 9 pls.

Root, F.M. (1927a) Note on the mosquito fauna of the Republic of Haiti. *American Journal of Hygiene*, 7 (4), 463–469.

Root, F.M. (1927b) Studies on Brazilian mosquitoes. II. *Chagasia fajardoi*. *American Journal of Hygiene*, 7 (4), 470–480, 4 pls.

Root, F.M. (1927c) Studies on Brazilian mosquitoes. III. The genus *Culex*. *American Journal of Hygiene*, 7 (5), 574–598, 6 pls.

Root, F.M. (1927d) Studies on Brazilian mosquitoes. IV. Notes on some Brazilian species of *Anopheles*. *American Journal of Hygiene*, 7 (5), 599–605, 2 pls.

Root, F.M. (1937) A new species of *Uranotaenia* (Diptera, Culicidae) from Haiti. *Journal of Parasitology*, 23 (1), 98–100.

Roper, R. (1914) An account of some *Anopheline* mosquitos found in British North Borneo, with description of a new species. *Bulletin of Entomological Research*, 5 (2), 137–147, 2 maps. DOI: 10.1017/S0007485300034003

Rosa-Freitas, M.G. (1989) *Anopheles (Nyssorhynchus) deaneorum*: a new species in the *albitarsis* complex (Diptera: Culicidae). *Memórias do Instituto Oswaldo Cruz*, 84 (4), 535–543. DOI: 10.1590/S0074-02761989000400011

Rosenberg, R. & Evenhuis, N.L. (1985) A new species of *Toxorhynchites* from Bangladesh (Diptera: Culicidae). *Mosquito Systematics*, 17 (1), 36–43.

Ross, E.S. (1943) The identity of *Aedes bimaculatus* (Coquillett) and a new subspecies of *Aedes fulvus* (Wiedemann) from the United States (Diptera, Culicidae). *Proceedings of the Entomological Society of Washington*, 45 (6), 143–151.

Rossi, G.C. (2015) Annotated checklist, distribution, and taxonomic bibliography of the mosquitoes (Insecta: Diptera: Culicidae) of Argentina. *Check List*, 11 (4), 1712. DOI: 10.15560/11.4.1712

Rossi, G.C. & Harbach, R.E. (2008) *Phytotelmatomyia*, a new Neotropical subgenus of *Culex* (Diptera: Culicidae). *Zootaxa*, 1879, 1–17. DOI: 10.11646/zootaxa.1879.1.1

Rossi, G.C., Laurito, M. & Almirón, W.R. (2008) Morphological description of the pupa and redescription of the adults and larva of *Culex (Culex) apicinus* Philippi (Diptera: Culicidae). *Zootaxa*, 1941, 31–42. DOI: 10.11646/zootaxa.1941.1.3

Rossi, P. (1790) *Fauna Etrusca: Sistens insecta quae in provinciis Florentina et Pisana praesertim collegit Petrus Rossius*. Vol. 2. Thomae Masi & Sociorum, Liburni, Italy. DOI: 10.5962/bhl.title.15771

Rozeboom, L.E. (1935) *Culex rooti*, a new *Culex* from Panama (Diptera: Culicidae). *Annals of the Entomological Society of America*, 28 (2), 251–253. DOI: 10.1093/aesa/28.2.251

Rozeboom, L.E. (1946) Three new *Finlayas* of the *dissimilis* group (Diptera: Culicidae) from the Philippine islands. *Journal of Parasitology*, 32 (6), 587–595.

Rozeboom, L.E. (1951) A new species of *Anopheles* from the Philippine islands (Diptera: Culicidae). *Journal of Parasitology*, 37 (5), 502–506.

Rozeboom, L.E. & Cabrera, B.D. (1964) Filariasis in Mountain Province, Luzon, Republic of the Philippines. *Journal of Medical Entomology*, 1 (1), 18–28. DOI: 10.1093/jmedent/1.1.18

Rozeboom, L.E. & Gabaldon, A. (1941) A summary of the "*tarsimaculatus*" complex of *Anopheles* (Diptera: Culicidae). *American Journal of Hygiene*, 33 (C), 88–100, 4 pls. DOI: 10.1093/oxfordjournals.aje.a118713

Rozeboom, L.E. & Kitzmiller, J.B. (1958) Hybridization and speciation in mosquitoes. *Annual Review of Entomology*, 3, 231–248. DOI: 10.1146/annurev.en.03.010158.001311

Rozeboom, L.E. & Knight, K.L. (1946) The *punctulatus* complex of *Anopheles* (Diptera: Culicidae). *Journal of Parasitology*, 32 (2), 95–131. DOI: 10.2307/3272584

Rozeboom, L.E. & Komp, W.H.W. (1948) Three new species of *Culex* (Diptera: Culicidae) from Colombia. *Journal of Parasitology*, 34 (5), 396–406. DOI: 10.2307/3273605

Rozeboom, L.E. & Komp, W.H.W. (1950a) A review of the species of *Culex* of the subgenus *Melanoconion* (Diptera, Culicidae). *Annals of the Entomological Society of America*, 43 (1), 75–114. DOI: 10.1093/aesa/43.1.75

Rozeboom, L.E. & Komp, W.H.W. (1950b) A new *Microculex, elongatus*, from Colombia, with notes on the subgenus (Diptera, Culicidae). *Proceedings of the Entomological Society of Washington*, 52 (3), 147–157.

Rozo-Lopez, P. & Mengual, X. (2015) Updated list of the mosquitoes of Colombia (Diptera: Culicidae). *Biodiversitiy Data Journal*, 3, e4567. DOI: 10.3897/BDJ.3.e4567

Rubio-Palis, Y., Guzmán, H., Pérez, E. & Estrada, Y. (2016) Revisión de la colección del género *Chagasia* (Diptera: Culicidae) del Museo Entomológico "Dr. Pablo Cova Garcia", Maracay, Venezuela. *Boletín de Malariología y Salud Ambiental*, 56 (1), 63–67.

Rudnick, A., Wallace, H.G., Furumizo, R.T., Burton, J.J.S. & Neill, W.A. (1986) Dengue virus ecology in Malaysia, further studies, 1976–1980. *Bulletin – Institute for Medical Research, Kuala Lumpur*, 23, 127–153.

Rueda, L.M. (2005) Two new species of *Anopheles (Anopheles)* Hyrcanus Group (Diptera: Culicidae) from the Republic of South Korea. *Zootaxa*, 941, 1–26. DOI: 10.11646/zootaxa.941.1.1

Rueda, L.M., Vongphayloth, K., Pecor, J.E., Hii, J., Deboun, M. & Brey, P.T. (2015) Mosquito faun of Lao People's Democratic Republic, with special emphasis on the adult and larval surveillance at Nakai District, Khammuane Province. *U.S. Army Medical Department Journal*, Jul–Sep, 25–32.

Ruiz, F., Linton, Y.-M., Ponsonby, D.J., Conn, J.E., Herrera, M., Quiñones, M.L., Vélez, I.D. Wilkerson, R.C. (2010) Molecular comparison of topotypic specimens confirms *Anopheles (Nyssorhynchus) dunhami* Causey (Diptera: Culicidae) in the Colombian Amazon. *Memorias do Instituto Oswaldo Cruz*, 105 (7), 899–903. DOI: 10.1590/S0074-02762010000700010

Ruiz, F., Quiñones, M.L., Erazo, H.F., Calle, D.A., J Alzate, J.F. & Linton, Y.-M. (2005) Molecular differentiation of *Anopheles (Nyssorhynchus) benarrochi* and *An. (N.) oswaldoi* from Southern Colombia. *Memórias do Instituto Oswaldo Cruz*, 100 (2), 155–160. DOI: 10.1590/S0074-02762005000200008

Ruiz-Lopez, F., Wilkerson, R.C., Conn, J.E., McKeon, S.N., Levin, D.M., Quiñones, M.L., Póvoa, M.M. & Linton, Y.-M. (2012) DNA barcoding reveals both known and novel taxa in the Albitarsis Group (*Anopheles*: *Nyssorhynchus)* of Neotropical malaria vectors. *Parasites & Vectors*, 5, 44. DOI: 10.1186/1756-3305-5-44

Ruiz-Lopez, F., Wilkerson, R.C., Ponsonby, D., Herrera, M., Sallum, M.A.M., Velez, I.D., Quiñones, M.L., Flores-Mendoza, C., Chadee, D.D., Alarcon, J., Alarcon-Ormasa, J. & Linton, Y.-M. (2013) Systematics of the Oswaldoi Complex (*Anopheles, Nyssorhynchus)* in South America. *Parasites & Vectors*, 6, 324. DOI: 10.1186/1756-3305-6-324

Russell, T.L. & Burkot, T.R. (2023) *A Guide to Mosquitoes in the Pacific*. Pacific Community, Noumea, New Caledonia.

Ruzzante, L., Reijnders, M.J.M.F. & Waterhouse, R.M. (2019) Of genes and genomes: mosquito evolution and diversity. *Trends in Parasitology*, 35 (1), 32–51. DOI: 10.1016/j.pt.2018.10.003

Sá, I.L.R. de, Hutchings, R.S.G., Hutchings, R.W. & Sallum, M.A.M. (2020) Revision of the Atratus Group of *Culex (Melanoconion)* (Diptera: Culicidae). *Parasites & Vectors*, 13, 269. DOI: 10.1186/s13071-020-3982-x

Sá, I.L.R. de, Hutchings, R.S.G., Hutchings, R.W. & Sallum, M.A.M. (2022) Revision of the Educator Group of *Culex (Melanoconion)* (Diptera, Culicidae). *Journal of Medical Entomology*, 59 (4), 1252–1290. DOI: 10.1093/jme/tjac051

Sabrosky, C.W. (1999) *Family-group Names in Diptera: An Annotated Catalog*. Backhuys Publishers, Leiden, Netherlands.

Sakakibara, M. (1959) On *Anopheles (Anopheles) omorii* n. sp. *Endemic Diseases Bulletin of Nagasaki University*, 1 (3), 288–295, 3 pls.

Salem, H.H. (1938) The mosquito fauna of Sinai Peninsula (Egypt) with a description of two new species. *Publications of the Egyptian University Faculty of Medicine*, 16, 1–31, 16 pls.

Sallum, M.A.M. & Forattini, O.P. (1996) Revision of the Spissipes Section of *Culex (Melanoconion)* (Diptera: Culicidae). *Journal of the American Mosquito Control Association*, 12 (3), 517–600.

Sallum, M.A.M. & Hutchings, R.S.G. (2003) Taxonomic studies on *Culex (Melanoconion) coppenamensis* Bonne-Wepster & Bonne (Diptera: Culicidae), and description of two new species from Brazil. *Memórias do Instituto Oswaldo Cruz*, 98 (5), 615–622. DOI: 10.1590/S0074-02762003000500006

Sallum, M.A.M., Hutchings, R.S.G., Leila, R. & Ferreira, R.L.M. (1997) *Culex gnomatos* a new species of the Spissipes Section of *Culex (Melanoconion)* (Diptera: Culicidae) from the Amazon region. *Memórias do Instituto Oswaldo Cruz*, 92 (2), 215–219. DOI: 10.1590/S0074-02761997000200014

Sallum, M.A.M., Obando, R.G., Carrejo, N. & Wilkerson, R.C. (2020a) Identification keys to the *Anopheles* mosquitoes of South America (Diptera: Culicidae). I. Introduction. *Parasites & Vectors*, 13, 583. DOI: 10.1186/s13071-020-04298-6

Sallum, M.A.M., Obando, R.G., Carrejo, N. & Wilkerson, R.C. (2020b) Identification keys to the *Anopheles* mosquitoes of South America (Diptera: Culicidae). II. Fourth-instar larvae. *Parasites & Vectors*, 13, 582. DOI: 10.1186/s13071-020-04299-5

Sallum, M.A.M., Obando, R.G., Carrejo, N. & Wilkerson, R.C. (2020c) Identification key to the *Anopheles* mosquitoes of South America (Diptera: Culicidae). III. Male genitalia. *Parasites & Vectors*, 13, 542. DOI: 10.1186/s13071-020-04300-1

Sallum, M.A.M., Obando, R.G., Carrejo, N. & Wilkerson, R.C. (2020d) Identification keys to the *Anopheles* mosquitoes of South America (Diptera: Culicidae). IV. Adult females. *Parasites & Vectors*, 13, 584. DOI: 10.1186/s13071-020-04301-0

Sallum, M.A.M., Peyton, E.L., Harrison, B.A. & Wilkerson, R.C. (2005a) Revision of the Leucosphyrus group of *Anopheles (Cellia)* (Diptera, Culicidae). *Revista Brasileira de Entomologia*, 49 (Supl. 1), 1–152. DOI: 10.1590/S0085-56262005000500001

Sallum, M.A.M., Peyton, E.L. & Wilkerson, R.C. (2005b) Six new species of the *Anopheles leucosphyrus* group, reinterpretation of *An. elegans* and vector implications. *Medical and Veterinary Entomology*, 19 (2), 158–199. DOI: 10.1111/j.0269-283X.2005.00551.x

Sallum, M.A.M., Schultz, T.R., Foster, P.G., Aronstein, K., Wirtz, R.A. & Wilkerson, R.C. (2002) Phylogeny of Anophelinae (Diptera: Culicidae) based on nuclear ribosomal and mitochondrial DNA sequences. *Systematic Entomology*, 27 (3), 361–382. DOI: 10.1046/j.1365-3113.2002.00182.x

Sallum, M.A.M., Schultz, T.R. & Wilkerson, R.C. (2000) Phylogeny of Anophelinae (Diptera Culicidae) based on morphological characters. *Annals of the Entomological Society of America*, 93 (4), 745–775. DOI: 10.1603/0013-8746(2000)093[0745:POADCB]2.0.CO;2

Sánchez-Covisa Villa, A., Rodríguez Rodríguez, J.A. & Guillén Llera, J.L. (1985) Estudio de la larva de cuarto estado de las especies del subgenero *Finlaya (Diptera: Culicidae)* de España Peninsular. *Suplemento ao Boletim da Sociedade Portuguesa de Entomologia*, 1 (Supl. 1), 439–448.

Sandosham, A.A. (1944) *Malaria in Malai: A Handbook for Anti-malaria Students, &c.* Syonan: Syonan Tokubetu-Si, Singapore.

Sant'Ana, D.C. & Sallum, M.A.M. (2017) *Anopheles (Nyssorhynchus) striatus*, a new species of the Strodei Subgoup (Diptera, Culicidae). *Revista Brasileira de Entomologia*, 61 (2), 136–145. DOI: 10.1016/j.rbe.2016.12.005

Sant'Ana, D.C. & Sallum, M.A.M. (2022a) A new species of the Nuneztovari Complex of *Nyssorhynchus* (Diptera: Culicidae) from the western Brazilian Amazon. *Zootaxa*, 5134 (2), 275–285. DOI: 10.11646/zootaxa.5134.2.6

Sant'Ana, D.C. & Sallum, M.A.M. (2022b) A new species of the Arthuri Complex of the Strodei Subgroup of *Nyssorhynchus* (Diptera: Culicidae). *Zootaxa*, 5175 (5), 559–569. DOI: 10.11646/zootaxa.5175.5.5

Sant'Ana, D.C. & Sallum, M.A.M. (2024) Revision of the Strodei Subgroup of *Nyssorhynchus* (Diptera: Culicidae), with descriptions of 2 new species. *Journal of Medical Entomology*, 61 (1), 87–109. DOI: 10.1093/jme/tjad149

Santos, V.S., Bridi, L.C. & Rafael, M.S. (2022) Phylogenetic relationships of the supercontig of sodium channel subunit I (NaV) in 17 species of *Anopheles* (Diptera: Culicidae). *Revista da Sociedade Brasileira de Medicina Tropical*, 55, e0701-2021. DOI: 10.1590/0037-8682-0701-2021

Saraiva, J.F. & Scarpassa, V.M. (2021) *Anopheles (Nyssorhynchus) tadei*: a new species of the Oswaldoi-konderi Complex (Diptera, Anophelinae) and its morphological and molecular distinctions from *An. konderi* sensu stricto. *Acta Tropica*, 221, 106004. DOI: 10.1016/j.actatropica.2021.106004

Sarala, K.S., Nutan, N., Vasantha, K., Dua, V.K., Malhotra, M.S., Yadav, R.S. & Sharma, V.P. (1994) Cytogenetic evidence for three sibling species in *Anopheles fluviatilis* (Diptera: Culicidae). *Annals of the Entomological Society of America*, 87 (1), 116–121. DOI: 10.1093/aesa/87.1.116

Sasa, M. & Nakahashi, Y. (1952) A new species of the *Aedes (Finlaya) niveus* subgroup of mosquito from Japan. *Japanese Journal of Experimental Medicine*, 22 (3), 257–265.

Sasa, M. & Takahasi, H. (1948) Descriptions of *Culex rubensis* n. sp. and the adult male of *Theobaldia kanayamensis* Yamada [in Japanese]. *Japanese Journal of Bacteriology*, 3 (2), 51–52.

Sasa, M., Kano, R. & Hayashi, S. (1950a) A new species of mosquito, *Aedes koreicoides n.* [*sic*] sp., from Hokkaido. *Japanese Journal of Experimental Medicine*, 20, 627–629.

Sasa, M., Kano, R. & Takahasi, H. (1950b) A revision of the adult Japanese mosquitoes of the genus *Aedes*, subgenus *Aedes*, with description of two new species. *Japanese Journal of Experimental Medicine*, 20, 631–640.

Satoto, T.B.T. (2001) *Cryptic species within* Anopheles barbirostris *van der Wulp, 1884, inferred from nuclear and mitochondrial gene sequence variation*. PhD Thesis, University of Liverpool, Liverpool, UK.

Say, T. (1823) Descriptions of dipterous insects of the United States. *Journal of the Academy of Natural Sciences of Philadelphia*, 3 (1), 9–54.

Scarpassa, V.M., Cunha-Machado, A.S. & Saraiva, J.F. (2016) Evidence of new species for malaria vector *Anopheles nuneztovari* sensu lato in the Brazilian Amazon region. *Malaria Journal*, 15, 205. DOI: 10.1186/s12936-016-1217-6

Scanlon, J.E. & Peyton, E.L. (1967) *Anopheles (Anopheles) tigertti*, a new species of the Aitkenii Group from Thailand. *Proceedings of the Entomological Society of Washington*, 69 (1), 18–23.

Schick, R.X. (1970a) Mosquito studies (Diptera, Culicidae). XX. The Terrens Group of *Aedes (Finlaya)*. *Contributions of the American Entomological Institute*, 5 (3), 1–158.

Schick, R.X. (1970b) Mosquito studies (Diptera, Culicidae). XXIII. Additions and corrections to the revision of the *Aedes terrens* group. *Contributions of the American Entomological Institute*, 7 (1), 13–40.

Schiner, J.R. (1864) *Fauna austriaca. Die Fliegen (Diptera). Nach der analytischen Methode bearbeite, mit der Charakteristik sämmtlicher europäischer Gattungen, der Beschreibung aller in Deutschland vorkommenden Arten und der Aufzählung aller bisher beschriebenen europäischen Arten. II Theil.* Carl Gerold's Sohn, Wien [Vienna], Austria. DOI: 10.5962/bhl.title.8525

Schiner, J.R. (1868) *Reise der Österreichischen Fregatte Novara um die Erde in den Jahren 1857, 1858, 1859 unter den Befehln des Commodore B. von Wüllerstorf-Urbair. Zoologischer Theil. Zweiter Band. 1. Abtheilung. B. 1. Diptera*. Kaiserlich-Königliche Hof- und Staatsdruckerei; in Commission bei K. Gerold's Sohn, Wien [Vienna], Austria. DOI: 10.5962/bhl.title.1597

Schmidt, E.R., Foley, D.H., Bugoro, H. & Bryan, J.H. (**2003**) A morphological study of the *Anopheles punctulatus* group (Diptera: Culicidae) in the Solomon Islands, with a description of *Anopheles (Cellia) irenicus* Schmidt, sp.n. *Bulletin of Entomological Research*, 93 (6), 515–526. DOI: 10.1079/BER2003267

Schmidt, E.R., Foley, D.H., Hartel, G.F., Williams, G.M. & Bryan, J.H. (**2001**) Descriptions of the *Anopheles (Cellia) farauti* complex of sibling species (Diptera: Culicidae) in Australia. *Bulletin of Entomological Research*, 91 (5), 389–410. DOI: 10.1079/BER2001113

Schrank, F. von P. (1776) *Beyträge zur Naturgeschichte*. Caspar Fritsch, Leipzig, Germany.

Schremmer, F. (1950) Morphologische und funktionelle Analyse der Mundteile und des Pharynx der Larve von *Anopheles maculipennis* Meig. *Österreichische Zoologische Zeitschrift*, 2, 173–222.

Schwetz, J. (1927) Synopsis des moustiques connus du Congo belge, avec quelques commentaires et considerations. *Revue Zoologique Africaine*, 15 (3), 271–319.

Scudder, S.H. (1890) *The tertiary insects of North America*. United States Geological Survey of the Territories. Vol. 13. Government Printing Office, Washington, DC, USA.

Sedaghat, M.M., Linton, Y.-M., Oshaghi, M.A., Vatandoost, H. & Harbach, R.E. (2003) The *Anopheles maculipennis* complex (Diptera: Culicidae) in Iran: molecular characterization and recognition of a new species. *Bulletin of Entomological Research*, 93 (6), 527–535. DOI: 10.1079/BER2003272

Séguy, E. (1921) Description d'un nouveau Moustique français du groupe de l'*Aedes Mariae* [*sic*] et synopsis des espèces de ce groupe. *Bulletin de la Société Entomologique de France*, 1921 (13), 192–195. DOI: 10.3406/bsef.1921.26829

Séguy, E. (1923) Remarques sur quelques moustiques. *Annales de la Société Entomologique de France*, 92 (1–2), 205–208. DOI: 10.1080/21686351.1923.12279945

Séguy, E. (1924) Les moustiques de l'Afrique Mineure de l'Égypte et de la Syrie étude comparative des moustiques des régions méditerranéennes, de l'Europe centrale et septentrionale leurs parasites suivi du catalogue des culicides néarctiques et paléarctiques. *Encyclopédie Entomologique Série A. Travaux Généraux*, 1, 1–257, 29 pls.

Séguy, E. (1927) Notes sur les moustiques de l'Afrique Mineure, de l'Égypte et de la Syrie. II. *Encyclopédie Entomologique Série B. Mémoires et Notes*, 4 (1), 27–28.

Senevet, G. (1936) Les moustiques de la Martinique. *Archives de l'Institut Pasteur d'Algérie*, 14 (2), 123–134.

Senevet, G. (1937) Les moustiques de la Guyane française (mission 1934). *Archives de l'Institut Pasteur d'Algérie*, 15 (3), 352–382.

Senevet, G. & Abonnenc, E. (1938) Quelques anophélinés de la Guyane française. *Archives de l'Institut Pasteur d'Algérie*, 16 (4), 486–512.

Senevet, G. & Abonnenc, E. (1939a) Les moustiques de la Guyane française – II. Le genre *Culex*. *Archives de l'Institut Pasteur d'Algérie*, 17 (1), 62–134.

Senevet, G. & Abonnenc, E. (1939b) Les moustiques de la Guyane Française – III. Les sabéthinés. *Archives de l'Institut Pasteur d'Algérie*, 17 (2), 247–281.

Senevet, G. & Abonnenc, E. (1941) Les moustiques de la Guyane Française. Le genre *Culex*. 2. Nouvelle espèce du sous-genre *Melanoconion*. *Archives de l'Institut Pasteur d'Algérie*, 19 (1), 41–44.

Senevet, G. & Prunnelle, M. (1927) Une nouvelle espèce d'anophèle en Algérie, *Anopheles marteri* n. sp. *Archives de l'Institut Pasteur d'Algérie*, 5 (4), 529–533.

Senevet, G., Chabelard, R. & Abonnenc, E. (1942) Les moustiques de la Guyane. III – Les sabéthinés (2). *Archives de l'Institut Pasteur d'Algérie*, 20 (4), 336–348.

Serafim, J., Jr & Davis, N.C. (1933) Distribution of *Aëdes (Taeniorhynchus) taeniorhynchus* (Wiedemann). *Aëdes (Taeniorhynchus) jacobinae*, new species. *Annals of the Entomological Society of America*, 26 (1), 13–19. DOI: 10.1093/aesa/26.1.13

Sergent, Ed. & Sergent, Ét. (1903) Observations sur les moustiques des environs d'Alger. *Annales de l'Institut Pasteur*, 17 (1), 60–67.

Service, M.W. (1958) A new *Anopheles* from the southern Cameroons, *Anopheles buxtoni* sp. nov. *Annals of Tropical Medicine and Parasitology*, 52 (2), 176–180. DOI: 10.1080/00034983.1958.11685858

Service, M.W. (1959a) A new species of the genus *Aëdes* from southern Nigeria. *Proceedings of the Royal Entomological Society of London, Series B: Taxonomy*, 28 (5–6), 73–76. DOI: 10.1111/j.1365-3113.1959.tb00132.x

Service, M.W. (1959b) A new culicine mosquito, *Aëdes (Aëdimorphus) lokojensis* sp. nov., from northern Nigeria. *Revue de zoologie et de botanique Africaines*, 59 (3–4), 243–248.

Service, M.Wss. (1960) A new species and variety of *Anopheles* from Nigeria, with notes on *Anopheles flavicosta* Edwards and *Aëdes (Aë.) wendyae* Service. *Proceedings of the Royal Entomological Society of London, Series B: Taxonomy*, 29 (7–8), 85–90. DOI: 10.1111/j.1365-3113.1960.tb01146.x

Service, M.W. (1970a) A new *Anopheles* (Dipt., Culicidae) from Nigeria and a note on *A. pretoriensis* (Theobald). *Bulletin of Entomological Research*, 59 (4), 647–649. DOI: 10.1017/S0007485300003631

Service, M.W. (1970b) Studies on the biology and taxonomy of *Aedes (Stegomyia) vittatus* (Bigot) (Diptera: Culicidae) in northern Nigeria. *Transactions of the Royal Entomological Society of London*, 122 (4), 101–143, 2 pls. DOI: 10.1111/j.1365-2311.1970.tb00529.x

Service, M.W. (1977) A new *Anopheles* (Dipt., Culicidae) from Nigeria together with a note on *An. brohieri* Edwards. *Annals of Tropical Medicine and Parasitology*, 71 (2), 233–236. DOI: 10.1080/00034983.1977.11687183

Service, M.W. (1990) *Handbook to the Afrotropical Toxorhynchitine and Culicine Mosquitoes, excepting* Aedes *and* Culex. British Museum (Natural History), London, UK.

Service, M.W. (1993a) *Mosquito Ecology: Field Sampling Methods*. 2nd edn. Elsevier Applied Science, London, UK.

Service, M.W. (1993b) Mosquitoes (Culicidae). In: Lane, R.P. & Crosskey, R.W. (eds), *Medical Insects and Arachnids*. Chapman & Hall, London, UK, pp. 120–240.

Shannon, R.C. (1931) On the classification of Brazilian Culicidae with special reference to those capable of harboring the yellow fever virus. *Proceedings of the Entomological Society of Washington*, 33 (6), 125–164.

Shannon, R.C. (1933) Correction. *Proceedings of the Entomological Society of Washington*, 35 (4), 58.

Shannon, R.C. & Cerqueira, N.L. (1943) *Psorophora lanei*, um novo mosquito do Brasil e da Bolivia (Diptera, Culicidae). *Revista de Entomologia*, 14, 135–137.

Shannon, R.C. & del Ponte, E. (1927) Cuatro notas sobre especies nuevas de Dípteros, Nematóceros, Hematófagos, o no, de la República Argentina. *Revista del Instituto Bacteriológico de Buenos Aires*, 4 (7), 724–736.

Shannon, R.C. & del Ponte, E. (1928) Los culicidos en la Argentina. *Revista del Instituto Bacteriológico de Buenos Aires* (for 1927), 5 (1), 29–140.

Shevchenko, A.K. & Prudkina, N.S. (1973) On morphology of genitals in mosquito males from the *Aedes* genus [in Russian]. *Vestnik Zoologii [Zoological Record] / Institut Zoologii, Akademiya Nauk Ukrainskoi SSR*, 6, 40–47.

Shidrawi, G.R. & Gillies, M.T. (1988) *Anopheles paltrinierii*, n. sp. (Culicidae: Diptera) from the Sultanate of Oman. *Mosquito Systematics* (for 1987), 19 (3), 201–211.

Shingarev, N.I. (1926) New information on Culicidae of U.S.S.R. [in Russian]. *Russkiĭ Zhurnal Tropicheskoĭ Meditsiny*, 4 (2), 47–48.

Shingarev, N.I. (1928) Notes on Culicidae. III [in Russian]. *Russkiĭ Zhurnal Tropicheskoĭ Meditsiny*, 1, 47–53.

Shriram, A.N., Ramaiah, K.D., Krishnamoorthy, K. & Sehgal, S.C. (2005) Diurnal pattern of human-biting activity and transmission of subperiodic *Wuchereria bancrofti* (Filariidea: Dipetalonematidae) by *Ochlerotatus niveus* (Diptera: Culicidae) on the Andaman and Nicobar islands of India. *American Journal of Tropical Medicine and Hygiene*, 72 (3), 273–277. DOI: 10.4269/ajtmh.2005.72.273

Shute, P.G. (1928) A new variety of *Culicella morsitans* (Theobald) (Diptera). *Entomologist*, 61, 186.

Silva do Nascimento, T.F. & Lourenço-de-Oliveira, R. (2002) *Anopheles halophylus*, a new species of the subgenus *Nyssorhynchus* (Diptera: Culicidae) from Brazil. *Memórias do Instituto Oswaldo Cruz*, 97 (6), 801–811. DOI: 10.1590/S0074-02762002000600010

Silva-do-Nascimento, T.F. & Lourenço-de-Oliveira, R. (2007) Diverse population dynamics of three *Anopheles* species belonging to the Triannulatus Complex (Diptera: Culicidae). *Memórias do Instituto Oswaldo Cruz*, 102 (8), 975–982. DOI: 10.1590/S0074-02762007000800013

Simon, C.M., Steffan, W.A., Moss, W.W. & Evenhuis, N.L. (1983) The genus *Toxorhynchites* (Diptera: Culicidae); numerical phylogenetic analysis of *Tx. splendens* and allies with phenetic comparisons. *Mosquito Systematics* (for 1982), 14 (4), 221–261.

Simaa, M.A.A., Harbach, R.E., Almalik, A.M.A., Ahmed, E.M., Esia, A.A., Mohamed, A.H. & Azrag, R.S. (2021) *Culex* mosquitoes (Diptera: Culicidae) recorded along the Nile River in northern Sudan, with a key for the identification of all species of the genus known to occur in the country. *Zootaxa*, 4963 (3), 401–411. DOI: 10.11646/zootaxa.4963.3.1

Singh, G. & Prakash, S. (2008) A new species of *Lutzia* (Diptera: Culicidae) comparable with *halifaxii* Theobald 1903 in a semi-arid zone of India, Chapter 24. In: Tyagi, B.K. (ed.), *Vector-borne Diseases: Epidemiology and Control*. Scientific Publishers (India), Jodhpur, India, pp. 195–204.

Sinton, J.A. & Covell, G. (1927) The relation of the morphology of the buccal cavity to the classification of anopheline mosquitoes. *Indian Journal of Medical Research*, 15, 301–308, 5 pls.

Sirivanakarn, S. (1968) The *Culex* subgenus *Lophoceraomyia in* New Guinea and Bismarck Archipelago (Diptera: Culicidae). *Pacific Insects Monographs*, 17, 75–186.

Sirivanakarn, S. (1969) Note on the change of a specific name of New Guinea *Culex (Lophoceraomyia)* (Diptera: Culicidae). *Journal of Medical Entomology*, 6 (2), 119. DOI: 10.1093/jmedent/6.2.119

Sirivanakarn, S. (1971) Contributions to the mosquito fauna of Southeast Asia. XI. Proposed reclassification of *Neoculex* Dyar based principally on the male terminalia. *Contributions of the American Entomological Institute*, 7 (3), 62–85.

Sirivanakarn, S. (1972) Contributions to the mosquito fauna of Southeast Asia. XIII. The genus *Culex*, subgenus *Eumelanomyia* Theobald in Southeast Asia and adjacent areas. *Contributions of the American Entomological Institute*, 8 (6), i, 1–86.

Sirivanakarn, S. (1973a) Descriptions of two new species of *Culex (Lophoceraomyia)* with notes on three other species from the Papuan Subregion (Diptera: Culicidae). *Journal of Medical Entomology*, 10 (2), 212–216. DOI: 10.1093/jmedent/10.2.212

Sirivanakarn, S. (1973b) Three new species of *Culex* subgenus *Culiciomyia* Theobald from Southeast Asia and a redescription of the type of *C. tricuspis* Edwards from Alor, Lesser Sunda Islands, Indonesia (Diptera: Culicidae). *Proceedings of the Entomological Society of Washington*, 75 (1), 112–124.

Sirivanakarn, S. (1975) A new species of *Culex (Eumelanomyia)* Theobald from Manus Island, Papua-New Guinea (Diptera: Culicidae). *Mosquito Systematics*, 7 (4), 394–400.

Sirivanakarn, S. (1976) Medical entomology studies – III. A revision of the subgenus *Culex* in the Oriental Region (Diptera: Culicidae). *Contributions of the American Entomological Institute*, 12 (2), iii, 1–272.

Sirivanakarn, S. (1977a) Medical entomology studies – VI. A revision of the subgenus *Lophoceraomyia* of the genus *Culex* in the Oriental Region (Diptera: Culicidae). *Contributions of the American Entomological Institute*, 13 (4), iii, 1–245.

Sirivanakarn, S. (1977b) Additional descriptions of three species of *Culex (Eumelanomyia)* with the description of a new species from peninsular Malaysia (Diptera: Culicidae). *Mosquito Systematics*, 9 (1), 73–87.

Sirivanakarn, S. (1977c) Redescription of four Oriental species of *Culex (Culiciomyia)* and the description of a new species from Thailand (Diptera: Culicidae). *Mosquito Systematics*, 9 (2), 93–111.

Sirivanakarn, S. (1977d) A new species of *Culex (Eumelanomyia)* from India with descriptions of pupae and larvae of *Cx. pluvialis* Barraud and *Cx. iphis* Barraud (Diptera: Culicidae). *Mosquito Systematics*, 9 (4), 537–547.

Sirivanakarn, S. (1978) The female cibarial armature of New World *Culex*, subgenus *Melanoconion* and related subgenera with notes on this character in subgenera *Culex, Lutzia* and *Neoculex* and genera *Galindomyia* and *Deinocerites* (Diptera: Culicidae). *Mosquito Systematics*, 10 (4), 474–492.

Sirivanakarn, S. (1979) A new species of *Culex (Melanoconion)* from Bolivia and Ecuador (Diptera: Culicidae). *Mosquito Systematics*, 11 (2), 135–138.

Sirivanakarn, S. (1983) A review of the systematics and a proposed scheme of internal classification of the New World subgenus *Melanoconion* of *Culex* (Diptera, Culicidae). *Mosquito Systematics* (for 1982), 14 (4), 265–333.

Sirivanakarn, S. & Belkin, J.N. (1980) The identity of *Culex (Melanoconion) taeniopus* Dyar and Knab and related species with notes on the synonymy and description of a new species (Diptera, Culicidae). *Mosquito Systematics*, 12 (1), 7–24.

Sirivanakarn, S. & Galindo, P. (1980) *Culex (Melanoconion) adamesi*, a new species from Panama (Diptera, Culicidae). *Mosquito Systematics*, 12 (1), 25–34.

Sirivanakarn, S. & Jakob, W.L. (1979) A new species of *Culex (Melanoconion)* from southern Brazil (Diptera: Culicidae). *Mosquito Systematics*, 11 (2), 139–143.

Sirivanakarn, S. & Jakob, W.L. (1982) *Culex (Melanoconion) sacchettae*, a new species from the state of São Paulo, Brazil (Diptera: Culicidae). *Mosquito Systematics* (for 1981), 13 (2), 191–194.

Sirivanakarn, S. & Kurihara, I. (1973) A new species of *Culex*, subgenus *Culiciomyia* Theobald from Ceram, Indonesia (Diptera: Culicidae). *Proceedings of the Entomological Society of Washington*, 75 (2), 220–224.

Sirivanakarn, S. & Ramalingam, S. (1976) A new species of *Culex (Eumelanomyia)* Theobald with notes on three other species from Malaysia (Diptera: Culicidae). *Mosquito Systematics*, 8 (2), 209–216.

Skuse, F.A.A. (1889) Diptera of Australia. Part V—The Culicidae. *Proceedings of the Linnean Society of New South Wales*, 3 (4), 1717–1764, 1 pl.

Skuse, F.A.A. (1895) The banded mosquito of Bengal. *Indian Museum Notes* (for 1894), 3 (5), 20.

Slooff, R. (1961) *Tripteroides (Rachisoura) tityae* (Diptera: Culicidae), a new species from Netherlands New Guinea. *Entomologische Berichten*, 21 (1), 22–23.

Slooff, R. (1963) *Uranotaenia gerdae*, a new species of the *tibialis* group, from West New Guinea (Diptera: Culicidae). *Annals of the Entomological Society of America*, 56 (1), 82–86. DOI: 10.1093/aesa/56.1.82

Snodgrass, R.E. (1959) The anatomical life of the mosquito. *Smithsonian Miscellaneous Collections*, 139 (8), iii, 1–87.

Snyman, L.P. (2021) Review of mosquitoes (Diptera: Culicidae) from the Afrotropical region using herpetofauna as hosts: conservation and epidemiological consequences of knowing little about many species. *Austral Entomology*, 60 (1), 163–171. DOI: 10.1111/aen.12501

Sobti, S.K. (1968) A new species of the *Anopheles funestus* complex (Diptera: Culicidae) from Zanzibar, United Republic of Tanzania. *Bulletin of the World Health Organization*, 38 (3), 481–483.

Soesilo, R. & van Slooten, J. (1931a) Verspreide aanteekeningen over Anophelinen in de Oost-Indischen Archipel. *Geneeskundig Tijdschrift voor Nederlandsch-Indië*, 71 (7–18), 1254–1259.

Soesilo, R. & van Slooten, J. (1931b) Miscellaneous notes on anopheline mosquitoes in the Dutch East Indies. *Mededeelingen van den Dienst Volksgezondheid in Nederlandsch-Indië*, 20, 124–128.

Soghigian, J., Andreadis, T.G. & Livdahl, T.P. (2017) From ground pools to treeholes: convergent evolution of habitat and phenotype in *Aedes* mosquitoes. *BMC Evolutionary Biology*, 17, 262. DOI: 10.1186/s12862-017-1092-y

Soghigian, J., Sither, C., Justi, S.A., Morinaga, G., Cassel, B.K., Vitek, C.J., Livdahl, T., Xia, S., Gloria-Soria, A., Powell, J.R., Zavortink, T., Hardy, C.M., Burkett-Cadena, N.D., Reeves, L.E., Wilkerson, R.C., Dunn, R.R., Yeates, D.K., Sallum, M.A., Byrd, B.D., Trautwein, M.D., Linton, Y.-M., Reiskind, M.H. & Wiegmann, B.M. (2023) Phylogenomics reveals the history of host use in mosquitoes. *Nature Communications*, 14, 6252. DOI: 10.1038/s41467-023-41764-y

Somboon, P. & Harbach, R.E. (2019) *Lutzia (Metalutzia) chiangmaiensis* n. sp. (Diptera: Culicidae), formal name for the Chiang Mai (CM) form of the genus *Lutzia* in Thailand. *Journal of Medical Entomology*, 56 (5), 1270–1274. DOI: 10.1093/jme/tjz072

Somboon, P., Namgay, R. & Harbach, R.E. (**1922c**) *Uranotaenia (Pseudoficalbia) bhutanensis* (Diptera: Culicidae), a new species from Bhutan. *Journal of Medical Entomology*, 59 (4), 1303–1307. DOI: 10.1093/jme/tjac061

Somboon, P., Phanitchakun, T., Namgay, R. & Harbach, R.E. (**2022b**) The *Anopheles lindesayi* species complex (Diptera: Culicidae) in Bhutan. *Journal of Medical Entomology*, 59 (4), 1236–1251. DOI: 10.1093/jme/tjac052

Somboon, P., Phanitchakun, T., Namgay, R. & Harbach, R.E. (**2020a**) Description of *Aedes (Hulecoeteomyia) bhutanensis* n. sp. (Diptera: Culicidae) from Bhutan. *Acta Tropica*, 203, 105280. DOI: 10.1016/j.actatropica.2019.105280

Somboon, P., Phanitchakun, T., Namgay, R. & Harbach, R.E. (**2022a**) The *Anopheles baileyi* species complex (Diptera: Culicidae: Anophelinae) in Bhutan. *Acta Tropica*, 226, 106241. DOI: 10.1016/j.actatropica.2021.106241

Somboon, P., Phanitchakun, T., Namgay, R., Wangdi, T., Pemo, D. & Harbach, R.E. (**2020b**) Molecular and morphological evidence of sibling species in *Anopheles baileyi* Edwards (Diptera: Culicidae) in Bhutan and Thailand. *Acta Tropica*, 209, 105549. DOI: 10.1016/j.actatropica.2020.105549

Somboon, P., Phanitchakun, T., Saingamsook, J., Namgay, R. & Harbach, R.E. (**2021a**) *Reinertia*, a new subgenus of the genus *Aedes* Meigen and its type species *Aedes (Reinertia) suffusus* Edwards (Diptera: Culicidae), newly recorded from Bhutan. *Journal of Medical Entomology*, 58 (3), 1138–1148. DOI: 10.1093/jme/tjaa277

Somboon, P., Phanitchakun, T., Saingamsook, J., Namgay, R. & Harbach, R.E. (**2021b**) *Culex (Culex) longitubus*, a new species of the Mimeticus Subgroup (Diptera: Culicidae) from Bhutan. *Journal of Medical Entomology*, 58 (6), 2198–2205. DOI: 10.1093/jme/tjab080

Somboon, P., Phanitchakun, T., Saingamsook, J., Yoshihide, M., Namgay, R. & Harbach, R.E. (**2021c**) *Culex bhutanensis*, a new species of the Mimeticus Subgroup of the nominotypical subgenus of the genus *Culex* Linnaeus (Diptera: Culicidae) from Bhutan. *Acta Tropica*, 217, 105868. DOI: 10.1016/j.actatropica.2021.105868

Somboon, P., Rory, A., Tsuda, Y., Takagi, M. & Harbach, R.E. (**2010**) Systematics of *Anopheles (Cellia) yaeyamaensis* sp. n., alias species E of the *An. minimus* complex in southeastern Asia (Diptera: Culicidae). *Zootaxa*, 2651, 43–51. DOI: 10.11646/zootaxa.2651.1.3

Somboon, P., Saeung, A., Saingamsook, J., Namgay, R. & Harbach, R.E. (**2024**) A new species of the *Anopheles barbirostris* complex (Diptera: Culicidae) from Bhutan. *Journal of Medical Entomology*, 61 (2), 377–388. DOI: 10.1093/jme/tjad161

Somboon, P., Saingamsook, J. & Harbach, R.E. (**2023c**) Problematic placement of species in subgenera of the genus *Topomyia* (Diptera: Culicidae: Sabethini) based on the length of foretarsomeres 2 and 3. *Zootaxa*, 5357 (1), 139–143. DOI: 10.11646/zootaxa.5357.1.7

Somboon, P., Thongwat, D. & Harbach, R.E. (2011) *Anopheles (Cellia) rampae* n. sp., alias chromosomal form K of the Oriental Maculatus Group (Diptera: Culicidae) in Southeast Asia. *Zootaxa*, 2810, 47–55. DOI: 10.11646/zootaxa.2810.1.5

Somboon, P., Wilai, P., Saeung, A., Saingamsook, J. & Harbach, R.E. (**2023a**) The identity of *Anopheles (Anopheles) barbirostris* species A3 of the Barbirostris Complex (Diptera: Culicidae). *Zootaxa*, 5353 (1), 096–100. DOI: 10.11646/zootaxa.5353.1.8

Somboon, P., Zhou, L., Phanitchakun, T., Saingamsook, J. & Harbach, R.E. (**2023b**) *Aedes yunnanensis* (Diptera: Culicidae): transferred from the subgenus *Hulecoeteomyia* to *Orohylomyia* subg. nov., with descriptions of the species and the new subgenus. *Journal of Medical Entomology*, 60 (5), 955–967. DOI: 10.1093/jme/tjad075

Speiser, P. (1909) 10. Diptera. 4. Orthorhapha. In: Sjöstedt, Y., *Wissenschaftliche ergebnisse der Schwedischen zoologischen expedition nach dem Kilimandjaro, dem Meru und den umgebenden Massaisteppen Deutsch-Ostafrikas 1905–1906*. 2. Band, Abteilung 8–14. P. Palmquists Aktiebolag, Stockholm, Sweden, pp. 31–112.

Stackelberg, A.A. (1927) The Culicidae of the USSR and the countries bordering [in Russian]. *Analytical Tables of the Fauna of the USSR*, 1, vi, 1–169.

Stackelberg, A.A. (1943) A new species of *Finlaya*, Theo. (Diptera, Culicidae) from Ussuri Land. *Bulletin of Entomological Research*, 34 (4), 311. DOI: 10.1017/S0007485300023816

Stanton, A.T. (1915) A new anopheline mosquito from Sumatra. *Bulletin of Entomological Research*, 5 (4), 373–375. DOI: 10.1017/S0007485300030364

Stanton, A.T. & Hacker, H.P. (1917) The *Anopheles* of Malaya—III. A new variety of *A. albotaeniatus*, Theo. *Bulletin of Entomological Research*, 7 (3), 273–275. DOI: 10.1017/S0007485300017612

Starkey, G.S. & Webb, J.E., Jr (1946) A new species of *Aedes* of the *Finlaya* group from Angaur Island. *Proceedings of the Entomological Society of Washington*, 48 (7), 179–182.

Statz, G. (1944) Neue Dipteren (Nematocera) aus dem Oberoligocän von Rott. II. Teil. V. Familie Culicidae (Stechmücken). *Palaeontographica*, 95 (A), 108–120, 6 pls.

Steffan, W.A. (1968) *Armigeres* of the Papuan Subregion (Diptera: Culicidae). *Journal of Medical Entomology*, 5 (2), 135–159. DOI: 10.1093/jmedent/5.2.135

Steffan, W.A. (1979) New *Culex (Acallyntrum)* from the Papuan Subregion (Diptera: Culicidae). *Journal of Medical Entomology*, 16 (2), 133–139. DOI: 10.1093/jmedent/16.2.133

Steffan, W.A. & Evenhuis, N.L. (1981) Biology of *Toxorhynchites*. *Annual Review of Entomology*, 26, 159–181. DOI: 10.1146/annurev.en.26.010181.001111

Steffan, W.A. & Evenhuis, N.L. (1982) A new species of *Toxorhynchites* from Papua New Guinea (Diptera: Culicidae). *Mosquito Systematics*, 14 (1), 1–13.

Steffan, W.A. & Evenhuis, N.L. (1985) Classification of the subgenus *Toxorhynchites* (Diptera: Culicidae) I. Australasian, eastern Palaearctic, and Oriental species-groups. *Journal of Medical Entomology*, 22 (4), 421–446. DOI: 10.1093/jmedent/22.4.421

Steffan, W.A., Evenhuis, N.L. & Manning, D.L. (1980) Annotated bibliography of *Toxorhynchites* (Diptera: Culicidae). *Journal of Medical Entomology Supplement*, 3, 1–140. DOI: 10.1093/jmedent/17.suppl3.1

Stegniy, V.N. & Kabanova, V.M. (1976) Cytoecological study of indigenous populations of the malaria mosquito in the territory of the U.S.S.R. I. Identification of a new species of *Anopheles* in the *maculipennis* complex by the cytodiagnostic method [in Russian]. *Meditsinskaya Parazitologiya i Parazitarnye Bolezni*, 45 (2), 192–198.

Stephens, J.F. (1825) Some observations on the British Tipulidae, together with descriptions of the species of *Culex* and *Anopheles* found in Britain. *Zoological Journal*, 1 (4), 448–457.

Stephens, J.F. (1828) Note on the foregoing paper, with a description of a new species of *Anopheles*. *Zoological Journal*, 3 (12), 502–504.

St John, O. (2007) *Phylogeny of the genus* Culex *(Diptera: Culicidae)*. MRes Thesis, Department of Life Sciences, Imperial College, London, UK.

Stoker, W.J. & Waktoedi Koesoemawinangoen, R. (1949) *Illustrated Map of the Anopheline Imagines of Indonesia (English Translation)*. Ministry of Health of Indonesia, Djakarta, Indonesia.

Stone, A. (1939) Two new *Aedes* from Guam (Diptera, Culicidae). *Proceedings of the Entomological Society of Washington*, 41 (5), 162–165.

Stone, A. (1944) Notes on the genus *Trichoprosopon* (Diptera, Culicidae). *Revista de Entomologia*, 15 (3), 335–341.

Stone, A. (1945) A new species of *Aedes* from Saipan and the larva of *Aedes pandani* (Diptera: Culicidae). *Proceedings of the Entomological Society of Washington*, 47 (3), 65–69.

Stone, A. (1948) A change of name in mosquitoes (Diptera, Culicidae). *Proceedings of the Entomological Society of Washington*, 50 (6), 161.

Stone, A. (1957) Corrections in the taxonomy and nomenclature of mosquitoes (Diptera, Culicidae). *Proceedings of the Entomological Society of Washington* (for 1956), 58 (6), 333–344.

Stone, A. (1961) A synoptic catalog of the mosquitoes of the world, supplement I (Diptera: Culicidae). *Proceedings of the Entomological Society of Washington*, 63 (1), 29–52.

Stone, A. (1966) A new species of *Coquillettidia* from Samoa (Diptera: Culicidae). *Proceedings of the Entomological Society of Washington*, 68 (4), 331–333.

Stone, A. (1967) A synoptic catalog of the mosquitoes of the world, supplement III (Diptera: Culicidae). *Proceedings of the Entomological Society of Washington*, 69 (3), 197–224.

Stone, A. (1970) A synoptic catalog of the mosquitoes of the world, supplement IV (Diptera: Culicidae). *Proceedings of the Entomological Society of Washington*, 72 (2), 137–171.

Stone, A. & Barreto, P. (1969) A new genus and species of mosquito from Colombia, *Galindomyia leei* (Diptera, Culicidae, Culicini). *Journal of Medical Entomology*, 6 (2), 143–146. DOI: 10.1093/jmedent/6.2.143

Stone, A. & Bohart, R.M. (1944) Studies on mosquitoes from the Philippine Islands and Australasia (Diptera: Culicidae). *Proceedings of the Entomological Society of Washington*, 46 (8), 205–225.

Stone, A. & Farner, D.S. (1945) Further notes on the *Aedes scutellaris* group (Diptera, Culicidae). *Proceedings of the Biological Society of Washington*, 58, 155–162.

Stone, A. & Hair, J.A. (1968) A new *Culex (Melanoconion)* from Florida (Diptera, Culicidae). *Mosquito News*, 28 (1), 39–41.

Stone, A. & Knight, K.L. (1956) A new name in Culicidae (Diptera). *Proceedings of the Entomological Society of Washington*, 58 (1), 34.

Stone, A. & Knight, K.L. (1957a) Type specimens of mosquitoes in the United States National Museum, V: the Sabethini (Diptera, Culicidae). *Journal of the Washington Academy of Sciences*, 47 (4), 117–126.

Stone, A. & Knight, K.L. (1957b) Proposed use of the plenary powers to validate the generic name *Toxorhynchites* Theobald as published in 1901 in the "Journal of Tropical Medicine" (Class Insecta, Order Diptera). *Bulletin of Zoological Nomenclature*, 13 (4), 128–131. DOI: 10.5962/bhl.part.3546

Stone, A. & Knight, K.L. (1958) Two new names in mosquitoes (Diptera, Culicidae). *Proceedings of the Entomological Society of Washington*, 60 (2), 69.

Stone, A. & Penn, G.H. (1947) A new species of *Culex* from New Guinea (Diptera, Culicidae). *Journal of the Washington Academy of Sciences*, 37 (3), 89–91.

Stone, A. & Penn, G.H. (1948) A new subgenus and two new species of the genus *Culex* L. (Diptera, Culicidae). *Proceedings of the Entomological Society of Washington*, 50 (5), 109–120.

Stone, A. & Rosen, L. (1952) A new species of *Aedes* from Tahiti (Diptera: Culicidae). *Proceedings of the Hawaiian Entomological Society*, 14 (3), 425–428.

Stone, A. & Rosen, L. (1953) A new species of *Culex* from the Marquesas Islands and the larva of *Culex atriceps* Edwards (Diptera: Culicidae). *Journal of the Washington Academy of Sciences*, 43 (11), 354–358.

Stone, A. & Thurman, E.B. (1958) *Armigeres (Armigeres) baisasi*, a new mosquito from the Philippine Islands (Diptera: Culicidae). *Journal of the Washington Academy of Sciences*, 48 (7), 240–243.

Stone, A., Knight, K.L. & Starcke, H. (1959) *A Synoptic Catalog of the Mosquitoes of the World (Diptera, Culicidae)*. The Thomas Say Foundation. Volume VI. Entomological Society of America, College Park, MD, USA.

Strickland, C. (1916) A new species of protanopheline from Malaya, *Myzorhynchus hunteri*. *Indian Journal of Medical Research*, 4 (2), 263–270.

Strickland, C. & Chowdhury, K.L. (1927a) *An illustrated key to the identification of the anopheline larvae of India, Ceylon, and Malaya, west of Wallace's Line with practical notes on their collection*. Thacker, Spink & Co., Calcutta & Simla, India.

Strickland, C. & Chowdhury, K.L. (1927b) A new species of anopheline *A. pseudojamesi* common in Bengal. *Indian Medical Gazette*, 62 (5), 240–243.

Strickland, E.H. (1911) Some new Culicidae from Western Australia, South Queensland, and Tasmania. *Entomologist*, 44 (576, 577, 578), 179–182, 201–204, 249–251. DOI: 10.5962/bhl.part.22058

Strickman, D. (1990) *Culex pseudostigmatosoma*, *Cx. yojoae* and *Cx. aquarius*: new Central American species in the subgenus *Culex* (Diptera: Culicidae). *Mosquito Systematics* (for 1989), 21 (3), 143–177.

Strickman, D. & Pratt, J. (1989) Redescription of *Cx. corniger* Theobald and elevation of *Culex (Culex) lactator* Dyar and Knab from synonymy based on specimens from Central America (Diptera: Culicidae). *Proceedings of the Entomological Society of Washington*, 91 (4), 551–574.

Su, L. & Zhang, Y. (1987) A new species of the subgenus *Ochlerotatus* Lynch Arribalzaga (Diptera: Culicidae) [in Chinese]. *Acta Entomologica Sinica*, 30 (4), 450–452.

Su, L. & Zhang, Y. (1988) A new species of *Aedes* (Diptera: Culicidae) [in Chinese]. *Entomotaxonomia*, 10 (1–2), 11–13.

Suguna, S.G., Rathinam, K.G., Rajavel, A.R. & Dhanda, V. (1994) Morphological and chromosomal descriptions of new species in the *Anopheles subpictus* complex. *Medical and Veterinary Entomology*, 8 (1), 88–94. DOI: 10.1111/j.1365-2915.1994.tb00392.x

Sukowati, S., Baimai, V., Harun, S., Dasuki, Y., Andris, H. & Efriwati, M. (1999) Isozyme evidence for three sibling species in the *Anopheles sundaicus* complex from Indonesia. *Medical and Veterinary Entomology*, 13 (4), 408–414. DOI: 10.1046/j.1365-2915.1999.00197.x

Sum, J.-S., Lee, W.-C., Amir, A., Braima, K.A., Jeffery, J., Abdul-Aziz, N.M., Fong, M.-Y. & Lau, Y.-L. (2014) Phylogenetic study of six species of *Anopheles* mosquitoes in Peninsular Malaysia based on inter-transcribed spacer region 2 (ITS2) of ribosomal DNA. *Parasites & Vectors*, 7, 309. DOI: 10.1186/1756-3305-7-309

Suman, D.S., De, S., Sharma, G., Chandra, K. & Banerjee, D. (2021) *Culex (Culex) gaugleri*, a new species (Diptera: Culicidae) from India. *Records of the Zoological Survey of India*, 121 (4), 429–439. DOI: 10.26515/rzsi/v121/i4/2021/166022

Suman, D.S., Sharma, G., De, S., Kumar, V., Veer, V., Chandra. K. & Banerjee, D. (2022) Description of *Culex (Culex) kodaikanalensis* sp. nov. (Diptera: Culicidae) from India with phyletic analysis using mitochondrial cytochrome oxidase I and 16S rRNA genes. *International Journal of Tropical Insect Science*, 42 (4), 2969–2978. DOI: 10.1007/s42690-022-00825-3

Sun, L., Li, T.-J., Fu, W.-B., Yan, Z.-T., Si, F.-L., Zhang, Y.-J., Mao, Q.-M., Demari-Silva, B. & Chen, B. (2019) The complete mt genomes of *Lutzia halifaxii*, *Lt. fuscanus* [*sic*] and *Culex pallidothorax* (Diptera: Culicidae) and comparative analysis of 16 *Culex* and *Lutzia* mt genome sequences. *Parasites & Vectors*, 12, 368. DOI: 10.1186/s13071-019-3625-2

Surcouf, J. (1912) Note sur les culicides. *Bulletin du Muséum national d'histoire naturelle*, 18 (2), 59–61.

Surcouf, J.M.R. & Gonzalez-Rincones, R. (1912) Diptères piqueurs et suceurs de sang. Actuellement connus, de la République do Venezuela. *Archives de Parasitologie*, 15 (2), 248–314.

Sutil O., E. & Pulido F., J. (1978) *Sabethes (Sabethes) carrilloi* sp. n. de Venezuela. *Boletin de la Direccion de Malariologia y Saneamiento Ambiental*, 18 (3), 199–204.

Sutil Oramas, E. & Pulido F., J. (1974) *Wyeomyia cova-garciai* n. sp. *Boletín informativo de la Dirección de Malariología y Saneamiento Ambiental*, 14 (1–2), 24–26.

Sutil Oramas, E., Pulido Florenzano, J. & Amarista Meneses, J.R. (1987) Dos especies nuevas de *Culex* de Venezuela (Diptera: Culicidae). *Boletin de la Direccion de Malariologia y Saneamiento Ambiental*, 27 (1–4), 81–85.

Swellengrebel, N.H. (1914) Een nieuwe Anopheline veer Deli: *Myzorhynchus Argyropus* [*sic*] n. sp. (*met autoreferaat.*). *Geneeskundig Tijdschrift voor Nederlandsch-Indië*, 54, 334–337, 1 pl.

Swellengrebel, N.H. & Rodenwaldt, E. (1932) *Die anophelen von Niederländisch-Ostindien*. 3te auflage. Gustav Fischer, Jena, Germany.

Swellengrebel, N.H. & Swellengrebel de Graaf, J.M.H. (1920) Addenda to bescription [*sic*] of larvea [*sic*] of Netherlands Indian anophelines. *Mededeelingen van den Burgerlijken Geneeskundigen Dienst in Nederlandsch-Indië* (for 1919), 9 (Add.), 1–3, 2 pls.

Szadziewski, R. (1998) New mosquitoes from Baltic amber (Diptera: Culicidae). *Polskie Pismo Entomologiczne*, 67 (3–4), 233–244.

Szadziewski, R. & Gilka, W. (2011) A new fossil mosquito, with notes on the morphology and taxonomy of other species reported from Eocene Baltic amber (Diptera: Culicidae). *Polskie Pismo Entomologiczne*, 80 (4), 765–777. DOI: 10.2478/v10200-011-0057-5

Szadziewski, R. & Szadziewska, M.M. (1985) *Culex erikae* sp. n. (*Diptera, Culicidae*) from the Baltic amber. *Polskie Pismo Entomologiczne*, 55, 513–518.

Szadziewski, R., Sontag, E. & Szwedo, J. (2019) Mosquitoes of the extant avian malaria vector *Coquillettidia* Dyar, 1905 from Eocene Baltic amber (Diptera: Culicidae). *Palaeoentomology*, 2 (6), 650–656. DOI: 10.11646/palaeoentomology.2.6.16

Taai, K. & Harbach, R.E. (2015) Systematics of the *Anopheles barbirostris* species complex (Diptera: Culicidae: Anophelinae) in Thailand. *Zoological Journal of the Linnean Society*, 174 (2), 244–264. DOI: 10.1111/zoj.12236

Takano, K.T., Nguyen, N.T.H., Nguyen, B.T.H., Sunahara, T., Yasunami, M., Nguyen, M.D. & Takagi, M. (2010) Partial mitochondrial DNA sequences suggest the existence of a cryptic species within the Leucosphyrus group [*sic*] of the genus *Anopheles* (Diptera: Culicidae), forest malaria vectors, in northern Vietnam. *Parasites & Vectors*, 3, 41. DOI: 10.1186/1756-3305-3-41

Talaga, S. & Duchemin, J.-B. (2023) Mosquitoes (Diptera: Culicidae) of the Amazonian savannas of French Guiana with a description of two new species. *Journal of Vector Ecology*, 49 (1), 15–27 (published 21 December 2023 for 2024). DOI: 10.52707/1081-1710-49.1.15

Talaga, S. & Gendrin, M. (2022) Three new species of *Culex (Melanoconion)* (Diptera: Culicidae) from French Guiana based on morphological and molecular data. *Zootaxa*, 5205 (2), 177–189. DOI: 10.11646/zootaxa.5205.2.5

Talaga, S., Dejean, A., Carinci, R., Gaborit, P., Dusfour, I. & Girod, R. (2015) Updated checklist of the mosquitoes (Diptera: Culicidae) of French Guiana. *Journal of Medical Entomology*, 52 (5), 770–782. DOI: 10.1093/jme/tjv109

Talaga, S., Duchemin, J.-B., Girod, R. & Dusfour, I. (**2020b**) The *Culex* mosquitoes (Diptera: Culicidae) of French Guiana: a comprehensive review with the description of three new species. *Journal of Medical Entomology*, 58 (1), 182–221. DOI: 10.1093/jme/tjaa205

Talaga, S., Gantier, J.-C. & Girod, R. (**2020a**) Mosquitoes (Diptera: Culicidae) originally described from French Guiana. *Zootaxa*, 4747 (2), 361–377. DOI: 10.11646/zootaxa.4747.2.8

Tanaka, K. (1998) A new species of mosquitoes of *Aedes (Ochlerotatus)* (Diptera, Culicidae) from Japan. *Japanese Journal of Systematic Entomology*, 4 (2), 215–225.

Tanaka, K. (2003) Studies on the pupal mosquitoes of Japan (9). Genus *Lutzia*, with establishment of two new subgenera, *Metalutzia* and *Insulalutzia* (Diptera, Culicidae). *Japanese Journal of Systematic Entomology*, 9 (2), 159–169.

Tanaka, K. (2004) Studies on the pupal mosquitoes of Japan (11) subgenera *Oculeomyia* (stat. nov.) and *Sirivanakarnius* (nov.) of the genus *Culex*, with a key of pupal mosquitoes from Ogasawara-guntô (Diptera: Culicidae). *Medical Entomology and Zoology*, 55 (3), 217–231. DOI: 10.7601/mez.55.217

Tanaka, K. (2014) Family Culicidae. In: Nakamura, T., Saigusa, T. & Suwa, M. (eds), *Catalogue of the Insects of Japan. Volume 8, Diptera. Part 1, Nematocera – Brchycera Aschiza*. Entomological Society of Japan, Fukuoka, Japan, pp. 181–201.

Tanaka, K. (2018) Taxonomic catalogue of the mosquitoes from Japan, with a key to Japanese genera and subgenera of the subtribe Aedina (Diptera: Culicidae). *Makunagi/Acta Dipterologica*, Supplement 3, 157 pp.

Tanaka, K. & Mizusawa, K. (1973) Two new species of the genus *Aedes (Neomacleaya)* from the Ryukyu Islands (Diptera, Culicidae). *Bulletin of the National Science Museum, Tokyo*, 16 (4), 625–638, 2 pls.

Tanaka, K., Mizusawa, K. & Saugstad, E.S. (1975a) Two new crab hole mosquitoes of the genus *Uranotaenia (Pseudoficalbia)* from the Ryukyu Islands (Diptera, Culicidae). *Mosquito Systematics*, 7 (1), 27–40.

Tanaka, K., Mizusawa, K. & Saugstad, E.S. (1975b) A new species of the genus *Aedes (Aedes)* from Japan, with synonymical notes on Japanese species of the subgenus *Aedes* (Diptera, Culicidae). *Mosquito Systematics*, 7 (3), 41–58, 174–177.

Tanaka, K., Mizusawa, K. & Saugstad, E.S. (1979) A revision of the adult and larval mosquitoes of Japan (including the Ryukyu Archipelago and the Ogasawara Islands) and Korea (Diptera: Culicidae). *Contributions of the American Entomological Institute*, 16, vii, 1–987.

Tantely, M.L., Le Goff, G., Boyer, S. & Fontenille, D. (2016) An updated checklist of mosquito species (Diptera: Culicidae) from Madagascar. *Parasite*, 23, 20. DOI: 10.1051/parasite/2016018

Taylor, B. (1972) A new species of *Aedes* from San Cristobal, British Solomon Islands Protectorate. *Journal of Medical Entomology*, 9 (4), 317–318. DOI: 10.1093/jmedent/9.4.317

Taylor, B. & Maffi, M. (1991) *Anopheles (Cellia) rennellensis*, a new species within the *punctulatus* complex of *Anopheles* (Diptera, Culicidae) from Rennell Island. In: Wolff, T. (ed.), *Natural History of Rennell Island, British Solomon Islands. Volume* 8 *(Zoology). Scientific Results of The Noona Dan Expedition (Rennell Section, 1962) and The Danish Rennell Expedition, 1965*. Danish Science Press, Copenhagen, Denmark, pp. 195–198.

Taylor, B. & Tenorio, J.A. (1974) *Aedes (Finlaya) maffii*, a new species of mosquito from the British Solomon Islands (Diptera: Culicidae). *Journal of Medical Entomology*, 11 (5), 577–581. DOI: 10.1093/jmedent/11.5.577

Taylor, F.H. (1912) Description of mosquitoes collected in the Northern Territory during the expedition, 1911. *Bulletin of the Northern Territory*, la, 25–34, 1 pl.

Taylor, F.H. (1914a) A revision of the *Culicidae* in the Macleay Museum, Sydney. *Proceedings of the Linnean Society of New South Wales* (for 1913), 38 (4), 747–760, 1 pl. DOI: 10.5962/bhl.part.13575

Taylor, F.H. (1914b) Contributions to a knowledge of Australian Culicidae. No. i. *Proceedings of the Linnean Society of New South Wales*, 39 (3), 454–468, 4 pls. DOI: 10.5962/bhl.part.2288

Taylor, F.H. (1914c) XXX. The *Culicidae* of Australia.—I. *Transactions of the Entomological Society of London*, 61 (4), 683–708, 4 pls. DOI: 10.1111/j.1365-2311.1914.tb01794.x

Taylor, F.H. (1914d) Culicidae from Papua. *Transactions of the Entomological Society of London*, 62 (1), 185–205, 2 pls. DOI: 10.1111/j.1365-2311.1914.tb01395.x

Taylor, F.H. (1915) Contributions to a knowledge of Australian Culicidae. No. ii. *Proceedings of the Linnean Society of New South Wales*, 40 (1), 176–184, 2 pls. DOI: 10.5962/bhl.part.18868

Taylor, F.H. (1916) Contributions to a knowledge of Australian *Culicidae*. No. iii. *Proceedings of the Linnean Society of New South Wales*, 41 (3), 564–574. DOI: 10.5962/bhl.part.15324

Taylor, F.H. (1919) Contributions to a knowledge of Australian Culicidae. No. iv. *Proceedings of the Linnean Society of New South Wales* (for 1918), 43 (4), 826–843, 4 pls.

Taylor, F.H. (1927) Two new species of Australian Culicidae (Diptera). *Bulletin of Entomological Research*, 18 (1), 67–69. DOI: 10.1017/S0007485300019696

Taylor, F.H. (1929) Notes on Australian Culicidae (Dipt.). *Bulletin of Entomological Research*, 20 (3), 271–277, 1 pl. DOI: 10.1017/S0007485300021258

Taylor, F.H. (1933) A new species of *Finlaya* (order Diptera: family Culicidae) from Papua. *Australian Zoologist*, 7 (5), 370.

Taylor, F.H. (1934) The Diptera of the Territory of New Guinea. I. Family Culicidae. *Proceedings of the Linnean Society of New South Wales*, 59 (3–4), 229–236.

Taylor, F.H. (1944) Contributions to a knowledge of Australian Culicidae. No. VII. *Proceedings of the Linnean Society of New South Wales*, 69 (3–4), 120–128.

Taylor, F.H. (1946) Contributions to a knowledge of Australian Culicidae. No. VIII. *Proceedings of the Linnean Society of New South Wales* (for 1945), 70 (5–6), 211–212.

Tenorio, J.A. (1975) *Bironella (Brugella) obscura*, a new species of mosquito from New Guinea (Diptera: Culicidae). *Journal of Medical Entomology* (for 1974), 11 (6), 734–738. DOI: 10.1093/jmedent/11.6.734

Tenorio, J.A. (1977) Revision of the genus *Bironella* (Diptera: Culicidae). *Journal of Medical Entomology*, 14 (3), 317–361. DOI: 10.1093/jmedent/14.3.317

Tewari, S.C. & Hiriyan, J. (1992a) Description of a new species of *Aedes (Rhinoskusea)* from South India. *Mosquito Systematics* (for 1991), 23 (2), 123–131.

Tewari, S.C. & Hiriyan, J. (1992b) Description of two new species of *Aedes (Diceromyia)* from South India (Diptera: Culicidae). *Mosquito Systematics*, 24 (2), 154–175.

Theobald, F.V. (1900a) *Report on the collections of mosquitoes (Culicidae) received at the British Museum (Natural History) from various parts of the world in connection with the investigation into the causes of malaria, conducted by the Colonial Office and the Royal Society*. British Museum (Natural History), London, UK. DOI: 10.5962/bhl.title.106823

Theobald, F.V. (1900b) A new *Anopheles (A. paludis)* from Sierra Leone. *Report of the Malaria Commission of the Royal Society*, 1, 75–76.

Theobald, F.V. (1901a) The classification of mosquitoes. *Journal of Tropical Medicine*, 4, 229–235 (July 1901).

Theobald, F.V. (1901b) *A monograph of the Culicidae or mosquitoes*. Vol. 1. British Museum (Natural History), London, UK. DOI: 10.5962/bhl.title.58067

Theobald, F.V. (1901c) *A monograph of the Culicidae or mosquitoes*. Vol. 2. British Museum (Natural History), London, UK. DOI: 10.5962/bhl.title.58067

Theobald, F.V. (1901d) Notes on a collection of mosquitoes from West Africa, and descriptions of new species. *Memoirs of the Liverpool School of Tropical Medicine*, 4 (Appendix), i–xiv, 3 pls.

Theobald, F.V. (1902a) A short description of the Culicidae of India, with descriptions of new species of *Anopheles*. *Proceedings of the Royal Society of London*, 69 (456), 367–394, 1 pl. DOI: 10.1098/rspl.1901.0122

Theobald, F.V. (1902b) The classification of the Anophelina. *Journal of Tropical Medicine*, 5, 181–183.

Theobald, F.V. (1903a) *A monograph of the Culicidae or mosquitoes*. Vol. 3. British Museum (Natural History), London, UK. DOI: 10.5962/bhl.title.58067

Theobald, F.V. (1903b) Report on a collection of mosquitoes and other flies from Equatorial East Africa and the Nile provinces of Uganda. *Reports of the Sleeping Sickness Commission of the Royal Society*, 7 (3), 33–42, map.

Theobald, F.V. (1903c) Report on a collection of mosquitoes or Culicidae, etc, from Gambia, and descriptions of new species. *Memoirs of the Liverpool School of Tropical Medicine*, 10 (Appendix), i–xi, 2 pls.

Theobald, F.V. (1903d) A new culicid from Senegal and notes on the species of mosquitoes, etc. *Memoirs of the Liverpool School of Tropical Medicine*, 11, Part 2, (Appendix), i–iii, 1 pl.

Theobald, F.V. (1903e) Two new Australian culicids. *Entomologist*, 36 (481), 154–157.

Theobald, F.V. (1903f) New Culicidae from the Federated Malay States. *Entomologist*, 36 (485), 256–259.

Theobald, F.V. (1903g) Two new Jamaican Culicidae. *Entomologist*, 36 (486), 281–283.

Theobald, F.V. (1904a) A new culicid genus from Uganda. *Journal of Tropical Medicine*, 7, 17–18.

Theobald, F.V. (1904b) The mosquitoes of Egypt, the Sudan and Abyssinia. In: Balfour, A. (director), *First Report of the Wellcome Research Laboratories at the Gordon Memorial College Khartoum*. Department of Education, Sudan Government, Khartoum, Sudan, pp. 62–83, 2 pls.

Theobald, F.V. (1904c) New Culicidae from the Federated Malay States. *Entomologist*, 37 (488, 489, 490, 491, 493, 495, 496), 12–15, 36–39, 77–78, 111–113, 163–165, 211–213, 236–239. DOI: 10.5962/bhl.part.2880

Theobald, F.V. (1905a) A new genus of Culicidae. *Entomologist*, 38 (501), 52–56.

Theobald, F.V. (1905b) New Culicidae from the west coast of Africa. *Entomologist*, 38 (503), 101–104, 154–158.

Theobald, F.V. (1905c) A new *Stegomyia* from the Transvaal. *Entomologist*, 38 (508), 224–225.

Theobald, F.V. (1905d) *The mosquitoes or Culicidae of Jamaica*. Institute of Jamaica, Kingston, Jamaica.

Theobald, F.V. (1905e) *Genera Insectorum. Diptera, Fam. Culicidae*. Fascicle 26. V. Verteneuil & L. Desmet, Bruxelles [Brussels], Belgium, 50 pp., 2 pls.

Theobald, F.V. (1905f) A catalogue of the Culicidae in the Hungarian National Museum, with descriptions of new genera and species. *Annales Musei Nationalis Hungarici*, 3 (1), 61–119, figure legends, 4 pls.

Theobald, F.V. (1905g) New Culicidae from India, Africa, British Guiana, and Australia. *Journal of Economic Biology*, 1, 17–36, 2 pls.

Theobald, F.V. (1905h) Some new mosquitoes from Ceylon. *Journal of the Bombay Natural History Society*, 16 (2), 237–250, 2 pls.

Theobald, F.V. (1905i) A new *Ficalbia* from West Africa. *Annals and Magazine of Natural History*, 15 (86), 199–200.

Theobald, F.V. (1906) Report on economic entomology. Part I. Second report on the mosquitoes or Culicidae of the Sudan. In: Balfour, A. (director), *Second Report of the Wellcome Research Laboratories at the Gordon Memorial College, Khartoum*. Department of Education, Sudan Government, Khartoum, Sudan, pp. 67–83.

Theobald, F.V. (1907) *A monograph of the Culicidae or mosquitoes*. Vol. 4. British Museum (Natural History), London, UK. DOI: 10.5962/bhl.title.58067

Theobald, F.V. (1908a) Notes on some Transvaal mosquitoes, including two new species and a new variety. *Entomologist*, 41 (540), 106–109. DOI: 10.5962/bhl.part.10161

Theobald, F.V. (1908b) First report on the collection of Culicidae and Corethridae in the Indian Museum, Calcutta, with descriptions of new genera and species. *Records of the Indian Museum*, 2 (3), 287–302.

Theobald, F.V. (1908c) New mosquitoes from the Sudan and list and synoptic table of all the known Sudanese species. In: Balfour, A. (director), *Third Report of the Wellcome Research Laboratories at the Gordon Memorial College, Khartoum*. Baillière, Tindall & Cox, London, UK, pp. 249–267, 2 pls.

Theobald, F.V. (1909) *Descriptions of the new mosquitoes collected by Dr. Graham in Ashanti*. Gold Coast Miscellaneous Publication No. 237. Great Britain Colonial Office, London, UK.

Theobald, F.V. (1910a) Second report on the collection of Culicidae in the Indian Museum, Calcutta, with descriptions of new genera and species. *Records of the Indian Museum*, 4 (1), 1–33, 3 pls.

Theobald, F.V. (1910b) The Culicidae of Fiji, including two new species. *Entomologist*, 43 (565), 155–159. DOI: 10.5962/bhl.part.28540

Theobald, F.V. (1910c) *A monograph of the Culicidae or mosquitoes*. Vol. 5. British Museum (Natural History), London, UK. DOI: 10.5962/bhl.title.58067

Theobald, F.V. (1911a) Uganda Culicidae. In: Theobald, F.V., *Novae Culicidae. Part 1*. Published by the author, Wye, Kent, UK, pp. 9–35.

Theobald, F.V. (1911b) The Culicidae or mosquitoes of the Transvaal. Report of the Division of Entomology of the Department of Agriculture, Union of South Africa, 1, 232–272, 1 pl.

Theobald, F.V. (1911c) A new genus and two new species of Culicidae from the Sudan. In: Balfour, A. (director), *Fourth Report of the Wellcome Tropical Research Laboratories at the Gordon Memorial College, Khartoum. Volume B.—General Science*. Baillière, Tindall & Cox, London, UK, pp. 151–15.

Theobald, F.V. (1911d) Culicidae of the R. Zool. Soc. "Natura Artis Magistra", Amsterdam and description of three new species. *Tijdschrift voor Entomologie*, 54, 233–240.

Theobald, F.V. (1912a) Three new Culicidae from the Transvaal. *Entomologist*, 45 (586), 92–93. DOI: 10.5962/bhl.part.9917

Theobald, F.V. (1912b) No. V.—Diptera, Culicidae. *Transactions of the Linnean Society of London. Second Series: Zoology*, 15 (1), 81–94, 1 pl. DOI: 10.1111/j.1096-3642.1912.tb00090.x

Theobald, F.V. (1913) New Culicidae from the Sudan. *Annals of Tropical Medicine and Parasitology*, 7 (4), 591–602. DOI: 10.1080/00034983.1913.11687629

Theobald, F.V. (1914) A new mosquito from Samoa. *Entomologist*, 47 (608), 36–37. DOI: 10.5962/bhl.part.15119

Thielman, A.C. & Hunter, F.F. (2007) A photographic key to the adult female mosquito species of Canada (Diptera: Culicidae). *Canadian Journal of Arthropod Identification*, 4, [ii], 116 pp. DOI: 10.3752/cjai.2007.04

Thompson, G.A. (1956) *Aedes stenei*, a new species of mosquito from Jamaica, B.W.I., with redescription of a closely related species (Diptera, Culicidae). *Mosquito News*, 16 (3), 196–200.

Thomson, C.G. (1869) Diptera. Species nova descripsit. In: Kongliga Svenska Vtenskaps Akademien (corporate author), *Kongliga svenska Fregatten Eugenies resa omkring jorden. Vetenskapliga Iakitagelser. II. Zoologi. 1. Insecta*. P. A. Norstedt & Soner, Stockholm. Sweden, pp. 443–614, 1 pl.

Thomson, F.W. (1903) Notes on the Culicidae of Dehra Dun. With a description of a new *Mansonia*, which mimics *Anopheles*. *Journal of Tropical Medicine*, 6, 314–315.

Thurman, D.C., Jr (1954) *Ayurakitia*, a new genus of mosquito from northern Thailand (Diptera: Culicidae). *Journal of the Washington Academy of Sciences*, 44 (6), 197–200.

Thurman, D.C. (1954) The discovery of *Paraedes (Udaya) argyrurus* Edwards, 1934, in Thailand (Diptera: Culicidae). *Mosquito News*, 14 (2), 83–86.

Thurman, D.C. (1955) *Culex (Culiciomyia) termi*, an unusual new mosquito from Thailand (Diptera, Culicidae). *Proceedings of the Entomological Society of Washington*, 57 (1), 17–23.

Thurman, D.C. & Thurman, E.B. (1958) Two new species in the genus *Armigeres* Theobald, 1901, from northern Thailand (Diptera: Culicidae). *Journal of the Washington Academy of Sciences*, 48 (6), 186–187.

Thurman, E.B. (1959) A contribution to a revision of the Culicidae of northern Thailand. *University of Maryland Agricultural Experimental Station Bulletin*, A-100, 1–182.

Thurman, E.B. & Winkler, E.C. (1950) A new species of mosquito in California, *Aedes (Ochlerotatus) bicristatus* (Diptera, Culicidae). *Proceedings of the Entomological Society of Washington*, 52 (5), 237–250.

Toma, T. & Higa, Y. (2004) A new species of *Ficalbia* (Diptera: Culicidae) from Iromote Island, Okinawa, Ryukyu Archipelago, Japan. *Medical Entomology and Zoology*, 55 (3), 195–199. DOI: 10.7601/mez.55.195

Toma, T. & Miyagi, I. (2002) *Armigeres (Armigeres) sembeli* (Diptera: Culicidae), a new species from Sulawesi, Indonesia and redescription of *Armigeres (Armigeres) giveni* Edwards. *Medical Entomology and Zoology*, 53 (4), 257–265. DOI: 10.7601/mez.53.257

Toma, T. & Miyagi, I. (2003) *Armigeres (Armigeres) laoensis* sp. nov. (Diptera: Culicidae) from Khammouane Province, Lao PDR. *Medical Entomology and Zoology*, 54 (2), 169–172. DOI: 10.7601/mez.54.169

Toma, T. & Mogi, M. (2003) *Topomyia (Suaymyia) miyagii* (Diptera: Culicidae): a new species from Flores Is., Indonesia. *Medical Entomology and Zoology*, 54 (1), 25–30. DOI: 10.7601/mez.54.25_1

Toma, L., Catalani, M., Antonella, A., Goffredo, M., Romi, R. & Di Luca, M. (2016) Finding of *Anopheles (Anopheles) hyrcanus* (Pallas, 1771) (Diptera, Culicidae) during the entomological surveillance for West Nile virus in Umbria, Italy. *Veterinaria Italiana*, 53 (3), 263–266. DOI: 10.12834/VetIt.668.3283.3

Toma, T., Miyagi, I. & Cabrera, B.D. (1984) *Culex (Culiciomyia) azurini*, a new crab hole mosquito species from the Philippines. *Mosquito Systematics*, 16 (2), 172–182.

Toma, T., Miyagi, I. & Syafruddin. (1995) Description of *Armigeres (Armigeres) alkatirii*, a new species from Sulawesi, Indonesia (Diptera: Culicidae). *Mosquito Systematics*, 27 (2), 110–117.

Toma, T., Miyagi, I. & Tamashiro, M. (2014) Blood meal identification and feeding habits of *Uranotaenia* species collected in the Ryukyu Archipelago. *Journal of the American Mosquito Control Association*, 30 (3), 215–218. DOI: 10.2987/14-6398R.1

Toma, T., Miyagi, I. & Tanaka, K. (1990) A new species of *Toxorhynchites* mosquito (Diptera: Culicidae) from the Ryukyu Archipelago, Japan. *Journal of Medical Entomology*, 27 (3), 344–355. DOI: 10.1093/jmedent/27.3.344

Torres Cañamares, F. (1951) Una nueva especie de *Aedes (Dipt. Cul.)*. *Eos: Revista Española de Entomología*, 27 (1), 79–92.

Torres-Gutierrez, C. & Sallum, M.A.M. (2015) Catalog of the subgenus *Melanoconion* of *Culex* (Diptera: Culicidae) for South America. *Zootaxa*, 4028 (1), 1–50. DOI: 10.11646/zootaxa.4028.1.1

Torres-Gutierrez, C., de Oliveira, T.M.P., Emerson, K.J., Bergo, E.S. & Sallum, M.A.M. (2018) Molecular phylogeny of *Culex* subgenus *Melanoconion* (Diptera: Culicidae) based on nuclear and mitochondrial protein-coding genes. *Royal Society Open Science*, 5, 171900. DOI: 10.1098/rsos.171900

Townson, H., Dyer, N., McAlister, E., Satoto, T.B.T., Boewono, D.T., Bangs, M.J. & Harbach, R.E. (2013) Systematics of *Anopheles barbirostris* van der Wulp and a sibling species of the Barbirostris Complex (Diptera: Culicidae) in eastern Java, Indonesia. *Systematic Entomology*, 38 (1), 180–191. DOI: 10.1111/j.1365-3113.2012.00653.x

Tsai, C. & Lien, J. (1950) A new species of *Aëdes (Finlaya)* found in Taiwan. *Journal of the Formosan Medical Association*, 49 (5), 177–183.

Tsuda, Y. (2019) *An Illustrated Book of the Mosquitoes of Japan: Adult Identification, Geographic Distribution and Ecological Note* [in Japanese]. Hokuryukan, Tokyo, Japan.

Tsukamoto, M. (1989) Two new mosquito species from a pitcher plant of Mt. Kinabalu, Sabah, Malaysia: *Culex rajah* and *Toxorhynchites rajah* (Diptera: Culicidae). *Japanese Journal of Tropical Medicine and Hygiene*, 17 (3), 215–228. DOI: 10.2149/tmh1973.17.215

Tung, L.H. (1955) On three new species of mosquitoes from Kiangsi Province, China [in Chinese]. *Acta Entomologica Sinica*, 5 (9), 309–315.

Turell, M.J., Dohm, D.J., Sardelis, M.R., O'Quinn, M.L., Andreadis, T.G. & Blow, J.A. (2005) An update on the potential of North American mosquitoes (Diptera: Culicidae) to transmit West Nile virus. *Journal of Medical Entomology*, 42 (1), 57–62. DOI: 10.1093/jmedent/42.1.57

Turell, M.J., O'Guinn, M.L., Dohm, D.J. & Jones, J.W. (2001) Vector competence of North American mosquitoes (Diptera: Culicidae) for West Nile virus. *Journal of Medical Entomology*, 38 (2), 130–134. DOI: 10.1603/0022-2585-38.2.130

Tuten, H.C., Bridges, W.C. & Adler, P.H. (2012) Comparative morphology of the pyloric armature of adult mosquitoes (Diptera: Culicidae). *Arthropod Structure & Development*, 41 (5), 475–481. DOI: 10.1016/j.asd.2012.04.002

Tyagi, B.K., Hiriyan, J., Tewari, S.C., Ayanar, K., Samuel, P.P., Arunachalam, N., Paramasivan, R., Krishnamoorthy, R., Dhananjeyan, K.J., Leo, S.V. & Rajendran, R. (2009) Description of a new species, *Anopheles pseudosundaicus* (Diptera: Culicidae) from Kerala, India. *Zootaxa*, 2219, 49–60. DOI: 10.11646/zootaxa.2219.1.5

Tyagi, B.K., Munirathinam, A., Krishnamoorthy, R., Baskaran, G., Govindarajan, R., Krishnamoorthi, R., Mariappan, R., Dhananjeyan, K.J. & Venkatesh, A. (**2015b**) A revision of genus *Toxorhynchites* Theobald, 1901, in the South-East Asian countries, with description of a new species *Toxorhynchites (Toxorhynchites) darjeelingensis* from West Bengal, India (Diptera, Culicidae). *Halteres*, 6, 13–32.

Tyagi, B.K., Munirathinam, A. & Venkatesh, A. (**2015a**) A catalogue of Indian mosquitoes. *International Journal of Mosquito Research*, 2 (2), 50–97.

Tyson, W.H. (1970a) Contributions to the mosquito fauna of Southeast Asia. VII. Genus *Aedeomyia* Theobald in Southeast Asia. *Contributions of the American Entomological Institute*, 6 (2), 1–27.

Tyson, W.H. (1970b) Contributions to the mosquito fauna of Southeast Asia. VIII. Genus *Aedes*, subgenus *Mucidus* Theobald in Southeast Asia. *Contributions of the American Entomological Institute*, 6 (2), 28–80.

Unti, O. (1941) Anofelinos do vale do Rio Paraíba, *Anopheles (Nyssorhynchus) strodei* Root, 1926 com a descrição de tres variedades novas. *Publica Serviço de Profilaxia da Malária*, 33, 3–18.

Urbanelli, S., Porretta, D., Mastrantonio, V., Bellini, R., Pieraccini, G., Romoli, R., Crasta, G. & Nascetti, G. (2014) Hybridization, natural selection, and evolution of reproductive isolation: a 25-year survey of an artificial sympatric area between two mosquito sibling species of the *Aedes mariae* complex. *Evolution*, 68 (10), 3030–3038. DOI: 10.1111/evo.12490

Valencia, J.D. (1973) Mosquito studies (Diptera, Culicidae). XXXI. A revision of the subgenus *Carrollia* of *Culex*. *Contributions of the American Entomological Institute*, 9 (4), 1–134.

Van As, J., Davies, A.J. & Smit, N.J. (2015) Life cycle of *Hepatozoon affluomaloti* sp. n. (Apicomplexa: Haemogregarinidae) in crag lizards (Sauria: Cordylidae) and in culicine mosquitoes from South Africa. *Folia Parasitologica*, 62, 008. DOI: 10.14411/fp.2015.008

van den Assem, J. (1959) Notes on New Guinean species of *Tripteroides*, subgenus *Rachisoura* (Diptera, Culicidae), with descriptions of two new species. *Tijdschrift voor Entomologie*, 102 (1), 35–55, 1 pl.

van der Beek, J.G., Dijkstra, K.D.B., Boerlijst, S.P., Busscher, L., Kok, M.L., Braks, MA.H., Schaffner, F., Davelaar, G.J., Henry, M., Hulshof, K., Leslie, T.E. & Schrama, M. (2020) Taxonomy, ecology and distribution of the mosquitoes (Diptera: Culicidae) of the Dutch Leeward Islands, with a key to the adults and fourth instar larvae. *Contributions to Zoology*, 89, 1–19. DOI: 10.1163/18759866-bja10005

van der Wulp, F.M. (1881) Diptera. In: Veth, P.J., 1881–92. *Midden-Sumatra. Reizen en onderzoekingen der Sumatra-expeditie, uitgerust door het Aardrijkskundig genootschap, 1877–1879, beschreven door de leden der expeditie, onder toezicht van Prof. P.J. Veth.* Natuurlijke Historie. Negende afdeeling. E.J. Brill, Leiden, Netherlands.

van der Wulp, F.M. (1884) Note XXXVIII. On exotic Diptera. Part 1. *Notes from the Leyden Museum*, 6 (4), 248–256.

van der Wulp, F.M. (1898) Dipteren aus Neu-Guinea in der Sammlung des Ungarischen National-Museums. *Természetrajzi Füzetek kiadja a Magyar nemzeti Muzeum*, 21, 409–426, 1 pl.

van Someren, E.C.C. (1946a) Ethiopian Culicidae: notes and descriptions of some new species and hitherto unknown larvae and pupae (Diptera). *Transactions of the Royal Entomological Society of London*, 96 (7), 109–124. DOI: 10.1111/j.1365-2311.1946.tb00446.x

van Someren, E.C.C. (1946b) Ethiopian Culicidae. Tribe Megarhinini—notes and descriptions. *Transactions of the Royal Entomological Society of London*, 97 (7), 177–186. DOI: 10.1111/j.1365-2311.1946.tb00373.x

van Someren, E.C.C. (1946c) Ethiopian Culicidae—descriptions of the adults and pupae of two new species and two subspecies of the genus *Aëdes* from East Africa (Diptera). *Proceedings of the Royal Entomological Society of London, Series B: Taxonomy*, 15 (1–2), 1–6. DOI: 10.1111/j.1365-3113.1946.tb00802.x

van Someren, E.C.C. (1947a) The description of a new mosquito from the Seychelles. *East African Medical Journal*, 24 (1), 29–35.

van Someren, E.C.C. (1947b) The description of a new *Anopheles* of the *Myzorhynchus* series from Madagascar with notes on its systematic position in relation to the Ethiopian species of this group. *East African Medical Journal*, 24 (1), 42–46.

van Someren, E.C.C. (1947c) Ethiopian Culicidae—the description of a new *Eretmapodites* Theobald. *Proceedings of the Royal Entomological Society of London, Series B: Taxonomy*, 16 (1–2), 13–15. DOI: 10.1111/j.1365-3113.1947.tb00840.x

van Someren, E.C.C. (1948) Ethiopian Culicidae: some new mosquitoes from Uganda. *Proceedings of the Royal Entomological Society of London, Series B: Taxonomy*, 17 (9–10), 128–132. DOI: 10.1111/j.1365-3113.1948.tb00902.x

van Someren, E.C.C. (1949a) Ethiopian Culicidae—descriptions of four new mosquitoes from Madagascar. *Proceedings of the Royal Entomological Society of London, Series B: Taxonomy*, 18 (1–2), 3–8. DOI: 10.1111/j.1365-3113.1949.tb01406.x

van Someren, E.C.C. (1949b) Ethiopian Culicidae—*Eretmapodites* Theobald: description of four new species of the *chrysogaster* group with notes on the five known species of this group. *Proceedings of the Royal Entomological Society of London, Series B: Taxonomy*, 18 (7–8), 119–129. DOI: 10.1111/j.1365-3113.1949.tb01433.x

van Someren, E.C.C. (1950a) Ethiopian Culicidae: description of a new *Aëdes* from East Africa and notes on two known *Aëdes* with which it might be confused. *Proceedings of the Royal Entomological Society of London, Series B: Taxonomy*, 19 (1–2), 7–10. DOI: 10.1111/j.1365-3113.1950.tb00905.x

van Someren, E.C.C. (1950b) Ethiopian Culicidae: a description of two new *Aëdes* of the subgenus *Aëdimorphus* Theobald, from Uganda. *Proceedings of the Royal Entomological Society of London, Series B: Taxonomy*, 19 (3–4), 39–41. DOI: 10.1111/j.1365-3113.1950.tb00921.x

van Someren, E.C.C. (1950c) Ethiopian Culicidae: a description of a new sub-species of *Aëdes (Aëdimorphus) tricholabis* Edwards. *Proceedings of the Royal Entomological Society of London, Series B: Taxonomy*, 19 (5–6), 67–69. DOI: 10.1111/j.1365-3113.1950.tb00933.x

van Someren, E.C.C. (1951) New Culicini from Kenya and Uganda. *Proceedings of the Royal Entomological Society of London, Series B: Taxonomy*, 20 (1–2), 1–9. DOI: 10.1111/j.1365-3113.1951.tb01005.x

van Someren, E.C.C. (1954a) New Culicini from the Kenya Coast. *Proceedings of the Royal Entomological Society of London, Series B: Taxonomy*, 23 (5–6), 103–109. DOI: 10.1111/j.1365-3113.1954.tb00110.x

van Someren, E.C.C. (1954b) Ethiopian Culicidae: descriptions of a new *Culex*, the female of *Eretmapodites tonsus* Edwards and the early stages of two *Aedes* of the subgenus *Banksinella* Theobald. *Proceedings of the Royal Entomological Society of London, Series B: Taxonomy*, 23 (7–8), 119–126. DOI: 10.1111/j.1365-3113.1954.tb00113.x

van Someren, E.C.C. (1956) Undescribed culicine larvae and pupae from Uganda. *Proceedings of the Royal Entomological Society of London Series B, Taxonomy*, 25 (1–2), 3–12. DOI: 10.1111/j.1365-3113.1956.tb01072.x

van Someren, E.C.C. (1962) Ethiopian Culicidae: three new *Aedes* from Tanganyika, with a description of the male of *Aedes usambara* Mattingly and the female of *Uranotaenia henrardi* Edwards. *Proceedings of the Royal Entomological Society of London, Series B: Taxonomy*, 31 (1–2), 19–26. DOI: 10.1111/j.1365-3113.1962.tb01164.x

van Someren, E.C.C. (1967) The female and early stages of *Culex (Culex) nakuruensis* Mattingly, with a description of a new subspecies of *Culex (Culex) shoae* Hamon & Ovazza. *Proceedings of the Royal Entomological Society of London, Series B: Taxonomy*, 36 (1–2), 11–16. DOI: 10.1111/j.1365-3113.1967.tb00528.x

van Someren, E.C.C. (1971) The description of a new *Aedes* mosquito of the subgenus *Skusea* from the Seychelles. *Journal of Entomology. Series B, Taxonomy*, 40 (1), 21–25. DOI: 10.1111/j.1365-3113.1971.tb00100.x

van Someren, E.C.C. & Hamon, J. (1964) Ethiopian Culicidae (Diptera). A new species of *Culex* from Tanganyika, the description of the pupa of *Aedes usambara* Mattingly and the early stages of *Eretmapodites tonsus* Edwards. *Journal of the Entomological Society of Southern Africa*, 27 (1), 78–85.

van Someren, E.C.C., Teesdale, C. & Furlong, M. (1955) The mosquitos of the Kenya Coast; records of occurrence, behaviour and habitat. *Bulletin of Entomological Research*, 46 (3), 463–493. DOI: 10.1017/S0007485300039481

van Thiel, P.H. (1927) Sur l'origine des variations de taille de l'*Anopheles maculipennis* dans les Pays-Bas. *Bulletin de la Société de Pathologie exotique*, 20, 366–390.

Vargas, D.L. (1979) *Anopheles (Anopheles) veruslanei* Vargas, n. sp. (Diptera: Culicidae). *Gaceta Medica de Mexico*, 115 (2), 87–88.

Vargas, L. (1941) *Anopheles (Anopheles) Gabaldoni* n. sp. del Estado de Tabasco, México. (Dipt. Culicidae). *Medicina: Revista Mexicana*, 21 (395), 385–394.

Vargas, L. (1943a) Los subgeneros americanos de *Anopheles* (Diptera, Culicidae). *Anopheles (Russellia) xelajuensisde Leon* [*sic*], 1938 n. subgn. y *Anopheles (Coelodiazesis) fausti* n. sp. *Revista del Instituto de Salubridad y Enfermedades Tropicales*, 4 (1), 57–77.

Vargas, L. (1943b) *Anopheles earlei* Vargas, 1942, n. sp. norteamericana del grupo *maculipennis*. *Boletín de la Oficina Sanitaria Panamericana*, 22 (1), 8–12.

Vargas, L. (1950) Los subgeneros de *Aëdes. Downsiomyia* n. subgen. (Diptera: Culicidae). *Revista del Instituto de Salubridad y Enfermedades Tropicales*, 11 (1), 61–69.

Vargas, L. (1953) *Megarhinus* de Norteamerica *(Diptera: Culicidae)* [*sic*]. *Revista del Instituto de Salubridad y Enfermedades Tropicales*, 13, 27–32.

Vargas, L. (1956) *Psorophora (Psorophora) stonei* n. s. *Revista del Instituto de Salubridad y Enfermedades Tropicales*, 16 (2), 15–16.

Vargas, L. & Diaz Nájera, A. (1961) Nueva especie de *Sabethes* de Mexico *(Diptera, Culicidae)*. *Revista del Instituto de Salubridad y Enfermedades Tropicales*, 21 (1–2), 73–77.

Vargas, L. & Downs, W.G. (1950) Tres especies nuevas de *Aedes* (Diptera, Culicidae). *Revista de la Sociedad Mexicana de Historia Natural*, 11 (1–4), 161–172, 5 pls.

Vargas, L. & Martínez Palacios, A. (1953) Descripcion de *Wyeomyia (Wyeomyia) stonei*, n. sp., y notas sobre otros Sabethini de Mexico. *Revista del Instituto de Salubridad y Enfermedades Tropicales*, 13 (4), 293–307.

Vattier, G. & Hamon, J. (1962) Description de la larve et de la nymphe de *Culex (Culiciomyia) gilliesi* Hamon et van Someren, 1961. Clef des larves du sous-genre *Culiciomyia* connues en Afrique au sud du Sahara. *Bulletin de la Société de Pathologie exotique*, 55 (2), 246–252.

Venhuis, W.G. (1932) Een nieuwe anopheline, *Anopheles meraukensis*. *Geneeskundig Tijdschrift voor Nederlandsch-Indië*, 72, 1040–1042, 1 pl.

Venhuis, W.G. (1933) Een nieuwe varieteit van de superspecies *An. punctulatus*. *Geneeskundig Tijdschrift voor Nederlandsch-Indië*, 73, 203–206.

Venhuis, W.G. (1940) Een nieuwe *Anopheles* larve uit Indochina, *Anopheles alongensis* nov. sp. *Geneeskundig Tijdschrift voor Nederlandsch-Indië*, 80, 173–178, 2 pls.

Ventrillon, E. (1905) Culicides nouveaux de Madagascar. *Archives de Parasitologie*, 9 (4), 441–450.

Ventrillon, E. (1906a) *Culex* nouveaux de Madagascar. *Bulletin du Muséum National d'Histoire Naturelle*, 12 (2), 100–106.

Ventrillon, E. (1906b) *Stegomyia cartroni*, culicide nouveau de Madagascar. *Bulletin du Muséum National d'Histoire Naturelle*, 12 (3), 143–145.

Vesgueiro, F.T., Demari-Silva, B., dos Santos Malafronte, R., Sallum, M.A.M. & Marrelli, M.T. (2011) Intragenomic variation in the second internal transcribed spacer of the ribosomal DNA of species of the genera *Culex* and *Lutzia* (Diptera: Culicidae). *Memórias do Instituto Oswaldo Cruz*, 106 (1), 1–8. DOI: 10.1590/S0074-02762011000100001

Vincke, I. & Leleup, N. (1949) Un nouvel Anophèle dendrophile du Katanga: *Anopheles Berghei* sp. nov. *Revue de zoologie et de botanique Africaines*, 42 (3–4), 248–249.

Vockeroth, J.R. (1954) Notes on northern species of *Aedes*, with descriptions of two new species (Diptera: Culicidae). *Canadian Entomologist*, 86 (3), 109–116. DOI: 10.4039/Ent86109-3

von Heyden, L. (1870) Fossile Dipteren aus der Braunkohle van Rott im Siebengebirge. *Palaeontographica*, 17 (6), 237–266, 2 pls.

von Humboldt, F.H.A. (1819) *Voyage aux régions équinoxiales du Nouveau Continent, fait en 1799, 1800, 1801, 1802, 1803 et 1804 par Al. de Humboldt et A. Bonpland; rédigé par Alexandre de Humboldt. Avec deux atlas, qui renferment, l'un les vues les cordillères et les monumens des peuples indigènes de l'amérique, et l'autre des cartes géographiques et physiques*. Tome Second. In: Humboldt, F.H.A. (ed.), *Voyage de Humboldt et Bonpland*. Premiére Partie. Relation historique, Paris, France.

von Röder, V. (1885) Dipteren von der Insel Portorico. *Stettiner Entomologische Zeitung*, 46 (7–9), 337–349.

Vu, T.P., Nguyen, D.M., Tran, D.H. & Nguyen, T.V. (1991) *Anopheles (Anopheles) cucphuongensis*: a new species from Vietnam (Diptera: Culicidae). *Mosquito Systematics* (for 1990), 22 (3), 145–148.

Vythilingam, I., Jeffery, J., Harbach, R.E., Chan, S.T., Tan, C.H. & Asmad, M. (2007) *Anopheles (Cellia) kokhani*, n. sp. (Diptera: Culicidae) from Kapit, Sarawak, East Malaysia. *Proceedings of the Entomological Society of Washington*, 109 (2), 463–468.

Waktoedi Koesoemawinangoen, R. (1954) *Anophelini di Indonesia*. Kjilid 1. Kementerian Kesehatam Republik Indonesia, Djakarta, Indonesia.

Walker, F. (1848) *List of the specimens of dipterous insects in the collection of the British Museum. Part I*. British Museum, London, UK.

Walker, F. (1856) *Insecta Saundersiana: or characters of undescribed insects in the collection of William Wilson Saunders, Esq., F.R.S., F.L.S., &c. Vol.* I. *Diptera*. John van Voorst, London, UK.

Walker, F. (1859) Catalogue of the dipterous insects collected in the Aru Islands by Mr. A. R. Wallace, with descriptions of new species. *Journal of the Proceedings of the Linnean Society of London Zoology*, 3, 77–131. DOI: 10.1111/j.1096-3642.1858.tb02413.x

Walker, F. (1860) Catalogue of the dipterous insects collected in Amboyna by Mr. A. R. Wallace, with descriptions of new species. *Journal of the Proceedings of the Linnean Society of London Zoology*, 5 (s1), 144–168. DOI: 10.1111/j.1096-3642.1861.tb00922.x

Walker, F. (1861) Catalogue of the dipterous insects collected at Dorey, New Guinea, by Mr. A. R. Wallace, with descriptions of new species. *Journal of the Proceedings of the Linnean Society of London Zoology*, 5 (19), 229–257. DOI: 10.1111/j.1096-3642.1861.tb02102.x

Walker, F. (1865) Descriptions of new species of the dipterous insects of New Guinea. *Journal of the Proceedings of the Linnean Society of London Zoology*, 89 (30–32), 102–130. DOI: 10.1111/j.1096-3642.1865.tb02425.x

Wang, H.-H. & Feng, C.-Y. (1964) Description of three new species of *Culex (Lophoceratomyia)* [in Chinese]. *Acta Zootaxonomica Sinica*, 1 (1), 37–41.

Wang, X.-X. (1984) Description of a new *Culex* species (Diptera: Culicidae) [in Chinese]. *Acta Entomologica Sinica*, 27 (2), 217–220.

Wang, Y., Zhou, X., Yang, D. & Rokas, A. (2015) A genome-scale investigation of incongruence in Culicidae mosquitoes. *Genome Biology and Evolution*, 7 (12), 3463–3471. DOI: 10.1093/gbe/evv235

Wanson, M. & Lebied, B. (1945) Un nouvel Anophèle cavernicole du Congo belge. *Anopheles (Myzomyia) Vanhoofi* spec. nov. *Revue de zoologie et de botanique Africaines*, 39 (1), 118–129.

Ward, R.A. (1984) Second supplement to "A catalog of the mosquitoes of the world" (Diptera: Culicidae). *Mosquito Systematics*, 16 (3), 227–270.

Ward, R.A. (1992) Third supplement to "A catalog of the mosquitoes of the world" (Diptera: Culicidae). *Mosquito Systematics*, 24 (3), 177–230.

Watson, M. (1910) A new anopheline from the Federated Malay States. *Annals of Tropical Medicine and Parasitology*, 4, 251–253. DOI: 10.1080/00034983.1910.11685715

Wattal, B.L. & Kalra, N.L. (1965) An entomological survey of Dehra Dun Valley (Uttar Pradesh). Part II: description of *Culex (Culiciomyia) ramakrishnii* sp. n., and allotype (♂) of *Uranotaenia maculipleura* Leicester, 1908. *Bulletin of the Indian Society for Malaria and Other Communicable Diseases*, 2 (4), 309–313.

Wattal, B.L., Bhatia, M.L. & Kalra, N.L. (1958) Some new records of culicines of Dehra Dun (Uttar Pradesh) with a description of a new variety. *Indian Journal of Malariology*, 12 (3), 217–230.

Webb, C., Doggett, S. & Russell, R. (2016) *A Guide to Mosquitoes of Australia*. CSIRO Publishing, Clayton South, Australia.

Wesenberg-Lund, C. (1921) Contributions to the biology of the Danish Culicidae. *Det Kongelige Danske Videnskabernes Selskabs Skrifter. Naturvidenskabelig og mathematisk Afdeling*, 7 (1), 1–210, 21 pls.

Westwood, D.O. (1835) Insectorum nonnullorum novum (ex ordine Dipterorum) descriptiones. *Annales de la Société Entomologique de France*, 4, 681–685. DOI: 10.1080/14786443508648642

Wharton, R.H. (1947) Notes on Australian mosquitoes (Diptera, Culicidae). Part VII. The genus *Harpagomyia*. *Proceedings of the Linnean Society of New South Wales*, 72 (1–2), 58–68.

Wharton, R.H. (1962) The biology of *Mansonia* mosquitoes in relation to the transmission of filariasis in Malaya. *Bulletin – Institute for Medical Research, Kuala Lumpur*, 11, 1–114, 6 pls.

White, G.B. (1974) Priority of *Ingramia* Edwards, 1912, over *Ravenalites* Doucet, 1957 as the name for a subgenus of *Mimomyia* Theobald, 1903 (Diptera: Culicidae). *Mosquito Systematics*, 6 (4), 239–242.

White, G.B. (1975) Notes on a catalogue of Culicidae of the Ethiopian Region. *Mosquito Systematics*, 7 (4), 303–344.

White, G.B. (1980) Family Culicidae. In: Crosskey, R.W. (ed.), *Catalogue of the Diptera of the Afrotropical Region*. British Museum (Natural History), London, UK, pp. 114–148.

White, G.B. (1985) *Anopheles bwambae* sp.n., a malaria vector in the Semliki Valley, Uganda, and its relationships with other sibling species of the *An. gambiae* complex (Diptera: Culicidae). *Systematic Entomology*, 10 (4), 501–522. DOI: 10.1111/j.1365-3113.1985.tb00155.x

White, G.B. & Faust, C. (2014) Appendix 4 – medical acarology and entomology. In: Farrar, J., Hotez, P., Junghanss, T., Kang, G., Lalloo, D. & White, N.J. (eds), *Manson's Tropical Diseases*. 23rd edn. Elsevier, Amsterdam, Netherlands, pp. 1258–1272.

Whitman, L. & Antunes, P.C.A. (1937) Studies on the capacity of various Brazilian mosquitoes, representing the genera *Psorophora, Aedes, Mansonia*, and *Culex*, to transmit yellow fever. *American Journal of Tropical Medicine*, 17 (6), 803–823. DOI: 10.4269/ajtmh.1937.s1-17.803

Wiedemann, C.R.G. (1819) Beschreibung neuer Zweiflügler aus Ostindien und Afrika. *Zoologisches Magazin*, 1 (3), 1–39.

Wiedemann, C.R.G. (1820) *Diptera exotica. Sectio I. antennis multiarticulatis*. Kiliae [Kiel], Germany.

Wiedemann, C.R.G. (1821) *Diptera exotica*. Pars I. Kiliae [Kiel], Germany. DOI: 10.5962/bhl.title.14404

Wiedemann, C.R.W. (1828) *Aussereuropäische zweiflügelige Insekten: als Fortsetzung des Meigenschen Werkes*. Vol. 1. Hamm, Germany. DOI: 10.5962/bhl.title.14603

Wijesundara, D.P. (1951) Notes on species of the subgenus *Aëdes* Meigen (sens. str.) in Ceylon, with descriptions of three new species. *Ceylon Journal of Science. Section B. Zoology and Geology*, 24 (3), 173–179, 2 pls.

Wilkerson, R.C. (1991) *Anopheles (Anopheles) calderoni* n.sp., a malaria vector of the Arribalzagia Series from Peru (Diptera: Culicidae). *Mosquito Systematics*, 23 (1), 25–38.

Wilkerson, R.C. & Linton, Y.-M. (2015) Elevation of *Pseudoskusea, Rusticoidus* and *Protomacleaya* to valid subgenera in the mosquito genus *Aedes* based on taxon naming criteria recently applied to other members of the Tribe Aedini (Diptera: Culicidae). *Parasites & Vectors*, 8, 668. DOI: 10.1186/s13071-015-1247-x

Wilkerson, R.C. & Peyton, E.L. (1990) Standardized nomenclature for the costal wing spots of the genus *Anopheles* and other spotted-wing mosquitoes (Diptera: Culicidae). *Journal of Medical Entomology*, 27 (2), 207–224. DOI: 10.1093/jmedent/27.2.207

Wilkerson, R.C. & Sallum, M.A.M. (1999) *Anopheles (Anopheles) forattinii:* a new species in Series Arribalzagia (Diptera: Culicidae). *Journal of Medical Entomology*, 36 (3), 345–354. DOI: 10.1093/jmedent/36.3.345

Wilkerson, R.C. & Strickman, D. (1990) Illustrated key to the female anopheline mosquitoes of Central America and Mexico. *Journal of the American Mosquito Control Association*, 6 (1), 7–34.

Wilkerson, R.C., Linton, Y.-M., Fonseca, D.M., Schultz, T.R., Price, D.C. & Strickman, D.A. (2015) Making mosquito taxonomy useful: a stable classification of tribe Aedini that balances utility with current knowledge of evolutionary relationships. *PLoS ONE*, 10 (7), e0133602. DOI: 10.1371/journal.pone.0133602

Wilkerson, R.C., Linton, Y.-M. & Strickman, D. (2021) *Mosquitoes of the World*. Volumes 1 and 2. Johns Hopkins University Press, Baltimore, MD, USA.

Wilkerson, R.C., Parsons, T.J., Klein, T.A., Gaffigan, T.V., Bergo, E. & Consolim, J. (1995) Diagnosis by random amplified polymorphic DNA polymerase chain reaction for four cryptic species related to *Anopheles (Nyssorhynchus) albitarsis* (Diptera: Culicidae) from Paraguay, Argentina, and Brazil. *Journal of Medical Entomology*, 32 (5), 697–704. DOI: 10.1093/jmedent/32.5.697

Wilkerson, R.C., Reinert, J.F. & Li, C. (2004) Ribosomal DNA ITS2 sequences differentiate six species in the *Anopheles crucians* complex (Diptera: Culicidae). *Journal of Medical Entomology*, 41 (3), 392–401. DOI: 10.1603/0022-2585-41.3.392

Wilkerson, R.C., Somboon, P. & Harbach, R.E. (2022) Reconsideration of the status of subspecies in the Japonicus Group of the subgenus *Hulecoeteomyia* Theobald of *Aedes* Meigen (Diptera: Culicidae). *Zootaxa*, 5162 (2), 198–200. DOI: 10.11646/zootaxa.5162.2.8

Williston, S.W. (1893) List of Diptera of the Death Valley Expedition. In: Riley, C.V., *Report on a small collection of insects made during the Death Valley Expedition* (pp. 235–268). [Series] *North American Fauna No. 7, The Death valley expedition. A biological survey of parts of California, Nevada, Arizona, and Utah*. Part 2. Government Printing Office, Washington, DC, USA, pp. 253–259.

Williston, S.W. (1896) On the Diptera of St. Vincent (West Indies). *Transactions of the Royal Entomological Society of London*, 44, 253–446, 7 pls. DOI: 10.1111/j.1365-2311.1896.tb00965.x

Williston, S.W. (1900) Culicidae. In: Osten Sacken, C.R., 1886–1901. *Biologia Centrali-Americana. Insecta. Diptera*. Volume 1. The Supplement [by S.W. Williston, J.M. Aldrich, W.M Wheeler and A.L. Melander]. R. H. Porter, London, UK, p. 224.

Wirth, W.W. (1948) Diptera. In: Usinger, R.L., La Rivers, I., Chandler, H.P. & Wirth, W.W., *Entomology 133: Biology of Aquatic and Littoral Insects*. Syllabus Series. University of California, Berkeley, CA, USA, pp. 86–136, 223–239.

Wolfs, J. (1945) *Culex (Neoculex) Wansoni* sp. n. *Annales de la Société Belge de Médecine Tropicale*, 25, 221–224, 2 pls.

Wolfs, J. (1947a) Un Culicide nouveau du Katanga, *Aedes (Banksinella) Bequaerti*, sp. n. *Revue de zoologie et de botanique Africaines*, 40 (1), 40–41.

Wolfs, J. (1947b) *Culex (Culex) bukavuensis* sp. n. *Revue de zoologie et de botanique Africaines*, 40 (2–3), 159–164.

Wolfs, J. (1947c) *Culex (Culex) Seldeslachtsi*, sp. nov. *Revue de zoologie et de botanique Africaines*, 40 (2–3), 224–228.

Wolfs, J. (1948a) *Taeniorhynchus (Coquillettidia) schoutedeni* n. sp. (Culic.). *Revue de zoologie et de botanique Africaines*, 40 (4), 244–247.

Wolfs, J. (1948b) *Taeniorhynchus (Coquillettidia) Vanoyei* sp. n. (Dipt. Culicidae). *Revue de zoologie et de botanique Africaines*, 41 (1), 83–86.

Wolfs, J. (1949) *Aedes (Stegomyia) apicoargenteus* var. *denderensis* var. n. (Culicidae). *Revue de zoologie et de botanique Africaines*, 42 (3–4), 190–192.

Wolfs, J. (1951) Un Culicide nouveau de la région éthiopienne: *Taeniorhynchus (Coquillettidia) karandalaensis* sp. n. *Revue de zoologie et de botanique Africaines*, 45 (1–2), 12–14.

Wolfs, J. (1958) Deux *Culicidae* nouveaux du Congo belge (Diptera Nematocera). *Revue de zoologie et de botanique Africaines*, 58 (3–4), 297–298.

Wood, D.M., Dang, P.T. & Ellis, R.A. (1979) *The insects and arachnids of Canada. Part 6. The mosquitoes of Canada. (Diptera: Culicidae)*. Canadian Government Publishing Centre, Supply and Services Canada, Hull, Quebec, Canada.

Woodhill, A.R. (1949) A new subspecies of *Aëdes (Stegomyia) scutellaris* Walker (Diptera, Culicidae) from northern Australia. *Proceedings of the Linnean Society of New South Wales*, 74 (3–4), 140–144.

Woodhill, A.R. & Lee, D.J. (1944) The subspecies of *Anopheles amictus* Edwards (Diptera, Culicidae). *Proceedings of the Linnean Society of New South Wales*, 69 (1–2), 62–66.

Worth, C.B. (1960) Description of a new species of the aedine subgenus *Neomelaniconion* from Tongaland, South Africa (Diptera: Culicidae). *Journal of the Entomological Society of Southern Africa*, 23 (2), 312–313.

Worth, C.B., Paterson, H.E. & de Meillon, B. (1961) The incidence of arthropod-borne viruses in a population of culicine mosquitoes in Tongaland, Union of South Africa (January, 1956, through April, 1960). *American Journal of Tropical Medicine and Hygiene*, 10 (4), 583–592. DOI: 10.4269/ajtmh.1961.10.583

Wu, S.-C. (1936) Further notes on the mosquitoes of Hangchow, Chekiang, with description of one new species. *Yearbook of the Bureau of Entomology of Chekiang Province* (for 1935), 5, 46–53.

Xavier, S.H. & da Silva Mattos, S. (1972) *Culex (Melanoconion) milwardi* sp. n. (Diptera, Culicidae). *Revista Brasileira de Biologia*, 32 (4), 569–571.

Xavier, S.H., da Silva, J.E. & da Silva Mattos, S. (1970a) *Culex (Melanoconion) olimpioi* sp. n. (Diptera, Culicidae). *Revista Brasileira de Malariologia e Doenças Tropicais*, 22 (1), 183–188.

Xavier, S.H., da Silva Mattos, S. & da Silva, J.E. (1970b) *Uranotaenia rachoui*, uma nova espécie do Brasil (Diptera, Culicidae). *Revista Brasileira de Malariologia e Doenças Tropicais*, 22 (2/4), 411–416.

Xu, J. & Luo, X. (1998) A new species of *Anopheles hyrcanus* group Nei Mongol Autonomous Region (Diptera: Culicidae). *Chinese Journal of Parasitology and Parasitic Diseases*, 16 (1), 45–52.

Yamada, M. (1937) A new species of *Anopheles* in Chosen (Korea). *Keijo Journal of Medicine*, 8, 237–255, 5 pls.

Yamada, S. (1917) On two new species of mosquitoes from Japan [in Japanese]. *Dōbutsugaku zasshi*, 29 (341), 61–72.

Yamada, S. (1918) A new species of *Anopheles* collected from Hokkaido [in Japanese]. *Eiseigaku Densenbyogaku Zasshi*, 13, 689–691.

Yamada, S. (1921) Description of ten new species of *Aedes* found in Japan, with notes on the relation between some of these mosquitoes and the larva of *Filaria bancrofti* Cobbold. *Annotationes Zoologicae Japonenses*, 10 (6), 45–81.

Yamada, S. (1924) A revision of the adult anopheline mosquitoes of Japan: systematic descriptions, their habits and their relations to human diseases, together with an account of three new species. *Scientific Reports from the Government Institute for Infectious Diseases, Tokyo Imperial University*, 3, 215–241.

Yamada, S. (1932) Culicidae. In: Esaki, T. *et al.*, *Nippon Konchu Zukan: Iconographia Insectorum Japonicorum*. Editio Prima. Hokuryukan, Tokyo, Japan, pp. 210–235.

Yamada, S. & Watanabe, P. (1918) A new species of Korean *Anopheles* [in Japanese]. *Japanese Journal of Experimental Medicine*, 2, 206–209.

Yamaguti, S. & LaCasse, W.J. (1952) *Culex (Culiciomyia) kyotoensis* n. sp. (Culicidae, Diptera) from Japan. Published by the authors, Kyoto, Japan.

Yamaguti, S. & Tamaboko, R. (1954) Two new species of *Aedes* (Culicidae, Diptera) from Japan. *Acta Medica Okayama* (for 1952), 8 (4), 414–422, 4 pls.

Yang, J.-L., Li, F.-H. & Chen, H.-B. (1993) A new species of *Culex* from Yunnan (Diptera: Culicidae) [in Chinese]. *Acta Zootaxonomica Sinica*, 18 (2), 225–228.

Yang, W., Zhao, S., Xie, Y., Liu, T., Kong, L., Guo, Y., Xie, Z., Liu, P. & Chen, X.-G. (2022) *Armigeres subalbatus* is a potential vector of Zika virus but not dengue virus. *Infectious Diseases of Poverty*, 11, 62. DOI: 10.1186/s40249-022-00990-0

Yao, Y.-C. & Wu, C.-C. (1944) A new anopheline mosquito, *Anopheles kweiyangensis*, sp. n. from Guiyang, Guizhou [in Chinese]. *National Medical Journal of China*, 30 (1), 1–10.

Yeo, H., Tan, C.-H., Chong, C.-S. & Lam-Phua, S.-G. (2019) Identification key to the genera of adult female mosquitoes (Diptera: Culicidae) of Singapore. *Zootaxa*, 4624 (3), 407–423. DOI: 10.11646/zootaxa.4624.3.9

Yee, D.A., Reyes-Tores, L.J., Dean, C., Seavo, N.A. & Zavortink, T.J. (2021) Mosquitoes (Diptera: Culicidae) on the islands of Puerto Rico and Vieques, USA. *Acta Tropica*, 220, 105959. DOI: 10.1016/j.actatropica.2021.105959

Yin, L.R. (1970) *Sensilla of fourth instar larvae of* Aedes aegypti *(L.), and a comparison with three other mosquito species*. MS Thesis, Department of Biology, University of Saskatchewan, Saskatoon, Canada.

Young, T.C.M. & Majid, S.A. (1928) A variety of Anopheles [*sic*] *karwari* collected in Coorg, S. India. *Indian Journal of Medical Research*, 16 (2), 469–471, 1 pl.

Zagaria, N. & Savioli, L. (2002) Elimination of lymphatic filariasis: a public-health challenge. *Annals of Tropical Medicine and Parasitology*, 96 (Suppl. 2), S3–S13.

Zavortink, T.J. (1968) Mosquito studies (Diptera, Culicidae). VIII. A prodrome of the genus *Orthopodomyia*. *Contributions of the American Entomological Institute*, 3 (2), 1–221.

Zavortink, T.J. (1969) Mosquito studies (Diptera, Culicidae). XV. A new species of treehole breeding *Anopheles* from the southwestern United States. *Contributions of the American Entomological Institute*, 4 (4), 27–38.

Zavortink, T.J. (1970a) Mosquito studies (Diptera, Culicidae) XVI. A new species of treehole breeding *Aedes (Ochlerotatus)* from southern California. *Contributions of the American Entomological Institute* (for 1969), 5 (1), 1–7.

Zavortink, T.J. (1970b) Mosquito studies (Diptera, Culicidae). XIX. The treehole *Anopheles* of the world. *Contributions of the American Entomological Institute*, 5 (2), 1–35.

Zavortink, T.J. (1970c) Mosquito studies (Diptera, Culicidae). XXII. A new subgenus and species of *Aedes* from Arizona. *Contributions of the American Entomological Institute*, 7 (1), 1–11.

Zavortink, T. (1971) Contributions to the mosquito fauna of Southeast Asia. IX. The genus *Orthopodomyia* Theobald in Southeast Asia. *Contributions of the American Entomological Institute*, 7 (3), 1–37.

Zavortink, T.J. (1972) Mosquito studies (Diptera, Culicidae). XXVIII. The New World species formerly placed in *Aedes (Finlaya)*. *Contributions of the American Entomological Institute*, 8 (3), 1–206.

Zavortink, T.J. (1973) Mosquito studies (Diptera, Culicidae). XXIX. A review of the subgenus *Kerteszia* of *Anopheles*. *Contributions of the American Entomological Institute*, 9 (3), 1–54.

Zavortink, T.J. (1979) Mosquito studies (Diptera, Culicidae). XXXV. The new sabethine genus *Johnbelkinia* and a preliminary reclassification of the composite genus *Trichoprosopon*. *Contributions of the American Entomological Institute*, 17 (1), 1–61.

Zavortink, T.J. (1981) Species complexes in the genus *Trichoprosopon*. *Mosquito Systematics*, 13 (1), 82–85.

Zavortink, T.J. (1986a) *Zinzala*, a new subgenus of *Wyeomyia* with two new species from pitcher-plants in Venezuela (Diptera, Culicidae, Sabethini). *Wasmann Journal of Biology* (for 1985), 43 (1–2), 46–59.

Zavortink, T.J. (1986b) The occurrence of *Runchomyia frontosa* in carnivorous bromeliads in Venezuela, with notes on the biology of its immatures (Diptera, Culicidae, Sabethini). *Wasmann Journal of Biology*, 44 (1–2), 127–129.

Zavortink, T.J. & O'Meara, G.F. (1999) *Culex (Micraedes) biscaynensis* n. sp. from Florida (Diptera: Culicidae). *Journal of the American Mosquito Control Association*, 15 (3), 263–270.

Zavortink, T.J. & Poinar, G.O., Jr (2000) *Anopheles (Nyssorhynchus) dominicanus* sp. n. (Diptera: Culicidae) from Dominican Amber. *Annals of the Entomological Society of America*, 93 (6), 1230–1235. DOI: 10.1603/0013-8746(2000)093[1230:ANDSND]2.0.CO;2

Zavortink, T.J. & Poinar, G.O., Jr (2008) *Toxorhynchites (Toxorhynchites) mexicanus*, n. sp. (Diptera: Culicidae) from Mexican amber: a New World species with Old World affinities. *Proceedings of the Entomological Society of Washington*, 110 (1), 116–125.

Zetterstedt, J.W. (1838) *Insecta Lapponica descripta*. Voss, Lipsiae [Leipzig], Germany.

Zhou, H., Zhu, G. & Lu, Y. (1999) One new species of the genus *Topomyia* (Diptera, Culicidae) [in Chinese]. *Acta Parasitology et Medica Entomologica Sinica*, 6 (2), 107–111.

Zsemlye, J.L., Hancock, R.G. & Foster, W.A. (2005) Analysis of a complex vertical copulatory-courtship display in the yellow fever vector *Sabethes chloropterus*. *Medical and Veterinary Entomology*, 19 (3), 276–285. DOI: 10.1111/j.1365-2915.2005.00570.x